EOCENE-OLIGOCENE CLIMATIC AND BIOTIC EVOLUTION

PRINCETON SERIES IN GEOLOGY AND PALEONTOLOGY
EDITED BY ALFRED G. FISCHER

EOCENE-OLIGOCENE CLIMATIC AND BIOTIC EVOLUTION

Edited by Donald R. Prothero and William A. Berggren

PRINCETON UNIVERSITY PRESS

Published by Princeton University Press, 41 William Street, Princeton, New Jersey 08540
In the United Kingdom: Princeton University Press, Oxford

Library of Congress Cataloging-in-Publication Data

Eocene-Oligocene climatic and biotic evolution / edited by Donald R. Prothero and William A. Berggren.
p. cm.—(Princeton series in geology and paleontology)
Includes index.
ISBN 0-691-08738-5 (alk. paper)—ISBN 0-691-02542-8 (pbk. : alk. paper).
1. Paleontology—Eocene. 2. Paleontology—Oligocene. 3. Geology, Stratigraphic—Eocene.
4. Geology, Stratigraphic—Oligocene.
5. Paleoclimatology. I. Prothero, Donald R.
II. Berggren, William A. III. Series.
QE737.E53 1992
560'.178—dc20 91-40143

The publisher would like to acknowledge Donald R. Prothero for providing the camera-ready copy from which this book was printed

Princeton University Press books are printed on acid-free paper, and meet the guidelines for permanence and durability of the Committee on Production Guidelines for Book Longevity of the Council on Library Resources

Printed in the United States of America

10 9 8 7 6 5 4 3 2 1

10 9 8 7 6 5 4 3 2 1
(Pbk.)

CONTENTS

PART III: THE MARINE RECORD

PART IV: THE TERRESTRIAL RECORD

LIST OF CONTRIBUTORS

John B. Anderson
Department of Geology and Geophysics
Rice University
Houston, TX 77251

Marie-Pierre Aubry
Centre Paléont. Strat. Paléoécol. (URA 11)
Université Claude Bernard
27-43 Blvd. du 11 Novembre
69622 Villeurbanne, Cedex, France
and
Woods Hole Oceanographic Institution
Woods Hole, MA 02543

Jack G. Baldauf
Department of Oceanography and
Ocean Drilling Program
Texas A&M University
College Station, TX 77845

Enriqueta Barrera
Department of Geological Sciences
University of Michigan
Ann Arbor, MI 48109-1063

Eric J. Barron
Earth System Science Center
Pennsylvania State University
512 Deike Building
University Park, PA 16802

Louis R. Bartek
Department of Geology and Geophysics
Rice University
Houston, TX 77251

Amanda Beecroft
Department of Geology and Geophysics
University of Adelaide
Adelaide, South Australia 5001 Australia

William A. Berggren
Department of Geology and Geophysics
Woods Hole Oceanographic Institute
Woods Hole, MA 02543

Thomas M. Bown
Branch of Paleontology and Stratigraphy
U.S. Geological Survey
Denver Federal Center
Box 25046
Denver, CO 80225

Henk Brinkhuis
Laboratory of Paleobotany and Palynology
Heidelberglaan 2
3584 CS Utrecht, The Netherlands

Burchard D. Carter
Department of Geology and Physics
Georgia Southwestern College
Americus, GA 31709

Scott Clay-Poole
Department of Botany KB-15
University of Washington
Seattle, WA 98195

Margaret E. Collinson
Division of Biosphere Sciences
King's College London
University of London
Campden Hill Road
London W8 7AH England

Stephen K. Donovan
Department of Geology
University of West Indies
Mona, Kingston 7, Jamaica

Robert J. Emry
Department of Paleobiology
National Museum of Natural History
Smithsonian Institution
Washington, DC 20560

Emmett Evanoff
Geology Section
University of Colorado Museum
Boulder, CO 80309-0315

R. Ewan Fordyce
Department of Geology
University of Otago
P.O. Box 56
Dunedin, New Zealand

Thor Hansén
Department of Geology
Western Washington University
Bellingham, WA 98225-5996

Jean-Louis Hartenberger
Laboratoire de Paléontologie
Institut des Sciences de l'Evolution
Université Montpellier II
Place Eugène-Bataillon
F34095 Montpellier Cedex 5 France

J. J. Hooker
Department of Palaeontology
Natural History Museum
Cromwell Road
London SW7 5BD England

J. Howard Hutchison
Museum of Paleontology
Department of Integrative Biology
University of California
Berkeley, CA 94720

Gerta Keller
Dept. Geological and Geophysical Sciences
Guyot Hall
Princeton University
Princeton, NJ 08544-1003

Dennis V. Kent
Lamont-Doherty Geological Observatory of Columbia University
Palisades, NY 10964

Robert H. Lander
Exxon Production Research Company
Houston, TX 77252-2189

Serge Legendre
Laboratoire de Paléontologie
Institut des Sciences de l'Evolution
Université Montpellier II
Place Eugène-Bataillon
F34095 Montpellier Cedex 5 France

Estella Leopold
Quaternary Research Center
Department of Geological Sciences
University of Washington
Seattle, WA 98195

Liu Gengwu
Quaternary Research Center
Department of Geological Sciences
University of Washington
Seattle, WA 98195

Spencer G. Lucas
New Mexico Museum of Natural History
P.O. Box 7010
Albuquerque, NM 87194-7010

Norman MacLeod
Dept. Geological and Geophysical Sciences
Guyot Hall
Princeton University
Princeton, NJ 08544-1003

Brian McGowran
Department of Geology and Geophysics
University of Adelaide
Adelaide, South Australia 5001 Australia

Michael L. McKinney
Department of Geological Sciences
University of Tennessee
Knoxville, TN 37996

Kenneth J. McNamara
Museum of Western Australia
Francis Street
Perth, Western Australia 6000 Australia

Kenneth G. Miller
Lamont-Doherty Geological Observatory of Columbia University
Palisades, NY 10964
and
Department of Geological Sciences
Rutgers University
New Brunswick, NJ 08903

Graham Moss
Department of Geology and Geophysics
University of Adelaide
Adelaide, South Australia 5001 Australia

John D. Obradovich
Branch of Isotope Geology
U.S. Geological Survey MS 963
Denver Federal Center
Box 25046
Denver, CO 80225

Donald R. Prothero
Department of Geology
Occidental College
Los Angeles, CA 90041

D. Tab Rasmussen
Department of Anthropology
Washington University
St. Louis, MO 63130

Gregory J. Retallack
Department of Geological Sciences
University of Oregon
Eugene, OR 97403-1272

Malcolm I. Ross
Department of Geology and Geophysics
Rice University
Houston, TX 77251

Elwyn L. Simons
Dept. of Biological Anthropology and Anatomy
Duke University
Durham, NC 80225

Lisa Cirbus Sloan
Department of Geological Sciences
University of Michigan
Ann Arbor, MI 48109-1063

Richard K. Stucky
Department of Earth Sciences
Denver Museum of Natural History
2001 Colorado Boulevard
Denver, CO 80205

Carl C. Swisher III
Geochronology Center
Institute of Human Origins
2453 Ridge Road
Berkeley, CA 94709

Ellen Thomas
Department of Earth Sciences
University of Cambridge
Downing Street
Cambridge CB2 3EQ England

Anne H. Walton
P.O. Box 8576
Austin, TX 78713

Wang Banyue
Institute of Vertebrate Paleontology
and Paleoanthropology
Academia Sinica
P.O. Box 643
Beijing 100044 People's Republic of China

Jack A. Wolfe
Branch of Paleontology and Stratigraphy
U.S. Geological Survey
Denver Federal Center
Box 25046
Denver, CO 80225

PREFACE

The transition from the Eocene to the Oligocene Epochs was the most significant event in earth history since the extinction of dinosaurs. From the warm, equable climates of the Eocene, the earth underwent significant changes. Global temperature cooled more than any time since the Mesozoic, and the first Antarctic ice sheets appeared. There were major extinctions in both plants and animals on the land and in the oceans. By the Oligocene, the earth had a much cooler, more temperate climate, with a much lower diversity of organisms. In essence, the Eocene-Oligocene transition was the change from the "greenhouse" world of the Cretaceous and early Cenozoic to the glaciated world of today.

This critical period of earth history has always fascinated scientists, but the recent interest in mass extinctions has increased the need to understand Eocene-Oligocene extinctions. Some have suggested that the Eocene-Oligocene extinctions were part of a periodic cycle that was also responsible for the Cretaceous-Tertiary extinctions. To test such hypotheses, there must be detailed studies of climatic information and organisms buried in Eocene-Oligocene sediments. Over the last decade, many such detailed studies have been undertaken, but very little of the information has been published. In marine geology, studies of the oceans around Antarctica have provided critical new evidence about the timing of Antarctic glaciation, and the oceanographic changes in the Southern Hemisphere. In terrestrial sections, new techniques of magnetostratigraphy and $^{40}Ar/^{39}Ar$ dating have revolutionized our correlation of terrestrial sections, and the fossils contained within them. To specialists, this unpublished information was known by informal networks and by word of mouth. There was little communication between specialties, nor publication of these data for the rest of the world to see.

For this reason, we thought that the best forum for such communication was an informal workshop. On August 1-5, 1989, we held a Penrose Conference on this subject in Rapid City, South Dakota. During that meeting, paleontologists and stratigraphers of both the marine and terrestrial persuasions were able to meet and exchange ideas. The meeting was extraordinarily fruitful, and it soon became apparent that an important synthesis was emerging. Consequently, another meeting was held, this time as a Theme Session at the Annual Meeting of the Geological Society of America in St. Louis, Missouri on November 6, 1989. After this theme session, we decided to organize the contributors and publish their research.

As the project grew, we learned of other scientists who had important information to contribute. Eventually, we were able to find authors representing nearly every major group of organisms in the ocean (from coccoliths, dinoflagellates, diatoms, foraminiferans, to molluscs, echinoids, and even whales) and on land (from soils, pollen, and plants, to land snails, reptiles, and land mammals). Since the effects were global, we have studies from both Southern and Northern Hemisphere oceans, and of land plants and animals in North America, Europe, Asia, and Africa. Besides the biological evidence, this book also includes evidence from seismology, stable isotopes, and sedimentology. There are also chapters on computer modeling of this critical period in climate history.

Finally, the radical changes in the late Paleogene time scale necessitated several chapters on the new chronology of the Eocene and Oligocene. In addition to the overviews of the marine and terrestrial timescales, there are several papers on critical new evidence for these correlations, particularly in North American terrestrial sections. Since the timing of events is critical to the discussion, these chapters open the book.

From the papers in this book, it is clear that the Eocene-Oligocene extinctions were much more complex than they are usually portrayed. Rather than a single abrupt "Terminal Eocene Event," there appear to be several events, the most of important of which occurred at the end of the middle Eocene. Although some of this complex story is summarized in the overview chapter, it is clear that we are just beginning to understand this critical and fascinating period of earth history.

Acknowledgements

The Penrose Conference would never have been possible without the efforts of two people. Dr. Philip R. Bjork of the Museum of Geology, South Dakota School of Mines and Technology, was the local convener of the conference. He engineered a logistical miracle in hosting 61 people, providing the audiovisual aids, running field trips to the Big Badlands, and organizing our closing night barbeque. Ms. Lois Elms coordinated the conference for the Geological Society of America, expertly handling all the planning and arrangements. The Center for the Study of the Evolution and Origin of Life at the University of California, Los Angeles, graciously provided funds for the participation of several of our foreign scientists. Their generosity made it a truly international meeting.

To speed up publication, this book was produced by Prothero as camera-ready copy on a Macintosh IIsi computer, using Microsoft Word 4.0 software and a LaserWriter II NT printer. Mr. Clifford R. Prothero helped with many of the page layouts. Mr. Steve King helped with software conversions, and Ms. Anita L. Argabright typed one of the chapters onto disk. Prothero was supported by NSF grant EAR87-08221 during the conferences and the production of this book.

EOCENE-OLIGOCENE CLIMATIC AND BIOTIC EVOLUTION: AN OVERVIEW

by William A. Berggren and Donald R. Prothero

ABSTRACT

The middle Eocene to early Oligocene witnessed major changes in global climate and ocean circulation which are reflected in significant turnovers in marine and terrestrial biota. The change from a thermospheric to thermohaline (psychrospheric) circulation is thought to reflect the establishment (and continued, if intermittent, presence) of ice on Antarctica, although a precise chronology of this conjunct evolution remains elusive due to conflicting interpretations of the isotopic record of carbon and oxygen. Paleoclimatic evidence clearly indicates global cooling and increased aridity during the late middle Eocene and again across the Eocene/Oligocene boundary and earliest Oligocene. The cooling during the middle Eocene has not been satisfactorily explained, although ice may have existed on the Antarctic continent by this time. The most significant cooling event and change in land floras occurred in the earliest Oligocene at about 33.5 Ma (in the revised chronology developed here). By the early Oligocene there were significant ice sheets on Antarctica, probably caused by the thermal isolation, and insulation, of the south polar region as cold currents flowed between the newly separated Antarctica and Australia.

The late middle Eocene - early Oligocene interval was the most significant episode of climatic change and extinction since the end of the Cretaceous, with the exception of the Paleocene/Eocene boundary event. However, the latter appears to have involved a relatively rapid reduction in global productivity, brief change from thermohaline to thermospheric circulation, and associated major extinction of deep-water benthic foraminifera superimposed on a longer-term trend toward increasing temperatures. Climatic changes at the Eocene-Oligocene transition are seen to have been the antithesis of those at the Paleocene/Eocene boundary, with sequential global extinctions superimposed on a gradually cooling (primarily at high latitudes) earth and increased eutrophication associated with increased biosiliceous productivity at high southern latitudes. In groups ranging from marine plankton to whales, and terrestrial pollen to land mammals, the major wave of extinctions took place at the end of the middle Eocene (Bartonian/Priabonian boundary). Highly diversified, warm-adapted species were the primary victims. Lesser extinctions took place during the late Eocene (mid-Priabonian), at the Eocene/Oligocene boundary (about 34 Ma in the chronology developed here), and in the early Oligocene (Rupelian). Extraterrestrial impacts occurred during the late Eocene (mid-Priabonian) but are not associated with any extinctions of consequence. Global biotic changes associated with the mid-late Paleogene climatic deterioration should be interpreted in the framework of endogenic (long-term changes in global circulation and associated restructuring of global environment and trophic resources) rather than exogenic (extraterrestrial) causes.

INTRODUCTION

The middle Eocene-early Oligocene was a critical period in earth history in that major changes in ocean circulation and global climate, reflected in significant turnovers in marine and terrestrial biotas, occurred over an approximately 10 m.y. interval. Because of its strategic importance intermediate between the globally warm (non-glaciated) Late Cretaceous and the cooler (glaciated) Neogene, and because of the flood of new information becoming available on mid-Paleogene geology, we organized a symposium devoted to the subject in August, 1989 (Prothero et al., 1990). The topic which forms the title of this book was vigorously discussed over the period of a five days.

At this point it may be appropriate to step outside the immediate topic of this book in order to put the symposium and the papers which fill this volume in perspective. There is a growing awareness in the Earth Sciences community of the direct link between global atmosphere - lithosphere - biosphere processes. We have only to recall the postulated, if not satisfactorily demonstrated, connection between core/mantle geodynamics, geomagnetic polarity reversals, modifications of ocean-continent geometries and resultant responses (feedback) in atmosphere-ocean circulation and climate change, and the associated responses in marine and terrestrial biotal geographies and turnovers. At the same time the information explosion in all branches of human knowledge renders it virtually impossible for the specialist (let alone the interested informed reader) to keep abreast of developments in areas to which an increasing number of specialists are bringing new information and insight at an ever increasing rate. Clearly there is a need for a symposium to bring together information dealing with the Eocene-Oligocene transition and timely publication of the contributed papers.

The chapters in this volume treat a broad spectrum of topics dealing with late Paleogene climate and ocean history and the responses engendered in marine and terrestrial biota. In this chapter we present an overview of these papers following the basic subdivisions adopted in the book itself.

A geochronologic framework is developed for the Paleogene and a consensus is seen to be emerging for a somewhat younger mid-Paleogene chronology than that suggested in some extant time scales with an Eocene/Oligocene boundary anchored at 34 Ma. Within this revised chronologic framework North American land mammal "ages" are recalibrated to the Global Polarity Time Scale (GPTS) and re-aligned with standard marine chronostratigraphies.

The papers dealing with micropaleontology and marine geology include empirical observations of the geological record (Antarctica), data on taxonomy and biogeography, stable isotopes and climatic modelling. Those dealing with marine invertebrates and marine and terrestrial vertebrates and plants contain biogeographic distributional information based on empirical observations from an essentially global stratigraphic database and provide a fascinating story on the intimate link between mid - late Paleogene marine and terrestrial ecosystems. While a consensus appears to exist that the Eocene "doubthouse" was, indeed, different from the Cretaceous "hothouse," on the one hand, and the Neogene "icehouse," on the other, there remain problems with the interpretation of observational data regarding the manner in which various investigators view the evolution of mid-Paleogene oceans and climate.

If we have raised the consciousness of the reader to the exciting world of the mid-Paleogene, provided evidence of the link which exists between all parts of a dynamic earth system and contributed in some way to the development of a global mid-Paleogene Historical Geology, we shall feel that we have accomplished the task we set ourselves in organizing the 1989 Penrose Conference and GSA Theme Session on the Eocene-Oligocene transition.

LATE PALEOGENE CHRONOSTRATIGRAPHY AND GEOCHRONOLOGY

The major nineteenth-century subdivisions of the Cenozoic (Paleocene through Pleistocene) were essentially biostratigraphic in connotation, whereas the lower-order units (stages) were essentially lithostratigraphic. The terms Paleogene and Neogene likewise were originally biostratigraphic in nature, connoting perceived temporal homogeneity in (predominantly) marine invertebrate faunas associated with the major subdivisions of the Cenozoic. Clarification and standardization of stratigraphic terminology and principles owes much to the tireless efforts of Hollis D. Hedberg (e.g., Hedberg, 1976) whose clear, didactic style lent authority to the enshrinement of the "holy trinity" of Litho-, Bio-, and Chronostratigraphy (for an up-to-date review of these concepts see Schoch, 1988).

Recent attempts to unify the global stratigraphic scale have crystallized around an agreement to standardize the major Phanerozoic units at their (lower) boundaries only, and at a single locality termed the boundary stratotype. Choice of a particular level is based upon a careful search for an appropriate horizon within a biostratigraphic sequence which does minimal injustice to existing practice, thus maintaining historical continuity with original definition and intent. The precise reference level for such a boundary (termed by some "The Golden Spike") is designated the Global Stratotype Section and Point (GSSP) which serves to denote that horizon where rock and time coincide (Harland et al., 1990), thus providing a clear philosophical distinction between bio- and chronostratigraphy. For while the boundary definition itself (a lithostratigraphic level) is (theoretically) globally isochronous (i.e., chronostratigraphic), the primary means (i.e., biostratigraphy) by which the boundary is recognized (i.e., correlated) away from the boundary stratotype is subject to varying degrees of asynchrony. Thus there is a need for a clear understanding of the differences between definition and correlation in discussing the denotation of chronostratigraphic units.

The Eocene/Oligocene boundary was recently stratotypified at the 19 m level in the Massignano section of the northern Apennines. It is denoted by the last occurence (LO) of the planktonic foraminiferal genus *Hantkenina* which occurs at a stratigraphic level in Magnetozone C13R, estimated to lie at a level 0.14 of the interpolated distance between the base of C13N and the top of C15N, with an estimated numerical age of 34.0 Ma (Berggren et al., this volume).

Why was this particular level chosen and is it consistent with historical definition and intent? A detailed analysis of this question is beyond the scope of this review but a brief examination of the background to the choice is pertinent to the discussion presented in some of the chapters in this volume (Brinkhuis; McGowran et al.). The passage of geologic time is recorded in the material (physical) rock record. However, the standard chronostratigraphic framework of the geologic record is based upon the evolution of biota in marine stratigraphic sections suitable for regional correlation. The standard subdivisions of the Cenozoic, for instance, were originally based on the stage of evolution of marine molluscs, but precise boundaries between these subdivisions were not specified. Modern usage dictates that the standard subdivisions of the geologic record be based upon marine successions with boundaries anchored at lithostratigraphic levels denoted by distinct biotic events capable of correlation away from the boundary stratotype section.

The Oligocene was created for a series of rocks temporally deposited between typical Lyellian Eocene and Miocene but it included rocks originally included in the Eocene by Lyell (1833). Lyell (1857) subsequently moved his Eocene/Miocene boundary to a level equivalent to Beyrich's Eocene/Oligocene boundary (see Berggren, 1971). Denoting the position of an Eocene/Oligocene boundary in Europe is difficult, due to the fact that a precise separation between the Eocene, Oligocene and Miocene series (in a modern chronostratigraphic sense) was not provided by the original nineteenth-century geologists, and to the nature of the upper Eocene-lower Oligocene depositional facies (shallow water marine-brackish carbonates in England; evaporitic and brackish water carbonates in the Paris Basin; within a sequence of essentially unfossiliferous sands in the Belgian Basin). In northwestern Europe, controversy surrounds usage of the Lattorfian Stage, which has been variously ascribed to the Eocene and Oligocene, despite its indequate representation in outcrop. The terms Priabonian (strato-typified in the Tethyan-Mediterranean region) and Rupelian (stratotypified in the Belgian Basin of northwestern Europe) have come to serve as the standard stages for the upper Eocene and lower Oligocene, respectively. Biotic elements useful for global correlation are notably absent or only sporadically present in the unconformity-ridden sections spanning the Eocene/Oligocene boundary of Europe precluding accurate global correlations. However, it has been possible to narrow down the biostratigraphic span within which a boundary stratotype level should be drawn by means of integrated studies involving various microfossil groups and cross-correlation with the standard magnetic polarity scale.

The delineation of an Eocene/Oligocene boundary has fallen primarily to micropaleontologists working with (predominantly) deep sea sediments, where the terminal part of the ranges of several taxa of planktonic foraminifera (*Turborotalia cerroazulensis, Globigerinatheka, Hantkenina*) has come to serve to distinguish Eocene from Oligocene. It was the LO of *Hantkenina* that was chosen as the definitive biostratigraphic criterion for characterization and correlation (not definition) of the boundary by IGCP Project 174 at the Massignano section in the Apennines.

Brinkhuis (this volume) observes that the FO of the dinoflagellate taxon *Glaphryocysta semitecta* (associated with the LO of hantkeninids in the Apennines) correlates with the base of the *Asterocyclina* Beds in the type Priabonian section, 15 m below the top of the (recently modified) type section (the Bryozoan Limestone unit). Accordingly he notes that the global LO of hantkeninids would correlate stratigraphically with the mid-part of the type Priabonian and questions whether the choice of the LO of *Hantkenina* was an appropriate criterion for Eocene/Oligocene boundary denotation. At the same time he records that the LO of the dinoflagellate taxon *Areosphaeridium diktyoplokum* coincides with the top of the type Priabonian section, with Chron C13N and with the T4.3/4.4 cycle boundary in the global sea level curve of Haq et al. (1988) and suggests that this level may be appropriately considered for the definitive Eocene/Oligocene boundary. However, in keeping with the precept that "base defines unit" and that the boundaries of higher level units in the stratigraphic hierarchy automatically encompass the boundaries in the next higher level (on the principle of coterminous boundaries), it will be seen that stratotypification of the Eocene/Oligocene boundary in the Massignano section automatically delimits (by decapitation) the upper level of the Priabonian Stage and the lower limit of the Rupelian Stage (by temporal extension) so that they become coterminous. The major $\delta^{18}O$ shift associated with a normal polarity event (C13N) is ~0.5 m.y. younger than the Eocene/Oligocene boundary itself, and both these events can serve as useful means of achieving *approximate* correlation with the boundary itself on a global scale (cf. McGowran, this volume, who has chosen these criteria for denoting the position of the Eocene/Oligocene boundary in South Australia). Realism, and a respect for the history of debate on chronostratigraphic boundaries would suggest, however, that we have not heard the last on the subject of the Eocene/Oligocene boundary.

Age estimates for the Eocene/Oligocene boundary have varied from 32-38 Ma over the past three decades and indeed extend to as old as 48 Ma if we wish to extend our collective memory back to the mid-1930s (Harland et al., 1990, p. 7, Figure 1.5). In the past decade, controversies have centered upon reliance on "low-temperature" glauconites versus whole rock K-Ar dates on "high-temperature" minerals (sanidines, biotites) and various problems of bio- and chronostratigraphic characterization of the boundary level itself on a regional and global scale (Berggren et al., this volume).

The basis for the age estimate of 36+ Ma for the Eocene/Oligocene boundary (Berggren et al., 1985) as opposed to somewhat younger estimates around 34 Ma which were being made by other investigators at the time, is reviewed in the papers by Berggren et al. (this volume) and Prothero and Swisher (this volume). The culprit seems to have been reliance at the time on earlier (30 year old) K-Ar dates on early Oligocene and early middle Eocene volcanics in terrestrial mid-continent stratigraphic sections, in which paleomagnetic stratigraphy was subsequently measured and correlated (incorrectly as it turns out) with the standard global polarity time scale. The advent of the single crystal laser fusion (SCLF) $^{40}Ar/^{39}Ar$ dating technique has led to a significant improvement in radioisotopic dating precision (and presumably accuracy; but the reader's attention is directed to the discussion in Berggren et al., this volume, on potential problems with disparities in radioisotopic dates owing to analytic disparities on interlab standards) and new data reviewed in these two papers are consistent with, and supportive of, an Eocene/ Oligocene boundary age estimate of 34 Ma at the boundary stratotype at Massignano. At the same time these dates suggest that any eventual revision to the Paleogene time scale will involve an approximately 2 m.y. younging for most of the mid-Paleocene to mid-Oligocene bio- and chronostratigraphic boundaries as well as magnetochrons.

The new $^{40}Ar/^{39}Ar$ dates have an even more profound effect on the terrestrial geochronology. In North America, the calibration of the land mammal "ages" has been shifted by more than 2 m.y. and two whole polarity chrons (Swisher and Prothero, 1990; Prothero and Swisher, this volume). As a consequence, North American terrestrial "late Eocene" is now middle Eocene, "early Oligocene" is now late Eocene, and "middle Oligocene" is now early Oligocene. The effects of this changing time scale on our conception of North American biotic change are outlined below.

THE RECORD OF GLOBAL CLIMATE CHANGE

The Paleogene may be viewed in a general sense as a time of transition during which the earth's climate gradually changed from an essentially uniformly warm Late Cretaceous to the cooler, more heterogeneous Neogene. Indeed, the Eocene has been recently characterized -- in more euphemistic terms -- the "doubthouse" period, intermediate between the late Mesozoic "hothouse" and the Neogene "icehouse" (Watkins and Mountain, eds., 1990). Yet in more specific terms, it has a more complicated "microclimate" of its own, with an anomalously warm global climate optimum spanning some 4-5 m.y. during the early Eocene. It is followed by a gradual, stepwise cooling during the remainder of the Paleogene, reflected in the biogeographic distribution of marine microplankton and in the stable isotope record as discussed in several papers in this volume. The major question facing paleoceanographers remains the timing of the initiation and subsequent history of ice on the Antarctic continent. While not solved at the time of this writing, several papers in this volume treat the problem in a forthright and comprehensive manner.

An understanding of the evolution of Paleogene climate and, in particular, the role of Antarctic glaciation in the evolution of global atmospheric and oceanic circulation, requires the integration of a global data set from a large variety of sources (i.al., empirical observations in the geologic record, faunal and floral distribution patterns of marine and terrestrial biota, stable isotope records, aeolian distribution patterns) and feedback obtained from climate models (Sloan and Barron, this volume). A comprehensive review of this story is beyond the scope of this summary, but a brief overview is presented in the context of the papers presented here, in an attempt to provide temporal constraints on the initiation of Antarctic glaciation.

Glaciation on the Antarctic continent may have developed by the evaporation and precipitation of warm, salty waters transported southwards from the equatorial Pacific flowing within the gradually (accelerating during the middle Eocene at about Chron C19 time) widening seaway formed from the separation of Australia and Antarctica (Bartek et al., this volume). The timing of glacial inception remains elusive, however. Drill-hole and seismic stratigraphic data from Antarctica indicate a glacial history (in the form of an ice-sheet grounding event in the Ross Sea concomitant with a shift from temperate/temperate glacial to polar glacial conditions) extending back at least to the late Oligocene (Bartek et al., this volume). The recent discovery by Birkenmajer and colleagues of glacial deposits ("Krakow" glaciation) from King Georges Island, West Shetlands, and of diamictites (of probable glacial origin) on the continental margin (Prydz Bay) of Antarctica by ODP Leg 119 point to an earlier inception of ice growth on the Antarctic continent. The former record dated at ~49 Ma (early middle Eocene) can probably be ascribed to mountain glaciation, but is seen to follow closely the major input of volcanic derived silica into the oceanic circulation system and the bipolar laterization associated with the global climatic optimum of the early Eocene. It also coincides conveniently with the early/middle Eocene major cooling indicated by the $\delta^{18}O$ record, steepening of temperature gradients, intensification of upwelling, and increasing fertility and productivity as reflected in the major increase in biosiliceous deposition in the ocean, the so called "silica burp" of McGowran (1989). The age of the diamictites remains equivocal. They have been interpreted by some as late Eocene in age but, as Miller (this volume) points out, the magnetobiochronology is weakly constrained by fossil evidence, and available data require only that they are stratigraphically lower, but not significantly older than, well-dated lower Oligocene strata above. The recent discovery of early Oligocene (Chron C13N) ice-rafted detritus on the East Antarctic margin of the Weddell Sea (ODP Leg 113) and on southern (ODP Leg 119) and central (ODP Leg 120) Kerguelen Plateau, associated with a strong positive (~ 1 ^{o}/oo) shift in δ^{18}O provides the oldest, unequivocal evidence of a continental ice sheet on Antarctica.

The Eocene-early Oligocene oxygen isotope record is generally interpreted as indicative of mid-high latitude cooling beginning in the early middle Eocene (Chrons C22N-C21N) with step-like increases of ~1°/oo at the early/middle Eocene (Chrons C22N-C21N), middle/late Eocene (Chrons C18N-C17N) and in the earliest Oligocene (Chron C13N), resulting in the establishment and evolution of a major temperature gradient between low and high latitudes and in bottom water temperature lowering from ~12°-13°C in the early Eocene to ~8°-9°C in the late Eocene and ~3°-4°C in the earliest Oligocene. Miller (this volume) and Thomas (this volume) review the difficulties inherent in reading the oxygen isotope record strictly as a temperature signal. They note the need for an ice volume correction factor, and conclude that while the intermittent presence of early Oligocene ice sheet(s) on Antarctica appears incontestable, the nature and extent of pre-Oligocene ice build-up on the Antarctic continent remains elusive. As they point out in their respective papers, reading Paleogene climate history from the stratigraphic record is complicated by such problems as:

1. How are global ice volume versus temperature and local salinity variations separated in the δ^{18}O record of foraminifera?

2. Were equatorial surface temperatures intermittently cooler (~18°C) during the Paleogene as suggested by some interpretations of the stable isotope record, or were they essentially stable and comparable to present day values (28°C), as suggested by other interpretations of the same stable isotope record and by the biotic record (Adams et al., 1990).

3. What was the mode of formation of deep water? Was ocean circulation primarily thermohaline and driven by deep waters formed at high latitudes, similar to today, or was it primarily thermospheric, driven by density differences caused by the formation at low latitudes of warm, saline bottom waters (WSBW)? And if so, did the reversal in circulation during the mid-Paleogene and the concomitant increase in δ^{18}O values in benthic foraminifera reflect the (mid/late Eocene) initiation of the psychrosphere, the deep, cold oceanic realm traditionally defined as having stable temperatures below 10°C?

These questions are not trivial, as Thomas (this volume) observes, because it may be necessary for those attempting to model the circulation and climate dynamics of the Paleogene ocean (Sloan and Barron, this volume) to consider the WSBW scenario to account for the efficient transport of heat to northern hemisphere high latitudes required by the subtropical biotic evidence, such as durophagous alligators, anguid and varanid lizards, dermopterans ("flying lemurs"), and paromomyid primates of the late early Eocene of Ellesmere Island in the Canadian Archipelago.

What changes might be expected in continental climate in terms of surface temperature and precipitation in response to the changes in sea-surface temperatures and continental topography that are known to have occurred during the Paleogene? Operating within the framework of increased latitudinal thermal gradients resulting from high-latitude surface water cooling (on the order of 7-10°C during the Paleogene) and increased continental tectonic uplift (Laramide, Andean and precursor Himalayan orogenies), Sloan and Barron (this volume) conducted a series of climate modeling sensitivity studies. Results revealed that: 1) increasing elevations resulted in lower surface temperatures nearby as well as at some distance from the elevated mountains; 2) changes in sea-surface temperature associated with heightened latitudinal surface water gradients had a more pronounded effect on simulated continental climate. For instance, high-latitude continental surface temperatures underwent large decreases in winter temperatures and only moderate decreases in summer, as well as mean annual, temperatures. Zonal precipitation decreased at the expense of increased precipitation in low latitudes as well. As the authors observe, the deterioration in higher-latitude climate (in the form of lower summer temperatures) and increased latitudinal surface water gradients mirrors conditions for the establishment and maintenance of ice at high latitudes derived from other recent studies (namely cool summers capable of retaining winter snow/ice accumulation). This suggests, in turn, that the initiation and continued growth of ice on Antarctica may have been the result of a gradual global cooling, coupled with uplift of continental regions far removed from Antarctica itself, and of heightened surface thermal gradients associated with the gradual, but inexorable, isolation and insulation of Antarctica from warmer low latitude surface waters by tectonic opening of passages to circum-Antarctic surface- and deep-water flow. Judging from the data presented and reviewed in several papers in this volume, this isolation may have been well under way, if not completed, by late Eocene-early Oligocene. While empirical observations of the geologic record and the modeling experiments decribed here appear to be supportive of the general scenario sketched above, the answer to the enigma of the early ice history of Antarctica may reside in the stable isotope record which as yet has not provided a clear and unequivocal solution to the Gordian Knot of Paleogene climate.

Kennett and Stott (1990), using stable isotope records from the Antarctic (Maud Rise), have argued for an early Paleogene two-layered ocean, *Proteus,* consisting of warm, saline deep waters formed at mid-low (Tethyan) latitudes overlain by cooler, intermediate waters of primarily mid-latitude origin. The late Paleogene ocean, *Proto-Oceanus,* was hypothesized to have been three-layered with cold, dense deep waters formed in Antarctic regions (proto-AABW), a warm, saline intermediate layer of low latitude origin, and surficial cool waters formed in the polar regions. The Neogene and modern ocean is characterized by a thermohaline circulation driven by deep waters formed at high latitudes.

Miller (this volume) observes that the source(s) of deep water during the middle and late Eocene is/are not apparent. Oceanic circulation was probably sluggish, according to the seismic stratigraphic evidence, and the regional distribution of unconformities, whereas the same evidence indicates vigorous circulation by earliest Oligocene time. Carbon and oxygen data suggest (at least) a brief pulse of Northern Component Water (NCW), together with Southern Component Water (SCW), i.e., a bipolar source for the formation of deep waters in the earliest Oligocene associated with the $1^{o}/oo$ increase in $\delta^{18}O$, implying a direct link between deep water and climate change. Using time series analysis of oxygen isotope records from the Maud Rise, he suggests that some of the deep waters in the region were formed in southern high latitudes , i.e., from an Antarctic "cold spigot," rather than from low latitude WSBW, while Thomas (this volume) argues against the *Proteus/Proto-Oceanus* model using the deep water benthic foraminiferal record from the Maud Rise. Her argument runs as follows: faunal turnover among taxa is stepwise and gradual and predates the change in $\delta^{18}O$. This turnover should have been (relatively) rapid in response to the (relatively) rapid (a few thousand years) change in deep-water circulation. The lack of correspondence in timing between faunal and isotopic records argues that the $\delta^{18}O$ increase across the Eocene/Oligocene boundary was due to ice volume, rather than simply to temperature. The interpretation of Thomas (this volume) is seen to be consistent with the observations of Kaminski (1987) of a sequential south (equatorial) to north decrease in diversity and extinction in the North Atlantic among deep water, low oxygen-adapted agglutinated (flysch type) benthic foraminifera. Diversity reduction and extinctions began in the abyssal realm at low latitudes in the late Eocene, proceeding to mid-latitudes and the bathyal realm in the earliest Oligocene while some forms survived into the Neogene in "refuges" in the Norwegian-Greenland Sea and North Sea. The reduction of these distinct forms was linked with the transition from a warm sluggishly circulating, low-oxygen deep-sea environment to a more oxygenated, thermohaline-driven circulation caused by bipolar cooling. In the next section we shall examine the response among other elements of the marine biota to the climatic/oceanographic changes that were taking place in the mid- to late Paleogene.

Finally, we may note that while several bolide impact events have been recorded in the late Eocene, they do not appear to coincide with extinction event(s) in the marine record (Miller et al., 1991), lending support to the claim by marine micropaleontologists and paleoceanographers that the late Eocene-early Oligocene biotic turnover should be ascribed to endogenic (long-term changes in paleocirculation and associated environmental responses) rather than to exogenic (extraterresrial) causes.

MARINE BIOTIC RESPONSE

In this section, we examine the response of marine and terrestrial biotas to the changing climatic conditions of the mid- to late Paleogene. In order to present a more comprehensive picture across the evolutionary spectrum, we include brief reviews of other groups not discussed in this volume.

The responses of the marine phytoplankton (calcareous nannoplankton, dinoflagellates and diatoms), zooplankton (planktonic foraminifera and radiolarians), and benthic microfaunas (deep-water and larger benthic foraminifera, ostracodes) to late Paleogene climatic deterioration was similar in all groups. It should be stressed that while little happened at the Eocene/Oligocene boundary, a major turnover is seen in all groups near the middle/late Eocene boundary. As pointed out by Boersma et al. (1987), Hallock et al. (1991) and Aubry (this volume), progressive intensification of cooling at high latitudes was the forcing agent of evolution in the planktonic and larger foraminifera and in the calcareous nannoplankton, and this can be generalized to all biotas.

While there is an agreement between authors in this volume concerning parallelism in biotic turnovers throughout the middle Eocene to early Oligocene as a result of decreasing temperatures, there is a strong discrepancy in the interpretation of the early Oligocene low diversity in calcareous groups. While Aubry (this volume) emphasizes that the early Oligocene extinctions may reflect a further step in the eutrophication of the euphotic zone rather than a sharp drop in temperature, Keller et al. (this volume) relate the lack of speciation in the planktonic foraminifera in the latest Eocene and early Oligocene to low surface-water productivity. In fact, while Boersma et al. (1987) showed that the middle late Eocene cooling was accompanied by eutrophication of the photic zone, Keller et al. (this volume) interpret a gradual decrease of the middle Eocene to Oligocene vertical $\delta^{13}C$ gradients between planktonic and benthic foraminifera as reflecting a progressive decrease in primary productivity. Contrary to Keller and colleagues, Thomas (this volume) observes an increase in $\delta^{13}C$ surface to deep water gradients in the middle Eocene, and suggests that an increase in surface productivity at this time correlates with the appearance of radiolarians and change in illite/smectite ratio in the southern ocean. Baldauf (this volume) points to the expansion of biosiliceous sediments in the southern high latitude in the earliest Oligocene, and to early Oligocene increased surface productivity. Early Oligocene low diversity in the planktonic and larger benthic foraminifera (Hallock et al., 1991) and in the calcareous nannoplankton (Aubry, this volume) is interpreted as reflecting a nutrient-rich euphotic zone, with loss of oligotrophic environments and maximum contraction of the trophic resource continuum (TRC). The $\delta^{13}C$ appears to have fluctuated rather largely during the middle Eocene, as indicated by the records at ODP Sites 689 and 690 (Kennett and Stott, 1990). These authors indicate that the $\delta^{13}C$ values increased by 0.5‰ to 1.5‰ across the Eocene/Oligocene boundary (suggesting high productivity), and decreased to 0.25‰ at 32 Ma (indicating low productivity; for comparison, a $\delta^{13}C$ peak of 3‰ occurs at the Paleocene/Eocene boundary). The differences in interpretation between various authors probably stems from the difference between productivity and fertility. Eutrophic waters are highly fertile, and high early Oligocene fertility is indicated by the expansion of biosiliceous facies at high southern latitudes. However, high fertility does not necessarily imply high productivity (Boersma and Premoli-Silva, pers. comm., and work in progress).

Cooling of the warm stable high latitudes in the early middle Eocene and subsequent ice build-up on Antarctica obviously involved a profound restructuring of oceanic circulation (see Miller, this volume), and consequently of biotic environments, and much of this still remains to be understood.

Calcareous nannofossils

The evolution of the calcareous nannoplankton in response to the late Paleogene climatic deterioration is discussed by Aubry (this volume). She shows that decrease in global diversity and increase in diachrony and provincialism result from progressive eutrophication and cooling from early middle Eocene to early Oligocene. The earliest evidence of important climatic change following the warm equitable early Eocene (Rea et al., 1990) is the reduction in regional diversity in the earliest middle Eocene in the southern high latitudes, with the loss of (inferred) K-mode selected taxa (e.g., discoasters) characteristic of oligotrophic tropical waters. This event which correlates with the ~52 Ma $\delta^{18}O$ event (see discussion above and Miller, this volume) was echoed by subsequent successions of evolutionary turnovers during the middle Eocene, leading eventually to a profound turnover near the middle/late Eocene boundary. As new lineages evolving in response to environmental changes co-existed with lineages which originated in the early Eocene, maximum diversity occurred in the middle Eocene (Zone NP15) ecotonal community. While there is no particular calcareous nannoplankton event associated with the Eocene/Oligocene boundary, a well marked event occurs in the latest Eocene when the two rosette-shaped discoasters (*D. barbadiensis* and *D. saipanensis*) became extinct. This event affected only low and middle latitudes, where the discoasters may constitute up to 30% of the assemblages. The most important event regarding the late Paleogene evolution of the calcareous nannoplankton occurred in the early Oligocene, mainly within Zone NP22. At this time, all but one of the long-ranging species which had evolved in the early Eocene, several of the long-ranging taxa which evolved in the middle Eocene in response to climatic deterioration, and all the short-ranging taxa which evolved in the late Eocene became extinct, bringing to 70% the reduction in diversity since the middle Eocene. Evolution of the Cenozoic calcareous nannoplankton has been essentially seen as temperature-controlled, but Aubry argues that the global early Eocene extinctions cannot be solely explained by a further drop in temperature. Rather, she proposes that they mainly result from a further step towards eutrophication as a result of increased mixing following the early Oligocene ice buildup. Early Oligocene eutrophication of the euphotic zone is indicated by the lithologic change in southern high latitudes from predominantly upper Eocene nannofossil oozes to lower Oligocene mixed diatom and nannofossil oozes.

Planktonic foraminifera

As pointed out by Aubry (this volume), evolutionary patterns similar to those described in the calcareous nannoplankton occur in the planktonic foraminifera (Boersma et al., 1987; Hallock et al., 1991). Boersma et al. (1987) described in great detail the changes in diversity and biogeography that the planktonic foraminifera exhibit in the Atlantic Ocean during the Eocene. They pointed to: 1) a biogeographic reorganization starting at the early/middle Eocene boundary and signaling the termination of early Eocene essentially stable conditions, and continuing through the middle Eocene until Biochron P11; 2) a peak in diversity in Zone P11 resulting from the co-occurrence of older early Eocene lineages with new lineages evolving in response to environmental changes; 3) a major (the largest of the Paleogene) turnover near the middle/late Eocene boundary (with "15 originations, 18 extinctions and 33 planktonic foraminiferal biotic overturn events" by the end of Biochron P14). Extinctions during Biochrons P13 and P14 affected "most of the older Eocene tropical and equatorial surface groups"; and 4) further extinction events in the late Eocene. Combining paleontologic evidence with stable isotopic records, Boersma et al. (1987) retraced the evolution of near surface and deep waters cooling, the correlative reduction/restriction of tropical surface water faunas and expansion of the temperate water taxa, and discussed the evolution of the Atlantic meridional and vertical thermal gradients. The Paleogene evolutionary and biogeographic history of the planktonic foraminifera has recently been interpreted in terms of changes in nutrient flux to surface waters by Hallock et al. (1991). Early Eocene K-strategists indicative of oligotrophic conditions were replaced by less specialized taxa characteristic of deeper and cooler waters as high latitude cooling intensified from middle Eocene to early Oligocene, causing increased rate of ocean overturn and nutrient return to surface resulting in early Oligocene eutrophy.

The pattern of middle Eocene-Oligocene faunal turnover in planktonic foraminifera is also linked with temperature and "productivity-nutrient supply" by Keller, MacLeod and Barrera (this volume), based on a comparative study of several Antarctic and low-latitude Indo-Pacific Deep Sea Drilling sites. The faunal turnover, with gradual extinction of over 80% of the fauna during an approximately 14 m.y. interval, involved predominantly warm-water surface dwellers which were replaced by cooler-water subsurface forms. As we have seen to be typical of other elements of the marine biota, two brief intervals of accelerated turnover occurred: at the middle/late Eocene and early/late Oligocene boundaries with no marked events associated with the Eocene/Oligocene boundary. Stressing the need for a species-level taxonomic database, the authors demonstrate that the faunal turnover across the Eocene-Oligocene transition is driven not so much by the accelerated rate of extinction in the late Eocene as by the lack of origination of new taxa in the dynamically evolving oligotrophic (but see above) environment of the early Oligocene. Major faunal turnovers are shown to be recognizable in depth stratified assemblage groups, the middle/late Eocene one involving the decline/extinction of warm, surface water forms (morozovellids, acarininids, among others) and a concomitant increase in cooler, surface- and intermediate-water forms. It is these latter forms (*"Turborotalia" ampliapertura*, *Pseudohastigerina* spp., among others) which survived into the Oligocene only to be eliminated by further reduction in depth-dependent habitats and reduced nutrient supply. The middle/late Eocene boundary turnover is linked with onset of major Antarctic glaciation (but see cautionary evaluation of the evidence by Miller, this volume) with a parallel decline in surface and intermediate water temperatures at high and low latitudes (cf. Matthews and Poore, 1980; Prentice and Matthews, 1988 who have interpreted stable isotope records to indicate low latitude temperatures to have remained stable at ~28°C throughout the Cenozoic). Keller and colleagues interpret the gradual decrease in $\delta^{13}C$ gradients between surface and intermediate dwelling planktonic and benthic forms in southern ocean profiles between middle Eocene to Oligocene to reflect a gradual decrease in primary marine productivity; they indicate that surface-water productivity remained low during the Oligocene, comparable to earlier findings at tropical North Pacific sites. This interpretation would appear to be at variance with the evidence cited by Aubry (this volume) and Baldauf (this volume) for high nutrient availability in the Antarctic region during the early Oligocene, as indicated by high biosiliceous productivity and the initiation of diatomaceous/nannofossil oozes, and with Hallock et al. (1991), who show that decrease in diversity during the middle Eocene to early Oligocene is linked to eutrophication and loss of oligotrophic habitats.

Benthic foraminifera

Lower bathyal benthic foraminifera from the Antarctic region (Maud Rise, Weddell Sea) exhibit a sequential, stepwise extinction pattern during the mid-Eocene to Oligocene, with accelerated turnovers in the middle Eocene (~Chrons C20-19), late Eocene (Chrons C16-15) and early Oligocene (Chrons C12-11) (Thomas, this volume). Each of the turnovers has a characteristic signature: 1) the first turnover is denoted by a decrease in diversity and increase in relative abundance of (inferred) epifaunal taxa (particularly uniserial lagenids and lenticulinids), the loss of heavily calcified buliminids and (the somewhat premature, relative to lower latitudes) last common occurrence of *Nuttalides truempyi;* 2) the second event is characterized by the loss of all buliminid taxa and the first appearance of *Turrilina alsatica;* and 3) the third event is characterized by a lowering of faunal diversity, a marked decrease in (inferred) epifaunal taxa, such as *T. alsatica,* and a corresponding increase in (inferred) epifaunal taxa and domination of deep water assemblages by *Nuttalides umbonifera* (which constitutes up to 70% of some of the younger early Oligocene assemblages). This is interpreted as a reponse to gradual cooling and increased corrosiveness of deep waters. Thomas observes that the gradual nature of the faunal turnover suggests that the formation of the psychrosphere was also gradual, reflecting the slow, sequential cooling of surface waters at high latitudes during the middle Eocene and Oligocene (see discussion above and ostracode evidence of Benson below). Finally, Thomas postulates a link between a possible increase in surface water productivity in the middle Eocene (based on the $\delta^{13}C$ on bulk carbonate, reflecting dissolved surface water carbonate values on planktonic organisms and not seen at the same magnitude in benthic organisms, suggesting an increased surface to deep water gradient at the time) and the initial appearance, albeit in low abundances, of radiolarians, a change in illite/smectite ratios indicating reduced hydrolysis during weathering and reflecting climatic cooling and/or aridity on Antarctica, and the first of the turnovers in the deep water benthic fauna cited above at about Chron C20 (mid-Eocene). In actual fact, if we are to get at the root of the problem of late Eocene-early Oligocene faunal response to climate and paleoceanographic changes, we shall probably have to begin with an investigation of the change in deep-water faunal communities which took place between the climatic optimum of the early Eocene (Chrons C24-C22) and the initial, cooling step of the early middle Eocene (Chrons C22-C21).

In a companion paper in this volume, McGowran and colleagues examine the effects of late Eocene-early Oligocene climate and sea-level changes upon neritic benthic foraminifera in two parallel stratigraphic sections from southern Australia. They note that first ocurrences (FOs) are associated with the marine transgressions while last occurrences (LOs) are associated with the Chinaman Gully regression, and that faunal similarity between the two sections was most marked during the earliest Oligocene transgression. The Chinaman Gully regression is correlated with the "Eocene/Oligocene boundary" type 1 sequence boundary TA4.3/TA4.4 (Haq et al., 1988) and the major positive $\delta^{18}O$ shift at the Chron C13N/R boundary. The changes in the neritic fauna are viewed as a "microcosm of global oceanic changes including the breakdown of a tendency toward water stratification at the onset of the psychrosphere". While recognizing that Paleogene climatic and oceanographic history is one of evolutionary change on a long-term scale, McGowran and colleagues view the Eocene/Oligocene boundary as a short, distinct, bipolar transenvironmental Terminal Eocene Event(s), superimposed upon a long-term trend toward decreasing global temperature.

An analysis of Gulf Coast Paleogene shallow water benthic foraminiferal faunas has shown increased extinction rates associated with the Lisbon-Cook Mountain/Jackson Group (middle/upper Eocene) boundary and with the Jackson/Vicksburg (Eocene/Oligocene) boundary (Gaskell, 1991). The former boundary is coincident with a major type 1 sequence boundary (TE2.3/TE3.1 in Haq et al., 1987=TA3.6/TA4.1 in Haq et al., 1988), the latter is situated within a condensed section representing a type 2 depositional sequence (TE.3.3 in Haq et al., 1987=TA4.3 in Haq et al., 1988); the contact may, in fact, be disconformable (Mancini and Tew, 1991). Gaskell (1991) makes the cogent observation that while faunal extinction associated with type 1 sequence boundaries may be due to habitat eradication by rapidly falling sea level, the same phenomenon associated with a relatively moderate amount and slow rate of sea level fall may be associated with climatic changes.

Larger benthic foraminifera thrived in the warm Eocene tropics. Massive shallow-water carbonates composed essentially of tests of the genus *Nummulites* occurred in a belt stretching from Indonesia to Gibraltar and delineating the former extent of the Tethyan seaway. They achieved their maximum latitudinal dispersion (New Zealand to Alaska) in the mid-late Eocene, attesting to relatively uniformly warm climates over a large part of the earth during this time. During the Oligocene taxonomic diversity and biogeographic distribution were reduced, reflecting the loss of warm, oligotrophic habitats associated with climatic cooling and more vigorous circulation and concomitant global sea-level changes near the Eocene/Oligocene boundary. A precise understanding of the timing of changes in the larger foraminiferal fauna is hampered by lack of planktonic biostratigraphic calibration, and by stratigraphic unconformities and disconformities, probably owing to sea level fluctuations, with the result that continuous sections across the boundary in the Indo-Pacific, European and Caribbean regions are (apparently) rare (Adams et al., 1986). Faunal turnover was primarily at the species level, but within the biostratigraphic resolution currently achievable, the extinction of the genera *Asterocyclina, Discocyclina, Fabiania, Pellatispira,* several taxa of reticulate *Nummulites* (*fabianii/retiatus* group), *Spiroclypeus vermicularis,* among others (Indo-Pacific letter--zonal boundary Tb/Tc) is viewed as relatively abrupt in the Indo-Pacific-Tethyan region, and has been associated with a level closely approximating the Eocene/Oligocene boundary as generally recognized. Early Oligocene faunas are characterized by the presence of striate *Nummulites* spp. (*fichteli* and/or *vascus*), and the appearance of *Cycloclypeus, Austrotrillina, Praerhapydionina, Fallotella* and (Indo-Pacific region) *Lepidocyclina* (*Eulepidina*) and *L.* (*Nephrolepidina*), among others. A faunal break of similar magnitude occurred in the Americas (Caribbean region in particular) but involved different taxa.

Deep water flysch-like agglutinated benthic foraminiferal faunas exhibit a gradual restriction to high latitudes during the Eocene (Kaminski, 1988). Mid-Eocene North Atlantic assemblages vary in diversity from about 25 taxa in the Norwegian-Greenland Sea to about 50 near Greenland (ODP Site 647). In the North Sea and Labrador Margin a series of LOs reflect basinal shallowing as sediment supply exceeded tectonic subsidence. *Spiroplectammina* and *Glomospira*, which disappeared on the Labrador Margin during the early Eocene, persisted until the Eocene/Oligocene boundary in the North Atlantic (DSDP Site 647) and the Norwegian-Greenland Sea. A sequential, diachronous south-to-north disappearance of *Spiroplectammina* occurred in the North Sea during the latest Eocene-earliest Oligocene. Flysch-type assemblages essentially disappeared at the Eocene/Oligocene boundary in Tethyan and North Atlantic basins, but the extinction pattern is one of gradual elimination of taxa. This suggested to Kaminski (1988) a causal relationship, with the production of well oxygenated waters of a southerly source, consequent deepening of the oceanic lysocline in much of the world ocean, and a sharp reduction in oceanic productivity. The Greenland-Scotland Ridge was said to have acted as a barrier to penetration of deep Atlantic waters into the Norwegian-Greenland Sea (and, by extension, the North Sea Basin) which served as a refuge for the persistence into the Neogene of a low-diversity flysch-type fauna.

Diatoms

Compiling an essentially global database on 120 selected marine diatoms from four geographic regions, Baldauf (this volume) notes four major floral turnover events from the middle Eocene through early Miocene. Two of them (late middle Eocene and latest Eocene-earliest Oligocene) are relevant in the context of this review. The first (middle Eocene) event is interpreted as a transition from typical middle Eocene to a late middle to late Eocene flora, said to represent a minor floral response to oceanographic changes and may reflect increased biogeographic provincialization associated with the increasing thermal isolation of Antarctica and decoupling of surface to bottom, and heightened, latitudinal thermal gradients. The second event spans the Eocene/Oligocene boundary and is characterized by a 45% floral turnover, with approximately the same number of incoming and departing taxa. The turnover occurred in all geographic regions and clear distinctions are seen between low, mid- and high latitude assemblages. In the low latitudes this event had previously been delineated as an increase in relative abundance from about 20% (latest Eocene) to 50-80% (earliest Oligocene) of species of the genus *Cestodiscus*. Similar faunal change is seen to occur in Labrador Sea and (with less stratigraphic resolution owing to a hiatus) Norwegian-Greenland Sea floras and is interpreted as a direct response to global climate cooling and reorganization of oceanic circulation. In the Southern Ocean, this event is said to correspond to, and reflect, the expansion of biosiliceous sedimentation into high southern latitudes and the development (if not initiation) of the Antarctic cryosphere. Baldauf concludes that this results from a decrease in vertical stratification of the water column and increased surface water productivity (cf. Keller et al., this volume, who interpret the carbon and oxygen isotope gradients in planktonic and benthic foraminifera to indicate decreasing productivity and thermal stratification during the late Eocene to early Oligocene).

Dinoflagellates

Marine dinoflagellates are treated by Brinkhuis (this volume). He notes that a detailed, nine-fold (local) zonal biostratigraphy has been developed for the Mediterranean region uppermost Eocene-lower Oligocene and calibrated to the magnetobiostratigraphic record of the Appenines. The invasion of the Mediterranean region by several low-latitude dinoflagellate taxa and a concomitant decrease in several taxa characteristic of low latitudes, is seen to reflect a lowering of surface water temperatures and salinities (SST) across the Eocene/Oligocene boundary. Two such events were noted, one in late Chron C13R and a second, more pronounced and somewhat longer, associated with Chron C13N1 and with the major positive peak in $\delta^{18}O$ (discussed above). Quantitatively based floral distribution patterns in the type (and reference) Priabonian sections are shown to be correlative with relative sea-level cycles TA4.1 through TA4.4 (Haq et al., 1988) and cooler floral intervals shown to be correlative to sea-level falls suggesting glacio-eustasy as a/the causal agency. In the early Oligocene, many dinoflagellate cyst taxa are reported to disappear. Diversity is particularly reduced among marginal marine forms, which suggests to Brinkhuis that marginal marine habitats may have suffered disruption during these relatively rapid sea level falls, particularly at the type 1 event associated with Chron C13N in the earliest Oligocene.

Radiolaria

Radiolaria appear to have been little affected by the late Eocene-early Oligocene events described here, although several biostratigraphic events have been recorded by Riedel and Sanfilippo (1986). Approximately three last occurrences (LO's; *Dictyoprora armadillo*, *Lophocyrtis jacchia*, *Lychnocanoma amphitrite*) and an evolutionary transition from *Lithocyclas arisoteles* to *L. angusta* coincide with the LO of rosette-shaped discoasters, which occurred >0.5 m.y. of the Eocene/Oligocene boundary (as currently delineated). Approximately 0.5 m.y. earlier, the LO of about 5 additional taxa occurred (*Calocyclas turris*, *Thyrsocyrtis bromia*, *T. rhizodon*, *Carpocanistrum azyx*, *Thyrsocyrtis tetracantha*) in low latitude oceanic sites associated, in Barbados, with a late Eocene mikrotektite layer situated within the *C. bandyca* Zone) near the Zone P15/16 boundary and dated at 35. 4 ± 0.6 Ma. At nearby North Atlantic ODP Site 612, *C. turris* and *T. bromia* appear to range somewhat higher into the *C. ornata* Zone and into P16-P17 levels (Miller et al., 1991).

Ostracodes

Working with a large data base from the first 14 legs of the R.V. *Glomar Challenger*, Benson (1975) recognized a single major change in the faunal composition of deep-water ostracodes between the late Cretaceous-Eocene and the Oligocene-Neogene. The origin of the modern psychrospheric fauna was shown to have occurred in the late Eocene over an approximately 4 m.y. interval (40-36 Ma in extant chronologies; about 38-34 Ma in the emerging chronology presented by Berggren et al., this volume). The faunal changes associated with this event were predominantly the appearance of new taxa through invasion of, and adaptation to, the newly structured two-layered ocean (Benson et al., 1984). The occurrence of taxa was noted to have been diachronous, moving from deeper to shallower waters and this pattern was considered to reflect threshold effects found in distantly stratified water mass structures. The high sample diversity and taxic abundances were viewed as the local response to mixing of waters of psychrospheric and thermospheric systems. A marked decrease in diversity (to pre-psychrospheric event levels) near the Eocene/Oligocene boundary was interpreted as indicating temporary cessation of mixing of these two water systems. Elements of these psychrospheric ostracode faunas occurred in the upper Eocene (Priabonian) of southern France, Italy and Moravia, persisted in several middle Miocene Mediterranean basins, and were finally eradicated by the late Neogene closure of Tethys following the collision of Africa and Europe.

The late Eocene-early Oligocene distribution of shallow-water ostracodes in the Gulf Coast and Caribbean-Central and northern South American region has been summarized by van den Bold (1986). As is the case with larger foraminifera, the stratigraphic record of this group suffers from lack of continuous sections across the Eocene/Oligocene boundary. In the Gulf Coast about 50% of the extant taxa cross the boundary. Several taxa, including *Haplocytheridea montgomeryensis*, disappear over the boundary interval and indeed, the disappearance of virtually all taxa of the genus *Haplocytheridea*, which was dominant during the Eocene, and its replacement by the related form *Hemicyprideis*, was said to be the main characteristic of ostracode faunal change. A larger percentage of forms were said to have become extinct in Mexican boundary sections, but the differences were ascribed to the presence of hiatuses.

No major event(s) in deep-sea ostracode faunas seem to be associated with the Eocene/Oligocene boundary in Barbados (van den Bold, 1986). Continental slope faunas were said to be dominated by members of the family Krithinae. Faunas at Barbados are rather low diversity and only a single taxon, *Krithe cancuenensis*, appears to have disappeared at the paleontologically determined Eocene/Oligocene boundary. Several other extinctions bracket the boundary in a sequential fashion. In deep-water faunas of Mexico, *Messinella ovata* was shown to have become extinct at the boundary.

Invertebrates

From microfossils, we now move up the trophic pyramid to megascopic marine organisms. The dominant invertebrate phylum of the Cenozoic has always been the Mollusca, especially the Classes Gastropoda and Bivalvia. Hansen (1987, this volume) shows that there was a middle Eocene peak of diversity in the Gulf Coast of North America, followed by a severe extinction of warm-water molluscs at the end of the middle Eocene. According to Hansen's (1987) calculations, 89% of the gastropod and 84% of the bivalve species became extinct at the Bartonian/Priabonian boundary. A lesser extinction occurred in the mid-Priabonian. The Gulf Coast record shows another extinction event at the end of the Eocene, when 97% of the surviving gastropods and 89% of the bivalves went extinct. Unfortunately, the Oligocene record of the Gulf Coast is represented by large unconformities. Hansen argues that these extinctions correlate better with cooling events, rather than sea-level changes (as argued by Dockery, 1986). As we saw in the microfossil record, the most severe extinctions affected warm-water forms at the end of the middle Eocene. Subsequent extinction events resulted in less severe effects, since the total diversity has declined at the end of the middle Eocene.

Hickman (1980) showed that warm-adapted species of molluscs began to decline in the Pacific Coast during the late Eocene, although her data do not resolve the details of the transition. According to Zinsmeister (1982, 1989), there was a abrupt decline of warm-water molluscs in New Zealand and Australia in the late Eocene. He attributes these extinctions to the abrupt Southern Ocean cooling triggered by the foundering of the South Tasman Rise, the beginning of Circum-Antarctic circulation, and the clear evidence of glaciation.

The second most important member of the Cenozoic shelly fauna is the echinoids. McKinney et al. (this volume) show that echinoids reached a peak of diversity of over 1000 species by the middle Eocene. There was a small to moderate decline in the middle Eocene (near the Lutetian/Bartonian boundary), but there does not seem to be a similar extinction at the end of the Bartonian. The biggest extinction was at the Terminal Eocene Event, when over 50% of the species became extinct. Diversity remained low throughout the Oligocene. McKinney et al. (this volume) point out that there is a strong correlation of these diversity declines with temperature changes. In addition, cold-adapted marsupiate echinoids appeared in Australia in the late Eocene, and managed to survive through the Oligocene.

Marine vertebrates

No updated systematic summary of Paleogene fish diversity is presently available. However, Sepkoski's compendium of marine families (1982) and genera (unpublished; summarized in Raup and Sepkoski, 1986) presents some data on the subject. According to Sepkoski (1982), one family of sharks (Orthacodontidae) and six families of bony fish (Pycnodontidae, Galkiniidae, Rhamphosidae, Blochiidae, Phyllodontidae, and Eotrigonodontidae) last appeared during the middle or late Eocene. However, Patterson (in Smith and Patterson, 1988, Appendix 2) checked many of these last occurrences. The "Orthacodontidae" are a polyphyletic group whose true extinction was in the Danian. The Pycnodontidae, on the other hand, are a monophyletic group of 15 genera whose last occurrence is documented in the Bartonian. The extinction of the Galkiniidae is late Jurassic, and the middle Eocene last appearance is apparently a misprint in Sepkoski (1982). The Rhamphosidae are monogeneric, last occurring in the middle Eocene. The Blochiidae include 2-3 genera and 7-8 species, which were last documented in the Bartonian. The Phyllodontidae are monophyletic, consisting of about 5 genera and 10 species, and were last recorded in the upper Eocene. The Eotrigonodontidae are a paraphyletic group; the middle Eocene "extinction" of this group is a pseudoextinction. Of the seven fish families listed by Sepkoski as having become extinct in the middle or late Eocene, only three are significant.

Smith and Patterson (1988, Figure 5) performed a similar analysis on Sepkoski's unpublished data base of marine genera. They showed that although there is a high level of generic extinction in the late Eocene, it is comparable to levels of extinction throughout the Eocene and also in the Cretaceous. According to these authors, the extinction peak at the end of the Eocene was exaggerated by the noise of paraphyletic and polyphyletic groups, wrongly dated taxa, monotypic taxa, and other pseudoextinctions.

Clearly, the data base on marine fish is insufficient at the present time to establish patterns of extinction during the Eocene-Oligocene transition. Although there are significant extinctions during this interval, the quality of the data is too poor (both in terms of systematics and stratigraphic resolution) to decide if there are significant fish extinctions concentrated at the Bartonian/Priabonian boundary, in the mid-Priabonian, or at the Terminal Eocene Event.

The top predators of the marine food pyramid in the Cenozoic are marine mammals. Since pinnipeds (seals, sea lions and walruses) do not appear until the late Oligocene, and sirenians (manatees and dugongs) are very rare in the Eocene or Oligocene, only the whales are useful to this discussion. Fordyce (this volume) summarizes the record for whales. There is a slight decrease in archaeocete whale diversity from the Bartonian to the Priabonian, and archaeocetes disappear at the end of the Eocene. In the early Oligocene, modern whales began to diversify, especially in the southern oceans. The first baleen whales are known from the earliest Oligocene, and the first toothed whales occur in the late early Oligocene. By the late Oligocene, there were as many as fifty species and more than eleven families of whales. Fordyce attributes this explosive diversification to increased food resources in the southern oceans as upwelling and deep bottom waters caused blooms of plankton, crustaceans, fish, and other elements of the food pyramid.

THE TERRESTRIAL BIOTIC RESPONSE

The dramatic difference between Eocene and Oligocene terrestrial biotas has been known for over a century. Stehlin (1909) was so impressed with the distinction that he called it "la Grande Coupure" ("the great break"). Several turn-of-the-century American paleontologists, such as Osborn (1910), commented upon the profound differences in northern hemisphere land mammals. As impressive as these differences were, however, their correlation with the marine record has always been problematic. Fossiliferous terrestrial deposits rarely interfinger with marine deposits during the later Eocene or Oligocene. Datable volcanic deposits occur in many North American sections, but not in Europe or Asia. In addition, the poor stratigraphic data on most fossil mammal collections resulted in lumping stratigraphic data into land mammal "ages" which typically span 2-5 million years. Such coarse resolution allowed Lillegraven (1972) and Savage and Russell (1983) to recognize general patterns of diversity and extinction, but prevented detailed correlation.

The recent developments of magnetic stratigraphy and $^{40}Ar/^{39}Ar$ dating have radically changed our understanding of the terrestrial record (Swisher and Prothero, 1990; Prothero and Swisher, this volume). Magnetic stratigraphy made it possible to correlate terrestrial sections directly to the magnetic polarity timescale. Since most pelagic marine sections and cores are routinely calibrated to the magnetic record, direct high-resolution correlations are possible for the first time. The development of $^{40}Ar/^{39}Ar$ dating has made it possible to obtain many more dates than before, with a higher degree of reliability, and also revealed that many older K-Ar dates were in error. Most of the papers in this volume have taken the new correlations into account, but there are exceptions, as discussed below.

North America

The most complete and well studied record of the terrestrial Eocene-Oligocene transition occurs in North America. Over 150 years of collecting and study of the abundant fossils from the Big Badlands of South Dakota, the Uinta Basin of Utah, and many correlative areas, has generated a large data base. Unfortunately, the correlations with the Lyellian stages and the marine record first codified by Wood et al. (1941) were in error (Swisher and Prothero, 1990; Prothero and Swisher, this volume). The Uintan land mammal "age" (long thought to be late Eocene) is now placed in the middle Eocene, and the Uintan/Duchesnean boundary appears to correspond to the middle/late Eocene (Bartonian/Priabonian) boundary. The Chadronian, long thought to be early Oligocene, now appears to be late Eocene, and the Chadronian/ Orellan boundary (at about 34 Ma) appears to correspond to the Eocene/ Oligocene boundary. The Orellan and Whitneyan, once considered middle and late Oligocene, are now placed in the early Oligocene. Since these discoveries are so recent, even faunal summaries like those of Prothero (1985, 1989) and Stucky (1990) are now out of date.

North America has yielded more climatic information, and data from more types of organisms, than any other continent. Retallack (1981; 1983a, b; 1990; this volume) has studied the climatic information obtained from ancient soil horizons, or paleosols, in the Clarno Formation of Oregon and the Big Badlands of South Dakota. The most striking feature is a long-term drying trend in the Badlands, from 1000 mm of annual precipitation in the Duchesnean (early late Eocene) to 500-900 mm in the Orellan (early Oligocene) to 350-450 mm in the early Arikareean (late Oligocene). Using the evidence of fossil root traces and soil profiles, Retallack suggests that late Eocene forests gave way to latest Eocene dry woodland, then early Oligocene wooded grassland with gallery woodland, and finally mid-Oligocene open grassland with trees only along watercourses.

Direct evidence of botanical change comes from other deposits which are not as highly oxidized as the deposits of the Big Badlands. Wolfe (1971, 1978, this volume; Wolfe and Hopkins, 1967; Wolfe and Poore, 1982) summarized the botanical record for North America. Radiometrically dated land floras are abundant from Alaska to the Mississippi embayment, and they show a consistent pattern. After a middle Eocene warm period, there is a severe cooling event of about 10° C around 41-40 Ma at the end of the middle Eocene. Another warming event spans the late Eocene, followed by a severe cooling of at least 13°C at about 32-33 Ma. Wolfe (1971) originally called this the "Oligocene deterioration," then in 1978 coined the term "Terminal Eocene Event" for this severe cooling, and other authors have since appropriated the term. Prothero (1985, 1989) correlated this event with the Chadronian-Orellan event (then considered "mid-Oligocene" based on erroneous dates). Wolfe (this volume) considers this deterioriation to be earliest Oligocene in age, and apparently it correlates with the striking early Oligocene oxygen isotope event described by Miller (this volume). Wolfe is particularly impressed with the rapidity and severity of this change, and suggests that fairly abrupt climatic changes are required.

In addition to the cooling, Wolfe (this volume) also argues that the early Oligocene was characterized by an increase in mean annual range of temperatures. In the middle Eocene of the Pacific Northwest, for example, the mean annual range of temperature was only 3-5°C, but during the Oligocene it was as high as 25°C. This was demonstrated by the striking change in the microthermal floras, typically found in middle latitude regions with small temperature fluctuations. These floras were typically coniferous before the deterioration and broadleaved deciduous afterwards. Due to the loss of the canopy of wind-pollinated conifers (under which the angiosperms must depend on insect pollination), the post-deterioration floras were dominated by wind-pollinated angiosperms. Most of the high-latitude flora went extinct, or survived only in refugia in southeast Asia.

Pollen data in the Gulf Coast marine section show a similar pattern (Frederiksen, 1988). Following a high diversity in the middle Eocene, there are major palynological extinctions at the end of the middle Eocene (Cook Mountain Formation in Texas-Louisiana, upper Lisbon Formation in Alabama, planktonic foram Zone P13-P14, nannofossil Zone NP16). There is another recovery during the late Eocene, followed by a severe cooling in the early Oligocene in nannofossil Zone NP21.

Leopold et al. (this volume) review the pollen data for western North America. The pattern is very similar to the floral data of Wolfe. Leopold et al. (this volume) are particularly struck by the absence of grass pollen during the evolution of Oligocene "savannas." Instead of grasses or herbaceous vegetation, they argue that Oligocene rangeland was dominated by woody vegetation. Leopold et al. (this volume) point out that grass fossils or their pollen are extremely rare in the late Oligocene and early Miocene, and do not become abundant until the mid-Miocene.

Retallack (1990) argues that there were Oligocene grasses, although they do not fossilize abundantly, and they were apparently dominated by non-grassy low-biomass scrub. He visualizes the Oligocene rangeland as a seasonally dry scrubby community similar to the *Ephedra*-saltbush communities of the Great Basin, or the bluebush-saltbush communities of central Australia. Based on the Badlands paleosols, Retallack (1990) thinks grasses may have been more common in the Oligocene than their fossils or pollen indicate. Thomasson (1985) contends that the mid-Miocene silicified grass fossil record is too diverse, and must have been preceded by an Oligocene evolutionary radiation of unsilicified, unfossilizable grasses. Silicification then evolved in the mid-Miocene as a response to heavy grazing by herbivorous mammals.

One of the best records of the Eocene/Oligocene (Chadronian/Orellan) boundary occurs in sections near Douglas, Wyoming (Evanoff et al., this volume). These sections clearly show a sedimentological change from moist Chadronian floodplains to semiarid landscapes with abundant wind-blown deposits during the Orellan. These same beds preserve an excellent record of climate-sensitive land snails. According to Evanoff, Chadronian land snails are large-shelled taxa similar to snails now found subtropical climates with seasonal precipitation, such as in the southern Rocky Mountains and central Mexican Plateau. Based on modern analogues, these snails indicate a mean annual temperature of 16.5°C and a mean annual precipitation of about 450 mm during the late Chadronian in eastern Wyoming. This is similar to the 500 mm estimate by Retallack (this volume) for the Big Badlands Orellan. By the Orellan, the large-shelled forms had been replaced by drought-tolerant small-shelled taxa indicative of a warm-temperate open woodland habitat with a pronounced dry season. Such snail faunas are today found in regions like southern California and northern Baja California.

The amphibian and reptile fauna shows a similar trend toward cooling and drying. According to Hutchison (1982, this volume), there were stepwise reductions in lower vertebrate diversity starting at the Bridgerian/Uintan boundary. Along with the changes in total numbers, the aquatic forms (especially salamanders, freshwater turtles, and crocodilians) steadily decline in the late Eocene, and by the Oligocene only terrestrial tortoises are common. This indicates a pronounced drying trend during the late Eocene. Along with the total diversity, the size of aquatic turtles closely mirrors the aridity, rather than temperature. The pronounced size reduction in freshwater turtles after the early Uintan, and a similiar size reduction in the crocodilians, again seems to indicate a drying trend. Crocodiles are absent by the Chadronian, but more cool-tolerant alligators persist until the early Orellan.

The most abundant fossils in the terrestrial sections are mammals. Unlike snails and reptiles, however, they are members of archaic groups whose ecological tolerances are not as apparent. Mammal diversity was first tabulated by Lillegraven (1972) and by Savage and Russell (1983), but with the advent of magnetic stratigraphy, much greater resolution is possible. More recent tabulations were done by Prothero (1985) and Stucky (1990), both of which are now outdated by the new $^{40}Ar/^{39}Ar$ dates. Stucky (this volume) brings the diversity calculations in line with the new chronology, although the qualitative trends discussed by Prothero (1985) and Stucky (1990) are still valid, regardless of the time scale changes.

Allowing for the adjustments of time scale, all of these studies show the same basic trend. The most significant changes in the fauna occurred at the end of the Uintan (middle-late Eocene transition) and at the end of the Chadronian (end of the Eocene). The extinctions at the Uintan/Duchesnean boundary are by far the most significant, marked by a loss of over 25% of the genera of land mammals. Some of these groups last found in the Uintan include mixodectids, microsyopids, taeniodonts, achaenodonts, uintatheres, nyctitheriids, anaptomorphine primates, sciuravid rodents, dichobunid artiodactyls, limnocyonid and "miacid" carnivorans, mesonychids, hyopsodonts, and isectolophid tapiroids. Most of the victims are members of archaic groups common in the early Eocene forests. They were adapted to arboreal life or to forest browsing, and drying climate and the breakup of the forest canopy had a profound effect on their survival.

In the Duchesnean, and especially at the beginning of the Chadronian, a number of new taxa immigrated to, or evolved in North America (Lucas, this volume; Emry, this volume). These are the members of the "White River Chronofauna" (Emry, 1981; Emry et al., 1987), which persisted with only minor changes until the early Miocene (late Arikareean). Many of these taxa were members of extant families which originated at this time, such as rabbits, dogs, camels, pocket gophers, squirrels, rhinos, and shrews. At the end of the Chadronian, a second minor wave of extinction wiped out the last vesitiges of archaic Eocene browsers and tree dwellers, including brontotheres, multituberculates, pantolestids, oromerycids, epoicotheres, and paramyid and cylindrodont rodents. They are replaced by some of the first hypsodont mammals adapted for grazing, such as the leptauchenine oreodonts and the eumyine cricetid rodents.

In summary, the North American record shows major changes at the end of the middle Eocene, and in the latest Eocene-earliest Oligocene. There is clear evidence of cooling, drying, and greater seasonality in the soils and vegetation. Land snails and reptiles adapted to these conditions are the main survivors by the early Oligocene. Among the mammals, archaic Eocene forest browsers and tree dwellers are replaced by members of a number of extant families adapted to more open scrubland and possibly grassland.

Europe

After North America, the best-studied terrestrial record is in Europe. Although Europe was an archipelago during much of the Eocene, with frequent interfingering of marine and nonmarine deposits, there are still great problems in correlations. Many of the mammalian faunas come from fissure fills or other isolated pockets, so determining stratigraphic superposition is difficult. Consequently, the evolutionary sequence is reconstructed primarily by stage of evolution, and the main localities are difficult to date by either radiometrics or magnetic stratigraphy.

There are severe correlation problems in placing the marine Eocene/Oligocene boundary with respect to the terrestrial sequence. The type Bartonian in southern England contains both mammals and marine faunas, so the middle Eocene is unambiguous. Hooker (1986, this volume) places the base of the Ludian transgression (Marnes à Pholadomyes of the Paris Basin) at the base of the Barton Clay bed J (upper tongue of the Barton Clay Formation). According to Aubry (1985), the Ludian transgression correlates with nannoplankton Zone NP18, and therefore with the Bartonian/Priabonian boundary. Higher in the English section in the Brockenhurst Bed (base of the Colwell Bay Member), late Priabonian nannofossils of Zones NP19/20 are also found.

Determining the Eocene/Oligocene boundary is much more difficult, especially because there have been many controversies about how to recognize it in marine sections (Berggren, 1971; Berggren and Van Couvering, 1978; Berggren et al., 1985). Traditionally, the "Grande Coupure" of Stehlin (1909) has been correlated with the Eocene/Oligocene boundary (Cavelier, 1979; Savage and Russell, 1983), although there have been dissenters (Lopez and Thaler, 1974; Sigé and Vianey-Liaud, 1979). Some of the earliest post-Grande Coupure faunas, Hoogbutsel and Hoeleden in Belgium, occur in the type area of the base of the upper Tongrian (Savage and Russell, 1983). They are found in a unit sandwiched between nannofossil Zones NP21 and NP22, making them early (but not earliest) Oligocene (Steurbaut, 1986; Gramann, 1990; Gullentops, 1990). These faunas are overlain by the Rupelian transgression. Fossil mammals from the Boom Clay, part of the type Rupelian Stage, also contain post-Grande

Coupure mammals (Savage and Russell, 1983). Thus, the earliest post-Grande Coupure mammals are not earliest Oligocene (which begins in nannofossil Zone NP21).

In addition, Hooker (this volume) points out that both the Rhine Graben and Mainz Basins of Germany produce a distinctly pre-Grande Coupure fauna associated with nannoplankton Zone NP22, which is early Oligocene. In England, the early Oligocene NP21/22 boundary lies between the Bembridge Oyster Bed and the *Nematura* band, and the earliest post-Grande Coupure fauna lies above the *Nematura* band. Thus, the Grande Coupure is not at the Eocene/Oligocene boundary, but in the early Oligocene, correlative with NP22 or the base of NP23. It appears to be 1-2 million years younger than the Eocene/Oligocene boundary (as recognized by the last occurrence of *Hantkenina*), but may correlate with the early Oligocene isotopic event (see Miller, this volume) and floral change (see Wolfe, this volume).

Collinson (this volume; see also Collinson et al., 1981; Collinson and Hooker, 1987) has reviewed the botanical data from Europe. Middle Eocene floras were predominantly tropical, but by the late Eocene, floras were dominated by subtropical evergreens. These include taxodiaceous swamps and reed marshes, with patches of woodland or forest. Early Oligocene floras include mixed deciduous/evergreen plants indicating a warm-temperate, seasonal climate. Reed marshes were particularly dominant during the Oligocene, although early Oligocene floras include serrate-margined, deciduous trees of the Betulaceae and Juglandaceae. Collinson (this volume) indicates that the main floral transition is difficult to date, but occurs below the MP21 (Hoogbutsel) level (i.e., before the Grande Coupure), and probably close to the Eocene/Oligocene boundary.

Pollen data from Europe (Collison, this volume; Hubbard and Boulter, 1982; Boulter and Hubbard, 1983; Boulter, 1984) also indicate that the transition is marked by the increase of temperate elements and conifer pollen, and loss of tropical and subtropical elements. Both pollen and macrofloral data indicate a cooling trend, although there is no obvious evidence of the drying trend seen in North America.

Rage (1986) reviewed the lower tetrapod record from Europe. Surprisingly, during the late Eocene, the diversity of reptiles and amphibians increased, with about 40 tropical taxa by the end of the Ludian. Only ten of these survived beyond the Grande Coupure. As in North America, aquatic taxa (include crocodiles and the aquatic snake family Paleopheidae) disappeared at the end of the Eocene.

The mammal record from Europe has received the most attention. As reviewed by Hooker (this volume; see also Collinson and Hooker, 1987) and Legendre and Hartenberger (this volume; see also Legendre, 1987, 1989; Hartenberger, 1986, 1987), the change in mammalian faunas is protracted throughout the later Eocene and Oligocene. Early and middle Eocene European faunas were dominated by archaic taxa, such as adapid primates, multituberculates, "insectivores," creodonts, archaic ungulates, tillodonts, and pantodonts. The artiodactyls and perissodactyls are mostly peculiar endemic forms that lived in isolation from other northern continents in the European archipelago. They include endemic perissodactyls (palaeotheres and lophiodonts) and a wide variety of unique artiodactyls (including xiphodonts, choeropotamids, cebochoerids, mixtotheriids, dacrytheriids, anoplotheriids, amphimerycids, and cainotheriids). There is a peak of turnover that may correspond to the Bartonian-Priabonian event (Hartenberger, 1986, Figure 4). At this event, most of the large perissodactyls go extinct and are replaced by species of *Palaeotherium*, with soft-browsing forms replaced by coarse-browsing forms. There was a drastic reduction in arboreal mammals (especially primates and apatemyids) as well as small mammals and insectivores, and an increase in large ground mammals and browsing herbivores (Collinson and Hooker, 1987). The rodent fauna changed from frugivorous pseudosciurids to browsing theridomyids.

The largest change, however, is at the Grande Coupure (between mammal levels MP20 and MP21), with as much as 60% of the fauna going extinct. As we have seen above, it appears that the Grande Coupure is an early Oligocene event, occurring perhaps 0.5 m.y. after the Eocene/Oligocene boundary (as recognized by the last occurrence of *Hantkenina*), and perhaps equivalent to the early Oligocene isotopic event. Post-Grande Coupure faunas are dominated by rabbits, theridomyid rodents, advanced carnivorans (mostly amphicyonids, mustelids, viverrids, procyonids, ursids, and nimravids), advanced artiodactyls (especially anthracotheres, leptomerycids, entelodonts, and tayassuids), and advanced perissodactyls (especially rhinocerotoids and chalicotheres). Arboreal mammals disappeared completely, and large mammals dominated. Granivorous rodents made their first appearance.

Almost all of these new taxa are immigrants from Asia, or from North America via Asia (Matthew, 1906; Stehlin, 1910; McKenna, 1983; Savage and Russell, 1983). The "Grande Coupure" is less a climatic event, and more of an example of sudden immigrant influx, and displacement of native endemic taxa (comparable to the Pliocene "Great American Interchange" between South and North America). However, this immigration may have had climatic triggers. Conventionally, this flood of Asian and North American taxa was thought to have crossed the Turgai Straits across the Ural region, which separated Asia and Europe in the Eocene (Vianey-Liaud, 1976). Lowered sea levels in the early Oligocene may have dried up these seaways and opened land passages (Plint, 1988). However, Heissig (1979) pointed out that some of these taxa had already migrated to the Balkan region in the middle Eocene. He argues for a more complex corridor system, with the Alpine uplift serving as the main corridor between eastern and western Europe during the Grande Coupure.

A plot of body size distributions, or "cenogram," reveals much about the community structure (Legendre, 1986, 1989). Middle and late Eocene communities are characterized by an abundance of large mammals, with a continuous unbroken distribution in sizes (Legendre and Hartenberger, this volume, Figure 26.6), which indicates humid tropical forest habitat. By the Oligocene, large-sized species are rare and medium-sized species are absent, drastically changing the shape of the cenogram. This distribution indicates a cooler, more arid and open environment, close to that of the modern savanna.

Thus, the patterns seen in Europe closely mirror those seen in North America, even though the taxa are different. The major differences appear to be that North America shows much greater evidence of drying during the Oligocene than does Europe. This is not surprising, since most North American data comes from the continental interior and behind the Rocky Mountain rain shadow, while European floras and faunas are from the coasts of islands. Both continents underwent major changes at the end of the middle Eocene, but the late Eocene-early Oligocene transition is very different in scale. The Grande Coupure in Europe is much bigger than the Chadronian-Orellan transition in North America. If the suggested correlations are correct, however, the Chadronian-Orellan transition (late in magnetic Chron C13R) matches the minor changes in the marine faunas at the Eocene/Oligocene boundary. If the Grande Coupure is really early Oligocene, 1-2 million years younger, then it corresponds to the major early Oligocene oxygen isotope event, and the dramatic change in North American floras in the early Oligocene (Wolfe, this volume).

Asia

Documentation and calibration of the Asian record is still in the preliminary stages. Although there has been much research on the mammalian faunas, particularly by the Chinese (summarized by Li and Ting, 1983), very few of these faunas have been radiometrically dated or correlated by magnetostratigraphy. Given the surprising realignment of the North American terrestrial biochronology due to new dates (Swisher and Prothero, 1990; Prothero and Swisher, this volume), it would not be surprising if the Asian record will soon have to be reinterpreted as well.

Wang (this volume) follows the conventional interpretation of early, middle, and late Oligocene in her discussion of Chinese mammalian faunas. These correlations are based in similarities of Asian mammals to those from North America or Europe. However, most of the mammals are endemics, so the similiarities are based on a few immigrational events. In examining Wang's correlations based on mammals, we believe that the Chinese "early Oligocene" is actually late Eocene, and the "middle Oligocene" is actually early Oligocene.

For example, Wang (this volume) points out that Chinese "early Oligocene" faunas are dominated by elements that we would consider middle or late Eocene elsewhere: pantodonts, mesonychids, archaic tapirs, *Amynodon*, *Amynodontopsis* (and several other amynodonts), the hyracodont *Forstercooperia*, and especially brontotheres, which go extinct at the end of the Eocene in both North America (Evanoff et al., this volume) and Europe (Lucas and Schoch, 1989). Wang claims that these taxa persisted longer in Asia, but it is simpler to assume that they come from late Eocene deposits. Her "early Oligocene" first occurrences include the rhinocerotoids *Ardynia*, *Cadurcodon*, plus *Bothriodon*, *Entelodon*, and other artiodactyls whose nearest relatives are late Eocene elsewhere. The primitive rhinocerotid, *Ronzotherium*, is early Oligocene in Europe, but its nearest relatives are Duchesnean and Chadronian (late Eocene) in North America (Prothero et al., 1989), so *Ronzotherium* is probably late Eocene (not "early Oligocene") in Asia.

Likewise, Wang's "early-middle Oligocene" transition is marked by the extinction of brontotheres and other archaic Eocene forms, which marks the Eocene/Oligocene boundary elsewhere. Chinese "middle Oligocene" faunas include many elements which are considered early Oligocene or post-Grande Coupure elsewhere, including abundant cricetid rodents (especially *Eumys* and *Eucricetodon*), the chalicothere *Schizotherium*, the cat-like *Nimravus*, the bear *Amphicynodon*, the leptomerycid *Prodremotherium*, the hyracodont rhino *Eggysodon*, and advanced rhinocerotids such as *Aprotodon* and "*Aceratherium*."

Corroborating this recorrelation of the Asian "middle Oligocene" is the only published radiometric date in this interval, from the Hsanda Gol Formation of Mongolia. Evernden et al. (1964, p. 193) gave dates of 31.3 and 32.0 Ma from a whole-rock analysis of a basalt. Clearly, these should be redated with $^{40}Ar/^{39}Ar$ methods, but taken at face value, these dates suggest an early Oligocene age. Dashzeveg (in Dashzeveg and Devyatkin, 1986; Dashzeveg, in prep.) reports a major change in Mongolian faunas between the Ergilin and Hsanda Gol levels. Like the late Eocene elsewhere, the Ergilin marks the last appearance of brontotheres, eomoropid chalicotheres, deperetellid and helaletid "tapiroids," and mesonychids, as well as the decline of amynodonts, entelodonts, and anthracotheres. Unfortunately, Dashzeveg follows the Chinese in calling the Ergilin "early Oligocene." The Hsanda Gol Formation, as we have just seen, has early Oligocene dates, and a typical early Oligocene pattern of diverse cricetid rodents and nimravid, ursid, and mustelid carnivorans. Contrary to Dashzeveg, the evidence seems clear that the Hsanda Gol is early, not middle Oligocene, and the major extinction in Mongolia was at the Eocene-Oligocene (Ergilin-Hsanda Gol) transition.

If we reinterpret the Asian "early Oligocene" as late Eocene, and the "middle Oligocene" as early Oligocene, then the patterns of extinctions in Asia closely resemble those of other continents. Wang (this volume) reports that a variety of archaic Eocene forms, including tillodonts, arctocyonids, helohyids, and eurymylids disappear at the end of the Sharamurunian land mammal "age." The Sharamurunian has traditionally been considered late Eocene, based on its similarites to the Uintan (especially in the miacid carnivores, uintatheres, mesonychids such as *Harpagolestes*, brontotheres such as *Dolichorhinus*, *Telmatherium* and *Manteoceras*, "tapiroids" such as *Colodon*, hyracodonts such as *Triplopus* and *Forstercooperia*, the diagnostic Uintan genus *Amynodon*, and chalicotheres like *Eomoropus*). Now that the Uintan is middle Eocene, it seems likely that at least part of the Sharamurunian is also middle Eocene. If so, then the 45 genera that go extinct at the end of the Chinese "late Eocene" actually go extinct at the middle-late Eocene transition, in good agreement with the major extinctions at the Uintan-Duchesnean transition in North America.

The extinction event at the end of the Houldjinian ("early Oligocene" of the Chinese, but late Eocene here) is comparable to the extinctions at the end of the Eocene elsewhere (Chadronian/Orellan transition in North America, Ludian/Stampian in Europe). All of the continents saw the extinction of brontotheres at the end of the Eocene, along with archaic "tapiroids" such as the Deperetellidae and Lophialetidae. In addition, this was the time of the extinction of Eocene holdovers such as the Mesonychidae, and Asian endemics such as the Anagalidae and Yuomyidae. Following the Houldjinian-Kekeamuan ("early-middle" Oligocene according to Wang, but Eocene/ Oligocene boundary here), there was a great diversification of hypsodont rodents (especially the cricetids), as also occurred in the Orellan of North America and after the Grande Coupure in Europe. Thus, the Asian record shows both a significant middle/late Eocene extinction, and a lesser extinction at the end of the Eocene.

Leopold et al. (this volume) summarize some of the botanical changes in Asia and the USSR during the Eocene and Oligocene. In the early Oligocene of the USSR, forest vegetation dominated with areas of woody savanna in Kazakhstan. By the early Miocene, grassy steppes and forest mosaics were widespread over Siberia and Kazakhstan. In China, there was a striking change in pollen between the middle and late Eocene. Since much of China (especially the Gobi Desert region) was arid in the middle Eocene, arid and subarid pollen suites dominated at that time. By the late Eocene, these subarid floras became restricted to northwestern China, while southeastern China became more mesic. Coastal areas had broad-leaved forests, cypress swamps, and even mangroves. A subtropical woody savanna developed in northwestern China during the late Eocene, lacking in grasses and herbaceous pollen (as reported for other continents). This northwest-southeast arid-humid gradient across China persisted into the early Oligocene, although with decreased diversity. The woody savanna persisted into the late Oligocene, with increased importance of salt bushes. In highland and coastal areas, the increase in temperate deciduous trees and conifers suggest late Oligocene cooling. By the Miocene, grasses and herbaceous vegetation began to dominate the savanna in China, as they did on other continents.

Other continents

Compared to North America and Eurasia, the continental record of Africa, South America, Australia, and Antarctica is much less well documented during the Eocene and Oligocene. However, there are adequate data for certain taxa at certain times in limited areas.

Rasmussen et al. (this volume) review the African record during the Eocene and Oligocene. The Fayum region of Egypt provides most of the data, and other, less tropical parts of the continent have yet to be adequately sampled. Rasmussen et al. (this volume) place the Eocene/ Oligocene boundary at the unconformity between Fayum Faunal Zone 2 and 3. They find no significant changes in the vegetation or fauna across the Eocene-Oligocene transition. However, Egypt was part of the tropical Tethyan belt during this interval. Apparently, the tropical Tethyan region was not as strongly affected by climatic changes as the higher latitudes.

South America has a good fossil record for much of the Cenozoic, but unfortunately not for the Eocene-Oligocene transition. Until recently, the Divisaderan land mammal "age" was placed in the late Eocene (about 41 Ma) and the Deseadan in the Oligocene, about 37-27 Ma (Marshall, 1985). However, recent dating and magnetostratigraphy has placed the Divisaderan and Deseadan in the latest Oligocene (28-21 Ma), and left an enormous gap representing the entire late Eocene-middle Oligocene record of South America (MacFadden et al., 1985). Novacek et al. (1989) and Flynn et al. (1990) reported a new fauna from Chile (radioisotopically dated at about 30 Ma) which may give us the first fossils from the early Oligocene of South America. This investigation is just beginning, so the implications for Eocene-Oligocene faunal changes in South American mammals are unclear.

Marshall and Cifelli (1989) summarized the pattern of diversity in South American land mammals using the old chronology. Despite the late Eocene-middle Oligocene gap, it is possible to compare the middle Eocene Mustersan faunas with the late Oligocene Divisaderan-Deseadan faunas. There is little difference in familial or generic diversity between the Mustersan and Divisadearan, but the Deseadan shows a great increase in generic diversity (from 34 to 83 genera), mostly due to the immigration of primates and rodents. However, since the Divisaderan is based on a limited mammalian sample from the Divisadero Largo fauna of Argentina, the real change is between the Mustersan and Deseadan. Other than this generic increase, there are no significant changes in extinction or origination rates from the Mustersan through the Deseadan.

Marshall and Cifelli (1989) point out that the Deseadan marks the beginning of Andean uplift, changing the vegetation from subtropical woodlands to seasonally arid savanna woodlands. This is shown by the extinction of many archaic browsing mammals (including the Sparnotheriodontidae, Pyrotheriidae, Colombitheriidae, Notopithecinae, Isotemnidae, and Oldfieldthomasiidae) and their replacement by hypsodont grazers (Hegetotheriidae, Interatheriidae, Mesotheriidae, Toxodontidae) during the Deseadan (Cifelli, 1985). Marshall and Cifelli do not address the possibility that the change to woodland-savannas and grazing mammals may also have had global climatic causes, as we have seen on other continents.

Pascual et al. (1985) argue that the changes between the Mustersan and Deseadan were among the most fundamental in South American history. As we have seen, it marked the end of nearly all the archaic groups, including not only the browsing ungulates, but also rodent-like marsupials (such as polydolopids and groeberiids), and primitive edentates. Many of these last occurrences represented the gigantic end-members of long-established lineages, such as the gigantic carnivorous marsupials, the Proborhyaeninae. By contrast, lineages with smaller species, such as the microbiotheriid and caenolestid marsupials, and smaller, less specialized marsupicarnivores, began to radiate in the Deseadan. As we have already seen, the most striking feature of the Deseadan is the addition of primates and rodents, and the appearance of mammals with high-crowned, grazing teeth.

Australia has no mammalian record for the Eocene-Oligocene transition. Until recently, the oldest known Cenozoic mammal fauna (from the Geilston Travertine of Tasmania) was slightly younger than 23 Ma, and therefore late Oligocene (Tedford et al., 1975). Godthelp (1991) has just described the Tinga Murra l.f., which he argues is Eocene. However, its "Eocene" age is based on doubtful comparisons to South American marsupials, so its true age is uncertain (M. Woodburne, pers. commun.). Most of the assemblage is composed of primitive dasyurids, so there is no indication of major turnover in Australian mammals.

However, Kemp (1978) has reviewed the Australian botanical record. Like the northern continents, middle Eocene floras in Australia were dominated by tropical rainforests. By the late Eocene, these floras began to decline in diversity. A major change took place at the Eocene/Oligocene boundary, when deepwater circulation between Antarctica and Australia commenced, leading to Antarctic glaciation. Oligocene plant assemblages are characterized by a low diversity of mostly cool-temperate plants tolerant of higher seasonality. The initiation of widespread brown coal swamps, however, suggests high rainfall in southeastern Australia, which was under the influence of rain-bearing westerly winds triggered by the circum-Antarctic current. Truswell and Harris (1982) showed that open forest habitats, with a more diverse herbaceous understory, began in the early Oligocene, as a consequence of increased aridity. According to Christophel (1990), Oligocene floras show a great increase in sclerophylly and the reduction of leaf size in many taxa, including conifers.

Although Antarctic land mammals are known from only a single taxon of polydolopid marsupial (Woodburne and Zinsmeister, 1984), land floras are better known. Middle Eocene floras and pollen from the La Meseta Formation on Seymour Island suggest a cool-temperate rain forest with large trees, similar to those found today in Tasmania, New Zealand, and southern South America (Case, 1988). If glaciers were present, they occurred mostly at higher elevations or on the Antarctic Peninsula, and did not prevent the spread of lush forests on the lowlands.

SUMMARY

The Eocene-Oligocene transition was gradual, spaced out over at least ten million years. The major extinction event took place at the middle-late Eocene transition (about 40-41 Ma), when the high diversity of animals and plants which favored warm climates and tropical forests were severely decimated. This apparent cooling may have resulted from incipient Antarctic glaciation, although the evidence is controversial. Lesser extinctions took place at the end of the Eocene (34 Ma), although some groups were more severely affected than others. In the earliest Oligocene (about 33.5 Ma), there was significant cooling and ice volume increase, resulting in major changes in land floras, and possibly the "Grande Coupure" immigration event in European land faunas. Major Antarctic glaciations marked the middle Oligocene (about 30 Ma), although most organisms were already cold-adapted survivors, and there were relatively few extinctions.

These extinctions were clearly related to changes in climate triggered by the thermal isolation of Antarctica as Australia drifted northward and allowed the development of deep cold bottom waters. Extraterrestrial impacts had no apparent effect, since they occurred in the middle of the late Eocene, when there were no extinctions of significance.

ACKNOWLEDGMENTS

We wish to thank the Penrose Conference participants, together with others who have generously contributed to this volume. The cooperative spirit which contributors have shown the convenors has rendered our task much easier in preparing this volume in a timely fashion. We would like to thank, in particular, Anne Boersma and Marie-Pierre Aubry for discussions on various aspects of the marine trophic resource continuum model, and Kenneth G. Miller and M.-P. Aubry for a critical review of the section dealing with marine micropaleontology and paleoceanography. Spencer Lucas critically read the sections on terrestrial organisms. Preparation of this review was supported by NSF Grants OCE-9101463 (WAB) and EAR87-08221 (DRP). This is Woods Hole Oceanographic Institution Contribution number 7793.

REFERENCES

Adams, C. G., Butterlin, J. and Samanta, B.K. 1986. Larger foraminifera and events at the Eocene/Oligocene boundary in the Indo-Pacific region. In Pomerol, C. and Premoli-Silva, I., eds., *Terminal Eocene Events*. Amsterdam, Elsevier, pp. 237-252.

Adams, C.G., Lee, D.E., and Rosen, B.R. 1990. Conflicting isotopic and biotic evidence for tropical sea-surface temperatures during the Tertiary. *Paleogeogr., Palaeoclimat., Palaeoecol.* 77 (1990): 289-313.

Aubry, M.P. 1985. Northwestern European Paleogene magnetostratigraphy, biostratigraphy, and paleogeography: calcareous nannofossil evidence. *Geology* 13 :198-202.

Benson, R.H. 1975. The origin of the psychrosphere as recorded in changes of deep sea ostracodes assemblages. *Lethaia* 8 : 69-83.

Benson, R.H., Chapman, R.E., and Deck, L.T. 1984. Paleoceanographic events and deep-sea ostracodes. *Science* 224 : 1334-1336.

Berggren, W.A. 1971. Tertiary boundaries, In Riedel, W.R. and Funnell, B.M., eds., *Marine Micropaleontology of the Oceans*, Cambridge, Cambridge Univ. Press, pp. 693-803.

Berggren, W.A., D.V. Kent, and J.J. Flynn, 1985. Paleogene geochronology and chronostratigraphy, In N.J. Snelling, ed., *The Chronology of the Geological Record. Mem. Geol. Soc. London* 10: 141-195.

Berggren, W.A., and Van Couvering, J.A. 1978. Biochronology. *Amer. Assoc. Petrol. Geol. Stud. Geol.* 6: 39-55.

Birkenmajer, K. 1987. Tertiary glacial and interglacial deposits, South Shetland Islands, Antarctica: Geochronology versus biostratigraphy (a progress report). *Bull. Polish Academy Sci. Earth Sci. 36*:133-145.

Boersma, A. and Premoli-Silva, I. 1991. Distribution of Paleogene planktonic foraminifera-analogies with the Recent? *Paleogeogr., Palaeoclimat., Palaeoecol.* 83: 29-48.

Boersma, A., Premoli-Silva, I. and Shackleton, N.J. 1987. Atlantic Eocene planktonic foraminiferal paleohydrographic indicators and stable isotope paleoceanography. *Paleoceanography* 2 (3): 287-331.

van den Bold, W.A. 1986. Distribution of Ostracoda at the Eo-Oligocene boundary in deep (Barbados) and shallow marine environment (Gulf of Mexico). In Pomerol, C. and Premoli-Silva, I., eds., *Terminal Eocene Events*. Amsterdam, Elsevier, pp. 259-263.

Boulter, M.C. 1984. Palaeobotanical evidence for land–surface temperature in the European Paleogene. In P.J. Brenchley, ed., *Fossils and Climate*. Chichester, John Wiley and Sons, pp. 35–47.

Boulter, M.C. and Hubbard, R.N.L.B. 1982. Objective paleoecological and biostratigraphic interpretation of Tertiary palynological data by multivariate statistical analysis. *Palynology* 6: 55-68.

Case, J.A. 1988. Paleogene floras from Seymour Island, Antarctic Peninsula. *Geol. Soc. Amer. Mem.* 169: 523-530.

Cavelier, C. 1979. La limite Eocène-Oligocène en Europe occidentale. *Sci. géol. Inst. Géol. Strasbourg, (Mém.)* 54 :1-280.

Christophel, D.C. 1990. The impact of mid-Tertiary climatic changes on the development of the modern Australian flora. *Geol. Soc. Amer. Abs. Prog.* 22 (7): A77.

Cifelli, R.L. 1985. South American ungulate evolution and extinction, In Stehli, F.G., and Webb, S.D., eds., *The Great American Biotic Interchange*, New York, Plenum Press, pp. 249-266.

Collinson, M.E. and Hooker, J.J. 1987. Vegetational and mammalian faunal changes in the early Tertiary of southern England. In Friis, E.M., Chaloner, W.G. and Crane, P.R., eds., *The Origins of Angiosperms and their Biological Consequences*: Cambridge, Cambridge University Press, pp. 259-304.

Collinson, M.E., Fowler, K. and Boulter, M.C. 1981. Floristic changes indicate a cooling climate in the Eocene of southern England. *Nature* 291, 315-317.

Dashvezeg, D. and Devyatkin, E.V. 1986. Eocene–Oligocene boundary in Mongolia.In Pomerol, C. and Premoli-Silva, I., eds., *Terminal Eocene Events*. Amsterdam, Elsevier, pp. 153–157.

Dockery, D.T. III. 1986. Punctuated succession of Paleogene mollusks in the northern Gulf Coastal plain. *Palaios* 1:582–589.

Emry, R.J. 1981. Additions to the mammalian fauna of the type Duchesnean, with comments on the status of the Duchesnean. *Jour. Paleont.* 55: 563-570.

Emry, R.J., P.R. Bjork, and L.S. Russell. 1987. The Chadronian. Orellan, and Whitneyan land mammal ages, In Woodburne, M.O., ed., *Cenozoic Mammals of North America, Geochronology and Biostratigraphy*, Berkeley, Univ. Calif. Press, pp. 118-152.

Evernden, J.F., D.E. Savage, G.H. Curtis, and G.T. James. 1964. Potassium-argon dates and the Cenozoic mammalian chronology of North America. *Amer. Jour. Sci.* 262: 145-198.

Flynn, J.J., and Wyss, A.R. 1990. New early Oligocene marsupials from the Andean Cordillera, Chile. *Jour. Vert. Paleo.* 10 (3): 22A (abstract).

Frederiksen, N.O. 1988. Sporomorph biostratigraphy, floral changes, and paleoclimatology, Eocene and earliest Oligocene of the eastern Gulf Coast. *U.S. Geol. Surv. Prof. Paper* 1448.

Gaskell, B.A., 1991. Extinction patterns in Paleogene benthic foraminiferal faunas: relationship to climate and sea level. *Palaios* 6: 2-16.

Godthelp, H. 1991. The Tinga Murra local fauna, Eocene mammals from Australia, In Vickers-Rich, P.V., Monaghan, J., Baird, R.J., and Rich, T.H., eds., *Vertebrate Palaeontology of Australasia*. Victoria, Monash Univ. Press (in press).

Gramann, F. 1990. Eocene/Oligocene boundary definitions and the sequence of strata in N.W. Germany. *Tert. Res.* 11 :73-82.

Gullentops, F. 1990. Sequence stratigraphy of the Tongrian and early Rupelian in the Belgian type area. *Tert. Res.* 11 :83-96.

Hallock, P. 1987. Fluctuations in the trophic resource continuum: a factor in global diversity cycles? *Paleoceanography* 2(5):457-471.

Hallock, P., Premoli-Silva, I., and Boersma, A. 1991. Similarities between planktonic and larger foraminiferal evolutionary trends through Paleogene paleoceanographic changes. *Palaeogeogr., Palaeoclimat., Paleoecol.* 83 (1991): 49-64.

Hansen, T.A. 1987. Extinction of late Eocene to Oligocene molluscs: Relationship to shelf area, temperature changes, and impact events. *Palaios* 2:69–75.

Haq, B.U., Hardenbol, J. and Vail, P.R. 1987. The chronology of fluctuating sea level since the Triassic. *Science* 235 : 1156-1167.

Haq, B.U., Hardenbol, J., and Vail, P.R. 1988. Mesozoic and Cenozoic chronostratigraphy and cycles of sea-level change. In Wilgus, C.K., Posamentier, H.R., Ross, C.A., and Kendall, C.G. St. C., eds., Sea Level Changes: An Integrated Approach, *Soc. Econ. Paleont. Min., Spec. Publ.* 42: 71-108.

Harland, W.B., Armstrong, R.L., Cox, A.V., Craig, L.E., Smith, A.G., and Smith, D.G. 1990. *A Geologic Time Scale: 1989*. Cambridge, Cambridge University Press, 263 pp.

Hartenberger, J.-L. 1986. Crises biologiques en milieu continental au cours du Paléogène: exemple des mammifères d'Europe. *Bull. Centr. Rech., Expl.-Prod. Elf-Aquitaine* 10 :489-500.

Hartenberger, J.-L. 1987. Modalités des extinctions et apparitions chez les mammifères du

Paléogène d'Europe. *Mem. Soc. Géol. France, N.S.* 150: 133-143.

Hartenberger, J.-L. 1988. Etudes sur la longevite des genres mammiferes fossils du Paleogene d'Europe. *C.R. Acad. Sci. Paris* 306: 1197–1204.

Hedberg, H.D., ed. 1976. *International Stratigraphic Guide: A Guide to Stratigraphic Classification, Terminology, and Procedure,* New York, John Wiley and Sons, 200 pp.

Heissig, K. 1979. Die hypothetische Rolle Sudosteuropas bei den Säugetierwanderungen im Eozän und Oligozän. *Neues Jb. Geol. Paläont. Monatsh.* 1979: 83-96.

Hickman, C.S. 1980. Paleogene marine gastropods of the Keasey Formation of Oregon. *Bull. Amer. Paleont.* 78: 1-112.

Hooker, J.J. 1986. Mammals from the Bartonian (middle/late Eocene) of the Hampshire Basin, southern England. *Bull. Brit. Mus. Nat. Hist. (Geol.)* 39: 191-478.

Hubbard, R.N.L.B. and Boulter, M.C. 1983. Reconstruction of Palaeogene climate from palynological evidence. *Nature* 301: 147-150.

Hutchison, J.H. 1982. Turtle, crocodilian and champsosaur diversity changes in the Cenozoic of the north–central region of the western United States. *Palaeogeogr., Palaeoclimat., Palaeoecol.* 37: 149–164.

Kaminski, M.A. 1987. Cenozoic deep-water agglutinated foraminifera in the North Atlantic. PhD Thesis, MIT/WHOI 88-3, 262 pp.

Kemp, E.M. 1978. Tertiary climatic evolution and vegetation history in the southeast Indian Ocean region. *Palaeogeogr., Palaeoclimat., Palaeoecol.* 24: 169-208.

Kennett, J. P., and Stott, L. D. 1990. Proteus and Proto-Oceanus: ancestral Paleogene oceans as revealed from Antarctic stable isotopic results; ODP Leg 113. In Barker, P. F., Kennett, J. P., et al., *Proceedings of the Ocean Drilling Program, Scientific Results,* 113: 865-880.

Legendre, S. 1986. Analysis of mammalian communities from the late Eocene and Oligocene of southern France. *Palaeovertebrata* 16: 191–212.

Legendre, S. 1987a. Concordance entre paléontologie continentale et les événements paléocéanographiques: exemple des faunes de mammifères du Paléogène du Quercy. *C.R. Acad. Sci. Paris,* Sér. 3, 304: 45-50.

Legendre, S. 1989. Les communautés de mammifères du Paléogène (Eocène supérieur et Oligocène) d'Europe occidentale: structures, milieux et évolution. *Münchner Geowiss. Abh., (A)*16: 1-110.

Li C.-K., andTing S.-Y. 1983. The Paleogene mammals of China. *Bull. Carnegie Mus. Nat. Hist.* 21: 1-93.

Lillegraven, J.A. 1972. Ordinal and familial diversity of Cenozoic mammals. *Taxon* 21: 261-274.

Lopez, N., and Thaler, L. 1974. Sur les plus ancien lagomorphe européen et la "Grande Coupure" Oligocene de Stehlin. *Palaeovertebrata* 6: 243-251.

Lucas, S.G., and Schoch, R.M. 1989. European brontotheres, In Prothero, D.R., and Schoch, R.M., eds., *The Evolution of Perissodactyls,* New York, Oxford Univ. Press, pp. 485-490.

Lyell, C. 1833. *Principles of Geology,* (1st Edition), vol. 3. London, John Murray, 398 pp.

Lyell, C. 1857. *Principles of Geology. Supplement to the Fifth Edition,* London, John Murray, 40 pp.

MacFadden, B.J., Campbell, K.E., Cifelli, R.L., Siles, O., Johnson, N.M., Naeser, C.W., and Zeitler, P.K. 1985. Magnetic polarity stratigraphy and mammalian faunas of the Deseadean (late Oligocene–early Miocene) Salla Beds of northern Bolivia. *Jour. Geol.* 93: 223–250.

Mancini, E.A. and Tew, B.H. 1991. Relationships of Paleogene stage and planktonic foraminiferal zone boundaries to lithostratigraphic and allostratigraphic contacts in the eastern Gulf Coastal Plain. *Jour. Foram. Res.* 21 (1): 48-66.

Marshall, L.G. 1985. Geochronology and land-mammal biochronology of the transamerican faunal interchange, In Stehli, F.G., and Webb, S.D., eds., *The Great American Biotic Interchange,* New York, Plenum Press, pp. 49-88.

Marshall, L.G., and Cifelli, R.L. 1989. Analysis of changing diversity patterns in Cenozoic land mammal age faunas, South America. *Palaeovertebrata,* 19: 169-210.

Matthew, W.D. 1906. Hypothetical outline of the continents in Tertiary times. *Bull. Amer. Mus. Nat. Hist.* 22: 353-383.

Matthews, R.K. and Poore, R.Z. 1980. Tertiary

$\delta^{18}O$ record and glacio-eustatic sea-level fluctuations. *Geology* 8 : 501-504.

McGowran, B. 1989. Silica burp in the Eocene ocean. *Geology* 17: 857-860.

McKenna, M.C. 1983. Holarctic landmass rearrangement, cosmic events, and Cenozoic terrrestrial organisms. *Ann. Missouri Bot. Garden* 70: 459-489.

Miller, K.G., Berggren, W.A., Zhang, J., and Palmer-Julson, A. 1991. Biostratigraphy and isotope stratigraphy of upper Eocene microtektites at Site 612: how many impacts? *Palaios* 6: 17-38.

Novacek, M.J., Wyss, A., Frassinetti, D., and Salinas, P. 1989. A new ?Eocene mammal fauna from the Andean main range. *Jour. Vert. Paleo.* 9 (3): 34A (abstract).

Osborn, H.F. 1910. *The Age of Mammals in Europe, Asia, and North America.* New York, MacMillan and Comp. 634 pp.

Pascual, R., Vucetich, M.G., Scillato-Yané, G.J., and Bond, M. 1985. Main pathways of mammalian diversification in South America, In Stehli, F.G., and Webb, S.D., eds., *The Great American Biotic Interchange,* New York, Plenum Press, pp. 219-248.

Plint, A.G. 1988. Global eustacy and the Eocene sequence in the Hampshire Basin, England. *Basin Res.* 1: 11-22.

Prentice, M.L. and Matthews, R.K. 1988. Cenozoic ice-volume history: development of a composite oxygen isotope record. *Geology* 16: 963-966.

Prothero, D.R. 1985. North American mammalian diversity and Eocene-Oligocene extinctions. *Paleobiology* 11(4): 389-405.

Prothero, D.R. 1989. Stepwise extinctions and climatic decline during the later Eocene and Oligocene, In S.K. Donovan, ed., *Mass Extinctions: Processes and Evidence.* New York, Columbia Univ. Press, pp. 211-234.

Prothero, D.R., Guérin, C., and Manning, E. 1989. The history of the Rhinocerotoidea, In Prothero, D.R., and Schoch, R.M., eds., *The Evolution of Perissodactyls,* New York, Oxford Univ. Press, pp. 322-340.

Prothero, D.R., W.A. Berggren, and P.R. Bjork. 1990. Penrose Conference Report: Late Eocene-Oligocene biotic and climatic evolution. *GSA News and Information* 12 (3): 74-75.

Rage, J.-C. 1986. The amphibians and reptiles of the Eocene-Oligocene transition in western Europe: an outline of the faunal alterations. In Pomerol, C. and Premoli-Silva, I., eds., *Terminal Eocene Events.* Amsterdam, Elsevier, pp. 309-310.

Raup, D.M., and Sepkoski, J.J. Jr. 1986. Periodic extinctions of families and genera. *Science* 231: 833-836.

Rea, D.K., Zachos, J.C., Owen, R.M., and Gingerich, P.D. 1990. Global change at the Paleocene-Eocene boundary: climatic and evolutionary consequences of tectonic events. *Palaeogeogr., Palaeoclimat., Palaeoecol.,* 79: 117-128.

Retallack, G.J. 1981. Preliminary observations on fossil soils in the Clarno Formation (Eocene to early Oligocene), near Clarno, Oregon. *Oregon Geol.* 43: 147–150.

Retallack, G.J. 1983a. Late Eocene and Oligocene paleosols from Badlands National Park, South Dakota. *Spec. Pap. Geol. Soc. Am.* 193: 82 pp.

Retallack, G.J. 1983b. A paleopedological approach to the interpretation of terrestrial sedimentary rocks: the mid–Tertiary fossil soils of Badlands National Park, South Dakota.*Geol. Soc. Amer. Bull.* 94: 823–840.

Retallack, G.J. 1990. *Soils of the past: an introduction to paleopedology,* London, Unwin–Hyman, 520 pp.

Riedel, W.R. and Sanfilippo, A. 1986. Radiolarian events and the Eocene-Oligocene boundary. In Pomerol, C. and Premoli-Silva, I., eds., *Terminal Eocene Events.* Amsterdam, Elsevier, pp. 253-257.

Savage, D.E., and Russell, D.E. 1983. *Mammalian Paleofaunas of the World.* Reading, Mass., Addison Wesley Publ. Company. 432 pp.

Schoch, R.M. 1989. *Stratigraphy: Principles and Methods.* New York, Van Nostrand-Reinhold Publ., 375pp.

Sepkoski, J.J., Jr. 1982. A compendium of fossil marine families. *Milw. Publ. Mus. Cont. Biol. Geol.* 51.

Shackleton, N.J. 1984. Oxygen isotope evidence for Cenozoic climate change. In P. Brenchley, ed., *Fossils and Climate,* John Wiley and Sons Ltd., pp. 27-33.

Sigé, B., and Vianey-Liaud, M. 1979. Impropriété de la Grande Coupure de Stehlin comme support d'une limite Eocène-

Oligocène. *Newsl. Stratigr.* 8: 79-82.

Smith, A.B., and Patterson, C. 1988. The influence of taxonomic method on the perceptions of patterns of evolution. *Evol. Biol.* 23: 127-216.

Stehlin, H.G. 1909. Remarques sur les faunules de mammifères des couches éocènes et oligocènes du Bassin de Paris. *Bull. Soc. Géol. Fr.* 9: 488-520.

Stehlin, H.G. 1910. Die Säugetiere des schweizerischen Eocaens 6. *Abh. Schweiz. Pal. Ges.* 26: 839-1164.

Steurbaut, E. 1986. Late middle Eocene to middle Oligocene calcareous nannoplankton from the Kallo Well, some boreholes and exposures in Belgium and a description of the Ruisbroek Sand Member. *Meded. Werkgr. Tert. Kwart. Geol.* 23(2): 49-83.

Stucky, R.K. 1990. Evolution of land mammal diversity in North America during the Cenozoic. *Curr. Mammal.* 2:375-432.

Swisher, C.C., III, and Prothero, D.R. 1990. Single-crystal $^{40}Ar/^{39}Ar$ dating of the Eocene-Oligocene transition in North America. *Science* 249: 760-762.

Tedford, R.H., Banks, M.R., Kemp, N.R., McDougall, I., and Sutherland, F.L. 1975. Recognition of the oldest known fossil marsupials from Australia. *Nature* 255: 141-142.

Thomasson, J.R. 1985. Miocene fossil grasses: possible adaptation in reproduction bracts (lemma and palea). *Ann. Missouri Bot. Garden* 72: 843-851.

Truswell, E.M., and Harris, W.K. 1982. The Cainozoic palaeobotanical record in arid Australia: fossil evidence for the origins of arid-adapted flora, In Barker, W.R., and Greenslade, P.J.M., eds., *Evolution of the Flora and Fauna of Arid Australia.* Adelaide, Peacock Publications, pp. 67-76.

Vianey-Liaud, M. 1976. L'évolution des Rongeurs à l'Oligocène en Europe occidentale. Thèse, Montpellier, France.

Watkins, J.S. and Mountain, G.S., eds. 1990. Role of ODP Drilling in the Investigation of Global Change in Sea Level. Report of a JOI/USSAC Workshop, El Paso, Texas, October 24-26, 1988, 70 p., 3 appendices. (Unpublished report, available from JOI/USSAC Office, Washington, D.C. or from the authors).

Wolfe, J. A. 1971. Tertiary climatic fluctuations and methods of analysis of Tertiary floras. *Palaeogeogr., Palaeoclimatol., Palaeoecol.* 9:27-57.

Wolfe, J.A. 1978. A paleobotanical interpretation of Tertiary climates in the Northern Hemisphere. *Amer. Scientist* 66: 694-703.

Wolfe, J. A. and Hopkins, D. M. 1967. Climatic changes recorded by Tertiary land floras in northwestern North America. In K. Hatai, ed., *Tertiary Correlation and Climatic Changes in the Pacific.* Tokyo, Sasaki Printing and Publishing. pp. 67-76.

Wolfe, J. A. and Poore, R. Z. 1982. Tertiary marine and nonmarine climatic trends. In W. Berger and J. C. Crowell, eds., *Climate in Earth History.* Washington, D.C., Natl. Acad. Sci.. pp. 154-158.

Wood, H.E., and others, 1941. Nomenclature and correlation of the North American continental Tertiary. *Geol. Soc. Amer. Bull.* 52: 1-48.

Woodburne, M.O., and Zinsmeister, W.J. 1984. The first land mammal from Antarctica and its biogeographic implications. *Jour. Paleont.* 58: 913-948.

Zinsmeister, W.J. 1982, Late Cretaceous-early Tertiary molluscan biogeography of southern circum-Pacific. *Jour. Paleont.* 56: 84-102.

Zinsmeister, W.J. 1989. A look at the dramatic change in molluscan diversity along the southern margin of the Pacific at the end of the Eocene. *Geol. Soc. Amer. Abs. Prog.* 21 (6): A88.

1. TOWARD A REVISED PALEOGENE GEOCHRONOLOGY

by William A. Berggren, Dennis V. Kent, John D. Obradovich, and Carl C. Swisher III

ABSTRACT

New information has become available that requires a revision of Paleogene chronology incorporated in most current Cenozoic time scales (e.g. Berggren et al., 1985a, b). Age estimates for the limits of the Paleogene (the Oligocene/Miocene and Cretaceous/Paleogene boundaries) have not changed appreciably and remain at about 24 Ma and about 66 Ma, respectively. However, new radioisotope data indicate that boundaries of subdivisions within the Paleogene are generally younger than previously estimated, for example, the Paleocene/Eocene and Eocene/Oligocene, by about 2 to 3 m.y. We review the current status of magnetobiostratigraphic correlations and new radioisotope data, with particular reference to late Eocene—early Oligocene geochronology and provide a reassessment of the age of the Eocene/Oligocene boundary as 34 Ma. We anticipate that with concurrent work on a fundamental revision of the geomagnetic polarity sequence, a comprehensive and detailed new time scale for the Cenozoic will soon be developed.

INTRODUCTION

It is six years since we published a time scale for the Paleogene (Berggren et al., 1985a, b). All available first-order correlations between Paleogene calcareous plankton and magnetostratigraphy were compiled in Berggren et al. (1985a). Since that time there have been several additions and refinements to this data set; those relevant to late Eocene – early Oligocene geochronology are shown in Tables 1.1 (planktonic foraminifera) and 1.2 (calcareous nannoplankton). Inasmuch as we view a time scale to consist of a data set which integrates information from biostratigraphy, radioisotopes, magnetostratigraphy and sea-floor anomalies (Aubry et al., 1988), we believe that a thorough evaluation of each of these disciplines is required before any basic revision(s) can be made to the extant scale. Moreover revisions to Paleogene geochronology are but part of a more comprehensive revision to the time scale for the Cenozoic Era in which we are currently engaged. Thus we view this contribution as an interim step on the way to a fundamental revision of Cenozoic geochronology.

We review below pertinent new Paleogene radioisotopic age data with particular emphasis on those bearing on the Eocene/Oligocene boundary and provide an assessment of the current status of Paleogene geochronology.

NEW RADIOISOTOPIC DATA

A compilation of (predominantly post-1985) radioisotopic data pertinent to Paleogene geochronology is presented in Tables 1.3, 1.5-1.7 and summarized and compared with the Paleogene time-scale of Berggren et al. (1985a, b) in Figure 1.1. A number of late Eocene – early Oligocene radioisotopic dates were generated from the Marche-Umbria Basin (Italy) in connection with the research for a potential Eocene/Oligocene boundary stratotype section and point (Nocchi et al., 1986; Premoli-Silva et al., 1988). We have compiled an annotated list of those dates relevant to the boundary (Table 1.3) considered reliable by the responsible authors (Montanari, 1988; Montanari et al., 1988; Odin, et al. 1988). Of particular relevance to Paleogene geochronology is the fact that the ages obtained in the northeastern Apennines are from volcanic ashes in stratigraphic sections having (in general) both magneto- and biostratigraphy. Comparison of the magnetic

TABLE 1.1. LATE EOCENE-EARLY OLIGOCENE MAGNETOBIOSTRATIGRAPHIC CORRELATIONS
PLANKTONIC FORAMINIFERA

NAME	FAD	LAD	PALEOMAGNETIC CHRON	REFS.	REMARKS / SOURCE
1. *Acarinina*		X	mid-C17N (ref 1 or top C18N (ref 2)	1, 2	
2. *Planorotalites*		X	top C18N	2	
3. *Porticulasphaera semiinvoluta*	X		top C18N	1, 2	= P14/P15 boundary of Blow (1979). Note brief overlap of *Truncorotaloides* and *P. semiinvoluta* (Blow, 1979) vs. brief, but distinct, separation of *Acarinina* and *P. semiinvoluta* in Nocchi *et al.* (1986, Fig. 1).
4. *Morozovella spinulosa*		X	base 17N	1	
5. (*Cribrohantkenina inflata*)	X		(C16N or basal part of C15R)	3, 4	= P15/P16 boundary of Blow (1979) and Berggren & Miller (1988). Occurs between FAD of *Isthmolithus recurvus* and LAD of *P. semiinvoluta* and *T. pomeroli* in Spain (Molina, 1986; Monechi, 1986; Molina *et al*, 1986). FAD *I. recurvus* is in C16N2 (ref. 1-4) and LAD of *T. pomeroli* and *P. semiinvoluta* occur in C15R (in Contessa Highway; ref. 1-3) or C16N2 (in Massignano section; ref. 4) which suggests that FAD of *C. inflata* corresponds approximately with C16N2. FAD not considered unequivocally determined (ref. 3). Brief overlap of *C. inflata* and *P. semiinvoluta* shown in upper part of *G. semiinvoluta* Zone (= lower Zone P16 of Blow, 1979) in ref. 5.
6. *Porticulasphaera semiinvoluta*		X	C15R (base)	1-3	Located in C16N2 in Massignano section (ref. 4).
7. *Turborotalia pomeroli*		X	C15R	1-3	Located in C16N, just above LAD of *P. semiinvoluta*, in Massignano section (ref. 4).
8. *Turborotalia cunialensis*	X		C15R	2-4	Just above (i.e., younger than) LAD of *P. semiinvoluta* (ref. 2, 3).
9. *Globigerapsis index*		X	C13R	1-4, 6	Essentially coincident with LAD *Discoaster saipanensis* and LAD of *D. barbadiensis* and with lower of 3 normal "events" in C13R in Contessa Highway section (ref. 2, 3). Occurs in younger part of C13R on Kerguelen Plateau (ODP Site 748) and thus appears to be reliable, globally synchronous datum.
10. *Cribrohantkenina inflata*		X	C13R	2-4	= P16/P17 boundary Blow (1979) and Berggren and Miller (1988) and located between LAD of *G. index* and *T. cunialensis* and *Hantkenina* in mid-part C13R at Massignano (ref. 4) and just below youngest of 3 normal "events" in C13R in Contessa Highway section (ref. 2).
11. *Turborotalia cerroazulensis* (incl. *T. cocoaensis* and *T. cunialensis*)		X	C13R	1-4	Associated with youngest of 3 normal "events" in C13R in Contessa Section (ref. 2, 3) and above normal "event" = C13N2 at Massignano (ref. 4).
12. *Hantkenina* spp.		X	C13R	1-4	Located just above LAD of *T. cerroazulensis - cunialensis* and youngest of 3 normal "events" in C13R in Contessa Highway section (ref. 2, 3) and above normal "event" = C13N2 at Massignano (ref. 4).
13. *Pseudohastigerina micra* & *P. danvillensis* (> 150 μm)		X	C13R	2-4	Coincides with LAD *Hantkenina*.
14. *Pseudohastigerina* spp. (*naguewichiensis* & *barbadoensis*) (< 150 μm)		X	C12R	1-3	

Refs:
1. Berggren et al. (1985a)
2. Nocchi et al. (1986) = Lowrie et al. (1982)
3. Premoli-Silva et al. (1988): Gubbio (Contessa Section)
4. Coccioni et al. (1988): Massignano Section (Ancona, Italy)
5. Toumarkine and Luterbacher (1985)
6. Berggren (1991)

( ) = second order correlation

TABLE 1.2. CALCAREOUS NANNOPLANKTON

NAME	FAD	LAD	PALEOMAGNETIC CHRON	REFS.	REMARKS / SOURCE
1. *Isthmolithus recurvus*	X		mid-C16N2	1-4	= CP15a/CP15b boundary; located at top of C16N2 at Contessa Highway Section (ref. 1) and mid-C16N2 at Massignano (ref. 4).
2. *Cribrocentrum reticulatum*		X	mid-C16N1	3, 4	
3. *Cyclococcolithina kingii* (= *C. protoannula)*		X	mid-C15N	3, 4	
4. *Discoaster saipanensis*		X	C13 (third normal event in C13R)	1-4	Coincident with LAD of *Globigerapsis index*
5. *Discoaster barbadiensis*		X	C13 (third normal event in C13R)	1-4	Coincident with LAD of *Globigerapsis index*
6. Acme *Ericsonia obruta*			lower C13N1	2-4	
7. *Ericsonia formosa*		X	C12R (lower part, ref. 1; or top C13N, ref. 2)	1-3	
8. *Isthmolithus recurvus*		X	C12R (lower part)	1-3	
9. *Reticulofenestra umbilica*		X	C12R (lower part)	1-3	

Refs: 1. Berggren et al., (1985a)
2. Nocchi et al.,(1986)
3. Premoli-Silva et al.,(1988)
4. Coccioni et al.,(1988)

CQ = Contessa Quarry; CH = Contessa Highway; MAS = Massignano; MCA = Monte Cagnero.
* = Rb-Sr age.
\+ = Supporting date, not directly tied to magnetobiostratigraphy; not included on Figure 1.1.
( ) = Approximate/estimated correlation.

1. Montanari et al. (1988)
2. Montanari (1988)
3. Odin et al. (1988)

TABLE 1.3. ISOTOPIC AGES RELEVANT TO THE EOCENE/OLIGOCENE BOUNDARY (ITALY)

SAMPLE	BIOSTRAT. LEVEL	PALEOMAG. CHRON	AGE in Ma.	REMARKS	REF.
1. +MCA/84-5	*Upper CP18*	*(base C9N)*	*28.0±0.7*	Approximate age of *28.0 Ma* considered acceptable pending 40 Ar/39 Ar laser-fusion probe analysis (Montanari et al., 1988).	*1,2*
2. CQ/GAR-274	Upper CP18	base C9N (= C9N.9)	28.1±0.3 27.8±0.2	Approximate age of 28.0 Ma considered acceptable pending 40 Ar/39 Ar laser-fusion probe analysis (Montanari et al., 1988).	1,2
3. +MCA/83-3	Lower CP18	(upper C12R)	31.7±0.6		1,2
4. CQ/BOB-247	Lower CP18	upper C12R (= C12R.12)	32.0±0.8	Approximately same biostratigraphic level as sample MCA/83-3 above.	1,2
5. MAS 84/1-14.7	base CP16a	mid-C13R (=C13R.62)	34.6±0.3	MAS 84/1-14.7 (Odin et al., 1988: 214) = MAS/85 -14.7 (Montanari et al., 1988: 206) = MAS/86 - 14.7 (Montanari, 1988: 218, Fig. 4B where it is listed as 34.3± 0.3 Ma).	1-3
6. MAS/84-2: 12.9	base CP15b; upper *T. cerroazulensis* Zone	base C13R (= C13R.82)	33.9±0.4 34.4±0.2*	MAS/84-2:12.9 (Montanari et al., 1988: 217) = MAS/86-12.9 on p. 218, Fig. 4A	1,2
7. MAS 84/1-7.2	CP15B	top C16N (=C16N.0)	35.3±0.7 36.3±0.4*	Same level as sample CQ/ETT-218 below.	1,2
8. CQ/ETT-218	mid-CP15B	top C16N (=C16N.0)	36.9±1.3 35.5±0.2*	Rb/Sr age (Montanari et al., 1988) considered more precise.	1,2
9a. CQ/CAT 210.5A	mid-CP15A	top C17N (=C17N.0)	36.3±0.3	Duplicate dates on same level equivalent to 153.5 m in CH section at top C17N.	1,2
9b. CQ/CAT 210.5B	mid-CP15A	top C17N (= C17N.0)	36.5±0.7		

TABLE 1.4. COMPARISON OF RECENT LATE EOCENE (C17N) – EARLY OLIGOCENE (C9N) AGE ESTIMATES OF MAGNETIC POLARITY CHRONS.

Magnetic Polarity Chron	Berggren et al. (1985 a,b)	Montanari (1988) Montanari et al. (1988) Odin et al. (1988)
	Ages in Ma	
C9N	28.15 - 29.21	~ 28 (base CN9)
C10N	29.73 - 30.33	–
C11N	31.23 - 32.06	–
C12N	32.46 - 32.90	32.0 ± 0.8 (uppermost C12R)
C13N	35.29 - 35.82	34.6 ± 0.8 (mid C13R 33.9 ± 0.4 and 34.4 ± 0.2 (base 13R)
C15N	37.24 - 37.68	–
C16N	38.10 - 39.24	35.3 ± 0.7 and 36.3 ± 0.4 (top of C16N) and 36.9 ± 1.3 and 35.5 ± 0.2 (top C16N)
C17N	39.53 - 41.11	36.3± 0.3 and 36.5 ± 0.7 (top C17N)

TABLE 1.5. RECALIBRATED NORTH AMERICAN LATE EOCENE-OLIGOCENE MAMMAL AGES (PROTHERO et al., 1982; SWISHER AND PROTHERO, 1990)

	Sample	Paleomagnetic Chron	Age in Ma (SEM) (Mean ages of multiple dates)	Remarks
1.	Roundhouse Rock Ash	C10R (Upper part)	28.59±0.32 b	Roundhouse Rock, Wildcat Ridge, Morrill Co., Nebraska.
2.	Nonpareil Ash	Lower C11N	30.05±0.09 b	Roundtop, Nebraska (just below Whitneyan/Arikareean boundary).
3	Upper Whitney Ash	C12N	30.58±0.18 b	Whitney ashes are located at Scottsbluff, Nebraska.
4.	Lower Whitney Ash	Upper C12R	31.85±0.01 b 31.81±0.03 a	"
5.	Persistent White Layer (=Glory Hole Ash)	Uppermost C13R	33.91±0.06 b	Located at Dilts Ranch, Wyoming.
6.	Lone Tree Ash J	C13R/C15N	34.48±0.08 b 34.72±0.04 a	Lone Tree Ash series located at Flagstaff Rim, Wyoming.
7.	Lone Tree Ash I	C15R	35.38±0.10 b	"
8.	Lone Tree Ash G.	C15R	35.57±0.06 b 35.72±0.03 a	"
9.	Lone Tree Ash F	basal C15R	35.72±0.11 b 35.81±0.04 a	"
10.	Lone Tree Ash B.	C16N (top)	35.92±0.01 b 35.97±0.22 a	"

$^{40}Ar/^{39}Ar$ ages based on monitor mineral MMhb-I at 520.4 Ma.
b - biotite; a - anorthoclase.

TABLE 1.6. MISCELLANEOUS EOCENE-OLIGOCENE ISOTOPIC AGES WITH AND WITHOUT MAGNETOBIOSTRATIGRAPHIC CONTROL.

SAMPLE	BIOSTRAT. LEVEL	PALEOMAG. CHRON	AGE in Ma	REMARKS/SOURCE
1. Iversen basalt, Point Arena, Calif.	Latest Zemorrian age; CP19b	(= C6CN)	23.8 wr	K-Ar age by Turner (1970), recalculated by Miller (1981) using new decay constants.
2. ODP Site 706 (basement basalt)[t]	CP16b or CP16c (=NP21)	C13N (~ C13N.5)	33.4± 0.5⁺ wr	Duncan and Hargraves (1990). Age estimate is a weighted mean from plateau and isochron age estimate of 4 samples; northern margin of Nazareth Bank, western subtropical Indian Ocean,13°06.85' S, 61°22.26' E, SE of Seychelles.
3. Society Ridge Core; base of 80.3-81 feet interval		(= C13R) (~C13R.75)	34.29±0.05⁺ b	Upper Yazoo Clay (D. Dockery, III, personal communication, 1990). Society Ridge Test Hole No. 1, Hinds Co., Miss., Sec. 24, T7N R1W; SE/4; NE/4, NW/4. Mean of $^{40}Ar/^{39}Ar$ dates by C. S.
4a. Upper bentonite	(top) NP19/NP20 (top) P16	(= C13R) (~C13R.75)	34.4±0.3⁺ s	Upper Yazoo Clay (undifferentiated); Satartia, 12-13 miles SSW Yazoo Miss.; SW 1/4, Sec. 31, TION, R3W (locality 3). Local. 3 is stratigraphically above the bentonite at local. 1 and 2 (see below). Dated by J.D.O.
4b. Upper bentonite	-"-	(= C13R) (~C13R.75)	34.31±0.07⁺ s	Same bentonite as 4a. Mean of 5 $^{40}Ar/^{39}Ar$ dates by C. S.
5. Lower bentonite	NP19/20	(= C13R) (~C13R.75)	34.9±0.3⁺ s	Yazoo Clay (Obradovich, pers. comm., 1989). Locality 1: SE 1/4, NW1/4, Sec. 9, T9N, R3W. Locality 2: SW 1/4, SW 1/4, Sec, 4, T9N., R3W. Localities 1 and 2 of Obradovich are of the same bentonite along unnamed creek. Dated by J.D.O.
6. North American tektites (Barbados)	*Turborotalia-cerroazulensis* Zone or youngest *G. semiinvoluta* (P15) Zone; NP20; base *Cryptoprora ornata* Zone	(=C15)	35.4±0.6⁺ g	Glass et al., (1986). Located about 26 m below Eocene/Oligocene boundary based on LAD of hantkeninids and *T.cerroazulensis*; correlative with Chron C15 (Miller et al., 1991; cf. Berggren et al., 1985; Nocchi et al., 1986).Note that earlier K-Ar ages (34.2 ± 0.6Ma) on North American tektite field (Zahringer, 1963) are about 1 m.y. younger.
7. DSDP Site 612 tektite	P15 (lower part); *C. ornata* Zone	(= C16N)	35.5±0.3 g	Spherule microtetikite layer just above middle/upper Eocene unconformity; Obradovich et al., 1989; dates of items 6 and 7 are essentially identical but tektites believed to be biostratigraphically separable, Miller, et al., 1991.
8. Bentonite in the Hurricane Lentil, Landrum Mbr., Cook Mountain Fm.	= *Ostrea sellaeformis* Zone (= NP16)	(=lower C18N to upper C20N)	42.0±0.8	Upper part Hurricane Lentil, lower part of Landrum Member, Cook Mountain Formation, East Bank Trinity River, Alabama Ferry locality, Houston Co., Texas (see Stenzel, AAPG, 1940, vol. 24, no. 9, p. 1663-1675). Dated by J. D. O.
9. Castle Hayne bentonite	CP13b (= NP15); P11 *Cubitostrea lisbonensis* Zone	(= lower C20R)	46.2±1.8 45.7±0.7*	Harris and Fullagar (1989). Sequence 1 of Castle Hayne Limestone.
10. DSDP Hole 516F[t]	P10 and NP15 (lower part)	C21N	46.8±0.5 b	Bryan and Duncan (1983). K-Ar determination on biotite from coarse sand in 516F - 76 - 4, 107-115 cm. Calcareous nannoplankton biostratigraphy (Barker, Carlson, Johnson, et al., 1983: 171, 248; Wei and Wise, 1989: 131) constrain this level correlated with Chron C21N (Berggren et al., 1983) to the lower (basal)) part of Zone NP15; see also Berggren et al. (1985).
11. ODP Site 713 (basement basalt)	CP13b (= NP15)	C20R	49.3±0.6⁺ wr	Duncan and Hargraves (1990). Age estimate is a weighted mean average from plateau and isochron age estimate of 5 samples. Northern edge of Chagos Bank, Central Indian Ocean, 04°11.58' S, 73°23.65' E.
12. Montagnais impact structure 200 km south of Halifax, N.S.	NP13; P9	(=C22)	50.5± 0.8⁺	Montagnais impact structure dated isotopically (Bottomley and York , 1988) and biostratigraphically (Aubry, et al., 1990). Additional data can be found in Jansa, et al. (1990).

TABLE 1.6 (continued)

13. Mo clay (-17 ash)	Upper part of *Wetzelliella (Apectodinium) hyperacantha* Zone (*Deflandrea oebisfeldensis* Acme)	(= C24R.33)	55.07±0.16⁺ s	Mo clay ash series at Ejerselev (Jutland) Denmark. Knox (1984) correlated DSDP Site 550 ash sequence (in lower Zone NP10) to Danish ash series from ash layer No. -17 to top of Fur Formation (= ash No. +140). Ash No. -17 thus lies at, or extremely close to, the Paleocene/Eocene boundary. Dated by J.D.O.

K/Ar ages unless otherwise indicated.
* = Rb/Sr age
+ = $^{40}Ar/^{39}Ar$ age
t = basement ages

(= C13R) : no magnetostratigraphy; second order correlation through biostratigraphy.
J. D. O = John D. Obradovich
C. S. = Carl Swisher (using MMhb-I at 520.4 Ma)
s - sanidine; b - biotite; g - glass; wr - basalt

TABLE 1.7. ISOTOPIC AGES RELEVANT TO THE CRETACEOUS/PALEOGENE BOUNDARY. ALL ARE LASER-FUSION $^{40}Ar/^{39}Ar$ AGES ON SANIDINES IN BENTONITES MADE BY CARL SWISHER (SWISHER AND DINGUS, UNPUBLISHED DATA) UNLESS OTHERWISE NOTED (SEE ITEMS 1 AND 6).

SAMPLE	BIOSTRATIGRAPHIC LEVEL	PALEOMAG.CHRON	AGE in Ma.	REMARKS/SOURCE
1. Z coal	c. 40 cm above lowest Paleocene palynoflora and Iridium anomaly; stratigraphically highest dinosaur occurrence is 3.0 m below base of Lerbekmo Z coal	(≅ C29R)	66.1±0.5	$^{40}Ar/^{39}Ar$ plateau age, same loc. as 2: Obradovich (1984). Recalculated from 66.0 Ma using 520.4 Ma for MMhb-I
2. Z coal	"	(≅C29R)	66.17±0.06	"Lerbekmo" Hill Creek locality; correlative with younger Z coal (item 1) (See also Lerbekmo et al., 1979). Mean of 7 dates.
3. Z coal	Bug Creek, Montana; correlative with younger Z Coal at Hell Creek.	(≅C29R)	66.14±0.01	Mean of 3 dates.
4a. Iridium bearing (lower) Z coal	3.1 m and 3.34 m above stratigraphically highest occurrence of K palynoflora and dinosaurs, respectively.	(=C29R)	66.22±0.08	Located just above iridium anomaly. Mean of 30 <u>single crystal</u> dates.
4b. Same as 4a.	"		66.26±0.06	Mean of 10 dates (multiple crystals).
5. Nevis Coal, Alberta	Within 2 m interval of palynomorphic break from (K) *Wodehouseia fimbriata* and (P) *W. spinata* Zones. Nevis coal is 4.5 m and 10.5 m above highest stratigraphic occurrence of *Triceratops* and *Tyrannosaurus* skeletons, respectively, and *Triceratops* is *ca.* 6 m below the K/P palynofloral break (Lerbekmo, et al., 1979).	C29R	66.00±0.05	Mean of 12 dates
6. Beloc, Haiti	Directly overlying uppermost Maastrichtian limestones of *Abathomphalus mayaroensis* Zone and *Micula murus* Zone and overlain by 5-cm basal Paleocene *Guembelitria cretacea* (PO) Zone. P α Zone forms (*i.al., Parvularugoglobigerina eugubina, Eoglobigerina eobulloides)* occur within 1.1 m above top of tektite bearing unit.	C29R	64.48±0.08# 65.2±0.1⁺	Weighted mean of 23 total fusion $^{40}Ar/^{39}Ar$ dates on single tektites using an age of 513.9 (#) or 520.4 (+) for MMhb-1 (Izett et al., 1991; see discussion in this paper). Age estimate of 64.5± 0.1 Ma is consistent with weighted mean average of 3 total laser fusion dates on sanidine crystals from the HS bentonite in Montana of 64.57 ± 0.23 Ma at terrestrial K/P boundary. For additional stratigraphic data see Sigurdsson et al. (1991) and Maurrrasse and Sen (1991). Dates are also consistent with a combined mean age of 64.68 ± 0.12 Ma for Z coal bentonite at Hell Creek, Montana (*cf.* item 3) and a correlative bentonite at Frenchman Valley, Saskatchewan (McWilliams et al., 1991).

TABLE 1.8. DATA FOR EOCENE/OLIGOCENE AGE ESTIMATE BASED ON INTERPOLATION FROM SOUTH ATLANTIC MAGNETIC ANOMALY DISTANCES (CANDE AND KENT, 1991)

ITEM	Chron	S. Atlantic Distance (Km)	Date (Ma)	Anomaly	S. Atlantic Distance (Km)
1. CQ/BOB-247	C12R.12	694.53	32.0	12	675.549
2. ODP 706	C13N.50	750.30	33.4	12R	687.792
3. MAS84/1-14.7	C13R.62	774.68	34.6	13	743.924
4, U. Yazoo.1	C13R.75	778.45	34.3	13R	756.732
5. U. Yazoo.2	C13R.75	778.45	34.4	E/O = 13R.14	760.790
6. U. Yazoo.3	C13R.75	778.45	34.3	15	785.690
7. U. Yazoo.4	C13R.75	778.45	34.9	15R	793.068
8. MAS84/2-12.9	C13R.82	780.48	33.9	16	803.444
9. NA tektites	C15N.00	785.700	35.4	16AR	808.159
10. MAS84/1-7.2	C16N.00	803.440	35.3	16B	812.224
11. CQ/ETT-218	C16N.00	803.440	35.5	16R	828.960

and biochronologically based numerical estimates for late Eocene to early Oligocene magnetic polarity chrons (Berggren et al., 1985a, b) with those based on the ages generated in the northeastern Apennines (Montanari, 1988; Montanari et al., 1988; Odin et al., 1988) (Table 1.4) reveals discrepancies between the two ranging from about 1 m.y. (Chron C9N) to 3 m.y. (Chron C16N - C17N) with younger values found in the data from the northeastern Apennines. These data suggested a need for a revision of the Eocene/Oligocene boundary (Berggren and Kent, 1988) from 36.6 Ma (Berggren et al., 1985a, b) to a value closer to 34 Ma.

Two Oligocene and one early middle Eocene calibration points were used to constrain the lower (predominantly Paleogene) part of the time scale of Berggren et al. (1985a, b). The Oligocene points were the Lone Tree Ashes J and B from Flagstaff Rim, Wyoming, dated respectively at 32.4 Ma and 34.6 Ma by Evernden et al. (1964) and correlated with Chron C12N and C13N, respectively, by Prothero et al. (1982). The supposedly long reversed magnetozone between these ashes (J and B) indicated by the K-Ar dates at Flagstaff Rim were interpreted as correlative with Chron C12R, consistent with the estimated duration of about 2.4 m.y. for Chron C12R on magnetic polarity chronologies. However, recent mean $^{40}Ar/^{39}Ar$ ages of 34.72 Ma (Ash J) and 35.97 Ma (Ash B) (Swisher and Prothero, 1990) (Table 1.5) made on single crystals of anorthoclase indicate that the stratigraphic section between these two ashes spans about half (1.25 m.y. vs. 2.2 m.y.) the time interval previously estimated based on the earlier K-Ar ages of Evernden et al. (1964), and that the paleomagnetic signature of the stratigraphic section spanning the 1.25 m.y. between these ashes (see items 6-10, Table 1.5) is to be correlated with Chron C15 (rather than C12; Prothero et al., 1982; or C13 as reinterpreted by Swisher and Prothero, 1990; see Prothero and Swisher, this volume). It should be noted that Montanari (1990) also reinterpreted the magnetostratigraphy of the High Plains sections based on the assumption that the earlier radioisotopic ages were correct and assigned the reversed interval between the two ashes to Chron C13R.

In similar fashion, newly obtained

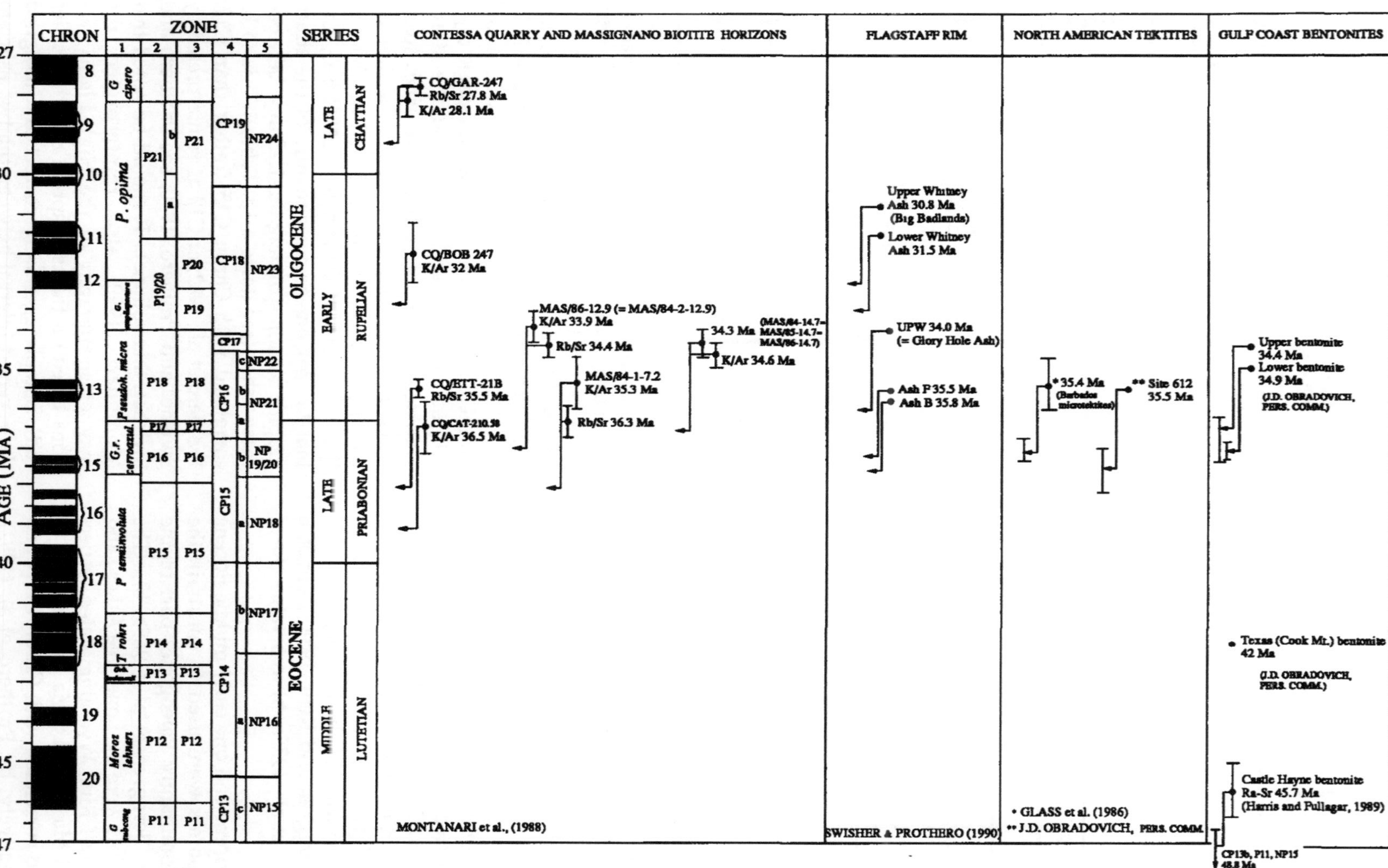

FIGURE 1.1. Magnetobiostratigraphic correlations of middle Eocene – Late Oligocene radioisotope data (Tables 1.3, 1.5, 1.6). Note that the radioisotopic data are plotted against the chronology at the left margin of the table which is that of Berggren et al. (1985 a, b). The arrows indicate the magnetobiostratigraphic position of these ages based on first or second order correlations (see text for further discussion)

$^{40}Ar/^{39}Ar$ ages on single crystals of biotite of 33.91 Ma on the Persistent White Layer (= Glory Hole Ash), just below the lower Orellan normal event and correlative with the terminal range of the titanotheres in eastern Wyoming, and of 30.58 Ma on the Upper Whitney Ash (within the upper Whitneyan normal event), suggest that the long Orellan-Whitneyan reversed interval with a duration of about 2.5 m.y. is consistent with its (re)interpretation as correlative with Chron C12R (Swisher and Prothero, 1990). The new interpretations of North American land mammal magnetobiochronology bring the terrestrial North American record more in line with the marine magnetobiochronology developed in the northern Apennines (Gubbio, Massignano), Italy (Premoli-Silva et al., 1988) and lead to a major realignment of NALM "ages" *vis à vis* marine chronostratigraphy (Swisher and Prothero, 1990).

There are a number of additional radioisotopic ages on Paleogene (primarily Eocene and Oligocene) stratigraphic levels, some of which have magnetostratigraphic, in addition to biostratigraphic, control (Table 1.6). We draw attention to several items of interest in the context of this paper.

1) A mean $^{40}Ar/^{39}Ar$ isochron age on basalt of 33.4 Ma on Chron C13N at Ocean Drilling Project (ODP) Site 706 (item 2, Table 1.6) should be compared with an age of 33.91 Ma on the Persistent White Layer (= Glory Hole Ash) in strata immediately below a normal polarity interval now identified as Chron C13N in Wyoming (item 5, Table 1.5).

2) Several single crystal $^{40}Ar/^{39}Ar$ ages between ~ 34 to 35 Ma in the upper part of the Yazoo Clay Formation of the Gulf Coast and within Zone NP19/20, and thus correlative with Chron C13R (items 3-5, Table 1.6), provide maximum ages for the Eocene/Oligocene boundary.

3) The age of 35.4 Ma on the North American tektite level at Barbados (item 6, Table 1.6) should be compared with ages of 33.9 Ma (K-Ar) and 34.4 Ma (Rb/Sr) at the base of Chron C13R in the Massignano Section (item 6, Table 1.3). The dated levels in both instances are within the *Turborotalia cerroazulensis* Zone, and again provide maximum ages for the Eocene/Oligocene boundary.

4) Ages of 46.2 Ma (K-Ar) and 45.7 Ma (Rb/Sr) on Zone NP15 in the lower Castle Hayne Formation (item 9, Table 1.6) of the Atlantic Coastal Plain and 46.8 Ma on the basal part of Zone NP15 and within a normal polarity interval identified as Chron C21N of Deep Sea Drilling Project (DSDP) Hole 516F (item 10, Table 1.6) support an age estimate of about 46^+ Ma for Chron C21N (cf. Berggren et al., 1985a, b where Chron C21N = 48.75 to 50.34 Ma).

The early middle Eocene calibration point in the (predominantly) Paleogene part of the time scale of Berggren et al. (1985a, b) was an age estimate of 49.5 Ma for the top of Chron C21N. This was based on an interpolation from radioisotopic (K-Ar) ages on lavas and tuffs stratigraphically bracketing the top of a normal magnetozone in continental beds in the western United States correlated to a normal magnetozone in the Ardath Shale (Zone P10) in San Diego, and thence to Anomaly 21 (Flynn, 1983, a, b; 1986).

Prothero and Swisher (this volume) and work in progress by Swisher indicate that the radioisotopic dates from Wyoming used by Flynn (1986) to calibrate Chron 21N may likely be anomalously old. New $^{40}Ar/^{39}Ar$ dates on single crystals of sanidine and biotite extracted from magnetic and biostratigraphic sections in Wyoming and Texas indicate an age of 46-45 Ma for Chron C21N. The Wyoming dates used by Flynn (1986) for the correlation of Chron C21N need to be redated by single crystal methods to determine if they are anomalously old as a result of contamination from older detrital minerals.

In view of the ages of ~ 46-47 Ma for the early Chron C20R to C21N interval cited above, we regard the estimates of 49.5 Ma for the top of Chron C21N (Flynn, 1983 a, b; 1986) and the age of 49.3 Ma on ODP Site 713 basement basalt (within Zone CP13B = NP15) immediately above Anomaly 21 (item 11, Table 1.6) as anomalous in the context of this discussion.

5) The $^{40}Ar/^{39}Ar$ age of 50.5 ± 0.8 Ma on the lowest melt horizon of the Montagnais impact

structure, biostratigraphically dated as Zone NP13 and P9, late early Eocene (item 12, Table 1.6), is seen to be about 2 m.y. younger than the estimated age, 53 Ma, for this level in the time scale of Berggren et al. (1985a, b). This age is consistent with those discussed under items 4 (above) and 6 (below) indicating that the Eocene part of a revised Paleogene time scale will require an approximately 2 m.y. younger shift in values.

6) The $^{40}Ar/^{39}Ar$ plateau age of 55.1 Ma on the -17 ash of the Mo Clay sequence in Denmark (item 13, Table 1.6) provides a much needed calibration point for the early Paleogene as it lies biostratigraphically very close to the Paleocene/Eocene boundary.

7) A large number of single-crystal laser-fusion $^{40}Ar/^{39}Ar$ ages have recently been made on the Hell Creek "Z" coal and the iridium bearing (lower) Z coal of Montana (Table 1.7). The ages range essentially between ~ 66.1 and 66.3 Ma and confirm previous estimates of ~ 66 Ma (e.g., Obradovich, 1984a; Obradovich et al., 1986; chronogram in Harland et al.,, 1990) for the Cretaceous/Paleogene boundary.

A recent study by Izett et al. (1991) reported three laser fusion $^{40}Ar/^{39}Ar$ dates on a sample split of the Baadsgaard et al. (1988) sanidine from the Hell Creek Z-Coal. These authors reported a mean age of 64.56 ± 0.16 Ma for the sanidine, an age approximately two percent younger than that reported here. Approximately one percent of this age difference can be explained by different ages used for the irradiation monitor mineral, the international standard Minnesota hornblende MMhb-I. The age of MMhb-I was originally published by Alexander et al. (1978) with an age of 519.4 Ma. This age was used by Obradovich (1984b) who used MMhb-I as a monitor mineral for the calibration of his $^{40}Ar/^{39}Ar$ plateau age of 66.0 ± 0.54 on sanidine from the Hell Creek Z-Coal. In 1987, the age of MMhb-I was modified to 520.4 Ma by Samson and Alexander (1987). As such, Obradovich's age for the Z-Coal sanidine would be about 0.2% older or 66.1 ± 0.54 Ma. The mean $^{40}Ar/^{39}Ar$ age based on single crystals of sanidine reported in this paper of 66.12 ± 0.14 Ma on sanidine also collected from the Hell Creek Z-Coal is also based on MMhb-I as a monitor mineral with a published age of 520.4 Ma. and is remarkably consistent with the age of 66.1 ± 0.54 Ma obtained by Obradovich.

Izett et al. (1991) report a weighted mean of three total fusion $^{40}Ar/^{39}Ar$ dates of 64.56 ±0.16 Ma. This age is based on an age of 513.9 Ma for MMhb-I, an age the Menlo Park lab has recently adopted for the MMhb-I interlaboratory standard based on an extensive set of potassium and argon measurements on MMhb-I and in-house standard SB-3. What is curious is that although the $^{40}Ar/^{39}Ar$ ages reported by Izett et al. for Hell Creek Z-Coal appear consistent with the K-Ar ages reported by Baadsgaard et al. (1988), they are based on an age of 513.9 Ma for MMhb-I (Izett et al., 1991) while the Australian National University (ANU) reports an age of 524.2 Ma for MMhb-I (Baadsgaard et al., 1988). If an age of 520.4 Ma is used for the Minnesota standard, the Menlo Park results become 65.4 Ma, an age still approximately 1 % younger than the ages adopted here. Obviously, the age of MMhb-I has not been fully resolved at the present time. As a result, we have chosen to report our ages here based on the published age of MMhb-I of 520.4 Ma as this age is most consistent with other data used in the calibration of the current time scale of Berggren et al. (1985a, b). If a younger or even older age for MMhb-I is adopted in the future, then all time scale calibration points will have to be evaluated together, not just those at the K/P boundary.

The other percent age difference for the K/P boundary is not as easily explained. Work in progress indicates that it may also be a bias in the absolute measurement of the $^{40}Ar/^{39}Ar$ of the standard. In one interlaboratory experiment, similar raw $^{40}Ar/^{39}Ar$ ratios were obtained on the K/P sanidine, but about a percent difference for the standard Fish Canyon whose age is based on K-Ar dates and, in part, on the age of MMhb-I. The actual cause of this discrepancy has not been agreed upon at this time, but work in progress will hopefully resolve this issue. We have chosen at this time to continue to use an age of approximately 66 Ma for the K/P boundary until this interlaboratory bias is resolved.

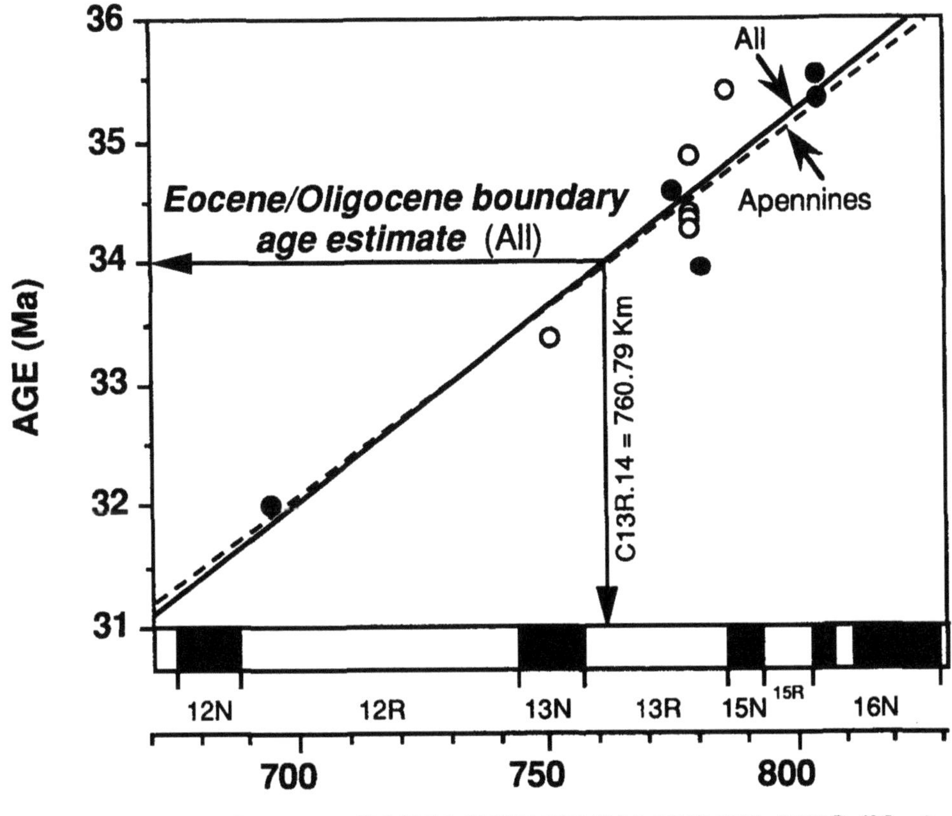

All (○, ● n=11): Age (Ma) = 9.677 + 0.031923*Distance (Km)

Apennines (● n=5): Age (Ma) = 10.837 + 0.030368*Distance (Km)

FIGURE 1.2. Estimated age of the Eocene/Oligocene boundary based on linear regression through 11 radioisotope age determinations plotted with respect to characteristic distance from ridge axis to correlative magnetic anomalies in the South Atlantic (Cande and Kent, 1991) according to magnetobiostratigraphic constraints (Table 1.8; see text for further discussion). Filled circles indicate subset of data from the Apennines (Table 1.3), open circles other data (Table 1.6).

THE EOCENE/OLIGOCENE BOUNDARY

The Eocene/Oligocene boundary has recently been stratotypified in the Massignano section (Umbro-Marche Apennines), near Ancona, Italy, where the boundary point is designated at the 19 meter level within C13R; this is 0.14 of the stratigraphic distance below the base of Chronozone C13N (=C13R.14) and corresponds to the extinction level of hantkeninids (Nocchi et al., 1986).

Data from Tables 1.3, 1.5, and 1.6 which are relevant to late Eocene–Oligocene geochronology are presented in Figure 1.1. The radioisotopic data are plotted directly to the chronologic scale at the left of the figure, whereas the arrows indicate the magnetobiostratigraphic position based on first or second order correlation. For example, the ages of 27.8 Ma and 28.1 Ma on Contessa Quarry biotites would correlate with early Chron C8R in the chronology of Berggren et al. (1985a, b). However, the dated level is within earliest Chron C9N and is seen to be about 1 m.y. younger (28 Ma vs. 29.2 Ma) than the age estimate of Berggren et al. (1985a, b). The other data are similarly plotted for each of the general areas discussed above.

Examination of the data in Tables 1.3, 1.5-1.7 and Figure 1.1 indicate that the Eocene/Oligocene boundary (= P17/P18 boundary, within Zone NP21, and within the upper

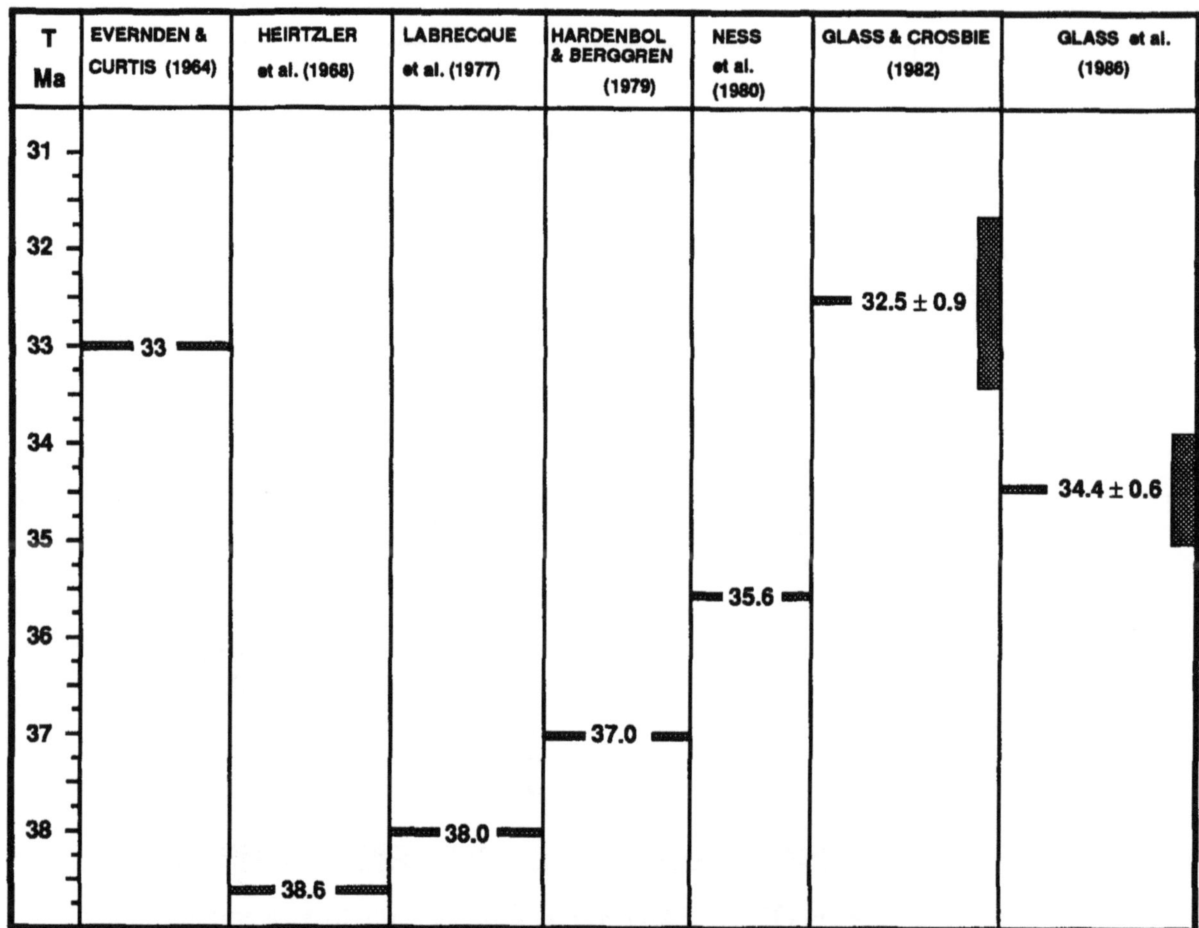

FIGURE 1.3. Historical vicissitude of the Eocene/Oligocene boundary.

third of Chron C13R (=C13R.14) in Massignano) is bracketed by:

1) an age of 33.4 Ma on Chron C13N at ODP Site 706 (item 2, Table 1.6).

2) ages of 34.28 Ma (item 3, Table 1.6), 34.4 Ma and 34.31 Ma (items 4a and 4b, Table 1.6), and 34.9 Ma (item 5, Table 1.6) on upper Yazoo Clay bentonites of the Gulf Coastal Plain which are correlative biostratigraphically with the lower part of Chron C13R.

An estimate of 34.0 Ma for the boundary (Figure 1.2) is obtained by a linear regression on a total of 11 dates from the Apennines, North American tektites in Barbados, and bentonites from the Upper Yazoo Clay (items 4-8, Table 1.3; items 2-6, Table 1.6) plotted according to magnetobiostratigraphic constraints with respect to characteristic distance from ridge axis to correlative magnetic anomalies in the South Atlantic as revised by Cande and Kent (submitted to *Journal of Geophysical Research*). This age estimate is based on the assumption that the Eocene/Oligocene boundary occurs at C13R.14, as indicated in the magnetobiostratigraphy of the Massignano section, and that seafloor spreading was uniform in the South Atlantic over just the short interval from about anomaly 12 to Anomaly 16. A very similar age estimate (33.9 Ma) for the Eocene/Oligocene boundary is obtained if the regression analysis is performed on only the 5 dates in this interval from the Apennines (items 4-8, Table 1.3).

The estimate of 34 Ma upon which most

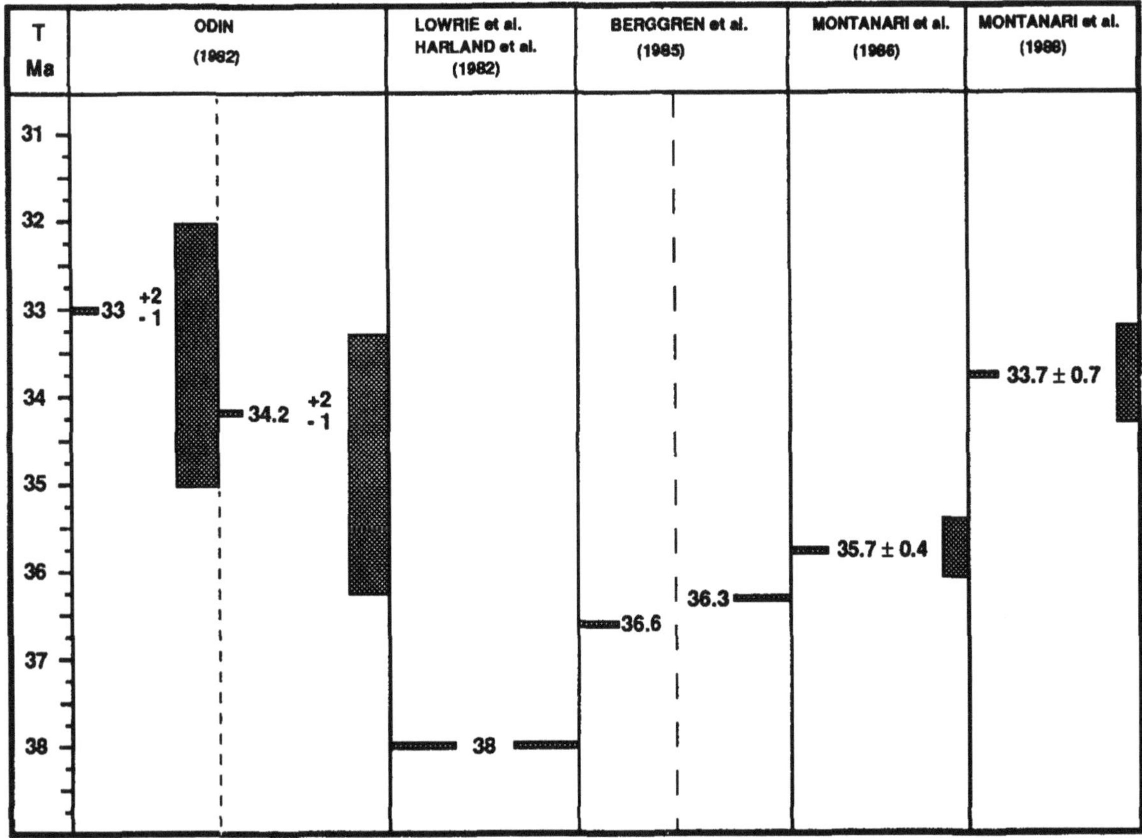

geochronologists now appear to agree should be contrasted with previous estimates of the age of the Eocene/Oligocene boundary (Figure 1.3) which ranged from < 33 Ma (Evernden et al., 1964; Glass and Crosbie, 1982) to > 38 Ma (Heirtzler et al., 1968; LaBrecque et al., 1977; Lowrie et al., 1982; Harland et al., 1982). In this context it is important to understand the nature of the methodologies used in obtaining these widely differing values. For example:

1) The estimate of 33 Ma by Evernden et al. (1964) was based on K-Ar ages on volcanics intercalated in mammal-bearing terrestrial sequences. However, because of the endemic nature of the North American fauna during this time interval, correlation with the standard European chronostratigraphic framework was based entirely upon stage of evolution of the mammalian faunas. Recent studies on these dated volcanics have shown that some of the earlier age estimates based on conventional K-Ar dates were too old due to contamination, while others were too young as a result of alteration (Swisher and Prothero, 1990).

2) The age estimates of 38.6 Ma, 38.0 Ma and 35.6 Ma by Heirtzler et al. (1968), LaBrecque et al. (1977), and Ness et al. (1980), respectively, were magnetochronologic estimates based on extrapolation and interpolation between direct ages or age estimates on selected magnetic polarity chrons or sea-floor anomalies. These age estimates for the Eocene/Oligocene boundary are for a level within the upper part of Chron C13R and may be found in the tables in Ness et al. (1980). The age of 38 Ma of Lowrie et al. (1982) and Harland et al. (1982) is essentially that of LaBrecque et al. (1977).

3) The age estimates of Glass and Crosbie (1982) and Glass et al. (1986) were based on sediment rate extrapolations to the biostratigraphically estimated level of the Eocene/Oligocene boundary in local sections based on K-Ar ages on the North American tektite strewn field. The age estimate of 34.4 Ma was based on a direct age of 35.4 Ma on the Barbados tektite

(see item 6, Table 1.6) which lies some 26 m below the biostratigraphically estimated position of the Eocene/Oligocene boundary.

4) The estimates of Odin (1982) are based on (predominantly) glauconite ages on upper Eocene – lower Oligocene sediments. The different age estimates reflect varying interpretations of the base of the Oligocene as equivalent to the Lattorfian/Bartonian stage boundary (34 $^{+2}/_{-1}$ Ma) or the Rupelian/Priabonian stage boundary (33 $^{+2}/_{-1}$ Ma).

5) The age of 36.6 Ma of Berggren et al. (1985a, b) was based on a linear regression through several calibration points which connected three linear segments of the sea-floor magnetic reversal sequence (see Berggren, 1986). The actual value for the Eocene/Oligocene boundary in Berggren et al. (1985a, b) should be 36.3 Ma in as much as the position of the boundary was plotted in the middle part of Chronozone C13R, rather than in the upper part of Chronozone C13R as now indicated by the Massignano section. This age estimate was based in part on the anomalously old K-Ar dates of Everden et al. (1964) (Swisher and Prothero, 1990).

6) The age estimates of 35.7 ± 0.4 Ma and 33.7 ± 0.7 Ma in Montanari et al. (1986, 1988, respectively) are based on a regression line through radioisotopic ages in the lower Oligocene and upper Eocene of the northeastern Apennines. The younger value resulted from the elimination of an age on a lower Oligocene (Chron C13N) stratigraphic level subsequently believed to be spurious. A recent estimate of 33.7 + 0.4 Ma of Odin et al. (1991) is in substantial agreement with the age of 34 Ma we derive here for the Eocene/Oligocene boundary.

CURRENT STATUS OF PALEOGENE GEOCHRONOLOGY

The following age estimates for boundaries of the major subdivisions of the Paleogene are suggested on the basis of our assessment of current information:

Oligocene/Miocene (Chattian/Aqutanian) boundary, correlated to Chron C6CN, at about 24 Ma (e.g., item 1, Table 1.6; chronogram of Harland et al., 1990);

Eocene/Oligocene (Priabonian/Rupelian) boundary, correlated to Chron C13R.14, at 34 Ma;

Paleocene/Eocene (Thanetian/Ypresian) boundary, correlated to Chron C24R.33, at about 55 Ma (item 12, Table 1.6);

Cretaceous/Paleogene (Maastrichtian/Paleocene) boundary, correlated to Chron C29R.3, at about 66 Ma (Table 1.7).

ACKNOWLEDGEMENTS

This paper grew out of a general review of the status of Paleogene geochronology presented by one of us (WAB) at the Penrose Conference (1989). We wish to thank M.-P. Aubry for her help in biostratigraphically dating several key samples with the aid of calcareous nannoplankton and discussions on various problems associated with stratigraphic interpretation, and K. G. Miller and S. Cande for thoughtful critiques of a draft version of this paper. This research was supported by NSF grants OCE-9101463 (WAB) and OCE-8811339 and OCE-9104447 (DVK). This is WHOI Contribution Number 7777 and L-DGO Contribution Number 4817.

REFERENCES

Alexander, E.C., Jr., Mickelson, G.M., and Lanphere, M.A. 1978. A new $^{40}Ar/^{39}Ar$ dating standard, In R.E. Zartman, ed., Short papers of the Fourth International Conference on Geochronology, Cosmochronology, Isotope geology 1978. *U. S. Geological Survey, Open-File Report* 78-701:6-8.

Aubry, M. P., Gradstein, F. M. and Jansa, L. F. 1990. The late Early Eocene Montagnais meteorite: No impact on biotic diversity: *Micropaleont.* 36(2): 164-172.

Baadsgaard, H., Lerbekmo, J.F and McDougall, I. J. F. 1988. A radiometric age for the Cretaceous-Tertiary boundary based on K-Ar, Rb-Sr, and U-Pb ages of bentonites from Alberta, Saskatchewan, and Montana, *Can. Jour. Earth Sci.* 25: 1088-1097.

Barker, P. F., Carlson, R. L., and Johnson, D. A. 1983. *Initial Reports of the Deep Sea Drilling Project*, 72.Washington, D. C., U. S. Government Printing Office: 1024 p.

Berggren, W. A. 1986. Geochronology of the Eocene/Oligocene boundary. In Pomerol, C. and Premoli-Silva, I., eds., *Terminal Eocene Events*, Amsterdam, Elsevier Science

Publishers, pp. 349-356

Berggren, W. A. 1991. Paleogene planktonic foraminifera magnetobiostratigraphy of the Southern Kerguelen Plateau (ODP Sites 747-749). In Wise, Jr., S. W., Schlich, R., et al., *Proc. ODP Sci. Results*, 120: College Station, TX (Ocean Drilling Program).

Berggren, W. A., Hamilton, N., Johnson, D. A. Pujol, C., Weiss, W., Cepek, P., and Gombos, Jr., A. M. 1983. Magnetobiostratigraphy of Deep Sea Drilling Project Leg 72, Sites 515-518, Rio Grande Rise (South Atlantic). In Barker, P. F., Carlson, R. L., and Johnson, D. A., et al. *Initial Reports of the Deep Sea Drilling Project*, 72: Washington, D.C., U.S. Government Printing Office: 939-948.

Berggren, W. A. and Kent, D. V. 1988. Late Paleogene time-scale: an update. *Geol. Soc. Amer., Abs. Prog.*, 21(6): A 86.

Berggren, W. A., Kent, D. V. and Flynn, J. J. 1985a. Paleogene geochronology and chronostratigraphy. In Snelling, H.J., ed., *The Chronology of the Geological Record, Geological Society Memoir* 10: 141-193.

Berggren, W. A., Kent, D. V., Flynn, J. J. and Van Couvering, J. A. 1985b. Cenozoic geochronology. *Geol. Soc. Amer. Bull.* 96: 1407-1418.

Bottomley, R., and York, D. 1988. Age measurements of the submarine Montagnais impact crater and the periodicty question: *Geophys. Res. Lett.* 14(12): 1409-1412.

Bryan, N. B. and Duncan, R. A. 1983. Age and provenance of clastic horizons from Hole 516F. In Barker, P. F., Carlson, R. L., and Johnson, D. A., et al. *Initial Reports of the Deep Sea Drilling Project*, 72: Washington (U.S. Government Printing Office): 475-477.

Cande, S.C. and Kent, D.V. 1991, A new geomagnetic polarity timescale for the late Cretaceous and Cenozoic. *Jour. Geophys. Res.*, in review.

Coccioni, R., Monaco, P., Monechi, S., Nocchi, M., and Parisi, G. 1988. Biostratigraphy of the Eocene–Oligocene boundary at Massignano (Ancona, Italy). In Premoli-Silva, I., Coccioni, R. and Montanari, A., eds., *The Eocene–Oligocene Boundary in the Marche-Umbria Basin (Italy)*, International Subcommission on Paleogene Stratigraphy, E/O Meeting, Ancona, 1987, Special Publication II. 1, Ancona: 59-74.

Duncan, R. A. and Hargraves, R. B. 1990. $^{40}Ar/^{39}Ar$ geochronology of basement rocks from the Mascarene Plateau, the Chagos Bank, and the Maldives Ridge. In Duncan, R. A., Backmann, J., Peterson, L. C. et al., *Proceedings ODP, Sci. Results*, 115: College Station, TX (Ocean Drilling Program): 43-51.

Evernden, J. F., Savage, D. E., Curtis, G. H., and James, G. T. 1964. Potassium-argon dates and the Cenozoic mammalian chronology of North America. *Amer. Jour. Sci.* 262: 145-198.

Flynn, J. J. 1983a. Correlation and geochronology of middle Eocene strata from the western United States. Ph.D. Dissertation, Columbia University, New York, 498 pages.

Flynn, J. J. 1983b. Correlation of middle Eocene marine and terrestrial strata. *Geol. Soc. Amer., Abs. Prog.* 15(6): 574-575.

Flynn, J. J. 1986. Correlation and geochronology of middle Eocene strata from the western United States. *Paleogeogr., Palaeoclimat., Palaeoecol.* 55(1986): 335-406.

Glass, B. P. and Crosbie, J. R. 1982. Age of the Eocene/Oligocene boundary based on extrapolation from North American microtektite layer. *Bull. Amer. Assoc. Petrol. Geol.* 66: 471-476.

Glass, B. P., Hall, C. D., and York, D. 1986. $^{40}Ar/^{39}Ar$ laser-probe dating of North American tektite fragments from Barbados and the age of the Eocene–Oligocene boundary. *Chem. Geol. (Isotope Geoscience Section)* 59: 181-186.

Hardenbol, J. and Berggren, W.A. 1978. A new Paleogene numerical time scale. *Amer. Assoc. Petrol Geol. Stud. Geol.* 6: 213-234.

Harland, W.B., Armstrong, R., Cox, A., Craig, L., Smith, A., and Smith, D. 1990, *A Geologic Time Scale.* Cambridge, Cambridge University Press, 263 pp.

Harland, W. B., Cox, A.V., Llewellyn, P.G., Pickton, C.A.G., Smith, A.G., and Walters, R. 1982, *A Geologic Time Scale.* Cambridge, Cambridge University Press, 131 pp.

Harris, N. B. and Fullagar, P. D. 1989. Comparison of Rb-Sr and K-Ar dates of middle Eocene bentonite and glauconite, southeastern Atlantic Coastal Plain. *Geol. Soc. Amer. Bull.* 101(4): 573-577.

Izett, G.A., Dalrymple, G.B., Snee, L.W. 1991. $^{40}Ar/^{39}Ar$ age of the Cretaceous -Tertiary boundary tektites from Haiti. *Science* 252: 1539-1542.

Jansa, L. F., Aubry, M.-P., and Gradstein, F. M. 1990. Comets and extinctions: cause and effect? *Geol. Soc, Amer. Spec. Paper* 247: 223-232.

Knox, R. W. O'B, 1984. Nannoplankton zonation and the Paleocene – Eocene boundary beds of NW Europe: an indirect correlation by means of volcanic ash layers. *Jour. Geol. Soc. London* 141: 993-999.

LaBrecque, J.L., Kent, D.V., and Cande, S.C. 1977, Revised magnetic polarity time scale for the Late Cretaceous and Cenozoic time. *Geology* 5: 330-335.

Lerbekmo, J.F., Evans, M.E., and Baadsgaard, H. 1979. Magnetostratigraphy, biostratigraphy, and geochronology of Cretaceous-Tertiary boundary sediments, Red Deer Valley. *Nature* 279 : 26-36.

Lowrie, W., Napoleone, G., Perch-Nielsen, K., Premoli-Silva, I., and Toumarkine, M. 1982. Paleogene magnetic stratigraphy in Umbrian pelagic carbonate rocks: the Contessa sections, Gubbio. *Geol. Soc. Amer. Bull.* 92: 414-32.

McWilliams, M.; Baksi, A.; and Baadsgaard, H. 1991. New $^{40}Ar/^{39}Ar$ ages from the K-T boundary bentonites in Montana and Saskatchewan. *EOS* 72(17): 301 (abstract).

Miller, K. G., Berggren, W. A., Zhang, J., and Palmer-Judson, A. 1991. Biostratigraphy and isotope stratigraphy of upper Eocene microtektites at Site 612: how many impacts? *Palaios* 6(1): 17-38.

Miller, P. 1981. Tertiary calcareous nannoplankton and benthic foraminifera biostratigraphy of the Point Arena, California. *Micropaleontology* 27(4): 419-443.

Molina, E. 1986. Description and biostratigraphy of the main reference section of the Eocene/Oligocene boundary in Spain: Fuente Caldera section. In Pomerol, C. and Premoli-Silva, I., eds., *Terminal Eocene Events*, Amsterdam, Elsevier Science Publishers, pp. 53-63.

Molina, E., Monaco, P., Nocchi, M. and Parisi, G. 1986. Biostratigraphic correlation between the Central Subbetic (Spain) and Umbro-Marchean) Italy) pelagic sequences at the Eocene/Oligocene boundary using foraminifera. In Pomerol, C. and Premoli-Silva, I., eds., *Terminal Eocene Events*, Amsterdam, Elsevier Science Publishers, pp. 75-85.

Monechi, S. 1986. Biostratigraphy of Fuente Caldera section by means of calcareous nannofossils. In Pomerol, C. and Premoli-Silva, I., eds., *Terminal Eocene Events*, Amsterdam, Elsevier Science Publishers, pp. 65-69.

Montanari, A. 1988. Geochemical characterization of volcanic biotites from the upper Eocene – upper Miocene pelagic sequence of the northeastern Apennines. In Premoli-Silva, I., Coccioni, R., and Montanari, A., eds., *The Eocene – Oligocene Boundary in the Marche-Umbria Basin (Italy). International Sub-commission Paleogene Stratigraphy, Eocene/Oligocene Boundary Meeting, Ancona, Oct. 1987, Special Publication*: 209-227.

Montanari, A. 1990. Geochronology of the terminal Eocene impacts; An update. *Geol. Soc. Amer. Spec. Paper* 247: 607-613.

Montanari, A., Deino, A. L. Drake, R. E., Turrin, B. D., DePaolo, D. J., Odin, G. S., Curtis, G. H., Alvarez, W., and Bice, D. 1988. Radioisotopic dating of the Eocene – Oligocene boundary in the pelagic sequences of the northeastern Apennines. In Premoli-Silva, I., Coccioni, R., and Montanari, A., eds., *The Eocene – Oligocene Boundary in the Marche-Umbria Basin (Italy). International Subcommission Paleogene Stratigraphy, Eocene/Oligocene Boundary Meeting, Ancona, Oct. 1987, Special Publication*: 195-208.

Ness, G., Levi, S., and Crouch, R. 1980. Marine magnetic anomaly timescales for the Cenozoic and Late Cretaceous: a precís, critique and synthesis. *Rev. Geophys. Space Phys.* 18(4): 753-770.

Nocchi, M., Parisi, G., Monaco, P., Monechi, S., Mandile, M., Napoleone, G., Ripepe, M., Orlando, M., Premoli-Silva, I., and Bice, D. M. 1986. The Eocene–Oligocene boundary in the Cambrian pelagic regression. In Pomerol, C. and Premoli-Silva, I., eds., *Terminal Eocene Events*, Amsterdam, Elsevier Science Publishers, pp. 25-40.

Obradovich, J. D. 1984a. What is the age of the Cretaceous – Tertiary boundary as recognized in continental deposits? *Geol. Soc. Amer. Abs. Prog.* 16(6): 612.

Obradovich, J.D. 1984b. An overview of the

measurement of geologic time and the paradox of geologic time scales. *Stratigraphy* 1:11-30.

Obradovich, J.D. and Cobban, W.A. 1975. A time-scale for the late Cretaceous of the western interior of North America, In W.G.E. Caldwell, ed., Cretaceous system in the western interior of North America. *Geol. Assoc. Canada Spec. Paper* 13: 31-54.

Obradovich, J.D., Sutter, J.F., and Kunk, M.J. 1986. Magnetic polarity chron tiepoints for the Cretaceous and early Tertiary. *Terra Cognita* 6: 140.

Odin, G., Guise, P., Rex, D. C., and Kreuzer, H. 1988. K-Ar and $^{39}Ar/^{40}Ar$ geochronology of late Eocene biotites from the northeastern Apennines. In Premoli-Silva, I., Coccioni, R., and Montanari, A., eds., *The Eocene – Oligocene Boundary in the Marche-Umbria Basin (Italy). International Sub-commission Paleogene Stratigraphy, Eocene/Oligocene Boundary Meeting, Ancona, Oct. 1987, Special Publication*: 239-245.

Odin, G.S., Montanari, A., Deino, A., Drake, R., Guise, P.G., Kreuzer, H., and Rex, D.C. 1991, Reliability of volcano-sedimentary biotite ages across the Eocene-Oligocene boundary (Apennines, Italy), *Chem. Geol. (Isotope Geoscience Section)* 86: 203-224.

Premoli-Silva, I., Coccioni, R., and Montanari, A., eds., 1988, *The Eocene – Oligocene Boundary in the Marche-Umbria Basin (Italy). International Sub-commission Paleogene Stratigraphy, Eocene/Oligocene Boundary Meeting, Ancona, Oct. 1987, Special Publication.*

Prothero, D. R., Denham, C. R., and Farmer, H. G., 1982. Oligocene calibration of the magnetic polarity time scale. *Geology* 10(11): 650-653.

Samson, S.D. and Alexander, E.C. Jr. 1987. Calibration of the interlaboratory $^{40}Ar/^{39}Ar$ standard, MMhb-I. *Chem. Geol. (Isotope Geoscience Section)* 66:27-34.

Swisher, C. C. and Prothero, D. R. 1990. Single-crystal $^{40}Ar/^{39}Ar$ dating of the Eocene – Oligocene transition in North America. *Science* 249: 760-762.

Toumarkine, M. and Luterbacher, H. 1985. Paleocene and Eocene planktic foraminifera. In Bolli, H. M., Saunders, J. B. and Perch-Nielsen, K. *Plankton Stratigraphy*, Cambridge, Cambridge Univ. Press, pp. 87-154.

Turner, D. C. 1970. Potassium-argon dating of Pacific coast Miocene foraminiferal stages. In Bandy, O. L., ed., Paleontologic zonation and radiometric dating. *Geol. Soc. Amer. Spec. Paper* 124: 91-129.

Wei, W. and Wise, Jr., S. W. 1989. Paleogene calcareous nannofossil magnetobiochronology: results from South Atlantic DSDP Site 516. *Marine Micropaleont.* 14: 119-152.

Zahringer, J. 1963. K-Ar measurements of tektites, In *Radioactive dating; Proceedings of a Symposium*, Athens, November 19-23, 1962: Vienna, I.A.E.A.: 289-305.

2. MAGNETOSTRATIGRAPHY AND GEOCHRONOLOGY OF THE TERRESTRIAL EOCENE-OLIGOCENE TRANSITION IN NORTH AMERICA

by Donald R. Prothero and Carl C. Swisher III

ABSTRACT

New $^{40}Ar/^{39}Ar$ dates and magnetic stratigraphy of Uintan through Whitneyan terrestrial sections in North America radically change long-held interpretations of the age and correlation of North American Land Mammal "Ages" with the Eocene - Oligocene timescale. Current data indicate that earliest Uintan (Shoshonian) faunas, as well as the Bridgerian/Uintan boundary occur at the base of magnetic Chron C20R, considered here to be about 46 Ma. Typical early Uintan faunas occur later in Chron C20R, and late Uintan faunas in Chron C20N to C18R (43-40 Ma). Earliest Duchesnean faunas first occur above the Lapoint Ash (dated at about 40 Ma, most likely within the later part of C18N). The Uintan/Duchesnean boundary would then fall within Chron C18N. The Duchesnean-Chadronian transition lies within Chron C16N at approximately 37 Ma. The Chadronian/Orellan boundary occurs near the top of Chron C13R, at about 33.9 Ma; the Orellan/Whitneyan boundary is in the middle of Chron C12R, just below a date of 31.8 Ma. The Whitneyan/Arikareean boundary occurs within Chron C11N, above a date of 30.05 Ma.

The major faunal discontinuity within the middle part of the Duchesne River Formation marks the Uintan/Duchesnean boundary in North America. This faunal transition would appear then to correlate with the major faunal turnover recorded across the marine middle/late Eocene boundary. A lesser faunal change at the Chadronian/Orellan boundary correlates approximately with the Eocene/Oligocene boundary at about 34 Ma.

INTRODUCTION

Since the first discovery of fossil vertebrates in the Big Badlands of South Dakota in 1846, terrestrial rocks and fossils of the North American Eocene-Oligocene transition have been extensively studied. The highly fossiliferous deposits of the White River Group in the Big Badlands of South Dakota (and in Nebraska, Wyoming, Colorado, and North Dakota) have been an irresistible magnet to both amateur and professional collectors (Figure 2.1). Early in this century, the Uinta Basin provided a window on the middle and late Eocene in North America. As a result, Wood et al. (1941) used fossils from these two areas to typify their Uintan, Duchesnean, Chadronian, Orellan, and Whitneyan North American Provincial Ages (now North American Land Mammal "Ages," or "NALMAs").

The last decade has seen great breakthroughs in stratigraphic and geochronologic study of these deposits. The first important new data came from extensive fossil collections made by Morris Skinner and colleagues at the Frick Laboratory beginning in the 1930s. Unlike most previous White River collections, these fossils were collected with detailed stratigraphic data; many specimens were zoned to the nearest foot from marker ashes. When these collections became available for study, they permitted the development of the first detailed biostratigraphy of the White River deposits and allowed detailed examination of the temporal sequence of mammalian evolution during the Eocene and Oligocene (Emry, 1973, this volume; Prothero, 1982). In recent years, a tremendous literature on the Uintan through Whitneyan interval has accumulated. The most recent reviews of the biochronology were published by Krishtalka et al. (1987) and Emry et al. (1987), although the systematics of the

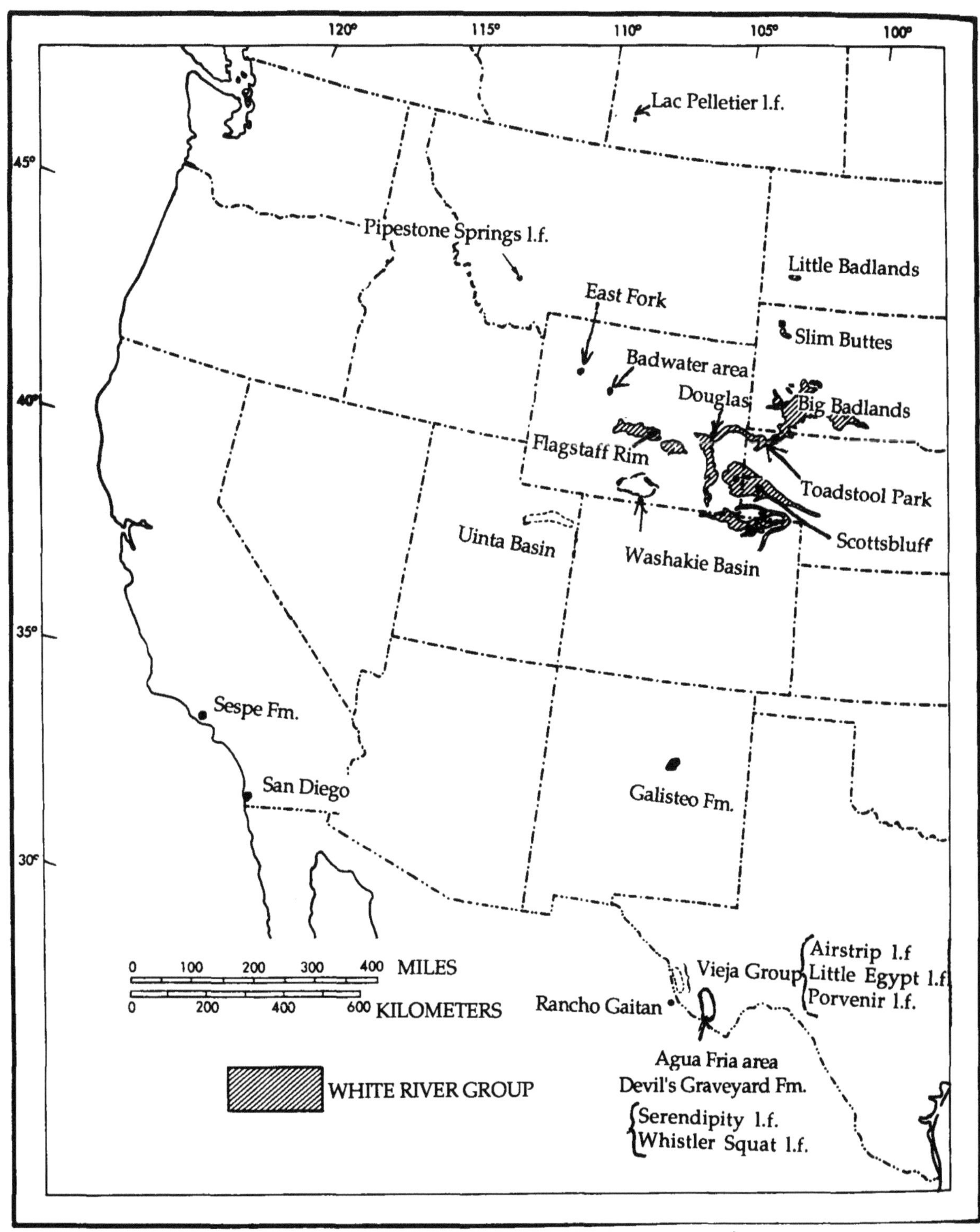

FIGURE 2.1. Index map, showing location of major terrestrial fossil localities of the Eocene-Oligocene transition in western North America. Modified from Emry et al. (1987, Figure 5.2).

mammals have not been thoroughly reviewed in 50 years (Scott, Jepsen, and Wood, 1940-1941). Such a detailed review of the fossil mammals is beyond the scope of the present paper, although much of the information is summarized by Stucky (this volume). The systematics and detailed biostratigraphy of the mammalian fauna will be published elsewhere (Prothero and Emry, in prep.).

Prior to the application of K-Ar dates in 1964, correlation of the North American Uintan through Whitneyan NALMAs with the global time scale rested almost entirely upon comparative stage of evolution of mammalian fossils. Little faunal interchange occurred with Europe and/or Asia during this period and interfingering marine/non-marine beds are scarce. Evernden et al. (1964) provided a series of dates from the Chadronian Flagstaff Rim section in Natrona County, Wyoming. The K-Ar dates indicated that the Chadronian represented a longer time interval than the preceding Duchesnean or succeeding Orellan and Whitneyan NALMAs. However, as no K-Ar dates existed on the European marine late Eocene or Oligocene, the correlation of the North American Chadronian within the European chronology was not altered.

A major breakthrough concerning Eocene-Oligocene intracontinental correlations came with the application of magnetic polarity stratigraphy to continental deposits. Soft, well-exposed badlands siltstones proved to be ideal for magnetic stratigraphy. Since 1976, magnetic sampling has covered most of the key sections (Prothero et al., 1982, 1983; Prothero, 1982, 1984, 1985a, 1985c, 1985d; Denham, 1984; Testarmata and Gose, 1979; Flynn, 1986). The combination of these detailed magnetostratigraphies calibrated with the K-Ar dates (provided by Evernden et al., 1964, and many other studies) culminated in the use of these dates for the calibration of the geologic time scale by Berggren et al. (1985). By magnetostratigraphic correlation, the Flagstaff Rim dates and the Bridgerian-Uintan dates cited in Flynn (1986) indicated an age of approximately 36.5 Ma for the Eocene/Oligocene boundary. However, shortly after the publication of the Berggren et al. (1985) time scale, detailed magnetostratigraphy and K-Ar dating of intercalated biotite horizons of the Massignano section in Italy called these correlations into question and indicated a much younger age of approximately 34 Ma for the Eocene/Oligocene boundary (Montanari et al., 1988).

The next refinement in the stratigraphy and geochronology of the North American sections occurred as a result of innovations in radioisotopic dating. With the advent of single-crystal laser-fusion $^{40}Ar/^{39}Ar$ dating, the accuracy of many of North American calibration points based on K-Ar dates have come into question. Rather than basing a calibration point on a single K-Ar date, multiple dates could be routinely obtained with laser-fusion $^{40}Ar/^{39}Ar$ dating. Furthermore, contaminant and altered crystals could be identified and avoided, yielding increased confidence in the geological accuracy of the dates. As a result, many new dates have recently become available for more precise calibration of the North American sequence. The new geochronology has necessitated not only a radical reassessment as to the duration of Uintan through Whitneyan NALMAs, but also of magnetic and biochronological correlations with the Lyellian epochs as well (Swisher and Prothero, 1990).

These breakthroughs in biostratigraphy, magnetostratigraphy, and geochronology have significantly changed the temporal framework of the entire late Paleogene North American sequence. Several generations of vertebrate paleontologists educated with the idea that "Uintan is late Eocene" and "Chadronian is early Oligocene" must now rethink the basis of these correlations. Even very recent studies (e.g., Prothero, 1989) are now completely out of date. In this paper, we present a summary of the present status of the North American chronology, although this interim report is liable to be revised by further detailed analyses.

GEOCHRONOLOGY

The intracontinental correlation of North American terrestrial deposits has been effectively accomplished by the use of fossil mammals. However, in many cases, fossil mammals do not occur uniformly throughout these deposits, and many deposits are unfossiliferous. Much to their credit, North American paleontologists have pieced together the North American record into a series of Land Mammal

"Ages" that now appear to cover much of Cenozoic time. However, numerous problems exist in the correlation of these mammal "Ages" with the global time scale. The Uintan through Whitneyan interval is particularly difficult in this regard, since it was period of isolation of the North American continent with little faunal exchange with Europe and Asia. The additional lack of fossiliferous intertonguing nonmarine /marine deposits during this period further hampered precise global correlations.

Recent advances in mass spectroscopy and the development of single-crystal laser-fusion (SCLF) $^{40}Ar/^{39}Ar$ dating has revolutionized the field of geochronology. The SCLF $^{40}Ar/^{39}Ar$ dating method is superior to conventional potassium-argon dating in the ability to date minute samples such as single crystals, obtain both potassium and argon measurements (or proxy measurements) from the same sample split at the same time by the same method, and determine multiple age components of contaminated tephra. Because such small samples are used, less "clean-up" time is required of the gases, resulting in the ability to obtain a date in approximately 20 minutes. Consequently, multiple dates can be routinely obtained, permitting increased precision and accuracy. As a result, SCLF $^{40}Ar/^{39}Ar$ dating is rapidly revising the age of many of the potassium-argon dates that have been in wide use for the last 25 years for the calibration of the geologic time scale.

We present new $^{40}Ar/^{39}Ar$ ages based on replicate dates on single crystals of sanidine, biotite, and plagioclase separated from key tephra collected from Duchesnean through Whitneyan sections (Tables 2.1 and 2.2). A summary of most of these dates were presented in Swisher and Prothero (1990). New dates on the Buckshot Ignimbrite and Bracks Rhyolite (from the Vieja Group, Trans-Pecos Texas) and the Lapoint Tuff (from the type Duchesne River Formation in the Uinta Basin, Utah) are presented for the first time. The status of the calibration points for Duchesnean through Whitneyan sections are briefly discussed below. In the next section these dates are incorporated into the context of a magnetobiochronology for the Uintan through Whitneyan NALMAs.

Duchesnean

The Lapoint Tuff occurs at the contact between the Dry Gulch Creek and Lapoint Members of the Duchesne River Formation (Figure 2.2) in eastern Utah (Andersen and Picard, 1972). The fauna from the Brennan Basin and Dry Gulch Creek Members below the tuff are considered late Uintan in Age, while that from the Lapoint Member is considered Duchesnean (Emry, 1981; Krishtalka et al., 1987).

A K-Ar date on biotite from the Lapoint Tuff was reported by McDowell et al. (1973) of 40.4 ± 0.8 Ma. A new sample of this tuff reported here yielded two new K-Ar dates (KA5856-01 and 5856-02) on biotite of 39.75 ± 0.7 Ma and 39.78 ± 0.7 Ma. Six $^{40}Ar/^{39}Ar$ dates made on single crystals of this same biotite yielded a consistent mean age of 39.74 ± 0.07 Ma (Tables 2.1-2.2). We consider the age of 39.74 ± 0.07 Ma. as the best estimate for the age of the Lapoint Tuff.

Current attempts by Swisher and Stucky (work in progress) to date a biotite-bearing unit overlying the Badwater 20 locality of the Hendry Ranch Member of the Wagonbed Formation in central Wyoming has resulted in questioning its use as a calibration point for the Duchesnean. A K-Ar date on biotite of 42.3 ± 1.4 Ma reported by Black (1969) has been used as a lower age limit of about 42 Ma for the Duchesnean NALMA (Krishtalka et al., 1987). Dating of the freshest-looking single crystals of this biotite fraction yielded a spectrum of $^{40}Ar/^{39}Ar$ ages, all older than 50 Ma, indicating the presence of detrital biotite. The further observation of muscovite in the dated sample indicates a granitic and/or metamorphic source of the unit. As a result, it would be virtually impossible to obtain a conventional K-Ar date on multiple biotite grains that could serve as an accurate calibration point for the Duchesnean NALMA. No younger age component has been discovered in this unit, indicating the unlikelihood of any primary volcanic event of Duchesnean age. Swisher and Stucky conclude that the Badwater 20 date reported by Black (1969) and Krishtalka et al. (1987) is based on a fortuitous averaging of older altered biotite grains and should therefore be rejected as a calibration point for the Duchesnean. Additional work is underway to determine if there is any younger age component, or whether the entire unit is detrital.

Duchesnean - Chadronian

The dating of the Duchesnean-Chadronian transition has been hampered, in part, because of the controversy surrounding the faunal definition of the beginning of the Chadronian (Emry et al., 1987). Most workers agree that the transition can be found in the Vieja sequence of west Texas, but disagreement continues over the placement of the Porvenir local fauna within the Duchesnean or Chadronian NALMA. Dates from the Bracks Rhyolite and Buckshot Ignimbrite, which bracket the Little Egypt and Porvenir local faunas (Figure 2.3), are widely used for the calibration of the Duchesnean-Chadronian transition (Emry et al., 1987). Two K-Ar dates on sanidine have been reported for the overlying Bracks Rhyolite of 37.8 Ma (Evernden et al., 1964) and 37.4 ± 1.2 Ma (Wilson et al., 1968), while the underlying Buckshot Ignimbrite yielded sanidine ages of 36.1 ± 2.3 Ma, 35.6 ± 2.0 Ma. and 39.6 ± 1.2 Ma. Four additional K-Ar dates on sanidine by Henry et al. (1986) indicate an average age of 37.23 ± 0.25 Ma for the Buckshot Ignimbrite. New $^{40}Ar/^{39}Ar$ dates made on single crystals of sanidine from these tuffs by Swisher and Henry (work in progress) yielded mean ages of 36.67 ± 0.04 Ma for the Bracks Rhyolite and 37.80 ± 0.06 Ma for the underlying Buckshot Ignimbrite. The single-crystal dates are considered here as the most accurate age estimates for the Vieja tephra.

Chadronian

Swisher and Prothero (1990) reported new $^{40}Ar/^{39}Ar$ dates on the ashes from the Flagstaff Rim (FSR) section in Wyoming. These samples were originally collected by Morris Skinner and dated by Evernden et al. (1964) by the K-Ar method. The new $^{40}Ar/^{39}Ar$ dates suggest that slight alteration of the biotite has resulted in small amounts of argon loss. While Swisher picked the freshest-looking biotite from these samples, slightly younger ages are noted in comparison with the anorthoclase dates from the same unit. To a greater extent this same phenomenon resulted in erroneously young ages in the bulk K-Ar dates of Evernden et al. (1964). We therefore, place higher confidence in the anorthoclase dates from FSR. As can be seen in Table 2.2, the lowermost dated ashes, Lone Tree ash B (LTB) and LTF, contain sufficient amounts of anorthoclase for conventional K-Ar dating. Evernden et al. 's (1964) K-Ar dates on anorthoclase are consistent with the $^{40}Ar/^{39}Ar$ dates presented here, although the single dates were unable to distinguish the age differences between LTB and LTF. The major change in the age of the Flagstaff Rim section occurs above LTB, where the biotite dates were as much as 2 million years younger than the new $^{40}Ar/^{39}Ar$ dates based on either anorthoclase or biotite. Evernden et al. (1964) did not publish anorthoclase dates for the upper ashes at Flagstaff Rim.

Chadronian - Orellan

Swisher and Prothero (1990) also presented single-crystal $^{40}Ar/^{39}Ar$ dates for the 5 Tuff of Evanoff et al. (this volume). It is also known as the "Glory Hole" Ash, near Douglas, Wyoming, and in Nebraska, the Persistent White Layer [PWL] of the Frick Laboratory, or the "Purple-White Layer" of Schultz and Stout (1955, 1961). This ash marks the boundary between the Chadron Formation and the overlying Orella Member of the Brule Formation. It contained a few detrital biotite crystals, indicating slight reworking. However, five consistent dates were obtained indicating an eruption age of 33.91 ± 0.06 Ma for the PWL. This dates the Chadronian/Orellan boundary, since the last appearance of Chadronian brontotheres, and the first appearance of several characteristic Orellan taxa, occurs about 6.4 m above this ash. No other Orellan calibration points have been reported.

Whitneyan

The Lower and Upper Whitney ashes are widespread marker beds in Nebraska used for the correlation of the Whitney Member of the Brule Formation (Schultz and Stout, 1955, 1961; Swinehart et al., 1985). Previous attempts to date these ashes by conventional K-Ar methods (Emry, pers. commun.) resulted in anomalously old dates due to the inclusion of older detrital biotite. Swisher and Prothero (1990) published mean $^{40}Ar/^{39}Ar$ dates on single crystals of anorthoclase, plagioclase, and biotite separated from the Lower Whitney Ash (LWA). The consistent dates on biotite and sanidine indicate a 31.8 Ma age for the LWA.

The sample from the Upper Whitney Ash (UWA) contained biotite and plagioclase. The plagioclase yielded a wide spectrum of ages and is considered too altered to be accurately dated. The biotite, however, yielded consistent single-crystal $^{40}Ar/^{39}Ar$ dates indicating an age of 30.58 ± 0.18 Ma for the UWA.

Whitneyan - Arikareean

Faunal continuity characterizes the Whitneyan-Arikareean transition (Tedford et al., 1985, 1987). In Nebraska, this transition occurs in the upper part of the Brule Fm., informally termed the "brown siltstone member" and in the lower part of the Gering Formation (Tedford et al., 1985). The boundary has not been clearly defined, but it is bracketed by two ashes, the Nonpareil Ash of the Brule Formation (Swinehart et al., 1985), and the Roundhouse Rock Ash of the Gering Formation. Swisher and Prothero (1990) indicate a mean $^{40}Ar/^{39}Ar$ date on single crystals of biotite of 30.05 ± 0.09 Ma for the Nonpareil Ash and 28.59 ± 0.32 Ma for the Roundhouse Rock Ash. The latter is consistent with a previous K-Ar date of 28.7 ± 0.7 Ma on the same ash (Obradovich et al., 1973).

MAGNETOSTRATIGRAPHY AND BIOSTRATIGRAPHY

Magnetic stratigraphy has made it possible to resolve many of these correlation problems (Prothero, 1988). Unlike other methods of correlation, magnetic stratigraphy offers the possibility of globally synchronous time correlation independent of facies or lithology. It is the only method that allows high-resolution correlation of the land sequences with the marine record. However, it also has its limitations. Magnetic signatures are only "normal" or "reversed," and by themselves are not indicative of a particular numerical age. This problem can be severe in continental sections. The apparent length of a particular magnetic polarity interval is strongly influenced by the sedimentation rate of the deposits. A long polarity interval in a stratigraphic section may in actuality represent a short period of time and thus will affect precise correlation of a local magnetostratigraphy to the magnetic polarity time scale (MPTS). Mammalian biostratigraphy has helped in some regards within North America, but has rarely been useful when correlating with other continents or the marine record. The numerical age of a particular reversal and its correlation with the MPTS is heavily dependent on radioisotopic dating. In most cases, magnetic correlations are only as good as the accuracy of the dates on which they were calibrated.

The new radioisotopic ages force a significant reinterpretation of the duration and correlation of Uintan through Whitneyan NALMAs with the the global timescale (Berggren et al., this volume). This, in conjunction with a revision in the calibration of the Paleogene time scale, necessitates a reinterpretation of all previously studied North American Eocene-Oligocene magnetic sections. The following is a status report of ongoing work on calibration of the Paleogene of North America. Due to space limitations, only a summary of the magnetic data is presented. Details of the magnetic procedures and rock magnetism of each of the sections are presented in Prothero et al. (1983), Prothero (1985a, 1985c) and Prothero (in prep.).

Uintan-Duchesnean

The Uintan and Duchesnean NALMAs were originally defined on mammalian fossils found in the Uinta Basin in northeastern Utah (Figure 2.1). However, Uintan and Duchesnean rocks have also been recorded in many other parts of the western United States and Canada, as reviewed by Krishtalka et al. (1987). Wood et al. (1941), correlated the Bridgerian Provincial Age (now North American Land Mammal "Age" or NALMA) with the middle Eocene, and the Uintan and Duchesnean with the late Eocene of Europe. Refinements in dating and biochronologic correlations of the middle and late Eocene in Europe, and of the North American sequence, has gradually pushed the Uintan into the middle Eocene (Berggren et al., 1985; Krishtalka et al., 1987). Ongoing work has seen the development of magnetostratigraphies covering many of the key sections pertinent to the calibration of the Uintan - Duchesnean interval.

Flynn (1986) provided the first magnetic data on sections spanning the Bridgerian/ Uintan boundary (Figure 2.2). He produced magnetic stratigraphies for sections in the East Fork and Washakie Basins of western

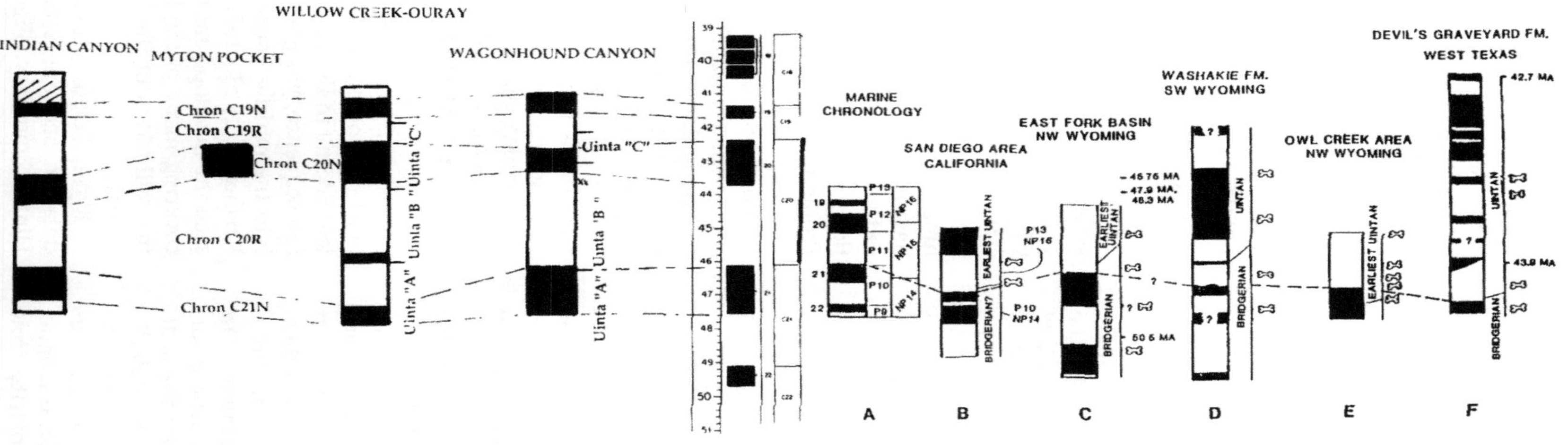

FIGURE 2.2. Correlation of the Uinta Basin magnetic pattern (left) and other Uintan magnetic sections from Flynn (1986) and Walton (this volume) with the magnetic polarity timescale of Berggren et al. (this volume).

Wyoming, and the San Diego area in southern California. Flynn noted that the Bridgerian/Uintan boundary consistently fell within a reversed interval which he identified as Chron C20R. Bridgerian faunas were also found in rocks of reversed and normal polarity below (interpreted to be Chrons C21N and C21R). Uintan faunas were found in the normal and reversed intervals identified as Chrons C20R and C20N.

The Uintan and Duchesnean NALMAs derive their names from the Uinta and Duchesne River Formations that crop out in the Uinta Basin of northeastern Utah. The Uinta Formation is stratigraphically complex, with the fluvial sandstones laterally interfingering with the lacustrine shales of the Green River Formation (Dane, 1954; Cashion, 1967). The lowest informal unit of the Uinta Formation, Uinta "A" of Osborn (1929), is poorly fossiliferous, although like the overlying Uinta "B" contains early Uintan faunas (Krishtalka et al., 1987); these two units were combined to form the Wagonhound Member of Wood (1934). Uinta "C" (the Myton Member of Wood, 1934) contains a late Uintan fauna, as do the overlying Brennan Basin and Dry Gulch Creek Members of the Duchesne River Formation (Andersen and Picard, 1972; Emry, 1981; Krishtalka et al., 1987).

Magnetostratigraphies are currently being developed for these important Uintan and Duchesnean sections (Prothero, in prep.). Prothero has collected paleomagnetic samples from four critical sections in the eastern, central, and western Uinta Basin (Figure 2.2). The easternmost section ran through Osborn's (1929) and Riggs' (1912) sections at Wagonhound Canyon, near Bonanza, Utah, and then up through the Coyote Basin to Kennedy Wash. Here, the lower Uinta Formation is almost entirely fluvial sandstones, and the Myton Member consists of floodplain mudstones. Most of the early collections of Uinta Basin mammals were made in this region (Osborn, 1929, Figure 63). Prothero's results indicate that the Uinta "A" is entirely of normal polarity in this section. A zone of reversed polarity began just above the base of Uinta "B" and extended to about 20 m below the "*Amynodon* sandstone" that marks the transition to Uinta "C." The Uinta "B-C" transition is in this normal interval, but the rest of Uinta "C" is reversed, as is the base of the lower Duchesne River Formation. The top of the section (still within the lower Duchesne River Formation) is of normal polarity.

The central section (following Kay, 1934) ran along Willow Creek to Ouray, Utah, and then up the Green River through Leota Bottom and Brennan Basin. This section spanned several important Carnegie Museum localities, including White River Pocket, and Leota, Thorne, and Skull Pass Quarries. In these sections, the lower Uinta Formation was predominately of reversed polarity up to just above White River Pocket. Lower Uinta "C" was of normal polarity, but the upper part was reversed. Another normal zone occurs near the base of the type section of the Brennan Basin Member of the Duchesne River Formation. A short parallel section was run through the important locality at Myton Pocket, seven miles southeast of Myton, Utah (Hamblin, 1987). This locality contains a Uinta "C" fauna like those of Leota Quarry, and both are of normal polarity.

The westernmost section was measured in Indian Canyon, southwest of Duchesne, Utah. Here the Uinta fluvial sandstones have been replaced with lacustrine shales, evaporites, and carbonates, called the "saline facies" and "sandstone and limestone facies" by Dane (1954). The magnetic section through Indian Canyon yielded a normal interval in the Green River Shale just above the Horse Bench Sandstone marker, another in the midst of the "saline facies" section, and a third near the top of the "sandstone and limestone facies."

In summary, there is a normal polarity interval in the lowermost Uinta Formation, another at the base of Myton Member, and a third within the Brennan Basin Member of the Duchesne River Formation. How does this polarity record correlate with other regions? Unfortunately, radioisotopic dates are ambiguous. The $^{40}Ar/^{39}Ar$ date of 47.4 ± 0.24 Ma in upper Uinta "B2" in Coyote Basin reported in an earlier abstract (Prothero and Swisher, 1990) appears to be due to detrital contamination. Mauger (1977) reported K-Ar dates of 43.1 ± 1.3 Ma (corrected for new constants) in the Green River Formation in Indian Canyon about 70 m above the Horse Bench Sandstone, and 42.8 ± 1.0 Ma for a tuff 19 m below the contact between the saline facies and limestone-sandstone facies of Dane (1954). Unfortunately, the reliability of Mauger's dates have been questioned (Krishtalka et al., 1977). Considering that the lower date falls in the early Uintan normal, and the upper date is in reversed rock laterally equivalent to the Uinta-Duchesne River contact (three polarity zones later), their small separation in ages is also suspicious. Swisher is currently resampling these ashes to date them with $^{40}Ar/^{39}Ar$ methods.

Flynn (1986) reported that the Bridgerian-Uintan faunal transition takes place early in Chron C20R (about 45.5 Ma, according to the time scale of Cande and Kent, in press). Bridgerian faunas occur at the beginning of Chron C20R in both the Washakie Basin (Fauna B, Flynn, 1986, Figure 9) and East Fork Basin (Fauna C, Flynn, 1986, Figure 9) of Wyoming. Earliest Uintan ("Shoshonian") faunas occur immediately above faunas these in both places. On this basis, we (Prothero, 1990; Prothero and Swisher, 1990) correlated the Uinta "A" normal with Chron C21N, and the Uinta "B" reversed with C20R. At the time, we followed conventional accounts that Uinta "A" was unfossiliferous, and therefore could be older than the *bona fide* early Uintan faunas of Uinta "B." Uinta "B" thus recorded the earliest Uintan fossils, and was correlative with the early Uintan faunas found elsewhere in Chron C20R.

However, the details of the biostratigraphy of the lower Uinta Formation clearly need more careful study. Krishtalka et al. (1987) reported that Uinta "A" yielded a sparse fauna of early Uintan fossils including *Amynodon, Triplopus, Metarhinus, Dolichorhinus,* and *Forstercooperia.* This was first reported by Osborn (1895, p. 75) as the fauna from "Horizon A, the *Telmatotherium megarhinum* beds." Osborn (1929, pp. 91-96) later subdivided his Uinta "A" into Uinta "A" plus Uinta "B1." The 350 feet of what Osborn (1895) called "Horizon B, the *Telmatotherium cornutum* beds" was redefined as Uinta "B2" in 1929. The supposed "Uinta A" fossils reported by Osborn (1895) came from Osborn's (1929) Uinta "B1," and Osborn was consistent in 1929 when he reported that Uinta "A" was barren. Similarly, Riggs (1912) reported that recollecting could not duplicate the supposed records of the brontotheres *Metarhinus* and *Sphenocoelus* from the base of Uinta "A." (Osborn, 1929, records these from Uinta "B1"). Kay (1934, 1957) indicated that Uinta "A" is barren except for one brontothere, *Sthenodectes priscus.* Krishtalka et al. (1987) report that that Uinta "A" perissodactyls could be found in the Carnegie Museum collections, but the detailed field and stratigraphic records on these specimens cannot be found to corroborate this (R. Stucky, pers. commun., 1991).

Our experience has shown that it is very difficult to distinguish Uinta "A" and "B" in the field. Outside the control of the measured sections, there is no way to tell one sandstone from the next. Osborn's changing definitions of Uinta "A" and "B" between 1895 and 1929 reflect this difficulty. It is very likely that specimens labeled "Uinta A" in museum collections of a century ago may be mislocated or mislabeled. Until museum collections and field records have been carefully re-examined, or new specimens have been found *in situ,* we regard the provenance of these supposed "Uinta A" fossils as questionable.

Uinta "B1" of Osborn (1929) is the first unit to produce unequivocal early Uintan fossils in the Uinta Basin. Stratigraphic documentation of the mammal fossils within this unit is inadequate at present, but the published records (e.g., Osborn, 1929, Figure 65) indicate that diagnostic Uintan taxa, such as *Triplopus, Amynodon, Eobasileus,* and *Dolichorhinus,* do not occur until midway through Uinta "B1" or higher. Only the brontotheres *Metarhinus* and *Rhadinorhinus* are documented from the lower half of Uinta "B1." Thus, it is possible that the lowermost part of Uinta "B1" is Bridgerian, as well. There is a specimen of the Bridgerian taxon *Hyrachyus eximius* (AMNH 1879) from the base of Uinta "B1." This is consistent with the interpretation that the Bridgerian-Uintan transition occurs early in Chron C20R, since these *bona fide* earliest Uintan fossils do not occur until well within the Uinta "B1" reversed interval (Figure 2.2). Similarly, Kay (1934, Plate XLVI) indicated that Uintan faunas first occurred midway through Uinta "B" in the Willow Creek-Ouray section of the central Uinta Basin, also midway through Chron C20R (Figure 2.2). It appears that Uinta "A" and the lowest part of Uinta "B" correlate with Chron C21N and the base of C20R, and that the

Bridgerian-Uintan transition occurs midway through Uinta "B."

The lower part of Uinta "C" also occurs in a normal polarity interval, which probably correlates with Chron C20N. Although the boundary between the Uinta and Duchesne River Formations occurs in the next reversed interval above (C19R), these sections do not span the transition between Uintan and Duchesnean faunas, since Uintan mammals are found in the lower two members of the Duchesne River Formation (Emry, 1981; Krishtalka et al., 1987). The higher parts of the Duchesne River Formation were sampled magnetically, but the results were discouraging, since these redbeds had a chemical overprinting from hematite that could not be removed.

The alternative to correlating the Uinta Basin sequence with Chrons C21N-C19N (Figure 2.2) is to shift the sequence up one complete polarity chron. In this interpretation, the Uinta "A" normal would be C20N, and Uinta "B" would correlated with C19R. However, this conflicts with late Uintan faunas in C20N in the Washakie Basin reported by Flynn (1986), and also the occurrence of late Uintan faunas in Chron C20N in the Devil's Graveyard Formation of Trans-Pecos Texas (Walton, this volume). The late Uintan Serendipity l.f. not only occurs just within C20R, but also lies beneath a K-Ar date of 42.7 ± 1.6 Ma. If this date is correct (it, too, should be rerun with $^{40}Ar/^{39}Ar$ methods), then the beginning of the late Uintan cannot be younger than this date, and the Uinta "A" normal could not be C20N. Clearly, much more dating and biostratigraphic work is needed to resolve these difficulties.

These correlations, if correct, would indicate that the radioisotopic dates used by Flynn (1986) to calibrate the age of C21N are too old, as are many of the K-Ar dates currently being used for the calibration of the Bridgerian-Uintan NALMAs (Krishtalka et al., 1987). Flynn estimated an age for the top of C21N at approximately 49.3 Ma by means of interpolation between dated horizons of 45.7 and 48.1Ma over 1500 feet above the polarity boundary, and 50.5 Ma from 950 feet below the top of C21N. Two independent studies in progress indicate that this age estimate may be too old. Swisher (work in progress) has discovered that contamination of detrital biotite appears a common problem with many of the Wyoming Eocene dates, resulting in erroneously old ages. $^{40}Ar/^{39}Ar$ dates on individual crystals of biotite from the Bridger B Tuff, for example, have resulted in a bimodal age distribution, suggesting abundant contamination of older detrital biotite. This work indicates that the K-Ar dates used for the calibration of the Bridgerian may be some 1.5 million years too old. A similar problem has been reported for many of Mauger's (1977) K-Ar dates for the Washakie Basin and the Uinta Basin, and work is underway to resolve the age of these important tuffs. It is premature to present an evaluation of each of the published Bridgerian-Uintan K-Ar dates, although almost every unit has reported dates of almost bimodal distribution. Although the older dates have been used by most paleontologists, it is likely that these are in fact too old due to contamination. These important units are currently being redated by single-crystal $^{40}Ar/^{39}Ar$ methods to separate the apparent multiple age components.

Support for a younger age for C21N also comes from dates on the Castle Hayne Formation of the Atlantic Coastal Plain, with Rb/Sr dates of 46.2 ± 1.8 Ma and 45.7 ± 0.7 Ma on this magnetic polarity chron. In addition, a date of 46.8 ± 0.5 Ma from DSDP Hole 516F also supports a younger C21N (see Berggren et al., this volume). This younger age estimate for C21N is supported by work in progress by Cande and Kent (in press). These workers have refit the Paleogene magnetic sea floor anomalies, refining their duration and age estimates. Their work indicates that the duration of C21N ranges from 48.0 Ma to 46.2 Ma (Kent, pers. commun.), approximately 1 to 2 million years younger than indicated by the Berggren et al. (1985) time scale. Similar adjustments were proposed for C19N -C22N.

These studies indicate that Flynn's (1986) paleomagnetic and biostratigraphic correlations and the ones suggested above are correct, although inconsistent with many published K-Ar dates associated with these units. Work is underway to resolve this discrepancy. Flynn (1986) showed that the normal polarity interval he identified as Chron C21N in the Ardath Shale in San Diego, California, contains a microfauna representing planktonic foraminiferal Zone P10 and nannofossil Zone NP14. Since these zones both include part of Chron C21N (Berggren et al., 1985), the marine stratigraphy is consistent with the magnetic stratigraphy, with the land mammals, and with the new dates, but not with the old K-Ar dates. The numerical age of C21N, however, is probably 46.2-48.0 (Kent, pers. commun), not 48.7-50.1 as reported in Berggren et al. (1985, Figure 5).

Walton (this volume) has sampled Bridgerian and Uintan volcaniclastics of the lower and middle members of the Devil's Graveyard Formation in Trans-Pecos Texas (Stevens et al., 1984; Wilson, 1986). Early Uintan faunas, such as Whistler Squat l.f., occur at the base of a reversed interval she identifies as C20R (Figure 2.2). Late Uintan faunas, such as Serendipity and Titanothere Hill, occur at the top of the reversed interval she identifies as C20R. According to her interpretation, the rest of the section spans C20N, and possibly C19N and C19R. The section terminated just beneath the Duchesnean Skyline and Cotter channels, and the upper Bandera Mesa Member of the Devil's Graveyard Formation has not been sampled.

Of particular interest is the polarity and age assignment of the Whistler Squat local fauna. The Whistler Squat l.f. is in sediments of reversed polarity, underlying a biotite-bearing tuff dated by K-Ar methods at 44.0 ± 0.9 Ma (McDowell, 1979). Rodents from Whistler Squat were considered by Wood (1973, 1974) to be late Bridgerian while the presence of *Amynodon* and other taxa indicate an early Uintan age (Krishtalka et al., 1987). This discrepancy in age assignment may reflect, in part, the mixture of Bridgerian and Uintan faunal elements noted by Flynn (1986) when defining the Shoshonian "subage" of the Uintan. Krishtalka et al. (1987) in an added note suggest a possible Shoshonian age assignment for the Whistler Squat l.f. If the reversed polarity interval containing the Whistler Squat l.f is C20R, it would be consistent with the polarity interpretation of all other Shoshonian faunas (Flynn, 1986). The date of 44 Ma from a unit overlying the Whistler Squat l.f. (also of reversed polarity) would then be consistent with the new, younger age estimate for the beginning of the Uintan NALMA.

Testarmata and Gose (1979) published a magnetic stratigraphy of the Duchesnean and early Chadronian portions of the Vieja Group in Trans-Pecos Texas, to the west of the Devil's Graveyard Formation (Wilson, 1977, 1984). Testarmata and Gose sampled only a portion of the Vieja section, and did not deal with the Uintan, nor most of the Chadronian localities. Their magnetic data were quite noisy, and produced numerous polarity intervals which have been frequently reinterpreted. For example, Testarmata and Gose (1979, Figure 7) originally correlated their section with Chrons C12N to C13R. The Duchesnean Porvenir l.f., and the ?Chadronian Little Egypt l.f., which occur in predominantly normal rocks between the Buckshot Ignimbrite and Bracks Rhyolite, were correlated with C13N. Prothero (in Prothero et al., 1982, 1983; Prothero, 1985a) reinterpreted the Vieja section to correlate with C13N to C15R, based on faunal similarities with the Flagstaff Rim section (see Wilson, 1986; Emry, this volume).

The new dates from Flagstaff Rim (Swisher and Prothero, 1990), along with the redating of the Bracks Rhyolite and Buckshot Ignimbrite (see above), force yet another reinterpretation. Since both the Duchesnean Porvenir l.f. and the Chadronian Little Egypt l.f. are located between these rhyolites, this would appear to date the Duchesnean-Chadronian transition. However, the Little Egypt l.f. contains very few truly diagnostic Chadronian taxa (Wilson, 1977, Tables 8 and 9). Only *Bathygenys* and *Merycoidodon* are found on the list of Chadronian first appearances given by Lucas (this volume). Whether or not the Little Egypt l.f. truly belongs in the Chadronian, most authors (Wilson, 1986; Krishtalka et al., 1987; Kelly, 1990; Lucas, this volume) agree that the Porvenir l.f. is latest Duchesnean.

Using the new dates presented here of 37.80 ± 0.06 Ma for the Buckshot Ignimbrite and 36.67 ± 0.04 Ma for the Bracks Rhyolite, the magnetic section of Testarmata and Gose (1979) probably correlate with Chrons C15N to C17N (Figure 2.3). The reversed interval containing the Bracks Rhyolite is probably Chron C16R. The Buckshot Ignimbrite apparently lies in reversed rock at the very base of the section, but its date suggests that it lies within the long normal of Chron C17N. Testarmata and Gose

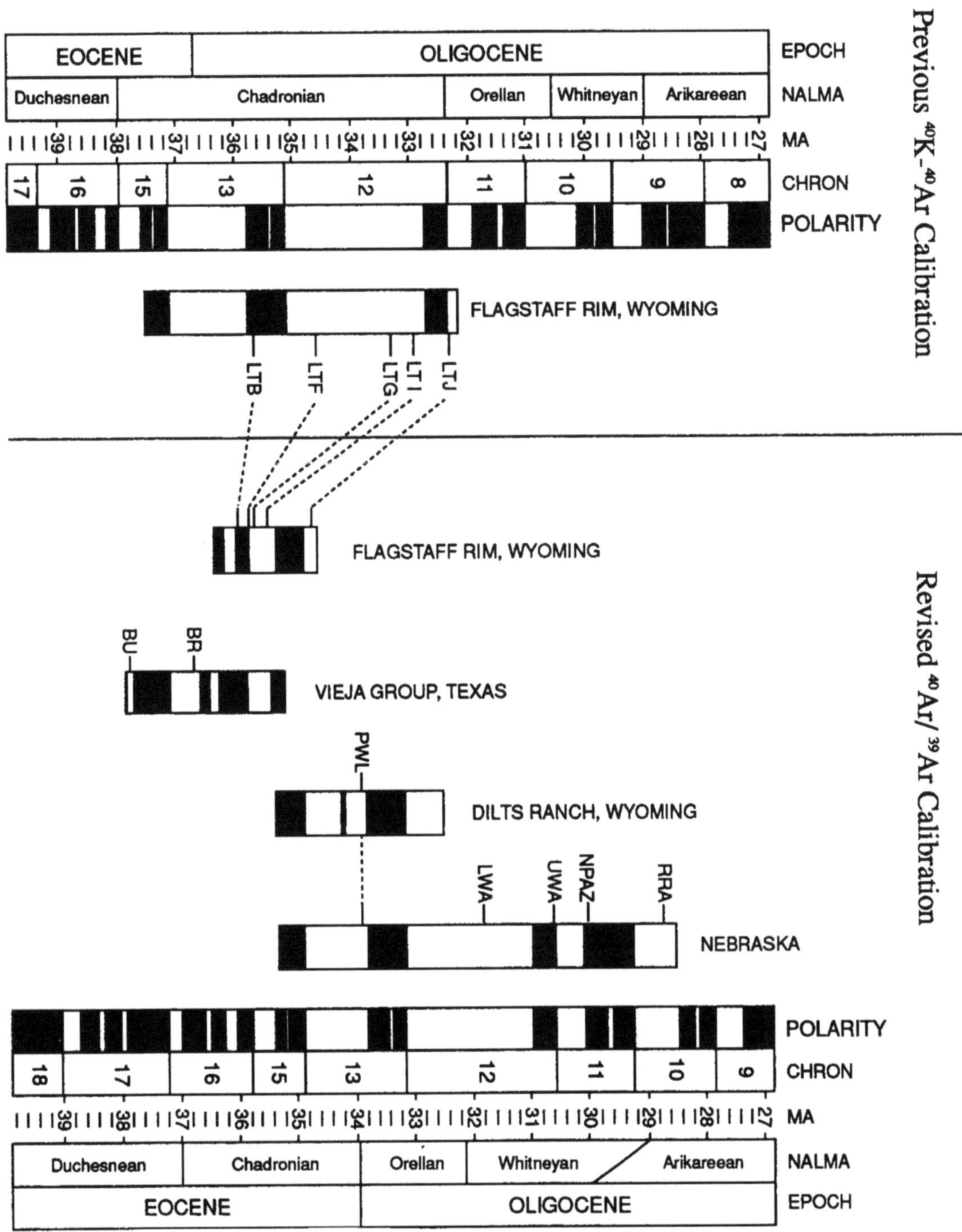

FIGURE 2.3. Correlation and calibration of the North American Duchesnean through Arikareean chronology in North America. Correlation based on K-Ar dates of Evernden et al. (1964) and timescale of Berggren et al. (1985) shown on left. New correlation based on ^{40}Ar/^{39}Ar dates of Swisher and Prothero (1990) shown on right. Magnetics of Flagstaff Rim based on Prothero (1985a; this volume, Fig. 2.4); Vieja Group from Testarmata and Gose (1979); Dilts Ranch from Evanoff et al. (this volume); Nebraska sections from Prothero (1985c; this volume, Figure 2.5 and 2.6). Abbreviations: BR, Bracks Rhyolite; BU, Buckshot Ignimbrite; LTB, LTF, LTG, LTI, LTJ, Lone Tree Gulch ashes from Flagstaff Rim, Wyoming (Emry, 1973); LWA, Lower Whitney Ash (Schultz and Stout, 1955, 1961; Swinehart et al., 1985); NPAZ, Nonpareil Ash Zone (Swinehart et al., 1985; Tedford et al., 1985); RRA, Roundhouse Rock Ash (Obradovich et al., 1973; Tedford et al., 1985); UWA, Upper Whitney Ash.

(1979) did not establish whether the Little Egypt and Porvenir local faunas lie in rocks of normal or reversed polarity, so they cannot be placed precisely within Chron C16R or Chron C17N. As of this writing, Prothero (in prep.) has sampled sections containing the late Uintan Candelaria l.f. It lies within a sequence of reversed-normal-reversed polarity that probably corresponds to C20R-C19R, based on its similarity to late Uintan faunas such as the Serendipity l.f. in Texas, and the Myton l.f. in Utah (which lie within Chron C20N). Clearly, much more work needs to be done in the Trans-Pecos Texas region to resolve the magnetostratigraphy.

Another region of Texas has yielded the first Eocene land mammals from the the North America coastal plain. The Casa Blanca l.f. in the Laredo Formation, Claiborne Group, Webb County, Texas yields a late Uintan assemblage of mammals in middle Eocene (upper nannoplankton Zone NP16) marine beds (Westgate, 1988, 1990). Although magnetic studies were inconclusive (Walton, 1986), the nannofossil information would correlate the Casa Blanca l.f., and possibly other late Uintan faunas such as Myton, Serendipity, Candelaria local faunas, with Chron C18R or C19N. This is consistent with the Uinta Basin results reported above (Figure 2.2).

Other Uintan and Duchesnean localities offer much promise for magnetostratigraphy. Many of these have been sampled by Prothero, and will be published shortly. These include the Sespe Formation in Ventura County (Kelley, 1990), and sections in San Diego County, California; the Galisteo Formation in New Mexico (Lucas, 1982); the Sage Creek and Dell beds in Beaverhead County, Montana (Tabrum et al., in press); and the Duchesnean Canyon Creek l.f., Natrona County, Wyoming. Sampling was also attempted in the Duchesnean Lac Pelletier locality in Saskatchewan (Storer, 1988, 1990), but the magnetic results were unstable.

Chadronian through Whitneyan

For a long time, it has been clear that the most complete record of the Chadronian occurs at Flagstaff Rim, Natrona County, Wyoming (Emry, 1973, this volume; Emry et al., 1987). This became the key section for the magnetic stratigraphy, since it has both superposed faunas with good stratigraphic data, and K-Ar dates (Evernden et al., 1964). However, the influence of those dates gave an erroneous notion of the age and duration of the Chadronian. With the new $^{40}Ar/^{39}Ar$ dates (Swisher and Prothero, 1990), the magnetic stratigraphy had to be radically rëinterpreted.

The original Flagstaff Rim magnetics were collected and run by Charles Denham and Harlow Farmer (Prothero, Denham, and Farmer, 1982, 1983; Denham, 1984). Denham's procedures were slightly different from those used by most other paleomagnetists. He collected only one sample per site, so site statistics could not be calculated, and individual anomalous overprints were hard to recognize. Denham did only limited thermal demagnetization on his samples, so the effects of a chemical overprint from high-coercivity iron hydroxide such as goethite could not be addressed. When Prothero resampled the lower, more problematic part of the section, he found that three samples per site and extensive thermal demagnetization changed the interpretation significantly (Prothero, 1985a). Since that time, the upper part of the section has also been completely resampled and treated with thermal demagnetization (Figure 2. 4); the results differ very little from the version presented in 1985.

Based on the K-Ar dates of Evernden et al. (1964), Prothero et al. (1982, 1983; Prothero, 1985a) considered the section to span Chrons C11R to C15N (Figure 2.3, left side). However, the new dates on these units reported by Swisher and Prothero (1990 and in this paper) clearly render that interpretation untenable, since they indicate that the dated parts of the section spans a much shorter interval, from about 34.5-37 Ma, not 32-37.5 Ma as suggested by the original K-Ar dates.

Swisher and Prothero (1990) correlated the middle and upper part of the Flagstaff Rim section with Chron C13R. The two normals in the lower part of the FSR section were correlated with Chrons C15N and C16N. However, John Obradovich (pers. commun.) has pointed out that the new dates of 35.9 ± 0.2 Ma for LTB and 35.7 ± 0.04 Ma for LTF suggest a correlation of the LTB normal interval with the upper part of Chron C16N (Figure 2.3). The normal at the

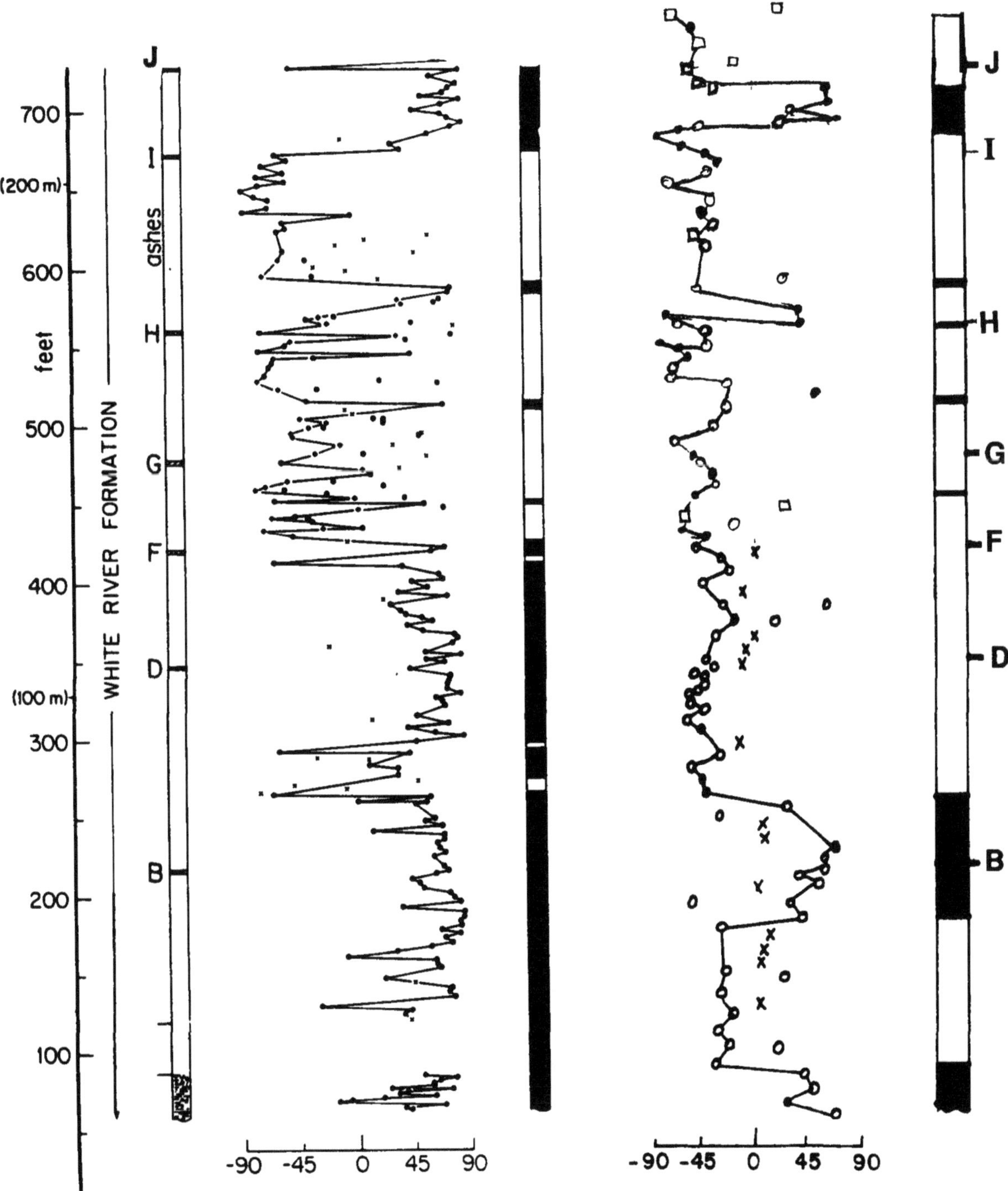

FIGURE 2.4. Magnetic stratigraphy of Flagstaff Rim, Natrona County, Wyoming (Emry, 1973). On left is interpretation of Denham (in Prothero, Denham and Farmer, 1982, 1983), which used only one sample per site and was not thermally demagnetized. On the right is the same section with three samples per site and extensive thermal demagnetization. Section below Ash F from Prothero (1985a, Figure 5); data above Ash F is new. The normal at the base is interpreted as C16N1, followed by C16R, C16N, C15R (from Ash D to Ash I), C13N, and C13R (around and above Ash J). Symbols as follows: solid circles are Class I (significant) sites; open squares are Class II (one sample lost) sites; open circles are Class III (one sample divergent) sites; X indicates indeterminate sites (terminology of Opdyke et al., 1977).

Toadstool Park–Roundtop

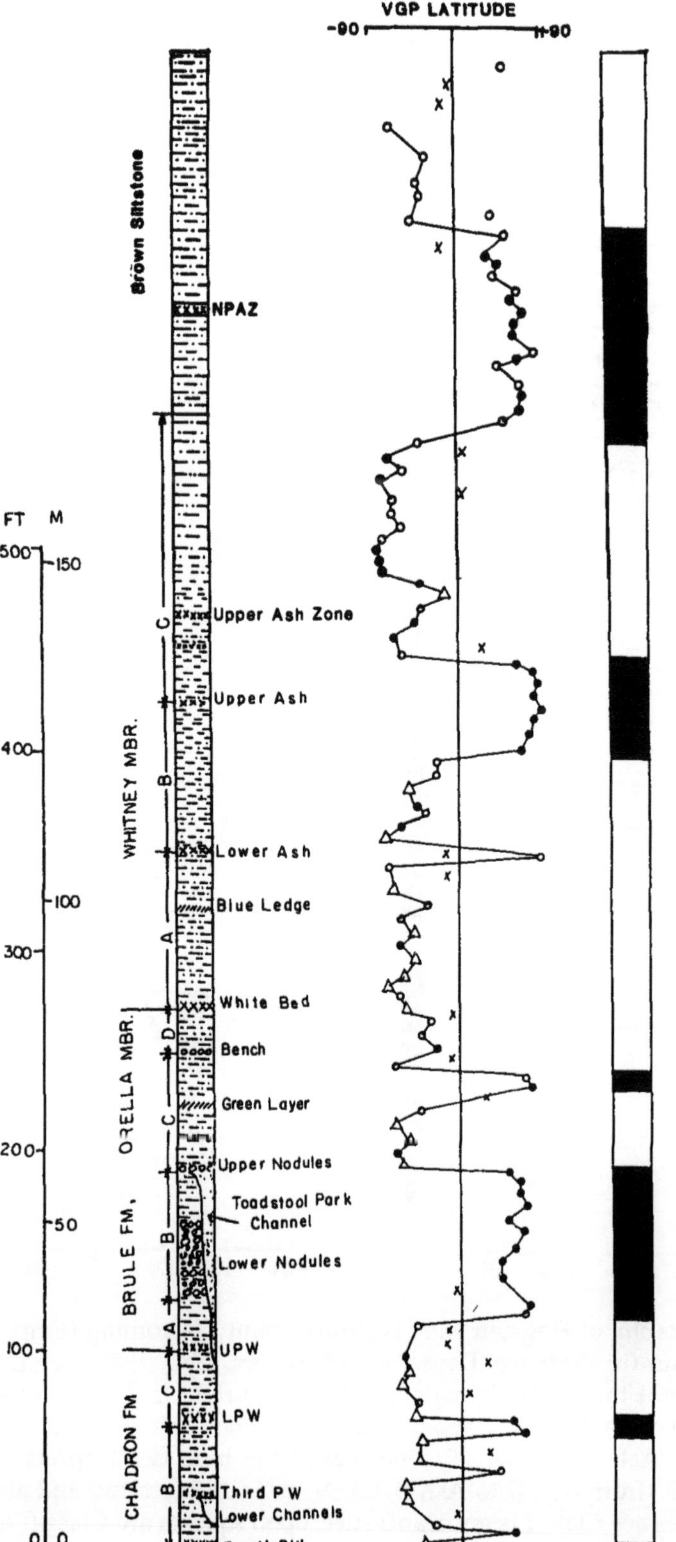

FIGURE 2.5. Magnetic stratigraphy of the Toadstool Park-Roundtop area, Sioux County, Nebraska. Stratigraphy after Schultz and Stout (1955, 1961) and Tedford et al. (1985). Symbols as in Figure 2.4, except that Class II sites are open triangles.

base of FSR section would thus correlate with the middle part of C16N, or possibly upper C17N, if the faunal information discussed by Emry (this volume) indicates a greater duration of time below ash B. If this interpretation is correct, then the long FSR reversal between LTF and LTJ correlates with C15R. The normal interval between Ash J ($^{40}Ar/^{39}Ar$ dated at 34.7 ± 0.04 Ma) and Ash I ($^{40}Ar/^{39}Ar$ dated at 35.4 ± 0.1 Ma) would be Chron C15N, and there are several meters of reversed section around and above Ash J which correlate with Chron C13R.

The new dates for the Chadronian also force reinterpretation of other previously published sections. The early Chadronian Ledge Creek section, about 20 km southeast of Flagstaff Rim, was once interpreted as spanning Chrons C12R to C15N (Prothero, 1985a, Figures 5, 8). The change in the Flagstaff Rim section suggests that Ledge Creek spans Chrons C13R to C16N. The early Chadronian Pipestone Springs l.f., in Boulder County, Montana, was originally interpreted to represent Chrons 12R-C13N (Prothero, 1984, 1985a, Figure 8). These revisions place it in Chrons C13R-C15N. The normal interval at the base of the Chadron Formation in Toadstool Park, Nebraska (Figures 2.3, 2.5) is probably Chron C15N.

Wang (this volume) disagrees with placing the Chadronian in the late Eocene. Most of the geochronological arguments she raises are discussed in this chapter. Wang argues that the presence of certain mammalian genera in both the Chadronian and the early Oligocene (Stampian) of Europe suggests their correlation. However, Europe was almost completely isolated from North America in the late Eocene, and shared very few genera in common. The appearance of Chadronian mammals in the early Oligocene of Europe simply indicates that they managed to immigrate to Europe after they appeared in North America. In fact, several genera (including *Eusmilus, Adjidaumo,* and *Plesispermophilus*) occur in the Orellan, so they are not diagnostic of the Chadronian. The rhinoceros *Trigonias* does not occur in Europe; it is based on a misidentification (Prothero, in prep.). Of Wang's list, only *Plesictis*, an extremely rare mustelid, is restricted to the Chadronian and the Stampian.

The only section which spans the Chadronian-Orellan transition with both radioisotopic dates and superposed faunas is the sequence near Douglas, Wyoming. As originally published (Prothero, 1985a), it was thought to span Chrons C10R to C13R. However, new dates and revised magnetic stratigraphy have shown that it spans Chrons C12R to C15N (Figure 2.3; Evanoff et al., this volume). The date of 33.9 Ma on the 5a Tuff (also known as the "Glory Hole Ash," and the "Persistent White Layer, " or "PWL" in other publications) is particularly critical, since it lies just 6.4 m below the last appearance of brontotheres, and therefore the Chadronian/Orellan boundary.

Besides the Douglas section, the most complete lower Oligocene sections are those in western Nebraska and eastern Wyoming. Sections along the Pine Ridge in Nebraska and Wyoming are well known for their large fossil collections and excellent stratigraphy tied to marker ashes. The longest of these is the classic section at Toadstool Park and Roundtop in Sioux County, Nebraska (Schultz and Stout, 1955, 1961). As published (Prothero et al., 1983, Figure 5; Prothero, 1985c, Figure 2), the section contained an interval of normal polarity in Orella B, and another around the Upper Whitney Ash. This section has since been extended up to the top (Figure 2.5), including previously unrecognized sections of the "brown siltstone member" of the Brule Formation (Swinehart et al., 1985; Tedford et al., 1985). Based on the current geochronology, the early Orellan normal interval is Chron C13N. The upper Whitney normal is Chron C12N, and the long reversed interval spanning the late Orellan and early Whitneyan is the unusually long Chron C12R (Figure 2.3).

Another critical section is the classic sequence at Scottsbluff, Nebraska (Figure 2.6). This section spans most of the Orella and Whitney members of the Brule Formation, up to an unconformity which cuts down from the overlying Gering Formation (Schultz and Stout, 1955, 1961). Like the Toadstool Park -Roundtop section, there is an early Orellan normal interval at the base of the section, and another normal interval encompassing the Upper Whitney Ash (Figure 2.6). This section is also important because it was the source of the dated samples of the Lower and Upper Whitney Ashes described above. The Arikareean portion of

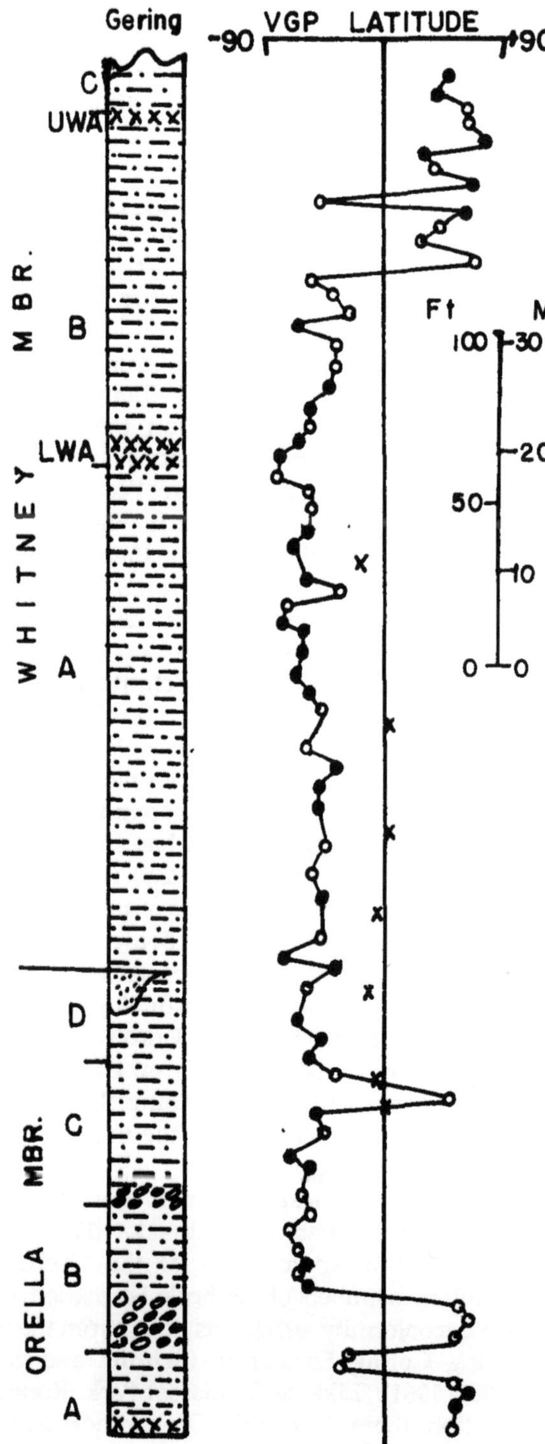

FIGURE 2.6. Magnetic stratigraphy of the White River Group at Scottsbluff National Monument, Scottsbluff County, Nebraska. Stratigraphy after Schultz and Stout (1955, 1961). Symbols as in Figure 2.4.

Scottsbluff (above the unconformity) has recently been sampled, and its interpretation will be described elsewhere (Swinehart et al., in press).

Finally, the classic sections at the Big Badlands of South Dakota were the original source of most early Oligocene fossils. These spectacular and well-exposed sections are actually less complete than the less famous sections found further to the south and west (Prothero, 1985c). The abundant Orellan fossils found in every rock shop come mainly from two restricted zones, the "lower nodules" and "upper nodules." Biostratigraphic comparisons (Prothero, 1982, 1985c; Stevens, in press) have shown that the earliest Orellan faunas found just above the PWL in the Douglas and Lusk areas of Wyoming do not occur in the Big Badlands. Instead, the faunas of the "lower nodules" are characteristic of the later early Orellan, and the Chadronian-Orellan transition is represented either by unconformities or by unfossiliferous rock in the Big Badlands.

The first published magnetic stratigraphy for the Big Badlands sections (Prothero et al., 1983; Denham, 1984; Prothero, 1985c) were partially based on sections collected by Denham that did not have three samples per site, and were not treated by thermal demagnetization. In 1986 and 1987, Prothero resampled all the important sections, and discovered that thermal demagnetization radically changed the polarity patterns (Figure 2.7). As suspected from the biostratigraphy, the early Orellan normal is very short, and tends to occur at the base of, or just below the "lower nodules" in most sections. Since these nodular bands are locally determined diagenetic features, it is not surprising that they are not the same age in every locality (Prothero, 1985d). The rest of the Orellan section (Scenic Member) is all reversed, as is the lower half of the Whitneyan section (Poleslide Member). Based on faunal correlation with the dated sections in Nebraska, this long Orellan-Whitneyan reversed interval is Chron C12R. As in the Nebraska sections, another short normal occurs high in the Poleslide Member; this appears to be Chron C12N. Finally, the uppermost Poleslide and lower half of the Sharps Formation are found in yet a third normal interval, which appears to be

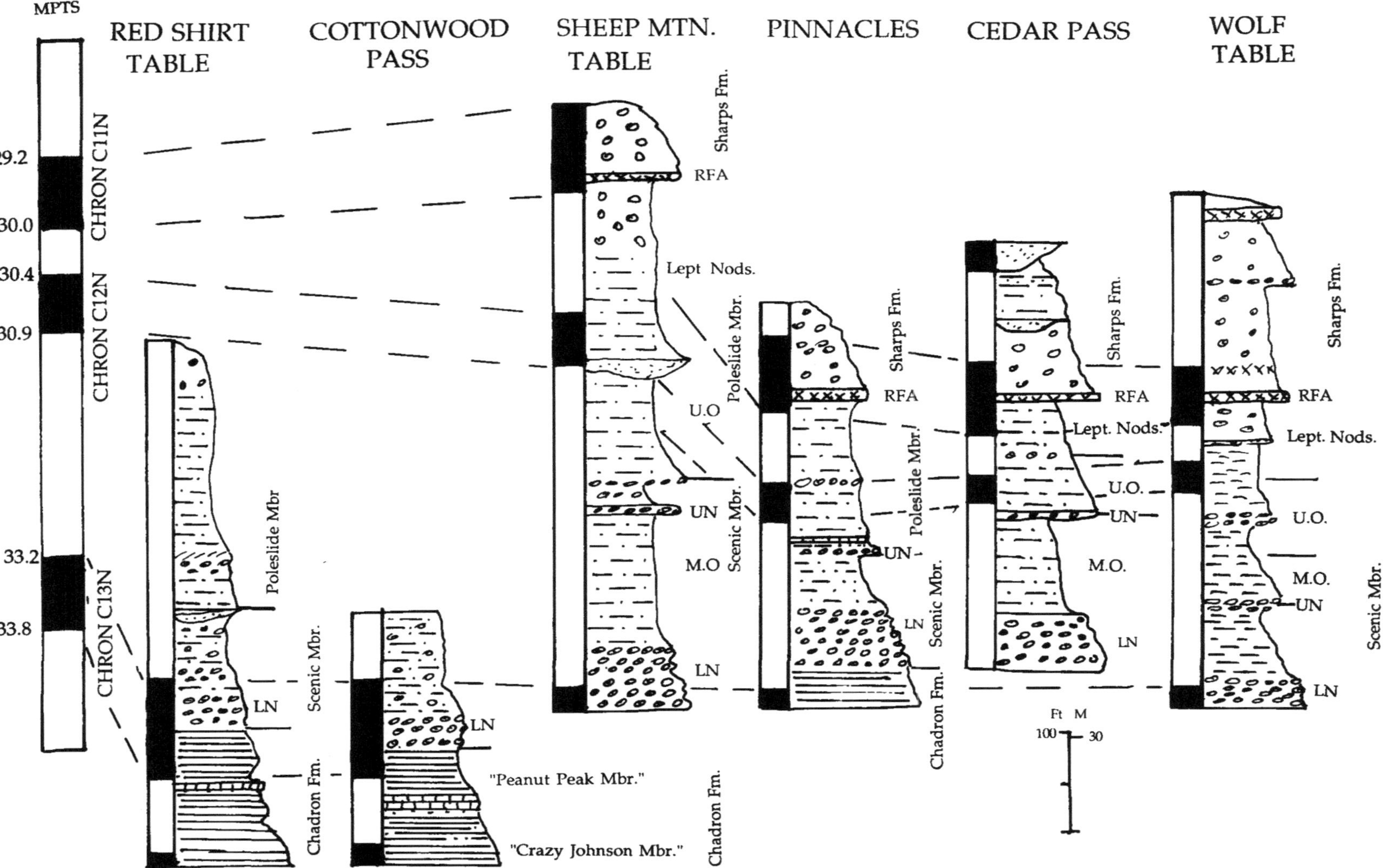

FIGURE 2.7. Revised magnetic stratigraphy of the White River Group in the Big Badlands National Park, South Dakota. Stratigraphy after Skinner (unpub. field notes) and Tedford (unpub. field notes). Abbreviations as follows: Lept. Nods., *Leptauchenia* nodules; LN, lower nodular zone; M.O., "Middle Oreodon Beds"; RFA, Rockyford Ash; UN, upper nodular zone; U.O., "Upper Oreodon Beds."

Chron C11N. This is in good agreement with the Nebraska sections, where the transition from the Whitney to the "brown siltstone" members occurs in Chron C11N (Figure 2.5). In both sections, a prominent ash occurs near the base of Chron C11N. Known as the Rockyford Ash in South Dakota, and the Nonpareil Ash Zone in Nebraska (Swinehart et al., 1985; Tedford et al., 1985), this ash has been dated at 30.05 Ma (see above).

IMPLICATIONS

The new correlations suggested by Swisher and Prothero (1990) and in this paper radically affect the interpretations of faunal changes. For example, the sequence and nature of changes in North American mammalian faunas discussed by Prothero (1985b, 1985d, 1989) are still the same, but their temporal spacing and correlation with global events are completely different. The major faunal changes that occur at the Uintan-Duchesnean transition, once thought to be within the late Eocene, are now at the middle/late Eocene boundary. The minor faunal change between the Duchesnean and Chadronian, once thought to be the Eocene-Oligocene transition, is now within the late Eocene. The extinctions at the Chadronian/Orellan boundary, once touted as the "Mid-Oligocene event" (Prothero, 1985b, 1985d), are now correlative with the Terminal Eocene Event of Wolfe (1978). There is no significant faunal event during the middle Oligocene in North America, as is also true of the marine record (Prothero, 1989; Berggren and Prothero, this volume).

The change in duration of the Chadronian from over 5 to less than 3 million years significantly affects the range compilations of Prothero (1985d) and Stucky (1990, this volume). For example, Prothero (1985d) and Stucky (1990) reported unusually low rates of origination and extinction for the Chadronian, since the rate calculation was based on a long duration of 5 million years. With a shorter Chadronian, these rates are much higher, and comparable to rates during other NALMAs (Stucky, this volume).

On the positive side, these new calibrations and correlations make a lot more sense. For example, Prothero (in Prothero et al., 1983, Prothero, 1985c) always obtained a relatively long reversed interval between the early Orellan normal and the late Whitneyan normal. The consistency of thickness of reversed rock is striking in sections ranging from North Dakota to Colorado. This long reversed interval seemed difficult to reconcile with correlation to Chron C10R, but makes much more sense if it is Chron C12R (Figure 2.3).

On a global scale, these correlations make even better sense. The conflicts between estimates of 36.5 Ma (Berggren et al., 1985; Aubry et al., 1988) and 32-34 Ma (Odin, 1982; Montanari et al., 1988) for the Eocene/Oligocene boundary now appear to be resolved (see Berggren et al., this volume). The most significant late Paleogene faunal event in the marine record is at the middle-late Eocene transition, and this now appears to correspond to the most significant late Paleogene event in North American land mammals, the Uintan-Duchesnean transition (Prothero, 1989; Prothero et al., 1990; Berggren and Prothero, this volume). Likewise, the minor changes at the Duchesnean-Chadronian transition are comparable to the marine record of the late Eocene, and the Chadronian-Orellan extinctions are comparable to marine turnover at the Terminal Eocene Event.

$^{40}Ar/^{39}Ar$ dating has greatly improved the magnetostratigraphic and faunal interpretations of the North American terrestrial record. Many of the controversies and arguments in correlation have apparently been resolved in such a way that the North American record is now very consistent with the global pattern of extinction and faunal change, although much work remains to be done with the middle Eocene. As can be seen elsewhere in this volume, that newly emerging global story can now shed light on the causes of floral and faunal change in North America.

ACKNOWLEDGEMENTS

We thank R.J. Emry, J.J. Flynn, J.A. Wilson, and M.O. Woodburne for their helpful comments on this chapter. Swisher thanks D. Savage, G. Curtis, R. Drake, and A. Deino for their help and support. Swisher was supported by NSF grant EAR89-16431 during preparation of the $^{40}Ar/^{39}Ar$ dates discussed here. Prothero was supported by NSF grant EAR87-08221, and grants from the Petroleum Research Fund of the American Chemical Society, during most of the magnetic sampling summarized here. Sampling would not have been possible without the help of the following people: Jeff Amato, Susan Briggs, Jill Bush, Jennifer Chean, Priscilla Duskin, Emmett Evanoff, Jim Finegan, John Foster, Jon Frenzel, Dana Gilchrist, Karen Gonzalez, Kecia Harris, Steve King, Allison Kozak, Rob Lander, Heidi Shlosar, and Annie Walton. Drs. Jim Swinehart, Richard Tedford, and Mike Voorhies were very helpful with the stratigraphy of the Nebraska sections, Bob Emry and Emmett Evanoff with the Wyoming sections, and Phil Bjork with the Badlands sections. Most of the stratigraphic groundwork, however, was laid by the late Dr. Morris F. Skinner, whose careful collection methods and stratigraphic work made this research feasible in the first place. We dedicate this work to his memory.

REFERENCES

Andersen, D.W., and Picard, M.D. 1972. Stratigraphy of the Duchesne River Formation (Eocene-Oligocene?), northern Uinta Basin, northeastern Utah. *Utah Geol. Min. Surv. Bull.* 97: 1-29.

Aubry, M.-P., Berggren, W.A., Kent, D.V., Flynn, J.J., Klitgord, K.D., Obradovich, J.D., and Prothero, D.R. 1989. Paleogene geochronology: an integrated approach. *Paleoceanography* 3 (6): 707-742.

Berggren, W.A., Kent, D.V., and Flynn, J.J. 1985. Paleogene geochronology and chronostratigraphy, In N.J. Snelling, ed., *The Chronology of the Geological Record. Mem. Geol. Soc. London* 10: 141-195.

Black, C.C. 1969. Fossil vertebrates from the late Eocene and Oligocene Badwater Creek area, Wyoming, and some regional correlations. *Wyo. Geol. Assoc. Guidebook, 21st Ann. Field Conf.*, pp. 43-48.

Cande, S.C. and Kent, D.V. 1991. A new geomagnetic polarity timescale for the late Cretaceous and Cenozoic. *Jour. Geophys. Res.*, in review.

Cashion, W.B. 1967. Geology and fuel resources of the Green River Formation, southeastern Uinta Basin, Utah and Colorado. *U.S. Geol. Surv. Prof. Paper* 548.

Dane, C.H. 1954. Stratigraphic and facies relationships of the upper part of the Green River Formation and lower part of the Uinta Formation in Duchesne, Uintah, and Wasatch Counties, Utah. *Amer. Assoc. Petrol. Geol. Bull.* 38: 405-425.

Denham, C.R. 1984. Statistical sedimentation and magnetic polarity stratigraphy, In W.A. Berggren and J.A. Van Couvering, eds., *Catastrophes and Earth History, The New Uniformitarianism.* Princeton, New Jersey, Princeton Univ. Press, pp. 101-112.

Emry, R.J. 1973. Stratigraphy and preliminary biostratigraphy of the Flagstaff Rim area, Natrona County, Wyoming. *Smithsonian Contrib. Paleobiol.* 18.

Emry, R.J. 1981. Additions to the mammalian fauna of the type Duchesnean, with comments on the status of the Duchesnean. *J. Paleont.* 55: 563-570.

Emry, R.J., P.R. Bjork, and L.S. Russell. 1987. The Chadronian. Orellan, and Whitneyan land mammal ages, In Woodburne, M.O., ed., *Cenozoic Mammals of North America, Geochronology and Biostratigraphy*, Univ. Calif. Press, Berkeley, Calif., pp. 118-152.

Evernden, J.F., D.E. Savage, G.H. Curtis, and G.T. James. 1964. Potassium-argon dates and the Cenozoic mammalian chronology of North America. *Amer. J. Sci.* 262: 145-198.

Flynn, J.J. 1986. Correlation and geochronology of middle Eocene strata from the western United States. *Palaeogeogr., Palaeoclimat., Palaeoecol.* 55: 335-406.

Hamblin, A.H. 1987. Paleogeography and paleoecology of the Myton Pocket, Uinta Basin, Utah (Uinta Formation--Upper Eocene). *Brigham Young Univ. Geology Stud.* 34: 33-60

Henry, C.D., McDowell, F.W., Price, J.G., and Smyth, R.C. 1986. Compilation of potassium-argon ages of Tertiary igneous rocks, Trans-Pecos Texas. *Bur. Econ. Geol. Univ. Texas Geol. Circ.* 86-2: 1-34.

Kay, J.L. 1934. The Tertiary formations of the Uinta Basin, Utah. *Ann. Carnegie Mus.* 23: 357-371.

Kelly, T.S. 1990. Biostratigraphy of Uintan and Duchesnean land mammal assemblages from the middle member of the Sespe Formation, Simi Valley, California. *Contrib. Science Nat. Hist. Mus. Los Angeles County* 419: 1-42.

Krishtalka, L., and others, 1987. Eocene (Wasatchian through Duchesnean) biochronology of North America, In Woodburne, M.O., ed., *Cenozoic Mammals of North America, Geochronology and Biostratigraphy*. Berkeley, Calif., Univ. Calif. Press, pp. 77-117.

Lucas, S.G. 1982. Vertebrate paleontology, stratigraphy, and biostratigraphy of the Eocene Galisteo Formation, north-central New Mexico. *New Mex. Bur. Mines. Min. Res. Circ.* 186: 1-34.

Mauger, R.L. 1977. K-Ar ages of biotites from tuffs in Eocene rocks of the Green River, Washakie, and Uinta Basins, Utah, Wyoming, and Colorado. *Contrib. Geol. Univ. Wyoming*, 15: 17-41.

McDowell, F.W. 1979. Potassium-argon dating in the Trans-Pecos Texas volcanic field. *Bur. Econ. Geol. Univ. Texas Guidebook* 19: 10-18.

McDowell, F.W., Wilson, J.A., and Clark, J. 1973. K-Ar dates for biotite from two paleontologically significant localities: Duchesne River Formation, Utah, and Chadron Formation, South Dakota. *Isochron/West* 7: 11-12.

Montanari, A., Deino, A., Drake, R., Turrin, B.D., DePaolo, D.J., Odin, G.S., Curtis, G.H., Alvarez, W., and Bice, D.M. 1988. Radioisotopic dating of the Eocene-Oligocene boundary in the pelagic sequence of the Northern Apennines, In I. Premoli-Silva, R. Coccini, and A. Montanari, eds., *The Eocene-Oligocene Boundary in the Marche-Umbria Basin (Italy)*, Int. Subcomm. Paleogene Strat., Ancona, Italy, pp. 195-208.

Obradovich, J.D., Izett, G.A., and Naeser, C.W. 1973. Radiometric ages of volcanic ash and pumice beds in the Gering Sandstone (earliest Miocene) of the Arikaree Group, southwestern Nebraska. *Geol. Soc. Amer. Abs. Prog.* 5: 499-600.

Odin, G.S. (ed.) 1982. *Numerical Dating in Stratigraphy*. John Wiley & Sons, New York.

Opdyke, N.D., Lindsay, E.H., Johnson, N.M., and Downs, T. 1977. The paleomagnetism and magnetic polarity stratigraphy of the mammal-bearing section of Anza-Borrego State Park, California. *Jour. Quat. Res.* 7: 316-329.

Osborn, H.F. 1895. Fossil mammals of the Uinta Basin. Expedition of 1894. *Bull. Amer. Mus. Nat. Hist.* 7: 71-105.

Osborn, H.F. 1929. The titanotheres of ancient Wyoming, Dakota, and Nebraska. *U. S. Geol. Surv. Monograph* 55: 1-953 (2 vols.).

Prothero, D.R. 1982. How isochronous are mammalian biostratigraphic events? *Proceedings of the 3rd North American Paleontological Convention*, 2: 405-409.

Prothero, D.R. 1984. Magnetostratigraphy of the Early Oligocene Pipestone Springs locality, Jefferson County, Montana. *Contributions to Geology of the University of Wyoming*, 23 (1): 33-36.

Prothero, D.R. 1985a. Chadronian (early Oligocene) magnetostratigraphy of eastern Wyoming: implications for the Eocene-Oligocene boundary. *Journal of Geology*, 93: 555-565.

Prothero, D.R. 1985a. Mid-Oligocene extinction event in North American land mammals. *Science*, 229: 550-551.

Prothero, D.R. 1985c. Correlation of the White River Group by magnetostratigraphy, In J.E. Martin, ed., *Fossiliferous Cenozoic deposits of western South Dakota and northwestern Nebraska. Dakoterra*, Museum of Geology, South Dakota School of Mines, 2(2): 265-276.

Prothero, D.R. 1985d. North American mammalian diversity and Eocene-Oligocene extinctions. *Paleobiology*, 11(4): 389-405.

Prothero, D.R. 1988. Mammals and magnetostratigraphy. *J. Geol. Educ.* 34(4): 227-236.

Prothero, D.R. 1989. Stepwise extinctions and climatic decline during the later Eocene and Oligocene, In S.K. Donovan, ed., *Mass Extinctions: Processes and Evidence*. New York, Columbia Univ. Press, pp. 211-234.

Prothero, D.R. 1990. Magnetostratigraphy of the middle Eocene Uinta Formation, Uinta Basin, Utah. *Jour. Vert. Paleo.* 10(3): 38A.

Prothero, D.R., W.A. Berggren, and P.R. Bjork, 1990. Penrose Conference Report: Late Eocene-Oligocene biotic and climatic evolution. *GSA News and Information* 12 (3): 74-75.

Prothero, D.R., Denham, C.R., and Farmer, H.G. 1982. Oligocene calibration of the magnetic polarity timescale. *Geology*, 10: 650-653.

Prothero, D.R., Denham, C.R., and Farmer, H.G. 1982. 1983. Magnetostratigraphy of the White River Group and its implications for Oligocene geochronology. *Palaeogeogr.*,

Palaeoclimat., Palaeoecol. 42: 151-166.

Prothero, D.R., and Swisher, C. C., III. 1990. Magnetostratigraphy and $^{40}Ar/^{39}Ar$ dating of the middle Eocene Uinta Formation, Utah. *Geol. Soc. Amer. Abs. Prog.* 22 (7): A364.

Riggs, E.S. 1912. New or little known titanotheres from the lower Uintah Formations. *Field Mus. Nat. Hist. Publ.* 159, v. IV, pp. 1-41.

Schultz, C.B., and Stout, T.M. 1955. Classification of the Oligocene sediments in Nebraska. *Bull. Univ. Neb. State Mus.* 4: 17-52.

Schultz, C.B., and Stout, T.M. 1961. Field conference on the Tertiary and Pleistocene of western Nebraska. *Spec. Publ. Univ. Neb. State Mus.* 2: 1-54.

Scott, W.B., Jepsen, G.L., and Wood, A.E. 1940-1941. The mammalian fauna of the White River Oligocene. *Trans. Amer. Phil. Soc.* 28, parts I-V.

Stevens, J.B., Stevens, M.S., and Wilson, J.A. 1984. Devil's Graveyard Formation (new), Eocene and Oligocene age, Trans-Pecos Texas. *Texas Mem. Mus. Bull.* 32: 1-21.

Steven, M.S. In press. Review of the Merycoidodontidae, In D.R. Prothero and R.J. Emry, eds., *The Terrestrial Eocene-Oligocene Transition in North America* (in press).

Storer, J.E. 1988. The rodents of the Lac Pelletier lower fauna, late Eocene (Duchesnean), of Saskatchewan. *J. Vert. Paleont.* 8:84-101.

Storer, J.E. 1990. Primates of the Lac Pelletier lower fauna (Eocene: Duchesnean), Saskatchewan. *Can. J. Earth Sci.* 27:520-524.

Stucky, R.K. 1990. Evolution of land mammal diversity in North America during the Cenozoic. *Curr. Mammal.* 2:375-432.

Swinehart, J.B., Souders, V.L., Degraw, H.M., and Diffendal, R.F., Jr. 1985, Cenozoic paleogeography of western Nebraska, In R.M. Flores and S. Kaplan, eds., *Cenozoic Paleogeography of the west central United States. Spec. Publ. Rocky Mtn. Sect. Soc. Econ. Paleont. Min.* pp. 209-229.

Swinehart, J., Tedford, R.H., Prothero, D.R., King, S.A., and Tierney, T. In Press, The Whitneyan-Arikareean transition in the High Plains, In D.R. Prothero and R.J. Emry, eds., *The Terrestrial Eocene-Oligocene Transition in North America* (in press).

Swisher, C.C., III, and Prothero, D.R. 1990. Single-crystal $^{40}Ar/^{39}Ar$ dating of the Eocene-Oligocene transition in North America. *Science* 249: 760-762.

Tabrum, A.R., Prothero, D.R., and Garcia, D. 1991. Eocene-Oligocene biostratigraphy and magnetostratigraphy of western Montana, In D.R. Prothero and R.J. Emry, *The Terrestrial Eocene-Oligocene Transition in North America* (in press).

Tedford, R.H., Swinehart, J.B., Hunt, R.M., and Voorhies, M.R. 1985. Uppermost White River and lowermost Arikaree rocks and faunas, White River Valley, northwestern Nebraska, and their correlation with South Dakota, In J.E. Martin, ed., *Fossiliferous Cenozoic deposits of western South Dakota and northwestern Nebraska. Dakoterra,* Museum of Geology, South Dakota School of Mines, 2: 335-352.

Tedford, R.H., and others, 1987. Faunal succession and biochronology of the Arikareean through Hemphillian interval (late Oligocene through earliest Pliocene Epochs) in North America, In Woodburne, M.O., ed., *Cenozoic Mammals of North America, Geochronology and Biostratigraphy,* Berkeley, Calif., Univ. Calif. Press, pp. 152-210.

Testarmata, M.M., and Gose, W.A. 1979. Magnetostratigraphy of the Eocene-Oligocene Vieja Group, Trans-Pecos Texas. *Bur. Econ. Geol. Univ. Texas Guidebook* 19: 55-66.

Walton, A. H. 1986b. Magnetostratigraphy of some Eocene vertebrate-bearing rocks in Texas. *Geol. Soc. Am. Abs. Prog. 99th Ann. Mtg*: 18(7): 781 (Abstr.).

Westgate, J.W. 1988. Biostratigraphic implications of the first Eocene land-mammal fauna fromthe North American coastal plain. *Geology* 16: 995-998.

Westgate, J.W. 1990. Uintan land mammals (excluding rodents) from an estuarine facies of the Laredo Formation (middle Eocene, Claiborne Group) of Webb County, Texas. *Jour. Paleont.* 64: 454-468.

Wilson, J.A. 1977. Stratigraphic occurrence and correlation of early Tertiary vertebrate faunas, Trans-Pecos Texas, Part 1: Vieja area. *Texas Mem. Mus. Bull.* 25: 1-42.

Wilson, J.A. 1980. Geochronology of the Trans-Pecos volcanic field. *New Mexico Geol. Soc.*

Guidebook, 31: 205-211.

Wilson, J.A. 1984. Vertebrate fossil faunas 49 to 36 million years ago and additions to the species of *Leptoreodon* found in Texas. *Jour. Vert. Paleont.* 4: 199-207.

Wilson, J.A. 1986. Stratigraphic occurrence and correlation of early Tertiary vertebrate faunas, Trans-Pecos Texas: Agua Fria-Green Valley areas. *Jour. Vert. Paleont.* 6: 350-373.

Wilson, J.A., P.C. Twiss, R.K. DeFord, and S.E. Clabaugh, 1968. Stratigraphic succession, potassium-argon dates, and vertebrate faunas, Rim Rock Country, Trans-Pecos Texas. *Amer. Jour. Sci.* 266: 590-604.

Wolfe, J.A. 1978. A paleobotanical interpretation of Tertiary climates in the Northern Hemisphere. *Amer. Scientist* 66: 694-703.

Wood, A.E. 1973. Eocene rodents, Pruett Formation, southwest Texas: their pertinence to the origin of South American Caviomorpha. *Pearce-Sellards Series* 20.

Wood, A.E. 1974. Early Tertiary vertebrate faunas, Vieja Group, Trans-Pecos Texas: Rodentia. *Texas Mem. Mus. Bull.* 21.

Wood, H.E. 1934. Revision of the Hyrachyidae. *Bull. Amer. Mus. Nat. Hist.* 67: 182-295.

Wood, H.E., and others, 1941. Nomenclature and correlation of the North American continental Tertiary. *Geol. Soc. Amer. Bull.* 52: 1-48.

TABLE 2.1. Summary of $^{40}Ar/^{39}Ar$ laser total fusion analyses.

Unit Name	Sample	40K-40Ar Published Age (Ma)*	L#	Mineral	40Ar/39Ar No. Dates	Mean Age (Ma)	SD (Ma)	SEM (Ma)
Arikareean								
Roundhouse Rock Ash	RRA	28.7 ± 0.7	1745	biotite	9	28.593	0.955	0.318
Whitneyan/Arikareean								
Nonpareil Ash	AJ86-36		2566	biotite	5	30.050	0.191	0.085
Whitneyan								
Upper Whitney Ash	#62		1584	biotite	11	30.580	0.610	0.184
Lower Whitney Ash	#61		1582	biotite	10	31.846	0.023	0.007
			1583	anorthoclase	4	31.811	0.050	0.025
			1621	plagioclase	2	31.673	0.219	0.155
Chadronian/Orellan								
Purple White Layer	PWL		1746	biotite	5	33.907	0.131	0.058
Chadronian								
Lone Tree Ash J	LTJ	32.4	1152	biotite	8	34.479	0.231	0.082
			1155	anorthoclase	12	34.720	0.126	0.036
Lone Tree Ash I	LTI		1148	biotite	8	35.375	0.294	0.104
Lone Tree Ash G	LTG	33.5	1147	biotite	9	35.574	0.165	0.055
			1157	anorthoclase	7	35.723	0.073	0.027
Lone Tree Ash F	LTF	34.6	1156	biotite	11	35.722	0.379	0.114
		36.6	1153	anorthoclase	5	35.813	0.085	0.038
Lone Tree Ash B	LTB	34.2	1146	biotite	12	35.919	0.344	0.099
		36.1	1151	anorthoclase	4	35.973	0.447	0.224
Duchesnean/Chadronian								
Brack's Rhyolite	H87-205		2552	sanidine	5	36.665	0.078	0.035
Buckshot Ignimbrite	BKST		2556	sanidine	5	37.798	0.150	0.060
Duchesnean								
La Point Ash	LP1		2554	biotite	6	39.742	0.171	0.070

* K-Ar analysis by Obradovich et al. (1973), all others by Evernden et al. (1964)
L # = Berkeley Geochronology Center lab number
s = standard deviation of analyses
SEM = standard error of the mean (Taylor, 1982)
All 40Ar/39Ar dates based on MMhb-l at 520.4 Ma

TABLE 2.2. Original data of $^{40}Ar/^{39}Ar$ laser total fusion analyses.

Lab Number	Unit Name	Mineral Dated	Ar 37/39	Ar 36/39	Ar 40*/39	% Rad.	Age (Ma)	SD (Ma)	SEM (Ma)
ARIKAREEAN									
Roundhouse Rock Ash									
1745-01	RRA	biotite	0.00739	0.002861	2.78403	76.7	27.512	0.332	
1745-02	RRA	biotite	0.00885	0.006735	2.93946	59.6	29.036	0.639	
1745-03	RRA	biotite	0.00822	0.001460	2.95362	87.2	29.174	0.399	
1745-04	RRA	biotite	0.01569	0.002260	2.83158	80.9	27.978	0.436	
1745-05	RRA	biotite	0.01099	0.002177	2.77137	81.1	27.388	0.412	
1745-06	RRA	biotite	0.01188	0.001877	2.83501	83.6	28.012	0.233	
1745-07	RRA	biotite	0.00698	0.007430	2.88300	56.8	28.482	0.494	
1745-08	RRA	biotite	0.00918	0.003711	2.99969	73.2	29.626	0.324	
1745-10	RRA	biotite	0.01282	0.031263	3.05066	24.8	30.125	1.011	
					Mean age	**=**	**28.593**	**0.955**	**0.318**
WHITNEYAN / ARIKAREEAN									
Nonpareil Ash									
2566-02	AJ86-36	biotite	0.00922	0.000497	0.95609	86.6	30.225	0.124	
2566-03	AJ86-36	biotite	0.01035	0.000327	0.95706	90.7	30.256	0.106	
2566-04	AJ86-36	biotite	0.01120	0.000244	0.94513	92.8	29.882	0.076	
2566-05	AJ86-36	biotite	0.00479	0.000262	0.94382	92.3	29.841	0.087	
2566-06	AJ86-36	biotite	0.00628	0.000206	0.94964	93.8	30.023	0.182	
					Mean age	**=**	**30.045**	**0.191**	**0.085**
2567-03	AJ86-36	plag	5.23086	0.002559	0.83774	71.5	26.511	0.281	
2567-10	AJ86-36	plag	3.42773	0.005045	0.89566	42.4	28.330	0.353	
2567-04	AJ86-36	plag	5.64202	0.002758	0.95219	72.6	30.103	0.351	
2567-02	AJ86-36	plag	5.75436	0.002570	0.95503	76.4	30.192	0.603	
2567-07	AJ86-36	plag	2.56354	0.001047	0.95580	90.3	30.216	0.162	
2567-06	AJ86-36	plag	6.68762	0.002898	0.97971	75.6	30.966	0.403	
2567-08	AJ86-36	plag	6.96063	0.003636	0.99689	66.0	31.504	0.541	
2567-01	AJ86-36	plag	6.71793	0.002334	1.00281	87.2	31.690	0.323	
2567-05	AJ86-36	plag	6.07960	0.003227	1.01122	68.6	31.953	0.468	
2567-09	AJ86-36	plag	6.48805	0.003545	1.01717	66.0	32.139	0.531	
					Mean age	**=**	**30.360**	**1.770**	**0.560**
WHITNEYAN									
Upper Whitney Ash									
1584-05	UWA	biotite	0.00967	0.001742	3.62576	87.5	31.291	0.104	
1584-11	UWA	biotite	0.01368	0.001951	3.50578	85.9	30.264	0.093	
1584-12	UWA	biotite	0.00908	0.001223	3.54891	90.7	30.633	0 130	
1584-14	UWA	biotite	0.09236	0.000472	3.64503	96.5	31.456	0.312	
1584-18	UWA	biotite	0.00446	0.001062	3.62112	92.0	31.251	0 169	
1584-19	UWA	biotite	0.02355	0.002795	3.44706	80.7	29.761	0.102	
1584-21	UWA	biotite	0.01386	0.002688	3.42695	81.2	29.589	0.143	
1584-22	UWA	biotite	0.01516	0.002180	3.49230	84.4	30.149	0.110	
1584-23	UWA	biotite	0.02992	0.001532	3.54864	88.7	30.631	0.124	
1584-24	UWA	biotite	0.04833	0.002513	3.53989	82.7	30.556	0.109	
1584-25	UWA	biotite	0.01263	0.001612	3.56848	88.2	30.801	0.162	
					Mean age	**=**	**30.580**	**0.610**	**0.184**
Lower Whitney Ash									
1582-01	LWA	biotite	0.01912	0.000312	3.69021	97.5	31.842	0.071	
1582-02	LWA	biotite	0.02801	0.001129	3.68655	91.7	31.811	0.093	
1582-03	LWA	biotite	0.02332	0.000314	3.69077	97.5	31.847	0.079	
1582-04	LWA	biotite	0.01876	0.000339	3.69069	97.3	31.846	0.059	
1582-05	LWA	biotite	0.01612	0.000274	3.69001	97.8	31.840	0.072	
1582-06	LWA	biotite	0.01939	0.000339	3.69075	97.3	31.847	0 059	
1582-07	LWA	biotite	0.02704	0.000460	3.69457	96.5	31 879	0.089	
1582-08	LWA	biotite	0.08109	0.000730	3.68804	94.6	31.824	0.082	
1582-09	LWA	biotite	0.03639	0.000445	3.69545	96.6	31.887	0.068	
1582-10	LWA	biotite	0.02303	0.000301	3.68953	97.6	31.836	0.085	
					Mean age	**=**	**31.846**	**0.023**	**0.007**

Lab Number	Unit Name	Mineral Dated	Ar 37/39	Ar 36/39	Ar 40*/39	% Rad.	Age (Ma)	SD (Ma)	SEM (Ma)
1583-01	LWA	sanidine	0.14783	0.000100	3.69020	99.5	31.842	0.060	
1583-02	LWA	sanidine	0.53810	0.000199	3.69206	99.5	31.858	0.058	
1583-04	LWA	sanidine	0.00738	0.000028	3.68492	99.7	31.797	0.054	
1583-05	LWA	sanidine	0.12494	0.000062	3.67909	99.7	31.747	0.055	
					Mean age	**=**	**31.811**	**0.050**	**0.025**
1583-03	LWA	plag	5.02252	0.001697	3.68855	97.6	31.828	0.361	
1583-06	LWA	plag	5.95862	0.002710	3.65227	92.0	31.518	0.340	
					Mean age	**=**	**31.673**	**0.219**	**0.155**
CHADRONIAN / ORELLAN									
Purple White Layer									
1746-01	PWL	biotite	0.11059	0.000191	3.43501	98.6	33.885	0.197	
1746-03	PWL	biotite	0.46109	0.000408	3.45083	97.6	34.040	0.204	
1746-06	PWL	biotite	0.09066	0.000175	3.42860	98.7	33.822	0.195	
1746-07	PWL	biotite	0.32825	0.000242	3.42101	98.7	33.748	0.207	
1746-11	PWL	biotite	0.01039	0.000136	3.45086	98.8	34.040	0.508	
					Mean age	**=**	**33.907**	**0.131**	**0.058**
CHADRONIAN									
Lone Tree Gulch Ash J									
1152-01	LTJ	biotite	0.05250	0.004193	4.37000	77.9	34.508	0.274	
1152-02	LTJ	biotite	0.03889	0.011822	4.34222	55.4	34.291	0.331	
1152-03	LTJ	biotite	0.01296	0.003465	4.32217	80.8	34.134	0.179	
1152-11	LTJ	biotite	0.01181	0.008022	4.33390	64.6	34.226	0.197	
1152-13	LTJ	biotite	0.00525	0.007249	4.36334	67.1	34.456	0.311	
1152-21	LTJ	biotite	0.07495	0.011160	4.39587	57.2	34.710	0.304	
1152-22	LTJ	biotite	0.10923	0.014907	4.40544	50.0	34.785	0.360	
1152-23	LTJ	biotite	0.04003	0.007926	4.39749	65.3	34.723	0.538	
					Mean age	**=**	**34.479**	**0.247**	**0.087**
1155-01	LTJ	sanidine	0.15701	-0.000005	4.42554	100.3	34.942	0.137	
1155-02	LTJ	sanidine	0.11554	0.000495	4.41950	97.0	34.895	0.068	
1155-04	LTJ	sanidine	0.16486	0.000383	4.36574	97.7	34.475	0.167	
1155-05	LTJ	sanidine	0.09153	0.000403	4.39042	97.5	34.668	0.094	
1155-11	LTJ	sanidine	0.14409	0.000472	4.38468	97.1	34.623	0.113	
1155-12	LTJ	sanidine	0.14357	0.000204	4.39810	98.9	34.728	0.087	
1155-13	LTJ	sanidine	0.09350	0.000638	4.38449	96.0	34.621	0.080	
1155-14	LTJ	sanidine	0.14544	0.000434	4.40083	97.4	34.749	0.072	
1155-21	LTJ	sanidine	0.26558	0.000720	4.37369	95.8	34.760	0.138	
1155-22	LTJ	sanidine	0.12235	0.004891	4.36146	75.2	34.664	0.149	
1155-23	LTJ	sanidine	0.17740	0.000427	4.37924	97.5	34.803	0.110	
1155-24	LTJ	sanidine	0.11402	0.000587	4.36768	96.3	34.712	0.139	
					Mean age	**=**	**34.720**	**0.126**	**0.036**
Lone Tree Gulch Ash I									
1148-01	LTI	biotite	0.00672	0.006322	4.49830	70.6	35.511	0.376	
1148-02	LTI	biotite	0.01146	0.000187	4.48345	98.8	35.395	0.222	
1148-03	LTI	biotite	0.01010	0.000573	4.43709	96.3	35.033	0.218	
1148-05	LTI	biotite	0.01399	0.000258	4.48521	98.3	35.409	0.240	
1148-11	LTI	biotite	0.01848	0.000626	4.48308	96.0	35.392	0.244	
1148-12	LTI	biotite	0.01530	0.001363	4.42446	91.6	34.934	0.246	
1148-13	LTI	biotite	0.00000	0.000689	4.54740	95.7	35.895	0.422	
1148-14	LTI	biotite	0.00440	0.000664	4.48792	95.8	35.430	0.380	
					Mean age	**=**	**35.375**	**0.294**	**0.104**
Lone Tree Gulch Ash G									
1147-01	LTG	biotite	0.03859	0.000658	4.49590	95.9	35.492	0.130	
1147-02	LTG	biotite	0.04201	0.000592	4.49243	96.3	35.465	0.114	
1147-03	LTG	biotite	0.01073	0.000556	4.49425	96.4	35.480	0.344	
1147-04	LTG	biotite	0.00938	0.001014	4.50879	93.7	35.593	0.109	

Lab Number	Unit Name	Mineral Dated	Ar 37/39	Ar 36/39	Ar 40*/39	% Rad.	Age (Ma)	SD (Ma)	SEM (Ma)
1147-11	LTG	biotite	0.00191	0.000508	4.55368	96.8	35.944	0.198	
1147-12	LTG	biotite	0.00595	0.000529	4.48195	96.6	35.383	0.270	
1147-13	LTG	biotite	0.00232	0.001745	4.52190	89.7	35.696	0.207	
1147-14	LTG	biotite	0.01089	0.000785	4.50401	95.1	35.556	0.142	
1147-15	LTG	biotite	0.00558	0.000402	4.50374	97.4	35.554	0.140	
					Mean age	**=**	**35.574**	**0.165**	**0.055**
1157-01	LTG	sanidine	0.01944	0.002203	4.50706	87.4	35.580	0.473	
1157-02B	LTG	sanidine	0.00750	0.000000	4.52919	100.0	35.753	0.070	
1157-03	LTG	sanidine	0.00740	0.000023	4.53308	99.8	35.783	0.085	
1157-04	LTG	sanidine	0.00759	0.000024	4.52105	99.8	35.689	0.067	
1157-05	LTG	sanidine	0.00622	0.000011	4.53452	99.9	35.794	0.070	
1157-11	LTG	sanidine	0.00846	0.000146	4.52565	99.0	35.725	0.062	
1157-14	LTG	sanidine	0.00758	0.000147	4.52747	99.0	35.739	0.120	
					Mean age	**=**	**35.723**	**0.073**	**0.027**
Lone Tree Gulch Ash F									
1156-01	LTF	biotite	0.03137	0.016414	4.54793	48.4	35.899	0.452	
1156-02	LTF	biotite	0.01861	0.004997	4.51240	75.3	35.621	0.298	
1156-03	LTF	biotite	0.10555	0.011865	4.61079	56.8	36.390	0.408	
1156-04	LTF	biotite	0.01784	0.050277	4.43961	23.0	35.052	1.090	
1156-05	LTF	biotite	0.05052	0.006143	4.44998	71.0	35.133	0.279	
1156-14	LTF	biotite	0.07397	0.001035	4.54755	93.8	35.896	0.097	
1156-15	LTF	biotite	0.00946	0.000424	4.54526	97.3	35.878	0.076	
1156-21	LTF	biotite	0.01426	0.007312	4.55712	67.8	35.971	0.311	
1156-22	LTF	biotite	0.02218	0.010452	4.54166	59.5	35.850	0.343	
1156-23	LTF	biotite	0.04719	0.018191	4.50994	45.6	35.602	0.471	
1156-24	LTF	biotite	0.01962	0.004192	4.51624	78.5	35.651	0.306	
					Mean age	**=**	**35.722**	**0.379**	**0.114**
1153-02	LTF	sanidine	0.38504	0.002334	4.52036	87.3	35.684	0.160	
1153-03	LTF	sanidine	0.09803	0.000389	4.54249	97.7	35.857	0.082	
1153-04	LTF	sanidine	0.11458	0.001033	4.54330	93.8	35.863	0.069	
1153-05	LTF	sanidine	0.19707	0.002684	4.54681	85.4	35.890	0.128	
1153-06	LTF	sanidine	0.42590	0.005950	4.53128	72.4	35.769	0.115	
					Mean age	**=**	**35.813**	**0.085**	**0.038**
Lone Tree Gulch Ash B									
1146-01	LTB	biotite	0.01420	0.002004	4.54412	88.5	35.869	0.076	
1146-02	LTB	biotite	0.00753	0.001760	4.57303	89.8	36.095	0.086	
1146-03	LTB	biotite	0.00315	0.002313	4.60883	87.1	36.375	0.214	
1146-05	LTB	biotite	0.02669	0.005810	4.52637	72.5	35.731	0.092	
1146-11	LTB	biotite	-0.00077	0.000370	4.58593	97.6	36.196	0.149	
1146-12	LTB	biotite	0.10535	0.004880	4.48100	75.7	35.376	0.161	
1146-13	LTB	biotite	0.00408	0.001740	4.54704	89.8	35.892	0.208	
1146-14	LTB	biotite	0.03280	0.000709	4.56814	95.6	36.057	0.133	
1146-15	LTB	biotite	0.02206	0.002074	4.53633	88.1	35.809	0.111	
1146-21	LTB	biotite	0.01036	0.005749	4.46348	72.4	35.239	0.159	
1146-22	LTB	biotite	-0.01714	0.008749	4.57906	63.9	36.142	0.204	
1146-23	LTB	biotite	0.01157	0.000998	4.59206	93.9	36.244	0.430	
					Mean age	**=**	**35.919**	**0.344**	**0.099**
1151-05	LTB	sanidine	0.43810	0.001107	4.48320	93.9	35.393	0.218	
1151-22	LTB	sanidine	0.41402	0.004504	4.54520	77.8	35.878	0.129	
1151-23	LTB	sanidine	0.46580	0.004869	4.58544	76.6	36.192	0.176	
1151-24	LTB	sanidine	0.36003	0.001383	4.61548	92.4	36.427	0.136	
					Mean age	**=**	**35.973**	**0.447**	**0.224**

Lab Number	Unit Name	Mineral Dated	Ar 37/39	Ar 36/39	Ar 40*/39	% Rad.	Age (Ma)	SD (Ma)	SEM (Ma)
DUCHESNEAN/CHADRONIAN									
Bracks Rhyolite									
2552-01	H87-205	sanidine	0.06295	0.000463	1.16164	89.7	36.657	0.204	
2552-02	H87-205	sanidine	0.05094	0.000062	1.16567	98.6	36.783	0.121	
2552-03	H87-205	sanidine	0.05987	0.000221	1.16023	94.9	36.613	0.133	
2552-04	H87-205	sanidine	0.05982	0.000211	1.16268	95.1	36.690	0.136	
2552-06	H87-205	sanidine	0.06189	0.000286	1.15920	93.4	36.581	0.206	
					Mean age	**=**	**36.665**	**0.078**	**0.035**
2552-05	H87-205	sanidine	0.06783	0.001367	1.21734	75.2	38.397	0.959	
Buckshot Ignimbrite									
2556-01	BKST	sanidine	0.03933	0.000788	1.20351	83.9	37.965	0.118	
2556-02	BKST	sanidine	0.04010	0.000286	1.20062	93.5	37.875	0.184	
2556-03	BKST	sanidine	0.04341	0.000157	1.19517	96.4	37.705	0.142	
2556-04	BKST	sanidine	0.03559	0.001028	1.20012	79.9	37.859	0.153	
2556-05	BKST	sanidine	0.04052	0.000100	1.19141	97.7	37.587	0.208	
					Mean age	**=**	**37.798**	**0.150**	**0.060**
DUCHESNEAN									
La Point Tuff									
2554-05	LP1	biotite	0.00186	0.000983	1.25371	81.1	39.531	0.606	
2554-11	LP1	biotite	0.00412	0.000817	1.25488	83.8	39.568	0.197	
2554-03	LP1	biotite	0.00454	0.000986	1.26082	81.1	39.753	0.552	
2554-14	LP1	biotite	0.00456	0.000843	1.26236	83.4	39.801	0.175	
2554-12	LP1	biotite	0.00000	0.001278	1.26247	76.9	39.804	0.252	
2554-13	LP1	biotite	0.01639	0.002163	1.26863	66.5	39.996	0.693	
					Mean age	**=**	**39.742**	**0.171**	**0.070**

3. MAGNETOSTRATIGRAPHY OF THE LOWER AND MIDDLE MEMBERS OF THE DEVIL'S GRAVEYARD FORMATION (MIDDLE EOCENE), TRANS-PECOS TEXAS

by Anne H. Walton

ABSTRACT

The lower and middle members of the Devil's Graveyard Formation in Trans-Pecos Texas include several localities with Bridgerian and Uintan (middle Eocene) vertebrate faunas. Paleomagnetic samples were collected from 150 meters of section in the Devil's Graveyard Formation and subjected to a rigorous program of stepwise thermal demagnetization and measurement. Rock magnetic studies show a strong remanence carried by detrital magnetite, with lesser (usually secondary) hematite, in most samples. Almost all samples revealed normal or reversed polarity though magnetic behavior was often complex.

The lower and middle members of the Devil's Graveyard Formation, constrained by both vertebrate biostratigraphy and three K-Ar dates, were probably deposited during Chrons C20R and C20N (by the timescale of Berggren et al., 1991). The West Texas results agree with the findings of Flynn (1983, 1986), which place the Bridgerian/ Uintan boundary, defined by the first appearance of *Amynodon*, at or near the boundary between Chrons C21N and C20R in Wyoming and southern California. West Texas retained a humid, equable climate through the Uintan while climate in the Northern Interior became progressively cooler and more arid. The persistence of some taxa considered "Bridgerian" through the late Uintan in Texas reflects intensifying environmental fragmentation.

INTRODUCTION

The Eocene North American Land Mammal "Ages" (NALMAs) were recognized and named in the Rocky Mountain basins of the Northern Interior. Precise correlation of these well-known faunas with those of other areas, such as the semitropical and tropical regions of North America or with other continents, could greatly refine the reconstruction of Eocene-Oligocene paleoclimates and biotic evolution. Eocene-Oligocene faunas from the boundary region between North America and Mesoamerica (*sensu lato*) are preserved in Texas. The Devil's Graveyard Formation (Stevens et al., 1984), in the Trans-Pecos volcanic field, includes four distinct vertebrate faunas ascribable to the Bridgerian to Chadronian NALMAs. The area also contains extrusive rocks which have yielded dates on high-temperature minerals by potassium-argon methods (Henry and McDowell, 1986; Henry et al., 1986) and, more recently, single-crystal $^{40}Ar/^{39}Ar$ dating (Swisher, pers. comm.). The contribution of magnetostratigraphy can further refine the understanding of the timing of deposition and faunal migrations.

STRATIGRAPHY OF THE DEVIL'S GRAVEYARD FORMATION

Trans-Pecos Texas is tectonically part of the southernmost Southern Rocky Mountains, with Laramide structures complicated by Tertiary vulcanism and Basin and Range faulting. Volcanic activity began in the middle Eocene. The volcanic contribution to fluvial and lacustrine systems (in the form of reworked volcanic sediment, ashfall tuffs, and eventually flows) generally increased with time through the Oligocene and gradually died out in the Miocene.

Eocene to early Oligocene fluvially-deposited volcaniclastic sediments exposed in Brewster and Presidio Counties, north and west of Big Bend National Park (Figure 3.1), were described by Stevens et al (1984) and named the

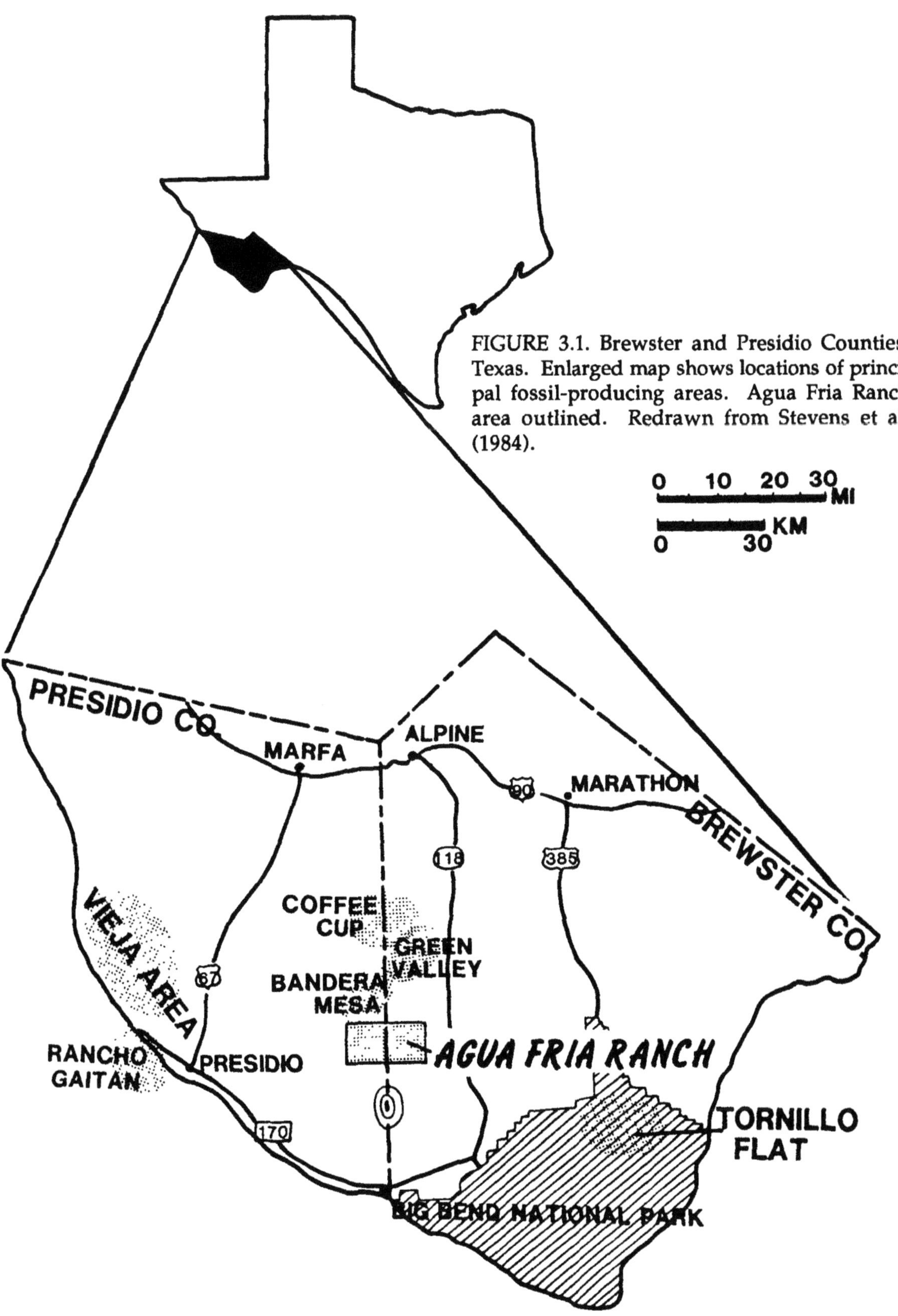

FIGURE 3.1. Brewster and Presidio Counties, Texas. Enlarged map shows locations of principal fossil-producing areas. Agua Fria Ranch area outlined. Redrawn from Stevens et al. (1984).

Devil's Graveyard Formation (DGF). This unit is often referred to in the older literature as the Pruett Formation or Duff Tuff. Sedimentologic evidence indicates most of the lower and middle members represent the distal facies of a volcanic apron from an unidentified volcanic source some tens of kilometers to the southwest, in Mexico (Runkel, 1990a).

The DGF is subdivided into three members: the unnamed lower and middle members, and the uppermost Bandera Mesa Member (Figure 3.2). An erosional unconformity separates the middle member from the Bandera Mesa Member. The section as a whole is characterized by massive nodular mudstones (Lithofacies H of Runkel, 1990a), interbedded with laterally-persistent fine-grained layers of variable thickness, and slightly meandering channel fill. Very few bedforms or other primary sedimentary structures have survived pedogenesis and diagenetic alteration. The rate of sediment accumulation was probably highly irregular, as is true for most continental sections and especially true for those in the vicinity of an intermittently active volcanic source.

Marker beds, source rocks for K-Ar dates, and local faunas are diagrammed in Figure 3.2. Lithologic description and more extensive discussion can be found in Stevens et al. (1984). The most reliable isotopic date presently available in the DGF (McDowell, pers. comm.) is from the lower orange clinoptilolite above the Whistler Squat Quarry tuff, with an age of 43.9 ± 0.9 Ma (McDowell, 1979).

VERTEBRATE BIOSTRATIGRAPHY

The faunas of the lower and middle members of the DGF are ascribable to the Uintan land mammal age though they differ in detail from the type faunas of the Northern Interior. The differences (summarized by Wilson, 1984, 1986) are probably controlled by contrasting environments, specifically to climate variation with latitude. Evidence from land snails (Roth, 1984) indicates a warm, equable, moist climate, similar to that of coastal southern Mexico today, prevailed during the Eocene in West Texas, even though faunal and paleosol evidence indicate a trend towards increasing aridity and seasonality in the Northern Interior at approximately the same time (Peterson and Chapin; Retallack; reported in Prothero et al., 1990). Busbey (1986) finds similar paleoclimatic indications from reptiles; the typically "Bridgerian" crocodilian *Pristichampsus* persists in West Texas through the lower and middle members of the DGF, long after its disappearance from faunas to the north. The Uintan climatic deterioration (Lillegraven, 1976) apparently did not affect the Texas Gulf Coast as strongly as the basins of the Northern Interior, restricting the range of some taxa. A microfauna of "Bridgerian" aspect persists late into the Uintan of West Texas (Walton, 1986a).

Faunal lists for Trans-Pecos Texas can be found in Wilson (1986). The lower member of the DGF (Figure 3.2) includes the Whistler Squat local fauna and the Basal Tertiary local fauna. The Basal Tertiary and Whistler Squat local faunas resemble those of the Tepee Trail Formation of Wyoming and the Friars and Mission Valley Formations of southern California in that *Amynodon* and *Leptoreodon* are found in association with characteristically "Bridgerian" elements such as the primates *Omomys* and *Microsyops*. The rodents also suggest a Bridgerian age, as observed by Wood (1973, 1974). The Whistler Squat local fauna is arguably ascribable to the Shoshonian (earliest Uintan) subage proposed by Flynn (1983, 1986), characterized by a mixture of "Bridgerian" and "Uintan" forms.

The Titanothere Hill, Serendipity, and Purple Bench faunas occur at or near the top of the lower member of the DGF; all are considered Uinta C equivalents (Wilson, 1984), or late Uintan, based on the presence of *Protoreodon petersoni* and its large mammal association, e.g. *Metamynodon mckinneyi*. Such genera as *Colodon* and *Toromeryx* are known from late Uintan and younger units elsewhere (Krishtalka et al., 1987). The micromammals, however, retain elements typical of the Bridgerian, such as *Omomys* and *Microsyops*, and the rodents *Mysops* and *Pauromys*.

The base of the Skyline Channels, separating the middle member from the Bandera Mesa Member, represents a temporal and faunal discontinuity. The Skyline local fauna is Duchesnean, based on the presence of *Amynodontopsis bodei* and *Hyaenodon* (Wilson, 1984, 1986).

Other Uintan faunas in Trans-Pecos Texas (Figure 3.1) occur in Big Bend National Park,

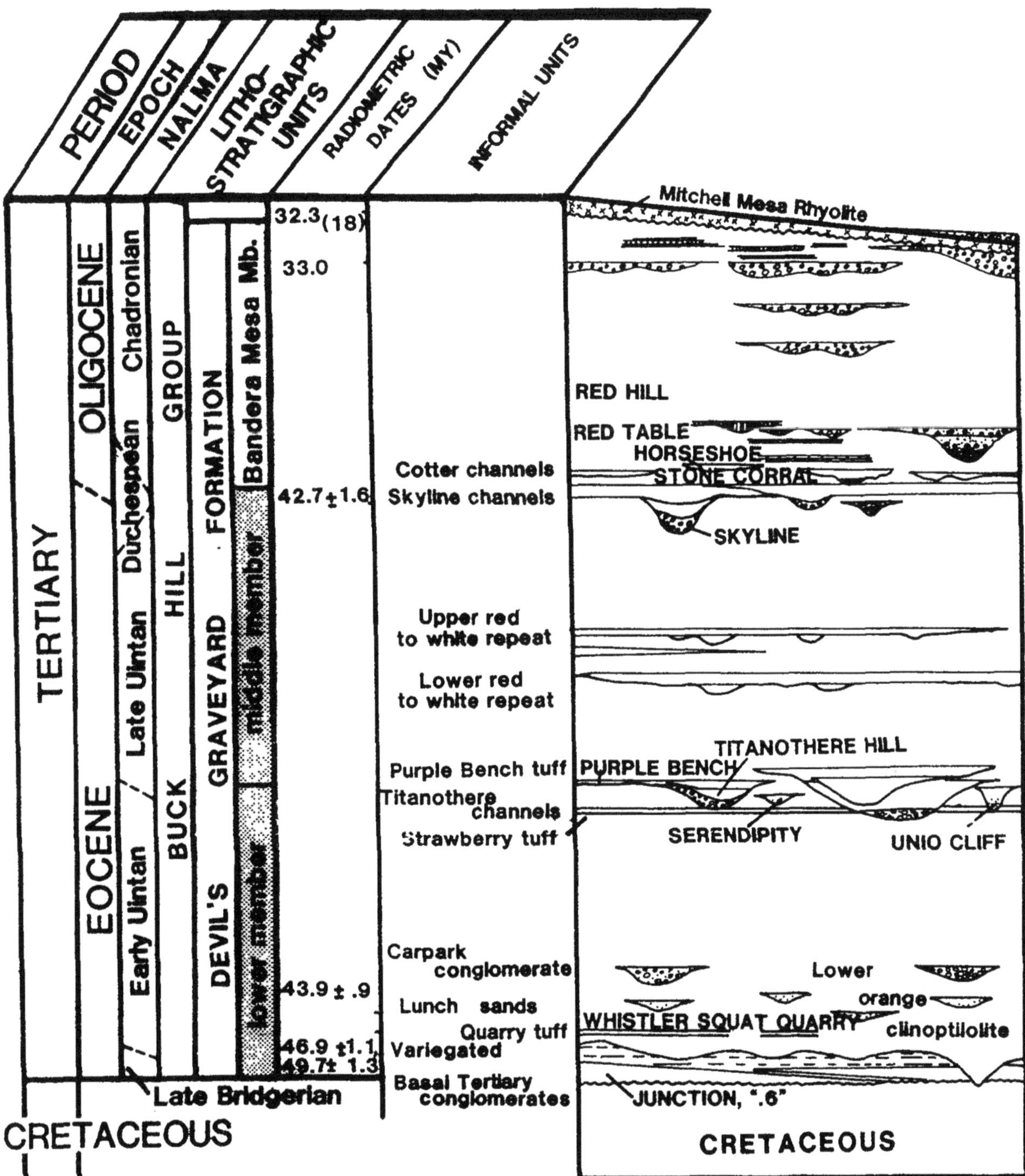

FIGURE 3.2. Diagrammatic stratigraphic section of the Devil's Graveyard Formation, showing fossil localities, sources of K-Ar dates, and informal units. The K-Ar date on the Mitchell Mesa Rhyolite is an average of 18 samples. Redrawn from Stevens et al. (1984).

southeast of the Agua Fria Ranch, and in the Vieja area, 100-120 kilometers to the northwest (the Candelaria local fauna; Wood, 1974; Wilson, 1986). A rich Uinta C-equivalent fauna from the Laredo Formation of the southern Rio Grande Valley (Westgate, 1988, 1990) also contains micromammal taxa commonly found in the Bridgerian elsewhere, such as *Pauromys* (Walton, in press). The Casa Blanca local fauna allows the correlation of the Serendipity and Candelaria local faunas of Trans-Pecos Texas and other Uinta C land mammal faunas with Gulf Coast marine units of the Cook Mountain and Laredo Formations belonging to Nannoplankton Zone 16. Uinta C faunas in North America are therefore correlated with the late Lutetian to early Bartonian ages in Europe (Westgate, 1988, 1990). Magnetostratigraphic studies in the Laredo Formation were inconclusive because of irregular magnetic behavior (Walton, 1986b).

SAMPLE COLLECTION AND EXPERIMENTAL PROCEDURE

Sections from twelve localities, encompassing the lower and middle members of the DGF, were sampled for paleomagnetic study in the Agua Fria Ranch. No more than two kilometers separate any two localities laterally. Vertically, there is considerable overlap between the sections at each locality; only the uppermost portion of the total section is not duplicated in at least two localities.

A total of 385 samples were collected.from 152 meters (500 feet) of composite section (detailed in Walton, 1986a). Stratigraphic density of sites (horizons) was favored over multiple sampling from individual sites. A minimum of two or as many as four samples were collected per site. With overlapping sections, as many as six samples were taken for some marker beds. Stratigraphic distance between sites ranged from a few cm to 30 m, but was usually about 1 to 1.5 m. Sampling emphasized fossil localities, marker beds, and long sections uninterrupted by channel sandstones whenever possible. Paleomagnetic sampling was confined to the well-exposed and relatively fossiliferous lower and middle members. The hiatus between the middle member and the Bandera Mesa Member, indicated by a deeply scoured contact and deposition of the overlying Skyline Channels, marks an important temporal and environmental discontinuity, so that the Bandera Mesa Member section should be considered separately from the rest of the DGF.

Most samples were collected using hand tools and oriented in place with a Brunton compass. Others were cored with a gasoline-powered drill and oriented with a Brunton compass fixed onto a custom-built brass orientation sleeve. Hand samples were later cored in the laboratory on a water-cooled drill press, and the orientation transferred to the core. Two or more cores were drilled from almost all the hand specimens. Small, fragile, or awkwardly shaped samples were sometimes cut into cubes and contained in plastic boxes, or reinforced using sodium silicate.

Samples were kept in a magnetically shielded room with a maximum ambient field of about one hundred gammas, and were never removed during the course of measurement. Samples were measured with a cryogenic magnetometer (Superconducting Technology, C-102 SRM) at the University of Texas at Austin. Every measurement was repeated to insure reproducibility of results.

Natural remanent magnetization (NRM) was measured for all samples. All samples were subjected to at least five and as many as twelve demagnetization steps in order to reveal the components of remanent magnetization. A Schonstedt GSD-1 was used for alternating field (AF) demagnetization, applied in three perpendicular directions. Thermal demagnetization (TD) was done in air and the samples were cooled in a shielded region with an ambient field of no more than 7 gammas. For the majority of sites, one sample was subjected to AF demagnetization, and the second sample to TD in order to reveal the components of remanent magnetization. Any additional samples received TD because the thermal method was more effective than AF demagnetization.

ROCK AND MINERAL MAGNETISM

Detrital grains of magnetite are visible in polished sections of several samples from the Devil's Graveyard Formation, and are often in the process of being altered to ilmenite, leucoxene, or hematite. Secondary hematite is probably responsible for the bright red pigment in beds such as the Variegated and the

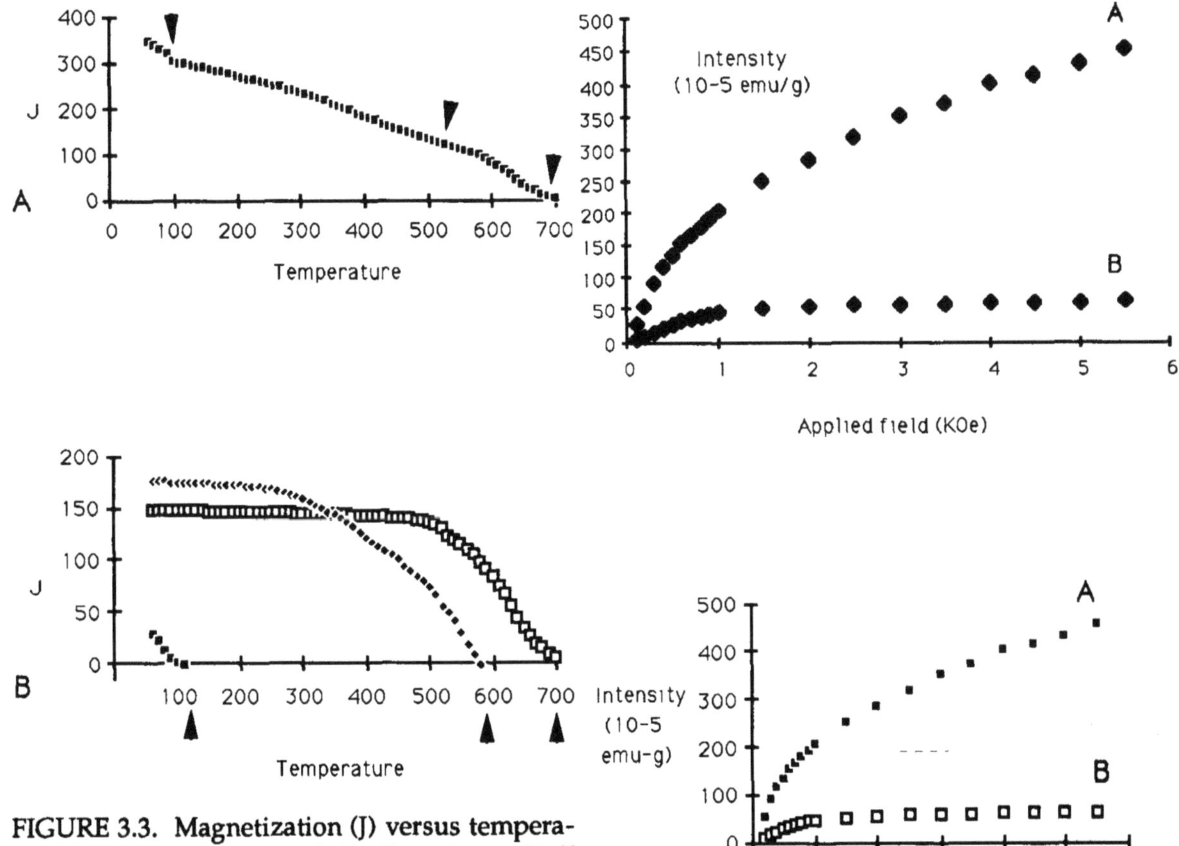

FIGURE 3.3. Magnetization (J) versus temperature (°C) for a sample of the Strawberry Tuff (TDG10-05). A. Three magnetic minerals contribute different Curie curves. Arrows mark flexure points of combined curves. B. Three curves separated. Arrows mark resultant Curie temperatures. The magnetite curve emerges with a Curie point of ~580°C.

FIGURE 3.4. Normalized IRM curves. A. Gray siltstone, sample TDG05-26. B. Bright red Strawberry Tuff from Titanothere Hill, sample TDG10-05.

Strawberry Tuff.

Thermomagnetic analysis was conducted in an oxygen-free atmosphere using a Cahn electrobalance (courtesy Brooks Elwood, University of Texas at Arlington). A sample of the bright red Strawberry Tuff from Titanothere Hill (TDG10-05) reveals a complex magnetic mineralogy (Figure 3.3). Inflection points in the Curie curve indicate three distinct magnetic phases: goethite (Tc = ~100°C), magnetite (Tc = 580°C), and hematite (Tc = 680°C).

The acquisition of isothermal remanent magnetization (IRM) is another means of identifying the carrier(s) of remanence. Demagnetized rock samples were subjected to increasing fields in an electromagnet up to a maximum field of 5.5 KOe. Figure 3.4 shows two extreme examples. Sample TDG05-26 saturates in a field of less than 3 KOe, indicating magnetite as the sole magnetic mineral. A sample from the Strawberry Tuff (TDG10-05), however, is not saturated by available magnetic fields. The steep initial rise is due to magnetite, and the increase above 3 KOe is from substantial amounts of hematite.

Directional behavior during demagnetization indicates that of the three magnetic minerals present, magnetite is predominant in most of the section and is probably of primary (volcanic) origin. Some hematite may also be primary, but in most cases is probably secondary. The third contributing mineral, goethite, is a typical weathering product. Consequently it is to be expected that the original

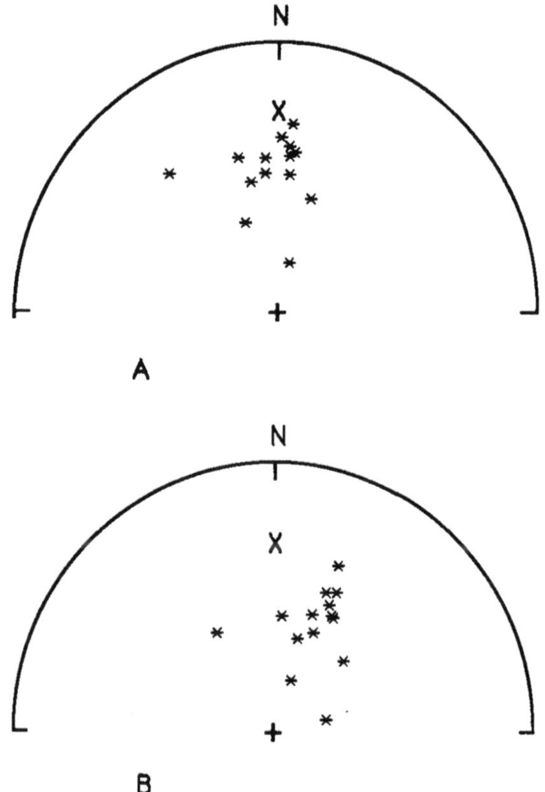

FIGURE 3.5. Stereo net plots of NRM directions for Titanothere Hill. Normal samples only. X marks direction of theoretical dipole field. Beds dip 20° to east. A. Without structural correction. B. With structural correction.

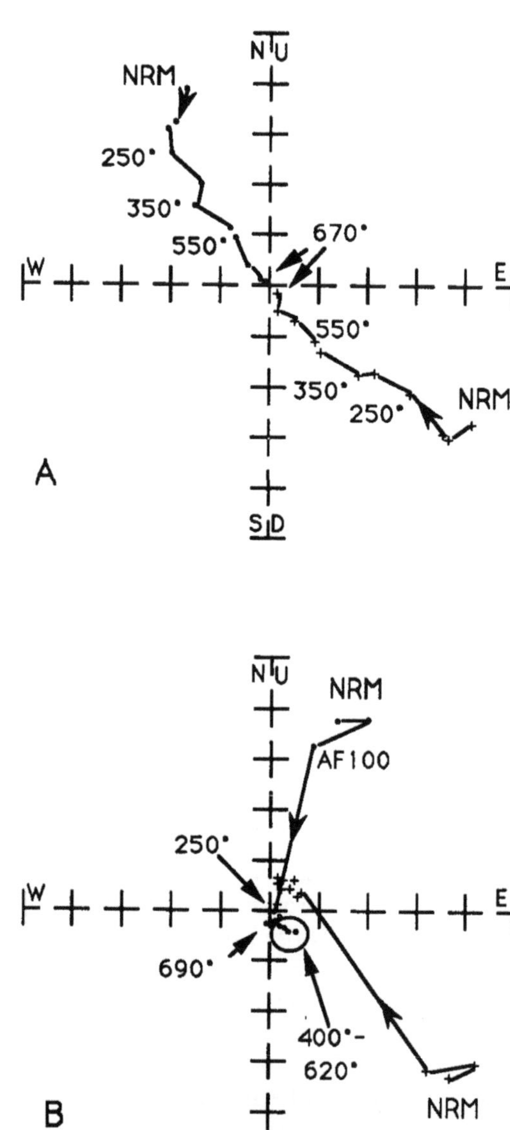

FIGURE 3.6. As-Zijderveld diagrams. Circles represent declination, crosses inclination. Distance from origin indicates intensity. A. Pink mudstone, sample TDG06-80A. Normal sample showing primary magnetite and hematite. B. Gray claystone, sample TDG10-02A. Reversed sample with normal overprint eliminated by heating to 250°C. Complex magnetic mineralogy indicated by heterogeneous clusters of remanence through successive demagnetization stages. TD temperatures indicated for declination only.

remanence in these samples is carried mainly by magnetite and is a detrital remanent magnetization, and that the samples will have a present field overprint. Some units, such as the bentonitic Variegated (discussed below), have a more complex magnetic mineralogical history, to the extent that the determination of remanence was inconclusive. It is unfortunate that irregular magnetic behavior was also characteristic of the heavily-sampled ashfall tuffs, such as the Whistler Squat Quarry Tuff.

DIRECTIONS OF MAGNETIZATION

The directions of the natural remanent magnetization (NRM) typically cluster around the present-day field direction. When NRM directions from the tilted beds at the Titanothere Hill section (normal polarity samples only) are

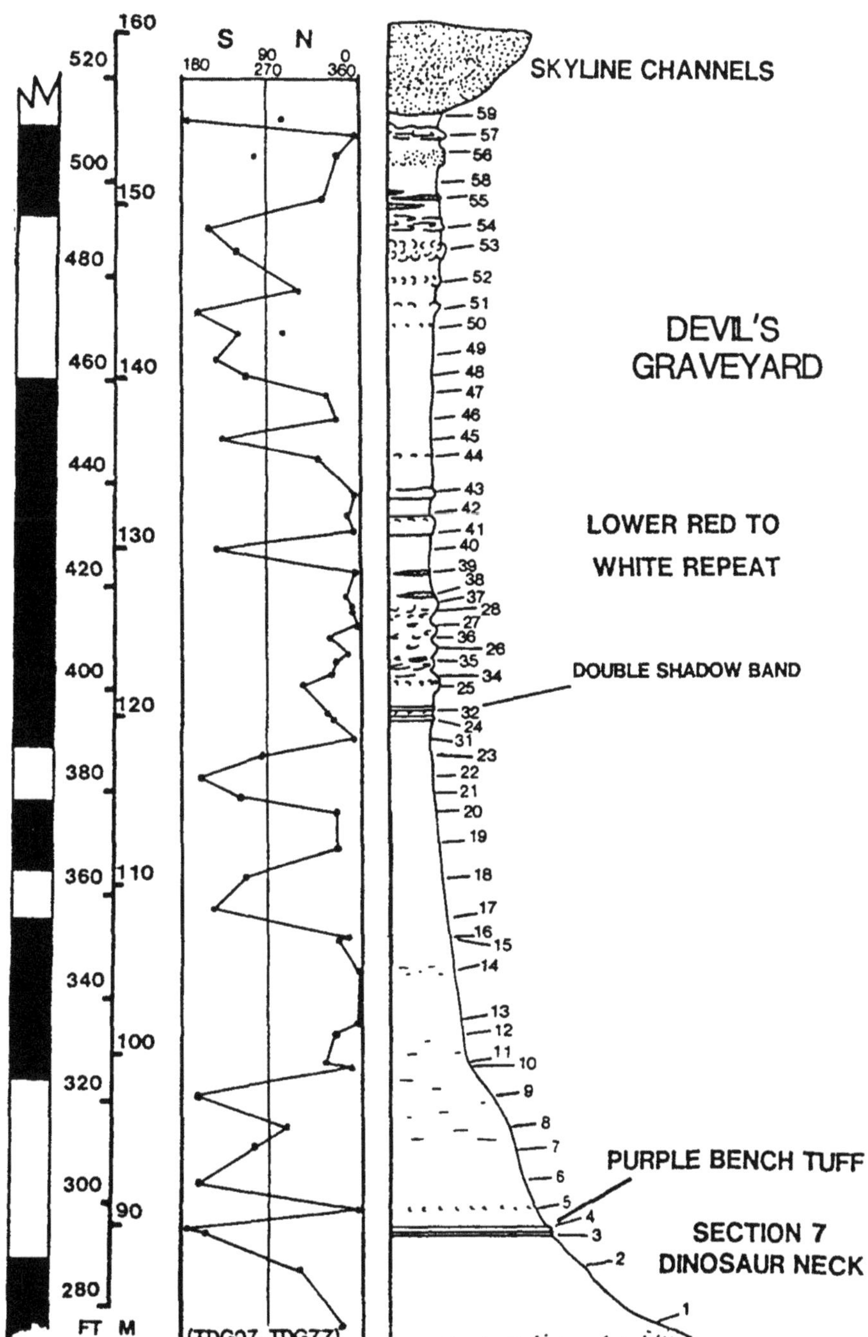

FIGURE 3.7. A representative sample section. A. Sampled sites, lithology, and stratigraphic markers. B. Plot of resulting mean site declinations. In cases where a sample subjected to AF demagnetization produced conspicuously different results from TD, both declinations are shown by points but TD results are preferred (e.g. sites 56 and 59) C. Stratigraphic level. D. Interpretation of magnetic reversal stratigraphy. Reversals shown when indicated by two or more sites.

compared before, and after, structural correction (Figure 3.5), it is clear that the overprint postdates Quaternary tilting.

The response to demagnetization is best shown in orthogonal vector projections (As-Zijderveld diagrams) which were also used to calculate the characteristic direction of magnetization. Figure 3.6 shows two examples. In the first case (Figure 3.6A), one component of magnetization is removed by heating to only 160°C and can be ascribed to goethite. The stable component is then revealed in the range from 200°C to 670°C, and is carried by magnetite and hematite. In this sample the hematite is probably primary, because its remanence is concordant with the direction carried by magnetite.

The second sample (Figure 3.6B) is reversely magnetized. Initial AF demagnetization has questionable effect, but heating to 250° erases a strong normal overprint ascribable to goethite. The remaining remanence is reversed, though at least two different directions of magnetization are revealed by overlapping clusters of points during demagnetization. The stable component is recognized in the range TD 400°-620°C, and is probably carried by both magnetite and hematite. After heating to 690° the sample retains less than 10% of its NRM intensity and the direction is probably random.

In most samples the polarity of magnetization could be determined with little ambiguity. The overall quality of the data, however, is not sufficient to permit the calculation of a meaningful paleomagnetic pole position supported by Fisher statistics. Samples subjected to AF demagnetization usually showed the same polarity as those demagnetized by TD, but directional behavior was frequently complex, because hematite is little affected by AF demagnetization. Inclination was especially inconsistent, as a result of inaccurate orientation during sample handling, swelling of clays exposed to humidity in the laboratory, or shallowing of the inclination induced by post-depositional compaction (Anson and Kodama, 1987). In contrast, declination is relatively consistent and was preferentially used as an indicator of paleomagnetic pole directions.

MAGNETOSTRATIGRAPHY

The magnetostratigraphic results for all twelve localities are shown in Walton, 1986a. Figure 3.7 shows one section in detail. The spacing between sites is typical for all sections. The plot of mean declination shows readily identifiable polarity in most cases. In this as in all the sections, there are several magnetozones documented by only one site. These short intervals may be real, but a magnetozone was assigned only when defined by at least two sites from one or more sections. Short intervals are difficult to trace even over small distances, and are not identifiable in the more compressed marine magnetic anomaly record.

Results from the twelve sections are summarized on the right side of Figure 3.8. The composite section was assembled using lithostratigraphic marker beds (such as the Strawberry Tuff) for correlation. In general, lithostratigraphy and magnetostratigraphy coincide. Small discrepancies are probably the result of time-trangressive, non-uniform deposition, and erosion.

The Variegated (Figure 3.2), a brightly pigmented bentonitic unit near the base of the lower member, is problematical. It is loosely consolidated, difficult to sample, hematized, and magnetically extremely viscous and erratic, as were all samples of swelling clay lithology. The few measurable samples appear to be normally magnetized. Based on the more reliable normal polarity of the underlying Basal Tertiary, the entire lower part of the lower member is portrayed as normal in Figure 3.8. The limits of this normal interval, however, are unknown.

Runkel (1988, 1990a, 1990b) demonstrated the lithostratigraphic and biostratigraphic equivalence of the lower member of the DGF to the lower part of the Canoe Formation in Big Bend National Park. Though not as densely fossiliferous as the DGF, the greater thickness of the lower Canoe Formation and mineralogical differences suggest magnetostratigraphic sampling in this area has potential for resolving the problems presented by the Variegated.

CORRELATION WITH THE MARINE POLARITY TIMESCALE

Figure 3.8 shows a proposed correlation of the lower and middle members of the Devil's Graveyard Formation (Figure 3.8B) with the marine polarity timescale of Berggren et al. (this volume) (Figure 3.8A). The correlation

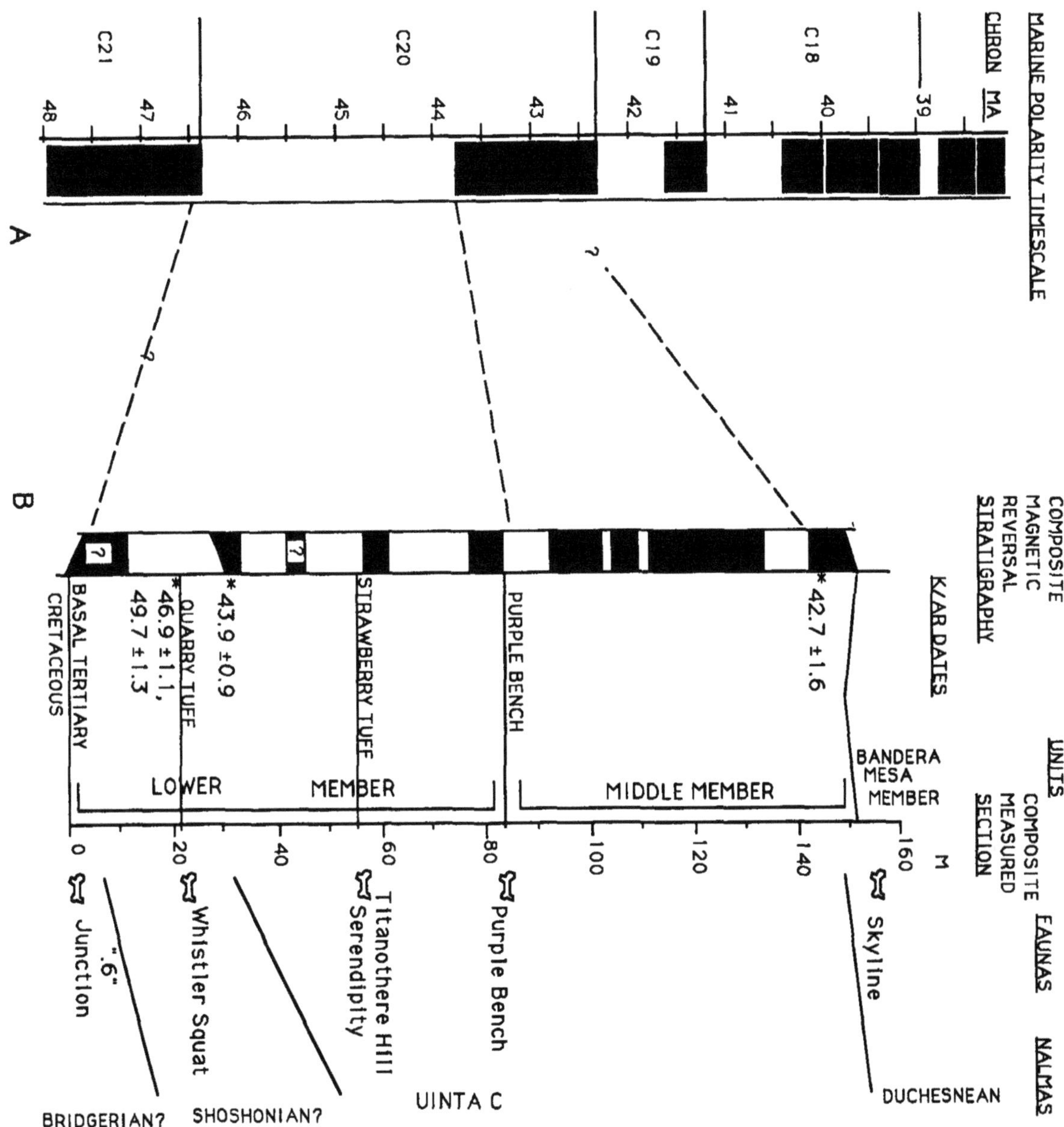

FIGURE 3.8. Proposed correlation of marine polarity timescale with Devil's Graveyard Formation. A. Timescale from Berggren et al., 1991. B. Composite magnetostratigraphic column for lower and middle members of the Devil's Graveyard Formation. Quarry Tuff, Strawberry Tuff, and Purple Bench Tuff used for lateral correlation. K-Ar dates, lithostratigraphic units, fossil-producing levels, and suggested NALMAs shown to right.

differs from that of Walton (1986a) because new calibrations of the marine timescale (Berggren et al., 1991) indicate Chrons C20 and C19 are younger than represented in the past (e.g. Berggren et al., 1985; Harland et al., 1982). The predominantly reversed lower member and the predominantly normal middle member are correlated with Chron C20R and C20N, respectively. This scheme requires no special assumptions about rate of sediment accumulation and is consistent with the most reliable of the K-Ar dates known from the section: 43.9 ± 0.9 Ma on the lower orange clinoptilolite (McDowell, 1979) and 42.7 ± 1.6 on an ash from beneath the Skyline Channels (Geochron Labs, in Stevens et al., 1984); and with the more reliable of two K-Ar dates on the Quarry Tuff, that of 46.9 ± 1.1 (McDowell, 1979).

If all of Chron C20R is preserved in the DGF section (as shown with query in Figure 3.8), the lower member was deposited at approximately 3 cm/1000 yr, a reasonable average rate for continental sections. An analogous rate of sediment accumulation for the entire section would suggest the Chron C20-C19 transition would be found several meters above the base of the Skyline Channels. The top of the middle member of the DGF would then fall within Chron C20, and is overlain discontinuously by the Bandera Mesa Member, so the upper limit on its age cannot be defined precisely.

Though the correlation presented in Figure 3.8 is believed to be the best solution, it would be misleading to claim it is the only solution. For example, acceptance of the younger limit for the date on the ash below the Skyline would place the top of the middle member within Chron C19 (Flynn, pers. comm.). There are many short reversals in the DGF section, so that the correlation with Chron C20R and C20N is at best a generalization. It is expected that more reversals should appear in continental sections than in even the best-preserved deep marine units, because more rapid (if irregular) deposition increases the probability that a short reversal will be sampled. The DGF section nevertheless seems extreme in the number and duration of short reversals.

IMPLICATIONS FOR FAUNAL AGES

If the correlation in Figure 3.8 is correct, the most precisely dated of the DGF faunas is that of Purple Bench, roughly coinciding with the Chron C20R-C20N transition (43.8 Ma by the timescale of Berggren et al., 1991). Other faunas are only datable as falling within Chron C20R (46.4-43.8 Ma).

The lowest part of the section is temporally and biostratigraphically ambiguous because of the complex magnetic behavior of the Variegated. The polarity of the Variegated is uncertain, so it is uncertain whether the Basal Tertiary is part of a normal interval of substantial length, or a short reversal. The Chron C21N-C20R transition shown near the base of the lower member is speculative. In the Agua Fria area there is no long section exposed below the Variegated where polarity could be corroborated.

If the Chron C21N-C20R occurs at or near its indicated position, the DGF could provide the means to test the isochrony of the Bridgerian-Uintan faunal transition between three widely-separated, ecologically distinct areas: the Northern Interior, the Pacific coast, and West Texas. Flynn (1983, 1986) has shown that the Bridgerian/Uintan boundary, taken as the first appearance of *Amynodon* as an immigrant from Asia, falls within Chron C20R in the Washakie Basin of Wyoming. In southern California, the boundary also lies within (or just below) Chron C20R. Jerskey's (1981) work in the Bridger Basin of Wyoming (Bridger Formation) also implies the Bridgerian/ Uintan boundary occurs in Chron C20R. In the Tepee Trail Formation of Wyoming, Flynn (1983, 1986) placed the boundary between Bridgerian and earliest Uintan faunas in the lowest part of Chron C20R in the type section in the East Fork Basin. In an abstract, Shive et al. (1980) suggest a normal polarity for beds spanning the first appearance of *Amynodon* (the lower Holy City beds; Eaton, 1985) in the Tepee Trail Formation in the Owl Creek area, so that the Bridgerian/ Uintan boundary would occur in the upper part of Chron C21N, rather than in lower Chron C21R as proposed by Flynn (1983, 1986). However, this work remains unpublished.

In Trans-Pecos Texas the Bridgerian/Uintan boundary, as recognized elsewhere, occurs below the first occurrence of *Amynodon* at Whistler Squat. However, the boundary cannot yet be confidently recognized in the Devil's Graveyard Formation as hoped by Walton (1986a) because the absence of diagnostic *Amynodon* from the Basal Tertiary may be accidental; the fauna is mainly composed of micromammals (Wilson, 1986). The close resemblance of the Basal Tertiary and the Whistler Squat faunas argues against any long span of time or major faunal turnover occurring between the two. Forty-one percent of the taxa found in the Whistler Squat assemblage also occur in the Basal Tertiary. If only small mammals (didelphids, insectivores, primates, and rodents) are considered, 78% of the fauna is held in common.

CONCLUSIONS

If the arrival of *Amynodon* occurred between Basal Tertiary and Whistler Squat time, and if the proposed magnetic reversal stratigraphy is correct, the Bridgerian/ Uintan boundary is roughly synchronous in such widely separated areas as Wyoming, southern California, and Texas. The Bridgerian-Uintan transition in Trans-Pecos Texas would coincide with the intitiation of local volcanism but not with substantial faunal turnover.

The most important implications of the improved dating of Uintan faunas in the DGF are paleobiogeographic. Faunal, geochemical, and paleosol evidence suggests the humid tropical forest prevailing in the Bridgerian of the Northern Interior degraded to mixed grasslands by the Orellan (Peterson and Chapin; Retallack; reported in Prothero et al., 1990). The Trans-Pecos and Laredo faunas and the Laredo flora indicate that tropicality endured on the Gulf Coast into the late Uintan (Westgate and Gee, 1990), so that genera such as *Pauromys* continued to diversify throughout the Uintan in the south even though they are not known from deposits younger than the Bridgerian in the north (Walton, in press). Climatic deterioration began earlier at high latitudes and did not affect subtropical latitudes simultaneously.

Duchesnean microvertebrates from Texas have much to offer in improving the resolution of latitudinal gradients in terrestrial climate change. The West Texas section is not continuous, so the patterns of faunal change from Uintan to Duchesnean faunas cannot be detailed. The absence of Uintan microvertebrates in Texas Duchesnean sections, so far as is known, suggests that though early and middle Eocene climate change may have been fairly gradual in North America, the Uintan-Duchesnean transition was almost as drastic in subtropical latitudes as it was in the Northern Interior.

ACKNOWLEDGMENTS

Funds for field work were provided by the University of Texas Geology Foundation. The Department of Geological Sciences assisted in providing liquid helium and lab time. Preparation of the manuscript was supported by a grant from the Owen Coates Fund. Brooks Elwood, at the University of Texas at Arlington, performed essential thermomagnetic analyses. Billy Pat McKinney, lessee of the Agua Fria Ranch, allowed access onto the property. There are too many people to thank individually for their West Texas hospitality.

Above all I am indebted to Drs. Wulf Gose and John A. Wilson of the University of Texas at Austin, advisors to the Master's thesis on which this paper is based. Margaret and Dr. Jim Stevens of Lamar University were invaluable help in the field and every other aspect of this project. The manuscript was reviewed thoughtfully by Drs. Jim Westgate, Tony Runkel, and John Flynn. For their help in the field I thank Arten Avakian, Carole Gee, Martin Sander, Jim and Karen Westgate, and Susan Witebsky.

REFERENCES

Anson, G. L. and Kodama, K. P. 1987. Compaction-induced inclination shallowing of the post-depositional remanent magnetization in a synthetic sediment. *Geophys. Jour. Roy. Astr. Soc.* 88:673-692.

Berggren, W. A., Kent, D. V., Flynn, J. J., and Van Couvering, J. A. 1985. Cenozoic geochronology. *Geol. Soc. Am. Bull.* 96:1407-1418.

Busbey, A. 1986. *Pristichampus* cf. *P. vorax* (Eusuchia: Pristichampsinae) from the Uintan of West Texas. *Jour. Vert. Paleo.* 6:101-103.

Eaton, J. G. 1985. Paleontology and correlation of the Eocene Teepee Trail and Wiggins Formations in the north fork of Owl Creek area, southeastern Absaroka Range, Hot Springs County, Wyoming. *Jour. Vert. Paleo.* 5:345-370.

Flynn, J. J. 1983. Correlation and geochronology

of Middle Eocene strata from the western United States. Ph.D. dissertation, Columbia University, 496 p.

Flynn, J. J. 1986. Correlation and geochronology of Middle Eocene strata from the western United States. *Palaeogeogr., Palaeoclimat., Palaeoecol.* 55:335-406.

Harland, W. B., Cox, A. V., Llewellyn, P. G., Pickton, C. A. G., Smith, A. G., and Walters, R. 1982. *A Geologic Time Scale.* Cambridge, Cambridge University Press, 131 pp.

Henry, C. D. and McDowell, F. W. 1986. Geochronology of the Tertiary volcanic field, Trans-Pecos Texas. In J. G. Price, C. D. Henry, and D. S. Barker, eds., *Igneous Geology of Trans-Pecos Texas.* Field Trip Guide, Geol. Soc. Am. 99th Ann. Meeting.

Henry, C.D., McDowell, F.W., Price, J. G., and Smyth, R. C. 1986. Compilation of potassium-argon ages of Tertiary igneous rocks, Trans-Pecos Texas. *Univ. Texas Bur. Econ. Geol. Circ.* 86-2:1-34.

Jerskey, R. G. 1981. A paleomagnetic study of the Bridger Formation, southern Green River Basin, Wyoming. M.S. thesis, University of Wisconsin at Milwaukee, 60 pp.

Krishtalka, L. K.; Stucky, R.; West, R. M.; McKenna, M. C.; Black, C. C.; Bown, T. M.; Dawson, M. R.; Flynn, J.J.; Golz, D. J.; Lillegraven, J. A.; and Turnbull, W. D. 1987. Eocene (Clarkforkian through Duchesnean) chronology of North America. In M. O. Woodburne, ed., *Cenozoic Mammals: Their Temporal Record, Biostratigraphy, and Biochronology,* Berkeley, Univ. California Press, 77-117.

Lillegraven, J. A. 1976. A biogeographical problem involving comparisons of later Eocene terrestrial vertebrate faunas of western North America. In J. Gray and A. Boucot, A., eds., *Historical Biogeography, Plate Tectonics, and the Changing Environment,* Corvallis, Oregon State University Press, 333-347.

McDowell, F. W. 1979. Potassium-argon dating in the Trans-Pecos volcanic field. *Univ. Texas Bur. Econ. Geol. Guidebook* 19:10-18.

Prothero, D. R.; Berggren, W. A.; and Bjork, P. R. 1990. Late Eocene-Oligocene climatic and biotic evolution. Penrose Conference report. *GSA News & Info.* March 1990: 74-75.

Roth, B. 1984. *Lysinoe* (Gastropoda: Pulmonata) and other land snails from Eocene-Oligocene of Trans-Pecos Texas, and their paleoclimatic significance. *The Veliger* 27:200-218.

Runkel, A. C. 1988. Stratigraphy, sedimentology, and vertebrate paleontology of Eocene rocks, Big Bend region, Texas. Ph.D. dissertation, University of Texas at Austin, 310 pp.

Runkel, A. C. 1990a. Lateral and temporal changes in volcanogenic sedimentation: analysis of two Eocene sedimentary aprons, Big Bend region, Texas. *Jour. Sed. Pet.* 60:747-760.

Runkel, A. C. 1990b. Stratigraphy and depositional history of Late Cretaceous and Paleogene rocks, Trans-Pecos Texas. South Texas Geol. Soc. Guidebook. *Amer. Assoc. Petrol. Geol. 74th Annual Convention*:117-146.

Shive, P. N., Sundell, K. A., and Rutledge, J. 1980. Magnetic polarity stratigraphy of Eocene volcaniclastic rocks from the Absaroka Mountains of Wyoming. *EOS* 61:945.

Stevens, J. B., Stevens, M. S., and Wilson, J. A. 1984. Devil's Graveyard Formation (new), Eocene and Oligocene age, Trans-Pecos Texas. *Texas Mem. Mus. Bull.* 32:1-21.

Walton, A. H. 1986a. Magnetostratigraphy and the ages of Bridgerian and Uintan faunas in the lower and middle members of the Devil's Graveyard Formation, Trans-Pecos Texas. M.A. thesis, University of Texas at Austin, 135 pp.

Walton, A. H. 1986b. Magnetostratigraphy of some Eocene vertebrate-bearing rocks in Texas. *Geol. Soc. Am. Abs. Prog.* 18 (7): 781 (Abstr.).

Walton, A. H. in press. New microrodents from the Uintan (Middle Eocene) of Texas. *Jour. Vert. Paleo.*

Westgate, J. W. 1988. Biostratigraphic implications of the first Eocene land-mammal fauna from the North American coastal plain. *Geology* 16:995-998.

Westgate, J. W. 1990. Uintan land mammals (excluding rodents) from an estuarine facies of the Laredo Formation (middle Eocene, Claiborne Group) of Webb County, Texas. *Jour. Paleont.* 64: 454-468.

Westgate, J. W. and Gee, C. T. 1990. Paleoecology of a middle Eocene mangrove biota (vertebrates, plants, and invertebrates) from southwest Texas. *Palaeogeogr. Palaeoclim. Palaeoecol.* 78:163-177.

Wilson, J.A. 1984. Vertebrate faunas 49 to 36 million years ago and additions to the species of *Leptoreodon* (Mammalia: Artiodactyla) found in Texas. *Jour. Vert. Paleo.* 4:199-207.

Wilson, J.A. 1986. Stratigraphic occurrence and correlation of Early Tertiary vertebrate faunas, Trans-Pecos Texas. Part 2: Agua Fria- Green Valley areas. *Jour. Vert. Paleo.* 6:350-373.

Wood, A. E. 1973. Eocene rodents, Pruett Formation, southwest Texas: their pertinence to the origin of the South American Caviomorpha. *Pearce-Sellards Ser.* 20: 41 pp.

Wood, A. E. 1974. Early Tertiary vertebrate faunas, Vieja Group, Trans-Pecos Texas: Rodentia. *Texas Mem. Mus. Bull.* 21: 1-112.

Wood, H. E., Chaney, R. W., Clark, J., Colbert, E., Jepsen, G. L., Reeside, J. B., and Stock, C. 1941. Nomenclature and correlation of the North American continental Tertiary. *Geol. Soc. Am. Bull.* 52:1-48.

4. REDEFINITION OF THE DUCHESNEAN LAND MAMMAL "AGE," LATE EOCENE OF WESTERN NORTH AMERICA

By Spencer G. Lucas

"*Duchesnean is probably the most hotly debated, and certainly the most weakly substantiated of the Nearctic land-mammal ages we adopt.*"-- Savage and Russell (1983, p. 120)

ABSTRACT

I redefine the Duchesnean land-mammal "age" (LMA) in western North America on the basis of the Halfway and Lapoint mammalian faunas of the Duchesne River Formation in northeastern Utah and their principal correlatives in South Dakota, New Mexico, California and Texas. The beginning of the Duchesnean and the beginning of the next youngest LMA, the Chadronian, are identified by numerous first appearances of land-mammal genera that include both immigrants from Asia and North American evolutionary originations. The oldest eubrontothere, *Duchesneodus*, is the index fossil of the Duchesnean LMA. Earlier arguments for recognizing Duchesnean as a "subage" of the Chadronian, or of the Uintan and Chadronian, are inconsistent with the definition and recognition of some other North American LMA's which are comparable to the Duchesnean in terms of duration, faunal distinctiveness and adequacy of the "type" fauna. Subdivision of the Duchesnean can be undertaken but is not formalized because of inadequate data. The Duchesnean LMA is late Eocene, about 37-42 Ma, a correlative of part of the Sharamurunian LMA of Asia and the European latest Bartonian-Priabonian.

INTRODUCTION

Wood et al. (1941) proposed a sequence of land-mammal "ages" to discriminate intervals of Cenozoic time in western North America. With the exception of Dragonian, a Paleocene LMA based on a fauna from Utah, all the LMA's of Wood et al. (1941) are still in use (Woodburne, 1987a). Yet, despite their durability, the validity of a few LMA's has been a source of contention. This is true of the Duchesnean, identified by Wood et al. (1941) as the youngest Eocene LMA. The Duchesnean LMA has either been questioned, abandoned, subdivided or defended. Here, I defend the validity of the Duchesnean LMA by redefining it.

As many workers have pointed out (e.g., Repenning, 1967; Tedford, 1970; Berggren and Van Couvering, 1974; Woodburne, 1977), the LMA's of Wood et al. (1941) are biochronological units, based on biochrons sensu Williams (1901), who coined the term to signify intervals of geologic time recognized by fossils, i.e., the interval of geologic time that corresponds to the duration of a taxon. As biochronological units, the LMA's are not tied to stratotypes and thus are not stage-ages. Instead, they are time intervals rooted in fossil assemblages. Although the North American LMA's were originally defined with reference to type faunas, "the ages are not necessarily coextensive with their types" (Wood et al., 1941, p. 6). Or, as Berggren and Van Couvering (1974, p. 7) well expressed it, "mammalian biochrons (Land Mammal Ages, etc.) also originate as local zones tied to reference sections and 'type faunas', even though they are commonly liberated from such earthly bondage almost at birth, and in many instances are created with inferred or abstract limits not observed in the type section itself." My goal in redefining the Duchesnean LMA is to demonstrate: (1) that its "type" fauna provides an adequate basis for recognizing a time interval (stage of mammalian evolution) between the Uintan and the Chadronian; (2) that principal correlative faunas of the Duchesnean "type" fauna further substantiate recognition of a time interval between the Uintan and Chadronian; and (3) that the beginning of the Duchesnean and of the Chadronian (= end of Duchesnean) can be defined by numerous first appearances, both of immigrants from Asia to North America and of evolutionary originations in North America.

In so doing, I do not make the distinction between "definition" and "characterization" of an LMA advocated by Woodburne (1977, 1987b). He has argued that we should "define" the beginnings of LMA's by the appearance of a single taxon and that "definitions that involve several taxa are certain to be ambiguous...and must actually be considered characterizations" (Woodburne, 1987b, p. 14). Thus, my definitions of the beginning of the Duchesnean and of the Chadronian would be termed "characterizations" by Woodburne.

The problems with single-taxon definitions of biochronologic boundaries have long been obvious -- inherent diachrony of the taxon's appearance over a broad area and the practical difficulties faced when the single taxon used for definition is absent in a local fauna. Given the sporadic occurrence of vertebrate fossils, the latter is a particularly large problem. Furthermore, multiple taxon events can be synchronous within typical levels of geochronological resolution. Woodburne's distinction between "definition" and "characterization" thus strikes me as a semantic way to differentiate single-taxon and multiple-taxon definitions of biochronological boundaries that need not be utilized in the multiple-taxon approach to boundary definition employed here.

HISTORY OF THE DUCHESNEAN

In reviewing the history of the Duchesnean LMA, I focus on articles that have addressed directly the definition and validity of this LMA. A large number of faunal and systematic studies have used Duchesnean as a valid LMA (e.g., Bjork, 1967; Golz, 1976; Prothero, 1985a) but were not directly concerned with its definition. Although these studies provide information important to this definition, I concentrate here on those articles in which discussions of the definition of the Duchesnean LMA have been most explicit.

The term Duchesnean has its roots in Peterson's (1931) proposal of the term "Duchesne Oligocene" (Formation) for the rock unit previously (Peterson and Kay, 1931) termed "Upper Uinta" in the Uinta Basin of northeastern Utah. Kay (1934) replaced Peterson's name with Duchesne River Formation because Keyes (1924) had already used the term Duchesne Limestone for a Jurassic rock-stratigraphic unit. Both Peterson and Kay identified the Duchesne River Formation as Oligocene, although they recognized its fauna as transitional between typical Uinta Eocene and Chadron faunas of the White River "series," the latter assigned by them an Oligocene age. Furthermore, Kay (1934) named three "horizons" of the Duchesne River Formation which he defined as rock-stratigraphic units - Randlett, Halfway and Lapoint (Figure 4.1). These horizons also corresponded to three stratigraphically successive faunas collected by Carnegie Museum crews working in the Uinta basin.

Wood et al. (1941) coined the term Duchesnean age as a "new provincial time term, based on the Duchesne River formation of northeastern Utah" (Figure 4.1). Although they listed California Institute of Technology Locality 150 in the Sespe Formation as a correlative, Wood et al. (1941) stated that "a faunal definition would be premature except to note the abundance of *Teleodus*." They then provided mammalian faunal lists for the Duchesne River Formation and the Sespe locality. In a further unusual step, Wood et al. (1941) discussed the Duchesnean/Chadronian boundary by identifying the Titus Canyon (California) and Yoder (Wyoming) mammal faunas as intermediate in age between the Duchesnean faunas and those of Pipestone Springs and Thompson Creek (Montana) which, in turn, they considered "probably slightly older than the Chadron Formation."

Scott (1945) monographed the mammalian faunas of the Duchesne River Formation and assigned them an Oligocene age. Simpson (1946), however, subsequently argued that the Duchesnean faunas are of Eocene, not Oligocene age (also see Simpson, 1933).

Kay's (1934) threefold division of the Duchesnean--Randlett, Halfway, Lapoint horizons/faunas--persisted until Gazin (1955, 1956, 1959) recommended removing the Randlett from the

Kay (1934)	Wood et al. (1941)	Gazin (1955)	Clark et al. (1967)	Andersen & Picard (1972)	Wilson (1978)	Wilson (1984, 1986)	Kelly (1990)	this paper
	CHADRONIAN AGE	CHADRONIAN	CHADRONIAN; DUCHESNEAN – Viejan: Vieja	Duchesne River F.: Starr Flat Mbr.	CHADRONIAN: local faunas Airstrip, Little Egypt	CHADRONIAN: local faunas Little Egypt, Rancho Gaitan	CHADRONIAN: local faunas	CHADRONIAN local faunas
Duchesne River F.: Lapoint horizon	DUCHESNEAN AGE: Sespe 150 locality	DUCHESNEAN: Lapoint	DUCHESNEAN – Lapointian: Lapoint	Duchesne River F.: Lapoint Member	CHADRONIAN: Porvenir	DUCHESNEAN: Porvenir, Lapoint	DUCHESNEAN – LATE: Porvenir, Lapoint	DUCHESNEAN: Porvenir, Tonque, Lapoint
Duchesne River F.: Halfway horizon	DUCHESNEAN AGE: Sespe 150 locality	DUCHESNEAN: Halfway	UINTAN: Halfway	Duchesne River F.: Dry Gulch Creek Member	UINTAN	DUCHESNEAN: Skyline, CIT 150, Badwater, 20 Wood Rodent	DUCHESNEAN – EARLY: Pearson Ranch	DUCHESNEAN: Slim Buttes, Skyline-Cotter, Pearson Ranch
Duchesne River F.: Randlett horizon	DUCHESNEAN AGE: Sespe 150 locality	UINTAN: Randlett	UINTAN: Randlett	Duchesne River F.: Brennan Basin Member	UINTAN	UINTAN: Candelaria	UINTAN – C	UINTAN
Uinta F: horizon C	UINTAN AGE	UINTAN: Uinta C	UINTAN	Uinta F.: Myton Member	UINTAN: Candelaria	UINTAN: Candelaria	UINTAN – C	UINTAN

FIGURE 4.1. Changing concepts of the Duchesnean LMA. Rock-stratigraphic/faunal subdivisions of the Duchesne River Formation in the Uinta basin (Kay, 1934; Andersen and Picard, 1972) are shown as well as differing schemes for subdividing late Uintan, Duchesnean and Chadronian time.

Duchesnean because of its faunal similarity to the Uintan Myton fauna (Figure 4.1). Black and Dawson (1966), however, included all three faunas in the Duchesnean.

Exclusion of both the Randlett and Halfway faunas from the Duchesnean was undertaken by Clark et al. (1967). They restricted the Duchesnean LMA to the Lapoint fauna in the Uinta basin and identified the Vieja fauna from Presidio County, Texas described by Stovall (1948) as Duchesnean, though younger than Lapoint. Clark et al. (1967) thus divided the Duchesnean into an earlier portion (their Lapointian), supposedly characterized by "*Teleodus*," *Epihippus*, *Diplobunops* and *Poabromylus* and a younger portion (their Viejan), supposedly characterized by "*Teleodus*," *Mesohippus v. viejaensis* and *Agriochoerus* (Figure 4.1).

Tedford (1970) rejected this subdivision, correctly pointing out that neither *Epihippus* nor *Diplobunops* is known from the Lapoint fauna and that the temporal succession of the Lapoint and Vieja faunas posited by Clark et al. (1967) was far from adequately substantiated. However, Tedford (1970, p. 692) did conclude (p. 692) that "the Lapoint fauna is still distinct, and older than those characteristic of the Chadronian Mammal 'Age,' and it deserves recognition as part of a revised Duchesnean."

At about the same time, Wilson et al. (1968) reported K-Ar dates and additional mammal fossils from the Vieja area that suggested Duchesnean faunas were older than 36-37 Ma. Particularly important was the first recognition of the Porvenir local fauna from the Chambers Tuff to encompass the fossils Stovall (1948) reported, as well as additional taxa.

Andersen and Picard (1972) presented a revised physical stratigraphy of the Duchesne River Formation that differed from that of Kay (1934) in creating four members (in ascending order): Brennan Basin, Dry Gulch Creek, Lapoint and Starr Flat (Figure 4.1). They also provided lists of the vertebrate fauna of each member.

Wilson (1978) first seriously questioned the validity of the Duchesnean (Figure 4.1). He stated that only one taxon (*Simimeryx minutus*) is unique to the Lapoint fauna and argued that the Porvenir local fauna is younger than the Lapoint fauna and of Chadronian age. According to Wilson, the first occurrence of *Mesohippus* marked the base of the Chadronian (see also Wood et al., 1941). Wilson thus believed that there is no Lapoint equivalent fauna in Texas between the Uintan Candelaria local fauna and the Chadronian Porvenir local fauna. Although Wilson (1978) did not recommend total abandonment of Duchesnean, he did not recognize it in Trans-Pecos Texas and clearly questioned its validity, concluding that (p. 38) "hopefully, a sufficiently distinctive fauna to warrant a mammalian 'age' called Duchesnean can be found."

More was written during the 1980s about the validity of the Duchesnean LMA than during the preceding forty years. Lucas et al. (1981; also see Lucas, 1982, 1983a, 1986b) recognized a valid Duchesnean LMA and attempted to define its beginning, as well as the beginning of the Chadronian, by immigration events, following the ideas of Repenning (1967). Emry (1981) added two taxa to the Lapoint fauna (Miacidae or Amphicyonidae, *Agriochoerus maximus*) and provided an updated faunal list. He noted the first appearance of many of the same immigrant taxa at the beginning of the Duchesnean as Lucas et al. (1981) and stated (Emry, 1981, p. 563) "these immigrant taxa are potentially useful in defining the beginning of a North American Land Mammal 'Age' which might better be called Chadronian, with the Duchesnean being its earliest subage." Emry et al. (1987) reiterated this conclusion, again identifying the Duchesnean land-mammal fauna as the beginning of the "White River chronofauna."

Savage and Russell (1983, p. 120) used the term Duchesnean but stated that "the faunas herein assigned to Duchesnean could be termed later or latest Uintan or "eo-Chadronian." They listed the Lapoint and Pearson Ranch faunas as Duchesnean and also noted that Krishtalka (1979) had described *Hyopsodus* from a Duchesnean fauna in Wyoming. They considered Duchesnean latest Eocene, a correlative of the European Headonian and identified Duchesnean as the *Eotylopus*-"*Teleodus*"-*Simimys* fauna.

Wilson (1984, 1986) agreed with Emry's arguments and considered the Duchesnean to be a "subage," but he divided it into Uintan and Chadronian "subages" (Figure 4.1). Wilson identified the early Duchesnean fauna as having taxa with Uintan affinities (*Hyopsodus*, *Harpagolestes* and *Diplobunops*), whereas the late Duchesnean is marked by the first appearance of *Hemipsalodon*, *Mesohippus*, *Toxotherium*, *Hyracodon primus*, *Brachyhyops* and *Merycoidodon*.

Krishtalka et al. (1987) tentatively retained the Duchesnean LMA. They stated that to resolve the issue of whether or not the Duchesnean should be a "subage" of the Chadronian (or Uintan and Chadronian), "additional faunas from this period as well as detailed stratigraphic collections are needed" (p. 90). Krishtalka et al. (1987, p. 90) also listed several Duchesnean faunas as well as "reported Duchesnean faunas that require further study."

Most recently, Kelly (1990) recognized a separate Duchesnean LMA. He divided the Duchesnean into earlier and later portions (Figure 4.1) along the lines proposed by Wilson (1984, 1986).

VALIDITY OF THE DUCHESNEAN AS A LAND MAMMAL "AGE"

The low taxonomic diversity and small number of specimens of fossil mammals from the Lapoint fauna of the Duchesne River Formation, usually considered the "type" fauna of the Duchesnean LMA, seems to be the main obstacle preventing recognition of the Duchesnean as a distinct LMA (e.g., Wilson et al., 1968; Emry, 1981; Wilson, 1984, 1986). Many taxa from the Lapoint and its putative correlatives (especially the Porvenir local fauna of Trans-Pecos Texas) also occur in Uintan and/or Chadronian horizons, and this has been used to support inclusion of the Duchesnean in the Chadronian as a "sub-age" (Wilson, 1978; Emry, 1981) or to divide the Duchesnean into late Uintan and early Chadronian "subages" (Wilson, 1984, 1986).

It is unfortunate that the Lapoint fauna (Figure 4.2) is based on so limited a collection of fossil mammals, but this fauna does contain temporally restricted and geographically widespread taxa (*Duchesneodus* is the best example) that allow its correlation with other mammalian faunas from a broad geographic area of western North America. I find it significant that the Tiffany local fauna, late Paleocene of southwestern Colorado, consists of 15 genera based on approximately 100 specimens (Simpson, 1935; Archibald et al., 1987). It is the "type" fauna of the Tiffanian LMA and little more diverse or specimen-rich than the Lapoint fauna, yet the validity of the Tiffanian has never been challenged because its "type" fauna is inadequate.

That many Duchesnean mammal taxa are found in Uintan and Chadronian horizons is essentially irrelevant to recognition of the Duchesnean as a distinct land-mammal "age." In fact, it is the norm for mammalian taxa, especially at the genus level, to crosscut LMA boundaries (Stucky, 1990). Thus, for example, data from Rose (1981) indicate that only 20-

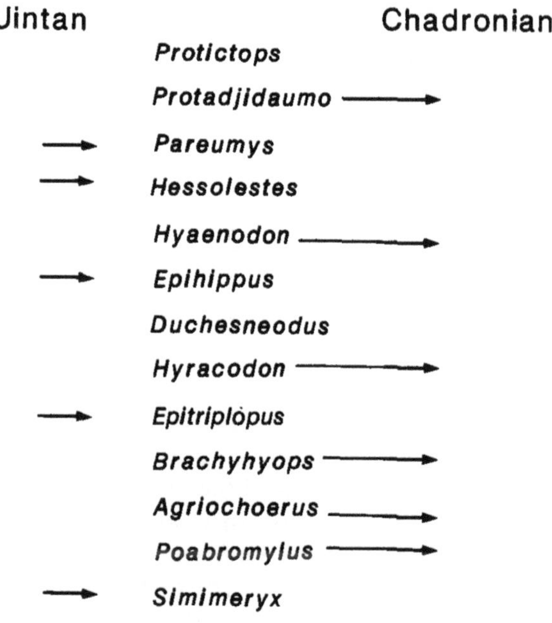

FIGURE 4.2. Genera of the Lapoint/Halfway fauna, "type" fauna of the Duchesnean LMA belong to three groups: (1) Duchesnean last appearances also known from the Uintan; (2) genera known only from the Duchesnean; and (3) Duchesnean first appearances also known from the Chadronian.

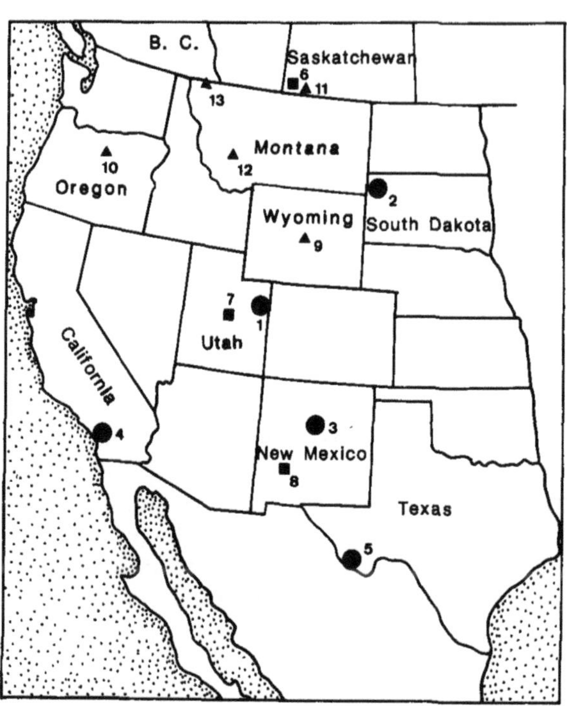

FIGURE 4.3. Location of Duchesnean local faunas, *Duchesneodus* occurrences and local faunas assigned a Duchesnean age by some workers. Dots are the Duchesnean "type" fauna and principal correlatives: 1. Lapoint/Halfway fauna; 2. Slim Buttes Formation; 3. Tonque local fauna; 4. Pearson Ranch local fauna; 5. Porvenir and Skyline-Cotter Channels local faunas. Squares indicate other occurrences of *Duchesneodus*: 6. Cypress Hills; 7. Green River Formation; 8. Black Range. Triangles denote other local faunas assigned a Duchesnean age by some authors: 9. Badwater; 10. Clarno Formation; 11. Lac Pelletier lower fauna; 12. Three Forks basin; 13. Kishenehn Formation.

25% of Clarkforkian land-mammal genera are restricted to that LMA. Yet the Clarkforkian is now regarded, without exception, as a valid LMA (Archibald et al., 1987; Krishtalka et al., 1987).

Emry (1981) and Emry et al. (1987) characterized the Duchesnean mammalian fauna as the beginning of the "White River chronofauna" that endured until the Arikareean. Although no precise definition of the "White River chronofauna" was offered, I assume it refers to the assemblages dominated by hyaenodontids, rabbits, rhinos, horses, oreodonts, camels, leptomerycids and, until the end of the Chadronian, eubrontotheres, characteristic of sediments of the White River Group. Emry (1981) and Emry et al. (1987) offered inclusion of the Duchesnean fauna in this chronofauna as a rationale for incorporating the Duchesnean into the Chadronian as a "subage."

However, identification of chronofaunas (*sensu* Olson, 1952) aims to discern ecological associations of relatively long duration, not the short-lived taxa and assemblages that are the groundmass of precise biochronologies. Chronofaunal schemes, like identification of the "White River chronofauna," divide time from the "top down" and produce units that are inherently difficult to subdivide further. Biochronologies are better built from the "bottom up," with short-lived taxa and assemblages used to discriminate short intervals of time and later aggregated into longer units based on the types of larger scale faunal turnovers that terminate chronofaunas. Thus, even though a significant turnover in the mammalian fauna took place at the Chadronian/Orellan boundary (Prothero, 1985a, b), it does not alter recognition of the "White River chronofauna," which persists until the Arikareean. And, no argument for reducing the Chadronian, Orellan or Whitneyan LMA's to "subage" status has been made, although all encompass portions of the duration of the "White River chronofauna."

Duchesnean is a relatively short interval of time, at most five million years (37-42 Ma). Its boundaries are readily defined and it has a characteristic mammalian fauna that can be recognized from Saskatchewan to California to the Big Bend of Texas. I thus redefine it below as a distinct LMA.

REDEFINITION OF THE DUCHESNEAN LMA

A revised definition of the Duchesnean LMA in the format of Wood et al. (1941) can be presented as follows:

Duchesnean land-mammal "age" -- A term proposed by Wood et al. (1941, p. 10), now based on the mammalian faunas (Halfway and Lapoint faunas: Figure 4.2) of the Dry Gulch Creek and Lapoint Members of the Duchesne River Formation, Uinta basin, northeastern Utah. The "type" fauna is hereafter referred to as the Halfway/Lapoint fauna.

Principal correlatives--(Figure 4.3) - Mammalian fauna from the Slim Buttes Formation, South Dakota (Bjork, 1967); Tonque local fauna, Galisteo Formation, New Mexico (Lucas, 1982); Pearson Ranch local fauna, Sespe Formation, California (Kelly, 1990); Porvenir local fauna, Chambers Tuff and Skyline-Cotter Channels local fauna, Devils Graveyard Formation, both Texas (Wilson, 1984, 1986 and references cited therein).

Index fossil -- *Duchesneodus*.

First appearance -- *Leptictis, Protadjidaumo, Ischyromys, Pseudocylindrodon, Ardynomys, Jaywilsonomys, Eutypomys, Adjidaumo, Aulolithomys, Yoderimys, Hemipsalodon, Hyaenodon, Daphoenus, Hyracodon, Menops, Mesohippus, Trigonias?, Subhyracodon?, Toxotherium, Brachyhyops, Archaeotherium, Agriochoerus, Poabromylus, Eotylopus, Aclistomycter, Heteromeryx, Hendryomeryx, Leptomeryx* and *Hypertragulus*.

Last appearance -- *Proterixoides, Simidectes, Sespedectes, Chumashius, Pareumys, Rapamys, Griphomys, Mytonomys, Ischyrotomus, Hessolestes, Harpagolestes, Amynodon, Triplopus?, Epihippus, Protoreodon, Leptoreodon, Simimeryx* and *Protylopus*.

Note that I have not used Wood et al.'s (1941) category "characteristic fossils" because it seems unnecessary when first and last appearances and index fossil(s) have been identified. Furthermore, most Duchesnean fossil assemblages consist of small numbers of fossils, so identifying abundant, characteristic forms is virtually impossible.

"Type" Fauna

As noted in the review of the history of the Duchesnean LMA, Wood et al. (1941) originally included three faunas in the Duchesnean -- Randlett, Halfway and Lapoint. Removal of the Randlett fauna because of its similarity to the late Uintan mammal fauna of the Myton Member of the Uinta Formation, first suggested by Gazin (1955), has long been accepted (e.g., Krishtalka et al., 1987). The Halfway "fauna" consists of two taxa, *Epihippus (Duchesnehippus) intermedius* and *Eosictis avinoffi*. As Wilson (1978) and Emry (1981) noted, *Epihippus* is the equid genus characteristic of the Uintan, though *E. intermedius* is larger than Uintan species and may fill the evolutionary gap between *Epihippus* and Chadronian *Mesohippus*. *Eosictis avinoffi* is based on a single right maxillary fragment with three teeth and part of a fourth originally identified, with reservations, by Scott (1945) as a felid. Bryant (1988), however, reevaluated this specimen and concluded that it is not a felid but probably an agriochoerid artiodactyl. It clearly is difficult to evaluate the precise age of the Halfway "fauna" because it is so small, but I tentatively include it in the Duchesnean because of *E. intermedius*. I thus combine the one genus from the Halfway "fauna" with the Lapoint fauna to arrive at the Halfway/Lapoint fauna, "type" fauna of the Duchesnean (Figure 4.2; Table 4.1).

As has already been emphasized, the Halfway/Lapoint fauna is of low diversity (13 mammalian species), most of which are only represented by one or two specimens. These taxa

TABLE 4.1. Mammalian genera known from the "type" fauna of the Duchesnean LMA and its principal correlatives. Uintan (U) and Chadronian (C) occurrences of the genera are also indicated. L/H = Lapoint/Halfway fauna, SB = Slim Buttes Formation, T = Tonque local fauna, P = Pearson Ranch local fauna, Po = Porvenir local fauna and S/C = Skyline-Cotter Channels local fauna. Based primarily on data in Bjork (1967), Wilson (1978, 1984, 1986), Emry (1981), Lucas (1982) and Kelly (1990).

GENUS	U	L/H	SB	T	P	Po	S/C	C
Insectivora:								
Centetodon	X					X		X
Leptictis						X		X
Apternodus	X					X		X
Proterixoides	X				X			
Protictops[1]		X						
Simidectes	X				X		X	
Sespedectes	X				X			
Primates:								
Chumashius	X				X			
Mahrgarita							X	
Rooneyia						X		
Rodentia:								
Pareumys	X	X			X			
Protadjidaumo		X						X
Microparamys	X				X	X		X
Rapamys	X				X			
Presbymys					X			
Griphomys	X				X			
Simimys	X				X	cf.		
Leptotomus	X					X	?	X
Mytonomys	X						X	
Ischyrotomus	X					X		
Manitsha	X					X		X
Ischyromys						X		X
Pseudocylindrodon						X		X
Ardynomys						X		X
Jaywilsonomys						X		X
Eutypomys						X		X
Adjidaumo						X		X
Viejadjidaumo						X		
Aulolithomys						X		X
Yoderimys						X		X
Creodonta:								
Hemipsalodon				X		X		X
Hyaenodon		X			X	X	X	X
Ischognathus		X						

Table 4.1. (continued)

GENUS	U	L/H	SB	T	P	Po	S/C	C
Carnivora:								
Miacis	X					X		X
Daphoenus						X		X
Condylarthra:								
Harpagolestes	X						X	
Hessolestes	X	X						
Perissodactyla:								
Duchesneodus		X	X	X	X	X	X	
Menops						X		X
Triplopus	X				?			
Colodon	X		?			X		X
Toxotherium						X	X	X
Amynodon	X			X				
Amynodontopsis			X		X		X	
Metamynodon	X					X		X
Hyracodon		X				X		X
Epitriplopus [2]	X	X						
Trigonias						?		X
Subhyracodon						?		X
Epihippus	X	X	X					
Haplohippus						X		
Mesohippus						X		X
Artiodactyla:								
Brachyhyops		X				X		X
Archaeotherium						X		X
Protoreodon	X			X	X	X	X	
Leptoreodon	X				X			
Hypertragulus							X	X
Agriochoerus		X				X	X	X
Poabromylus		X		X		X		X
Simimeryx	X	X			X			
Protylopus	X				X			
Eotylopus					X	X	?	X
Aclistomycter						X	X	X
Heteromeryx						X		X
Hidrosotherium						X		
Hendryomeryx						X	?	X

[1] I tentatively retain the name *Protictops* here despite Lillegraven et al.'s (1981, p. 46) claim that its type and only species, *P. alticuspidens*, is a *nomen nudum*. Lillegraven et al. based this claim on their inability to locate the holotype of *P. alticuspidens*. However, Peterson's (1934, p. 374-375) introduction of the name conforms to Article 13 of the Code of Zoological Nomenclature (ICZN, 1985), so *P. alticuspidens* is not a *nomen nudum*. If *Protictops* is a synonym of *Centetodon*, then another taxon from the Lapoint fauna would continue from the Uintan through the Chadronian.

[2] I follow Radinsky (1967, p. 30) in considering *Mesamynodon* Peterson, 1931 to be a junior subjective synonym of *Epitriplopus* Wood, 1927.

form three groups (Figure 4.2): (1) those that are also known from Uintan faunas -- *Pareumys, Hessolestes, Epitriplopus, Epihippus* and *Simimeryx*; (2) those that are also known from the Chadronian -- *Protadjidaumo, Hyaenodon, Hyracodon, Brachyhyops, Agriochoerus* and *Poabromylus*; and (3) those known only from the Duchesnean as here redefined -- *Protictops* and *Duchesneodus* (note that *Protictops* is known only from a single specimen). The three-component nature of the Halfway/Lapoint fauna is characteristic of Duchesnean faunas as a whole (see below). Although the Halfway/Lapoint fauna may not be an ideal "type" fauna because of its relatively low diversity and small number of collected specimens, it nevertheless is unambiguously post-Uintan and pre-Chadronian, as these terms are used here, and easily correlated with its principal correlative faunas as detailed below.

Principal Correlatives

Faunas with *Duchesneodus*, the index fossil of the Duchesnean LMA, are readily correlated with the Lapoint/Halfway fauna as its principal correlatives. These faunas share other taxa with the Lapoint/Halfway that further support the correlation. They also contain taxa not known from the Lapoint/Halfway fauna that are unique to Duchesnean time as well as other taxa that have their first or last appearance during the Duchesnean but are also known from the Uintan or Chadronian, respectively. The Lapoint/Halfway fauna thus can be seen as the hub of a network of correlated faunas formed by it and its principal correlatives. This produces faunal characterization of the Duchesnean much the same way as Shaw (1964) developed a composite standard for correlation by combining the stratigraphic ranges of taxa from different sections. Shaw (1964, p. 189) argued that "the combined information [from two or more sections] is more likely to be representative of the fauna as a whole than is the information from either of the consituent sections alone." And, I similarly argue that the Duchesnean fauna composed of the type fauna and its principal correlatives (Table 4.1) is more representative of the Duchesnean mammal fauna than is the Lapoint/Halfway fauna alone. The key to this characterization is whether or not the Lapoint/Halfway fauna can be correlated with confidence with its principal correlatives. This is the case, although as discussed below, it may be slightly younger than some of these correlatives, as has been argued by Wilson (1984, 1986) and Kelly (1990).

The mammalian fauna of the Slim Buttes Formation in Harding County, South Dakota (Bjork, 1967) includes *Duchesneodus* as well as *Epihippus intermedius, Colodon* and *Amynodontopsis. Amynodontopsis* is restricted to the Duchesnean, but *Colodon* is a long-ranging form also known from Uintan and Chadronian faunas. The Slim Buttes artiodactyls -- an agriochoerid and a leptotraguline -- are of no precise biochronological significance.

The Tonque local fauna of the Galisteo Formation in north-central New Mexico (Lucas, 1982) also includes *Duchesneodus*. Duchesnean first occurrences are *Poabromylus* and *Hemipsalodon. Amynodon* and *Protoreodon* from the Tonque local fauna are Duchesnean last occurrences.

The Pearson Ranch local fauna from the Sespe Formation in Ventura County, California, has just been reviewed by Kelly (1990). Duchesnean last appearances are *Simidectes, Chumashius, Rapamys, Pareumys, Triplopus?, Protoreodon, Leptoreodon* and *Protylopus*. First appearances are *Hyaenodon* and *Eotylopus*. Taxa restricted to the Duchesnean are *Presbymys, Duchesneodus* and *Amynodontopsis*.

The Porvenir and Skyline-Cotter Channels local faunas from Trans-Pecos Texas have been listed recently by Wilson (1986). Porvenir is the most diverse Duchesnean fauna, consisting of 43 genera. Duchesnean first appearances not seen in the other principal correlatives are *Leptictis, Pseudocylindrodon, Ardynomys, Jaywilsonomys, Adjidaumo, Viejadjidaumo, Aulolithomys, Yoderimys, Eutypomys, Daphoenus, Mesohippus, Menops* (=*Menodus sensu* Osborn, 1929: Mader, 1989), *Trigonias?, Subhyracodon?, Toxotherium, Hypertragulus, Aclistomycter, Heteromeryx* and *Leptomeryx*. Taxa unique to the Porvenir, and therefore possibly unique to the Duchesnean, are *Rooneyia, Ischognathus* and *Haplohippus*.

The Skyline-Cotter Channels local fauna is much less diverse than Porvenir (Wilson and Stevens, 1986). Other than *Mahgarita, Simidectes, Mytonomys, Amynodontopsis, Hypertragulus* and *Harpagolestes*, Skyline-Cotter is a depauperate subset of Porvenir. Correlation of Porvenir and Skyline-Cotter with one or more of the other Duchesnean principal correlatives is well substantiated (e.g., Wilson, 1984, 1986). Radiometric ages suggest Porvenir is older than 37.8 Ma (McDowell, 1979; Krishtalka et al., 1987), which is older than the beginning of the Chadronian as here defined (Swisher and Prothero, 1990). There is some biochronological basis for a younger age assignment of Porvenir to the post-Duchesnean Chadronian as advocated by Emry et al. (1987, Figure 5.3). However, Porvenir includes *Duchesneodus*, and I assign it a Duchesnean age as the probable youngest Duchesnean local fauna.

The Duchesnean principal correlatives produce a composite standard of 67 mammalian genera (Table 4.1). Of these 67, 10 (15%) are known only from the Duchesnean, whereas 19 (28%) are holdovers from the Uintan whose last appearance is in the Duchesnean. The largest group of genera are the 29 (43%) that first appear in the Duchesnean and continue on into the Chadronian, whereas only 9 genera (13%) pass from the Uintan into the Chadronian.

Index Fossil

Wood et al. (1941, p. 8) used the term index fossils for those "known only from deposits of the age in question." This is consistent with definition of an index fossil as "one which identifies and dates the strata or succession of strata in which it lies" (Shimer and Shrock, 1944, p. 1). But, I prefer more rigorously defined index fossils according to criteria generally used by invertebrate paleontologists. As Weller (1960, p. 549) put it, "an ideal index fossil should be (1) easily recognized and easily distinguished from all others, (2) restricted to a narrow stratigraphic zone, (3) extensively distributed geographically, (4) [of an organism that was] adaptable to a wide range of environments so that specimens occur in many different types of rock, and (5) represented by abundant specimens." However, Weller's fourth factor, adaptability to a wide range of environments, is an underlying reason for the wide geographic range of an index fossil, so it should not stand alone as a criterion.

Duchesneodus (=*Teleodus* of most previous authors: see Lucas and Schoch, 1982) is the only Duchesnean mammal that comes close to meeting these criteria, and I list it as the only Duchesnean index fossil. However, *Amynodontopsis* is restricted to and known from three Duchesnean faunas. Further sampling may substantiate it as an additional Duchesnean index fossil.

Besides *Duchesneodus* and *Amynodontopsis*, eight genera are known only from the Duchesnean "type" or principal correlative faunas and would have been listed by Wood et al. (1941) as index fossils: *Protictops, Rooneyia, Mahrgarita, Presbymys, Viejadjidaumo, Ischognathus, Haplohippus* and *Hidrosotherium*. But, these taxa have such restricted geographic distributions, are known from so few specimens and/or are of such doubtful taxonomic status (e.g., *Ischognathus, Hidrosotherium*), that they are of little or no biochronological utility at present.

It is significant that Wood et al. (1941) essentially considered "*Teleodus*" to be a Duchesnean index fossil and equated (plate I) the Duchesnean with a "*Teleodus* faunal zone." Further collecting and taxonomic study has upheld their judgment (Lucas and Schoch, 1989). I believe this supports the viability of large-mammal biochrons (usually called "zones") for subdividing Cenozoic time, a method that has a tradition in North America dating back to Marsh (1891).

First and Last Appearances

Duchesnean first and last appearances are culled from the list of genera (Table 4.1) provided by the type fauna and the principal correlatives. These are genera that first appear during the Duchesnean and continue into the Chadronian and genera known from the Uintan which last occur during the Duchesnean. Other genera might be added to one or the other list (for example, *Hyopsodus* as a last occurrence) based on other faunas of possible Duchesnean age. But, because of uncertainties in correlating these faunas (see below), I am reluctant to do this at present.

BEGINNING OF THE CHADRONIAN

The main problem when defining the Duchesnean land-mammal "age" is not definition of the Uintan/Duchesnean boundary or identifying a mammal fauna characteristic of the Duchesnean. It is definition of the Duchesnean/Chadronian boundary, a problem recognized, but not satisfactorily solved by Wood et al. (1941, p. 11; also see Emry et al., 1987 for a discussion of the failure of Wood et al.'s definition).

Following the ideas of Repenning (1967), Lucas et al. (1981) attempted to identify immigrant taxa that could be used to define the Duchesnean/Chadronian boundary: the anthracothere *Bothriodon*, the carnivores *Mustelavus, Palaeogale, Hoplophoneus* and *Parictis* and the manid *Patriomanis*. These taxa, however, are rare in early Chadronian faunas, and, as Wilson (1984) noted, none of them are known from what clearly are early Chadronian faunas in Trans-Pecos Texas. These immigrant taxa, by themselves, thus do not provide a practical definition of the beginning of the Chadronian.

Nevertheless, additional taxa, evolutionary first occurrences in western North America, can be used to define the Duchesnean/Chadronian boundary. Combining immigrants and evolutionary originations produces a substantial list of first appearances that define the beginning of the Chadronian. These are: Sciuridae (*Protosciurus*), *Palaeolagus*, Nimravidae (*Dinictis, Hoplophoneus*), *Daphoenictis, Daphoenocyon, Penetrigonias, Stibarus*, Tayassuidae (*Perchoerus*), Anthracotheriidae (*Heptacodon, Bothriodon*), *Bathygenys, Merycoidodon, Poebrotherium, Montanatylopus, Pseudoprotoceras* and *Hypisodus*.

These first appearances provide a sufficient basis for identifying as early Chadronian a wide geographic range of faunas usually identified as early Chadronian (Emry et al., 1987). These include: Little Egypt local fauna, Trans-Pecos Texas (Wilson, 1978); Rancho Gaitan local fauna, Coahuila, Mexico (Ferrusquia-Villafranca, 1969); Yoder local fauna, eastern Wyoming (Kihm, 1987); Flagstaff Rim fauna below Ash A, Natrona County, Wyoming (Emry, 1973; written commun., 1989; this volume); fauna from the Big Sand Draw Sandstone Lentil, Fremont County, Wyoming (Emry, 1975); fauna of the Ahearn Member of the Chadron Formation, South Dakota (Clark et al., 1967); and the South Fork local fauna, Cypress Hills, Saskatchewan (Storer, 1984).

Three taxa that have long been considered Chadronian first appearances (Wood et al., 1941; Emry et al., 1987) that do not appear on my list are *Leptictis, Archaeotherium* and *Leptomeryx*. These taxa first appear in the Porvenir local fauna, though admittedly from what may be the youngest strata from which this local fauna is derived, the Blue Cliff horizon of the Chambers Tuff (Wilson, 1978). It might be tempting to assign this horizon to the early Chadronian and assign underlying strata of the Chambers Tuff, and hence the remainder of the Porvenir local fauna, to the Duchesnean. However, *Duchesneodus* is well documented from the Blue Cliff horizon (Lucas and Schoch, 1989), as is *Protoreodon*. Furthermore, the *Leptomeryx* specimens from the Blue Cliff horizon may be *Hendryomeryx* (cf. Black, 1978), and *Archaeotherium* from this horizon is based on a single, fragmentary specimen that may pertain to *Brachyhyops*. I thus include the Blue Cliff horizon in the Duchesnean and posit an apparent pre-Chadronian first appearance of three otherwise characteristically Chadronian genera.

SUBDIVIDING THE DUCHESNEAN

In principle, subdivision of the Duchesnean is a laudable goal. However, I question whether the present state of knowledge will allow formal subdivision of the Duchesnean. I am particularly troubled by our dim understanding of facial and biogeographic factors that controlled the distribution of Duchesnean mammals, the limited sampling and study thus far undertaken on some Duchesnean units and the rarity of some biochronologically significant Duchesnean mammal taxa.

According to Wilson (1984, 1986) and Kelly (1990), earlier Duchesnean faunas can be recognized by the possession of *Leptoreodon* and/or *Amynodontopsis*, whereas later Duchesnean faunas lack these taxa and include *Brachyhyops, Hyracodon* and *Poabromylus*. *Hyaenodon* and *Duchesneodus*, according to Wilson and Kelly, occur in both early and late Duchesnean faunas. This scheme thus identifies Lapoint and Porvenir as late Duchesnean and Skyline-Cotter and Pearson Ranch as early Duchesnean.

In a similar vein, Slim Buttes has *Amynodontopsis* as well as *Epihippus*, so it and the Halfway occurrence of *Epihippus* arguably are early Duchesnean. Tonque has *Poabromylus*, so might be considered early Duchesnean, although *Amynodon*, a characteristically Uintan taxon is present.

Porvenir has a large number of Duchesnean first appearances, 26 of 43 genera (60%). On that basis alone, it is tempting to argue it is late Duchesnean. The other Duchesnean faunas, except Lapoint, have Duchesnean last appearances equal to or greater than the number of Duchesnean first appearances. Furthermore, radioisotopic dates suggest Porvenir is older than about 37 Ma, Tonque is also older than about 38 Ma and that Halfway is older than 40.6 Ma (Kautz et al., 1981; Krishtalka et al., 1987). However, *Brachyhyops*, a supposed late Duchesnean taxon, is known from the Baca Formation in west-central New Mexico which underlies lower Datil Group volcanics that yield K-Ar ages as old as about 40 Ma (Lucas, 1983a; Cather et al., 1987). Moreover, many Uintan holdovers (Duchesnean last appearances) are based on a single specimen (e.g., *Harpagolestes* in the Skyline-Cotter Channels local fauna). Finally, all of the Duchesnean faunas except Porvenir are of low diversity with many taxa based on one or a few specimens. Collection of these Duchesnean faunas has generally moved only a short distance up the rarefaction curve, and a few new discoveries could significantly modify their composition.

I thus see the Duchesnean record as currently inadequate to provide a basis for formal subdivision of the LMA. But, I do agree with Wilson (1984, 1986) and Kelly (1990) that Lapoint and Porvenir (and Tonque) probably are younger than the remaining Duchesnean faunas (Figure 4.1).

OTHER POSSIBLE DUCHESNEAN OCCURRENCES/LOCAL FAUNAS

Isolated occurrences of *Duchesneodus* outside of the principal correlative faunas are of Duchesnean age. Some local faunas that lack *Duchesneodus* have also been assigned a Duchesnean age (Savage and Russell, 1983; Krishtalka et al., 1987). Here, I review briefly the age assignments of these occurrences and local faunas (Figure 4.3).

Green River Formation, central Utah

Nelson et al. (1980) described a skull of *Duchesneodus* from the Green River Formation at Spring City Cuesta, Sanpete County, Utah. This indicates a Duchesnean horizon.

Cypress Hills *Duchesneodus*, Saskatchewan

The type and only known specimen of *Duchesneodus primitivus* is from an uncertain locality and horizon in the Cypress Hills of Saskatchewan (Lambe, 1908; L. S. Russell, written commun., 1986; Lucas and Schoch, 1989). It indicates a Duchesnean horizon but cannot be related stratigraphically to the mammalian local faunas from the Cypress Hills.

Black Range, New Mexico

A lower jaw of *Duchesneodus* is a Duchesnean fossil from the Rubio Peak Formation in the Black Range, Catron County, New Mexico (Lucas, 1983a). At nearby localities in the Rubio Peak Formation, a small Chadronian land-mammal fauna (Lucas, 1986a, b) consists of *Jaywilsonomys*, *Parvitragulus* and *Montanatylopus*. This early Chadronian fauna cannot be stratigraphically related to the Duchesnean locality because of structural problems and a lack of precise locality data on the Duchesnean fossil. The Kneeling Nun Tuff, which overlies the Rubio Peak Formation in the northern Black Range (Harrison, 1986), is an ignimbrite with a well-established Ar/Ar age of 35.28 Ma (McIntosh et al., 1986).

Baca Formation, New Mexico

The Baca Formation has produced the following mammals (Schrodt, 1980; Schiebout and Schrodt, 1981; Lucas, 1983a, b): *Hyaenodon*, *Diplacodon*, *Brachyhyops*, *Protoreodon* and an agriochoerid, protoceratid and camelid. Some of these taxa suggest Uintan horizons (especially *Diplacodon*), whereas others (*Hyaenodon*, *Brachyhyops*) indicate a Duchesnean age. No stratigraphic organization of Baca mammal localities exists, and it is likely that most are of Uintan age with only those near Veteado Mountain being Duchesnean (Lucas, 1984). The oldest K-Ar age from overlying volcanics is 39.6 ± 1.5 Ma, whereas the oldest Ar/Ar age is 37.02 ± 0.15 Ma (Cather et al., 1987).

Badwater Localities, Wyoming

Badwater locality 20 in the Hendry Ranch

Member of the Wagonbed Formation, Wind River basin, Wyoming, has long been considered Duchesnean, as have the nearly stratigraphically equivalent "Wood" and "Rodent" localities (e.g., Krishtalka et al., 1987). These localities are just above a tuff with a K-Ar age of 42.3 ± 1.4 Ma (Black, 1969; Krishtalka et al., 1987; but see Prothero and Swisher, this volume).

Poabromylus and an amphicyonid are Duchesnean first appearances at locality 20, which also includes *Apternodus, Pareumys, Pseudocylindrodon, Diplobunops, Leptotragulus, Hendryomeryx* and a characteristically Bridgerian-Uintan brontothere, *Telmatherium* (Black, 1969, 1970, 1971, 1974, 1978, 1979; Krishtalka and Setoguchi, 1977; Dawson, 1980). The "Rodent" locality includes *Leptotomus, Pseudocylindrodon, Colodon* and *Hendryomeryx*, whereas the "Wood" locality includes *Leptotomus, Ischyromys, Microparamys, Pseudocylindrodon, Epihippus, Diplobunops* and *Hendryomeryx* (Wood, 1949; Black, 1970, 1971, 1974, 1978, 1979). On the strength of the two Duchesnean first appearances at locality 20, I consider these localities Duchesnean and note they may be the oldest Duchesnean localities known.

Clarno Formation, Oregon

The Hancock Quarry in the upper Clarno Formation of north-central Oregon has produced a fauna of large land mammals that includes: *Hemipsalodon, Haplohippus, Epihippus, Protitanops, Colodon, Procadurcodon, Telataceras, Heptacodon* and *Diplobunops* (Krishtalka et al., 1987; Hanson, 1989, and in press). *Procadurcodon* and *Telataceras* also occur in the Uglov Svita near Artem in the Primorskyi Territory of the eastern U.S.S.R. (Russell and Zhai, 1987). *Colodon* occurs in the Hosan Formation of Kokaido Province, North Korea (Russell and Zhai, 1987). *Haplohippus* is a Porvenir taxon, but *Epihippus* (supposedly *E. gracilis*) and *Diplobunops* suggest an Uintan age. *Heptacodon* is a Chadronian first appearance.

A Duchesnean or early Chadronian age assignment for the Hancock Quarry mammals has generally been advocated (Emry et al., 1987; Krishtalka et al., 1987), and either has some evidence to support it. Perhaps it is worth arguing that the Hancock Quarry is in a different faunal province than are the Duchesnean mammal faunas, a province that encompasses far eastern Asia and the Pacific Northwest. Within that province, the Hancock Quarry mammals pertain to the later portion of the Sharamurunian LMA of Li and Ting (1983).

Lac Pelletier Local Fauna, Saskatchewan

Storer (1987, 1988) described the rodents and gave a preliminary list of other taxa in the Lac Pelletier local fauna from the Cypress Hills Formation in southwestern Saskatchewan. According to Storer (1987, Table 1), the Lac Pelletier local fauna contains many taxa typically seen in Uintan faunas and the Duchesnean last appearance *Protoreodon*. But, it also includes the Chadronian first appearances *Heptacodon* and *Pseudoprotoceras*. I thus tentatively consider the Lac Pelletier local fauna early Chadronian pending a complete description of the fauna. However, it is tempting to wonder whether Lambe's *Duchesneodus* from the Cypress Hills came from the same stratigraphic level as the Lac Pelletier local fauna.

Three Forks Basin, Montana

Krishtalka et al. (1987) tentatively considered the Shoddy Springs local fauna from the Climbing Arrow Formation, Three Forks basin, southwestern Montana (Robinson, 1963; Robinson et al., 1957) to be Duchesnean. However, taxa thus far listed or described from this local fauna (Black, 1967; Krishtalka and Black, 1975; Krishtalka, 1979) are consistent with an Uintan age, e.g., *Ectypodus, Hyopsodus, Dilophodon* and *Protoreodon*. Both Krishtalka (1979) and Krishtalka et al. (1987), however, allude to unpublished specimens from the Shoddy Springs local fauna as the basis for the Duchesnean age assignment, i. e., "an age comparable to that of the Duchesnean Badwater locality 20" (Krishtalka, 1979, p. 389).

Kishenehn Formation, Montana and British Columbia

Savage and Russell (1983, Figure 4-34) listed the Kishenehn Formation of Montana-British Columbia as Duchesnean, following the suggestion of Russell (1954). Emry et al. (1987, Figure 5.3), however, listed it as early Chadronian. The only published Kishenehn mammals are from British Columbia (Russell, 1954): *Peratherium, Thylacaelurus, Desmatolagus, Pseudocylindrodon, Protadjidaumo, Paradjidaumo* and *Leptotragulus*? M. R. Dawson (oral commun., 1989) reports *Pseudocylindrodon, Adjidaumo, Paradjidaumo, Leptomeryx* and *Merycoidodon*? from the Kishenehn Formation along the north fork of the Flathead River in Glacier National Park, Montana.

None of the reported Kishenehn mammals mandate a Duchesnean age, whereas if *Merycoidodon* is present, it indicates a Chadronian age. I thus see no reason to assign a Duchesnean age to the Kishenehn mammals and instead consider them Chadronian.

CORRELATION OF DUCHESNEAN WITH OTHER CHRONOLOGIES

Earlier arguments about the Eocene or Oligocene age of the Duchesnean noted above have been put to rest by geochronometric and biochronologic evidence of the Eocene age of the Duchesnean assisted by an upward shift of the Eocene/Oligocene boundary during the past 50 years. Radiometric dates suggest the Duchesnean may have a duration of as much as 5 million years, from about 37 Ma to as old as 42 Ma (Emry et al., 1987; Krishtalka et al., 1987; Swisher and Prothero, 1990; but see Prothero and Swisher, this volume). This means that Duchesnean faunas are no younger than part of Chron C16 and no older than Chron C18, although no convincing magnetostratigraphy of rocks containing Duchesnean mammals is currently available.

Discussion above of the age of the Clarno mammals suggests that Duchesnean is a correlative of the later part of the Asian Sharmurunian LMA of Li and Ting (1983). No direct mammal-based correlation of Duchesnean with the standard marine stage-ages of Europe can be made. However, the radioisotopic ages indicate the Duchesnean is equivalent to the latest Bartonian and to most or all of the Priabonian (Aubry et al., 1988; Swisher and Prothero, 1990).

ACKNOWLEDEGMENTS

I thank R. J. Emry, J. Storer and J. A. Wilson for their reviews of an earlier version of this article.

REFERENCES

Andersen, D. W. and Picard, M. D. 1972. Stratigraphy of the Duchesne River Formation (Eocene-Oligocene), northern Uinta Basin, northeastern Utah. *Bull. Utah Geol. Min. Surv.* 97: 1-29.

Archibald, J. D., Clemens, W. A., Gingerich, P. D., Krause, D. W., Lindsay, E. H. and Rose, K. D. 1987. First North American land mammal ages of the Cenozoic Era. In M. O. Woodburne, ed., *Cenozoic Mammals of North America Geochronology and Biostratigraphy*. Berkeley, University of California Press, 24-76.

Aubry, M.-P., Berggren, W. A., Kent, D. V., Flynn, J. J., Klitgord, K. D., Obradovich, J. D. and Prothero, D. R. 1988. Paleogene geochronology: an integrated approach: *Paleoceanog.* 3: 707-742.

Berggren, W. A. and Van Couvering, J. A. 1974. The late Neogene: biostratigraphy, geochronology and paleoclimatology of the last 15 million years. *Develop. Paleont. Strat.* 2: 1-216.

Bjork, P. R. 1967. Latest Eocene vertebrates from northwestern South Dakota. *Jour. Paleont.* 41: 227-236.

Black, C. C. 1967. Middle and late Eocene mammal communities: a major discrepancy. *Science* 156: 62-64.

Black, C. C. 1969. Fossil vertebrates from the late Eocene and Oligocene, Badwater Creek area, Wyoming, and some regional correlations. In J. A. Barlow, ed., *Wyoming Geological Association 21st Field Conference Guidebook 1969 Symposium on Tertiary Rocks of Wyoming*. Casper, Wyoming Geological Association, 43-47.

Black, C. C. 1970. Paleontology and geology of the Badwater Creek area, central Wyoming. Part 5. The cylindrodont rodents. *Ann. Carnegie Mus.* 41: 201-214.

Black, C. C. 1971. Paleontology and geology of the Badwater Creek area, central Wyoming. Part 7. Rodents of the Family Ischyromyidae. *Ann. Carnegie Mus.* 43: 179-217.

Black, C. C. 1974. Paleontology and geology of the Badwater Creek area, central Wyoming. Pt. 9. Additions to the cylindrodont rodents from the late Eocene. *Ann. Carnegie Mus.* 45: 151-160.

Black, C. C. 1978. Paleontology and geology of

the Badwater Creek area, central Wyoming Part 14. The artiodactyls. *Ann. Carnegie Mus.* 47: 223-259.

Black, C. C. 1979. Paleontology and geology of the Badwater Creek area, central Wyoming. Part 19. Perissodactyla. *Ann. Carnegie Mus.* 48: 391-401.

Black, C. C. and Dawson, M. R. 1966. A review of late Eocene mammalian faunas from North America. *Amer. Jour. Sci.* 264: 321-349.

Bryant, H. N. 1988. The anatomy, phylogenetic relationships and systematics of the Nimravidae (Mammalia: Carnivora). Ph.D. Dissertation, University of Toronto, 414 pp.

Cather, S. M., McIntosh, W. C. and Chapin, C. E. 1987. Stratigraphy, age, and rates of deposition of the Datil Group (upper Eocene-lower Oligocene), west-central New Mexico. *New Mex. Geol.* 9: 50-54.

Clark, J., Beerbower, J. R. and Kietzke, K. K. 1967. Oligocene sedimentation, stratigraphy, paleoecology and paleoclimatology in the Big Badlands of South Dakota. *Fieldiana: Geol. Mem.* 5: 1-158.

Dawson, M. R. 1980. Geology and paleontology of the Badwater Creek area, central Wyoming. Pt. 20: The late Eocene Creodonta and Carnivora. *Ann. Carnegie Mus.* 49: 79-91.

Emry, R. J. 1973. Stratigraphy and preliminary biostratigraphy of the Flagstaff Rim area, Natrona County, Wyoming. *Smithson. Contrib. Paleobiol.* 18: 1-43.

Emry, R. J. 1975. Revised Tertiary stratigraphy and paleontology of the Western Beaver Divide, Fremont County, Wyoming. *Smithson. Contrib. Paleobiol.* 25: 1-20.

Emry, R. J. 1981. Additions to the mammalian fauna of the type Duchesnean, with comments on the status of the Duchesnean "age." *Jour. Paleont.* 55: 563-570.

Emry, R. J., Bjork, P. R. and Russell, L. S. 1987. The Chadronian, Orellan, and Whitneyan North American land mammal ages. In M. O. Woodburne, ed., *Cenozoic Mammals of North America Geochronology and Biostratigraphy*. Berkeley, University of California Press, 118-152.

Ferrusquia-Villafranca, I. 1969. Rancho Gaitan local fauna, early Chadronian, northeastern Chihuahua. *Soc. Geol. Mex. Bol.* 30(2): 99-138.

Gazin, C. L. 1955. A review of the upper Eocene Artiodactyla of North America. *Smithson. Misc. Coll.* 128(8): 1-96.

Gazin, C. L. 1956. The geology and vertebrate paleontology of upper Eocene strata in the northeastern part of the Wind River Basin, Wyoming, Part 2. The mammalian fauna of the Badwater area. *Smithsonian Misc. Coll.* 131(8): 1-35.

Gazin, C. L. 1959. Paleontological exploration and dating of the early Tertiary deposits adjacent to the Uinta Mountains. *Inter-mountain Assoc. Petrol. Geol. Guidebook* 10: 131-138.

Golz, D. J. 1976. Eocene Artiodactyla of southern California. *Nat. Hist. Mus. Los Angeles Co. Sci. Bull.* 26: 1-85.

Hanson, C. B. 1989. *Telataceras radinskyi*, a new primitive rhinocerotid from the late Eocene Clarno Formation of Oregon. In D. R. Prothero and R. M. Schoch, eds., *The Evolution of Perissodactyls*, Oxford University Press, New York, 379-398.

Hanson, C. B. (in press). Vertebrate faunas, biostratigraphy, and geochronology of the type Clarno Formation, north-central Oregon. In D.R. Prothero and R.J. Emry, eds., *The Terrestrial Eocene-Oligocene Transition in North America.* (in press)

Harrison, R. W. 1986. General geology of Chloride mining district, Sierra and Catron Counties, New Mexico. In R. E. Clemons, W. E. King and G. H. Mack, eds., *Truth or Consequences Region*, Socorro, New Mexico Geological Society, 265-272.

International Commission on Zoological Nomenclature. 1985. *International Code of Zoological Nomenclature Third Edition.*, London, International Trust for Zoological Nomenclature, 338 pp.

Kautz, P. F., Ingersoll, R. V., Baldridge, W. S., Damon, P. E. and Shafiqullah, M. 1981. Geology of the Espinaso Formation (Oligocene), north-central New Mexico. *Geol. Soc. Amer. Bull.* 92: 2318-2400.

Kay, J. L. 1934. The Tertiary formations of the Uinta basin, Utah. *Ann. Carnegie Mus.* 23: 357-371.

Kelly, T. S. 1990. Biostratigraphy of Uintan and Duchesnean land mammal assemblages from the middle member of the Sespe Formation, Simi Valley, California. *Contrib. Sci.*

Nat. Hist. Mus. Los Angeles Co. 419: 1-42.

Keyes, C. R. 1924. Grand staircase of Utah. *Pan-Amer. Geol.* 41: 36.

Kihm, A. J. 1987. Mammalian paleontology and geology of the Yoder Member, Chadron Formation, east-central Wyoming. *Dakoterra* 3: 28-45.

Krishtalka, L. 1979. Paleontology and geology of the Badwater Creek area, central Wyoming. Part 18. Review of late Eocene *Hyopsodus*. *Ann. Carnegie Mus.* 48: 377-389.

Krishtalka, L. and Black, C. C. 1975. Paleontology and geology of the Badwater Creek area, central Wyoming. Part 12. Description and review of late Eocene Multituberculata from Wyoming and Montana. *Ann. Carnegie Mus.* 45: 287-297.

Krishtalka, L. and Setoguchi, T. 1977. Paleontology and geology of the Badwater Creek area, central Wyoming. Pt. 13: The late Eocene Insectivora and Dermoptera. *Ann. Carnegie Mus.* 46: 71-99.

Krishtalka, L., Stucky, R. K., West, R. M., McKenna, M. C., Black, C. C., Bown, T. M., Dawson, M. R., Golz, D. J., Flynn, J. J., Lillegraven, J. A. and Turnbull, W. D. 1987. Eocene (Wasatchian through Duchesnean) biochronology of North America. In M. O. Woodburne, ed., *Cenozoic Mammals of North America: Geochronology and Biostratigraphy*. Berkeley, University of California Press, 77-117

Lambe, L. M. 1908. The Vertebrata of the Oligocene of the Cypress Hills, Saskatchewan. *Contribs. Canadian Paleont.* 3: 1-64.

Li, C. and Ting, S. 1983. The Paleogene mammals of China. *Bull. Carnegie Mus. Nat. Hist.* 21: 1-93.

Lillegraven, J. A., McKenna, M. C. and Krishtalka, L. 1981. Evolutionary relationships of middle Eocene and younger species of *Centetodon* (Mammalia, Insectivora, Geolabididae) with a description of the dentition of *Ankylodon* (Adapisoricidae). *Univ. Wyoming Pub. Geol. Sci.* 45: 1-115.

Lucas, S. G. 1982. Vertebrate paleontology, stratigraphy and biostratigraphy of Eocene Galisteo Formation, north-central New Mexico. *N. Mex. Bur. Mines Min. Res. Circular* 186: 1-34.

Lucas, S. G. 1983a. The Baca Formation and the Eocene/ Oligocene boundary in New Mexico. In C. E. Chapin and J. F. Callender, eds., *Socorro Region II*. Socorro, New Mexico Geological Society, 187-192.

Lucas, S. G. 1983b. *Protitanotherium* (Mammalia, Perissodactyla) from the Eocene Baca Formation, west-central New Mexico. *N. Mex. Jour. Sci.* 23: 39-47.

Lucas, S. G. 1984. Correlation of Eocene rocks of the northern Rio Grande rift and adjacent areas: implications for Laramide tectonics. In W. S. Baldridge, P. W. Dickerson, R. E. Riecker and J. Zidek, eds., *Rio Grande Rift: Northern New Mexico*. Socorro, New Mexico Geological Society, 123-128.

Lucas, S. G. 1986a. The first Oligocene mammal from New Mexico. *Jour. Paleont.* 60: 1274-1276.

Lucas, S. G. 1986b. Oligocene mammals from the Black Range, southwestern New Mexico. In R. E. Clemons, W. E. King and G. H. Mack, eds., *Truth or Consequences Region*, Socorro, New Mexico Geological Society, 261-263.

Lucas, S. G. and Schoch, R. M. 1982. *Duchesneodus*, a new name for some titanotheres (Perissodactyla, Brontotheriidae) from the late Eocene of western North America. *Jour. Paleont.* 56: 1018-1023.

Lucas, S. G. and Schoch, R. M. 1989. Taxonomy of *Duchesneodus* (Brontotheriidae) from the late Eocene of North America. In D. R. Prothero and R. M. Schoch, eds., *The Evolution of Perissodactyls*, New York, Oxford, 490-503.

Lucas, S. G., Schoch, R.M., Manning, E. and Tsentas, C. 1981. The Eocene biostratigraphy of New Mexico. *Geol. Soc. Amer. Bull.* 92: 951-967.

Mader, B. J. 1989. The Brontotheriidae: a systematic revision and preliminary phylogeny of North American genera. In D. R. Prothero and R. M. Schoch, eds., *The Evolution of Perissodactyls*, New York, Oxford, 458-484.

Marsh, O. C. 1891. Geologic horizons as determined by vertebrate fossils. *Am. Jour. Sci.* 42: 336-338.

McDowell, F. W. 1979. Potassium-argon dating in the Trans-Pecos volcanic field. *Bur. Econ. Geol. Univ. Texas Guidebook* 19: 10-18.

McIntosh, W. C., Sutter, J. F., Chapin, C. E., Osburn, G. R. and Ratte, J. C. 1986. A stratigraphic framework for the eastern Mogollon-Datil volcanic field based on paleomagnetism and high precision Ar/Ar dating of ignimbrites--a progress report. In R. E. Clemons, W. E. King and G. H. Mack, eds., *Truth or Consequences Region*, Socorro, New Mexico Geological Society, 183-195.

Nelson, M. E., Madsen, J.H., Jr. and Stokes, W. L. 1980. A titanothere from the Green River Formation, central Utah: *Teleodus uintensis* (Perissodactyla: Brontotheriidae). *Contrib. Geol. Univ. Wyo.* 18: 127-134.

Olson, E. C. 1952. The evolution of a Permian vertebrate chronofauna. *Evolution* 6: 181-196.

Osborn, H. F. 1929. The titanotheres of ancient Wyoming, Dakota, and Nebraska. *U. S. Geol. Surv. Monog.* 55: 1-953.

Peterson, O. A. 1931. New species from the Oligocene of the Uinta. *Ann. Carnegie Mus.* 21: 61-78.

Peterson, O. A. 1934. List of species and description of new material from the Duchesne River Oligocene, Uinta basin, Utah. *Ann. Carnegie Mus.* 23: 373-389.

Peterson, O. A. and Kay, J. L. 1931. The upper Uinta Formation of northeastern Utah. *Ann. Carnegie Mus.* 20: 293-306.

Prothero, D. R. 1985a. North American mammalian diversity and Eocene-Oligocene extinctions. *Paleobiol.* 11: 389-405.

Prothero, D. R. 1985b. Mid-Oligocene extinction event in North American land mammals. *Science* 229: 550-551.

Radinsky, L. B. 1967. A review of the rhinocerotoid family Hyracodontidae. *Bull. Amer. Mus. Nat. Hist.* 136: 1-45.

Repenning, C. A. 1967. Palearctic-Nearctic mammalian dispersal in the late Cenozoic. In D. M. Hopkins, ed., *The Bering Land Bridge*. Stanford, Stanford University Press, 288-311.

Robinson, G. D. 1963. Geology of the Three Forks quadrangle Montana. *U.S. Geol. Surv. Prof. Pap.* 370: 1-143.

Robinson, G. D., Lewis, E. and Taylor, D. W. 1957. Eocene continental deposits in Three Forks Basin, Montana. *Geol. Soc. Amer. Bull.* 68: 1786.

Rose, K. D. 1981. The Clarkforkian land-mammal age and mammalian faunal composition across the Paleocene-Eocene boundary. *Univ. Mich. Pap. Paleont.* 26: 1-197.

Russell, D. E. and Zhai, R. 1987. The Paleogene of Asia: mammals and stratigraphy. *Mem. Mus. Nat. d'Hist. Nat.* (C) 51: 1-488.

Russell, L. S. 1954. Mammalian fauna of the Kishenehn Formation, southeastern British Columbia. *Ann. Rept. Nat. Mus. Canada* 132: 92-111.

Savage, D. E. and Russell, D. E. 1983. *Mammalian Paleofaunas of the World*. Reading, Addison-Wesley Publishing Co., 432 pp.

Schiebout, J. A. and Schrodt, A. K. 1981. Vertebrate paleontology of the lower Tertiary Baca Formation of western New Mexico. *Geol. Soc. Amer. Bull.* 92: 976-979.

Schrodt, A. K. 1980. Depositional environments, provenance, and vertebrate paleontology of the Eocene-Oligocene Baca Formation, Catron County, New Mexico. Unpublished M.S. thesis, Baton Rouge, Louisiana State University, 174 pp.

Scott, W. B. 1945. The Mammalia of the Duchesne River Oligocene. *Trans. Amer. Phil. Soc.* 34: 209-253.

Shaw, A. B. 1964. *Time in Stratigraphy*. New York, McGraw-Hill, 365 pp.

Shimer, H. W. and Shrock, R. R. 1944. *Index Fossils of North America*. Cambridge, M.I.T. Press, 837 pp.

Simpson, G. G. 1933. Glossary and correlation charts of North American Tertiary mammal-bearing formations. *Bull. Amer. Mus. Nat. Hist.* 67: 79-121.

Simpson, G. G. 1935. The Tiffany fauna, upper Paleocene I--Multituberculata, Marsupialia, Insectivora, and ?Chiroptera. *Amer. Mus. Novit.* 795: 1-19.

Simpson, G. G. 1946. The Duchesnean fauna and the Eocene-Oligocene boundary. *Amer. Jour. Sci.* 244: 52-57.

Storer, J. E. 1984. Fossil mammals of the Southfork local fauna (early Chadronian) of Saskatchewan. *Can Jour. Earth Sci.* 21: 1400-1405.

Storer, J. E. 1987. Dental evolution and radiation of Eocene and early Oligocene Eomyidae (Mammalia, Rodentia) of North America, with new material from the Duchesnean of Saskatchewan. *Dakoterra* 3: 108-117.

Storer, J. E. 1988. The rodents of the Lac Pelletier lower fauna, late Eocene (Duchesnean) of Saskatchewan. *Jour. Vert. Paleont.* 8: 84-101.

Stovall, J. W. 1948. Chadron vertebrate fossils from below the rim rock of Presidio County, Texas. *Am. Jour. Sci.* 246: 78-95.

Stucky, R. K. 1990. Evolution of land mammal diversity in North America during the Cenozoic. *Current Mammal.* 2: 375-432.

Swisher, C. C. III and Prothero, D. R. 1990. Single-crystal $^{40}Ar/^{39}Ar$ dating of the Eocene-Oligocene transition in North America. *Science* 249: 760-762.

Tedford, R. H. 1970. Principles and practices of mammalian geochronology in North America. *Proc. N. Amer. Paleont. Conv.* 2F: 666-703.

Weller, J. M. 1960. *Stratigraphic Principles and Practice.* New York, Harper and Brothers, 725 pp.

Williams, H. S. 1901. The discrimination of time-values in geology. *Jour. Geol.* 9: 570-585.

Wilson, J. A. 1978. Stratigraphic occurrence and correlation of early Tertiary vertebrate faunas, Trans-Pecos Texas, Part 1: Vieja area. *Texas Mem. Mus. Bull.* 25: 1-42.

Wilson, J. A. 1984. Vertebrate faunas 49 to 36 million years ago and additions to the species of *Leptoreodon* (Mammalia: Artiodactyla) found in Texas. *Jour. Vert. Paleont.* 4: 199-207.

Wilson, J. A. 1986. Stratigraphic occurrence and correlation of early Tertiary vertebrate faunas, Trans-Pecos Texas: Agua Fria-Green Valley areas. *Jour. Vert. Paleont.* 6: 350-373.

Wilson, J. A., and Stevens, M. S. 1986. Fossil vertebrates from the latest Eocene, Skyline Channels, Trans-Pecos Texas. *Contrib. Geol. Univ. Wyo. Spec. Pap.* 3: 221-235.

Wilson, J. A., Twiss, P. C., DeFord, R. K. and Clabaugh, S. E. 1968. Stratigraphic succession, potassium-argon dates and vertebrate faunas, Vieja Group, Rim Rock Country, Trans-Pecos Texas. *Amer. Jour. Sci.* 266: 590-604.

Wood, A. E. 1949. Small mammals from the uppermost Eocene (Duchesnean) near Badwater, Wyoming. *Jour. Paleont.* 23: 556-565.

Wood, H. E. II, Chaney, R. W., Clark, J., Colbert, E. H., Jepsen, G. L., Reeside, J. B., Jr. and Stock, C. 1941. Nomenclature and correlation of the North American continental Tertiary. *Geol. Soc. Amer. Bull.* 52: 1-48.

Woodburne, M. O. 1977. Definition and characterization in mammalian chronostratigraphy. *Jour. Paleont.* 51: 220-234.

Woodburne, M. O. ed. 1987a. *Cenozoic Mammals of North America: Geochronology and Biostratigraphy.* Berkeley, University of California Press, 336 pp.

Woodburne, M. O. 1987b. Principles, classification, and recommendations. In M. O. Woodburne, ed., *Cenozoic Mammals of North America: Geochronology and Biostratigraphy.* Berkeley, University of California Press, 9-17.

5. MAMMALIAN RANGE ZONES IN THE CHADRONIAN WHITE RIVER FORMATION AT FLAGSTAFF RIM, WYOMING

By Robert J. Emry

ABSTRACT

When local range zones are plotted for mammalian taxa in the Chadronian White River Formation at Flagstaff Rim in Wyoming, there seem to be no particular stratigraphic levels where first or last appearances of taxa are concentrated that are not also due in part to the mode of preservation. A minor faunal event, marked by the apparently coincident last appearances of *Protadjidaumo, Toxotherium*, and *Parvitragulus*, could indicate the end of early Chadronian time. Gross sedimentological characteristics agree with faunal change in suggesting that more time is represented by the thin, lower, undated, part of the section than by the thicker, upper part that has dated volcanic tuffs. Within the late Eocene to early Oligocene interval in North America, an important faunal event, which includes the first appearance of several immigrant taxa, occurs at the Uintan–Duchesnean transition. Following this, throughout the Oligocene, first appearances are predominantly the products of evolution in North America, making the limits of land mammal ages less objectively definable.

INTRODUCTION

The White River Formation at Flagstaff Rim in central Wyoming was recognized as a potentially very important section for developing a Chadronian mammalian biostratigraphy (Emry, 1973) that might be useful for resolving time intervals within the Chadronian NALMA. Its potential derives from several attributes that are not all present in a single continuous section elsewhere: it is thick compared to most Chadronian sections, is abundantly fossiliferous through much of its extent, appears to represent much of Chadronian time, and has a series of volcanic ash beds for which radiometric ages had been determined (Evernden et al., 1964). These same characteristics made the section an important one for magnetostratigraphic analysis (Prothero, Denham and Farmer, 1982, 1983; Prothero, 1985). The radiometric dates have not only provided calibration points for the magnetic polarity chronology, but through intercontinental magnetostratigraphic correlation, might also be implicated in calibrating the epoch chronology. More reliable age determinations, recently obtained with new techniques, appear to date the end of the Chadronian at about 34 Ma, and correlate this with the Eocene–Oligocene boundary (Swisher and Prothero, 1990).

Because the Flagstaff Rim section has become increasingly important, it seems worthwhile to compile what is known about the distribution of fossil mammalian taxa within the section. The succession of mammalian species through Chadronian time should provide the basis for establishing a range–zone biostratigraphy that would provide greater resolution in correlating other more–limited Chadronian faunas. The distribution of species, and the evolutionary change in lineages through time within this interval, can also play an important role in interpreting the magnetostratigraphic data. The identification of specific magnetozones depends to some extent on the relative length (in time terms) of magnetozones in a given series (for example, compare Prothero, Denham, and Farmer, 1982, 1983, with Prothero, 1985, with Swisher and Prothero, 1990, and finally with Prothero and Swisher, this volume). And it cannot be assumed, *a priori*, that the amount of time is in direct proportion to the thickness of rock section representing each normal or reversed polarity

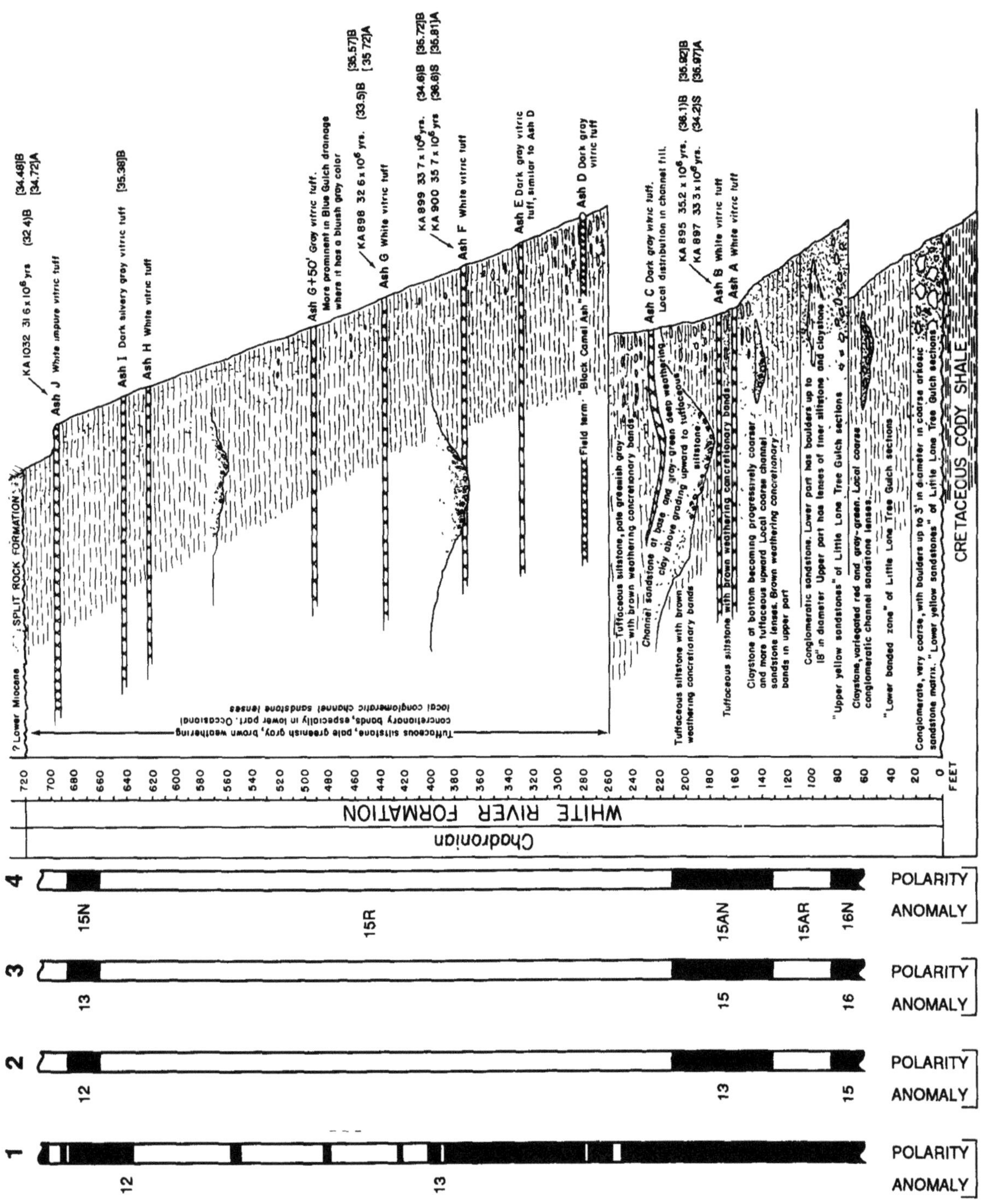

FIGURE 5.1. Generalized section used for stratigraphic documentation of fossil specimens collected from the White River Formation in the Flagstaff Rim area (Emry, 1973). Radioisotopic dates indicated are K-Ar determinations by Evernden et al. (1964), corrected (in parentheses) according to the new constants (Dalrymple, 1979) and, in brackets, the new $^{40}Ar/^{39}Ar$ dates (Swisher and Prothero, 1990). A, anorthoclase; B, biotite; S, sanidine. Magnetostratigraphic columns show: (1), original interpretations by Prothero et al. (1982, 1983); (2), revision resulting from resampling and reanalyzing (Prothero, 1985); (3), reinterpretation based on new age determinations (Swisher and Prothero, 1990), and (4) further reinterpretation (Prothero and Swisher, this volume).

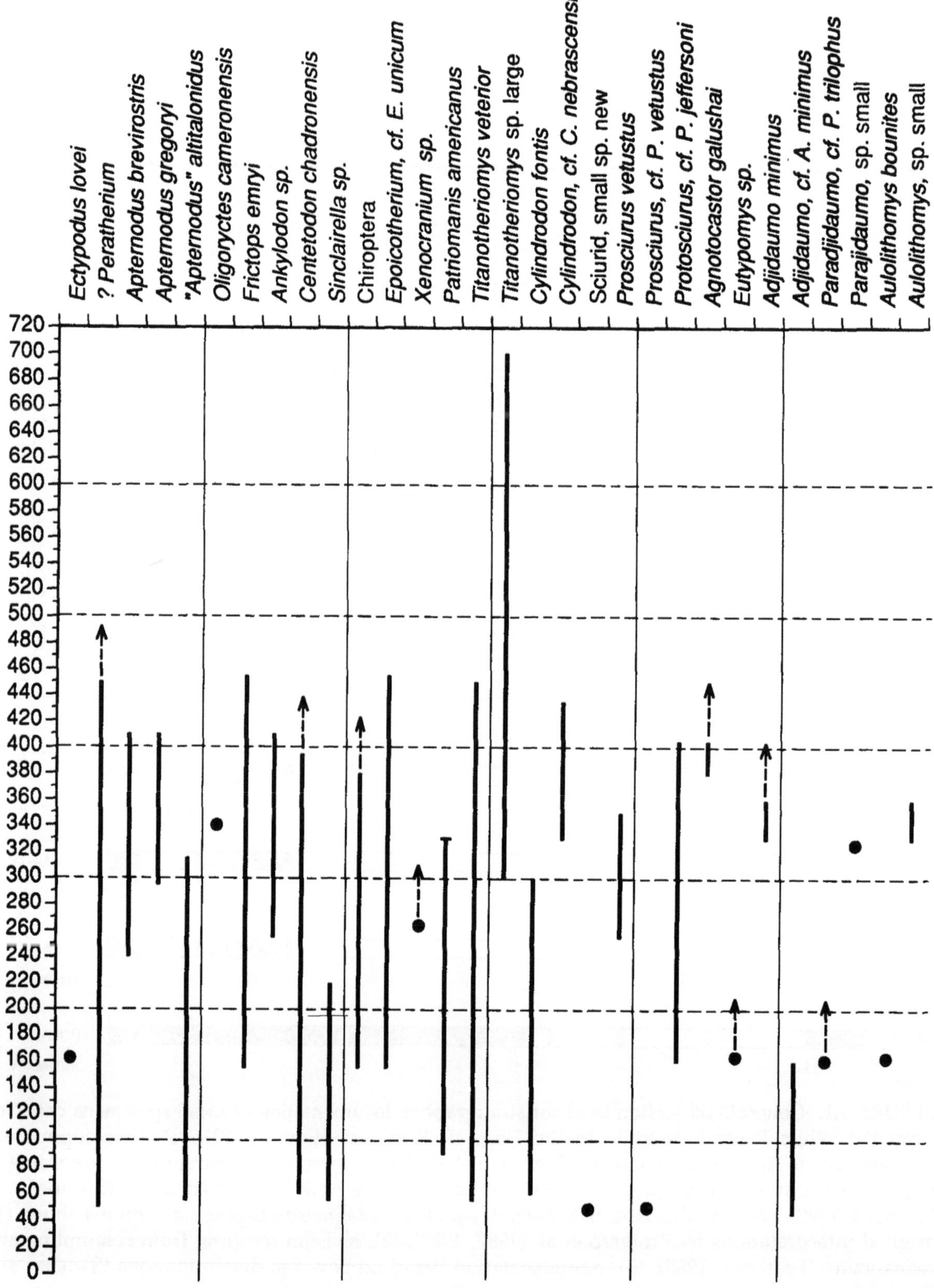
Ectypodus lovei
? Peratherium
Apternodus brevirostris
Apternodus gregoryi
"Apternodus" altitalonidus
Oligoryctes cameronensis
Frictops emryi
Ankylodon sp.
Centetodon chadronensis
Sinclairella sp.
Chiroptera
Epoicotherium, cf. E. unicum
Xenocranium sp.
Patriomanis americanus
Titanotheriomys veterior
Titanotheriomys sp. large
Cylindrodon fontis
Cylindrodon, cf. C. nebrascensis
Sciurid, small sp. new
Prosciurus vetustus
Prosciurus, cf. P. vetustus
Protosciurus, cf. P. jeffersoni
Agnotocastor galushai
Eutypomys sp.
Adjidaumo minimus
Adjidaumo, cf. A. minimus
Paradjidaumo, cf. P. trilophus
Parajidaumo, sp. small
Aulolithomys bounites
Aulolithomys, sp. small
720
700
680
660
640
620
600
580
560
540
520
500
480
460
440
420
400
380
360
340
320
300
280
260
240
220
200
180
160
140
120
100
80
60
40
20
0

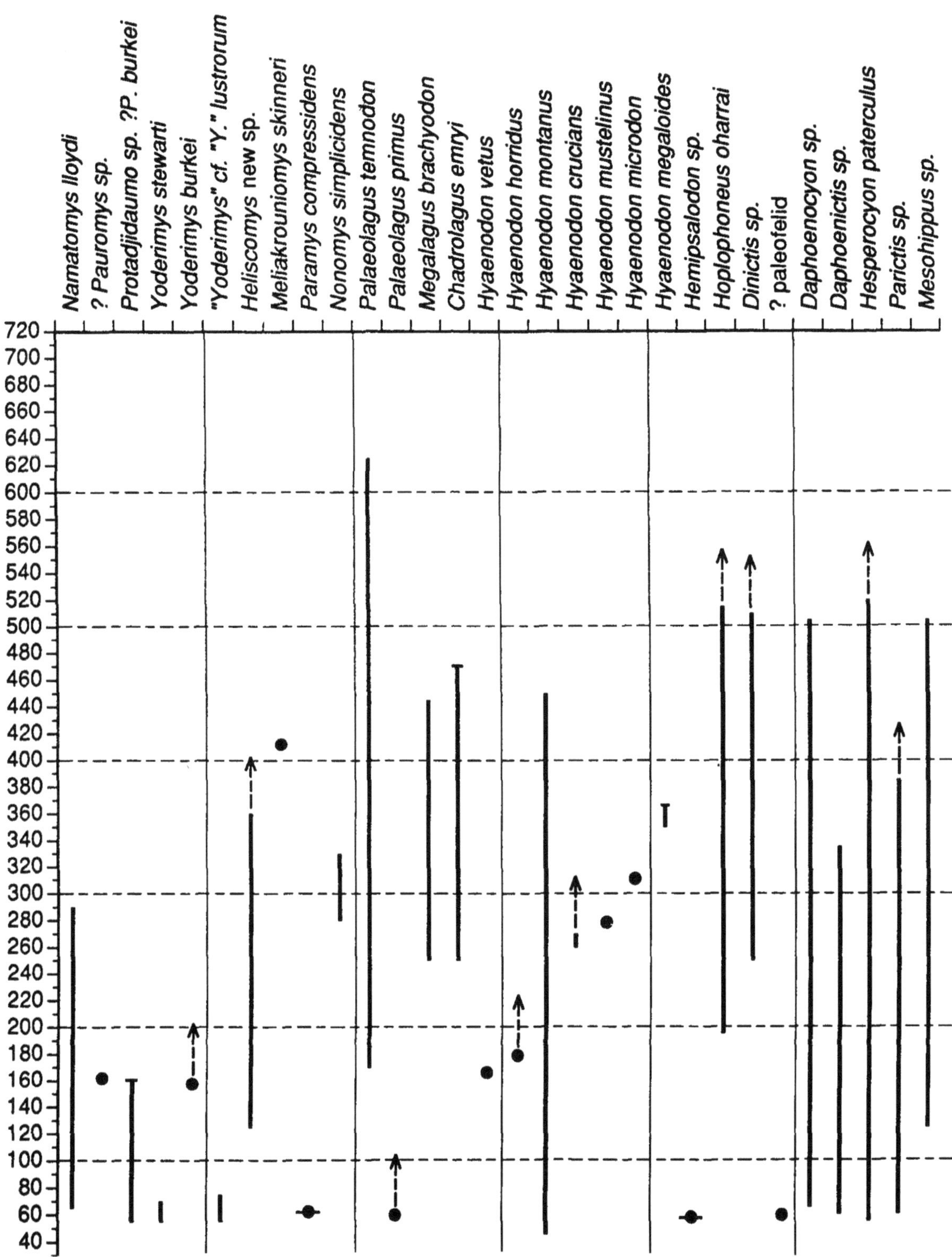

FIGURE 5.2. Local observed ranges of mammalian taxa in the Flagstaff Rim sequence. Footage corresponds to that seen in Figure 5.1. Arrow indicates that taxon is known to occur later elsewhere. Horizontal bar at end of range indicates taxon is not known to occur later elsewhere, nor are descendants known. Dots indicate taxa known from single specimens, or, if more than one specimen, all occur at one quarry site.

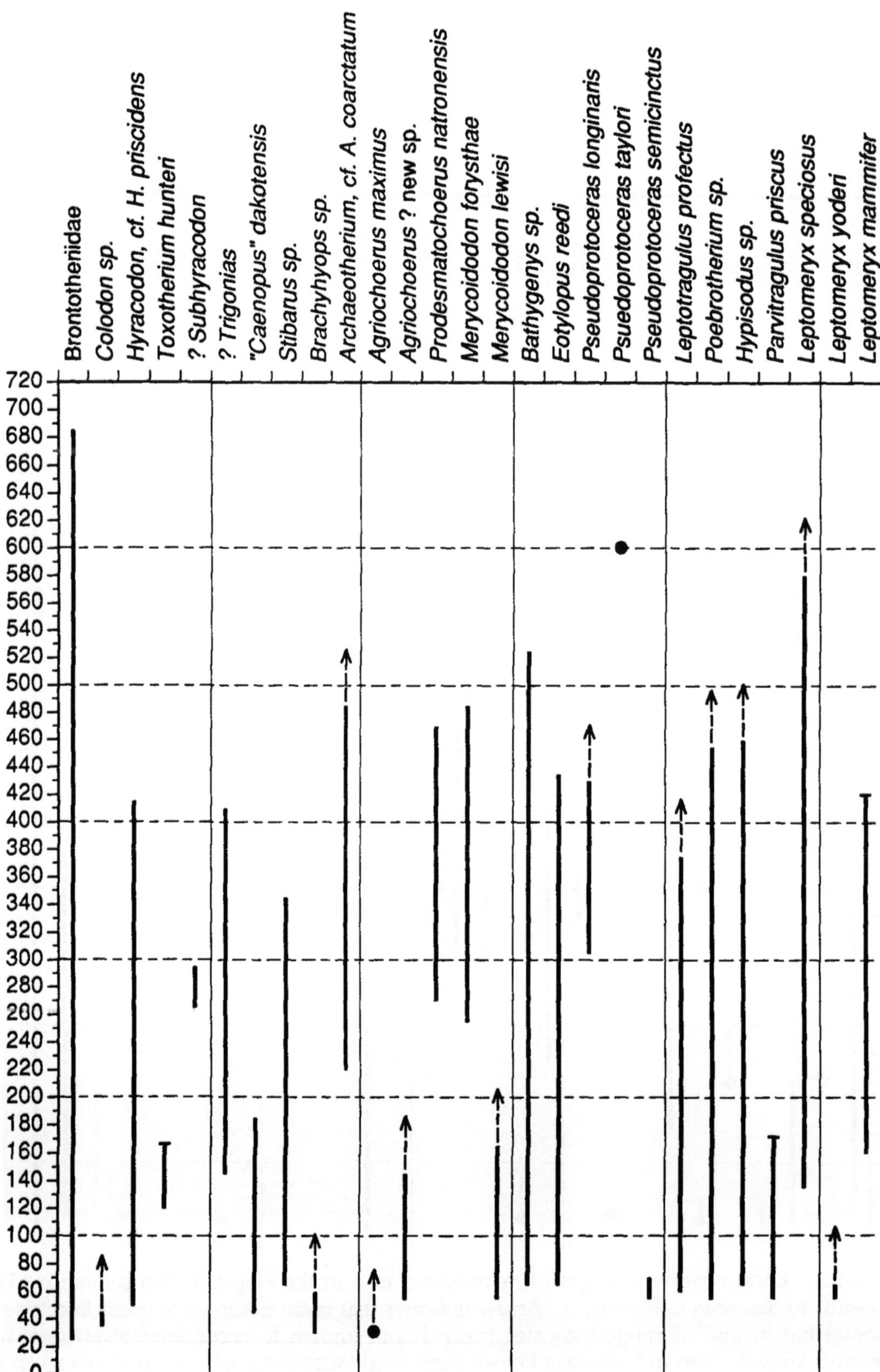
Brontotheriidae
Colodon sp.
Hyracodon, cf. H. priscidens
Toxotherium hunteri
? Subhyracodon
? Trigonias
"Caenopus" dakotensis
Stibarus sp.
Brachyhyops sp.
Archaeotherium, cf. A. coarctatum
Agriochoerus maximus
Agriochoerus ? new sp.
Prodesmatochoerus natronensis
Merycoidodon forysthae
Merycoidodon lewisi
Bathygenys sp.
Eotylopus reedi
Pseudoprotoceras longinaris
Psuedoprotoceras taylori
Pseudoprotoceras semicinctus
Leptotragulus profectus
Poebrotherium sp.
Hypisodus sp.
Parvitragulus priscus
Leptomeryx speciosus
Leptomeryx yoderi
Leptomeryx mammifer
720
700
680
660
640
620
600
580
560
540
520
500
480
460
440
420
400
380
360
340
320
300
280
260
240
220
200
180
160
140
120
100
80
60
40
20
0

interval, unless it is also assumed that the rate of sedimentation remained uniform throughout the sequence being sampled. The possibility of changing rates of sedimentation must also be considered when interpreting the paleomagnetic data. The pattern and degree of change seen in the fossils through time, though certainly also subject to rate changes, nevertheless provides an independent check in evaluating the significance of changing sedimentation rates.

The gross sedimentological characteristics of the White River sequence at Flagstaff Rim suggest that the rate of deposition may have accelerated at about the Ash B level, or about 53 m (175 feet) above the base of the Lone Tree Gulch section (see Figure 5.1). Below that level, a greater proportion of the sediments are locally derived clastic material, the finer sediments are generally brightly colored by oxidized iron, and volcanic components are weathered and altered to a much greater degree than higher in the section. Above the Ash B level, the sediments are predominantly composed of air–transported volcaniclastics, with the minor amount of locally derived coarser clastics usually confined to well–defined stream channel sandstones. The volcanic material appears quite fresh and unaltered, and at certain levels is relatively pure ash, usually white or silvery white, but with beds of dark gray to black ash as well. The more distinct of these ash beds are the ones labelled Ashes A through J (Figure 1), and are the ones radiometrically dated (Evernden et al., 1964; Swisher and Prothero, 1990). It appears that a great influx of volcanic material began at about the Ash B level, and that the part of the section above this level accumulated at a much greater rate.

MAMMALIAN LOCAL RANGE ZONES

When the observed local ranges of fossil mammalian taxa from Flagstaff Rim are plotted against the section (Figure 5.2) it becomes obvious that the pattern of appearances is to a large degree an artifact of preservation. The upper limits of many local ranges, for example, simply reflect the relatively unfossiliferous nature of the upper part of the section. And what appear to be significant occurrence events at about 18 m (60 feet) and about 50 m (about 160 feet) actually indicate the stratigraphic positions of two large quarry samples that yielded much of the fossil material known from the lower part of the section. Among these, the apparently coincident last occurrences of *Protadjidaumo, Toxotherium,* and *Parvitragulus* at the 60 m level probably do indicate an important minor faunal event that might be used to define the end of early Chadronian. From the base of the section up to about 60 m (200 feet), or approximately to the Ash B level, most fossils occur in several small quarry concentrations, with fewer numbers of single random occurrences. In the middle part of the section, from about 60 to 136 m (about 200 to 450 feet) most fossils occur as single specimens, apparently randomly distributed, with a few small quarry concentrations. The upper third of the section is very sparsely fossiliferous, with occasional isolated specimens, usually of the more common taxa. Occasional large bone and tooth fragments indicate that brontotheres occur to within a few feet of the top of the local section, confirming that the section does not continue beyond Chadronian. The apparent unfossiliferous nature of the upper part of the section is a function, at least in part, of topography; this part of the section makes up the steeper part of the escarpment, which results in an outcrop area that is not only smaller, but much more difficult to prospect.

On the range chart (Figure 5.2) the vertical bars for each taxon link the lowest and highest observed occurrences (this chart is calibrated in feet, rather than meters, because the original section was described in feet and the thousands of specimens collected subsequently are tied to that measured section). Single occurrences are represented by a dot; these may indicate a taxon represented by a single specimen, or in some instances indicate several specimens all known from a single quarry. An arrow point on the top end of a range bar indicates that the taxon, or another that could reasonably be derived from it, is known to occur later elsewhere. A horizontal line across the top of a range bar indicates that neither that taxon nor another that might conceivably be derived from it is known to occur later—in other words, these could represent extinctions in the real sense. The larger number of taxon range bars end neither with arrows nor horizontal lines; these are taxa whose species level relationships are

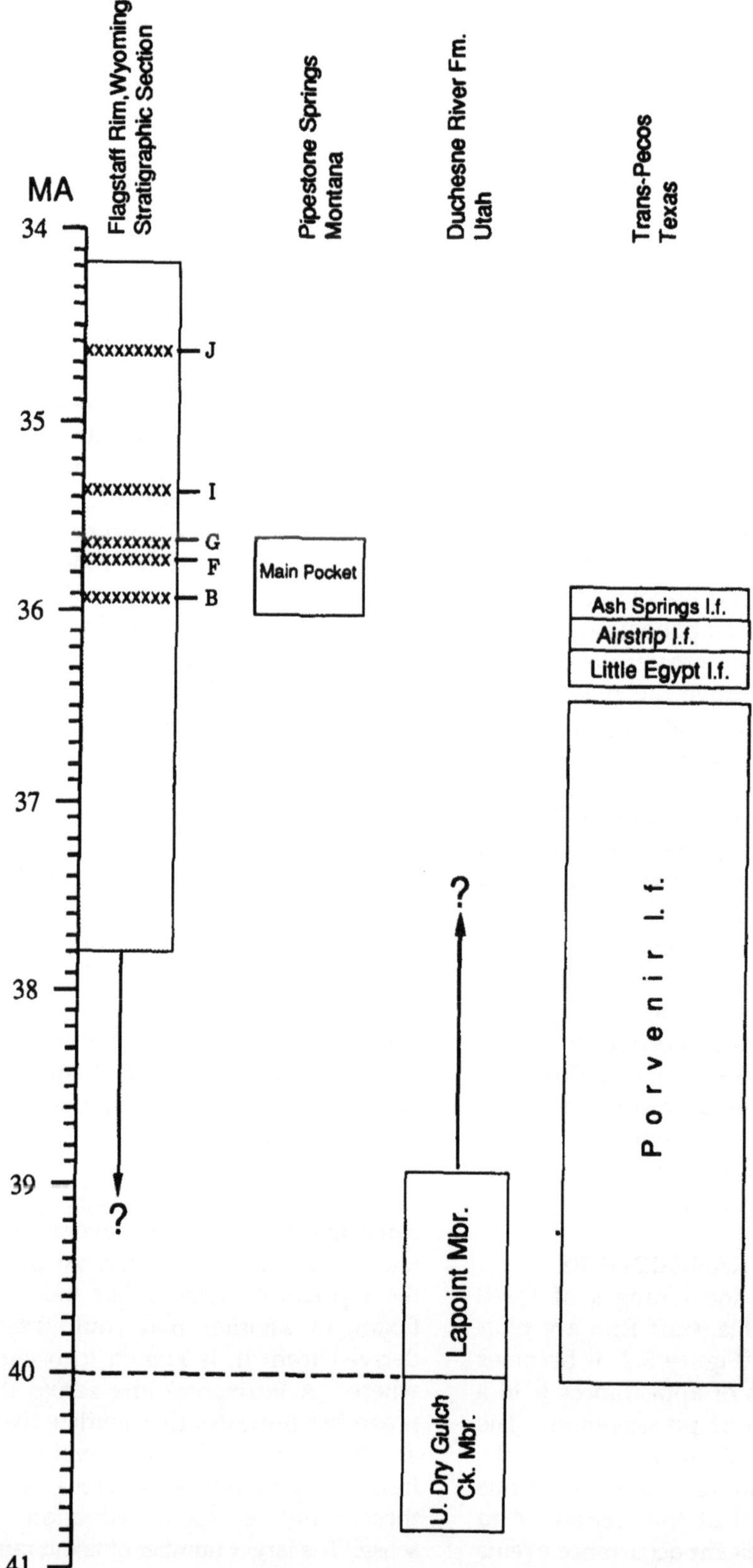

FIGURE 5.3. Some Duchesnean and Chadronian correlations suggested by faunal information in conjunction with new radiometric dates. Arrows indicate uncertain position of boundary with respect to numerical scale.

too poorly understood to allow a confident guess as to what part of their total range zone may be represented by these local occurrences.

DISCUSSION

As a generality, most Chadronian mammalian genera continue into the Orellan, and some continue even into Whitneyan and later. In some cases these genera are represented by a single species that continues through, apparently unchanged, but more commonly, different species names are applied to morphologically distinct parts of a single evolving lineage.

Where more than one species of a genus occurs in the Chadronian, it is sometimes problematic which, if any of those known, is the most likely source for a related Orellan species. An example is seen in the small ungulate genus *Leptomeryx*, one of the genera occurring most abundantly in the Flagstaff Rim sequence, where it is represented by at least three species and two co–occurring lineages. Up to about 136 m (450 feet) on the local section, both *L. mammifer* and the smaller species, *L. speciosus*, are very abundant. Above 136 m (450 feet), *L. mammifer* is absent, while *L. speciosus* continues into higher levels, and may be the only reasonable source for *L. evansi* of the Orellan. Because *L. mammifer* is so abundant up to about 136 m (450 feet) in the section, and it is not known to occur later anywhere else, its disappearance here can be reasonably interpreted as a real extinction

This range chart (Figure 5.2) does show changes in the faunal composition through the section, which allows some preliminary correlations with other important Chadronian units.

Figure 5.3 is not intended to establish correlations, but to indicate some preliminary correlations that might be suggested strictly on the basis of faunal data. On this diagram the part of the Flagstaff Rim section above ash B is tied to the numerical scale according to the new $^{40}Ar/^{39}Ar$ dates (Swisher and Prothero, 1990); the lower part of this section (below ash B) has no isotope dates, so its downward extent, relative to the numerical scale is not really known (the arrow and question mark indicate this uncertainty). However, the amount of faunal change observed (changes in faunal composition as well as morphologic change in individual lineages) suggest that more time may be represented by the part of the section below ash B than by the ash B to G interval. The data on *Leptomeryx*, for example, support this conclusion. The Lapoint is tied to the numerical scale by the date at the base of the Lapoint Member (recalculated with new constants from McDowell et al., 1973); its upward extent is likewise not dated and is also indicated by a questioned arrow. The other suggested correlations are based on paleontologic evidence.

The fauna of the "main pocket" at Pipestone Springs, Montana, for example (Tabrum and Fields, 1980), is essentially duplicated by the fauna occurring from the Ash B to Ash G levels of Flagstaff Rim. This correlation is consistent with magnetostratigraphic information (Prothero, 1984). Some previously accepted correlations of other faunas, however, appear to be inconsistent with the Flagstaff Rim faunal range chart (Figure 5.2) and the new Flagstaff Rim dates (Swisher and Prothero, 1990). In the Trans–Pecos Texas sequence, for example, the Ash Springs local fauna could not be correlated lithostratigraphically with the radioisotopically dated sequence in the Big Bend area, but Wilson (1977), based on the "evolutionary position of taxa," considered the Ash Springs l.f. to be younger than the Brite Ignimbrite (31.4 Ma). But the Ash Springs l.f. contains *Toxotherium* and *Parvitragulus*, both of which occur also in the Flagstaff Rim sequence, but only below Ash B, which is approximately 35.9 Ma, according to Swisher and Prothero (1990). This suggests either that *Toxotherium* and *Parvitragulus* survived in Texas for several million years after their last known occurrence in Wyoming, or that the Ash Springs l.f. is actually older than the Brite Ignimbrite, and that the Little Egypt, Airstrip, and Ash Springs local faunas (see Wilson, 1977, and references therein) may all be nearer the age of the Buckshot Ignimbrite (36.8 Ma), which underlies the Little Egypt l.f. The Porvenir local fauna of Trans–Pecos Texas has *Brachyhyops* in its lower levels and *Archaeotherium* in its upper levels (Wilson, 1971). In the Flagstaff Rim sequence, *Brachyhyops* occurs near the base and *Archaeotherium* does not appear until about Ash B level. This suggests the correlation shown in Figure 5.3, where the Porvenir local fauna is extended upward to correlate with the *Archaeotherium*-

bearing level at Flagstaff Rim.

The lowest levels at Flagstaff Rim have several taxa that occur in the Lapoint Duchesnean (*Hemipsalodon, Brachyhyops, Protadjidaumo*, for example), and would be called Duchesnean if one wishes to recognize Duchesnean by the presence of an assemblage that includes these taxa. As I have argued previously (Emry, 1981), it might be possible to recognize a "Duchesnean fauna," but defining a Duchesnean NALMA that is practical requires that it also have recognizable limits. And a definition of Duchesnean based on such a limited assemblage relies heavily on gaps in the record for boundaries. The first occurrence of entelodonts (*Brachyhyops*) in North America, for example, is surely an important faunal event, useful in defining the beginning of Duchesnean time. But the last occurrence of *Brachyhyops* to signal the end of Duchesnean, or, alternatively,the first appearance of *Archaeotherium* to indicate the beginning of Chadronian, are equally surely of less importance, as faunal events, for recognizing a NALMA boundary. A definition based on these occurrences would effectively reduce to a subjective taxonomic decision; is the entelodont present in a fauna of questionable age really a large *Brachyhyops* or a small *Archaeotherium*? This problem is not merely potential, but realized; J. Storer (*in litt.*) suggests that the entelodont reported as *Archaeotherium* from the upper part of the Porvenir local fauna (mentioned above) might be more appropriately referred to the larger *Brachyhyops* species, *B. viensis*.

Surely there is a substantial interval, probably 4 million years or more, between the latest typical Uintan faunas and the first undisputed Chadronian faunas. The fauna of this Duchesnean interval is not as well known as either that of the Uintan or Chadronian, but what we do know of it indicates that the more prominent faunal event, including the first appearance in North America of several immigrant taxa, is at the Uintan/Duchesnean transition. Thereafter, until the Arikareean, first appearances of taxa are predominantly those produced by evolution within North America, rather than immigration. This naturally limits the stability, and objectivity in recognizing, the NALMA boundaries during this interval.

ACKNOWLEDGEMENTS

Thanks are due to William Korth for assistance in identifying some of the rodents, to Dan Chaney for help in drafting the illustrations, and to John Storer and Spencer Lucas for constructive criticism of the manuscript.

REFERENCES

Emry, R. J. 1973. Stratigraphy and preliminary biostratigraphy of the Flagstaff Rim area, Natrona County, Wyoming. *Smithsonian Contrib. to Paleobiol.* 18:1–42.

Emry, R. J. 1981. Additions to the mammalian fauna of the type Duchesnean, with comments on the status of the Duchesnean "Age." *Jour. Paleontol.* 55:563–570.

Dalrymple, G.B. 1979. Critical tables for conversion of K-Ar ages from old to new constants. *Geology* 7: 558-560.

Evernden, J. R., Savage, D. E., Curtis, G. H., and James, G. T. 1964. Potassium–argon dates and the Cenozoic mammalian chronology of North America. *Am. Jour. Sci.* 262:145–198

McDowell, F. W., Wilson, J. A., and Clark, J. 1973. K–Ar dates for biotite for two paleontologically significant localities: Duchesne River Formation, Utah and Chadron Formation, South Dakota. *Isochron/West* 7:11–12.

Prothero, D. R. 1984. Magnetostratigraphy of the early Oligocene Pipestone Springs locality, Jefferson County, Montana. *Contrib. Geol., Univ. Wyoming*, 23:33–36.

Prothero, D.R., 1985. Chadronian (early Oligocene) magnetostratigraphy of eastern Wyoming: implications for the Eocene-Oligocene boundary. *Jour. Geology*, 93: 555-565.

Prothero, D.R., Denham, C. R., and Farmer, H. G. 1982. Oligocene calibration of the magnetic polarity time scale. *Geology* 10: 650–653.

Prothero, D.R., Denham, C. R., and Farmer, H. G. 1983. Magnetostratigraphy of the White River Group and its implications for Oligocene geochronology. *Palaeogeog., Palaeoclimat., Palaeoecol.* 42:151–166.

Swisher, C. C., III, and Prothero, D. R. 1990. Single–crystal $^{40}Ar/^{39}Ar$ dating of the Eocene–Oligocene transition in North America. *Science* 249:760–762.

Tabrum, A. R., and Fields, R. W. 1980. Revised mammalian faunal list for the Pipestone

Springs Local Fauna (Chadronian, early Oligocene) Jefferson County, Montana. *Northwest Geol.* 9:45–51.

Wilson, J. A. 1971. Early Tertiary vertebrate faunas, Vieja Group, Trans-Pecos Texas: Entelodontidae. *Pearce-Sellards Series* 17:1–17.

Wilson, J. A. 1977. Stratigraphic occurrence and correlation of early Tertiary vertebrate faunas, Trans-Pecos Texas. *Texas Mem. Mus. Bull.* 25:1–42.

6. EOCENE-OLIGOCENE CLIMATIC CHANGE IN NORTH AMERICA: THE WHITE RIVER FORMATION NEAR DOUGLAS, EAST-CENTRAL WYOMING

By Emmett Evanoff, Donald R. Prothero, and R.H. Lander

ABSTRACT

The White River Formation exposed near the town of Douglas in east-central Wyoming records a change from a moist subtropical climate in the latest Eocene to a semiarid warm temperate climate in the early Oligocene. These climates are indicated by the sedimentology and nonmarine gastropods (primarily land snails) of the formation. The area also contains a dated volcanic tuff and a magnetostratigraphic record that constrain the age of the boundary between the Chadronian and Orellan land mammal "ages." This land mammal "age" boundary is at or near the Eocene/Oligocene boundary.

INTRODUCTION

The White River Group, a fine-grained volcaniclastic sequence in the western United States, records faunal, depositional, and paleoclimatic changes across the Eocene/Oligocene boundary. It contains the most complete late Eocene and early Oligocene vertebrate record in North America, and somewhat less well known invertebrate, sedimentologic, and volcaniclastic records. The White River Group is an unconformity-bounded nonmarine sequence (*sensu* Hanneman, 1988, 1989) that extends across the northern Great Plains and central Rocky Mountains (Figure 6.1). It is treated as a formation in Colorado, Wyoming, and North Dakota, but is considered a group in Nebraska and South Dakota, where it includes the Chadron and Brule Formations. Four areas of White River exposures have received intensive study: the Big Badlands of South Dakota (Wanless, 1923; Clark, 1937; Clark and others, 1967; Retallack, 1983), the Pine Ridge of northwest Nebraska (Darton, 1899; Schultz and Stout, 1955; Harvey, 1960), northeast Colorado (Galbreath, 1953), and Flagstaff Rim in central Wyoming (Emry, 1973). Other areas in the Great Plains and Rocky Mountains with thick sections of White River rocks have received less attention but are nevertheless important for regional correlations and paleoenvironmental reconstructions.

An important but relatively unstudied area of White River exposures occurs southeast of the town of Douglas in Converse County, east-central Wyoming (Figure 6.1). Here, the White River Formation is exposed in 40 square kilometers of badland outcrops in two areas separated by the North Platte River (Figure 6.2). The oldest White River rocks are exposed in the southern or "Dilts Ranch" area. The upper half of the formation is best exposed in the northern area (the "Wulff Ranch" and "Morton Ranch" areas of Figure 6.2). Both areas are situated topographically on the southern margin of the Powder River basin, but are structually south of the northern boundary fault of the Laramie Mountains (Denson and Horn, 1975; Blackstone, 1988). The Precambrian core of the Laramie Mountains is 22 km to the southwest, but locally the formation overlies strongly folded Cretaceous rocks. The White River has been slightly folded and cut by normal faults on the east side of the Dilts Ranch area.

The 230 m thick exposures in the Douglas area contain abundant terrestrial fossils and 13 volcanic tuff beds, including a dated tuff near the Eocene/Oligocene boundary (Swisher and Prothero, 1990). This dated tuff and detailed magnetostratigraphic analysis (Prothero, 1982a, 1985) constrain the age of a major change in mammal faunas (the boundary between the Chadronian and Orellan land mammal ages) and provide a temporal framework for compar-

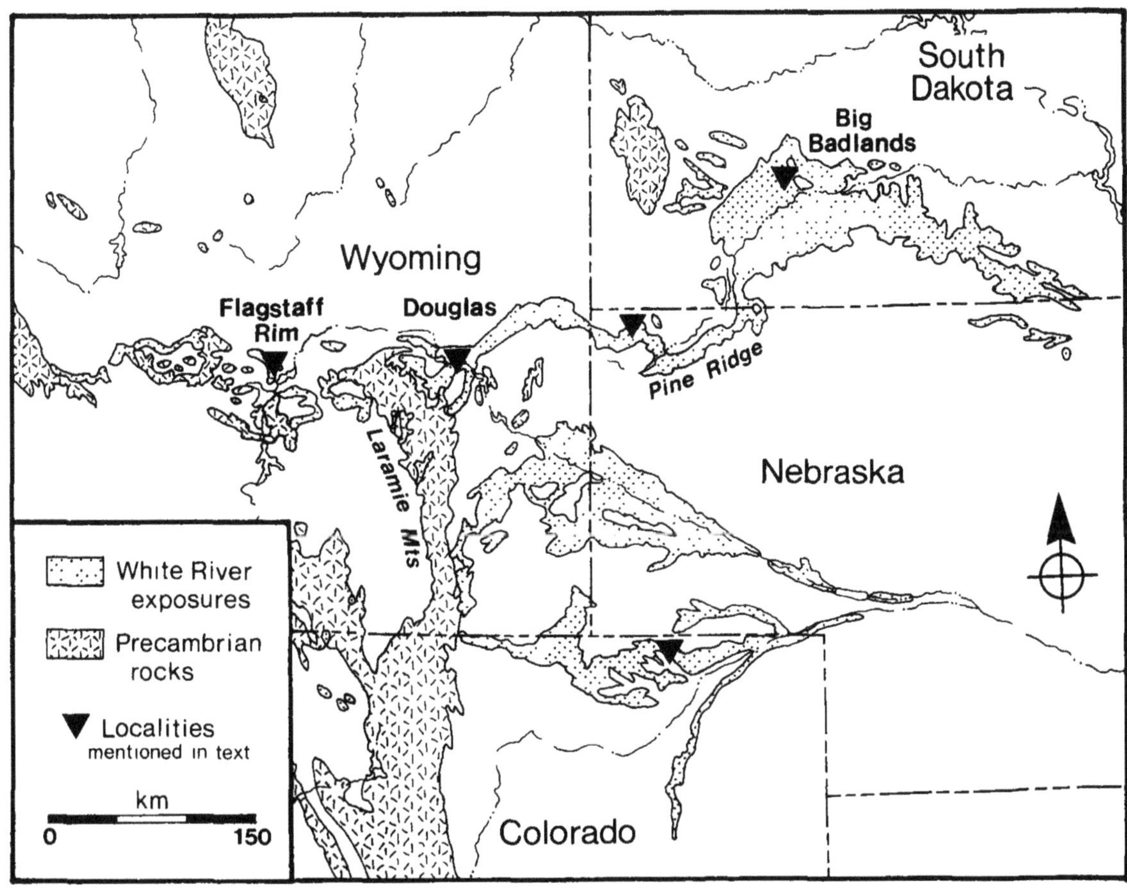

FIGURE 6.1. Distribution of the White River deposits, Precambrian cores of Laramide uplifts, and localities mentioned in the text. Base map from King and Beikman (1974), with additional information from Tweto (1979) and Love and Christiansen (1985), and Burchett (1986).

ing local events with regional and worldwide events. The sedimentology of the White River Formation records a progressive increase in aridity, as indicated by a shift from fluvial to dryland loess deposition. The fossil land snail fauna in the area also indicates increased aridity and a minor drop in temperature across the boundary.

STRATIGRAPHY

The White River Formation in the Douglas area is divided into the the Chadron and the Brule Members (Evanoff, 1990). These units are separated by the basal contact of a widespread, thick, white tuff bed (5 tuff of Evanoff, 1990; the "Glory Hole ash," purplish white layer, or persistent white layer of Prothero, 1982a, 1982b, 1985). The basal Chadron Member is characterized by green to brown clayey mudstones, brown to tan (typically nodular) sandy mudstones, thin sheet sandstones, and numerous thick ribbon sandstones. The overlying Brule Member is characterized by brown to tan nodular sandy mudstones, massive tan sandy siltstones, rare ribbon sandstones, and rare conglomeratic sheet sandstones. Use of the names Chadron and Brule for members in the Douglas area reflects: 1) the subtle lithologic differences between the "upper" and "lower" parts of the White River Formation; and 2) lithologic similarities between the Douglas area sequence and the Chadron and Brule formations in northwest Nebraska. However, the White River sequence in the Douglas area cannot logically be divided into formations, because the two members cannot always be distinguished in

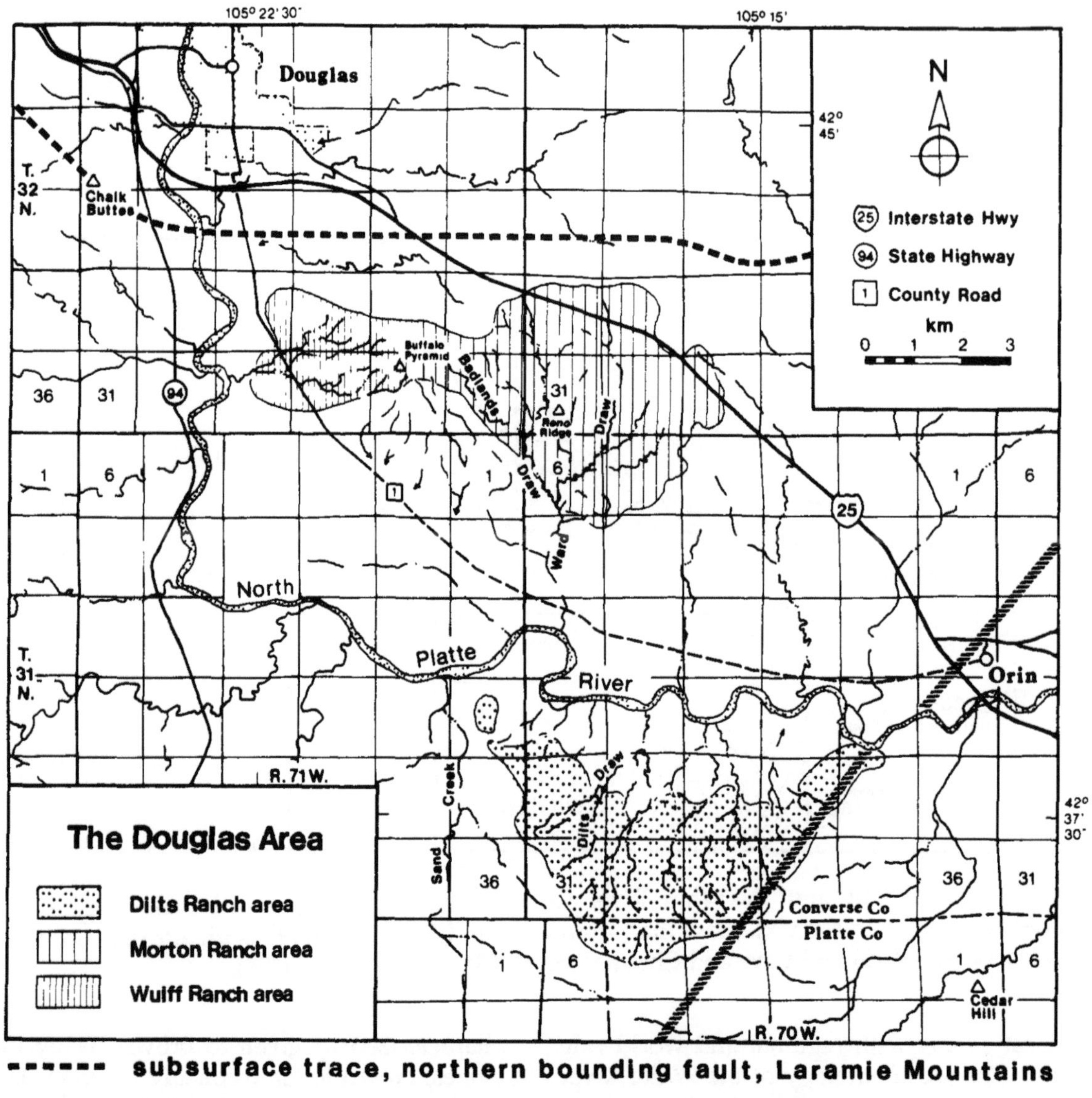

FIGURE 6.2. Distribution of White River Formation outcrops, fold and fault zones, and geographic features in the Douglas area. The northern exposures include the Morton Ranch and Wulff Ranch areas.

isolated outcrops and are not as distinct as their equivalent formations in Nebraska.

Tuff beds are prominent and typically can be traced throughout the area. The tuffs range in mean thickness from 0.5 m to 1.05 m, although some pinch out. Locally the tuffs contain depositional fabrics although in most places the fabric is disrupted by numerous root traces and burrows. Brule tuffs are glass-rich whereas most Chadron tuffs are wholly altered to opal-CT, clinoptilolite (a silica-rich zeolite), or smectite (Lander, 1985, 1987, 1991). White-colored Brule tuffs are generally characterized by rhyolitic glass and plagioclase feldspar compositions ranging from An37 to An45. Altered Chadron tuffs were also probably rhyolitic,

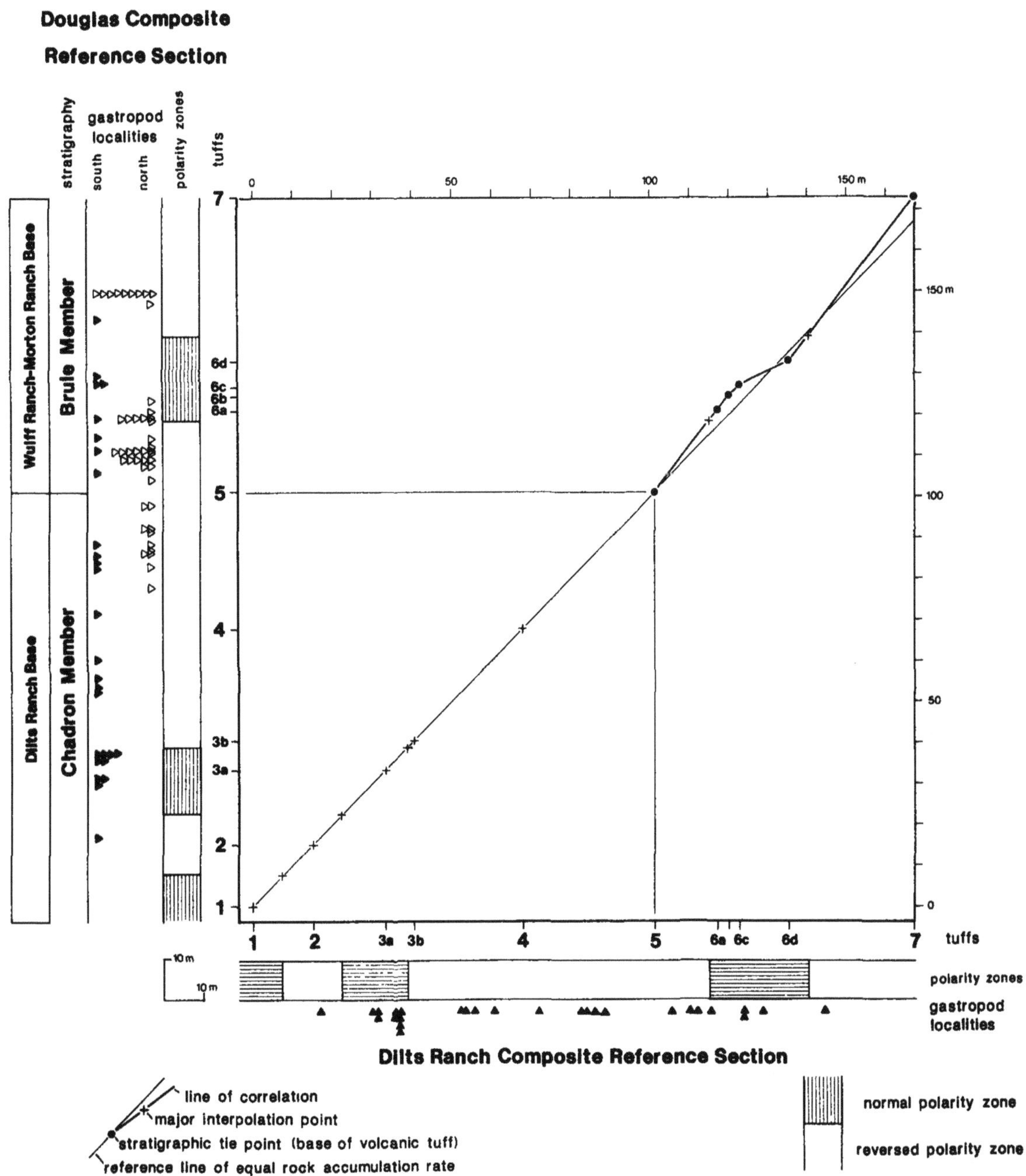

FIGURE 6.3. Method of correlating features within the White River Formation in the Douglas area. In this example, the stratigraphic positions of fossil gastropod localities and polarity event boundaries are projected by graphic correlation from the X-axis to the Y-axis (the Douglas area composite reference section) through the line of correlation. The mean stratigraphic position of tuffs in the two sections determines the position of the line of correlation. Evanoff (1990) describes the derivation of the Douglas area composite reference section, and Edwards (1984) discusses the graphic correlation procedure and its underlying assumptions.

based on similar plagioclase feldspar compositions (Lander, 1991). Several Brule tuffs are dark gray and contain quartz-latite glass and considerably more calcic plagioclase (An65-85). Each tuff has a unique stratigraphic position, and some can be distinguished by associated overlying or underlying strata. The tuff beds are chronostratigraphic markers from which the positions of fossil localities, physical event horizons, and magnetopolarity zones can be determined with great accuracy (Figure 6.3).

A radioisotopic date from the 5 tuff by single-crystal laser fusion $^{40}Ar/^{39}Ar$ analysis of biotite is 33.9 ± 0.1 Ma (Swisher and Prothero, 1990). This date is well within the age range of the Eocene/Oligocene boundary as determined from radioisotopic age dates of tuffs within pelagic limestones and marls that were continuously deposited across the Eocene/Oligocene boundary in the northeast Apennines (Montanari and others, 1988). This age date also constrains the magnetic polarity normal interval above the 5 tuff to C13N. Using the age estimates of 33.7 Ma for the Eocene/Oligocene boundary (Montanari and others, 1988) and 23.7 Ma for the Oligocene/Miocene boundary (Berggren and others, 1985), the age of the base of C13N is herein interpolated as approximately 33.1 Ma. The base of C13N is 13.7 m above the 5 tuff, so the position of the Eocene/Oligocene boundary is 3.4 m (± 8.6 m) above the 5 tuff (Figure 6.4) by simple interpolation assuming a constant depositional rate. The range of possible positions of the Eocene/Oligocene boundary includes the boundary between the Chadronian and Orellan land mammal ages (Figure 6.5), and a major transition in land snail faunas near the Chadron-Brule boundary. The radioisotopic date of the 5 tuff and the position of the C13R/C13N boundary in the Douglas area indicates the Chadron/Orellan boundary is at or very near the Eocene/Oligocene boundary.

The magnetic polarity of the White River Formation in the study area continues to be refined. The original sequence of polarity zones (Prothero, 1982a, 1985) was done with limited thermal demagnetization (in the case of the Wulff Ranch and Reno Ranch south of Tower sections--Fig. 6.6). All of these sections have been reinterpreted by using more extensive thermal demagnetization, using the tuffs as marker beds, and reanalyzing problematic intervals (Figure 6.6). A new section was run in the Reno Ranch East area to corroborate the pattern north of the river. Samples were thermally demagnetized in several steps, and the results obtained between 300° and 400°C gave the most consistent results. AF demagnetization and IRM acquisition studies were also conducted. They showed that the remanence is carried by a low coercivity mineral which magnetically saturates at 1 KOe, and therefore is probably some form of titanomagnetite. Means of the significant normal and reversed sites passed a reversal test. The revised polarity sequence has been interpolated onto the Douglas composite reference section by using the positions of tuffs as tie points (Figure 6.7). Three normal polarity zones and three reversed polarity zones are present.

Vertebrate fossils are abundant in the lower two-thirds of the White River Formation and include mammals of the Chadronian and Orellan land mammal ages. No unequivocal Whitneyan mammals are known from the area. Detailed vertebrate biostratigraphic data is available only for mammals of the upper Chadron Member and lower half of the Brule Member (Figure 6.5). Based on original faunal definition of the Chadronian Land Mammal "Age" (Wood and others, 1941), the highest occurrence of brontotheres marks the end of the Chadronian. However, this criterion has come under question because rare brontotheres can co-occur with taxa thought to reflect the Orellan Land Mammal "Age," and the absence of brontotheres in local lower Brule sequences does not necessarily indicate a non-Chadronian age (Emry and others, 1987). The Chadronian-Orellan boundary is currently being re-evaluated (Korth, 1989), but until more diagnostic criteria are accepted, the highest occurrence of

FIGURE 6.4. (opposite page) Physical features and events as indicated by the White River Formation in the Douglas area. Stippled area in the dominant processes column represents the transition between the fluvial- and eolian-dominated intervals; black areas in the magnetostratigraphy column are zones of normal polarity; and vertical lines between the Eocene and Oligocene indicate the range of possible positions of the series boundary.

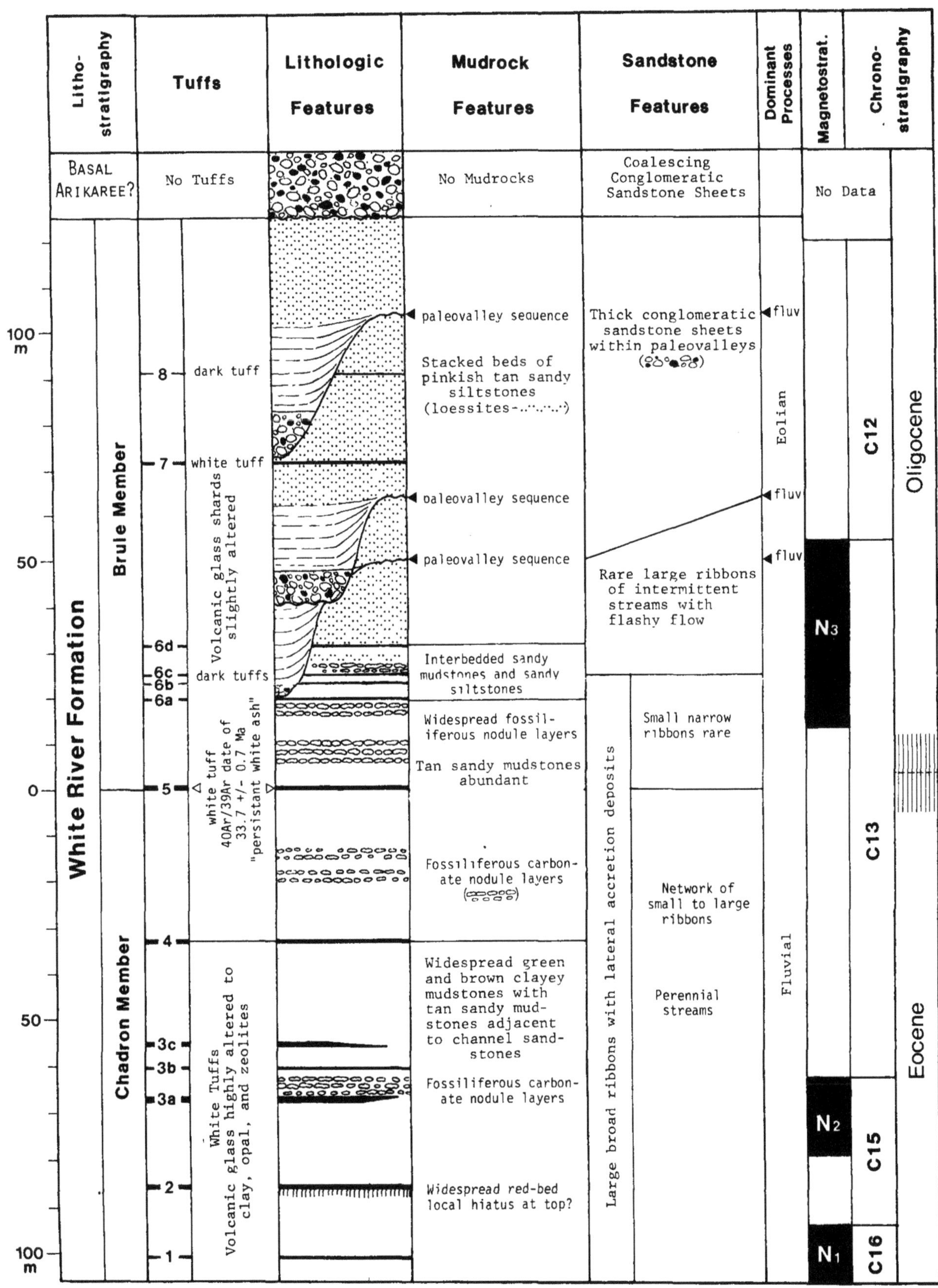

Litho-stratigraphy
Tuffs
Lithologic Features
Mudrock Features
Sandstone Features
Dominant Processes
Magnetostrat.
Chrono-stratigraphy
BASAL ARIKAREE?
No Tuffs
No Mudrocks
Coalescing Conglomeratic Sandstone Sheets
No Data
100 m
50
0
50
100 m
White River Formation
Brule Member
Chadron Member
8 dark tuff
7 white tuff
Volcanic glass shards slightly altered
6d
6c
6b
6a
dark tuffs
5 white tuff 40Ar/39Ar date of 33.7 +/- 0.7 Ma "persistant white ash"
4
3c
3b
3a
2
1
White Tuffs Volcanic glass highly altered to clay, opal, and zeolites
paleovalley sequence
Stacked beds of pinkish tan sandy siltstones (loessites)
paleovalley sequence
paleovalley sequence
Interbedded sandy mudstones and sandy siltstones
Widespread fossiliferous nodule layers
Tan sandy mudstones abundant
Fossiliferous carbonate nodule layers
Widespread green and brown clayey mudstones with tan sandy mudstones adjacent to channel sandstones
Fossiliferous carbonate nodule layers
Widespread red-bed local hiatus at top?
Thick conglomeratic sandstone sheets within paleovalleys
Rare large ribbons of intermittent streams with flashy flow
Large broad ribbons with lateral accretion deposits
Small narrow ribbons rare
Network of small to large ribbons
Perennial streams
fluv
fluv
fluv
Eolian
Fluvial
N3
N2
N1
C12
C13
C15
C16
Oligocene
Eocene

brontotheres (6.4 m above the 5 tuff) marks the Chadronian/Orellan boundary in the study area. The highest occurrence of the eomyid rodent *Yoderimys* sp. (D. G. Kron, personal communication, 1985), the lowest occurrences of the rabbit *Litolagus molidens* Dawson, 1958, and the cricetid rodent *Scottimus viduus* Korth, 1981, are at or near this boundary.

LITHOLOGIES AND DEPOSITIONAL ENVIRONMENTS

The White River Formation is composed primarily of tuffaceous mudrocks. The fine-grained tuffaceous clasts in the mudrocks and tuff beds were derived from contemporary volcanic centers in Colorado and the Great Basin, several hundred kilometers to the south and west (Mutschler and others, 1987). These volcaniclasts were transported to the area primarily by winds and secondarily by local streams. Plagioclase compositions and smectite Fe and Ti abundances in mudrocks indicate that the bulk of pyroclastic material was dacite to latite in composition. Coarse, gravelly sandstone and thin tuff beds are volumetrically minor lithologic components. The coarse sandstones are composed of arkosic detritus that was transported by streams flowing from the Laramie Mountains (Evanoff, 1990). The contrasting sizes of fine volcaniclastic and coarse epiclastic sediment greatly enhance the differences in lithofacies.

Sandstones and conglomerates

Sandstone and conglomerate beds record changes in the stream systems during deposition of the White River Formation. Chadron sandstones are dominated by ribbon bodies of various sizes arranged in complex networks. Large Chadron ribbons have low sinuosities and widths ranging from 10 m to 60 m. Stratification in these ribbons includes basal conglomeratic sands arranged in trough crossbed sets, and upper beds inclined normal or oblique to original stream flow that represent lateral accretion deposits. These large Chadron ribbons are the channel deposits of single-channel, low-sinuosity, perennial streams. Smaller ribbons (less than 10 m wide) are either nonstratified or have few trough crossbed sets, and are normal or oblique to larger sandstone bodies. These small ribbons represent crevasse and flood channels. Chadron streams were part of a complex network of low-sinuosity, laterally migrating channels that shifted by avulsion.

The Brule Member has fewer sandstone bodies than the Chadron Member. Small ribbon sandstones are almost absent in all of the Brule. The few large ribbons in the lower Brule (between the 5 and 6c tuffs) resemble the large Chadron ribbons in internal stratification. The large ribbons of the middle Brule (between the 6c and 7 tuffs) have complex internal stratification dominated by alternating, thin to very thin beds of sandstone and sandy mudstone. The thin sandstone beds have abundant parting lineation and the sandy mudstones contain mudcracks. The stratification and geometry of these ribbons indicate shallow, unconfined, and flashy flow in channels of low-sinuosity, intermittent streams.

Large conglomeratic sheet sandstones occur within very large cut-and-fills representing paleovalley sequences in the middle and upper Brule Member. These sheet sandstones are very thick (as much as 11 m), wide (as much as 875 m), and elongate with low sinuosities. They contain large cobbles and boulders, the largest clasts of all sandstone bodies in the area. Stratification in these sheets is characterized by large lenticular packages as much as 5.5 m thick of trough and tabular crossbed sets and the absence of lateral accretion deposits. These sheet sandstones were deposited by powerful, low sinuosity streams which transported very coarse sediment.

FIGURE 6.5. (opposite page) Ranges of White River mammal and gastropod taxa in the Douglas area. The range data for mammalian taxa are derived primarily from the collections of Skinner (1958-1959), as presented in Prothero (1982a, Figure 24), with modifications from unpublished data by D. G. Kron (1991, written communication). Solid bars are ranges determined by Skinner; open bars are range modifications by Kron. Gastropod ranges are from Evanoff (1990); diamonds represent isolated stratigraphic occurrences. The stippled area shown in the tephrochronology column represents the transition between the fluvial- and eolian-dominated intervals.

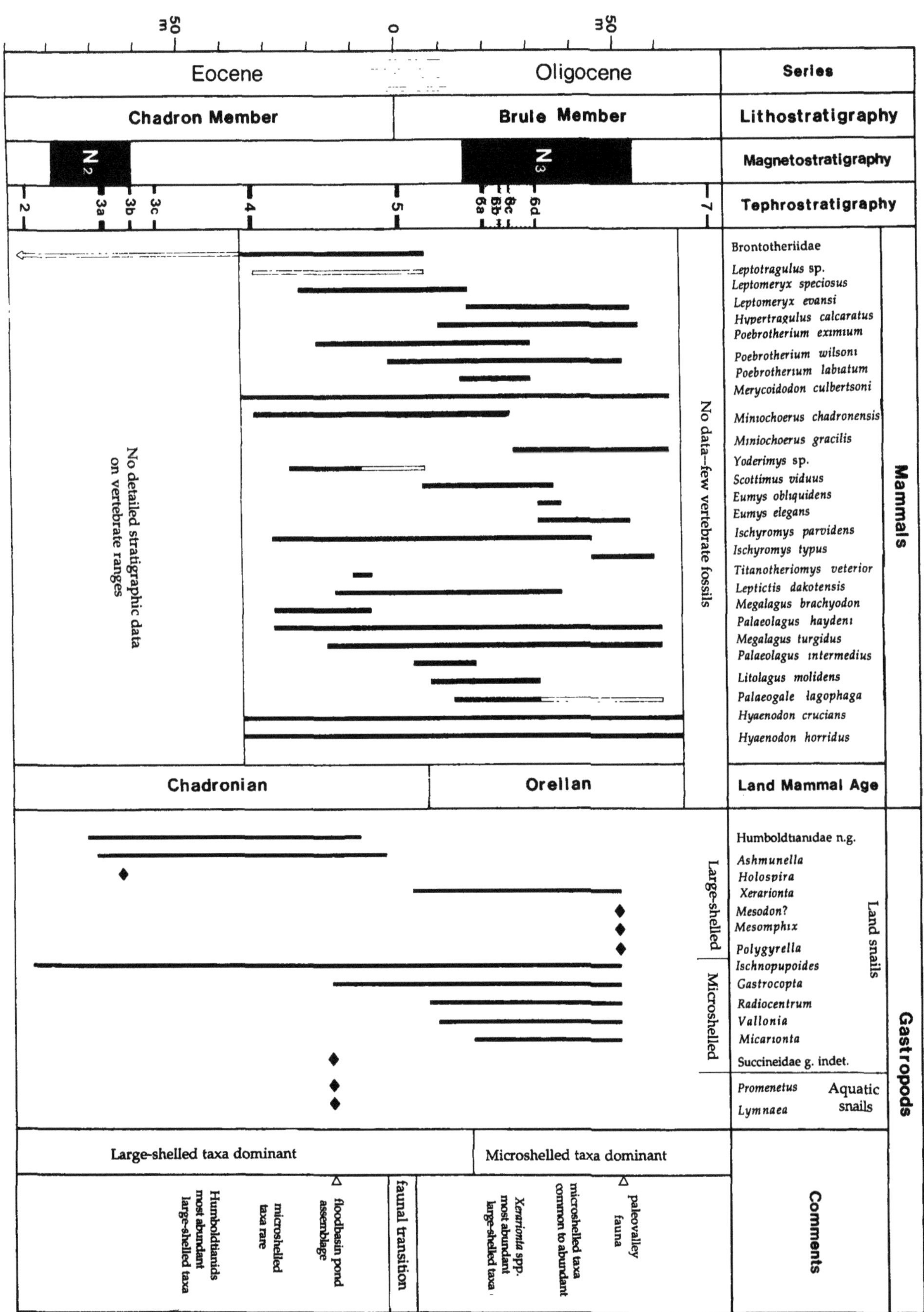
50 m
0
50 m
Eocene
Oligocene
Series
Chadron Member
Brule Member
Lithostratigraphy
N2
N3
Magnetostratigraphy
2
3a
3b
3c
4
5
6a
6b
6c
6d
7
Tephrostratigraphy
No detailed stratigraphic data on vertebrate ranges
No data–few vertebrate fossils
Brontotheriidae
Leptotragulus sp.
Leptomeryx speciosus
Leptomeryx evansi
Hypertragulus calcaratus
Poebrotherium eximium
Poebrotherium wilsoni
Poebrotherium labiatum
Merycoidodon culbertsoni
Miniochoerus chadronensis
Miniochoerus gracilis
Yoderimys sp.
Scottimus viduus
Eumys obliquidens
Eumys elegans
Ischyromys parvidens
Ischyromys typus
Titanotheriomys veterior
Leptictis dakotensis
Megalagus brachyodon
Palaeolagus haydeni
Megalagus turgidus
Palaeolagus intermedius
Litolagus molidens
Palaeogale lagophaga
Hyaenodon crucians
Hyaenodon horridus
Mammals
Chadronian
Orellan
Land Mammal Age
Large-shelled
Microshelled
Humboldtianidae n.g.
Ashmunella
Holospira
Xerarionta
Mesodon?
Mesomphix
Polygyrella
Ischnopupoides
Gastrocopta
Radiocentrum
Vallonia
Micarionta
Succineidae g. indet.
Land snails
Promenetus
Lymnaea
Aquatic snails
Gastropods
Large-shelled taxa dominant
Microshelled taxa dominant
floodbasin pond assemblage
microshelled taxa rare
Humboldtianids most abundant large-shelled taxa
faunal transition
paleovalley fauna
microshelled taxa common to abundant
Xerarionta spp. most abundant large-shelled taxa
Comments

Mudrocks

Mudrocks of the White River Formation in the Douglas area indicate a shift from dominantly fluvial to eolian depositional processes. The mudrocks include clay-rich and sand-rich mudstones deposited in a fluvial system, and sandy siltstones deposited as loess in an eolian system.

The Chadron and lower Brule Members are characterized by widespread horizontal beds of structureless, commonly bioturbated mudstones typically less than 1 m thick. Where preserved, volcanic glass shards in these mudstones are fragmented. Clayey mudstones, which are most abundant in the Chadron Member, are so altered that no shards are preserved. Mudstones in the Douglas area are fluvially reworked volcaniclastic dust deposits, that were subsequently bioturbated and altered by weathering and diagenesis.

One of the most prominent feature of mudstones in the Douglas area is the abundance of carbonate nodules. Individual nodules are spherical to oblate in shape and range from 0.1 to over 2 m thick. Nodules may be isolated, stacked into vertical strings, or laterally linked into layers. Carbonate nodules are best developed in sandy mudstones but are present in any of the other lithologies. Some nodule layers contain abundant fossils including rhizoliths, invertebrate burrows, pupa cases, fecal pellets, shells of land snails, and reptile and mammal bones. Most of the fossiliferous nodule layers in the study area do not extend laterally for more than a few hundred meters. However, some lower Brule fossiliferous nodule layers extend for several kilometers, making prominent marker beds that reflect large areas of landsurface stability. The fossiliferous nodule layers are not evenly distributed throughout the White River Formation, but cluster into stratigraphic bundles (Figure 6.4). Lander and others (1992, in press) have suggested that these carbonate nodules formed at the groundwater table shortly after deposition.

The middle and upper Brule Member are characterized by very thick beds of homogeneous sandy siltstones. Glass shards in the siltstones are only slightly altered and unbroken, retaining fine needle-like terminations and bubble chambers. Siltstone beds are typically more than 1 m thick, internally structureless, and arranged in stacked sequences up to 28 m thick. Fossils are well preserved, but are rare and scattered. These siltstones represent ancient volcaniclastic loess deposits, as indicated by their homogeneous fine-grained lithology, abundance of delicate angular glass shards, lack of stratification within beds, sheet geometry, and general lack of associated channel deposits.

The middle and upper parts of the Brule Member also contain three very large cut-and-fill features that represent ancient valleys (Figure 6.4). These paleovalleys are characterized by scours, at least 20 m deep, filled with very large conglomeratic sandstone sheets, and medium to thick beds of alternating muddy sandstone and mudstone. The sides of the paleovalleys are marked by inclined nodular mudstone beds derived from mass movement of the surrounding sandy siltstones. The paleovalley fills were formed from a combination of fluvial, eolian, and mass movement processes (Evanoff, 1990).

GASTROPOD FAUNAS

Fossil nonmarine gastropods are abundant in the area and occur from the middle part of the Chadron Member to the middle Brule Member (Figure 6.5). Of 75 known fossil gastropod lo-

---

FIGURE 6.6. (opposite page) Revised magnetic polarity stratigraphy of the Douglas area, after Prothero (1982a, p. 18; 1985, Figure 6). Sections located as follows: "Wulff Ranch" - center, NW1/4 sec. 36, T. 32 N., R. 71 W., Irvine, WY (1949) 7 1/2 minute quadrangle; "Reno Ranch East" - W1/2N W1/4 sec. 32, and W1/2 sec. 29, T. 32 N., R. 70 W., Irvine, WY (1949) 7 1/2 minute quadrangle; "Reno Ranch S of Tower: - SW1/4 sec. 31, T. 32 N., R. 70 W., Irvine, WY (1949) 7 1/2 minute quadrangle; Dilts Ranch Composite - SE1/4 sec. 29, NE1/4 sec. 32, and E1/2 sec. 33, T. 31 N., R. 70 W., Irvine, WY (1949) and Dilts Ranch, WY (1949) 7 1/2 minute quadrangles; "Tripyramid"- SW1/4 NW1/4 NW1/4 sec. 30, T. 31 N., R. 70 W., Irvine, WY (1949) 7 1/2 minute quadrangle; "Sphinx" - SW1/4 SW1/4 NE1/4 sec. 24, T. 31 N., R. 71 W., Irvine, WY (1949) 7 1/2 minute quadrangle. The "Wulff Ranch", "Reno Ranch East", and "Reno Ranch S of Tower" sections are in the Morton Ranch area; the Dilts Ranch Composite, "Tripyramid", and "Sphinx" sections are in the Dilts Ranch area.

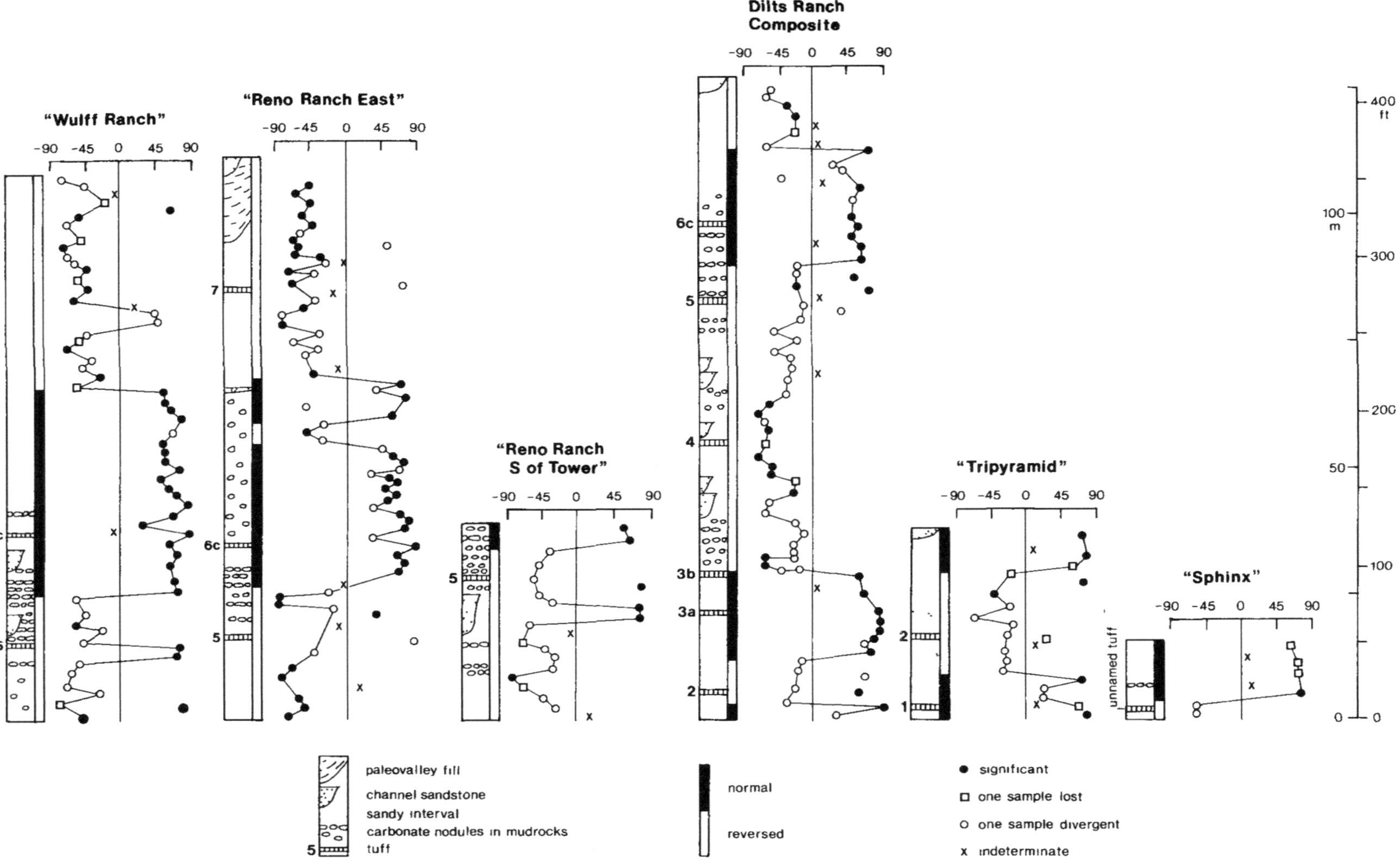
"Wulff Ranch"
"Reno Ranch East"
"Reno Ranch S of Tower"
Dilts Ranch Composite
"Tripyramid"
"Sphinx"
-90 -45 0 45 90
6c
5
7
4
3b
3a
2
1
unnamed tuff
400 ft
300
200
100
0
100 m
50
paleovalley fill
channel sandstone
sandy interval
carbonate nodules in mudrocks
tuff
normal
reversed
significant
one sample lost
one sample divergent
indeterminate

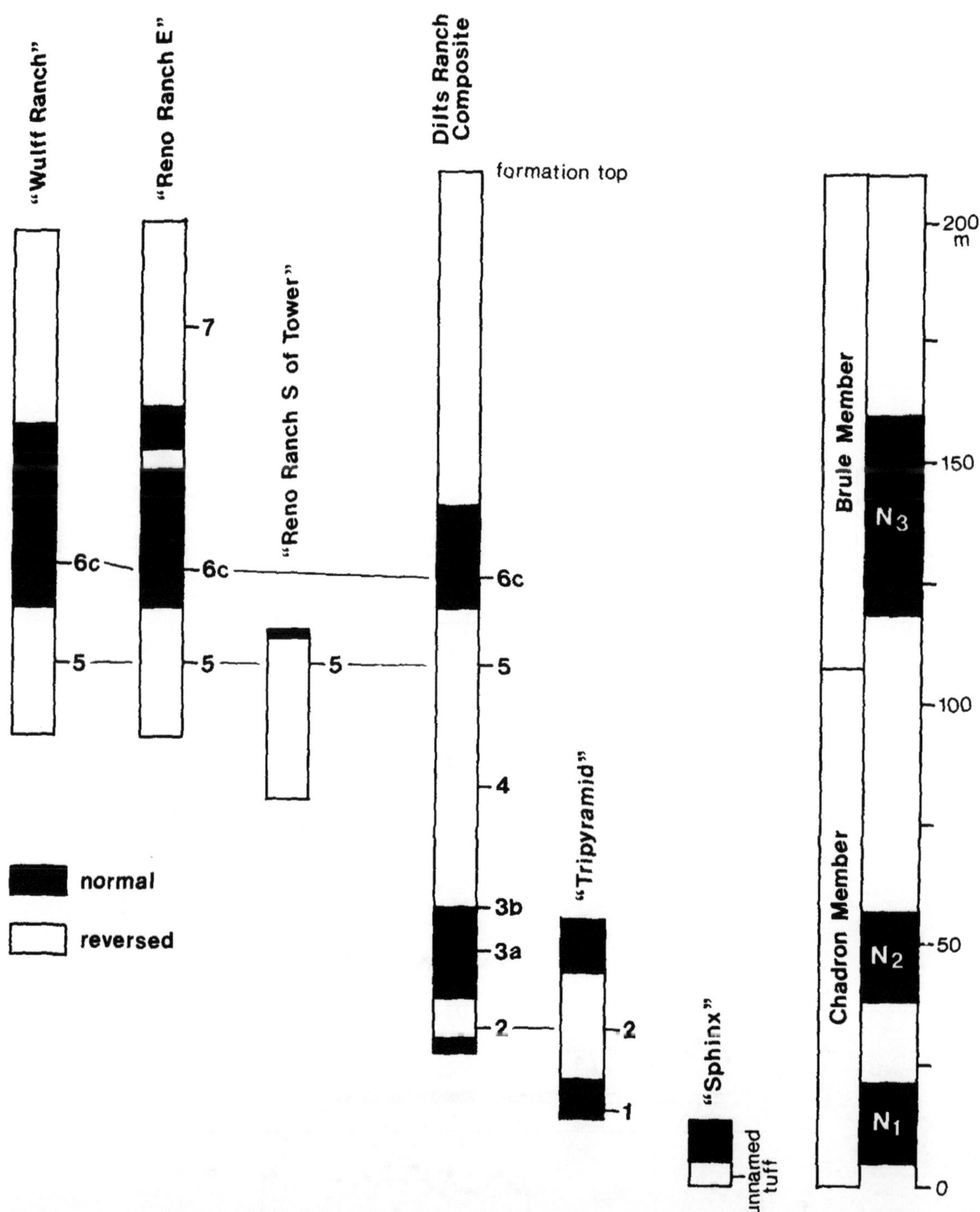

FIGURE 6.7. Revised magnetostratigraphy of the White River Formation in the Douglas area, based on correlations of tuffs and reanalysis of problematic intervals. According to the interpretation of Swisher and Prothero (1990) and Prothero and Swisher (this volume), the N_1 normal interval is Chron C16N, N_2 is Chron C15N, and N_3 is Chron C13N. The 5 ash is dated at 33.9 Ma, and lies just beneath the Chadronian-Orellan transition.

calities in the area, only one contains freshwater snails (Pulmonata: Basommatophora), indicating that habitats suitable for aquatic snails were rare. The other 74 localities contain only land snails (Pulmonata: Stylommatophora). All the taxa can be referred to modern families, and in most cases to modern genera (Evanoff, 1990). In general the taxa can be separated into groups by shell size: large-shelled taxa with diameters larger than 10 mm, and microshelled taxa with smaller diameters (Figure 6.5). The habitat and climatic implications of the fossil land snails are inferred from the habitats and climatic parameters of related modern analog taxa.

Land snail faunas of the Chadron Member are almost exclusively large-shelled forms (Humboldtianidae n. gen., *Ashmunella, Holospira*) which are now found in the central Mexican Plateau and the southern-most Rocky Mountains. The land snail fauna of the lower and middle Brule Member includes taxa now living in southern California and northern Baja California (*Xerarionta, Micarionta*, and *Radiocentrum*), and in the modern Rocky Mountains (*Ischnopupoides, Gastrocopta, Vallonia*, and *Radiocentrum*). Brule faunas were initially dominated by large-shelled forms, but microshell forms became much more abundant in the upper part of the lower Brule. The oldest paleovalley sequence of the middle Brule has the most diverse land snail fauna, and is dominated by large-shelled forms. This paleovalley fauna contains taxa typical of the lower and middle Brule, mixed with taxa now found in the Eastern United States (*Mesodon*?, *Mesomphix, Omphalina*) and in the northwest Rocky Mountains (*Polygyrella*).

The distributions of modern analogue taxa of the fossil land snails provide estimates of ancient temperatures and precipitation. For example, the distributions of modern *Humboldtiana, Holospira*, and *Ashmunella* indicate a mean annual temperature of 16.5°C and a mean annual precipitation of 450 mm during deposition of the Chadron Member (Evanoff, 1990). The mean annual temperature estimate is not far from the temperature estimate of 14°C from the correlative Florissant flora of Colorado (Meyer, 1986). Work is currently in progress to determine additional temperatue and precipitation estimates for the land snail faunas.

CLIMATIC IMPLICATIONS

The White River Formation in the Douglas area records a paleoclimatic shift toward drier conditions from the late Eocene to the early Oligocene. This increased drying is indicated by both the rocks and land snail faunas.

The most important depositional change in the area was the shift from a fluvial depositional system in the Chadron and lower Brule members to a predominantly eolian depositional system in the middle and upper Brule. This transition represents a paleoclimatic change from moist to dry conditions. The fluvial system which deposited the Chadron Member became less active during deposition of the lower Brule, as indicated by the decrease in channel sandstones. Eventually this fluvial system was not able to rework the volcaniclastic dust that continued to fall in the area, and thick beds of loess accumulated to form the middle and upper Brule Member. Dry conditions during loess deposition are indicated by rare channel sandstones of intermittent streams that were subject to flashy flow. The lack of extensive fluvial reworking and weathering of the loessites indicates that these dust deposits accumulated under dry, probably semiarid conditions, based on modern conditions of dryland loess accumulation (Pye and Tsoar, 1987; Pye, 1987). The alternation between loess deposition and paleovalley cuts and fills by streams suggests alternating dry and wet conditions during deposition of the middle and upper Brule Member.

The fossil gastropods mirror the paleoclimatic changes indicated by the rocks. Chadronian snails indicate a woodland habitat in a subtropical climate with seasonal precipitation. Snails of the lower Brule Member indicate open woodland habitats in a warm temperate climate with a pronounced dry season. The shift in the lower Brule from a fauna dominated by large-shelled taxa to a fauna dominated by microshelled taxa indicates a reduction in the amount of leaf-litter shelter, suggesting a shift from an open woodland to a bushland environment. The microshell-dominant fauna persists into the loessites of the middle Brule, suggesting that bushlands were the primary trap for dust, as they are in modern dryland sites (Gerson and Amit, 1987). The snail fauna of the lowest paleovalley indicates dense woodlands with moisture available for longer periods of time throughout the year. Whether this increase in moisture reflects increased precipitation or local environmental conditions within the valley is uncertain. In summary, the land snail faunas of the Douglas area suggest overall drying during deposition of the White River Formation, with only a minor decrease in temperature near the Eocene/Oligocene boundary.

The climatic changes indicated by the White River Formation near Douglas have also been recognized at other areas in the region. Increased drying from the late Eocene (Chadronian) through the middle Oligocene (Whitneyan) has been recognized in the sequences at the Big Badlands (Clark and others, 1967; Retallack, 1983, 1986) and northwest Nebraska (Harvey, 1960). Temperature indicators in these areas suggest a minor cooling trend from subtropical to warm temperate climates across the Chadronian/Orellan boundary. This is especially evident from the distribution of large, non-burrowing tortoises (*Hesperotestudo, Stylemys*) across the boundary. These large tortoises are abundant in Chadron and Brule rocks throughout the region and could not have withstood average winter temperatures below 13° C (Hutchison, 1982).

The timing of climatic events may not be the same over the entire region. For example, the fluvial-to-eolian transition occurs near Ash I (late Chadronian) at Flagstaff Rim; at the top of the lower Brule Member (end of the early Orellan) at Douglas; near the top of the Orella Member (end of the late Orellan) at northwest Nebraska; and in the upper Poleslide Member (late Whitneyan) in the central Big Badlands (Evanoff, unpublished field data). This trend suggests a west-to-east progression of dryland deposition, but this hypothesis needs to be tested with additional sedimentologic and stratigraphic studies. The stratigraphy, sedimentology, and paleomalacology of the Douglas area provide a valuable reference for such paleoclimatic work.

ACKNOWLEDGEMENTS

Evanoff's research was supported by funds from the University of Colorado Museum, the Geological Society of America, Sigma Xi, and the University of Colorado Department of Geological Sciences. Evanoff was assisted in the field by L. Ivy, D. Cunningham, G. Cunningham, and M. Gillam. Prothero's research was supported by NSF grant EAR87-08221, and a grant from the Petroleum Research Fund. D. Kron was very helpful in introducing Prothero to the stratigraphy of Dilts Ranch. The following people helped with the magnetic sampling: K. Gonzalez, H. Shlosar (1979); A. Kozak, A. Walton (1983); D. Gilchrist, K. Harris, A. Kozak (1986); J. Amato, J. Chean, J. Finnegan (1988). Prothero thanks J. Kirschvink for graciously allowing the use of his paleomagnetics lab. R.J. Emry, D.L. Hanneman, and J.B. Swinehart reviewed the manuscript, and made helpful suggestions. D. Kron provided comments on the stratigraphic ranges of vertebrates given in Figure 6.5. We also thank the landowners and lessees in the Douglas area for allowing access to their property.

REFERENCES

Blackstone, D. L. Jr. 1988. Thrust faulting: southern margin Powder River Basin, Wyoming. *Wyo. Geol. Assoc. 39th Ann. Field Conf. Guidebook*, pp. 35-44.

Berggren, W. A., Kent, D. V., Flynn, J. J., and Van Couvering, J. A. 1985. Cenozoic geochronology. *Geol. Soc. Amer. Bull.* 96: 1407-1418.

Burchett, R.R. 1986. Geologic bedrock map of Nebraska. *Nebraska Geol. Surv. Map*, scale 1:1,000,000.

Clark, J. 1937. The stratigraphy and paleontology of the Chadron Formation in the Big Badlands of South Dakota. *Ann. Carnegie Mus.* 25: 261-351.

Clark, J., Beerbower, J. R., and Kietzke, K. K. 1967. Oligocene sedimentation in the Big Badlands of South Dakota. *Fieldiana Geol. Mem.* 5: 1-158.

Darton, N. H. 1899. Preliminary report on the geology and water resources of Nebraska west of the one hundred and third meridian. *U.S. Geological Survey 19th Annual Report*, 4: 721-785.

Denson, N. M., and Horn, G. H. 1975. Geologic and structure map of the southern part of the Powder River Basin, Converse, Niobrara, and Natrona counties, Wyoming. *U.S. Geol. Survey Misc. Inv. Series Map* I-877, 2 sheets, scale 1:125,000.

Edwards, L. E. 1984. Insights on why graphic correlation (Shaw's method) works. *Jour. Geol.* 92: 583-597.

Emry, R. J. 1973. Stratigraphy and preliminary biostratigraphy of the Flagstaff Rim area, Natrona County, Wyoming. *Smithsonian Contrib. Paleobiol.* 18: 1-43.

Emry, R. J., Bjork, P. R., and Russell, L. S. 1987. The Chadronian, Orellan, and Whitneyan North American Land Mammal Ages, In Woodburne, M. O., ed., *Cenozoic mammals of North America.* Berkeley, University of California Press, pp. 118-152.

Evanoff, E. 1990. Late Eocene and early Oligocene paleoclimates as indicated by the sedimentology and nonmarine gastropods of the White River Formation near Douglas, Wyoming [Ph.D. thesis]. Boulder, University of Colorado, 440 pp.

Galbreath, E. C. 1953. A contribution to the Tertiary geology and paleontology of northeastern Colorado. *Univ. Kansas Paleont. Contrib., Vertebrata* 4: 1-120.

Gerson, R., and Amit, R. 1987. Rates and modes of dust accretion and deposition in an arid region - the Negev, Israel, In Frostick, L., and Reid, I., eds., *Desert sediments: ancient and modern. Geol. Soc. London Spec. Publ.* 35: 157-169.

Hanneman, D. L. 1988. Depositional systems of Cenozoic basins, southwestern Montana. *Geol. Soc. Amer. Abst. Prog.* 20 (7): A315-A316.

Hanneman, D. L. 1989. Cenozoic basin evolution in a part of southwestern Montana [Ph.D. thesis]. Missoula, University of Montana, 347 pp.

Harvey, C. 1960. Stratigraphy, sedimentation, and environment of the White River Group of the Oligocene of northern Sioux County, Nebraska [Ph.D. thesis]. Lincoln, University of Nebraska, 151 pp.

Hutchison, J.H. 1982. Turtle, crocodilian, and champsosaur diversity changes in the Cenozoic of the north-central region of western United States. *Palaeogeogr., Palaeoclimat., Palaeoecol.* 37: 149-164.

King, P. B., and Beikman,H. M. 1974. *Geologic map of the United States.* U.S. Geological Survey, 3 sheets, scale 1:2,500,000.

Korth, W. W. 1989. Stratigraphic occurrence of rodents and lagomorphs in the Orella Member, Brule Formation (Oligocene), northwestern Nebraska. *Univ. Wyoming Contrib. Geol.* 27: 15-20.

Lander, R. H. 1985. Origin, pedogenesis, and post-pedogenic alteration of White River marker beds [Undergraduate thesis]. Galesburg, Illinois, Knox College, 102 pp.

Lander, R. H. 1987. Zeolitic alteration of tuffaceous White River Group sediments (Oligocene, western U.S.A.) *Proc. 24th Ann. Mtg. Clay Min. Soc.* Socorro, New Mexico, p. 84 (abs.).

Lander, R. H. 1991, White River Group diagenesis. [Ph.D. thesis]. Urbana, University of Illinois, 143 pp.

Lander, R. H., Evanoff, E., Anderson, T. F., and Hay, R. L. 1992. Soil or groundwater origin for White River Group calcrete? *Geol. Soc. Amer. Bull.*, in press.

Love, J. D., and Christiansen, A. C. 1985. *Geologic map of Wyoming.* U.S. Geological Survey, 3 sheets, scale 1:500,000.

Meyer, H.W. 1986. An evaluation of the methods for estimating paleoaltitudes using Tertiary floras from the Rio Grande Rift vicinity, New Mexico and Colorado. [Ph.D. thesis]. Berkeley, University of California, 230 pp.

Montanari, A., Deino, A. L., Drake, R. E., Turrin, B. D., DePaolo, D. J., Odin, G. S., Curtis, G. H., Alvarez, W., and Bice, D. M. 1988. Radioisotopic dating of the Eocene-Oligocene boundary in the pelagic sequence of the northeast Apennines, In Premoli-Silva, I., Coccioni, R., and Montanari, A., eds., *The Eocene-Oligocene boundary in the Marche-Umbria Basin (Italy).* Ancona, Italy, International Union of Geological Sciences, Commission on Stratigraphy, pp. 195-208.

Mutschler, F. E., Larson, E. E., and Bruce, R. M. 1987. Laramide and younger magmatism in Colorado - new petrologic and tectonic variations on old themes. *Colo. Sch. Mines Quart.* 82: 1-47.

Prothero, D. R. 1982a. Medial Oligocene magnetostratigraphy and mammalian biostratigraphy: testing the isochroneity of mammalian biostratigraphic events (Ph.D. dissertation). New York, Columbia University, 284 pp.

Prothero, D. R. 1982b. How isochronous are mammalian biostratigraphic events? *3rd N. Amer. Paleont. Conv., Proc.* 2: 405-409.

Prothero, D. R. 1985. Chadronian (early Oligocene) magnetostratigraphy of eastern Wyoming: implications for the age of the Eocene-Oligocene boundary. *Jour. Geol.* 93: 555-565.

Pye, K. 1987. *Aeolian Dust and Dust Deposits.* London, Academic Press, 334 pp.

Pye, K., and Tsoar, H. 1987. The mechanics and geological implications of dust transport and deposition in deserts with particular reference to loess formation and dune sand diagenesis in the northern Negev, Israel, In Frostick, L., and Reid, I., eds., *Desert sediments: ancient and modern. Geol. Soc. London Spec. Publ.* 35: 139-156.

Retallack, G. J. 1983. Late Eocene and Oligocene paleosols from Badlands National Park, South Dakota. *Geol. Soc. Amer. Spec. Paper* 193, 82 pp.

Retallack, G. J. 1986. Fossil soils as grounds for interpreting long-term controls on ancient rivers. *Jour. Sed. Petrol.* 56: 1-18.

Schultz, C. B., and Stout, T. M. 1955. Classification of Oligocene sediments in Nebraska. *Univ. Neb. State Museum Bull.* 4: 17-52.

Skinner, M. F. 1958-1959. Unpublished field notebook volume 6: New York, American Museum of Natural History, 100 p.

Swisher, C. C., III, and Prothero, D. R. 1990. Laser-fusion $^{40}Ar/^{39}Ar$ dating of the Eocene-Oligocene transition in North America. *Science* 249: 760-762.

Tweto, O. 1979. *Geologic map of Colorado.* U.S. Geological Survey, 2 sheets, scale 1:500,000.

Wanless, H. R. 1923. The stratigraphy of the White River beds of South Dakota. *Proc. Amer. Phil. Soc.* 62: 190-269.

Wood, H. E., Chaney, R. W., Clark, John, Colbert, E. H., Jepsen, G. L., Reeside, J. B., Jr., and Stock, C. 1941. Nomenclature and correlation of the North American continental Tertiary. *Geol. Soc. Amer. Bull.* 52: 1-48.

7. EVIDENCE FROM THE ANTARCTIC CONTINENTAL MARGIN OF LATE PALEOGENE ICE SHEETS: A MANIFESTATION OF PLATE REORGANIZATION AND SYNCHRONOUS CHANGES IN ATMOSPHERIC CIRCULATION OVER THE EMERGING SOUTHERN OCEAN?

by Louis R. Bartek , Lisa Cirbus Sloan, John B. Anderson, and Malcolm I. Ross

ABSTRACT

A review of the literature and an analysis of Deep Freeze 87, Polar Duke 90 and published data indicates that the first major ice sheet grounding event occurred in the Ross Sea during late Oligocene time. Correlation of a pronounced unconformity in seismic data, to drill cores indicates a shift from temperate/ temperate-glacial to polar-glacial conditions in late Oligocene time. This unconformity is present over an extensive portion of the Ross Sea (at least 100,000 km^2) and suggests that the extent of this late Oligocene grounding event may have been comparable with that of Wisconsinan Glacial Maximum. The onset of this glacial activity in Antarctica is significant because it may have had tremendous impact on global paleoclimate, eustacy, and biotic evolution.

In the recent past, researchers have inferred, from indirect evidence (ice-rafted debris fluctuations, oxygen isotope curves, and deep sea hiatuses), that the Cenozoic ice sheets did not form on Antarctica until Miocene time, and it has been suggested that an ice sheet did not ground on the Ross Sea continental shelf until late Pliocene/Pleistocene time. They also inferred that Antarctica was relatively warm until Miocene time, when its climate supposedly cooled and ice sheets formed in response to the initiation of deep circum-polar oceanic circulation following the opening of Drake Passage (22 Ma.). However, the correlations described above, along with reinterpretations of some of the indirect evidence indicate that continental scale ice sheets were developing by late Paleogene time rather than during the Miocene. Identification of glacial deposits in Paleogene outcrops (on the Antarctic Peninsula) and drill cores from the continental margin (e.g., Leg 119 in Prydz Bay), corroborates an "early" Cenozoic initiation of continental scale glaciation in Antarctica. In light of this evidence, a mechanism that is not related to the the opening of Drake Passage and the initiation of deep circum-polar oceanic circulation is presented.

Antarctica has been cold enough to form ice sheets ever since it moved into a polar position, and the limiting factor on ice sheet development was the availability of significant quantities of moisture to form an ice sheet. Results from parametrically modelled simulations of Southern Ocean paleoclimates suggest that the development of the East Antarctic ice sheet may have been triggered by the opening of a seaway between Antarctica and Australia during late Cretaceous/early Paleogene time. The development of a seaway between the two continents decreased the continentality of the region and resulted in onshore transport of moist air over Antarctica. Transport of moisture over the cold Antarctic continental landmass resulted in precipitation and the formation of the Antarctic ice sheets by at least late Paleogene time. Widespread erosion of the continental shelf and shelf overdeepening occurred when the ice advanced onto the Ross Sea continental shelf during late Oligocene time. It is hypothesized that metastable, marine-based ice sheets have waxed and waned on the Ross Sea continental shelf since the Oligocene grounding event. These results are at least in part supported by results from an atmospheric general circulation model.

INTRODUCTION

The onset of extreme glacial conditions in Antarctica and the ensuing glacial history of the continent have been the objects of controversy for a number of years. The controversy arises from the fragmented nature of the available data bases. Ninety-seven percent of the continent is shrouded in ice, so researchers are often faced with disjointed bits of information that are available in only one part of the continent (or ocean basin) and that frequently represent only a small segment of the stratigraphic record from that geographic area. Another factor that has contributed to the controversy is the variety of data bases that are being studied. There are researchers analyzing stratigraphic hiatuses, oxygen isotope, and ice-rafted debris (IRD) records from the deep sea. Others are studying the biostratigraphy, sedimentology, and seismic stratigraphy of cores and seismic data collected from the continental shelf of Antarctica, and there is yet another group that is trying to decipher the glacial history of Antarctica by analyzing the Cenozoic rocks that are exposed on the continent itself. Many of the deep sea data bases are thought to be proxies for Antarctice ice volume, but they are several steps removed from the continent, and the systematics of these proxies are still not completely understood (Anderson and Arthur, 1983; Frakes, 1983; Anderson, 1986). Interpretations of the glacial history of Antarctica based solely upon the deep sea record may be flawed by our poor understanding of the complex relationships between variations in the parameters that are measured in deep sea cores and fluctuations in Antarctic ice volume. Therefore, this paper focuses primarily upon: 1) interpreting the early history of the Antarctic ice sheets from the data bases that are available from the continent and the seismic stratigraphy of cores and seismic data collected from the Antarctic continental margin, and 2) presenting a mechanism for the formation of extensive ice sheets prior to the initiation of a deep circum-Antarctic current. An important caveat to this paper is that our knowledge of the stratigraphy of the Antarctic margin is still rather limited due to extensive ice coverage and paucity of long cores in the older portion of the Cenozoic stratigraphic record. More core from the Antarctic margin is needed to fully test the hypotheses that are presented in this paper.

PALEOGENE ICE SHEETS IN ANTARCTICA: EVIDENCE PRO AND CON

Recent drilling on the Antarctic margin and analysis of outcrop stratigraphy have provided direct evidence of glacial conditions in Antarctica by late Paleogene time. Tills discovered in the Ocean Drilling Program (ODP) Leg 119 cores from the Prydz Bay region indicate that glacial conditions existed on the East Antarctic continental shelf since middle Eocene-Oligocene time (Barron et al., 1989). Widepread Paleogene glacial activity in Antarctica is also suggested by: 1) Birkenmajer's (1987a, b; Birkenmajer et al., 1987), and Porebski and Gradzinski 's (1987) documentation of tills in the South Shetland Islands of West Antarctica, which are capped by andesitic lavas that have been K-Ar dated at more than 49.4 Ma; 2) LeMasurier and Rex's (1982) interpretation of an ice cap on Marie Byrd Land based upon the presence of extensive Oligocene hyaloclastite deposits; and 3) the contention of Stump et al. (1980) of development of an East Antarctic ice sheet prior to 19 Ma based upon the discovery of two large subglacially erupted volcanoes in the Transantarctic Mountains.

The stratigraphy of the Ross Sea continental margin is another important indicator of the onset extensive glacial conditions in Antarctica. It is an area that is roughly the size of Texas and has been (Barrett, 1975), and presently is, the recipient of a significant porportion (nearly a third of the ice sheet drainage of Antarctica) of East and West Antarctic ice sheet drainage. Evidence of widespread glacial erosion and subglacial deposition on the Ross Sea margin are therefore indicators of large scale ice sheet development in Antarctica. In the Ross Sea sector of Antarctica widespread glacial conditions during Paleogene time are indicated by the

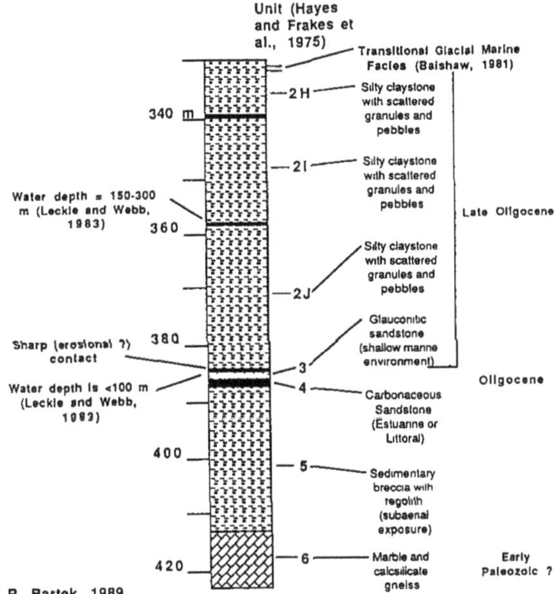

FIGURE 7. 1. This figure presents an expanded view of DSDP Site 270 (from 330-422.5 m). It is a compilation of data from Hayes and Frakes et al. (1975); Balshaw (1981); and Leckie and Webb (1983) and demonstrates that the change in sedimentation styles recorded at DSDP Site 270 is not merely a manifestation of normal basin subsidence.

stratigraphy of the McMurdo Sound Sediment and Tectonic Studies (MSSTS)-1, Cenozoic Investigations of the Ross Sea (CIROS)-1 and Deep Sea Drilling Project (DSDP) Leg 28 cores and seismic stratigraphic relationships. In the southwestern corner of the Ross Sea, at the CIROS-1 site (Figure 7.2), Upper Oligocene tills and striated and faceted pebbles within Lower Oligocene sediment indicate Paleogene glaciation in this portion of the Ross Sea (Barrett et al., 1989; and Hambrey et al., 1989). In the central Ross Sea, at DSDP Site 270, the transition from temperate-littoral conditions when carbonaceous sand and greensand (units 3 and 4 of Hayes and Frakes et al., 1975) were deposited (late Oligocene) to cold, deep-water (≈500 meters) conditions when glacial marine sediments (units 2j-2a of Hayes and Frakes et al., 1975) were deposited (late Oligocene to Recent) indicates Paleogene glacial activity in this region (Figure 7.1). Hayes and Frakes et al. (1975) state that the core and seismic data from the central Ross Sea indicate a gradual deepening of Site 270 and the central Ross Sea from subaerial exposure under more temperate climatic conditions during Oligocene time, when the breccia that has been interpreted as a talus deposit (unit 5) was deposited, to a depth of 500 meters under cold climatic conditions when the glacial marine sediments (units 2j-2a) were deposited (late Oligocene to Recent). A similar climatic change has been change documented in the Oligocene deposits recovered at CIROS-1 (Hambrey et al., 1989; and Harwood et al., 1989). It is important to note that the results of the early investigations of the glacial history, based solely on the older data sets, suggested to Hayes and Frakes and others (1975) that the grounding of an ice sheet in the central Ross Sea did not occur until about 4.5 to 5.0 million years ago. However, the possibility that ice was grounded on the central and eastern Ross Sea continental shelf before Pliocene time must be given strong consideration in light of the evidence supplied by subsequent investigations.

Correlations from the drill cores to seismic data are very important because they help to establish the regional scale of the glacial events. Unfortunately, subsurface conditions and structural complexity make it difficult to tie seismic data to the excellent climatic and chronologic data bases of CIROS-1 and MSSTS-1. Therefore, the focus of this paper will be on the central and eastern portions of the Ross Sea continental margin where correlations between core and seismic data facilitate a demonstration of the regional extent of Paleogene and Neogene glacial events in Antarctica.

There are two lines of evidence that support the possibility of a grounded ice sheet on the central and eastern Ross Sea continental shelf in Paleogene time: 1) rapid changes in sedimentologic and paleontologic characteristics of sediment across a mid- [used in this paper to refer to latest early Oligocene through earliest late Oligocene time] to late Oligocene unconformity, and 2) seismic evidence of erosive surfaces that lie too far below the sea floor to correlate to the Pliocene event. The shift from deposition under shallow water continental shelf conditions, as indicated by the glauconitic greensand

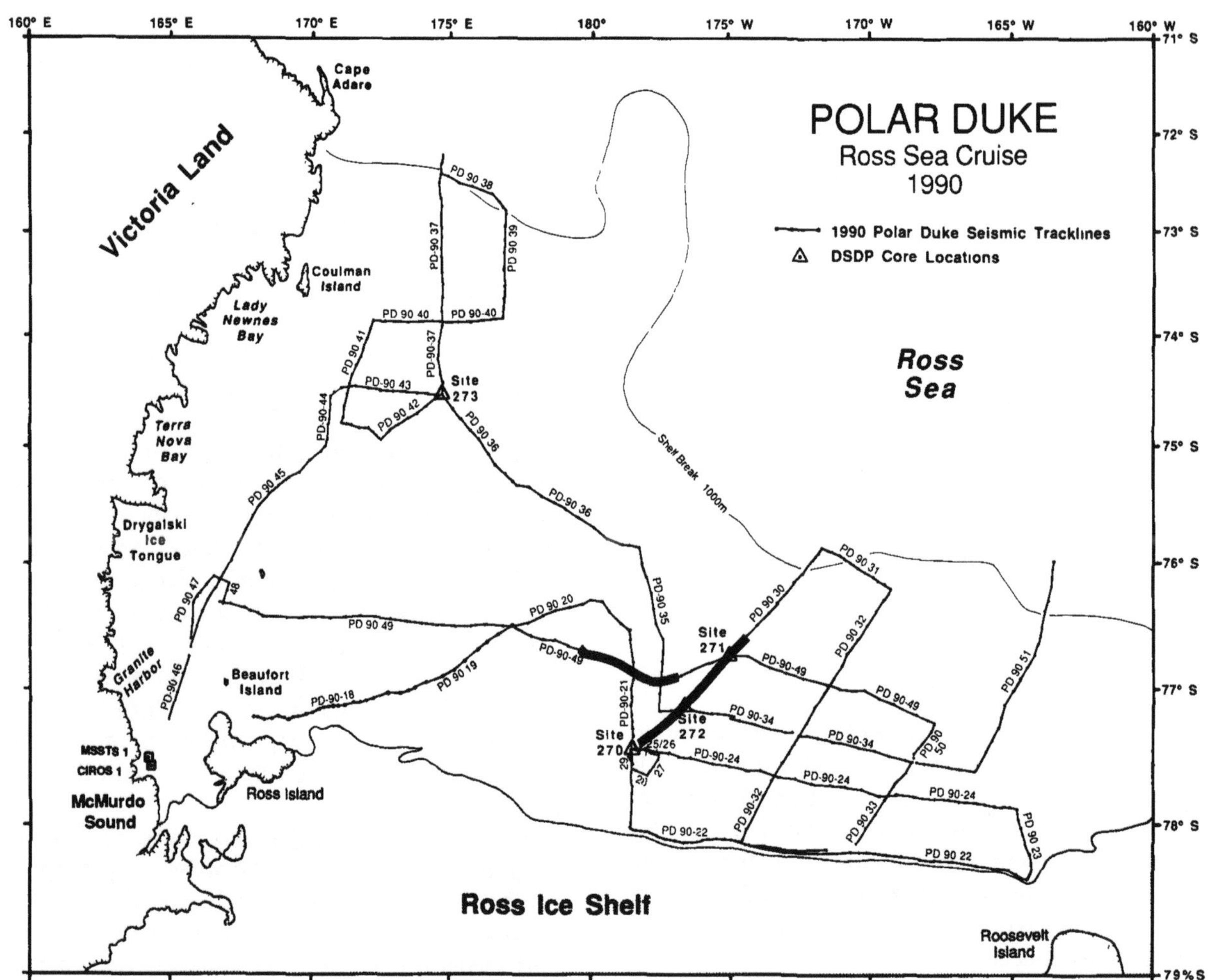

FIGURE 7. 2. Location of Polar Duke 90 (PD-90) seismic tracklines and DSDP core locations in the Ross Sea, Antarctica. The heavier lines along PD-90-30 and PD-90-49 indicate the location of the profiles that are displayed in Figures 7.3 and 7.4.

deposited in the basal portion of Site 270, to deposition under glacial conditions, as indicated by the diamictites that unconformably overlie the greensand, without the presence of a transitional facies or interbedding between the two units, suggests an erosive contact between the greensand and the overlying diamictites (Figure 7.1). In addition to the dramatic sedimentologic change that has been documented between units 3 and 2-I at Site 270 by Hayes and Frakes et al. (1975), Leckie and Webb (1983) have documented a shift from benthic foraminiferal assemblages that are commonly associated with a shallow water inner shelf setting, to deep water assemblages typical of an outer shelf or upper slope environment (150-300 m). The change in foraminiferal assemblages occurs over a stratigraphic thickness of less than 30 meters (385-379 m) and there may be a hiatus of 2 million years at the contact between the greensand and the basal diamictite (Allis, 1975; Hayes and Frakes et al., 1975; and McDougall, 1976). It has been argued that these changes are merely a manifestation of subsidence (Leckie and Webb, 1983). However, the sharp contact between the shallow water greensands and the overlying glacial diamictites, the increase in water depth, and the

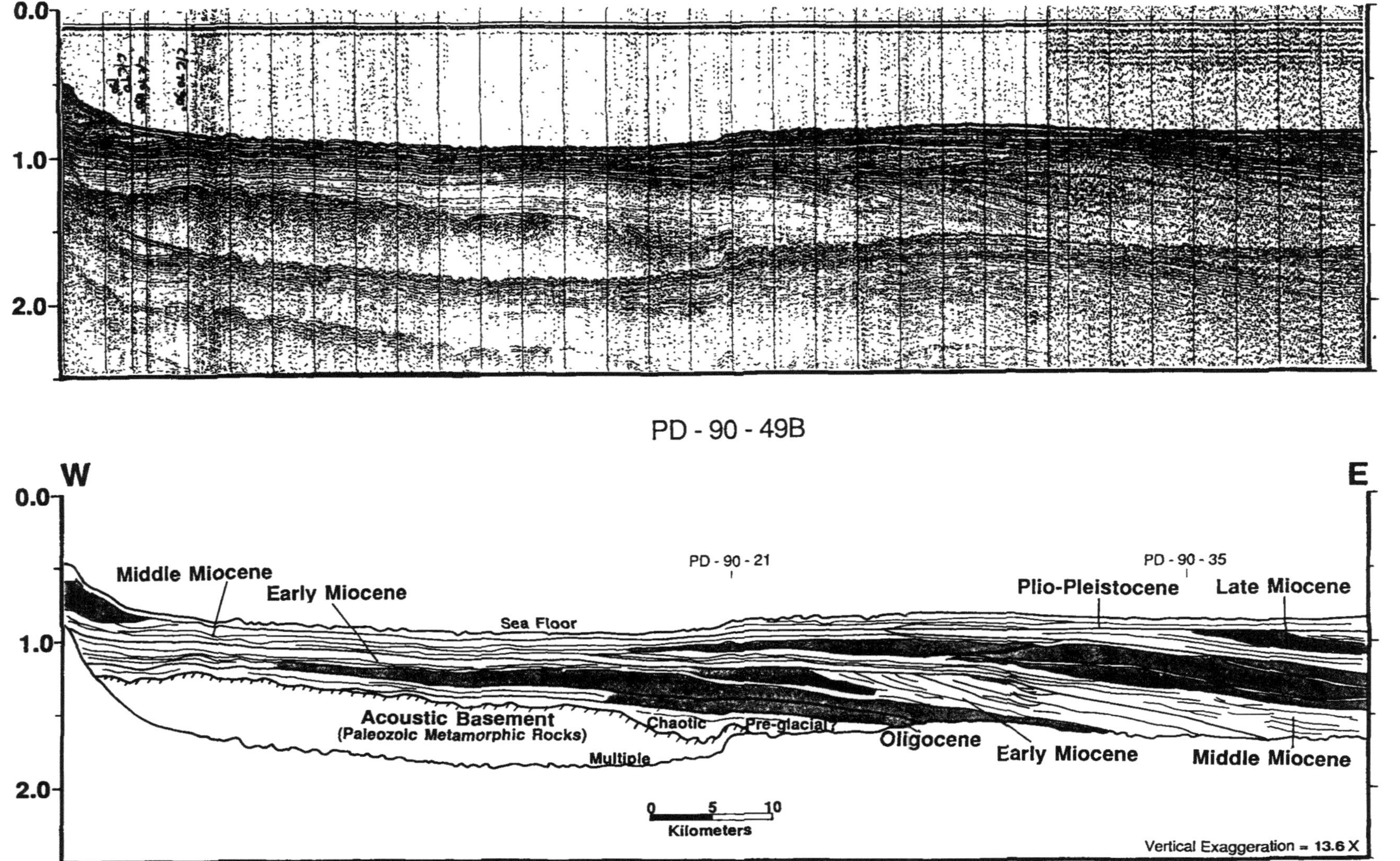

FIGURE 7. 3. Portion of seismic line PD-90-49 and line drawing illustrating the presence of subglacial seismic facies as old as late Oligocene. The location of the profile is indicated by the heavier line along PD-90-49 in Figure 7.2. Note the massive intervals and the cross cutting relationships of the Neogene strata in this strike section. These are features that are similar to the till tongue stratigraphy described by King and Fader (1986).

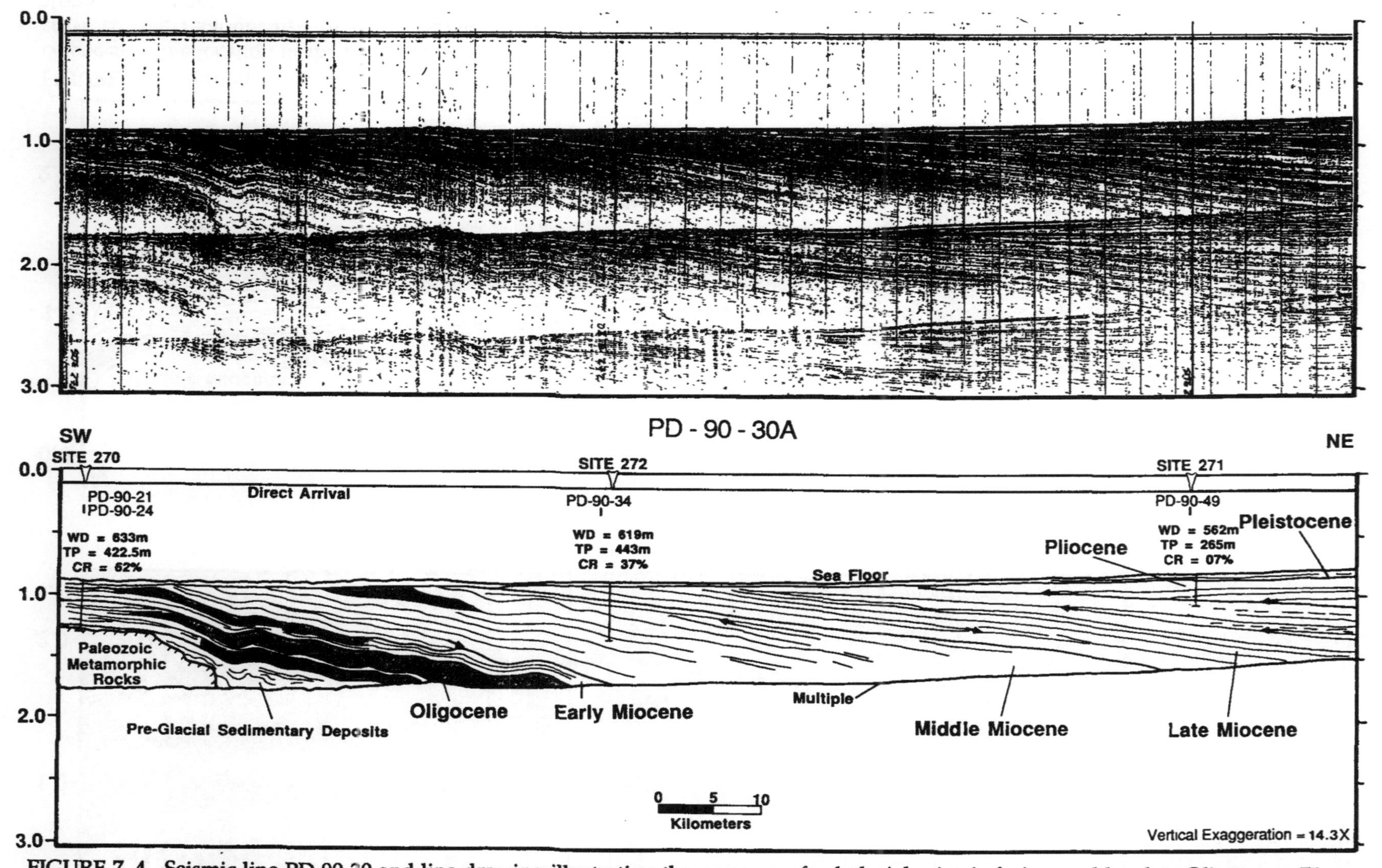

FIGURE 7. 4. Seismic line PD-90-30 and line drawing illustrating the presence of subglacial seismic facies as old as late Oligocene. The location of the profile is indicated by the heavier line along PD-90-30 in Figure 7.2. The multiple obscures a portion of the section, but one can see the aggradational and progradational patterns of the margin in this area.

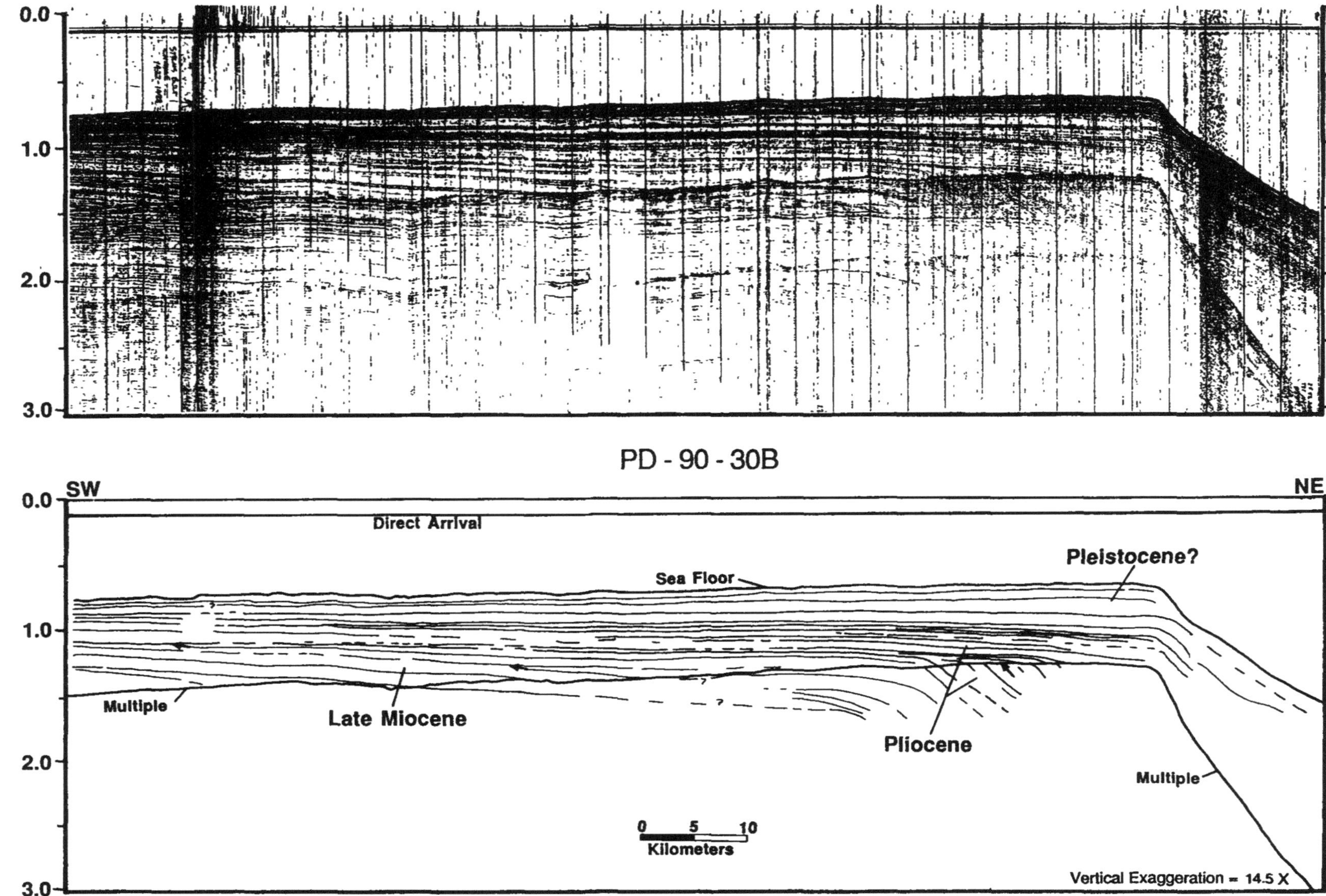
0.0
1.0
2.0
3.0
PD - 90 - 30B
SW
NE
Direct Arrival
Sea Floor
Pleistocene?
Multiple
Late Miocene
Pliocene
Multiple
0 5 10
Kilometers
Vertical Exaggeration = 14.5 X

stratigraphic hiatus, may also imply grounding of an ice sheet and coincident removal of up to 250 meters of sediment by glacial erosion during mid- to late Oligocene time in the central Ross Sea.

Further evidence of a large scale, glacial-erosional event is found in seismic data from the central and eastern Ross Sea. Recent seismic studies on the Canadian shelf (King and Fader, 1986), in the North Sea (King et al., 1987) and at Ice Stream B in Antarctica (Alley et al., 1989) provide important information about the nature of seismic facies associated with both sub-ice sheet and glacial marine sequences. These results provide a rough framework for seismic studies in Antarctica aimed at deciphering the history of glaciation on the continental shelf in areas, such as the Ross Sea, where drill sites provide needed chronostratigraphic information. Velocity analysis of the strata that infill the Central and Eastern Basins of the Ross Sea via sonobuoy refraction surveys (Hayes and Frakes et al., 1975; Hinz and Block, 1983; and Sato et al., 1984) facilitates correlation of these surfaces to DSDP sites 270-272.

A preliminary analysis of high-resolution seismic data collected during Leg 2 of the voyage of *R/V Polar Duke* in the Ross Sea indicates that an ice sheet was grounded on the Ross Sea continental shelf by at least late Oligocene time (Bartek and Anderson, 1990; Anderson and Bartek, in press). The first occurrence in upper Oligocene strata of glacial erosional surfaces and till tongues (Figures 7.2 and 7.3) that are similar to those described by King and Fader (1986) and King et al. (1987) indicates that the grounding event occurred by late Paleogene time. The cut and fill stratigraphy that is seen in the strike section displayed in Figure 7.3 indicates that the ice sheets waxed and waned following the late Paleogene grounding event (Bartek et al., in press). The width of the scours associated with the till tongues is comparable with the width of modern ice streams. The massive tongue-like units in Figure 7.3 are at least 70 to 80 kilometers wide; modern ice streams have widths on the order of 80 to 100 kilometers (Alley et al., 1989). The widths of the trough-like scours on the Ross Sea continental shelf also exceeds the widths of incisions carved by fluvial entrenched valley systems. Incisions carved by fluvial systems are typically 10 to 20 kilometers wide (Berryhill, 1986; Thomas and Anderson, 1989).

The lateral extent of the mid- to late Oligocene ice sheet grounding event is indicated by the work of Hinz and Block (1983), and Barrett et al. (1989). Hinz and Block (1983) correlate an unconformity (U6) to the mid- to late Oligocene boundary between the preglacial greensands and the overlying glacial deposits at site 270 and they mapped it over an extensive portion (≈100,000 km^2) of the Central and Eastern Basins of the Ross Sea (Figure 5 in Hinz and Block, 1983). A mid-Oligocene unconformity has also been identified in the western Ross Sea by Harwood et al. (1989) at the CIROS-1 site. The lateral extent of this unconformity in the western Ross Sea is still unresolved. The U6 unconformity is interpreted by Hinz and Block (1983) as a product of erosion associated with a paleoceanographic event rather than ice sheet grounding. Hinz and Block (1983) suggest that the grounding of ice sheets did not play an important role in producing the numerous unconformities that are present in the central Ross Sea until unconformity U1 time (Pliocene?). However, the recently acquired high resolution seismic data suggest that the extensive shelf aggradation and progradation that followed the mid- to late Oligocene grounding event (Figure 7.4) was also intimately associated with the waxing and waning of meta-stable, marine-based ice sheets on the Ross Sea continental shelf (Figure 7.3). The ice grounding stratigraphy appears to have been produced by two mechanisms. The till tongue-like stratigraphy, which appears to be the most prevalent, may have been produced by the buoyancy line migration mechanism discussed by King and Fader (1986). In this model, tongues of till are deposited on the shelf as wet-based ice sheets wax and wane. Data from the eastern side of the Eastern Basin of the Ross Sea also show evidence of the subglacial delta stratigraphy discussed by Alley et al. (1989) and indicate that subglacial deltaic deposition may also be an important mechanism producing the ice sheet grounding stratigraphy of the Ross Sea.

Other evidence supporting the idea of pre-Pliocene ice sheet grounding events in the Central Ross Sea is found in Balshaw's (1981) analysis of DSDP Leg 28 cores from the Ross Sea. Balshaw analyzed the grain size distributions of the massive and stratified diamictites that comprise unit 2 of Sites 270-273. She classified a sample from approximately 331 meters sub-bottom at Site 270 as transitional glacial marine (Figure 7.1). Transitional glacial marine facies are deposited at or near the grounding lines of ice sheets (Anderson et al., 1983) and the presence of such a unit at 331 meters sub-bottom (late Oligocene - Early Miocene) at Site 270 indicates that an ice sheet was grounded on the continental shelf in the central Ross Sea before Pliocene time. It also must be something more than mere serendipity that one of the largest Cenozoic sea level falls (Haq et al., 1987) occurs during the late Oligocene. The presence of this unconformity and eustatic fall has also been documented in the southwestern Ross Sea by Barrett et al. (1989) at CIROS-1. The sudden change of lithologies across the late Oligocene unconformity at Site 270, the presence of extensive late Oligocene glacial-erosional unconformity, and the large fall in eustacy all support the concept of an expansion of Antarctic ice sheets and an extensive ice sheet grounding event on the Ross Sea continental shelf in mid- to late Oligocene time.

Corroborating evidence for Paleogene ice sheets can be found in: 1) Margolis and Kennett's (1971) discovery of IRD in Eocene deep sea sediments from the southern Pacific; 2) Breza and others' (1989) detection of IRD in lower Oligocene sediment from the Kerguelen Plateau; 3) re-interpretations of the oxygen isotope record by Matthews and Poore (1980), Keigwin and Corliss (1986), Miller and others (1987), and Prentice and Matthews (1988) which indicate that ice volume has effected the $\delta^{18}O$ record since at least the Eocene/Oligocene boundary; 4) the eustatic sea level curves published by Haq and others (1987) which indicate that sea level has been fluctuating at rates that are associated with glacial waxing and waning during the Cenozoic and late Mesozoic rather than the lower rates that are normally associated with tectonic activity alone; 5) the numerical modelling of Cretaceous and Eocene climates by Barron et al. (1981), Barron (1987), and Sloan and Barron (1990) which indicates that it is difficult to produce an equable, globally above-freezing climate (as is usually inferred for the Cretaceous and early Paleogene) at middle and high latitudes, with known forcing mechanisms of reasonable magnitude.

As stated in an earlier section, the interpreted glacial history of Antarctica is based upon diverse data bases and is very controversial. Authors who have argued against the possibility of major pre-Miocene Antarctic glaciation (Shackleton and Kennett, 1975; Kennett, 1977; Savin et al., 1975; Mercer, 1983; Kennett, 1983; and Barker and Kennett et al., 1988) have done so based upon the oxygen isotope record, the presence of terrestrial palynomorphs, paleobiogeography, the presence of smectite and kaolinite in deep sea sediments, and plate tectonic constraints on paleocirculation of the southern ocean. Based upon analyses of oxygen isotope data from DSDP cores from the South Tasman Rise and the Campbell Plateau, Shackleton and Kennett (1975) argued that there has been Antarctic glacial activity and formation of cold bottom water in the southern Ocean since late Eocene time. However, they believe that the formation of a large ice sheet in Antarctica did not occur until early middle Miocene. They base the timing of the ice sheet formation upon a large positive shift in the oxygen isotope record that occurs in early middle Miocene time. Shackleton and Kennett also argue that the ice sheet has been stable since that time. Maintenance of low bottom water temperatures, as indicated by oxygen isotope data, while the temperatures of late Miocene surface water rose significantly, suggests to Shackleton and Kennett that cold bottom water was continually produced in the presence of a stable ice sheet.

It has also been argued that the Oligocene palynomorphs recovered by Kemp and Barrett (1975) from glacial marine sediments in the Ross Sea preclude the possibility of an Oligocene ice cap in Antarctica (Kennett, 1977). However, the recovery of marine fauna and flora (Webb et al., 1984) as well as fossil wood (Webb and Harwood, 1987) from the Sirius Group indicates that the volume of Antarctic ice sheets may have fluctuated dramatically, and that it may indeed be possible to reforest the once glaciated landscape. The fossil wood also suggests that there may have been enclaves where vegetation was able to endure the harsh conditions of the glacial maxima. Although this possibility seems less likely during intervals when ice was able to expand all the way out to the shelf edge.

Paleobiogeographic arguments for a Neogene timing of Antarctic ice sheet formation center upon the location of siliceous and calcareous sediment belts around Antarctica. It is believed that the expansion of the siliceous sediment belt is proxy for the cooling of the waters around Antarctica and associated ice sheet development. However, as Kennett (1977) points out, "it is not known whether this sedimentary boundary reflects the same water mass boundary or temperature changes for the Cenozoic as it does at the present-day Antarctic Convergence". Additionally, the ancient boundary separates diatomaceous ooze from calcareous nannofossil ooze, while the modern oceanic boundary separates diatomaceous ooze from calcareous foraminiferal ooze (Kennett, 1977). Thus, inferences made about glacial conditions in Antarctica from biogeography are based upon a data base whose systematics are not fully understood.

Barker and Kennett and others (1988) have maintained that the presence of smectite in cores from the Maud Rise (Site 690) indicates that chemical, rather than physical, weathering was the dominant denuding process occurring in East Antarctica during Eocene time. It is inferred that the presence of smectite in Eocene sediments at Site 690 indicates that conditions on East Antarctica were warm. However, one may also conclude that the presence of smectite in Eocene sediments at Site 690 is merely the result of erosion of smectitic clays (of earlier origin) from the continent, possibly by glacial processes, and delivery to the site during Eocene time. The other results from ODP Leg 113 are also ambiguous because of the poor recovery of Tertiary sequences from the continental margin of East Antarctica, and the dependence of the glacial history reconstructions presented by Barker and Kennett and others (1988) on the quantity of IRD in cores and biogeography.

The primary argument involving plate tectonics and a Neogene timing for the growth of ice sheets in Antarctica revolves around the role that continental landmasses played in blocking deep circumpolar flow (Kennett, 1977; Mercer, 1978). Kennett has contended that ice sheets did not develop on Antarctica until middle Miocene time. He has stated that the ice sheets were able to form only after the initiation of deep circumpolar flow. It is believed that circumpolar flow was initiated following the development of a deep water passage between Antarctica, Australia (and the South Tasman Rise) and New Zealand (and the Campbell Plateau). Kennett's arguments were based upon the geological and geophysical data and the plate tectonic reconstructions that were available in 1977 (Weissel and Hayes, 1972; McGowran, 1973; Deighton et al., 1976; Christoffel and Falconer, 1972; Hayes and Ringis, 1973; Andrews et al., 1975; Hampton, 1975; Kennett et al., 1975; Herron and Tucholke, 1976; Tucholke and Houtz, 1976; and Barker and Burrell, 1976). These data bases and interpretations suggested to Kennett (1977) that the Drake Passage opened and was no longer a barrier to deep circumpolar flow by mid- to late Oligocene time (Barker and Burrell, 1976) and that the South Tasman Rise was a barrier to circumpolar circulation until some time during the Oligocene (Weissel and Hayes, 1972). Kennett concludes that the dramatic shift to heavier oxygen isotope values during middle Miocene time is a manifestation of the first large scale glaciation of Antarctica. He indicates that the glaciation was caused by the establishment of a circum-Antarctic current and subsequent thermal isolation of the continent.

The work by Barron and Larsen, et al. (1989), Stump et al. (1980), LeMasurier (1972), Barrett et al. (1989), the reinterpretation of the oxygen isotope records, and the evidence presented in this paper for an Oligocene ice sheet grounding event in the Ross Sea indicate that the initiation of circumpolar flow was not the driving force behind development of polar ice caps in Antarctica, for the ice sheets had developed before the initiation of circumpolar flow. Thus, it is clear that the argument for an initial formation of ice sheets in Antarctica during middle Miocene time, solely as a result of coeval development of deep circumpolar flow, is tenuous. However, it is evident that plate reorganization did play an important role in the development of ice sheets in Antarctica, for Antarctica has been in a polar position since the late Cretaceous (McElhinny, 1973; Lowrie and Hayes, 1975), and yet the earliest evidence of glacial conditions and ice sheet development is presently limited to Eocene time.

ATMOSPHERIC CIRCULATION AND ICE SHEET FORMATION

The breakup of Gondwana and the separation of Australia and New Zealand from Antarctica may have played a key role in the development of Antarctic ice sheets by reducing the continentality of the region. The importance of increased availability of moisture in the development of large-scale glacial conditions has been recognized by a number of authors (Ewing and Donn, 1956 and 1958; Donn and Ewing, 1966; Crowell, and Frakes, 1970; Fairbridge, 1973; and Frakes and Kemp, 1973). In this paper we apply this concept to the role that the opening of a seaway between Antarctica and Australia had upon the initial formation of ice sheets in Antarctica. The work of Cande and Mutter (1982) indicates that plate reorganization was responsible for the development of a seaway between Antarctica and Australia during early Paleogene time. Cande and Mutter's (1982) and Veveers' (1986) reevaluations of the paleomagnetic data from the south Pacific indicates a much earlier opening of a deep water passage between Antarctica, Australia, and New Zealand. Cande and Mutter (1982) indicate that the magnetic anomalies that were originally identified as anomalies 19 to 22 may actually be anomalies 20 to 34. Thus, the initiation of sea floor spreading between Antarctica and Australia has been pushed back some 30 million years from early Eocene to midlate Cretaceous time. It is possible that the opening of a seaway between Antarctica and Austro-New Zealand landmasses had a dramatic effect upon atmospheric circulation which, in turn, led to formation of ice sheets on Antarctica. Parametric modelling of atmospheric circulation and version 0AV6 of the Community Climate Model (CCM) located at the National Center for Atmospheric Research (NCAR) were utilized in an initial test of this hypothesis. These modelling experiments are referred to as inital tests because the current versions of these models will be vastly improved by better coupling to more realistic ocean and sea-ice models, improved mesoscale cyclonic simulations, and by acquisition of larger Antarctic meteorological data bases (Bromich, pers. commun.). In this initial test of the hypothesis, parametric modelling of atmospheric circulation was conducted for intervals just before (Cenomanian), and after (Lutetian) the break-up of the Antarctic and Austro-New Zealand landmasses. Grid spacing on these climate simulations is 3 degrees (≈111 km at the equator) and the axis of rotation of the earth was varied from -23.5 degrees for austral winter simulations to 23.5 degrees for austral summer simulations. Surface atmospheric pressures range from 1000 mbar to 1030 mbar in the simulations. Results from an independent atmospheric general circulation model (GCM) investigation (see Sloan and Barron, this volume) of Eocene conditions are cited along with the results of the parametric modelling where the GCM results have a bearing upon the parametric modelling conclusions. The CCM was used to produce a series of Eocene climate simulations under January and July conditions. For a more complete description of the model boundary conditions and results see Sloan and Barron (this volume).

Figures 7.5-7.8 are the products of parametric modelling (algorithms that apply the rules that govern today's climate and atmospheric circulation to plate tectonic reconstructions of the past) (Scotese and Summerhayes, 1986) of surface atmospheric circulation in the Southern Ocean region before the formation of a seaway between Antarctica and the Austro-New Zealand landmasses (Cenomanian: 97.5-91.0 Ma), and after the formation of the seaway (Lutetian: 52.0-43.6 Ma). The figures depict summer and winter atmospheric circulation using two different thermal gradients. The geometry of the Subpolar Low (Polar Front) is modified according to the season. During the summer the distribution of the Subpolar Low around the Polar High is essentially symmetric (Figures 7.5 and 7.7) and during the winter, when the equator-to-pole thermal gradient is highest, Rossby wave formation intensifies and the Subpolar Low acquires a tri- or quadra-lobate geometry (Musk,1988) (Figure 7.6 and 7.8). Air flowing through these waves converges and diverges. Air of the eastern side of a trough diverges and induces convergence at the surface, which in turn causes surface air to rise (Musk, 1988). The rising surface air creates surface

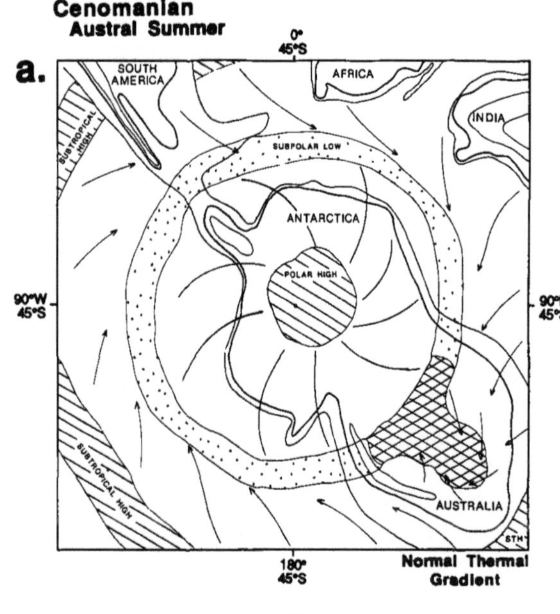

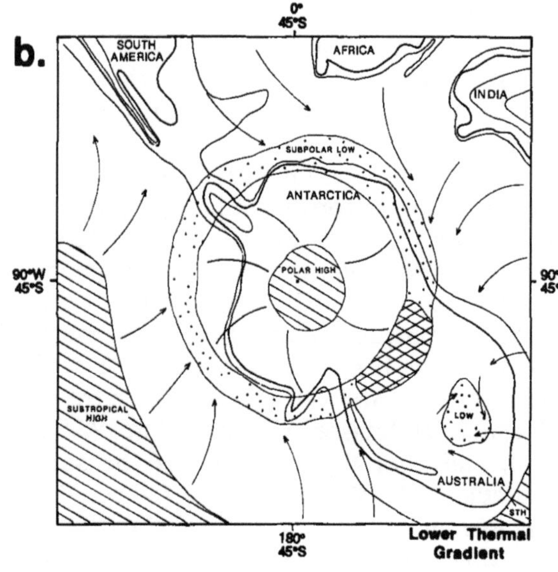

FIGURE 7. 5. This figure presents a parametric model of atmospheric circulation during the austral summer over the Southern Ocean region before the break-up of the Antarctica/ Australia landmass (Late Cretaceous). Two thermal gradients were used in this experiment. Figure 7.5a illustrates surface atmospheric pressure distribution and circulation with today's ("normal") thermal gradient and Figure 7.5b depicts circulation and pressure distribution with a lower thermal gradient. In both cases the Subpolar Low is symmetrically distributed around the Polar High. Note that the low pressure centers over Antarctica are relatively small in areal extent and that the winds along the Antarctic-Australia margin are derived from a large continental landmass and are therefore probably dry. However, with a lower gradient, the low impinges on the coast of Antarctica and may have produced precipitation on Antarctica. A key to abbreviations and symbols used in Figures 7.5 - 7.8 is provided below.

MAD. MADAGASCAR

N.Z. NEW ZEALAND

COASTLINE

CONTINENTAL MARGIN

STH SUBTROPICAL HIGH

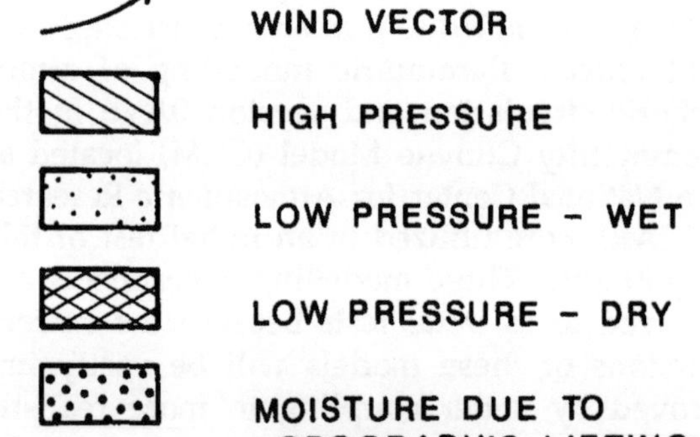

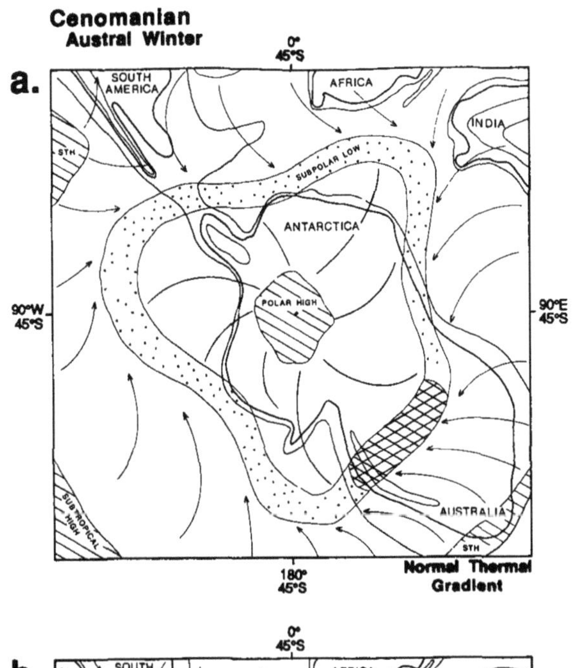

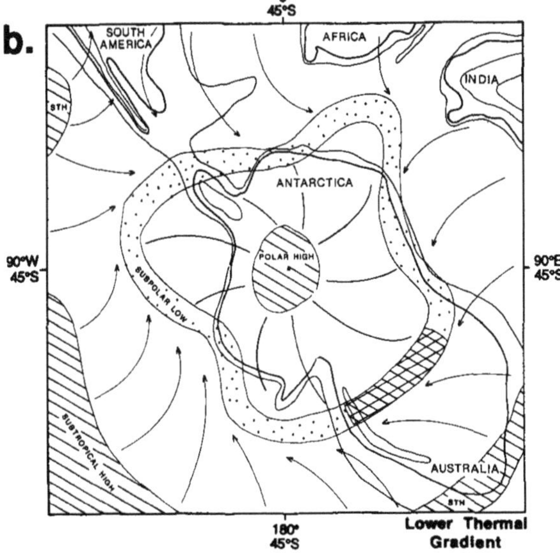

FIGURE 7. 6. This is a parametric model of atmospheric circulation during the austral winter over the Southern Ocean region before the break-up of the Antarctica/Australia landmass (Cenomanian). Two thermal gradients were used in this experiment. Figure 7.6a illustrates surface atmospheric pressure distribution and circulation with today's ("normal") thermal gradient and Figure 7.6b depicts circulation and pressure distribution with a lower thermal gradient. In both cases Rossby Waves have been sketched in and deform the Subpolar Low into a multi-lobate geometry. Rossby wave intensification during the winter enhances wave cyclone formation. Precipitation is strongly associated with wave cyclones (Ahrens, 1988). Wave cyclones transport moisture onto the continent as they move southward from the Polar Front (Carroll, 1986). Thus, the winter circulation patterns appear to be the most conducive to ice sheet formation during Cenomanian time.

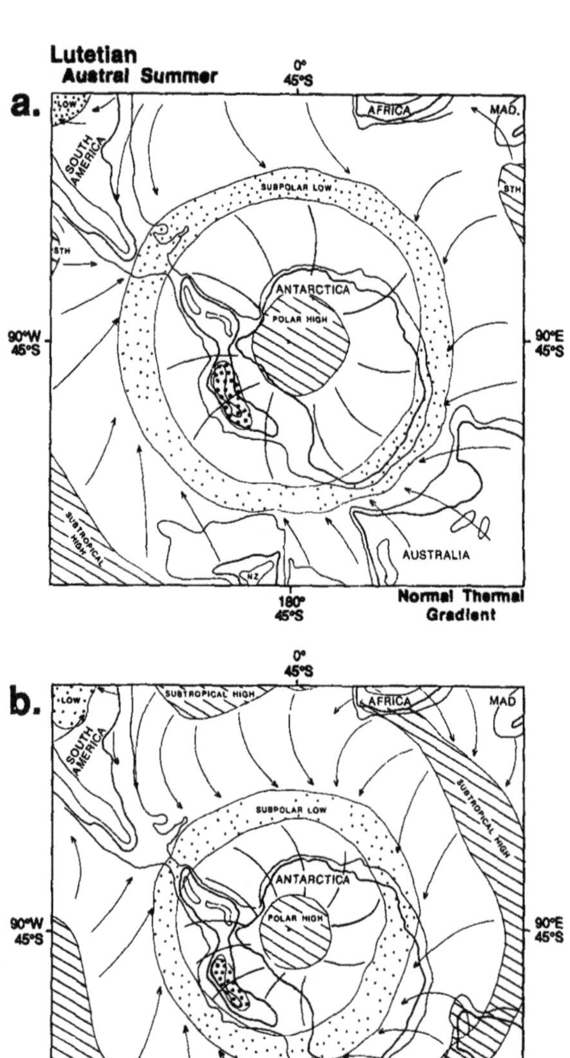

FIGURE 7. 7. This is a parametric model of atmospheric circulation during the austral summer over the Southern Ocean region after the break-up of the Antarctica/Australia landmass (middle Eocene). Two thermal gradients were used in this experiment. Figure 7.7a illustrates surface atmospheric pressure distribution and circulation with today's ("normal") thermal gradient and Figure 7.7b depicts circulation and pressure distribution with a lower thermal gradient. Note that the low pressure centers over Antarctica are now relatively large in areal extent and that the winds approaching Antarctica travel over a broad seaway and are therefore moisture laden relative to the pre-break-up situation. The change in atmospheric circulation associated with the formation of a seaway between Antarctica and Australia may have led to increased precipitation on a cold continent and, therefore, to the development of Eocene or older ice sheets in Antarctica. Also note that the lower thermal gradient scenario (Figure 7.7b) is more conducive to increased precipitation on Antarctica (a continent in a polar position and presumably in the dark for much of the year during Eocene time) and probably results in ice sheet development. This situation casts doubt upon the warm, equable, Eocene climates proposed by paleoceanographers.

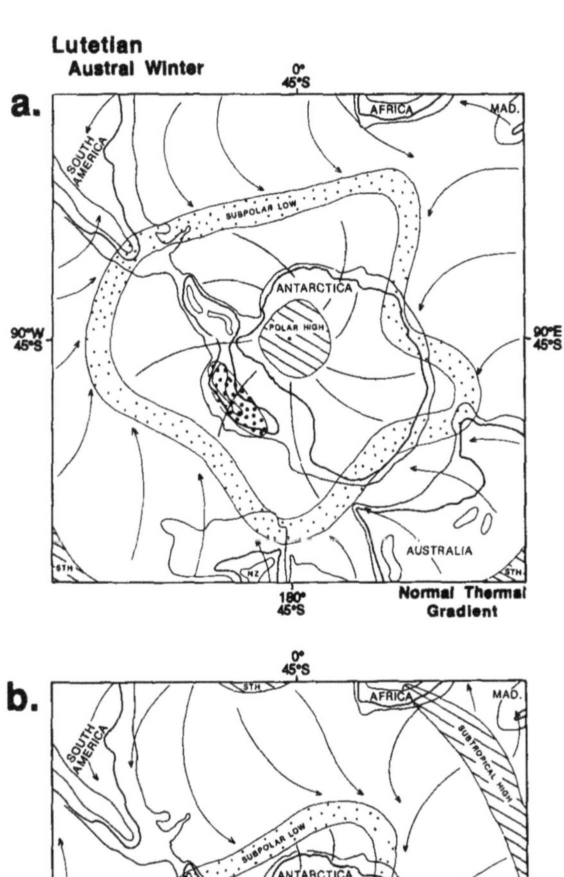

FIGURE 7. 8. This figure presents a parametric model of atmospheric circulation during the austral winter over the Southern Ocean region after the break-up of the Antarctica/Australia landmass (middle Eocene). Two thermal gradients were used in this experiment. Figure 7.8a illustrates the surface atmospheric pressure distribution and circulation with today's ("normal") thermal gradient and Figure 7.8b depicts circulation and pressure distribution with a lower thermal gradient. In both cases Rossby Waves have been sketched in and deform the Subpolar Low into a multi-lobate geometry. Rossby wave intensification during the winter enhances wave cyclone formation. Precipitation is strongly associated with wave cyclones (Ahrens, 1988). Wave cyclones transport moisture onto the continent as they move southward from the Polar Front (Carroll, 1986). Thus, the winter circulation patterns appear to be the most conducive to ice sheet formation during middle Eocene time.

depressions, cloud development, and precipitation (Musk, 1988). Rossby wave intensification during the winter also enhances wave cyclone formation. Precipitation is strongly associated with wave cyclones (Ahrens, 1988). Wave cyclones transport moisture onto the continent as they move southward from the Polar Front (Carroll, 1986). Thus, the winter circulation patterns appear to be the most conducive to ice sheet formation.

Paleoceanographers argue that the Eocene equator-to-pole thermal gradient was less than one half its present value (Shackleton and Boersma, 1981; Boersma and Shackleton, 1987). However, Tertiary glacial deposits recovered from the Antarctic continental margin (LeMasurier, 1972; Birkenmajer, 1987a; Barron and Larsen, et al., 1989; Barrett et al., 1989) suggest that the thermal gradient may have been the same as it is today. The thermal gradient exerts tremendous control on the position of the Polar Front, and hence the ability for moisture to be transported onto Antarctica. Today the winter equator-to-pole thermal gradient in the Northern Hemisphere can be as high as 70° C, while the summer gradient is usually one half that value (Musk, 1988). Musk (1988) indicates, that during the Northern Hemisphere summer, the polar front shifts nearly 20° north of its winter position as a result of the decrease in the equator-to-pole thermal gradient. Since the thermal gradient strongly controls the position of the polar front, and there is controversy over the nature of the Paleogene and Cretaceous gradients, two different thermal gradients were used in the models. In this study the position of the polar front was determined by using a modern thermal gradient for one set of experiments, and it was conservatively shifted 5° poleward to simulate a low thermal gradient situation.

The Cenomanian reconstructions (Figures 7.5 and 7.6) indicate that a Polar High was present over Antarctica and that the continent was surrounded by a Subpolar Low. The Subpolar Low is broad region of surface low pressure. It is the product of convergence of northward flowing cold polar air and southward flowing mild air that originated at the Subtropical High. At the convergence of the these two air masses, the surface air rises and produces a zone of low pressure at the surface (Ahrens, 1988). As the air rises in the Subpolar Low, it cools and condensation occurs. If the air that rises in the low has been transported over a broad expanse of water, and therefore has a high moisture content, precipitation will occur. During the Cenomanian it appears as though significant quantities of moisture may have been advected onto Antarctica in the coastal area along the Indian Ocean Sector of Antarctica and Marie Byrd Land (Figure 7.5) by atmospheric circulation under low thermal gradients and during the normal gradient winter (Figures 7.5a, 7.6a, b). Atmospheric circulation under a normal thermal gradient during the Cenomanian austral summer (Figure 7.5a) positioned the Subpolar Low too far north of the continent for precipitation in the Low to have any effect on continental ice sheet development. Therefore, it appears as though a portion of Antarctica received a significant quantity of precipitation during Cenomanian time, regardless of the thermal gradient. In fact, it appears as though a lower thermal gradient enhances ice sheet development.

The concept of ice caps in Antarctica/ Australia during late Mesozoic time is supported by: 1) recalibrated oxygen isotope curves, which indicate ice volume effects back into the late Cretaceous (Prentice and Matthews, 1988), 2) Late Cretaceous, globally-synchronous, high frequency shifts in coastal onlap (Haq et al., 1987), 3) oxygen isotope data from carbonate cements in concretions and 4) IRD from lower Cretaceous deposits in Australia. The oxygen isotope data from the Cretaceous concretions in Australia (in the Otway and Gippsland Basins) indicate that the mean annual temperature of the early Cretaceous at paleolatitudes of 75 - 80° S was around -2 ± 5° C with seasonal extremes of -17°C (Rich et al., 1988 ; Gregory et al., 1989). The observation of IRD within the Bulldog Shale in the Eromanga Basin of Australia suggests the presence of high-latitude ice near sea level (Frakes and Francis, 1988). These data suggest that the climate may have been cool enough to initiate glacier growth (Oerlemans, 1982) by early Cretaceous time. Perhaps the recovery of Cretaceous and Paleogene sediments from drilling on the Enderby Land and the Queen Maud Land margins or through the Antarctic ice sheet will establish whether there were late Mesozoic ice sheets in Antarctica. However, modelling results suggest that more of the continent was able to receive moisture after the formation of a large seaway between Antarctica and the Austro-New Zealand landmass (Lutetian time?). Therefore, it is presumed that large ice sheets did not develop on Antarctica until Australia and New Zealand rifted form Antarctica.

Figures 7.7 and 7.8 illustrate models of the atmospheric circulation over the Southern Ocean region during the middle Eocene austral summer and austral winter, under both low and normal thermal gradients. A deep water passage between Antarctica and South America had not yet formed at this time (Barker and Burrell, 1982), but there was broad seaway present between Antarctica and Australia and New Zealand (Cande and Mutter, 1982). Formation of the seaway appears to have modified atmospheric circulation over the Southern Ocean. The Polar High was still situated over the center of Antarctica, and the continent was still surrounded by the Subpolar Low (Figures 7.7 and 7.8), but once the seaway between Antarctica and the Austro-New Zealand landmasses formed, moisture was advected onto the Wilkes Land margin throughout the year (Figures 7.7 and 7.8). Additionally, atmospheric circulation after the break-up, and under a low thermal gradient, appears to have resulted in more extensive coverage of East Antarctica by the Subpolar Low (Figure 7.7b). During middle Eocene time, air masses traveled south from the Subtropical High, over the broad seaway to the Subpolar Low that lay over East Antarctica (Figure 7.7). Convergence with southward flowing Polar air masses at the Subtropical Low caused the moist air that originated at the subtropical high to rise. Condensation of the moisture in the air that originated at the Subtropical High produced precipitation over an extensive area of the cold Antarctic landmass and presumably led to the formation of ice sheets on Antarctica at least by middle Eocene time. The upper Eocene glacial deposits recovered in Prydz Bay (Barron and Larsen, et al., 1989) may be a manifestation of the atmospheric circulation patterns that developed after the formation of a seaway between Antarctica, Australia and New Zealand.

Yet another artifact of tectonic reorganization is the formation of a continental seaway between Marie Byrd Land and East Antarctica. Presumably, air flowing away from the Polar High crossed this body of water, and absorbed moisture. As the air flowed over the mountains of Marie Byrd Land, the moisture in the air condensed during orographic lifting, and precipitation fell upon Marie Byrd Land (Figures 7.7 and 7.8). Examination of Figure 7.8 reveals that the Subpolar Low lies just off of the coast of Marie Byrd Land. Wave cyclones initiated at the Subpolar Low during the austral winter may have transported enough moisture onto the continent as well. The orographic process combined with the advection of moisture from the Polar Front to produce an initially higher volume of moisture on Marie Byrd Land. This may have resulted in the formation of an ice cap on Marie Byrd Land during Oligocene time (LeMasurier and Rex, 1982).

The Antarctic Peninsula's more northerly position may have prevented significant glacial development in that area until Oligocene time. The middle Eocene Krakow glaciation (Birkenmajer, 1987) is considered to have been a local alpine glaciation, as deposits of this age on nearby Seymour Island lack evidence of glaciation. An important caveat to this statement is that there are a number of unconfomities (perhaps even glacially carved) in the middle Eocene of Seymour Island, so caution must be exercised when comparing the stratigraphies from King George Island and Seymour Island. Moisture transported over the Antarctic Peninsula by wave cyclones during the austral winter was probably responsible for the glacial conditions and the glacial deposits that have been documented on the South Shetland Islands (Antarctic Peninsula) by Birkenmajer and his colleagues (1987).

The discussion in the preceding paragraphs suggests that small glaciers and ice caps may have formed on the Antarctic/Australian landmass during late Cretaceous time. However, modelling indicates that the interval following an opening of a seaway between Antarctica and the Austro-New Zealand landmass was probably the time that large ice sheets formed on Antarctica. The low thermal gradients proposed by paleoceanographers for Eocene time are not difficult to reconcile with the presence of glacial deposits on the Antarctic margin. Antarctica has been in a polar position since Cretaceous time, and hence has been

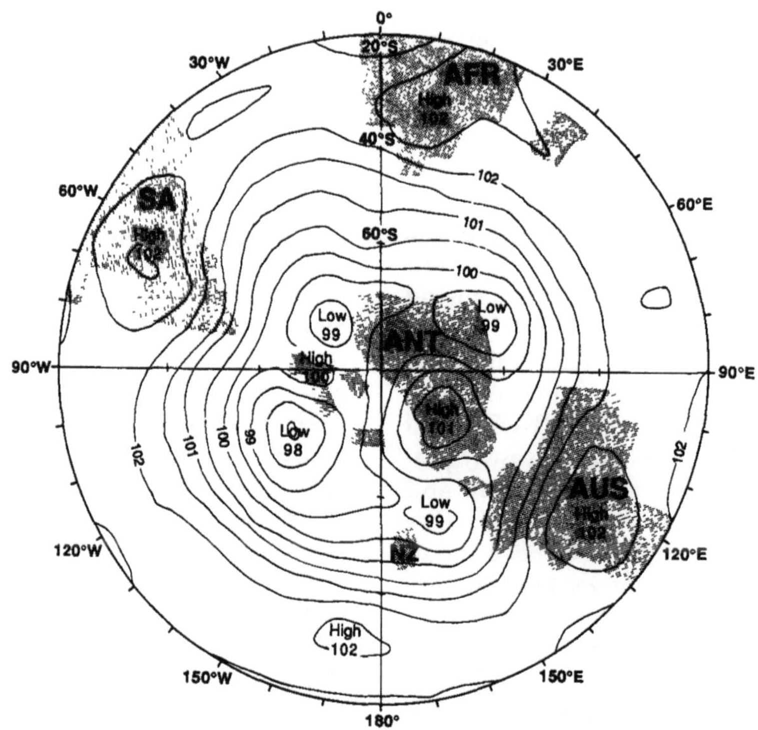

FIGURE 7. 9. This is a GCM simulation of Eocene atmospheric pressure distribution at sea level over the Southern Ocean region during the austral winter. The posted values along the isobars are mb (10^{-1}) and the areas marked with H's and L's correspond to areas of high and low pressure. Note the multi-lobate geometry of the isobars over Antarctica.

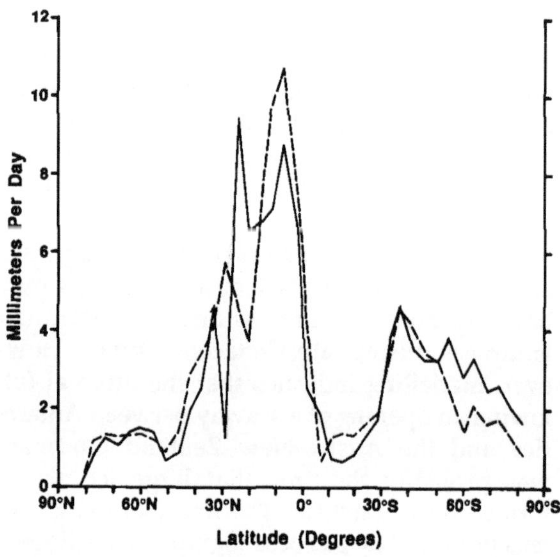

FIGURE 7.10. Mean latitudinal precipitation rates (mm/day) over land in Eocene time for average July (austral winter) conditions with both reduced and steep meridional ocean SST gradients, and it indicates that precipitation is more abundant in the low gradient situation. Solid line is the reduced gradient and the dashed line is the steep gradient.

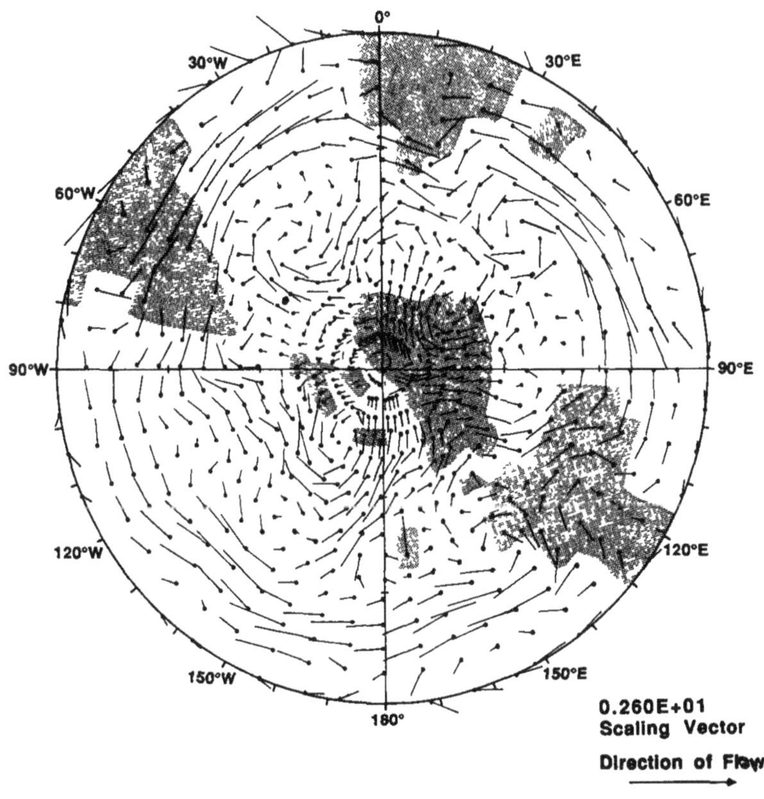

FIGURE 7.12. GCM simulation of precipitation distribution during the austral winter over the Southern Ocean region in Eocene time. The posted values are precipitation in mm/day and the areas marked with H's and L's correspond to areas of high and low precipitation. Note the relatively high level of precipitation (> 1 mm/day) generated over much of East Antarctica by the moisture bearing onshore winds that are illustrated in Figure 7.11.

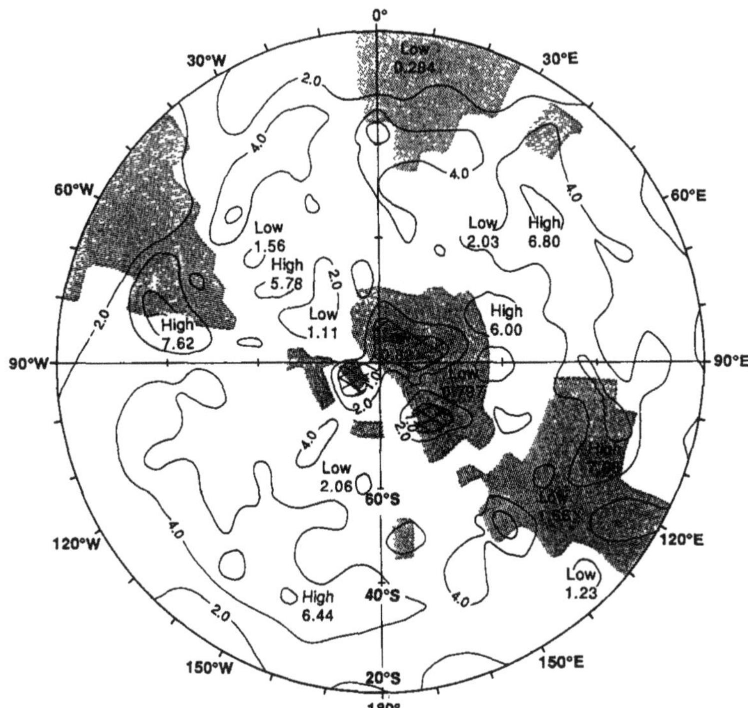

FIGURE 7.11. GCM simulation of Eocene atmospheric circulation over the Southern Ocean region at the 900 mb surface during the austral winter. The arrows are scaled wind vectors. The length of the vector below the diagram is scaled to 2.6 ms^{-1}. Note the onshore winds over much of East Antarctica.

subjected to extended periods of low solar insolation. It is therefore difficult to imagine that the continent was much warmer than it is today. However, plant fossils recovered from Neogene deposits in Antarctica (Askin and Markgraf, 1986; Webb and Harwood, 1987) indicate that there were relatively warm intervals interspersed between the glacial periods. Warmer atmospheric conditions on the continent and slightly cooler equatorial conditions would have produced a lower equator to pole thermal gradient, but modelling results indicate that such a situation would result in increased onshore transport of moist air and precipitation on a continent that was still relatively cold. Such a situation may have triggered formation of ice sheets (Figures 7.7 and 7.8). Thus, lower thermal gradients, at least in terms of precipitation, may actually enhance ice sheet development rather than retard it.

GCM experiments (see Sloan and Barron, this volume) also suggest that a reduced meridional thermal gradient would have enhanced ice sheet development in Antarctica. The GCM simulations suggest that the winter circulation patterns were the most conducive to ice sheet formation. The intensification of Rossby Wave formation during the winter, as illustrated by the development of a tri- or quadra-lobate geometry along the Polar Front in Figure 7.9, may have caused increased wave cyclone formation and resulted in more advection of moisture onto the continent as the wave cyclones moved southward from the Polar Front during the winter. Simulations with both reduced and steep meridional ocean sea surface temperature (SST) gradients (for descriptions of gradients see Sloan and Barron, this volume) indicate that precipitation is more abundant over Antarctica in the low gradient situation (Figure 7.10). Figures 7.11 and 7.12 illustrate the distribution of near surface winds and precipitation for the GCM simulation of austral winter (July) with a reduced gradient. The patterns displayed in these figures are similar to the parametric simulation of these conditions in that the GCM experiment also suggests that winds traveling across the seaway between Antarctica and Australia advected moisture onto the continent. The GCM experiments suggest that there was an abundant supply of moisture in the atmosphere in the low gradient situation. In fact the simulation suggests that there was nearly twice as much precipitation as there is today. Schwerdtfeger (1984) indicates that much of East Antarctica receives less than 10 cm of precipitation per year. The GCM simulation for the reduced gradient case suggests that Antarctica may have received 9 cm in the winter months alone (see areas with at least ≈1 mm/day in Figure 7.12). The relatively high precipitation rates of the reduced gradient scenario may represent the early stages of the development of the Antarctic ice sheets. A shift to a meridional thermal gradient that was more like the modern gradient (i.e. steeper), and reduced precipitation in Antarctica (as illustrated by Figure 7.14) may have occurred when the climate in Antarctica cooled further, due to the increases elevation and albedo that were associated with the growth of the ice sheets.

GCM generated surface temperature maps (Figure 7.13), for average July conditions, also indicate that it was cold enough to permit snow accumulation in Antarctica. This is true even for the reduced gradient case, where the simulation indicates that surface temperatures were cool enough, except for right at the continental margin, for snow accumulation (Figure 7.13). Both models indicate that in spite of warm ocean surface temperatures near the poles, winter continental surface temperatures were below freezing. These conclusions are also supported by other climate model results (Ogelsby, 1989; Sloan and Barron, 1990).

Another critical factor in the establishment and growth of continental-scale ice sheets that must also be considered is summer surface temperatures. As discussed above, both parametric and GCM models suggest that there was relatively abundant precipitation over a large portion of Antarctica in late Paleogene time. Moreover, GCM simulations and limited paleoclimatic data (e.g. Frakes and Francis, 1988; Rich et al., 1988; and Gregory et al., 1989) indicate that winter surface temperatures in the Southern Ocean region, an area the was occupied by a portion of Gondwana in late Mesozoic and early Paleogene time, were low enough to permit snow/ice accumulation (Oerlemans, 1982) at this time (Figure 7.13). However, the simulations of conditions during the austral summer generate surface temperatures that are

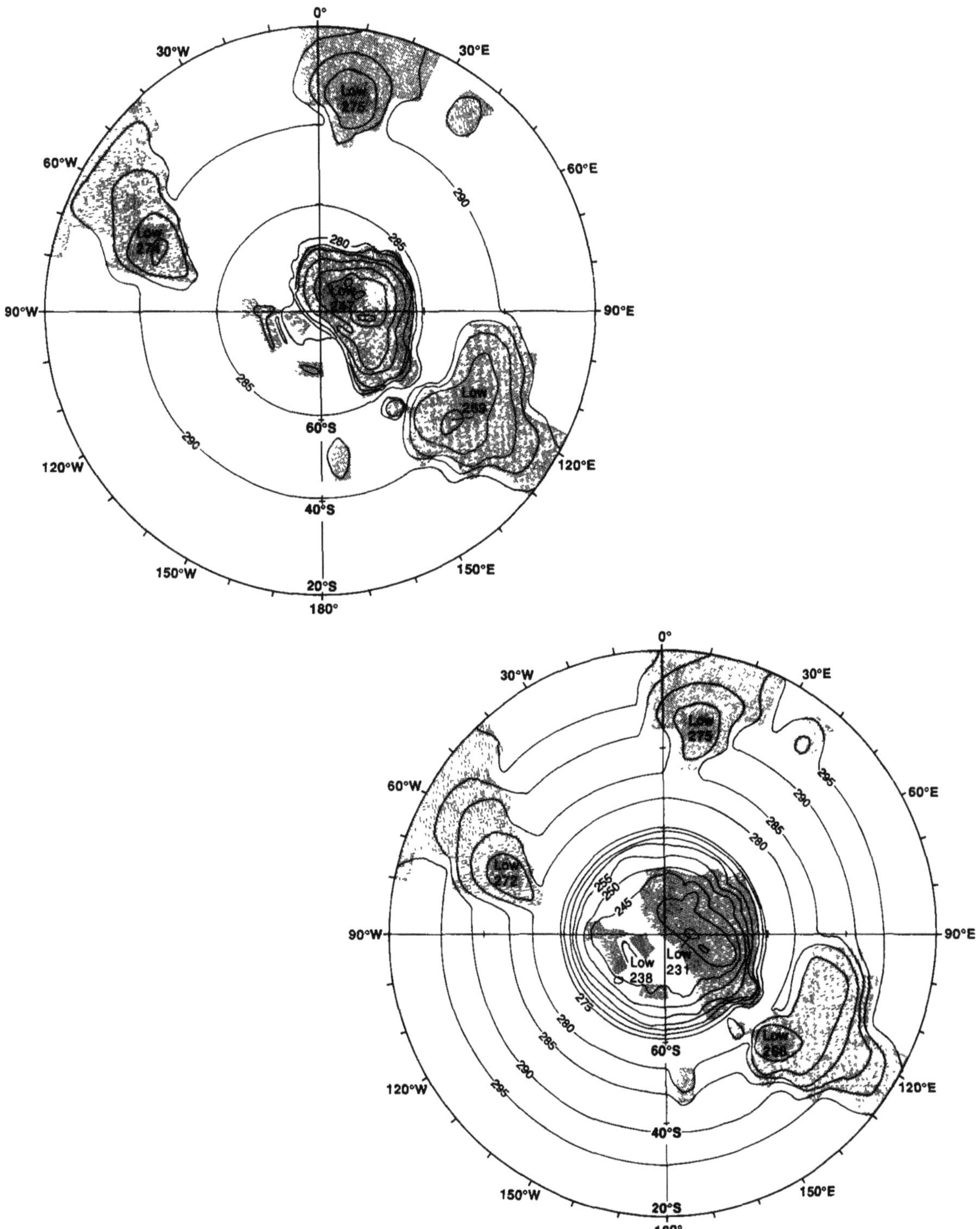

FIGURE 7.13. GCM simulation of Eocene surface temperature distributions generated by shallow and steep meridional thermal gradients during the austral winter. The posted values are degrees Kelvin and the areas marked with H's and L's correspond to high and low temperatures. Contour interval is 5° K. Note that the temperatures in Antarctica are below freezing (273° K) in both simulations.

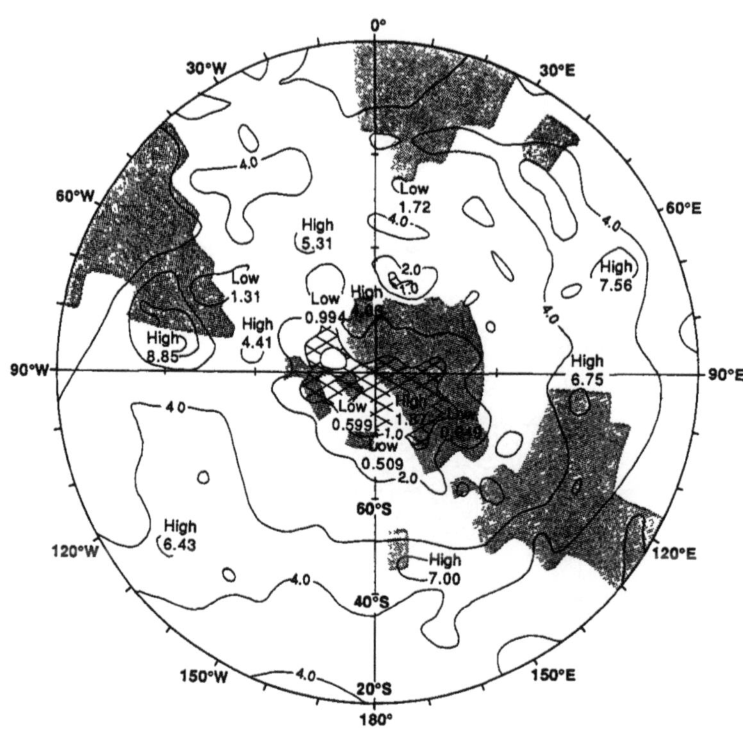

FIGURE 7.14. GCM simulation of precipitation distribution during the austral winter over the Southern Ocean region in Eocene time. The posted values are precipitation in mm/day and the areas marked with H's and L's correspond to areas of high and low precipitation. Note that the precipitation generated over East Antarctica with the steep thermal gradient is less than that produced by the shallow thermal gradient of Figure 7.11. The areal distributions of Figures 7.11 and 7.14 show the same trend that is illustrated in the latitudinal precipitation chart of Figure 7.10.

above freezing (Sloan and Barron, 1990). Therefore, the growth of ice sheets in Antarctica is a summer mass balance problem. In order to produce substantial ice, summer continental surface temperatures would have had to have been low enough to prevent melting of all of the snow and ice that accumulated during the preceding winters (Crowley et al., 1987).

Unfortunately, existing model results do not provide good estimates of summer temperatures for Antarctica during the Eocene and Oligocene. Austral summer surface temperatures from the GCM simulations are high enough to melt ice, but there are several factors that contribute to the (artificially high) temperatures of the model. Low continental elevations (less than or equal to 500 m) were specified for Antarctica in the GCM experiments. The ice sheets on Antarctica have raised present day elevations, over a large portion of Antarctica, to between 2 to 3 kilometers above sea level (Stuiver et al, 1981). Simulations utilizing higher elevations may be more realistic given the evidence for late Paleogene ice sheets in Antarctica (Barrett et al., 1986; Bartek et al., in press; Barron and Larsen et al., 1989). Higher elevations would result in significantly lower surface temperatures (Olgelsby, 1989; and Sloan and Barron, 1990). For an increase in elevation of 2 kilometers, a lapse rate of 6.5° C/km (Sloan and Barron, 1990) would decrease summer surface temperatures by as much as 13° C. Another factor that may have contributed to artifically high summer surface temperatures in the GCM simulations is a specified albedo that was too low. In the simulations presented here, an albedo for ice-free conditions was utilized. Ice-albedo feedback will reduce the summer surface temperatures. Therefore, given the necessary winter climatic conditions over Antarctica, and sufficient elevation, there were areas in which ice was sustained through the summer. The preservation of ice through the summer would eventually have been enhanced by the increased albedo of the ice. Positive feedback of this system would have continued as Antarctica became more isolated and the surrounding ocean temperatures decreased. Cooling ocean surface temperatures would have permitted ice to spread onto the continental margins To test these hypotheses, more simulations are needed to quantify how much of an effect ice-albedo would have on summer surface temperatures.

NEGATIVE FEEDBACK AND ICE SHEET DISINTEGRATION

There is a large body of data that indicates that Antarctic ice sheet volume fluctuated dramatically during Cenozoic time. Cenozoic ice sheet disintegration episodes are indicated by the presence of Eocene, upper Oligocene, upper Miocene and Pliocene marine flora and fauna that were discovered in the Sirius Group by Webb and others (1984), Askin and Markgraf (1986), and Harwood (1986). The marine flora and fauna recovered from the Sirius Group indicate that marine basins existed in areas that are now covered by the East Antarctic Ice Sheet. High frequency fluctuations in the oxygen isotope record (Miller et al., 1987; Prentice and Matthews, 1988) and in coastal onlap (Haq et al., 1987) corroborate the concept of an unstable Antarctic ice sheet (Harwood, 1985, 1986b).

The circulation models presented in the preceding paragraphs may explain the formation of ice sheets on Antarctica prior to the initiation of circum-polar flow, but they do not explain deglaciation. Albedo effects and collapse of marine-based ice sheets may have led to the high frequency ice volume fluctuations that are thought to have occurred in the Cenozoic. Albedo effects may have led to further cooling of the region and caused intensification of the Polar High. Intensification of the Polar High may have caused the Subpolar Low to shift further north. This would have greatly retarded transportation of moist air onto the continent, and provided negative feedback to the system. Thus, the ice sheet moved into a metastable condition because of the negative feedback that it provided to the system that nourished it. In this state, the ice sheet may have been susceptible to processes that led to the multiple deglaciations that have been recorded by microfossils in the Sirius Group.

Mechanisms for rapid ice sheet disintegration have been proposed by Hughes (1973), Clark and Lingle (1977), and Thomas and Bentley (1978), and conceivably may have resulted in the most recent rapid deglaciation of Antarctic continental shelf that has been documented in the Ross Sea by Anderson and others (1980) and Kellogg and others (1979). It has been proposed that marine-based ice sheets are capable of catastrophic disintegration (Hughes, 1973). Marine-based ice sheets are based below sea level and buttressed on topographic highs (pinning points). These ice sheets are inherently unstable because they are based below sea level, and changes in the relative position of the base of the ice sheet and the pinning points can cause the marine-based ice sheet to become unpinned and result in rapid retreat of the ice sheet. Changes in the relative position of the pinning point and the base of the ice sheet can arise from eustatic fluctuations, isostatic depression of the sea floor, or erosion. The marine flora and fauna from the Sirius Group indicate that there are marine basins below the East Antarctic Ice Sheet. Prior to the significant uplift of the Transantarctic Mountains, these marine basins may have been occupied by unstable marine-based ice sheets (Harwood, 1985). Thus, the initiation of large scale glaciation of Antarctica, and the high frequency deglaciations may be explained by the circulation models presented in the preceding paragraphs, negative feedback from albedo, and the collapse of marine-based ice sheets.

The formation of Antarctic ice sheets following the ice sheet collapses may be linked to variations in surface water temperature of the ocean surrounding Antarctica. Rapid growth of ice sheets may only occur during intervals when climatic conditions generate coastal low pressure systems and onshore winds over relatively "warm" surface waters. This model may help explain the paradox of a $\delta^{18}O$ record from the deep sea of warm Eocene water temperatures, and the glacial history recorded in the ODP Leg 119 cores from the Prydz Bay area of Antarctica which seem to indicate extensive glacial conditions in Antarctica during Eocene-Oligocene time.

CONCLUSIONS

A dramatic shift in the mode of sedimentation during late Oligocene time has been recorded in CIROS-1 and DSDP Leg 28 cores from the Ross Sea. An angular unconformity observed in seismic records from throughout the Ross Sea is correlated to this shift in sedimentation style. Based upon the extensive and irregular nature of the angular unconformity, and its correlation to the dramatic change in the lithostratigraphy of cores from the Ross Sea, it is interpreted as a glacial erosional unconformity. The angular unconformity and the change in sedimentation style recorded in the CIROS-1 and DSDP Leg 28 cores is believed to mark the grounding of an ice sheet on the Ross Sea continental shelf, and a shift to subpolar or possibly even polar glacial conditions in the Ross Sea region during late Oligocene time.

The interpretation of an Oligocene ice sheet grounding event on the Ross Sea continental shelf corroborates other data that suggest extensive glaciation in Antarctica. The formation of the Paleogene ice sheets prior to the initiation of deep circumpolar flow indicates that thermal isolation of the continent was not necessarily the triggering event of extensive Cenozoic glaciation of Antarctica. It is suggested that glaciation of Antarctica is linked to the development of a seaway between Antarctica and the Austro-New Zealand landmasses. The formation of a seaway resulted in the development of seasonal low pressure systems along the Antarctic coast. The low pressure systems may have produced onshore winds that transported moist air on to the cold Antarctic landmass during Eocene time reducing the region's aridity. Thus, ice sheet formation during late Eocene time may have been the result of the seasonal precipitation associated with the development of the low pressure systems and onshore winds as a seaway formed between Antarctica and the Austro-New Zealand landmasses.

The documentation of the presence of a late Oligocene ice sheet, and the presence of Eocene, late Oligocene, late Miocene and Pliocene marine microfossils and Pliocene wood (Webb and Harwood, 1987) in the Sirius Group suggest that the ice sheets in Antarctica were not stable (Webb et al., 1984), as has been the contention of other workers. The marine microfossils were derived from interior basins that are now situated below the East Antarctic Ice Sheet and their presence in the Sirius Group suggests that the late Oligocene ice sheet underwent a severe reduction in volume, possibly even disappeared, and then reestablished itself. It appears as though ice sheets (marine) formed on Antarctica during late Eocene/early Oligocene time and have waxed and waned since that time.

ACKNOWLEDGEMENTS

NSF Grant 8818523 to John B. Anderson funded a portion of this investigation. We wish to thank Arthur A. Few of the Space Physics department at Rice University for his assistance with the parametric climate simulations and David Harwood for his insightful review of this paper. We also wish to thank Eric J. Barron for assistance to LCS. Modelling study by LCS was completed at the National Center for Atmospheric Research, funded by NSF grant ATM-8804084 to Eric J. Barron. Stephanie Staples-Shipp assisted us with drafting.

REFERENCES

Ahrens, C.D. 1988. *Meteorology Today: An Introduction to Weather, Climate, and the Environment*. New York, West Publishing Co., 582 pp.

Alley, R.B., Blankenship, D.D., Rooney, S.T., and Bentley, C.R. 1989. Sedimentation beneath ice shelves--the view from ice stream B. *Marine Geology* 85: 101-120.

Allis, R.G., Barrett, P.J., and Christoffel, D.A. 1975. A paleomagnetic stratigraphy for Oligocene and Early Miocene marine glacial sediments at Site 270, Antarctica. In D.E. Hayes and L.A. Frakes, *Init. Rept. Deep Sea Drilling Project*, Washington D. C., U. S. Government Printing Office, 28: 879-884.

Anderson, J.B. 1986. Critical evaluation of some criteria used to infer Antarctica's glacial and climatic history from deep-sea sediments, *South Afr. Jour. Sci.* 82: 503-505.

Anderson, J.B., and Bartek, L.R., in press. High resolution seismic record of Antarctic glacial history, In *Antarctic Sciences: Geology and Biology, Proceedings of the Second Symposium on Antarctic Science*, Seoul, South Korea, Korean Ocean Research and Development Institute.

Anderson, J.B., Kurtz, D.D., Domack, E.W., Balshaw, K.M. 1980. Glacial and glacial marine sediments of the Antarctic continental shelf. *Jour. Geol.* 88: 399-414.

Anderson, J.B., Brake, C., Myers, N., Wright, R. 1983. Development of a polar glacial-marine sedimentation model from Antarctic Quaternary deposits and glaciological information. In B.F. Molnia, ed., *Glacial Marine Sedimentation*, New York, Plenum Press, pp. 233-264.

Anderson, T.F., and Arthur, M.A. 1983. Stable isotopes of oxygen and carbon and their application to sedimentologic and paleoenvironmental problems, In *Stable Isotopes in Sedimentary Geology, SEPM Short Course No.* 10, Dallas, 1983, p. 1-1--1-151.

Andrews, P.B., Gostin, V.A., Hampton, M.A., Margolis, S.V., and Ovenshine, A.T. 1975. Synthesis--Sediments of southwest Pacific Ocean, southwest Indian Ocean, and South Tasman Sea, In J.P. Kennett, R.E. Houtz, et al., *Init. Rept. Deep Sea Drilling Project*, Washington D.C., U.S. Government Printing Office, 29: 147-1154.

Askin, R.A., and Markgraf, V. 1986. Palynomorphs from the Sirius Formation, Dominion Range, Antarctica, *Antarctic Jour. U.S.* 21 (5): 34-35.

Balshaw, K.M. 1981. Antarctic glacial chronology reflected in the Oligocene through Pliocene sedimentary section in the Ross Sea, Unpublished Ph.D. Dissertation, Rice University, 140 pp.

Barker, P.F., and Burrell, J. 1976. The opening of Drake Passage. In *Proceedings of the Joint Oceanographic Assembly, Food and Agricultural Organization of the United Nations*, Rome, p. 103.

Barker, P.F. and Burrell, J. 1982. The influence upon Southern Ocean circulation, sedimentation, and climate of the opening of Drake Passage, In C. Craddock, ed., *Antarctic Geoscience*. Madison, Univ. Wisconsin Press, p. 377-385.

Barker , P.F. and Kennett, J.P. et al. 1988. *Proceedings of the Ocean Drilling Program, Volume 113 Initial Reports, Weddell Sea, Antarctica*, College Station, Texas, 785 pp.

Barrett, P.J. 1975. Characteristics of pebbles from Cenozoic marine glacial sediments in the Ross Sea (DSDP Sites 270-274) and the South Indian Ocean (Site 268), In D.E. Hayes, L.A. Frakes, et al., *Init. Rept. Deep Sea Drilling Project*, Washington D. C., U. S. Government Printing Office, 28: 769-784.

Barrett, P.J., ed. 1986. Antarctic Cenozoic history from the MSSTS-1 drillhole, McMurdo Sound, *Science and Information Publishing Centre, Dept. of Scientific and Industrial Research Bulletin* 237, Wellington, New Zealand, pp. 174..

Barrett, P.J., Hambrey, M.J., Harwood, D.M., Pyne, A.R., and Webb, P.N. 1989. Synthesis, In P.J. Barrett, ed., Antarctic Cenozoic History from the CIROS-1 Drillhole, McMurdo Sound, *Science and Information Publishing Centre, Dept. of Scientific and Industrial Research Bulletin* 245, Wellington, New Zealand, pp. 241-251.

Barron, E.J. 1987, Eocene equator-to-pole surface ocean temperatures: A significant climate problem? *Paleoceanography* 2: 729-739.

Barron, E.J., Thompson, S.L., and Schneider, S.H. 1981. An ice-free Cretaceous? Results from climate model simulations. *Science* 212: 501-508.

Barron, J., Larsen, B., et al., 1989. *Proceedings of the Ocean Drilling Program, Volume 119 Initial Reports, Kerguelen Plateau - Prydz Bay, Antarctica*, College Station, Texas, pp. 1-939.

Bartek, L.R., Vail, P.R., Anderson, J.B., Emmet, P.A., and Wu., S. submitted. The effect of Cenozoic ice sheet fluctuations on the stratigraphic signature of the Neogene, In S. Cloetingh and D.L. Sahagian, eds., Causes and Consequences of Long Term Sea Level Change, *Journal of Geophysical Research.*

Bartek, L.R., and Anderson, J.B. 1990. Neogene stratigraphy of the Ross Sea continental shelf: revelations from Leg 2 of the 1990 Ross Sea Expedition of the R/V Polar Duke, In A.K. Cooper and P.N. Webb (conveners) International Workshop on Antarctic Offshore Seismic Stratigraphy (ANTOSTRAT): Overview and Extended Abstracts, *U.S. Geological Survey Open-file Report* 90-309, p. 53-62.

Berryhill, H.L. Jr. 1986. The continental shelf off South Texas, In H.L. Berryhill Jr., ed. Late Quaternary facies and structure,

Northern Gulf of Mexico. *Amer. Assoc. Petrol. Geol. Stud. Geol.* 23: 11-79.

Birkenmajer, K. 1987a. Oligocene-Miocene glacio-marine sequences of King George Island (South Shetland Islands), Antarctica. *Palaeontologia Polonica.* 47: 9-37.

Birkenmajer, K. 1987b. Tertiary glaciation in the South Shetland Islands, West Antarctica: Evaluation of data, Abst., *5th International Symposium on Antarctic Earth Sciences,* Cambridge, p.16.

Birkenmajer, K., Gazdzicki, A., Gradzinski, R., Kreuzer, H., Porebski, S.J., and Tokarski, A.K. 1987. Origin and age of Pectinid-bearing conglomerate (Tertiary) on King George Island, West Antarctica. *5th International Symposium on Antarctic Earth Sciences,* Cambridge, p. 17.

Boersma, A., and Shackleton, N.J. 1987. Atlantic Eocene planktonic foraminiferal paleohydrographic indicators and stable isotope paleoceanography. *Paleoceanography* 2: 287-331.

Breza, J., Wise, S.W., Zachos, J.C., and the The 120 Shipboard Party. 1989. Lower Oligocene ice rafted debris at 58°S on the Kerguelen Plateau: The 'Smoking Gun' for the existence of an early Oligocene ice sheet on East Antarctica, *Third International Conference on Paleoceanography,* Cambridge, Blackwell Scientific Publications, Abstract Volume, p. 24.

Cande, S.C. and Mutter, J.C. 1982. A revised identification of the oldest sea-floor spreading anomalies between Australia and Antarctica. *Earth Planet. Sci. Lett.* 58: 151-160.

Carroll, J.J. 1986. Processes determining regional thermal and precipitation climatology. *South Afr. Jour. Sci.* 82: 493-497.

Christoffel, D., and Falconer, R. 1972. Marine magnetic measurements in the southwest Pacific Ocean and the identification of new tectonic features, In D.E. Hayes, ed., *Antarctic Oceanology II, The Australian-New Zealand Sector, American Geophysical Union.* Washington D.C., Antarctic Research Series, 19: 197-209.

Clark, J.A., and Lingle, C.S. 1977. Future sea-level changes due to West Antarctic ice sheet fluctuation. *Nature* 269: 206-209.

Cooper, A.K., Davey, F.J., and Behrendt, J.C. 1987. Seismic stratigraphy and structure of the Victoria Land Basin, western Ross Sea, Antarctica, In A.K. Cooper and F.J. Davey,, eds., *The Antarctic Continental Margin: Geology and Geophysics of the Western Ross Sea, Circum-Pacific Council for Energy and Mineral Resources Earth Science Series,* v. 5B, Houston, Texas, p. 27-65.

Crowell, J.C. and Frakes, L.A. 1970. Phanerozoic glaciation and the causes of ice ages. *Amer. Jour. Sci.* 268: 193-224.

Crowley, T.J., Short, D.A., Mengel, J.G., and North, G.R. 1987. The role of seasonality in the evolution of climate during the last 100 million years. *Science* 231: 579-584.

Deighton, I., Falvey, D.A., and Taylor, D.J. 1976. Depositional environments and geotectonic framework: Southern Australia continental margin, *Australian Petroleum Exploration Association Journal* 16: 25-36.

Donn, W.L., and Ewing, M. 1966. The theory of ice ages III, *Science* 152: 1706-1712.

Ewing, M., and Donn, W.L. 1956. The theory of ice ages I, *Science* 123: 1061-1066.

Ewing, M., and Donn, W.L. 1958. The theory of ice ages II, *Science* 127: 1159-1162.

Fairbridge, R.W. 1973. Glaciation and plate migration, In D.H. Tarling and S.K. Runcorn, eds., *Implications of Continental Drift to the Earth Sciences.* Academic Press, New York, 1: 501-515.

Frakes, L.A. 1984. Problems in Antarctic marine geology: a review, In R.L. Oliver, P.R. James, and J.B. Jago, eds., *Antarctic Earth Science,* Canberra, A. C. T., Australian Academy of Science, pp. 375-378.

Frakes, L.A., and Francis, J.E. 1988. A guide to Phanerozoic cold polar climates from high-latitude ice-rafting in the Cretaceous, *Nature* 333: 547-549.

Frakes, L.A., and Kemp, E.M. 1973. Paleogene continental positions and evolution of climate, In D.H. Tarling and S.K. Runcorn, eds., *Implications of Continental Drift to the Earth Sciences,* New York, Academic Press, 1: 535-558.

Gregory, R.T., Douthitt, C.B., Duddy, I.R., Rich, P.V., and Rich, T.H. 1989. Oxygen isotopic composition of carbonate concretions from the lower Cretaceous of Victoria, Australia: implications for the evolution of meteoric waters on the Australian continent in a paleopolar environment, *Earth Planet. Sci. Lett.* 92: 27-42.

Hambrey, M.J., Barrett, P.J., and Robinson, P.H. 1989. Stratigraphy, In P.J. Barrett, ed., *Antarctic Cenozoic History from the CIROS-1 Drillhole, McMurdo Sound, Science and Information Publishing Centre, Dept. of Scientific and Industrial Research Bulletin* 245: 23-48. Wellington, New Zealand.

Hampton, M.A. 1975. Detrital and biogenic sediment trends at DSDP Sites 280 and 281, and evolution of middle Cenozoic currents, In J.P. Kennett, R.E. Houtz, et al., *Init. Rept. Deep Sea Drilling Project*, Washington D.C., U.S. Government Printing Office. 29: 1071-1076.

Harwood, D.M. 1985. Late Neogene climatic fluctuations in the high-southern latitudes: implications of a warm Gauss and deglaciated Antarctic continent, *South Afr. Jour. Sci.* 81: 239-241.

Harwood, D.M. 1986a. Recycled siliceous microfossils from the Sirius Formation, *Antarctic Jour. U.S.* 21 (5): 101-103.

Harwood, D.M. 1986b. Diatom biostratigraphy and paleoecology with a Cenozoic history of Antarctic ice sheets, unpublished Ph.D. dissertation, Ohio State University, Columbus, Ohio, pp. 592.

Harwood, D.M., Barrett, P.J., Edwards, A.R., Rieck, H.J., and Webb, P.-N. 1989. Biostratigraphy and chronology, In P.J. Barrett, ed., *Antarctic Cenozoic History from the CIROS-1 Drillhole, McMurdo Sound, Science and Information Publishing Centre, Dept. of Scientific and Industrial Research Bulletin* 245: 231-240, Wellington, New Zealand.

Haq, B.L., and Hardenbol, J., and Vail, P.R. 1987. Chronology of fluctuating sea levels since the Triassic. *Science* 235: 1156-1167.

Hayes, D.E., and Frakes, L.A., et al. 1975. *Init. Rept. Deep Sea Drilling Project*, Washington D. C., U. S. Government Printing Office, 28: 1-1017.

Hayes, D.E., and Ringis, J. 1973. Sea floor spreading in the Tasman Sea. *Nature* 243: 454-458.

Herron, E.M., and Tucholke, B.E. 1976. Seafloor magnetic patterns and basement structure in the southeastern Pacific, In C.D. Hollister, C. Craddock, et al., *Initial Reports of the Deep Sea Drilling Project*, Washington D. C., U. S. Government Printing Office, 35: 263-278.

Hinz, K., and Block, M. 1983. Results of geophysical investigations in the Weddell Sea and in the Ross Sea, Antarctica, *Proceedings 11th World Petroleum Congress, London*, PD 2, Westchester, New York, John Wiley & Sons, Ltd., pp. 79-91.

Hughes, T. 1973. Is the West Antarctic Ice Sheet disintegrating? *Jour. Geophys. Res.* 78: 7884-7910.

Keigwin, L.D., and Corliss, B.H. 1986. Stable isotopes in late middle Eocene to Oligocene foraminifera. *Geol. Soc. Amer. Bull.* 97: 335-345.

Kellogg, T.B., Truesdale, R.S., and Osterman, L.E. 1979. Late Quaternary extent of the West Antarctic ice sheet: New evidence from Ross Sea cores. *Geology* 7: 249-253.

Kemp, E.M., and Barrett, P.J. 1975. Antarctic glaciations and early Tertiary vegetation. *Nature* 258: 507-508.

Kennett, J.P. 1977. Cenozoic evolution of Antarctic glaciation, the circum-Antarctic ocean, and their impact on global paleoceanography, *Jour. Geophys. Res.* 82: 3843-3860.

Kennett, J.P. 1983. Paleoceanography: Global ocean evolution, *Rev. Geophys. Space Phys.* 21: 1258-1274.

Kennett, J.P., Houtz, R.E., Andrews, P.B., Edwards, A.R., Gostin, V.A., Hahos, M., Hampton, M.A., Jenkins, D.G., Margolis, S.V., Ovenshine, A.T., and Perch-Nielsen, K. 1975. Cenozoic paleoceanography in the southwest Pacific Ocean, Antarctic glaciation and the development of the circum-Antarctic current, In J.P. Kennett, R.E. Houtz, et al., *Init. Rept. Deep Sea Drilling Project*, Washington D. C., U. S. Government Printing Office, 29: 1155-1169.

King, L.H., and Fader, G. 1986. Wisconsinan glaciation of the continental shelf, southeast Atlantic Canada. *Geol. Surv. Canada Bull.* 363: 1-72.

King, L.H., Rokoengen, K., and Gunleiksrud, T. 1987. Quaternary seismostratigraphy of the Mid Norwegian shelf, 65°-67°30' N -- A till tongue stratigraphy. *Continental Shelf and Petroleum Technology Research Institute Publication* 114, Trondheim, Norway, 58 pp.

Leckie, R.M., and Webb, P.N. 1983. Late

Oligocene-early Miocene glacial record of the Ross Sea, Antarctica: Evidence from DSDP Site 270. *Geology* 11: 578-582.

LeMasurier, W.E. 1972. Volcanic record of Cenozoic glacial history of Marie Byrd Land, In R.J. Adie, ed., *Antarctic Geology and Geophysics*. Universitetsforlaget, Oslo, pp. 251-259.

LeMasurier, W.E., and Rex, D.C. 1982. Volcanic record of Cenozoic glacial history in Marie Byrd Land and western Ellsworth Land: Revised chronology and evaluation of tectonic factors, In C. Craddock, ed., *Antarctic Geoscience,* Madison, Univ. Wisconsin Press, pp. 725-734.

Lowrie, W., and Hayes, D.E. 1975. Magnetic properties of oceanic basalt samples, In D.E. Hayes and L.A. Frakes, eds., *Init. Rept. Deep Sea Drilling Project,* Washington D. C., U. S. Government Printing Office, 28: 869-878.

Margolis, S.V., and Kennett, J.P. 1971. Cenozoic paleoglacial record in Cenozoic sediments of the Southern Ocean, *Amer. Jour. Sci.* 271: 1-36.

Matthews, R.K., and Poore, R.Z. 1980. Tertiary $\delta^{18}O$ record and glacio-eustatic sea-level fluctuations. *Geology* 8: 501-504.

McElhinny, M.W. 1973. *Paleomagnetism and Plate Tectonics*, New York, Cambridge University Press, 358 pp.

McGowran, B. 1973. Rifting and drift of Australia and the migration of mammals. *Science* 180: 759-761.

Mercer, J.H. 1978. Glacial development and temperature trends in the Antarctic and in South America, In E.M. van Zinderen Bakkar, ed., *Antarctic Glacial History and World Paleoenvironments*. Rotterdam, Balkema, p. 73-93.

Mercer, J.H. 1983. Cenozoic glaciation in the Southern Hemisphere, *Ann. Rev. Earth Planet. Sci.* 11: 99-132.

Miller, K.G., Fairbanks, R.G., and Mountain, G.S. 1987. Tertiary oxygen isotope synthesis, sea level history, and continental margin erosion, *Paleoceanog.* 2: 1-19.

Musk, L.F. 1988. *Weather Systems*. New York, Cambridge University Press, 160 pp.

Oerlemans, J. 1982. A model of the Antarctic Ice Sheet. *Nature* 297: 550-553.

Oglesby, R.J. 1989. A GCM study of Antarctic glaciation.*Climate Dynamics* 3: 135-156.

Porebski, S., and Gradzinski, R. 1987. Depositional history of the Polenez Cove Formation (Oligocene), King George Island, West Antarctica: A record of continental glaciation, shallow-marine sedimentation and contemporaneous volcanism. *Studia Geologica Polonica* 93: 7-74.

Prentice, M.L., and Matthews, R.K. 1988. Cenozoic ice-volume history: Development of a composite oxygen isotope record, *Geology* 16: 963-966.

Rich, P.V., Rich, T.H., Wagstaff, B.E., McEwen Mason, J., Douthitt, C.B., Gregory, R.T., and Felton, E.A. 1988. Evidence for low temperatures and biologic diversity in Cretaceous high latitudes of Australia. *Science* 242: 1403-1406.

Sato, S., Asakura, N., Saki, T., Oikawa, N., Kaneda, Y. 1984. Preliminary results of geological and geophysical survey in the Ross Sea and in the Dumont D'urville Sea, off Antarctica, *Memoirs of National Institute of Polar Research Special Issue* 33, National Institute of Polar Research, Tokyo, pp. 66-92.

Savin, S.M., Douglas, R.G., Stehli, F.G. 1975. Tertiary marine paleotemperatures, *Geol. Soc. Amer. Bull.* 86: 1499-1510.

Schwerdtfeger, W. 1984. *Weather and Climate of the Antarctic*. New York, Elsevier Science Publishers, pp. 261.

Scotese, C.R., and Summerhayes, C.P. 1986. Computer model of paleoclimate predicts coastal upwelling in the Mesozoic and Cenozoic.*Geobyte* 1: 28-42.

Shackleton, N.J., and Kennett, J.P. 1975. Paleotemperature history of the Cenozoic and the initiation of Antarctic glaciation: oxygen and carbon isotope analysis in DSDP Sites, 277, 279, and 281, In J.P. Kennett, R.E. Houtz, et al., *Init. Rept. Deep Sea Drilling Project,* Washington D.C., U.S. Government Printing Office, 29: 743-755.

Shackleton, N.J., and Boersma, A. 1981. The climate of the Eocene ocean. *Jour. Geol. Soc. London* 138: 153-157.

Sloan, L.C., and Barron, E.J. 1990. "Equable" climates during Earth history. *Geology* 18: 489-492.

Stuiver, M., Denton, G.H., Hughes, T.J., and Fastook, J.L. 1981. History of the marine ice sheet in West Antarctica during the last glaciation: A working hypothesis, In G.H. Denton and T.J. Hughes, eds., *The Last Great Ice Sheets*, New York, John Wiley and Sons, pp. 319-436.

Stump, E., Sheridan, M.F., Borg, S.G., and Sutter, J.F. 1980. Early Miocene subglacial basalts, the East Antarctic Ice Sheet, and uplift of the Transantarctic Mountains. *Science* 207: 757-759.

Thomas, M.A., and Anderson, J.B. 1989. Glacial eustatic controls on seismic sequences and parasequences of the Trinity/Sabin incised valley, Texas Continental Shelf. *Gulf Coast Assoc. Geol. Soc. Trans.* 39: 563-569.

Thomas, R.H., and Bentley, C.R. 1978. A model for Holocene retreat of the West Antarctic Ice Sheet, *Quat. Res.* 2: 150-170.

Tucholke, B.E., and Houtz, R.E. 1976. Sedimentary framework of the Bellingshausen Basin from seismic profiler data, In C.D. Hollister, C. Craddock, et al., *Init. Rept. Deep Sea Drilling Project,* Washington D. C., U.S. Government Printing Office, 35: 197-227.

Veevers, J.J. 1986. Breakup of Australia and Antarctica estimated as mid-Cretaceous (95 ± 5 Ma) from magnetic and seismic data at the continental margin. *Earth Planet. Sci. Lett.* 77: 91-99.

Webb, P.N., and Harwood, D.M. 1987. Terrestrial flora of the Sirius Formation: Its significance for Late Cenozoic glacial history. *Antarctic Jour. U.S. (Review Issue)* 22: 7-11.

Webb, P.N., Harwood, D.M., McKelvey, B.C., Mercer, J.H., and Stott, L.D. 1984. Cenozoic marine sedimentation and ice-volume variation on the East Antarctic craton. *Geology* 12: 287-291.

Weissel, J.K., and Hayes, D.E. 1972. Magnetic anomalies in the southeast Indian Ocean, In D.E. Hayes, ed., *Antarctic Oceanology II, The Australian-New Zealand Sector,* American Geophysical Union, Washington D.C., Antarctic Research Series 19: 234-249.

8. MIDDLE EOCENE TO OLIGOCENE STABLE ISOTOPES, CLIMATE, AND DEEP-WATER HISTORY: THE TERMINAL EOCENE EVENT?

By Kenneth G. Miller

ABSTRACT

The "Terminal Eocene Event" in the marine realm is comprised of a sequence of events that record the climate transition from peak early Eocene warmth to cold, glaciated Oligocene conditions. These events began with a benthic foraminiferal $\delta^{18}O$ increase of ~1.0 ‰ near the early/middle Eocene boundary (Chron C22N-C21N; estimated age 52-49 Ma using time scale of Berggren et al., 1985). Another 1.0 ‰ benthic foraminiferal $\delta^{18}O$ increase occurred near the middle/late Eocene boundary. This latter event has not been directly tied to magnetostratigraphy, and the timing is not well constrained (Chron C18N to early Chron C17N; estimated age 42-41 Ma). These $\delta^{18}O$ increases may be attributed to two deep-water coolings of ~5°C each, although it is possible that they could reflect some ice growth. An earliest Oligocene $\delta^{18}O$ increase of 1.0 ‰ occurred in benthic and most planktonic foraminiferal records in the Atlantic, Pacific, and Indian Oceans. This event has been directly tied to Chron C13N; it post-dates the planktonic foraminiferal extinctions used to recognize the Eocene/Oligocene boundary. At least 0.3-0.4 ‰ of the increase resulted from Antarctic glaciation and a glacioeustatic lowering of greater than 30 m.

The sources of middle to late Eocene deep and bottom waters are not clear. Seismic stratigraphic evidence and the distribution of hiatuses indicates generally sluggish circulation at this time. Carbon isotope evidence for deep-water sources is equivocal because of poor Pacific Ocean sections. Seismic stratigraphic evidence and the development of widespread hiatuses indicate vigorous deep-water circulation during the early Oligocene both from the Arctic/North Atlantic and Antarctic. Carbon and oxygen isotope data are consistent with a short pulse of Northern Component Water (NCW) during the earliest Oligocene (ca. 35.5-34.5 Ma) together with input of cold Southern Component Water (SCW). This deep-water change is associated with the earliest Oligocene $\delta^{18}O$ increase, suggesting a link between deep water and climate changes.

INTRODUCTION

First introduced by Wolfe (1978), the concept of the "Terminal Eocene Event" has been used to describe climatic, faunal, and sedimentological changes in the marine and terrestrial realms which occurred near the end of the Eocene (Van Couvering et al., 1981). The synchrony and causal relationships among these events are still debatable. The nature and timing of the middle Eocene-Oligocene biotic and climatic changes are significant in view of impact-related extinctions because this is one of the predicted periodic mass extinctions (Raup and Sepkoski, 1982). In contrast to the biotic synthesis of Raup and Sepkoski (1982), more detailed stratigraphic studies have established that there was not a mass extinction event in either plankton or deep water benthos at the end of the Eocene (Keller et al., 1983, this volume; Corliss et al., 1984; Snyder et al., 1984; Miller et al., 1985a; Thomas, 1985, this volume; Fenner, 1986; Keller, 1986a, b; Aubry, this volume; among others). The prevailing view of these authors is that there was a sequence of faunal abundance changes and first/last occurrences during the late middle Eocene to early Oligocene rather than a catastrophic turnover.

Climate has been cited as the cause of these biotic changes, yet interpretations of late Paleogene climate have been debatable. The

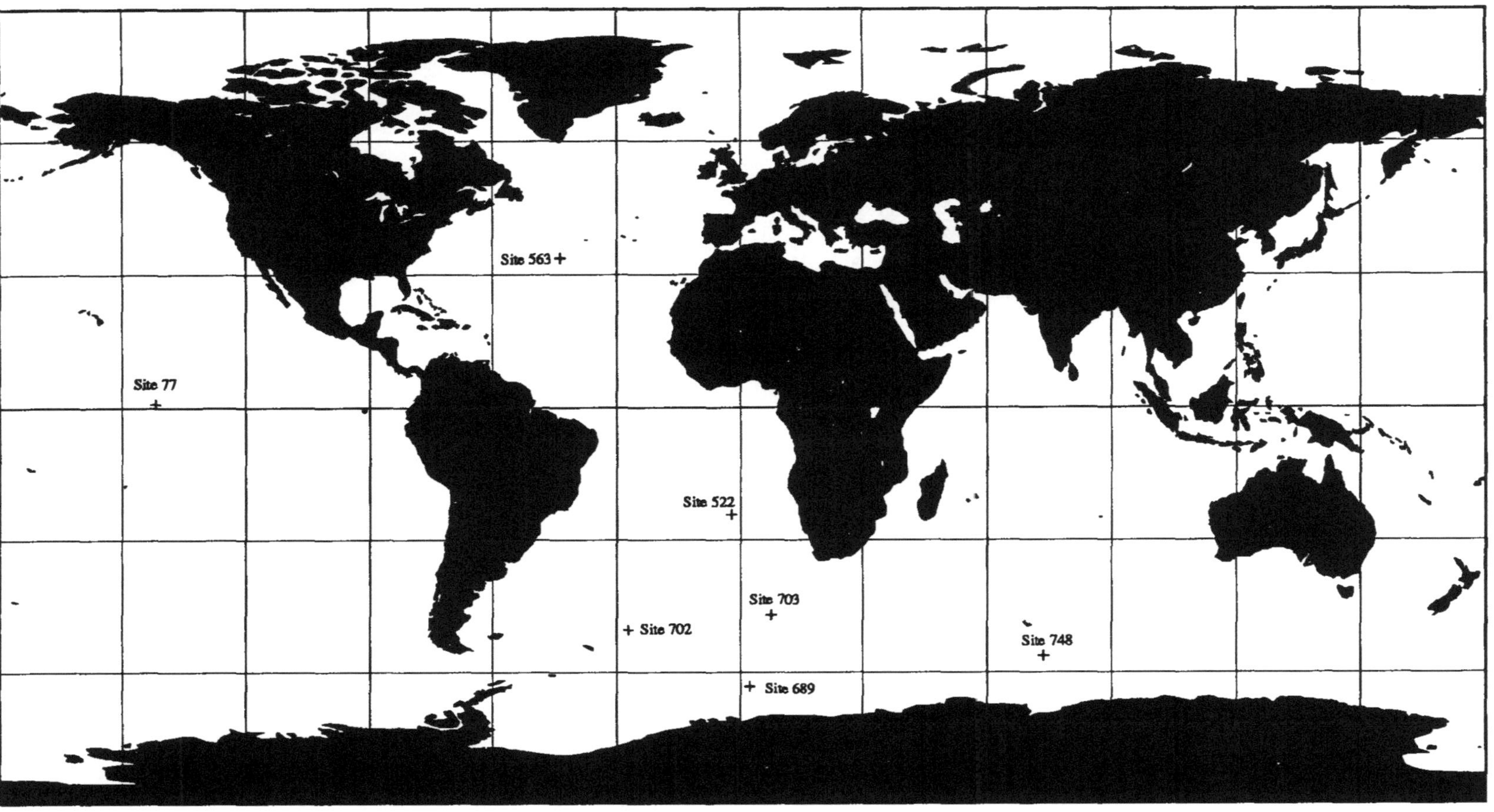

FIGURE 8.1. Global location map showing sites discussed in text.

reason is that marine paleoclimatic reconstructions rely primarily on foraminiferal $\delta^{18}O$ records which record not only ocean temperature changes, but also global ice-volume and local salinity variations. In addition, the late Paleogene foraminiferal isotope record suffers from inadequate sections and poor geographic coverage.

Our understanding of the amplitude and timing of late Paleogene $\delta^{18}O$ fluctuations has depended largely on Atlantic locations (e.g., Deep Sea Drilling Project (DSDP) Legs 73, 74, 80, and 82; Shackleton et al., 1984; Oberhansli et al., 1984; Poore and Matthews, 1984; Miller et al., 1985a, b; Oberhansli and Toumarkine, 1985; Keigwin and Corliss, 1986; Miller et al., 1987a, 1988). Miller et al. (1987a) synthesized available Atlantic and Pacific benthic foraminiferal $\delta^{18}O$ data. No appropriate data were available for the middle-late Eocene from the Pacific, while Southern Ocean records were limited. Ocean Drilling Program (ODP) Legs 113, 114, 119, and 120 recovered several Paleogene sections which allow the evaluation of $\delta^{18}O$ fluctuations in the Southern Ocean. I use recently published and new data from these sections to examine the timing of late Paleogene $\delta^{18}O$ fluctuations, discuss their possible interpretations, and evaluate implications of the $\delta^{18}O$ and $\delta^{13}C$ records to deep- and bottom-water sources.

METHODS

I used published isotope data from Atlantic, Pacific, and Southern Ocean locations (Figure 8.1; Table 8.1; Kennett and Stott, 1990; Katz and Miller, 1991; Zachos et al., 1991) together with new benthic foraminiferal isotope data from ODP Site 703 (Table 8.2; Figure 8.2). All isotopic analyses were derived from *Cibicidoides* spp., which secrete their tests constantly offset from $\delta^{18}O$ equilibrium and which reflect the distribution of $\delta^{13}C$ of SCO_2 (e.g., Shackleton and Opdyke, 1973; Graham et al., 1981). *Nuttallides truempyi* is often used as a substitute for *Cibicidoides* spp. for lower Paleogene sections (Miller et al., 1987b; Katz and Miller, 1991). Although Shackleton et al. (1984) determined that there were systematic offsets between *Cibicidoides* and *N. truempyi* in $\delta^{13}C$ and $\delta^{18}O$ composition, Kennett and Stott (1990) and Katz and Miller (1991) suggested that the offsets between these taxa may have changed slightly through time. Because of this, the comparisons shown here are restricted to *Cibicidoides* data.

Samples were obtained for Site 703 at approximately one per section (1.5 m; Figure 8.2). Deposition was apparently continuous for the interval examined except for a distinct "middle" Oligocene hiatus (ca. 30-25 Ma; Table 8.2; Figure 8.2). My sampling interval for the lower Oligocene (ca. 36-35 Ma) varies from approximately 0.1-0.3 m.y. (Table 8.2).

Samples examined for benthic foraminiferal isotope analyses were washed with sodium metaphosphate (5.5 g/liter) and/or hydrogen peroxide (3% solution) in tap water through a 63-mm sieve and air-dried. Benthic foraminifers were ultrasonically cleaned for 5-10 s and roasted at 370°C in a vacuum. Isotope measurements were made using a Carousel-48 automatic carbonate preparation device attached to a Finnigan MAT 251 (Table 8.2). Replicate samples yielded mean $\delta^{18}O$ differences of 0.236 ‰ and mean $\delta^{13}C$ differences of 0.133 ‰, respectively (Table 8.2). These differences probably overestimate variability because duplicate analyses were performed on levels which displayed large offsets from adjacent samples.

The stable isotope records were correlated using magnetostratigraphy where possible. Atlantic Sites 563 and 522 have excellent magnetostratigraphic control, and age

TABLE 8.1. Depths and references for sites considered here.

Site	Present depth	Paleodepth	Reference
Atlantic			
522	4441 m	3000 m	Miller et al. (1988)
563	3796 m	2300 m	Miller et al. (1985)
Pacific			
77	4291 m	2700 m	Keigwin & Keller (1984) Miller & Thomas (1985)
Southern			
689	2080 m	1650 m	Kennett and Stott (1990)
702B	3083 m	2000 m	Katz and Miller (1991)
703	1796 m	1050-1450 m	this study
748	1291 m	900-1200 m	Zachos et al. (1991)

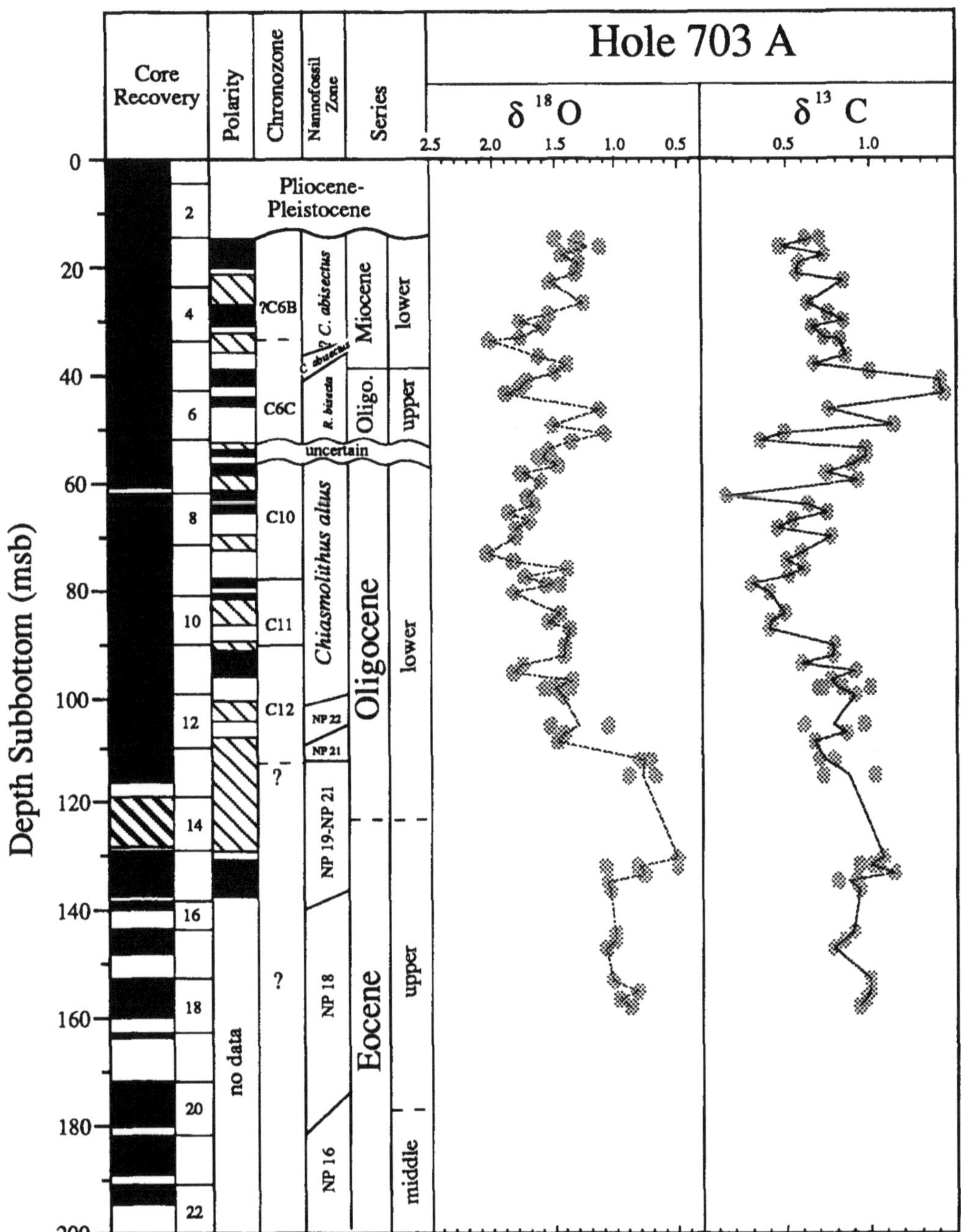

FIGURE 8.2. Magnetostratigraphy, nannofossil biostratigraphy, $\delta^{18}O$ data, and $\delta^{13}C$ data for Ocean Drilling Project Site 703. Polarity interpretations are after Clement and Hailwood (1991); nannofossil biostratigraphy is after Crux (1991). Age model was derived by linearly interpolating between top *C. abisectus* Zone, 34.25 meters subbottom (msb), 23.20 Ma; top *R. bisecta* 39.01 msb, 23.70 Ma; extrapolate sedimentation rate to 57.39 msb, 25.60 Ma; last occurrence *Chiloguembelina* spp., 57.40 msb, 30.00 Ma; base Chronozone C12n, 95.65 msb, 32.90 Ma; $\delta^{18}O$ increase, 108.38 msb, 35.8 Ma; top Zone NP18, 138.msb, 37.8 Ma; top Zone NP16, 177.5 msb, 42.3 Ma.

TABLE 8.2. Stable isotope data, Site 703

Sample	Depth, msb	Age (Ma)	$\delta^{18}O$	$\delta^{13}C$
3-1, 65-69	15.05	21.18	1.307	0.704
			1.507	0.612
3-2, 65-69	16.55	21.34	1.137	0.478
			1.326	0.463
3-3, 65-69	18.05	21.50	1.448	0.734
3-4, 65-69	19.55	21.66	1.306	0.572
3-5, 65-69	21.05	21.81	1.343	0.559
3-6, 65-69	22.55	21.97	1.539	0.843
3, CC	23.20	22.04	2.771*	-0.348*
4-3, 8-12	26.98	22.44	1.277	0.624
4-4, 8-12	28.48	22.59	1.543	0.751
4-5, 8-12	29.98	22.75	1.785	0.841
4-6, 8-12	31.40	22.90	1.595	0.656
4-7, 8-12	32.98	23.07	1.779	0.728
5-1, 30-34	33.70	23.14	2.033	0.825
5-3, 30-34	36.70	23.46	1.642	0.858
5-4, 30-34	38.20	23.61	1.407	0.662
5-5, 30-34	39.70	23.77	1.510	0.997
5-6, 30-34	41.20	23.93	1.736	1.410
5-7, 30-34	42.70	24.08	1.778	1.404
6-1, 72-76	43.62	24.18	1.899	1.438
6-3, 72-76	46.62	24.49	1.155	0.757
6-5, 72-76	49.62	24.80	1.520	1.141
6-6, 72-76	51.12	24.95	1.103	0.490
			0.071*	0.961*
6-7, 25-29	52.60	25.10	1.369	0.350
7-1, 132-136	53.72	25.22	1.553	0.970
7-2, 132-136	55.22	25.38	1.550	0.959
			1.644	0.977
7-3, 132-136	56.72	25.53	1.477	0.893
7-4, 132-136	58.22	30.06	1.784	0.737
			1.766	0.745
7-5, 132-136	59.72	30.18	1.625	0.917
8-1, 90-94	62.80	30.41	1.730	0.137
8-2, 90-94	64.30	30.52	1.673	0.631
8-3, 90-94	65.80	30.64	1.887	0.742
8-4, 90-94	67.30	30.75	1.709	0.538
8-5, 90-94	68.80	30.86	1.825	0.447
8-6, 90-94	70.30	30.98	1.815	0.769
9-2, 12-16	73.02	31.18	2.049	0.589
9-3, 12-16	74.52	31.30	1.839	0.495
9-4, 12-16	76.02	31.41	1.402	0.599
9-5, 12-16	77.52	31.53	1.754	0.514
9-6, 12-16	79.02	31.64	1.471	0.291
			1.574	0.301
9-7, 12-16	80.52	31.75	1.846	0.392

TABLE 8.2 (cont.). Stable isotope data, Site 703

Sample	Depth, msb	Age (Ma)	$\delta^{18}O$	$\delta^{13}C$
10-3, 88-92	84.78	32.08	1.467	0.485
10-4, 88-92	86.28	32.19	1.569	0.414
10-5, 88-92	87.78	32.30	1.385	0.401
11-1, 44-48	90.84	32.54	1.434	0.785
11-2, 44-48	92.34	32.65	1.419	0.776
11-3, 44-48	93.84	32.76	1.759	0.584
11-4, 44-48	95.34	32.88	1.851	0.911
11-5, 44-48	96.84	33.14	1.371	0.764
11-6, 44-48	98.34	33.44	1.590	0.712
			1.382	1.004
			1.534	0.695
			1.512	0.838
11-7, 48-52	99.88	33.75	1.439	0.910
12-4, 98-102	105.38	34.85	1.550	0.604
			1.096	0.958
12-5, 98-102	106.88	35.15	1.429	0.856
12-6, 98-102	108.38	35.45	1.479	0.663
13-2, 108-112	111.98	35.98	0.833	0.693
			0.746	0.778
13-4, 108-112	114.98	36.27	0.914	0.718
	114.98	36.27	0.716	1.032
15-2, 10-14	130.00	37.72	0.520	1.079
15-3, 10-14	131.50	37.86	0.519	1.015
			0.850	1.049
			1.116	0.933
15-4, 10-14	133.00	38.00	0.780	1.142
15-5, 10-14	134.50	38.15	1.095	0.814
			1.083	0.923
15-6, 10-14	136.00	38.29	1.074	0.939
17-1, 76-80	144.16	39.08	1.042	0.902
17-2, 76-80	145.66	39.23	1.033	0.848
17-3, 76-80	147.16	39.37	1.103	0.784
18-1, 60-64	153.00	39.93	1.056	0.999
18-2, 60-64	155.00	40.13	0.840	1.003
18-3, 60-64	156.50	40.27	0.997	0.969
18-4, 60-64	158.00	40.42	0.900	0.934

* Data are considered outliers and are not plotted on Figure 8.2. Age model was derived by linearly interpolating between top *C. abisectus* Zone, 34.25 meters subbottom (msb), 23.20 Ma; top *R. bisecta* 39.01 msb, 23.70 Ma; extrapolate sedimentation rate to 57.39 msb, 25.60 Ma; last occurrence *Chiloguembelina* spp., 57.40 msb, 30.00 Ma; base Chronozone C12n, 95.65 msb, 32.90 Ma; $\delta^{18}O$ increase, 108.38 msb, 35.8 Ma; top Zone NP18, 138.msb, 37.8 Ma; top Zone NP16, 177.5 msb, 42.3 Ma.

estimates were obtained by magnetochronology (Miller et al., 1985b, 1988). The magnetostratigraphic records at Southern Ocean Sites 689, 702, 703, and 748 are more ambiguous, and age estimates were based on magnetobiochronology (Stott and Kennett, 1990; Speiss, 1990; Katz and Miller, 1991; Zachos et al., 1991; Berggren, 1991; Figure 8.2). Age models for Site 77 were based on biostratigraphy (Keigwin and Keller, 1984). Because the earliest Oligocene $\delta^{18}O$ increase is globally synchronous among locations, it provides an excellent correlation point (e.g., Miller et al., 1991a); isotope records were "tuned" to the early Oligocene $\delta^{18}O$ increase by aligning the maximum earliest Oligocene $\delta^{18}O$ values (i.e., the 35.8 Ma Oi1 event of Miller et al., 1991a).

The ages of biostratigraphic events, magnetochrons, and geological boundaries were assigned using the Geomagnetic Polarity Time Scale (GPTS) of Berggren et al. (1985). It is clear that this time scale must be changed for the Eocene-early Oligocene (Montanari et al., 1988; Berggren and Kent, 1989; Swisher and Prothero, 1990; Berggren et al., this volume). However, most correlations were based on first-order calibrations of isotopes to magnetobiostratigraphy. Any future recalibration of the age of the GPTS will not compromise isotope-polarity correlations, and new age estimates may be readily obtained. For example, the age of the Eocene/Oligocene boundary is approximately 34 Ma (Premoli-Silva et al., 1988; Berggren et al., this volume), rather than the 36.6 Ma estimated by Berggren et al. (1985). Changing the age of the boundary does not change the correlation of the earliest Oligocene $\delta^{18}O$ increase with earliest Oligocene Chron C13N (Oberhänsli et al., 1984; Miller et al., 1988; see discussion below); only the numerical age estimate of Chron C13N and the associated oxygen isotope increase will change.

OXYGEN ISOTOPES, CHRONOLOGY, AND CLIMATE

The lowest benthic foraminiferal $\delta^{18}O$ values of the Cenozoic were attained in the early Eocene at intermediate- and deep-water locations throughout the world ocean (e.g., Shackleton and Kennett, 1975; Savin et al., 1975; Miller et al., 1987a). Based on these low values and the presence of thermophilic taxa in high latitudes, Miller et al. (1987a) inferred that the early Eocene was an ice-free, "greenhouse world." The general Eocene to earliest Oligocene $\delta^{18}O$ increase represented a transition from the "greenhouse world" to a glaciated "ice house world" (Miller et al., 1991a). This general increase occurred as three distinct steps:

1) An increase of approximately 1.0 ‰ began near the early/middle Eocene boundary in the Atlantic (Figure 1 in Miller et al., 1987a), Pacific (Figure 3b in Savin, 1977; Figure 5 in Miller et al., 1987b), and Southern Oceans (Figures 8.1, 8.3). The timing and amplitude of the $\delta^{18}O$ increase were not constrained well in any ocean until recently. The records at South Atlantic Sites 525 and 527 suggest that the increase began in Chron C21N (~50-49 Ma; Shackleton et al., 1984; time scale of Berggren et al., 1985). In contrast, there were hints from Pacific Site 577 that it began in Chron C22N (~52 Ma; Miller et al., 1987b). Southern Ocean Site 702 provides the best chronology for this event, which indicates that the increase began during Chron C22N (52.6-52 Ma) and continued through Chron C21 (ca. 49 Ma; Figure 8.3; Katz and Miller, 1991). Thus, the more complete record at Site 702 agrees with records from Sites 577 (where the increase began in Chron C22N) and 527 (where the increase continued through Chron C21).

2) A 1.0 ‰ $\delta^{18}O$ increase occurred near the middle/late Eocene boundary in the Atlantic (Oberhansli et al., 1984; Keigwin and Corliss, 1986; Miller et al., 1987) and Southern Oceans (Figure 8.3; Shackleton and Kennett, 1975; Kennett and Stott, 1990; Katz and Miller, 1991; this study). The timing of this increase is not constrained well. Magnetostratigraphic data for this interval are poor at most locations; this interval is also difficult to interpret magnetostratigraphically because polarities were dominated by long normal intervals. Based on biostratigraphy at Site 702, the increase occurred in the latest middle Eocene (Figure 8.3; ca. 42-41 Ma; Katz and Miller, 1991).

3) A 1.0-1.5 ‰ $\delta^{18}O$ increase occurred throughout the Atlantic, Pacific, Indian, and Southern Oceans in the earliest Oligocene (ca. 35.8 Ma; Shackleton and Kennett, 1975; Savin et al., 1975; Kennett and Shackleton, 1976; Keigwin, 1980; Corliss et al., 1984; Miller et al., 1987; this study; Figure 8.2). The age estimate

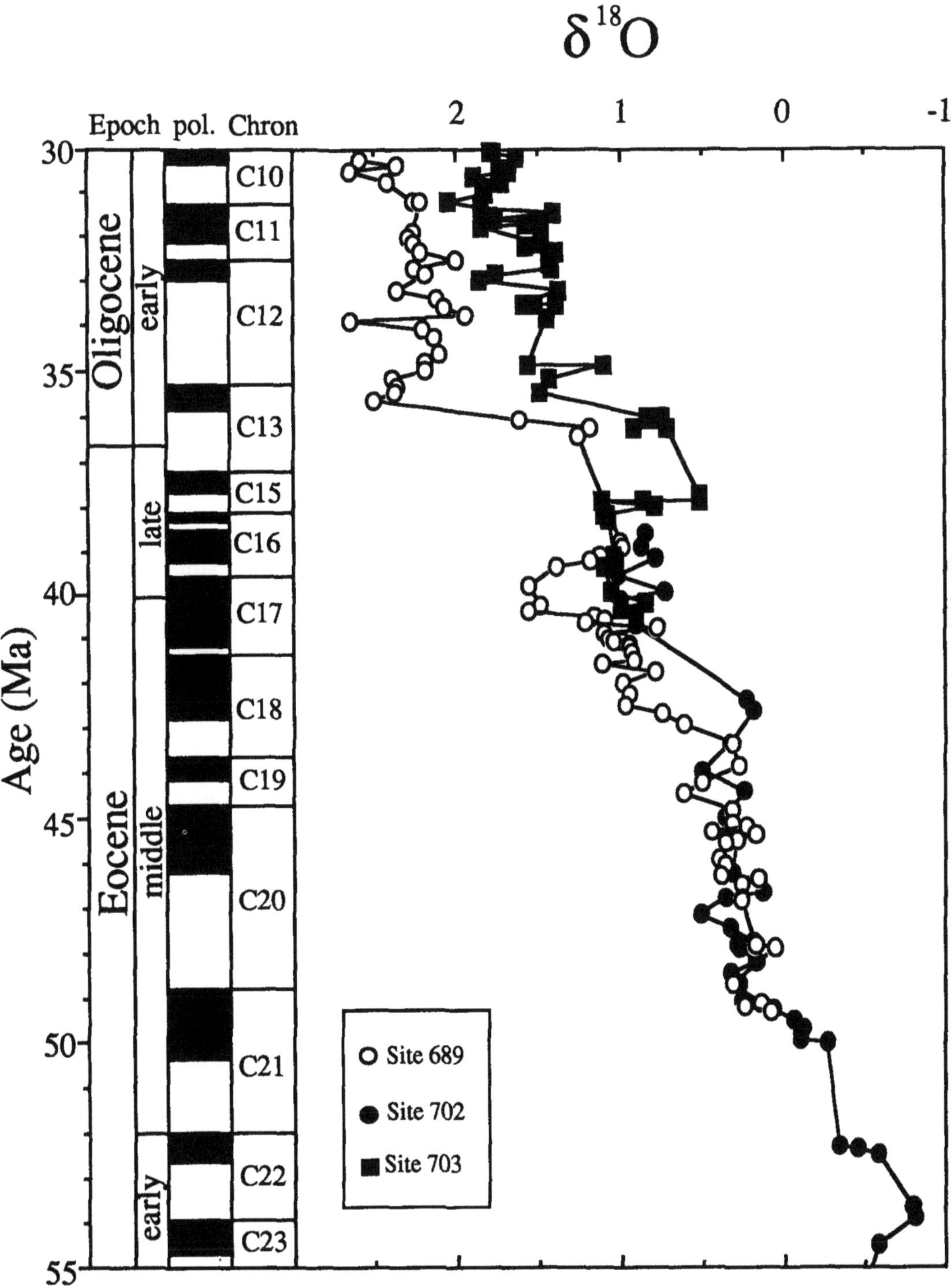

FIGURE 8.3. Late early Eocene to early Oligocene $\delta^{18}O$ data for the taxon *Cibicidoides* from Southern Ocean Sites 689, 702 , and 703. Data sources are listed in Table 8.1.

for this increase is derived from magnetochronology at South Atlantic Site 522 (Tauxe et al., 1984), where it occurred in Chron C13N (Oberhansli et al., 1984; Miller et al., 1988). At Southern Ocean Sites 689 and 703, a 1.0 ‰ $\delta^{18}O$ increase also occurred in the earliest Oligocene (Figures 8.2-8.4). In low latitudes, this increase clearly post-dates the last occurrence of *Hantkenina* spp., a taxon used to recognize the Eocene/Oligocene boundary (Keigwin, 1980; Corliss et al., 1984; Oberhansli et al., 1984; Miller et al., 1985a, 1988; among others). The Eocene/Oligocene boundary stratotype ("golden spike") was placed in Chronozone C13R at the level of the LO of *Hantkenina* in the Massignano section (Premoli-Silva et al., 1988). Although the Massignano section is diagenetically altered and not suitable for stable isotope studies, magnetobiostratigraphic correlations between Massignano and DSDP/ODP sections establish that the $\delta^{18}O$ increase clearly post dates the Eocene/Oligocene boundary.

Oxygen isotope data are equivocal as to the presence of Eocene ice, and it is not clear if the $\delta^{18}O$ increases near the early/middle and middle/late Eocene boundaries represent some ice growth. There seems to be a consensus that the early Eocene was ice free (Miller et al., 1987a, 1991a; Prentice and Matthews, 1988; Kennett and Barker, 1990). Large ice sheets existed during part of the early Oligocene, although these ice sheets were probably transient (e.g., Miller et al., 1987a, 1991; Kennett and Barker, 1990). In contrast, the evidence for middle-late Eocene ice sheets is equivocal, and it is not clear if large ice sheets existed during this "doubt house world" (Miller et al., 1991a). Benthic foraminiferal $\delta^{18}O$ values were less than 1.8 ‰ (in *Cibicidoides*) during the Eocene, and only exceeded this value during the Oligocene-Recent. While values greater than 1.8 ‰ (in *Cibicidoides*) require a glacial world[2] (Miller and Fairbanks, 1983, 1985; Miller et al., 1987a, 1991a), values less than this are equivocal as to the presence or absence of ice sheets. Covariance between the benthic and low-latitude planktonic $\delta^{18}O$ records provides the best isotope evidence for ice growth (Shackleton and Opdyke, 1973; Matthews and Poore, 1980; Miller et al., 1987a, 1991a). However, suitable low latitude planktonic isotope records are lacking across the critical early/middle and middle/late Eocene transitions.

Interpretations of inferred Eocene glacial sediments are also uncertain. Birkenmajer (1987) found 49 Ma (earliest middle Eocene) tills from King Georges Island, West Shetlands. Since this is a tectonically active region, these areally limited tills may be ascribed to mountain glaciation. Massive diamictites recovered from Prydz Bay, eastern Antarctica provide the best hint of possible late Eocene ice sheets (Barron et al., 1989). These sediments are typical deposits of a large ice sheet, and cannot simply be ascribed to mountain glaciation. The age of these deposits is not well known. Approximately 300 m of glacial sediments underlie lowermost Oligocene diamictites, and Barron et al. (1989) have speculated that these are Eocene deposits. Stratigraphic control of these inferred Eocene sediments is poor, and superposition only requires that they be lowermost Oligocene or older.

Circumstantial evidence for late Eocene glaciation is provided by the Sr-isotope record. By assuming that the major late middle Eocene (ca. 42-40 Ma) $\delta^{18}O$ increase was a glacial event, Miller et al. (1991b) were able to reconcile the $\delta^{18}O$ and $^{87}Sr/^{86}Sr$ records. Sr-isotope values began to increase soon after the late middle Eocene $\delta^{18}O$ increase, consistent with glacial erosion of the Antarctic craton and increased continental Sr-flux (Miller et al., 1991b). However, causes of the Sr-isotope increase are speculative, and the increase in Sr-isotope ratios could be attributed to tectonic rather than climate changes.

The earliest Oligocene $\delta^{18}O$ increase was caused by an increase in Antarctic ice volume and a decrease in bottom-water temperatures. The increase has been linked with a pulse of ice rafted detritus (IRD) at Southern Ocean Site 748 (Zachos et al., 1991) and widespread deposition of glaciomarine sediments around Antarctica (see summary of Miller et al., 1991a). The glaciomarine sediment evidence confirms the glacial significance of this event,

[2]Assuming an ice-free world and $\delta^{18}O$ values greater than 1.8 ‰ would require deep-water temperatures to be colder than today's, an assumption incompatible with ice-free conditions (Miller and Fairbanks, 1983, 1985).

but does not provide evidence for its amplitude.

Covariance between planktonic and benthic foraminiferal $\delta^{18}O$ records suggest that at least 0.3-1.0 ‰ of the earliest Oligocene increase is attributable to ice growth. Estimates of covariance ideally require equatorial non-upwelling planktonic isotope $\delta^{18}O$ records (Shackleton and Opdyke, 1973), although Crowley and Matthews (1983) suggested that subtropical planktonic $\delta^{18}O$ records may be used in place of tropical records. Low-latitude planktonic $\delta^{18}O$ data are scarce for the earliest Oligocene $\delta^{18}O$ increase. The only low-latitude record available is from Philippine Sea Site 292, where the planktonic increase is only 0.3‰; however, this record is limited by discontinuous coring (Keigwin, 1980). The earliest Oligocene increase is nearly 1.0 ‰ in subtropical planktonic foraminiferal $\delta^{18}O$ records (e.g., Site 522; Oberhansli et al., 1984). Keigwin and Corliss (1986) and Zachos et al. (1991) have shown that the mean $\delta^{18}O$ values of surface-dwelling planktonic foraminifera increased by 0.3-0.4 ‰ between the late Eocene and the early Oligocene. Their reconstructions provide reasonable lower limits for the global δ_w increase, although the averaging of values for each time interval underestimates the amplitude present in time series. Using the Pleistocene $\delta^{18}O$/sea-level calibration (0.11‰/ 10 m; Fairbanks and Matthews, 1978), this increase represents 30 m to 90 m of glacio-eustatic lowering (assuming 0.3 and 1.0‰ increases, respectively). Considering the maximum $\delta^{18}O$ of freezing ice, an outside limit for pre-Pleistocene $\delta^{18}O$/sea-level calibration was estimated to be 0.055 ‰/10 m (Miller et al., 1987a). If this calibration is used, the $\delta^{18}O$ may represent as much as 180 m of glacio-eustatic lowering; because this calibration represents an extreme limit, actual glacio-eustatic falls were probably closer to those estimated using the Pleistocene calibration (Miller et al., 1991a).

Benthic foraminiferal $\delta^{18}O$ values decreased in the early Oligocene to about 1.5‰, and increased by ~0.8-1.0 ‰ again in the late early Oligocene (ca. 31.5 Ma; Miller et al., 1991a). This "middle Oligocene" increase also has been correlated with deposition of glaciomarine sediments (Miller et al., 1991a). Equatorial planktonic isotope records are limited for this "middle" Oligocene interval, and the amount of ice growth associated with this event remains uncertain (Miller et al., 1991a).

DEEP WATER

Deep-water circulation plays an important role in climate modulation through redistribution of heat and salt and its control of atmospheric CO_2. Pleistocene deep-water changes have been documented using benthic foraminiferal $\delta^{18}O$ and $\delta^{13}C$ isotopes as tracers (e.g., Boyle and Keigwin, 1982; Curry and Lohmann, 1982; Shackleton et al., 1983; Mix and Fairbanks, 1985; Oppo and Fairbanks, 1987). Two strategies have been employed to isolate Pleistocene deep-water $\delta^{13}C$ changes from global budget and local effects: 1) synoptic interbasinal $\delta^{13}C$ comparisons; and 2) synoptic $\delta^{13}C$ comparisons over a wide bathymetric range within a small region. These Pleistocene carbon isotope strategies have been applied to Miocene-Pliocene deep-water circulation changes because Miocene-Pliocene ocean basin configurations were similar to the present (e.g., Miller and Fairbanks, 1983, 1985; Wright et al., 1991). Basinal configurations and gateways were much different during the Paleogene (e.g., Berggren and Hollister, 1974, 1977; Berggren, 1982). Nevertheless, interbasinal and vertical $\delta^{18}O$ and $\delta^{13}C$ comparisons demonstrate that these isotopes provide good tracers of Paleogene deep-water circulation (Miller and Fairbanks, 1985; Miller and Katz, 1987; Miller et al., 1987b; Kennett, and Stott 1990; Barrera and Huber, 1991; Zachos et al., 1991). For example, Southern Ocean locations had higher $\delta^{13}C$ values than the Pacific during much of the late Paleocene and early Eocene. This suggests that the Southern Ocean was proximal to a source of deep water (Miller et al., 1987b; Katz and Miller, 1991).

Carbon isotope reconstructions require comparisons of the regional signal with the global reservoir; Pacific deep water provides a monitor of the global deep-water reservoir (e.g., Miller and Fairbanks, 1985). Middle to late Eocene deep-water sources remain enigmatic because Pacific sections are poorly represented over this interval. Seismic stratigraphic evidence and the distribution of hiatuses suggests generally sluggish circulation at this time (Moore et al., 1978; Miller and

Tucholke, 1983; Mountain and Miller, in prep.), although the absence of widespread anoxia indicates that bottom waters remained oxygenated (e.g., Tucholke and Vogt, 1979). Following a late Paleocene erosional pulse in the Pacific, Indian (Moore et al., 1978), and Atlantic (Mountain and Miller, in prep.), the deep-sea record of the early and middle Eocene was relatively continuous (Moore et al., 1978), although Keller et al. (1987) documented a late middle Eocene hiatus. The poor late Eocene record in the deep sea (Moore et al., 1978) reflects erosion by swiftly flowing earliest Oligocene currents (see below).

There is evidence of dramatic latest Eocene to earliest Oligocene deep-water changes. Seismic stratigraphic evidence and the distribution of hiatuses suggest that the high latitudes were a significant source of vigorously circulating deep water by the earliest Oligocene. A pulse of Southern Component Water (SCW, analogous to Antarctic Bottom Water) from Antarctica caused widespread erosion in the Southern Ocean during the earliest Oligocene (e.g., Kennett, 1977). A similar pulse of Northern Component Water (NCW; analogous to North Atlantic Deep Water) resulted in erosion of seismic reflectors A^u and R4 in the North Atlantic during the earliest Oligocene (Miller and Tucholke, 1983; Mountain and Tucholke, 1985). Widespread erosion during the earliest Oligocene indicates strongly circulating deep and bottom waters (Moore et al., 1978; Miller and Tucholke, 1983; Mountain and Tucholke, 1985; Keller et al., 1987). The distribution of unconformities in the Southern Ocean (Kennett, 1977) and northern North Atlantic (Miller and Tucholke, 1983) suggest bipolar sources.

Preliminary $\delta^{18}O$ and $\delta^{13}C$ reconstructions (Figures 8.4, 8.5) indicate that a pulse of cool, nutrient-depleted NCW and cold, moderate nutrient SCW occurred in the earliest Oligocene. From ca. 35.5-34.5 Ma, the North Atlantic had the highest $\delta^{13}C$ values (see also Miller and Fairbanks (1985), the Southern Ocean had intermediate $\delta^{13}C$ values, and the Pacific had the lowest values (Figure 8.5). This is the same pattern observed for the late Pleistocene interglacials (Oppo and Fairbanks, 1987) and late Miocene (Wright et al., 1991). The highest $\delta^{18}O$ values were recorded adjacent to Antarctica (Figure 8.4; Site 689), with intermediate $\delta^{18}O$ values in the Atlantic, and lowest $\delta^{18}O$ values in the Pacific (Site 77) and Subantarctic (Site 703). These data indicate that the coldest waters were adjacent to Antarctica, while the highest $\delta^{13}C$ values were in the North Atlantic.

Oxygen isotope data show that there was a "cold spigot" (high $\delta^{18}O$ values) of SCW adjacent to Antarctica during the early Oligocene (Figure 8.4, Site 689). The relative influence of SCW may have been less during this early Oligocene interval than it is in the modern ocean. Its influence on $\delta^{18}O$ values is not apparent in the Atlantic and Pacific sites. The coldest waters (highest $\delta^{18}O$ values) may have mixed rapidly as they moved away from the continent, because data from the earliest Oligocene at Site 703 (Subantarctic sector of the Atlantic) shows Pacific-like $\delta^{18}O$ values. However, I caution that the lowermost Oligocene Site 703 record may suffer from some mixing due to drilling; the lower part of Core 12 through Core 13 was somewhat disturbed, and Core 14 was empty (Figure 8.2; Ciesielski et al., 1988). Therefore, the results from these cores may suffer from some vertical homogenization, and must be considered suspect. Nevertheless, the scenario presented here is not dependent on Subantarctic (Sites 703 and 748) and South Atlantic (Site 522) locations (all shown as unconnected symbols, Figure 8.5). Rather, this scenario depends on the quality and correlations of the Antarctic (Site 689), Pacific (Site 77), and North Atlantic (Sites 563) records (solid lines, Figure 8.5).

The stable isotope reconstructions shown in Figures 8.4 and 8.5 are reasonable, but the available data lack sufficient resolution and stratigraphic control to document this apparently rapid (~0.5 m.y.), transient pulse. The quality and correlations of the critical Antarctic, Pacific, and North Atlantic records need to be verified. These reconstructions require additional isotope data and better stratigraphic correlations to confirm the scenario presented here. Nevertheless, the isotope data are consistent with the scenario developed from seismic stratigraphic studies (Miller and Tucholke, 1983) and from the study of deep-sea hiatuses (Kennett, 1977; Miller and Tucholke, 1983; Keller et al., 1987). In

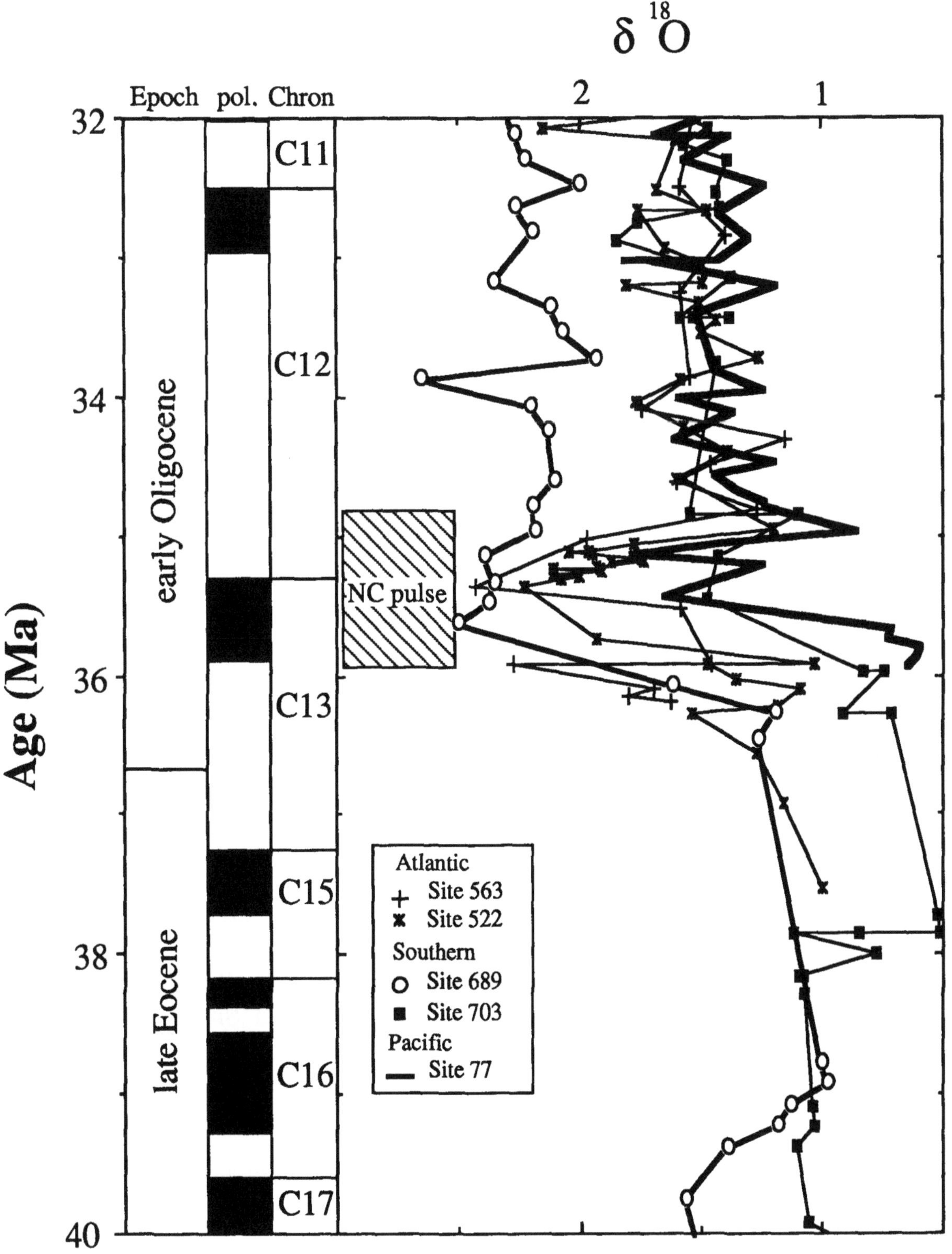

FIGURE 8.4. Summary of late Eocene to earliest Oligocene $\delta^{18}O$ data for *Cibicidoides* from the Atlantic, Pacific, and Southern Oceans. Data sources are listed in Table 8.1.

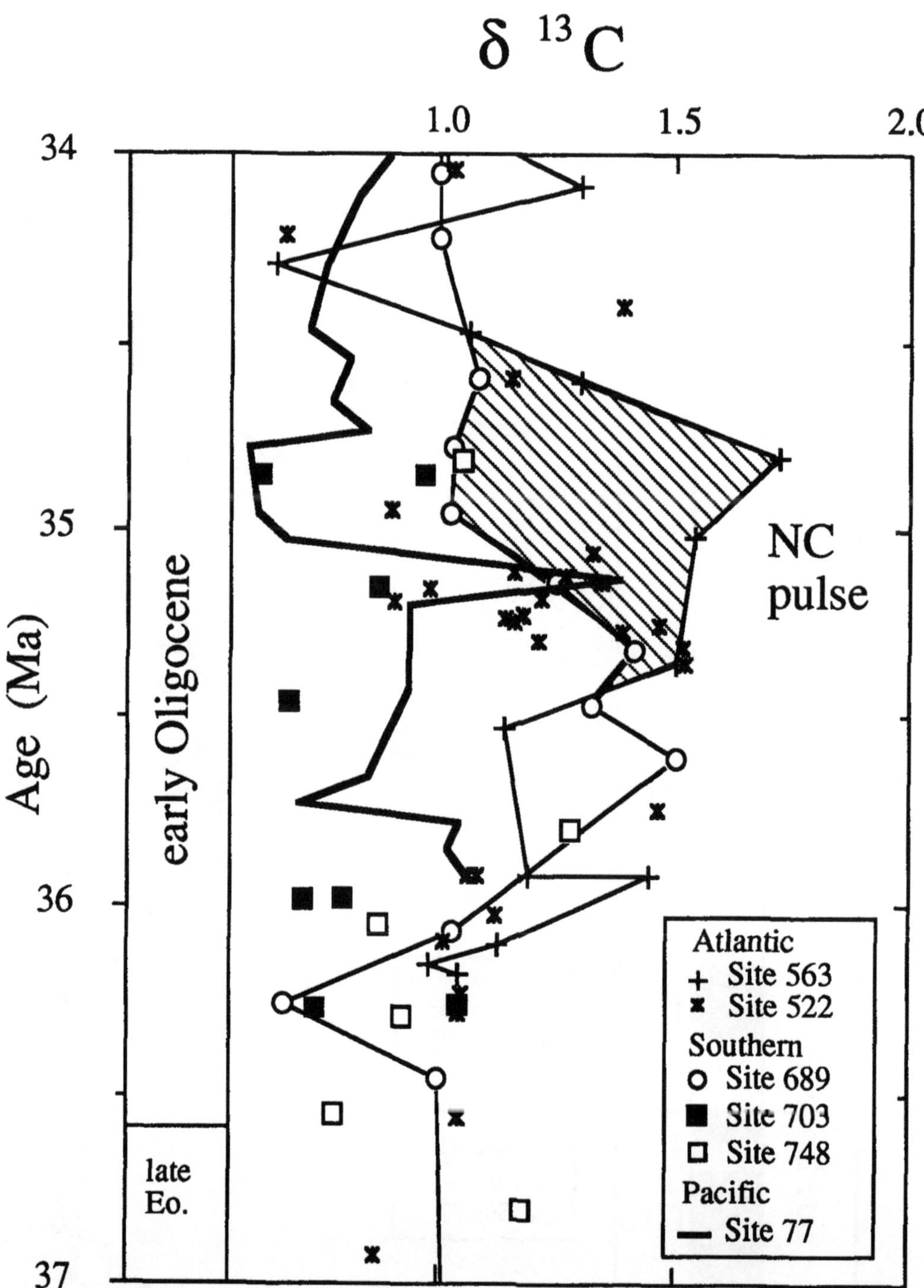

FIGURE 8.5. Summary of latest Eocene to earliest Oligocene $\delta^{13}C$ data for *Cibicidoides* from the Atlantic, Pacific, and Southern Oceans. Data sources are listed in Table 8.1.

addition, the increased ventilation of the deep-sea implied by stable isotope data, seismic stratigraphy, and hiatus distribution is supported by a large drop in the Calcite Compensation Depth (CCD) near the end of the Eocene (e.g., van Andel et al., 1975).

Carbon isotope records from all oceans converged for most of the Oligocene (e.g., Miller and Katz, 1987; Zachos et al., 1991). The absence of detectable basin-basin $\delta^{13}C$ differences during much of the Oligocene is puzzling. The low Oligocene $\delta^{13}C$ differences could be explained by a single deep-water source: the Antarctic. Deposition resumed in the Atlantic during the Oligocene, suggesting that NCW may have ceased (Miller and Tucholke, 1983; Miller et al., 1985b). Zachos et al. (1989) speculated that low $\delta^{13}C$ differences resulted from reduced deep-water connections among basins. However, carbon isotopes may not be the best monitor for the Oligocene ocean. The amplitude of $\delta^{13}C$ differences resulting from basin to basin fractionation is dependent on mean ocean nutrient levels (Boyle, 1986). The Oligocene may have been an oligotrophic ocean with much lower mean ocean nutrient levels than today. This would explain the apparent similarity between basins, because the difference would be within the noise of the $\delta^{13}C$ tracer and uncertainties in correlations. This scenario is supported by the low Oligocene surface to deep-water $\delta^{13}C$ differences (e.g., Miller and Fairbanks, 1985). Such low surface to deep-water differences would be expected for lower mean ocean nutrient levels (Broecker and Peng, 1982).

Recent studies of Antarctic (Maude Rise Sites 689 and 690; Kennett and Stott, 1990) and Subantarctic (Sites 699 and 703; Mead and Hodell, in prep.) locations have suggested that warm saline deep water (WSDW; Brass et al., 1982) influenced high southern latitudes during the Paleogene. Evidence for this consists of a reversed $\delta^{18}O$ gradient between deeper and shallower sites (i.e., lower $\delta^{18}O$ values at deeper sites), especially during the Oligocene. Kennett and Stott (1990) noted lower $\delta^{18}O$ values at Site 690 (~2250 m paleodepth) than at Site 689 (~1650 m paleodepth) during the late Eocene to Oligocene. Mead and Hodell (in prep.) similarly note slightly lower $\delta^{18}O$ values at Site 699 (> 3000 m paleodepth) than at Site 703 (~1000-1500 m paleodepth) during the late Eocene and early Oligocene. Mead and Hodell (in prep.) attributed the warm saline deep water not to low latitude source (i.e., the classic WSDW of Brass et al., 1982), but to an analog of NADW which today transports a tremendous amount of heat and salt (e.g., Worthington, 1970).

My time-series reconstructions suggest a different interpretation of the source of southern ocean waters. Southern Ocean Site 689 had distinctly higher early Oligocene $\delta^{18}O$ values than the rest of the world ocean (Figure 8.4). Southern Ocean Site 690 (not shown) would plot slightly lower in $\delta^{18}O$ values than Site 689, but would always be heavier than Subantarctic, Pacific, and Atlantic locations. This suggests that the sources for the waters at Sites 689 and 690 were from the Antarctic. NCW is observed as a water mass of intermediate $\delta^{18}O$ values (Figure 8.4) but high $\delta^{13}C$ values (Figure 8.5) during the earliest Oligocene. Mixing yielded a Southern Ocean with intermediate $\delta^{13}C$ values and a NCW with intermediate $\delta^{18}O$ values. It is quite clear from Figure 8.4 that NCW could not have been the source of the highest $\delta^{18}O$ values; rather, these values seem restricted to the Antarctic "cold spigot."

CONCLUSIONS AND FUTURE WORK

Benthic foraminiferal $\delta^{18}O$ values increased by nearly 3 ‰ during the middle Eocene to the early Oligocene, constituting the largest climate change of the Cenozoic. This overall increase must be ascribed both to deep-water cooling and to development of ice sheets. The timing of ice sheet development within this interval remains controversial. I conclude that large ice sheets existed in Antarctica during the earliest Oligocene, but that the middle to late Eocene record of ice sheets is debatable.

The studies of Shackleton and Kennett (1975) and Savin et al. (1975) established that Cenozoic marine climates changed in an apparently step-like fashion (e.g., Figure 8.3; see Berger, 1982 for discussion of the importance of these rapid climatic steps). Kennett and Shackleton (176) established that the earliest Oligocene $\delta^{18}O$ increase was fairly rapid (<100 k.y.). In contrast, the available data indicate that early middle and late middle Eocene $\delta^{18}O$

increases each apparently occurred within about a 3 m.y. interval (Figure 8.3). However, the actual nature of these large changes (including the earliest Oligocene increase) are not known, and many "Milankovitch scale" (10-100 k.y.) cycles may be embedded in these long-term signals. More detailed studies (5 k.y. sampling interval) of these important climate transitions are needed.

Deep-water history of the Eocene remains uncertain. To address deep-water history for this interval we need unaltered, complete middle to upper Eocene Pacific and North Atlantic sections. Seismic stratigraphic evidence and hiatus distribution indicate that the earliest Oligocene was an interval of strongly circulating deep-water from both northern (Arctic/ North Atlantic) and southern (Antarctic) sources. Stable isotope data indicate that a pulse of cool, nutrient-depleted NCW and cold, moderate nutrient SCW occurred in the earliest Oligocene. These data are consistent with the deep-water circulation scenario, although more detailed data are needed to confirm this.

ACKNOWLEDGMENTS

This study could not have taken place without M.E. Katz and J.D. Wright, who measured the Site 703 isotope data. I thank them and R.G. Fairbanks (L-DGO) for use of the Finnigan MAT-251 mass spectrometer, M.E. Katz and J.D. Wright for comments, and W.A. Berggren, A.N. Brower, and J. Zachos for reviews. This work was supported by NSF grants OCE88-17563 and OCE90-19569. Samples were supplied by the ODP. This is L-DGO contribution number 4818.

REFERENCES

Barrera, E. and Huber, B. T. 1991. Paleogene and early Neogene oceanography of the southern Indian Ocean: Leg 119, foraminifer stable isotope results. *Proc. Ocean Drill. Prog. Sci. Results* 119: in press.

Barron, J. A. et al. 1989. *Proc. Ocean Drill. Prog. Initial Repts.* 119: 942 pp.

Berger, W.H. 1982. Deep-sea stratigraphy: Cenozoic climate steps and the search for chemo-climatic feedback. In G. Einsele and A. Seilacher, eds., *Cyclic and Event Stratification,* Berlin, Springer Verlag, pp. 121-157.

Berggren, W. A. 1982. Role of ocean gateways in climatic change. In *Climate in Earth History, Studies in Geophysics,* National Academy Press: 118-125.

Berggren, W.A. 1991. Paleogene planktonic foraminifer magnetobiostratigraphy of the southern Kerguelen Plateau (ODP Sites 747-749). *Proc. Ocean Drilling Prog.,* 120, Pt. B: in press.

Berggren, W. A. and Hollister, C. D. 1974. Paleogeography, paleobiogeography and the history of circulation in the Atlantic Ocean. In W. E. Hay, ed., *Studies in Paleo-oceanography, SEPM Spec. Publ.* 20:126-186.

Berggren, W. A. and Hollister, C. D. 1977. Plate tectonics and paleocirculation: commotion in the ocean. *Tectonophysics* 38:11-48.

Berggren, W. A. and Kent, D. V. 1989. Late Paleogene time scale: An update. *Abstr. with Prog GSA Ann. Mtg*: 21(7): A86.

Berggren, W. A., Kent, D. V. and Flynn, J. J. 1985. Paleogene geochronology and chronostratigraphy. In N. J. Snelling, ed., *The Chronology of the Geological Record, Geol, Soc. London, Mem.* 10: 141-195.

Birkenmajer, K. 1987. Tertiary glacial and interglacial deposits, South Shetland Islands, Antarctica: Geochronology versus biostratigraphy (a progress report). *Bull. Polish Academy Sci. Earth Sci.* 36:133-145.

Boyle, E. A. 1986. Paired Cd and carbon isotope data in benthic foraminifera: implication for changes in ocean phosphorous, ocean circulation and atmospheric carbon dioxide. *Geochim. Cosmochim. Acta* 50:256-276.

Boyle, E. A., and Keigwin, L. D. 1982. Deep circulation of the North Atlantic over the last 200,000 years: Geochemical evidence. *Science* 218:784-787.

Brass, G. W., Southam, J. R., and Peterson, W. H. 1982. Warm saline bottom water in the ancient ocean. *Nature* 296: 620-623.

Broecker, W.S., and Peng, T.-H. 1982. *Tracers in the Sea.* Palisades, N.Y., Eldigio Press, 690 pp.

Ciesielski, P.F. et al. 1988. *Proceedings of the Ocean Drilling Program. Initial Reports,* vol. 114, Pt. A, Ocean Drilling Program, College Station, TX, 815 pp.

Corliss, B. H. Aubry, M.-P. Berggren, W. A., Fenner, J. M., Keigwin, L. D., and Keller, G. 1984. The Eocene/Oligocene boundary event in the deep sea. *Science* 226:806-810.

Crowley, T. C. and Matthews, R. K. 1983. Isotope-plankton comparisons in a late Quaternary core with a stable temperature history. *Geology* 11:275-279.

Curry, W. B. and Lohmann, G. P. 1982. Carbon isotopic changes in benthic foraminifera from the western South Atlantic: Reconstruction of glacial abyssal circulation patterns. *Quat. Res.* 18:218-235.

Fairbanks, R. G. and Matthews, R. K. 1978. The marine oxygen isotopic record in Pleistocene coral, Barbados, West Indies. *Quat. Res.* 10: 81-196.

Fenner, J. 1986. Information from diatom analysis concerning the Eocene-Oligocene boundary. In Ch. Pomerol and I. Premoli-Silva, eds., *Terminal Eocene Events.* Amsterdam, Elsevier, pp. 25-40.

Graham, D. W., Corliss, B. H., Bender, M. L., and Keigwin, L. D. 1981. Carbon and oxygen isotopic disequilibria of Recent benthic foraminifera. *Mar. Micropaleo.* 6: 483-497.

Katz, M. E. and Miller, K.G. 1991. Early Paleogene benthic foraminiferal assemblage and stable isotope composition in the Southern Ocean, Ocean Drilling Program Leg 114. *Proc. Ocean Drilling Prog.*, 114, Pt. B: in press.

Keigwin, L. D. 1980. Paleoceanographic change in the Pacific at the Eocene-Oligocene boundary. *Nature* 287:722-725.

Keigwin, L. D. and Keller, G. 1984. Middle Oligocene climate change from equatorial Pacific DSDP Site 77. *Geology* 12:16-19.

Keigwin, L. D. and Corliss, B. H. 1986. Stable isotopes in Eocene/Oligocene foraminifera. *Geol. Soc. Am. Bull.* 97:335-345.

Keller, G. 1986a. Late Eocene impact events and stepwise mass extinctions. In C. Pomerol, and I. Premoli-Silva, eds., *Terminal Eocene Events,* Amsterdam, Elsevier, pp. 403-412.

Keller, G. 1986b. Stepwise mass extinctions and impact events. *Mar. Micropaleo.* 10:267-293.

Keller, G., D'Hondt, S., and Vallier, T. L. 1983. Multiple microtektites horizon in upper Eocene marine sediments: no evidence for mass extinctions. *Science* 221:150-152.

Keller, G., Herbert, T., Dorsey, R., D'Hondt, S., Johnsson, M., Chi, W.R., 1987. Global distribution of late Paleogene hiatuses. *Geology* 15: 199-203.

Kennett, J. P. 1977. Cenozoic evolution of Antarctic glaciation, the Circum-Antarctic Ocean, and their impact on global paleoceanography. *J. Geophys. Res.* 82: 3843-3860.

Kennett, J. P. and Stott, L. D. 1990. Proteus and Proto-Oceanus: Paleogene oceans as revealed from Antarctic stable isotopic results. *Proc. Ocean Drilling Prog.* 113, Pt. B: 865-880.

Kennett, J. P. and Barker, P. F. 1990. Latest Cretaceous to Cenozoic climate and oceanographic developments in the Weddell Sea, Antarctica: An ocean drilling perspective. *Proc. Ocean Drilling Prog.* 113, Pt. B:937-960.

Matthews, R. K. and Poore, R. Z. 1980. Tertiary $\delta^{18}O$ record and glacio-eustatic sea-level fluctuations. *Geology* 8:501-504.

Mead, G. A., and Hodell, D. A. 1991. Late Eocene to early Oligocene vertical oxygen isotope gradients in the South Atlantic: Implications for warms saline deep water, in prep.

Miller, K. G., and Fairbanks, R. G. 1983. Evidence for Oligocene-Middle Miocene abyssal circulation changes in the western North Atlantic. *Nature* 306:250-253.

Miller, K. G. and Fairbanks, R. G. 1985. Oligocene to Miocene global carbon isotope cycles and abyssal circulation changes. In E.T. Sundquist and W.S. Broecker, eds., *The Carbon Cycle and Atmospheric CO_2: Natural Variations Archean to Present, Amer. Geophys. Union, Geophys. Monograph* 32:469-486.

Miller, K. G., and Katz, M. E. 1987. Oligocene to Miocene benthic foraminiferal and abyssal circulation changes in the North Atlantic. *Micropaleontology* 33:97-149.

Miller, K. G. and Thomas, E. 1985. Late Eocene to Oligocene benthic foraminiferal isotopic record, Site 574, equatorial Pacific. *Initial Rept. Deep Sea Drilling Proj.* 85:771-777.

Miller, K. G. and Tucholke, B. E. 1983. Development of Cenozoic abyssal circulation south of the Greenland-Scotland Ridge. In M. H. P. Bott, S. Saxov, M. Talwani, and J. Thiede, eds., *Structure and Development of the Greenland-Scotland Ridge,* Plenum Press, New York: 549-589.

Miller, K. G., Curry, W. B. and Ostermann, D. R. 1985a. Late Paleogene (Eocene to Oligocene) benthic foraminiferal oceanography of the Goban Spur region, Deep Sea Drilling Project Leg 80. *Initial Rep. Deep*

Sea Drill. Proj., 80:505-538.

Miller, K. G, Aubry, M.-P., Khan, J., Melillo, A. J., Kent, D. V., and Berggren, W. A. 1985b. Oligocene to Miocene biostratigraphy, magnetostratigraphy, and isotopic stratigraphy of the western North Atlantic. *Geology* 13:257-261.

Miller, K. G., Fairbanks, R. G. and Mountain, G. S. 1987a. Tertiary oxygen isotope synthesis, sea-level history, and continental margin erosion. *Paleoceanography* 1:1-19.

Miller, K. G., Feigenson, M. D., Kent, D. V., and Olsson, R. K. 1988. Oligocene stable isotope ($^{87}Sr/^{86}Sr$, $\delta^{18}O$, $\delta^{13}C$) standard section, Deep Sea Drilling Project Site 522. *Paleoceanography* 3:223-233.

Miller, K. G., Feigenson, M. D., Wright, J. D., and Clement, B. M. 1991b. Miocene isotope reference Section, Deep Sea Drilling Project Site 608: An evaluation of isotope and biostratigraphic resolution. *Paleoceanography* 6: 33-52.

Miller, K. G., Janecek, T. R., Katz, M. E., and Keil, D. 1987b. Abyssal circulation and benthic foraminiferal changes near the Paleocene/Eocene boundary. *Paleoceanography* 2: 741-761.

Miller, K. G., Wright, J. D., and Fairbanks, R. G. 1991a. Unlocking the Ice House: Oligocene-Miocene oxygen isotopes, eustasy, and margin erosion. *J. Geophys. Res.* 96: 6829-6848.

Mix, A. C., and Fairbanks, R. G. 1985. North Atlantic surface-ocean control of Pleistocene deep-ocean circulation. *Earth Planet. Sci. Lett.* 73: 231-243.

Montanari, A., et al. 1988. Radioisotopic dating of the Eocene-Oligocene boundary in the pelagic sequence of the northeastern Apennines. In I. Premoli-Silva, R. Coccioni, and A. Montanari, *The Eocene-Oligocene Boundary in the Marche-Umbria Basin (Italy)*, Ancona (Italy), pp. 195-208.

Moore, T. C., van Andel, Tj. H., Sancetta, C., and Pisias, N. 1978. Cenozoic hiatuses in pelagic sediments. *Micropaleontology* 24: 113-138.

Mountain, G. S., and Tucholke, B. E. 1985. Mesozoic and Cenozoic geology of the U.S. Atlantic continental slope and rise. In C.W. Poag, ed., *Geologic Evolution of the United States Atlantic Margin*. New York, Van Nostrand Reinhold, pp. 292-341.

Oberhänsli, H., and Toumarkine, M. 1985. The Paleogene oxygen and carbon isotope history of Sites 522, 523, and 524 from the central South Atlantic. In K. J. Hsü and H. J. Weissert, eds., *South Atlantic Paleoceanography*, Oxford, Oxford Univ. Press, pp. 124-147.

Oberhänsli, H., McKenzie, J., Toumarkine, M., and Weissert, H. 1984. A paleoclimatic and paleoceanographic record of the Paleogene in the central South Atlantic (Leg 73, Sites 522, 523, and 524). *Initial Rep. Deep Sea Drill. Proj.* 73:737-747.

Oppo, D., and Fairbanks, R. G. 1987. Variability in deep and intermediate water circulation during the past 25,000 years: Northern hemisphere modulation of the southern ocean. *Earth Planet. Sci. Lett.* 86: 1-15.

Raup, D.M., and Sepkoski, J.J. 1982. Mass extinctions in the marine fossil record. *Science* 215:1501-1503.

Poore, R. Z., and Matthews, R. K. 1984. Late Eocene-Oligocene oxygen and carbon isotope record from South Atlantic Ocean DSDP Site 522. *Initial Rep. Deep Sea Drill. Proj.* 73: 725-735.

Premoli-Silva, I., Coccioni, R., and Montanari, A., eds. 1988. The Eocene-Oligocene boundary in the Marche-Umbria Basin (Italy), *Internat. Subcomm. Paleogene Strat., Special Publ.* Ancona: 268 pp.

Prentice, M. L., and Matthews, R. K. 1988. Cenozoic ice-volume history: Development of a composite oxygen isotope record. *Geology* 17:963-966.

Savin, S.M. 1977. The history of the earth's surface temperature during the past 100 million years. *Ann. Rev. Earth Planet. Sci.* 5: 319-355.

Savin, S. M., Douglas, R. G. and Stehli, F. G. 1975. Tertiary marine paleotemperatures, *Geol. Soc. Am. Bull.* 86: 1499-1510.

Shackleton, N. J., and Kennett, J. P. 1975. Paleotemperature history of the Cenozoic and initiation of Antarctic glaciation: Oxygen and carbon isotopic analyses in DSDP Sites 277, 279, and 281. *Initial Rep. Deep Sea Drill. Proj.* 29: 743-755.

Shackleton, N. J., and Opdyke, N. D. 1973. Oxygen isotope and paleomagnetic stratigraphy of Equatorial Pacific core V28-238:

Oxygen isotope temperatures and ice volumes on a 10^5 year and 10^6 year scale. *Quat. Res.* 3: 39-55.

Shackleton, N. J., Imbrie, J. and Hall, M. A. 1983. Oxygen and carbon isotope record of East Pacific core V19-30: Implications for the formation of deep water in the late Pleistocene North Atlantic. *Earth Planet. Sci. Lett.* 65: 233-244.

Shackleton, N. J., Hall, M. A., and Boersma, A. 1984. Oxygen and carbon isotope data from Leg 74 foraminifers. *Initial Rep. Deep Sea Drill. Proj.* 74: 599-612.

Snyder, S.W., Muller, C. and Miller, K.G. 1984. Biostratigraphy and paleoceanography across the Eocene/Oligocene boundary at Site 549. *Geology* 12:112-115.

Speiss, V. 1990. Cenozoic magnetostratigraphy of Leg 113 drill sites, Maud Rise, Weddell Sea, Antarctica. *Proc. Ocean Drilling Prog.* 113, Pt. B:261-315.

Stott, L. D., and Kennett, J. P. 1990. Antarctic Paleogene planktonic foraminiferal biostratigraphy: ODP Leg 113, Sites 689 and 690. *Proc. Ocean Drilling Prog.* 113, Pt. B:549-569.

Swisher, C. C. and Prothero, D. R. 1990. Single-crystal $^{40}Ar/^{39}Ar$ dating of the Eocene-Oligocene transition in North America. *Science* 249: 760-762.

Tauxe, L., Tucker, P., Peterson, N. P., and LaBrecque, J. L. 1984. Magnetostratigraphy of Leg 73 sediments. *Initial Rep. Deep Sea Drill. Proj.* 73: 609-612.

Thomas, E. 1985. Late Eocene to Recent deep-sea benthic foraminifers from the central equatorial Pacific Ocean. *Initial Rep. Deep Sea Drill. Proj.* 85: 655-694.

Tucholke, B.E., and Vogt, P.R. 1979. Western North Atlantic: Sedimentary evolution and aspects of tectonic history. *Initial Rep. Deep Sea Drill. Proj.* 43: 791-825.

van Andel,Tj.H., Heath, G.R., and Moore, T.C. 1975. Cenozoic tectonics, sedimentation, and paleoceanography of the central equatorial Pacific. *Geol. Soc. Amer. Mem.* 143.

Van Couvering, J. A., Aubry, M.-P., Berggren, W. A., Bujak, J. P. and Wiesser, T. 1981. The terminal Eocene event and the Polish connection. *Palaeogeogr. Palaeoclimatol. Palaeoecol.* 36:321-362.

Wolfe, J.A. 1978. A paleobotanical interpretation of Tertiary climates in the Northern Hemisphere. *Amer. Scientist* 66:694-703.

Worthington, L. V. 1970. The Norwegian Sea as a mediterranean basin, *Deep-Sea Res.* 17: 77-84.

Wright, J. D., Miller, K. G. and Fairbanks, R. G. 1991. Evolution of modern deep-water circulation: Evidence from the late Miocene Southern Ocean. *Paleoceanography*: in press.

Zachos, J., Berggren, W. A., Aubry, M.-P., and Mackenson, A. 1991. Eocene-Oligocene climatic and abyssal circulation history of the southern Indian Ocean. *Proc. Ocean Drilling Prog., Sci. Rep.* 120:in press.

Zachos, J., et al. 1989. Paleogene climatic and circulation changes of the Southern Oceans: Inferences from stable isotope data. *Third Int. Conf. Paleoceanography*, Cambridge.

9. LATE EOCENE AND EARLY OLIGOCENE IN SOUTHERN AUSTRALIA: LOCAL NERITIC SIGNALS OF GLOBAL OCEANIC CHANGES

By Brian McGowran, Graham Moss, and Amanda Beecroft

ABSTRACT

Foraminiferal profiles have been established for two parallel sections through most of the upper Eocene and lower Oligocene in southern Australia. The sections are from the St.Vincent and Otway Basins, chosen to contrast restricted neritic conditions (low plankton numbers) with open conditions (relatively high plankton numbers).

Taxic patterns display a consistency between the Eocene and the Oligocene and between the two basins. Thus: species' first appearances attain their highest numbers at marine transgressions and their last appearances are most numerous at a marine regression (the Chinaman Gully Formation). The Simpson coefficient of similarity shows that interbasin taxic similarity is highest at the earliest Oligocene transgression. The greatest turnover in foraminiferal species is at the Chinaman Gully. Biofacies profiles are based on the relative abundances of higher taxa. They too show the greatest change at the Chinaman Gully, and they too show stronger interbasinal similarities than contrasts.

The regression is correlated with the major shift in oceanic $\delta^{18}O$ profiles, inferred to be the cooling event, and with the inferred fall in sealevel. Our local neritic evidence supports the notion of a short, sharp, bipolar, transenvironmental Terminal Eocene Event rather than merely a conclusion to protracted late Eocene change. (This TEE postdates the oceanic microspherule horizons and may be earliest Oligocene.) Whilst the local neritic changes can be explained in large part by a ventilating of the immediate benthic environment at the Chinaman Gully, it is reasonable to view them as a microcosm of global oceanic changes including the breakdown of a tendency toward water stratification at the onset of the psychrosphere. Furthermore, biofacies changes are both crisper, and mark much more clearly the maximum flooding surface, in the late Eocene chemical facies than in the Oligocene bryozoal siliciclastics. We are tempted to suggest that that too is of more pervasive importance.

INTRODUCTION

The increased scrutiny of "mass extinctions" and of other events in the stratigraphic record has enforced a much more detailed and quantified plotting of the fossil record itself than was deemed necessary until quite recently. Apart from tracing the iridium-enriched layer and associated phenomena at the Mesozoic/Cenozoic boundary, most of that scrutiny has been invested in oceanic sections. And for good reason: the microfossil record is richer; control is better with hydraulic piston cores; correlation and age determination are tighter in the presence of magnetostratigraphy and of diverse oceanic microplanktonic assemblages.

There is not a great amount of modern research on neritic benthic foraminifera that addresses specifically such interesting and important questions as: What does the record tell us about the Terminal Eocene Event (TEE)? Do we see a TEE when we respond to the analysis of systematic monographs that indicates a late Eocene mass extinction by inspecting critical sections in detail? Is there a link between the putative extraterrestrial impact (microspherules etc.) and enhanced extinction rates? Perhaps most importantly:

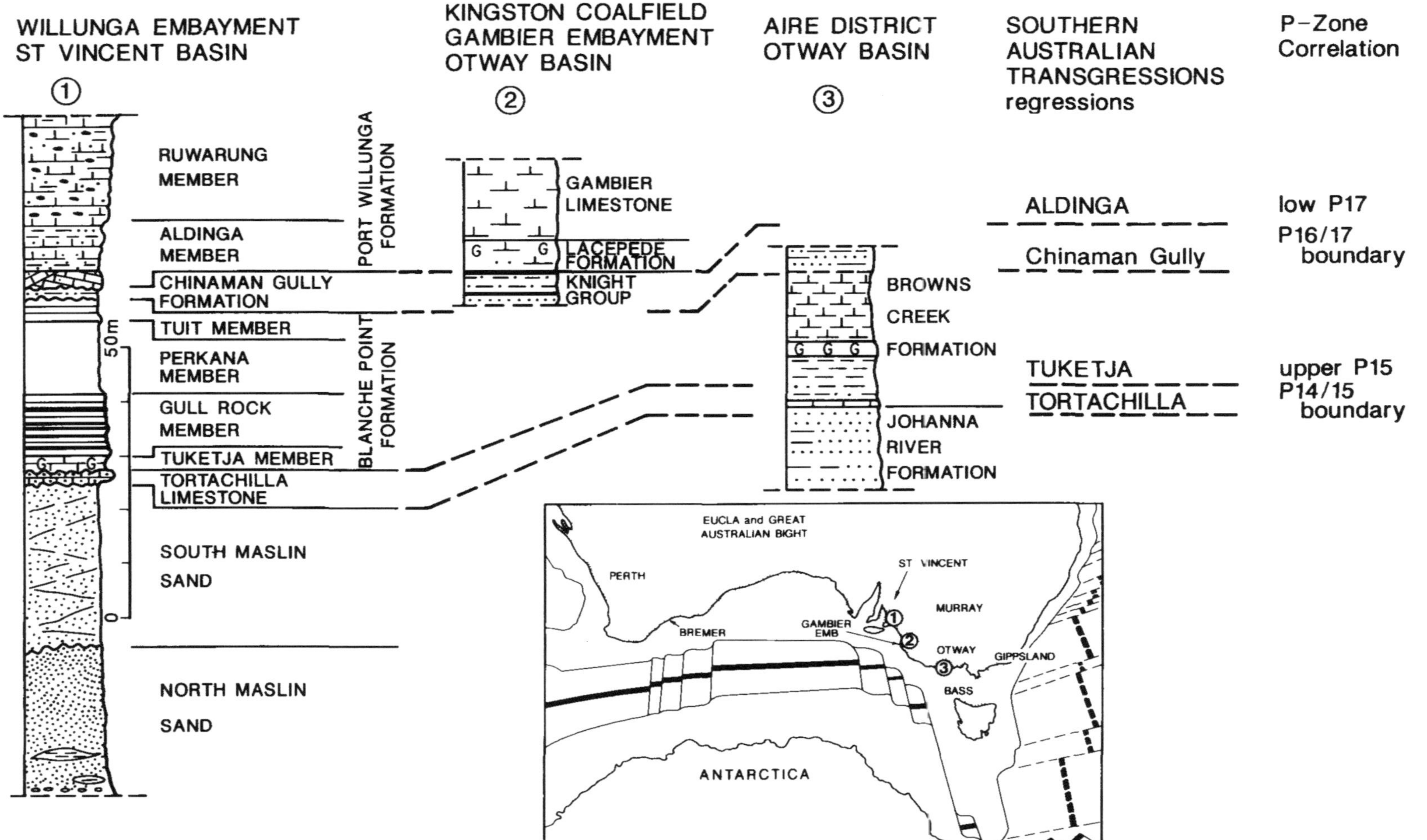

FIGURE 9.1a. The sections studied are from the St.Vincent Basin (1) and from the Otway Basin at its western end (2) and at its eastern end (3). Note that the St.Vincent Basin is behind Kangaroo Island, a Delamerian (Caledonian) massif that inhibited oceanic influence. The outcropping section (1) is 200 km from the shelf edge.

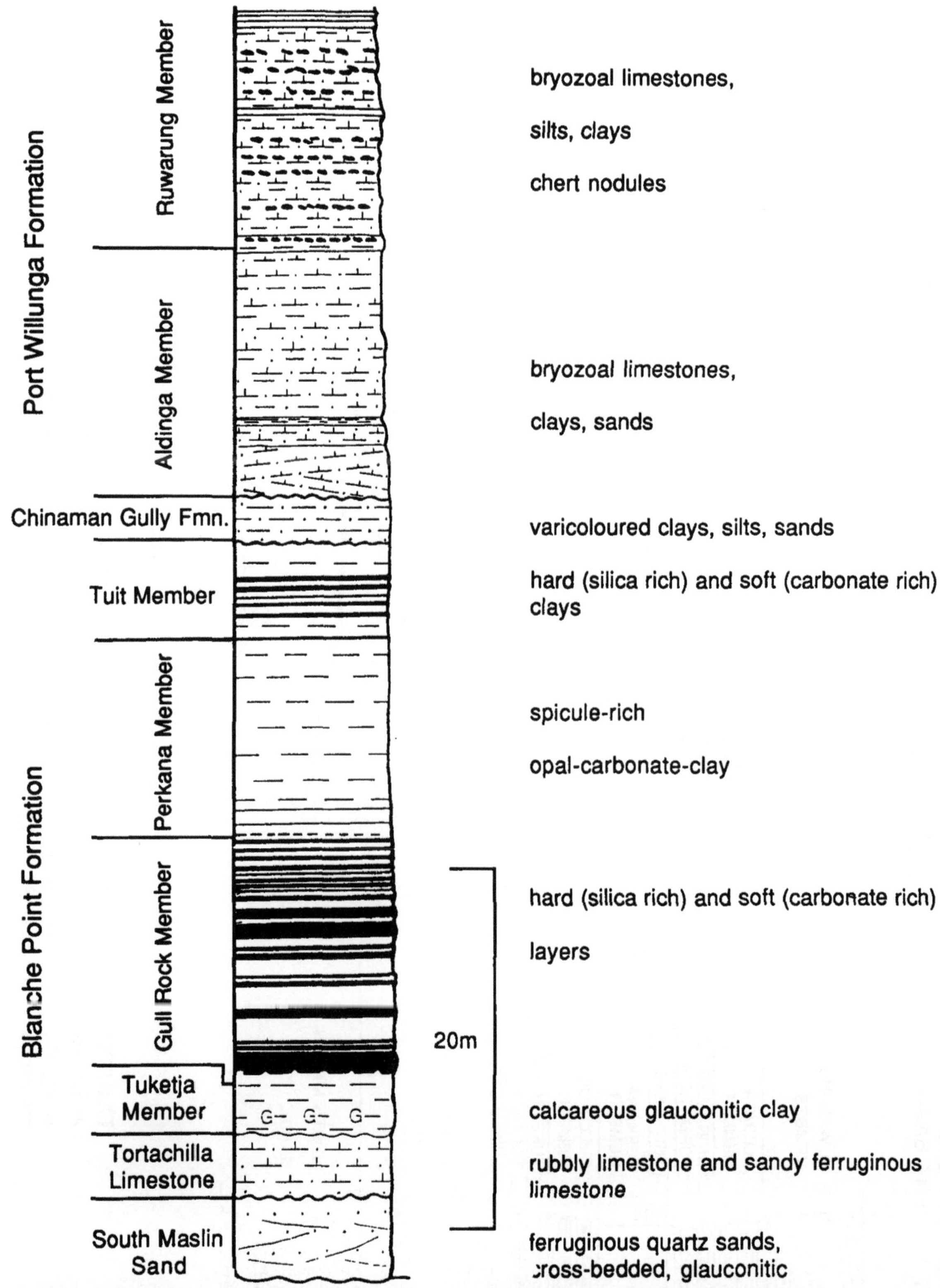

FIGURE 9.1b. The lithostratigraphic section in the Willunga Embayment [upper part of (1) in Figure 9.1a], based on Mack (1989), Moss (1989), and other sources. The lower part of this outcropping section is illustrated by McGowran (1990a, Figure 1) and the upper part by Lindsay and McGowran (1986, Figure 1).

can we establish better diplomatic relationships between the neritic and oceanic realms?

Some of those rhetorical questions bear upon the geohistorical record and its possible interaction with the biohistorical record. It has been said many times that the fossil record of the foraminifera, so rich and detailed and eminently applicable to biostratigraphy and biofacies, has contributed pathetically little to the progress of macroevolutionary research. That is changing--more in the pelagic realm--but paleobiological or evolutionary analysis for its own sake, as something more than a spinoff from applied research, is still rather slow.

In this paper we summarize and extend previous work on the patterns of relative abundances--biofacies--and of species turnover, or taxic change from the late Eocene into the early Oligocene. These neritic communities lived at about 55°S, or some 20° further south than their present location.

Taxic change in neritic communities only rarely can be separated into cladogenetic and migrational components, but at least we can make a beginning with the census, as it were. We emphasize always the comparative approach. Thus, we compare patterns across the critical stratigraphic boundaries and between basins of contrasting neritic environments. We indulge in a little extrapolation from the local-neritic to the global-oceanic. It all depends on correlation, but our most basic conclusion is that the evidence even from this mid-high latitude continental shelf suggests that we have barely begun to exploit the fertile interrelationships and feedbacks between the oceans and their margins at crucial times such as the Eocene-Oligocene transition.

STRATIGRAPHY

The location of the studied sections is shown in Figure 9.1. A more general study showed that the later Eocene marine stratigraphic record in southern Australia is essentially one of four rapid, far-reaching transgressions . In ascending order, they are the Wilson Bluff, Tortachilla, Tuketja, and Aldinga transgressions (McGowran, 1989a). Three of the names are taken from rock units in the St.Vincent Basin, as shown here. The St.Vincent Basin section is focused in Figure 9.1a and Table 9.1 displays the broad outlines of the units and their characteristics and some sequence-stratigraphic horizons, anticipating discussion on correlation, below.

The four members of the Blanche Point Formation are coloured grey-green to almost black (upper Gull Rock; Tuit) and their mineralogy is dominated by opaline silica, smectite, clinoptilolite, carbonate, glauconite; they are remarkably impoverished in kaolinite and detrital quartz (Jones and Fitzgerald, 1984, 1987). Richly fossiliferous in several horizons, the Blanche Point nevertheless is high in faunal dominance (especially the infaunal snail *Spirocolpus* and the shrimp burrow *Thalassinoides*) and low in diversity. In contrast, the bracketing sediments are well oxidized, yellow-red-brown in colour, and of quartz-bryozoal facies of extratropical aspect. That contrast is the most significant generalization to be made about this section and correlates closely with the microfaunal changes. The Blanche Point is bounded by an emergent unconformity below and a nonmarine unit above. The prominent banding in the opal-CT-rich Gull Rock is grouped into five bundles probably recording the Milankovitch 10^5 years band (McGowran, 1990a, Figure 1; Mack,1989).

The Otway Basin section is composite, as shown in Figure 9.1. The sands, clays and marls of the Aire district are succeeded by the glauconitic calcilutites and calcarenites of the Kingston coalfield, the two sections being linked confidently by a regressive to nonmarine interval, identified and referred to henceforth as the Chinaman Gully regression. Correlations between the sheltered St.Vincent Basin and the Otway Basin, facing the nascent Southern Ocean more openly, are shown in all the comparative profiles here.

SPECIES OVERTURN

We have updated and extended a previous study of benthic species overturn (McGowran and Beecroft, 1986b; McGowran, 1987). Figure 9.2 presents a census of the incomings and outgoings of about 200 species from the Tortachilla transgression to well into the Oligocene. The data are shown visually as three bar graphs of species numbers for samples compounded into 3- or 5-meter lots. The graphs respectively are the standing simple diversity, the species

TABLE 9.1. Summary of stratigraphy, biofacies, sequence events, Willunga Embayment of St.Vincent Basin, as discussed in text. In the Tortachilla Limestone and the Aldinga Member, the benthic foraminiferal associations of *Halkyardia, Linderina,*and *Maslinella* indicate warm-water incursions at this mid-paleolatitude.

formation member	minerals macrofauna	benthic foram. association	comments
	Willungan/Janjukian Stage boundary regression and sequence boundary		
Ruwarung Member	as below also cherty	*Cibicides* elphidiids	sequence strat. obscured
	maximum flooding surface		
Aldinga Member	quartz bryozoal echinoidal	*Cibicides* *Halkyardia-* *Linderina-* *Maslinella*	restoration of neritic carbonate facies
	Aldinga transgression		
Chinaman Gully Formn	detrital	--	essentially nonmarine
	Chinaman Gully regression sequence boundary		
Tuit Member	banded opal-CT rich *Spirocolpus* *Limopsis*	*Uvigerina* *Sphaeroidina* *Cerobertina*	last grey-green (black) facies
			[sequence stratigraphy obscure]
Perkana Member	poorly bedded opal-A rich spongolite	*Bolivina* *Sphaeroidina* *Uvigerinella*	
	maximum flooding surface		
Gull Rock Member	banded opal CT-rich *Spirocolpus* *Thalassinoides*	*Cibicides* *Uvigerina*	very dark at top
Tuketja Member	glauconitic marl, mollusc-rich	*Cibicides* *Uvigerina*	first grey-green facies
	Tuketja transgression hiatus and hard ground; sequence boundary		
Tortachilla Limestone	shelly fauna abundant	*Cibicides* *Halkyardia-* *Linderina-* *Maslinella*	neritic carbonate facies
	maximum flooding surface		
South Maslin Sand	detrital fossils sparse	--	"estuarine"
	Tortachilla transgression sequence boundary		
North Maslin Sand	detrital oxbow plant assemblages	--	nonmarine

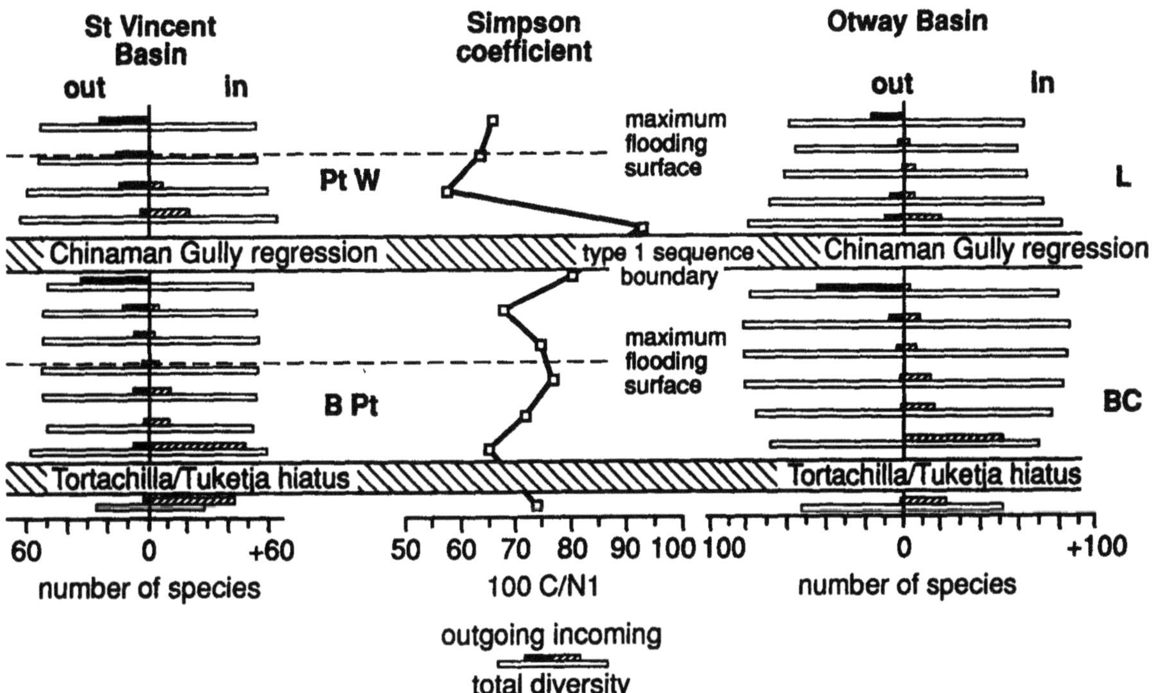

FIGURE 9.2. Summary of taxic change in the benthic foraminiferal assemblages. For abbreviations, see lithostratigraphic names in Figure 9.1.

appearing for the first in the respective sections, and the species appearing for the last time. Of course, to begin, we have to plot incomings against something that went before; and our only option was the rather unsatisfactory chalky bryozoal facies of the Wilson Bluff Limestone in the Eucla Basin to the west (McGowran and Beecroft, 1986b). The number of species incoming at the Tortachilla probably is exaggerated. Similarly, we had to truncate the top of the study so that outgoings had a chance of meaning something more than being merely the top of the available sections. The final departures may be exaggerated a little. But neither caveat significantly distorts the conclusions to be drawn here.

It is immediately apparent that species turnover varies more than does total diversity. Incomings are highest at the transgressions--the Tortachilla, Tuketja and Aldinga transgressions. Outgoings are low in number at first but build up toward the Chinaman Gully regression, drop sharply, then build up again. Since almost all species carry over from the Tortachilla to the Tuketja, we infer that the intervening regression had much less impact on the communities than did the later Chinaman Gully, even though there was a pronounced wave of newcomers at the Tuketja.

In contrast, the peak in taxic turnover at the Chinaman Gully is composed of both arrivals and departures. Clearly, that event had a major impact on the benthic communities.

Equally striking in this taxic pattern is the parallelism between the two successions. It was, accordingly, of some interest to ask whether the parallel patterns extended to parallel impacts at the specific level: do the same species tend to come in or go out at the same times? To test this we applied the Simpson coefficient of similarity to the profiles (Figure 9.2). It measures *c*, species in common, against *n1*, the smaller of the two diversities (Simpson, 1960). The results show that the similarity is most marked at the Chinaman Gully event, especially at the time of the Aldinga restocking, but that it drops immediately to its lowest value. Thus the wave of replacements lowered endemism but the effect did not persist. Again, the earlier Tortachilla/Tuketja event failed to engender so much drama.

We asked also whether there was any variation in pattern among the higher taxa of benthic foraminifera. Figure 9.3 suggests that there is no marked partitioning of the main pattern among the taxa. The most species-rich orders, the Nodosariida, Buliminida and Uvigerinida, are similar to the main game. Finally, Figure 9.4 converts the taxic data into rates (Wei and Kennett, 1986).

BIOFACIES CHANGES ACROSS THE EOCENE/OLIGOCENE BOUNDARY

First we present profiles of the relative abundances of selected taxa through the Blanche Point Formation (Figure 9.5). The taxa range in category from genus to suborder; note too the considerable variation in the scales of percentage abundances among the profiles.

It is clear, as has been emphasized already (McGowran and Beecroft, 1986a), that the major changes apparent on simple visual inspection are consonant with the lithostratigraphic boundaries, which in turn follow the major mineralogical changes (Table 9.1). Thus, the Uvigerinidae rise sharply at the onset of the accumulation of grey-green sediments (Tuketja Member), fluctuate spectacularly in counterpoint with the Cibicididae in the Gull Rock and Tuit Members, but remain relatively low through the Perkana Member. The Bolivinidae tend to replace the Uvigerinidae at the Gull Rock/Perkana contact just as *Sphaeroidina* and *Cassidulina* begin to make their run, but retreat to low numbers again in the Tuit Member.

Inspection of the other profiles supports the close match between biofacies and lithofacies and the importance of the lithostratigraphic boundaries.

Those profiles are extended above the Chinaman Gully Formation and compared with profiles from the Otway Basin in Figures 9.6a,b,c, and 9.7. The comparison with the Otway Basin requires correlation, which is achieved mostly by using the events labelled *a-f* in Figure 9.7. (Note that these correlations require that the respective last peaks in *Globigerinatheka index*, labelled *(i)*, *(ii)*, are not coeval. The last peak in the Blanche Point matches the penultimate peak in the other section.) The essential contrast underlying all of these comparisons between the two basins is illustrated in the profiles for the plankton (Globigerinida), which is always much more common in the Otway Basin profile.

The dominant higher taxa overall are the Uvigerinidae and the Cibicididae. In the St.Vincent Basin there is a sustained and pronounced decrease in uvigerinid numbers across the Chinaman Gully. That is not so clear in the composite Otway Basin profile, where they are common enough but never so dominant in the richer, open-neritic Browns Creek assemblages and where the Lacepede Formation is glauconitic; these factors combine to dampen the contrast. But the Cibicididae sustain a marked increase in both profiles. The most noteworthy change otherwise is the incoming in substantial numbers of the Elphidiidae. Both that event and the disappearance of the Miliolida in significant numbers are diachronous---taking the Chinaman Gully horizon as the control---probably reflecting the facies/environmental differences between the Aldinga and the Lacepede.

The main conclusions that we draw from these profiles are as follows:

(i) The incoming in sustained numbers of the Cibicididae and the Elphidiidae, together with the big drop in one profile in the Uvigerinidae, signals the facies change that is also seen in the change from grey-green to yellow-brown sediments. Even the Globigerinida, for all their contrasts, reflect that change.

(ii) The dramatic changes seen at lithological/mineralogical boundaries within the grey-green facies are not matched in intensity in the succeeding regime. The changes emphasized in the Blanche Point Formation (Figure 9.5) are not repeated or mimicked in the Port Willunga Formation.

(iii) On the whole, the parallels between the profiles from the two basins, easily contrasted by their planktonic counts, are more impressive than are their differences. Perhaps that generalization is brought home best in the event correlations achieved among both the plankton and the benthos and illustrated in Figure 9.7.

CORRELATION AND AGE DETERMINATION

It is one thing to identify the Chinaman

St Vincent Basin foraminiferal turnover

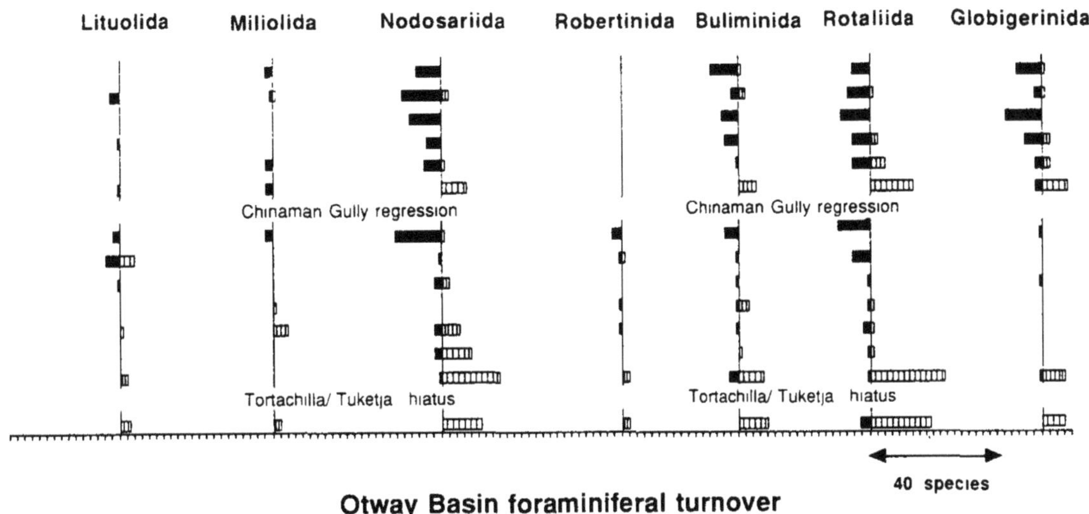

Otway Basin foraminiferal turnover

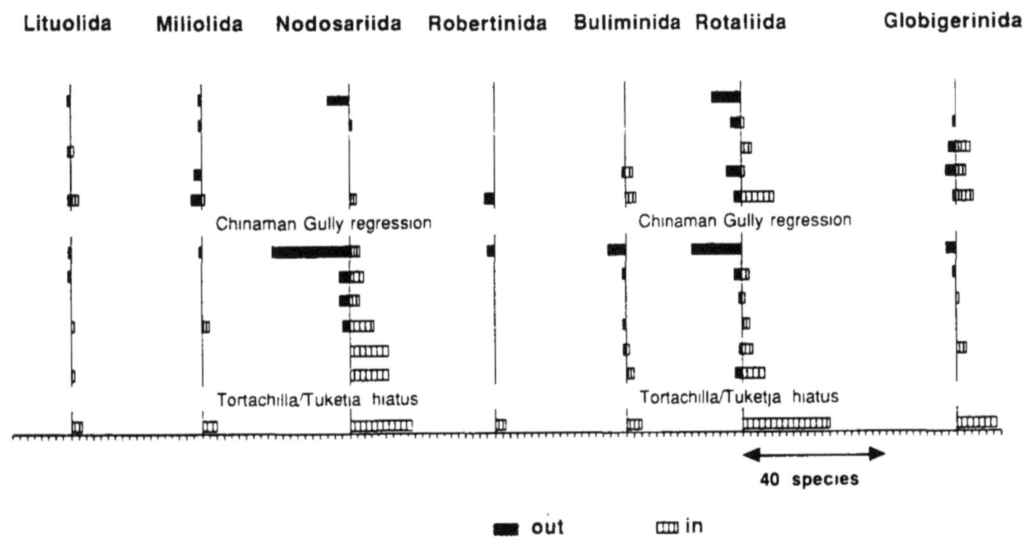

FIGURE 9.3. Foraminiferal turnover in the St.Vincent and Otway sections, by taxonomic order, with the plankton added.

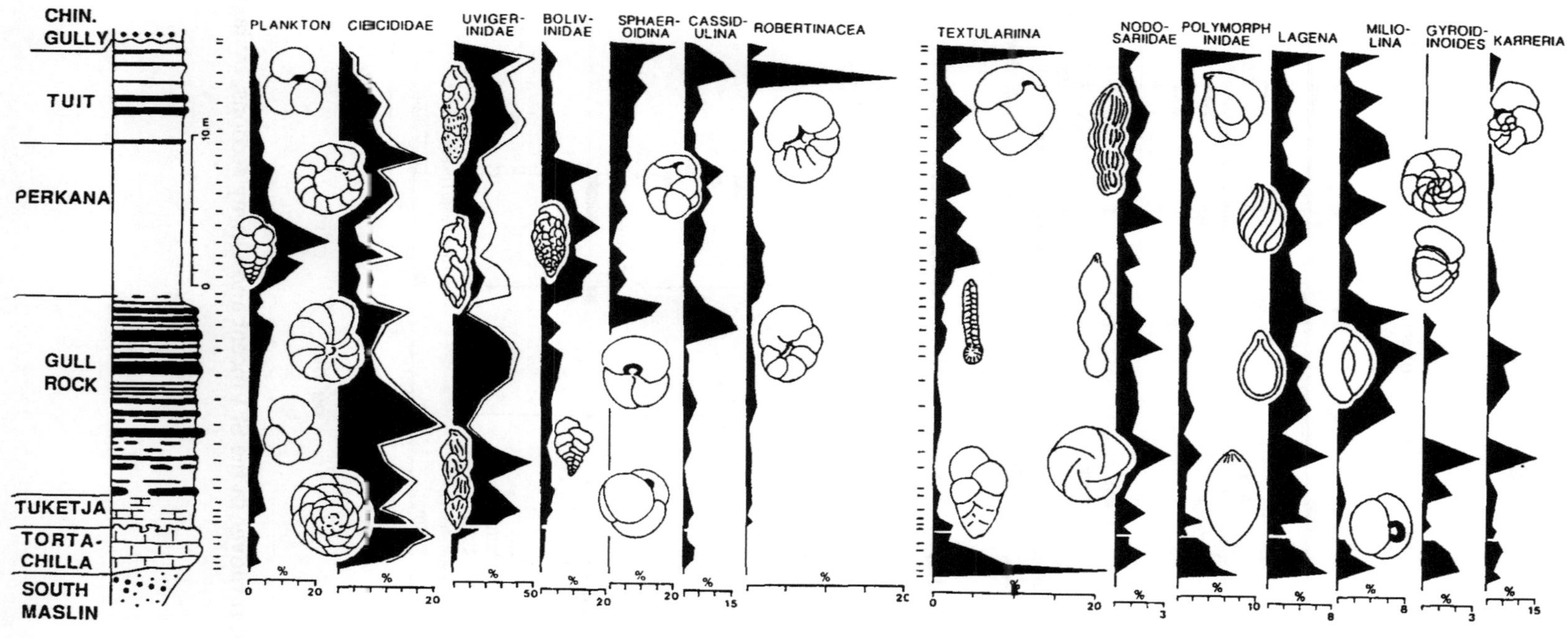

FIGURE 9.5. Foraminiferal biofacies changes in the Tortachilla Limestone and Blanche Point Formation, based on selected taxa of various ranks, as discussed by McGowran and Beecroft (1986). Note the different percentage scales.

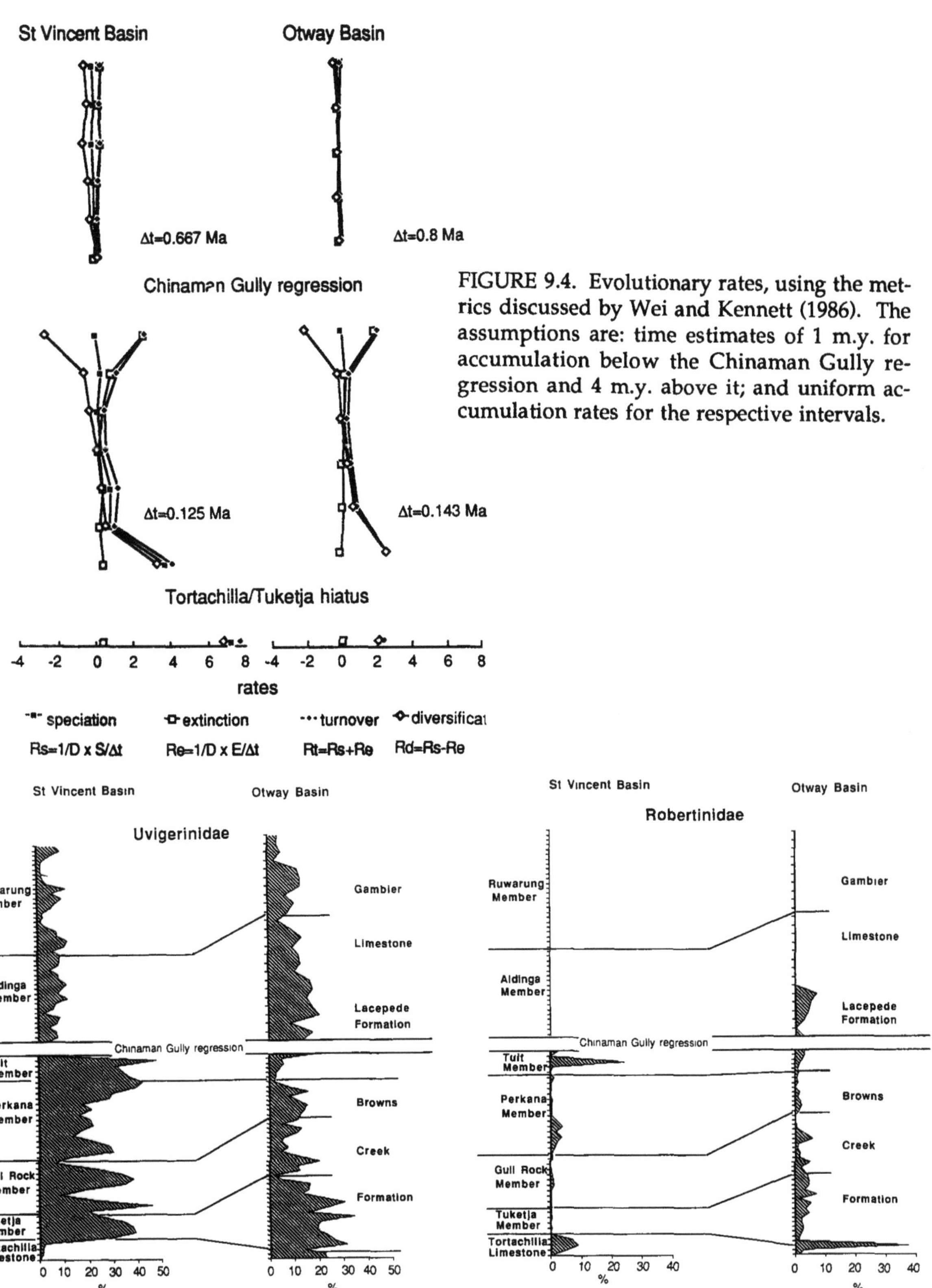

FIGURE 9.4. Evolutionary rates, using the metrics discussed by Wei and Kennett (1986). The assumptions are: time estimates of 1 m.y. for accumulation below the Chinaman Gully regression and 4 m.y. above it; and uniform accumulation rates for the respective intervals.

FIGURES 9.6a, 9.6b, 9.6c. Comparative profiles of relative abundances, St.Vincent and Otway Basin sections. One genus and seven higher taxa of benthic foraminifera are profiled, plus the plankton (Globigerinidae). Lines are correlation ties; ticks are sampling points.

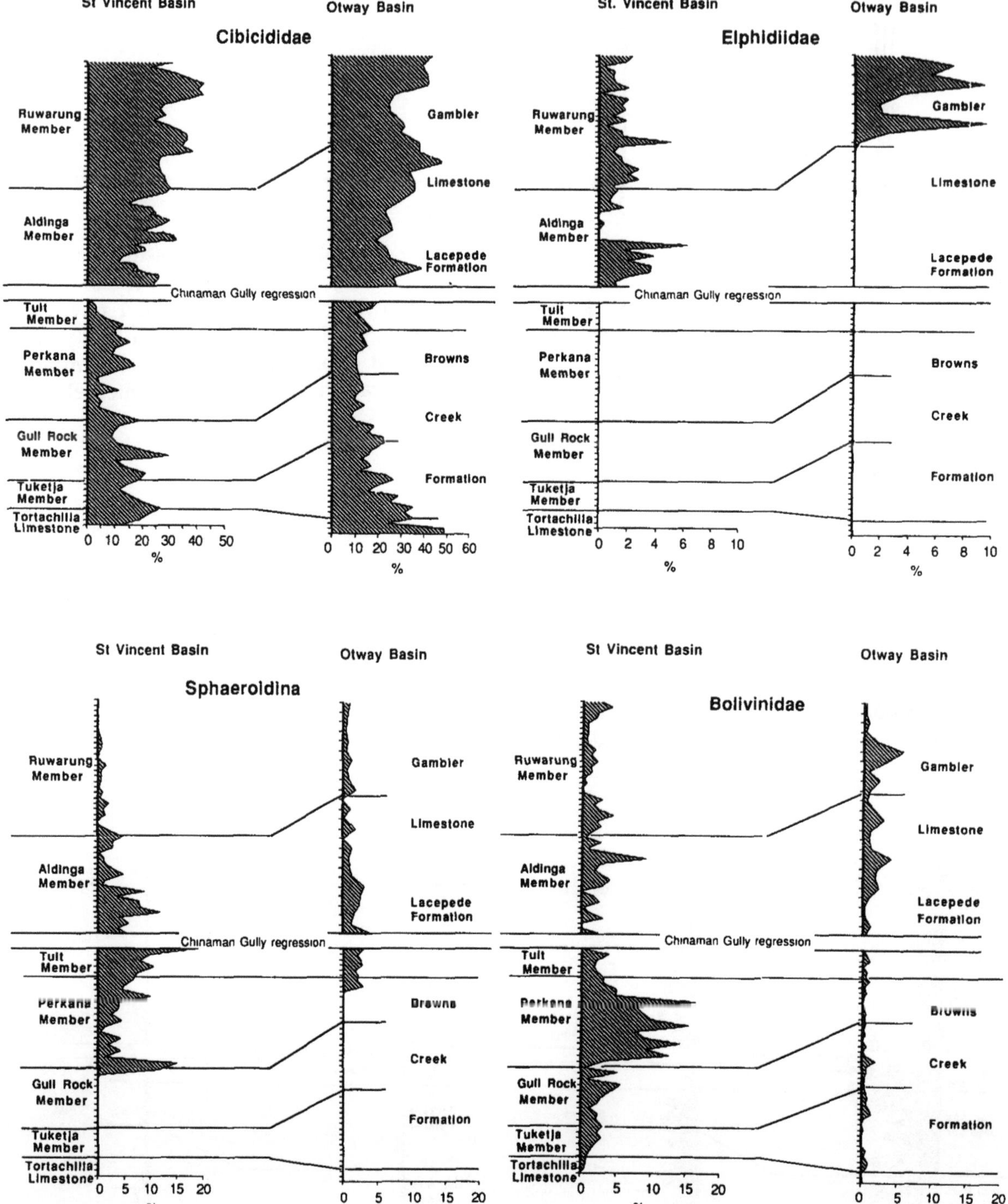
St Vincent Basin
Otway Basin
Cibicididae
Ruwarung Member
Aldinga Member
Gambler
Limestone
Lacepede Formation
Chinaman Gully regression
Tuit Member
Perkana Member
Gull Rock Member
Tuketja Member
Tortachilla Limestone
Browns
Creek
Formation
%
St. Vincent Basin
Elphidiidae
Sphaeroidina
Bolivinidae

FIGURE 9.6b

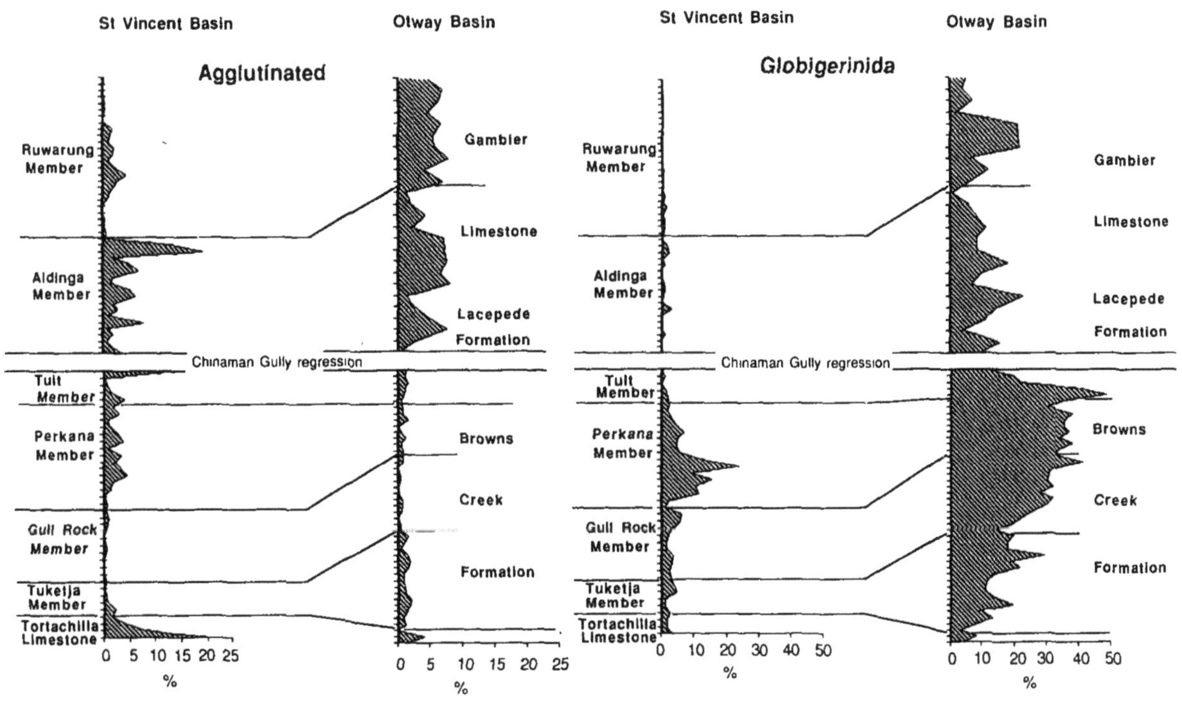
St Vincent Basin
Otway Basin
Agglutinated
Ruwarung Member
Aldinga Member
Chinaman Gully regression
Tuit Member
Perkana Member
Gull Rock Member
Tuketja Member
Tortachilla Limestone
Gambier
Limestone
Lacepede
Formation
Browns
Creek
Formation
0 5 10 15 20 25
%
St Vincent Basin
Otway Basin
Globigerinida
0 10 20 30 40 50
%

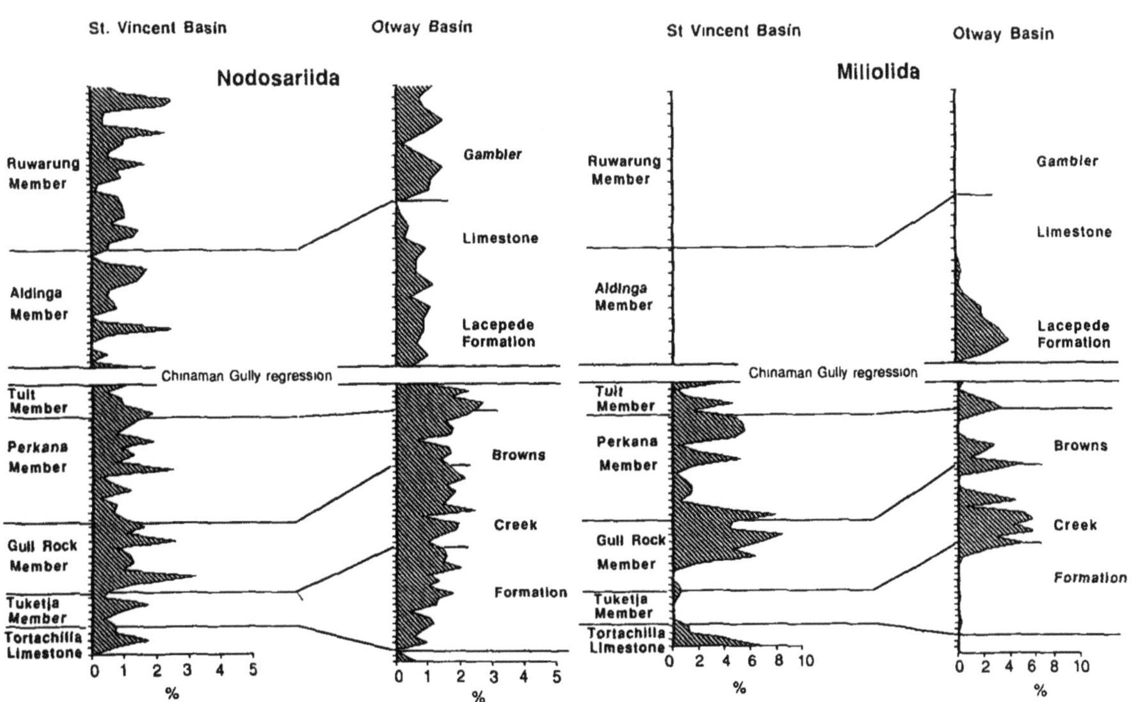
St. Vincent Basin
Otway Basin
Nodosariida
Ruwarung Member
Aldinga Member
Chinaman Gully regression
Tuit Member
Perkana Member
Gull Rock Member
Tuketja Member
Tortachilla Limestone
Gambier
Limestone
Lacepede Formation
Browns
Creek
Formation
0 1 2 3 4 5
%
St Vincent Basin
Otway Basin
Miliolida
0 2 4 6 8 10
%

FIGURE 9.6c

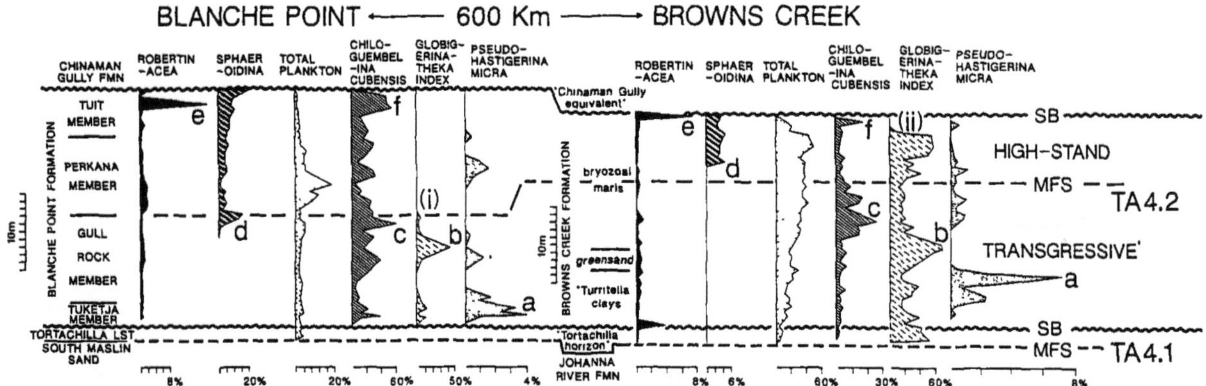

FIGURE 9.7. Event correlation between the Tortachilla-Blanche Point and Browns Creek Formations. Sequence stratigraphy is also shown (SB, sequence boundary; MFS, maximum flooding surface). Note that if relative abundance events (a) to (f) are approximately correlated, as seems likely, then the last occurrences of *Globigerinatheka index* at (i) and (ii) are allochronous.

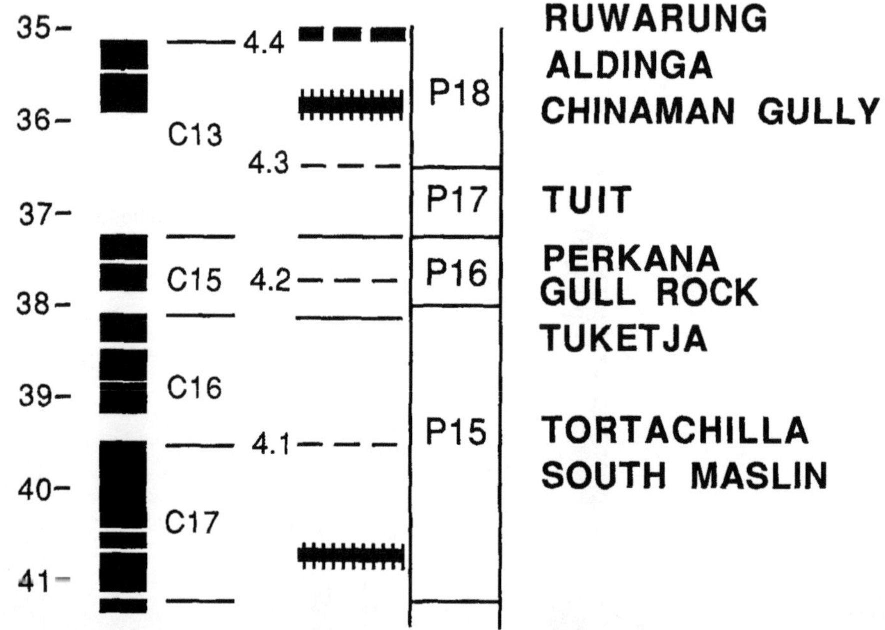

FIGURE 9.8. Correlations of P-zones, chrons, and sequence-stratigraphic events with the rock units of the St.Vincent Basin, scaled to the geochronology of Aubry et al.(1988). Those authors place the Eocene/Oligocene boundary within Zone P17 and hence quite low in Chron C13R.

Gully regression unambiguously as a particularly strong cluster of paleobiological events; it is quite another to show that it correlates with a major global exogenic change. As ever, the problems of correlation here boil down to a lack of biostratigraphic events in common between these mid-high latitude neritic facies and the lower latitude, deeper water facies including proposed boundary stratotypes. We have considered these problems repeatedly (Lindsay, 1985; Lindsay and McGowran, 1986; McGowran and Beecroft, 1986a,b; McGowran, 1987, 1989a, 1990b). However, it is necessary to outline a correlation once again, particularly since we have changed our collective mind on a rather small point---but a point that makes all the difference in whether there is a satisfying fit between the local and the global story or a niggling misfit.

But first: correlation with what? The integrated geochronology and chronostratigraphy of this part of the Palaeogene are in a state of flux (Swisher and Prothero, 1990; Berggren, personal communication, 1989). It is not all that critical that the numerical calibration of the Eocene-Oligocene is unsettled so long as there is a firm cross-correlation between biostratigraphy, sequence stratigraphy and magnetostratigraphy; but that blessed state eludes us as yet. Compare, for example, the different correlations of chrons, zones and sequence events, whilst disregarding the numerical estimates, displayed in Figures 9.8-9.10.

Table 9.2 lists local events that might play some part in the identification of the Eocene/Oligocene boundary. For a long time we have taken top *Subbotina linaperta*, carefully distinguished morphotypically from *Subbotina angiporoides* (Lindsay, 1985), as the prime marker, Lindsay having shown also that *S. linaperta* extends slightly but meaningfully beyond the last occurrence of *Globigerinatheka index* in New Zealand. That was at least consistent with correlating the last bulge in the abundance profile in *G. index* (horizon (ii) in Figure 9.7) with about the top of Zone P16. It suggested that the *Halkyardia-Linderina-Maslinella* "warm-water" benthic assemblage of the extensive Aldinga transgression (McGowran, 1989a) was the final warm climatic pulse of the late Eocene before the global oceanic cooling. Thus, the Chinaman Gully event at about the Zone P16/P17 boundary slightly but significantly preceded that cooling (McGowran, 1987, Figure 8), reinforcing the notion of the TEE as a drawn-out affair. All of this was, of course, a fragile and tentative model because of the lack of bioevents that would permit the necessary direct correlations. A markedly different correlation is presented by Waghorn (1989), whose overriding precept is that the distribution of calcareous nannofossils in these neritic facies is to be read literally---something that cannot be indulged in the interpretation of foraminiferal events, at any rate. The inevitable upshot is that coccolith events are taken by Waghorn as isochronous, planktonic foraminiferal events as diachronous.

Problems emerged with the application of sequence-stratigraphic concepts. Perhaps the

TABLE 9.2. Events pertinent to the recognition of the Eocene/Oligocene boundary in southern Australia (Lindsay and McGowran, 1986; McGowran, 1990b).

(ix)	TOP	*Subbotina linaperta*
		[ALDINGA/RUWARUNG boundary]
(viii)	BASE	*Cassigerinella chipolensis* (in presence of *C. winniana*)
(vii)	TOP	*Tenuitella aculeata* (sporadic at top of range)
(vi)	BASE	*Guembelitria triseriata* (sporadic at base of range)
(v)	BASE	*Turborotalia ampliapertura*
(iv)	TOP TOP	*Globigerinatheka index* (rare) *Tenuitella insolita* (rare)
(iii)		*Halkyardia-Linderina-Maslinella* benthic "warm" assemblage
[(ii)	CHINAMAN GULLY REGRESSION]	
(i)	TOP TOP	*Globigerinatheka index* (abundant) *Tenuitella insolita* (common)

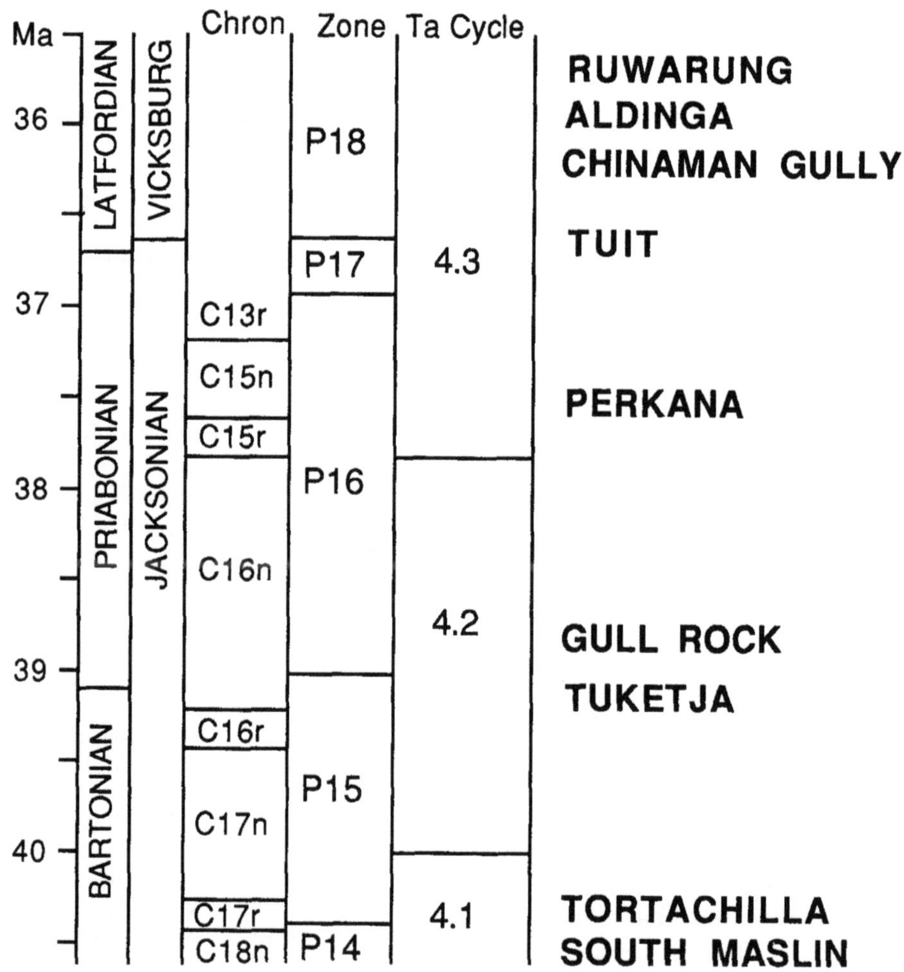

FIGURE 9.9. Correlations as for Figure 9.8, but using the correlations and scaling by Hazel (1989).

most striking element in the global pattern for this part of the record (Haq et al., 1987) is the emphatic maximum flooding surface (MFS) and transgression in the early Oligocene--MFS 4.4 preceded by a type 1 sequence boundary (SB 4.3/4.4) and sharp regression at about the Zone P17/P18 boundary. A search for SB4.3/4.4 and MFS 4.4 in the Port Willunga Formation yielded remarkably little evidence of either (Moss, 1989)--so long as the Eocene/Oligocene boundary is held at the last *S. linaperta* at the Aldinga/Ruwarung boundary. If we entertain the notion of the Chinaman Gully as the local manifestation of SB4.3/4.4, then the Aldinga transgression becomes the transgressive systems tract (TST) of 4.4. That would seem to be a strong confirmation of the magnitude ascribed by Haq et al. (1987) to cycle 4.4. The *Halkyardia-Linderina-Maslinella* assemblage at a high paleolatitude then postdates the major cooling event (see below) but accords with the expected match between major transgression and climatic amelioration.

Ironically, this correlation with its implication that the Chinaman Gully is at or very close to the Eocene/Oligocene boundary takes us back to the correlations of the early 1960s (Ludbrook, 1963; Wade, 1964). Those correlations were made under the influence of the Eames et al. (1963) pattern in which very little Oligocene was recognized anywhere away from the stratotypes. However, Lindsay's (1967) splendid corrective study still stands except that the lower Oligocene includes the Aldinga

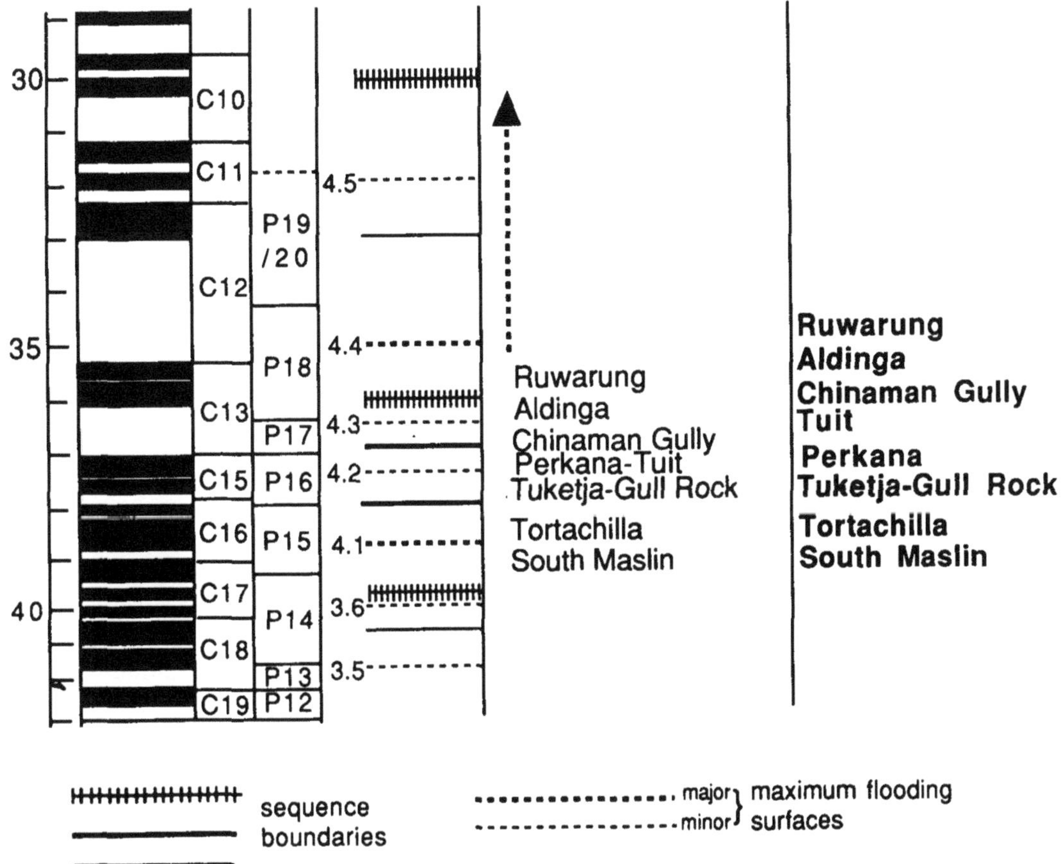

FIGURE 9.10. Previous (centre) and present (right) correlations, as discussed in the text, of the St.Vincent Basin lithostratigraphy with the geochronology and sequence stratigraphy of Haq et al. (1987). In contrast to Figure 9.8, in this version the top of Zone P17 is at the Chron C13R/C13N boundary; and the same horizon is the 4.3/4.4 sequence boundary and also the Eocene/Oligocene boundary.

Member as well as the Ruwarung member of the Port Willunga Formation.

Correlations within the late Eocene are not changed dramatically. The composite succession of local planktonic foraminiferal events (McGowran, 1989a,1990a) indicates that the Tortachilla transgression with last *Acarinina collactea* and first *Tenuitella insolita* and *Cassigerinella winniana* is coeval with low Zone P15. The Tortachilla/Tuketja unconformable contact is an emergent hard ground corresponding to SB4.1/4.2. The Tuketja Member with its *Hantkenina* ingression (shown as the *Pseudohastigerina* peak at event *a* in Figure 9.7) and the Gull Rock Member are TST4.2. The Gull Rock/Tuketja contact is at MFS4.2; the upper part of the Gull Rock shows pronounced darkening in colour, preservation of *Thalassinoides* burrows and other indications of condensed sedimentation (Mack, 1989). We do not clearly see, as yet, how the facies changes in the Perkana and Tuit Members fit the minor sequence 4.3. In terms of best fit, that is a trivial point compared with the fit of the Aldinga transgression with TST 4.4.

The obscurity of sequence-stratigraphic events in the Ruwarung Member (Moss, 1989) accords well with the minor ranking of both the SB4.4/4.5 and MFS4.5. The top of the Ruwarung becomes sandy and is shallowing upward. This is the Willungan/Janjukian local stage boundary (Lindsay, 1985; see Figure11). Being as it is close to the last occurrence of *Cassigerinella winniana* and *Chiloguembelina*

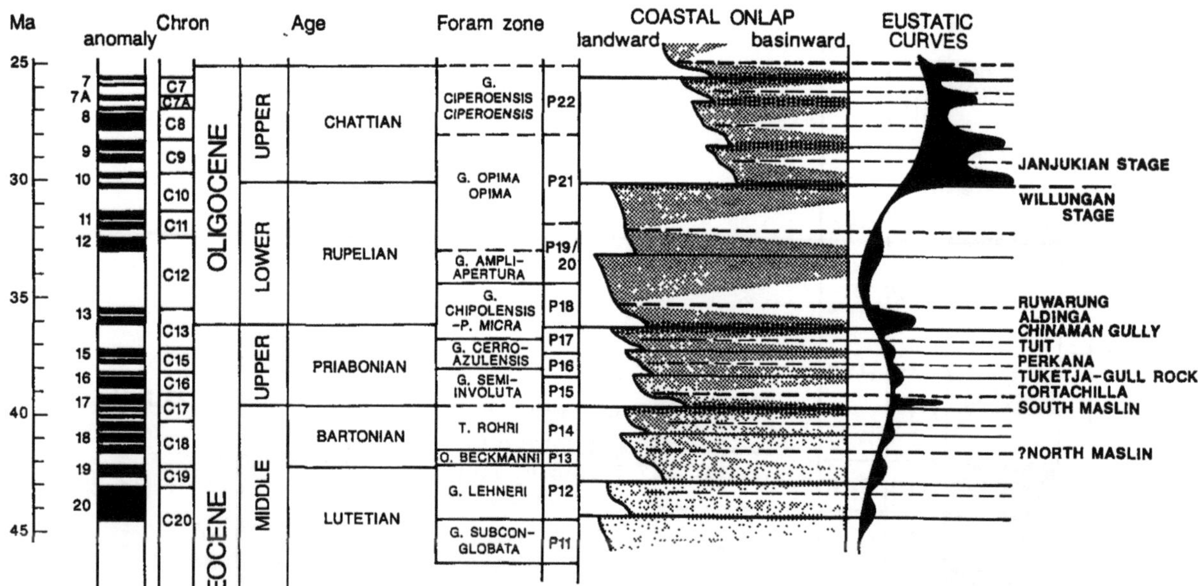

FIGURE 9.11. A summary chart. Using the Haq et al. (1987) time scale (as in Figure 9.10), sequence stratigraphy and sealevel curves, these correlations show the Chinaman Gully regression at (i) a type 1 sequence boundary, (2) the chron C13R/C13N boundary which is the level of the great $\delta^{18}O$ shift, and (3) the Eocene/Oligocene boundary. The peak in taxic turnover is at the Chinaman Gully which clearly is a reasonable Terminal Event Event, *sensu stricto*--but one that on cogent argument is in the earliest Oligocene (Premoli Silva et al., 1988). A TEE, *sensu lato* would begin at the previous type 1 sequence boundary in the vicinity of the middle/late Eocene boundary, itself a major event (Prothero et al., 1990) although not displaying the same degree of change in the neritic assemblages studied here. The mid-Oligocene event correlates in southern Australia with the Willungan/Janjukian Stage boundary (Lindsay, 1985).

cubensis and approximating the Zone P21a/21b boundary, the local stage boundary must also be close to the lower/ upper Oligocene boundary and the major sequence event at the supercycle TA4/TB1 boundary.

We conclude that the Chinaman Gully Formation correlates with the type 1 SB4.3/4.4. The old and new correlations are shown in Figure 9.10. SB4.3/4.4, in turn, is correlated by Haq et al. (1987) with about the Chron C13N/C13R boundary. Independently of sequence stratigraphic correlations, Shackleton (1986b) has correlated the spectacular oxygen isotopic shift in both surface and bottom water profiles with the Chron C13N/13R boundary. Oberhänsli and Toumarkine (1985) identify the shift at the C13R3 to C13N2 boundary. We have, then a correlation of the Chinaman Gully regression both with the sharpest putative fall in global sealevel in this part of the stratigraphic record, and with "probably the most rapid cooling demonstrated in the geological record" (Shackleton, 1986a).

In the Haq et al. (1987) correlation (Figures 9.10, 9.11), this is the Eocene/ Oligocene boundary; in the Aubry et al. (1989) correlation (Figure 9.8), it is 200,000-300,000 years above that boundary. The proposal for a boundary stratotype at Massignano includes a recommendation that the extinction of *Hantkenina* and its relatives be the P17/P18 planktonic foraminiferal zonal boundary and that this horizon in turn is the most reliable event for defining the Eocene/Oligocene boundary (Premoli-Silva et al., 1988). The last *Hantkenina* are found in chron C13R, always just above the highest normal magnetic event within C13R--clearly below the C13R/13N boundary. That recommen-dation accords more closely with the Aubry et al. correlations. It would follow that the TEE *sensu stricto*, as discussed below, is earliest Oligocene in age, and so is the Chinaman Gully regression.

THE TERMINAL EOCENE EVENT IN PALEOGENE BIOGEOHISTORY

It is probably the majority view now that global exogenic systems do not change gradually but, instead, tend to lurch from one state to the next. By that we mean that the time taken in the actual change is considerably less than the time of occupation of the respective states. The transformation of the essentially greenhouse world of the Cretaceous to the essentially icehouse world of the Neogene took place in a series of steps, as outlined by Berger (1982) in perhaps the most influential of the discussions on this topic. The most celebrated of those steps that came after the extinctions and the strangelove ocean at the end of the Cretaceous is the Terminal Eocene Event. Questions about the TEE that are still interesting and important include: Was it a short, sharp, pervasive event? Can diverse signals be correlated well enough to justify the appellation "event", or is it a fuzzy collection of phenomena spread through two or three million years? Either way, did it really impact severely on the biosphere?

First, however, we place the TEE in a global framework in the Palaeogene. Table 9.3 is constructed explicitly on the precept that the world moves rapidly from one mode to the next. There is plenty of evidence for that. The successional global tectonic regimes are based on discussion in McGowran (1989a); the identification of Chrons C24r and C19 as times of concentrated crustal change are well established. McGowran (1989a, 1990a) and especially Rea et al. (1990) have argued that those changes had a rapid effect on the environment (sealevel, climatic factors) and thence on the biosphere.

The global ocean recovered in due course from the strangelove state at the end of the Cretaceous (Hsü, 1986). It is inferred that productivity collapsed, on the evidence of $\delta^{13}C$ time series, due to either an extraterrestrial impact, or Deccan volcanism, or both. In contrast to the Deccan, the next major volcanic episode, the Thulean, was associated with a range of oceanic crustal patterns and effects at Chron C24R (e.g. Aubry et al., 1989; McGowran, 1989a, b; Rea et al.,1990).

The isotopically recognized oceanic cooling across the early/middle Eocene boundary is established well enough and matched cogently enough with other evidence to deserve the sobriquet "major cooling I". It terminates the episode of reversed oceanic circulation of the early Eocene. However, the middle-late Eocene constitutes an intermediate condition between that reversal and the psychrospheric situation of the Oligocene. Although the Antarctic "Krakow glaciation" (Birkenmajer et al., 1986) can be made consistent with the relatively rapid cooling of the ocean, we do not see the influence of the cryosphere in the Eocene. Miller et al. (1987) concluded that such influence developed later, in the Oligocene when glacioeustasy became important. $\delta^{18}O$ profiles show a continued drop in bottom water temperature through the later Eocene, whereas surface water values hold steadier. Thus, there is an increasing oceanic gradient, notwithstanding the diverse neritic and continental evidence for a part-return to the early Eocene warm conditions as from Chron C19 or Khirthar times onward into the late Eocene (McGowran, 1989a, and references therein). There is also evidence for oceanic stratification in the late Eocene (Boersma et al., 1987; McGowran, 1987,1989c). Boersma et al. suggest that the cause of stratification was the formation of dense bottom water by polar cooling.

The 40 Ma event includes prominent changes in planktonic foraminiferal communities, the greatest change in oceanic benthic foraminiferal communities (Tjalsma and Lohmann, 1982; Corliss and Keigwin, 1986) since the overturn in response to the Chron C24r happenings, and a great change in the oceanic ostracode record which is identified as the origin of the psychrosphere (Benson et al., 1985). At the supercycle TA3/TA4 boundary, this is inferred to have been the sharpest fall in sealevel for ten million years (Haq et al., 1987). Is this the real beginning of the TEE? Some would say that the 40 Ma event was more significant than the TEE, *sensu stricto*, itself (Prothero et al., 1990). It is approximately the beginning of the interval for which records of marine animal genera are conflated to yield the end-Eocene major extinction peak, the first such event since

TABLE 9.3. Palaeogene biogeohistory. Sources are cited in the text.

m.y	time scale	global tectonics	global ocean	magic numbers	evolution:pelagic & neritic foraminifera
30			cryosphere influential	30 Ma event	
	early Oligocene	HIMALAYAN tectonic regime "new"	efficient mixing psychrosphere		opportunists dominant
		crustal	MAJOR COOLING II stratification	Terminal Eocene event	K- extinctions: neritic and pelagic
	late Eocene	configurat			
40		-ion		40 Ma event	
	middle Eocene	Chron 19	cryosphere not influential	Chron 19 event	partial restoration: neritic K - forms
					oligotrophic habitats diminished
		interregnum			
50		INDIAN-ASIA	MAJOR COOLING I reversed oceanic circulation	cooling	K -forms:
	early Eocene	collision			maximum
			weakened atmospheric circulation		diversity and
		Chron 24r		Chron 24r event	turnover
	late Paleocene				
60		LARAMIDE tectonic regime "old"			K- selection reestablished
	early Paleocene	crustal configurat			opportunist survivors
		-ion	strangelove ocean	mass extinction	extinctions of Cretaceous biotas

the end of the Mesozoic (e.g. Sepkoski, 1986).

Following the end-Mesozoic extinctions, planktonic and larger benthic foraminiferal faunas recovered in Palaeogene evolutionary radiations. Using the ecologic notions of *r*- and *K*-selection related to trophic resource gradients in the photic zone, as developed especially by Hallock (1987), Premoli Silva et al. (1991) relate the fluctuating fortunes of those faunas to the Palaeogene paleoceanographic events, as outlined here in Table 9.3. Opportunists flourished in the Danian, to some degree in the Lutetian, and in the Rupelian. Between-times, niche specialization and diversification occurred mostly (it is inferred) through the extensive development of symbiotic relationships with phytoprotists. The faunal changes among the warm-water large benthic foraminifera give the Indo-Pacific letter-zonal boundaries Ta3/Tb and Tb/Tc at or close to the 40 Ma event and to the TEE respectively, thus coevally with extinctions among the planktonic K-forms. Reduced oceanic mixing--enhanced tendency toward stratification--leads to oligotrophy and niche specialization; cooling stimulates mixing and promotes eutrophy and opportunism. This cause and effect in the context of trophic resource gradients in the photic zone is richly heuristic. Perhaps it is the most cogent link found yet in the old, old search for such links between climatic change and organic evolution in the marine realm.

The TEE in the continental realm has derived from Stehlin's "Grande Coupure," a faunal overturn and replacement by migration among the larger mammals in western European assemblages (e.g. Russell and Tobien, 1986). Modern studies of taxic and transformational evolutionary change seem both to sharpen and to focus the event as an event and to show a cogent correlation with the (marine) $\delta^{18}O$ shift (Legendre, 1987). In the neritic warm-water environment, a very cautious analysis of turnover among the large benthic foraminifera alluded to above indicates a sharp event in the Indo-Pacific and American provinces at the same time, i.e. Zone P17 (Adams et al., 1986). Both sets of data are more impressive than the evidence of the calcareous plankton,which boils down to the disappearance of a very few lineages (of which much is made: Hut et al., 1986). It is much more important that the planktonic foraminiferal evidence supports the notion of vulnerability among the K-strategists, as applauded above.

We assess the TEE as follows:

(i) The TEE is a sharp event perceived coevally in different parts of the biosphere, not merely part of a continuum of change from the middle/late Eocene boundary onwards.

(ii) The biospheric events can be correlated variously in direct and in indirect ways with the major sequence boundary and sudden kick in the putative sealevel curve, on the one hand, and with the great $\delta^{18}O$ shift in the oceanic surface and bottom water profiles, on the other.

(iii) There is evidence of extraterrestrial impacts, but it is concentrated in Zone P16 as multiple microtektite horizons (e.g. Keller, 1986; Hazel, 1989). Not only are the horizons too old; they record a singularly unimpressive impact on the biosphere; and they are quite superfluous to convincing explanations of the TEE.

THE SOUTHERN AUSTRALIAN NERITIC RECORD AND THE TERMINAL EOCENE EVENT

Our investigation of the neritic benthic foraminiferal record in southern Australia indicates the following:

(i) The main biofacies changes in the sheltered St. Vincent Basin section match the change from grey-green-black sediments rich in opaline silica to a yellow-brown quartz-bryozoan facies. That match of biota with lithology is enhanced by the sharply lowered dominance by the Uvigerinidae and other groups expected to flourish where environmental stress--lowered levels of oxygen--occurs. The change occurs at a brief but very well-marked regression, the Chinaman Gully event.

(ii) Although the change is marked somewhat less clearly across the Chinaman Gully horizon (identified by correlation) in the Otway Basin composite section where plankton numbers are much higher, the parallels between the contrasting neritic sections are more impressive than are the differences.

(iii) The taxic overturn among species is most marked at the same level. Species packing is greatest at the transgressions and loss of species is greatest at the regression.

(iv) Again, the similarities in taxic overturn between the contrasting profiles are more impressive than are their differences. The degree of mutual taxic similarity reached a maximum in the recovery immediately after the disruptive regression. There is no apparent concentration of taxic response in some higher foraminiferal taxa at the expense of others.

The assessment of these patterns presents us with a problem of autocyclic *vis-a-vis* allocyclic causation. It is clear enough that the Chinaman Gully marks a change in the ventilation of the marine environment of the St.Vincent Basin. The underlying Blanche Point Formation is very rich in opaline silica. Whether released during the breakdown of volcanogenic ash (Jones and Fitzgerald, 1984) or from another source, that silica could be trapped long enough to be fixed biologically in the St.Vincent basin but was flushed out to sea in the more open Otway Basin (McGowran and Beecroft, 1986a). The ciculatory situation was not restored when marine conditions were restored with the Aldinga transgression.

That is essentially an *ad hoc*, autocyclic explanation. The parallels subsequently established with the Otway Basin (absence of silica-rich chemical sediments notwithstanding) made the explanation seem a little parochial (McGowran and Beecroft, 1986b; McGowran,1987). Then it was realized that the local evidence for temporary water stratification over the continental shelf, especially blooms of *Chiloguembelina* (event *c* in Figure 9. 7) , had coeval counterparts in the open ocean (McGowran, 1987; their significance has been traced by Boersma et al., 1989). Finally, the opal-rich Blanche Point Formation, tucked behind Kangaroo Island 200 km from the shelf edge, has *coeval* silica-rich counterparts in New Zealand and in Peru on the other side of the Pacific, demanding some substantial component of shared explanation (McGowran, 1989c). Thus has grown the realization that allocyclic thinking is required. The ventilating of the local environment at Chinaman Gully time may not actually be incorrect but is not sufficient explanation.

It seems possible then that the combined changes in neritic lithofacies, biofacies and taxic assemblages constitute a microcosm of the global oceanic situation. This study begins at the 40 Ma event and concludes at the 30 Ma event, and so has nothing to say about either. We find much less happening at the Tortachilla/Tuketja unconformity than at the Chinaman Gully regression. On present correlations the latter is coeval with the great $\delta^{18}O$ shift and the major Ta 4.3/4.4 sequence boundary. The collective "event" found at the Chinaman Gully supports the view that the TEE is real, sharp, bipolar, and manifested in the neritic realm. Furthermore, it signals a change from a greenhouse world, with an ocean prone to chemical sedimentation (cf. Berger, 1979), to an icehouse world less so prone. (But, as foreshadowed above, it may be within the earliest Oligocene.)

We might speculate further about responses in the benthic foraminiferal community to those changes in the neritic realm that yield sequence-stratigraphic configurations. Perhaps those responses are more volatile in late Eocene-type environments than in those of the early Oligocene. We do not find anything in the Oligocene microfaunal or lithofacies changes to compare in magnitude with the late Eocene changes summarized in Figure 9.5. Extrapolating from that sriking contrast, can we predict that it will be easier microfaunally to recognize maximum flooding surfaces and condensed sections in the greenhouse parts of the stratigraphic column than in the icehouse intervals? Did the biosphere become more robust whilst coping with the psychrospheric revolution? Such questions are testable. We believe that they are worth testing.

ACKNOWLEDGMENTS

We thank D.R. Prothero for his initiative, patience and enthusiasm at all stages, from the arrangement of the conference to the final production of this book. The manuscript was reviewed by W.A. Berggren and R.K. Olsson. The research was supported by grants from the Australian Research Council and by an Australian Postgraduate Research Award. Sherry Proferes drafted most of the figures. Of the numerous people whom we have shown the Eocene-Oligocene succession in the Willunga Subbasin, it is Peter Vail who particularly stimulated our current thinking.

REFERENCES

Adams, C.G., Butterlin, J., and Samanta, B.K. 1986. Larger foraminifera and events at the Eocene/Oligocene boundary in the Indo-West Pacific region. In Ch. Pomerol and I. Premoli-Silva, eds., *Terminal Eocene Events*, New York, Elsevier, pp. 237-252.

Aubry, M.-P., Berggren, W.A., Kent, D.V., Flynn, J.J., Klitgord, K.D., Obradovich, J.D., and Prothero, D.R. 1989. Paleogene geochronology: an integrated approach. *Paleoceanography* 3: 707-742.

Benson, R.H., Chapman, R.E., and Deck, L.T. 1985. Evidence from the ostracodes of the major events in the South Atlantic and world-wide over the past 80 million years. In K.J. Hsü and H. Weissert, eds., *South Atlantic Paleoceanography*, New York, Cambridge University Press, pp. 325-350.

Berger, W.H. 1979. Impact of deep-sea drilling on paleoceanography. In M. Talwani, W W. Hay, and W.B.F. Ryan, eds., *Deep Drilling results in the Atlantic Ocean, Continental Margins and Paleoenvironment*. Washington, D.C., American Geophysical Union, Maurice Ewing Series, 3: 297-314.

Berger, W.H. 1982. Deep-sea stratigraphy: Cenozoic climatic steps and the search for chemo-climatic feedback. In G. Einsele and A. Seilacher, eds., *Cyclic and Event Stratification*, Berlin, Springer-Verlag, pp. 121-157.

Birkenmajer, K., Delitala, M.C., Narebski, W., Nicoletti, M., and Petrucciani, C. 1986. Geochronology of Tertiary island-arc volcanics and glacigenic deposits, King George Island, South Shetland Islands (West Antarctica). *Bull. Polish Acad. Sci., Earth Sci.* 34: 257-272.

Boersma, A. and Premoli-Silva, I. 1989. Atlantic Paleogene biserial heterohelicid foraminifera and oxygen minima. *Paleoceanography* 4: 271-286.

Boersma, A., Premoli-Silva, I., and Shackleton, N.J. 1987. Atlantic Eocene planktonic foraminiferal paleohydrographic indicators and stable isotopic paleoceanography. *Paleoceanography* 2: 287-331.

Corliss, B.H. and Keigwin, L.D., Jr. 1986. Eocene-Oligocene paleoceanography. In K.J. Hsü , ed., *Mesozoic and Cenozoic Oceans*, Washington, D.C., American Geophysical Union, Geodynamics Series 15: 101-118.

Eames, F.E., Banner, F.T., Blow, W.H., and Clarke, W.J. 1962. *Fundamentals of Mid-Tertiary Stratigraphical Correlation*, Cambridge, Cambridge University Press.

Hallock, P. 1987. Fluctuations in the trophic resource continuum: a factor in global diversity cycles? *Paleoceanography* 2: 457-471.

Haq, B.U., Hardenbol, J., and Vail, P.R. 1987. Chronology of fluctuating sea levels since the Triassic. *Science* 235: 1156-1167.

Hazel, J.E. 1989. Chronostratigraphy of upper Eocene microspherules. *Palaios* 4: 318-329.

Hsü, K.J. 1986. Cretaceous/Tertiary boundary event. In K.J. Hsü, ed., *Mesozoic and Cenozoic Oceans*, American Geophysical Union, Geodynamics Series 15: 75-84.

Hut, P., Alvarez, W., Elder, W.P., Hansen, T., Kauffman, E.G., Keller, G., Shoemaker, E.M., and Weissman, P.R. 1987. Comet showers as a cause of mass extinction. *Nature* 329: 118-126.

Jones, J.B. and Fitzgerald, M.J. 1984. Extensive volcanism associated with the separation of Australia and Antarctica. *Science* 226: 346-348.

Jones, J.B. and Fitzgerald, M.J. 1987. An unusual and characteristic sedimentary mineral suite associated with the evolution of passive margins. *Sedimentary Geology* 52: 45-63.

Keller, G. 1986. Stepwise mass extinctions and impact events: Late Eocene to Early Oligocene. *Marine Micropaleontology* 10: 267-294.

Legendre, S. 1987. Mammalian faunas as paleotemperature indicators: concordance between oceanic and paleontological evidence. *Evolutionary Theory* 8: 77-86.

Lindsay, J.M. 1967. Foraminifera and stratigraphy of the type section of Port Willunga Beds, Aldinga Bay, South Australia. *Transactions of the Royal Society of South Australia* 91: 93-110

Lindsay, J.M. 1985. Aspects of South Australian foraminiferal biostratigraphy, with emphasis on studies of *Massilina* and *Subbotina*. In J.M. Lindsay, ed., *Stratigraphy, Palaeontology, Malacology*, Adelaide, Spec. Publ. No. 5: 187-214, Department of Mines and Energy, D.J. Woolman

Government Printer.

Lindsay, J.M. and McGowran, B. 1986. Eocene/Oligocene boundary, Adelaide region, South Australia. In Ch. Pomerol and I. Premoli-Silva, eds., *Terminal Eocene Events*, New York, Elsevier, pp. 165-173.

Ludbrook, N.H. 1963. Correlation of Tertiary rocks of South Australia. *Transactions of the Royal Society of South Australia* 87: 5-15.

Mack, D.A. 1989. Sequence stratigraphy and chronostratigraphic analysis of the Blanche Point Formation, St. Vincent Basin, South Australia. Unpubl. thesis, The University of Adelaide.

McGowran, B. 1987. Late Eocene perturbations: foraminiferal biofacies and evolutionary overturn, southern Australia. *Paleoceanography* 2: 715-727.

McGowran, B. 1989a. The later Eocene transgressions in southern Australia. *Alcheringa* 13: 45-68.

McGowran, B. 1989b. The silica burp in the Eocene ocean. *Geology* 17: 857-860.

McGowran, B. 1989c. Comment and reply on "Late Eocene diatomite from the Peruvian coastal desert, coastal upwelling in the eastern Pacific, and Pacific circulation before the Terminal Eocene Event." *Geology* 17: 957-958.

McGowran, B. 1990a. Fifty million years ago. *American Scientist* 78: 30-39.

McGowran, B. 1990b. Maastrichtian and early Cainozoic, southern Australia: foraminiferal biostratigraphy. In M.A.J. Williams, ed., *The Cainozoic of the Australian Region*. Geol. Soc. Australia, Spec. Publ. (in press).

McGowran, B. and Beecroft, A. 1986a. Foraminiferal biofacies in a silica-rich neritic sediment, Late Eocene, South Australia. *Palaeogeog., Palaeoclimat., Palaeoecol.* 52: 321-345.

McGowran, B. and Beecroft, A. 1986b. Neritic, southern extratropical foraminifera and the Terminal Eocene Event. *Palaeogeog., Palaeoclimat., Palaeoecol.* 55: 23-34.

Miller, K.G., Fairbanks, R.G., and Mountain, G.S. 1987. Tertiary oxygen isotope synthesis, sea level history, and continental margin erosion. *Paleoceanography* 2: 1-19.

Moss, G.D. 1989. The Port Willunga Formation: Eocene/Oligocene boundary stratigraphy and foraminiferal turnover. Unpubl. thesis, The University of Adelaide.

Oberhänsli, H. and Toumarkine, M. 1985. The Paleogene oxygen and carbon isotope history of sites 522, 523 and 524 from the central South Atlantic. In K.J. Hsü and H. Weissert, eds., *South Atlantic Paleoceanography*, New York, Cambridge University Press, pp. 125-147.

Premoli-Silva, I., Coccioni, R., and Montanari, A., eds. 1988. *The Eocene-Oligocene boundary in the Marche-Umbria basin (Italy)*. International Subcommission on Paleogene Stratigraphy, International Union of Geological Sciences, Eocene-Oligocene meeting, Ancona (Italy), Spec. Publ.

Premoli-Silva, I., Hallock, P., and Boersma, A. (ms) Similarities between planktonic and larger foraminiferal evolutionary trends through the Paleogene paleoceanographic changes. *Palaeogeog., Palaeoclimat., Palaeoecol.* (in press).

Prothero D.R., Berggren, W.A., Bjork, P.R. 1990. Penrose Conference Report: Late Eocene-Oligocene climatic and biotic evolution. *Geol. Soc. Amer. News and Information*, March 1990, pp. 74-75.

Rea, D.K., Zachos, J.C., Owen R.M., and Gingerich, P.D. 1990. Global change at the Paleocene-Eocene boundary: climatic and evolutionary consequences of tectonic events. *Palaeogeogr., Palaeoclimat., Palaeocol.* 79: 117-128.

Russell, D.E. and Tobien, H. 1986. Mammalian evidence concerning the Eocene-Oligocene transition in Europe, North America and Asia. In Ch. Pomerol and I. Premoli-Silva, eds., *Terminal Eocene Events*, New York, Elsevier, pp. 299-307.

Shackleton,N.J. 1986a. Preface. *Palaeogeog., Palaeoclimat., Palaeoecol.* 57: 1-2.

Shackleton, N.J. 1986b. Paleogene stable isotope events. *Palaeogeog., Palaeoclimat., Palaeoecol.* 57: 91-102.

Simpson, G.G. 1960. Notes on the measurement of faunal resemblance. *American Journal of Science* 258A: 300-311.

Sepkoski, J.J.,Jr., 1986. Global bioevents and the question of periodicity. In O.H. Walliser, ed., *Global Bio-events*, Berlin, Springer-Verlag, pp. 47-61.

Swisher, C.C. III and Prothero, D.R. 1990.

Single-crystal $^{40}Ar/^{39}Ar$ dating of the Eocene-Oligocene transition in North America. *Science* 249: 760-762.

Tjalsma, R.C. and Lohman, G.P. 1982. Paleocene-Eocene bathyal and abyssal benthic foraminifera from the Atlantic Ocean. *Micropaleontology Spec. Publ.* 4: 1-90.

Waghorn, D.B. 1989. Middle Tertiary calcareous nannofossils from the Aire district, Victoria; a comparison with equivalent assemblages in South Australia and New Zealand. *Marine Micropaleontology* 14: 237-255.

10. PALEOGENE CLIMATIC EVOLUTION: A CLIMATE MODEL INVESTIGATION OF THE INFLUENCE OF CONTINENTAL ELEVATION AND SEA-SURFACE TEMPERATURE UPON CONTINENTAL CLIMATE

by Lisa Cirbus Sloan and Eric J. Barron

ABSTRACT

An atmospheric general circulation model has been used to perform a series of sensitivity experiments to examine the effects of continental elevation and sea surface temperature upon continental climate during the Paleogene. Reconstructed geography of approximately 40 Ma was used for all experiments. Increasing the height of mountain ranges in the simulations produced drier continental conditions leeward of the mountains and wetter conditions windward of the mountains. Surface temperatures generally decreased in areas which experienced increased elevations, and surface temperatures at locations away from the altered mountains were also affected. Changing sea surface temperatures in the model from a distribution characterized by a low pole-to-equator temperature gradient to one with a steeper gradient which included lower high-latitude ocean surface temperatures had more widespread effects upon simulated continental climate. Among the most important responses to sea surface temperatures, continental surface temperatures at high latitudes experienced large decreases in winter values and small decreases in summer values. Surface temperatures also responded with a decrease in mean annual temperatures in many high latitude regions with the change to a steeper gradient ocean surface temperature distribution. Additionally, zonal precipitation intensified at low latitudes and decreased at high latitudes. Overall, results of the sensitivity experiments indicate that the steepening latitudinal ocean surface temperature gradient, with the associated ocean surface temperature distribution, and increasing tectonic uplift which occurred through the Paleogene played major roles in continental climatic changes.

INTRODUCTION

The Paleogene was a time during which global climate underwent great transitions. In an extreme sense, the transition can be represented by the change from the "globally warm" late Cretaceous to the glaciated Pleistocene. In between these extremes a general cooling took place (e.g., Savin, 1977, 1982). According to a variety of sources from the marine record, a key time of change in this long-term transition was during the Eocene and Oligocene (Savin, 1977; Keller, 1983; Keigwin and Corliss, 1986; Oberhansli and Hsü, 1986; Shackleton, 1986; Boersma et al., 1987; Rea et al., 1990). There is also evidence of large changes in continental climate conditions during this time (e.g., Chaney, 1940; Leopold and MacGinitie, 1972; Kemp, 1978; Wolfe and Poore, 1982; Wolfe, 1980, 1985; Retallack, 1986).

According to marine and terrestrial evidence from a wide variety of sources, the early Eocene was the warmest time interval of the Cenozoic (e.g., Savin, 1977; Haq et al., 1977; Kemp, 1978; McKenna, 1980, 1983; Estes and Hutchinson, 1980; Boersma et al., 1987). Warm marine conditions are indicated by oxygen isotope ratios of planktonic and benthic foraminifera recovered from many locations (e.g., Savin et al., 1975; Shackleton, 1986; Boersma et al., 1987; Stott and Kennett, 1989; Zachos et al., 1989). Warm continental climate is indicated by the presence of forests at high latitudes in both hemispheres during the early and middle Eocene

(Frakes and Kemp, 1973; Axelrod, 1984; Wolfe, 1980, 1985; Case, 1988), and the presence of early Eocene-age mammals and reptiles at latitudes poleward of 70° in the Northern Hemisphere (Dawson et al., 1976; Estes and Hutchinson, 1980; McKenna, 1980, 1983). The common denominator for nearly all of these paleoclimatic interpretations is the condition of a low latitudinal thermal gradient existing during this time. After the early Eocene, global cooling took place, predominantly at high latitudes. This is demonstrated by interpreted surface and deep water marine temperatures and by changing terrestrial biotic assemblage characteristics.

On a very long time scale (~100 million years), factors such as geography, atmospheric CO_2, and solar radiation levels have had a major impact upon climate (Barron and Washington, 1984; Barron, 1985; Crowley et al., 1987; Kasting, 1987). However, with the possible exception of geography (especially from a bathymetric point of view), during the Paleogene the changes in these factors are somewhat limited.

Within the Paleogene, two of the most noticeable changes in large-scale physical elements of the earth were (1) increasing tectonic uplift on the continents, and (2) changing sea surface temperature (SST) distributions. The Laramide orogeny was taking place in North America with many western regions of the continent experiencing uplift or extensive volcanism (Haun and Kent, 1965; Tweto, 1975; Axelrod, 1981). At approximately the same time in South America, volcanic activity and tectonic uplift were occurring during the Andean orogeny (Noble et al., 1979; Coira et al., 1982). South central Asia experienced volcanic activity and early uplift which culminated in the uplift of the Tibetan plateau in the later Cenozoic (Achache et al., 1984; Besse and Courtillot, 1988; Ruddiman et al., 1989).

Changes in ocean surface temperatures during the Paleogene are characterized by a steepening meridional ocean surface temperature gradient (Shackleton and Boersma, 1981; Keigwin and Corliss, 1986; Barron, 1987; Boersma et al., 1987). The steeper gradient is primarily the result of cooling of high latitude surface waters. In fact, high latitude surface water temperatures are estimated to have decreased by as much as 7-10°C during the Paleogene (Shackleton and Kennett, 1975; Savin, 1977; Boersma et al., 1987).

An important question arises from these recorded changing conditions. What effect did the changes in surface conditions, both in marine temperatures and continental topography, have on continental climate? Responses by the climate system would be expected to have an impact upon existing and evolving biota. This paper presents results from a series of climate modeling sensitivity studies in which the response of continental climate characteristics of surface temperature and precipitation to changes in sea surface temperatures and continental topography are examined. An atmospheric general circulation model is used to simulate climatic conditions that occur in conjunction with various combinations of sea-surface temperature and topography characteristics. January and July climates are simulated to produce estimates of the response of seasonal extremes to changing global conditions. Climatic responses for North America are given special attention because of the abundant paleoclimatic data of Eocene and Oligocene age from that continent.

Description of the Model

For this study version 0AV6 of the atmospheric general circulation model known as the Community Climate Model (CCM) from the National Center for Atmospheric Research (NCAR) was used. The CCM is a numerical model that simulates the three-dimensional behavior of the atmosphere based on thermodynamic and hydrodynamic laws. It is a spectral model based on the climate model described by Bourke et al. (1977) and McAvaney et al. (1978).

Vertical space in the model is divided into nine layers representing a height of 24 km. The horizontal dimension is represented by a global grid of approximately 4.5° latitude by 7.5° longitude resolution, producing a 40 by 48 element array. Land surface temperatures are calculated by the model from surface energy fluxes. The model version used for this study has no active ocean component, but instead employs a fixed-temperature ocean. The ocean is represented by surface temperatures that are constant throughout the duration of the

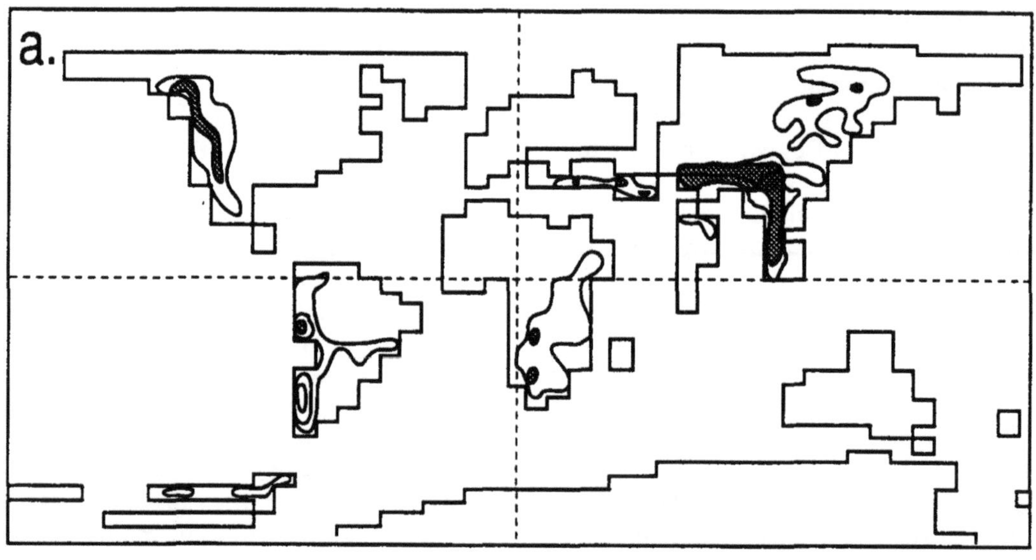

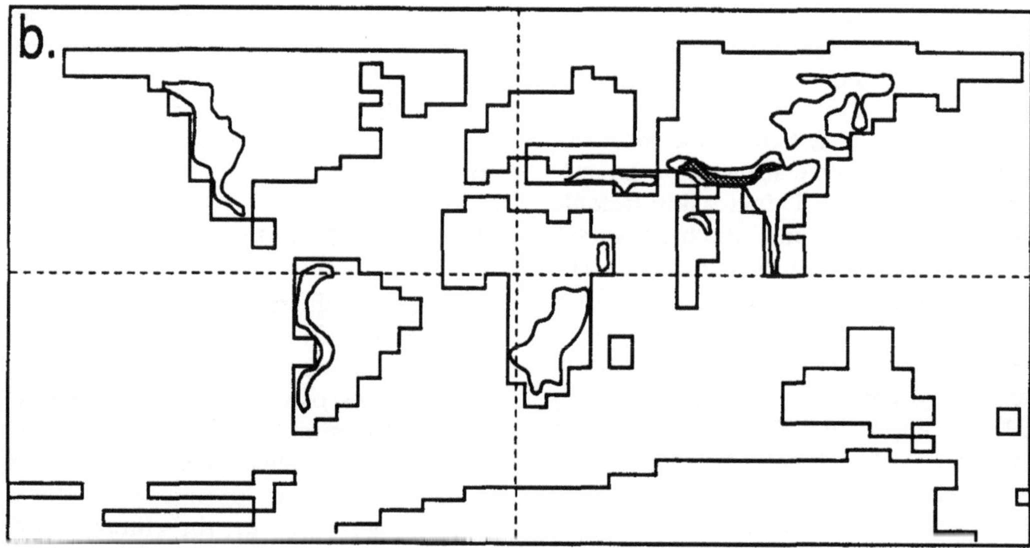

FIGURE 10.1. Land-sea distribution incorporated into all climate model simulations of this study at model resolution, shown with specified continental topography. (a) Full-height topography from Barron (1985) , (b) low mountain topography. Reconstruction is for approximately 40 Ma. Contour interval is 500 m, with elevations greater than 1 km shaded.

simulation, and no heat storage or transfer by the oceans can occur. This idealized system produces an adequate representation of first-order climate responses while allowing rapid, computationally inexpensive, solution.

Clouds are predicted in the model, using the Ramanathan et al. (1983) radiation-cloudiness formulation. The model version includes the surface hydrology scheme of Washington and Williamson (1977) in which model-derived precipitation and evaporation rates are used to predict changes in soil moisture and snow cover. Fixed solar radiation distributions for either January or July conditions are incorporated into the simulations. A more complete description of the model and comparison of model results with observations can be found in Pitcher et al. (1983).

Description of Sensitivity Experiments

A series of climate model sensitivity studies were carried out to evaluate the impact of sea surface temperature distribution and continental topography upon Eocene climate. As a sensitivity study, only one boundary condition is changed between each simulation. When the results of two simulations are compared, any differences between the climates can be attributed to the single factor varied between cases. All of the climate model experiments documented here incorporate paleogeography of approximately 40 Ma (Barron et al., 1981; Barron, 1985). While this geographic distribution is inexact for any specific stage of the Eocene or Oligocene, at model resolution it provides a representative geography with which to examine the climate of the Paleogene (Figure 10.1). The present day level of atmospheric CO_2 was specified for all simulations, and no ice was defined on the continents or oceans. Continental vegetation descriptions and albedo were held constant for all simulations.

Two different continental topographic data sets were incorporated into the climate model: the continental elevation data set of Barron (1985), which is referred to as the "mountain" case, and an elevation data set in which mountain elevations were reduced, referred to as the "low mountain" case. Both elevation distributions are shown in Figure 10.1. For the "low mountain" case, all continental elevations greater than or equal to 1 km were reduced by one-half (at model grid space resolution) and the surrounding topography was smoothed, where necessary, to preserve the original topographic expression. For example, maximum elevation reduction occurs in the region of the proto-Himalayas, with elevations reduced from approximately 2.5 km to 1.25 km, and the proto-Rockies vary from 1 km to 500 m between the two elevation cases. Results from the climate model simulations incorporating these elevational boundary conditions provide indication of the first order climatic change produced by rising continental elevations during the middle and late Paleogene. Because minimum surface temperatures for North America are of specific interest for Eocene climate reconstructions, climate simulations for each of these continental descriptions were produced for average January conditions. Differences between the results for these two cases are due to orographic influence upon climate simulated for average January conditions.

Two ocean surface temperature distributions were specified in the Eocene climate model simulations. These distributions can be thought of as extreme endmember ocean surface temperature distributions for the Paleogene. The first SST distribution is defined by a pole-to-equator ocean surface temperature gradient that is reduced relative to present observed temperatures. This distribution is referred to as the "low gradient" distribution. The gradient also has tropical surface temperatures that are 2-3°C lower than are presently observed (Figure 10.2). The second SST distribution is based upon the observed global mean annual ocean surface temperature gradient and is referred to as the "steep gradient" distribution (Figure 10.2). Both gradients are constrained to be symmetrical about the equator. SST distributions within the model for each case are specified as constant across all model latitudes to compensate for uncertainties in ocean surface temperature structure during the Paleogene. Climate model simulations incorporating each SST distribution were carried out for January and for July solar radiation conditions in order to produce an estimate of the seasonal extremes that would occur with each distribution of ocean surface temperatures. For each season, differences between climate model results for the cases incorporating the different SST distributions will be

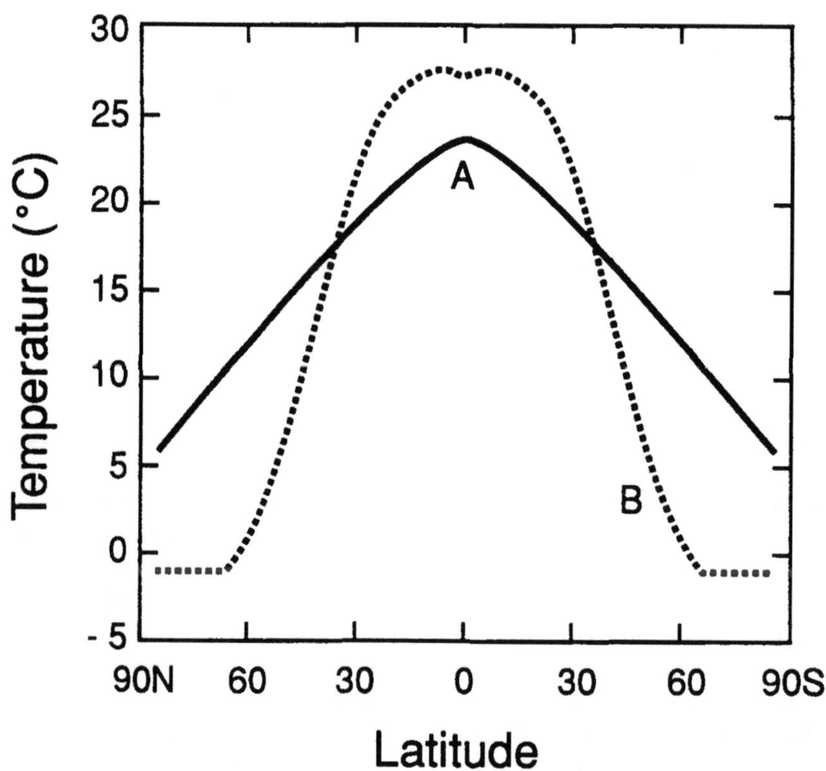

FIGURE 10.2. Meridional sea surface temperature gradients used to define the sea surface temperature distributions incorporated into the climate model simulations. (a) "low gradient" distribution, (b) "steep gradient" distribution.

TABLE 10.1. **Eocene Climate Simulations: Critical Boundary Conditions**

1. MOUNTAIN JANUARY:	40 Ma geography, "low-gradient" SST distribution, full-height continental elevations, January solar radiation
2. LOW MOUNTAIN JANUARY:	40 Ma geography, "low-gradient" SST distribution, reduced-height continental elevations, January solar radiation
3. LOW GRADIENT SST - JANUARY:	40 Ma geography, full-height continental elevations, "low-gradient" SST distribution, January solar radiation
4. LOW GRADIENT SST - JULY:	40 Ma geography, full-height continental elevations, "low-gradient" SST distribution, July solar radiation
5. STEEP GRADIENT SST - JANUARY:	40 Ma geography, full-height continental elevations, "steep-gradient" SST distribution, January solar radiation
6. STEEP GRADIENT SST - JULY:	40 Ma geography, full-height continental elevations, "steep-gradient" SST distribution, July solar radiation

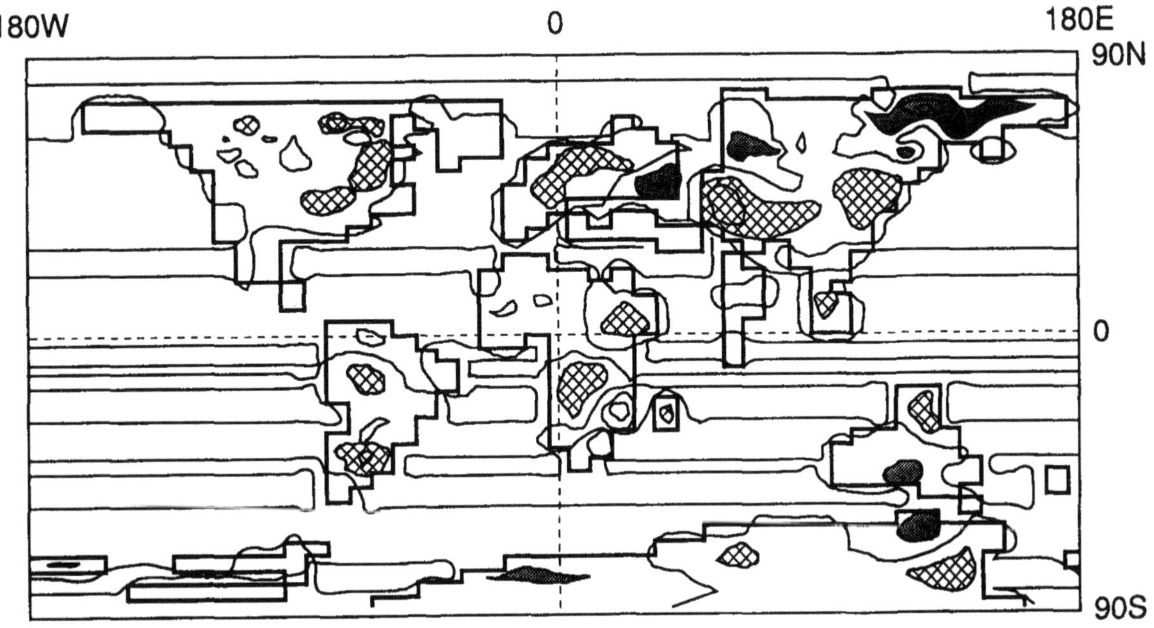

FIGURE 10.3. January surface temperature difference, "mountain" case minus "low mountain" case. Contour interval is 3°K; stippled areas indicate positive temperature difference; hatched areas indicate negative temperature difference.

attributed to the altered SST distribution.

The climate simulations are summarized in Table 10.1. A comparison of results from cases 1 and 2 (Table 10.1) will provide an estimate of Eocene continental climate sensitivity to topography. Intercomparison of the January results or the July results from the two SST distribution cases (e.g., case 3 to case 5, case 4 to case 6), provides an estimate of the seasonally extreme climatic response to the SST distributions that represent idealized Paleogene ocean surface conditions. Comparison of the results of case 3 to case 4 and of case 5 to case 6 will define the approximate mean annual response of climatic conditions to the specified SST distributions.

MODEL RESULTS

Effects of Mountain Elevations upon Continental Climate

In the first sensitivity analysis, average January climatic conditions were produced with each topographic distribution. The responses of continental surface temperatures and precipitation for the two simulations are compared, and the results are presented below as difference fields for each climate characteristic.

Surface Temperature

Surface temperature differences for the "mountain" case minus the "low mountain" case show substantial temperature responses to elevation in many regions. However, the temperature variations balance to produce a global mean surface temperature difference of only 0.2°K, with the slight global warming occurring in the "low mountain" case. The areal distribution of temperature differences indicates that many regional-scale temperature variations occurred in areas both with and without elevation modifications (Figure 10.3). In regions where elevations were increased (e.g., the proto-Rockies, Andes, and Himalayas) surface temperatures responded with a temperature decrease in the range of 3-8°K. This occurs due to temperature-atmospheric lapse rate relationships, but could also be caused by feedback between infrared opacity and elevation (Barron and Washington, 1984). There are also changes in surface temperature values (1-8°K range) between the "mountain" and "low mountain" cases that occur in areas where little or no elevational adjustments were carried out (e.g., eastern Asia, eastern North America, Europe, and

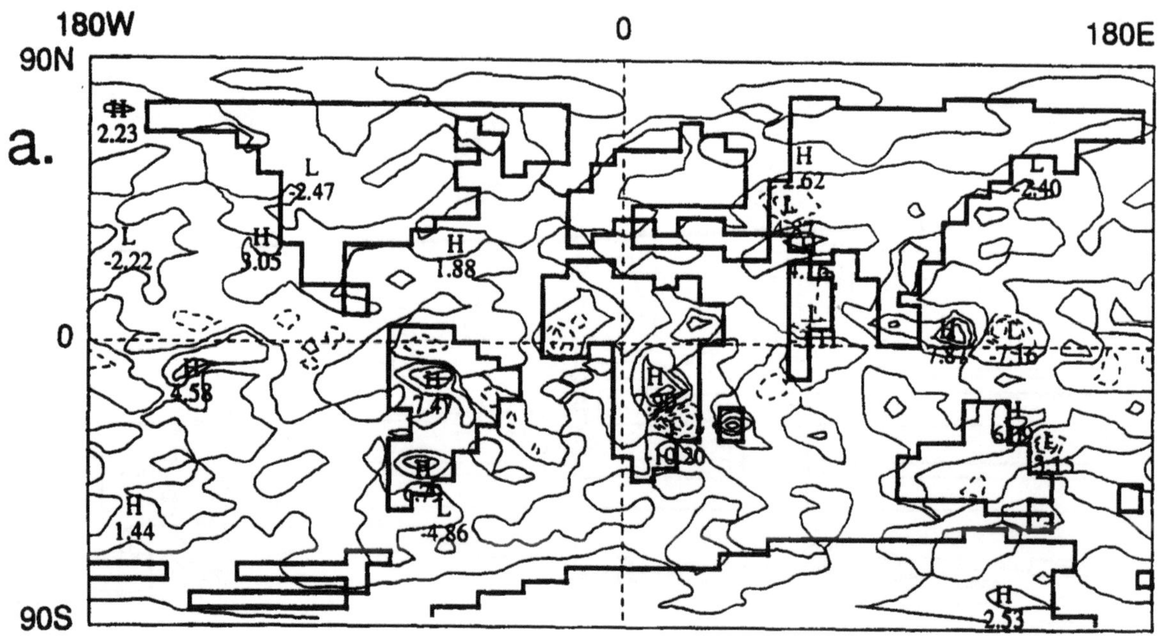

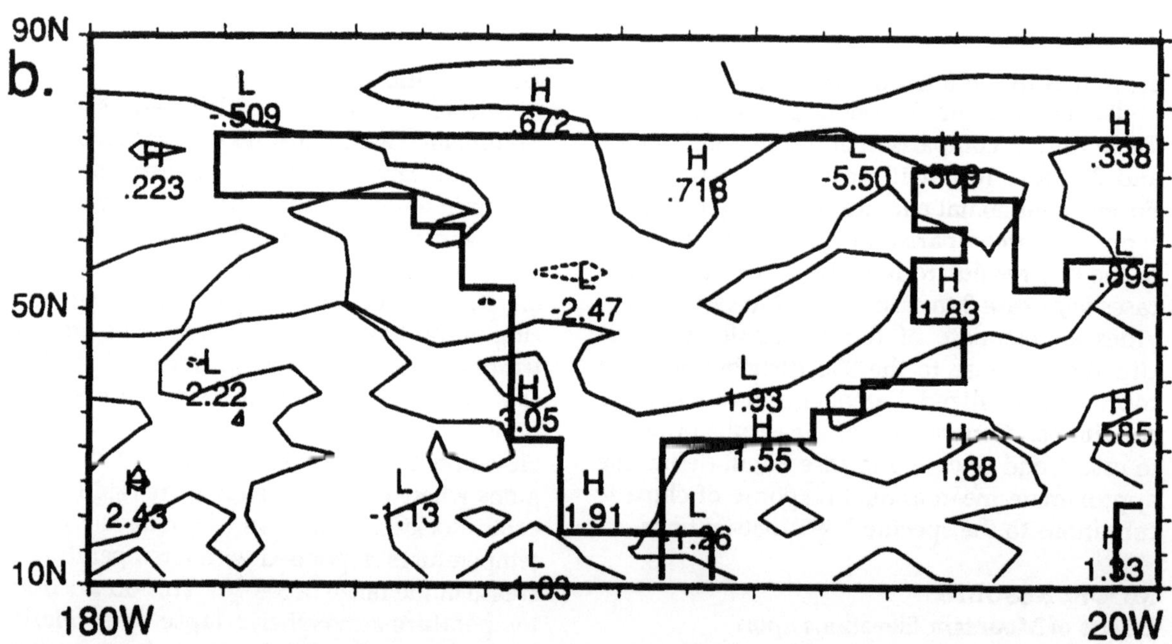

FIGURE 10.4. Precipitation difference, "mountain" case minus "low mountain" case. Contour interval is 2 mm/ day. (a) Global, (b) North America (contour interval 1 mm/day).

northern Australia). Changes in these areas are primarily the result of altered atmospheric circulation patterns between the two cases.

Results of this study indicate that mountain elevations can influence surface temperatures even in areas removed from the mountains, a result also found in late Cenozoic climate simulations of Ruddiman and Kutzbach (1989). Tectonic influence upon continental climate in the Paleogene cannot be considered as limited to the tectonically active regions alone, but instead could have had wider-reaching effects. The results also indicate that an elevation change of 500 - 1000 m is enough to affect surface temperature distributions (at model resolution); however, the climatic perturbation caused by the small change in topography of this study is much less than that generated by the larger-scale, late Cenozoic, uplift modeled by Kutzbach et al. (1989). Based on these results, it is suggested that for the Paleogene, rising mountains produced cooler temperatures in the areas of orogenic uplift, and also produced surface temperature changes in regions not experiencing tectonic uplift.

Precipitation

Varying the height of mountain elevations on the Eocene continents had the greatest effect upon precipitation distributions on the windward sides of mountain ranges, most noticeably over North America (Figure 10.4). For example, to the west of the Rocky Mountains the "mountain" case experiences an average of 3 mm/day more precipitation than the "low mountain" case. The higher elevations of the "mountain" case force upward motion of eastward-traveling air parcels which triggers precipitation on the westward sides of the mountains. There is then a decrease in precipitable moisture that is available for the continental interior. In the "low mountain" case this rain shadow effect is absent. There is a less pronounced precipitation maximum on the western sides of mountains, and precipitation is distributed more evenly about both sides of the lowered mountains. The concept of orographically-triggered continental aridity has been supported by other climate modeling studies (Manabe and Terpstra, 1974; Manabe and Broccoli, 1990; Ruddiman et al. 1989; Kutzbach and Ruddiman, 1989), and has also been supported in North Americaby paleosol records from the Bighorn Basin and the Badlands region (Bown and Kraus, 1981; Retallack, 1986). Although it is detectable, the precipitation response to the specified mountain elevations is not large (compare Figure 10.5 to Manabe and Broccoli, 1990, Figure 2, p. 193). This may be in part due to the fact that precipitation maxima are not best defined in January, and in part due to the relatively low mountain elevations specified in both simulations.

The implications of these results for continental precipitation during the Paleogene is that orographically-induced drying of continents with rising mountains could have begun when the mountains were as low as or lower than 1000 m in elevation, although at this height, large-scale continental aridity was not yet completely established. This agrees with the conclusions of Manabe and Broccoli (1990) and Ruddiman et al. (1989) and Kutzbach and Ruddiman (1989) whch suggest that aridity of midlatitudinal continental interiors became firmly established within the last 10 million years when mountain elevations were much greater.

Effects of Sea Surface Temperature Distribution upon Continental Climate

For the second sensitivity analysis, climate simulations incorporating each SST distribution were carried out for January and for July conditions. The response of several aspects of surface temperatures characteristics and of precipitation to the ocean surface temperatures are discussed in the following section.

Surface Temperature

January surface temperature differences for the "low gradient" SST distribution minus the "steep gradient" SST distribution (refer to Figure 10.2) are shown in Figure 10.5, and July surface temperature differences are shown in Figure 10.6. Both distributions both show several important changes in continental surface temperatures as the ocean surface temperatures were modified to reflect a steeper meridional gradient. First, the response of continental surface temperatures was greatest at middle and high latitude continental margins. The largest temperature changes (up to 35° K) between the cases in each simulated month occur in regions

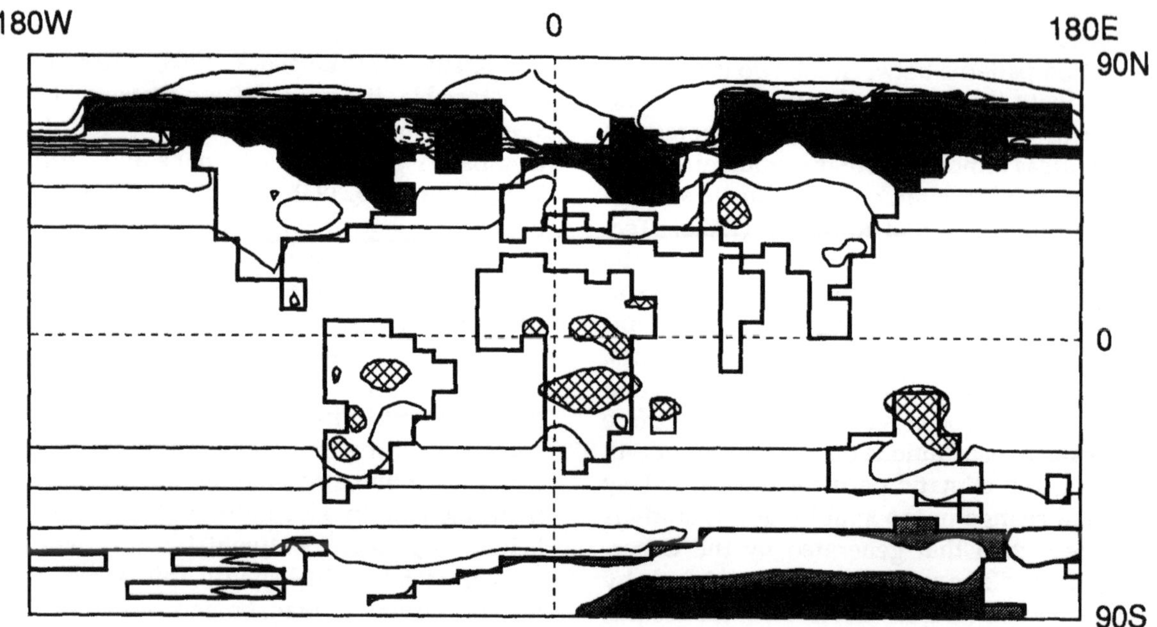

FIGURE 10.5. January surface temperature difference, "low gradient" SST case minus "steep gradient" SST case. Contour interval is 3°K. Stippled areas indicate positive difference, hatched areas indicate negative differences.

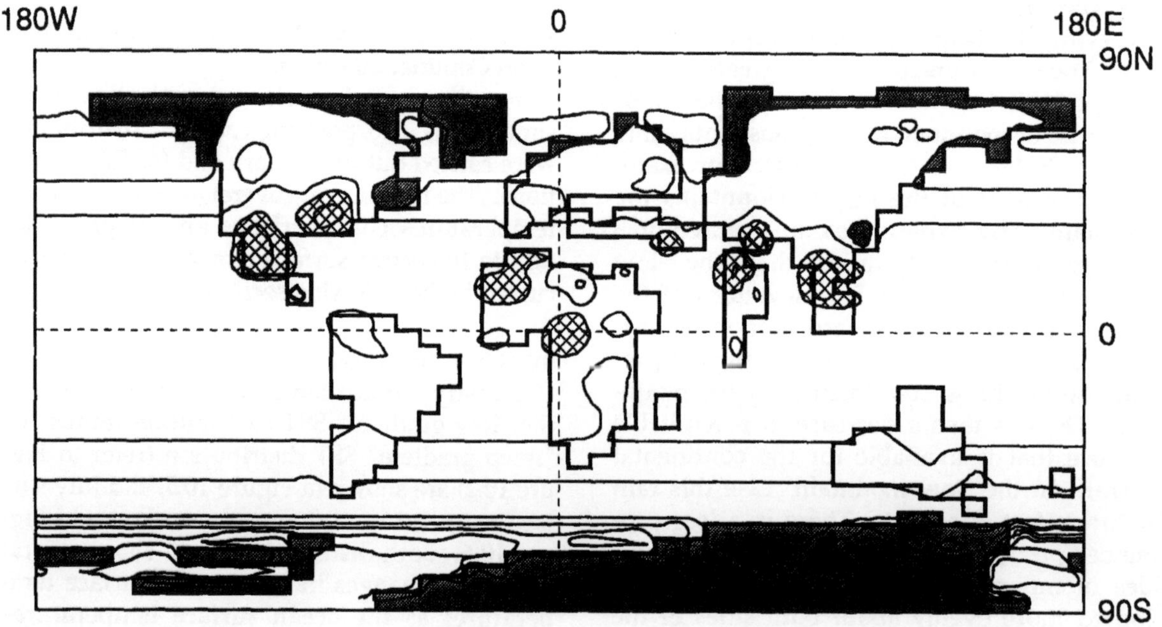

FIGURE 10.6. July surface temperature difference, "low gradient" SST case minus "steep gradient" SST case. Contour interval is 3°K. Stippled areas indicate positive difference, hatched areas indicate negative differences.

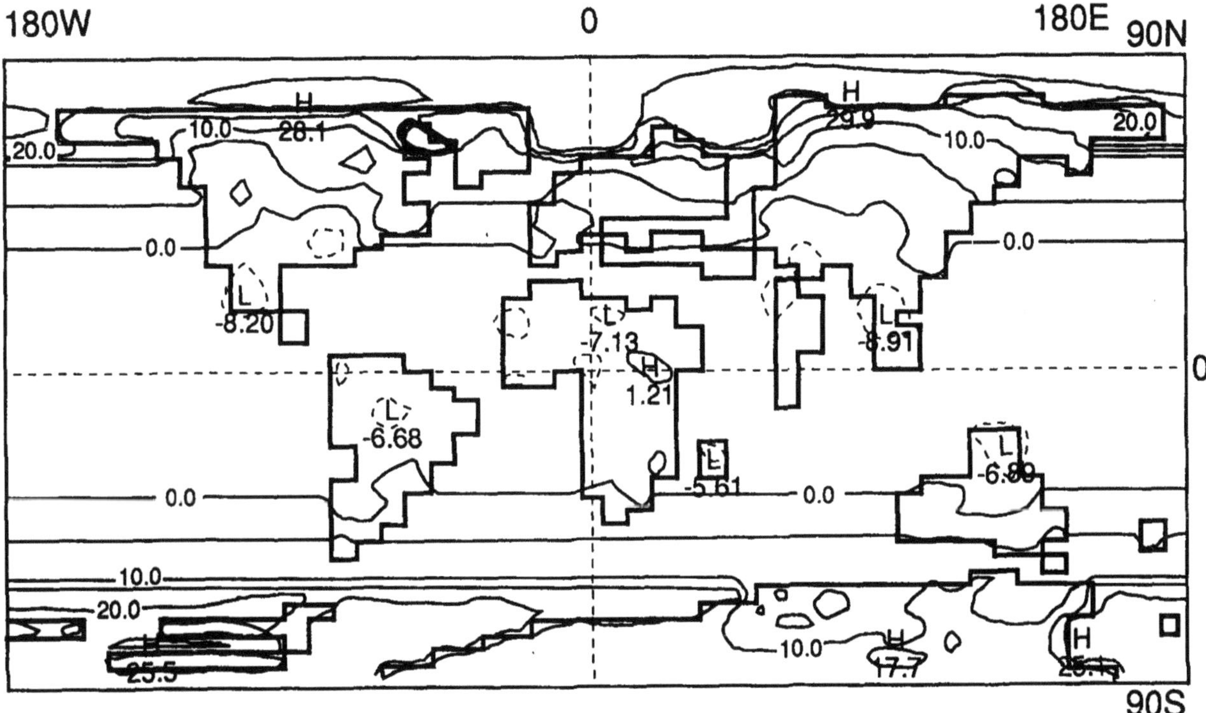

FIGURE 10.7. Estimated mean annual surface temperature difference, "low gradient" SST case minus "steep gradient" SST case. Contour interval is 5°K. Solid contours indicate positive difference or no difference, dashed contours indicate negative difference.

poleward of 60° latitude. Continental surface temperatures in these regions are clearly responding to surrounding ocean surface temperatures.

Second, minimum (winter) surface temperatures were more sensitive to SST distribution than were maximum (summer) temperatures. In the winter hemispheres, continental surface temperatures cool by up to 35°C when the ocean surface temperatures were changed from the "low gradient" to the "steep gradient" distribution. In the summer hemispheres the high-latitude regions experience a decrease in maximum summer temperatures by up to 9°C. This suggests an interesting implication for the eventual establishment of ice at high latitudes. An important condition for ice initiation and stablization to take place is that a cool summer in which some snow/ice will survive must follow a winter of snow accumulation (Crowley et al., 1987). While the summer temperatures produced by these simulations would not sustain snow, the trend of decreasing summer temperatures at high latitudes is certainly suggestive of such conditions later in the Tertiary.

Third, the average global surface temperature difference between each January simulation and between each July simulation is within 1.1°C, with the warmer average occurring in the "low gradient" case each time. This demonstrates that while the average global surface temperature response to SST values is small, the regional response of continental surface temperatures can be quite large and variable.

Finally, an important feature of the results is that the large change in specified ocean surface temperatures for the simulations resulted in a very small surface temperature response by continental interiors. These results and others (Schneider et al., 1985; Sloan and Barron, 1990) suggest that continental interior surface temperatures are not affected by ocean surface temperatures but are primarily controlled by solar insolation levels. Thus through the Paleogene continental interior temperatures were

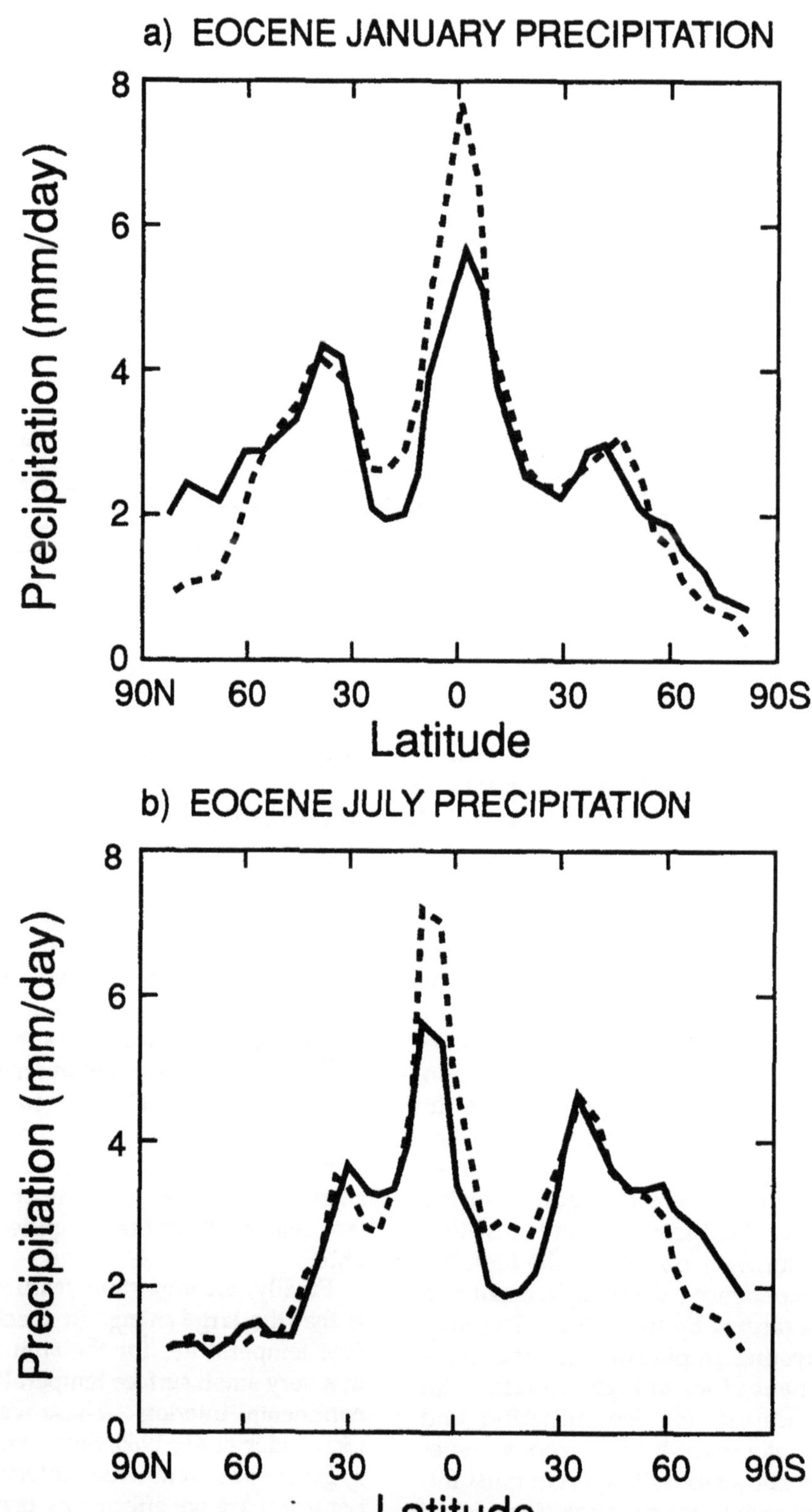

FIGURE 10.8. Mean zonal precipitation (mm/day); (a) January, (b) July. Solid lines indicate the "low gradient" SST case, dashed lines indicate the "steep gradient" SST case.

probably less influenced by cooling high latitude SSTs and a steepening meridional ocean surface temperature gradient than they were by continental elevation changes.

Mean Annual Surface Temperature

For each SST distribution, the mean values calculated from January and July continental surface temperature results were used as estimates of mean annual surface temperatures. A map of the differences between those values is presented in Figure 10.7. As expected from the January and July surface temperature results described above, mean annual temperatures at high latitudes were most sensitive to the specified SST distributions. The "steep gradient" ocean surface temperature distribution resulted in significantly reduced mean annual surface temperatures at high latitudes. Mean annual surface temperatures at high latitudes are as much as 20°K lower in the steep gradient case. Similarly, because of the specified tropical SSTs for the low and steep gradient SST cases, there is a slight increase in mean annual temperature at low latitudes for the steep gradient case. If the ocean surface temperature distributions used in this study reflect actual tropical ocean surface temperatures during the middle and late Paleogene, paleoclimatic evidence of tropical continental warming would be a significant discovery.

Precipitation

Global precipitation responded to the variation in SST distributions in a similar manner for both seasons. As a general trend, the "low gradient" SST distribution produced average latitudinal precipitation patterns that were more evenly distributed across latitudes than did the SST distribution with a steeper meridional temperature gradient (Figure 10.8). That is, the low gradient cases produced more precipitation at high latitudes and less at low latitudes than the steep gradient cases. The areal response of the precipitation distributions to the specified SSTs varied extensively, in part because precipitation is by nature highly variable. Using North America an an example, however, for the January simulations precipitation decreases over the northwest margin but slightly increases over the east-central margin from the "low gradient" to the "steep gradient" SST distribution. Differences between the cases are more pronounced in the July simulations (Figure 10.9). From the "low gradient" case to the "steep gradient" case there is a 3-7 mm/day average precipitation decrease over the southwestern margin of North America and an increase over the southeast margin of the continent. The trend of a drier southwestern North American margin (at least for January and July) is in agreement with Eocene precipitation records interpreted from paleosol records in southwestern California (Peterson and Abbott, 1979), although model results indicate that the rainy season occurred in summer, while Peterson and Abbott (1979, p. 83) estimate the rainy season as possibly winter.

DISCUSSION

Sensitivity studies have been completed which provide an estimate of the first order effects of sea surface temperature and of continental elevations upon climate during the Paleogene. Estimated tectonic uplift shows some effect upon continental precipitation patterns, suggesting the beginnings of midcontinental aridity establishment. Continental surface temperatures are also influenced by rising mountains, including in areas away from the actual uplift. Rising mountains probably played a greater role in determining continental interior surface temperatures than did the ocean surface temperatures, most certainly at regional spatial scale. Climatic response to the specified SST distributions was limited in large part to coastal regions of the continents. High latitude coastal regions responded most strongly to the change in SST distribution, from one governed by a low meridional surface gradient to a distribution controled by a steep meridional surface gradient. Winter continental surface temperatures in these locations cooled by up to 30°K and summer temperatures cooled by up to 10°K with the change from the "low gradient" SST distribution to the "steep gradient" distribution.

The model results suggest that the conditions of continental elevations and changing ocean surface temperatures were largely responsible for the Paleogene climate change that is documented by a variety of continental evidence. On a time scale of 10 million years or less, these conditions were probably the dominant factors

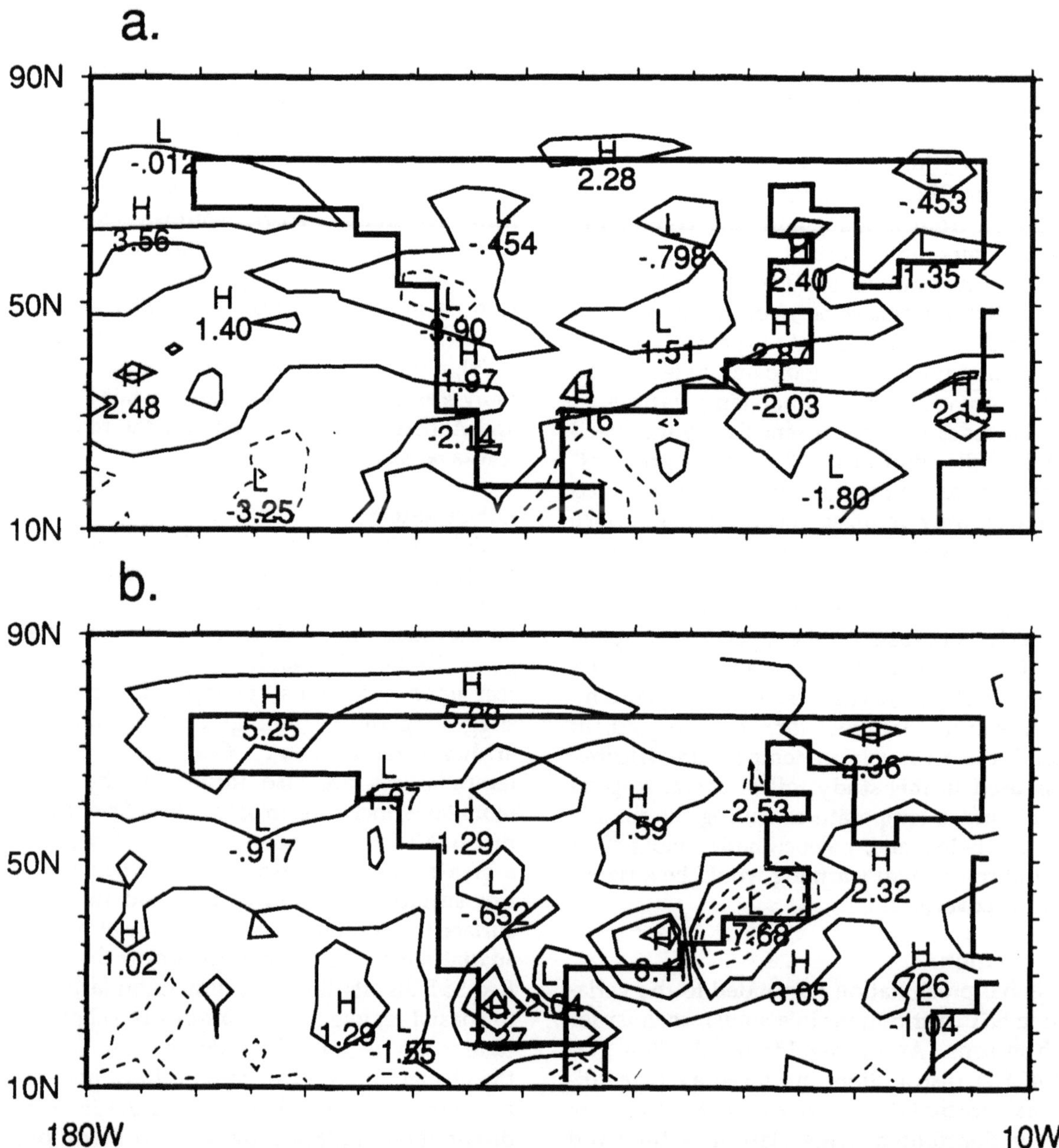

FIGURE 10.9. Precipitation difference for North America, "low gradient" SST case minus "steep gradient" SST case. (a) January, (b) July. Contour interval 2 mm/day.

in climate change during the Paleogene.

ACKNOWLEDGEMENTS

The climate modeling research documented here was supported by National Science Foundation Grant ATM-8804084 to EJB. Computer time was provided by the National Center for Atmospheric Research, which is funded by the National Science Foundation. This manuscript was written in part while LCS was visiting the Climate and Global Dynamics Division at the National Center for Atmospheric Research, and in part while LCS was funded by NASA grant NAGW-176 to J.C.G. Walker. We thank D. Rea for a helpful review and A. Modahl for typing the manuscript.

REFERENCES

Achache, J., Courtillot, V., and Zhoa, Y.X. 1984. Paleogeographic and tectonic evolution of Southern Tibet since middle Cretaceous times: new paleomagnetic data and synthesis. *Jour. Geophys. Res.* 89: 331-339.

Axelrod, D.I. 1981. Role of volcanism in climate and evolution. *Geol. Soc. Amer. Special Paper* 185, 59 p.

Axelrod, D.I. 1984. An interpretation of Cretaceous and Tertiary biota in polar regions. *Palaeogeog., Palaeoclimat., Palaeoecol.* 45: 105-147.

Barron, E.J. 1985. Explanations of the Tertiary global cooling trend. *Palaeogeog., Palaeoclimat., Palaeoecol.* 50: 729-739.

Barron, E.J. 1987. Eocene equator-to-pole surface ocean temperatures: A significant climate problem? *Paleoceanography* 2: 729-739.

Barron, E.J., Harrison, C.G.A., Sloan, J.L., II, and Hay, W.W. 1981. Paleogeography 180 million years ago to the present. *Eclogae Geologicae Helvetiae* 74: 443-470.

Barron, E.J., and Washington, W.M. 1984. The role of geographic variables in explaining paleoclimates: Results from Cretaceous climate model sensitivity studies. *Jour. Geophys. Res.* 89: 1267-1279.

Besse, J., and Courtillot, V. 1988. Paleogeographic maps of the continents bordering the Indian Ocean since the Early Jurassic. *Jour. Geophys. Res.* 93: 11791-11808.

Boersma, A., Premoli-Silva, I., and Shackleton, N.J. 1987. Atlantic Eocene planktonic foraminiferal paleohydrographic indicators and stable isotope paleoceanography. *Paleoceanography* 2: 287-331.

Bourke, W., McAvaney, B., Puri, K., and Thurling, R. 1977. Global modeling ofatmospheric flow by spectral methods. In Chang, J., ed., *Methods in computational physics. General circulation models of the atmosphere.* New York, Academic Press, 17: 267-324.

Bown, T.M., and Kraus, M.J. 1981. Lower Eocene alluvial paleosols (Willwood Formation, northwest Wyoming, U.S.A.) and their significance for paleoecology, paleoclimatology, and basin analysis. *Palaeogeog., Palaeoclimat., Palaeoecol.* 34: 1-30.

Case, J.A. 1988. Paleogene floras from Seymour Island, Antarctic Peninsula. In Feldmann, R.M., and Woodburne, M.O., eds., *Geology and paleontology of Seymour Island, Antarctic Peninsula. Geol. Soc. Amer. Memoir* 169: 523-530.

Chaney, R.W. 1940. Tertiary forests and continental history. *Geol. Soc. Amer. Bull.* 51: 469-488.

Coira, B., Davidson, J., Mpodozis, C., and Ramos, V. 1982. Tectonic and magmatic evolution of the Andes of Northern Argentina and Chile. *Earth-Science Reviews* 18: 303-332.

Crowley, T.J., Mengel, J.G., and Short, D.A. 1987. Gondwanaland's seasonal cycle. *Nature* 329: 803-807.

Dawson, M.R., West, R.M., Langston, W., Jr., and Hutchinson, J.H. 1976. Paleogene terrestrial vertebrates: Northernmost occurrence, Ellesmere Island, Canada. *Science* 192: 781-782.

Estes, R., and Hutchinson, J.H. 1980. Eocene lower vertebrates from Ellesmere Island, Canadian Arctic Archipelago. *Palaeogeog., Palaeoclimat., Palaeoecol.* 30: 325-347.

Frakes, L.A., and Kemp, E.M. 1973. Palaeogene continental positions and evolution of climate. In Tarling, D.H., and Runcorn, S.K., eds., *Implications of continental drift to the Earth sciences,* New York, Academic Press, 1: 541-559.

Haq, B.U., Premoli-Silva, I., and Lohmann, G.P. 1977. Calcareous plankton paleobiogeographic evidence for major climatic fluctuations in the early Cenozoic Atlantic Ocean:*Jour. Geophys. Res.* 82: 3861-3876.

Haun, J.D., and Kent, H.C. 1965. Geologic history of Rocky Mountain region. *Amer. Assoc. Petrol. Geol. Bull.* 49: 1781-1800.

Kasting, J.F. 1987. Theoretical constraints on oxygen and carbon dioxide concentrations in the Precambrian atmosphere. *Precambrian Research* 34: 205-229.

Keigwin, L.D., and Corliss, B.H. 1986. Stable isotopes in late middle Eocene to Oligocene foraminifera. *Geol. Soc. Amer. Bull.* 97: 335-345.

Keller, G. 1983. Paleoclimatic analyses of middle Eocene through Oligocene planktic foraminiferal faunas. *Palaeogeog., Palaeoclimat., Palaeoecol.* 43: 73-94.

Kemp, E.M. 1978. Tertiary climatic evolution and vegetation history in the southeast Indian Ocean region. *Palaeogeog., Palaeoclimat., Palaeoecol.* 24: 169-208.

Kutzbach, J.E., Guetter, P.J., Ruddiman, W.F., and Prell, W.L. 1989. Sensitivity of climate to Late Cenozoic uplift in southern Asia and the American West: Numerical experiments. *Jour. Geophys. Res.* 94: 18393-18407.

Leopold, E.B., and MacGinitie, H.D. 1972. Development and affinities of Tertiary floras in the Rocky Mountains. In Graham, A., ed., *Floristics and paleofloristics of Asia and Eastern North America*. Amsterdam, Elsevier, pp. 147-200.

Manabe, S., and Terpstra, T.B. 1974. The effects of mountains on the general circulation of the atmosphere as identified by numerical experiments. *Jour. Atm. Sci.* 31: 3-42.

Manabe, S., and Broccoli, A.J. 1990. Mountains and arid climates of middle latitudes. *Science* 247: 192-195.

McAvaney, B.J., Bourke, W., and Puri, K. 1978. A global spectral model for simulation of the general circulation. *Jour. Atm. Sci.* 35: 1528-1557.

McKenna, M.C. 1980. Eocene paleolatitude, climate and mammals of Ellesmere Island. *Palaeog., Palaeoclimat., Palaeoecol.* 30: 349-362.

McKenna, M.C. 1983. Cenozoic paleogeography of North Atlantic land bridges. In Bott, M.H.P., Saxov, S., Talwani, M., and Theide, J., eds., *Structure and development of the Greenland-Scotland Ridge. NATO Conference Series IV, Marine Sciences*. New York, Plenum Press, 8: 351-400.

Noble, D.C., McKee, E.H., and Megard, F. 1979. Early Tertiary "Incaic" tectonism, uplift, and volcanic activity, Andes of central Peru. *Geol. Soc. Amer. Bull.* 90: 903-907.

Oberhänsli, H., and Hsü, K.G. 1986. Paleocene-Eocene paleoceanography. In Hsü, K.J., ed., *Mesozoic and Cenozoic Oceans. American Geophysical Union Geodynamics Series*, 15: 85-100.

Peterson, G.L., and Abbott, P.L. 1979. Mid-Eocene climatic change, southwestern California and northwestern Baja California. *Palaeogeog., Palaeoclimat., Palaeoecol.* 26: 73-87.

Pitcher, E.J., Malone, R.C., Ramanathan, V., Blackmon, M.L., Puri, K., and Bourke, W. 1983. January and July simulations with a spectral general circulation model. *Jour. Atm. Sci.* 40: 580-604.

Ramanathan, V., Pitcher, E.J., Malone, R.C., and Blackmon, M.L. 1983. The response of a spectral general circulation model to refinements in radiative processes. *Jour. Atm. Sci.* 40: 605-630.

Rea, D.K., Zachos, J.C., Owen, R.M., and Gingerich, P.D. 1990. Global change at the Paleocene-Eocene boundary: Climatic and evolutionary consequences of tectonic events. *Palaeogeog., Palaeoclimat., Palaeoecol.* 79: 117-128.

Retallack, G.J. 1986. Fossil soils as ground for interpreting long-term controls on ancient rivers. *Jour. Sed. Petrol.* 56: 1-18.

Ruddiman, W. F., and Kutzbach, J.E. 1989. Forcing of Late Cenozoic Northern Hemisphere climate by plateau uplift in southern Asia and the American West. *Jour. Geophys. Res.* 94: 18409-18427.

Ruddiman, W.F., Prell, W.L., and Raymo, M.E. 1989. Late Cenozoic uplift in southern Asia and the American West: Rationale for general circulation modeling experiments. *Jour. Geophys. Res.* 94: 18379-18391.

Savin, S.M. 1977. The history of the Earth's surface temperature during the past 100 million years. *Ann. Rev. Earth Planet. Sci.* 5: 319-355.

Savin, S.M. 1982. Stable isotopes in climatic reconstructions. In *Climate in Earth history, National Research Council Studies in Geophysics*. Washington, D.C., National Academy Press, p. 164-171.

Savin, S.M., Douglas, R.G., and Stehli, F.G.

1975. Tertiary marine paleotemperatures. *Geol. Soc. Amer. Bull.* 86: 1499-1510.

Schneider, S.H., Thompson, S.L., and Barron, E.J. 1985. Mid-Cretaceous continental surface temperatures: Are high CO_2 concentrations needed to simulate above freezing winter conditions? In Sundquist, E.T., and Broecker, W.S., eds., *The carbon cycle and atmospheric CO_2. Natural variations Archean to present. American Geophysical Union Monograph* 32: 554-560.

Shackleton, N.J. 1986. Paleogene stable isotope events. *Palaeoceanog., Palaeoclimat., Palaeoecol.* 57: 91-102.

Shackleton, N.J., and Boersma, A. 1981. The climate of the Eocene ocean. *Geol. Soc. London Jour.* 138: 153-157.

Sloan, L. Cirbus, and Barron, E.J. 1990. "Equable" climates in Earth history? *Geology* 18: 489-492.

Stott, L.D., and Kennett, J.P. 1989. Evolution of the Antarctic seasonal thermoclimate and seasonality during the late Paleogene: Inferences from the stable isotopic composition of planktonic foraminifers. *Geol. Soc. Amer. Abst. Prog.* 21(7): A87.

Tweto, O. 1975. Laramide (Late Cretaceous-Early Tertiary) Orogeny in the Southern Rocky Mountains. *U.S. Geological Survey Memoir* 144: 1-24.

Washington, W.M., and Williamson, D.L. 1977. A description of the NCAR global circulation models. In Chang, J., ed., *Methods in computational physics. General circulation models of the atmosphere.* New York, Academic Press, 17: 111-172.

Wolfe, J.A. 1980. Tertiary climates and floristic relationships at high latitudes in the Northern Hemisphere. *Palaeogeog., Palaeoclimat., Palaeoecol.* 30: 313-323.

Wolfe, J.A. 1985. Distributions of major vegetational types during the Tertiary. In Sundquist, E.T., and Broecker, W.S., eds., The carbon cycle and atmospheric CO_2. Natural variations Archean to present. *Amer. Geophys. Union Geophys. Monog.* 32: 357-376.

Wolfe, J.A., and Poore, R.Z. 1982. Tertiary marine and nonmarine climatic trends. In *Climate in Earth history. National Research Council Studies in Geophysics.* Washington, D.C., National Academy Press, pp. 154-158.

Zachos, J.C., Lohmann, K.C., and Barrera, E. 1989. Eocene and Oligocene climatic events of the southern Indian Ocean: Results of ODP drilling on Kerguelen Plateau. *Geol. Soc. Amer. Abst. Prog.* 21(7): A87.

11. EOCENE-OLIGOCENE FAUNAL TURNOVER IN PLANKTIC FORAMINIFERA, AND ANTARCTIC GLACIATION

by Gerta Keller, Norman MacLeod and Enriqueta Barrera

ABSTRACT

Low-latitude planktic foraminiferal populations experienced a major faunal turnover between the late middle to early late Oligocene. This faunal turnover involved over 80% of planktic foraminiferal species and took place quasi-continuously over an interval of approximately 14 m.y. The overwhelming majority of species becoming extinct during this interval were surface-dwelling forms that were ecologically replaced by more cold-tolerant subsurface-dwelling species as the thermal contrast between surface and subsurface (> 400 m) marine pelagic habitats diminished. Within this 14 m.y. interval of widespread ecological reorganization of the planktic foraminiferal faunas, two subintervals stand out as being characterized by brief, but markedly intensified turnover; these are the middle/late Eocene and the early/late Oligocene. Contrary to previous reports, there was no major faunal change across the Eocene/Oligocene (E/O) boundary.

Stable isotope records and glacio-marine sediments from high latitude southern ocean ODP Legs 113 and 119 provide evidence of major glaciation on East Antarctica during the late middle to late Eocene and early Oligocene with glaciation persisting into the late Oligocene. Carbon and oxygen isotopic gradients for planktic and benthic foraminiferal species reflect decreasing surface productivity and thermal stratification during this time. The remarkably close correspondence between these stable isotope records and planktic foraminiferal turnovers strongly suggests that changes in climate and productivity were the primary driving forces behind the gradual decline and eventual extinction of the Eocene planktic foraminiferal fauna.

INTRODUCTION

Contrary to some previous reports, the so-called Eocene-Oligocene (E/O) mass extinction among planktic foraminifera was not centered at this boundary, but rather was part of a long-term trend that began during the middle Eocene and continued into the late Oligocene (Steineck, 1971; Keller, 1983a,b, 1985, 1986; Corliss et al., 1984; Molina et al., 1988), an interval spanning over 14 m.y. Within this interval, accelerated faunal turnovers can be recognized near the middle/late Eocene and early/late Oligocene boundaries. These faunal turnovers include the successive extinction of tropical and subtropical faunas and their replacement by cool and temperate assemblages of species (Kennett, 1977, 1978; McGowran, 1978; Keller, 1983a,b, 1985, 1986).

Correlated with these faunal changes was a major shift in the earth's climate. Stable isotopic records indicate a significant cooling (~1.0 per mil) beginning in the early middle Eocene (magnetochron C22 to C21; 52.6 to 49 Ma, Katz and Miller, 1991) and a major cooling (~2.0 per mil) near the middle/ late Eocene boundary (Oberhansli et al., 1984; Keigwin and Corliss, 1986; Miller et al., 1987; Prentice and Matthews, 1988; Kennett and Stott, 1990; Barrera and Huber, 1991). A further drop in temperature began near the E/O boundary and reached a global minimum at the base of Chron 13N (Keigwin and Keller, 1984; Oberhänsli et al., 1984; Keigwin and Corliss, 1986; Miller et al., 1987). Cooler surface and bottom water temperatures continued through most of the Oligocene reaching maximum low temperatures at about 29 Ma and followed by a short warming during the latest Oligocene (Keigwin and Keller, 1984).

The late Paleogene cooling trend has generally been interpreted to be a consequence of the development of circum-Antarctic circulation made possible by the northward movement of Tasmania and Australia by middle Eocene time (Kennett et al., 1975; Shackleton and Kennett, 1975). Until recently, the middle to late Eocene increases in $\delta^{18}O$ values have been attributed solely to high latitude cooling that was devoid of major Antarctic glaciation. In contrast, the early Oligocene isotopic shift has been interpreted as a combination of cooling and the first major glaciation on Antarctica (Matthews and Poore, 1980; Keigwin and Keller, 1984; Miller and Thomas, 1985; Keigwin and Corliss, 1986; Miller et al., 1987). However, evidence of extensive early to late Oligocene glaciation has recently been discovered in glacio-marine sediments in the Ross Sea, Antarctica (Barrett, 1989; Bartek et al., this volume), as well as in ice-rafted debris from early Oligocene sediments from the Kerguelen Plateau (Site 744, Barrera and Huber, 1991). Prentice and Matthews (1988) have used a composite low to mid-latitude planktic and benthic oxygen isotope record to argue that substantial ice volume (equal to the present) has existed since about 40 Ma. These observations are consistent with major Antarctic glaciation as early as 42 Ma; this in turn suggests the possibility of a causal link between changes in marine planktic habitats brought about by substantial global cooling and planktic foraminiferal faunal turnovers.

In this paper we examine the middle Eocene to Oligocene record of planktic foraminiferal faunal turnover beginning with the source of the common misconception of a mass extinction at the E/O boundary. After using taxic and morphotypic data to describe the major outlines of this faunal turnover, we then turn to a more detailed consideration of abundance variations within the dominant foraminiferal species in order to document specific intervals of increased environmental stress on these foraminiferal populations. The effect of this environmental stress can most clearly be seen in terms of changes in the isotopically determined depth habitat preferences of individual species. Finally, we discuss and correlate these faunal turnover data with the stable isotope records of high and low latitude stratigraphic sequences in order to demonstrate how changes in marine productivity, thermal stratification and climate all contributed to the terminal decline in Paleogene planktic foraminiferal populations.

THE EOCENE-OLIGOCENE MASS EXTINCTION: A TAXIC EVALUATION

Micropaleontologists have long regarded the late Eocene through early Oligocene as a time of widespread extinction in the planktic foraminiferal fauna. Evidence for this view has traditionally come from two sources: i) the planktic foraminiferal classifications of Loeblich and Tappan (1964) and Blow (1979), and ii) Cifelli's (1969) analysis of the iterative evolution of major planktic foraminiferal morphotypes. One striking aspect of the systematic data used to characterize extinction patterns in these studies is the almost universal employment of rather coarse taxonomic groupings and stratigraphic intervals. These data typically consist of the observed stratigraphic ranges of either families or genera assessed to the nearest inclusive epoch (e.g., Harland et al., 1967; Sepkoski, 1982). This data set tends to distort the true extinction record in favor of abrupt changes. More recently, this picture of a sudden and pervasive late Paleogene biotic crisis in the deep sea has been given additional support through the discovery of a late Eocene deep-sea iridium anomaly (Ganapathy, 1982; Glass et al., 1982; Keller et al., 1987), the observation of multiple late Eocene microtektite and microspherule-bearing layers in both continental sections and deep sea cores (Keller et al., 1983; 1987; Miller et al., 1991; Hazel, 1989), and the extinction periodicity analyses of Raup and Sepkoski (1984, 1986; see also Sepkoski and Raup, 1985).

Since 1964 there have been three major Paleogene planktic foraminiferal classifications: Loeblich and Tappan (1964), Blow (1979), and Loeblich and Tappan (1988). While only a handful of new planktic foraminiferal species have been described during this interval, the supraspecific taxonomy has more than doubled, rising from four Eocene-Oligocene families representing seventeen genera (Loeblich and Tappan, 1964) to ten Eocene-Oligocene families representing thirty-nine genera (Loeblich and Tappan, 1988). The demographic effects of this ongoing taxonomic revision can be seen by plotting the total number of planktic foraminiferal families and genera present in the six Eocene-

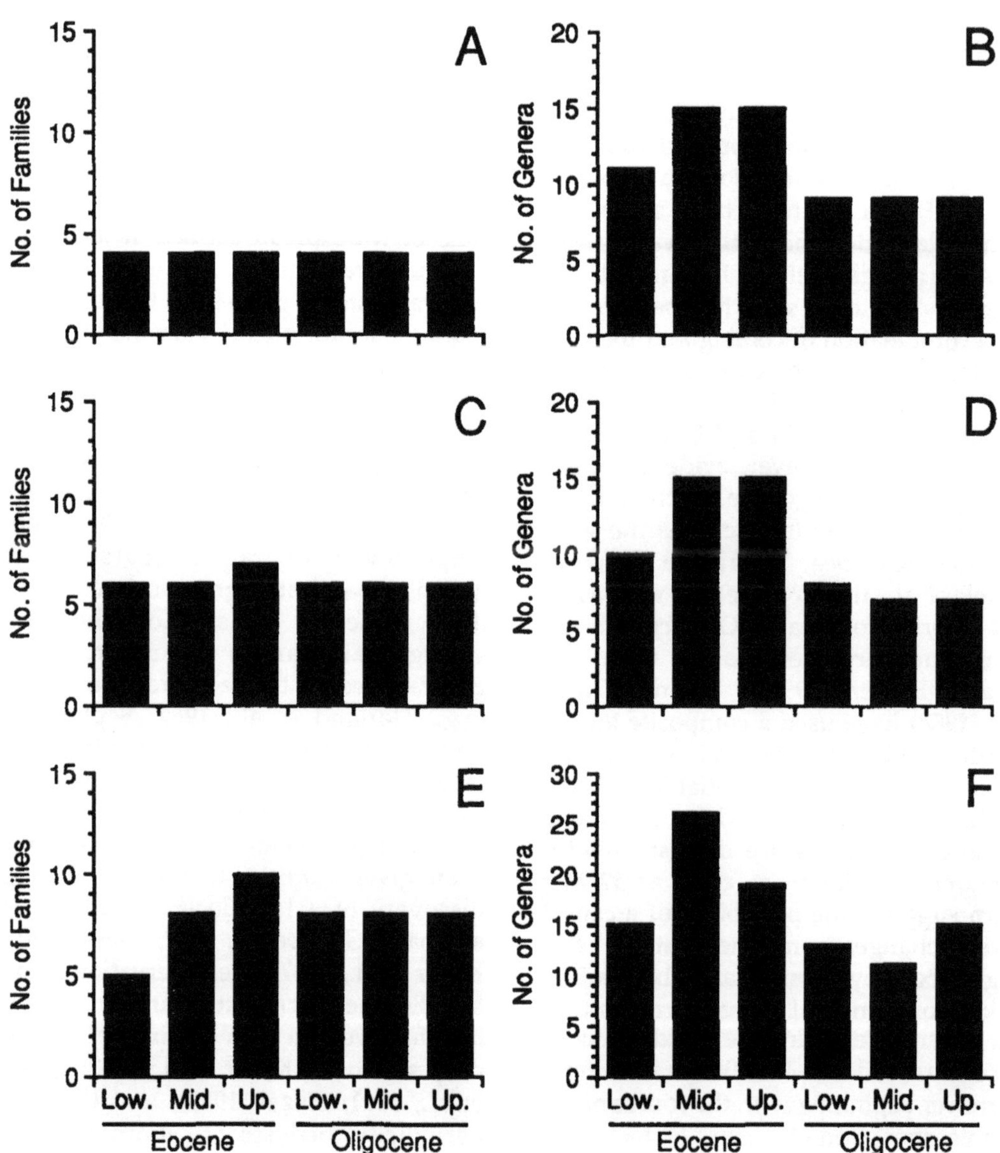

FIGURE 11.1. Histograms of planktic foraminiferal taxic (families and genera) richness for Eocene and Oligocene time based on Loeblich and Tappan (1964) [A-B], Blow (1979) [C-D], and Loeblich and Tappan (1988) [E-F].

Oligocene epochs for each classification (Figure 11.1). In essence, these diagrams describe two different taxic richness patterns in what are, for the most part, the same group of species.

The classifications of Loeblich and Tappan (1964) and Blow (1979) show similar patterns through the Eocene-Oligocene interval with little or no change at the family-level (Figures 11.1A and 11.1C), but a sudden rise in generic richness at the lower/ middle Eocene boundary followed by an equally sudden drop at the E/O boundary (Figures 11.1B and 11.1D). In contrast, Loeblich and Tappan's most recent classification (1988) exhibits a progressive increase in the Eocene family richness, followed by a slight decline at the E/O boundary. Unlike the previous classifications of Loeblich and Tappan (1964) and Blow (1979), however, Loeblich and Tappan's generic data exhibits an abrupt taxic richness *increase* in the middle Eocene followed by a more or less continuous decline into the middle Oligocene that, in turn, passes into another generic richness increase in the Upper Oligocene (Figures 11.1E and 11.1F). In addition, all three classifications exhibit marked discrepancies between taxic richness patterns assessed at different levels within the taxonomic hierarchy.

On the basis of these discrepancies, one is forced to conclude that: i) late Paleogene supraspecific taxic richness patterns are strongly influenced by the classification scheme employed; and ii) temporal trends in taxic richness observed at one level in the taxonomic hierarchy may not necessarily be representative of different levels. While the Loeblich and Tappan (1964) and Blow (1979) classifications have traditionally provided the main empirical support for recognizing the E/O boundary as a time of sudden and widespread change in planktic foraminiferal faunas, the greatly expanded generic classification of Loeblich and Tappan (1988) shows a gradual, rather than sudden, Eocene-Oligocene faunal turnover that is much more reminiscent of preliminary species-level data (e.g., Corliss et al., 1984). As a result of these very obvious sources of bias within various supraspecific planktic foraminiferal classifications, any rigorous evaluation of the Eocene-Oligocene planktic foraminiferal extinction record must employ species-level taxonomic data that have been resolved to a considerably more refined stratigraphic scale than that of nearest inclusive stage.

Fortunately, a large part of these species-level taxonomic/high resolution stratigraphic data have been summarized by Toumarkine and Luterbacher (1985) and Bolli and Saunders (1985). These biostratigraphic compendia list 109 species and subspecies of planktic foraminifera along with their associated stratigraphic ranges to (at least) the nearest inclusive biozone. Although these data do not represent an exhaustive census of all Eocene-Oligocene planktic foraminifera, they do include not only the majority of such taxa, but essentially all of the common, well-studied forms whose stratigraphic ranges have been reliably established. Moreover, those species and subspecies excluded from this dataset are predominately longer-ranging forms that are of little use in high-resolution biostratigraphic analyses and, if anything, would tend to obscure the very faunal turnover patterns we seek to study.

In order to evaluate these species- and subspecies-level data, taxic richness estimates [corrected for intrazonal turnover via Harper's (1975) index] were made for each of seventeen constituent Eocene and Oligocene planktic foraminiferal biozones as shown in Figure 11.2A. Starting at comparatively low values in the early Eocene, planktic foraminiferal taxic richness increased rapidly reaching a peak in the middle part of the middle Eocene, after which it underwent a progressive decline that extended through the Oligocene where it returned to early Eocene values. These patterns remain unchanged regardless of whether or not subspecies (light shading in Figure 11.2A) are excluded from consideration. Both species and species + subspecies data are also uncorrelated with variations in the temporal duration of the Eocene-Oligocene biozones (MacLeod, in press).

The taxic richness data illustrated in Figure 11.2A represent the dynamic interplay between rates of extinction and origination. Separate tabulations of these extinction and origination data (Figures 11.2B and 11.2C) reveal several interesting aspects of this extended faunal transition. For instance, elevated numbers of species and subspecies extinctions were not confined to the two late Eocene biozones, both of which contain unambiguously impact-related debris layers. Instead, relatively high levels of planktic foraminiferal extinctions occurred throughout an interval from the middle Eocene to the early Oligocene (Figure 11.2B). But, while the elevated levels of species originations that characterize the early/middle Eocene boundary remain high throughout most of the middle and late Eocene, this relatively high late Eocene planktic foraminiferal origination rate undergoes a precipitous decline at the E/O boundary (Figure 11.2C). It is this failure of taxic origination in the early Oligocene rather than the acceleration of late Eocene extinction rates (for which there is no evidence) that drives the planktic foraminiferal faunal turnover at the E/O boundary. Irrespective of this highly provocative and previously unanticipated decline in species originations across the E/O boundary, the taxic richness data as a whole provide strong evidence in favor of the gradual Eocene-Oligocene planktic foraminiferal turnover

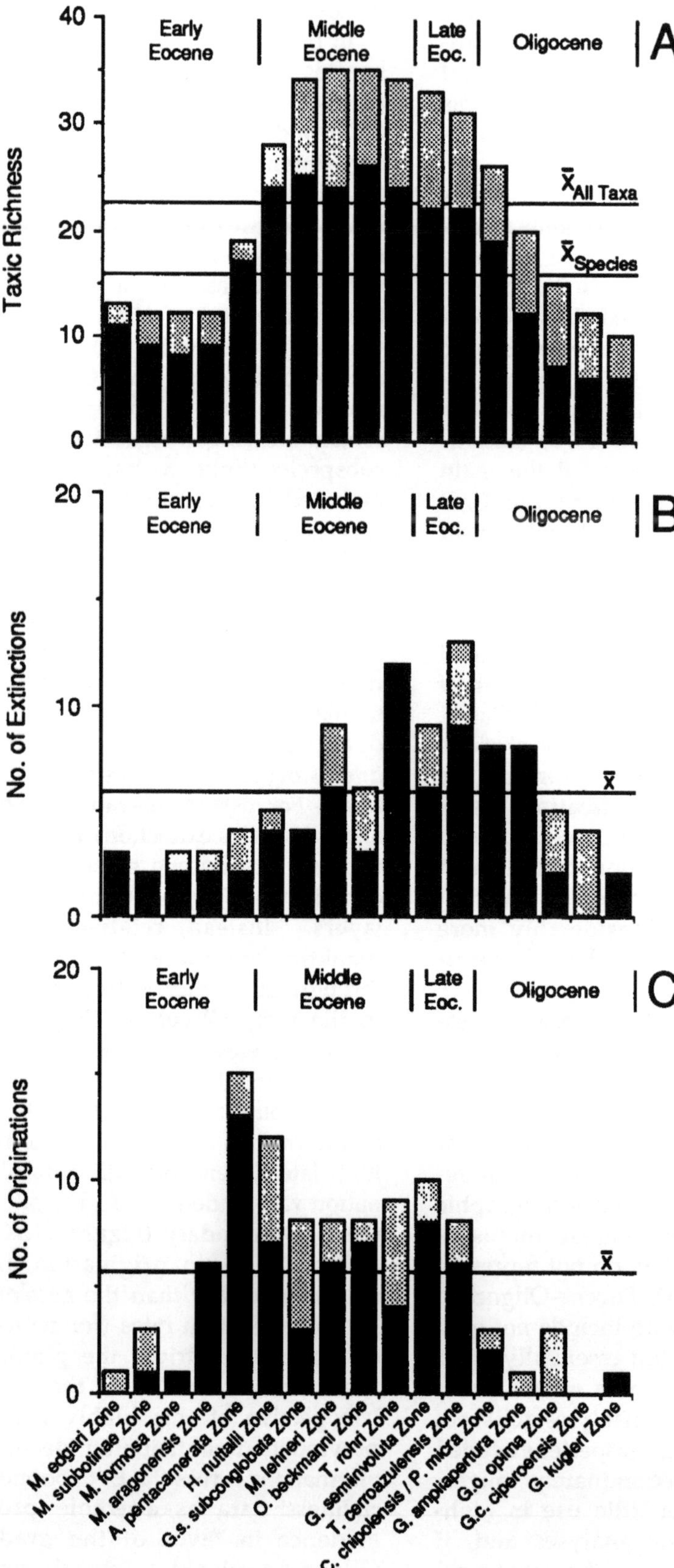

FIGURE 11.2. Histograms of estimated planktonic foraminiferal taxic (species and subspecies) richness (A), numbers of taxic extinction (B) and originations (C) in each of the seventeen Eocene and Oligocene planktic foraminiferal biozones. Black bars = species data, stippled pattern = subspecies data. Data from Toumarkine and Luterbacher (1985) and Bolli and Saunders (1985).

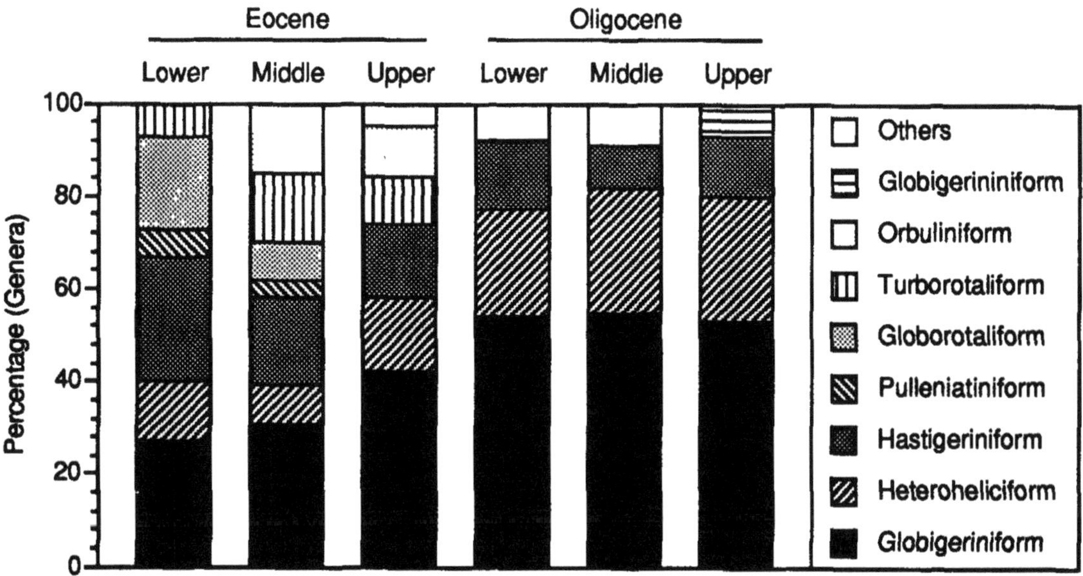

FIGURE 11.3. Morphotypic richness patterns for planktic foraminiferal faunas in each of the six Eocene-Oligocene epochs. Morphotypic classification based on Cifelli (1969): i) globigeriniform -- test consisting of a series of round chambers arranged in a trochospiral coil with a single umbilical aperture; ii) orbuliniform -- globigerine tests in which the final chamber envelops the previous coils; iii) turborotaliform -- globigerine tests with angular chambers and an acute periphery; iv) globorotaliform--globigerine tests with angular chambers and a true peripheral keel; v) pulleniatiniform -- globigerine tests in which axis of coiling migrates to a more umbilical position during ontogeny; vi) hastigeriniform -- planispirally coiled, bilaterally symmetrical tests; vii) globigeriniform -- globigerine tests that exhibit supplementary apertures on the spiral side; viii) heteroheliciform --tests with biserially or triserially arranged chambers. See text for discussion.

model reflected in the generic classification of Loeblich and Tappan (1988).

Was the Eocene-Oligocene faunal turnover the result of a preferential elimination of specific morphotypes that reflect either ecological similarities or functional equivalences? In his classic paper on iterative evolution in planktic foraminifera, Cifelli (1969) used seven morphotypic groupings of taxa to summarize the evolutionary history of planktic foraminifera. Based on changes in the relative abundance of these morphotypes he concluded that while both orbuliniform and globorotaliiform species went extinct before the beginning of the late Eocene, the hastigeriniforms continued to diversify into the late Eocene. Similarly, turborotaliiform morphotypes remained present throughout the Eocene but became extinct by the end of that epoch. By the late Oligocene, however, "all [morphotypes] except the globigerin[iforms] had disappeared" (Cifelli, 1969, p. 159).

In order to update Cifelli's (1969) analysis and determine the extent to which his characterization of the Eocene-Oligocene faunal transition remains valid, we have expanded Cifelli's original morphotypic classification to include a category for species exhibiting either biserial or triserial chamber arrangements (heteroheliciforms) along with an additional category for forms (e.g., *Cassigerinella*) that cannot be fit into any of the aforementioned groups. Results of our revised morphotypic characterization are shown in Figure 11.3.

While Cifelli's (1969) morphotypic analysis appears to be largely consistent with contemporary data (despite the revisions in our taxonomic view of the Eocene-Oligocene planktic foraminiferal transition), our results extend those of Cifelli in several important respects. Although Cifelli (1969) described the Oligocene planktic foraminiferal fauna as being exclusively globigeriniform in character, our expanded morphotypic classification shows that both globigeriniform and heteroheliciform taxa not only dominate the Oligocene planktic foraminiferal assemblage (at least in terms of numbers of genera), but that these morphotypes were also major constituents of the Eocene fauna. It is also important to note that regardless of the low taxic richness values that characterize in early Eocene planktic foraminiferal faunas (Figure 11.2), they nevertheless exhibit the highest morphotypic richnesses observed in the late Paleogene.

Cifelli (1969) concluded that the replacement of the morphotypically diverse middle and late Eocene planktic foraminiferal fauna by (what he described as) a nearly monotypic Oligocene assemblage (of globigeriniform species) was brought about by a reduction in the thermal differentiation of marine surface waters. This inference was based on the ecological preferences of morphologically similar Recent planktic foraminiferal species. Since that time, a large number of isotopic, taxonomic, and biogeographic data have supported the general model of late Paleogene climatic cooling. In the following sections we analyze the Eocene-Oligocene faunal transition based on current relative abundance data and the grouping of species according to their depth habitat preferences in order to explore the relationship between these patterns of taxonomic/morphotypic faunal turnover and physical changes in marine planktic habitats. This analysis will both refine the timing of the major planktic foraminiferal faunal turnover(s) and document the evolutionary/ecological fates of individual species groups in greater detail.

MAJOR FAUNAL TURNOVERS

Planktic foraminiferal assemblages are generally dominated by five to six species which may constitute up to 80% of the total number of individuals present. Changes in the relative abundance of these dominant species reflect changes in the physical and biological character of the local environment. Quantitative analysis of microfossil assemblages can therefore be used to examine the relationship between environmental change and planktic foraminiferal turnovers.

Relative species abundance data for planktic foraminifera from DSDP sites in low and middle latitude Pacific, Atlantic and Indian Oceans as well as onshore marine sections in Spain and the U.S. Gulf Coast have been published for over 20 middle Eocene to Oligocene sequences (Keller, 1983a, 1985, 1986; Keller et al., 1987; Molina et al., 1988). Although individual species abundances vary among these sections, the overall trends in their biostratigraphy and extinction record are similar. Figure 11.4 illustrates this record for Indian Ocean DSDP Site 219. Site 219, as well as most other deep-sea sections, have short hiatuses or nondeposition events in the middle Eocene (*M. lehneri*/*T. rohri* Zone, *O. beckmanni* Zone missing), at the middle/late Eocene boundary (*T. rohri*/*G. semiinvoluta* Zone) and in the late Eocene (*G. semiinvoluta*/*T. cerroazulensis* Zone, Keller, 1983a, 1986; Keller et al., 1987).

Among the dominant species there was a gradual replacement of warm-tolerant tropical species by successively more temperate and cold-tolerant taxa. For example, in Site 219 the middle Eocene *G. subconglobata* Zone is dominated by *Acarinina bullbrooki*, *A. planodorsalis* and *A. broedermanni*, all of which terminally decline at the top of this zone and into the following *M. lehneri* Zone. In the succeeding *T. rohri* Zone, *T. rohri* and *Globigerinatheka* spp. (excluding *G. semiinvoluta*) dominate and terminally decline in abundance at the top of the zone. A notably cooler water temperate faunal assemblage dominates the early late Eocene *G. semiinvoluta* Zone (*Globorotaloides carcoselleensis*, *Globigerinatheka semiinvoluta*, *Pseudohastigerina micra*; Figure 11.4) only to be replaced by a still colder water fauna in the late Eocene *T. cerroazulensis* Zone dominated by *Subbotina angiporoides*, *S. linaperta*, and *Globigerina ouachitaensis*. The appearance of successively more cold-tolerant elements within the dominant planktic foraminiferal fauna through the middle and late Eocene results from the migration of cooler water taxa into lower latitudes and coincides with global climatic cooling.

At the Site 219 E/O boundary, however, a significant change in planktic foraminiferal species diversity is conspicuously absent. Only *Turborotalia cerroazulensis*, which constitutes less than 8% of the total fauna, disappears at this boundary (Figure 11.4). Overall, the

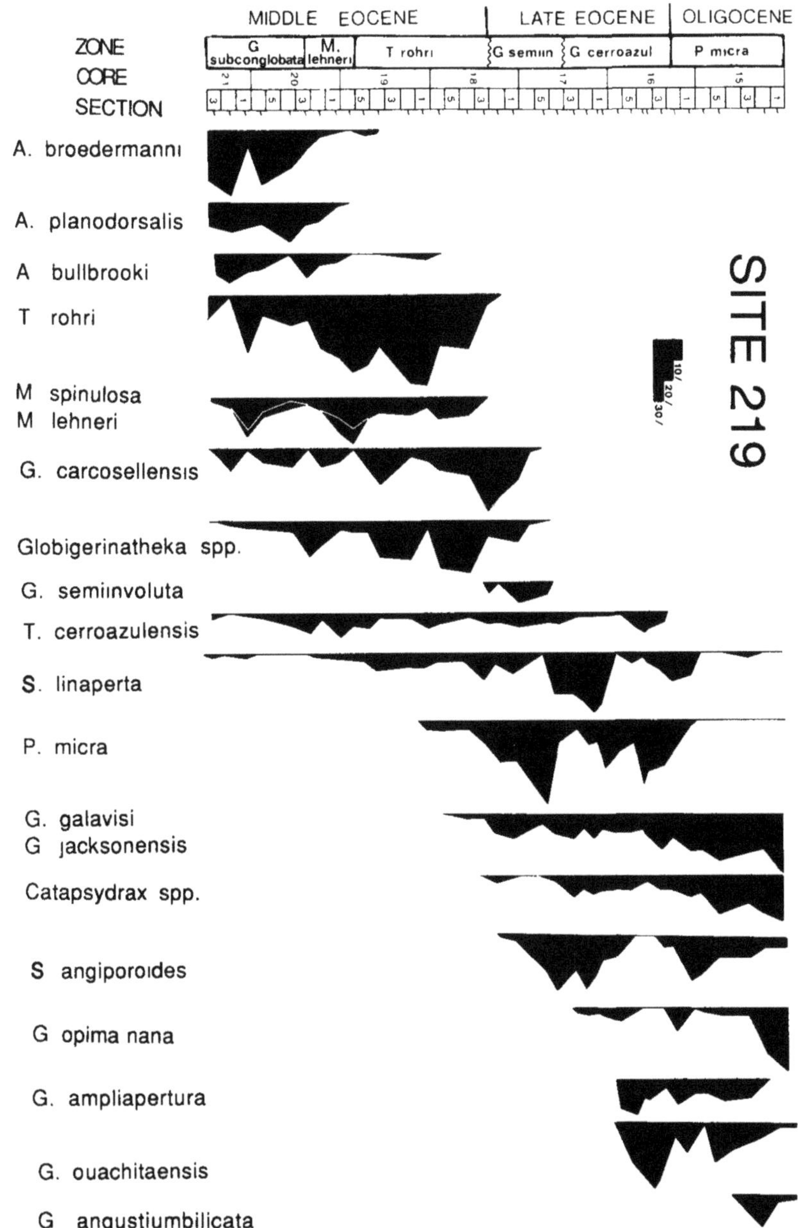

FIGURE 11.4. Abundance distribution of dominant planktic foraminiferal species in Indian Ocean DSDP Site 219. Wavy lines mark hiatuses, tick marks in core-sections indicate sample positions.

dominant cool-temperate taxa that appeared during the latest Eocene continued across the boundary into the early Oligocene (Figures 11.4, 11.5; Keller, 1983a, 1985, 1986); suggesting a simple continuation of the late Eocene cooling trend. The next major faunal turnover begins in the late early Oligocene near the *G. ampliapertura/ G. opima* Zone boundary and continues into *G. opima* Zone (Figure 11.5). In this interval, most of the surviving Eocene taxa (e.g., *Pseudohastigerina micra, Globigerina ampliapertura, Subbotina angiporoides, S. linaperta, S. utilisindex, Subbotina galavisi, G. jacksonensis)* decline in abundance and disappear. At the same time *Globorotalia opima* appears and predominates in the zone bearing its name,

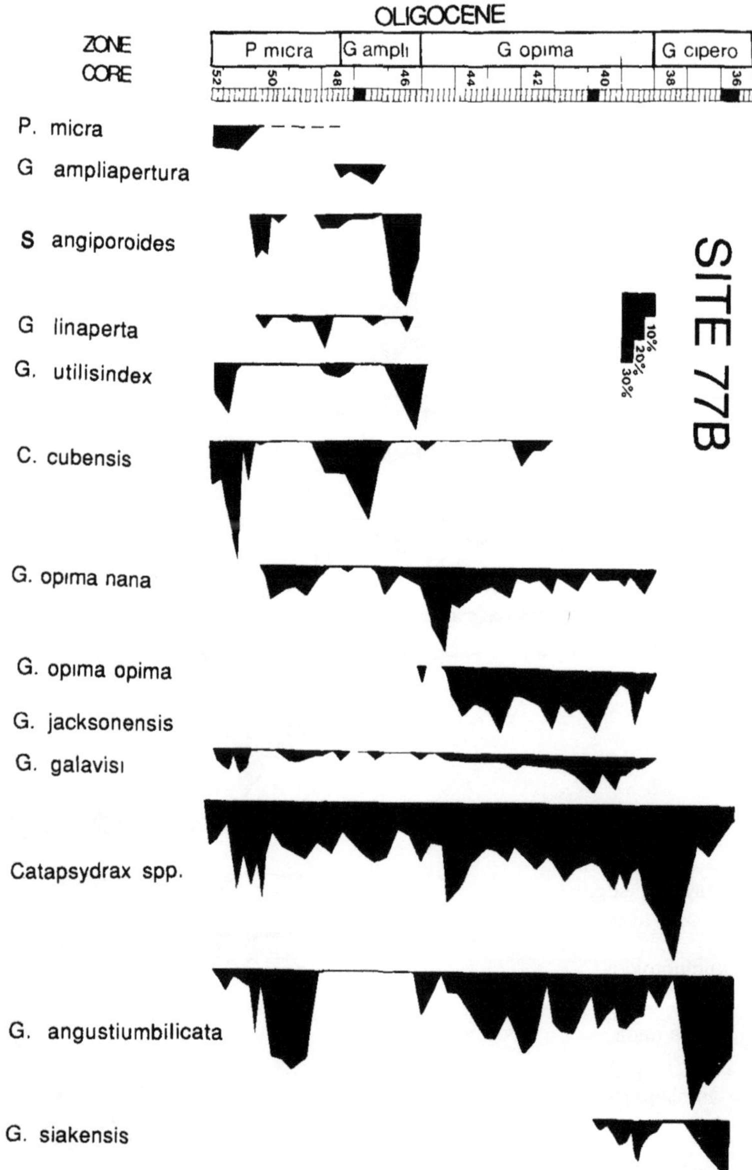

FIGURE 11.5. Abundance distribution of dominant planktic foraminiferal species in east equatorial Pacific DSDP Site 77B.

followed by increased abundance of *Tenuitella angustiumbilicata,* and the appearance of *Globorotalia siakensis, G. pseudocontinuosa, Globigerina ciperoensis,* and *G. angulisuturalis* (Figure 11.5; Keller, 1983a,b; 1984; 1985). These faunal turnovers represent global climatic and oceanographic changes as well as a major evolutionary transition from a relatively warm-tolerant and diverse middle Eocene fauna to a cold-tolerant, low diversity Oligocene assemblage.

DEPTH STRATIFICATION OF PLANKTIC FORAMINIFERA

Planktic foraminifera live in the upper portion of the oceanic water column with the highest density of individuals in the upper 100 m and few individuals below 400 m depth (Bé and Tolderlund, 1971; Bé et al., 1971). Within this interval, species are depth stratified with many species remaining at relatively constant depths while others migrate downward during their life cycle (Lidz et al., 1968; Berger, 1969, Bé and Hemleben, 1970; Bé and Tolderlund; Williams et al., 1979; Fairbanks et al., 1980, 1982). The relative depth habitat of individual fossil species can be determined from oxygen and carbon isotope ranking. This depth ranking is typically based on the relative values of $\delta^{18}O$, with the lower $\delta^{18}O$ values characterizing species that predominantly inhabit warmer surface waters while higher $\delta^{18}O$ values indicate species living either at greater depths or within cooler water masses. Although some recent spinose taxa have been observed to calcify their tests out of equilibrium with the surrounding sea water (Fairbanks, 1980, 1982), the stable isotopic depth ranking of Recent species exhibits a general agreement with the vertical stratification observed from plankton tow studies (Williams et al., 1979; Williams and Healy-Williams, 1980) despite the presence of seasonal temperature variations, the non-equilibrium isotopic fractionation of some foraminiferal species (e.g., *Cibicidoides*), and other vital effects. Consequently, the stable isotopic ranking of species can be used to subdivide the standing foraminiferal fauna into species groups which predominantly inhabit surface, intermediate and deep water sectors within the upper part of the oceanic water column.

These data, when coupled with assessments of relative abundance changes for individual species inhabiting each depth sector, can, in turn, be used to infer changes in the water mass stratification and the physical expansion or contraction of particular depth-related habitats. Physical oceanographic changes such as these are presumably driven by external factors, such as variations in temperature, salinity, and density. Thus, once species have been isotopically ranked, it is then possible to construct vertical temperature profiles of marine surface waters and construct a detailed paleoecological description of species extinctions and originations in terms of the historical record of the occupation of depth-related planktic habitats.

In the present study, stable isotopic ranking of planktic foraminifera in middle Eocene through Oligocene sections was used to understand the oceanographic and evolutionary changes underlying planktic foraminiferal faunal turnovers. Preliminary stable isotopic depth rankings for most common species have been provided by Saito and Donk (1974), Douglas and Savin (1973; 1975; 1978) Boersma and Shackleton (1981), Boersma and Premoli-Silva (1983, 1989), Shackleton et al. (1985), Keigwin and Corliss (1986), Corliss and Keigwin (1986) and our own unpublished data. Based on this composite data set, we have compiled a preliminary depth stratification of species grouped into surface, intermediate and deep water dwellers (Table 11.1; Keller and MacLeod, in press). Although further isotopic ranking of species is still necessary to refine the relative positions of species in the water column, the data presented here are sufficient to demonstrate the potential of this tool for understanding the general pattern of planktic foraminiferal evolution and species diversity and their relationship to changing climatic and oceanographic regimes.

By analogy with present oceanic conditions, we have assumed that the isotopically lightest species inhabited the surface (mixed) layer that constitutes the uppermost 100 m of the water column (Table 11.1). Intermediate dwellers are inferred to have occupied the interval below the surface (mixed) layer and at or just below the permanent (as opposed to the seasonal) thermocline that resides between 100 m to 250 m below the surface. Deep water dwellers are inferred to have lived below the permanent thermocline (Berger, 1969; Bé and Tolderlund, 1971; Williams et al., 1979; Williams and Healy-Williams, 1980; Fairbanks et al., 1980, 1982).

Middle Eocene-Oligocene depth stratification

The relative abundance of dominant species groups in surface, intermediate and deep water habitats is illustrated in Figure 11.6 based on data from the equatorial Pacific DSDP Site 77B for the Oligocene interval and Indian Ocean DSDP Site 219 for the Eocene interval. Within each depth habitat, species have been grouped by common temporal trends. For example, species that evolved during the same time span are grouped together and followed

TABLE 11.1. Relative depth ranking of middle Eocene to Oligocene planktic foraminifera based on oxygen isotope variations.

Surface

Middle Eocene	Late Eocene	Oligocene
Acarinina bullbrooki (●)	Chiloguembelina cubensis (◎)	Globigerina ampliapertura (◎)
A. broedermanni (●)	Truncoroaloides rohri (●)	Pseudohastigerina barbadoensis (◎)
A, planodorsalis (●)	Globigerinatheka semiinvoluta (●)	Globigerina officinalis
Muricoglobigerina senni (●)	G. index (●)	G. ouachitaensis
	Turborotalia cerroazulensis (●)	G. angustumbilicata
	T. cunialensis (●)	Globorotalia siakensis
	Hantkenina alabamaensis (●)	G. pseudokugleri
	Pseudohastogerina barbadoensis (◎)	G. mendacis
	Globigerina ampliapertura	
	G. officinalis	
	G. ouachitaensis	
	G. angustiumbilicata	

Intermediate

Middle Eocene	Late Eocene	Oligocene
Morozovella lehneri (●)	Subbotina linaperta	Subbotina linaperta
M. spinulosa (●)	S. angiporoides	S. angiporoides
M. aragonensis (●)	S. utilisindex	S. utilisindex
Truncorotaloides rohri (●)	Globigerina eocaena	Globigerina eocaena
Globigerinatheka semiinvoluta (●)	G. galavisi	G. galavisi
G. index (●)	G. euapertura	G. praebulloides
	G. medizzai (●)	Pseudohastogerina micra
	Pseudohastigerina micra	Globoquadrina dehiscens
	Globorotaloides carcosellensis (●)	Globorotalia opima opima
		G. opima nana

Deep

Middle Eocene	Late Eocene	Oligocene
Globigerinatheka spp.	Globoquadrina venezuelana	Globoquadrina venezuelana
	G. pseudovenezuelana	G. pseudovenezuelana
	G. tripartita	G. tripartita
	Catapsydrax spp.	G. praedehiscens
		Catapsydrax spp.

●Species included in solid (black) pattern on Figure 11.6.
◎Species included in dark stipple pattern of surface and intermediate columns of Figure 11.6.

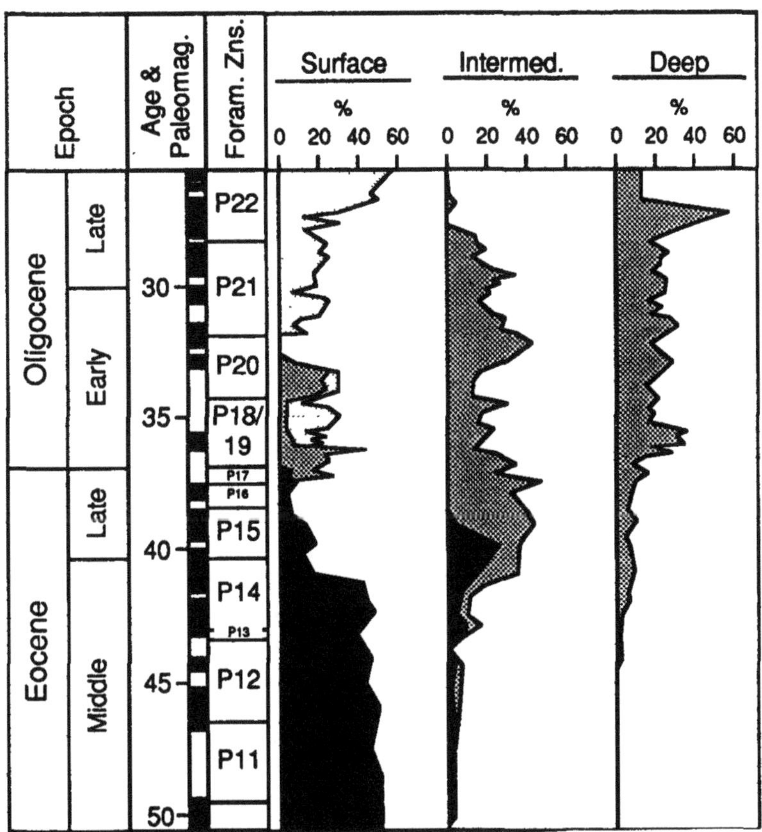

FIGURE 11.6. Relative depth stratification of the predominant constituents of the middle Eocene to Oligocene planktic foraminiferal fauna. Relative abundance of species grouped into surface-, intermediate- and deep-dwelling forms based on relative oxygen isotope ranking. Eocene data from DSDP Site 219, Oligocene data from DSDP Site 77B. Time scale based on Berggren et al. (1985). Faunal changes within each group are illustrated by different stipple patterns. Species included in the solid (black) pattern are marked by (O) in Table 11.1. Species included in the dark stipple area of the surface column are marked by a (O) in Table 11.1. Species listed in other stipple patterns as listed in Table 11.1 for surface-, intermediate- and deep-dwelling forms. See text for discussion.

through to their extinction with different shading patterns being used to delineate these temporally-related groups of taxa. In this way three major species groups can be identified in the surface (mixed) layer, but only two and one group in the intermediate and deeper sectors respectively. These results indicate that evolutionary diversification preferentially took place within the surface layer and secondarily at the intermediate depths of the thermocline. Our data also suggests that little evolutionary activity took place below the permanent thermocline (see also Hart, 1980).

In addition, Figure 11.6 illustrates that the transition from one faunal assemblage to another was generally a gradual process that occurred over a period 10^5 to 10^6 yrs. This transition pattern was typically characterized by a sharp terminal decline in numerical abundance and the lingering of these numerically rare species during the rise to dominance of the succeeding faunal group. Where abrupt faunal changes are observed, this is generally due to artifacts of the stratigraphic record such as short hiatuses and carbonate dissolution (Keller, 1983a, 1985, 1986; Keller et al., 1986; MacLeod and Keller, in press).

The major faunal turnovers noted in the

species abundance plots (Figures 11.4, 11.5) discussed earlier can be recognized in the depth stratified assemblage groups of Figure 11.6. The middle to late Eocene faunal turnover was characterized by the decline and eventual extinction of the warm surface water fauna (solid pattern) and a correlative increase in the cooler water intermediate fauna (dark stipple, Figure 11.6). A new, but for the most part short-lived, cooler surface water fauna (e.g., *G. ampliapertura*, *T. angustiumbilicata*, *Chiloguembelina cubensis*) evolved in the latest Eocene and disappeared again in the late early Oligocene (dark stipple), except for *T. angustiumbilicata* that predominates throughout the Oligocene. This surface water fauna was accompanied by a temporary decrease in the intermediate-dwelling species group (Figure 11.6). For the most part, *T. angustiumbilicata* was the dominant surface dweller (except for a short interval of increased carbonate dissolution in the *G. opima* Zone (Keller, 1983a), and intermediate and deep dwellers were abundant during the Oligocene at Site 77B. In the latest Oligocene (*G. ciperoensis* Zone) the surface group increased dramatically, partly due to the evolution of several new species (e.g., *Globorotalia siakensis*, *G. mayeri*, *G. kugleri*, *Globigerina ciperoensis*, *G. angulisuturalis*, Keller, 1983b), and was accompanied by a correlative decrease in the numbers of intermediate and deeper dwelling species.

Middle to late Eocene faunal turnover

Where and how in this long term climatic cooling trend did the so-called late Eocene-Oligocene mass extinctions occur? Figure 11.6 illustrates the presence of a major middle to late Eocene faunal turnover with its climax near the middle/ late Eocene boundary and a second faunal turnover of lesser magnitude during the late early Oligocene. Although this turnover cannot be characterized as a mass extinction, the numbers of taxa evolving and becoming extinct during the middle to late Eocene clearly identify this interval as marking a major ecological crisis for planktic foraminiferal populations. Figure 11.7 details this faunal turnover and shows that the relative abundance of all taxa becoming extinct at DSDP Site 219 constitute over 80% of the middle Eocene fauna. The decline in numerical abundance of surface dwellers, that began in the *G. subconglobata* Zone (Figure 11.4), continued through the *M. lehneri* Zone, accelerating into the upper part of the *T. rohri* Zone where it resulted in the extinction of several surface-dwelling species (e.g., *Truncorotaloides rohri*, *Morozovella lehneri* and *M. spinulosa*; Figures 11.4, 11.7). The surviving surface dwellers (e.g., *Globigerinatheka* spp., *Turborotalia cerroazulensis* s.l.) and one intermediate-dwelling species (*Globorotaloides carcoselleensis*) declined rapidly near the top of the *G. semiinvoluta* Zone. Since few surface-dwelling species survived to the E/O boundary (Figures 11.4, 11.7), the faunal turnover appears to have affected warm-tolerant surface dwellers first, followed successively by more cold-tolerant taxa. In contrast, only one intermediate-dwelling species disappeared during this interval, whereas the remaining intermediate-dwelling taxa survived well into the Oligocene. This middle to late Eocene faunal decline was accompanied by the rise to dominance of a new fauna, strikingly different in both taxonomic and ecologic composition. This new fauna consists primarily of intermediate-dwelling forms with a few deep-dwelling species whose number increased throughout the late Eocene and Oligocene. Only near the E/O boundary did a new surface-dwelling species group evolve (Figure 11.7). This interval of accelerated faunal turnover spanned approximately 4 m.y.

What caused this increase in the relative faunal turnover rate? Based on available isotopic data, this event most likely reflects major changes in the temperature-depth gradient between the ocean surface and the permanent thermocline that presumably decreased as a result of global cooling (Savin et al., 1975; Savin and Yeh, 1981; Barrera et al., 1990). Accompanying changes in the isotopic composition of planktic foraminifera during this interval indicate that this cooling trend reduced the vertical heterogeneity of the upper marine water column largely through surface water cooling which, in turn, would have been accompanied by a reduction in depth-related planktic habitats as well as a contraction of tropical biogeographic provinces. Increased vertical and latitudinal thermal gradients were likely accompanied by decreased oceanic circulation, diminished upwelling, and reduced nutrient

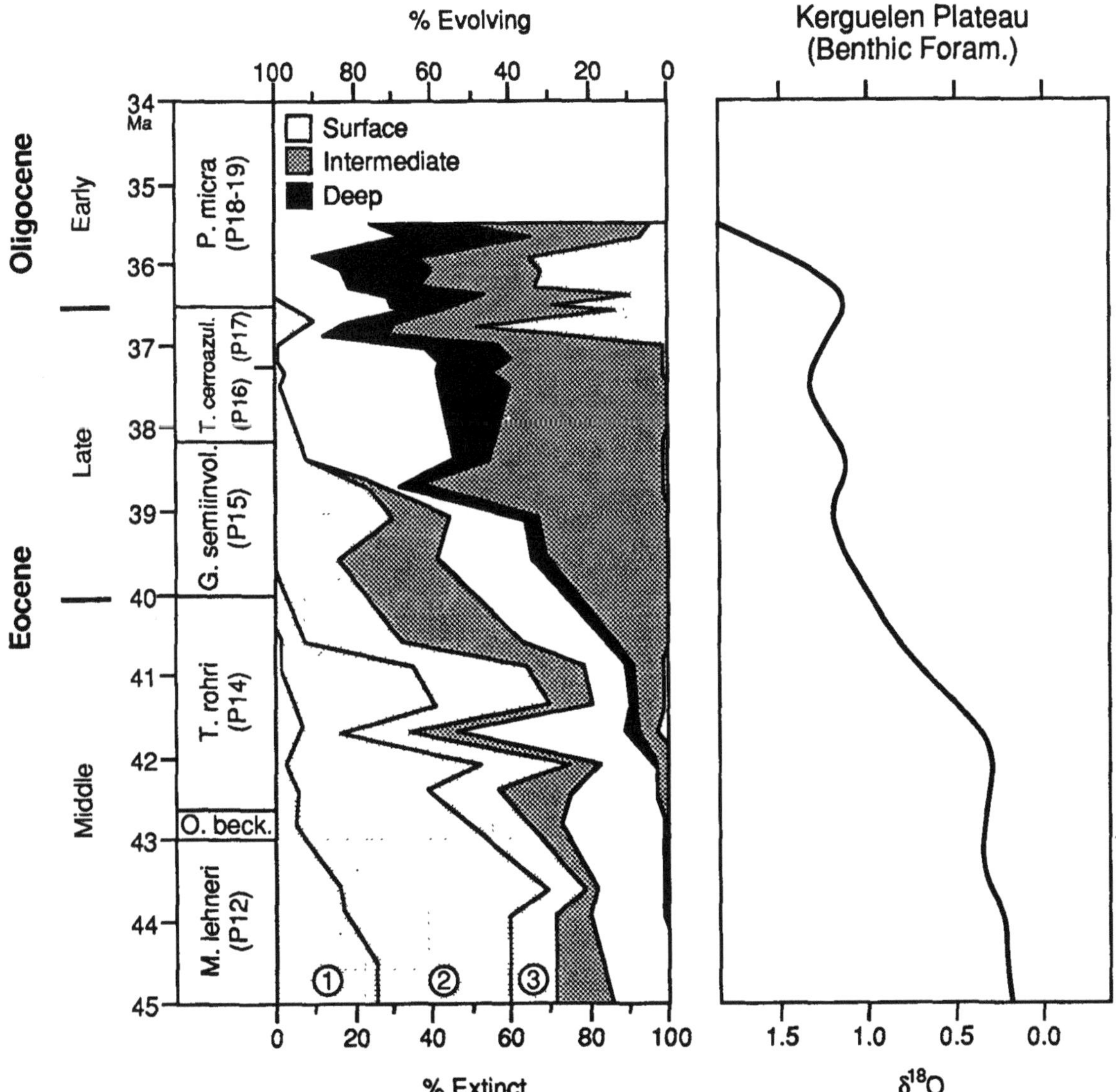

FIGURE 11.7. Middle Eocene to early Oligocene faunal turnover at Indian Ocean DSDP Site 219 based on the relative abundance of species going extinct and of species evolving in surface, intermediate and deep water sectors as determined from oxygen isotopic ranking. White zone in the center of the figure represents the relative proportion of the (undifferentiated) planktic foraminiferal fauna persisting throughout the interval. Benthic foraminiferal (*Cibicidoides*) $\delta^{18}O$ curve for ODP Sites 689 and 690; see Figure 11.11. Note that during the late middle to late Eocene, extinction predominantly takes place within surface-dwelling assemblage, including a successive pattern of disappearance among the following groups: 1. *Acarinina broedermanni, A. planodorsalis, A. bullbrooki;* 2. *Truncorotaloides rohri, Morozovella lehneri, M. spinulosa;* 3. *Globigerinatheka* spp. Time scale based on Berggren et al. (1985).

resources (an inference supported by the southern high latitude $\delta^{13}C$ record as discussed below). Our data thus provide empirical support for Cifelli (1969) and Lipps (1970, 1986) prior attribution of the middle to late Eocene faunal turnover as primarily resulting from habitat destruction, brought on by global cooling, via altered oceanographic patterns.

Oligocene faunal turnover

The second interval characterized by accelerated rates of faunal turnover spans the early/late Oligocene boundary, a time period characterized by the decline and eventual extinction of nearly all surviving Eocene species. The numerical abundance of these taxa and that of the evolving species is illustrated in Figure 11.8 for Site 77B along with the benthic $\delta^{18}O$ record (Keigwin and Keller, 1984). In this faunal turnover the species becoming extinct consisted primarily of surface-dwelling forms that evolved during the latest Eocene (e.g., *Globigerina ampliapertura*, *Pseudohastigerina spp.*) and the intermediate-dwelling Eocene survivors (e.g., *Subbotina linaperta*, *S. angiporoides*, *S. utilisindex*, *Globorotalia opima nana*, *Subbotina galavisi*, *G. jacksonensis*). These taxa begin their decline at approximately 30.5 Ma and disappear altogether between 28 to 29 Ma. This decline and eventual extinction was accompanied by a major global cooling (1.0 per mil for $\delta^{18}O$) that reached its maximum at 29 Ma (see Savin et al., 1975; Savin, 1977; Keigwin and Keller, 1984; Figure 11.8). Among the species that originate during this time interval was the relatively short-lived intermediate-dwelling form *Globorotalia opima* (restricted to *G. opima* Zone) and small globigerine and globorotaliid surface dwellers (e.g., *Globorotalia siakensis*, *G. mayeri*, *G. kugleri*, *G. mendacis*, *Tenuitella angulisuturalis*; Keller, 1983b). These taxa first appear in the late *G. opima* and *G. ciperoensis* Zones and survive well into the early Miocene. The evolution and diversification of these surface dwellers was accompanied by a significant global warming (1.0 per mil for $\delta^{18}O$; see Savin et al., 1977; Shackleton et al. 1984; Keigwin and Keller, 1984; Oberhänsli, 1986; Barrera and Huber, 1991). *Tenuitella augustiumbilicata* constitutes a relatively unaffected dominant faunal component of this faunal turnover in surface waters whereas *Catapsydrax* and *Globoquadrina* predominate in deeper waters (Keller, 1983a). The intermediate-dwelling fauna was much reduced during the Oligocene with the exception of a short interval near the *G. ampliapertura*/*G. opima* Zone boundary (Figure 11.8).

These changes in the depth stratification of Oligocene planktic foraminifera imply that the middle to late Eocene cooling trend continued through the early late Oligocene. Moreover, the Oligocene was much cooler than the late Eocene as indicated by; i) stable isotope data (Figures 11.9, 11.10; Savin et al., 1975; Shackleton et al., 1984; Barrera and Huber, 1991); ii) increased abundances of cold-tolerant deep-dwelling species; and iii) a correlative decrease in both intermediate- and surface-dwelling groups (Figure 11.6). The elimination of the late Eocene survivors at both surface and intermediate depths near the early/late Oligocene boundary and in the early late Oligocene, along with the near absence of newly evolving species until the late Oligocene, indicates a further reduction in depth-related habitats since the late Eocene (Figures 11.5, 11.6, 11.8).

Stable isotopic data of Site 77B suggest that this early/ late Oligocene faunal turnover was the result of renewed global cooling that was probably augmented by Antarctic glaciation during the Paleogene. As before, this global cooling would have been expected to reduce vertical and latitudinal thermal gradients, upwelling, and oceanic circulation, all of which would have resulted in habitat destruction for planktic biotas. In the late Oligocene this cooling trend was reversed, however, with low latitude climatic warming, and the diversification of surface-dwelling forms, indicating an increase in the thermal differentiation of marine surface waters, a deepening of the permanent thermocline, and (probably) increased upwelling and nutrient supply.

EOCENE TO OLIGOCENE ANTARCTIC COOLING

The general trends of the late Paleogene paleoclimatic history are well understood (as discussed earlier), whereas relatively high resolution stable isotopic records have only recently been recovered from sediments in the southern

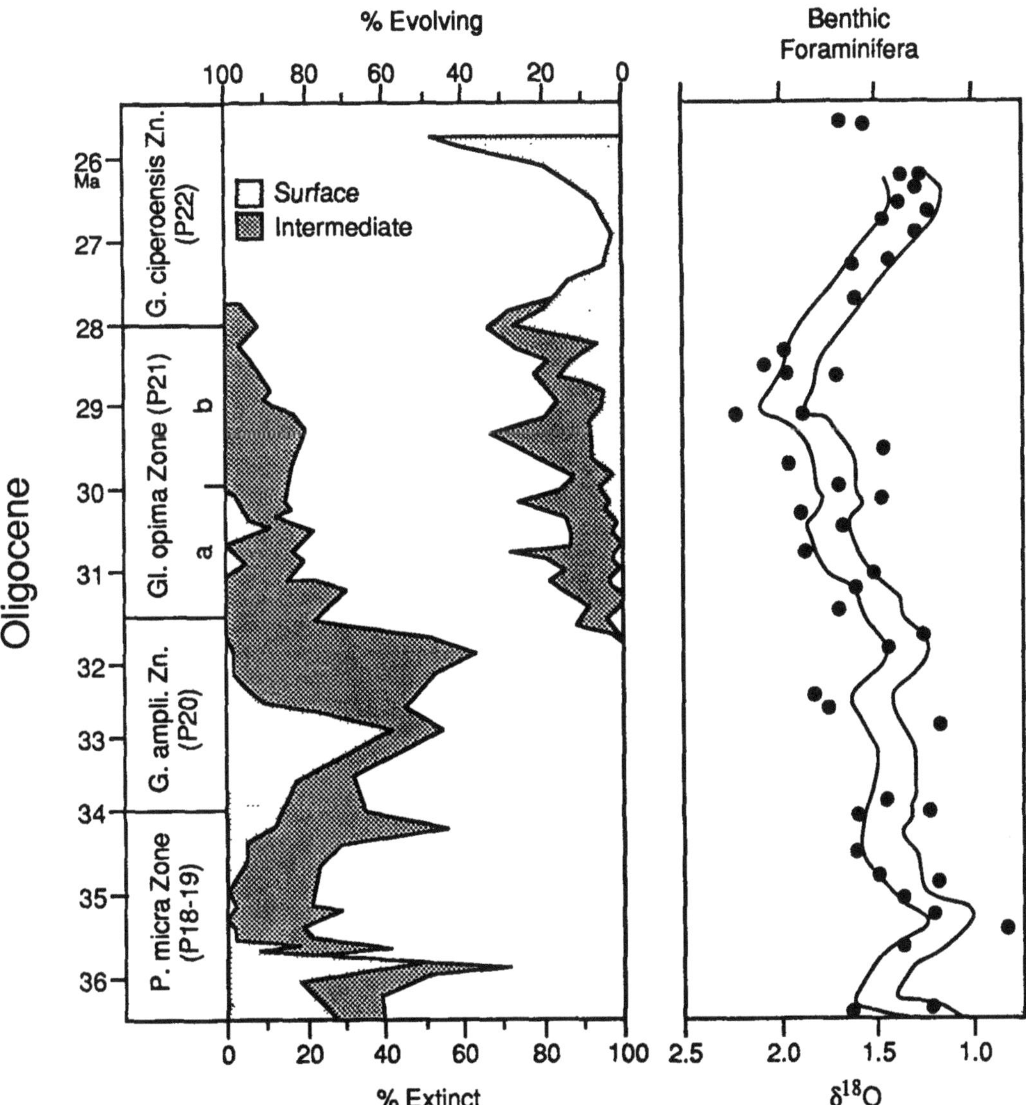

FIGURE 11.8. Oligocene faunal turnover in east equatorial Pacific DSDP Site 77B based on relative abundances of species going extinct and of species evolving in surface, intermediate and deep water sectors as determined by oxygen isotope ranking of species. White zone in the center of the figure represents the relative proportion of the (undifferentiated) planktic foraminiferal fauna persisting throughout the interval. Benthic foraminiferal (*Cibicidoides*) $\delta^{18}O$ curve for Site 77B from Keigwin and Keller (1984). Time scale based on Berggren et al. (1985).

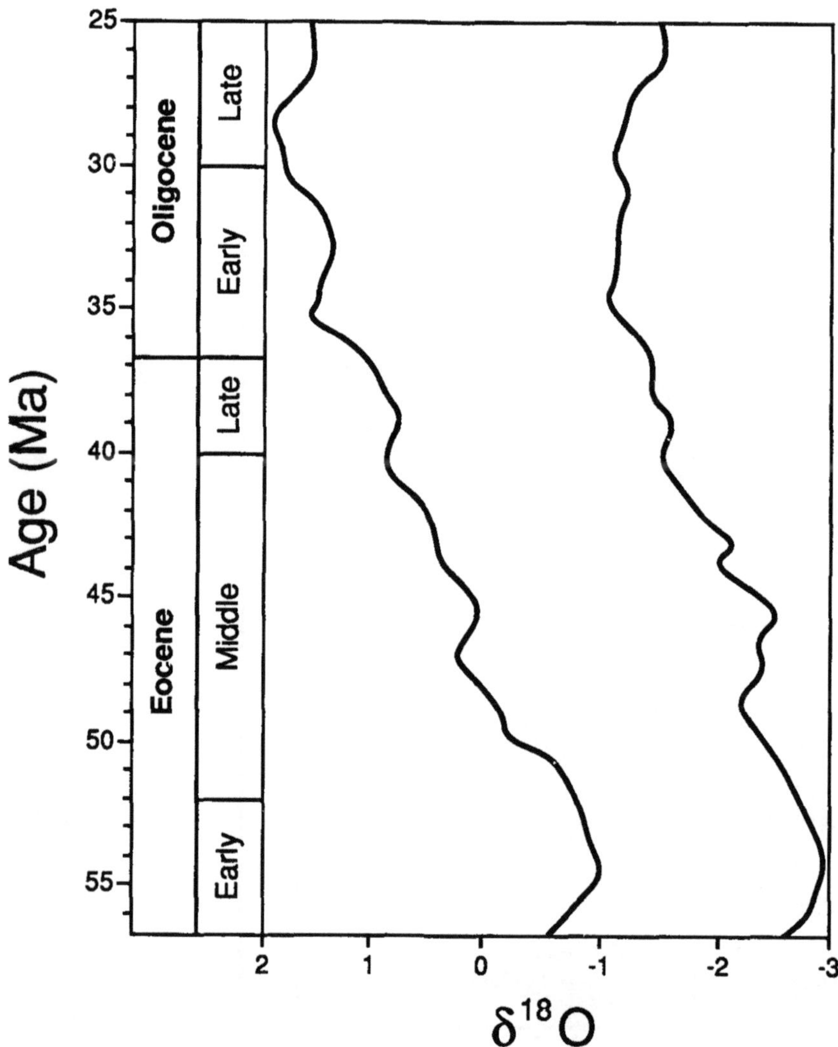

FIGURE 11.9. Composite oxygen isotope curves for tropical shallow-dwelling planktic foraminifers and benthic foraminifers as modified from Prentice and Matthews, 1988. Isotopic data from Poore and Mathews (1984) and Shackleton et al. (1984). Time scale based on Berggren et al. (1985).

oceans near Antarctica on ODP Legs 113 and 119. These high latitude isotopic records together with low and mid-latitude records have yielded the necessary climatic data that document the nature of the late Paleogene cooling and its relationship to the planktic foraminiferal turnovers.

Figure 11.9 presents the composite $\delta^{18}O$ record for planktic and benthic foraminifera from low and mid-latitudes as modified from Prentice and Matthews (1988). This $\delta^{18}O$ record most closely reflects the prevailing temperature conditions at Sites 219 and 77B during the time of the middle Eocene to Oligocene faunal turnovers. Unfortunately, no corresponding composite $\delta^{13}C$ record is available at this time. In Figures 11.10 and 11.11 we illustrate the $\delta^{18}O$ and $\delta^{13}C$ records from the Kerguélen Plateau (Sites 738, 744) and Maude Rise (Sites 689, 690). These data are based on Stott et al. (1990) and Barrera and Huber's (1991) analyses of the planktic foraminiferal taxa *Acarinina* and

Chiloguembelina (reflecting near-surface water conditions), *Subbotina* and *Globorotaloides suteri* (reflecting conditions at or below the thermocline), and the benthic foraminifer *Cibicidoides* spp.. (recording water conditions on the ocean floor [= intermediate water at these locations]). Short-time fluctuations in the raw isotopic data have been removed by interpolating the data to a constant temporal spacing (500,000 years) and then applying a smoothing filter (Sheppard's five-term filter; see Whittaker and Robinson, 1929) to remove any remaining single point anomalies. This data smoothing procedure is preferable to the more commonly employed equally-weighted average filters in that it efficiently represents the overall trend of the isotopic time series while simultaneously minimizing the distortion brought about through the temporally forward or backward migration of local extrema. The isotopic records of these three groups reflect variations in surface productivity along with the thermal differentiation of marine surface waters during the Eocene-Oligocene Antarctic climate cooling trend.

CLIMATE COOLING-$\delta^{18}O$ RECORD

The $\delta^{18}O$ records of both the Maude Rise (ODP Sites 689, 690) and Kerguelen Plateau (Sites 738, 744) as well as the composite low latitude record, show very similar temperature trends for benthic foraminiferal species that resided within the intermediate to deep water mass. Maximum Cenozoic warming occurred between 53 Ma and 55 Ma (Figures 11.9-11.11). Thereafter, intermediate water temperatures decreased by about 12°C-15°C (3.5 per mil) in both high and low latitudes reaching a maximum low in the early Oligocene. At the same time high latitude surface water temperatures also cooled by 12°C-15°C (3.5 per mil) whereas low latitude temperatures cooled by 6°C-8°C (1.5-2.0 per mil). The majority of this cooling occurred during the middle and late Eocene. Although in its early stages this cooling trend appears to be gradual, marked drops in intermediate water temperatures of about 0.5 per mil can be identified between 52-50 Ma (early/middle Eocene boundary) and between 47-48 Ma. Thereafter, two pronounced increases in $\delta^{18}O$ values of intermediate waters, (each averaging 1.0 to 1.5 per mil in high latitude) span the middle/late Eocene boundary (39-43 Ma) and the E/O boundary (35-37 Ma). These major high latitude cooling episodes are associated with ice-rafted debris in the Kerguelen Plateau sites both above the E/O boundary and near the middle/late Eocene boundary (dated at 34.5-36.5 Ma, and about 42.0 Ma, Barrera et al., 1990; Barrera and Huber, 1991; Ehrmann, 1991). Such glacio-marine sediments provide strong evidence for a major episode of East Antarctic glaciation beginning during the late middle Eocene and intensifying throughout the Oligocene. Based on the composite low latitude $\delta^{18}O$ record, Prentice and Matthews (1988) have concluded that glaciation began at about 40 Ma though their interpretation of these data differs from ours in that they assume that low latitude temperatures remained constant throughout the interval.

Comparison of the low latitude faunal turnover of Indian Ocean Site 219 (paleolatitude ~2°S) with the high latitude benthic $\delta^{18}O$ record from Kerguelen Plateau sites, shows a close correlation between the onset of major global cooling and Antarctic glaciation at about 42 Ma and the onset of the major middle to late Eocene faunal turnover (Figure 11.7). In fact, the faunal turnover (largely a record of the progressive elimination of warm-tolerant surface-dwelling species) parallels the decline in surface and intermediate water temperatures at both high and low latitude sites (Figures 11.9, 11.10). Moreover, the disappearance of this warm surface water fauna that was predominately composed of globorotaliform morphotypes, was nearly complete by 38 Ma; thus coinciding to the end of the first major cool event. The second major cooling in the early Oligocene corresponds to the elimination of surface-dwelling taxa that evolved during the latest Eocene cooling (Figures 11.7, 11.8), and the onset of cold climatic conditions also correlate well with the early/late Oligocene faunal turnover.

$\delta^{18}O$ AND $\delta^{13}C$ GRADIENTS

The $\delta^{18}O$ gradients between planktic foraminifers in surface, near-surface and benthic foraminifers reflect thermal stratification of the water column, whereas $\delta^{13}C$ gradients generally reflect changes in surface water productivity. Figures 11.10 and 11.11 illustrate

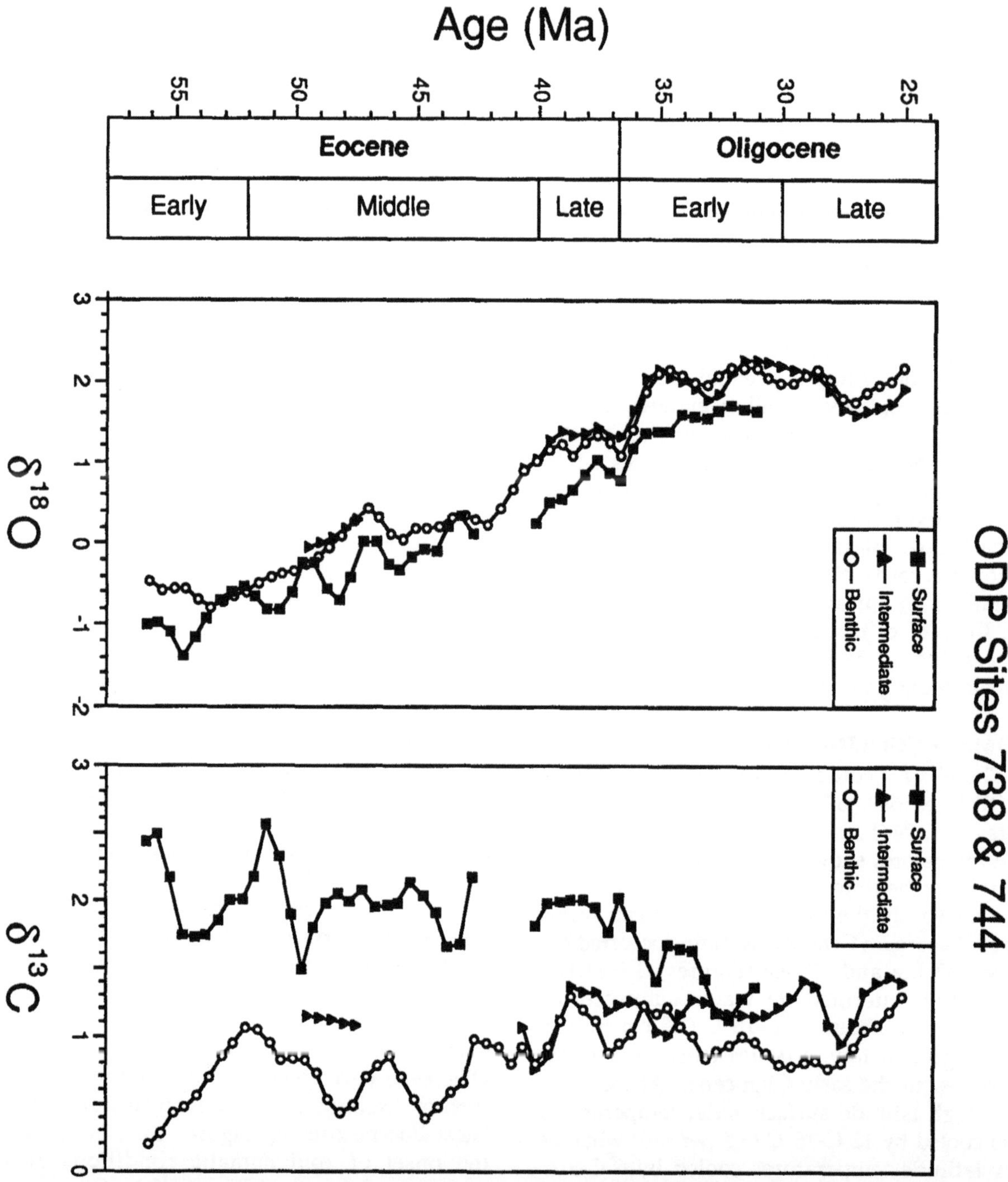

FIGURE 11.10. Eocene-Oligocene oxygen and carbon isotope curves for surface- (*Acarinina* spp. [from 43-52 Ma] and *Chiloguembelina* spp. [from 30-40 Ma]), and intermediate- (*Globorotaloides suteri*) dwelling planktic foraminifera along with benthic foraminifera (*Cibicidoides* spp.) from Kerguelen Plateau, ODP Sites 738, 744, southern Indian Ocean. Data from Barrera and Huber, in press. Time scale based on Berggren et al. (1985). See text for discussion.

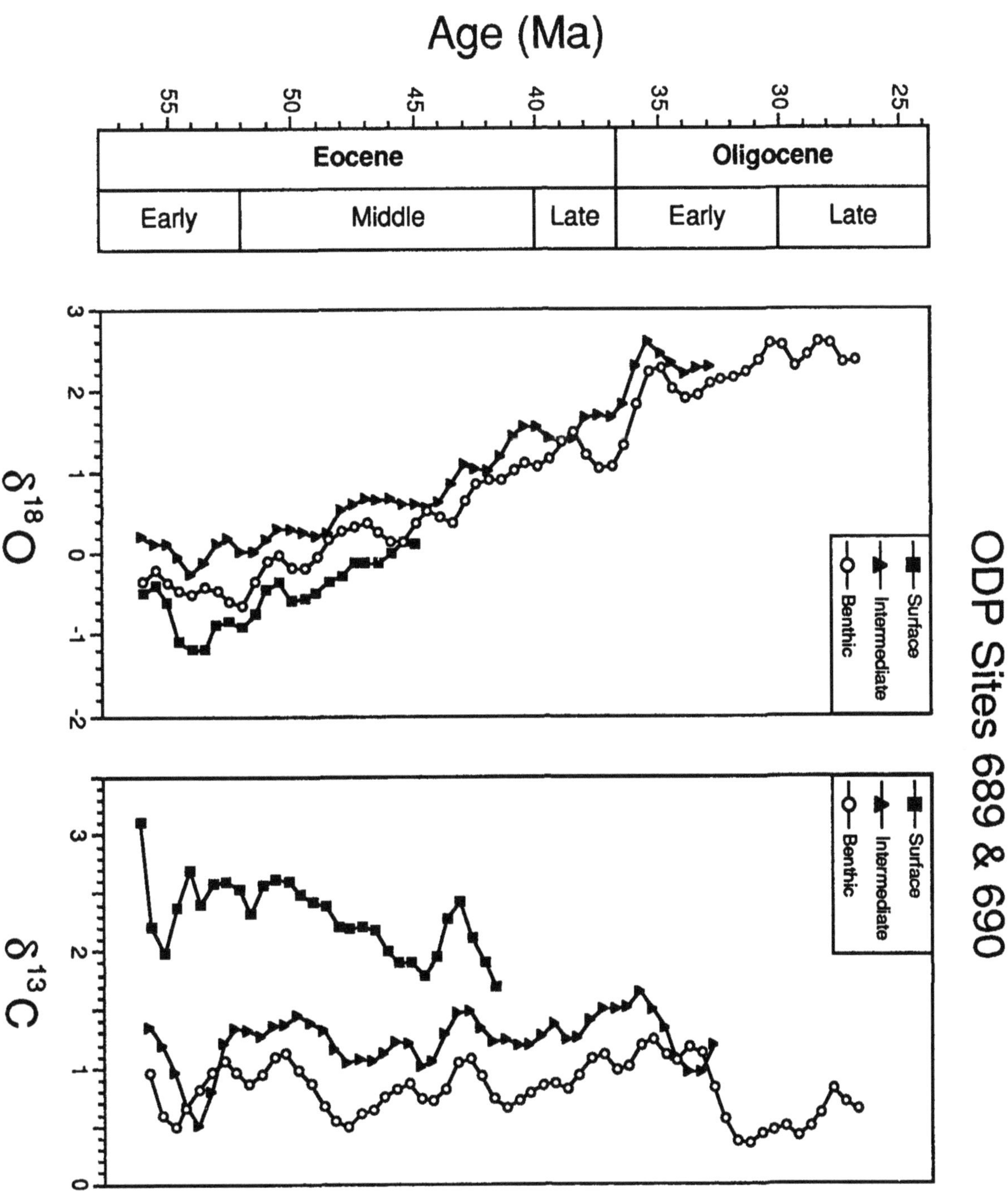

FIGURE 11.11. Eocene-Oligocene oxygen and carbon isotope curves for surface- (*Acarinina* spp.) and intermediate- (*Subbotina* spp.) dwelling planktic foraminifera and benthic foraminifera (*Cibicidoides* spp.) from Maude Rise, ODP Sites 689, 690, Wedell Sea, Antarctica. Data from Barrera and Huber, in press. Time scale based on Berggren et al. (1985). See text for discussion.

these gradients for the Kerguelen Plateau (Sites 738, 744) and Maude Rise (Sites 689, 690). It is interesting to note that at both Sites 738 and 744, the $\delta^{18}O$ values of *Globorotaloides suteri* are nearly the same as those of *Cibicidoides* spp., suggesting a very shallow thermal gradient. This implies that water temperatures between the base of the thermocline and the ocean floor differed by only a few degrees, considering a disequilibrium fractionation of *Cibicidoides* of about 0.5 per mil (paleodepth between 1500 m and 1800 m during middle Eocene to Oligocene time, Barrera and Huber, 1991). Surface water $\delta^{18}O$ values (*Acarinina*), however, average about 0.5 per mil lighter and are more variable reflecting both warmer overall surface temperatures and the presence of substantial climatic fluctuations. Similarly, higher sea surface temperatures than those below the thermocline are observed at Sites 689 and 690 on Maude Rise (Figure 11.11). But, as opposed to the Kerguelen Plateau sites, $\delta^{18}O$ values of *Subbotina* spp. average 0.2 to 0.5 per mil heavier than for *Cibicidoides* spp. This implies that bottom water temperatures were similar, or only about 1°C lower than just below the thermocline. High latitude southern ocean surface to deep thermal gradients show no significant changes between middle Eocene and Oligocene time. Low latitude surface to deep thermal gradients also remain essentially unchanged through the Eocene but increase slightly during the Oligocene (Figure 11.9).

In contrast, $\delta^{13}C$ gradients between surface, near-surface (below thermocline) and the ocean floor show significant changes between Eocene and Oligocene time. Unlike the $\delta^{18}O$ gradients, the $\delta^{13}C$ gradient patterns between Maude Rise and Kerguelen Plateau sites appear to be very similar. The $\delta^{13}C$ difference is greatest between surface water (*Acarinina, Chiloguembelina*) and the ocean floor (*Cibicidoides*), but decreases steadily between early Eocene and Oligocene. During the early Eocene this difference averages about 1.5 per mil, decreasing to 1.0 per mil during the middle Eocene, to 0.5 per mil during the late Eocene and then declining to nearly 0.0 in the early Oligocene (Figures 11.10, 11.11). The progressive decrease in this $\delta^{13}C$ gradient is primarily due to the steadily decreasing $\delta^{13}C$ values of surface dwellers and secondarily to a slight increase of about 0.5 per mil in *Cibicidoides* during the late Eocene. Both benthic (*Cibicidoides*) and subsurface (*Subbotina, Globorotaloides suteri*) values remain relatively stable through the Eocene, although they mirror fluctuations observed in the surface $\delta^{13}C$ record and decrease in the early Oligocene.

If formed under equilibrium conditions, $\delta^{13}C$ ratios of foraminifers reflect the isotopic composition of dissolved CO_2 in ambient sea water. These values vary primarily with the level of surface water productivity. Because ^{12}C is preferentially removed during photosynthesis, surface waters are enriched in the heavier ^{13}C and bottom waters are enriched in the lighter ^{12}C through the return of this isotopic species to the ocean via oxidation at depth (Deuser and Hunt, 1969; Kroopnick, 1974). The gradual decrease in the vertical $\delta^{13}C$ gradient between surface, and subsurface and benthic foraminifers observed between middle Eocene and Oligocene time therefore may reflect a progressive decrease in primary marine productivity. Within this general trend, a major drop in surface productivity parallels the latest Eocene cooling (37-38 Ma) and ends in the early Oligocene (34 Ma) when surface productivity was only marginally higher than that of the subsurface planktic foraminifers (*Subbotina* spp., *Globorotaloides suteri*). Surface water productivity appears to have remained low through the early Oligocene. A similar decrease in the values for surface waters from the middle Eocene through late Oligocene has been observed from tropical North Pacific sites (Savin et al., 1975).

DISCUSSION AND CONCLUSIONS

The remarkable correspondence between middle Eocene to Oligocene planktic foraminiferal turnovers with major Antarctic glaciation accompanied by steadily declining global surface water temperatures in both high and low latitudes (Figures 11.9-11.11) and decreasing surface productivity suggest that changes in climate and productivity are the primary forces driving planktic foraminiferal faunal turnovers during this interval. Each of the two subintervals within this period that are characterized by elevated rates of faunal change (Figures 11.7, 11.8) and associated with times of accelerated temperature and productivity declines. If declining productivity-nutrient supply was the primary driving force behind these biotic patterns, random extinctions resulting from increased inter-specific competition within all depth-related habitats would be expected. But, if changes in the planktic habitat itself, brought about by a decrease in the temperature-density differentiation of marine surface waters, was the primary driving force, the resulting extinctions should exhibit a marked ecological structure. During the middle Eocene to Oligocene cooling trend, in particular, a decrease in warm-tolerant surface-dwelling species accompanied by an increase in more cold-tolerant deep-dwelling taxa would be expected.

The latter was indeed the case during the middle to late Eocene when virtually all warm surface dwellers were successively eliminated and more cold-tolerant intermediate-dwelling forms increased in relative abundance (Figure 11.7) [*Note*: the obvious discrepancy between our interpretation and that of Poore and Matthews (1984) will be addressed by Barrera et al., in prep.]. Regrettably, the role of lower surface productivity in this faunal turnover remains unclear at this time. It is possible that the composite $\delta^{13}C$ data reflect less efficient utilization of available nutrient resources due to surface water habitat reduction (note decreased diversity and numerical abundance of planktic foraminifers), rather than reduced nutrient supply per se. Despite this uncertainty, however, the middle to late Eocene faunal turnover appears to have been driven primarily by surface cooling and associated changes in the temperature-density differentiation of the surface waters and secondarily by the productivity-nutrient relationship.

In contrast, the early to late Oligocene faunal turnover occurred at a time of already cool and relatively stable surface water stratification, but greatly reduced productivity-nutrient supply. In fact, surface productivity at this time decreased to nearly the same levels as intermediate-dwelling forms (e.g., *Subbotina* spp., *Globorotaloides suteri*, Figures 11.9, 11.10). If available nutrient supply were a controlling factor, one would expect increased competition primarily in the photic zone (surface layer) and thermocline (intermediate) layer. This appears to have been the case. Species extinctions occurred in surface and intermediate layers seemingly at random, although it is interesting to note that only Eocene survivor taxa actually disappeared. The preferential extinction of Eocene survivors during this interval suggests strongly implies elimination through competition for a limited nutrient supply.

Our study indicates that evolutionary turnover in the middle Eocene through Oligocene planktic foraminiferal fauna occurred primarily in surface and secondarily in intermediate water groups, whereas deep-dwelling forms remained essentially unaffected. Morphologic diversity was highest among surface dwellers (which included globorotaliform, turborotaliform and pulleniatiniform morphotypes--Figure 11.3, Table 11.1), lower among the intermediate-dwelling species group (including heteroheliciform, hastigeriniform and globigeriniform morphotypes) and lowest among deep dwellers (predominately globigeriniform morphotypes). This pattern of decreasing morphologic diversity with depth may be related to the high density of phytoplankton above the thermocline thereby promoting the elaboration of specialized morphologies to either actively capture or more efficiently graze different types of prey. Phytoplankton are known to be a primary food source for many modern planktic foraminiferal species including juveniles of omnivorous and carnivorous taxa which are too small to effectively prey upon zooplankton (Hemleben et al., 1989). In addition, greater variability in test morphology may also be related to the high physical water density in the thermocline layer that may offer greater freedom in terms of test shape variation (e.g., less stringent selection pressure on morphological attributes that affect an individual's ability to maintain its position within the water column). In contrast, the low morphologic variability of predominantly globigeriniform test shapes below the thermocline layer may reflect a more omnivorous or scavenging trophic pattern that coincides with reduced nutrient supply, along with higher selection pressures on morphological attributes that contribute to the organism's ability to maintain its position in the water column.

Our relative depth classification of Paleogene planktic foraminiferal morphotypes contrasts strongly with that of Hart (1980).

Whereas Hart (1980) attempted to predict the depth distribution of Cretaceous through Recent genera based on the observed distribution of Recent morphological analogues along with some preliminary oxygen isotopic data from middle Cretaceous species, our results are based on the empirically assessed relative isotopic ranking of individual Paleogene species. These data reveal the existence of previously unanticipated levels of intra-generic heterogeneity within the stable isotopic biochemistry of these Paleogene faunas. This implies that their ecology, distribution, and trophic structure may have differed substantially from that of their Recent counterparts (see also Shackleton et al., 1985). In addition, this observed intra-generic ecological heterogeneity calls into question the conclusions of studies (e.g., Stanley et al., 1988) that attempt to use Hart's (1980) genus/ biocharacter-based depth classification to account for differences in the evolutionary histories of major planktic foraminiferal clades.

In summary, abundant faunal and isotopic data strongly suggests that changes in climate and productivity are the primary driving force in the Eocene to Oligocene faunal turnovers. There is no evidence of catastrophic extinctions, or sudden species abundance changes, directly associated with any of the three well documented microspherule layers in the *G. semiinvoluta* and *T. cerroazulensis* Zones (Keller et al., 1983, 1987; Keller, 1986; Glass and Burns, 1987; Miller et al., 1991). Nor is there evidence that these extraterrestrial impact events triggered the climatic change, since climate cooling precedes the emplacement of impact-related debris.

ACKNOWLEDGEMENTS

We would like to thank J. C. Ingle, J. Lipps, and C. Benjamini who reviewed a preliminary version of this paper and provided many thoughtful comments. This research was supported by NSF Grants OCE-2088079 and OCE-8800049 (G. K.), BSR-8708563 (N. M.) and OCE-8800049, DPP-9096290 (E. B.), the Princeton University Tuttle Fund (G. K. and N. M.) and the Michigan Society of Fellows (N. M.).

REFERENCES

Barrera, E., Lohmann, K. C., and Barron, J. 1990. Stable isotope and sedimentologic evidence for late middle Eocene to early Oligocene glaciation in East Antarctica: Results from ODP Leg 119, southern Indian Ocean, *Trans. Amer. Geophys. Union* 71(43):1398.

Barrera, E. and Huber, B. T. 1991. Paleogene and early Neogene oceanography of the southern Indian Ocean: Leg 119 Foraminifer stable isotope results. In J. Barron, et al., eds., *Proc. Ocean Drilling Project, Leg* 119, College Station, Texas, in press.

Barrett, P. J. 1989. Antarctic Cenozoic history from the CIROS-1 drill hole, McMurdo Sound. *New Zealand. Dept. Sci. Ind. Res. Misc. Bull.* 245.

Bé, A. W. H. and Hemleben, C. 1970. Calcification in a living planktonic foraminifer, *Globigerinoides sacculifer* (Brady). *Lamont-Doherty Geol. Obs. Contrib.* 1399:221-234.

Bé, A. W. H. and Tolderlund, D. S. 1971. Distribution and ecology of living planktonic foraminifera in surface waters of the Atlantic and Indian Oceans. In B. M Funnell and W. R. Riedel, eds., *Micropaleontology of Oceans.* London, Cambridge University Press, pp. 105-149.

Bé, A. W. H., Vilks, G., and Lott, L. 1971.Winter distribution of planktonic foraminifera between the Grand Banks and the Caribbean. *Micropaleontology* 17:31-42.

Berger, W. H. 1969. Ecologic patterns of living planktonic Foraminifera. *Deep Sea Res.* 16:1-24.

Berggren, W. A., Kent, D. V., and Van Couvering, J. A. 1985. Paleogene geochronology and chronostratigraphy. In N. J. Snelling, ed., *The Chronology of the Geologic Record. Geol. Soc. London, Memoir* 10: 141-195.

Blow, W. H. 1979. *The Cainozoic Globigerinida.* Leiden, E. J. Brill (3 vols.), 1413pp.

Boersma, A. and Shackleton, N. J. 1981. Oxygen and carbon-isotope variations and planktonic Foraminifer depth habitats, Late Cretaceous to Paleocene, Central Pacific, Deep Sea Drilling Project Sites 463 and 465. In J. Theide et al., eds., *Init. Rept. Deep Sea Drill. Proj.* Washington, D.C., U.S. Government Printing Office, 62: 515-526.

Boersma, A. and Premoli-Silva, I. 1983. Paleocene planktonic foraminiferal biogeography and the paleoceanography of the Atlantic Ocean. *Micropaleontol.* 29(4):355-381.

Boersma, A. and Premoli-Silva, I. 1989. Atlantic Paleogene biserial heterohelicid foraminifera and oxygen minima. *Paleoceanog.* 4(3):271-286.

Bolli, H. M. and Saunders, J. B. 1985. Oligocene to Holocene low latitude planktic foraminifera. In H. M. Bolli, J. B. Saunders, K. Perch-Nielsen, eds., *Plankton Stratigraphy,* London, Cambridge University Press, pp. 155-262.

Cifelli, R. 1969. Radiation of Cenozoic planktonic Foraminifera. *Syst. Zool.* 18:154-168.

Corliss, B. H. and Keigwin, L. D. Jr. 1986. Eocene-Oligocene paleoceanography. *Mesozoic and Cenozoic Oceans. Amer. Geophys. Union, Geodynamics Series* 15:101-118.

Corliss, B. H., Aubry, M.-P., Berggren, W. A., Fenner, J. M., Keigwin, L. D. Jr., and Keller, G. 1984. The Eocene/Oligocene boundary event in the deep sea. *Science* 226:806-810.

Deuser, W. G. and Hunt, J. M. 1969. Stable isotope ratios of dissolved inorganic carbon in the Atlantic. *Deep Sea Res.* 16:221-225.

Douglas, R. G. and Savin, S. M. 1973. Oxygen and carbon isotope analyses of Cretaceous and Tertiary foraminifera from the central North Pacific. In E. L. Winterer, et al., eds., *Init. Rept. Deep Sea Drill. Proj.* Washington, D.C., U.S. Government Printing Office, 17: 591-605.

Douglas, R. G. and Savin, S. M. 1975. Oxygen and carbon isotope analyses of Tertiary and Cretaceous microfossils from Shatsky Rise and other sites in the North Pacific Ocean. In R. L. Larson, et al., eds., *Init. Rept. Deep Sea Drill. Proj.* Washington, D.C., U.S. Government Printing Office, 32: 590-592.

Douglas, R. G. and Savin, S. M. 1978. Oxygen isotope evidence for the depth stratification of Tertiary and Cretaceous planktic foraminifera. *Mar. Micropaleontol.* 3:175-196.

Ehrmann, W. U. in press. Implications of sediment composition on the southern Kerguelen Plateau for paleoclimate and depositional environment. In J. Barron, et al., eds., *Proceedings of the Ocean Drilling Project, Leg 119,* College Station, Texas, in press.

Fairbanks, R. G., Wiebe, P. H., and Bé, A. H. W. 1980. Vertical distribution and isotopic composition of living planktonic foraminifera in the Western North Atlantic. *Science* 207: 61-63.

Fairbanks, R. G., Sverdlove, M., Free, R., Wiebe, P. H., and Bé, A. W. H. 1982. Phenotypic variation and oxygen isotope ratios in recent planktonic foraminifera. *Jour. Foram. Res.* 2:55-67.

Ganapathy, R. 1982. Evidence for a major meteorite impact on Earth 34 million years ago: Implications for Eocene Extinctions. *Science* 216: 885-886.

Glass, B. P. and Burns, C. A. 1987. Late Eocene crystal-bearing spherules: Two layers or one? *Meteoritics* 24: 209-218.

Glass, B. P., DuBois, D. L., and Ganapathy, R. 1982. Relationship between an iridium anomaly and the North American microtektite layer in Core RC9-58 from the Caribbean Sea. *Jour. Geophys. Res.* 87: 425-428.

Harland, W.B., et al. 1967. *The Fossil Record.* London, Geol. Soc. London, 828 pp.

Harper, C. W. Jr. 1975. Standing diversity of fossil groups in successive intervals of geologic time: a new measure. *Jour. Paleontol.* 49:752-757.

Hart, M. B. 1980. A water depth model for the evolution of the planktonic Foraminiferida. *Nature* 286:252-254.

Hazel, J. E. 1989. Chronostratigraphy of upper Eocene microspherules. *Palaios* 4:318-329.

Hemleben, C., Spindler, M., and Anderson, O. R. 1989. *Modern Planktonic Foraminifera,* New York, Springer-Verlag, 363pp.

Katz, M. E and Miller, K. G. 1991. Early Paleogene benthic foraminiferal assemblage and stable isotope composition in the southern ocean, In P. F. Ciesielski, et al., eds., *Proceedings of the Ocean Drilling Project, Leg 114,* Texas, College Station, pp. 481-514.

Keigwin, L. D. Jr. and Corliss, B. H. 1986. Stable isotopes in late middle Eocene to Oligocene foraminifera. *Geol. Soc. Am. Bull.* 97:335-345.

Keigwin, L. D. Jr. and Keller, G. 1984. Middle Oligocene cooling from equatorial Pacific DSDP Site 77B. *Geology* 12:16-19.

Keller, G. 1983a. Biochronology and paleoclimatic implications of middle Eocene to Oligocene planktonic foraminiferal faunas. *Mar. Micropaleontol.* 7:463-486.

Keller, G. 1983b. The Paleogene/Neogene boundary in the equatorial Pacific Ocean. *Rev. Ital. Paleontol. Stratig.* 89(4):529-545.

Keller, G. 1985. Eocene and Oligocene stratigraphy and erosional unconformities in the Gulf of Mexico and Gulf Coast. *Jour. Paleontol.* 59(4):882-903.

Keller, G. 1986. Stepwise mass extinctions and impact events: Late Eocene to early Oligocene. *Mar. Micropaleontol.* 10:267-293.

Keller, G. D'Hondt, S., and Vallier, T. L. 1983. Multiple microtektite horizons in Upper Eocene marine sediments: no evidence for mass extinctions. *Science* 221:150-152.

Keller, G., Herbert, T., Dorsey, R., D'Hondt, S., Johnsson, M., and Chi, W. R. 1987. Global distribution of late Paleogene hiatuses. *Geology* 15:199-203.

Keller, G., D'Hondt, S. L., Orth, C. J., Gilmore, J. S., OLiver, P. Q., Shoemaker, E. M., and Molina, E. 1987. Late Eocene impact microspherules: Stratigraphy, age and geochemistry. *Meteoritics* 22(1):25-60.

Keller, G. and MacLeod, N. in press. Faunal Turnover and Depth Stratification: Their relationships to climate and productivity events in the Eocene to Miocene pelagic realm, *Science and Reports of the Tohoku University*, Sendai, Japan.

Kennett, J. P., Hontz, R. E., Andrews, P. B., Edwards, A. R., Gostin, V. A., Hajos, M., Hampton, M. A., Jenkins, D. G., Margolis, S. V., Ovenshine, A. T., and Perch-Nielsen, K. 1975. Cenozoic paleoceanography in the southwest Pacific Ocean, Antarctic glaciation and the development of the circum-Antarctic current. In Kennett, J. P. Homtz, R. E. et al., eds., *Init. Rept. Deep Sea Drill. Proj.* 29, Washington, D.C., U.S. Government Printing Office, pp. 1155-1170.

Kennett, J. P. 1977. Cenozoic evolution of Antarctic glaciation, the circum-Antarctic Ocean and their impact on global paleoceanography, *Jour. Geophys. Res.* 82(27): 3843-3859.

Kennett, J. P. 1978. The development of planktonic biogeography in the southern ocean during the Cenozoic. *Mar. Micropaleontol.* 3:301-345.

Kennett, J. P. and Stott, L. D. 1990. Proteus and Proto-Oceanus: Paleogene oceans as revealed from Antarctic stable isotope results. In P. Barker, et al., eds., *Proceedings of the Ocean Drilling Project, Leg 113*, College Station, Texas, pp. 549-569.

Kroopnick, P. 1974. The dissolved O_2-CO_2-^{13}C system in the eastern equatorial Pacific, *Deep Sea Res.* 21:211-227.

Lidz, B., Kehn, H. and Miller, H. 1968. Depth habitats of pelagic foraminifera during the Pleistocene. *Nature* 217:245-247.

Lipps, J. H. 1970. Plankton evolution. *Evolution* 24:1-22.

Lipps, J. H. 1986. Extinction dynamics in pelagic ecosystems. In D. K. Elliot, ed., *Dynamics of Extinction*, New York, John Wiley & Sons, Inc., pp. 89-103.

Loeblich, A. R. Jr. and Tappan, H. 1964. Foraminiferal classification and evolution, *Jour. Geol. Soc. India* 5:6-40.

Loeblich, A. R., Jr. and Tappan, H. 1988. *Foraminiferal Genera and Their Classification*, New York, Van Nostrand Reinhold Company, (2 vols.), 970pp.

MacLeod, N. in press. Effects of late Eocene impacts on planktic foraminifera. In V. L. Sharpton and P. Ward, eds., Proceedings of the Conference on Global Catastrophes in Earth History: An Interdisciplinary Conference on Impacts, Volcanism, and Mass Mortality, *Geol. Soc. Amer. Special Paper* 247.

MacLeod, N. and Keller, G. in press. How complete are Cretaceous/ Tertiary boundary sections? A chronostratigraphic estimate based on graphic correlation. *Geol. Soc. Am. Bull.*

Matthews, R. K. and Poore, R. Z. 1980. Tertiary $\delta^{18}O$ record and glacio-eustatic sea-level fluctuations. *Geology* 8:501-504.

McGowran, B. 1978. Stratigraphic record of early Tertiary oceanic and continental events in the Indian Ocean region. *Mar. Geol.* 26:1-39.

Miller, K. G. and Thomas, E. 1985. Late Eocene to Oligocene benthic foraminiferal isotope record, Site 574 equatorial Pacific. In L. Mayer, et al., eds., *Init. Rept. Deep Sea Drill. Proj.* 85, Washington, D.C., U.S. Government Printing Office, pp. 771-777.

Miller, K. G. Fairbanks, R. G., and Mountain, G. S. 1987. Tertiary oxygen isotope synthesis, sea level history, and continental margin erosion. *Paleoceanog.* 2(1):1-19.

Miller, G. K., Berggren, W. A., Jhang, J., and Palmer-Judson, A. A. 1991. Biostratigraphy and isotope stratigraphy of upper Eocene microtektites at Site 612: How many impacts? *Palaios* 6: 17-38.

Molina, E., Keller, G., and Madile, M. 1988. late Eocene to Oligocene Events: Molino de Cobo, Betic Cordillera, Spain. *Rev. Española de Micropaleontol.* 20(3):491-514.

Oberhänsli, H. 1986. Latest Cretaceous-early Neogene oxygen and carbon isotopic record at DSDP Sites in the Indian Ocean. *Mar. Micropaleontol.* 10:91-115.

Oberhänsli, H., McKenzie, J., Toumarkine, M., and Weissart, H. 1984. A paleoclimatic and paleoceanographic record of the Paleogene in the Central South Atlantic (Leg 73, Sites 522, 523, 524). In K. J. Hsü, et al., eds., *Init. Rept. Deep Sea Drill. Proj.* Washington, D.C., U.S. Government Printing Office, 73: 737-748.

Poore, R. Z. and Mathews, R. K. 1984. Oxygen isotope ranking of late Eocene and Oligocene planktonic foraminifera: Implications for Oligocene sea-surface temperatures and global ice-volume. *Mar. Micropaleontol.* 9: 111-134.

Prentice, M. L. and Matthews, R. K. 1988. Cenozoic ice-volume history: Development of a composite oxygen isotope record. *Geology* 16: 963-966.

Raup, D. M. and Sepkoski, J. J. Jr. 1984. Periodicity of extinctions in the geologic past. *Proc. Nat. Acad. Sci.* 81:801-805.

Raup, D. M. and Sepkoski, J. J. Jr. 1986. Periodic extinction of families and genera. *Science* 231: 833-836.

Saito, T. and Van Donk, J. 1974. Oxygen and carbon isotope measurements of late Cretaceous and early Tertiary foraminifera. *Micropaleontol.* 20:152-177.

Savin, S. M. 1977. The history of the earth's surface temperature during the past 100 million years. *Ann. Rev. Earth Planet. Sci.* 5: 319-355.

Savin, S. M., Douglas, R. G., and Stehli, F. G. 1975. Tertiary marine paleotemperatures. *Geol. Soc. Amer. Bull.* 86:1499-1510.

Savin, S. M. and Yeh, H.-W. 1981. Stable isotopes in ocean sediment. In C. Emiliani, ed., *The Oceanic Lithosphere: The Sea* (vol. 7), New York, Wiley Interscience, pp.1521-1554.

Sepkoski, J. J. Jr. 1982. A compendium of fossil marine families. *Contrib. Milwaukee Publ. Mus.* 51: 1-125.

Sepkoski, J. J. Jr. and Raup, D. M. 1985. Periodicity in marine mass extinctions, In D. K Elliot, ed., *Dynamics of Extinction*; New York, John Wiley & Sons, Inc., pp. 3-36.

Shackleton, N. J., Corfield, R. M., and Hall, M. A. 1985. Stable isotope data and the ontogeny of Paleocene planktonic foraminifera. *Jour. Foram. Res.* 15(4):321-336.

Shackleton, N. J., and Kennett, J. P. 1975. Paleotemperature history of the Cenozoic and the initiation of Antarctic glaciation: oxygen, and carbon isotope analyses in DSDP Sites 277, 279, and 281. In Kennett, J. P. Homtz, R. E. et al., eds., *Init. Rept. Deep Sea Drill. Proj.* Washington, D.C., U.S. Government Printing Office, 29: 599-612.

Shackleton, N. J., Corfield, R. M., and Hall, M. A. 1985. Stable isotope data and the ontogeny of Paleocene planktonic foraminifera. *Jour, Foram. Res.* 15(4):321-336.

Stanley, S. M., Wetmore, K. L., and Kennett, J. P. 1988. Macroevolutionary differences between two major clades of Neogene planktonic Foraminifera. *Paleobiology* 14(3):235-249.

Steineck, P. L 1971. Middle Eocene refrigeration – new evidence from California planktonic foraminiferal assemblages. *Lethaia* 4:125-129.

Stott, L. D., Kennett, J. P., Shackleton, N. J., and Corfield, R. M. 1990. The evolution of Antarctic surface waters during the Paleogene: Inferences from stable isotopic composition of planktonic foraminifers, ODP Leg 113. In P. Barker, et al., eds., *Proceedings of the Ocean Drilling Project, Leg 113*, Texas, College Station, pp. 849-863.

Toumarkine, M. and Luterbacher, H. 1985. Paleocene and Eocene planktic foraminifera, In H. M. Bolli, J. B. Saunders and K. Perch-Nielsen, eds., *Plankton Stratigraphy,* London, Cambridge University Press, pp. 155-262.

Williams, D. R., Bé, A. W. H., and Fairbanks, R. 1979. Seasonal oxygen isotopic variations in living planktonic foraminifera off Bermuda. *Science* 206:447-449.

Williams, D. R and Healy-Williams, N. 1980. Oxygen isotopic-hydrographic relationships among recent planktonic foraminifera from the Indian Ocean. *Nature* 283:848-852.

Whittaker, E. T and Robinson, G. 1929. *The*

Calculus of Observations, 2nd ed., Glasglow, Blackie and Sone Ltd. 395pp.

12. MIDDLE EOCENE - LATE OLIGOCENE BATHYAL BENTHIC FORAMINIFERA (WEDDELL SEA): FAUNAL CHANGES AND IMPLICATIONS FOR OCEAN CIRCULATION

By Ellen Thomas

ABSTRACT

Lower bathyal benthic foraminiferal faunas from Maud Rise (Weddell Sea, Antarctica) underwent gradual, but stepped extinctions from middle Eocene through Oligocene, with steps at about 46.4-44.6 Ma, 40-37 Ma, and 34-31.5 Ma. Faunal changes at these high latitudes encompassed decreasing diversity and increasing relative abundance of epifaunal species, in combination with loss of large, heavily calcified *Bulimina* species (first step), followed by the disappearance of all *Bulimina* species and the appearance of *Turrilina alsatica* (second step), followed by a strong decrease in abundance of *T. alsatica*, resulting in faunas dominated by *Nuttallides umbonifera* (third step). These faunal changes may reflect the reaction of the fauna to gradual cooling (and thus a gradual increase in corrosiveness) of the high-latitude lower bathyal waters. The gradual nature of the faunal changes in lower bathyal benthic foraminiferal assemblages and the absence of catastrophic extinctions (on Maud Rise, as well as worldwide) suggest that the psychrosphere was established as a result of gradual cooling of surface waters at high latitudes and that lower bathyal waters were formed by cooling and sinking at high latitudes during the middle Eocene through Oligocene.

INTRODUCTION

Dramatic climate changes occurred in the Cenozoic, especially from middle Eocene through early Oligocene (e.g., Shackleton and Kennett, 1975; Kennett and Shackleton, 1976; Kennett, 1977; Savin, 1977; Shackleton and Boersma, 1981; Mercer, 1983; Shackleton, 1986; Miller et al., 1987a; Kennett and Barker, 1990; Thomas, 1989, 1990a; Webb, 1990; Barron et al., 1991; Zachos et al., 1991, Wise et al., 1991). Deep waters in the world's oceans and surface waters at high latitudes cooled strongly after the very warm early Eocene: early Eocene surface water temperatures at high latitudes were estimated to have been about 15-17°C (Stott et al., 1990). At some time during the middle Eocene - early Oligocene the psychrosphere was established (Benson, 1975), as well as at least partial ice sheets on eastern Antarctica (Kennett and Shackleton, 1976; Keigwin and Keller, 1984; Miller and Thomas, 1985; Miller et al., 1987a; Kennett and Barker, 1990; Barron et al., 1991; Wise et al, 1991). The extent and nature of these ice sheets (whether they were true continental ice sheets, temperate ice sheets or upland and coastal glaciers) is under considerable discussion, even after the large increase in knowledge resulting from Ocean Drilling Program (ODP) Legs 113, 114, 119, 120 at high southern latitudes (Kennett and Barker, 1990; Barron et al., 1991; Wise et al., 1991).

There is no agreement on the interpretation of the oxygen isotopic records of benthic foraminifers, especially in how far this record demonstrates establishment of continental ice sheets (the ice-volume effect), and in how far it represents cooling of deep waters in the oceans (e.g., Matthews and Poore, 1980; Poore and Matthews, 1984; Keigwin and Corliss, 1986; Miller et al., 1987a; Prentice and Matthews, 1988; Wise et al., 1991; Zachos et al., 1991; Oberhaensli et al., 1991). Therefore it is not clear whether equatorial surface waters were cooler than now during these warm periods (e.g., Shackleton, 1984), or remained essentially at the same temperature throughout the

Cenozoic, as indicated by the distribution of tropical biota such as hermatypic corals, mangroves and larger foraminifera (Adams et al., 1990). There also is no agreement on whether the flat Eocene isotopic gradients in planktonic foraminifera from low to high latitudes in the Atlantic Ocean reflect very low to flat temperature gradients (Shackleton and Boersma, 1981; Shackleton, 1984; Keigwin and Corliss, 1986; Boersma et al., 1987), or are influenced by fresh-water influx at higher latitudes (e.g., Wise et al., 1991). In addition, there is no agreement regarding modes of deep-water formation in the oceans during the Paleogene: did deep oceanic waters (intermediate and bottom water masses) predominantly form at high latitudes after cooling and sinking, as they do today (Manabe and Bryan, 1985; Barrera et al., 1987; Katz and Miller, 1991), or did they predominantly form at low latitudes by evaporation for at least part and possibly all of this period (Chamberlain, 1906; Shackleton and Boersma, 1981; Brass et al., 1982; Hay, 1989)? Was the circulation of the deep and possibly intermediate water masses of the oceans thus essentially reversed (from dominant formation at high, to dominant formation at low latitudes) in all or part of the Paleogene (e.g., Kennett and Stott, 1990)? Did a final reversal from such a halothermal circulation system to the present system cause the origination of the psychrosphere? Oceans "running the reverse" from the modern circulation pattern (i.e., sinking at low latitudes) might be required to model satisfactorily the high heat transfer from low to high latitudes required to maintain the warm Eocene climate at high latitudes (Barron, 1985, 1987). Carbon and oxygen isotopic data and benthic foraminiferal data, however, do not unequivocally point to the existence of such an evaporation-driven deep water circulation, and may be read to indicate an overall Paleocene ocean circulation dominated by deep-water formation at high latitudes, although short (<0.5 m.y.) periods of high-volume formation of warm, salty bottom waters might have occurred (Barrera et al., 1987; Miller et al., 1987b; Thomas, 1989; 1990a; Kennett and Stott, in prep.; Katz and Miller, 1991; Zachos et al., 1991).

Recent deep-water benthic foraminiferal faunas reflect the physicochemical properties of water masses, and faunal patterns thus reflect global deep oceanic circulation (see e.g., Douglas and Woodruff, 1981, and Culver, 1987, for a review). Many studies of bathyal and upper abyssal benthic foraminifers from the Atlantic and Pacific Oceans have demonstrated that there were no catastrophically sudden extinctions at the end of the Eocene: there were extinctions over a period of several millions of years from the middle Eocene into the early Oligocene in benthic foraminiferal faunas as well as in ostracode faunas (e.g., Corliss, 1981; Tjalsma and Lohmann, 1983; Miller, 1983; Miller et al., 1984; Boersma, 1984; 1985; Corliss and Keigwin, 1986; Berggren and Miller, 1989; Boltovskoy, 1980; Boltovskoy and Boltovskoy, 1988; 1989; reviews in Douglas and Woodruff, 1981; and in Culver, 1987).

Benthic foraminiferal faunas from high latitudes might be expected to be a good source of information on deep-water formational processes: at these high latitudes deep waters are not only cool or cold, but also "young", i.e., these waters were in equilibrium with the atmosphere only a short time before reaching the site, and thus rich in oxygen, poor in nutrients and CO_2 (a short period means short compared to the turnover time of the oceans, around 1000 years; Gascard, 1990). Deep waters formed by evaporation at subtropical latitudes are relatively warm, and they must have travelled from sub-equatorial latitudes to the high latitudes, thus they are "old" (i.e., have not been in contact with the atmosphere for at least several hundreds of years, and are enriched in CO_2 and nutrients, depleted in oxygen). Thomas (1989; 1990a, 1990b) argued that bathyal benthic foraminiferal faunal changes at Sites 689 and 690 on the Maud Rise (Weddell Sea, Antarctica) indicate that bathyal waters at these sites formed at high latitudes during the Maestrichtian and early Paleocene, with at least one, possibly several short (less than 0.5 m.y.) periods of reversal of ocean circulation at the end of the Paleocene and in the earliest Eocene, during which warm salty deep waters bathed the Maud Rise sites (in agreement with conclusions by Katz and Miller, 1991, for faunas from ODP Leg 114 Sites in the southernmost Atlantic Ocean). The period of most intense deep-water formation at low latitudes was said to be at the end of the Paleocene, and the cause of the worldwide extinction of deep-water benthic foraminifers, the most severe and sudden extinction event of benthic foraminifers in the entire Cenozoic (Schnitker, 1979; Tjalsma and Lohmann, 1983; Miller et al., 1987; Boltovskoy and Boltovskoy, 1988; 1989; Katz and Miller, 1991; Mackensen and Berggren, 1991; Nomura, 1991; Kaiho, 1991).

In this paper I present information on middle Eocene through upper Oligocene lower bathyal benthic foraminiferal faunas from Ocean Drilling Program Site 689 (paleodepth about 1500-2000 m) on Maud Rise (Antarctica), in order to evaluate the response of benthic foraminiferal faunas to changing climate and deep-water circulation.

MATERIALS AND METHODS

Ocean Drilling Program Sites 689 (64°31.009' S, 03°05.996' E, present water depth 2080 m) and 690 (65°9.629'S, 1°12.296'E, present water depth 2914 m) were drilled on Maud Rise, an aseismic ridge on the eastern entrance of the Weddell Sea (Barker, Kennett et al., 1988; Figure 12.1). Site 689 is on the northeastern side of Maud Rise, Site 690 is 116 km to the southwest, on the southwestern flank. The sediments are lowermost Maestrichtian through Pleistocene biogenic oozes (Figure 12.2; see also Thomas et al., 1990). Core recovery was good and deformation minimal over most of the drilled section (Barker, Kennett et al., 1988). The Upper Cretaceous through lower middle Eocene consists of chalks and calcareous oozes. Upper middle Eocene through lowermost Miocene are mixed siliceous-calcareous biogenic oozes, with a gradual increase in the siliceous component up-section. Middle Miocene and younger sediments are dominantly siliceous oozes. Calcium carbonate microfossils shows signs of dissolution from the uppermost middle Eocene upwards (Barker, Kennett et al., 1988; Diester-Haass, 1991).

Sites 689 and 690 are on an aseismic ridge, so that backtracking following Parsons and Sclater's (1977) methods for sites on normal oceanic crust is not justified. Backtracking methods may be valid even for sites on oceanic plateaus, however, although the original depth of the site cannot be assumed to have been at the depth of average ridge-crest (Detrick et al., 1977; Coffin, 1991). Backtracking of Site 689 resulted in an estimate of 1300-1700 m paleodepth for the middle Eocene-Oligocene, using data on sediment density after Barker, Kennett et al. (1988), and the correction for sediment loading after Crough (1983). This paleodepth agrees with the depth estimates of 1000-1500 m from benthic foraminiferal data (Thomas, 1990a).

Biostratigraphic records of calcareous nannofossils and planktonic foraminifers combined with magnetostratigraphy show that there are several unconformities at both sites (Figure 12.2). At Site 690 large parts of the upper and uppermost middle Eocene are not represented in the sediments (Thomas et al., 1990). At both sites there are unconformities across the Oligocene/Miocene boundary, at Site 689 corresponding to the lower part of paleomagnetic Chronozone C6 and the upper part of C7 (20.45-25.60Ma). Another unconformity at Site 689 covers most of the lower Eocene and parts of the lowermost middle Eocene (lower Chronozone C21 through upper Chronozone C24, about 51-56Ma). The section appears to be complete (within the stratigraphic resolution) between these two unconformities, i.e., an interval corresponding to the time between roughly 51 and 27 Ma (lower middle Eocene through lower upper Oligocene). All data are presented versus sub-bottom depth (with nannofossil zonation after Pospichal and Wise, 1990; Wei and Wise, 1990). Data were not plotted versus numerical age because of the possible revisions of the age of the Eocene/Oligocene boundary (Berggren et al., this volume); numerical ages follow Berggren et al. (1985). All numerical ages were derived by extrapolation between paleomagnetic tie points listed in Thomas et al. (1990).

Data are presented from 93 samples between 68.92 and 202.41 mbsf (26.80 - 50.99 Ma, see Thomas et al., 1990), i.e., from the middle Eocene through Oligocene section at Site 689. Preliminary data on the faunas were published by Thomas (1989; 1990a), but data over the middle Eocene-Oligocene in these papers were limited to a sample resolution of 2 samples per core (9.6 m). Sample resolution for this paper was 1.5m, resulting in an average time-resolution of between 0.25 and 0.30 m.y.

Samples (15 cm^3) were dried at 75°C, soaked in Calgon, and washed over a sieve with openings of 63 μm; residues were dried at 75°C.

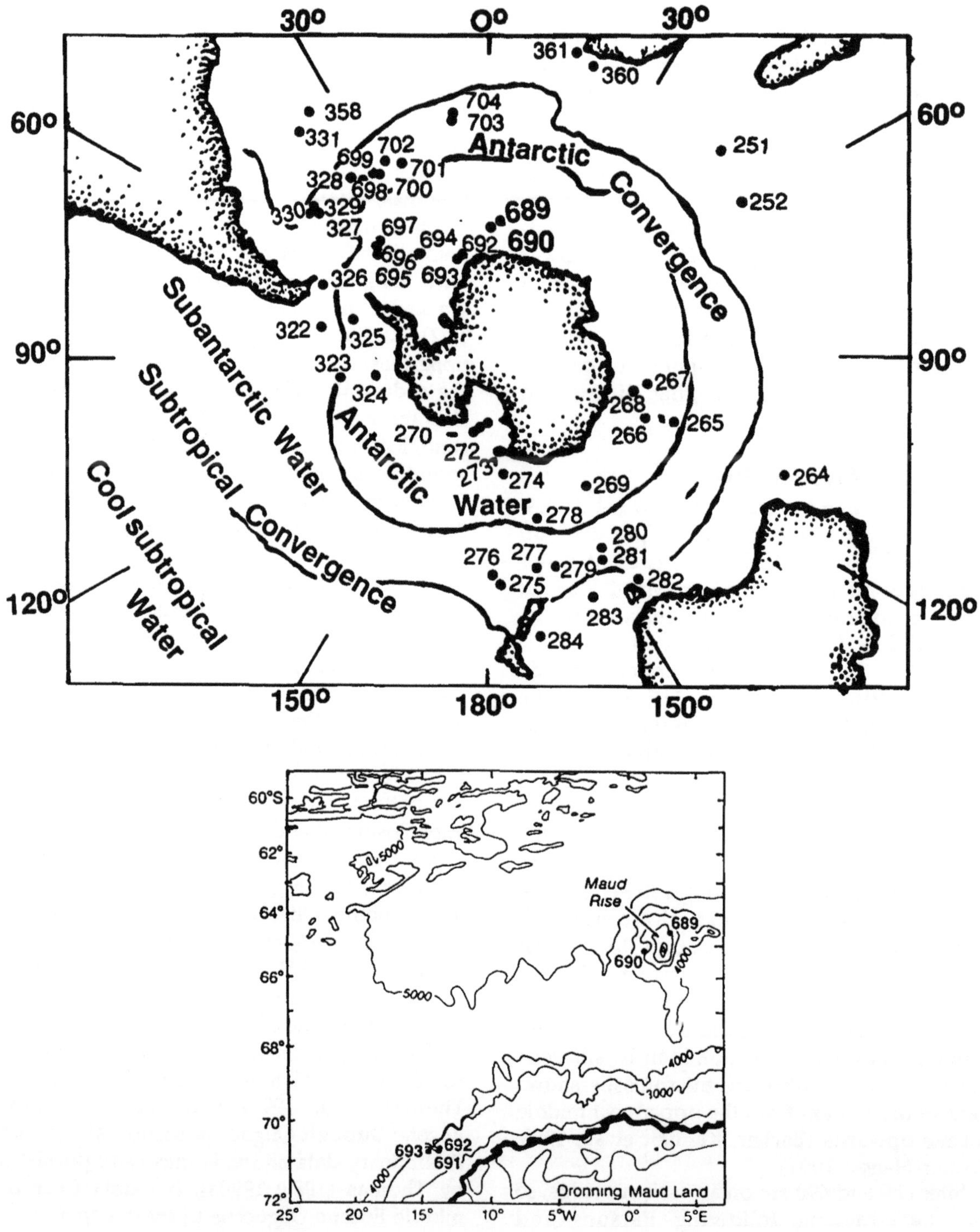

FIGURE 12.1. Location of Sites 689 and 690 on Maud Rise.

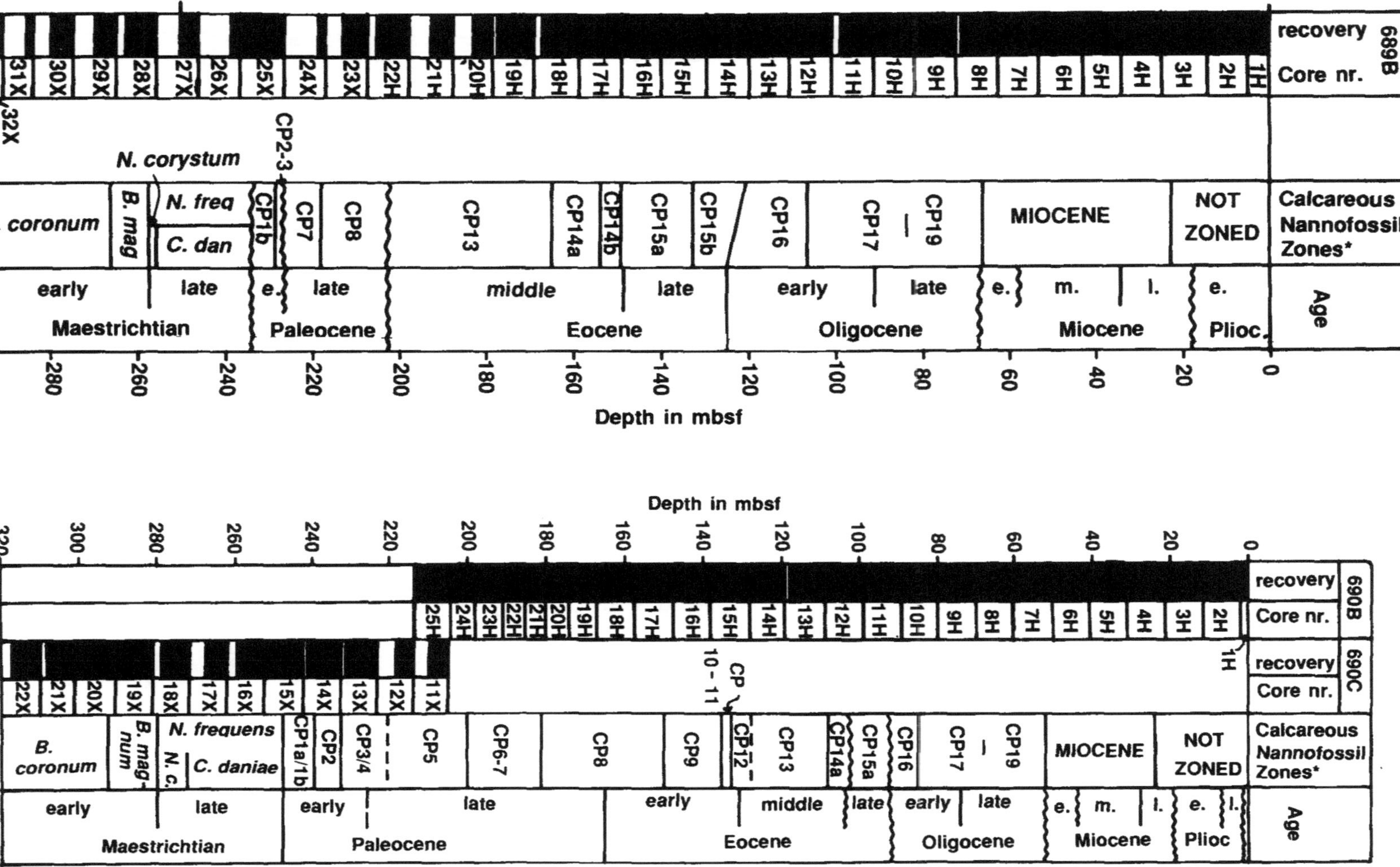

FIGURE 12.2. Core recovery and stratigraphy at Maud Rise, Sites 689 (left) and 690 (right). Calcareous nannofossil zones after Pospichal and Wise (1990) and Wei and Wise (1990).

Specimens were studied from the greater than 63 μm size fraction in order to ensure representation of smaller species (Thomas, 1985; Schroeder et al., 1987). At the beginning of the study rarefaction curves were drawn for several samples (plots of numbers of species versus number of specimens; Thomas, 1985). Rarefaction curves become parallel to the specimen axis at about 270 specimens, thus 300 specimens or more were picked from the sample or a split of the sample. There is no correlation between number of specimens and number of species for the studied samples, demonstrating that the number of species counted per sample does not reflect the number of counted specimens. Taxonomy is discussed in Thomas (1990a), and largely follows van Morkhoven et al. (1986).

RESULTS

Cretaceous through Neogene benthic foraminiferal faunas at Sites 689 and 690 were subdivided into eight assemblages by Thomas (1989; 1990a), using first and last appearances as discussed below (Figure 12.3). The interval studied in this paper contains Thomas' benthic foraminiferal assemblage 2 (upper boundary in the Miocene, lower boundary in the uppermost Eocene, Chronozone C16N, 38.5Ma), assemblage 3 (uppermost Eocene, 38.5Ma, through upper middle Eocene, lower part of Chronozone C20N, 46Ma) and assemblage 4 (upper middle Eocene, 46Ma through middle lower Eocene, 54.5 Ma). The latter was subdivided into two sub-assemblages, 4A and 4B, with a boundary close to the middle/lower Eocene boundary (Chronozone C21/C22 boundary) at about 52Ma.

Within the time-resolution available, the boundaries between the Maud Rise assemblages are largely coeval with the abyssal benthic foraminiferal zonal boundaries proposed by Berggren and Miller (1989), although most of their zonal marker species were not present, very rare, or had different ranges at the Maud Rise sites (Figure 12.3). One exception to this agreement might be that assemblage 2B at Maud Rise appears to agree better with bathyal zone BB5 than with abyssal zone AB8 (as discussed further below), whereas lower bathyal faunas usually show larger resemblance to abysssal than to middle bathyal assemblages. Another exception to the age-agreement between the Maud Rise benthic foraminiferal assemblages and the zones in Berggren and Miller (1989) might be the assemblage boundary (2/3) estimated to occur at 38.5Ma at Site 689, i.e., earlier than the boundary between Zones AB7 and AB8 in Berggren and Miller (1989), which was placed at the Eocene/Oligocene boundary. At Maud Rise this assemblage boundary occurs between 130.31 and 131.81 mbsf (Thomas, 1990a; but see below for discussion), and thus occurred before the globally recognized rapid increase in $\delta^{18}O$ values in benthic foraminifera in the lowermost Oligocene (paleomagnetic Chronozone C13R), which at Site 689 is recorded at about 120 mbsf (Thomas et al., 1990; and see below). The benthic faunal boundary occurs below the LAD of the planktonic foraminifer *Globigerinatheka index*, which occurs between 128.82 and 129.92 mbsf at Site 689; this LAD was estimated to have occurred at about 38.4Ma in Chronozone C16N (Stott and Kennett, 1990). A different interpretation of the paleomagnetic record is possible (Spiess, 1990), however, which would place this LAD close to the Eocene/Oligocene boundary (Berggren, 1991). In this paper I followed the interpretation of Stott and Kennett (1990), which is in agreement with the interpretation of nannofossil events in Wei and Wise (1990) as shown in Thomas et al. (1990).

All counts are presented in the appendix. Ranges of the more common species are listed (Figure 12.4), and relative abundances of the stratigraphically and paleoecologically most useful taxa provided (Figure 12.5). Diversity (expressed as number of species per 300 specimens), and relative abundance of epifaunal, infaunal and cylindrical taxa (subdivision of taxa in groups according to Thomas, 1990a; see Table 12.1, following Corliss and Chen, 1988), as well as the number of first and last appearances are shown in Figure 12.6, together with the $CaCO_3$ content of the sediments (O'Connell, 1990), the oxygen and carbon isotopic values of *Cibicidoides* spp. of Kennett and Stott (1990), and the bulk carbonate isotopic data from Shackleton and Hall (1990). First and last appearances are strictly local events; for most benthic foraminiferal taxa datum levels have not been globally determined. Thus first and last appearances may be originations and extinctions, or immigrations and emigrations,

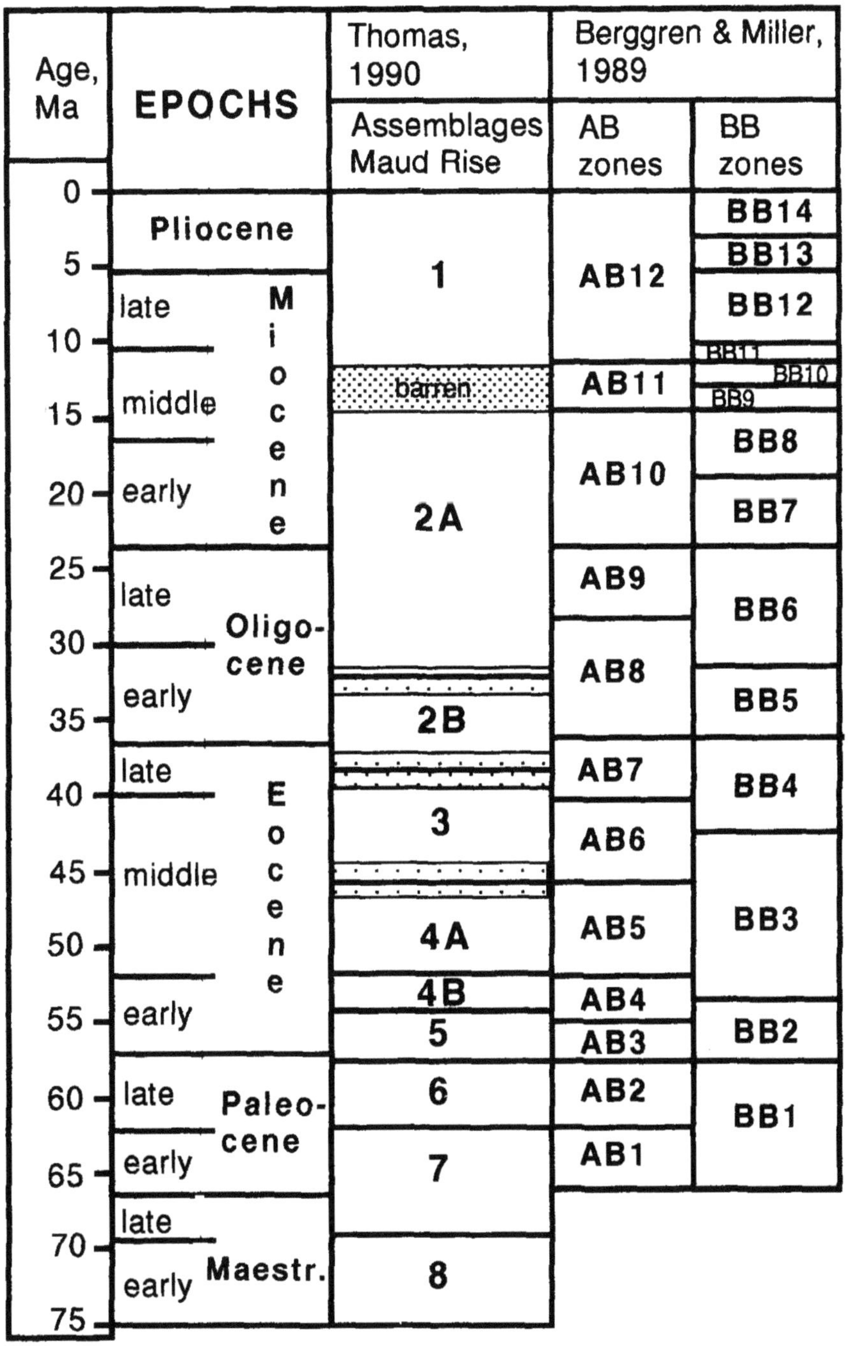

Age, Ma	EPOCHS	Thomas, 1990: Assemblages Maud Rise	Berggren & Miller, 1989: AB zones	Berggren & Miller, 1989: BB zones
0, 5, 10, 15, 20, 25, 30, 35, 40, 45, 50, 55, 60, 65, 70, 75	Pliocene; Miocene: late, middle, early; Oligocene: late, early; Eocene: late, middle, early; Paleocene: late, early; Maestr.: late, early	1, barren, 2A, 2B, 3, 4A, 4B, 5, 6, 7, 8	AB12, AB11, AB10, AB9, AB8, AB7, AB6, AB5, AB4, AB3, AB2, AB1	BB14, BB13, BB12, BB11, BB10, BB9, BB8, BB7, BB6, BB5, BB4, BB3, BB2, BB1

FIGURE 12.3. Benthic foraminiferal assemblages as recognized on Maud Rise at Sites 689 and 690, after Thomas (1990a). The assemblages are compared with the benthic foraminiferal zones for the bathyal regions (BB zones) and abyssal regions (AB zones) as defined by Berggren and Miller (1989).

FIGURE 12.4. Ranges of most common species at Site 689, versus sub-bottom depth, arranged by highest appearance.

very rare	(0 - 4 %)
rare	(5 - 9 %)
few	(10-24 %)
common	(25-49 %)
abundant	(> 50 %)

Species Location Index: index numbers are the columns in which species appear:

1	*Anomalina spissiformis*	33	*Gyroidinoides mediceus*
2	*Anomalinoides pseudogrosserugosa*	40	*Gyroidinoides planulatus*
35	*Anomalinoides semicribrata*	11	*Gyroidinoides soldanii*
3	*Astrononion pusillum*	12	*Lenticulina* spp.
4	*Astrononion umbilicatulum*	36	*Nonion havanense*
34	*Bigenerina nodosaria*	45	*Nuttallides truempyi*
5	*Bolivina decussata*	13	*Nuttallides umbonifera*
38	*Bolivina huneri*	14	*Oridorsalis tener*
48	*Bulimina callahani*	15	*Oridorsalis umbonatus*
37	*Bulimina elongata*	16	*Pleurostomella* spp.
46	*Bulimina ovula*	17	*Pullenia bulloides*
49	*Bulimina semicostata*	18	*Pullenia quinqueloba*
43	*Bulimina simplex*	19	*Pullenia salisburyi*
6	*Cibicidoides mundulus*	20	*Reussella tortuosa*
42	*Cibicidoides praemundulus*	47	*Siphogenerinoides eleganta*
44	*Cibicidoides trincherasensis*	21	*Stilostomella aculeata*
7	*Eggerella bradyi*	22	*Stilostomella annulifera*
8	*Eilohedra weddellensis*	32	*Stilostomella consobrina*
9	*Epistominella exigua*	23	*Stilostomella subspinosa*
39	*Fursenkoina bradyi*	24	*Turrilina alsatica*
41	*Fursenkoina fusiformis*	25	unilocular species
29	*Globocassidulina subglobosa*	26	uniserial lagenids
30	*Gyroidinoides acutus*	27	*Uvigerina graciliformis*
31	*Gyroidinoides girardanus*	28	*Uvigerina peregrina* group
10	*Gyroidinoides lamarckianus*		

the latter geographically as well as depth-wise (vertical).

Assemblage 4 (lower to middle middle Eocene)

The assemblage occurs at Site 689 from Sample 689B-22X-4, 40-42cm (202.41m; just above an unconformity between calcareous nannofossil zones CP8 and CP12). Assemblage 4 is characterized by a high diversity (36 to 58 species per 300 specimens, average 44.5 ± 5.6 species). Characteristic species are *Bulimina semicostata, Bulimina trinitatensis,* and *Bulimina callahani,* all large, ornamented, heavily calcified buliminid species; *Bulimina simplex,* a small, smooth-walled buliminid species; and *Siphogenerinoides eleganta.* Uniserial lagenid species and lenticulinids are common and diverse. *Stilostomella* spp., especially *S. subspinosa,* are common in many samples (Figures 12.4, 12.5, 12.6). Other commonly present species are *Oridorsalis umbonatus,* and *Gyroidinoides* spp.; *Cibicidoides praemundulus* is usually present and common to rare, as is *Nuttallides umbonifera.* The relative abundance of *Nuttallides truempyi* fluctuates strongly (Figure 12.5), but the species is most common in the upper part of the assemblage.

The lower boundary of the assemblage was taken at the first appearance (FA) of *Siphogenerinoides eleganta,* but its age could not be

01 *Anomalina spissiformis*
02 *Anomalinoides pseudogrosserugosus*
03 *Astrononion umbilicatulum*
04 *Astrononion pusillum*
05 *Bolivina decussata*
06 *Cibicidoides mundulus*
07 *Eggerella bradyi*
08 *Eilohedra weddellensis*
09 *Epistominella exigua*
10 *Gyroidinoides lamarckianus*
11 *Gyroidinoides soldanii*
12 *Lenticulina spp*
13 *Nuttallides umbonifera*
14 *Oridorsalis tener*
15 *Oridorsalis umbonatus*
16 *Pleurostomella spp*
17 *Pullenia bulloides*
18 *Pullenia quinqueloba*
19 *Pullenia salisburyi*
20 *Reussella tortuosa*
21 *Stilostomella aculeata*
22 *Stilostomella annulifera*
23 *Stilostomella subspinosa*
24 *Turrilina alsatica*
25 unilocular taxa
26 uniserial lagenids
27 *Uvigerina graciliformis*
28 *Uvigerina peregrina group*
29 *Globocassidulina subglobosa*
30 *Gyroidinoides acutus*
31 *Gyroidinoides girardanus*
32 *Stilostomella consobrina*
33 *Gyroidinoides mediceus*
34 *Bigenerina nodosaria*
35 *Anomalinoides semicribratus*
36 *Nonion havanense*
37 *Bulimina elongata*
38 *Bolivina huneri*
39 *Fursenkoina bradyi*
40 *Gyroidinoides planulatus*
41 *Fursenkoina fusiformis*
42 *Cibicidoides praemundulus*
43 *Bulimina simplex*
44 *Cibicidoides trincherasensis*
45 *Nuttallides truempyi*
46 *Bulimina ovula*
47 *Siphogenerinoides eleganta*
48 *Bulimina callahani*
49 *Bulimina semicostata*

SAMPLE	DEPTH, MBSF
08H-5, 040-043cm	068 92
08H-6, 041-043cm	070 42
08H, CC	071 93
09H-1, 040-042cm	072 51
09H-2, 040-042cm	074 01
09H-3, 040-042cm	075 51
09H-4, 040-042cm	077.01
09H-5, 040-042cm	078 51
09H-6, 040-042cm	080 01
09H, CC	080 66
10H-1, 039-043cm	082 11
10H-2, 039-043cm	083 61
10H-3, 039-043cm	085 11
10H-4, 039-043cm	086 61
10H-5, 039-043cm	088 11
10H-6, 039-043cm	089 61
10H, CC	090 49
11H-1, 040-044cm	091 72
11H-2, 040-044cm	093 22
11H-3, 040-044cm	094 72
11H-4, 040-044cm	096 22
11H-5, 040-044cm	097 72
11H, CC	098 96
12H-1, 040-042cm	101 41
12H-2, 040-042cm	102 91
12H-3, 040-042cm	104 41
12H-4, 040-042cm	105 91
12H-5, 040-042cm	107 41
12H-6, 040-042cm	108 91
12H, CC	110 60
13H-1, 040-042cm	111 01
13H-2, 040-042cm	112 51
13H-3, 040-043cm	114 01
13H-4, 040-042cm	115 51
13H-5, 040-042cm	117 01
13H-6, 040-042cm	118 51
13H, CC	120 20
14H-1, 043-045cm	120 64
14H-2, 041-043cm	122 12
14H-3, 041-043cm	123 62
14H-4, 041-043cm	125 12
14H-5, 041-043cm	126 62
14H-6, 041-043cm	128 12
14H CC	129 76
15H-1 040-043cm	130 32
15H-2, 040-043cm	131 82
15H-3 040-043cm	133 32
15H-4, 040-043cm	134 82
15H-5, 040-043cm	136 32
15H-6, 040-043cm	137 82
15H, CC	139 40
16H-1, 040-042cm	139 81
16H-2, 040-042cm	141 31
16H-3 040-042cm	142 81
16H-4 040-042cm	144 31
16H-5, 040-042cm	145 81
16H-6, 040-042cm	147 31
16H-7 040-042cm	148 81
16H CC	149 09
17H-1, 040-042cm	149 51
17H-2, 040-042cm	151 01
17H-3, 040-042cm	152 51
17H 4 040-042cm	154 01
17H-5, 040-042cm	155 51
17H-6 040-042cm	157 01
17H-7, 040-042cm	158 51
17H CC	158 80
18H-1 039-041cm	159 20
18H-2 039-041cm	160 70
18H-3 039-041cm	162 20
18H-4 039-041cm	163 70
18H 5 039-041cm	165 20
18H CC	166 85
19H 1 039-042cm	167 31
19H-2 039-042cm	168 81
19H-3 039-042cm	170 31
19H-4 039-042cm	171 81
19H-5 039-042cm	173 31
19H-6 039-042cm	174 81
19H CC	177 05
20H 1 040-044cm	178 52
20H-2 040 044cm	180 02
20H 3 040 044cm	181 52
20H 4 040 044cm	183 02
20H CC	184 46
21H 1 040-044cm	188 22
21H 2 040-044cm	189 72
21H 3 040 044cm	191 22
21H CC	192 28
22X 1 040 042cm	197 91
22X 2 040 042cm	199 41
22X 3 040-042cm	200 91
22X 4 040 042cm	202 41

ascertained unequivocally because of stratigraphic problems at Site 689 as well as at Site 690. Thomas (1990a) stated that the upper boundary of the assemblage occurred between samples 689B-19H-2, 40-42 cm (168.68mbsf, 45.55 Ma) and 689B-19H-1, 40-42 cm (167.31 mbsf, 45.14 Ma); at Site 690 the upper boundary occurred at an unconformity. At Site 689 this upper boundary is difficult to locate precisely, however, because the changes in the benthic foraminiferal faunas occur gradually, from Sample 689B-19H-5, 39-41 cm (173.31 mbsf; 46.64 Ma) through Sample 689B-18H-5, 39-41 mbsf (165.20 mbsf; 44.57 Ma; Figures 12.4, 12.5; Appendix). These gradual changes include the last appearances of *S. eleganta* and all large, heavily calcified buliminids, the last appearance of *B. simplex*, the last common appearance of *N. truempyi*, and a decrease in diversity, largely as the result of a loss of diversity of uniserial lagenids and lenticulinids. The first common occurrence of *Bulimina elongata*, a buliminid species that is typical of the next younger assemblage, is within the interval of assemblage 4, in sample 689B-19H-5, 39-41cm.

Assemblage 3 (middle Eocene - upper Eocene)

The assemblage occurs at Site 689 above the interval of gradual change at the upper part of assemblage 4 (173.31-165.20 mbsf; 46.64-44.57 mbsf). The assemblage is characterized by a lower diversity than the older assemblage 4 (30 to 49 species, average 40.9 ± 5.1 species). Typically common to abundant species are *Bulimina elongata*, a small, smooth-walled buliminid, with strongly fluctuating relative abundance (Figure 12.5), *Nuttallides umbonifera* (above sample 689B-15H-5, 40-42 cm; 136.32 mbsf, 39.58 Ma), *Globocassidulina subglobosa*, *Stilostomella* spp., *Bolivina huneri* (below sample 689B-17H-1, 40-42 cm, 149.51 mbsf, 40.88 Ma), and both *Cibicidoides praemundulus* and *Cibicidoides mundulus* over most of the assemblage interval.

Thomas (1990a) located the upper boundary of assemblage 3 between samples 689B-15H-2, 40-42cm (131.82 mbsf; 38.80 Ma) and 689B-15H-1, 40-42cm (130.31 mbsf, 38.50 Ma). The assemblage boundary is difficult to determine precisely because of the gradual nature of faunal change. The upper boundary was chosen at the first appearance of *Turrilina alsatica* and the last common appearance of *Bulimina elongata*. The former species, however, occurs sporadically from sample 689B-16H-1, 40-42 cm (139.81 mbsf, 39.93 Ma) up, with a first common appearance in Sample 689B-14H-3, 41-43 cm (123.62 mbsf, 37.09 Ma). The latter species fluctuates strongly in relative abundance, with its last common continuous appearance in sample 689B-15H-2, 40-43 cm (131.82 mbsf, 38.80 Ma), but another common appearance (above several samples without specimens) in sample 689B-14H-4, 41-43 cm (125.12 mbsf, 37.69 Ma). Thus there is an interval of gradual changes in the benthic foraminiferal faunas, from 139.81 mbsf (39.93 Ma) through 123.62 mbsf (37.09 Ma). This gradual faunal change encompasses a decrease in relative abundance of infaunal species (largely represented by *B. elongata*, Figure 12.6), and an increase in epifaunal and cylindrical taxa.

Assemblage 2 (upper Eocene - lower Miocene)

Assemblage 2 occurs above the interval of gradual faunal change through the top of the interval studied for this paper. The assemblage is characterized by much lower diversity than the older assemblages (22 through 46 species per 300 specimens, average 32.9 ± 6.5 species). Thomas (1989, 1990a) did not subdivide the assemblage, although she noticed that *T. alsatica* is common in the lower part of the assemblage at Sites 689 and 690. In this study, I propose a subdivision in a lower sub-assemblage 2B (characterized by the common to abundant occurrence of *T. alsatica*), and an upper sub-assemblage 2A (*T. alsatica* rare or absent). The boundary between these sub-assemblages cannot be located precisely (Figures 12.4, 12.5): the last common continuous occurrence of the species is in sample 689B-12H,CC (110.60 mbsf, 33.80 Ma), but after a few samples without the species the last common occurrence is in sample 689B-12H-1, 40-42 cm (101.41 mbsf; 31.54 Ma).

A strong (although fluctuating) decrease in diversity occurs in sub-assemblage 2B, between the lowest sample in the assemblage (15H-1, 40-43cm, 130.32 mbsf, 38.5Ma) and sample 689B-14H-1, 43-45cm, 120.64 mbsf, 36.16Ma). This decrease in diversity starts below the strong, sudden increase in $\delta^{18}O$-values in *Cibicidoides* spp. and bulk carbonate (Figure 12.6),

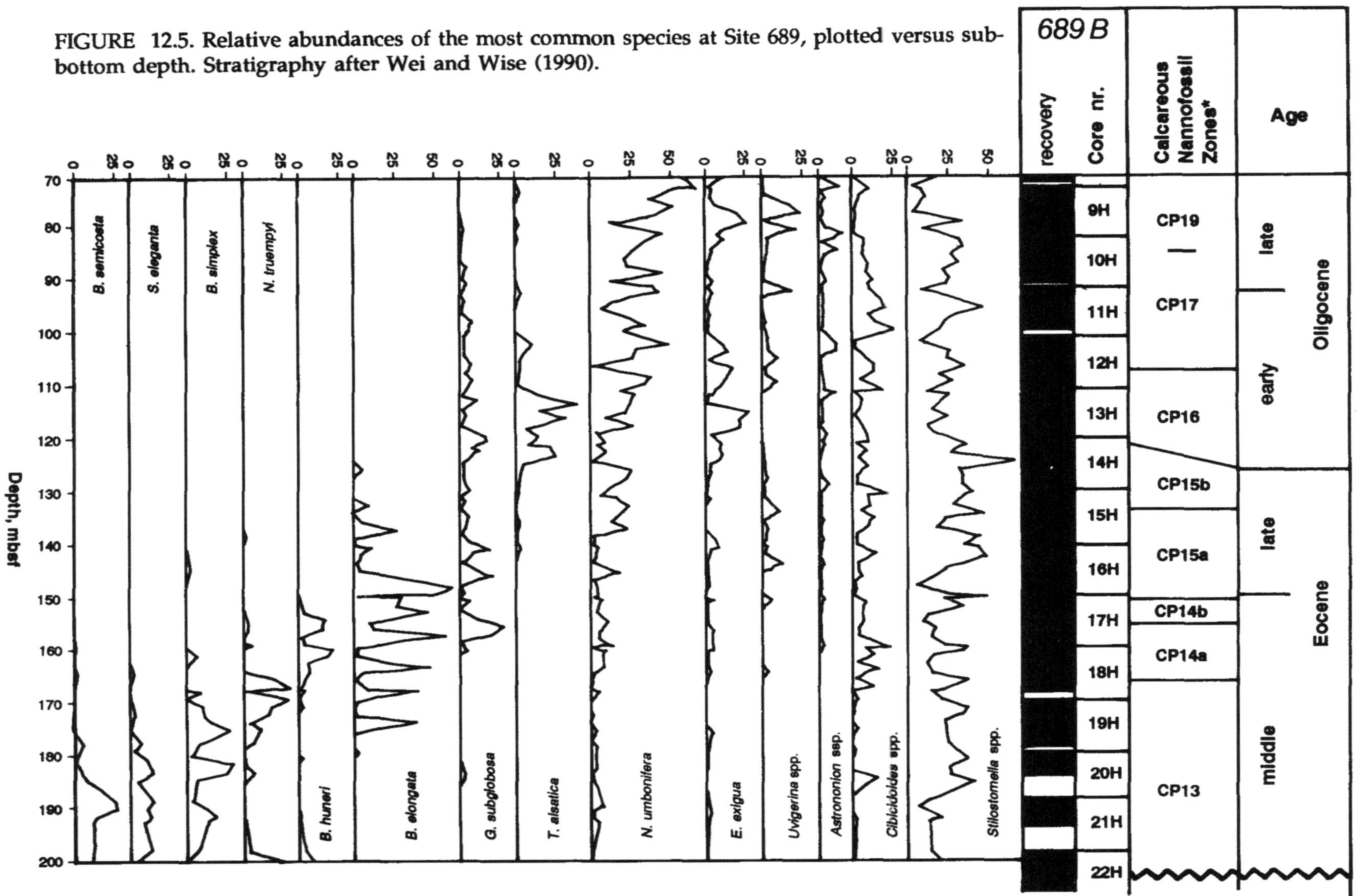

FIGURE 12.5. Relative abundances of the most common species at Site 689, plotted versus sub-bottom depth. Stratigraphy after Wei and Wise (1990).

which occurs between 118.45 mbsf (35.6 Ma) and 121.30 mbsf (36.5 Ma; Kennett and Stott, 1990).

Sub-assemblage 2B is characterized by the common presence of *T. alsatica*, common *Nuttallides umbonifera* and *Stilostomella* spp., common to rare *Epistominella exigua*, *C. mundulus*, and *G. subglobosa*. Sub-assemblage 2A lacks *T. alsatica*, but otherwise resembles sub-assemblage 2B. In sub-assemblage 2A there are a few intervals where the possibly infaunal taxa *Uvigerina* spp. and *Astrononion pusillum* are common (Figure 12.5). Of the *Stilostomella* species, *S. annulifera* is common to abundant especially in sub-assemblage 2B. Within assemblage 2 the diversity decreases upwards, while the relative abundance of epifaunal species (mainly *N. umbonifera*, which reaches up to 70% of the total fauna in the upper part of assemblage 2) increases and that of infaunal species (largely *T. alsatica*) decreases.

SPECIES RANGES AND FAUNAL EVENTS

In the middle Eocene through late Oligocene, major changes occurred in the bathyal benthic foraminiferal faunas of Maud Rise (Antarctica). In the first place, the diversity decreased from an average of 45 to an average of 30 species, i.e., a species loss of about 33%, largely as a result of the local disappearance of uniserial lagenids and lenticulid species. In the second place, large, ornamented and heavily calcified buliminid species disappeared during the later part of the middle Eocene, followed by the last disappearance of smaller buliminid species in the late middle to early late Eocene. In the third place, trochospiral (possibly epifaunal) taxa such as *N. umbonifera* and *E. exigua* increased in relative abundance. Thus a middle Eocene, diverse, buliminid (possibly infaunal species)-rich fauna was gradually replaced by a much less diverse fauna dominated by trochospiral (possibly epifaunal) taxa.

Cumulative plots of first and last appearances (Figure 12.6; FAs and LAs; local, not necessarily global events) show alternation of stable intervals and intervals with many last appearances. Such intervals with many last appearances occur between 130 and 140 mbsf (in the transition interval of assemblages 2 and 3, 37-40 Ma), and between 160 and 170 mbsf (43-46 Ma; in and just above the transition interval between assemblages 3 and 4, 44.5-46.6 Ma). Another interval with an increased number of last appearances occurs between 100 and 105 mbsf, in the transition interval of sub-assemblages 2A and 2B (31-33 Ma). The large number of last appearances above 80 mbsf is not an artifact of end-of-studied-section: these last appearances occur even when data for the higher part of the section are considered (Thomas, 1990a), and result from strong dissolution in the higher part of the section (see also $CaCO_3$-values in Figure 12.6; Diester-Haass, 1991).

First appearances are concentrated in the lower part of the section because of the presence of an unconformity at 203 mbsf. An interval with relatively many first appearances occurs roughly between 150 and 160 mbsf (41-43 Ma), another around 140 mbsf (about 40 Ma). The uppermost Eocene and the Oligocene are characterized by very few first appearances (Figure 12.6).

Many authors have described similar faunal patterns for the Eocene: a loss of buliminid species (e.g., Miller, 1983; Site 549 in Bay of Biscay; Douglas and Woodruff, 1981, Pacific; Boersma, 1986, Atlantic Ocean; Boltovskoy and Boltovskoy, 1989, Walvis Ridge, Southern Atlantic; Mueller-Merz and Oberhaensli, 1991, Walvis Ridge, Southern Atlantic), replacement of *N. truempyi* by *N. umbonifera* , and increase in relative abundance of *N. umbonifera*. Boersma (1981) described that *N. umbonifera* increased its depth-range after the earliest Oligocene.

Many specific FAs and LAs, however, occurred at different times at the Maud Rise sites than recorded elsewhere. The last common appearance of *N. truempyi* is commonly recorded elsewhere (e.g., Miller, 1983) near the middle/upper Eocene boundary, but occurs within the middle Eocene (close to the nannofossil zonal boundary between CP13 and CP14a) at Site 689. The LA of *B. semicostata* occurs in the lowermost Oligocene according to van Morkhoven et al. (1986), but in the middle Eocene at the Maud Rise sites.

These two taxa are easily recognized, and the differences in timing are not the result of taxonomic confusion. Differences in opinion on the timing of the FA of *Nuttallides umbonifera* may, however, be an artifact of study methods. This FA is commonly recorded in the lowermost

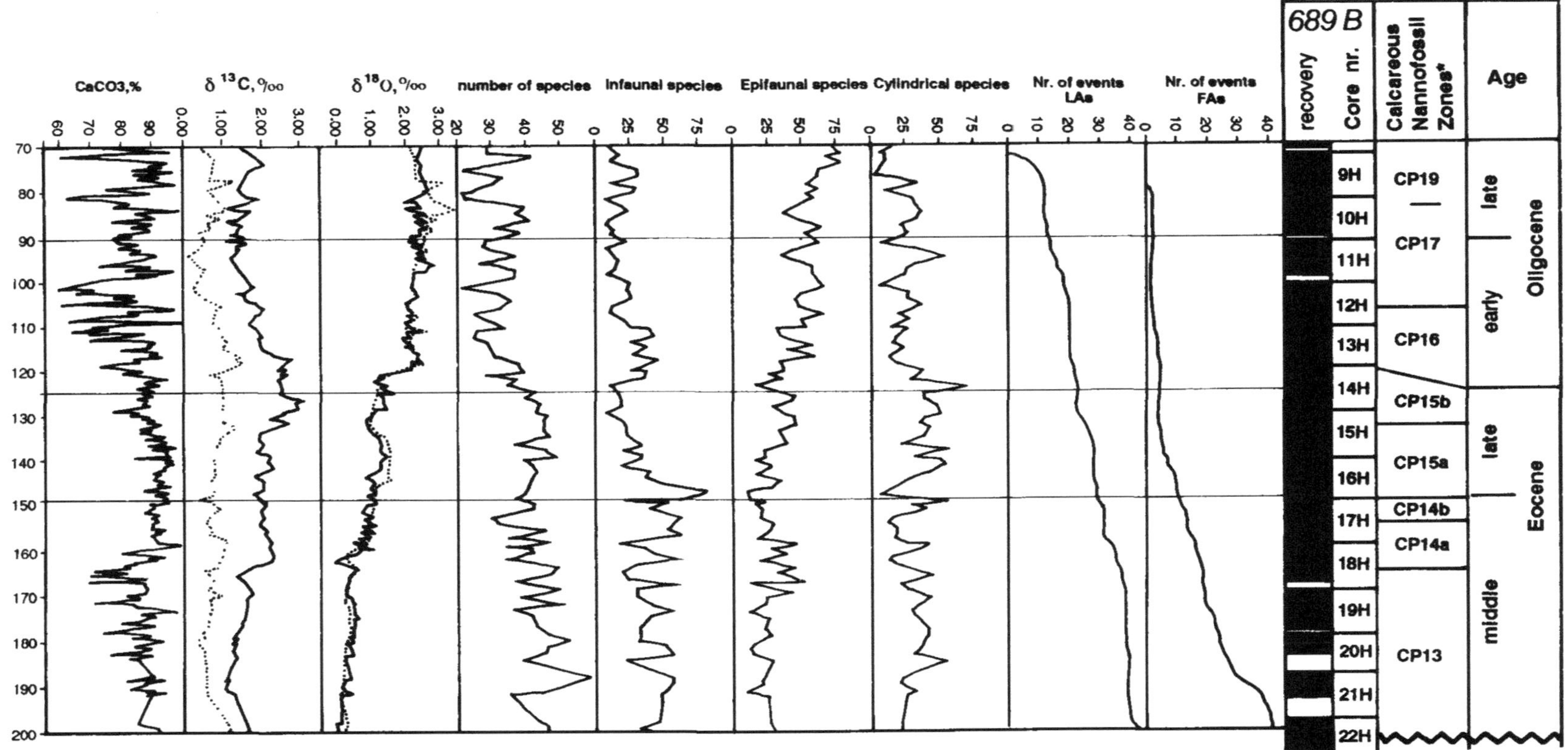

FIGURE 12.6. $CaCO_3$ contents of the sediments after O'Connell (1990), carbon and oxygen isotopic data after Kennett and Stott (1990) for *Cibicidoides* spp. (dotted line), and for bulk sediment after Shackleton and Hall (1990; heavy line), combined with diversity expressed as number of benthic foraminiferal species, relative abundance of infaunal, epifaunal and cylindrical morphogroups (see Table 12.1), and cumulative plots of first and last appearances. Stratigraphy after Wei and Wise (1990).

Oligocene/uppermost Eocene (Corliss, 1981; Tjalsma and Lohmann, 1983; Miller, 1983), but in the Maud Rise sites it occurs from the lower Eocene. This may be a result of difference in size fractions studied by different authors: the species is small when it first occurs, and becomes larger in the lowermost Oligocene (a common trend in benthic foraminifera; Boltovskoy, 1988). The earliest occurrences of the species can thus easily be missed when the larger than 125 or 150 μm fraction is studied. *Nuttallides umbonifera* was recorded from at least the middle Eocene on in South Atlantic sites (Walvis Ridge; Mueller-Merz and Oberhaensli, 1991), where the authors studied the size fraction larger than 100 μm. *Nuttallides umbonifera* is commonly described as having an acme in the lowermost Oligocene (Miller, 1983; Corliss, 1981; Kaminski et al., 1989), but at Site 689 the species appears to have its first increase in relative abundance lower, in the uppermost Eocene (136.32mbsf, about 39.6Ma; see also Corliss, 1981).

Very interesting is the occurrence of assemblage 2B with common *T. alsatica*: this assemblage occurs at Site 689 as well as at the deeper (upper abyssal) Site 690, but appears to correlate well with the bathyal *T. alsatica* BB5 Zone in Berggren and Miller (1989; duration: P18-P20; Figure 12.3), instead of with their abyssal zone AB8 (duration: P18-21). All other Paleogene assemblages at the Maud Rise sites appear to correlate much better with the abyssal zones of Berggren and Miller (1989), as is common for lower bathyal sites. A similar *T. alsatica* acme zone was observed to occur at about the same time (as far as can be said from the time-resolution available) in the Indian Ocean (Schroeder-Adams, 1991), and at high northern latitudes (Kaminski et al., 1989). The latter authors describe lower Oligocene calcareous benthic faunas with *T. alsatica* above Eocene agglutinated faunas at abyssal ODP Site 747 in the Southern Labrador Sea. A *T. alsatica* Zone was not observed by Thomas (1985) in abyssal sites in the equatorial Pacific, however; possibly the species occurs over a greater depth-range at higher latitudes (both North and South) than in the equatorial regions. At Site 689 the acme of *T. alsatica* begins at 123.62 mbsf (37.09 Ma), i.e., below the increase in $\delta^{18}O$-values of *Cibicidoides* spp., at 122.12-118.51 mbsf.

DISCUSSION

Benthic foraminiferal faunal changes from middle Eocene through late Oligocene at lower bathyal Site 689 were gradual, but stepped. There were periods of faunal change alternating with stable periods, but the "periods of change" were several millions of years long: the transition between assemblages 4 and 3 lasted from about 46.7 to 44.6 Ma (2.1 m.y.), the transition from assemblage 3 to 2 lasted from about 40 to about 37 Ma (3 m.y.), and that from sub-assemblage 2B to 2A from about 34 to 31.5 Ma (2.5 m.y.). The duration of these periods of faunal change is comparable with that of the well-known middle Miocene faunal change of deep-sea benthic foraminifera (Thomas, 1985, 1986a, b; Thomas and Vincent, 1987, 1988; Miller and Katz, 1987; Boltovskoy and Boltovskoy, 1989).

During the first period of faunal change (middle middle Eocene, 46.7-44.6 Ma) large, heavily calcified *Bulimina* species and many lagenids disappeared (at least, locally); during the second period (late Eocene, 40-47 Ma) the small, smooth species *Bulimina elongata* disappeared to be replaced by *Nuttallides umbonifera* and *Turrilina alsatica*, and in the last period of faunal change (late early Oligocene, 34-31.5 Ma) *T. alsatica* became extremely rare, while *N. umbonifera* increased in relative abundance. Over the whole period the overall diversity decreased, and the relative abundance of triserial (probably infaunal) species decreased.

The timing of faunal changes at Site 689 resembles that described from many sites worldwide: there were stepped, gradual changes during the middle to late Eocene. No clear, sudden extinction event occurred at the end of the Eocene, or at the time of the sudden increase in $\delta^{18}O$-values in the earliest Oligocene (Corliss, 1981; Miller, 1983; Douglas and Woodruff, 1981; Corliss et al., 1984; Keigwin and Corliss; 1986; Corliss and Keigwin, 1986; Boltovskoy and Boltovskoy, 1988, 1989; Kaminski et al., 1989; Kaihu, 1991; Mackensen and Berggren, 1991; Mueller-Merz and Oberhaensli, 1991). The overall pattern of faunal change also appears to be valid worldwide (*ibidem*): in the middle Eocene (to a lesser degree in the late Eocene), lower bathyal to upper abyssal faunas contained common to abundant elements (such as *Bulimina* species) that, in terms of modern faunas, would be considered more typical for an upper to upper-middle bathyal environment. During the middle and late Eocene the deeper faunas gradually lost these components, and from the Oligocene on the lower bathyal - upper abyssal faunas typically contained common to abundant *Nuttallides umbonifera* and *Epistominella exigua*.

There is no clear and simple correlation between the benthic foraminiferal faunal records and the records of carbon and oxygen isotopic values in benthic and planktonic foraminifers and bulk samples (Figure 12.6; see also Stott et al., 1990; Kennett and Stott, 1990; Shackleton and Hall, 1990), and between faunal records and the record of $CaCO_3$-content of the sediments (O'Connell, 1990). Conspicuously, there is no correlation between the lower Oligocene sudden increase in $\delta^{18}O$-values of benthic foraminifera and faunal change: in fact, the faunal change from assemblage 3 to 2 occurred clearly before the isotopic change at Site 689. Many authors recorded faunal changes which correspond in character to the changes at Site 689, but which occurred slightly later (at the Eocene/Oligocene boundary, i.e., still before the rapid increase in $\delta^{18}O$ in the earliest Oligocene; e.g., Berggren and Miller, 1989; Miller, 1983; Kaminski et al., 1989). This apparent difference in timing may result from differences in stratigraphic resolution (many studies were done at lower resolution than this study). On the other hand, the difference in timing may result from problems in biostratigraphic correlation at the extremely high-latitude Site 689; the LAD of the planktonic foraminifer *Globigerinatheka index*, for instance, appears to be much earlier at Site 689 than at other high latitude sites (compare Stott and Kennett, 1990 with Berggren, 1991). The initiation of benthic foraminiferal faunal changes, however, always pre-dates the isotopic change, although some changes in relative abundance have been reported to be coeval with the isotopic change (migration of *N. umbonifera* to lesser depths; Boersma, 1985; Katz and Miller, in prep.).

This lack of correspondence in timing of faunal and isotopic events supports the hypothesis that the global increase in $\delta^{18}O$ was at least partially, and probably largely caused by ice-volume increase, not water-temperature decrease (Zachos et al., 1991). The fact that faunal change was gradual and occurred before the isotopic shift suggests that the isotopic change may indeed have been the result of a "threshold effect" (e.g., Kennett and Shackleton, 1976; Kennett, 1977). First there was gradual environmental change (cooling) until temperatures of surface waters at high latitudes reached a threshold-value, then a sudden increase in ice-volume followed (driven by positive feed-back in the system, such as increased albedo from larger ice-volume). The faunas reacted to the gradual environmental change, i.e., the gradual cooling of the waters at lower bathyal depths in the middle Eocene, when decreasing diversity mirrored increasing $\delta^{18}O$ values of benthic foraminifera (Figure 12.6; see also Kennett and Stott, 1990; Figure 5 in Kennett and Barker, 1990); they have no reason to react to ice-volume increases on land.

At Site 689 the gradual increase in $\delta^{18}O$ values started at about 164 mbsf (about 44.5 Ma), slightly higher than faunal change from assemblage 4 to 3 (173-165 mbsf). This time in the middle Eocene was apparently a time of many, and complex, environmental changes at Site 689. The $\delta^{13}C$ data on bulk carbonate and *Cibicidoides* spp. (Figure 12.6; Shackleton and Hall, 1990; Kennett and Stott, 1990) show some indication that a change in surface-ocean productivity might have occurred just before the cooling: an increase in bulk-values (probably representing surface-water dissolved carbonate values) occurs at about 165 mbsf, and is not accompanied by a change in benthic values of the same magnitude; thus there may have been an increase in the surface-deep water gradients at that time. Such an increase can be explained by an increase in productivity (Shackleton, 1986; Miller and Fairbanks, 1985). There is additional evidence that there may have been an increase in surface-ocean productivity at that time: at about 160 mbsf radiolarians and diatoms first appeared (although they are rare; Thomas, unpublished data), and there is a strong and sudden increase in the abundance of radiolarians at 150 mbsf (Diester-Haass, 1991). Thus an increase in surface productivity might have occurred in the middle Eocene before a cooling event, just as has been proposed for the early Miocene "carbon excursion," pre-dating the middle Miocene increase in oxygen isotopic values (Berger and Vincent, 1985; Thomas and Vincent, 1987). It is not clear whether this middle Eocene event occurred worldwide, or was a local, high-latitude occurrence.

A change in clay-mineral content occurs at the same levels where radiolarians and diatoms first appear in the sediments. At that level illite/smectite ratios increase, possibly indicating climatic changes on the Antarctic mainland, notably decreased hydrolysis during weathering, as a result of increased aridity or cooling (Robert and Maillot, 1990). Thus there is considerable evidence of cooling (on land), coeval with increased biosiliceous productivity in the surface waters, directly followed by cooling of oceanic surface and deep-water, starting at some time in the middle Eocene (46 Ma). The ultimate cause of these climatic and biotic changes is not clear, but it might be sought in the plate-tectonic motion of Australia northward from Antarctica, causing surface currents of colder water to reach the coasts of Maud Rise (Kennett, 1977; Kemp, 1978; Mercer, 1983; Webb, 1990).

The faunal changes may well represent the reaction of the lower bathyal faunas to cooling of the deep waters, and thus increased solubility of calcite (i.e., more difficulty in abstracting calcite by foraminifera): the disappearance of large, heavily calcified species may result from lesser availability of calcite. The strong increase in relative abundance of *N. umbonifera* probably reflects increased $CaCO_3$-corrosivity (Bremer and Lohmann, 1981).

One of the most important characteristics of benthic foraminiferal faunal change from middle Eocene through early Oligocene is its gradual nature: all changes occurred over several millions of years, in contrast with the benthic extinction at the end of the Paleocene, which occurred over less than 4000 years (Thomas, 1990a; 1990b; in preparation; Kennett and Stott, 1991). Faunal changes were gradual worldwide, not just on Maud Rise (references quoted above; see also Culver, 1987 for an overview). This strongly argues against the hypothesis that an overall change in direction of deep ocean circulation occurred at any time during this interval, and against the model of development of ocean circulation as proposed by Kennett and Stott (1990). These authors proposed that Maud Rise was bathed in Warm Salty Bottom Waters (WSBW) during most of the Eocene (at least since 46 Ma), and that proto-Antarctic Bottom Waters filled the basins below the WSBW mass from the Oligocene on (pushing the WSBW to lesser depths). In this model, Site 689 was within a watermass that can be seen as a precursor of Antarctic Intermediate Water (AAIW), generated by sinking from the surface at high latitudes, whereas Site 690 was within WSBW during most of the Paleogene. The similarity between the benthic foraminiferal faunas at Sites 689 and 690 throughout the Paleogene (Thomas, 1990a) argues against the presence of two greatly different watermasses at the two sites. In addition, a circulation change must have been quick (a few thousand years at most) on a geologic time scale, as a result of the short circulation time of deep water in the oceans (Gascard, 1990). It is inconceivable that such a sudden, rapid, major change in deep and intermediate water circulation should not be reflected in large and sudden faunal changes in benthic foraminifera, in at least a few locations in the world's oceans. The observed gradual (although stepped) changes are much more likely the result of gradual cooling of deep (bathyal) waters that formed by sinking at high latitudes during the full period of middle Eocene through Oligocene. The gradual changes in the faunas then reflect gradual cooling of surface waters at high latitudes, resulting in a gradual establishment of the psychrosphere, instead of a sudden shift in the location of dominant formation of cold deep waters.

This fauna-based view of deep-water circulation can, as yet, not be made to agree with the presence of a "reversed isotope gradient" at Sites 689 and 690 (Kennett and Stott, 1990), and at the more northerly Sites 699 and 703 (Mead and Hodell, 1991). The explanation of the oxygen isotopic record, however, may be complex because salinity as well as temperature are involved, and stable density stratification may be salinity- as well as temperature-controlled (e.g., Railsback et al., 1989; Railsback, 1990). For instance, it is not easy to predict whether calcium carbonate deposited in a

WSBW mass is actually isotopically heavier or lighter than calcium carbonate deposited below a cold water mass: the WSBW is formed as a result of evaporation, leading to heavier isotopic values of the remaining water (Craig and Gordon, 1965; Railsback et al., 1989; Railsback, 1990). In addition, Site 689 is not only shallower than Site 690, but also on the opposite flank of Maud Rise. The current pattern around the rise is presently complex, with rising, less cold watermasses from greater depths causing the largest polynya in the Antarctic region at the surface (Comiso and Gordon, 1987). Therefore the evidence from benthic faunal composition should be taken into account while interpreting the isotopic records.

CONCLUSIONS

1. Lower bathyal benthic foraminiferal faunas from Maud Rise (Antarctica) show gradual, stepped faunal changes from middle Eocene through Oligocene, with steps at 46.6-44.6 Ma, 40-37 Ma, and 34-31.5 Ma (numerical ages after Berggren et al., 1985).

2. The faunal changes at Site 689 can be summarized as follows: (1) decreasing diversity and increasing relative abundance of epifaunal species, combined with the loss of large, heavily calcified *Bulimina* species; (2) the disappearance of all *Bulimina* species and the appearance of *Turrilina alsatica*; (3) the disappearance of *T. alsatica*, resulting in faunas dominated by *Nuttallides umbonifera*.

3. These faunal changes may well reflect a gradual cooling (and thus a gradual increase in corrosiveness) of the high-latitude lower bathyal waters. The gradual nature of the lower bathyal faunal changes and the absence of catastrophic extinctions (on Maud Rise, and worldwide) suggests that the psychrosphere was established as a result of the cooling of surface waters at high latitudes, while lower bathyal waters formed by cooling and sinking at high latitudes during the full interval of middle Eocene and Oligocene. Warm, salty deep waters might have been present during this time in the deepest basins, but did not reach up into the lower bathyal reaches of Maud Rise.

ACKNOWLEDGEMENTS

I thank Bill Berggren and Don Prothero for asking me to participate in the meetings on this interesting subject, and Ken Miller for a constructive review of the paper. The research was partially funded by a grant from USSAC (Leg 113). This is University of Cambridge contribution #1893.

REFERENCES

Adams, C. G., Lee, D. E., and Rosen, B. R. 1990. Conflicting isotopic and biotic evidence for tropical sea-surface temperature during the Tertiary. *Palaeogeogr., Palaeoclimat., Palaeoecol.* 77: 289-313

Barker, P. F., Kennett, J. P., et al. 1988. *Proc. ODP, Init. Repts. 113*: College Station, TX, Ocean Drilling Program: 785 pp.

Barrera, E., Huber, B., Savin, S. M., and Webb, P. N. 1987. Antarctic marine temperatures: late Campanian through early Paleocene. *Paleoceanography* 2: 21-48

Barron, E. J. 1985. Explanations of the Tertiary global cooling trend. *Palaeogeogr., Palaeoclim., Palaeoecol.* 50: 45-61

Barron, E. J. 1987. Eocene equator-to-pole surface ocean temperatures: a significant climate problem? *Paleoceanography* 2: 729-740

Barron, J., Larsen, B. L., and Baldauf, J. 1991. In press. Evidence for late Eocene-early Oligocene Antarctic glaciation and observations on late Neogene glacial history of Antarctica: results from ODP Leg 119. In Barron, J., Larsen, B. L., eds., *Proc. ODP, Sci. Results* 119: College Station, TX, Ocean Drilling Program.

Benson, R. H. 1975. The origin of the psychrosphere as recorded in changes of deep-sea ostracode assemblages. *Lethaia* 8: 69-83

Berger, W. H., and Vincent, E. 1985. Carbon dioxide and polar cooling in the Miocene: the Monterey hypothesis. In E. T. Sundquist and W. S. Broecker, eds., *The Carbon Cycle and Atmospheric CO_2: natural variations, Archean to Present. Geophysical Monographs* 32:455-468

Berggren, W. A., Kent, D. V., Flynn, J. J., and Van Couvering, J. A. 1985. Cenozoic geochronology. *Geol. Soc. Am. Bull.* 96: 1407-1418

Berggren, W. A., and Miller, K. G. 1989. Cenozoic bathyal and abyssal calcareous benthic foraminiferal zonation. *Micropaleont.*

35: 308-320

Berggren W. A. 1991. Paleogene planktonic foraminifer magnetobiostratigraphy of the southern Kerguelen Plateau. In Schlich, R., Wise, S. W. Jr., et al., *Proc. ODP, Sci. Results 120;* College Station, TX, Ocean Drilling Program.

Bremer, M., and Lohman, P. 1982. Evidence for primary control of the distribution of certain Atlantic Ocean benthonic foraminifera by degree of carbonate saturation. *Deep Sea Res.* 29: 987-998

Boersma, A., 1984. Oligocene and other Tertiary benthic foraminifera from a depth traverse down Walvis Ridge, Deep Sea Drilling Project Leg 74. In Hay, W. W., Sibuet, J. C., et al., *Init. Rept. Deep Sea Drill. Proj.* Washington, U. S. Government Printing Office, 75: 1273-1300

Boersma, A. 1985. Oligocene benthic foraminifers from North Atlantic sites: benthic foraminifers as water-mass indexes in the North and South Atlantic. In Bougault, H., Cande, S. C., et al., *Init. Rept. Deep Sea Drill. Proj.* Washington, U. S. Government Printing Office, 82: 611-628

Boersma, A. 1986. Eocene-Oligocene Atlantic paleooceanography, using benthic Foraminifera. In Ch. Pomerol and I. Premoli-Silva, eds., *Terminal Eocene Events.* Amsterdam, Elsevier, pp. 225-236

Boersma, A., Premoli-Silva, I., and Shackleton, N. J. 1987. Atlantic Eocene planktonic foraminiferal paleohydrographic indicators and stable isotope stratigraphy. *Paleoceanography* 2: 287-331

Boltovskoy, E. 1980. On the benthonic bathyal-abyssal foraminifera as stratigraphic guide fossils. *J. Foram. Res.* 10: 163-172

Boltovskoy, E. 1988. Size change in phylogeny of Foraminifera. *Lethaia* 21: 375-382

Boltovskoy, E., and Boltovskoy, D. 1988. Cenozoic deep-sea benthic foraminifera: faunal turnovers and paleobiographic differences. *Rev. Micropaleont.* 31: 67-84

Boltovskoy, E., and Boltovskoy, D. 1989. Paleocene-Pleistocene benthic foraminiferal evidence of major paleoceanographic events in the eastern South Atlantic (DSDP Site 525, Walvis Ridge). *Marine Micropaleont.* 14: 283-316

Brass, G. W., Southam, J. R., and Peterson, W. H. 1982. Warm saline bottom water in the ancient ocean. *Nature* 296: 620-623

Chamberlin, T. C. 1906. On a possible reversal of deep-sea circulation and its influence on geologic climates. *J. Geol.* 14: 363-373

Coffin, M. F. 1991. Subsidence of the Kerguelen Plateau: the Atlantic concept. In Schlich, R., Wise, S. W. Jr., et al., *Proc. ODP, Sci. Results* 120; College Station, TX, Ocean Drilling Program.

Corliss, B. H. 1981. Deep-sea benthonic foraminiferal faunal turnover near the Eocene/Oligocene boundary. *Marine Micropaleont.* 6: 367-384

Corliss, B. H., and Chen, C. 1988. Morphotype patterns of Norwegian deep-sea benthic foraminifera and ecological implications. *Geology* 16: 716-719

Corliss, B. H., and Keigwin, L. D. Jr. 1986. Eocene-Oligocene paleoceanography. In K. Hsu, ed., *Mesozoic and Cenozoic Oceans, Amer. Geophys. Union, Geodynamics Series* 15:101-118

Corliss, B. H., Aubry, M.-P., Berggren, W. A., Fenner, J. M., Keigwin, L. D. Jr., and Keller, G. 1984. The Eocene/Oligocene boundary in the deep sea. *Science* 226: 806-810

Comiso, J. C., and Gordon, A. L. 1987. Recurring polynyas over the Cosmonaut Sea and the Maud Rise. *J. Geophys. Res.* 92: 2819-2833

Craig, H., and Gordon, L. I. 1965. Deuterium and Oxygen 18 variations in the ocean and the marine atmosphere. In *Stable Isotopes in Oceanographic Studies and Paleotemperatures*, E. Tongiorgi, ed., Consiglio Nazionale delle Ricerche, Laboratorio di Geologia Nucleare, Pisa, Italy: 1-122

Crough, S. T. 1983. The correction for sediment loading on the seafloor. *J. Geophys. Res.* 88: 6449-6454

Culver, S. J. 1987. Foraminifera. In Lipps, J. R., ed., *Fossil Prokaryotes and Protists. University of Tennessee Studies in Geology* 18: 169-212

Detrick, R. S., Sclater, J. G., and Thiede, J. 1977. The subsidence of aseismic ridges. *Earth Planet. Sci. Lett.* 34: 185-196

Diester-Haass, L. 1991. Eocene/Oligocene paleoceanography in the Antarctic Ocean, Atlantic sector (Maud Rise, ODP Leg 113, Site 689 and 690).

Douglas, R. G., and Woodruff, F. 1981. Deep sea

benthic foraminifera. In Emiliani, C., ed. *The Sea; The Oceanic Lithosphere* 7: 1233-1327

Gascard, J.-C. 1990. Deep Convection and Deep-water Formation. *EOS*, 71 (49): 1837-1839

Hay, W. W. 1989. Paleoceanography: a review for GSA Centennial. *Geol. Soc. Am. Bull.* 100: 1934-1956

Kaminski, M. A., Gradstein, F. M., and Berggren, W. A. 1989. Paleogene benthic foraminifer biostratigraphy and paleoecology at Site 647, Southern Labrador Sea. In Srivastava, S. P., Arthur, M., Clement, B., et al., *Proc. ODP, Sci. Results*, College Station, TX, Ocean Drilling Program, 105: 705-756

Katz, M. R., and Miller, K. G. 1991. Early Paleogene benthic foraminiferal assemblage and stable isotope composition in the Southern Ocean, ODP Leg 114. In Ciesielski, P. F., and Kristofferson, Y., et al., *Proc. ODP, Sci. Results* 114: College Station TX, Ocean Drilling Program.

Keigwin, L. D. Jr., and Corliss, B. H. 1986. Stable isotopes in late middle Eocene through Oligocene foraminifera. *Geol. Soc. Am. Bull.* 97: 335-345

Keigwin, L. D. Jr., and Keller, G. 1984. Middle Oligocene cooling from Equatorial Pacific DSDP Site 77B. *Geology* 12: 16-19

Kemp, E. M. 1978. Tertiary climatic evolution and vegetation history in the Southeast Indian Ocean region. *Palaeogeog., Palaeoclim., Palaeoecol.* 24: 169-208

Kennett, J. P. 1977. Cenozoic evolution of Antarctic glaciation, the circum-Antarctic Ocean, and their impact on global paleoceanography. *J. Geophys. Res.* 82: 3843-3860

Kennett, J. P., and Barker, P. F. 1990. Latest Cretaceous to Cenozoic climate and oceanographic developments in the Weddell Sea, Antarctica: an ocean-drilling perspective. In Barker, P. F., Kennett, J. P., et al., *Proc. ODP, Sci. Results*, College Station, TX, Ocean Drilling Program, 113: 937-962

Kennett, J. P., and Shacleton, N. J. 1976.Oxygen isotope evidence for the development of the psychrosphere 38 myr. ago. *Nature* 260: 513-515

Kennett, J. P., and Stott, L. D. 1990. Proteus and Proto-Oceanus: ancestral Paleogene oceans as revealed from Antarctic stable isotopic results. In Barker, P. F., Kennett, J. P., et al., *Proc. ODP, Sci. Results*, College Station, TX, Ocean Drilling Program, 113: 865-880

Kennett, J. P., and Stott, L. D. 1991. Terminal Paleocene deep-sea benthic crisis: sharp deep sea warming and paleoceanographic changes in Antarctica. *Nature* (in press)

Kaihu, K. 1991. Global changes of Paleogene aerobic/anaerobic benthic foraminifera and deep-sea circulation. *Palaeogeog., Palaeoclim., Palaeoecol.* 83: (in press)

Mackensen, A., and Berggren, W. A. 1991. Paleogene benthic foraminifers from the Southern Indian Ocean (Kerguelen Plateau): biostratigraphy and paleoecology. In Wise, S. W. Jr., Schlich, R., et al., *Proc. ODP, Sci. Results*, College Station, TX, Ocean Drilling Program, 120:

Manabe, S., and Bryan, K. 1985. CO_2-induced change in a coupled ocean-atmosphere system and its paleoclimatic implications. *J. Geophys. Res.* 90: 11689-11707

Matthews, R. K., and Poore, R. Z. 1980. Tertiary $\delta^{18}O$ record and glacioeustatic sea-level fluctuations. *Geology* 8: 501-504

Mead, G. A., and Hodell, D. A. 1991. Late Eocene to early Oligocene vertical oxygen isotope gradients in the South Atlantic: implications for warm saline deep water.

Mercer, J. H. 1983. Cenozoic glaciation in the southern hemisphere. *Ann. Rev. Earth Planet. Sci.* 11: 99-132

Miller, K. G. 1983. Eocene-Oligocene paleoceanography in the deep Bay of Biscay. *Marine Micropaleont.* 7: 403-440

Miller, K. G., Curry, W. B., and Ostermann, D. R. 1984. Late Paleogene (Eocene to Oligocene) benthic foraminiferal oceanography of the Goban Spur Region, DSDP Leg 80. In De Graciansky, P. G., Poag, C. W., et al., *Init. Rept. Deep Sea Drill. Proj.* 80: 505-538

Miller, K. G., and Fairbanks, R. G. 1985. Oligocene-Miocene global carbon and abyssal circulation changes. In Sundquist, E., and Broecker, W. S., eds., *The carbon cycle and atmospheric CO_2: natural variations, Archean to Present. Amer. Geophys. Union Monog. Series* 32: 469-486

Miller, K. G., Fairbanks, R. G., and Mountain, G. S. 1987a. Tertiary isotope synthesis, sea level history, and continental margin erosion. *Paleoceanography* 2: 1-20

Miller, K. G., Janecek, T. R., Katz, M. R., and Keil, D. J. 1987b. Abyssal circulation and benthic foraminiferal changes near the Paleocene/Eocene boundary. *Paleoceanography* 2: 741-761

Miller, K. G., and Katz, M. Oligocene to Miocene benthic foraminiferal and abyssal circulation changes in the North Atlantic. *Micropalaeontology* 33; 97-149

Miller, K. G., and Thomas, E. 1985. Late Eocene to Oligocene benthic foraminiferal isotopic record, Site 574 Equatorial Pacific. In L. A. Mayer, F. Theyer, et al., *Initial Reports of the Deep Sea Drilling Project* 85: 771-777

Mueller-Merz, E., and Oberhaensli, H. 1991. Eocene bathyal and abyssal benthic foraminifera from a South Atlantic transect at 20°-30°S. *Palaeogeog., Palaeo-clim., Palaeoecol.* 83:(in press).

Nomura, R. 1991. Paleoceanography of upper Maestrichtian to Eocene benthic foraminiferal assemblages at ODP Sites 752, 753 and 754, eastern Indian Ocean. In Peirce, J., Weissel, J., et al., *Proc. ODP, Sci. Results*, College Station, TX, Ocean Drilling Program, 121: (in press)

Oberhaensli, H., Mueller-Merz, E., and Oberhaensli, R. 1991. Eocene paleoceanographic Evolution at 20°-30°S in the Atlantic Ocean. *Palaeogeog., Palaeoclimat., Palaeoecol.* 83: (in press)

O'Connell, S. 1990. Variations in Upper Cretaceous and Cenozoic calcium carbonate percentages, Maud Rise, Weddell Sea, Antarctica. In Barker, P. F., Kennett, J. P., et al., *Proc. ODP, Sci. Results*, College Station, TX, Ocean Drilling Program, 113: 971-984

Parsons, B., and Sclater, J. G. 1977. An analysis of the variation of ocean floor bathymetry and heat flow with age. *J. Geophys. Res.* 82: 803-827

Poore, R. Z., and Matthews, R. K. 1984. Late Eocene - Oligocene oxygen and carbon isotopic record from the South Atlantic DSDP Site 522. In Hsu, K. J., LaBrecque, J. L., et al., *Init. Rept. Deep Sea Drill. Proj.* Washington, U. S. Government Printing Office, 73: 725-735

Pospichal, J. J., and Wise, S. W., Jr. 1990. Paleocene to middle Eocene calcareous nannofossils of ODP Sites 689 and 690, Maud Rise, Weddell Sea. In Barker, P. F., Kennett, J. P., et al., *Proc. ODP, Sci. Results*, College Station, TX, Ocean Drilling Program, 113: 613-638

Prentice, M. L., and Matthews, R. K. 1988. Cenozoic ice-volume history: development of a composite oxygen isiotope record. *Geology* 16: 963-966

Railsback, L. B. 1990. Influence of changing dep ocean circulation on the Phanerozoic oxygen isotopic record. *Geoch. Cosmoch. Acta* 54: 1501-1509

Railsback, L. B., Anderson, T. F., Ackerly, S. C., and Cisne, J. L. 1989. Paleoceanographic modeling of temperature-salinity profiles from stable isotope data. *Paleoceanography* 4: 585-591

Robert, C., and Maillot, H. 1990. Paleoenvironments in the Weddell Sea area and Antarctic climates as deduced from clay mineral associations and geochemical data, ODP Leg 113. In Barker, P. F., Kennett, J. P., et al., *Proc. ODP, Sci. Results*, College Station, TX, Ocean Drilling Program, 113: 51-70

Savin, S. M. 1977. The history of the Earth's surface temperature during the past 100 million years. *Ann. Rev. Earth Planet. Sci.* 5: 319-344

Schnitker, D. 1979. Cenozoic deep water benthic foraminifers, Bay of Biscay. In Montadert, L, Roberts, D. G., et al., *Init. Rept. Deep Sea Drill. Proj.*, Washington, U. S. Government Printing Office, 48: 377-414

Schroeder-Adams, C.J., 1991. Eocene to Recent benthic foraminifer assemblages from the Kerguelen Plateau (Southern Indian Ocean). In Barron, J., Larsen, B., et al., *Proc. ODP, Sci. Results*, College Station, TX, Ocean Drilling Program, 119,

Schroeder, C. J., Scott, D. B., and Medioli, F. S. 1987. Can smaller benthic foraminifera be ignored in paleoenvironmental analyses? *J. Foram. Res.* 17: 101-105

Shackleton, N. J. 1984. Oxygen isotopic evidence for Cenozoic climate change. In P. Brenchley, ed., *Fossils and Climate* (Wiley, Chichester), 27-34

Shackleton, N. J. 1986. Paleogene stable isotope events. *Palaeogeogr., Palaeoclimat., Palaeoecol.* 57: 91-102

Shackleton, N. J. 1987. The carbon isotope history of the Cenozoic. In Brooks, J., and Fleet, A. J., (eds.), *Petroleum Source Rocks.* London, Blackwell, pp. 427-438

Shackleton, N. J., and Boersma, A. 1981. The climate of the Eocene ocean. *J. Geol. Soc. London* 138: 153-157

Shackleton, N. J., and Hall, M. A. 1990. Carbon isotope stratigraphy of bulk sediments, ODP Sites 689 and 690, Maud Rise, Antarctica. In Barker, P. F., Kennett, J. P., et al., *Proc. ODP, Sci. Results* College Station, TX, Ocean Drilling Program, 113: 985-989

Shackleton, N. J., and Kennett, J. P. 1975. Palaeotemperature history of the Cenozoic and the initiation of Antarctic glaciation: oxygen and carbon isotope analyses in DSDP Sites 277, 279 and 281. In Kennett, J. P., Houtz, K. E., et al., *Init. Rept. Deep Sea Drill. Proj.* Washington, D.C., U. S. Government Printing Office, 29: 743-755.

Spiess, V. 1990. Cenozoic magnetostratigraphy of Leg 113 drill sites, Maud Rise, Weddell Sea, Antarctica. In Barker, P. F., Kennett, J. P., et al., *Proc. ODP, Sci. Results,* College Station, TX, Ocean Drilling Program, 113: 261-315

Stott, L. D., Kennett, J. P., Shackleton, N. J., and Corfield, R. M. 1990. The evolution of Antarctic surface waters during the Paleogene: inferences from the stable isotopic composition of planktonic foraminifera, ODP Leg 113. In Barker, P. F., Kennett, J. P., et al., *Proc. ODP, Sci. Results,* College Station, TX, Ocean Drilling Program, 113: 849-864

Stott, L. D., and Kennett, J. P. 1990. Antarctic Paleogene planktonic foraminiferal biostratigraphy: ODP Leg 113, Sites 689 and 690. In Barker, P. F., Kennett, J. P., et al., *Proc. ODP, Sci. Results.* College Station, TX, Ocean Drilling Program, 113: 549-570

Thomas, E. 1985. Late Eocene to Recent deep-sea benthic foraminifers from the central equatorial Pacific Ocean. In L. A. Mayer, F. Theyer, et al., *Init. Rept. Deep Sea Drill. Proj.* 85: 655-679

Thomas, E. 1986a. Changes in composition of Neogene benthic foraminiferal faunas in equatorial Pacific and north Atlantic. *Palaeogeogr., Palaeoclimat., Palaeoecol.* 53: 47-61

Thomas, E. 1986b. Early to middle Miocene benthic foraminiferal faunas from DSDP Sites 608 and 610, North Atlantic. In Summerhayes, C. P., and Shackleton, N. J., (eds.), *North Atlantic Palaeoceanography.* London, Blackwell, pp. 205-218

Thomas, E., and Vincent, E. 1987. Major changes in benthic foraminifera in the equatorial Pacific before the middle Miocene polar cooling. *Geology* 15: 1035-1039

Thomas, E., and Vincent, E. 1988. Early to middle Miocene deep-sea benthic foraminifera in the equatorial Pacific. *Revue de Paleobiologie, Special Volume* 2 (BENTHOS '86): 583-588

Thomas, E. 1989. Development of Cenozoic deep-sea benthic foraminiferal faunas in Antarctic waters. In Crame, J. A. (ed.), *Origins and evolution of Antarctic biota. Geol. Soc. London Spec. Publ.* 47: 283-296

Thomas, E. 1990a. Late Cretaceous through Neogene deep-sea benthic foraminifers, Maud Rise, Weddell Sea, Antarctica. In Barker, P. F., Kennett, J. P., et al., *Proc. ODP, Sci. Results* College Station, TX, Ocean Drilling Program, 113: 571-594

Thomas, E. 1990b. Late Cretaceous-early Eocene mass extinctions in the deep sea. In Global Catastrophes, *Geol. Soc. Amer. Spec. Publ.* 247.

Thomas, E., Barrerra, E., Hamilton, N., Huber, B. T., Kennett, J. P., O'Connell, S., Pospichal, J. J., Spiess, V., Stott, L. D., Wei, W., and Wise, S. W. Jr. 1990. Upper Cretaceous - Paleogene stratigraphy of Sites 689 and 690, Maud Rise (Antarctica). In Barker, P. F., Kennett, J. P., et al., *Proc. ODP, Sci. Results.* College Station, TX, Ocean Drilling Program. 113: 901-914

Tjalsma, R. C., and Lohman, G. P. 1983. Paleocene-Eocene bathyal and abyssal benthic foraminifera from the Atlantic Ocean. *Micropaleontology Spec. Publ.* 94 pp.

van Morkhoven, F. P. C., Berggren, W. A., and Edwards, A. S. 1986. *Cenozoic cosmopolitan deep-water benthic foraminifera.* Elf-Aquitaine (Pau, France): 421 pp.

Webb, P.-N. 1990. The Cenozoic history of Antarctica and its global impact. *Antarctic Sci.* 2: 3-21

Wei, W., and Wise, S. W. Jr. 1990. Middle Eocene to Pleistocene calcareous nannofossils recovered by ODP Leg 113 in the Weddell Sea. In Barker, P. F., Kennett, J. P., et al., *Proc. ODP, Sci. Results*. College Station, TX, Ocean Drilling Program, 113: 639-666

Wise, S. W. Jr., Breza, J. R., Harwood, D. M., Wei, W., and Zachos, J. C. 1991. Paleogene glacial history of Antarctica in the light of ODP Leg 120 drilling Results. In Schlich, R., Wise, S. W. Jr., et al., *Proc. ODP, Sci. Results* College Station, TX, Ocean Drilling Program, 120.

Zachos, J., Berggren, W. A., and Aubry, M.-P. 1991. Eocene-Oligocene climatic and abyssal circulation history of the southern Indian Ocean. In Schlich, R., Wise, S. W. Jr., et al., *Proc. ODP, Sci. Results* College Station, TX, Ocean Drilling Program, 120.

TABLE 12.1A

	ABYSSAMINA QUADRATA	AGGLUTINANTS, OTHER	ALABAMINA DISSONATA	ALLOMORPHINA TRIGONA	ANGULOGERINA EARLANDI	ANOMALINA SPISSIFORMIS	ANOMALINOIDES ACUTUS	ANOMALINOIDES PSEUDOGROSSE	ANOMALINOIDES SEMICRIBRAT	ARAGONIA ARAGONENSIS	ASTRONONION PUSILLUM	ASTRONONION UMBILICATULUM	BIGENERINA NODOSARIA	BOLIVINA DECUSSATA	BOLIVINA HUNERI	BOLIVINA PSEUDOPUNCTATA	BOLIVINOIDES SP. AFF DELI	BULIMINA ALAZANENSIS	BULIMINA CALLAHANI	BULIMINA CF. SEMICOSTATA	BULIMINA ELONGATA	BULIMINA MACILENTA	BULIMINA MICROCOSTATA
068.92	...	...	...	...	...	1	...	2	...	...	7	3	...	1	...	...	...	...	...	...	...	1	...
070.42	...	...	...	...	...	...	...	...	...	...	1	1	...	...	...	...	...	...	...	...	...	...	...
071.93	...	...	...	...	...	...	...	1	...	...	24	...	...	...	...	1	...	...	...	...	...	...	...
072.51	...	...	...	...	...	...	...	1	...	...	2	...	1	16	...	...	...	...	...	...	...	...	...
074.01	...	...	...	...	...	...	2	3	...	...	3	6	...	2	...	...	...	...	...	...	...	...	...
075.51	...	...	...	1	...	...	...	...	1	...	1	...	...	14	...	...	...	...	...	...	...	...	5
077.01	...	...	...	1	...	...	...	...	...	...	5	...	3	3	...	...	...	...	...	...	...	...	...
078.51	...	...	...	...	...	...	3	9	...	...	...	2	...	7	...	...	...	...	...	...	...	...	...
080.01	...	...	...	...	...	...	...	5	...	...	13	1	1	2	...	...	...	...	...	...	...	...	1
080.66	...	...	...	...	9	...	...	...	...	...	31	...	3	...	...	...	...	...	...	...	...	...	...
082.11	...	...	...	...	...	...	...	1	...	...	3	1	12	...	...	...	...	...	...	...	...	...	...
083.61	...	...	...	...	...	...	1	...	...	...	35	...	2	...	...	...	...	...	...	...	1	3	2
085.11	...	...	...	...	...	...	...	...	...	...	7	...	5	...	2	...	...	...	...	...	...	...	43
086.61	...	...	...	...	...	...	...	1	...	...	2	...	9	...	...	...	...	...	...	...	...	...	...
088.11	...	...	...	...	...	...	...	...	...	...	...	2	12	...	...	...	...	...	...	...	...	...	1
089.61	...	...	...	3	...	...	...	4	...	...	4	6	10	...	...	...	...	...	...	...	1	...	...
090.49	...	...	...	...	...	...	...	7	...	...	3	...	12	...	...	...	...	...	...	...	...	...	1
091.72	...	...	...	...	...	...	...	...	...	...	4	...	6	...	...	...	...	...	...	...	...	...	...
093.22	...	...	...	1	...	...	...	...	1	...	5	...	6	...	...	...	...	...	...	...	...	...	...
094.72	...	...	...	...	...	...	...	1	...	...	5	...	...	...	...	...	...	...	...	...	...	...	...
096.22	...	...	...	...	...	...	...	...	...	...	7	...	3	...	...	...	...	...	...	...	...	...	...
097.72	...	...	...	...	...	...	...	1	...	...	3	...	...	...	...	...	...	...	...	...	1	...	...
098.96	...	...	...	...	...	...	...	1	...	...	3	2	...	...	...	...	...	...	...	...	...	...	...
101.41	...	...	...	1	...	...	...	...	...	...	...	31	...	...	...	...	...	...	...	...	...	...	...
102.91	...	...	...	1	...	...	...	...	...	...	32	...	...	...	...	...	...	...	...	...	...	...	...
104.41	...	...	...	...	...	...	...	...	...	...	1	...	...	...	...	...	...	...	...	...	...	...	1
105.91	...	...	...	...	...	...	...	17	...	...	...	2	2	...	...	...	...	...	...	...	...	...	...
107.41	...	...	...	5	...	...	...	...	...	...	1	2	1	...	...	...	...	...	...	...	...	...	...
108.91	...	...	...	1	...	1	...	1	...	...	3	...	1	...	...	...	...	...	...	...	...	...	...
110.60	...	...	...	1	...	...	...	...	...	...	5	...	8	...	...	...	...	...	...	...	...	...	...
111.01	...	...	...	2	...	...	...	...	...	...	29	...	...	...	...	...	...	...	...	...	...	...	...
112.51	...	...	...	1	...	3	...	...	...	...	3	...	1	...	...	...	...	...	...	...	...	...	...
114.01	...	...	...	1	...	...	...	1	...	...	...	...	...	...	1	...	...	...	...	...	...	...	...
115.51	...	...	...	...	...	...	1	...	...	...	...	4	...	...	...	...	...	...	...	...	...	...	...
117.01	...	...	...	4	...	1	...	...	...	...	...	2	3	...	...	...	...	...	...	...	...	...	...
118.51	...	...	...	1	...	9	...	2	...	...	4	10	5	...	...	...	...	...	...	...	...	...	...
120.20	...	...	...	...	...	1	...	...	2	...	3	...	1	...	...	...	...	...	...	...	...	...	...
120.64	...	...	...	3	...	4	...	...	...	...	1	5	...	...	...	...	...	...	...	...	...	...	...
122.12	...	...	...	2	...	4	...	...	...	...	...	1	...	...	3	...	...	...	...	...	...	...	...
123.62	...	...	...	...	...	5	...	...	...	...	...	...	2	...	...	...	...	...	...	...	...	...	...
125.12	...	...	...	3	...	1	...	...	...	...	...	...	...	...	...	...	...	...	...	...	17	...	...
126.62	...	...	...	2	...	6	1	...	...	...	...	1	...	...	...	...	...	...	...	...	...	...	...
128.12	...	1	...	2	...	...	...	...	2	...	2	12	...	...	...	...	...	...	...	...	...	...	...
129.76	...	...	...	...	...	...	...	...	1	...	...	...	...	...	...	...	...	...	...	...	...	...	...
130.32	...	...	...	...	...	2	...	...	...	...	...	1	2	...	...	...	...	...	...	...	1	1	...
131.82	...	...	...	...	...	1	...	...	4	...	...	...	1	...	...	...	...	...	...	...	29	...	...
133.32	...	...	...	1	...	1	...	...	...	...	...	...	...	...	1	...	...	...	...	...	1	...	...
134.82	...	...	...	...	...	2	...	...	1	...	5	...	...	...	...	...	...	...	...	...	15	3	...
136.32	...	...	...	1	...	...	...	...	...	...	1	...	...	...	...	...	...	...	...	...	84	...	...
137.82	...	...	...	2	...	...	...	...	...	...	1	2	...	...	...	...	...	...	...	...	2	...	...
139.40	...	...	...	2	...	5	...	...	...	...	...	...	...	...	...	...	...	...	...	...	5	...	...
139.81	...	...	...	1	...	...	...	...	...	...	2	...	...	...	...	...	...	...	...	...	40	...	...
141.31	...	...	...	...	...	...	3	...	...	...	...	...	...	...	...	...	...	...	...	...	17	...	...
142.81	...	...	...	...	...	...	...	...	...	...	6	...	...	...	...	...	...	...	...	...	7	1	...
144.31	...	...	...	1	...	...	1	...	...	...	1	...	...	...	...	...	...	...	...	...	14	...	...
145.81	...	...	...	...	...	2	...	...	...	...	1	...	...	...	...	...	...	...	...	...	117	...	...
147.31	...	...	...	...	...	...	...	...	...	...	...	...	...	...	...	...	...	...	...	...	203	1	...
148.81	...	...	...	1	...	...	...	...	...	...	...	...	...	...	...	...	...	...	...	...	148	2	...
149.09	...	...	...	...	...	8	1	...	...	...	4	1	...	...	...	...	...	...	...	...	10	2	...
149.51	...	...	...	...	...	...	...	...	...	...	4	...	...	...	...	...	...	...	...	...	96	...	...
151.01	...	...	...	1	...	...	2	...	...	...	...	...	...	...	5	...	...	...	...	...	86	1	...
152.51	...	...	...	...	...	...	...	...	...	...	...	...	...	...	10	...	...	...	...	...	169	4	...
154.01	...	...	...	...	...	1	3	...	...	...	...	...	...	...	49	...	...	...	...	...	33	1	...
155.51	...	...	...	...	...	1	1	...	...	...	...	...	...	...	46	...	...	...	...	...	39	3	...
157.01	...	...	...	...	...	...	...	...	...	...	...	...	...	...	7	...	...	...	...	...	171	3	...
158.51	...	...	...	4	...	...	4	...	...	...	4	...	...	...	11	...	...	...	...	3	9	2	...
158.80	...	...	...	...	...	...	5	...	...	...	...	...	...	...	10	...	...	...	...	2	10	...	...
159.20	...	...	...	...	...	...	...	...	1	...	...	...	...	...	69	...	...	...	...	...	7	4	...
160.70	...	...	...	...	...	4	...	...	2	...	...	...	...	...	59	...	...	...	...	...	8	4	...
162.20	...	...	...	1	...	1	4	...	...	...	1	...	...	...	20	...	...	...	...	...	159	...	...
163.70	...	...	...	...	...	1	...	...	2	...	...	...	...	...	22	...	...	...	1	6	15	...	...
165.20	...	...	...	...	...	...	...	...	4	...	...	...	...	...	12	...	...	...	1	1	...	...	...
166.85	...	...	...	...	...	...	...	...	2	...	...	...	...	...	4	...	...	...	3	...	25	...	...
167.31	...	...	...	...	...	1	2	...	2	...	...	...	...	...	13	...	...	...	11	1	123	...	...
168.81	...	...	...	...	...	9	...	...	3	...	...	...	...	...	2	...	...	...	14	...	8	...	...
170.31	...	...	...	...	...	...	...	...	3	...	...	...	3	...	7	...	...	...	7	...	...	...	...
171.81	...	...	...	1	...	4	...	...	3	...	...	...	1	...	3	...	...	...	45	...	13	...	...
173.31	...	...	...	...	...	3	...	...	5	...	...	...	...	...	1	...	...	...	...	2	128	...	...
174.81	...	...	...	...	...	4	...	...	...	...	...	1	1	...	...	...	...	...	...	...	5	...	...
177.05	...	...	4	...	...	1	...	...	...	...	...	1	...	...	1	...	...	...	1	...	...	...	...
178.52	...	...	...	...	...	...	...	...	5	...	...	1	...	...	...	...	...	...	3	...	5	6	...
180.02	...	...	...	...	...	2	...	...	3	...	...	...	...	...	5	...	...	...	14	...	...	4	...
181.52	...	...	...	1	...	2	...	...	...	...	...	...	...	...	...	...	...	...	1	...	...	2	...
183.02	...	...	...	...	...	...	1	...	1	...	...	1	...	...	3	...	...	...	2	...	...	...	...
184.46	...	...	...	...	...	3	...	...	1	...	...	...	...	...	1	...	...	2	1	...	...	...	...
188.22	...	...	...	1	...	2	1	...	6	...	...	...	...	...	1	...	...	...	1	...	...	1	...
189.72	...	...	...	...	...	2	1	...	...	...	...	3	...	...	3	...	...	...	2	...	...	...	...
191.22	...	...	...	...	...	...	...	...	...	...	...	2	...	...	1	...	...	...	...	...	...	...	...
192.28	...	...	...	...	...	...	1	...	1	...	...	...	...	...	...	...	...	...	1	...	...	...	...
197.91	...	...	...	...	...	1	...	...	18	...	...	...	...	...	14	...	...	...	...	...	...	...	...
199.41	...	...	...	...	...	...	...	...	[illegible]	...	...	1	...	...	23	...	3	...	...	...	...	...	...
200.91	...	...	...	...	...	...	...	...	8	1	...	1	...	...	46	...	3	...	...	...	...	...	...
202.41	2	...	...	...	...	7	...	...	1	...	...	...	...	...	...	...	...	...	...	...	...	...	...

TABLE 12.1B

BULIMINA MIDWAYENSIS	BULIMINA OVULA	BULIMINA SEMICOSTATA	BULIMINA SIMPLEX	BULIMINA TRINITATENSIS	BULIMINA VELASCOENSIS	BULIMINELLA BEAUMONTI	BULIMINELLA GRATA	CASSIDULINA SPP.	CHILOSTOMELLA OVOIDEA	CIBICIDES LOBATULUS	CIBICIDES VARIABILIS	CIBICIDOIDES BRADYI	CIBICIDOIDES GRIMSDALEI	CIBICIDOIDES HAVANENSIS	CIBICIDOIDES MUNDULUS	CIBICIDOIDES PERLUCIDUS	CIBICIDOIDES PRAEMUNDULUS	CIBICIDOIDES ROBERTSONIANUS	CIBICIDOIDES TRINCHERASENSIS	EGGERELLA BRADYI	EILOHEDRA WEDDELLENSIS	EOUVIGERINA GRACILIS	EPISTOMINELLA EXIGUA	EPONIDES SP
...	...	...	...	...	...	...	...	...	1	...	...	...	...	...	16	...	...	...	...	2	5	...	20	...
...	...	...	...	...	...	...	...	1	3	...	...	...	...	...	1	...	...	...	...	2	8	...	28	...
...	...	...	...	...	...	...	...	...	...	...	1	...	...	...	13	...	...	...	...	...	...	...	1	...
...	...	...	...	...	...	...	...	1	1	...	1	1	...	...	29	...	...	...	...	2	1	...	6	...
...	...	...	...	...	...	...	...	...	...	1	1	...	...	...	11	...	...	...	...	2	44	...	8	...
...	...	...	...	...	...	...	...	...	...	...	4	...	...	...	...	...	...	...	...	2	5	...	25	...
...	...	...	...	...	...	...	...	...	...	1	2	...	...	...	3	...	...	...	...	...	15	...	60	...
...	...	...	...	...	...	...	...	...	...	...	...	...	...	...	4	...	...	...	...	...	14	...	78	...
...	...	...	...	...	...	...	...	...	...	...	...	...	...	...	4	...	...	...	...	...	8	...	34	...
...	...	...	...	...	...	...	...	1	...	...	...	...	...	...	...	...	...	...	...	2	...	...	16	...
...	...	...	...	...	...	...	...	...	...	...	...	...	...	...	23	...	...	...	...	6	4	...	35	...
...	...	...	...	...	...	...	...	...	...	...	1	...	...	...	19	...	...	...	...	2	4	...	15	...
...	...	...	...	...	...	...	...	...	...	...	...	...	...	...	18	...	...	...	...	...	1	...	7	...
...	...	...	...	...	...	...	...	...	...	1	1	...	...	...	26	...	...	...	...	10	11	...	8	...
...	...	...	...	...	...	...	...	...	...	...	...	...	...	...	25	...	...	...	...	3	6	...	1	...
...	...	...	...	...	...	...	...	...	...	...	...	...	...	...	39	...	...	...	...	6	14	...	10	...
...	...	...	...	...	...	...	...	...	...	...	...	...	...	...	24	...	...	...	...	...	...	...	4	...
...	...	...	...	...	...	...	...	...	...	2	...	...	...	...	31	...	...	...	...	8	13	...	11	...
...	...	...	...	...	...	...	...	...	...	1	...	...	...	...	59	...	...	...	...	4	3	...	10	...
...	...	...	...	...	...	...	...	...	...	...	...	...	...	...	62	...	...	...	...	3	2	...	4	...
...	...	...	...	...	...	...	...	...	...	...	...	...	...	...	29	...	...	...	...	10	19	...	9	...
...	...	...	...	...	...	...	...	...	...	...	...	...	...	...	43	...	...	...	...	3	15	...	6	...
...	...	...	...	...	...	...	...	...	...	...	...	...	...	...	63	...	...	...	...	...	2	...	3	...
...	...	...	...	...	...	...	...	...	...	...	...	...	...	...	4	...	...	...	...	3	11	...	38	...
...	...	...	...	...	...	...	...	...	...	...	1	...	...	...	4	...	...	...	...	...	5	...	50	...
...	...	...	...	...	...	...	...	...	...	...	...	...	...	...	20	2	...	...	...	...	6	...	8	...
...	...	...	...	...	...	...	...	...	...	...	...	...	...	...	41	...	...	...	...	2	21	...	53	...
...	...	...	...	...	...	...	...	...	...	...	...	...	...	...	33	1	...	...	...	...	4	...	41	...
...	...	...	...	...	...	...	...	...	...	...	...	1	...	...	15	...	...	...	...	...	...	...	28	...
...	...	...	...	...	...	...	...	...	2	...	...	...	...	...	40	...	...	...	...	1	2	...	12	...
...	...	...	...	...	...	...	...	...	...	...	1	...	...	...	6	...	...	...	...	6	1	...	6	...
...	...	...	...	...	...	...	...	...	...	...	...	...	...	...	18	2	...	...	...	3	1	...	3	...
...	...	...	...	...	...	...	...	...	...	...	...	...	...	...	14	...	...	...	...	...	3	...	86	...
...	...	...	...	...	...	...	...	...	...	...	...	...	...	...	5	...	...	...	...	...	3	...	66	...
...	...	...	...	...	...	...	...	...	...	...	...	...	1	...	15	1	...	...	...	...	1	...	68	...
...	...	...	...	...	...	...	...	...	...	...	1	...	...	...	24	...	1	...	...	1	11	...	13	...
...	...	...	...	...	...	...	...	...	2	2	...	...	...	...	17	...	10	...	...	...	...	...	31	...
...	...	...	...	...	...	...	...	...	...	...	...	...	...	...	8	...	...	...	...	1	7	...	30	...
...	...	...	...	...	...	...	...	2	...	...	...	...	...	...	27	...	2	...	...	...	...	...	35	...
...	...	...	...	...	...	...	...	...	...	...	...	...	...	...	8	...	6	...	...	...	...	...	11	...
...	...	...	...	...	...	...	...	1	...	...	...	1	...	...	22	...	5	...	...	...	...	...	12	...
...	...	...	...	...	...	...	...	...	...	...	...	3	...	...	...	...	11	...	...	3	...	...	6	...
...	...	...	...	...	...	...	...	...	...	...	...	...	...	...	...	...	5	...	...	5	...	...	5	...
...	...	...	...	...	...	...	...	...	...	...	...	...	...	...	19	...	46	1	...	...	...	...	1	...
...	...	...	...	...	...	...	...	...	...	...	...	...	...	...	11	...	18	1	...	4	...	...	9	...
...	...	...	1	...	...	...	3	...	...	...	...	...	...	...	12	...	20	...	...	...	...	...	...	...
...	...	...	...	...	...	...	...	...	...	...	...	...	...	...	13	...	3	...	...	...	1	...	3	...
...	...	...	...	...	...	...	...	...	...	...	...	...	...	...	26	...	14	...	...	...	...	...	3	...
...	...	...	...	...	...	...	...	...	...	...	...	...	...	...	28	...	1	...	1	...	...	...	2	...
...	...	...	...	...	...	...	...	...	...	...	...	...	...	...	5	...	1	...	2	3	1	...	14	...
...	...	...	...	...	...	...	...	...	...	...	...	...	...	...	...	1	7	...	...	1	...	...	23	...
...	...	...	1	...	...	...	...	...	...	...	...	...	...	...	1	...	7	...	...	...	...	...	11	...
...	...	...	...	...	...	...	...	...	...	...	1	...	...	3	31	...	6	...	...	1	1	...	3	...
...	...	...	...	...	...	...	...	...	...	...	1	...	...	...	12	...	2	...	...	1	...	...	3	...
...	...	...	3	...	...	...	...	...	...	...	...	...	...	...	3	...	6	...	...	...	2	...	6	...
...	...	...	...	...	...	...	...	...	...	...	...	...	...	...	...	2	9	...	4	7	3	...	4	...
...	...	...	...	...	...	...	...	...	...	...	...	...	...	...	...	...	3	...	...	6	...	...	2	...
...	...	...	1	...	...	...	...	...	...	...	...	...	...	...	...	1	4	...	1	1	1	...	1	...
...	...	...	...	...	...	...	...	...	...	...	...	...	...	...	...	2	2	...	3	...	2	...	12	...
...	...	...	1	...	...	...	...	...	...	...	2	...	...	...	...	...	...	...	3	...	1	...	1	...
...	...	...	1	...	...	...	...	...	...	...	1	...	...	...	...	...	20	...	7	1	...	...	1	...
...	...	...	...	...	...	...	...	...	...	...	...	...	...	...	...	...	20	...	3	...	...	...	2	...
...	...	...	...	...	...	...	...	...	...	...	...	...	...	...	...	...	14	...	3	...	...	...	7	...
...	...	...	1	...	...	...	...	...	...	...	...	...	...	...	...	1	9	...	3	...	1	...	15	...
...	...	...	...	...	...	...	...	...	...	...	...	...	...	...	...	...	9	...	1	...	...	...	15	...
...	...	...	3	...	...	...	...	...	...	...	...	...	...	...	...	7	38	...	...	...	1	...	14	...
...	2	...	...	...	...	...	...	...	...	...	...	...	...	...	...	...	63	...	6	...	...	...	6	...
...	...	...	2	...	...	...	...	...	...	...	...	...	...	...	...	1	28	...	...	...	4	...	3	...
...	4	...	23	...	...	...	...	...	...	...	...	...	...	...	...	...	40	...	...	...	...	...	1	...
...	...	...	8	...	...	...	...	...	1	...	...	...	...	...	...	...	12	...	...	...	...	...	5	...
...	...	...	...	...	...	...	...	...	...	...	...	...	...	8	...	...	38	...	...	...	...	8	5	...
...	...	1	1	...	...	...	...	...	...	...	...	...	...	...	...	1	...	...	6	...	...	...	...	...
...	...	...	...	...	...	...	...	...	...	...	...	...	...	...	...	...	28	...	10	...	...	...	...	...
...	2	...	32	...	...	...	...	...	...	...	...	...	...	...	...	...	1	...	...	...	...	...	...	...
...	8	1	3	...	...	...	...	...	...	...	...	...	...	...	...	3	3	...	...	...	...	...	...	...
2	4	...	34	...	...	...	...	...	...	...	...	...	...	...	...	...	...	...	4	...	...	...	...	...
7	8	...	56	...	...	...	...	...	...	...	...	...	...	...	...	...	1	...	...	...	...	...	1	...
...	1	...	53	...	...	...	...	...	...	...	...	...	...	...	...	...	...	...	...	...	...	...	...	...
...	3	1	86	...	...	...	...	...	...	...	...	...	...	...	...	...	7	...	4	...	...	...	8	...
...	...	16	25	...	...	...	...	...	...	...	...	...	...	...	...	...	...	...	...	...	...	...	3	2
...	...	8	15	...	...	...	...	...	...	...	...	...	...	...	...	...	5	...	...	...	...	...	3	...
...	...	3	8	...	...	...	...	...	...	...	...	...	...	1	...	...	2	...	...	...	...	...	1	...
...	...	2	85	...	...	...	...	...	...	...	...	...	...	...	...	...	...	...	...	...	...	1	1	...
...	...	12	74	...	...	...	...	...	...	...	...	...	...	2	...	...	...	...	...	...	...	1	1	...
...	...	21	16	...	...	...	...	...	...	...	...	...	...	...	...	...	44	...	...	...	...	...	...	...
...	1	79	9	...	...	...	...	...	...	...	...	...	...	...	...	...	1	...	...	...	1	...	2	...
...	...	81	38	...	...	...	...	...	...	...	...	...	...	...	...	...	...	...	...	...	1	...	5	...
...	...	39	58	...	...	...	...	...	...	...	...	...	...	...	...	...	5	...	...	...	...	...	1	...
...	2	23	32	...	...	...	...	...	...	...	...	...	...	...	...	...	7	...	...	...	...	...	...	...
...	...	35	20	7	...	...	...	...	...	...	...	...	...	...	...	...	6	...	...	...	1	...	...	...
...	...	31	3	...	1	...	...	...	...	...	...	...	...	...	...	...	2	...	...	...	...	...	...	...
...	4	2	24	...	1	1	...	...	...	...	...	...	...	...	...	...	2	...	...	...	...	...	...	...
...	1	...	62	4	...	1	...	...	...	...	...	...	...	...	...	...	11	...	...	...	...	...	...	...

TABLE 12.1C

EPONIDES TUMIDULUS	FRANCESITA ADVENA	FURSENKOINA BRADYI	FURSENKOINA DAVISI	FURSENKOINA FUSIFORMIS	FURSENKOINA SP., REGULAR	FURSENKOINA SP., THIN	GAUDRYINA LAEVIGATA	GLOBOBULIMINA OVATA	GLOBOCASSIDULINA SUBGLOBOS	GRAVELLINA NARIVAENSIS	GYROIDINOIDES ACUTUS	GYROIDINOIDES DEPRESSUS	GYROIDINOIDES GIRARDANUS	GYROIDINOIDES LAMARCKIANUS	GYROIDINOIDES MEDICEUS	GYROIDINOIDES PLANULATUS	GYROIDINOIDES SOLDANII	HANZAWAIA AMMOPHILA	HERONALLENIA LINGULATA	KARRERIELLA BRADYI	KARRERIELLA SP., TAILED & LO	KARRERIELLA SUBGLABRA	LENTICULINA SPP	MARTINOTTIELLA SPP
...	...	...	...	...	...	...	...	...	...	...	...	...	...	1	...	...	3	...	...	1	...	...	2	...
...	...	...	...	...	...	...	...	...	2	...	2	...	2	...	...	...	4	...	...	1	...	...	1	...
...	...	...	...	...	...	...	...	...	...	...	...	...	...	2	1	...	4	...	...	...	...	...	1	...
...	3	...	11	...	...	...	...	...	...	...	2	...	4	...	...	...	4	...	...	1	...	...	2	...
...	...	...	2	...	...	...	...	...	1	...	3	...	10	...	...	...	14	...	...	...	...	...	...	...
...	...	...	...	...	...	...	...	...	...	...	...	...	...	...	...	...	2	...	...	...	...	...	1	...
...	1	...	...	...	...	...	...	...	1	...	...	...	3	...	1	...	3	...	1	1	...	...	...	...
...	...	...	...	...	...	...	...	...	4	...	8	...	...	2	...	...	10	...	...	...	...	...	...	...
...	...	...	...	...	...	...	...	...	2	...	3	...	2	...	...	...	5	...	...	1	...	...	1	...
...	...	...	...	...	...	...	...	...	1	...	3	...	...	...	...	...	2	...	...	...	...	...	...	...
...	...	...	...	...	...	...	...	...	...	...	4	...	...	...	...	...	7	...	...	...	...	...	...	...
...	1	...	...	...	...	...	...	...	1	...	6	...	...	...	...	...	8	...	...	1	...	...	...	...
...	1	...	...	...	...	...	...	...	1	...	5	...	...	...	6	...	...	...	...	...	...	...	2	...
...	...	1	...	...	...	...	...	...	12	...	5	...	2	...	...	...	11	...	...	1	...	...	2	...
...	...	...	...	...	...	...	...	...	3	...	3	...	3	...	...	2	5	...	...	2	...	...	2	...
...	1	...	...	...	...	...	...	...	10	...	13	...	...	...	13	...	5	...	...	2	...	...	1	...
...	1	...	...	...	...	...	...	...	3	...	8	...	...	...	2	...	1	...	...	3	...	...	...	...
...	...	...	...	...	...	...	...	...	2	...	6	...	...	...	...	...	3	...	...	...	...	...	...	...
...	...	...	...	...	...	...	...	...	11	...	...	...	...	...	1	...	...	...	...	...	...	...	1	...
...	3	...	...	...	...	...	...	...	4	...	2	...	4	...	...	...	3	...	...	...	...	...	...	...
...	3	...	...	...	...	...	...	...	23	1	2	...	...	...	3	...	4	...	1	...	...	...	...	...
...	...	...	...	...	...	...	...	...	22	...	...	...	1	...	4	...	3	...	1	...	...	...	...	2
1	...	...	...	1	...	...	...	...	7	...	...	...	5	...	8	...	1	...	...	...	...	2	1	...
...	...	...	...	...	...	...	...	...	7	...	1	...	...	...	2	...	...	...	1	...	...	...	...	...
...	...	1	...	...	...	...	...	...	7	...	1	...	...	...	...	...	1	...	...	...	...	...	1	...
...	...	...	...	...	...	...	...	...	18	...	...	...	1	1	4	...	...	...	1	...	...	...	...	...
...	1	...	...	...	...	...	...	...	6	...	2	...	...	2	...	...	9	...	1	...	...	3	...	...
...	3	...	...	...	...	...	...	...	19	...	1	...	1	...	...	...	7	...	...	...	...	...	2	...
...	...	...	...	...	...	...	...	...	5	...	3	...	...	...	...	...	5	...	...	...	...	...	...	...
...	...	2	...	5	...	...	...	...	3	...	2	...	...	3	11	...	...	...	...	1	...	1	...	...
...	1	...	...	...	...	...	...	...	30	...	1	...	...	1	3	...	...	...	...	...	...	...	...	...
...	2	...	...	...	...	...	...	...	3	...	...	...	...	1	1	...	...	...	...	...	...	2	2	...
...	...	...	...	3	...	...	...	...	9	...	1	...	...	1	2	...	3	...	...	...	...	...	...	...
...	...	1	...	...	...	...	...	...	2	...	1	...	...	3	5	...	...	...	...	...	...	...	1	...
...	...	...	...	...	...	...	...	...	49	...	5	...	...	4	2	6	...	...	...	1	...	...	...	...
...	...	...	...	7	...	...	...	...	53	...	2	...	7	1	7	...	2	...	...	3	...	...	1	...
1	...	...	...	1	...	...	...	...	27	...	2	...	...	7	3	...	...	...	...	1	...	1	...	...
...	...	...	...	4	...	...	...	...	39	...	2	...	...	3	6	2	...	...	...	...	...	...	...	...
...	...	...	...	6	...	...	...	...	15	...	1	...	2	...	...	...	...	...	...	...	...	...	...	...
...	...	...	...	6	...	...	...	...	1	...	2	...	1	...	3	...	...	...	...	...	...	2	1	...
...	...	1	...	...	...	...	...	...	4	2	...	...	...	6	4	...	...	...	1	2	...	...	1	...
...	...	5	...	4	...	...	...	...	6	...	...	...	...	15	9	1	...	...	...	2	...	...	1	...
...	...	...	...	...	...	1	...	...	15	...	...	...	...	...	10	...	...	...	1	...	...	...	1	...
...	...	...	...	1	...	...	...	...	2	1	...	...	3	3	5	...	3	...	...	...	...	5	...	...
...	...	2	...	4	...	...	...	...	8	...	1	...	5	1	6	...	...	...	...	...	...	3	2	...
...	...	...	...	...	...	...	...	...	1	...	1	...	...	1	2	...	...	...	...	1	...	1	7	...
...	...	...	...	2	...	...	...	...	13	...	2	...	1	...	8	...	...	...	...	...	...	1	...	...
...	...	1	...	...	...	...	...	...	12	...	2	...	...	1	5	...	...	...	...	3	...	...	...	...
...	...	...	...	4	...	...	...	...	2	...	...	...	...	1	5	...	...	...	1	...	...	2	...	1
...	...	4	4	...	...	...	...	...	13	...	1	...	...	8	3	...	...	...	...	2	...	...	...	...
...	...	2	...	1	...	...	...	1	56	...	2	...	1	4	3	...	...	...	...	...	...	4	2	...
...	...	...	...	3	...	2	...	...	24	...	1	...	...	5	...	...	...	...	...	...	...	...	...	...
...	...	...	...	10	...	...	...	...	6	...	...	...	...	...	...	1	6	...	1	...	...	2	1	...
...	...	...	...	17	...	...	...	...	34	...	...	...	1	4	3	...	...	...	...	...	...	...	...	...
...	...	...	...	7	...	...	...	...	64	...	11	...	2	...	3	...	...	...	1	...	...	2	1	...
...	...	4	...	6	...	...	...	...	...	...	...	...	1	...	...	...	7	2	1	...	...	4	1	...
...	...	6	...	5	...	1	...	...	12	...	...	...	2	...	1	2	4	...	...	...	...	1	...	...
...	...	12	...	...	...	1	...	...	1	...	1	...	1	...	1	...	...	...	...	...	...	1	...	...
...	...	11	...	4	...	...	...	...	19	...	...	...	6	...	...	1	...	...	...	...	...	1	...	...
...	...	6	7	...	...	...	...	...	16	...	...	...	1	...	1	...	1	...	...	...	...	1	...	...
...	...	...	...	4	...	...	...	...	3	...	...	...	...	...	...	...	3	...	...	...	...	1	...	...
...	...	...	...	2	3	1	...	...	43	...	1	...	...	...	1	...	...	...	...	...	...	...	1	...
...	...	...	...	1	...	...	...	...	93	...	...	...	...	...	...	11	1	...	...	...	...	...	2	...
...	...	1	...	...	1	...	...	...	63	1	6	...	...	...	...	6	1	...	1	...	...	...	...	...
...	...	...	...	...	...	...	...	...	1	...	...	...	...	...	...	3	3	...	...	...	...	1	2	...
...	...	5	5	...	...	...	...	...	13	...	1	...	...	...	...	9	1	...	...	...	1	1	2	...
...	...	4	...	3	...	...	...	...	11	...	...	...	3	...	...	6	...	1	...	...	1	...	...	...
...	...	11	...	3	...	...	...	...	...	...	...	...	5	...	...	2	3	1	...	...	...	1	2	...
...	...	12	...	4	...	...	...	...	2	1	...	...	...	...	1	...	7	...	...	...	3	3	10	...
1	...	1	...	...	...	...	...	...	4	...	...	...	1	...	...	1	...	...	...	...	...	...	3	...
...	...	...	...	...	...	...	...	...	2	...	...	...	5	...	...	4	...	17	...	...	...	4	16	...
...	...	2	...	1	...	...	...	...	3	...	...	...	...	...	...	...	1	...	...	...	2	3	4	...
...	...	4	...	8	...	...	...	...	2	...	...	...	...	...	...	2	1	1	...	...	...	1	2	...
...	...	...	...	1	...	...	1	...	...	...	...	...	...	...	...	...	...	...	1	...	...	...	9	...
...	...	...	...	5	...	...	...	...	...	...	...	...	2	...	2	...	...	...	1	...	1	...	7	...
...	...	...	...	2	...	...	...	...	1	...	...	...	1	1	...	2	...	...	...	...	...	...	6	...
...	...	5	...	...	...	...	...	...	1	...	...	...	2	...	...	...	...	...	1	...	...	...	4	...
...	...	...	...	3	...	...	...	...	...	...	...	...	...	...	...	...	...	...	...	...	1	...	2	...
...	...	...	...	1	...	...	...	...	...	...	...	...	...	...	...	...	1	...	1	...	...	...	3	...
...	...	2	...	1	...	...	...	...	2	...	...	...	5	...	...	2	3	...	...	1	4	3	8	...
...	...	...	...	...	...	...	...	...	2	...	...	...	9	...	...	2	...	...	1	...	2	2	4	...
...	...	5	...	2	...	...	...	...	2	...	1	...	2	1	...	1	...	9	...	...	2	1	...	...
...	...	1	...	2	...	...	...	...	9	...	...	...	...	...	...	6	...	1	1	...	...	...	...	...
...	...	1	...	...	...	...	...	...	10	...	...	...	...	...	6	3	...	...	1	...	...	...	...	...
...	...	3	...	...	...	...	...	...	...	...	...	...	...	...	...	2	1	...	...	...	...	...	3	...
...	...	5	...	1	...	...	...	...	1	1	...	...	...	...	...	4	4	...	1	...	...	...	8	...
...	...	...	...	5	...	...	...	...	2	...	...	...	4	...	...	2	8	3	...	...	2	...	6	...
...	...	...	...	7	...	...	2	...	...	...	...	...	...	...	...	...	2	...	...	...	...	5	13	...
...	...	...	...	...	4	...	...	...	...	...	...	...	...	...	...	7	...	...	2	...	...	...	14	...
...	...	...	...	...	7	...	...	...	...	...	1	...	...	...	...	5	...	...	...	...	...	1	8	...
...	...	...	...	...	4	...	...	...	...	...	...	...	...	...	1	5	...	...	...	...	...	...	8	...
...	...	...	...	...	...	5	...	...	...	1	...	...	...	...	...	11	...	...	1	...	...	...	25	...
...	...	...	...	...	4	...	...	...	...	1	...	1	...	...	...	1	...	...	...	...	...	...	13	...

TABLE 12.1D

NONION HAVANENSE	NONIONELLA LABRADORICA	NONIONELLA ROBUSTA	NUTTALLIDES SP., HIGH	NUTTALLIDES TRUEMPYI	NUTTALLIDES UMBONIFERA	OPHTHALMIDIUM PUSILLUM	ORIDORSALIS TENER	ORIDORSALIS UMBONATUS	OSANGULARIA INTERRUPTA	PLEUROSTOMELLA SPP.	PLEUROSTOMELLIDS, OTHER	POLYMORPHINIDS	PSEUDOPARRELLA SP.	PULLENIA BULLOIDES	PULLENIA QUADRILOBA	PULLENIA QUINQUELOBA	PULLENIA SALISBURYI	PULLENIA SUBCARINATA	PYRAMIDINA RUDITA	QUADRIMORPHINA ALLOMORPHI	QUINQUELOCULINA SPP.	REUSSELLA TORTUOSA	SIGMOILINA TENUIS	SIPHOGENERINOIDES BREVISPI	SIPHOGENERINOIDES ELEGANTA
...	...	...	...	...	109	...	6	2	...	2	...	...	...	1	...	3	2	...	...	...	...	47	...	...	...
...	...	...	...	...	178	...	1	...	...	1	...	...	...	2	...	...	...	...	...	...	...	18	...	...	...
...	...	...	...	...	131	...	3	4	...	6	...	...	...	1	2	...	1	...	...	...	1	1	...	...	...
...	3	...	...	...	147	...	6	1	...	4	...	...	...	3	3	2	...	...	...	...	3	7	...	...	...
...	...	...	...	...	103	...	22	12	...	2	...	...	1	2	...	...	2	...	...	...	...	5	...	...	...
...	...	...	...	...	159	...	1	1	...	...	...	...	...	5	...	5	...	...	...	...	1	11	...	...	...
...	...	...	...	...	104	...	...	...	...	1	2	...	1	2	5	...	3	...	...	...	...	1	...	...	...
1	...	...	...	...	32	...	3	6	...	6	...	...	...	2	2	...	2	...	...	...	...	1	...	...	...
...	...	...	...	...	130	...	...	2	...	2	...	...	...	2	2	...	...	...	...	...	...	...	...	...	...
...	...	...	...	...	99	...	1	...	...	4	...	...	...	...	...	...	1	...	...	...	...	...	...	...	...
...	...	...	...	...	98	...	2	8	...	2	...	...	...	...	...	3	...	1	...	...	...	...	...	...	...
...	...	...	...	...	79	...	7	10	...	2	...	...	...	...	...	2	...	...	...	...	1	...	...	...	...
2	...	...	...	...	60	...	2	16	...	9	...	...	...	...	...	2	3	4	...	...	...	...	...	...	...
...	1	...	...	...	67	...	5	12	...	3	1	...	1	...	3	...	3	...	...	...	...	...	...	...	...
...	...	...	...	...	138	...	4	18	...	1	...	...	...	5	...	3	...	3	...	...	...	...	...	...	...
...	...	...	...	...	38	...	9	15	...	4	...	...	5	...	...	2	5	...	...	...	2	4	...	...	...
...	...	...	...	...	101	...	10	4	...	...	...	...	...	...	1	2	...	1	...	...	...	...	...	...	...
...	...	...	...	...	124	...	...	9	...	1	...	...	1	...	3	1	3	...	...	...	...	...	...	...	...
...	1	...	...	...	49	...	...	30	...	6	1	...	...	...	1	...	...	...	...	...	...	...	...	...	...
...	...	...	...	...	15	...	...	10	...	13	1	...	...	...	...	5	2	3	...	...	1	...	...	...	...
...	...	...	...	...	64	...	...	2	...	4	...	...	8	...	5	...	6	1	...	...	2	...	...	...	...
...	...	...	...	...	100	...	2	4	...	8	1	...	...	...	1	1	4	...	...	...	2	5	...	...	...
...	...	...	...	...	49	...	14	5	...	6	1	...	...	...	1	2	2	...	...	...	...	...	...	...	...
...	...	...	...	...	150	...	...	5	...	1	...	...	2	...	...	...	...	...	...	...	...	...	...	...	...
...	...	...	...	...	93	...	2	3	...	6	...	...	6	3	1	...	2	...	...	...	2	4	...	...	...
...	...	...	...	...	77	...	8	5	...	5	...	...	...	2	...	4	2	...	...	...	1	16	...	...	...
...	...	...	...	...	8	...	...	...	...	5	...	...	...	2	2	3	...	...	...	...	1	21	...	...	...
...	...	...	...	...	110	...	1	...	...	3	...	...	2	1	...	...	...	...	...	...	...	...	...	...	...
...	...	...	...	...	105	...	...	2	...	2	...	...	1	10	9	5	4	...	...	...	...	...	...	...	...
...	...	...	...	...	45	...	...	7	...	4	...	1	...	...	7	5	...	...	...	...	...	...	...	...	...
...	...	...	...	...	81	...	...	1	...	4	...	1	2	3	1	...	1	...	...	...	...	...	...	...	...
2	...	...	...	...	77	...	...	...	...	3	...	...	...	...	...	5	...	...	...	...	...	...	...	...	...
2	...	...	...	...	74	...	...	1	...	1	...	...	2	...	...	...	5	...	...	...	...	...	...	...	...
...	...	...	...	...	29	...	1	7	...	2	...	...	...	3	...	...	2	...	...	...	...	1	...	...	...
...	...	...	...	...	84	...	...	4	...	1	...	...	...	...	...	1	4	...	...	...	1	...	...	...	...
2	1	...	...	...	16	...	...	4	...	2	2	...	5	14	...	...	2	...	...	...	1	...	...	...	...
3	2	...	...	...	27	...	...	13	...	4	3	...	...	11	...	2	2	1	...	...	...	...	...	...	...
...	2	...	...	...	14	...	...	10	...	2	...	1	...	...	...	...	1	...	...	...	...	...	...	...	...
2	...	...	...	...	32	...	...	14	1	7	1	1	...	...	1	3	2	...	...	...	...	...	...	...	...
3	...	...	...	...	8	...	...	1	1	5	...	3	...	...	...	...	3	3	...	...	...	...	...	...	...
1	...	...	...	...	79	...	...	11	...	10	2	4	...	...	...	...	3	...	...	...	...	...	...	...	...
...	1	...	...	...	75	...	...	5	...	10	1	...	...	6	...	3	6	10	...	...	...	...	...	...	...
...	...	...	...	...	49	...	...	17	...	25	1	1	...	4	...	3	...	...	...	...	1	...	...	...	...
...	...	...	...	...	25	...	...	8	1	19	...	1	...	...	...	2	2	6	...	...	...	...	...	...	...
1	...	...	...	...	53	...	...	25	...	5	2	3	...	2	...	2	...	4	...	...	...	...	...	...	...
1	...	...	...	...	77	...	...	8	7	4	6	2	...	...	...	...	...	...	...	...	1	...	...	...	...
2	...	...	...	...	57	2	...	6	...	11	...	2	...	4	...	5	3	1	...	...	...	...	...	...	...
4	...	3	...	...	43	5	...	5	...	6	5	1	...	9	...	...	3	...	...	...	3	...	...	...	...
...	...	...	...	...	70	4	...	3	...	3	...	...	...	1	...	...	2	4	...	...	1	...	1	...	...
1	...	...	...	2	8	2	...	22	...	1	4	...	...	6	3	2	3	...	...	...	1	...	...	...	...
1	3	...	...	...	7	...	...	16	...	3	...	3	...	2	1	10	4	2	...	...	...	...	...	...	...
...	1	...	...	...	16	3	...	5	...	7	4	1	...	1	3	2	2	...	...	...	1	...	...	...	...
...	1	...	...	...	8	6	...	10	...	8	...	...	...	2	...	3	...	...	...	...	1	...	...	...	...
2	...	...	...	...	8	1	1	20	...	11	1	...	...	1	3	4	4	...	...	...	...	1	...	...	...
4	...	...	...	...	52	4	...	13	...	5	1	...	...	2	...	2	...	...	...	...	...	...	...	...	...
11	...	...	...	...	10	8	...	25	...	3	...	1	...	2	...	3	1	...	...	...	...	...	2	...	...
7	...	...	...	...	11	3	...	4	...	...	...	3	...	...	2	2	2	3	...	...	1	...	...	...	...
5	...	...	...	...	15	...	...	11	...	5	2	...	...	...	2	...	...	...	...	...	...	...	...	...	...
...	...	...	...	...	13	4	...	15	...	6	2	1	...	...	3	4	...	...	...	...	...	...	...	...	...
5	2	...	...	1	23	1	...	15	...	9	...	3	1	3	3	...	...	2	...	...	...	...	...	...	...
6	1	...	...	1	16	1	...	13	...	6	...	1	...	5	...	3	1	1	...	...	...	...	...	...	...
...	...	...	...	2	19	8	...	16	...	1	1	1	...	...	2	...	...	...	...	...	...	...	...	...	...
1	...	...	...	6	36	...	...	3	...	1	...	...	...	...	...	1	...	3	...	...	...	...	...	...	...
...	...	...	...	7	19	1	...	4	...	4	1	1	...	2	...	2	...	1	...	...	...	...	...	...	...
1	...	...	...	1	18	1	...	1	...	3	1	...	...	1	...	2	1	1	...	...	...	...	...	...	...
1	1	...	...	4	46	...	...	11	...	5	...	...	...	7	1	...	...	2	...	...	...	...	...	...	...
...	...	...	...	16	11	...	...	9	2	6	1	...	...	1	...	1	3	6	...	...	...	...	...	...	...
2	...	...	...	...	30	...	...	4	2	4	1	1	...	6	...	1	...	...	...	...	...	...	...	...	...
4	...	3	...	1	29	...	...	30	1	...	3	...	...	6	2	...	...	2	...	...	...	...	...	...	1
3	...	...	...	...	32	...	...	4	5	1	...	...	...	...	...	...	...	...	...	...	...	...	...	...	...
11	...	3	...	3	10	...	...	30	2	6	3	1	...	7	2	4	...	...	...	...	...	...	...	...	4
13	...	...	...	60	6	...	...	14	...	14	3	2	...	...	...	1	2	...	...	...	...	...	...	...	3
14	...	...	...	94	4	...	...	13	...	4	2	...	...	1	...	1	4	...	...	...	...	...	...	...	...
16	...	...	...	3	18	...	...	8	...	14	1	1	...	...	...	1	2	2	...	...	...	...	...	...	...
30	...	1	...	86	4	...	...	7	1	10	...	1	...	8	...	3	...	...	...	...	...	...	...	...	1
16	...	...	...	53	...	...	...	9	...	13	...	1	...	1	...	...	4	8	...	...	...	...	...	...	2
16	...	...	1	74	8	...	...	17	...	11	...	1	...	3	2	...	3	4	...	...	...	...	...	...	11
5	...	...	...	13	1	...	...	14	...	6	1	2	...	1	...	...	2	2	...	...	...	...	...	...	1
13	...	...	...	33	13	...	...	6	...	12	...	2	...	...	...	3	4	...	...	...	...	...	...	...	1
2	...	...	...	17	1	...	...	10	...	3	...	2	...	1	...	2	6	7	...	...	...	...	...	...	19
11	...	...	...	...	13	...	...	36	...	13	3	1	...	1	...	...	5	17	...	...	...	...	...	...	6
6	...	1	...	...	11	...	...	19	...	14	...	...	...	...	...	...	3	8	1	...	...	...	...	...	26
1	...	3	...	6	15	...	...	5	...	13	1	...	...	2	...	...	2	...	...	1	...	...	...	...	40
3	...	3	...	20	3	...	...	5	...	7	2	...	...	...	...	...	1	8	...	...	...	...	...	...	41
1	...	3	...	2	5	...	...	17	6	4	3	1	...	...	...	1	...	3	...	...	...	...	...	...	13
6	...	...	...	1	23	...	...	7	...	14	2	2	...	1	...	5	2	5	3	...	...	...	...	...	41
2	...	2	...	1	5	...	...	21	...	16	2	3	...	...	...	1	...	...	...	...	...	...	...	...	29
4	...	...	...	...	6	...	...	2	...	11	1	3	...	1	...	3	...	...	...	...	...	...	...	...	37
11	...	...	...	1	12	...	...	13	...	18	2	...	...	3	2	3	7	...	...	...	...	...	...	...	15
15	...	...	...	13	8	...	...	25	...	14	...	...	...	3	...	1	...	3	...	...	...	...	...	...	38
6	...	...	...	52	4	...	...	12	...	7	1	2	...	1	...	1	3	5	...	1	...	...	...	...	14
7	...	...	...	19	5	...	...	24	...	8	...	1	...	4	2	2	...	10	...	...	...	...	...	...	31
11	...	...	...	4	12	...	...	25	...	3	1	...	...	2	...	1	6	6	...	...	...	...	...	6	12

TABLE 12.1E

SIPHOTEXTULARIA SPP.	SPIROPLECTAMMINA SPECTABIL	STAINFORTHIA COMPLANATA	STILOSTOMELLA ACULEATA	STILOSTOMELLA ANNULIFERA	STILOSTOMELLA CONSOBRINA	STILOSTOMELLA SP., SMOOTH	STILOSTOMELLA SUBSPINOSA	TAPPANINA SELMENSIS	TEXTULARIA ALABAMENSIS	TEXTULARIA SP.	TRITAXIA GLOBULIFERA	TROCHAMMINA GLOBIGERINIFOR	TURRILINA ALSATICA	TURRILINA BREVISPIRA	UNILOCULAR SPECIES	UNISERIAL LAGENIDS	UVIGERINA GRACILIFORMIS	UVIGERINA PEREGRINA GROUP	VALVULINERIA CAMERATA	VALVULINERIA LAEVIGATA	VULVULINA ADVENA	VULVULINA MEXICANA	VULVULINA SPINOSA	
...	...	...	5	7	...	...	24	...	...	...	...	...	1	...	2	4	22	4	...	...	...	...	...	068.92
...	...	...	4	2	1	...	38	...	...	...	...	...	...	...	2	3	1	...	...	...	...	...	...	070.42
...	...	...	1	...	...	...	3	...	...	...	...	...	1	...	2	2	...	...	...	...	...	...	...	071.93
...	...	3	2	2	...	...	15	...	...	...	...	...	2	...	...	9	1	...	...	...	...	...	...	072.51
...	...	...	...	...	...	...	27	...	...	...	...	2	...	...	3	1	3	1	...	...	...	...	...	074.01
...	...	...	...	...	...	...	17	...	...	...	...	...	...	...	1	1	54	...	...	...	...	...	...	075.51
...	...	...	...	4	...	...	3	...	...	...	...	...	...	...	4	1	23	52	...	...	...	...	...	077.01
...	...	...	1	96	...	...	2	...	...	...	...	2	2	...	3	4	4	...	...	...	...	...	...	078.51
...	...	...	...	22	...	...	4	...	...	...	...	...	...	...	1	...	56	10	...	...	...	...	...	080.01
...	...	...	2	8	...	...	22	...	...	...	...	...	1	...	...	4	18	...	...	...	...	...	...	080.66
...	...	...	...	88	...	...	3	...	...	...	...	...	...	...	2	3	7	1	...	...	...	...	...	082.11
...	...	...	2	76	6	...	20	...	...	...	...	...	...	...	8	4	6	2	...	...	...	...	...	083.61
...	...	...	1	21	61	...	19	...	...	...	...	...	...	...	6	6	...	4	...	...	...	...	...	085.11
...	...	...	...	9	62	...	22	...	...	...	...	...	1	...	13	2	...	2	...	...	...	...	...	086.61
...	...	...	...	32	32	...	1	...	...	...	...	...	...	...	8	3	...	...	...	...	...	...	...	088.11
...	...	...	...	78	...	...	2	...	...	...	...	...	...	...	9	7	...	...	...	...	...	...	...	089.61
...	1	...	2	28	...	...	34	...	...	...	...	1	2	...	...	4	...	19	...	...	...	...	...	070.49
...	...	...	...	6	10	...	6	...	...	...	...	...	4	...	4	1	...	53	...	...	...	...	...	091.72
...	...	...	...	34	65	...	6	...	...	...	...	...	...	...	7	5	...	...	...	...	...	...	...	093.22
...	...	...	...	87	48	...	3	...	...	...	...	...	...	...	11	6	...	...	...	...	...	...	4	094.72
...	...	...	...	60	...	...	33	...	...	...	...	...	...	...	6	4	...	...	...	...	...	...	...	096.22
...	...	...	1	34	13	...	21	...	...	...	...	...	...	...	5	2	3	...	...	...	...	...	...	097.72
...	...	...	...	10	1	...	35	...	...	...	...	...	1	...	4	6	1	...	...	...	...	...	...	098.96
...	...	...	...	...	15	...	3	...	...	...	...	...	32	...	2	1	8	...	...	...	...	...	1	101.41
...	...	...	4	36	6	...	38	...	...	...	...	...	17	...	1	1	7	1	1	...	...	...	...	102.91
...	...	...	1	11	14	...	44	...	...	...	...	...	3	...	6	6	21	4	...	...	...	...	...	104.41
...	...	...	8	19	34	...	45	...	...	...	...	...	...	...	3	6	1	...	...	...	...	...	...	105.91
...	...	...	...	4	43	...	10	...	...	...	...	...	...	...	...	2	...	3	...	...	...	...	...	107.41
...	...	...	6	29	40	...	9	...	...	...	...	...	3	...	...	5	11	15	...	...	...	...	...	108.91
...	...	...	3	1	22	...	2	...	...	...	...	...	28	...	2	3	...	...	2	...	...	...	...	110.60
...	...	...	...	6	71	...	...	...	...	...	...	...	53	...	1	2	...	...	...	...	...	...	...	111.01
...	...	...	33	...	28	...	...	...	...	...	...	...	118	...	3	...	...	...	...	...	...	...	...	112.51
...	...	...	1	...	43	...	...	...	...	...	...	...	49	...	11	6	...	...	...	...	...	...	...	114.01
...	...	...	2	68	...	...	...	...	...	...	...	...	93	...	9	2	...	...	1	...	...	...	...	115.51
...	...	...	3	1	34	...	1	...	...	...	...	...	22	...	5	1	...	...	...	...	...	...	...	117.01
...	...	...	...	...	26	...	28	...	...	...	...	...	44	...	4	1	...	...	...	...	...	...	...	118.51
...	...	...	3	...	3	...	103	...	...	...	...	...	23	...	4	4	...	...	...	...	...	...	...	120.20
...	...	...	...	...	30	...	88	...	...	...	...	...	65	...	2	1	...	...	...	...	...	...	1	120.64
...	...	...	...	21	51	...	20	...	...	...	...	...	78	...	6	1	...	1	...	...	...	...	2	122.12
...	...	...	4	...	25	...	168	...	...	...	...	...	11	...	4	7	...	1	...	1	...	...	4	123.62
...	...	...	8	...	2	...	88	...	...	...	...	...	7	...	5	8	...	3	1	...	...	...	1	125.12
...	...	...	65	4	4	...	35	...	...	...	...	...	4	...	4	6	...	8	1	...	...	...	...	126.62
...	...	...	...	4	9	...	96	...	...	2	...	...	4	...	16	11	...	...	...	...	...	...	...	128.12
...	...	...	29	6	13	...	69	...	...	...	...	...	1	...	7	22	...	1	...	...	...	...	1	129.76
...	...	...	46	5	6	...	48	...	...	...	...	...	...	...	8	6	1	1	...	...	...	...	...	130.32
...	...	...	1	34	15	...	39	...	...	...	...	...	...	...	3	11	4	14	...	...	1	...	...	131.82
...	...	...	16	7	54	...	32	...	...	...	...	...	...	...	7	8	3	26	...	...	...	...	1	133.32
...	1	...	8	4	2	...	59	...	...	...	...	...	6	...	13	41	...	...	...	...	...	...	...	134.82
...	...	...	6	...	5	...	47	...	...	...	...	...	2	...	4	14	...	5	...	...	...	...	...	136.32
...	1	...	63	...	...	...	97	...	...	...	...	...	...	...	12	12	...	7	...	...	...	...	...	137.82
...	...	...	17	2	17	...	72	...	...	...	...	...	6	...	9	13	...	1	...	...	...	...	...	139.40
...	...	...	91	...	11	...	32	...	...	...	...	...	7	...	6	15	...	9	...	...	...	...	...	139.81
...	...	...	77	...	5	...	63	...	...	...	...	...	...	...	8	19	...	3	...	...	...	...	...	141.31
...	...	...	72	9	4	...	33	...	...	...	...	...	...	...	12	13	...	35	...	...	...	...	...	142.81
...	...	...	47	12	...	...	10	...	...	...	...	...	...	...	13	9	...	...	...	...	...	...	...	144.31
...	...	...	38	...	3	...	3	...	...	...	...	...	...	...	12	8	...	...	...	...	...	...	...	145.81
...	...	...	3	...	7	...	5	...	...	...	...	...	...	...	11	11	...	...	...	...	...	...	...	147.31
...	...	...	16	...	...	...	47	...	...	...	...	...	...	...	6	11	...	...	...	...	...	...	...	148.81
...	...	...	2	...	7	...	141	...	...	...	...	...	...	...	4	9	...	7	...	...	...	...	...	149.09
...	...	...	1	2	13	...	58	...	...	...	...	...	...	...	6	13	...	16	1	...	...	...	...	149.51
...	...	...	6	...	3	...	95	...	...	...	...	...	...	...	3	9	...	3	1	...	...	...	...	151.01
...	...	...	...	6	3	...	45	...	...	...	...	...	...	...	4	8	...	...	...	...	...	...	...	152.51
...	...	...	22	6	...	...	8	...	...	...	...	...	...	...	6	3	...	...	...	...	...	...	1	154.01
...	...	...	34	8	2	...	11	...	...	...	...	...	...	...	9	5	...	...	...	...	...	...	1	155.51
...	...	...	...	8	3	...	35	...	...	...	...	...	...	...	4	10	...	...	...	...	...	...	...	157.01
...	...	...	4	1	...	...	68	...	...	...	...	...	...	...	9	9	...	...	...	...	...	...	...	158.51
...	...	...	6	...	...	...	111	...	...	...	...	...	...	...	3	12	...	...	...	...	...	...	...	158.80
...	...	...	5	...	12	...	100	...	...	...	...	...	...	...	4	6	...	...	2	...	...	...	...	159.20
...	...	...	1	...	3	...	45	...	...	...	...	...	...	...	4	2	...	...	1	...	...	...	...	160.70
...	...	...	8	...	...	...	27	...	...	...	...	...	...	...	7	5	...	...	...	...	...	...	2	162.20
...	...	...	8	...	7	...	32	...	...	...	...	...	...	...	2	18	...	6	...	...	...	...	1	163.70
...	...	...	...	...	34	...	73	...	...	1	...	...	...	...	15	8	...	...	...	...	...	...	...	165.20
...	...	...	...	4	...	...	64	...	...	...	...	...	...	...	6	10	...	...	...	...	...	...	...	166.85
...	...	...	7	5	11	...	26	...	...	...	...	...	...	...	8	10	...	...	...	...	...	...	1	167.31
...	...	...	7	2	14	...	38	...	...	...	...	...	...	2	9	12	...	...	...	...	1	...	...	168.81
...	...	...	6	2	...	...	115	...	...	...	...	...	...	...	7	5	...	...	...	...	...	...	...	170.31
...	...	...	4	5	11	...	135	...	...	...	...	...	...	1	12	5	...	...	...	...	...	...	...	171.81
...	...	...	58	12	10	...	3	...	...	...	...	...	...	...	4	17	...	...	...	...	...	...	...	173.31
...	...	...	1	3	6	...	68	...	...	...	...	...	...	...	10	20	...	...	...	...	...	...	...	174.81
...	...	...	21	...	41	...	49	...	...	...	...	...	...	3	3	16	...	...	...	...	...	...	...	177.05
...	...	...	11	...	28	...	53	...	...	...	...	...	...	2	15	20	...	...	...	...	...	...	...	178.52
...	...	...	14	...	76	...	26	...	...	...	...	...	...	2	12	26	...	...	...	...	...	...	...	180.02
...	...	...	8	...	9	...	63	...	...	...	...	...	...	2	12	20	...	...	...	...	...	...	...	181.52
...	...	...	9	...	12	...	58	2	...	...	...	...	...	2	8	8	...	...	...	...	...	...	...	183.02
...	1	...	...	...	68	...	53	...	...	...	...	...	...	...	...	22	...	...	...	...	...	...	...	184.46
...	...	...	4	...	15	...	24	...	...	...	...	...	...	3	14	17	...	...	...	...	...	...	1	188.22
...	...	...	1	...	10	...	14	...	...	...	9	...	...	...	9	23	...	...	...	...	...	...	...	189.72
...	...	...	7	...	8	...	51	...	...	...	...	...	...	...	14	25	...	...	...	...	...	11	...	191.22
...	...	...	...	...	...	...	32	...	...	...	...	...	...	...	4	12	...	...	...	...	...	...	...	192.28
...	8	...	...	...	...	47	...	...	1	...	...	...	...	...	2	15	...	...	...	...	...	...	...	197.91
3	4	...	2	...	...	54	4	1	1	...	...	...	...	...	2	27	...	...	...	...	...	...	...	199.41
...	...	...	2	18	...	30	...	1	3	...	...	...	...	2	3	11	...	...	...	...	...	...	...	200.91
...	...	...	8	...	8	30	...	20	1	...	...	...	...	5	3	15	...	...	...	...	...	...	...	202.41

13. LATE PALEOGENE CALCAREOUS NANNOPLANKTON EVOLUTION: A TALE OF CLIMATIC DETERIORATION

by Marie-Pierre Aubry

ABSTRACT

Diversity changes and patterns of diachrony and provincialism exhibited by the late Paleogene calcareous nannoplankton are analyzed. The latest Eocene (~37 to 36.3 Ma), often regarded as a time of major extinctions, witnessed only a weak change in diversity compared with the profound turnover that occurred near the middle/late Eocene boundary (~40 Ma) and the global extinctions in the early Oligocene (~35.2 to 34. 5 Ma). The intensity of changes in the calcareous nannoplankton varied with latitude, the major change at high latitudes occurring near the early/middle Eocene boundary (~52 Ma). There is an excellent correlation between the timing of changes in the calcareous nannoplankton and the timing of cooling events as inferred from isotopic studies. There is also a remarkable parallelism in the middle Eocene to early Oligocene evolution of the calcareous nannoplankton and planktonic foraminifera. This is used to show that in addition to decreasing temperatures, eutrophication was a determinant agent in evolutionary turnovers and extinctions in the calcareous nannoplankton whose late Paleogene evolution reflects expansions and contractions of the TRC (trophic resources continuum).

INTRODUCTION

Calcareous nannofossils are calcitic particles representative of the Phylum Haptophyta (Margulis and Schwartz, 1988) in the Kingdom Protoctista (Copeland, 1956). The coccolithophorids which belong to the order Prymnesiida (Hibberd and Leedale, 1985) constitute a major component of the modern phytoplankton and have probably maintained this position since the Jurassic (see Tappan and Loeblich, 1971; Berger, 1976). The calcareous nannoplankton are particularly sensitive indicators of oceanographic/climatic changes which have occurred during the last 150 million years. For instance, changes in their biogeographic patterns have been established on a broad scale allowing the delineation of trends in temperature during the Cenozoic (e.g., Haq and Lohmann, 1976a; Haq, 1980). High resolution studies of abundance fluctuations of selected species have revealed latitudinal shifts of water masses (Beaufort and Aubry, 1991) and fluctuating patterns of seasonality (Beaufort, 1991) during the Neogene. Calcareous nannofossils thus constitute an essential element for reconstructing the evolution of the biosphere.

IGCP Project 174, "Geological events at the Eocene/Oligocene boundary," has fostered numerous studies on calcareous nannofossils in upper Eocene and lower Oligocene deep sea- and land-based sections, while other studies were prompted by the challenging claim that mass extinction occurred at the Eocene/Oligocene boundary (Raup and Sepkoski, 1984, 1986) linked to the impact of a bolide (Ganapathy, 1982; Alvarez et al., 1982). Consequently, a large documentation is available regarding the stratigraphic distribution of calcareous nannofossil species in upper Eocene-lower Oligocene sections (e.g., Saunders et al., 1984; Nocchi et al., 1988a; Backman, 1987; Bybell, 1982; Coccioni et al., 1988), abundance patterns exhibited by the most common species particularly across the boundary (e.g., Backman, 1987; Madile and Monechi, 1991), and biogeographic distribution of characteristic assemblages (e.g., Haq et al., 1977; Wei and Wise, 1990a).

The objective of this paper is not to review and discuss these various studies which have sometimes yielded contradictory results. Rather, it is an attempt at outlining characters that the late Paleogene calcareous nannoplankton exhibit, which have been little discussed although they are most significant for understanding the middle Eocene to late Oligocene oceanographic/climatic evolution of our planet. In particular, emphasis is placed on changes in diversity that the calcareous nannoplankton exhibited during the late Paleogene.

CHANGES IN CALCAREOUS NANNOFOSSIL DIVERSITY: EVIDENCE FOR GLOBAL COOLING DURING THE MIDDLE EOCENE TO EARLY OLIGOCENE

Stratigraphic/temporal framework

The extinction of the two last representatives of the rosette-shaped discoasters, *Discoaster barbadiensis* and *D. saipanensis*, has long been regarded as one of the most striking event at the Eocene/Oligocene boundary, as recognized by Gartner (in Cita et al., 1970) in DSDP Site 10. The extinction of *D. saipanensis* was also used by Martini (1970) to characterize the Eocene/Oligocene boundary, and correlation was suggested by this author (Martini, 1971) between the planktonic foraminiferal and calcareous nannofossil zonal boundaries P17/P18 and NP20/NP21 (respectively Blow, 1969; Martini, 1970). Subsequent studies showed, however, that the extinction level of *D. saipanensis* is somewhat older than the P17/P18 planktonic foraminiferal zonal boundary and it follows from the recent approval by the Paleogene Subcommission on Stratigraphy of the use of the last hantkeninids to characterize the Eocene/ Oligocene boundary (Nocchi et al., 1988b) that the extinction of the rosette-shaped discoasters occurred prior to the Eocene/Oligocene boundary. Thus, for the purpose of stratigraphic correlations, Zone NP21 (Martini, 1970, 1971) and Subzone CP16a (Okada and Bukry, 1980) straddle the Eocene/Oligocene boundary. Also, because the FAD of *Sphenolithus pseudoradians* which defines the base of Zone NP20 has been shown to be unreliable (see discussion in Aubry, 1983a), Zones NP19 and NP20 have been formally regrouped as Zone NP19-20 (Aubry, 1983a), equivalent to Subzone CP15b (Okada and Bukry, 1980; = *Isthmolithus recurvus* Subzone, Bukry, 1975).

The discussion which follows is mainly centered on the middle/middle Eocene to lower/early Oligocene stratigraphic/temporal interval. In terms of calcareous nannofossil stratigraphy, it mainly covers the interval from Zone NP14 to Zone NP23 following Martini's zonal scheme (1971), and Zone CP12 to Zone CP18 following the nomenclatural scheme of Okada and Bukry (1980). Although under revision (see Berggren et al., this volume) the Paleogene time scale of Berggren et al. (1985) and Aubry et al. (1988) is used because it provides the best means of correlation presently available. Thus the interval discussed here is estimated to extend between ~52 Ma and ~32 Ma.

Diversity change near the Eocene/Oligocene boundary

The extinction of the rosette-shaped discoasters (*sensu* Aubry, 1984; = the *Heliodiscoaster mohleri* group of Theodoridis, 1984) has been regarded as a dramatic late Eocene event. While this appears to be true on quantitative grounds, it is only a minor event in terms of diversity change. Rosette-shaped discoasters may constitute up to 30% of the components of late Eocene assemblages; however, their high abundance is the product of only two species, *Discoaster barbadiensis* and *D. saipanensis*, the former becoming extinct (a few thousand years) prior to the latter. In the calcareous nannoplankton, the extinction of these two taxa, and the closely preceding extinction of *Reticulofenestra reticulata* account for what has been dubbed the late Eocene mass extinction. *Discoaster binodosus* is a third species in the genus *Discoaster* which may have become extinct close to the Eocene/Oligocene boundary, but its upper range is poorly documented due to its scarcity in upper Eocene (?lower Oligocene) deposits. The small change in diversity in the calcareous nannoplankton, as in the Foraminifera and planktonic diatoms, near the Eocene/Oligocene boundary was illustrated by Corliss et al. (1984).

While it is generally thought that the latest Eocene was the time of pronounced extinctions in the Paleogene (Haq, 1981), calcareous nannofossil evidence indicates 1) that a larger number of extinctions occurred in the early Oligocene than in the late Eocene, and 2) that a more pronounced turnover occurred near the middle/late Eocene boundary than near the Eocene/Oligocene boundary.

Diversity change during the early Oligocene

The extinctions of *D. saipanensis* and *D. barbadiensis* occurred about 0.4 m.y. prior to the Eocene/Oligocene boundary. A larger wave of extinctions occurred, however, in the early Oligocene, ~1.6 m.y. after the extinction of the rosette-shaped discoasters. Within a ~0.5 million-year-long interval which started ~1.2 m.y. after the Eocene/Oligocene boundary (mainly within Zone NP22), at least 12 species became extinct, among them *Calcidiscus protoannulus*, *Ericsonia formosa*, *Clausicoccus fenestratus*, *C. subdistichus*, *Hayella situliformis*, *Helicosphaera reticulata*, *Isthmolithus recurvus*, *Peritrachelina joidesa*, *Reticulofenestra dictyoda*, *R. hillae*, *R. oamaruensis*, and *R. umbilicus*. These were followed by further extinctions during Biochron NP23 (e.g., *Bramletteius serraculoides*, *Discoaster tani nodifer*, *D. tani*, *Helicosphaera compacta*, *Lanternithus minutus*). As speciation was very reduced during this time (perhaps only 2 new species, *S. distentus* and *H. perch-nielseniae*) this resulted in "mid" Oligocene calcareous nannofossil assemblages of very low diversity. Indeed, the early Oligocene is characterized by the lowest rates of evolution for the entire Paleogene (see Haq, 1973, Figure 3).

Diversity change near the middle/late Eocene boundary

In contrast to the isolated extinctions which occurred near the Eocene/Oligocene boundary, there is a sharp turnover in the composition of the calcareous nannoplankton at the middle/late Eocene boundary as equated with the NP17/NP18 biochronal boundary. Whereas the assemblages of Zone NP19-20 differ from those of Zone NP21 essentially in the absence of the rosette-shaped discoaters, assemblages of Zone NP18 (lower upper Eocene) differ from those of Zone NP17 (upper middle Eocene) by both the loss of several species in different genera and the appearance of new taxa. From a compilation of the range charts given for various deep sea sites (in particular, in Thierstein, 1974; Edwards and Perch-Nielsen, 1975; Perch-Nielsen, 1977; Wise, 1983; Wei and Wise, 1989, 1990b; Aubry, 1991 and unpublished data) and for various other sections (in particular, in Gartner, 1971, Saunders et al., 1984, Aubry, 1986 and unpublished data) it appears that among the taxa common in oceanic assemblages, at least five species (*Sphenolithus obtusus*, *Pseudotriquetrorhabdulus inversus*, *Cruciplacolithus delus*, *Chiasmolithus expansus* and *C. grandis*) disappear close to the top of Zone NP17, and at least four new taxa (*Chiasmolithus oamaruensis*, *C. altus*, *Peritrachelina joidesa* and *Hayella situliformis*) appear. In addition, the FAD of *Helicosphaera reticulata* may occur close to the base of Zone NP18 although the species has been reported from Zone NP17 (Bybell, 1975). The stratigraphic range of taxa restricted to shallow marine environments is difficult to establish because of the low number of sections studied which span the upper Eocene and lower Oligocene, and because of the presence of unconformities. However, it seems clearly established that *Orthozygus aureus* first occurs in Zone NP18 (see Báldi-Beke, 1971, Table 4). Using an estimate of 30 calcareous nannofossil species present in late middle and late Eocene oceanic assemblages, the turnover at the middle/late Eocene boundary affected one-third of the specific composition of assemblages while the extinctions near the Eocene/Oligocene boundary affected only a tenth of it. The greater change in diversity at the middle/late Eocene boundary than at the Eocene/Oligocene boundary is also apparent from the histogram by Tappan giving the number of calcareous nannofossil species by stage for the last 200 m.y. (Tappan, 1979, Figure 3; compare also the different order of magnitude of change in calcareous nannofossil diversity at the Eocene/Oligocene boundary and at the Cretaceous/ Paleocene boundary).

The middle Eocene to early Oligocene evolutionary trend

The changes in specific composition which occurred in the calcareous nannoplankton in the earliest Oligocene (between ~35.2 to 34.5 Ma), late Eocene (at ~36.7 Ma) and at the middle/late Eocene boundary (at ~ 40 Ma) are part of a broader evolutionary trend which initiated in the early middle Eocene (~52 Ma; although perhaps as early as the late early Eocene) and extended into the early Oligocene (~32 Ma) (Haq, 1973). The middle/late Eocene turnover and the late Eocene extinctions should be seen as steps in this trend (Aubry, 1983b). A compilation of Paleogene calcareous nannofossil data from deep sea and epicontinental sections shows maximum diversity values (~120 species) during the early middle Eocene (Zones NP 14-15) and minimum diversity values (~37 species) in the early Oligocene. There are difficulties in estimating changes in diversity in the calcareous nannoplankton through species counts. Calcareous nannofossils are morphotypes and the paleontologic concept of a calcareous nannofossil species is far removed from the biologic concept of species in the Coccolithophoridae (Aubry, 1989). Also, heterogeneous species concepts applied by various authors may introduce strong biases. This is particularly true for the representatives of the genus *Reticulofenestra* (inclusive of *Cribrocentrum, Cyclicargolithus, Dictyococcites*). Haq (1971) gives slightly different values than above, and the differences may be related to different taxonomic concepts. In addition, not all species in the calcareous nannoplankton may have fossil remains (in particular those secreting holococcoliths, a type of coccolith strongly susceptible to dissolution (Roth and Berger, 1975). Finally, in genera such as *Braarudosphaera, Micrantholithus, Pemma, Pontosphaera, Scyphosphaera*, the range of most individual species is poorly known and ambiguous stratigraphic ranges may introduce strong biases. Even with the best taxonomic scheme available (in which the paleontologic concept of species would be very close to the biologic concept), a curve of specific diversity would remain somewhat illusory. Thus the species numbers given above cannot be taken at face value as indicative of total diversity. Yet, these estimates suggest that the calcareous nannoplankton experienced a reduction in diversity of approximately 70% between the middle Eocene and the early Oligocene, i.e., in 8 to 10 m.y.

Figure 13.1 shows the ranges of the calcareous nannofossil species which contribute most significantly to middle Eocene through late Oligocene assemblages. Taxa which are always rare are omitted because their stratigraphic range is difficult to establish, and because the species which form the bulk of assemblages are those which yield the most paleoceanographic/paleoclimatic significance (Beaufort, 1991). Also, for the sake of clarity, few of the taxa restricted to Zones NP15 and NP16 are included in Figure 13.1, and category 2 is represented only by the taxa with the longest ranges (species such as *Toweius magnicrassus, T. gammation, T. callosus* which evolved in the early Eocene and disappeared in biochrons NP14 and NP15 are not represented). For a more complete representation of middle Eocene diversity the reader is referred to Perch-Nielsen (1985, in particular Figures 19, 30 and 69). Middle Eocene to late Oligocene assemblages comprise taxa which can be grouped into six categories. Short-ranging taxa restricted to the middle Eocene constitute the first category (Figure 13.1, bold solid line). In the second category (Figure 13.1, bold dashed line with double dots) are long-ranging taxa, most of which evolved in the earliest Eocene, which became extinct during the middle Eocene. The third category (Figure 13.1, double dashed and dotted line) contains similar taxa but which became extinct in the latest Eocene and during the Oligocene. The fourth category (Figure 13.1, bold dotted line) is formed of long-ranging taxa which evolved during the middle Eocene and, except for rare exceptions such as *Reticulofenestra floridana* and *Discoaster deflandrei*, disappeared progressively between the latest Eocene and the latest Oligocene. *Reticulofenestra dictyoda*, the stem species of the *Reticulofenestra* group, which evolved in the late early Eocene, belongs to this category. The bulk of the late Eocene and Oligocene assemblages includes taxa in the last two categories. The fifth category (Figure 13.1, long and short dashed line) comprises taxa which evolved in the late Eocene, mainly in the earliest late Eocene (early Biochron NP18) and became extinct in the early Oligocene. Most taxa in the last category (Figure 13.1, short dashed line) evolved during the late early and late Oligocene, but some evolved as early as earliest Oligocene (e.g., *H. recta* and *S. distentus* which occur in Biozone NP21, respectively from the Parathetys Basin, and the Gulf of Mexico, Caribbean area and central Pacific; Saunders et al., 1984; Aubry, 1988 and unpublished data). The middle middle to early late Eocene is thus a transitional interval, between a time of very high diversity (the late early and early middle Eocene, see

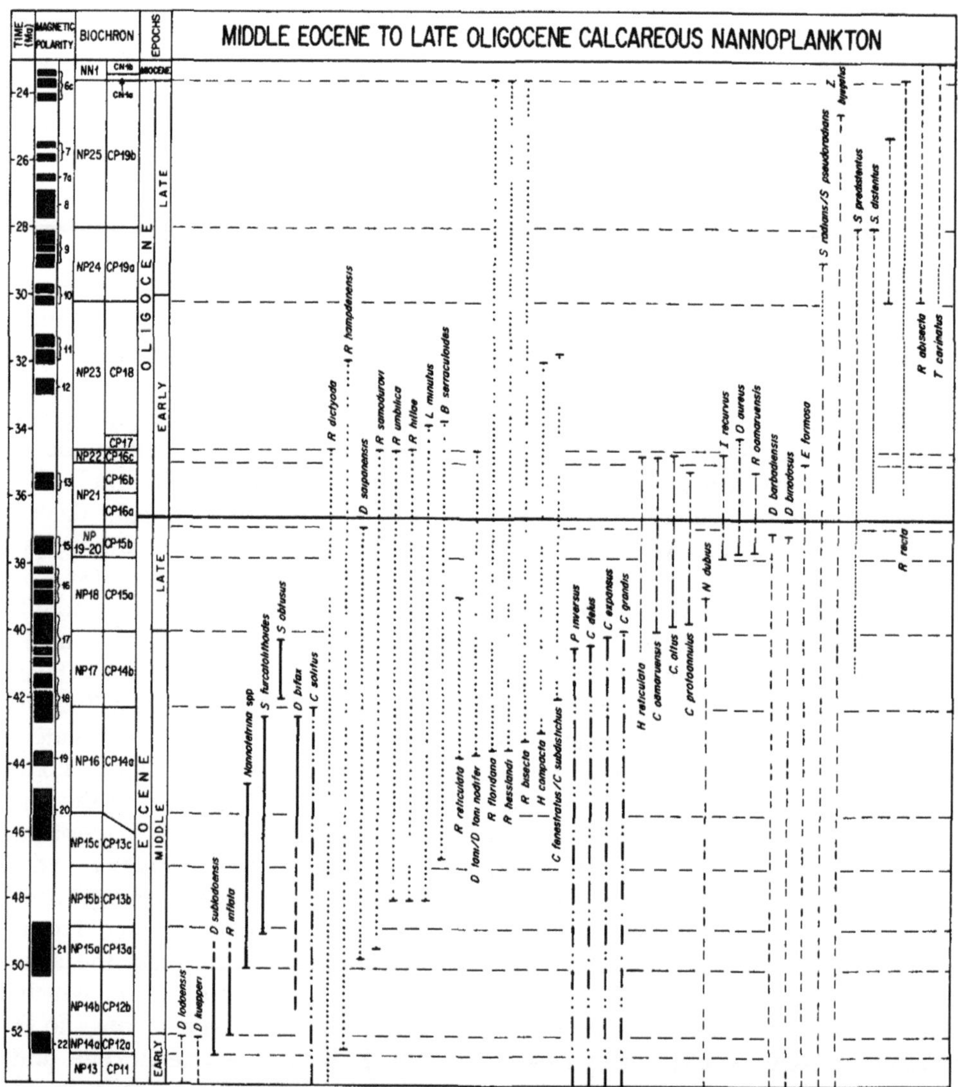

FIGURE 13.1. Range of some of the calcareous nannofossil species which contribute most significantly to middle Eocene through late Oligocene assemblages. See text for further explanation.Correlations between appearance/extinction events and magnetochrons are approximate as such direct correlations are not available for all taxa.

Haq, 1973, Figure 3) and a time of reduced diversity (the late Eocene and early Oligocene). In turn, the late Eocene to early Oligocene is an interval of increasingly reduced diversity. Figures 2 and 3 in Beckmann et al. (1981), which illustrate the early Eocene to early Miocene ranges of selected calcareous nannofossil species, also reflect this middle Eocene turnover with high evolutionary rates and the late Eocene-early Oligocene diversity reduction with very low evolutionary rates.

Three genera are strongly implicated in the middle Eocene to early Oligocene evolution. In the genus *Chiasmolithus* there was a replacement near the middle/late Eocene boundary when the last *(C. grandis)* but one *(C. titus)* of the 12 early and middle Eocene species became extinct and when 2 new species (*C. oamaruensis* and *C. altus)* evolved. In the genus *Discoaster*, sharp impoverishment occurred among the rosette-shaped discoasters as a result of both extinction and lack of speciation later than biochron NP15. *Discoaster barbadiensis* and *D. saipanensis* are best regarded as two long-ranging species (respectively NP11-NP19-20 and NP15-NP19-20), the only two taxa which ranged later than Biochron NP16. In contrast to both *Chiasmolithus* and *Discoaster* which suffered substantial extinctions during the middle Eocene, the genus *Reticulofenestra* (inclusive of *Cribrocentrum*, *Cyclicargolithus* and *Dictyococcites*) underwent diversification at this time, and it is contended that evolution in this genus was closely related to the middle Eocene to early Oligocene oceanographic/climatic evolution.

Specific taxonomy in the genus *Reticulofenestra* is bewildering and unsettled, mostly because intraspecific variability has rarely been considered in the erection of new species so that end members in a taxon may have been regarded as separate taxa. Broad intraspecific variability is suspected in the genus *Reticulofenestra* as indicated by the morphologic analysis of *R. hampdenensis* (Edwards, 1973a) and the morphometric study of *R. umbilicus* (Backman and Hermelin, 1986). This renders almost impossible a precise account of the late Paleogene evolution of the genus. Nevertheless, the appearance of its first representative, *R. dictyoda*, in the late early Eocene (during Biochron NP13), the appearance of *R. umbilicus* s.l. in the early middle Eocene (during Biochron NP15), the subsequent middle Eocene diversification of the genus and its progressive impoverishment during the Oligocene are definite facts (see Perch-Nielsen, 1985, Figure 58) which strongly suggest a close relationship between the evolution of the genus and late Paleogene oceanographic/climatic evolution.

Discussion

Haq (1971, 1973) and Bukry (1978) showed that in the calcareous nannoplankton a parallel relationship exists between specific diversity and temperature throughout the Cenozoic, higher diversity being associated with warmer intervals, lower diversity with colder times.

The decline in diversity that the calcareous nannoplankton underwent from middle Eocene to early Oligocene is thus a clear indication of global climatic cooling. How progressive this cooling was is unclear, however, because the response of the calcareous nannoplankton to climatic deterioration may not have been linear. If it was, the data suggest a stepwise cooling beginning in the middle middle Eocene, intensifying in the later middle Eocene with cooling pulses near the middle/late Eocene boundary, in the latest Eocene, and in the early Oligocene.

LATE PALEOGENE CALCAREOUS NANNOPLANKTON BIOGEOGRAPHY: EVIDENCE FOR INCREASING LATITUDINAL DIFFERENTIATION.

Calcareous nannofossil biogeographic studies were pioneered by Haq and Lohmann (1976a) who showed that the distribution of calcareous nannofossil assemblages, like that of the modern Coccolithophoridae (McIntyre and Bé, 1967; Okada and Honjo, 1973) is primarily a function of latitude, hence of climate, and that climatic evolution could be inferred from the equatorwards/polewards expansion/migration of biogeographically significant assemblages. One of the important finding of these authors was the invasion of the mid-latitudes by the Eocene high-latitude assemblages between ~49 and 30 Ma (age estimates adjusted to the time scale of Berggren et al., 1985) suggestive of a middle Eocene cooling that was confirmed by subsequent paleobiogeographic (Haq et al., 1977) and isotopic (Shackleton and Kennett, 1975; Boersma and Shackleton, 1977) studies. No other large-scale quantitative study of the translatitudinal paleobiogeographic evolution of the calcareous nannoplankton has since been published which would establish more precisely the timing of the expansion/contraction of the characteristic assemblages in the three major oceans during the Paleogene.

The available quantitative studies can be advantageously supplemented with qualitative and semi-quantitative information regarding diachrony and provincialism which, together, contain ample evidence for strong latitudinal differentiation during the late Paleogene. One of the best known examples of diachronous ranges among the Cenozoic calcareous nannofossils is the last occurrence of *Discoaster saipanensis* which has been shown to have been time transgressive from mid- to low latitudes (Cavelier, 1975). In addition, the restriction of *Isthmolithus recurvus* to mid- and high-latitudes is a well known instance of late Eocene provincialism.

Diachrony and provincialism are closely related and represent the distinct expressions of a single phenomenon, biogeographic differentiation. Diachrony is essentially involved with

AGE		BIOCHRON (NP)	NORTHERN HEMISPHERE								
			DSDP SITE 343	DSDP SITE 33g	DSDP SITE 407	DSDP SITE 336	DSDP SITE 403	DSDP SITE 405	DSDP SITE 406	DSDP SITE 13g	DSDP SITE 183
OLIGOCENE	LATE	25	—	13	20	13	13	—	14	—	—
		24	—		15			—	11	—	—
	EARLY	23	—	—	—	—	—	—	—	—	—
		22	9	—	—	—	—	—	·8	10	10
		21		—	—	—	—	—	—		
EOCENE	LATE	19 - 20		—	—	16	—	—	14		
		18	—	—	—	—	—	—	13	5	
	MIDDLE	17	—	—	—	—	—	—	—	—	6
		16	—	—	—	—	14	—	—	12	10
		15	—	—	—	—	—	—	—	—	—
		14	—	—	—	—	—	36	23	—	—
	EARLY	13	—	13	—	—	—	38	—	12 species recorded from zone NP12[(1)]	21 species recorded from zone NP12[(1)]

TABLE 13.1: Diversity of early Eocene to late Oligocene calcareous nannofossil assemblages from high latitudes DSDP and ODP sites from the North Atlantic (336, 338, 349, 403, 405-407), North Pacific (183, 192), South Pacific (277, 278), South atlantic (689, 690, 693) and Indian (748, 749) Oceans; arranged from highest to lowest latitude, left to right for the northern hemisphere, roght to left for the southern hemisphere. Present latitude of the sites given in Table 13.2. Eocene latitude essentially the same except for DSDP Sites 277 and 278 which were located at ~65°S (Kennett, Houtz, et al., 1975a). Data from Aubry (1991), Bukry (1973a), Edwards and Perch-Nielsen (1975), Martini (1979), Müller (1976), Pospichal and Wise (1990), Steinmetz (1979), Wei and Wise (1990b), Worsley (1973), and Aubry (unpublished data). Note the poor stratigraphic record as a result of

SOUTHERN HEMISPHERE									BIOCHRON (NP)	AGE	
DSDP SITE	DSDP SITE	DSDP SITE 277	DSDP SITE 278	ODP SITE 748	ODP SITE 749	ODP SITE 68g	ODP SITE 690	ODP SITE 693			
		8	10	10	7	8	6	2	25	LATE	OLIGOCENE
									24		
									23	EARLY	
		22	—	16	—	12	8	—	22		
			—				9	—	21		
		17	---		10			—	19 - 20	LATE	EOCENE
		21	—	20	15	13	16	—	18		
			—	—	13	11	—	—	17	MIDDLE	
			—	27	25	12[3] 16[4]	15	—	16		
		19	—	16	15	20	—	—	15		
		16	—	22	34	19	24	—	14		
		21	—		26	—		—	13	EARLY	

unconformities, in particular for the middle Eocene, at sites located in the northern hemisphere compared with that at sites located in the southern hemisphere. Diversity of early Eocene to late Oligocene assemblages from DSDP Sites 363 and 522, and of late Eocene to late Oligocene assemblages from DSDP Site 292 are given for comparison. Eocene latitude of DSDP Sites 363 and 522 similar to present (Sclater et al., 1977b). DSDP Site 292 was located near the Equator in the late Eocene (Sclater et al., 1977a). (): data for DSDP Site 522; *: from Proto Decima et al. (1978); +: from Ellis (1975) [1]; from Bukry (1973a, Table 5); [2] from Worsley (1973, Table 1); [3]: upper part of Zone NP16, Core 12, Wei and Wise (1990); [4]: lower part of Zone NP16, Cores 17, 18, Pospichal and Wise (1990).

the differential stratigraphic range of discrete taxa in different paleoenvironments (e.g., epicontinental versus oceanic, low-latitude versus mid- or high-latitude). Provincialism is more concerned with the frequency of taxa in different environments. Diachrony and provincialism are two factors which contribute to obscure middle Eocene to upper Oligocene translatitudinal correlations (Worsley and Jorgens, 1974; Perch-Nielsen in Beckmann et al., 1981). However, while they impede stratigraphic resolution, they provide important information on late Paleogene climatic evolution.

The discussion which follows is based on a comparative analysis of the early Eocene to late Oligocene calcareous nannofossil assemblages from Deep Sea Drilling Project (DSDP) and Ocean Drilling Project (ODP) sites and from various land-based sections. As will be shown, there is a considerable difference in the intensity of diachrony and provincialism between the northern Atlantic and the southern hemisphere. The extent of the differences between assemblages from high northern and high southern latitudes is, however, extremely difficult to evaluate. High latitude sites in the North Atlantic Ocean have yielded highly discontinous stratigraphic records in particular for the middle Eocene (Table 13.1). Direct documentation of the calcareous nannofossil assemblages from the northern Pacific Ocean is limited to two lower Eocene-lower Oligocene sections recovered from DSDP Sites 183 and 192 (Worsley, 1973; Bukry, 1973a), sites thought to have experienced no plate motion since the early Eocene (Scholl and Creager, 1973; but see alternative interpretation in Worsley and Davies, 1979). It is somewhat supplemented by valuable, but scanty information from dredged samples from the eastern Gulf of Alaska continental slope (Poore and Bukry, 1979), and from outcrops in Oregon (Bukry and Snavely, 1988) and Washington (Worsley and Crecelius, 1972; Armentrout and Worsley, 1980).

Diachrony

The diachronous nature of biostratigraphic datum events may be difficult to establish when diachrony is of small amplitude. Small amplitude diachrony (i.e., <1 m.y.) may be more apparent than real. It may be the result of different taxonomic concepts, of unrecognized unconformities in the stratigraphic record, and/or, in the case of last occurrence/appearance datums, of possible reworking. Thus the diachrony discussed here is not of the kind illustrated by Wei and Wise (1989). Instead, large amplitude diachrony (>>1 m.y.) is often readily demonstrated and probably more significant with regard to paleoceanography/paleoclimatology than is small amplitude diachrony. However, in the case of taxa which show strong provincialism and may be very rare outside their biogeographic province, the extent of the diachrony may be difficult to establish. In particular, discontinuous occurrences in the upper stratigraphic range of a taxon may be difficult to distinguish from occurrences related to reworking. Figure 13.2 attempts to illustrate not only the strong diachrony in the last and first occurrences of selected taxa between low and high latitudes, but also between the northern Atlantic and southern high latitudes.

It appears from Figure 13.2 that: 1) the regional disappearances of the warm/temperate water taxa occurred earlier at high than at low latitudes; and 2) also notably earlier in high southern latitudes than in the northern Atlantic. Particularly remarkable is the diachrony shown by the last occurrence of *Ericsonia formosa*, a species which became extinct in the earliest Oligocene. Its highest occurrence defines the NP21/NP22 zonal boundary, which has been shown to be a remarkably reliable datum event at low-latitude (Wei and Wise, 1989). At high southern latitudes, the last occurrence of *E. formosa* appears to constitute a well-defined horizon in the middle Eocene Zone NP16, as shown by its distribution at ODP Sites 689 (~64°S) and 690 (~65°S) (see Pospichal and Wise, 1990, Tables 1 and 3; and Wei and Wise, 1990b, Tables 1 and 2), at ODP Sites 748 (~58°S) and 749 (~58°S) (see Aubry, 1991, Tables 3 and 5), and at DSDP Site 277 (~65°S) (see Edwards and Perch-Nielsen, 1975, Table 4c, = *Ericsonia alternans*). At northern high-latitude DSDP Site 192 (NO Pacific, ~53°N), the last occurrence of *E. formosa* appears to lie in Zone NP16 (see Worsley, 1973, table 2B). Yet, at DSDP Site 336 (Norwegian-Greenland Sea) with an Eocene latitude of ~63°N, *E. formosa* is present in sediments assignable to the upper Eocene Zone NP19-20

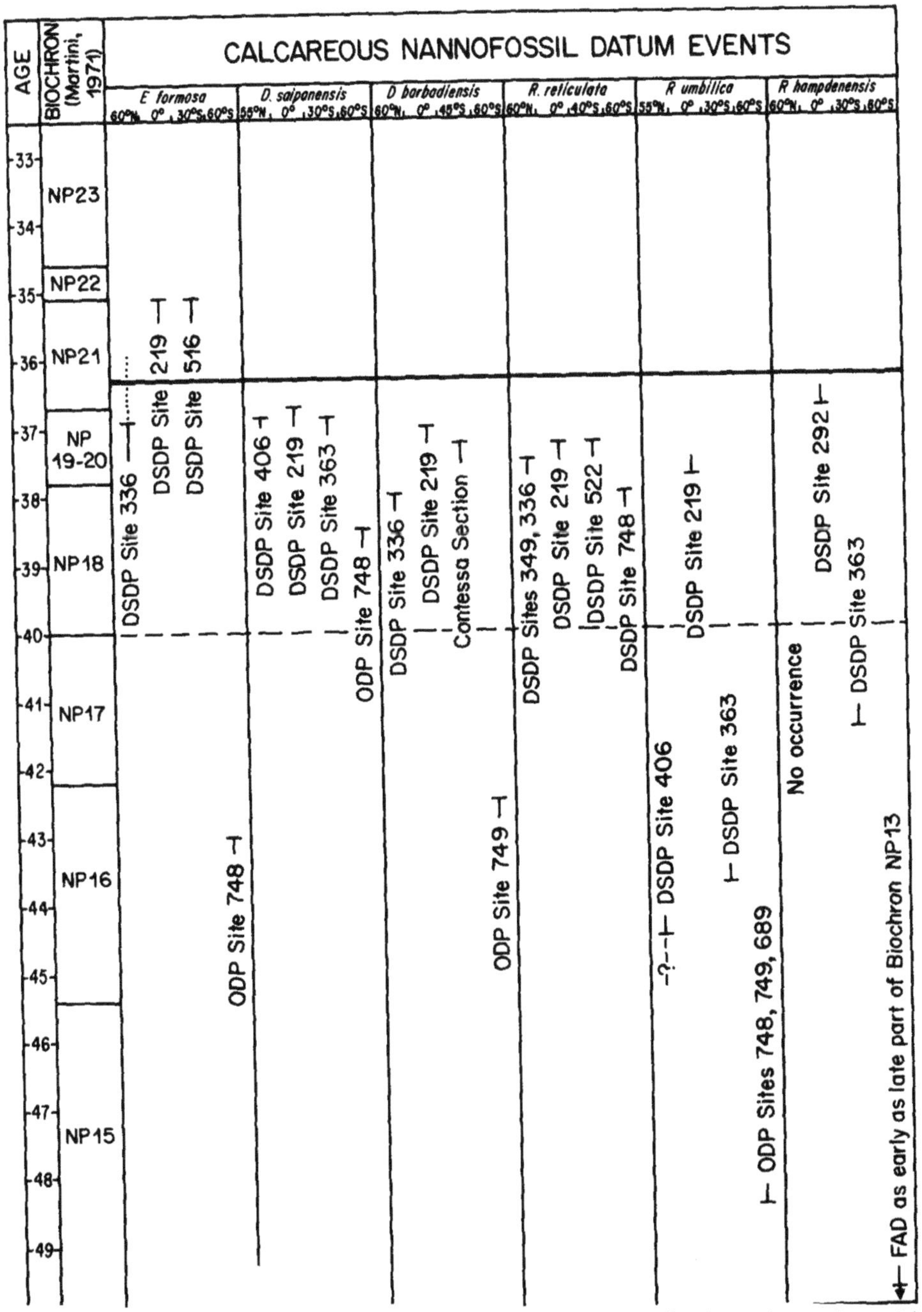

FIGURE 13.2. Diachrony in the highest and lowest occurrences of selected species with respect to latitude. Data for DSDP Sites 336, 349 and 406 are from Müller (1976; 1979); for DSDP Sites 219, 277, 292, 363, 511 and 516 respectively from Boudreaux (1974), Ellis (1975), Edwards and Perch-Nielsen (1975), Proto Decima et al. (1978), Wise (1983), and Wei and Wise (1989), with additional pers. obs.; for ODP Sites 689, 748 and 749 respectively from Wei and Wise (1990b) and Aubry (1991). Data for the Contessa Section are from Lowrie et al. (1982). A number of DSDP sites from equatorial, tropical, temperate and high-latitude regions are chosen to illustrate diachrony. Numerous other sites (e.g., DSDP Sites 94, 253, 277, 548, 366) would yield the same information. The paleolatitudes for DSDP Sites 219, 292 are taken from respectively Sclater et al. (1977a) and (1971), for DSDP Sites 363 and 516 from Sclater et al. (1977b), for DSDP Sites 511 and 522 from Smith and Briden (1977), for ODP Sites 748 and 749 from Shipboard Scientific Party (1989), and for the Contessa region from Lowrie et al. (1982).

characterized by the co-occurrence of *R. reticulata* and *Isthmolithus recurvus* (see Müller, 1976, Table 3).

Although not as marked, diachrony of the last occurrence of *R. reticulata* between low and high latitudes on the one hand, and between the northern Atlantic and high southern latitudes on the other, can be clearly demonstrated as well. At low latitude, *R. reticulata* became extinct in the latest Eocene, slightly prior to *D. barbadiensis*. At Norwegian-Greenland Sea DSDP Site 336 (~63°N), its last occurrence is also located in Zone NP19-20 (Müller, 1976), as it is in South Atlantic DSDP Site 522 (~40°S). At higher southern latitudes (ODP Sites 689 (~63°S), 690 (~65°S), 748 (~58°S) and 749 (~58°S) (see respectively Wei and Wise, 1990b, Tables 1 and 2; Aubry, 1991, Tables 3 and 5), its last occurrence is in Zone NP18. However, it should be noted that at southwestern Pacific DSDP Site 277 and at southern Atlantic DSDP Site 511, both with an Eocene paleolatitude >60°S, *R. reticulata* occurs in Zone NP19-20 (Edwards and Perch-Nielsen, 1975; Wise, 1983; Aubry, pers. obs.).

More tentative, perhaps, is the determination of the extent of the diachrony of the last occurrences of *D. saipanensis* and *D. barbadiensis*, because of the scarcity and discontinuous occurrence of these species in high-latitude assemblages. At low-latitude sites, the last occurrence of *D. saipanensis* is in the uppermost Eocene, slightly below the Eocene/Oligocene boundary. At ODP Sites 748 and 749, the last occurrence of *D. saipanensis* is within upper Eocene Zone NP18 (see Aubry, 1991). Yet, at DSDP Site 406 (~55°N) and 511 (~60°S), *D. saipanensis* occurs in Zone NP19-20. More marked is the diachrony of the last occurrence of *D. barbadiensis* which at low-latitude is in Zone NP19-20, sligthly below that of *D. saipanensis* and above that of *R. reticulata*. At DSDP Site 336 (~63°N), the last occurrence of *D. barbadiensis* appears to be located in the upper part of Zone NP18. Yet, at ODP Site 749 where the species is associated with *Chiasmolithus solitus* and *R. reticulata*, it is in the upper part of Zone NP16, while at ODP Site 689 (~64°S), it is located below the first occurrence of *R. reticulata*, possibly in the lower part of Zone NP16 or in Zone NP15 (see Pospichal and Wise, 1990, Table 1). At ODP Site 690, located 1° further South, the last occurrence of *D. barbadiensis* may be as low as Zone NP14 (see Pospichal and Wise, 1990, Table 3).

Although not as clearly settled as the pattern of diachrony of last occurrences, it seems that: 1) the first occurrences of cold/temperate water taxa happened later at low than at high latitudes and 2) also happened later in the high northern latitudes than in the high southern ones. This pattern is however difficult to establish, first because of insufficient data from northern high latitudes due to the incompleteness of the stratigraphic record, second, because the taxa which are involved (species of the genus *Reticulofenestra*) are poorly defined taxonomically so that data from different sources may be difficult to compare. At southern high-latitude ODP Sites 689 (see Pospichal and Wise, 1990, Table 1, =*R. daviesi*) and 748 and 749 (see Aubry, 1991, Tables 4 and 5) *R. hampdenensis* first occurs in Zone NP14, while at DSDP Site 277, its lowest occurrence was recorded in Zone NP15 (see Edwards and Perch-Nielsen, 1975, Table 4c); at DSDP Site 363 (~30°S), its lowest occurrence is located in mid-Zone NP16 (Aubry, unpublished) from where it is also reported at DSDP Site 516 (Wei and Wise, 1989). At DSDP Site 292 (~3°S), its lowest occurrence is in Zone NP19-20 (Aubry, unpublished). *Reticulofenestra hampdenensis* seems to be restricted to the southern hemisphere. The pattern of diachrony for *R. umbilicus* is more tentative because of taxonomic inconsistencies, and it will require morphometric analysis to resolve it. Yet, it is clear that the lower occurrence of large placoliths of *R. umbilicus* at tropical and equatorial sites is in the upper Eocene Zone NP19-20.

It follows from this discussion that the last occurrences of *E. formosa*, *D. saipanensis*, *D. barbadiensis* and *R. reticulata* in tropical and equatorial regions correspond to the extinctions (true LAD's) of the respective taxa while their last occurrences in temperate and high latitudes correspond to the local disappearances of the taxa. Conversely, the first occurrence of *R. hampdenensis* and that of *R. umbilicus* at high latitudes correspond to the evolutionary appearances of these taxa (true FAD's) while their first occurrences at lower latitudes are local geographic appearances. These opposite patterns of diachrony in the first and last occurrences of, respectively, cold and warm water species imply that warm water masses contracted towards equatorial regions during the Eocene while cold water masses expanded towards the tropics. Particularly significant is the latitudinal progression of *R. hampdenensis* which appears to have evolved in the early middle Eocene in southern high latitudes and migrated progressively towards the equator that it reached only in the late Eocene.

The pattern of differential diachrony between the southern high latitudes and the northern Atlantic Ocean is remarkable. At equivalent latitudes warm-water taxa which had disappeared from the southern high latitudes were persisting in the northern Atlantic. This could imply that the intensity of the middle Eocene to early Oligocene cooling of water masses was not the same in both hemispheres. The data from the northern Pacific Ocean are tenuous, but the fact that *E. formosa* and *D. barbadiensis* have their last occurrence in Zone NP16 at DSDP Site 192 (see Worsley, 1973, Table 2B) as they do in the southern ocean suggests that the symmetrical pattern of diachrony between both hemispheres is strongly modified in the North Atlantic by the transport of warm waters from the tropical regions to the high latitudes. This implies that the Gulf Stream has been an efficient transport system since at least the middle Eocene which is consistent with the Paleogene erosional history of this western boundary current (Pinet et al., 1981). Similarly, the occurrences of *D. saipanensis* and *R. reticulata* in the latest Eocene at DSDP Site 511 (and high diversity of middle Eocene calcareous nannofossil assemblages, Wise et al., 1985) suggest that a proto-Brazil current system may have reached the Falkland Plateau in the late Eocene, possibly confirming that a thermal gradient decreasing to the east was maintained throughout the Eocene in the South Atlantic (Boersma et al., 1987).

Provincialism

The rarity, or absence, of *Isthmolithus recurvus* and *Chiasmolithus oamaruensis* in low-latitude deposits and that of *D. barbadiensis* and *D. saipanensis* in high-latitude deposits have long been acknowledged as impediments to upper Eocene stratigraphic resolution (Martini and Worsley, 1971; Edwards, 1973b; Roth, 1973). Haq and Lohmann (1976b) noted that species of the genus *Chiasmolithus* were abundant in early Oligocene assemblages from southern high latitudes while they were rare in the northern latitudes. While increased provincialism in the Oligocene served for developing new techniques of stratigraphic correlation (Worsley and Jorgens, 1974), its use for paleoceanographic/paleoclimatic reconstruction has remained limited to the establishment of the *Discoaster/Chiasmolithus* ratio as an indicator of Eocene temperature based on provincialism in the genera *Discoaster* (warm water temperature) and *Chiasmolithus* (cold water temperature) (Table 13.3; Bukry, 1973b).

The calcareous nannoplankton have exhibited latitudinal provincialism throughout the Cenozoic, as early as early Paleocene (Perch-Nielsen, 1981; Perch-Nielsen et al., 1982), and even when the planet experienced the warmest temperatures in the latest Paleocene and early Eocene (Frakes, 1979). Provincialism was then considerably attenuated, as testified by the similarity in specific composition of early Eocene calcareous nannofossil assemblages from the Kerguelen Plateau in the southern Indian Ocean (Aubry, 1991) and the London Clay and lower Bracklesham Beds (Aubry, 1983a, 1986). In contrast, Blechschmidt and Worsley (1975, 1977) noted an increase in provincialism during the Oligocene.

A satisfactory latitudinal coverage is now available to study provincialism in the oceans of the southern hemisphere. For the northern hemisphere, the situation is different (Tables 13.1 and 13.2). In the Pacific Ocean, the northern high latitudes are represented by a meager 2 data points as already mentioned above. While the Paleogene sections recovered from DSDP Sites 183 and 192 are extremely valuable for the insight they provide, they lack stratigraphic continuity and yield scarce calcareous nannofossil assemblages (Worsley, 1973). In the Northern Atlantic, latitudinal coverage is more satisfactory, but the Paleogene sections recovered are riddled with unconformities so that little is known of the middle Eocene. Also, because of the strong influence that the Gulf Stream already had on the northern Atlantic biogeographic province(s) as discussed above, the Paleogene history of provincialism in the

Table 13.2. Latitude of DSDP and ODP sites considered in Table 13.1

Northern Hemisphere		Southern Hemisphere	
Site	Latitude	Site	Latitude
DSDP Site 292	15°49.11'		
		DSDP Site 363	19° 38.75'
		DSDP Site 522	26° 6.843'
		DSDP Site 277	52° 13.43'
DSDP Site 183	52°34.30'		
DSDP Site 192	53° 00.57'		
DSDP Site 406	55° 15.50'		
DSDP Site 405	55° 20.18'		
DSDP Site 403	56° 08.31'		
		DSDP Site 278	56° 33.42'
		ODP Site 748	58° 26.45'
		ODP Site 749	58° 43.03'
DSDP Site 336	63° 21.06'		
DSDP Site 407	63° 56.32'		
		ODP Site 689	64° 31.009'
		ODP Site 690	65° 9.6'
DSDP Site 338	67° 47.11'		
DSDP Site 349	69° 12.41'		
		ODP Site 693	70° 49.82'

TABLE 13.3. Variations in the frequency of *Discoaster* and *Chiasmolithus* in early and middle Eocene assemblages as a function of latitude. Frequency expressed as a percentage of the taxon in a total count of 300 to 500 specimens in the genera *Discoaster* and *Chiasmolithus*. Data from Bukry (1973b).

Age		Lat.	Site	*Discoaster*	*Chiasmolithus*
Middle Eocene	NP17	Low	DSDP Site 44 (19°N)	98%	2%
	NP14	High	DSDP Site 112 (54°N)	33%	67%
Early Eocene	NP13	Low	DSDP Site 210 (14°S)	76%	24%
		High	DSDP Site 207A (37°S)	46%	54%

TABLE 13.4: Variation in the frequency of *Discoaster* and *Chiasmolithus* across the early/middle Eocene boundary in the southern high-latitudes. Frequency expressed as the percentage of the taxon in a total count of 200 specimens in the genera *Discoaster* and *Chiasmolithus*. Upper four cells: from ODP Site 749 (58° 43.03' S), Southern Kerguelen Plateau. Lower four cells: from ODP Site 748 (58° 26.45' S), Southern Kerguelen Plateau. While data from ODP Site 749 suggest a progressive decrease in the *Discoaster*/*Chiasmolithus* ratio during the early Eocene, data from ODP Site 748 suggest early Eocene fluctuations of this ratio. The differences in the ratio obtained for the two sites reflect, in part, differential preservation. However, they also suggest variations in the composition of early Eocene assemblages as a result of climatic/oceanographic instability.

Sample	Age		*Discoaster*	*Chiasmolithus*
749B – 12X – cc	Middle Eocene	NP14b	8%	92%
749B – 13X – cc	early Eocene	NP13	10.5%	89.5%
749C – 7R – 1			18%	82%
749C – 7R – 2			24%	76%
748C – 2R – cc	Middle Eocene	NP15 – NP16	traces	≅ 100%
748C – 10R – 1 70 – 71	early Eocene	NP13	32%	68%
748C – 10R – 2 70 – 71			15%	85%
748C – 10R – 3 70 – 71			23%	77%

TABLE 13.5. Variations in the frequency of the rosette-shaped discoasters, *Reticulofenestra hampdenensis*, and a group comprising *Reticulofenestra bisecta* + *R. hesslandi* + *R. umbilica* across the Eocene/Oligocene boundary as a function of latitude. Frequency expressed as the percentage of the taxon in the total assemblage, established from a count of 300 specimens.

DSDP Sites	Age	Rosette-shaped discoasters	*R. bisecta* + *R. hesslandii* + *R. umbilicus s.l.*	*Reticulofenestra hampdenensis*
219 (≅ 3°S)	e. Oligocene NP 21 – NP 22	—	25%	0%
	l. Eocene NP 19 – 20	25%	5%	0%
292 (≅ 0°)	e. Oligocene NP 21 – NP 22	—	25%	5%
	l. Eocene NP 19 – 20	20%	12%	0%
363 (≅ 30°S)	e. Oligocene NP 21 – NP 22	—	------insufficient data---------	
	l. Eocene NP 19 – 20	15%	25%	10%
277 (≅ 60°S)	e. Oligocene NP 21 – NP 22	—	12%	45%
	l. Eocene NP 19 – NP 20	0.1%	30%	30%

North Atlantic Ocean would probably not apply to the North Pacific where the calcareous nannoplankton are likely to have had a biogeographic history similar to that in the southern hemisphere. The biogeographic history of the calcareous nannoplankton in the southern hemisphere is thus considered to be representative of the climatic evolution of the planet during the late Paleogene.

Provincialism in the southern hemisphere

As shown by the comparison of the calcareous nannofossil assemblages recovered from the numerous mid and low-latitude DSDP and ODP sites with the calcareous nannofossil assemblages recovered from DSDP Site 277 (~65°S Eocene latitude, Kennett, Houtz, et al., 1975a) on the Campbell Plateau in the southwestern Pacific Ocean, from ODP Sites 689 and 690 (~65°S Eocene latitude, Kennett and Barker, 1990) on Maud Rise in the Weddell Sea and from ODP Sites 748 and 749 (~58°S Eocene latitude, Shipboard Scientific Party, 1989) on Kerguelen Plateau in the southern Indian Ocean (respectively, Edwards and Perch-Nielsen, 1975; Pospichal and Wise, 1990; Wei and Wise, 1990b; Aubry, 1991), provincialism in the southern hemisphere accentuated in a stepwise fashion from middle Eocene to Oligocene. The most pronounced step seems to have been at the early/middle Eocene boundary (Pospichal and Wise, 1990; Aubry, 1991) at which time there was a sharp reduction in diversity at the southern high latitudes, with a loss of warm water taxa (e.g., *Sphenolithus radians*), a net decrease in the abundance of discoasters, and the establishment of specific dominance by representatives of the cold and temperate water genera *Chiasmolithus* and *Reticulofenestra* (Table 13.4). In addition, taxa which appeared during the early middle Eocene at low and mid latitudes did not migrate to southern high latitudes (e.g., *Rhabdosphaera inflata*, *Chiasmolithus gigas*) or were extremely sporadic (e.g., *Discoaster sublodoensis*, *Nannotetrina fulgens* and related species).

The definite turnover that occurs at mid- and low latitudes around the middle/late Eocene boundary is not as strongly marked at southern high latitudes because many of the species affected did not occur at high latitudes or were rare. However, the local successive disappearances of additional warm water taxa (i.e., *Ericsonia formosa* and *Discoaster barbadiensis*) in Zone NP16 are indicative of further climatic/oceanographic evolution at southern high-latitude during the middle Eocene. It should be noted in passing that the last occurrence of *D. barbadiensis* seems to constitute a well defined horizon in Zone NP16 in the southern Indian Ocean as it does (in the mid Eocene 6/7, Bortonian Stage, see Edwards, 1971, Figure 5) in the southwest Pacific and in New Zealand (Edwards, 1973b).

There is no increase in provincialism at the Eocene/Oligocene boundary; in fact this boundary cannot be delineated at southern high latitudes on the basis of calcareous nannofossils alone. While there is a striking (although somewhat superficial, as discussed above) difference between late Eocene and early Oligocene assemblages from low and mid latitudes, there is no significant difference between late Eocene and early Oligocene assemblages at southern high latitudes (Figures 13.3 to 13.5; Table 13.5; Aubry, 1983b; 1991, and unpublished).

The most remarkable difference between high- and low-latitude calcareous nannofossil assemblages in the southern hemisphere resides undoubtly in the increased latitudinal provincialism in the Oligocene (Aubry, 1991 and in progress). During the middle and late Eocene, the calcareous nannoplankton shows a continuous trend towards decreasing diversity at both low and high latitudes. This trend reversed at low latitudes during the late early Oligocene when speciation lead to increasing diversity in particular in the warm water genera *Sphenolithus* and *Helicosphaera*, and also in the cold to temperate genus *Reticulofenestra*. However, no such reversal occurred during the Oligocene at southern high latitudes where calcareous nannoplankton diversity further decreased as a consequence of both lack of speciation and lack of migration of the low and mid-latitude taxa into the high latitudes. While late Oligocene assemblages from low latitudes may include over 25 species, late Oligocene assemblages in the southern high latitudes are restricted to less than 10 species, 4 of which being extremely abundant (i.e., *Chiasmolithus altus*, *Coccolithus pelagicus*, *Reticulofenestra floridana* and *R. hampdenensis*). No rebound occurred at southern high latitudes before the late early Miocene as shown by moderately diversified early Miocene calcareous nannofossil assemblages at ODP Site 748 (Aubry, unpublished).

Provincialism in the North Atlantic Ocean

It is not possible to reconstruct the development of provincialism in the North Atlantic because northern high-latitude sites have recovered almost no middle Eocene sediments. However, many have contained Oligocene calcareous sediments, and this reveals a striking difference between the late Oligocene calcareous nannofossil assemblages in the southern high latitudes and in the northern Atlantic. While the trend towards decreasing diversity continued throughout the late Paleogene at southern high latitudes, this trend reversed during the late Oligocene in the Northern Atlantic. In Table 13.2 the number of taxa reported from upper Paleogene sediments recovered from high-latitude sites are given in an attempt to compare changes in regional diversity through time. It is clear that the database is insufficient to allow comparison for other zonal intervals than those of the Oligocene. There are

FIGURE 13.3. (following pages) Middle Eocene to early Oligocene abundance patterns in the calcareous nannoplankton from DSDP Site 219. Late Eocene latitude was ~3°S (Sclater et al., 1977a). Planktonic foraminiferal stratigraphy from Corliss et al. (1984). Note the abundance of the rosette-shaped discoasters in the late Eocene assemblage, resulting in a sharp contrast in quantitative composition between assemblages across the Eocene/Oligocene boundary. Note also the difference in composition with DSDP Site 292, from similar latitude. Although specific composition is similar at both sites, with high abundance of rosette-shaped discoasters in the late Eocene and their replacement by species of the genus *Reticulofenestra* in the early Oligocene, the proportions between taxa differ greatly between both sites, in particular in the Oligocene. Abscissa gives the percentage of species in the whole assemblage, as estimated from counts of 300 specimens.

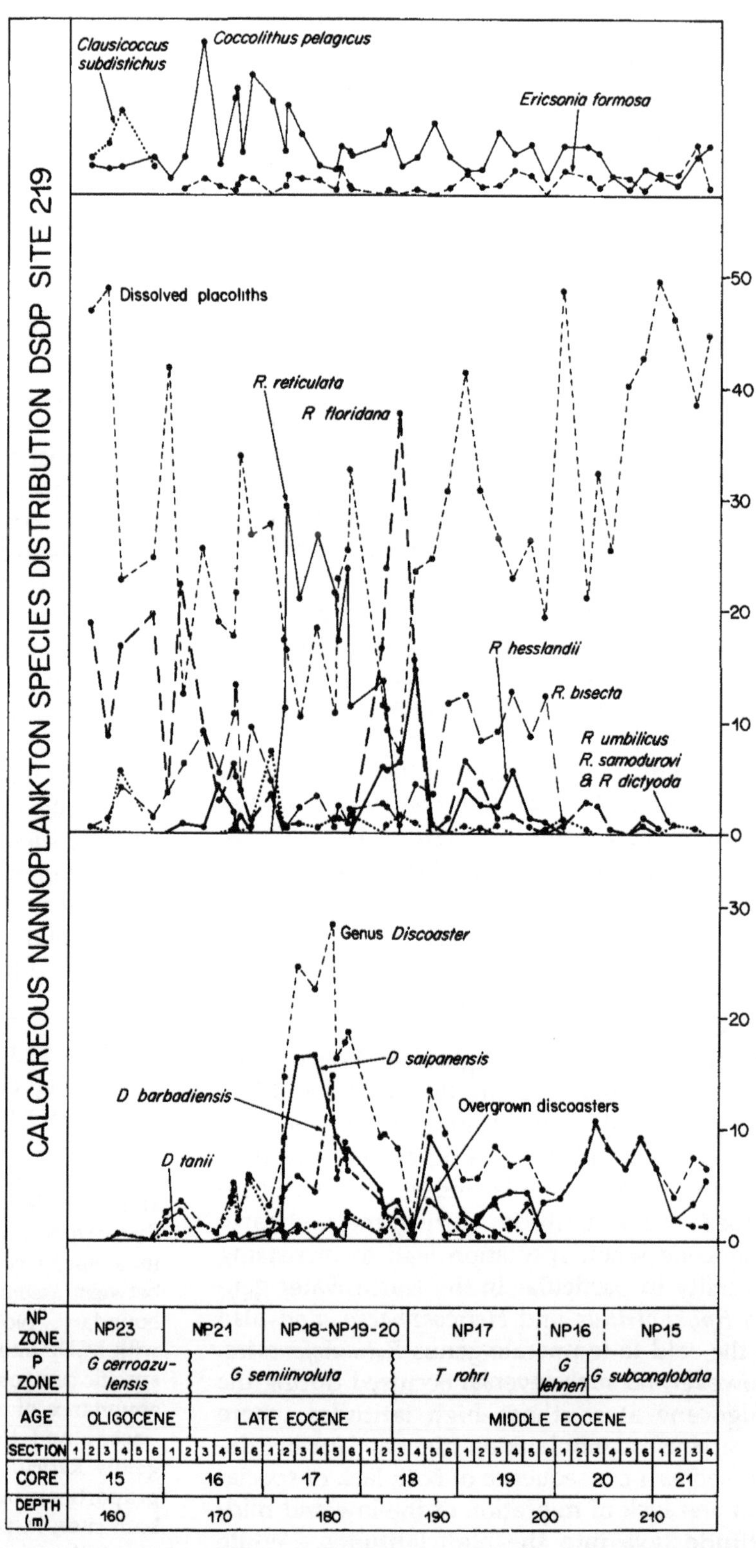
CALCAREOUS NANNOPLANKTON SPECIES DISTRIBUTION DSDP SITE 219
Clausicoccus subdistichus
Coccolithus pelagicus
Ericsonia formosa
Dissolved placoliths
R. reticulata
R floridana
R hesslandii
R. bisecta
R umbilicus
R. samodurovi
& R dictyoda
Genus Discoaster
D saipanensis
D barbadiensis
Overgrown discoasters
D tanii
NP ZONE
NP23
NP21
NP18-NP19-20
NP17
NP16
NP15
P ZONE
G cerroazu-lensis
G semiinvoluta
T rohri
G lehneri
G subconglobata
AGE
OLIGOCENE
LATE EOCENE
MIDDLE EOCENE
SECTION
CORE
DEPTH (m)

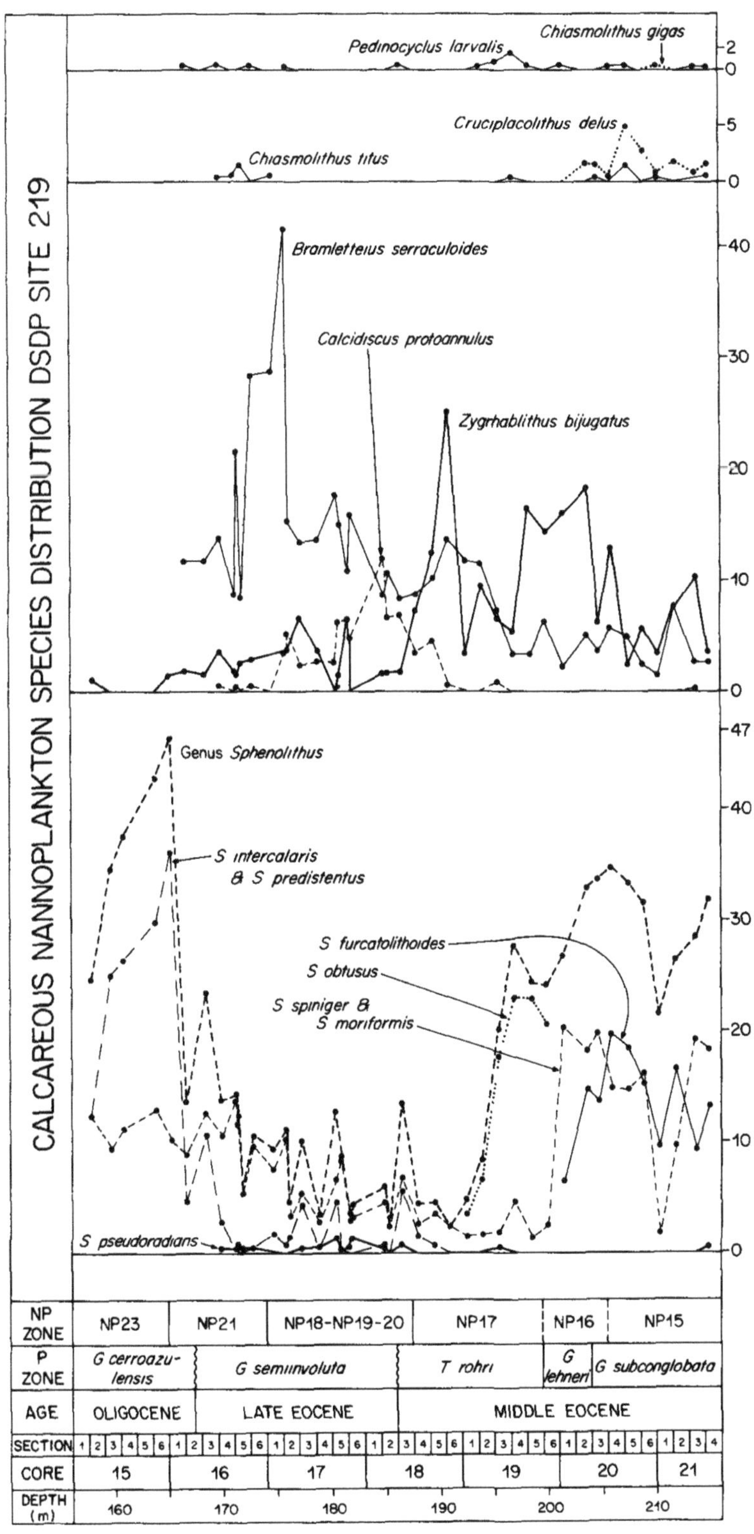
CALCAREOUS NANNOPLANKTON SPECIES DISTRIBUTION DSDP SITE 219
Chiasmolithus gigas
Pedinocyclus larvalis
Cruciplacolithus delus
Chiasmolithus titus
Bramletteius serraculoides
Calcidiscus protoannulus
Zygrhablithus bijugatus
Genus Sphenolithus
S intercalaris & S predistentus
S furcatolithoides
S obtusus
S spiniger & S moriformis
S pseudoradians
NP ZONE
NP23
NP21
NP18-NP19-20
NP17
NP16
NP15
P ZONE
G cerroazulensis
G semiinvoluta
T rohri
G lehneri
G subconglobata
AGE
OLIGOCENE
LATE EOCENE
MIDDLE EOCENE
SECTION
CORE
DEPTH (m)

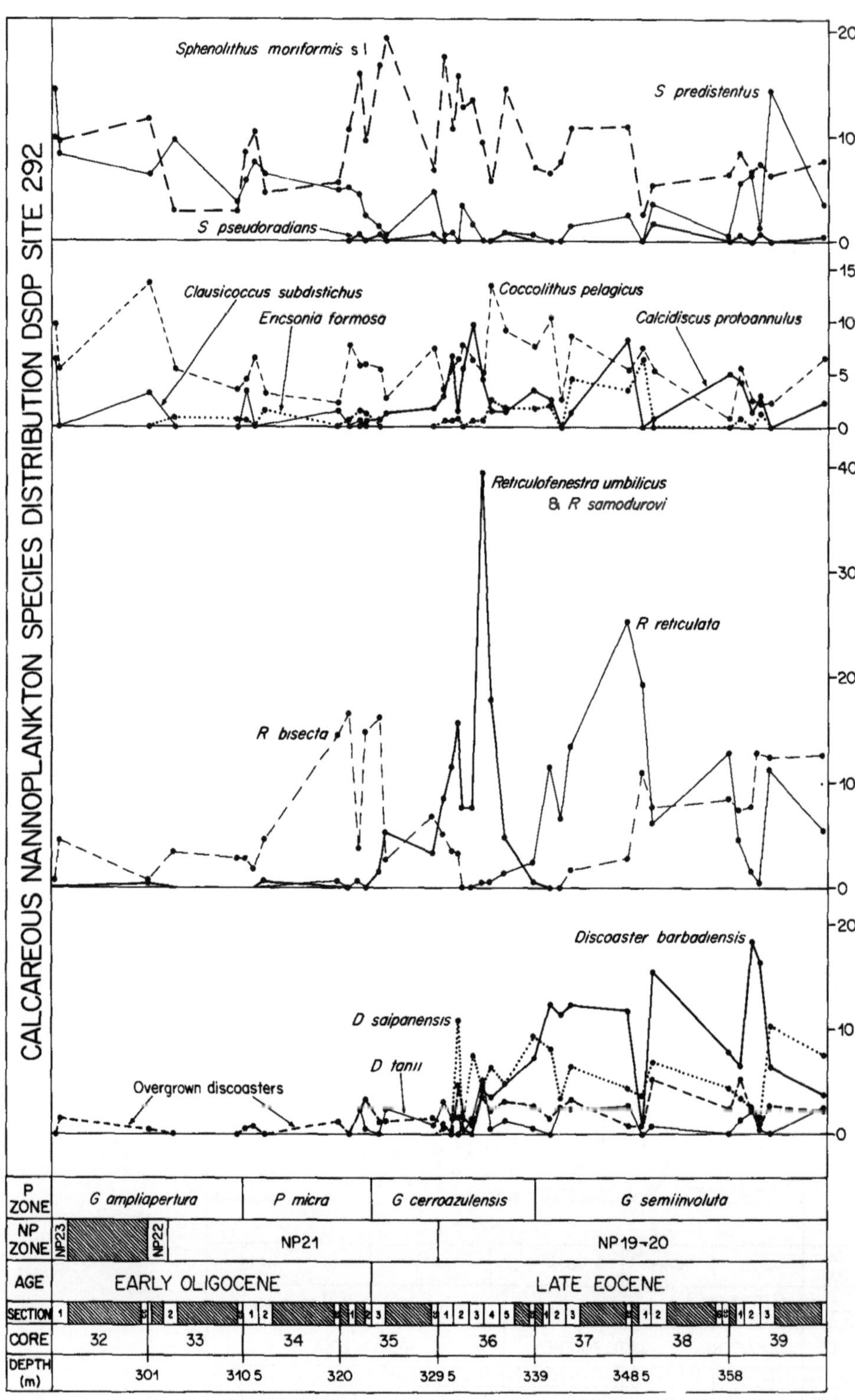

FIGURE 13.4. Late Eocene to early Oligocene abundance patterns in the calcareous nannoplankton from DSDP Site 292. This site was near the equator in the late Eocene (Sclater et al., 1971). Planktonic foraminiferal stratigraphy from Corliss et al. (1984). See Fig 3 for further explanation.

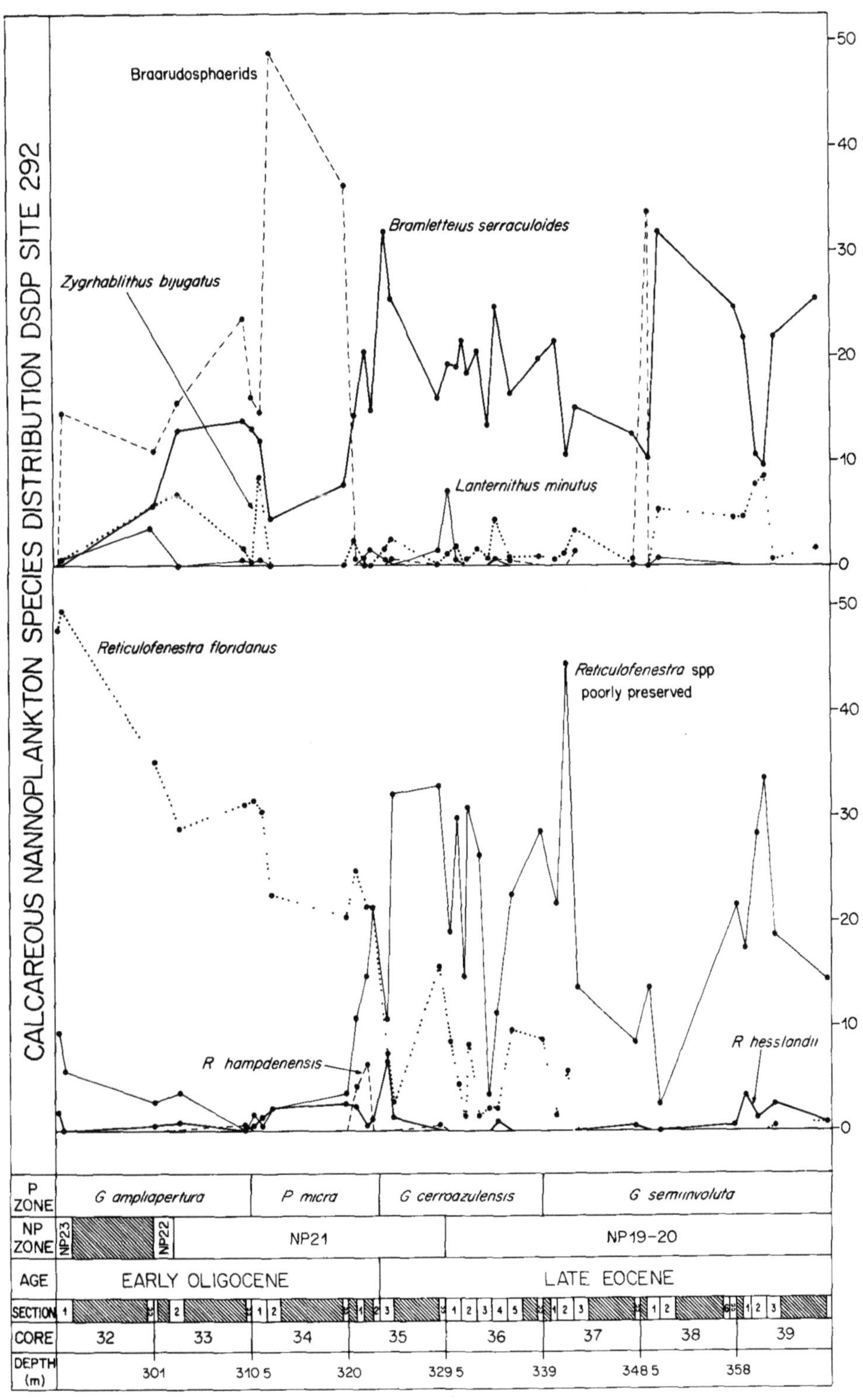
CALCAREOUS NANNOPLANKTON SPECIES DISTRIBUTION DSDP SITE 292
Braarudosphaerids
Bramletteius serraculoides
Zygrhablithus bijugatus
Lanternithus minutus
Reticulofenestra floridanus
Reticulofenestra spp poorly preserved
R hampdenensis
R hesslandii
50
40
30
20
10
0
P ZONE
G ampliapertura
P micra
G cerroazulensis
G semiinvoluta
NP ZONE
NP23
NP22
NP21
NP19-20
AGE
EARLY OLIGOCENE
LATE EOCENE
SECTION
CORE
32
33
34
35
36
37
38
39
DEPTH (m)
301
310 5
320
329 5
339
348 5
358

difficulties associated with comparing inventories from different sources. In addition to the fact that different authors may use different taxonomic concepts, inventories at different sites may not be as thorough particularly when assemblages are scarce. Moreover, preservation which influences greatly the composition of assemblages (Roth and Berger, 1975) was poor at many sites, and this is reflected by the variable number of species in assemblages of the same age recovered from adjacent sites (e.g., compare 36 species with 23 species in Zone NP14 respectively at DSDP Sites 405 and 406). Thus, a detailed site-to-site comparison would be meaningless. However, it appears worthwhile to compare the number of species in the late Oligocene assemblages from the southern high latitudes and the northern Atlantic. Diversity values (as expressed by the total number of species recorded) are as high as 20 in the latter, but reach a maximum of 10 in the former. As coccoliths are abundant and well-preserved in the Oligocene sediments recovered from ODP Sites 748 and 749, low diversity at these sites must be genuine. The similar values observed at all southern high-latitude sites support this view. Thus, during the Oligocene, while provincialism was increasing in the southern ocean, it was decreasing in the North Atlantic. This can only be explained by the flow of the Gulf Stream bringing warm waters from tropical regions into the northern latitudes, thus reversing the effect of the middle Eocene-early Oligocene global cooling.

Discussion

Diachrony and provincialism concur in indicating that strong biogeographic differentiation existed during the late Paleogene. Although the data discussed here may suggest that provincialism is linked to latitude, this is probably true only on a broad scale in as much as water masses may primarily follow a latitudinal trajectory. However, in detail, the distribution of species of the calcareous nannoplankton is not related to latitude but strictly to that of water masses (Burns, 1973; Bukry, 1978, 1980; Winter, 1985), itself driven by climatic forcing and tectonic evolution. The late Paleogene history of provincialism and diachrony shown by the calcareous nannoplankton in the North Atlantic is linked to the flow of the Gulf Stream which attenuated, delayed and eventually reversed the effects at high latitudes of the global climatic deterioration. In contrast, cold waters of the California current flowing southward on the Washington and Oregon continental margin accentuated and precipitated the effects of this climatic deterioration in temperate regions, as indicated by the calcareous nannofossil assemblages from the Oregon Coast Range (~45°N) (Bukry and Snavely, 1988).

From the biogeographic contrasts described in the southern hemisphere, which best reflects the late Paleogene climatic deterioration, it is apparent that the middle Eocene to early Oligocene cooling did not have the same effect on the calcareous nannoplankton simultaneously at all latitudes, but that its effect was latitudinally transgressive from high to low latitudes. While at mid- and low latitudes, the middle/late Eocene boundary turnover appears to be the most prominent event of the middle Eocene-early Oligocene evolutionary trend, the sharpest event in this trend at high-latitude appears to be the early/middle Eocene boundary decrease in diversity. This high-latitude cooling event (as inferred from the calcareous nannoplankton) was reflected at lower latitudes by rapid successive middle Eocene turnovers suggestive of climatic instability. Further high-latitude cooling eventually led to the middle/late Eocene boundary turnover, the extinction of the rosette-shaped discoasters in the latest Eocene, and the early Oligocene extinction of the *Reticulofenestra umbilicus* group, while in the mean time high-latitude calcareous nannoplankton species invaded tropical and equatorial regions.

The little evidence available from high-

FIGURE 13.5. (opposite page) Late Eocene to early Oligocene abundance patterns in the calcareous nannoplankton from DSDP Site 277. Late Eocene latitude was ~65°S (Kennett, Houtz, et al., 1975a). Planktonic foraminiferal stratigraphy from Corliss et al. (1984). Note the rare occurrences of the rosette-shaped discoasters, and the absence of variations in the quantitative composition of assemblages across the Eocene/Oligocene boundary. Note also the lower diversity of assemblages compared with the low-latitude sites.

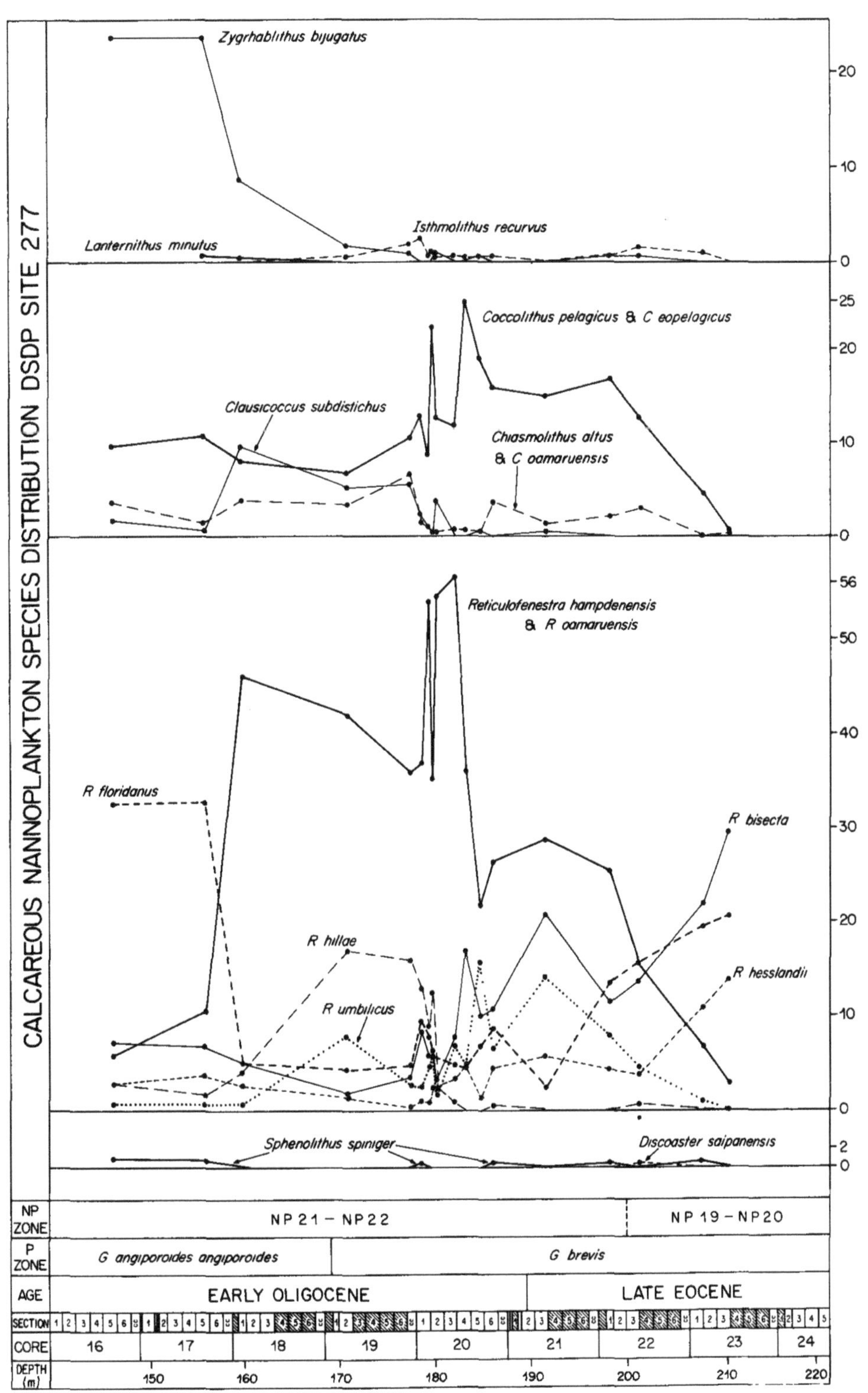
CALCAREOUS NANNOPLANKTON SPECIES DISTRIBUTION DSDP SITE 277
Zygrhablithus bijugatus
Isthmolithus recurvus
Lanternithus minutus
Coccolithus pelagicus & C eopelagicus
Clausicoccus subdistichus
Chiasmolithus altus
& C oamaruensis
Reticulofenestra hampdenensis
& R oamaruensis
R floridanus
R bisecta
R hillae
R hesslandii
R umbilicus
Sphenolithus spiniger
Discoaster saipanensis
NP ZONE
NP 21 – NP 22
NP 19 – NP 20
P ZONE
G angiporoides angiporoides
G brevis
AGE
EARLY OLIGOCENE
LATE EOCENE
SECTION
CORE
16
17
18
19
20
21
22
23
24
DEPTH (m)
150
160
170
180
190
200
210
220

latitude sections in the North Pacific and from regions affected by the cold California current suggests the same sequence of events in the northern hemisphere. Early Eocene assemblages (Biochron NP13) from the eastern Gulf of Alaska continental slope (~58°N) indicate warm water conditions (Poore and Bukry, 1979). A reduction in the abundance of warm water taxa occurs in the calcareous nannofossil assemblages from the Oregon Coast Range (~45°N) near the early/middle Eocene boundary (Bukry and Snavely, 1988), which suggests that northern high latitudes became cooler at the same time as southern high latitudes. In the middle Eocene Kellogg Shale of northern California, warm water species are replaced by cold water taxa around the NP15/NP16 zonal boundary (Barron et al., 1984). It has been shown above that the last occurrences of *E. formosa* and *D. barbadiensis* in the northern Pacific (~53°N) appear to occur at the same time as at southern high latitudes. Worsley and Crecelius (1972) reported on the dearth of discoasters and the abundance of species of *Chiasmolithus* and *Reticulofenestra* in middle Eocene and Oligocene assemblages from Olympic Peninsula, Washington (~48°N), an indication that the California Current was already transporting cold waters toward temperate regions. Reinterpretation of the calcareous nannofossil stratigraphy (Armentrout and Worsley, 1980) of the Willapa River, Pe Ell Doty, Mid Fork Satsop River and Canyon River sections, which expose middle Eocene (Zone NP16) formations unconformably overlain by the Lincoln Creek Formation, which extends from upper Eocene or lower Oligocene to upper lower Oligocene or younger (Zone NP19-21?- Zone NP22, Zone NP23 and perhaps younger), shows low middle Eocene diversity (16 species, no discoasters) and a drop in diversity at the NP22/NP23 zonal boundary from 21 species to 7, suggesting very cold waters in the northern Pacific in the late early Oligocene.

Thus, the "Terminal Eocene Event" stems from a high-latitude (polar) cooling event around the early/middle Eocene boundary, and the early Oligocene extinction of *R. umbilicus* (and of many other taxa) is the outcome of global climatic deterioration as a result of intensified high-latitude cooling. There is evidence that the high-latitude cooling did not proceed at regular pace. If the biogeographic distribution of taxa is primarily related to water temperature (McIntyre and Bé, 1967), the stratigraphically sporadic occurrences of warm water taxa (e.g., *D. sublodoensis, Nannotetrina fulgens* and related) in the southern Indian Ocean (Kerguelen Plateau, Aubry, 1991), in the Weddell Sea (Maud Rise, Pospichal and Wise, 1990) and in the southwest Pacific (Campbell Plateau, Edwards and Perch-Nielsen, 1975) suggest repeated migrations of warm/cold water masses as a result of climatic oscillations (in a pattern similar to that described by Beaufort and Aubry, 1991). Similarly, Haq and Lohmann (1976a) deduced from the southward migration of warm water assemblages that a warming event occurred in the late Eocene.

The calcareous nannoplankton indicate that climatic deterioration was decisive in isolating a cold water mass in the southern ocean. It is not possible to determine the timing of this isolation from the calcareous nannoplankton alone. Isolation may have resulted from the early Oligocene ice build-up (Shackleton and Kennett, 1975; Miller et al., 1987, 1991b). However, planktonic foraminiferal records suggest that the Antarctic was thermally isolated from warm water boundary current systems as early as late middle Eocene (planktonic foraminiferal Zones P13-P14) (Boersma et al., 1987). It is clear that isolation was effective in the late early Oligocene as it prevented that an Oligocene rebound in the southern high-latitude calcareous nannoplankton echoed the late early Oligocene recovery of the nannoplankton at lower latitudes. There are no direct data concerning possible recovery of the calcareous nannoplankton in the later Oligocene at high latitudes in the North Pacific Ocean. The very low diversity assemblages which occur in the Lincoln Creek Formation of southwestern Washington is the only relevant but indirect evidence of a cold water mass in the northern Pacific, equivalent to that in the southern hemisphere.

LATE PALEOGENE CALCAREOUS NANNOPLANKTON EVOLUTION AND BIOGEOGRAPHY: PRODUCTS OF MIDDLE EOCENE TO EARLY OLIGOCENE CLIMATIC DETERIORATION

Quantitative studies of the Paleogene biogeography of the calcareous nannoplankton in the Atlantic Ocean (Haq and Lohmann, 1976a; Haq et al., 1977) revealed three intervals of cooling during the late Paleogene, as indicated by the invasion of mid and/or low latitudes by high-latitude assemblages. These are: 1) the middle Eocene, between ~49 and 40 Ma (age adjusted to the time scale of Berggren et al., 1985), when high-latitude assemblages invaded the middle latitudes; 2) the early Oligocene, between ~ 36 and 34.8 Ma (age adjusted to the time scale of Berggren et al., 1985), when the high-latitude assemblages spread toward the lower latitudes; and 3) the latest early and the earliest late Oligocene, between ~32 and 29 Ma (age adjusted to the time scale of Berggren et al., 1985), when the warm water assemblages became restricted to the low latitudes. These conclusions are in agreement with those drawn from the present analysis of diachrony and provincialism. The difference in the dating of the onset of the middle Eocene cooling is due to the fact that no section from a latitude higher than 50°S was available at the time of those studies. As I have shown, the northern Atlantic, even at a latitude as high as 60°N, is not a faithful recorder of the climatic evolution of the planet because of the imprint of the Gulf Stream. The fact that high-latitude assemblages invaded the low latitudes in the early Oligocene suggests a cooling stronger than that in the middle Eocene (but see below). I have shown that the early Oligocene was the time of lowest evolutionary rates for the late Paleogene. The restriction of warm water assemblages to low latitudes suggests further cooling in the latest early Oligocene. I have shown that this is a time of increased provincialism, and consequently stronger latitudinal gradients. Calcareous nannofossil assemblages of extremely low diversity but high specific dominance and of unchanging specific composition throughout the late early and late Oligocene (from ~34.6 to 25.6 Ma, and extending beyond) suggest that temperatures in the southern high latitudes remained very cold throughout the Oligocene and support the hypothesis of a permanent, albeit small, ice sheet on the Antarctic continent (Zachos et al., 1991), although most authors believe that Oligocene ice sheets were transient (Miller et al., 1987, 1991b; Kennett and Barker, 1990).

Temperature

All evidence from the calcareous nannoplankton suggests that the late Paleogene climatic deterioration resulted from an initial cooling of the high latitudes near the early/middle Eocene boundary, whose effects progressed toward the middle and low latitudes as it intensified in a step-wise fashion, first during the middle Eocene, then in the latest middle Eocene, and in the earliest Oligocene when the temperatures of the surface waters in the low latitudes presumably reached their minimum. The timing of the initial cooling and that of its subsequent intensifications established independently through isotopic studies (see Miller, this volume) correspond remarkably well with the timings inferred from evolutionary changes in diversity and biogeography. High-latitude cooling at ~52 Ma (Katz and Miller, 1991) seems to have been immediately reflected by a change in diversity of the calcareous nannoplankton in the northern and southern oceans, in particular by the loss or decrease in abundance of the warm water taxa. Cooling at ~40 Ma (timing poorly constrained, see Miller, this volume) coincides with a major turnover, with the loss of many long-ranging species (see Figure 13.1). Cooling at 35.8 Ma (Miller et al., 1988; Zachos et al., 1991) was followed by a wave of extinctions which eliminated most temperate water taxa. The gradual middle Eocene increase of ~1.5‰ in $\delta^{18}O$ values, between ~52 and ~40 Ma (Miller et al., 1987; Miller, this volume) correlates with the progressive middle Eocene withdrawal of warm water taxa from the southern high latitudes.

The relationships between thermal gradients and provincialism/biogeography of temperature-significant assemblages is not a simple one. It can be confidently stated from calcareous nannofossil evidence that the thermal gradients from pole to equator during the early Eocene were the weakest for the whole Eocene, as was provincialism. This is in agreement with other lines of evidence which led to the suggestion that weak early Eocene thermal gradients may have resulted from enhanced sea floor thermal activity (Rea et al., 1990). As cooling proceeded during the middle Eocene, thermal gradients and provincialism first increased in parallel, resulting in the progressive disappearance of the warm water taxa from the high latitudes and the equatorward migration of cold and temperate water assemblages. However, increasing cooling led to the extinction of the warm-water taxa so that provincialism tended to decrease. The withdrawal of the warm-water taxa from high latitudes and their subsequent extinction at low-latitude suggest that cooling at high latitudes induced delayed cooling at low latitudes, so that the pole to equator thermal gradients, although stronger than in the early Eocene, may have remained reduced. This situation would be in agreement with that described by Shackleton and Boersma (1981).

Stott et al. (1990) have estimated the Eocene surface-water temperatures of the Weddell Sea from isotopic values recorded by the shallow dwelling planktonic foraminifera acarininids. Surface water temperatures during the austral summer season dropped from 15-17°C in the early Eocene to 15°C in the earliest middle Eocene, at ~52 Ma, and to 12 to 13°C in the middle Eocene, at ~45 Ma, approximately at the time when both *E. formosa* and *D. barbadiensis* disappeared from the high latitudes. These estimates are comparable with those of early Eocene highs ~16°C for near-surface waters in the southeastern Atlantic and 18 to 21°C in the southwestern Atlantic, with a decrease to ~13°C in the late Eocene for both sectors of the South Atlantic (Boersma et al., 1987). If it is assumed that ecologic preferences of species do not change through time and that middle Eocene to early Oligocene extinctions were primarily temperature-controlled, the extinction of *E. formosa* and *D. barbadiensis* would have happened when the surface water temperatures in the low latitudes reached a threshold of ~12 to 13°C. These estimates, which would imply that the low-latitude regions were colder during the early Oligocene than the high latitudes during the early Eocene, are unacceptably low (Adams et al., 1990), and beyond the estimated temperature extremes for the Eocene (Shackleton, 1984; Matthews, 1984). While temperature is the apparent major agent of migration and extinction in the calcareous nannoplankton during the middle Eocene to early Oligocene, other directly or indirectly related factors must have played a more direct though less apparent role in the late Paleogene evolution of the calcareous nannoplankton.

Trophic resources

Temperature has been regarded as the most important factor controlling the distribution of the calcareous nannoplankton following the characterization of five temperature-controlled assemblages of living Coccolithophoridae in the Atlantic Ocean by McIntyre and Bé (1967). Little is known of the influence of other ecologic parameters on the distribution and abundance of the Coccolithophoridae (see review in Tappan, 1980), but Birkenes and Braarud (1952) remarked that high nutrient supply was a prerequisite for the "*Coccolithus huxleyi* -summers". The role of nutrient concentrations on the differential abundance of living species in the calcareous nannoplankton is becoming appreciated (Fincham and Winter, 1989; Kleijne et al., 1989). Also, it has been shown that surface water fertility controlled the biogeographic patterns exhibited by the calcareous nannoplankton during the middle Cretaceous (Roth, 1989; Roth and Krumback, 1986). Apart from a few taxa whose distribution is thought to be controlled by salinity and turbidity (e.g. *Braarudosphaera bigelowii*, *Scyphosphaera* spp.; Martini, 1965; Müller in Doebl et al., 1976), most Cenozoic calcareous nannofossil species are currently regarded as temperature-significant (e.g., Bukry, 1973b; Wei and Wise, 1990a). However, it may be that trophic resources contributed a greater role than anticipated in the geographic distribution of the Paleogene Coccolithophoridae.

Fluctuations in nutrient availibility during the Paleogene have been delineated based on the relationships between planktonic and larger benthic foraminiferal assemblages and oceanic paleochemistry as indicated by oxygen and carbon isotopes (Boersma et al., 1987; Hallock et al., 1991). Expanded oligotrophy in euphotic waters throughout the open ocean and margins during the early Eocene was followed by increasing eutrophication and loss of oligotrophic habitats during the middle and late Eocene, resulting in a maximum contraction of the trophic resources continuum (TRC; Hallock, 1987) in the early Oligocene. The frequence of at least some of the low-latitude taxa (i.e., the discoasters) at high latitudes during the early Eocene may be related as much to oligotrophic conditions as to warmth, and their extinction may have been determined by a loss of their oligotrophic habitat in the late Eocene rather than by cooling alone. Similarly, the numerous early Oligocene extinctions may be the consequence of further eutrophication in addition to increased cooling. It is difficult to separate the role of temperature and nutrient concentrations in the late Paleogene evolution of the calcareous nannoplankton, first because cooling was accompanied by eutrophication as a result of intensified circulation and mixing (Boersma et al., 1987), second because both temperature and nutrient concentrations (the latter in part through determining the depth of the euphotic zone) probably regulate simultaneously species growth and distribution in the Coccolithophoridae. However, it is possible that discoasters were K-mode specialists, adapted to oligotrophic environments. Wei and Wise (1990a) pointed to the fact that a relationship between *Discoaster* abundance and warm waters was not straight-forward. Discoasters are abundant in tropical gyres but less common in oceanic areas which may be warmer but receive abundant nutrients, and in epicontinental seas. Their lower abun-dance at high fertility equatorial DSDP Site 366 (Fenner, 1986) than at mid-latitude DSDP Sites 522 and 516 located within a gyre supports an inferred adaptation of *Discoaster* to oligotrophy.

Unlike diatoms which require abundant nutrient supplies, calcareous nannoplankton (Coccolithophoridae) dominate the phytoplankton under oligotrophic conditions (Smayda, 1980). The link between high diversity and oligotrophy is shown by the fact that communities of Coccolithophoridae in nutrient-rich marginal seas have much lower diversity than in the open ocean (Okada and Honjo, 1975). In permitting photosynthesis throughout a thick euphotic zone, oligotrophy promotes species stratification and high diversity (Hallock, 1987). This is well illustrated by the vertical distribution of the communities of Coccolithophoridae in the subtropical gyres of the Pacific Ocean from surface waters down to 200 m (Okada and Honjo, 1973). While it is often thought that the calcareous nannoplankton inhabit the superficial waters, their depth distribution is generally similar to that of the planktonic foraminifera (mainly 0 to 200 m), and there is a marked stratification with presence and abundance of species characteristic of various levels. For instance, in the central Pacific Ocean, *Florisphaera profunda* and *Thorosphaera flabellata* are restricted to depths between 100 and 200 m, being abundant only between 150 and 200 m. In addition, *F. profunda* is more abundant at 150 m while *T. flabellata* is more abundant at 200 m (Okada and Honjo, 1973). Oligotrophic conditions in the early Eocene would have favoured vertical stratification and expansion of the deep-dwelling Coccolithophoridae species (e.g., the discoasters) into the warm high-latitude regions.

Discussion

If the late Paleogene evolution of the calcareous nannoplankton was linked to trophic levels (i.e., fertility), the middle Eocene to early Oligocene turnover described in Figure 13.1 could reflect the combined history of oceanic TRC and temperature. The main steps in this turnover are strikingly similar to those described for the planktonic foraminifera (Boersma et al., 1987). As cooling and eutrophy increased, the long-ranging species (category 2) which had evolved in the early Eocene when temperature and oligotrophy were high, progressively disappeared. The turnover was completed by the end of the middle Eocene. However, originations first exceeded extinctions and diversity reached its peak in the middle Eocene (Zone NP15) as an ecotonal community had developed (see Hallock et al., 1991). Short-ranging species (category 1; e.g., *Nannotetrina* spp., *Sphenolithus furcatolithoides, Chiasmolithus gigas, Discoaster bifax, D. mirus, D. wemmelensis*) which were probably K-mode selective and contributed to oligotrophic assemblages evolved at the same time as long-ranging taxa (category 4) which form the late Eocene and early Oligocene mesotrophic assemblages. It is remarkable that the calcareous nannoplankton show the same evolution as the planktonic foraminifera which reached peak diversity in Zone P11 (Berggren, 1969; Boersma et al., 1987) and in which "most of older Eocene tropical and equatorial surface water groups had disappeared" by the end of latest middle Eocene Zone P14 (Boersma et al., 1987, p. 322).

During the middle Eocene, taxa in category 2 were replaced by taxa (category 4) which would tolerate cooler temperatures but also higher trophic levels. The *Reticulofenestra* group is likely to have evolved in response to associated cooling and eutrophication. Yet, the first representatives of the genus, *R. dictyoda* and a form referred to as *R. umbilica* by Perch-Nielsen (1972, Plate 9, Figure 6) but shown to be a separate taxon (Backman and Hermelin, 1986), seem to have evolved when oligotrophy was the highest and temperatures were the warmest. Both first occur in sediments assigned to Zone NP13. In addition to the fact that the latter form appears to be restricted to the high latitudes (Perch-Nielsen, 1972), climatic fluctuations occurred toward the end of the early Eocene (Boersma et al., 1987), which may have been accompanied by slight fluctuations in fertility. The early Oligocene extinctions at ~34.5 to 35. 2 Ma, in particular that of *R. umbilica* and closely related forms, probably reflect a further step towards eutrophy rather than an abrupt cooling in surface waters as suggested by Wei (in press). Since diatoms require high nutrient levels for growth, diatom-oozes are good indicators of eutrophic conditions. In the southern oceans a lithologic change is seen at the Eocene/Oligocene boundary or in the early Oligocene. At DSDP Site 277 lowermost Oligocene "foram-rich nanno oozes" are overlain by lower Oligocene "diatom and spicule-bearing foram-rich nanno ooze" (Kennett, Houtz, et al., 1975b). At ODP sites 689 and 690, lower Oligocene "radiolarian-bearing diatom nannofossil oozes" succeed upper Eocene nannofossil oozes (Shipboard Scientific Party, 1988). At ODP Sites 748 and 749 upper Eocene nannofossil oozes underlie lower Oligocene "siliceous nannofossil oozes" with radiolarians and diatoms (Site 748) and "nannofossil oozes enriched with diatoms" (Shipboard Scientific Party, 1989). This change from calcareous oozes to mixed siliceous and calcareous oozes at the Eocene/Oligocene boundary or in the earliest Oligocene is suggestive of an increase in nutrient concentrations in the euphotic zone at high-latitude. At DSDP Site 511 the early Oligocene and latest Eocene is represented by "nannofossil diatomaceous oozes" and "muddy diatomaceous oozes" (Ludwig, Krasheninnikov, et al., 1983). This suggests that eutrophic conditions had developed at this site by late Eocene, perhaps as a result of upwelling at the boundary between the cold waters of the higher latitudes and the warmer waters of the proto-Brazil current. Since the Eocene record at this site is truncated by an unconformity, the extent of the diatomaceous facies in older Eocene levels is unknown. It should be noted that the first occurrence of diatomaceous sediments in the Weddell Sea during the late Eocene may not directly result from further cooling (Kennett and Barker, 1990) but more likely from cooling-induced eutrophy.

CONCLUSIONS

The sequence of evolutionary and biogeographic events in the calcareous nannoplankton can be linked to the late Paleogene climatic deterioration (Figure 13.6). Parallelism between the late early Eocene to Oligocene evolution in the calcareous nannoplankton and planktonic foraminifera is also shown. As evolution in the planktonic foraminifera has been shown to reflect a progressive eutrophication as well as cooling during the late Paleogene, I conclude that the similar evolution of the calcareous nannoplankton not only reflects temperature changes but also a contraction of the TRC. As a result, biogeographic gradients cannot be regarded as solely reflecting temperature gradients as indicated by Wei and Wise (1990a).

The response of the nannoplankton to climatic deterioration has been two-fold. It has consisted of a reduction of regional diversity. This is what happened initially, for instance, when cooling was minimal and the TRC slightly contracted. Reduction in regional diversity also took place at high and mid latitudes throughout the middle and late Eocene as a result of the combined effect of decreasing temperatures and increased eutrophication (resulting in further contraction of the TRC). However, in mid- and low latitudes, a peak in diversity occurred in the middle

Eocene, resulting from the co-existence of older lineages (categories 2 and 3) with taxa (categories 1 and 4) which evolved in response to the progressive environmental changes. It has also consisted of evolutionary turnovers and, supposedly when temperatures were lowest and when further eutrophication had eliminated (deeper) habitats, of extinction without further speciation. Extinction first affected the warm water taxa (e.g., *Sphenolithus radians*) and those most adapted to oligotrophic conditions (i.e., the early Eocene species of *Discoaster*), mainly forms which evolved in the early Eocene (see Figure 13.1), subsequently the more temperate species and those adapted to mesotrophic conditions. It should be noted that extinctions affected only taxa in the low and middle latitudes during the middle and late Eocene (except in the northern Atlantic where data are biased by the effect of the Gulf Stream). In contrast, the early Oligocene extinctions affected all latitudes. At this time, the last but one of the long-ranging species (category 3) which evolved in the early Eocene, several of the long-ranging taxa (category 4) which evolved in the middle Eocene in response to climatic deterioration, and all the short-ranging taxa (category 5) which evolved in the late Eocene became extinct. In particular, the *Reticulofenestra umbilicus* group which extended from polar regions to the equator became extinct. While there may be some diachrony in this event (Wei and Wise, 1990a; Aubry, 1991) which has yet to be established satisfactorily, it seems that this is the only time in the Paleogene when an extinction event occurred globally. Because of its global nature, it would be difficult to assume that it simply reflects an increased global cooling. The *R. umbilicus* group developed during the middle Eocene and was probably adapted to the prevalent middle and late Eocene mesotrophic conditions. Its extinction is likely linked to early Oligocene change in global circulation and fertility (Fenner, 1986; Boersma et al., 1987) in addition to a temperature decrease, and may be related to a loss of its habitat through increased eutrophication. This suggests that the $\delta^{18}O$ peak at 35.8 Ma is more likely related to ice growth (resulting in intensified upwelling and enrichment in nutrients of the euphotic zone) than to a sharp drop in temperatures at southern high latitudes. Thus, the late early Oligocene (early Biochron NP23; with the coolest temperatures and the maximum contraction of the TRC) can be seen as the antithesis of the latest Paleocene-earliest Eocene when warm temperatures, sluggish oceanic circulation and maximum extension of the TRC allowed warm water taxa to penetrate into the high latitudes.

The habitats and life strategies of the various Paleogene calcareous nannoplankton species remain to be established. This is a difficult endeavour. Because of their minute size, individual species cannot be isolated for direct measurement of their isotopic values which would indicate their depth range in the euphotic zone. Most of the living coccolithophorids in the open ocean appear to be extremely K-selected, and only a few species (e.g., *Gephyrocapsa oceanica*, *Emiliania huxleyi*) have characteristics of r-strategists (Kilham and Soltau Kilham, 1980). I have suggested that the early *Discoaster* species were deep dwelling K-mode specialists. This is based on the assumption that species adapted to the deeper part of the euphotic zone under highly oligotrophic conditions (expanded TRC) would be the first affected by eutrophication (see Hallock, 1987, Fig. 5). Although there is no obvious relationship between morphology/ structure of the coccoliths in the living Coccolithophoridae, it can be speculated that the robust, thick discoasters with one or two heavy stem(s) or knob(s) characteristic of the early Eocene were adapted for living in the deeper part of the euphotic zone. I have also suggested that *R. umbilicus* and related forms were adapted to mesotrophic conditions. It is likely that species which evolved in the middle and late Eocene and occupied the niches vacated as forms adapted to more oligotrophic conditions became extinct, were also adapted to mesotrophic conditions. *Reticulofenestra reticulata*, often regarded as a warm water taxon, may have been adapted to mesotrophic conditions, which would explain its abundance in middle and early late Eocene assemblages from high latitudes and in middle and late Eocene assemblages at oceanic sites close to a continental source of nutrients (DSDP Site 363, Gulf of Mexico) and in epicontinental basins. *Chiasmolithus altus*, *C. oamaruensis*, *R.*

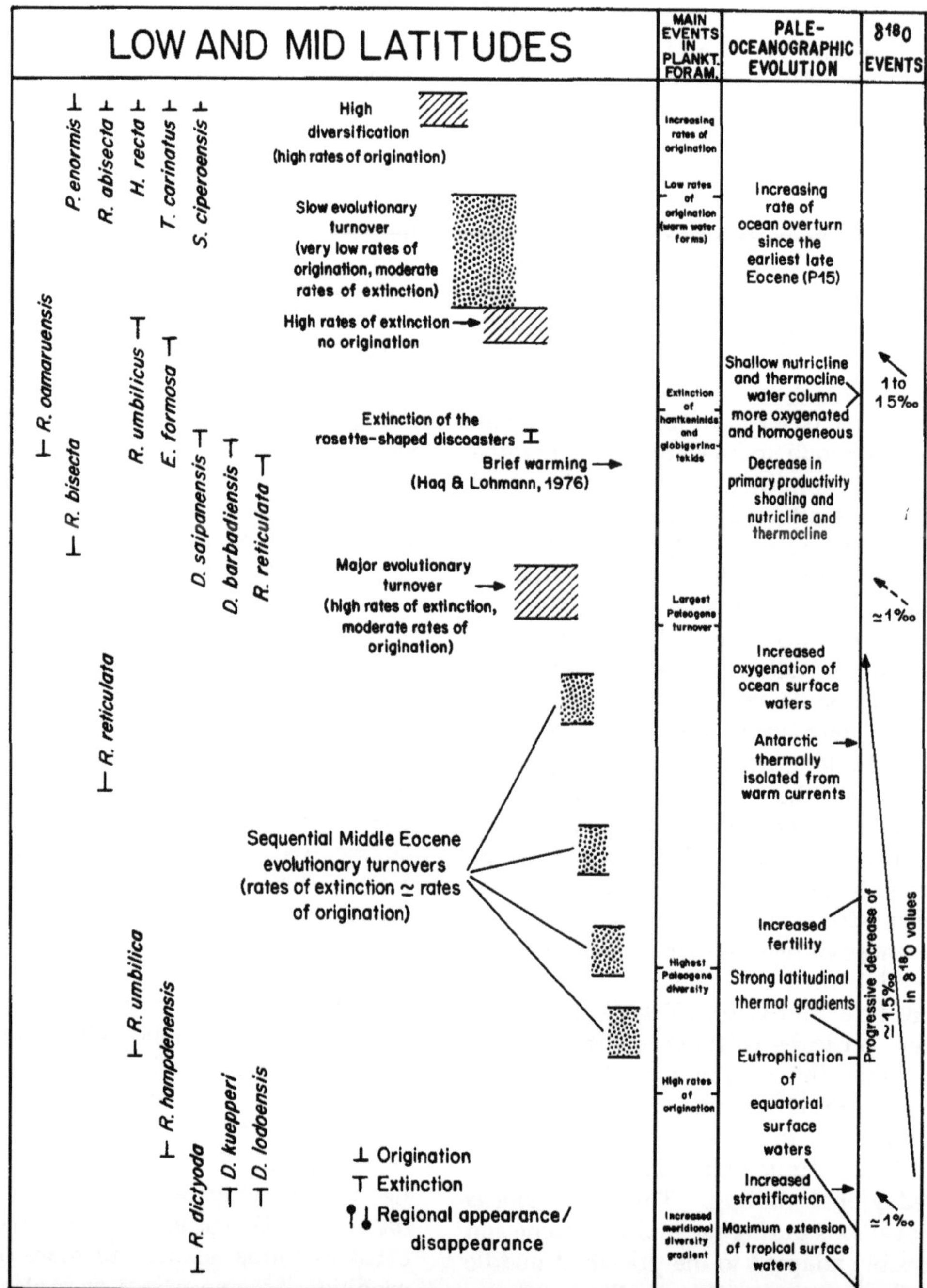

FIGURE 13.6. Chronology of the events which occurred in the calcareous nannoplankton in response to early middle Eocene to early Oligocene climatic deterioration. Magnetobiochronologic framework from Berggren et al. (1985). Oxygen isotope chronology from Corliss and Keigwin (1986), Katz and Miller (1991) and Zachos et al. (1991). Paleoceanographic evolution and chronology of events in the planktonic foraminifera from Boersma et al. (1987) and Hallock et al. (1991).

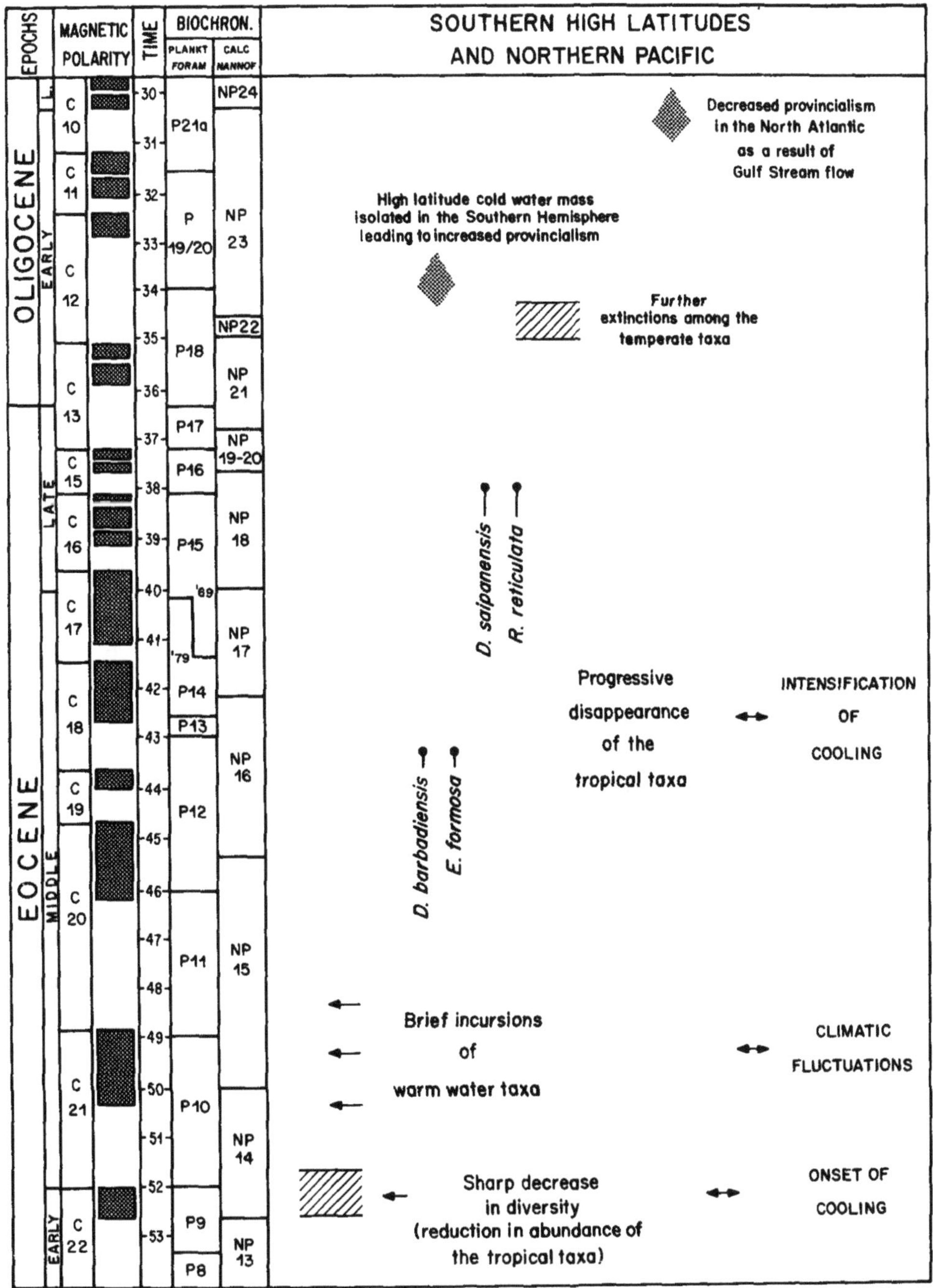
EPOCHS
MAGNETIC POLARITY
TIME
BIOCHRON.
PLANKT FORAM
CALC NANNOF
SOUTHERN HIGH LATITUDES AND NORTHERN PACIFIC
OLIGOCENE
EOCENE
L.
EARLY
LATE
MIDDLE
EARLY
C 10
C 11
C 12
C 13
C 15
C 16
C 17
C 18
C 19
C 20
C 21
C 22
P21a
P 19/20
P18
P17
P16
P15
'89
'79
P14
P13
P12
P11
P10
P9
P8
NP24
NP 23
NP22
NP 21
NP 19-20
NP 18
NP 17
NP 16
NP 15
NP 14
NP 13
Decreased provincialism in the North Atlantic as a result of Gulf Stream flow
High latitude cold water mass isolated in the Southern Hemisphere leading to increased provincialism
Further extinctions among the temperate taxa
D. saipanensis
R. reticulata
Progressive disappearance of the tropical taxa
INTENSIFICATION OF COOLING
D. barbadiensis
E. formosa
Brief incursions of warm water taxa
CLIMATIC FLUCTUATIONS
Sharp decrease in diversity (reduction in abundance of the tropical taxa)
ONSET OF COOLING

floridana and *R. hampdenensis* which are very abundant in late Eocene and Oligocene assemblages from southern high latitudes may be regarded as opportunists, adapted to more eutrophic conditions and able to form massive blooms (Aubry, in progress).

EPILOGUE

This study has shown the great sensitivity of the calcareous nannoplankton to climatic fluctuations. Temperature and fertility fluctuations have driven changes in species distribution, extinctions and speciation throughout the late Paleogene. The (weak) extinction peak that has been shown to occur in the paleontologic record at the Eocene/Oligocene boundary (Raup and Sepkoski, 1984, 1986; Sepkoski, 1990) and the late Eocene occurrence of impact-related microtektites (see review in Miller et al., 1991a) appear to be more coincidental than related. This study shows that the times of greater extinctions should not necessarily be linked to impact events as has been suggested (Alvarez, 1986; Davis et al., 1984; Raup, 1990), but should be viewed as the outcome of long-term paleoclimatic/paleoceanographic change.

This study supports the view that climatic changes served as major agents of extinction (Stanley, 1984). I have shown that during the middle to late Eocene extinctions in the calcareous nannoplankton affected the progressively refrigerated mid and low-latitude regions while high-latitude species migrated equatorwards. While temperature is often regarded as the dominant cause of extinction, other directly or indirectly related parameters such as trophic resources may have been more determinant although less apparent agents of extinction. Global early Oligocene extinctions in the calcareous nannoplankton may not primarily reflect a further drop in temperature, but mainly a threshold in eutrophication as a result of ice build-up and ocean mixing. It is interesting to note that most of the tropical species which evolved in the late Oligocene do not stem from high-latitude taxa, but evolved from few warm-water remnants, as shown by the early late Oligocene radiations in the genera *Sphenolithus* and *Helicosphaera*.

ACKNOWLEDGEMENTS

I am grateful to the conveners (Drs. D. Prothero and W. A. Berggren) of the Penrose Conference on "Eocene-Oligocene Climatic and Biotic Evolution," Rapid City, South Dakota, August 1989, for inviting me to present this paper. I am thankful to A. Boersma, who drew my attention to the trophic resource continuum concept and for fruitful discussions on its implications for Paleogene biogeography, to J. C. Steinmetz and K. G. Miller for reviewing the manuscript, and to Jack Cook (W.H.O.I.) for preparing the figures. This is Woods Hole Oceanographic Institution Contribution number 7792.

REFERENCES

Adams, C. G., Lee, D. E., Rosen, B. R. 1990. Conflicting isotopic and biotic evidence for tropical sea-surface tempeartures during the Tertiary. *Palaeogeogr., Palaeoclimat., Paleoecol.* 77: 289-313.

Alvarez, W. 1986. Toward a theory of impact crises. *EOS* 67: 649, 653-655, 658.

Alvarez, W., Asaro, F., Michel, H. V., and Alvarez, L. 1982. Iridium anomaly approximately synchronous with terminal Eocene extinctions. *Science* 216: 886-888.

Armentrout, J. M,. and Worsley, T. R. 1980. Tertiary calcareous nannofossils from southwestern Washington. *Newslett. Stratig.* 9: 58-71

Aubry, M.-P. 1983a. Biostratigraphie du Paléogène épicontinental de l'Europe du Nord-Ouest. Etude fondée sur les nannofossiles calcaires. *Docum. Lab. Géol. Lyon,* 89, 317 p.

Aubry, M.-P. 1983b. Late Eocene to early Oligocene calcareous nannoplankton biostratigraphy and biogeography. *Amer. Assoc. Petrol. Geol. Bull.* 67:415.

Aubry, M.-P. 1984. *Handbook of Cenozoic calcareous nannoplankton, Book 1: Ortholithae (Discoasters).* New York, Micropaleontology Press, 1, 266p.

Aubry, M.-P. 1986. Paleogene calcareous nannoplankton biostratigraphy of northwestern Europe. *Palaeogeogr., Palaeoclimat., Palaeoecol.* 55: 267-334.

Aubry, M.-P. 1988. Phylogeny of the Cenozoic calcareous nannoplankton genus *Helicosphaera. Paleobiology* 14(1): 64-80.

Aubry, M.-P. 1989. Phylogenetically based calcareous nannofossil taxonomy: Implications for the interpretation of geological events. In Crux, J. A., and Van Heck, S., eds., *Nannofossils and their applications*, London, Ellis Horwood Ltd., pp. 21-40.

Aubry, M.-P. 1991. Paleogene calcareous nannofossils from the Kerguelen Plateau, Leg 120. In Wise, S. W., Jr., Schlich, R. et al., *Proc. Ocean Drilling Program, Scientific Results*, 120: (in press)

Aubry, M.-P., Berggren, W. A., Kent, D. V., Flynn, J. J., Klitgord, K. D., Obradovich, J. D., and Prothero, D. 1988. Paleogene geochronology: An integrated approach. *Paleoceanography*, 3: 707-747.

Backman, J. 1987. Quantitative calcareous nannofossil biochronology of middle Eocene through early Oligocene sediment from DSDP Sites 522 and 523. *Abh. Geol. B.-A.* 39: 21-32.

Backman, J., and Hermelin, O. 1986. Morphometry of the Eocene nannofossil *Reticulofenestra umbilicus* lineage and its biochronological consequences. *Palaeogeogr., Palaeoclimat., Palaeoecol.* 57: 103-116.

Báldi-Beke, M. 1971. The Eocene nannoplankton of the Bakony Mountains, Hungary. *Ann. Inst. Geol. Publ. Hung.* 54: 13-39.

Barron, J. A., Bukry, D., Poore, R. Z. 1984. Correlation of the middle Eocene Kellogg Shale of northern California. *Micropaleontology*, 30: 138-170.

Beaufort, L. 1991. Dynamique du nannoplankton calcaire au cours du Néogène: Implications climatiques et océanographiques. *Docum. Lab. Géol. Lyon*, 151 p., in press.

Beaufort, L., and Aubry, M.-P. 1991. Paleoceanographic implications of a 17-m.y.-long record of high-latitude Miocene calcareous nannoplankton fluctuations. In Wise, S. W., Jr., Schlich, R. et al., *Proc. Ocean Drilling Program, Scientific Results*, 120: (in press)

Beckmann, J.-P., Bolli, H. M., Perch-Nielsen, K., Proto Decima, F., Saunders, J. B., and Toumarkine, M. 1981. Major calcareous nannofossil and foraminiferal events between the middle Eocene and the early Miocene. *Palaeogeogr., Palaeoclimat., Palaeoecol.* 36: 155-190.

Berger, W. H. 1976. Biogenous deep sea sediments: Production, preservation and interpretation. In Riley, J. P., and Chester, R., eds., *Chemical Oceanography*, 5, 2nd edit., London, Academic Press, pp. 265-388.

Berggren, W. A. 1969. Rates of evolution in some Cenozoic planktonic foraminifera. *Micropaleontology* 15: 351-365.

Berggren, W. A., Kent, D. V., and Flynn, J. J. 1985. Paleogene geochronology and chronostratigraphy. In Snelling, N. J., ed., *The chronology of the geological record, Geol. Soc. London, Mem.* 10: 141-195.

Birkenes, E., and Braarud, T. 1952. Phytoplankton in the Oslo fjord during a "*Coccolithus huxleyi*-summer". *Avh. Norske vidensk. Akad. Oslo*, I Math. Nat. Kl, 2: 1-23.

Blechschmidt, G. and Worsley, T. R. 1975. Inter-ocean provincialism of Oligocene calcareous nannofossils. *Geol. Soc. Amer. Abst. prog.* 7: 1002.

Blechschmidt, G. and Worsley, T. R. 1977. Oligocene calcareous nannofossil distribution in deep sea. *Bull. Amer. Assoc. Petrol. Geol.* 61: 769.

Blow, W. H. 1969. Late middle Eocene to Recent planktonic foraminiferal biostratigraphy. In Bronnimann, P. R., and Renz, H. H., Eds, *Proceedings of the First International Conference on Planktonic microfossils, Geneva, 1976*, Leiden, E. J. Brill, 1: 199-421.

Boersma, A., and Shackleton, N. J. 1977. Tertiary oxygen and carbon isotope stratigraphy, site 357 (mid latitude South Atlantic Ocean). In Supko, P. R., Perch-Nielsen, K., et al., *Init. Rept. Deep Sea Drill. Proj.* Washington, U.S. Government Printing Office, 39: 911-924.

Boersma, A., Premoli-Silva, I., and Shackleton, N. 1987. Atlantic Eocene planktonic foraminiferal paleohydrographic indicators and stable isotope paleoceanography. *Paleoceanography*, 2: 287-331.

Boudreaux, J. E. 1974. Calcareous nannoplankton ranges, Deep Sea Drilling Project Leg 23. In Whitmarsh, R. B., Weser, O. E., Ross, D. A., et al., *Init. Rept. Deep Sea Drill. Proj.* Washington, U.S. Government Printing Office, 23: 1073-1090.

Bukry, D. 1973a. Coccoliths and silicoflagellates from Deep Sea Drilling Project Leg 19, North Pacific Ocean and Bering Sea. In Creager, J. S., and Scholl, D. W., et al., *Initial Reports of the Deep Sea Drilling Project*, Washington, U.S. Government Printing

Office, 19: 857-867.

Bukry, D. 1973b. Coccolith and silicoflagellate stratigraphy, Tasman Sea and southwestern Pacific Ocean, Deep Sea Drilling Project Leg 21. In Burns, R. E., Andrews, J. E. et al., *Init. Rept. Deep Sea Drill. Proj.* Washington, U.S. Government Printing Office, 21: 885-893.

Bukry, D. 1975. Coccolith and silicoflagellate stratigraphy, northwestern Pacific Ocean, Deep Sea Drilling Project Leg 32. In Larson, R. L., Moberly, R. et al., *Init. Rept. Deep Sea Drill. Proj.* Washington, U.S. Government Printing Office, 32: 677-701.

Bukry, D. 1978. Biostratigraphy of Cenozoic marine sediment by calcareous nannofossils. *Micropaleontology* 24: 44-60.

Bukry, D. 1980. Coccolith stratigraphy, tropical eastern Pacific Ocean, Deep Sea Drilling Project Leg 54. In Rosendahl, B. R., Hekinian, R., et al., *Init. Rept. Deep Sea Drill. Proj.* Washington, U.S. Government Printing Office, 54: 535-543.

Bukry, D. and Snavely, P. D. 1988. Coccolith zonation for Paleogene strata in the Oregon Coast range. In Filewicz, M. V., and Squires, R. L., eds., *Paleogene stratigraphy, West Coast of North America, Pacific Section, S.E.P.M. West Coast Paleogene Symposium,* 58: 251-263.

Burns, D. A. 1973. The latitudinal distribution and significance of calcareous nannofossils in the bottom sediments of the south-west Pacific Ocean (lat. 15-55°S) around New Zealand. In *Oceanography of the South Pacific 1972,* Fraser, R., comp., New Zealand National Commission for UNESCO, p. 221-228.

Bybell, L. 1975. Middle Eocene calcareous nannofossils at Little Stave Creek, Alabama. *Tulane Stud. Geol. Paleontol.* 11: 177-252.

Bybell, L. 1982. Late Eocene to early Oligocene calcareous nannofossils in Alabama and Missisipi. *Trans. Gulf Coast Assoc. Geol. Soc.* 32: 295-302.

Cavelier, C. 1975. Le diachronisme de la Zone à *Ericsonia subdisticha* (nannoplancton) et la position de la limite Eocène/Oligocène en Europe et en Amérique du Nord. *Bull. Bur. Rech. géol. min.* (2), sec. IV, 3, 201-225.

Cita, M. B., Nigrini, C., and Gartner, S. 1970. Biostratigraphy. In Peterson, M. N. A., et al., *Init. Rept. Deep Sea Drill. Proj.* Washington, U.S. Government Printing Office, 2: 391-411.

Coccioni, R., Monaco., P., Monechi, S., Nocchi, M., and Parisi, G. 1988. Biostratigraphy of the Eocene-Oligocene boundary at Massignano (Ancona, Italy). In Coccioni, R., et al., *Int. Subcomm. Paleog. strat., E/O Meeting, Ancona, October 1987, spec. Publ.* II.1, 59-96.

Copeland, H. F. 1956. *The classification of lower organisms.* Palo Alto, Calif., Pacific Books, 302 p.

Corliss, B. H., and Keigwin, L. D. 1986. Eocene-Oligocene paleoceanography. In Hsü, K. J., ed., *Mesozoic and Cenozoic oceans, Amer. Geophys. Union, Geodynamics Ser.* 15: 101-118.

Corliss, B. H., Aubry, M.-P., Berggren, W. A., Fenner, J., Keigwin, L. D., Jr., and Keller, G. 1984. The Eocene/Oligocene boundary in the deep sea. *Science* 226: 806-810.

Davis, M., Hut, P., and Muller, R. A. 1984. Extinction of species by periodic comet showers. *Nature* 308: 715-717.

Doebl, F., Müller, C., Schuler, M., Sittler, C., and Weiler, H. 1976. Les marnes à foraminifères et les schistes à poissons de Bremmelbach (Bas-Rhin). Etudes sédimentologiques et micropaléontologiques. Reconstitution du milieu au début du Rupélien dans le fossé rhénan. *Sci. Géol. Bull.* 29: 285-320.

Edwards, A. R. 1971. A calcareous nannoplankton zonation of the New Zealand Paleogene. In Farinacci, A., ed., *Proceedings of the Second Conference on Planktonic Microfossils, Roma, 1970,* Rome, Ediz. Tecnoscienza, 1: 381-419.

Edwards, A. R. 1973a. Key species of New Zealand calcareous nannofossils. *New Zealand Jour. Geol. Geophys.* 16: 68-89.

Edwards, A. R. 1973b. Calcareous nannofossils from the southwest Pacific, Deep Sea Drilling Project, Leg 21. In Burns, R. E., Andrews, J. E., et al., *Init. Rept. Deep Sea Drill. Proj.* Washington, U.S. Government Printing Office, 21: 641-691.

Edwards, A. R. and Perch-Nielsen, K. 1975. Calcareous nannofossils from the southern southwest Pacific, Deep Sea Drilling Project, Leg 29. In Kennett, J. P., Houtz, R. E., et al., *Init. Rept. Deep Sea Drill. Proj.* Washington, U.S. Government Printing Office, 21: 469-539.

Ellis, C. H. 1975. Calcareous nannofossil biostratigraphy - Leg 31, DSDP. In Karig, D. E., Ingle, J. C., Jr., et al., *Init. Rept. Deep Sea Drill. Proj.*Washington, U.S. Government Printing Office, 31: 655-676.

Fenner, J. 1986. Infomation from diatom analysis concerning the Eocene-Oligocene boundary. In Pomerol, C., and Premoli-Silva, I., *Terminal Eocene Events,* New York, Elsevier, pp. 283-287.

Fincham, M. J., and Winter, A. 1989. Paleoceanographic interpretations of coccoliths and oxygen-isotopes from the sediment surface of the southwest Indian Ocean. *Mar. Micropaleontol.* 13: 325-351.

Frakes, L. A. 1979. *Climates throughout geological time,* New York, Elsevier, 310p.

Ganapathy, R. 1982. Evidence for a major meteorite impact on the Earth 34 million years ago: Implication for Eocene extinctions. *Science* 216: 885-886.

Gartner, S. 1971. Calcareous nannofossils from the Joides Blake Plateau cores, and revision of Paleogene nannofossil zonation. *Tulane Stud. Geol. Paleontol.* 8: 101-121.

Hallock, P. 1987. Fluctuations in the trophic resource continuum: A factor in global diversity cycles? *Paleoceanography* 2: 457-471.

Hallock, P., Premoli-Silva, I., and Boersma, A. 1991. Similarities between planktonic and larger foraminiferal evolutionary trends through Paleogene paleoceanographic changes. *Palaeogeogr., Palaeoclimat., Palaeoecol.* 83: 49-64.

Haq, B. U. 1971. Paleogene calcareous nannoflora. Part IV: Paleogene nannoplankton biostratigraphy and evolutionary rates in Cenozoic calcareous nannoplankton. *Stockholm Contr. Geol.* 24:129-158.

Haq, B. U. 1973. Transgressions, climatic change and the diversity of calcareous nannoplankton. *Mar. Geol.* 15:M25-M30.

Haq, B. U. 1980. Biogeographic history of the Miocene calcareous nannoplankton and paleoceanography of the Atlantic Ocean. *Micropaleontology* 26: 414-443.

Haq, B. U. 1981. Paleogene paleoceanography: Early Cenozoic oceans revisited. *Oceanol. Acta,* Actes 26e congrès International de Géologie, Colloque Géologie des océans, Paris, 7-17 juillet 1980, 71-82.

Haq, B. U., and Lohmann, G. P. 1976a. Early Cenozoic calcareous nannoplankton biogeography of the Atlantic Ocean. *Mar. Micropaleontol.* 1: 119-194.

Haq, B. U., and Lohmann, G. P. 1976b. Remarks on the Oligocene calcareous nannoplankton biogeography of the Norwegian Sea (DSDP Leg 38). In Talwani, M., Udintsev, G., et al., *Init. Rept. Deep Sea Drill. Proj.* Washington, U.S. Government Printing Office, 38: 141-145.

Haq, B. U., Premoli-Silva, I., and Lohmann, G. P. 1977. Calcareous plankton paleobiogeographic evidence for major climatic fluctuations in the early Cenozoic Atlantic Ocean. *Jour. Geophys. Res.* 82: 3861-3876.

Hibberd, D. J., and Leedale, G. F. 1985. Order 7. Prymnesiida. In Lee, J. J., ed., *Illustrated Guide to the Protozoa,* Soc. Protozoologists, pp. 74-87.

Katz, M. E., and Miller, K. G. 1991. Early Paleogene benthic foraminiferal assemblages and stable isotopes in the southern ocean. In Cieselski, P. F., Kristoffersen, Y., et al., 1991. *Proceedings of the Ocean Drilling Program, Scientific Results,* 114: 481-512.

Kennett, J. P., Houtz, R. E., et al. 1975a. Cenozoic paleoceanography in the southwest Pacific Ocean, Antarctic glaciation, and the development of the circum-Antarctic current. In Kennett, J. P., Houtz, R. E., et al., *Init. Rept. Deep Sea Drill. Proj.* Washington, U.S. Government Printing Office), 29: 1155-1169.

Kennett, J. P., Houtz, R. E., et al., 1975b. *Init. Rept. Deep Sea Drill. Proj.* Washington, U.S. Government Printing Office, 29: 45-120.

Kennett, J. P., and Barker, P. F. 1990. Latest Cretaceous to Cenozoic climate and oceanographic developments in the Weddell Sea, Antarctica: An ocean-drilling perspective. In Barker, P. F., Kennett, J. P., et al., *Proceedings of the Ocean Drilling Program, Scientific Results,* 113: 937-960.

Kilham, P., and Soltau Kilham, S. 1980. The evolutionary ecology of phytoplankton. In Morris, I., ed., *The physiological ecology of phytoplankton. Studies in Ecology* 7: 493-570.

Kleijne, A., Kroon, D., and Zevenboom, W. 1989. Phytoplankton and foraminiferal frequencies in the northern Indian Ocean and Red Sea surface waters. *Netherlands Jour.*

Sea Res. 24: 531-539.

Lowrie, W., Alvarez, W.,Napoleone, G., Perch-Nielsen, K., Premoli-Silva, I., and Toumarkine M. 1982. Paleogene magnetic stratigraphy in Umbrian pelagic carbonate rocks: The Contessa sections, Gubbio. *Geol. Soc. Amer. Bull.* 93: 411-432.

Ludwig, W. J., Krasheninnikov, V. A., et al. 1983. *Init. Rept. Deep Sea Drill. Proj.* Washington, U.S. Government Printing Office, 29: 21-109.

Madile, M., and Monechi, S. 1991. Late Eocene to early Oligocene calcareous nannofossil assemblages from sites 699 and 703, subantarctic South Atlantic Ocean, In Ciesielski, P. F., Kristoffersen, Y., et al., *Proceedings of the Ocean Drilling Program, Scientific results,* 114: 179-192.

Margulis, L., and Schwartz, K. V. 1988. *Five kingdoms,* New York, W.H. Freeman and Co., 2nd edit., 376 p.

Martini, E. 1965. Mid-Tertiary calcareous nannoplankton from Pacific deep-sea cores. *Proc. 17th Symposium Colston Res. Soc.* Colston Papers 17: 393-411.

Martini, E. 1970. Standard Paleogene calcareous nannofossil zonation. *Nature,* 226: 560-561.

Martini, E. 1971. Standard Tertiary and Quaternary calcareous nannoplankton zonation. In Farinacci, A., ed., *Proceedings of the Second Conference on Planktonic Microfossils, Roma, 1970,* Rome, Ediz. Tecnoscienza, 2: 739-785.

Martini, E. 1979. Calcareous nannoplankton and silicoflagellate biostratigraphy at Reykjanes ridge, northeastern North atlantic (DSDP Leg 49, Sites 407 and 409). In Luyendyk, B. P., Cann, J. R., et al., *Init. Rept. Deep Sea Drill. Proj.* Washington, U.S. Government Printing Office), 49: 533-549.

Martini, E. and Worsley, T. 1971. Tertiary calcareous nannoplankton from the western equatorial Pacific. In Winterer, E. L., et al., *Init. Rept. Deep Sea Drill. Proj.* Washington, U.S. Government Printing Office, 7(2): 1471-1507.

Matthews, R. K. 1984. Oxygen isotope record of ice-volume history: 100 million years of glacio-eustatic sea-level fluctuation. In Schlee, J. S., ed., Interregional unconformities and hydrocarbon accumulation. *Amer. Assoc. Petrol. Geol. Mem.* 36: 97-107.

McIntyre, A., and Bé, A. H. W. 1967. Modern Coccolithophoridae of the Atlantic Ocean. *Deep Sea Res.* 14: 561-597.

Miller, K. G., Feigenson, M. D., Kent, D. V., and Olsson, R. K. 1988. Oligocene stable isotope ($^{87}Sr/^{86}Sr$, $\delta^{18}O$, $\delta^{13}C$) standard section, Deep Sea Drilling project Site 522. *Paleoceanography* 3: 223-233.

Miller, K. G., Fairbanks, R. G., and Mountain, G. S. 1987. Tertiary oxygen isotope synthesis, sea level history, and continental margin erosion. *Paleoceanography* 2: 1-19.

Miller, K. G., Berggren, W. A., Zhang, J., and Palmer-Judson, A. A. 1991a. Biostratigraphy and isotope stratigraphy of upper Eocene microtektites at Site 612: How many impacts? *Palaios* 6: 17-38.

Miller, K. G., Wright, J. D., and Fairbanks, R. G. 1991b. Unlocking the ice House: Oligocene-Miocene oxygen isotopes, eustasy, and margin erosion. *Jour. Geophys. Res.* 96B: 6829-6848

Müller, C. 1976. Tertiary and Quaternary calcareous nannoplankton in the Norwegian-Greenland sea, DSDP, Leg 38. In Talwani, M., Udinstev, G. et al., *Init. Rept. Deep Sea Drill. Proj.* Washington, U.S. Government Printing Office, 38: 823-841.

Müller, C. 1979. Calcareous nannofossils from the North Atlantic, Leg 48. In Montadert, L., Roberts, D. G. et al., *Init. Rept. Deep Sea Drill. Proj.* Washington, U.S. Government Printing Office, 48: 589-639.

Nocchi, M., Parisi, G., Monaco, P., Monechi., S., and Madile, M. 1988a. Eocene and early Oligocene micropaleontology and paleoenvironments in SE Umbria, Italy. *Palaeogeogr., Palaeoclimat., Palaeoecol.* 67: 181-244.

Nocchi, M., Monechi, S., Coccioni, R., Madile, M., Monaco, P., Orlando, M., Parisi, G., and Premoli-Silva, I. 1988b. The extinction of Hantkeninidae as a marker for recognizing the Eocene/Oligocene boundary: A proposal. In Coccioni, R., et al., *Int. Subcomm. Paleog. strat., E/O Meeting, Ancona, October 1987,* spec. Publ., V.1, 249-252.

Okada, H., and Bukry, D. 1980. Supplementary modification and introduction of code numbers to the low-latitude coccolith biostratigraphic zonation (Bukry, 1973; 1975). *Mar. Micropaleontol.* 5: 321-325.

Okada, H., and Honjo, S. 1973. The distribution of oceanic coccolithophorids in the Pacific. *Deep Sea Res.* 20: 355-374.

Okada, H., and Honjo, S. 1975. Distribution of Coccolithophores in marginal seas along the western Pacific Ocean and in the Red Sea. *Mar. Biol.* 31: 271-285.

Perch-Nielsen, K. 1972. Remarks on Late Cretaceous to Pleistocene coccoliths from the North Atlantic. In Laughton, A. S., Berggren, W. A. et al., *Init. Rept. Deep Sea Drill. Proj.* Washington, U.S. Government Printing Office, 12: 1003-1071.

Perch-Nielsen, K. 1977. Albian to Pleistocene calcareous nannofossils from the wsetern South Atlantic, DSDP Leg 39. In Supko, P. R., Perch-Nielsen, K., et al., *Init. Rept. Deep Sea Drill. Proj.* Washington, U.S. Government Printing Office, 39: 699-823.

Perch-Nielsen, K. 1981. Nouvelles observations sur les nannofossiles calcaires à la limite Crétacé-Tertiaire près de El Kef (Tunisie). *Cah. Micropaléontol.* 3:25-36.

Perch-Nielsen, K. 1985. Cenozoic calcareous nannofossils. In Bolli, H. M., Saunders, J. B., and Perch-Nielsen, K., eds., *Plankton Stratigraphy,* Cambridge, Cambridge Univ. Press, pp. 427-554.

Perch-Nielsen, K., McKenzie, J., and He, Q. 1982. Biostratigraphy and isotope stratigraphy and the 'catastrophic' extinction of the calcareous nannoplankton at the Cretaceous/Tertiary boundary. *Geol. Soc. Amer. Spec. Paper* 190: 353-371.

Pinet, P. R., Popenoe, P., and Nelligan, D. F. 1981. Gulf Stream: Reconstruction of Cenozoic flow patterns over the Blake Plateau. *Geology* 9: 266-270.

Poore, R. Z., and Bukry, D. 1979. Preliminary report on Eocene calcareous plankton from the eastern Gulf of Alaska continental slope. *Geol. Surv. Circ.* 804B: 141-143.

Pospichal, J. J., and Wise, S. W. 1990. Paleocene to middle Eocene calcareous nannofossils of ODP Sites 689 and 690, Maud Rise, Weddell sea. In Barker, P. F., Kennett, J. P., et al., *Proceedings of the Ocean Drilling Program, Scientific results* 113: 613-638.

Proto Decima, F., Medizza, F., and Todesco, L. 1978. Southeastern Atlantic Leg 40 calcareous nannofossils. In Bolli, H. M., Ryan, W. B. F., et al., *Init. Rept. Deep Sea Drill. Proj.* Washington, U.S. Government Printing Office, 40: 571-634.

Raup, D. M. 1990. Impact as a general cause of extinction; a feasibilty test. *Geol. Soc. Amer. Spec. Pap.* 247: 27-32.

Raup, D. M., and Sepkoski, J. J., Jr. 1984. Periodicity of extinctions in the geologic past. *Proc. Natl. Acad. Sci. USA* 81: 805-801.

Raup, D. M., and Sepkoski, J. J., Jr. 1986. Periodic extinctions of families and genera. *Science* 231: 833-836.

Rea, D. K., Zachos, J. C., Owen, R. M., and Gringerich, P. D. 1990. Global change at the Paleocene/Eocene boundary: Climatic and evolutionary consequences of tectonic events. *Palaeogeogr., Palaeoclimat., Palaeoecol.* 79: 117-128.

Roth, P. H. 1973. Calcareous nannofossils - Leg 17, Deep Sea Drilling Project. In Winter, E. L., Ewing, J. L., et al., *Initial Reports of the Deep Sea Drilling Project,* Washington, U.S. Government Printing Office, 17: 695-795.

Roth, P. H. 1989. Ocean circulation and calcareous nannoplankton evolution during the Jurassic and the Cretaceous. *Palaeogeogr., Palaeoclimat., Palaeoecol.* 74: 111-126.

Roth, P. H., and Berger, W. H. 1975. Distribution and dissolution of coccoliths in the south and central Pacific. In Sliter, W. V., Bé, A. W. H., and Berger, W. H., eds., *Dissolution of deep-sea carbonates, Cushman Foundation for Foraminiferal Research, Spec. Publ.* 13: 87-113.

Roth, P. H., and Krumbach, K. R. 1986. Middle Cretaceous calcareous nannofossil biogeography and preservation in the Atlantic and Indian Oceans: Implications for paleoceanography. *Mar. Micropaleontol.* 10: 235-266.

Saunders, J. B., Bernouilli, D., Müller-Merz, E., Oberhänsli, H., Perch-Nielsen, K., Riedel, W. R., Sanfilippo, A., and Torrini, R. Jr. 1984. Stratigraphy of the late middle Eocene to early Oligocene in the Bath cliff section, Barbados, West Indies. *Micropaleontology* 30: 390-425.

Sclater, J. G., Anderson, R. N., and Bell, M. L. 1971. Elevation of ridges and evolution of the central eastern Pacific. *Jour. Geophys. Res.* 76: 7888-7915.

Sclater, J. G., Abbott, D. and Thiede, J. 1977a. Paleobathymetry and sediments of the Indian Ocean. In Heirtzler, J. R., Bolli, H. M., Davies, T. A., Saunders, J. B., and Sclater, J., *Indian Ocean geology and biostratigraphy,* Washington, D.C., Amer. Geophys. Union, pp. 25-59.

Sclater, J. G., Hellinger, S., and Tapscott, C. 1977b. The paleobathymetry of the Atlantic ocean from the Jurassic to the Present. *J. Geol.* 85: 509-552.

Scholl, D. W.,Creager, J. S. 1973. Geological synthesis of Leg 19 (DSDP) results; Far North Pacific, and Aleutien ridge, and Bering Sea. In Creager, J. S., and Scholl, D. W., et al., *Init. Rept. Deep Sea Drill. Proj.* Washington, U.S. Government Printing Office, 19: 897-913.

Sepkoski, J. J. 1990. The taxonomic nature of periodic extinction. *Geol. Soc. Amer. Spec. Pap.* 247: 33-44.

Shackleton, N. J. 1984. Oxygen isotope evidence for Cenozoic climatic change. In Brenchley, P., ed., *Fossils and climate.* Chichester, Wiley, pp. 27-34.

Shackleton, N. J., and Boersma, A. 1981. The climate of the Eocene ocean. *J. Geol. Soc. London* 138: 153-157.

Shackleton, N. J., and Kennett, J. P. 1975. Paleotemperature history of the Cenozoic and the initiation of Antarctic glaciation: Oxygen and carbon isotope analyses in DSDP Sites 277, 279 and 281. In Kennett, J. P., Houtz, R. E., et al., *Init. Rept. Deep Sea Drill. Proj.* Washington, U.S. Government Printing Office, 29: 743-755.

Shipboard Scientific Party, 1988. Sites 698 and 690. In Barker, K., Kennett, J. P., et al., *Proceedings of the Ocean Drilling Program, Initial Reports* 113: 89-181, 183-292.

Shipboard Scientific Party, 1989. Sites 748 and 749. In Schlich, R., Wise, S. W., et al., *Proceedings of the Ocean Drilling Program, Initial Reports* 120: 157-235, 237-274, 439-568.

Smayda, T. J. 1980. Phytoplankton species succession. In Morris, I., ed., *The physiological ecology of phytoplankton. Studies in Ecology* 7: 493-570.

Smith, A. G., and Briden, J. C. 1977. *Mesozoic and Cenozoic paleocontinental maps.* Cambridge, Cambridge University Press.

Steinmetz, J. C. 1979. Calcareous nannofossils from the North Atlantic Ocean, Leg 49, Deep sea Drilling Project. In Luyendyk, B. P., Cann, J. R., et al., *Init. Rept. Deep Sea Drill. Proj.* Washington, U.S. Government Printing Office, 49: 519-531.

Stanley, S. M. 1984. Temperature and biotic crises in the marine realm. *Geology* 12: 205-208.

Stott, L. D., Kennett, J. P., Shakleton, N. J., and Corfield, R. M. 1990. The evolution of Antarctic surface waters during the Paleogene: Inferences from the stable isotopic composition of planktonic foraminifers, ODP Leg 113. In Barker, P. F., Kennett, J. P., et al., *Proceedings of the Ocean Drilling Program, Scientific Results,* 113: 849-863.

Tappan, H. 1979. Protistan evolution and extinction at the Cretaceous/Tertiary boundary. In Christensen, W. K., and Birkelund, T., eds., *Cretaceous-Tertiary Boundary Events symposium,* II. Proceedings, 13-21.

Tappan, H. 1980. *The Paleobiology of Plant Protists.* New York, W.H. Freeman and Co., 1028 p.

Tappan H. and Loeblich, A. R., Jr. 1971. Geobiologic implications of fossil phytoplankton evolution and time space distribution. In Kosanke, R. and Cross, A. T., eds., *Symposium on palynology of the Late Cretaceous and early Tertiary, Geol. Soc. Amer. Spec. paper* 127: 247-340.

Theodoridis, S. 1984. Calcareous nannofossil biozonation of the Miocene and revision of the helicoliths and discoasters. *Utrecht Micropaleontol. Bull.* 32, 271 p.

Thierstein, H. R. 1974. Calcareous nannoplankton - Leg 26, Deep Sea Drilling Project. In Davies, T. A., Luyendyk, B. P., et al., *Initial Reports of the Deep Sea Drilling Project,* Washington, U.S. Government Printing Office, 26: 619-668.

Wei, W. 1991. Evidence for an earliest Oligocene abrupt cooling in the surface waters of the Southern Ocean. *Geology,* in press.

Wei, W., and Wise, S. W., Jr., 1989. Paleogene calcareous nannofossil magnetochronology: Results from South Atlantic DSDP Site 516. *Mar. Micropaleontol.* 14: 119-152.

Wei, W., and Wise, S. W., Jr. 1990a. Biogeographic gradients of middle Eocene-Oligocene calcareous nannoplankton in the South Atlantic Ocean. *Palaeogeogr., Palae-*

oclimatol., Palaeoecol. 79: 29-61.

Wei, W., and Wise, S. W., Jr. 1990b. Middle Eocene to Pleistocene calcareous nannofossils recovered by ocean drilling Program Leg 113 in the Weddell Sea. In Barker, P. F., Kennett, J. P., et al., *Proceedings of the Ocean Drilling Program, Scientific Results*, 113: 639-666.

Winter, A. 1985. Distribution of living coccolithophores in the California current system, southern California borderland. *Mar. Micropaleontol.* 9: 383-393.

Wise, S. W. 1983. Mesozoic and Cenozoic calcareous nannofossils recovered by Deep Sea Drilling Project Leg 71 in the Falkland Plateau Region, Southwest Atlantic Ocean. In Ludwig, W. J., Krasheninikov, V. A., et al., *Init. Rept. Deep Sea Drill. Proj.* Washington, U.S. Government Printing Office, 71: 481-550.

Wise, S. W., Jr., Gombos, A. M., and Muza, J. P. 1985. Cenozoic evolution of polar water masses, Souhtwest Atlantic Ocean. In Hsü, K. J., and Weissert, H. J., eds., *South Atlantic paleoceanography*, Cambridge, Cambridge University Press, pp. 283-324.

Worsley, T. R. 1973. Calcareous nannofossils: Leg 19 of the Deep Sea Drilling Project. In Creager, J. S., and Scholl, D. W., et al., *Init. Rept. Deep Sea Drill. Proj.* Washington, U.S. Government Printing Office, 19: 741-750.

Worsley, T. R., and Crecelius, E. 1972. Paleogene calcareous nannofossils from the Olympic Peninsula, Washington. *Geol. Soc. Amer. Bull.* 83: 2859-2862.

Worsley, T. R., and Davies, T. A. 1979. Cenozoic sedimentation in the Pacific Ocean: Steps toward a quantitative evaluation. *Jour. Sed. Petrol.* 49: 1131-1146.

Worsley, T. R., and Jorgens, M. L. 1974. Oligocene calcareous nannofossil provinces. In Ross, C. A., ed., *Paleogeographic provinces and provinciality, Soc. Econ. Paleontol. Mineral. Spec. Publ.* 21: 85-108.

Zachos, J. C., Berggren, W. A., Aubry, M.-P., and Mackensen, A. 1991. Eocene-Oligocene paleoclimatic variations of the southern Indian Ocean. In Wise, S. W., Jr., Schlich, R. et al., *Proceedings of the Ocean Drilling Program, Scientific Results*, 120: (in press).

14. MIDDLE EOCENE THROUGH EARLY MIOCENE DIATOM FLORAL TURNOVER

By Jack G. Baldauf

ABSTRACT

A database of occurrence data from the Southern Ocean, the low-latitude Atlantic and Pacific, the Labrador Sea and the Norwegian-Greenland Sea was developed to determine the spatial and temporal response of the diatom flora to oceanographic and climatic changes during the Paleogene and early Neogene. This database provides insight on periods of floral turnover. Four events, exhibiting greater than a 30 percent turnover of the diatom flora, are identified and include the middle Eocene (41-43 Ma), the latest Eocene-earliest Oligocene (~35-37 Ma), the latest Oligocene-earliest Miocene (25-23 Ma), and the late early Miocene (17-19 Ma).

Each period of floral turnover (number of first occurrences vs the number of last occurrences) differs in the magnitude (~30-80 percent of the flora) of turnover, in the number of species disappearing or appearing, and in the geographic extent. The floral changes identified in part reflect species adaptation to increased provincialism resulting from a gradual intensification of latitudinal thermal gradients.

INTRODUCTION

Numerous studies have examined diatoms present in Paleogene and lower Neogene sediments. These include studies by Baldauf and Monjanel (1989), Benda (1972), Cleve-Euler (1941), Fenner (1976, 1977, 1981, 1984a, 1984b, 1985, 1986), Fourtanier (in press), Gombos (1977, 1982, 1983, 1984, and 1987), Gombos and Ciesielski (1983), Grunow (1866), Heiberg (1863), Kanaya (1957), Schrader and Fenner (1976), and Schulz (1927), among others. The majority of these studies document the diatom assemblage from stratigraphic sequences or individual samples for specific biostratigraphic or taxonomic purposes. Few studies have addressed the evolutionary response of the Paleogene and early Neogene diatom flora to changing oceanographic and climatic conditions such as; the development of the cryosphere, change from circum low latitude to circum-polar circulation, intensification of thermal latitudinal gradients or increased thermalhaline circulation (see Barron et al. in press; Shackleton and Kennett 1975, Barron 1987; Miller et al., 1987, among others).

Studies that present a partial perspective on floral turnover include those of Jouse (1978) and Fenner (1981, 1985, 1986). Jouse (1978) examined the stratigraphic range of 50 diatom genera to establish a generic level biostratigraphy. In doing so, she recognized distinct diatom assemblages (at the generic level) which she grouped into temporal intervals. Three of these intervals (Stages II-IV) are Paleogene or early Neogene in age.

Stage II (Jouse, 1978; Paleocene to middle Eocene) is characterized by a period of steady-state, with relatively insignificant appearances of new genera. The assemblage is typified by genera such as *Actinoptychus, Hemiaulus, Sceptroneis, Stephanopyxis, Triceratium* and *Trinacria*. Stage III (Jouse, 1978; late Eocene to late early Oligocene) is characterized by the decrease in dominance of genera in Stage II and the appearance of several new genera including *Cestodiscus* and *Rhizosolenia*. Stage IV (Jouse, 1978; late Oligocene to early Miocene) is characterized by reorganization of the diatom flora with the introduction of several new genera including; *Bogorovia, Synedra, Thalassionema,* and *Thalassiothrix*.

Fenner (1981, 1985, 1986) elaborated on the work of Jouse (1978) by examining the diatom assemblages at the species rather than the genus level. In doing so, Fenner (1985) defined four intervals (early/middle Eocene boundary, late middle Eocene, Eocene/Oligocene

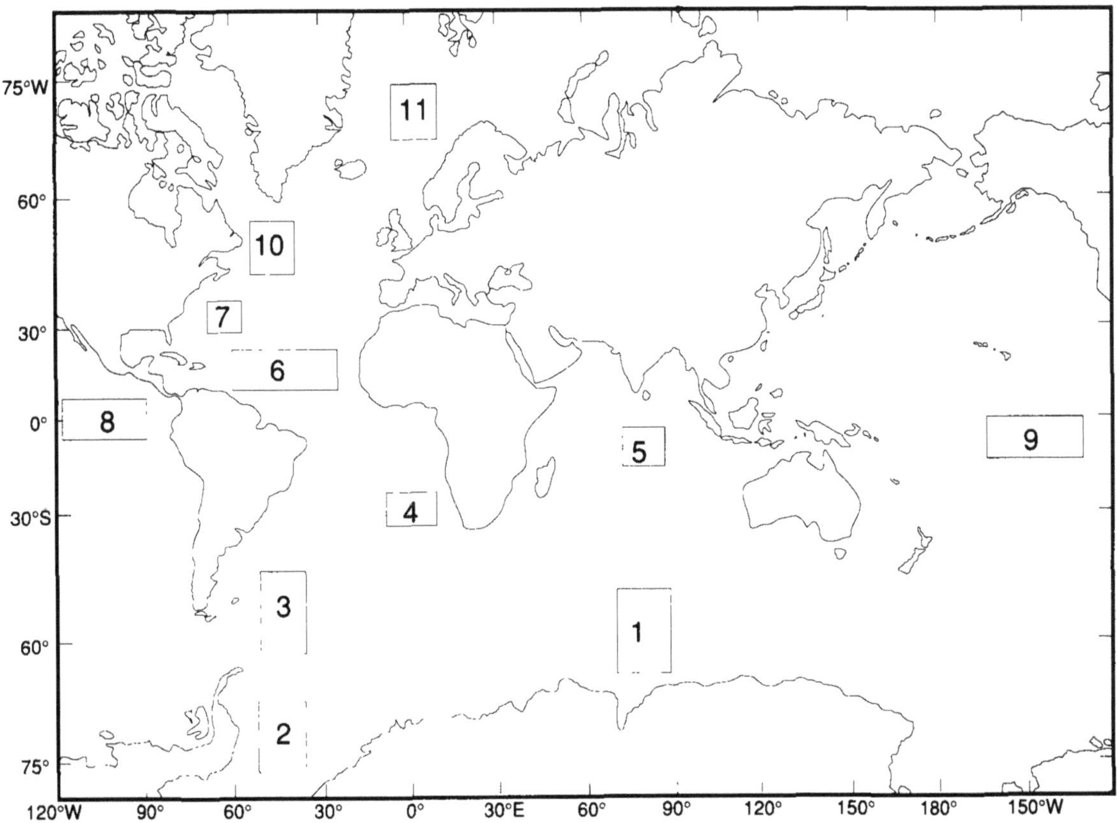

FIGURE 14.1. Location of selected studies of Paleogene diatoms from DSDP and ODP drill holes. 1=Baldaulf and Barron (in press) and Harwood and Maruyama (in press); 2=Gersonde and Burckle (in press); 3=Fenner (in press), Gombos (1976, 1983), and Gombos and Ciesielski (1983); 4=Gombos (1984); 5=Fountainer (in press); 6=Fenner (1977, 1981, 1984a, 1984b, 1987); 7=Gombos (1987); 8=Fenner (1984a, 1984b, 1987), Barron (1983); 9=Fenner (1984a, 1984b, 1987); 10=Baldauf and Monjanel (1989); 11=Schrader and Fenner (1976) and Fenner (1987).

boundary, and late early Oligocene) of floral turnover during the Paleogene. The early/middle Eocene interval is characterized by a gradual replacement of the Paleocene and early Eocene diatom assemblage with numerous new species. This gradual floral replacement continues until the end of the middle Eocene when previously dominant species disappear representing Fenner's second interval of floral turnover.

The third interval of floral change, as identified by Fenner (1985) approximates the Eocene/Oligocene boundary. This interval corresponds to a dramatic increase in abundance and geographic distribution of several species of *Cestodiscus*, similar to that previously described by Jouse (1978). Fenner (1985) estimates a 50% increase (relative abundance) in *Cestodiscus* in the low-latitudes at this time. The fourth and final turnover of the Paleogene diatom floral identified by Fenner (1985) occurs during the late early Oligocene. This interval is characterized by the disappearance of species typical of the Eocene and the appearance of precursors to the Neogene assemblages such as *Synedra, Thalassionema* and *Thalassiothrix*.

Although the studies of Jouse (1978) and Fenner (1981, 1985, 1986) provide a preliminary interpretation of floral evolution of Paleogene and early Neogene diatoms, they do not provide a temporal or spatial framework with the resolution necessary to establish the response of the flora to changing oceanographic and climatic conditions. The purpose of this paper is to summarize the current knowledge of middle Eocene through early Miocene diatoms in an effort to document the magnitude and geographic extent of floral turnovers in the Paleogene and early Neogene.

DATA BASE

Four geographic regions (Southern Ocean, low-latitude Atlantic and Pacific Ocean, Labrador Sea, and Norwegian-Greenland Sea) were identified to examine the spatial and temporal response of the diatom flora to oceanographic and climatic changes during the Paleogene and early Neogene. The database presented in this manuscript is compiled from previous work by Baldauf and Monjanel (1989), Baldauf and Barron (in press), Barron (1983) Fenner (1977, 1981, and 1985), Harwood and Maruyama (in press), and Schrader and Fenner (1976), as well as unpublished work by the author (see Figure 14.1).

Using a database which incorporates results from several independent workers may introduce a degree of bias as a result of differing taxonomic concepts between individuals. This potential biased was minimize by; 1) limiting the number of studies incorporated into the database; 2) including, for each data set only those species with well defined species concepts; and 3) avoiding, for each data set the use of forms with taxonomic uncertainty, such as *Goniothecium* sp. or resting spores such as *Xanthiopyxis* or *Pseudopyxilla*. Therefore the database used here represents a subset of the diatom flora recorded from the Paleogene and lower Neogene sedimentary record. It should be noted that each dataset represents a particular geographic region and therefore a specific depositional environment.

The temporal range of taxa presented in this manuscripts is based on the zonal concepts of Baldauf and Barron (in press) for the Southern Ocean, Baldauf and Monjanel (1989), Barron (1983), and Fenner (1985) for the low to middle latitudes, and Schrader and Fenner (1976) and Fenner (1985) for the Norwegian-Greenland Sea. Calibration of these zonal schemes to a geochronology is as follows. The southern ocean diatom zonation is calibrated directly with magnetostratigraphy (Baldauf and Barron, in press); the low to middle latitude zonations are in part directly calibrated to a magnetostratigraphy (Barron 1983, 1985) and indirectly calibrated to a chronology via calcareous nannofossil zonations (Fenner 1985; Baldauf and Monjanel 1989); the Norwegian-Greenland Sea diatom zonation is indirectly calibrated to a geochronology via calcareous microfossil zonations (Fenner 1985). The geochronology of Berggren et al. (1985) is used in this manuscript.

Quantitative data presented for Oligocene sediments from Ocean Drilling Program (ODP) Site 744 on the Kerguelen Plateau is based on work by the author. Sample preparation follows a modified version of Laws (1981) and counting procedures follows that of Schrader and Gersonde (1978).

The middle Eocene through early Miocene database use for this study is shown in Figures

TABLE 14.1. Occurrence of stratigraphic sequences for each two million year time slices used in this study. Southern Ocean Site 744 is located in 2307 m of water on the Kerguelen Plateau; Low latitude Sites 71, 149 and 366 are located in 4419, 3972, and 2853 m of water, respectively; Labrador Sea Site 647 is positioned in 3869 m of water; and Norwegian-Greenland Sea Site 338 is located in 1315 m of water; += sequences present; \=hiatus present, and -=no sequence available.

AGE	744	71, 149, 366	647	338
16-18	+	+	-	+
18-20	+	+	-	+
20-22	+	+	-	+
22-24	\	+	-	+
24-26	+	-	-	+
26-28	+	-	-	+
28-30	\	-	+	+
30-32	+	+	+	+
32-34	+	+	+	+
34-36	+	+	+	\
36-38	+	+	+	\
38-40	+	+	-	\
40-42	+	+	-	+
42-44	+	-	-	+
44-46	+	+	-	-

14.2-14.6. Figure 14.2 is a composite diagram showing the temporal range of about 120 selected species from the four geographic regions (see Figures 14.3-14.6). Initial examination of the database shows a limited data set for the early middle Eocene with data available only from the equatorial Atlantic. The database improves slightly for the remainder of the Eocene with data available from the equatorial Atlantic and in part from the Southern Ocean and the Norwegian-Greenland Sea. The database is good for the Oligocene and early Miocene with data sets available from the Southern Ocean, low latitude Atlantic and Pacific, and the Norwegian-Greenland Sea. Data is also incorporated for the early Oligocene of the Labrador Sea (see Table 14.1 for summary).

DISCUSSION

The diatom assemblage present in the Paleogene and early Neogene world's oceans, based on a represented sub-sample (Figure 14.2) consists of both temporally long ranging and geographically wide spread species (such as *Coscinodiscus marginatus*, *Actinoptychus senarius*, *Stephanopyxis turris*, and *Hemiaulus polycystinorum*), as well as species which are spatially and temporally restricted (such as *Sceptroneis pupa*, *Lisitzinia ornata*, *Hemiaulus gondolaformis*, and *Brightwellia spiralis*). However, it should be noted that the temporally long ranging forms are in part based on a species concept that is board and that incorporates a wide degree of morphological variation (for an example, see Baldauf and Barron, 1982 for discussion of *Coscinodiscus marginatus*).

General observations of Figure 14.2 suggest that the middle Eocene to early Miocene diatom flora can be divided into five distinct assemblages based on differences in floral composition. These include a middle Eocene assemblage, middle to latest Eocene assemblage, earliest Oligocene to late Oligocene assemblage, late Oligocene to early Miocene assemblage, and late early Miocene to middle Miocene assemblage. Results of a cluster analysis for each 2 m.y. time interval (Figure 14.7) indicates that the time intervals can be grouped into similar groupings with periods of floral turnover correspond to intervals representing the *Brightwellia imperfecta* Zone (41-43 Ma), the *Coscinodiscus excavatus* and lower *Cestodiscus reticulatus* Zones (~35-37 Ma); the *Rocella gelida* Zone (21-24.5 Ma), and the *Crucidenticula nicobarica* Zone (17-19 Ma).

The magnitude of floral turnover is unique for each period with regards to both the percentage of floral change and the percentage of species being introduced into, versus those being removed from the assemblage. Figure 14.8 represents the number of bioevents as a percentage of the diatom assemblage for each 2 m.y. time-slice.

The gradual increase in the percentage of the flora affected by turnover (20-33% for Eocene-Oligocene, 45-80?% for the Miocene) suggests that the mechanisms responsible for floral change had further reaching consequences during the Miocene than during the Eocene.

Event 1 (middle Eocene, 41-43 Ma)

Event 1 is tentatively assigned to the upper portion of the *Hemiaulus alatus*, the *Brightwellia imperfecta* and lowermost portion of the *Asterolampra marylandica* Zones (Zones 13-11 on Figures 14.2-6). This interval corresponds to the uppermost portion of calcareous nannofossil Zone NP16 and the lowermost portion of Zone NP17. This event is recognized in the equatorial Atlantic (Figure 14.4) and is considered tentative because data representing the interval prior to this event is limited. Within the database presented here, middle Eocene data older than this event is based on one equatorial Atlantic site (Site 149; Figure 14.4). Although the data set from the Norwegian-Greenland Sea approximates the interval corresponding to the event, it lacks consistent data for the interval prior to this event.

Event 1 occurs slightly after (~1-2 Ma) the change in the diatom assemblage defined by Fenner (1985) at the base of the *Triceratium kanayae* Zone. Additional quantitative analysis is required to determine the relationship between these two events.

Although limited, the data available suggests that this event represents a transition from a characteristic middle Eocene flora to a late middle Eocene-upper Eocene flora. Prior to this event the low-latitude assemblage is dominated by specimens of *Actinoptychus senarius*, *Parlaria sulcata*, *Pyxilla reticulata* (several different forms of *Pyxilla* were grouped under

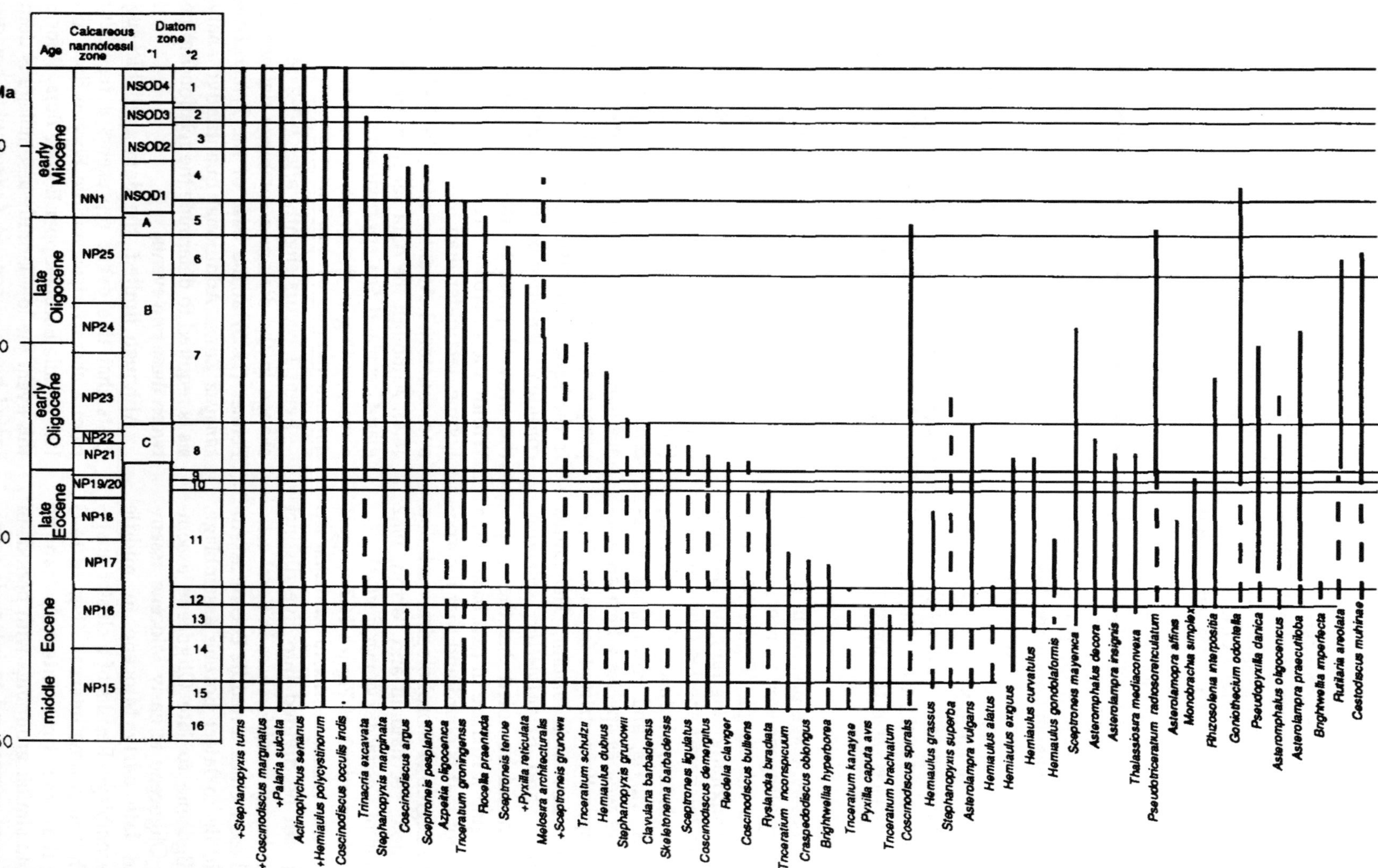

FIGURE 14.2a. Temporal distribution of selected diatom species for the world ocean's based on data presented in Figures 14.3-14.6; *1-Diatom zonation of Baldauf and Barron (in press); A=*Rocella gelida* Zone; B=*Rocella vigilans* Zone; C=*Rhizosolenia oligocenica* Zone; *2-Composite diatom zonation based on Barron (1985) and Fenner (1984, 1985); 1=*Crucidenticula nicobarica*; 2=*Triceratium pileus*; 3=*Craspedodiscus elegans*; 4=*Rossiela paleacea*; 5=*Rocella gelida*; 6=*Bogorovia veniamini*; 7=*Rocella vigilans*; 8=*Cestodiscus reticulatus*; 9=*Coscinodiscus excavatus*; 10=*Baxterium brunii*; 11=*Asterolampra marylandica*; 12=*Brightwellia imperfecta*; 13=*Hemiaulus alatus*; 14=*Pyxilla caput avis*; 15=*Triceratium kanayae*; 16=*Craspedodiscus oblongus*; +=indicates taxonomic grouping.

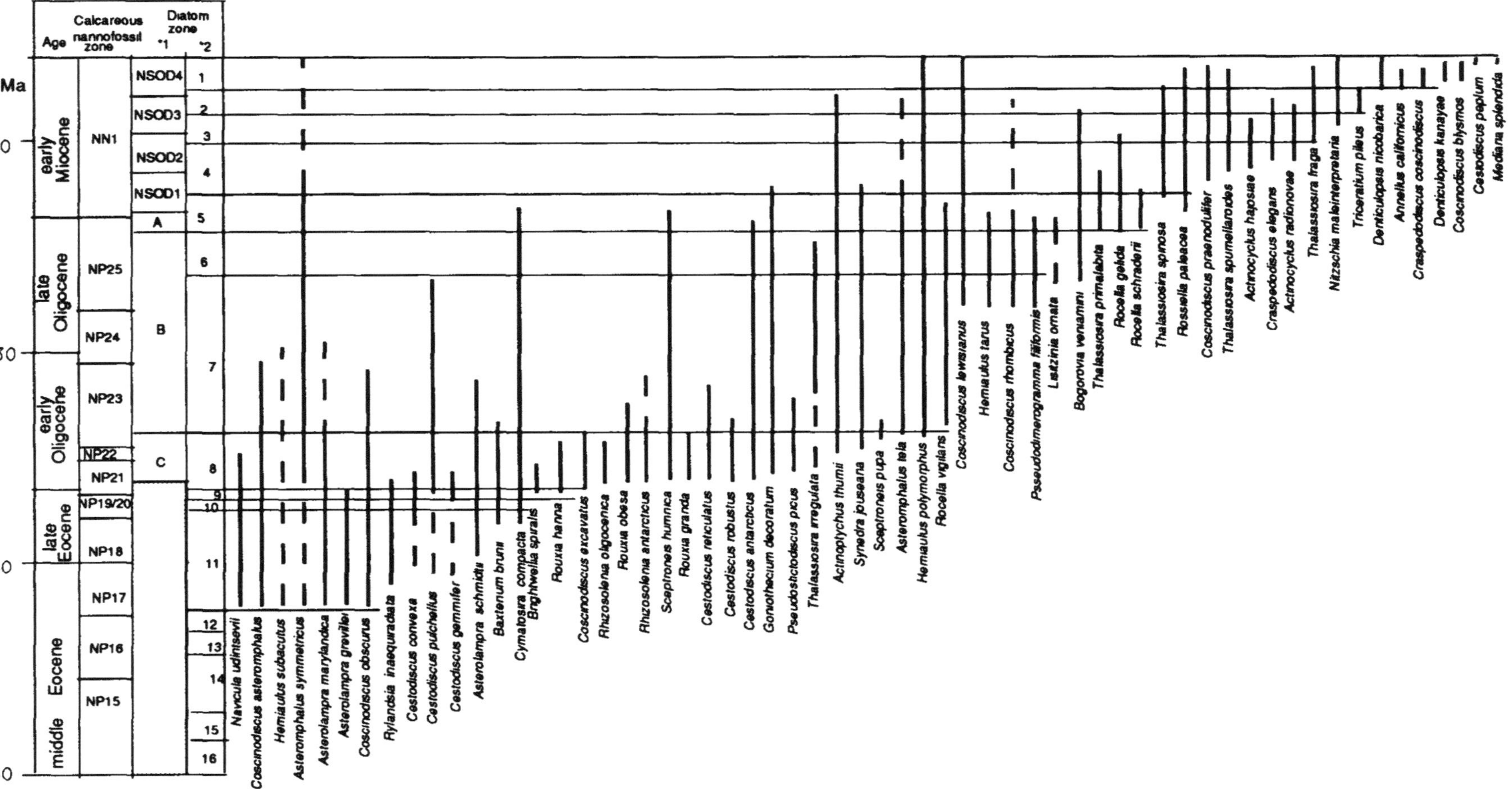

FIGURE 14.2b. Temporal distribution of selected diatom species for the world ocean's based on data presented in Figures 14.3-14.6; *1-Diatom zonation of Baldauf and Barron (in press); A=*Rocella gelida* Zone; B=*Rocella vigilans* Zone; C=*Rhizosolenia oligocenica* Zone; *2-Composite diatom zonation based on Barron (1985) and Fenner (1984, 1985); 1=*Crucidenticula nicobarica*; 2=*Triceratium pileus*; 3=*Craspedodiscus elegans*; 4=*Rossiela paleacea*; 5=*Rocella gelida*; 6=*Bogorovia veniamini*; 7=*Rocella vigilans*; 8=*Cestodiscus reticulatus*; 9=*Coscinodiscus excavatus*; 10=*Baxterium brunii*; 11=*Asterolampra marylandica*; 12=*Brightwellia imperfecta*; 13=*Hemiaulus alatus*; 14=*Pyxilla caput avis*; 15=*Triceratium kanayae*; 16=*Craspedodiscus oblongus*; +=indicates taxonomic grouping.

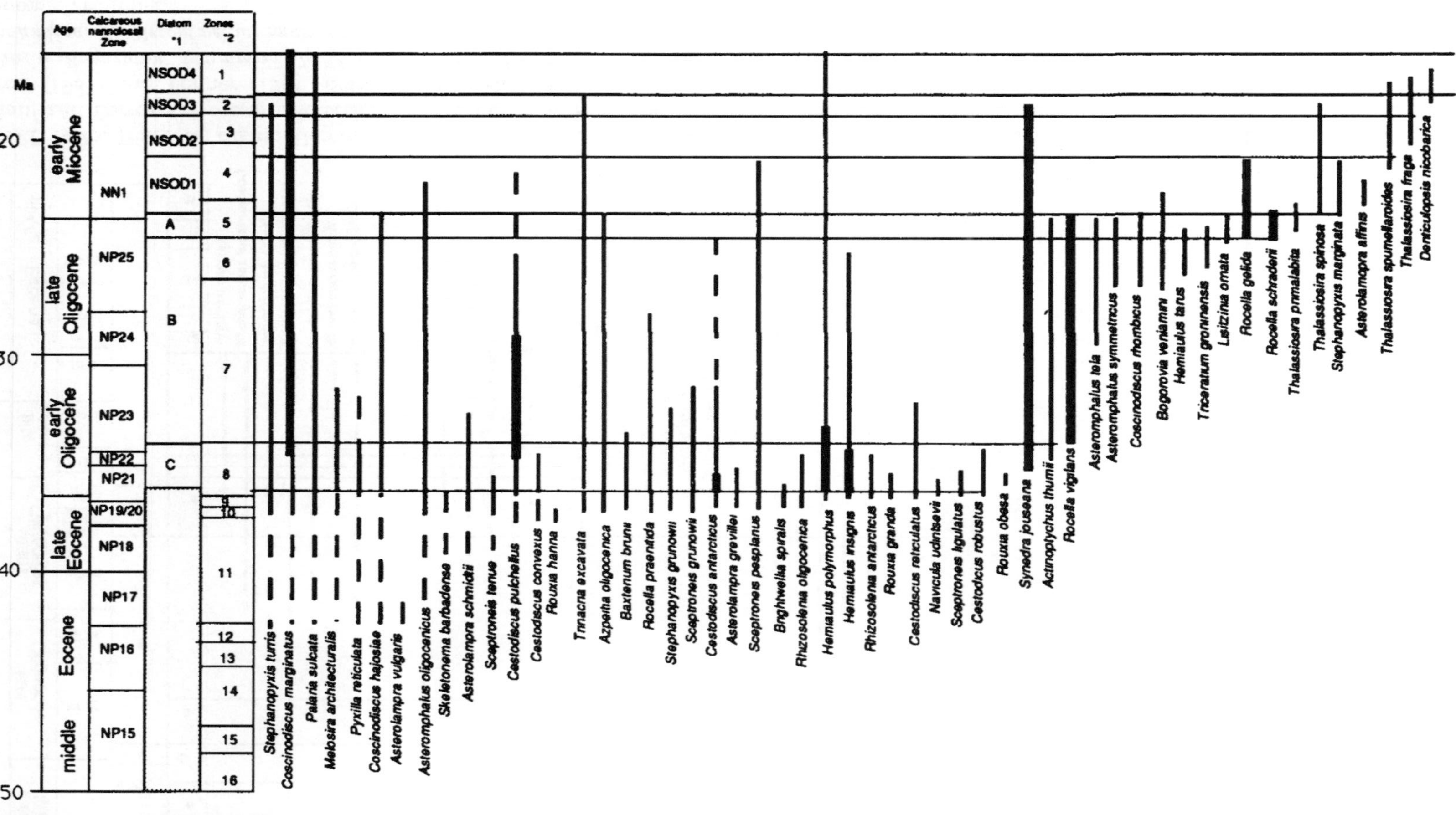

FIGURE 14.3. Temporal distribution of selected diatom species for the Southern Ocean. Modified from Baldauf and Barron (in press) and Harwood and Maruyama (in press); *1-Diatom zonation of Baldauf and Barron (in press); A=*Rocella gelida* Zone; B=*Rocella vigilans* Zone; C=*Rhizosolenia oligocenica* Zone; *2-Composite diatom zonation based on Barron (1985) and Fenner (1984, 1985); 1=*Crucidenticula nicobarica*; 2=*Triceratium pileus*; 3=*Craspedodiscus elegans*; 4=*Rossiela paleacea*; 5=*Rocella gelida*; 6=*Bogorovia veniamini*; 7=*Rocella vigilans*; 8=*Cestodiscus reticulatus*; 9=*Coscinodiscus excavatus*; 10=*Baxterium brunii*; 11=*Asterolampra marylandica*; 12=*Brightwellia imperfecta*; 13=*Hemiaulus alatus*; 14=*Pyxilla caput avis*; 15=*Triceratium kanayae*; 16=*Craspedodiscus oblongus*.

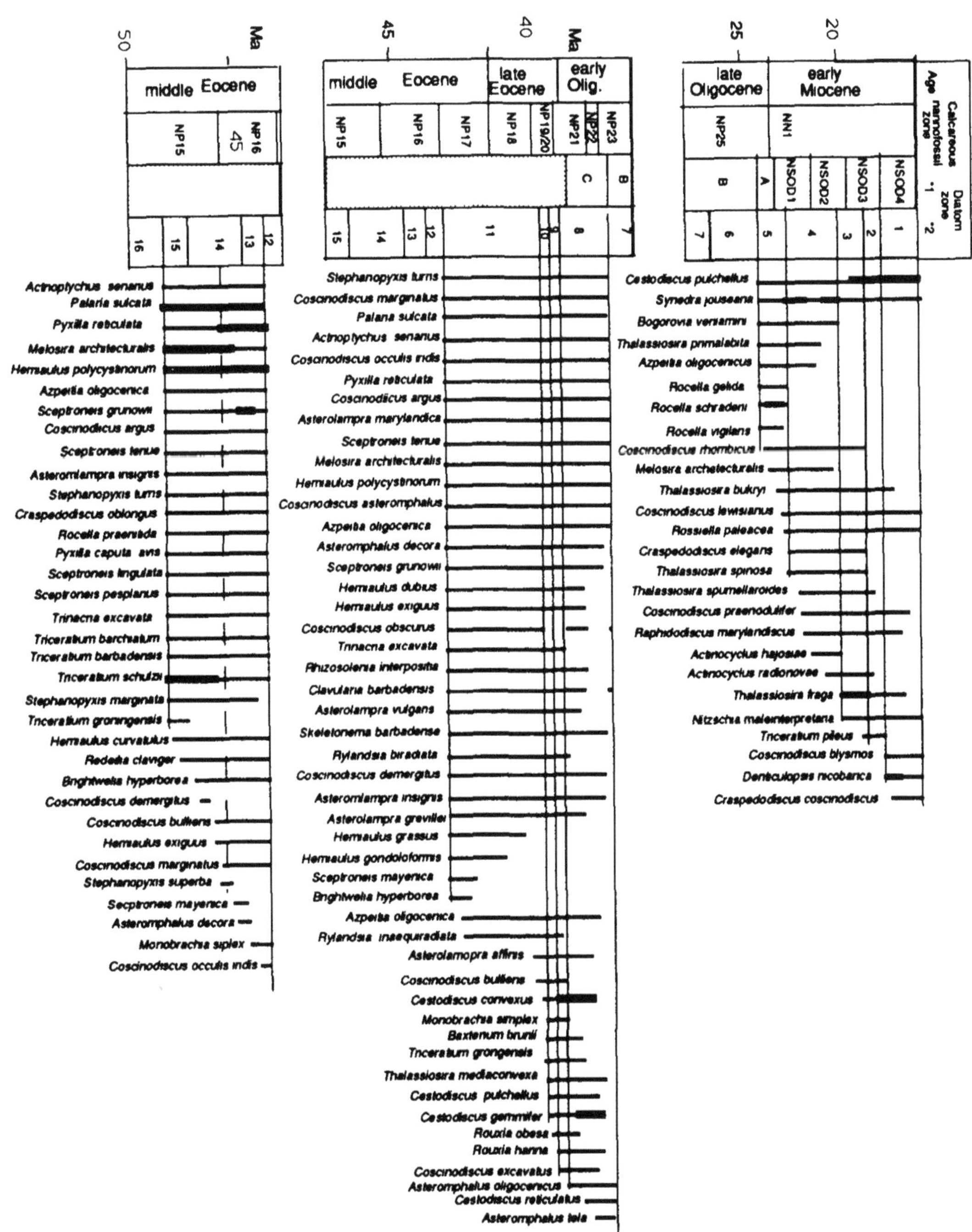

FIGURE 14.4. Temporal distribution of selected diatom species for the low latitudes Atlantic and Pacific. Modified from Fenner (1984, 1985) *1-Diatom zonation of Baldauf and Barron (in press); B=*Rocella vigilans* Zone; C=*Rhizosolenia oligocenica* Zone; *2-Composite diatom zonation based on Barron (1985) and Fenner (1984, 1985); 1=*Crucidenticula nicobarica;* 2=*Triceratium pileus;* 3=*Craspedodiscus elegans;* 4=*Rossiela paleacea;* 5=*Rocella gelida;* 6=*Bogorovia veniamini;* 7=*Rocella vigilans;* 8=*Cestodiscus reticulatus;* 9=*Coscinodiscus excavatus;* 10=*Baxterium brunii;* 11=*Asterolampra marylandica;* 12=*Brightwellia imperfecta;* 13=*Hemiaulus alatus;* 14=*Pyxilla caput avis;* 15=*Triceratium kanayae;* 16=*Craspedodiscus oblongus.*

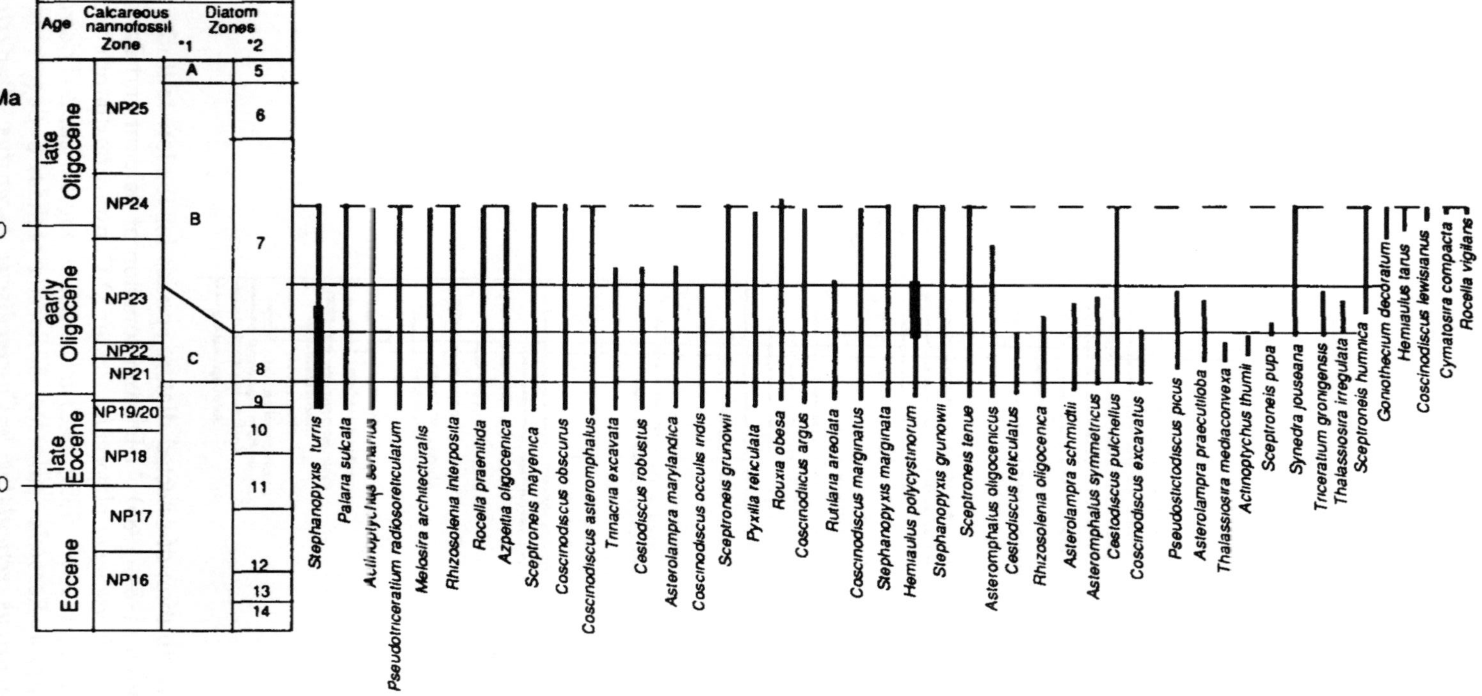

FIGURE 14.5. Temporal distribution of selected diatom species for the Labrador Sea. Modified from Baldauf and Monjanel (1989); *1-Diatom zonation of Baldauf and Monjanel (1989); A=*Rocella gelida* Zone; B=*Rocella vigilans - Cestodiscus reticulatus*; C=*Cestodiscus reticulatus*; *2-Composite diatom zonation based on Barron (1985) and Fenner (1984, 1985); 1=*Crucidenticula nicobarica*; 2=*Triceratium pileus*; 3=*Craspedodiscus elegans*; 4=*Rossiela paleacea*; 5=*Rocella gelida*; 6=*Bogorovia veniamini*; 7=*Rocella vigilans*; 8=*Cestodiscus reticulatus*; 9=*Coscinodiscus excavatus*; 10=*Baxterium brunii*; 11=*Asterolampra marylandica*; 12=*Brightwellia imperfecta*; 13=*Hemiaulus alatus*; 14=*Pyxilla caput avis*; 15=*Triceratium kanayae*; 16=*Craspedodiscus oblongus*.

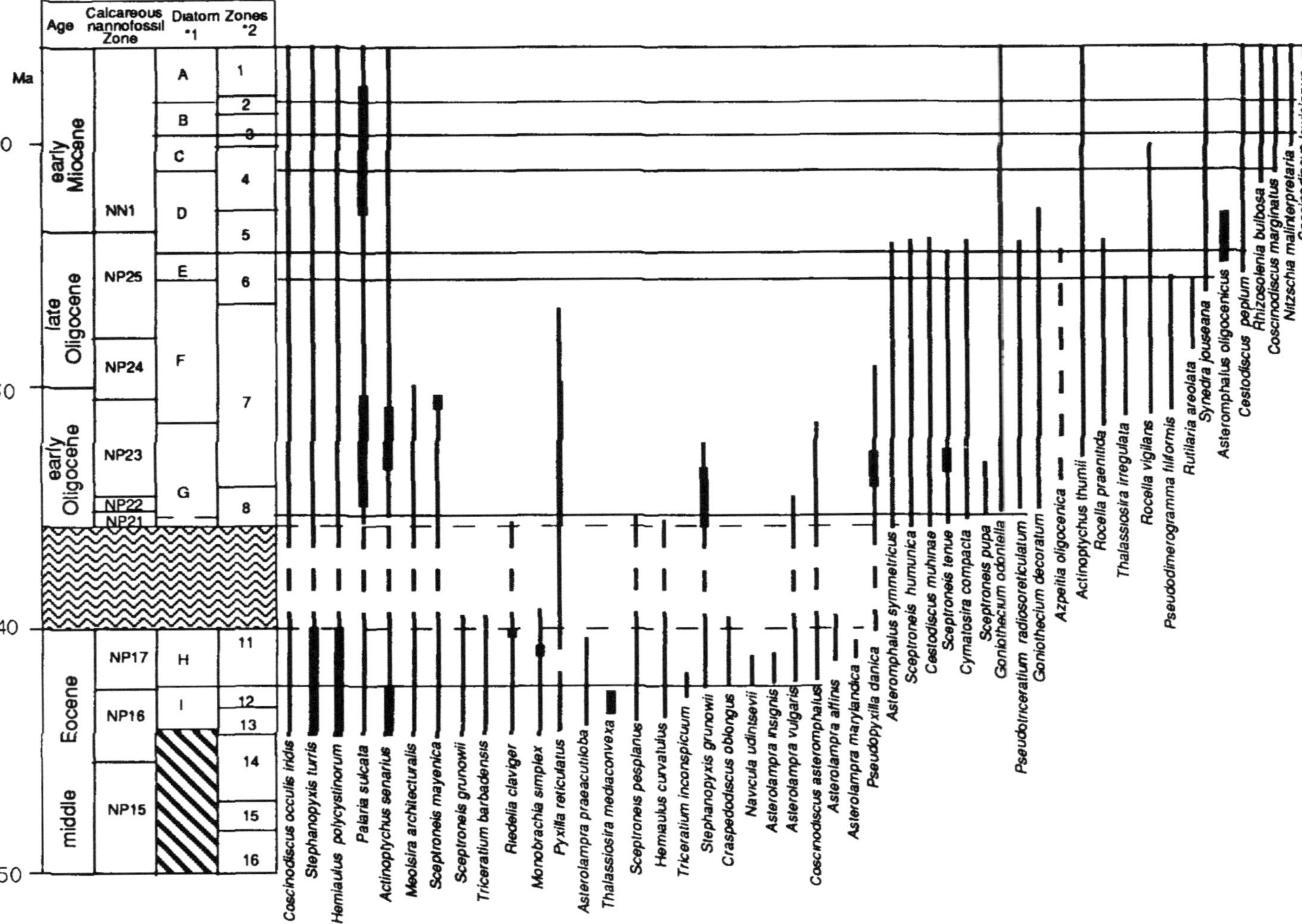

FIGURE 14.6. Temporal distribution of selected diatom species for the Norwegian-Greenland Sea. Modified from Schrader and Fenner (1976) and Fenner (1985). *1-Diatom zonation of Schrader and Fenner (1976) and Fenner (1985); A=*Thalassiosira fraga* ; B=*Nitzschia maleinterpretaria*; C=*Rocella vigilans*; D=*Rhizosolenia norwegica* to *Rocella praenitida*; E=*Thalassiosira irregulata*; F=*Pseudodiogramma filiformis* ; G=*Sceptroneis pupa* ; H=*Craspedodiscus oblongus*; I=*Trinacria* excavata; *2-Composite diatom zonation based on Barron (1985) and Fenner (1984, 1985); 1=*Crucidenticula nicobarica*; 2=*Triceratium pileus*; 3=*Craspedodiscus elegans*; 4=*Rossiela paleacea*; 5=*Rocella gelida*; 6=*Bogorovia veniamini*; 7=*Rocella vigilans*; 8=*Cestodiscus reticulatus*; 9=*Coscinodiscus excavatus*; 10=*Baxterium brunii*; 11=*Asterolampra marylandica*; 12=*Brightwellia imperfecta*; 13=*Hemiaulus alatus*; 14=*Pyxilla caput avis*; 15=*Triceratium kanayae*; 16=*Craspedodiscus oblongus*.

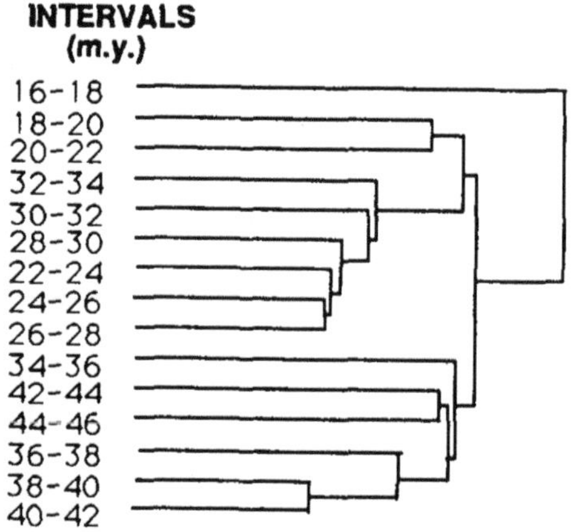

FIGURE 14.7. Cluster analysis of the diatom flora for each 2 m.y. interval based on the dataset in Figure 14.2. Intervals of floral change are; 17-19 Ma; 21-23 Ma; Ma; 34?-36; and Ma; 41-43 Ma.

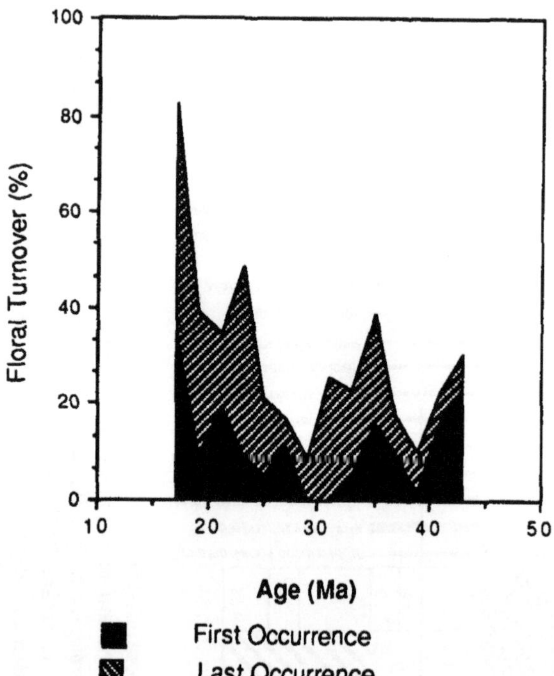

FIGURE 14.8. Turnover of the diatom flora for each 2 m.y. interval as indicated by the number of first and last occurrence for each time interval. Data is based on data set shown in Figure 14.2.

this species), *Melosira architecturalis, Hemiaulus polycystinorum, Triceratium kanayae* and to a lesser extent *Sceptroneis grunowii.* Results form various age equivalent continental sequences generally show a similar dominant assemblage (see Kanaya, 1957; Barron et al., 1984 for examples). As suggested by the data in Figure 14.8, event 1 may represent an expansion of the diatom flora with numerous species being introduced and only a few species being removed. In the low latitudes, species introduced include forms such as *Asterolampra decora, Asterolampra marylandica, Brightwellia imperfecta, Hemiaulus subacutus, Monobrachia simplex,* and *Rutilaria areolata.* Species removed from the assemblage include those such as *Brightwellia hyperborea, Hemiaulus gondolaformis, Craspedodiscus oblongus,* and *Pyxilla caputa avis.*

The data set for the Norwegian-Greenland Sea includes an interval that approximates this event. Because of poor biosiliceous preservation, the characteristics of the assemblage prior to this event are unclear. The Norwegian-Greenland Sea assemblage is similar to that for the low-latitude including species such as *Stephanopyxis turris, Actinoptychus senarius, Hemiaulus polycystinorum, Monobrachia simplex, and Asterolampra vulgaris,* yet this assemblage differs from that in the low latitude by incorporating species such as *Thalassiosira mediaconvexa* and lacking typical low latitude species such as *Hemiaulus gondolaformis.*

Event 1 represents a minor floral response to oceanographic change, the increase in the number of new species with minimal expense of the established flora may represent niche expansion resulting from increased floral provincialism. The timing of event 1 (~41-43 Ma) approximates that of the increased isolation of Antarctica from warm currents and a decoupling of near-surface and bottom temperatures resulting in strengthened latitudinal thermal gradients (Boersma et al., 1987; Barron and Baldauf,1989). Additional support of increased niche development and provincialism is the expansion of biosiliceous sediments in coastal upwelling settings along the Pacific Rim (Barron and Baldauf, 1989).

Event 2 (latest Eocene-earliest Oligocene, 35-37 Ma)

Event 2 corresponds to an interval representing the *Coscinodiscus excavatus* and lower portion of the *Cestodiscus reticulatus* Zones (Zones 9 and 8 on Figures 14.2-14.6). This interval corresponds to the lower portion of calcareous nannofossil Zone NP21 and approximates the Eocene/Oligocene boundary. This event is recognized in the Southern Ocean, equatorial Atlantic, and in part in the Labrador Sea and the Norwegian-Greenland Sea (Figures 14.3-14.6). This event is equivalent to that previously noted by Jouse (1978) and Fenner (1985) and represents about a 45% turnover in the diatom flora (based on this database) with approximately equal number of new species replacing previously established species (Figure 14.8).

Floral turnover during this time occurs in all geographic regions with clear floral distinctions observed between the low to middle and high latitude assemblages. In the Southern Ocean (Figure 14.3) this event is characterized by the increased dominance of *Hemiaulus polymorphus* and *Hemiaulus insignis*, as well as the first occurrence of numerous species such as *Cestodiscus antarctica*, *Cestodiscus robustus*, *Baxterium brunii*, *Brightwellia spiralis*, *Rhizosolenia oligocenica*, and *Rouxia granda*.

In the low latitudes (Figure 14.4) this event is characterized by the introduction and increase in abundance of numerous species of *Cestodiscus*. (*Cestodiscus convexus*, *C. gemminifer*, and *C. reticulatus*). Fenner (1985) estimates that in the low latitudes *Cestodiscus* increases from a relative abundance of 20% in the latest Eocene to 50-80% of the assemblage in the earliest Oligocene. In addition, similar to the Southern Ocean assemblage, new species such as *Baxterium brunii*, *Cestodiscus pulchellus*, *Coscinodiscus excavatus*, *Rouxia hanna*, and *Rouxia obesa* are introduced into the flora and forms such as *Hemiaulus exiguus*, *Trinacria excavata*, *Asterolampra vulgaris*, *Rylandia biradiata*, *R. inaequiradiata* disappear.

The data set from the Labrador Sea represents an interval equivalent to event 2. Poor biosiliceous preservation limits any knowledge about the diatom assemblage prior to this event for this region. The occurrence of several forms of *Cestodiscus* (*Cestodiscus robustus*, *C. reticulatus*, and *C. pulchellus*) with *Coscinodiscus excavatus*, and *Rouxia obesa* suggests that the diatom flora in this region of the Labrador Sea is predominantly similar to the assemblage observed from the lower latitudes. However, the occurrence of forms such as *Sceptroneis pupa*, *Thalassiosira mediaconvexa*, *Sceptroneis mayenica*, and *Sceptroneis tenue* also suggests the assemblage has affinities with that in the Norwegian-Greenland Sea and represents a transitional assemblage between the low and high latitudes.

Recognition of this event in the Norwegian-Greenland Sea is hampered by a hiatus which separates middle Eocene from the lowermost Oligocene sediments (Fenner, 1985). The Norwegian-Greenland Sea assemblage differs above and below this hiatus suggesting a change in the floral composition. The assemblage approximating event 2 is characterized by a dominance of *Palaria sulcata*, *Stephanopyxis grunowii* and the introduction of numerous forms such as *Cestodiscus muhinae*, *Cymatosira compacta*, *Sceptroneis pupa*, and *Sceptroneis humnica*.

Event 2 (35-37 Ma) most likely reflects a floral response to oceanographic and atmospheric reorganization resulting from climatic cooling. Numerous studies (Barron et al., in press; Baldauf and Barron, 1990; Kennett 1978 among others) have presented evidence for early Oligocene changes in ocean circulation and ice build up on east Antarctica. This event in the Southern Ocean corresponds to the expansion of biosiliceous sediments in to the southern high latitudes (see Baldauf and Barron, 1990).

Figure 14.9 compares the diatom floral (Baldauf unpublished data) and the oxygen isotope (Barrera and Huber, in press) results for ODP Site 744 on the Kerguelen Plateau. The diatom assemblage shows an abrupt increase in abundance (valves per gram sediment) in the sedimentary record commencing about 36-37 Ma. Below this increase diatoms were not observed. Above this initial increase diatoms maintain values between 2-8 $x10^4$ valves/gram until about 35.8 Ma where they decrease to values less than 1.5×10^4 valves/gram and remain relatively constant through the remainder of the Oligocene and lower Miocene.

The diatom assemblage at this site can be divided into four components. A "cosmopolitan" component including species such as *Synedra*

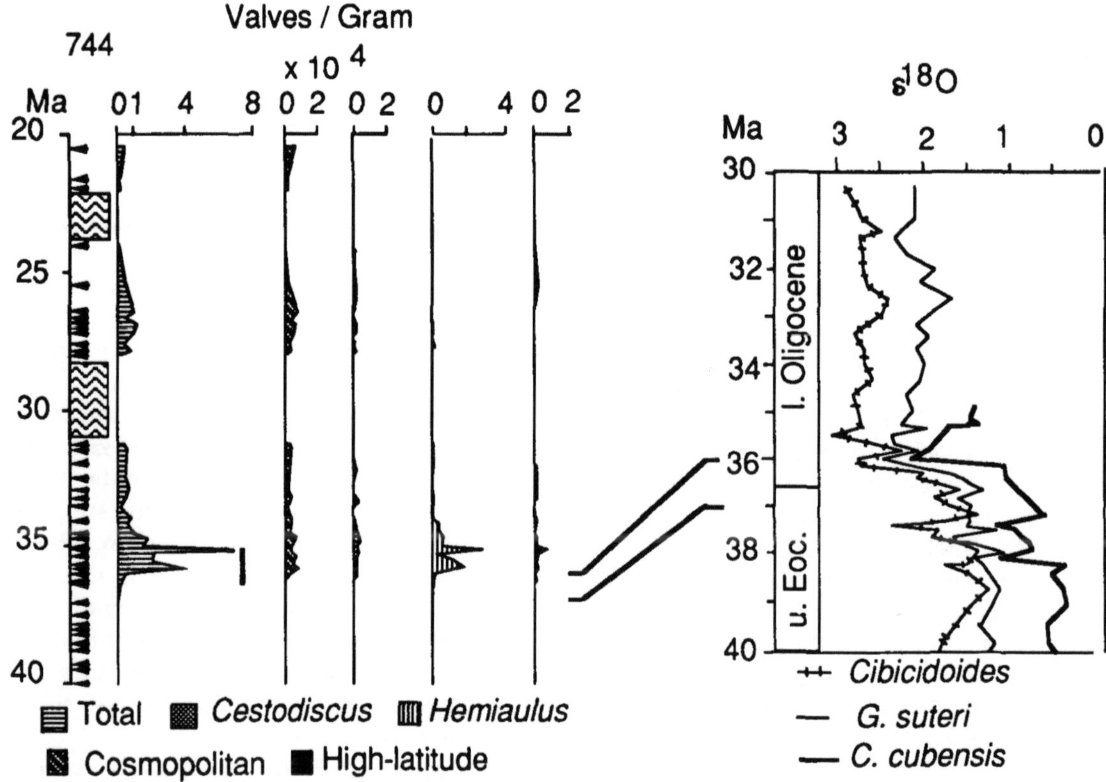

FIGURE 14.9. Comparison of the abundance of specific components of the diatom assemblage with oxygen isotope reults of Berrea and Huber (in press). Data shows corrected value. Arrow along left axis represent samples quantitatively examined, wavy lines indicate stratigraphic breaks in the record.

jouseana, Thalassiosira fraga, Stephanopyxis turris, and *Coscinodiscus marginatus;* a *Cestodiscus* component characterized by species of *Cestodiscus* such as *C. convexus, C. parmula, C. pulchellus, C. reticulatus,* and *C. robustus;* a *Hemiaulus* component typified by *Hemiaulus polycystinorum, H. insignis,* and *H. pungens;* and a high latitudes component typified by *Rhizosolenia oligocenica, Cestodiscus antarcticus,* and *Rhizosolenia antarctica*

Although detailed biogeographic studies are still required for Paleogene diatoms, previous studies have shown species of *Cestodiscus* to dominate the assemblage in the low to middle latitudes (Fenner, 1981; Baldauf and Monjanel, 1989). The occurrence of numerous species of *Cestodiscus* and the presence of forms restricted to the high southern latitudes suggests the development of partial, but weak latitudinal barriers to floral communication at this time.

The temporal agreement between floral turnover, an introduction of a high latitude species component, and the expansion of biosiliceous sediments into the southern high latitudes suggest that event 2 represents at least in part a floral response to oceanographic reorganization. This is further supported by the oxygen isotope record of Barrera and Huber (in press) which shows an increase in isotope values (of equal magnitude) in both planktonic (surface and near surface) and benthic foraminifera between 36 and 37 Ma (Figure 14.9). Comparison between the oxygen isotope record and the floral composition suggest that the expansion of biosiliceous sediments into the high southern latitudes resulted from increased surface water productivity and particle flux to the sea floor.

Event 3 (latest Oligocene to earliest Miocene; 23-25 Ma)

Event 3 is placed in an interval approximating the upper portion of the *Rocella vigilans* and the *Rocella gelida* Zones (Zones 6 and 5 on Figures 14.2-14.6). This interval corresponds to calcareous nannofossil Zones NP25 and NN1. Event 3 is characterized by an abrupt turnover of the diatom assemblage (>40%) and the replacement of the dominant Oligocene assemblage (species such as *Rocella vigilans, Rutilaria areolata, Asteromphalus oligocenicus, Goniothecium odontella, Cymatosira compacta,* and *Triceratium groningensis*) with a transitional Oligocene-Miocene assemblage (species such as *Bogorovia veniamini, Synedra jouseana, Rocella gelida, Rocella schraderi, and Thalassiosira primlabata*). This event is recognized in the Southern Ocean, in the low latitudes, and in the Norwegian-Greenland Sea (Figures 14.3, 14.4, and 14.6).

In the Southern Ocean this event is characterized by the last appearance of species such as *Rocella vigilans, Asterolampra tela,* and *Hemiaulus tarus* and the first occurrence of species such as *Rocella gelida, Thalassiosira prilamabiata and Thalassiosira spinosa.* In addition, species such as *Lizistinia ornata* and *Rocella schraderii* have short temporal ranges approximating this interval.

Similar floral turnover is observed in the low latitudes with in part the same species disappearing from the assemblage. However the first and/or last occurrence of specific species between these latitudes are often not contemporaneous. For example, the first occurrence of *Rocella gelida* has an estimated age of 24.5 Ma in the low latitudes (Barron, 1985) and an age of 27.5 Ma in the southern high latitude (Baldauf and Barron , in press). In the Norwegian-Greenland Sea, event 3 is characterized by the disappearances of species such as *Cestodiscus muhinae, Cymatosira compacta, Rocella praenitida,* and *Asteromphalus oligocenica,* and the first occurrences of species such as *Synedra jouseana* and *Cestodiscus peplum.*

Event 3 is characterized by the gradual introduction of new species into the diatom flora at the expense of the established population. The introduction of species several which are common in geographically restricted regions of high surface water productivity commences about 5-7 Ma prior to Event 3. The initial introduction of such species is recognized by Fenner (1985) and Barron and Baldauf (1989) and occurs with the *Rocella vigilans* Zone. The data presented in Figure 14.8 implies that there is a continued gradual disappearance of species following Event 2 (35-37 Ma) with a minimal number of new species introduced through the late Oligocene. In fact the data used here suggests few if any species were introduced between 29-31 Ma. Commencing at about 29 Ma there is a gradual increase (to ~15%) in the number of species introduced which continues until Event 3. Event 3 represents the culmination in this replacement process with the establishment of a Oligocene-Neogene transitional assemblage well established by 21-23 Ma. The mechanism(s) responsible are at present unknown but may be related to the increased decoupling of surface waters from intermediate and deep waters resulting in the gradual geographic restriction of surface water productivity.

Event 4 (late early Miocene; 17 -19 Ma)

Event 4 is tentatively defined here as an interval that approximates the *Crucidenticula nicobarica* and *Triceratium pileus* Zones. This event correlates with calcareous nannofossil zones CN2-CN3. This event is tentative as a result of the limited data set available from the four geographic regions and the minimal data available for the interval slightly younger than this event. Event 4 represents a replacement of the Oligocene-Miocene transitional flora with a characteristic Miocene assemblage. Floral turnover is high with about 80% of the assemblage affected. Thirty percent of the assemblage represents newly introduced species while 50% of the assemblage disappears. Although these values may be biased as a result of the database, this event represents the most significant floral turnover during the Paleogene and early Neogene.

This event is recognized in the Southern Ocean, the low latitudes, and the Norwegian-Greenland Sea. Characteristic of this event is the disappearance of species such as *Synedra jouseana, Thalassiosira irregulata, Thalassiosira spinosa, Rossielia paleacea, Thalassiosira. spumellaroides, Thalassiosira fraga* and *Coscinodiscus rhombicus.* In addition, numerous species such as *Medaria splendia Cruci-*

denticula nicobarica, Actinocyclus ingens, Annelus californicus and *Craspedodiscus coscinodiscus* are introduced into the assemblage. Furthermore, the assemblage exhibits an increase in dominance of species such as *Thalassionema nitzschioides* and *Thalassiothrix longissima,* typical in regions of surface water productivity.

Age constraint of the current database suggest that Event 4 commences at about 17-19 Ma. Because younger data is not included in this data set, the duration of this event is unclear. Evidence summarized in Barron and Baldauf (1989) suggest that the floral turnover may continue until about 15 Ma.

Similar to the previous events, Event 4 represents a floral response to reorganization of ocean circulation. The timing of this event (~17 Ma) corresponds to a decline in biosiliceous sediments in the Caribbean and equatorial Atlantic and the initial increase of biosiliceous sediments in the equatorial Pacific. The "switch" in biosiliceous sediments from the Atlantic to the Pacific continues until 15 Ma by which time the low and middle latitude Atlantic is nearly devoid of and the Pacific enriched with such sediments (Keller and Barron, 1982; Barron and Baldauf, 1989; Baldauf and Barron, 1990). This silica transition has been suggested by numerous authors (see above) to reflect the documented oceanographic and climatic changes which are of similar age (further development of the Antarctic cryosphere, increased intermediate and bottom formation/circulation). Coinciding with this switch in biosiliceous sedimentation is a change in faunal composition of benthic foraminifera which is interpreted by Thomas and Vincent (1987) to represent increased variability of corrosiveness of bottom waters, possibly resulting from enhanced productivity in surface waters. As such, Event 4 represents a floral response to further ocean reorganization and intensification of vertical and latitudinal gradients.

CONCLUSION

A subset of about 120 species, representing the Paleogene and early Neogene diatom flora from the Southern Ocean; the low latitude Atlantic and Pacific, the Labrador Sea, and the Norwegian-Greenland Sea provides the means to identify periods of floral turnover. Four events are recognized and include the middle Eocene, the latest Eocene to earliest Oligocene, the latest Oligocene - earliest Miocene and the late early Miocene. In general these events reflect species adaptation to oceanographic reorganization in response to intensification of latitudinal and vertical gradients, gradual development of the cryosphere and tectonic changes. Changes in the diatom flora during this time interval also correspond with intervals of floral reorganization of biosiliceous sedimentation patterns and reflect a transition from species characteristic of diffuse regions of high surface water productivity to apparently more opportunistic and competitive species within the geographically restricted regions of moderate to high surface water productivity.

The middle Eocene event represents about a 30% turnover in the diatom assemblage consisting of mainly new species being introduced into the assemblage possibly as a result of increased provincialism. This event is tentative because of the limited data available for this time interval.

The latest Eocene to earliest Oligocene event represents about a 40% turnover for the assemblage equally partitioned between the number of first occurrences and the number of last occurrences. This event corresponds to expansion of biosiliceous sediment in the southern high latitude and most likely represents a floral response to development of the Antarctic cryosphere and further intensification of floral provincialism.

The late Oligocene to early Miocene event represents a turnover of about 50% of the flora and is characterized by dominance of species disappearances This event is characterized by a transition from an Eocene assemblage to the occurrence of an Oligocene to early Miocene transitional assemblage.

The final event occurring during the late early Miocene is the most dramatic with approximately 80% turnover in the diatom assemblage. This event is characterized by the increased dominance of species (such as *Thalassionema nitzschioides* and *Thalassiothrix longissima*) at the expense of the Oligocene-early Miocene transitional assemblage. Event 4 corresponding to increased high latitude cooling and changes in the biosiliceous deposition patterns, reflects floral response to ocean reorganization.

REFERENCES

Baldauf, J. G. and Barron, J. A. 1982. Diatom biostratigraphy and paleoecology of the type section of the Luisian Stage, central California. *Micropaleontology* 28:59-84.

Baldauf, J. G. and Monjanel, A. 1989. An Oligocene diatom biostratigraphy for the-Labrador Sea: DSDP Site 112 and ODP Hole 647A. In S. P. Srivastava, M. Arthur, B. Clement et al., *Proceedings of the Ocean Drilling Program Scientific Results* 105:323-347.

Baldauf, J. G. and Barron, J. A. 1990. Evolution of biosiliceous sedimentation patterns - Eocene through Quaternary: Paleoceanographic response to polar cooling. In U. Bleil, J. Thiede, eds., *Geological history of the Polar Oceans: Arctic versus Antarctic,* Kluwer Academic Publishers, 575-607.

Baldauf, J. G. and Barron, J. A. in press. Diatom biostratigraphy: Kerguelen Plateau and Prydz Bay regions of the Southern Ocean. In J. A. Barron, B. Larsen, et al., *Proceedings of the Ocean Drilling Program Scientific Results* 119.

Barrera, E. and Huber, B. in press. In J.A. Barron, B. Larsen, et al., *Proceedings of the Ocean Drilling Program Scientific Results* 119.

Barron, Eric J. 1987. Eocene equator-to-pole surface ocean temperatures: A significant climate problem? *Paleoceanog.* 2: 729-739.

Barron, J. A. 1983. Latest Oligocene through early middle Miocene diatom biostratigraphy of the eastern tropical Pacific. *Marine Micropaleontology* 7:487-515.

Barron, J. A. 1985. Late Eocene to Holocene diatom biostratigraphy of the eastern equatorial Pacific Ocean, DSDP Leg 85. In L. F. Mayer, F. Theyer, et al., *Init. Rept. Deep Sea Drill. Proj.* 85:413-456.

Barron, J. A. and Baldauf, J. G. 1989. Tertiary cooling steps and paleoproductivity as reflected by diatoms and biosiliceous sediments. In W. H. Berger, V. S. Smetacek, G. Wafer, eds., *Productivity of the Oceans present and past.* New York, Wiley, pp. 341-354.

Barron, J. A., Larsen, B., and Baldauf, J. G. in press. Evidence for late Eocene-early Oligocene Antarctic glaciation and observations on late Neogene glacial history of Antarctica: Results from ODP Leg 119. In J. A. Barron, B. Larsen, et al., *Proceedings of the Ocean Drilling Program Scientific Results* 119.

Barron J. A., Poore, R., and Bukry, D. 1984. Correlation of the middle Eocene Kellogg Shale of northern California. *Micropaleontology* 20:138-170.

Benda, L. 1972. The diatoms of the Moler Formation of Denmark. *Nova Hedw., Beihft.* 39:251-266.

Berggren, W., Kent, D. V., and Flynn, J. 1985. Jurassic to Paleogene: Part 2 Paleogene geochronology and chronostratigraphy. In N. J. Snelling, ed., *The Chronology of the Geological Record,* Oxford, Blackwell Scientific Publications, pp. 141-195.

Boersma, A., Premoli-Silva, I., and Shackleton, N. J. 1987. Atlantic Eocene planktonic foraminiferal paleohydrographic indicators and stable isotope paleogeography. *Paleoceanog.* 2:287-331.

Cleve-Euler, A. 1941. Alttertiare Diatomeen und Silicoflagellaten im inneren Schwedesn. *Paleontographica* XCII:165-209.

Fenner, J. 1984. Eocene-Oligocene planktic diatom stratigraphy in the low latitudes and high southern latitudes. *Micropaleontology* 30:319-342.

Fenner, J. 1984. Middle Eocene to Oligocene planktonic diatom stratigraphy from Deep Sea Drilling Sites in the South Atlantic Equatorial Pacific and Indian Oceans. In W. W. Hay, J. C. Sibuet, et al., *Init. Rept. Deep Sea Drill. Proj.* 75:1245-1271.

Fenner, J. 1977. Cenozoic diatom biostratigraphy of the equatorial and southern Atlantic Ocean. In P. R. Supko, K. Perch-Nielsen, et al., *Init. Rept. Deep Sea Drill. Proj.* 39:491-622.

Fenner, J. 1981. Diatoms in the Eocene and Oligocene sediments off northwest Africa, their stratigraphic and paleoceanographic occurrences. Dissertation zur Erlangung des Doktorgrades der Maathematisch-Naturwissenschaftlichen Fakultat der Christian-Albrechts-Universitat zu Kiel, 230p.

Fenner, J. 1986. Information from diatom analysis concerning the Eocene-Oligocene boundary. In C. H. Pomerol, I. Premoli-Silva, eds., *Terminal Eocene Events.* Amsterdam, Elsevier, pp. 283-288.

Fourtanier, E. in press. Paleocene and Eocene diatom biostratigraphy and taxonomy of eastern Indian Ocean Site 752, ODP Leg 121. In J. Peirce, J. Weissel et al., *Proceedings of the Ocean Drilling Program Scientific Results* 121.

Gombos, A. M. Jr. 1977.Paleogene and Neogene diatoms from the Falkland Plateau and Malvinas outer Basin: Leg 36 DSDP. In P. F. Baker, I. W. D. Dalziel et al., *Init. Rept. Deep Sea Drill. Proj.* 36:575-687.

Gombos, A .M. Jr. 1980. The early history of the diatom family *Asterolampraceae*. *Bacillaria* 3:227-272.

Gombos, A .M. Jr. 1982. Early and middle Eocene diatom Evolutionary Events. *Bacillaria* 5:225-242.

Gombos, A .M. Jr. 1983. Survey of diatoms in the upper Oligocene and lower Miocene in Holes 515B and 516F. In P. F. Barker, R. L., Carlson, D. A. Johnson, et al., *Init. Rept. Deep Sea Drill. Proj.* 71:793-804.

Gombos, A .M. Jr. 1984. Late Paleocene diatoms in Cape Basin. In K. J. Hsü, J. L. La Brecque et al., *Init. Rept. Deep Sea Drill. Proj.* 73: 495-511.

Gombos, A .M. Jr. 1987. Middle Eocene diatoms from the North Atlantic, Deep Sea Drilling Project Site 605. In J. E. van Hinte, S. Wise Jr. et al., *Init. Rept. Deep Sea Drill. Proj.* 92:793-799.

Gombos, A. M. Jr. and Ciesielski, P. 1983. Late Eocene to early Miocene diatoms from the southwest Atlantic In Ludwig V. A. Krasheninikov et al., *Init. Rept. Deep Sea Drill. Proj.* 71:583-634.

Grunow, A. 1866. Maler aus Jutland eingesandt von Th. Jensen, analysiert von A. Grunow (uber *Sceptroneis*? n. sp. Grun.), *Hedwigia* 5: 145-146.

Harwood, D. M., and Maruyama, T. in press. Middle Eocene to Pleistocene diatom biostratigraphy of the Southern Ocean sediments from the Kerguelen Plateau, ODP Leg 120. In R. Schlich., S. Wise Jr. et al., S., *Proceedings of the Ocean Drilling Program Scientific Results*, 120.

Heiberg, P. A. C., 1863. Conspectus criticus diatomacearum Danicarum: Kjobenhavn (Wilhelm Priors Ferlag), p. 136.

Kanaya, T. 1957. Eocene diatom assemblages from the Kellogg and Sidney Shales, Mt. Diablo Area, California. *Tohoku Univ., Sendai, Japan, Sci. Rept. ser. 2 (Geol.)* 28:1-124.

Keller, G. and Barron, J. A. 1983. Paleoceanographic implications of Miocene deep-sea hiatuses. *Geol. Soc. Amer. Bull.* 94:590-613.

Kennett, J.P. 1978. The development of planktonic biogeography in the Southern Ocean during the Cenozoic. *Mar. Micropaleo.* 3:301-345.

Jouse, A. P. 1978. Diatom biostratigraphy on the generic level. *Micropaleontology* 24:316-326.

Laws, R. A. 1983. Preparing strewn slides for quantitative microscopical analysis: A test using calibrated microsphere. *Micropaleontology* 28:60-65.

Miller, K. G., Fairbanks, R. G., and Mountain, G. S. 1987. Tertiary oxygen isotope synthesis, sea-level history, and continental margin erosion, *Paleocean.* 1: 1-19.

Schrader, H. J. and Fenner, J. 1976. Norwegian Sea Cenozoic diatom biostratigraphy and taxonomy. In M. Talwani, G. Udintsev et al., *Init. Rept. Deep Sea Drill. Proj.* 38:921-1100.

Schrader, H. J. and Gerseonde, R. 1978. Diatoms and silicoflagellates. *Utrecht Micropal. Bull.* 17:129-176.

Schulz, P. 1927. Diatomeen aus norddeutschen Basalttuffen und Tuffgeschieben. *Zeit. Geschieb* 3: 66-78, 113-126.

Shackleton, N., and Kennett, J. P. 1975. Paleotemperature history of the Cenozoic and the initiation of Antarctic glaciation: Oxygen and carbon isotope analyses in DSDP sites 277, 279, 281, *Init. Rept. Deep Sea Drill. Proj.* 29: 801-807.

Thomas, E. and Vincent, E. 1987. Equatorial Pacific deep-sea benthic foraminifera; faunal changes before the middle Miocene polar cooling. *Geology* 15:1035-1039.

15. LATE PALEOGENE DINOFLAGELLATE CYSTS WITH SPECIAL REFERENCE TO THE EOCENE/OLIGOCENE BOUNDARY

by Henk Brinkhuis

ABSTRACT

This paper aims to summarize developments of Eocene-Oligocene dinoflagellate cyst research of the past decades and to present an inventory of events around the Eocene/Oligocene (E/O) boundary. Furthermore, the potential of dinoflagellate cyst palynology as a tool for the recognition, timing and understanding of paleo-environmental changes across the E/O boundary is discussed. It is shown that dinoflagellate cysts are particularly suited to correlate marginal marine sediments with time-equivalent more offshore deposited strata. This is demonstrated on the basis of results of studies recently carried out on sections straddling the E/O boundary in central and northeastern Italy, including the E/O boundary stratotype section and the type section of the Priabonian Stage.

INTRODUCTION

The number of detailed reports on Paleogene dinoflagellate cyst assemblages have increased progressively during the last decades, stimulated by their application to oil and gas exploration. The accumulation of knowledge on their qualitative distribution now allows detailed stratigraphic correlations, especially in former marginal marine areas. Well over 400 species have been recognized in the late Eocene, an all-time high in the geologic record of dinoflagellate cysts (Williams and Bujak, 1977a). Despite the successful application of dinoflagellate cysts in Eocene and Oligocene biostratigraphy, only a few attempts have been made to integrate these microfossils in multidisciplinary research aimed at specific problems, e.g., the recognition of the Eocene/Oligocene (E/O) boundary and the understanding of the nature of the distinct global events which occur at or near this boundary (see Pomerol and Premoli-Silva, 1986).

EOCENE-OLIGOCENE DINOFLAGELLATE CYSTS

Mid- to high latitude Eocene and, to a lesser degree, Oligocene dinoflagellate cysts are relatively well-known on a global scale through a large variety of papers (see recent overviews by e.g., Williams and Bujak, 1985; Head and Norris, 1989; Brinkhuis and Biffi, in prep.). Most information, however, is available from northwest Europe (see e.g., Châteauneuf, 1986; Costa and Manum, 1988). In contrast, Eocene-Oligocene assemblages from lower latitudes are still poorly documented, available studies being derived mainly from the Mediterranean area, India or Australia (see Brinkhuis and Biffi, in prep., for an overview).

After some decades of pioneer work on dinoflagellate cyst occurrences by various authors, Williams (1977) established a world-wide Mesozoic-Cenozoic dinoflagellate cyst zonation scheme. This scheme, however, does not have a particularly high stratigraphic resolution and was evidently based on limited available data. After this attempt no other world-wide 'standard' dinoflagellate cyst zonation scheme has been established. Existing zonation schemes have been erected on a local or regional basis.

Regional dinoflagellate cyst zonations of Eocene-Oligocene deposits are essentially based on marginal marine sequences from northwest Europe, the most important ones being those of Costa and Downie, (1976), Bujak et al. (1980), Châteauneuf and Gruas-Cavagnetto (1978) and Châteauneuf (1980; see also the review by Head and Norris, 1989). These schemes make use of shallow marine forms such as the

AGE			AREA	NW EUROPE	NW EUROPE	ENGLAND	GERMANY	GERMANY	FRANCE	FRANCE	SWISS ALPS	ITALY
				COSTA AND DOWNIE, 1976 EMEND BUJAK, 1979	COSTA AND MANUM, 1988	BUJAK ET AL 1980	BENEDEK AND MULLER 1974	BENEDEK AND MULLER, 1976	CHATEAUNEUF AND GRUAS-CAVAGNETTO 1978	CHATEAUNEUF, 1980	JAN DU CHENE 1977	BIFFI AND MANUM 1988
OLIGOCENE	L	P 22	NP 25		D 15		HOMOTRYBLIUM FLORIPES	HOMOTRYBLIUM FLORIPES				DO 3 (DO 3b; DO 3a)
		P 21	NP 24				OPERCULODINIUM XANTHIUM	CORDOSPHAERIDIUM XANTHIUM				DO 2
	E	P 19/20	NP 23	WETZELIELLA GOCHTII	D 14		PHTANOPERIDINIUM COMATUM	HYSTRICHOGONYAULAX COREOIDES	W 14 WETZELIELLA GOCHTII	CHIROPTERIDIUM PARTISPINATUM; DAPSILIDINIUM SIMPLEX; WETZELIELLA GOCHTII		DO 1
		P 18	NP 22					WETZELIELLA SYMMETRICA; PENTADINIUM LATICINCTUM		PHTANOPERIDINIUM AMOENUM		
			NP 21		D 13			CHIROPTERIDIUM ASPINATUM				
EOCENE	L	P 17	NP 20	RHOMBODINIUM PERFORATUM	D 12				W 13 KISSELOVIA CLATHRATA SUBSP ANGULOSA	KISSELOVIA CLATHRATA SUBSP ANGULOSE	RHOMBODINIUM PERFORATUM	DE
		P 16	NP 19									
		P 15	NP 18									
	M	P 14	NP 17	GOCHTODINIUM SIMPLEX; RHOMBODINIUM POROSUM	D 11	POLYSPHAERIDIUM CONGREGATUM; HOMOTRYBLIUM VARIABILE; RHOMBODINIUM POROSUM			W 12 RHOMBODINIUM PERFORATUM; W 11 RHOMBODINIUM POROSUM	RHOMBODINIUM POROSUM		
		P 13	NP 16	RHOMBODINIUM DRACO	D 10: D 10b	AREOSPHAERIDIUM FENESTRATUM; HETERAULACYSTA POROSA			W 10 RHOMBODINIUM DRACO	RHOMBODINIUM INTERMEDIUM & AREOSPHAERIDIUM DIKTYOPLOKUS		
		P 12										
		P 11	NP 15	KISSELOVIA COLEOTHRYPTA	D 10a	GLAPHYROCYSTA INTRICATA; AREOSPHAERIDIUM ARCUATUM; PHTANOPERIDINIUM COMATUM			W 9 WETZELIELLA AFF W ARTICULATA	SYSTEMATOPHORA PLACACANTHA	WETZELIELLA ARTICULATA	
		P 10	NP 14			PENTADINIUM LATICINCTUM			W 8 KISSELOVIA FASCIATA			
	E	P 9	NP 13		D 9	HOMOTRYBLIUM ABBREVIATUM			W 7 KISSELOVIA COLEOTHRYPTA SUBSP ROTUNDA		WETZELIELLA CF W LUNARIS	
		P 8									WETZELIELLA HORRIDA	
		P 7	NP 12		D 8	KISSELOVIA RETICULATA			W 6 KISSELOVIA COLEOTHRYPTA		KISSELOVIA COLEOTHRYPTA	
		P 6	NP 11	DRACODINIUM VARIELONGITUDUM	D 7 (D 7b; D 7a)	EATONICYSTA URSULAE			W5 DRACODINIUM VARIELONGITUDUM		WILSONIDINIUM TABULATUM	
			NP 10	DRACODINIUM SIMILE	D 6 (D 6b; D 6a)	DEFLANDREA PHOSPHORITICA			W4 DRACODINIUM SIMILE; W3 WETZELIELLA MECKELFELDENSIS		WETZELIELLA MECKELFELDENSIS	

FIGURE 15.1. Overview of Eocene-Oligocene dinoflagellate cyst zonation schemes (after Williams and Bujak, 1985; Brinkhuis and Biffi, in prep.).

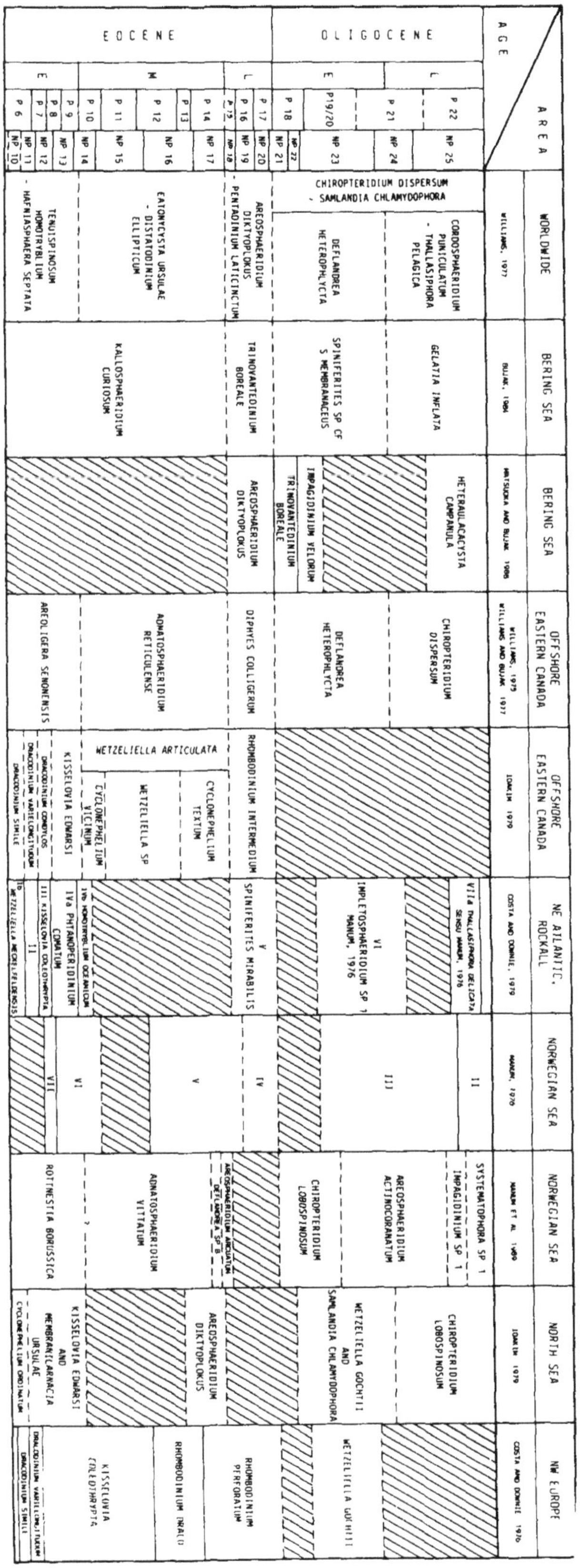

Wetzeliellaceae and the *Phtanoperidinium* lineages. Although the schemes from northwest Europe are largely comparable and allow correlations within this area, they are often of limited correlative value across larger distances, because consistent occurrences of these dinoflagellate cysts are bound to extremely marginal marine environments. Moreover, many of the *Wetzeliella* and *Phtanoperidinium* species employed in these zonation schemes only occur in northwest Europe. Subsequently, therefore, many mid- to high latitude regional zonations were defined for other areas, e.g., for offshore eastern Canada (Williams and Bujak, 1977b, Ioakim, 1979), for the North Atlantic (Costa and Downie, 1979), for the northern Pacific (Bujak, 1984; Matsuoka and Bujak, 1988), and for the northern North Sea and Norwegian Sea (Manum, 1976; Ioakim, 1979; Manum et al., 1989). Unfortunately, correlations between the established zones remain problematic. For the purpose of IGCP project 124 ('The northwest European Tertiary Basin'), a new Paleogene northwest European dinoflagellate cyst zonation scheme was constructed, based on concurrent ranges of many taxa rather than on one single group. The project resulted in the erection of 20 Cenozoic biozones, coded D1 to D20 (Costa and Manum, 1988; see Figure 15.1). However, this scheme was somewhat opportunistically pieced together using index-species from mid- and higher latitudes without well-documented calibration against other microfossil distributions or magnetostratigraphy. Nevertheless, Köthe (1990) successfully applied the D-coded zonal scheme in her study of Paleogene sediments of northwest Germany, with only minor modifications.

Biffi and Manum (1988) defined a new uppermost Eocene and three Oligocene zones on the basis of sections from central Italy, because the assemblages they recovered differ markedly from those described earlier from other areas. This zonation represents the first comprehensive scheme to be applied in lower latitudes. An overview of the existing Eocene to Oligocene zonation schemes in presented in Figure 15.1.

THE EOCENE/OLIGOCENE BOUNDARY IN TERMS OF DINOFLAGELLATE CYSTS

As a result of the application of different

criteria, a precise definition of the Eocene/Oligocene boundary is still controversial (see discussions in Berggren et al., 1985; Pomerol and Premoli-Silva, 1986). However, in terms of marine biostratigraphy the boundary is generally considered to correspond to the last occurrence (LO) of *Hantkenina* and morphologically related planktonic foraminiferal genera (e.g., Beckmann et al., 1981; Poore et al., 1982; Nocchi et al., 1988a, b). The extinction of the hantkeninids now equates to the planktonic foraminiferal P17/P18 Zonal boundary (Blow, 1969; redefined by Nocchi et al., 1988a, b). The bioevent occurs at a level just above the last short-term normal "event" within Chron C13R and is believed to be reasonably synchronous on a global scale (e.g., Poore et al., 1982; papers in Premoli-Silva et al., 1988). This concept is followed in the present paper. However, because most published late Eocene-earliest Oligocene dinoflagellate studies have poor to no control via calcareous plankton and/or magnetostratigraphic data, the location of the E/O boundary is poorly constrained in many cases. Moreover, existing zonation schemes (Figure 15.1) are mainly based on sequences where the E/O transition interval is either missing, condensed or developed in restricted or continental facies. In other cases, even the approximate location of the E/O boundary is unclear. Also the application of the zonal scheme developed by Costa and Manum (1988) for northwest Europe is problematic with regard to the E/O boundary. In their scheme, the E/O boundary corresponds to the boundary between zones D12 and D13, defined by the LO of *Areosphaeridium diktyoplokus*. Unfortunately, in northwest Europe the E/O transition is characterized by unconformities and continental deposits due to significant sea level fluctuations (see discussions in e.g., Châteauneuf, 1980; Berggren et al., 1985). The E/O boundary is therefore difficult to recognize in this area and the corresponding dinoflagellate event to discriminate between the latest Eocene and earliest Oligocene used by Costa and Manum is poorly constrained.

Only recently a dinoflagellate zonal scheme across the E/O boundary on the basis of complete, otherwise well-documented pelagic sections in central Italy was developed by Brinkhuis and Biffi (in prep.). On the basis of over 45 events in the dinoflagellate record, nine dinoflagellate zones were defined across the Eocene-Oligocene transition, calibrated against calcareous plankton and paleomagnetic data previously provided by Premoli-Silva and coworkers (see papers in Premoli-Silva et al., 1988). The relatively high degree of resolution reached with the nine dinoflagellate zones spanning only four to five calcareous plankton zones can not be compared to the rather large-scaled zones defined previously (see Figure 15.2). The authors emphasized the local nature of the zonal scheme; only few of the events used by them can be traced outside the Mediterranean area.

Events often claimed crucial for recognizing the E/O boundary in dinoflagellate studies and zonation schemes are the LOs of *Areosphaeridium diktyoplokus, Diphyes colligerum* and *Heteraulacacysta leptalea* at the end of the Eocene, and the first occurrences (FOs) of *Areoligera semicirculata, Reticulatosphaera actinocoronata, Wetzeliella gochtii* and *Phtanoperidinium amoenum* in the early Oligocene (see e.g. Morgenroth, 1966, Liengjarern et al., 1980; Châteauneuf, 1980; Benedek, 1986; Gruas-Cavagnetto and Barbin, 1988; Biffi and Manum, 1988; Head and Norris, 1989). Especially the LO of *Areosphaeridium diktyoplokus* and the FO of *Wetzeliella gochtii* are frequently emphasized as being indicative for the E/O junction. It is remarkable that even with regard to the ranges of these few species, extensive debates concerning their 'true' last or first occurrences continue in the literature (see discussion by Brinkhuis and Biffi, in prep.). In the Mediterranean area, where first order correlation to calcareous plankton and magnetostratigraphic records could be established, *A. diktyoplokus* is consistently present in the earliest Oligocene, its LO being calibrated against Chron C13N1 (Brinkhuis and Biffi, in prep.). *Wetzeliella gochtii* s.s. was not recorded by these authors. However, a morphologically similar form (*W.* sp. cf. *W. gochtii*) first appears shortly above the LO of *A. diktyoplokus*. Evidently, both events occur during the early Oligocene and are not associated with the E/O boundary horizon, if this boundary is linked to the LO of the hantkeninids.

The LO of *Diphyes colligerum* has been suggested by Williams (1975, 1977) to indicate the E/O boundary, based on material from offshore

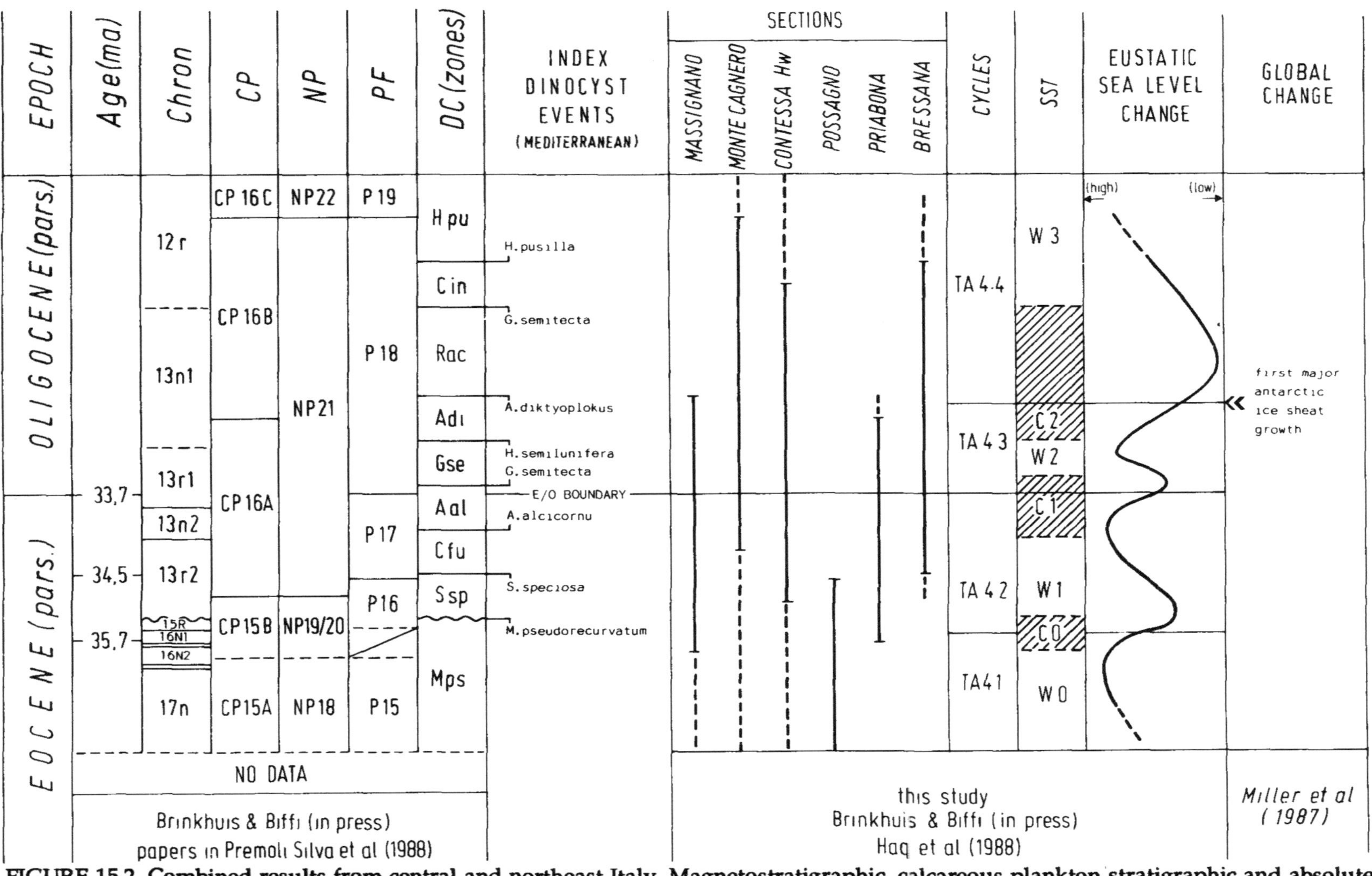

FIGURE 15.2. Combined results from central and northeast Italy. Magnetostratigraphic, calcareous plankton stratigraphic and absolute age interpretations from papers *in* Premoli Silva et al., 1988. Dinoflagellate cyst biostratigraphy, sequence stratigraphy and recognition of relatively cool (C0-C2) and warm (W0-W3) intervals from Brinkhuis and Biffi (in prep.) and Brinkhuis (in prep.; see also Figures 15.3 and 15.4). Please note that although the central Italian successions are complete, not all intervals contain palynomorphs. Hence, the gap in the stratigraphic record indicated is due to palynologically barren samples.

eastern Canada and a world-wide literature review of occurrences. The LO of *Heteraulacacysta leptalea* and the FO of *Areoligera semicirculata* near the E/O boundary are less well-constrained, being based on few data from northwest Europe (e.g., Eaton, 1976; Bujak et al., 1980) and Italy (Biffi, pers. comm. in Nocchi et al., 1986) or Germany (Morgenroth, 1966) and Belgium (Weyns, 1970) respectively. The data of Brinkhuis and Biffi (in prep.) and Brinkhuis (in prep.) from the Mediterranean suggest that *Areoligera semicirculata* has its FO in the early Oligocene and *Heteraulacacysta leptalea* and *Diphyes colligerum* to have their LOs in the same interval. *Reticulatosphaera actinocoronata* has been reported to have its FO in the early Oligocene of northwest Europe (e.g., as "*Cleistosphaeridium actinocoronatum*" by Benedek, 1972, and as "*Impletosphaeridium* sp. I" by Manum, 1976). However, in the Mediterranean, this taxon is already consistently present in the latest Eocene (Brinkhuis and Biffi, in prep.). The FO of *Phtanoperidinium amoenum* is used in parts of northwest Europe and in the Gulf of Mexico to be indicative for the earliest Oligocene (e.g., Drugg and Loeblich, 1967; Châteauneuf, 1980; Liengjarern et al., 1980). In contrast, Manum (1976) claimed late Eocene occurrences of the species. Furthermore, although neither Biffi and Manum (1988) nor Brinkhuis and Biffi (in prep.) recorded this species from the Oligocene of Italy, Gruas-Cavagnetto and Barbin (1988) reported the species to occur in the early Oligocene of northeast Italy. A subsequent study by Brinkhuis (in prep.) in the area could not corroborate this report. A recent paper by Médus and Pairis (1990) mentions the occurrence of *P. amoenum* in early Oligocene sediments from southeast France.

Other events frequently mentioned in the literature associated with the E/O transition are the LOs of *Melitasphaeridium pseudorecurvatum*, *Schematophora speciosa* and *Hemiplacophora semilunifera* (e.g., Haq et al., 1988; Head and Norris, 1989). These events indeed occur in the late Eocene to early Oligocene and are useful for correlation across large distances, but are not tied to the E/O boundary horizon (Brinkhuis and Biffi, in prep.). Hence, it seems that there is little evidence that one or more of the above mentioned events, often used for discriminating the latest Eocene from the earliest Oligocene, are specifically bound to the E/O boundary horizon, defined by the LO of the hantkeninids.

The recent studies in the Mediterranean show the introduction of taxa well-known from the middle and even early Eocene of higher latitudes closely around and at the E/O boundary horizon, e.g., the subsequent FOs of *Corrudinium incompositum*, *Achomosphaera alcicornu* s.s., *Glaphyrocysta semitecta*, *Gelatia inflata* and *Impagidinium pallidum*. Other taxa, like *Rottnestia borussica*, *Achilleodinium biformoides* and *Palaeocystodinium golzowense* are not known to occur consistently in the early late Eocene of Italy, but return close to the E/O boundary horizon. Brinkhuis and Biffi (in prep.) concluded therefore that assemblages containing e.g., *Achomosphaera alcicornu* s.s. and *Achilleodinium biformoides* characterize the latest Eocene in the Mediterranean area, whereas the incoming of *Glaphyrocysta semitecta* marks the base of the Oligocene, following correlations with the calcareous plankton and paleomagnetic records. Based on other reports, including information from Armenia, U.S.S.R. (Zaporcek, 1989), the authors argued that the incoming of *G. semitecta* during earliest Oligocene times can be traced across the Mediterranean area. However, since this event is the result of migration of the species to lower latitudes during this time, it cannot be employed to characterize the E/O boundary on a global scale.

In summary, comparison of all available information, including the recent results from Italy, indicates that the LO of *Areosphaeridium diktyoplokus* represents the most significant, globally recognizable dinoflagellate event associated with the E/O boundary. In fact, most authors use this event to be indicative of the E/O boundary. Data from the Mediterranean, however, show the event to occur above the LO of the hantkeninids, during the early part of C13N.

GLOBAL CHANGE ACROSS THE EOCENE/OLIGOCENE BOUNDARY: EVIDENCE FROM THE DINOFLAGELLATE RECORD.

During the last decade, attention was drawn to the occurrence of several distinct large-scale climatic, biotic and other events at or near the

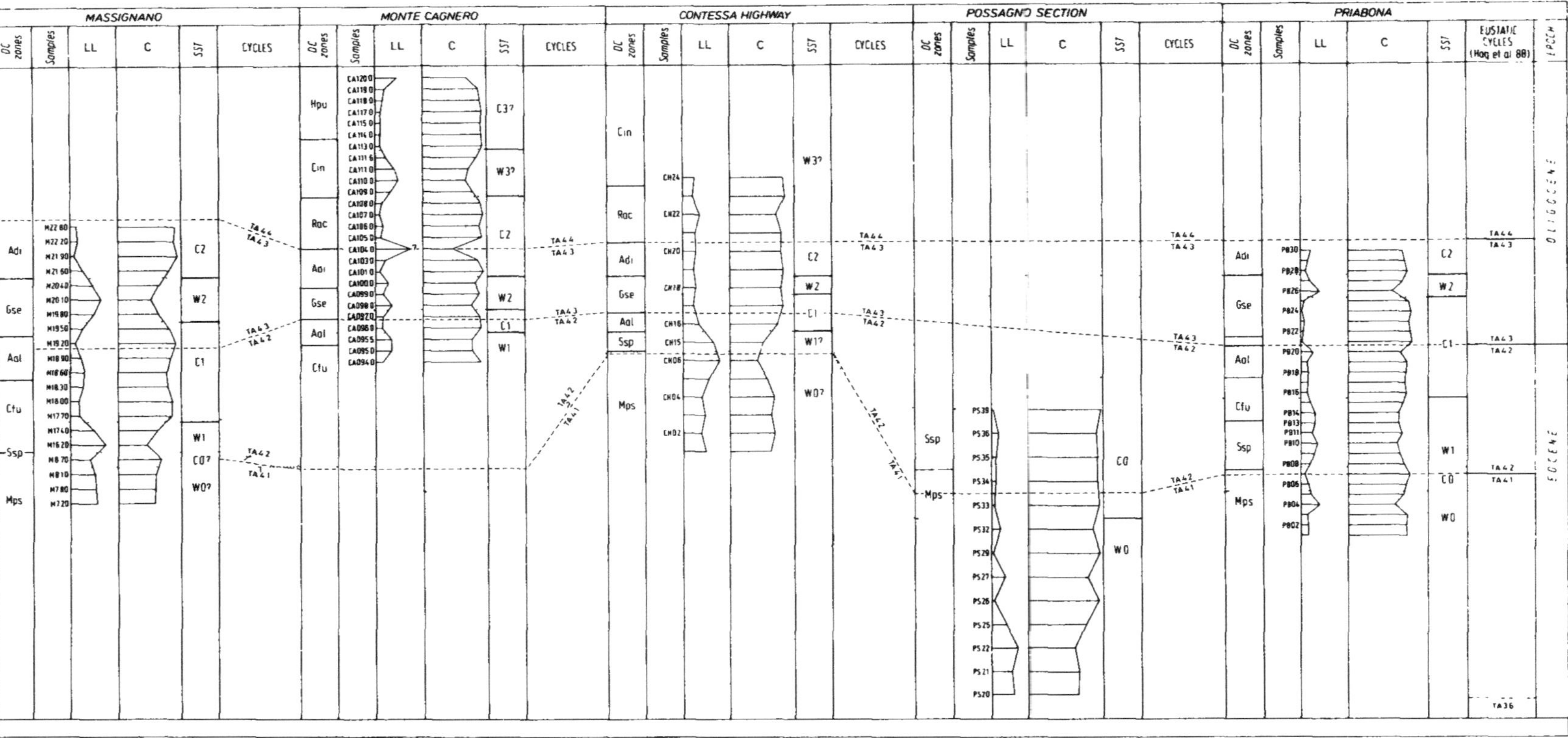

FIGURE 15.3. Quantitative distribution patterns of lower latitude (LL) versus cosmopolitan (C) taxa within samples from sections from the Priabonian type-area, correlation to the results from the central Italian sections of Brinkhuis and Biffi (in prep.) and recognition of relatively cool (C0-C2) and warm (W0-W3) intervals of sea surface temperature (SST; after Brinkhuis, in prep.).

E/O boundary. As was mentioned earlier, few dinoflagellate studies have specifically been undertaken in this respect and the potential of dinoflagellate cyst palynology in the interval remains poorly explored. One of the few examples is the integrated study by Van Couvering et al. (1981) of the E/O transition in the Carpathians. In that particular case, however, dinoflagellate cysts were only employed for general biostratigraphic purposes rather than for recognition and interpretation of paleoenvironmental changes. Châteauneuf (1986), contributing to IGCP Project 174 ('Terminal Eocene Events'), summarized trends in late Eocene-early Oligocene dinoflagellate cysts and, among others, concluded that early Oligocene assemblages are poorly diversified in comparison with the high-diversity of the Eocene dinoflagellate cyst assemblages. The reason for this pronounced decrease of species-diversity was not further discussed by Châteauneuf. In their review of all contributions to IGCP Project 174, Pomerol and Premoli-Silva (1986) indirectly linked this phenomena to the pronounced drop of sea surface temperature (SST) in the early Oligocene. They considered the cooling-event the most "...generalized event at the E/O boundary, which culminated in a drop of 5° C in all realms, some 300 Kyr later than the major extinction amongst planktonic foraminifera.." (i.e. the LO of the hantkeninids).

Brinkhuis and Biffi (in prep.) link the invasion of the Mediterranean by several higher latitude dinoflagellate cyst taxa and the significant decrease in relative abundances of lower latitude taxa to SST lowering across the E/O boundary. Moreover, these authors show that this phenomenon not only occurs around the E/O boundary, calibrated against C13R1, but also, and even more prominently so, in slightly younger sediments, calibrated against C13N1. They conclude that the dinoflagellate record indicates two subsequent cooling events during the latest Eocene to early Oligocene relative to maximum warming in the late Eocene (see Figures 15.2 and 15.3). Furthermore, Brinkhuis and Biffi suggest that the second early Oligocene cooling (during C13N1) was of greater magnitude and duration than the first cooler episode falling in C13R1. This 'second', more pronounced cooling corresponds to the event mentioned earlier by e.g., Pomerol and Premoli-Silva (1986) which they suggested to have occurred some 300 Kyr after the LO of the hantkeninids. Many more authors mentioned a distinct cooling event to have occurred in the early Oligocene. For example, the study by Miller et al. (1987) of stable oxygen isotope patterns in benthic foraminifers from a great number of DSDP and ODP sites indicated significant cooling on a world-wide scale during the interval calibrated against C13N. These authors linked the event to earliest extensive Antarctic glaciation and also suggested it to have caused a major glacio-eustatic sea level fall (Figure 15.2).

Brinkhuis and Biffi (in prep.) interpret shifts in the quantitative composition of successive assemblages also in terms of changes in apparent sea level. The interpretations are mainly based on dinoflagellate cyst distribution in Recent sediments, published e.g., by Wall et al. (1977). Two subsequent sea level lowerings were postulated to have occurred during latest Eocene to early Oligocene times, each of which coinciding with relatively cooler SSTs (see Figure 15.2).

The dinoflagellate cyst zones defined on the basis of the pelagic central Italian sections have been recognized in the time-equivalent nearshore successions from the Priabonian type-area (Brinkhuis, in prep.). Evidence from the Priabonian type-area is in line with the inferred inception of cooling events during the latest Eocene to early Oligocene relative to maximum warming in the late Eocene (Figure 15.3). In addition, some evidence for (less pronounced) early late Eocene SST fluctuations is available on the basis of the dinoflagellate cyst succession at Priabona. The quantitative dinoflagellate cyst distributions recorded from Priabonian type- and reference sections are also interpreted in terms of sea level changes and can be correlated to cycles TA4.1 through 4.4 of Haq et al. (1988). Correlation of the investigated sections with the use of the boundaries between these cycles support the proposed biostratigraphic correlation (Figure 15.4). Combining data from central and northeast Italy, more evidence that the periods of sea level lowering correlate to relatively cool intervals became available. This information supports glacio-eustacy as the cause for the sea level fluctuations.

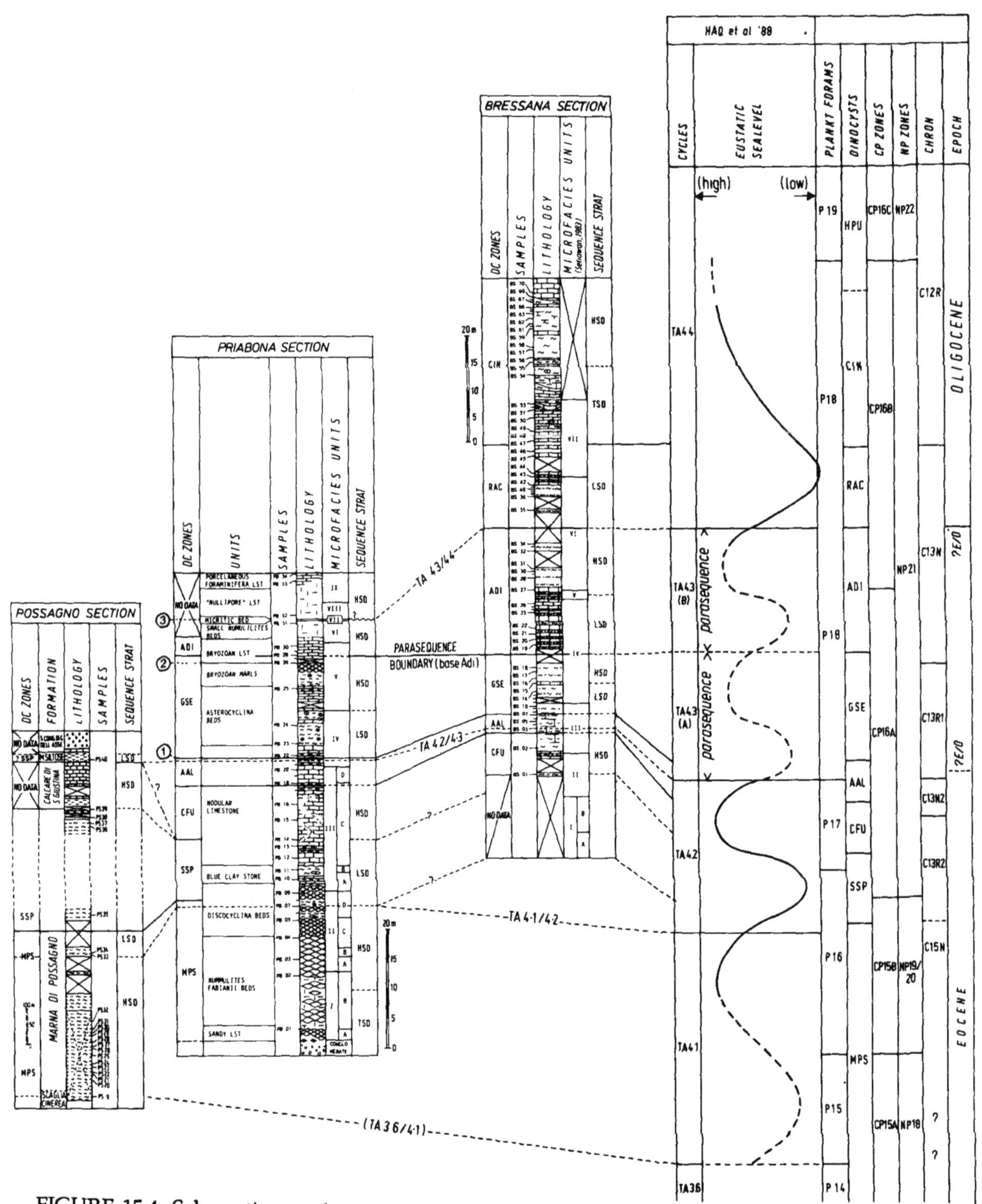

FIGURE 15.4. Schematic correlation of the sections from the Priabonian type-area with estimated eustatic sea level fluctuations and preliminary sequence stratigraphic interpretation on the basis of dinoflagellate cyst distribution, combined with results from central Italy (after Brinkhuis and Biffi, in prep.; Brinkhuis, in prep.). Numbers 1-3 in the Priabona section indicate (1) FO of *G. semitecta*/LO of hantkeninids, (2) base of Bryozoan limestones, i.e. top of the type-section following Hardenbol (1968), and (3) proposed top of the type section of Barbin and Bignot (1986).

Also in the Mediterranean many dinoflagellate cyst species disappear in the course of the early Oligocene, corroborating earlier reports world-wide. It may be noted that, in particular, presumed marginal marine species exhibit a decline in diversity. This may indicate that specifically marginal marine habitats became disrupted or no longer existed, leading to the impoverished and poorly diversified dinoflagellate cyst assemblages characteristic of the late early Oligocene. The array of climatic perturbations and relatively rapid sea level changes across the E/O boundary may well be considered responsable for this, apparently global, phenomenon. In addition, a marked change in tectonic and sedimentary setting during the late early Oligocene may have had additional effects in the Mediterranean area (Brinkhuis and Biffi, in prep.).

In summary, quantitative and qualitative dinoflagellate cyst events across the E/O boundary from both offshore as well as marginal marine settings in the Mediterranean allow the recognition of successive warmer and cooler intervals, in addition to apparent sea level fluctuations. Late Eocene warm conditions and sea level highstand became rapidly replaced by short-term cooler conditions and lowstand across the E/O boundary horizon (TA4.2/4.3 cycle boundary of Haq et al., 1988). After a short-term warmer interval and highstand during the earliest Oligocene, a second cooling and lowstand episode occurred (associated with the TA4.3/4.4 cycle boundary). This second interval of lowered SST was of greater magnitude and duration than the one at the E/O horizon and corresponds to the globally recognized cooling event falling within C13N. These results emphasize the importance and potential of dinoflagellate cyst palynology as a tool for the recognition of environmental changes, especially in marginal marine areas.

DINOFLAGELLATES AND LATE EOCENE-EARLY OLIGOCENE CHRONOSTRATIGRAPHY

Animated discussions with regard to the chronostratigraphic positions of the various late Eocene-early Oligocene stages, viz, the Bartonian (England), Priabonian (Italy), Ludian and Stampanian (France), Rupelian (Belgium) and especially the Lattorfian (Germany), still continue today (see e.g., Berggren et al., 1985). It is clear that selection of any particular suite of these stages leads to problems in biostratigraphic correlation, as they all have been defined from completly different sedimentary basins. This holds especially for Priabona, with its location in the Mediterranean area. Dinoflagellates have been studied from most of these stages, viz., from the Bartonian (e.g., Bujak et al., 1980), Priabonian (Gruas-Cavagnetto, 1974; Gruas-Cavagnetto and Barbin, 1988), Ludian and Stampian (e.g., Châteauneuf, 1980) and Rupelian (Weyns, 1970), but the results could only partly contribute to a better assessment of their relative and absolute positions (see review by Châteauneuf, 1986).

Recent studies in Italy resulted in a better understanding of the chronostratigraphic position of the Priabonian Stage. Combining the wealth of biostratigraphic and paleomagnetic data obtained from the pelagic Eocene-Oligocene sections in central Italy, including the recently proposed E/O boundary stratotype section at Massignano (Odin and Montanari, 1988), with data gathered from the sections from the Priabonian type-area, it was shown that the E/O boundary occurs in the middle part of the section and that the boundary is correlative with the TA4.2/4.3 cycle boundary of Haq et al (1988), falling within C13R1 and within an interval of lowered SST (see Figures 2-4; Brinkhuis, in prep.). This is primarily because the FO of *Glaphyrocysta semitecta*, an event calibrated against a horizon just above the LO of the hantkeninids in central Italy, occurs at the base of the *Asterocyclina* beds, some fifteen meters below the top of the type section (i.e., the base of the Bryozoan limestone unit, see Figure 15.4). Also other dinoflagellate events occur in the Priabona section in an order very similar to that recorded in central Italy, confirming the correlation as depicted in Figure 15.4 (Brinkhuis, in prep.).

The discovery that the LO of the hantkeninids is correlative with a horizon in the middle part of the Priabonian type-section raises the question of the validity of this event as the E/O boundary criterion. This discrepancy is a result of the decision of the International Colloquium on the Eocene in Paris (1968) to formally accept the section at Priabona as the type section of the Priabonian Stage, its limits defined by basalts at the base and Bryozoan limestones at the top (see e.g., Hardenbol, 1968; Cita, 1969; Setiawan, 1983). However, since (1) the E/O boundary stratotype has recently been defined at Massignano (Odin and Montanari, 1988) and (2) the boundary itself is characterized by the LO of the hantkeninids, the results from Priabona suggest that the Priabonian Stage must be decapitated in its upper part and the Rupelian Stage lowered so that its base corresponds to the base of the Oligocene as defined by the boundary stratotype (Berggren, pers. commun.).

Notwithstanding this, it is interesting to note that Barbin and Bignot (1986) proposed to redefine the top of the Priabonian type-section to a horizon some three meters *above* the top of the Bryozoan limestone unit. On the basis of the dinoflagellate cyst distribution, Brinkhuis (in prep.) suggests that the horizon proposed by Barbin and Bignot most probably correponds to the early part of Chron C13N1, the TA4.3/4.4 cycle boundary, and the LO of *Areosphaeridium diktyoplokus* (Figure 15.4). It therefore appears that if the stage and the type is accepted sensu Barbin and Bignot (1986), the LO of *Areosphaeridium diktyoplokus* does indeed represent the E/O junction, and provides the best biostratigraphic criterion for recognition of the boundary. However, the underlying Bryozoan beds are at present still considered to represent the lowest Oligocene at Priabona. Dinoflagellate events associated with this unit are the LO of *Hemiplacophora semilunifera* and the FO of *Areoligera semicirculata*. (Figure 15.4; see also discussion in Brinkhuis and Visscher, in prep.).

The chronostratigraphic position of the basal deposits of the Priabonian type section also forms a continuing debate in the literature (see e.g., Setiawan, 1983; Berggren et al., 1985; Barbin, 1986). Bio- and sequence stratigraphic interpretation of the dinoflagellate cyst distribution suggests that the basal deposits at Priabona belong to the transgressive part of cycle TA4.1 of Haq et al. (1988), correlative to the *Melitasphaeridium pseudorecurvatum* Zone of Brinkhuis and Biffi (in prep.). This assessment suggests correlation to the early part of zones P16, NP19/20 and CP15B, corresponding to an early Priabonian (s.l.) age (Figure 15.4).

ACKNOWLEDGEMENTS

Comments from W.A. Berggren, L.E. Edwards, H. Visscher and W.J. Zachariasse substantially improved the manuscript. Stimulating discussions with I. Premoli-Silva, L.E. Stover and H. Leereveld are gratefully acknowledged. Drawings were made by T. Schipper and H.R. Rypkema. The author acknowledges support from The Netherlands Foundation for Earth Science Research (AWON) and financial aid from The Netherlands Organization for the Advancement of Scientific Research (NWO) as well as from the LPP Foundation.

REFERENCES

Barbin, V., 1986. Le Priabonien dans sa région-type (Vicentin, Italie du Nord). Stratigraphie; Micropaléontologie; Essai d'intégration dans l'échelle chronostratigraphique. *Thesis University P. and M. Curie,* Paris, France, 227p.

Barbin, V., and Bignot, G. 1986. New proposal for an Eocene-Oligocene boundary according to microfacies from the Priabonian-type section. In Pomerol, C., and Premoli-Silva, I., eds., *Terminal Eocene Events.* Amsterdam, Elsevier, pp. 49-52.

Beckmann, J.P., Bolli, H.M., Perch-Nielsen, K., Proto-Decima, F., Saunders, J.B., and Toumarkine, M. 1981. Major calcareous nannofossil and foraminiferal events between the Middle Eocene and Early Miocene. *Palaeogeogr., Palaeoclimat., Palaeoecol.* 36 (3-4): 155-190.

Benedek, P.N. 1972. Phytoplankton aus dem Mittel- und Ober-Oligozän von Tönesberg (Niederrheingebiet). *Palaeontographica* Abt. B, 137: 1-71.

Benedek, P.N. 1986. Ergebnisse der Phytoplankton Untersuchungen aus dem Nordwestdeutschen Tertiär. In Tobien, H., ed., *Nordwestdeutschland im Tertiär; Beiträge zur regionalen geologie der Erde,* Gebr. Borntr ger, Berlin-Stuttgart, 18: 157-185.

Benedek, P.N., and Müller, C. 1974. Nannoplankton-Phytoplankton-Korrelation im Mittel- und Ober-Oligozän von NW-Deutschland. *N. Jb. Geol. Paläont., Mh.* 7: 385-397.

Benedek, P.N., and Müller, C. 1976. Die Grenze

Unter-/Mittel-Oligozän am Doberg bei Bünde/Westfalen. I. Phyto- und Nannoplankton. *N. Jb. Geol. Paläont., Mh.* 9: 129-144.

Berggren, W.A., Kent, D.V. and Flynn, J.J. 1985. Jurassic to Paleogene: Part 2: Paleogene geochronology and chronostratigraphy. In N.J. Snelling, ed., *The Chronology of the Geological Record. Geol. Soc. Mem.* 10: 141-189.

Biffi, U. and Manum S.B. 1988. Late Eocene-Early Miocene dinoflagellate cyst biostratigraphy from the Marche Region (Central Italy). *Boll. Soc. Pal. Ital.* 27 (2): 163-212.

Blow, W.H. 1969. Late middle Eocene to Recent planktonic foraminiferal biostratigraphy. In Brönnimans, R., Renz, N.H. , eds.. *Proceedings of the First International Conference on Planktonic Microfossils,* Geneva, 1967, 1, E.J. Brill, Leiden: 199-421.

Brinkhuis, H., in prep. Late Eocene to early Oligocene dinoflagellate cysts from the the Priabonian type-area (northeast Italy); biostratigraphy and palaeoenvironmental interpretation. *Palaeogeogr., Palaeoclimatol., Palaeoecol.*

Brinkhuis, H., and Biffi, U. in prep. High-resolution dinoflagellate cyst stratigraphy of the Eocene/Oligocene transition in central Italy. *Mar. Micropaleontol.*

Brinkhuis, H., and Visscher, H. in prep. The chronostratigraphic position of the Priabonian stage. *Newsl. Strat.*

Bujak, J.P. 1979. Proposed phylogeny of the dinoflagellates *Rhombodinium* and *Gochtodinium. Micropaleontology* 25: 308-324.

Bujak, J.P. 1984. Cenozoic dinoflagellate cysts and acritarchs from the Bering Sea and northern Pacific, DSDP Leg 19. *Micropaleontology* 30 (2): 180-212.

Bujak, J.P., Downie, C., Eaton, G.L., and Williams, G.L. 1980. Dinoflagellate cysts and acritarchs from the Eocene of southern England. *Palaeont. Assoc., Spec. Pap. Palaeont.* 24, 100p.

Châteauneuf, J.-J. 1980. Palynostratigraphie et Paleoclimatologie de l'Éocene superieur et de de l'Oligocéne du Bassin de Paris. *Mém. Bur. Rech. Géol. Min.* 116: 1-360.

Châteauneuf, J.-J. 1986. Evolution of the microflora and dinocysts at the Eocene-Oligocene boundary in western Europe. In Pomerol, C., and Premoli-Silva, I., eds., *Terminal Eocene Events.* Amsterdam, Elsevier, pp. 289-294.

Châteauneuf, J.-J., and Gruas-Cavagnetto, C. 1978. Les zones de Wetzeliellaceae (Dinophyceae) du Bassin de Paris. *Bull. Bur. Rech. Géol. Min.* 2, IV (2): 59-93.

Cita, M.B. 1969. Le Paléocène et l'Éocène de l'Italy du Nord. In Colloque sur l'Éocène. *Mém. Bur. Rech. Géol. Min.* 69: 417-429.

Costa, L.I., and Downie, C. 1976. The distribution of the dinoflagellate *Wetzeliella* in the Palaeogene of northwestern Europe. *Palaeontology* 19: 591-614.

Costa, L.I., and Downie, C. 1979. Cenozoic dinocyst stratigraphy of Sites 403 to 406 (Rockall Plateau), IPOD, Leg 48. In L. Montadert, D.G. Roberts, et al., eds., *Init. Rept. Deep Sea Drill. Proj.,* Washington D.C., U.S. Government Printing Office, 48: 513-529.

Costa, L.I., and Manum, S.B., et al. 1988. The description of the interregional zonation of the Paleogene (D 1 - D 15) and the Miocene (D 16 - D 20). In Vinken, K.L., Renier, P.K., et al., eds., *The Northwest European Tertiary Basin,* Results of the International Geological Correlation Programme, Project No 124. *Geol. Jahrb.* A (100): 321-342.

Drugg, W.S., and Loeblich, A.R. 1967. Some Eocene and Oligocene phytoplankton from the Gulf Coast, U.S.A. *Tulane Studies in Geology* 5: 181-194.

Eaton, G.L. 1976. Dinoflagellate cysts from the Bracklesham Beds (Eocene) of the Isle of Wight, southern England. *Bull. British Mus. Geol.* 26: 227-332.

Gruas-Cavagnetto, C. 1974. La palynoflore et le microplancton du Priabonien de sa localité type (prov. Vicenza, Italie). *Bull. Soc. Géol. France,* 7 (16): 86-90.

Gruas-Cavagnetto, C., and Barbin, V. 1988. Les dinoflagellés du Priabonien stratotypique (Vicentin, Italie); mise en evidence du passage Eocène/Oligocène. *Rev. Paléobiol.* 7 (1): 163-198.

Haq, B.U., Hardenbol, J., and Vail, P.R., et al. 1988. Mesozoic and Cenozoic chronostratigraphy and cycles of sea level change. In Wilgus, C. K., Hastings, B. S., et al., eds., *Sea Level Changes; An Integrated Approach. Soc. Econ. Paleontol. Mineral., Spe-*

cial Publ. 42: 71-108.

Hardenbol, J. 1968. The Priabonian type section (a preliminary note). In Colloque sur l'Éocène. *Mém. Bur. Rech. Géol. Min.* 58: 629-635.

Head, M.J., and Norris, G. 1989. Palynology and dinocyst stratigraphy of the Eocene and Oligocene in ODP-Leg 105, Hole 647A, Labrador Sea. In Srivastava, S. P., Arthur, M., Clement, B., et al., eds., *Proc. ODP, Sci. Results,* College Station, Texas, Ocean Drilling Program, 105: 515-550.

Ioakim, C. 1979. Étude comparative des dinoflagellés du Tertiaire inférieur de la Mer du Labrador et de la Mer du Nord. Thesis University Pierre et Marie Curie, Paris.

Jan du Chêne, R.E. 1977. Palynostratigraphic (Maastrichtien-Eocène inférieur) des Flysches du Schlieren (Canton d'Obwald, Suisse centrale). *Rev. Micropaleontol.* 20: 147-156.

Köthe, A. 1990. Paleogene dinoflagellates from northwest Germany. *Geol. Jahrb.* 118. 111p.

Liengjarern, M., Costa, L.I., and Downie, C. 1980. Dinoflagellate cysts from the Upper Eocene-Lower Oligocene of the Isle of Wight. *Palaeontology* 23: 475-499.

Manum, S.B. 1976. Dinocysts in Tertiary Norwegian-Greenland Sea sediments (Deep Sea Drilling Project Leg 38), with observations on palynomorphs and palynodebris in relation to environment. In Talwani, M. et al., eds., *Init. Rept. Deep Sea Drill. Proj.,* Washington D.C., U.S. Government Printing Office, 38: 879-919.

Manum, S.B., Boulter, M.C., Gunnarsdottir, H., Rangnes, K., and Scholze, A. 1989. Palynology of the Eocene to Miocene sedimentary sequence of ODP Leg 104 (Norwegian Sea). In Edholm, O., Thiede, J., et al., eds., *Proc. ODP, Init. Repts.,* College Station, Texas, Ocean Drilling Program, 104: 611-662.

Matsuoka, K., and Bujak, J.P. 1988. Cenozoic dinoflagellate cysts from the Navarin Basin, Norton Sound and St. George Basin, Bering Sea. *Bull. Fac. Liberal Arts, Nagasaki University, Nat. Sci.* 29: 1-147.

Médus, J., and Pairis, J-.L. 1990. Reworked pollen assemblages and the Eocene-Oligocene boundary in the Paleogene of the western external French Alps. *Palaeogeogr., Palaeoclimat., Palaeoecol.* 81 (1-2): 59-78.

Miller, K.G., Fairbanks, R.G., and Mountain, G.S. 1987a. Tertiary oxygen isotope synthesis, sea level history and continental margin erosion. *Paleoceanography* 2 (1): 1-19.

Morgenroth, P. 1966. Neue in organischer Substanz erhaltene Mikrofossilien des Oligozäns. *N. Jb. Geol. Paläont. Abh.* 127: 1-12.

Nocchi, M., Parisi, G., Monaco, P., Monechi, S., Madile, M., Napoleone, G., Ripepe, M., Orlando, M., Premoli-Silva, I., and Bice, D. 1986. The Eocene-Oligocene boundary in the Umbrian pelagic sequences (Italy). In Pomerol, C., and Premoli-Silva, I., eds., *Terminal Eocene Events.* Amsterdam, Elsevier, pp 25-40.

Nocchi, M., Parisi, G., Monaco, P., Monechi, S., and Madile, M. 1988a. Eocene and Early Oligocene micropaleontology and paleoenvironments in SE Umbria, Italy. *Palaeogeogr., Palaeoclimat., Palaeoecol.* 67: 181-244.

Nocchi, M., Monechi, S., Coccioni, R., Madile, M., Monaco, P., Orlando, M., Parisi, G., and Premoli-Silva, I. 1988b. The extinction of Hantkeninidae as a marker for recognizing the Eocene-Oligocene boundary: a proposal. In Premoli-Silva, I., Coccioni, R., and Montanari, A., eds., *The Eocene-Oligocene Boundary in the Marche-Umbria Basin (Italy).* International Union of Geological Sciences Commission on Stratigraphy; International Subcommission on Paleogene Stratigraphy Report (Ancona, Italy, 1988), pp. 249-252.

Odin, G.S., and Montanari, A. 1988. The Eocene-Oligocene boundary at Massignano (Ancona, Italy): a potential boundary stratotype. In Premoli-Silva, I., Coccioni, R., and Montanari, A., eds., *The Eocene-Oligocene Boundary in the Marche-Umbria Basin (Italy).* International Union of Geological Sciences Commision on Stratigraphy; International Subcommission on Paleogene Stratigraphy Report (Ancona, Italy, 1988), pp. 253-263.

Pomerol, C., and Premoli-Silva, I. 1986. The Eocene-Oligocene transition: events and boundary. In Pomerol, C., and Premoli-Silva, I., eds., *Terminal Eocene Events.* Amsterdam, Elsevier, pp. 1-24.

Poore, R.Z., Tauxe, L., Percival, S.F., and LaBrecque, J.L. 1982. Late Eocene-Oligocene magnetostratigraphy and biostratigraphy at South Atlantic DSDP Site 522. *Geology*

10: 508-511.

Premoli-Silva, I., Coccioni, R., and Montanari, A., eds., 1988. *The Eocene-Oligocene Boundary in the Marche-Umbria Basin (Italy)*. International Union of Geological Sciences Commision on Stratigraphy; International Subcommission on Paleogene Stratigraphy Report (Ancona, Italy, 1988), 268pp.

Setiawan, J.R. 1983. Foraminifera and microfacies of the type Priabonian. *Utrecht Micropal. Bull.* 29, 161pp

Van Couvering, J.A., Aubry, M.-P., Berggren, W.A., Bujak, J.P., Naeser, C.W., and Wieser, T. 1981. The terminal Eocene event and the Polish connection. *Palaeogeogr., Palaeoclimatol., Palaeoecol.* 36: 321-362.

Wall, D., Dale, B., Lohmann, G.P. and Smith, W.K. 1977. The environmental and climatic distribution of dinoflagellate cysts in modern marine sediments from regions in the North and South Atlantic oceans and adjacent seas. *Mar. Micropaleontol.* 2: 121-200.

Weyns, W. 1970. Dinophycées et acritarches des "Sables de Grimmertingen" dans leur localité-type, et les problèmes stratigraphiques de Tongrien. *Bull. Soc. Belge Géol. Paléontol. Hydr.* 79 (3-4): 247-268.

Williams, G.L. 1975. Dinoflagellate and spore stratigraphy of the Mesozoic-Cenozoic, offshore Eastern Canada. *Geol. Surv. Can. Pap.* 74-30, 2: 107-161.

Williams, G.L. 1977. Dinocysts: their classification, biostratigraphy and palaeoecology. In A.T.S. Ramsay, ed., *Oceanic Micropalaeontology*. London, Academic Press, pp. 1231-1325.

Williams, G.L., and Bujak, J.P. 1977a. Distribution patterns of some North Atlantic Cenozoic dinoflagellate cysts. *Mar. Micropaleontol.* 2: 223-233.

Williams, G.L., and Bujak, J.P. 1977b. Cenozoic palynostratigraphy of offshore Eastern Canada. *Am. Ass. Strat. Palynol., Contrib. Series* 5A: 14-47.

Williams, G.L., and Bujak, J.P. 1985. Mesozoic and Cenozoic dinoflagellates. In Perch-Nielsen, K., et al., eds. *Plankton stratigraphy*, Cambridge, Cambridge Univ. Press, pp. 847-964.

Zaporcek, N.J. 1989. Upper Eocene and Lower Oligocene Palynocomplexes and phytoplankton from Borehole no. 1., Landzar (Armenia, U.S.S.R.). In *Phanerozoic Paleoflora and stratigraphy*. Edit. Soviet Acad. Sci., Moscow, 1989, 85-103 (in Russian).

16. THE PATTERNS AND CAUSES OF MOLLUSCAN EXTINCTION ACROSS THE EOCENE/ OLIGOCENE BOUNDARY

By Thor Hansen

ABSTRACT

The late Eocene extinctions of marine molluscs in the U.S. Gulf Coast were spread out over the entire late Eocene interval. Selective extinction of warm water taxa and the general coincidence of molluscan and planktonic foraminiferal extinctions suggest cooling as the direct cause. Resolution of the precise pattern of the molluscan extinctions in the latest Eocene awaits a detailed analysis of this crucial interval.

INTRODUCTION

Although not in the "first tier" of extinctions in terms of magnitude, the Eocene–Oligocene extinction affected a broad range of organisms including planktonic foraminifera and benthic molluscs in the marine realm. Several prominent Eocene molluscan genera and subgenera that were geographically wide-spread and resilient through earlier changes in temperature and shoreline position, disappear at or near the Eocene/Oligocene boundary. These taxa include the bivalves *Yoldia (Calorhadia)*, *Venericardia (Venericor)*, *Pachecoa*, and the gastropods *Calyptraphorous*, *Pseudoliva*, and *Lapparia*. *Athleta*, a prominent Eocene gastropod, became extinct in North America but continued into the Oligocene in Europe (Dockery, 1984). Study of the Eocene/ Oligocene boundary also bears on the impact theory of extinctions because it is one of the few extinction horizons, besides the Cretaceous–Tertiary, near which there is physical evidence of extraterrestrial impact (Ganapathy, 1982; Alvarez et al., 1982).

In this report, I will document the landscape molluscan diversity pattern of the Gulf Coast not only for the late Eocene–early Oligocene, but for the entire Paleocene–early Oligocene interval. Such scope is necessary to understand both the environmental controls on diversity and the historical effect of the Cretaceous–Tertiary extinction. I will then discuss the two competing theories for the controls of that pattern and for the Eocene–Oligocene extinction in particular: changes in temperature, and shelf area.

NATURE OF THE RECORD

The ideal collecting ground for studying an extinction event should include (1) faunas that are well preserved, (2) faunas that are taxonomically well known, (3) a rock sequence that is continuous and free of large facies changes, and (4) tight biostratigraphic control. The early Tertiary deposits of the U.S. Gulf Coast come as close as any in the world to satisfying these criteria. The Paleogene of the Gulf Coast contains a beautifully preserved molluscan fauna where shell material is commonly preserved in such a state that original color patterns are present, and the matrix can be wet-sieved with nearly complete shell recovery. The molluscan systematics have been well studied, including a general tabulation and synonymy of the entire Paleocene–Eocene fauna (Palmer and Brann, 1965–1966) and a modern systematic treatment of the late Eocene and early Oligocene faunas (Dockery, 1977, 1982; MacNeil and Dockery, 1984). There are facies changes around the Eocene/Oligocene boundary. The upper Eocene Yazoo Formation was probably deposited in significantly deeper water than its surrounding formations, but the entire late middle Eocene to Oligocene interval is composed of marine deposits without a major hiatus or nonfossiliferous nonmarine unit. Finally, because of the economic value of Gulf Coast oil, this rock sequence has been extensively studied biostratigraphically and is relatively easy to correlate to the International Paleogene Time Scale.

PATTERNS OF DIVERSITY

A tabulation of molluscan diversity for the early Tertiary of the U.S. Gulf Coast was made using the sources cited in Hansen (1988). In this context diversity will refer to landscape or gamma species diversity as opposed to within-habitat or alpha diversity (Whittaker, 1970). In the Kincaid and Clayton Formations (Figure 16.1), diversity was relatively low after the Cretaceous–Tertiary (K–T) extinctions. The early Paleocene diversity level of around 100 species was an 80% reduction from the Late Cretaceous gamma diversity of the Gulf Coast (Hansen, 1988). Diversity can then be seen to fall and rise in three successive cycles, each valley being extremely low and the peaks successively higher through the late middle Eocene. The Paleocene and lower Eocene valleys correspond to times of widespread nonmarine deposits (the Naheola and Halls Summit Formations in the Paleocene and the Meridian, Carrizo, Reklaw and Queen City Formations in the lower Eocene) and the fossiliferous marine deposits that are present at these horizons (the Upper Wills Point Formation and the Tallahatta and Cane River Formations) contain relatively poorly preserved fossils. The diversity peaks correspond to times when fossiliferous marine deposits were widespread; the Clayton–Kincaid Formations, the Sabinetown and Bashi Member of the Hatchitigbee Formations, and the Cook Mountain–Upper Lisbon Formations (Figure 16.1). The Gosport–Cockfield–Yegua horizon is an exception to this rule in that the lone marine unit, the Gosport, produces a molluscan assemblage as diverse as the underlying Cook Mountain–Upper Lisbon interval. Although I think the diversity valleys represent times of lowered diversity along the Gulf Coast, I do not believe diversity was as low as the numbers suggest. If they were, each drop would represent an extinction horizon equivalent in magnitude to the Eocene/Oligocene boundary, which is not reasonable considering the lack of generic level extinctions during these intervals. On the other hand, I believe the diversity peaks and the flanks of the peaks *are* roughly representative of the gamma diversity levels of the Gulf Coast at their times of deposition. Each of the widespread marine horizons that make up these levels contain numerous fossil localities across the Gulf Coast that have been collected for many years. One recent monograph based on extensive new collecting of the Moodys Branch Formation (upper Eocene, see Figure 16.1) did not add substantially to the number of new species described for that unit (Dockery, 1977). This suggests that the number of species discoveries in the well-preserved horizons has plateaued.

CORRELATIONS WITH THE DIVERSITY PATTERN

If we assume then that the diversity curve in Figure 16.1 is fairly accurate except in the bottoms of the valleys, with what does this diversity trend correlate? The two most commonly cited correlates of benthic invertebrate diversity changes are temperature (Stanley, 1984) and shoreline changes (Hallam, 1981, 1989). While the latter argument has not been advanced specifically for the early Tertiary Gulf Coast, it is a popular idea as a general cause for mass extinctions and diversity fluctuations.

Shoreline Changes

Inspection of the shoreline curve (a trend line derived from the middle of the Mississippi embayment) in Figure 16.2 shows little correspondence with diversity. This lack of correspondence is particularly well documented for the late Eocene–Oligocene interval. The diversity peak of the Upper Lisbon Formation and its equivalents occurs during a widespread transgression. This is followed by a regression and subsequent transgression which deposits the Gosport Formation. The transgression during the Gosport interval was much less extensive than the Upper Lisbon marine incursion but diversity remained the same. The upper Eocene Moodys Branch Formation represented a further transgression and increase in shelf area yet diversity levels declined. Paleogeographic reconstructions of actual shelf area confirm this shoreline trend. The determination of the position of the continental shelf edge during the early Tertiary (Winker, 1984) has recently allowed the quantification of shelf areas between

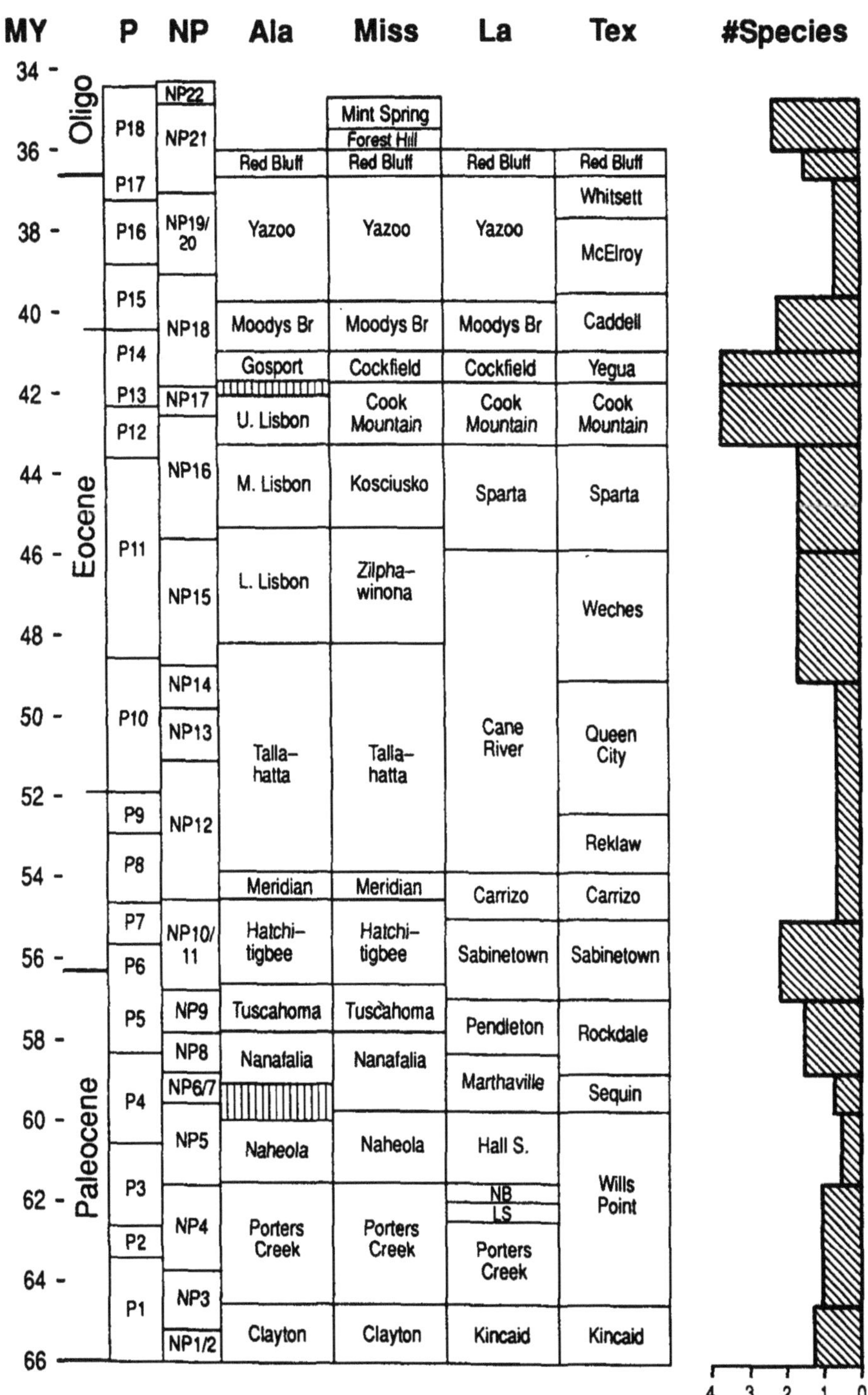

FIGURE 16.1. Correlation chart of early Tertiary formations of the Gulf Coast with microfossil biostratigraphy and molluscan gamma diversity. MY, million years before present; P, planktonic foraminiferal zonation; NP, nannofossil zonation; Ala, Alabama; Miss, Mississippi; La, Louisiana; Tex, Texas; #Species, gamma diversity, scale graduated in hundreds of species. Correlations taken from Hansen (1980) and Hazel et al. (1984).

transgressive and regressive maxima. Paleogeographic reconstructions of the Upper Lisbon–Cook Mountain, and Gosport–Cockfield–Yegua shows that the areal extent of the Gosport sea was roughly half that of the Upper Lisbon (Hansen, 1987a,b), yet the molluscan diversities were the same.

Temperature Change

Diversities of Gulf Coast molluscs show a much closer correlation with temperature than with shelf area (Figure 16.2 and McKinney and Oyen, 1989). The diversity peaks of the early Eocene and late middle Eocene correspond nicely with temperature highs, the troughs of the middle Paleocene and lower middle Eocene correlate with temperature lows. The difference in effect on diversity between temperature and shelf area is especially apparent in the Upper Lisbon to Moodys Branch interval around the middle to late Eocene boundary. The Upper Lisbon diversity peak corresponds to a temperature high and maximum transgression. The subsequent loss of shelf area during the Gosport Formation (while temperature is still high) produces no loss of diversity but the temperature drop at the mid–late Eocene boundary is accompanied by a diversity decline even though shelf area increases. The mid–late Eocene boundary is cited as the first of several cooling episodes which caused a decline in planktonic foraminiferal diversity (Keller, 1983).

DISCUSSION

Shoreline Changes

There are two possible mechanisms for extinction of marine benthic organisms by shoreline regression, 1) change in habitat area per se causing greater competition for space and 2) change in area causing a loss of certain habitats (this second factor is dependent on environmental heterogeneity). As can be seen in Figure 16.2 and from the above discussion, shelf area per se as a control of diversity trends in early Tertiary molluscs is not supported. There is also no reason to believe that loss of habitat would have been a significant factor in these diversity fluctuations. The early Tertiary Gulf Coast did not contain significant endemic perched faunas that would be vulnerable to loss of habitat during a regression, as has been the case during times of widespread epeiric seaways, e.g., the Ordovician (Berry and Boucot, 1973; Sheehan, 1973). The early Tertiary marine shelf assemblages occupied a marginal marine band roughly parallel to the present-day shoreline. Shoreline changes in such a setting cause changes in area but do not cause elimination of habitat unless the shelf is completely emergent.

Dockery (1986, 1987, 1988) argued that during regressions prodeltaic mud blanketed the Gulf shelf and eliminated habitat for molluscs that favored clear water, stable substrate environments. He went on to cite this as a major factor in Gulf Coast molluscan diversity trends. There is no question that most benthic molluscs prefer clear water and stable substrates and that this is a strong control on within habitat (alpha) diversities. But how much of a role did these factors play in gamma diversity trends for the Gulf Coast and species extinctions? For mud blanketing to cause a species extinction, it would need to completely cover the species habitat throughout its biogeographic range. It has not been demonstrated that this occurred frequently in the early Tertiary Gulf Coast. If mud blanketing was a first order control on diversities and extinctions one would expect mud tolerant taxa to preferentially survive these events. Yet the extinction percentages of deposit–feeding nuculanid bivalves and the mud-tolerant neogastropod families Turridae, Fasciolariidae and Buccinidae were as high as that of the suspension–feeding bivalves and as high as the average extinction rate overall.

Temperature Changes

The temperature curve correlates fairly well with the observed molluscan diversity pattern (Figure 16.2). Temperature as a cause for this pattern is suggested by an analysis of the temperature preferences of the constituent species (Figure 16.2). Cool-tolerant genera such as *Turritella, Calyptraea,* and *Crepidula,* and species of the family Naticidae, form a larger percentage of the fauna during low temperature intervals. Alternatively, the percentage of warm–water taxa increases during temperature highs (Hansen, 1987a,b; 1988a,b). In addition, the molluscan extinctions at the mid–late Eocene boundary correlate well with the first

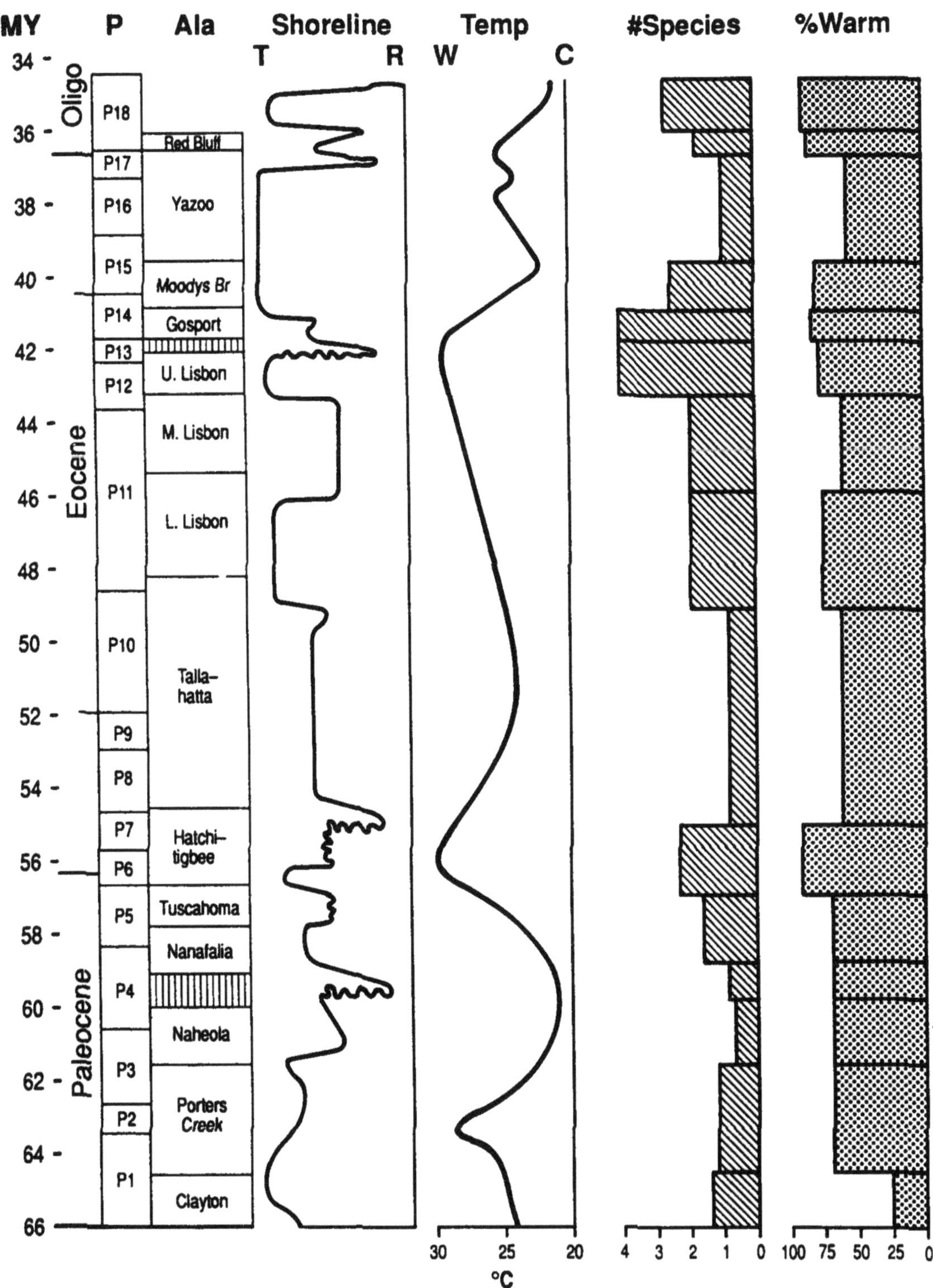

FIGURE 16.2. Correlation of early Tertiary formations with shoreline trends, temperature, molluscan gamma diversity and percentage of warm water species. See Figure 16.1 for abbreviations not listed here. T, transgression; R, regression; W, warm; C, cold. Shoreline trend taken from Dockery (1986) with emendation of the curve around the Gosport Formation interval from Baum and Vail (1988). Temperature curve, taken from Wolfe and Poore (1984), is derived from southeast North American terrestrial floras.

major pulse of late Eocene planktonic foraminiferal extinctions of the North Atlantic (Keller, 1983). The coincidence of benthic mollusc and planktonic foraminiferal extinctions at this time is strong evidence for temperature as a common cause.

Note that although the temperature peaks of Figure 16.2 are of roughly equal amplitude, the molluscan diversity peaks climb from a low of about a hundred species in the Paleocene to about 400 species in the late middle Eocene. I believe this trend of increasing diversity reflects rebound from the Cretaceous–Tertiary extinction. The K–T extinction eliminated on the order of 80% of molluscan species, reducing the Cretaceous landscape diversity of the Gulf Coast from around 500 species in the Late Cretaceous to little over 100 in the early Paleocene (Hansen, 1988). A line connecting the Paleocene, lower Eocene and middle Eocene diversity peaks, delineates a radiation rate that is exactly that predicted by Stanley (1979) for the maximum rate of radiation of the Bivalvia. This is a doubling rate of 11 million years. This suggests that early Tertiary molluscan diversity can be modeled by a combination of climate and temporal distance from the K–T extinction. The stratigraphic position of the diversity peaks was determined by the paleoclimate (warm spells produced high diversity) and the amplitude of the peaks depended on the time elapsed since the K–T extinction. The Eocene–Oligocene extinction "reset" the clock and was also the beginning of a general cooling trend, so the early Oligocene diversity peak is lower than that of the middle Eocene. The importance of radiation rate and temporal distance from the K–T boundary is confirmed by analysis of the species richness of the early Tertiary of the Paris Basin where molluscan diversity doubles at the same rate as in the U.S. Gulf Coast. As in the U.S. example, diversity also declines in the late Eocene (Dockery, 1984).

Impacts

The importance of extraterrestrial impacts on the late Eocene extinctions, either as a direct cause or as a climate forcing mechanism, is difficult to assess. Although the drawn-out period of extinction during the late Eocene appears to argue against impact as a cause, the relationship is not so easy to dismiss because there appears to have been an extended interval of impacts during the late Eocene. There are at least three (Keller et al., 1983) and possibly several more (Byerly et al., 1988; Hazel, 1989) microtektite layers in the Eocene/Oligocene boundary interval. The lack of precise correspondence between the tektite layers and foraminiferal extinctions led Keller et al. (1983) to discount the importance of impacts in the late Eocene extinctions. Still, the fact that the two largest extinctions of the Cenozoic, the Cretaceous–Tertiary and the Eocene–Oligocene, are associated with evidence of impacts is a remarkable coincidence (Hut et al., 1987). As yet, microtektites have not been found in macrofossiliferous marine shelf deposits, so the direct relationship between late Eocene molluscan extinctions and impacts cannot be explored.

THE REST OF THE STORY

Up to the mid point of the late Eocene there is a good correlation between temperature and molluscan diversity in the Gulf Coast. Temperatures dropped rapidly at the mid–late Eocene boundary and so did the number of molluscan species. What happened at the Eocene/Oligocene boundary? Did the number of species continue to decline in the late Eocene to a low at the Eocene/Oligocene boundary? In this scenario, the extinction at the epoch boundary would simply be the culmination of events occurring throughout the late Eocene (possibly in a stepwise fashion; Hansen, 1987a,b; 1988b). Or were there relatively few species extinctions through most of the late Eocene with another abrupt and significant extinction at the Eocene/Oligocene boundary itself (Dockery, 1986) The short answer to this question is: We do not know. There are no outcrops in North America where we can obtain a reliable molluscan species count immediately below the Eocene/Oligocene boundary. The richly fossiliferous shallow marine sediments of the late Eocene Moodys Branch Formation grade upward into the Yazoo Formation which represents a very fine-grained deep shelf environment (not the optimal environment for molluscs) and contains sparse and poorly preserved molluscan fossils. Lower Oligocene deposits contain a fairly diverse shallower water assemblage, the taxonomic composition of which reflects the major extinctions that have been ascribed to an Eocene/Oligocene boundary event. A literal reading of the fossil record suggests that many extinctions occurred near the Moodys Branch–Yazoo boundary but this could reflect facies control. However there is one outcrop of the Yazoo Formation that may provide a clue to this puzzle. The clay quarry near Cynthia, Mississippi exposes about 30 m of the upper 25% of the Yazoo Formation. The fossils here, unlike those exposed in the rest of the Yazoo Formation, are very well-preserved and easily washed out with sieves. Although this outcrop does not span the Eocene/Oligocene boundary itself, it is stratigraphically above most of the late Eocene foraminiferal extinctions and microtektite horizons (Byerly et al., 1988) and may provide important information as to what was still alive in the waning stages of the late Eocene. We are currently working on this section, in conjunction with David Dockery of the Mississippi Bureau of Geology, to try to pin down the latest Eocene molluscan extinction pattern.

CONCLUSIONS

The Eocene–Oligocene extinction of molluscs was stretched out over the entire late Eocene interval, with accelerated extinction rates starting at least by the mid-late Eocene boundary. The molluscan extinctions at the beginning of the late Eocene are best explained by climate change and are approximately synchronous with major extinctions of planktonic foraminifera. Resolution of the pattern of the latest Eocene molluscan extinctions awaits an adequate sampling of the molluscan fauna of this critical interval.

ACKNOWLEDGEMENTS

I thank C. Hickman and S. Stanley for reviewing the manuscript.

REFERENCES

Alvarez, W., Asaro, F., Michel, H.V., and Alvarez, L.W. 1982. Iridium anomaly approximately synchronous with terminal Eocene extinctions. *Science* 216:886–888.

Baum, G.R. and Vail, P.R. 1988. Sequence stratigraphic concepts applied to Paleogene outcrops, Gulf and Atlantic basins.In Sea-level changes—An integrated approach. *SEPM Special Publication* 42: 309–327.

Berry, W.B.N. and Boucot, A.J. 1973. Glacio-eustatic control of late Ordovician–early Silurian platform sedimentation and faunal changes. *Bull. Geol. Soc. Amer.* 84: 275–284.

Byerly, G.R., Hazel, J.E., and McCabe, C. 1988. A new late Eocene microspherule layer in central Mississippi. *Mississippi Geology* 8(4):1–5.

Dockery, D.T., III. 1977. Mollusca of the Moodys Branch Formation, Mississippi. *Miss. Geol. Surv. Bull.* 120: 1-212.

Dockery, D.T., III. 1982. Lower Oligocene Bivalvia of the Vicksburg Group in Mississippi. *Miss. Bur. of Geol. Bull.*123: 1-261.

Dockery, D.T., III. 1984. Crisis events for Paleogene molluscan faunas in the southeastern United States. *Mississippi Geology* 5(2):1–7.

Dockery, D.T., III. 1986. Punctuated succession of Paleogene mollusks in the northern Gulf Coastal plain. *Palaios* 1:582–589.

Dockery, D.T., III. 1987. Eocene–Oligocene molluscan extinctions: Comment. *Palaios* 2:620–621.

Dockery, D.T., III. 1988. Molluscan extinction rates in question. *Nature* 331:123.

Ganapathy, R. 1982. Evidence for a major meteorite impact on the Earth 34 million years ago: Implication for Eocene extinctions. *Science* 216:885–886.

Hallam, A. 1981. *Facies Interpretation and the Stratigraphic Record*. New York, W.H. Freeman and Co., 291 p.

Hallam, A. 1989. The case for sea-level change as a dominant causal factor in mass extinction of marine invertebrates. *Phil. Trans. R. Soc. Lond.* B325:437–455.

Hansen, T.A. 1987a. Extinction of late Eocene to Oligocene molluscs: Relationship to shelf area, temperature changes, and impact events. *Palaios* 2:69–75.

Hansen, T.A. 1987b. Eocene–Oligocene molluscan extinctions: Reply. *Palaios* 2: 621–622.

Hansen, T.A. 1988a. Early Tertiary radiation of marine molluscs and the long-term effects of the Cretaceous–Tertiary extinction. *Paleobiology* 14:37–51.

Hansen, T.A. 1988b. Molluscan extinction rates in question: Reply. *Nature* 331:123.

Hazel, J.E. 1989. Chronostratigraphy of upper Eocene microspherules. *Palaios* 4:318–329.

Hazel, J.E., Edwards, L.E., and Bybell, L.M. 1984. Significant unconformities and the hiatuses represented by them in the Paleogene of the Atlantic and Gulf Coastal Province. In Schlee, J.S., ed., Interregional unconformities and hydrocarbon accumulation. *Amer. Assoc. Pet. Geol. Memoir* 36:59–66.

Hut, P., Alvarez, W., Elder, W., Hansen, T.A., Kauffman, E.G., Keller, G., Shoemaker, E.M., and Weissman, P.R. 1987. Comet showers as a cause of mass extinctions. *Nature* 329:118–126.

Keller, G. 1983. Biochronology and paleoclimatic implications of middle Eocene to Oligocene planktic foraminiferal faunas. *Marine Micropaleontology*. 7:463–486.

Keller, G., D'Hondt, S., and Vallier, T.L. 1983. Multiple microtektite horizons in upper Eocene marine sediments: No evidence for mass extinctions. *Science* 221:150–152.

MacNeil, F.S., and Dockery, D.T., III. 1984. lower Oligocene Gastropoda, Scaphopoda, and Cephalopoda of the Vicksburg Group in Mississippi. *Miss. Bur. Geol. Bull.* 124, 416 p.

McKinney, M.L. and Oyen, C.W. 1989. Causation and nonrandomness in biological and geological time series: Termperature as a proximal control of extinction and diversity. *Palaios* 4:3–15.

Palmer, K.V.W. and Brann, D.C. 1965, 1966. Catalogue of the Paleocene and Eocene Mollusca of the southern and eastern United States. *Bull. Amer. Paleo.* 48, v. 218, 1027 p.

Sheehan, P.M. 1973. The relation of late Ordovician glaciation to the Ordovician–Silurian changeover in North American brachiopod faunas. *Lethaia* 6:147–154.

Stanley, S.M. 1979. *Macroevolution*. San Francisco, W.H. Freeman,332 p.

Stanley, S.M. 1984. Temperature and biotic crises in the marine realm. *Geology* 12:205–208.

Winker, C.D. 1984. Clastic shelf margins of the post–Comanchean Gulf of Mexico: Implications for deep–water sedimentation. In Characteristics of gulf basin deep–water sediments and their exploration potential: *Fifth Annual Research Conference, Gulf Coast Section, Soc. of Econ. Paleo. and Min. Foundation*, Austin, Texas. pp. 109–120.

Whittaker, R.H. 1970. *Communities and ecosystems*. New York, Macmillan, 162p.

Wolfe, J.A. and R.Z. Poore. 1982. Tertiary marine and nonmarine climatic trends. In Berger, W.H. and Crowell, J.C., eds. *Climate in Earth History*. Washington, D.C., National Academy of Sciences, pp. 154–158.

17. EVOLUTION OF PALEOGENE ECHINOIDS: A GLOBAL AND REGIONAL VIEW

by Michael L. McKinney, Kenneth J. McNamara, Burchard D. Carter, and Stephen K. Donovan

ABSTRACT

Echinoid global diversity shows an Eocene peak bounded by much lower diversities in the Paleocene and Oligocene. Regional, stage-level diversity patterns from the United States, Caribbean, Indo-Pacific, and Mediterranean are presented for a more refined view. Most of these regions imply that the Eocene peak occurred rather early in the middle Eocene, followed by a small to moderate diversity loss, possibly at the Lutetian/Bartonian boundary. A second, much larger loss occurred at the Eocene/Oligocene boundary, with major reductions of over 50% in nearly all orders. A key deviation from this regional pattern is North Africa, which shows a diversity peak in the early Eocene, possibly because preservation is better.

Regional diversities remained very low throughout most of the Oligocene, indicating prolonged environmental deterioration. Echinoids provide two lines of evidence that this deterioration was related to climatic cooling: 1) a strong correlation between diversity loss and global temperature; and 2) the appearance of marsupiate echinoids in the late Eocene of Australia, indicating the onset of cool climate.

INTRODUCTION

Echinoids provide much information about biotic and climatic evolution in the early Cenozoic. The evolutionary patterns of the group itself are of great interest because echinoids underwent a profound evolutionary radiation, from less than 200 known species in the Paleocene to well over 1000 species in the Eocene. They were then greatly decimated by the events toward the end of the Eocene and well into the Oligocene. Numerous orders within the Echinoidea show distinct trends as biotic interactions occurred, such as the rise of clypeasteroids at the expense of the cassiduloids. In addition, echinoids provide excellent independent clues to the causes of the biotic crises that occurred toward the end of the Eocene. For instance, marsupiate (larvae-brooding) echinoids are much more common in cold environments today and their appearance in Australia in late Eocene time clearly marks the beginning of cooling water temperatures. Finally, echinoids are among the best of all fossil groups for studying evolutionary patterns in that their true abundance is matched by exceptional tendencies for preservation. Unlike many marine groups, which secrete aragonite, echinoids produce calcitic tests that do not recrystallize. Irregular echinoid tests are particularly durable.

In this paper, we briefly review Eocene-Oligocene evolutionary patterns in the echinoids from a global and regional perspective. The regional patterns are admittedly limited to those regions that are among the best known and reflect our collective expertise. Continued work in poorly studied areas such as South and Central America are essential for a complete reconstruction of events. We begin with an overview of global patterns, followed by discussions of: 1) southeastern United States; 2) Caribbean; 3) Indo-Pacific; and 4) Mediterranean.

GLOBAL PATTERNS

Basic Patterns

Figure 17.1 shows the basic outline of regular and irregular echinoid evolution during the Cenozoic. It is based on Kier (1977), so it includes species described to 1970. The very low species diversity during the Paleocene is followed by a pronounced rise in the Eocene.

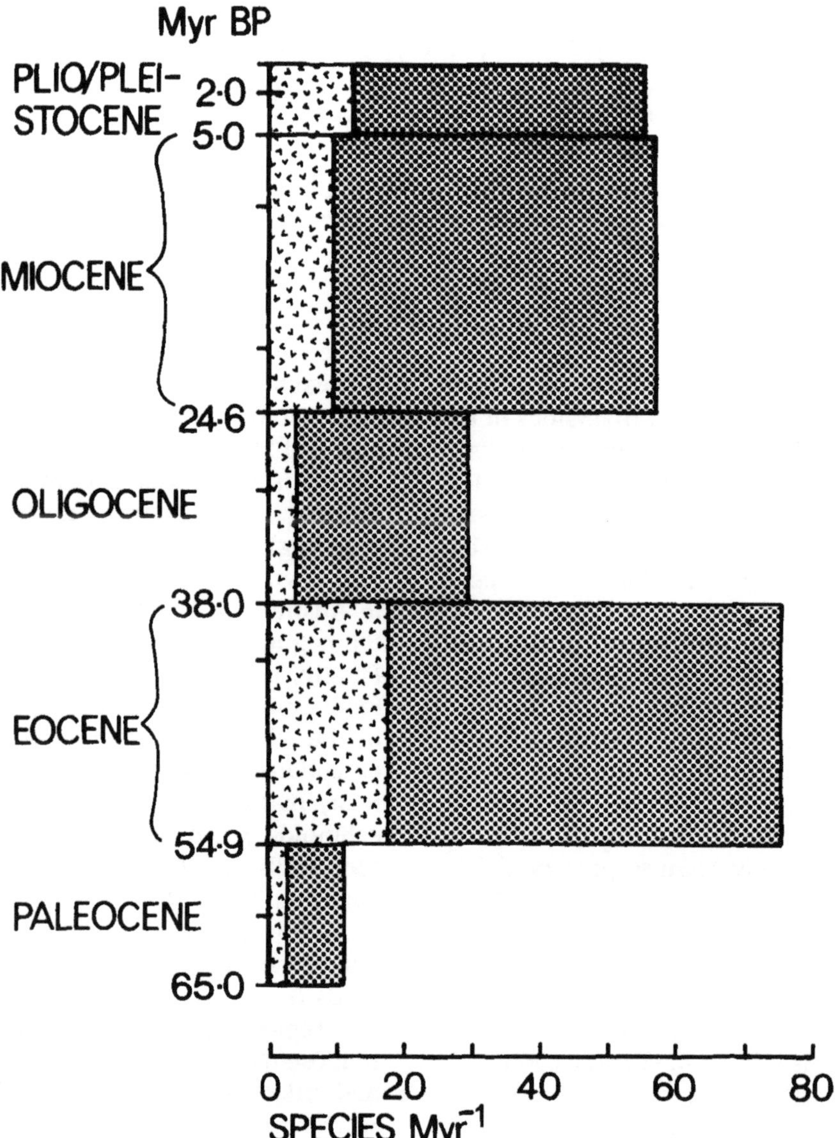

FIGURE 17.1. Worldwide stratigraphic distribution of fossil echinoid species in the Cenozoic; based on data from Kier (1977, Table 1).

The Oligocene is a time of sustained low diversity, with generally increasing diversity thereafter. Any explanation for the Terminal Eocene "events" (Prothero, 1989) must account for this pattern.

A generic view of the global echinoid pattern is seen in Figure 17.2, based on data from Kier (1974) and McKinney and Oyen (1989). Both originations and extinctions are high during the Eocene with a large decrease in both in the Oligocene. Of 173 Eocene genera, 98 genera (57%) were lost during the Eocene and 124 (72%) first appear. Of 101 Oligocene genera, only 21 (21%) became extinct and 26 (26%) first appear. Equally telling of a major extinction during the Eocene is that 81 (47%) of the Eocene genera are limited to the Eocene while only 7 (7%) of the Oligocene genera are limited to the Oligocene. The more detailed view provided later by regional data show that the originations tended to occur in the earlier parts of the Eocene and the extinctions in the later parts.

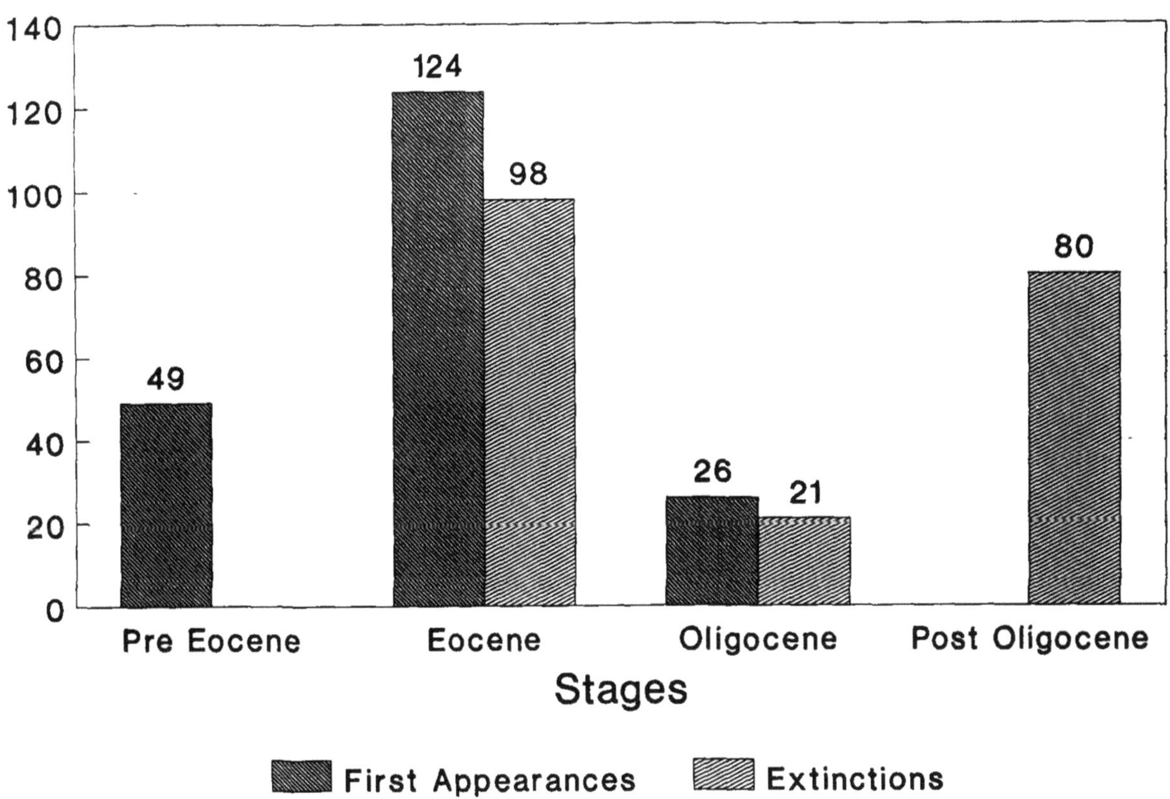

FIGURE 17.2. Global originations and extinctions of echinoid genera. Data from Kier (1974) and McKinney and Oyen (1989).

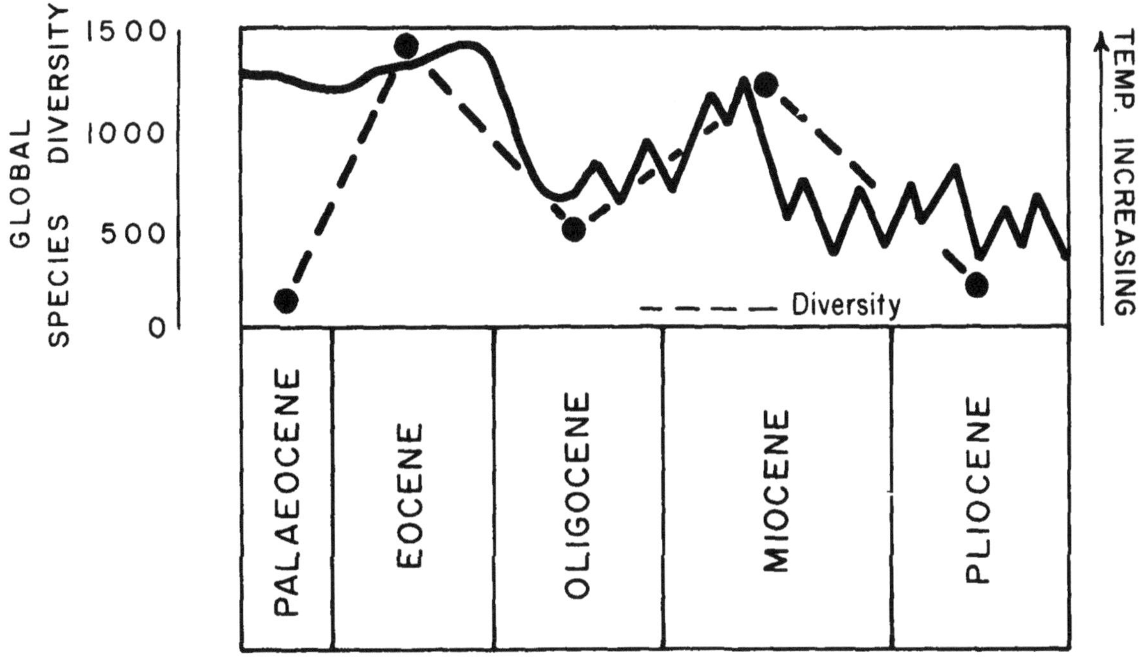

FIGURE 17.3. Comparison of global echinoid diversity (data from Table 17.1) with global temperature (data from Frakes, 1979).

Refined View

Table 17.1 shows a tabulation of species diversity by order for species described up to 1970 (from McKinney and Oyen, 1989). The overall diversity of the Eocene (1351 species) is greater than that of any other epoch. (It is not standardized for epoch duration but this does not change the general pattern much--see Figure 17.1). Kier (1974) tabulated only 909 Eocene species described up to 1925 so we can see that over 400 species have been described since then. Oligocene diversity is only about one-fourth that of the Eocene. The Miocene saw a major re-expansion to over 1000 species.

The ordinal-level patterns seen in Table 17.1 show that some orders show a much larger diversity loss in the Oligocene than others. The cassiduloids and holectypoids are most notable in this respect, with percentage losses far greater than the other groups. Statistical tests, such as the chi-square and test of proportions, confirm that the cassiduloid and holectypoid losses are far out of proportion relative to the other groups. Furthermore, neither group fully recovers. This type of extinction selectivity is one of the major clues that biotic data can provide about extinction causation, as discussed next.

Causation: the Global View

The most recent attempt to use echinoids to acount for Eocene-Oligocene events was by McKinney and Oyen (1989). They concluded, on the basis of rigorous statistical analysis, that global temperature was a significantly better predictor of diversity than was global sea level. Specifically, comparison of global Cenozoic species diversity (Table 17.1) with three sea level curves showed a poor correlation. In contrast, global echinoid diversity seems to be much more closely associated with the abiotic control of global temperature as illustrated in Figure 17.3. An important deviation to the "tracking" of diversity with temperature is in the Paleocene, where high temperatures are not associated with the expected high diversity. (This is also largely a time of high sea level, so that is not a factor either.) We suggest that this low diversity is a "lag effect," reflecting the relatively long time necessary for echinoid diversification to rebound to higher levels following the end-Cretaceous mass extinction. In any case, these global data are obviously coarse and it is at the more detailed regional level that the importance of temperature is more evident. Our basic conclusion in the context of the focus of this book will be that at least a large part of the diversity drop by the Oligocene was caused by climatic cooling promulgated by middle and late Eocene events in the Antarctic area and the initiation of the Circum-Antarctic current, as discussed in detail elsewhere in this book.

While temperature is seen as a controlling abiotic influence on diversity, we hasten to add that biotic interactions and properties must also be considered. This is seen most clearly in the ordinal level selectivity noted above. Specifically, one interpretation of the disproportionate demise of the cassiduloids and holectypoids might be that they contained a larger number of warm-water species. However it is also true that the expansion of some of the other groups may have occurred at the expense of the cassiduloids and holectypoids via competitive exclusion. For example, Mooi (1990) has elegantly described how clypeasteroids arose in the early Cenozoic, evolving substantial modifications (such as the proliferation of tube feet) that allowed them to exploit new environments and old environments more effectively than older groups. There is little doubt that this accounts for the expansion and success of clypeasteroids but it is unclear to what extent it was bought at the direct exclusion of cassiduloids and holectypoids. Thus, we must continue to ask whether the Eocene-Oligocene physical events alone accounted for the disproportionate loss in the cassiduloids and holectypoids at that time or whether biotic interactions were important too. Proving that competition causes extinction is a very difficult task (outside the scope of this brief overview), as shown by the continued and protracted debates on this topic with other taxa (e.g., bivalves replacing brachiopods).

North American Echinoids

This section will focus on the fossil echinoids of the southeastern United States, which has some of the most echinoid-rich deposits in the world. McKinney and Oyen (1989) concluded from rigorous statistical analysis that echinoid

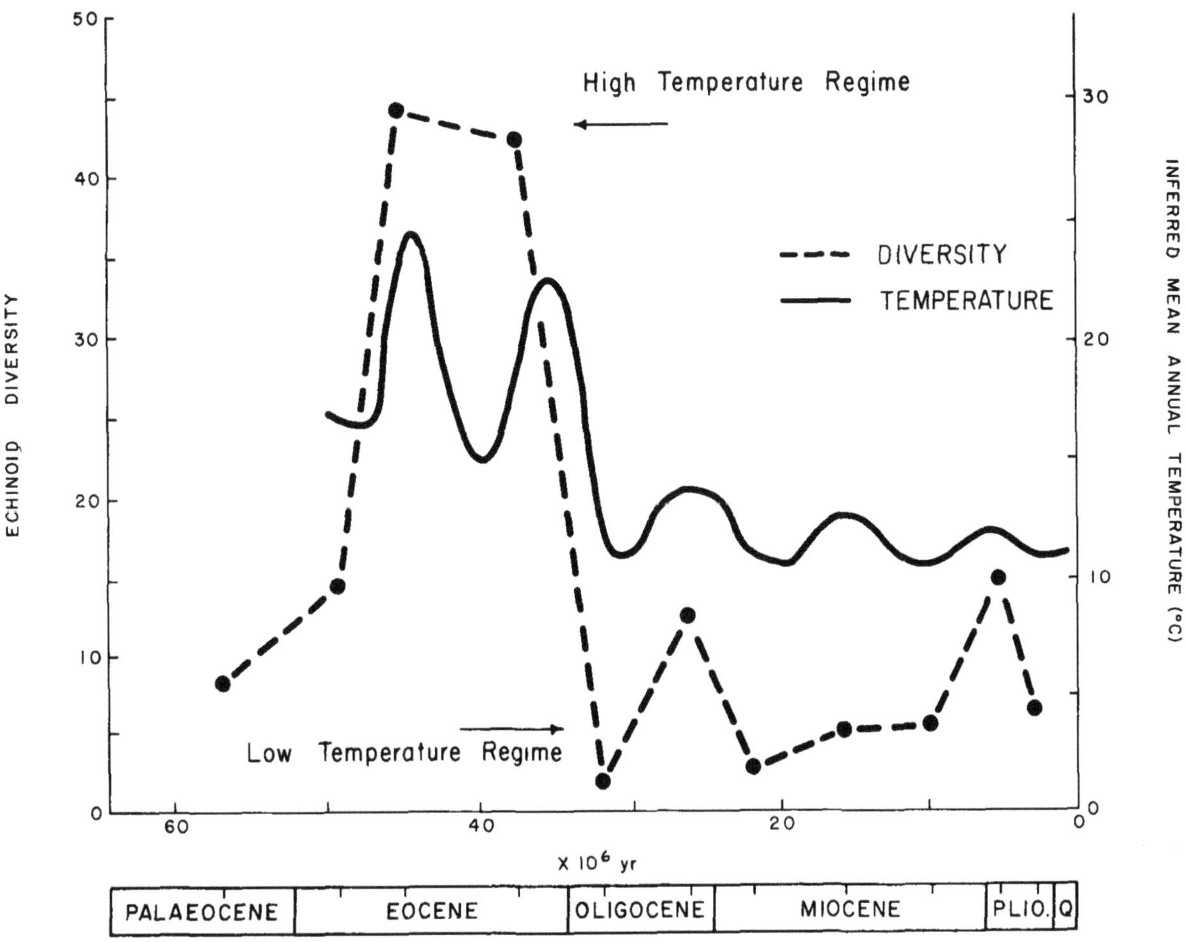

FIGURE 17.4. U.S. Coastal Plain echinoid diversity (from Table 17.2) compared to mean annual temperature of North America through time. Temperature curve is based on leaf emargination studies by Wolfe (1978).

species diversity in this area was associated with temperature. This is graphically shown in Figure 17.4, where a middle Eocene diversity peak decreases slightly in the late Eocene, followed by a radical decrease in the early Oligocene. Thereafter, there are rises in diversity, but they never again approach those highs of the Eocene. The similarity of echinoid diversity to mean annual temperature of North America is visible, and a statistical analysis reveals a very high correlation of 0.9 between the two curves (McKinney and Oyen, 1989).

Similar comparisons of diversity to sea level curves showed significantly less correlation. Most relevant to Eocene-Oligocene events is that this stage level data showed that sea level does not decrease in the early Oligocene while diversity drops radically. Indeed, there is an inverse relationship because diversity rose considerably in the late Oligocene even though sea level decreased according to three different sea level curves (Figure 6 in McKinney and Oyen, 1989). These findings support the assertions of Hansen (1987) that molluscan Paleogene diversity in this area was more closely related to temperature. (This is in contrast to Dockery (1986), who argued for sea level control of Paleogene molluscan diversity in the area.) This is important independent evidence in that mollusks are mainly recorded in the Paleogene nearshore clastic units. They are much more poorly preserved than echinoids in the extensive Paleogene carbonate units of Florida, Georgia, and elsewhere in the region.

TABLE 17.1. Global echinoid diversity as of 1970 (McKinney and Oyen, 1989).

ORDER	Paleocene	Eocene	Oligocene	Miocene	Pliocene	Pleistocene-Recent	Total
Cidaroida	14	111	31	50	15	237	458
Temnopleuroida	0	41	23	78	24	169	335
Echinoida	1	10	5	53	23	84	176
Phymosomatoida	7	88	10	0	1	2	108
Diadematoida	2	2	1	8	0	63	76
Echinothurioida	2	0	0	0	0	67	69
Arbacioida	3	19	4	8	2	30	66
Salenioida	6	9	1	4	0	21	41
Pedinoida	0	9	0	0	0	15	24
Hemicidaroida	0	2	0	0	0	0	2
Total Regulars	35	291	75	201	65	688	1355
Irregulars:							
Spatangoida	35	374	163	348	90	304	1314
Clypeasteroida	1	185	109	441	104	209	1049
Cassiduloida	28	431	61	169	13	39	741
Holasteroida	17	4	0	3	0	32	56
Holectypoida	5	65	5	8	1	4	88
Pourtelasesoida	5	0	0	0	0	3	8
Neolampadoida	0	1	0	1	0	1	3
Total Irregulars	91	1060	338	970	208	592	3259
TOTAL ECHINOIDS	126	1351	413	1171	273	1280	4614

TABLE 17.2. Species diversity of Cenozoic echinoids from the Gulf Coast area (McKinney and Oyen, 1989).

		Eocene			Oligocene		Miocene			Pliocene			
GROUP	Pal	L	M	U	L	U	L	M	U	L	U	Pl-Rc	TOTAL
Holectypoids	1	1	0	1	0	0	0	0	0	0	0	0	3
Holasteroids	0	2	0	0	0	0	0	0	0	0	0	0	2
Oligopygoids	0	0	1	4	0	0	0	0	0	0	0	0	5
Cassiduloids	1	2	6	5	0	2	0	0	1	1	0	1	19
Clyeasteroids	0	0	18	13	0	3	0	3	1	6	3	9	56
Spatangoids	3	2	10	15	1	4	1	2	2	4	1	8	53
Regulars	3	8	8	4	0	4	0	0	1	5	2	10	45
TOTALS	8	15	43	42	1	13	1	5	5	16	6	28	183

TABLE 17.3. Times of first appearance of some echinoid genera in the Indo-West Pacific region.

Genus	Oman	India	Australia	New Zealand
Eupatagus	M. Eocene	Oligocene	Lt. Eocene	E. Oligocene
Breynia	E. Oligocene	E. Miocene	M. Miocene	-
Brissus	E. Oligocene	-	E. Miocene	Lt. Oligocene
Meoma	-	M. Eocene	Lt. Eocene	E. Oligocene
Opissaster	Lt. Eocene	-	-	E. Oligocene
Cyclaster	Senonian	-	Lt. Eocene	E. Oligocee
Echinocyamus	M. Eocene	Oligocene	Pliocene	-
Sismondia	E. Eocene	-	M. Miocene	?
Orthlophus	-	-	Lt. Eocene	? Oligocene

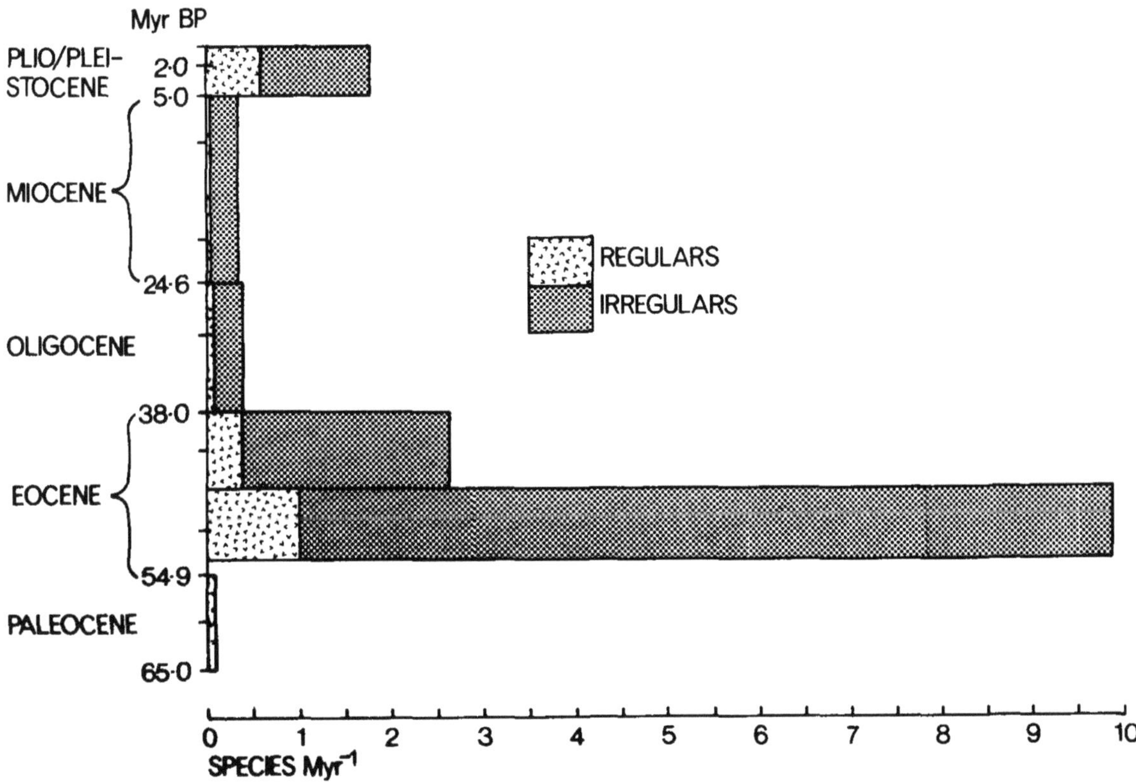

FIGURE 17.5. Temporal distribution of fossil echinoid species in the Cenozoic of Jamaica, including unpublished records. The Eocene record is divided into three unequal parts; pre-depositional (early early Eocene; no echinoids known), Yellow Limestone Group (mid early to mid middle Eocene), and early formations of the White Limestone Supergroup (mid middle to latest Eocene).

Table 17.2 shows echinoid diversity of the southeastern United States when broken down into ordinal-level groups (with regulars grouped together). The basic pattern is similar to the global patterns of Table 17.1, with deviations from a much smaller sample size. The main advantage is the stage-level resolution, lacking in Table 17.1, revealing that the lower Oligocene contained only one echinoid, a spatangoid, *Cyclaster drewryensis*, of the Red Bluff Limestone.

CARIBBEAN ECHINOIDS

Many taxonomic studies of Caribbean fossil echinoids have been published since the mid 1800s, but data and comments on the stratigraphy have often been poor. Indeed, many species are known only from unique specimens, and may lack stratigraphic context entirely. For example, although Donovan (1988) was able to incorporate most of Arnold and Clark's (1927, 1934) species into a preliminary biostratigraphy of Jamaica's echinoids, about 20% of the *circa* 130 species have yet to be stratigraphically determined. Any assessment of echinoid faunal dynamics during the Caribbean Paleogene will thus be constrained, although broad patterns are apparent. The Jamaican echinoid fauna is currently the best-studied in the Caribbean, so it is used as the basis of this section. However, it is supplemented by published data on other parts of the Caribbean, especially Cutress (1980), Kier (1967, 1984), and Mooi (1989).

Mid middle Eocene decline and Oligocene Scarcity

About 75% of the stratigraphically-determined echinoids in Jamaica are of Eocene age. About 70 species of echinoids from the

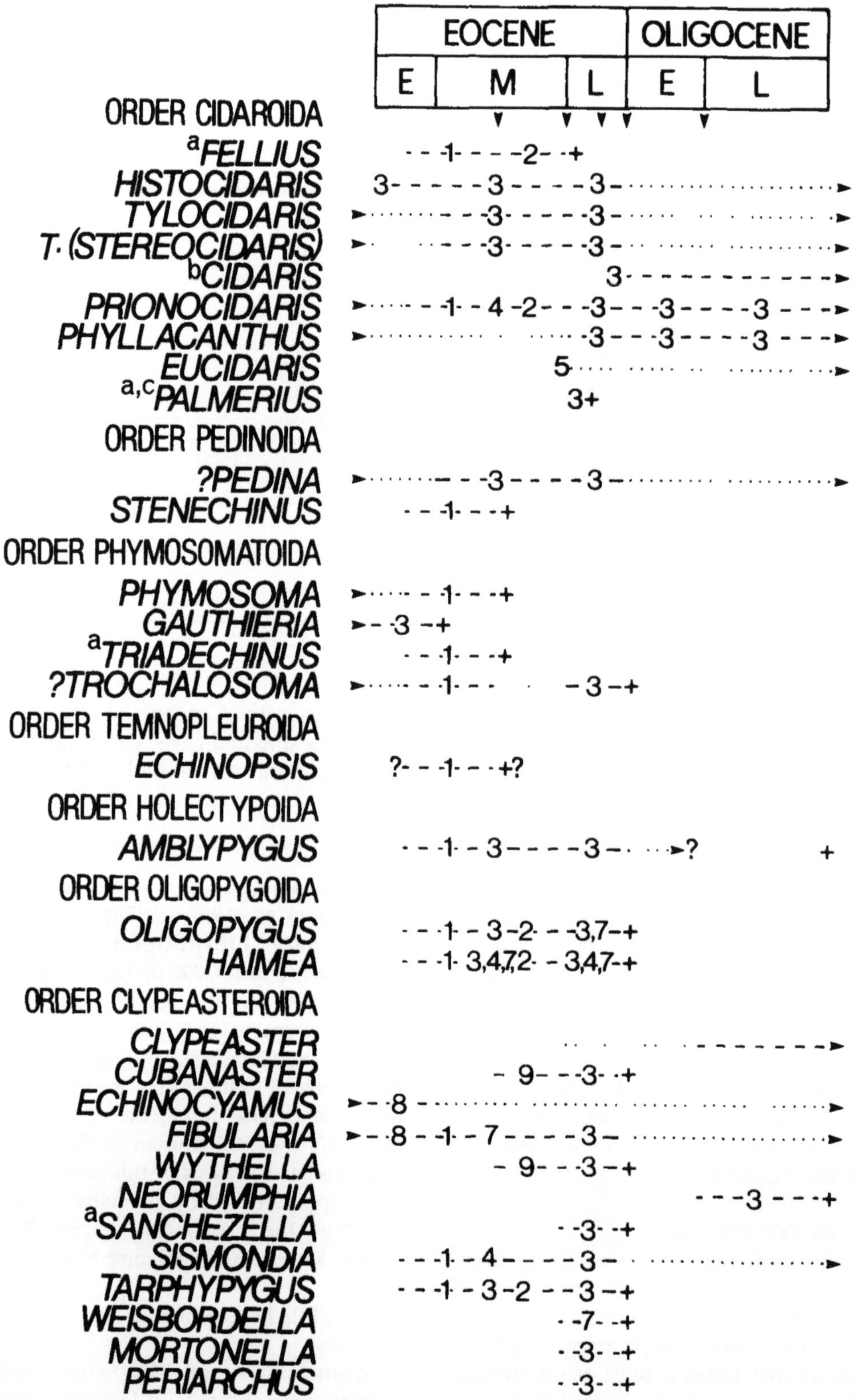

	EOCENE			OLIGOCENE	
	E	M	L	E	L

Taxon	Range
ORDER CIDAROIDA	
[a]*FELLIUS*	- - -1- - - -2- - +
HISTOCIDARIS	3- - - - - 3 - - - - 3 - · · · · · · · · · · · · · · · · · · · ►
TYLOCIDARIS	► · · · · · · · - - 3 - - - - 3 - · · · · · · · · · · · · · · · ►
T. (STEREOCIDARIS)	► · · · · - - - 3 - - - - 3 - · · · · · · · · · · · · · · · ►
[b]*CIDARIS*	3 - - - - - - - - - - - ►
PRIONOCIDARIS	► · · · · - - 1 - - 4 - 2 - - - 3 - - - 3 - - - - 3 - - - ►
PHYLLACANTHUS	► · · · · · · · · · · · · · · · · · 3 - - - 3 - - - - 3 - - - ►
EUCIDARIS	5 · · · · · · · · · · · · · · · · · ►
[a,c]*PALMERIUS*	3+
ORDER PEDINOIDA	
?PEDINA	► · · · · · · - - - 3 - - - - 3 - · · · · · · · · · · · · · · · ►
STENECHINUS	- - -1- - - +
ORDER PHYMOSOMATOIDA	
PHYMOSOMA	► · · · · - - 1 - - - +
GAUTHIERIA	► - - 3 - +
[a]*TRIADECHINUS*	- - -1- - - +
?TROCHALOSOMA	► · · · · - - 1 - - - - 3 - +
ORDER TEMNOPLEUROIDA	
ECHINOPSIS	?- - -1- - - +?
ORDER HOLECTYPOIDA	
AMBLYPYGUS	- - -1- - 3 - - - - 3 - · · · ►? +
ORDER OLIGOPYGOIDA	
OLIGOPYGUS	- - -1- - 3 - 2 - - - 3,7 - +
HAIMEA	- - -1- 3,4,7,2 - - 3,4,7 - +
ORDER CLYPEASTEROIDA	
CLYPEASTER	· · · · · · - - - - - - - - ►
CUBANASTER	- 9 - - - 3 - - +
ECHINOCYAMUS	► - - 8 - · ►
FIBULARIA	► - - 8 - - 1 - - 7 - - - - 3 - · · · · · · · · · · · · · · · ►
WYTHELLA	- 9 - - - 3 - +
NEORUMPHIA	- - - 3 - - - +
[a]*SANCHEZELLA*	- - 3 - - +
SISMONDIA	- - -1- - 4 - - - - 3 - · · · · · · · · · · · · · · · · ►
TARPHYPYGUS	- - -1- - 3 - 2 - - 3 - +
WEISBORDELLA	- - 7 - - +
MORTONELLA	- - 3 - - +
PERIARCHUS	- - 3 - - +

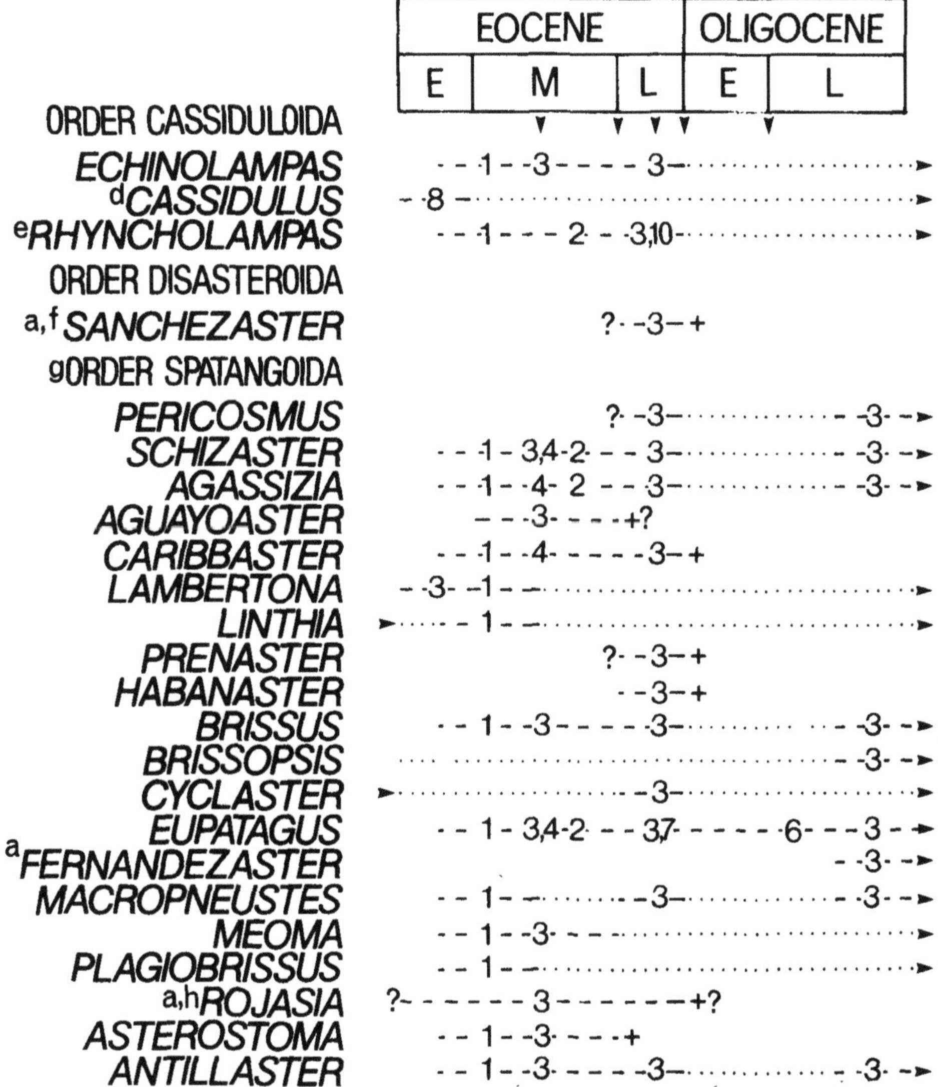

FIGURE 17.6. The stratigraphic distribution of 56 genera of Eocene and/or Oligocene echinoids in the Caribbean region. Dashed lines indicate maximum stratigraphic ranges of taxa within stratigraphic divisions, such as the late Eocene (for example, see *Habanaster*), or lithologic units, such as the Yellow Limestone Group of Jamaica (for example, see *Triadechinus*). The maximum lengths of these lines do not imply that these are the true stratigraphic distributions of these taxa, rather that they are the maximum possible known ranges of the various genera. Dotted lines indicate gaps in known ranges or parts of ranges where the taxa are recorded from outside the Caribbean. Arrows indicate extensions of ranges beyond the Eocene and Oligocene. Arrows at the top of the figure indicate the times of known Eocene and Oligocene extinctions (Prothero, 1989). Key: E=early; M=middle; L=late: 1=Yellow Limestone Group of Jamaica (mid early to mid middle Eocene); 2=Swanswick Formation of Jamaica (mid to late middle Eocene); Cuba; 4=St. Bartholomew; 5=Jamaica, unit not defined; 6=Brown's Town Formation of Jamaica (Oligocene); 7=Trinidad; 8=Barbados; 9=Claremont Formation of Jamaica (mid middle to mid late Eocene): a=monospecific; b=*sensu* Cutress, that is, *Cidaris* + *Stylocidaris*; c=based on radioles only; d=the horizon(s) of Jamaican *Cassidulus* is unknown; e=*Rhyncholampas matleyi* (Hawkins) may be an *Eurhodia* (R. Mooi, personal communication; S. Suter, written communication); f=Horizon given as Eocene by Kier (1984, p 15); g=a number of taxa in Kier (1984) have a given range of 'Oligocene-Miocene', which is shown herein as late Oligocene and younger; h=Eocene, precise range unknown (Kier, 1984, p.124): +=extinction.

Jamaican Cenozoic have been collected from the Yellow Limestone Group (mid-early to mid-middle Eocene), In contrast, echinoids from the overlying White Limestone Supergroup (mid middle Eocene to early late Miocene) are poorly known: only about 15 species are known from the Eocene of post-Yellow Limestone rocks, mainly from one locality (Donovan et al., 1989). Only 5 species are known from the Jamaican Oligocene. This information is graphically shown in Figure 17.5, where echinoid species diversity undergoes an abrupt decline after the mid middle Eocene and a further, smaller decline into the Oligocene.

However, it seems likely that at least some of the apparent decline in post-mid middle Eocene echinoids is artifactual. Macrofossils in the Yellow Limestone are generally well-preserved, locally abundant, and easily collected from the soft, friable limestone and clastics. In contrast, the White Limestone is generally well-cemented and sometimes dolomitized. This not only diminishes preservation but makes it less attractive to collectors. In spite of the possible artifactual component, the basic patterns are probably real when compared with other patterns globally and in other regions at this time (e.g., Table 17.1). Also, there is evidence from the literature on other parts of the Caribbean. None of the Trinidadian or Venezuelan taxa described by Cooke (1961) were considered to be Oligocene. Similarly, of the 24 species of Caribbean fossil cidaroids recorded by Cutress (1980, Table 16) only three are considered as Oligocene.

In summary, two basic patterns emerge:

1) an apparent decline in diversity in the mid middle Eocene (at least in Jamaica), and

2) a Caribbean-wide scarcity of Oligocene echinoids. The mid middle Eocene decline correlates with the Lutetian-Bartonian event recognized by Prothero (1989). The lack of Oligocene echinoids likewise points to regional echinoid decimation by the Terminal Eocene crisis documented by Prothero (1989), with no rebound to former diversity. This last is indicative of middle Eocene onset of adverse conditions that worsened in the late Eocene, mimicking the pattern seen globally and in the southeastern United States above.

A Refined View

The stratigraphic distribution of Eocene and Oligocene Caribbean echinoids is shown in Figure 17.6. Genera have been used because species are often known only from a single horizon or locality. Even with genera, precise ranges are usually unknown. Therefore available data is insufficient to determine the effect (if any) of the late Priabonian event (Prothero, 1989, p. 220). Similarly, few Caribbean taxa have been clearly recorded from the early Oligocene, so recognition of the mid-Oligocene event is not possible. Of the 56 genera shown in Figure 17.6, 16 of these (over 25%) show an apparent extinction in the Caribbean Paleogene. These genera (*Histocidaris, Tylocidaris, Stereocidaris, Eucidaris, Pedina, Echinocyamus, Fibularia, Lambertona, Linthia, Cyclaster, Meoma, Plagiobrissus*) are all known from the Eocene but not the Oligocene of the Caribbean. However, all of these genera are present in the Oligocene or later strata elsewhere, or post-Oligocene strata of the Caribbean. Therefore, extreme caution is needed in analysing apparent extinctions in such groups. Another problem is the possibility of taxonomic pseudoextinction (Patterson and Smith, 1987).

Four genera (*Stenechinus, Phymosoma, Triadechinus*, and possibly *Echinopsis*) appear to go extinct at the Lutetian-Bartonian (mid middle Eocene) event. These are all Yellow Limestone taxa and were perhaps poorly adapted to the range of substrates available in the succeeding, "purer" carbonate environment of the White Limestone. None of these genera is known from more than two species in the Yellow Limestone and this restricted diversity may have played a role in their decline. Indeed, *Stenechinus* and *Triadenchinus* are only known from the Yellow Limestone. It is significant that a fifth genus, *Fellius*, would have been included with these "end Yellow Limestone" extinctions had it not been tentatively identified from the Swanswick Formation (Donovan et al., 1989). Three genera (at most) disappear at the Bartonian-Priabonian (end middle Eocene) event: *Fellius, Asterostoma, Aquayoaster*. These are all taxa limited to Jamaica and/or Cuba, the most diverse being *Asterostoma* with three known species (Kier, 1984).

The Terminal Eocene Event seems to have been especially severe on the basis of this evidence, leading to the demise of 15 genera (Figure 17.6): *?Trochalosoma, Oligopygus, Haimea, Cubanaster, 'Jacksonaster,' Sanchezella, Tarphypyqus, Weisbordella, Mortonella, Periarchus, Sanchezaster, Caribbaster, Prenaster, Harbanaster, Rojasia.* These include a number of clypeasteroids and spatangoids that are only known from the late Eocene, mainly of Cuba. While some of these 15 may have been extensively studied (Kier, 1967; McKinney, 1984) disappeared at this time.

In summary, the Lutetian/Bartonian and Bartonian/Priabonian boundaries show little diversity loss of echinoid genera. Loss seems to be limited to only a few endemic genera of limited species diversity. This is in contrast to the significant loss at the species level at the Lutetian/Bartonian boundary. Most profound of all was the Terminal Eocene Event, which caused major losses at both the species and genus level.

Caribbean Echinoids: General Conclusions

1) The mid middle Eocene (Lutetian/ Bartonian boundary) was a time of major diversity reduction at the species level. However, at the generic level, the translated effects were minimal, being limited to a few endemic genera containing few species.

2) Following this, there is little evidence for significant species or generic diversity loss at the Bartonian/Priabonian boundary or any time up to the Terminal Eocene.

3) At the Terminal Eocene loss was followed by a long period of sustained low diversity, throughout the Oligocene and Miocene, indicating no return to the former, benign environmental conditions.

INDO-WEST PACIFIC ECHINOIDS

This section combines data from Ali's (1983) large data set with updated information (collected by K.J.M.). These data show that 64 echinoid genera have been described from the Eocene of the Indo-West Pacific, but only 36 from the Oligocene. Twenty-two of these Oligocene genera also occur in the Eocene. From these data, it seems that the Eocene-Oligocene transition saw drastic extinctions, with 66% of the genera disappearing, and only 41% of the Oligocene genera having originated after the Eocene/Oligocene boundary. Such data would appear to support a major mass extinction of echinoids at the Eocene/Oligocene boundary (i.e., a Terminal Eocene Event).

A Refined View

To assess the validity of this apparent event, diversity changes need to be studied at much finer time scales. In the Indo-West Pacific region, the best studied fossil echinoid fauna is from southern Australia (Philip, 1963a,b, 1964, 1965, 1969; Philip and Foster, 1971; Foster and Philip, 1976a,b, 1978; McNamara and Ah Yee, 1989; McNamara et al., 1986; Kruse and Philip, 1985). The richly fossiliferous deposits in this area extend from the late Eocene to the Miocene, within a number of basins with intercorrelatable strata.

The echinoids are concentrated at horizons that correspond to a series of marine transgressions (McGowran and Beecroft, 1986). These are the Tortachilla (T), Aldinga (A), Ruwarung (R), Janjukian (J), and Longfordian (L) (see McGowran et al., this volume). The stratigraphic ranges of all known echinoid genera from the southern Australian late Paleogene (late Eocene-Oligocene) are shown in Figure 17.7. These occur within the five horizons, with the T and A horizons being late Eocene, R and J are Oligocene, and L is Miocene. T is in planktonic foraminiferal zone 15, A is zone 19/20 and J is zone 21/22 (Figure 17.7).

Figure 17.8 summarizes the diversity changes in these fauna. An important feature is that extinction levels were highest before the Terminal Eocene: after the T transgression, 18% of late Paleogene genera of southern Australia became extinct. Furthermore, extinction rate declined to zero at the Eocene-Oligocene boundary, and increasing thereafter. These rates reflect true extinctions in the southern Australian region in that they are based on last known Cenozoic records. Levels of origination follow a similar pattern to extinctions (Figure 17.8). They are highest in the T transgression (horizon), with 32% of late Paleogene genera appearing, declining rapidly to 8% in the latest late Eocene (when only *Meoma*, *Eupatagus*, and *Paradoxechinus* appear in the Australian record). Origination rates then increase through the Oligocene to the late

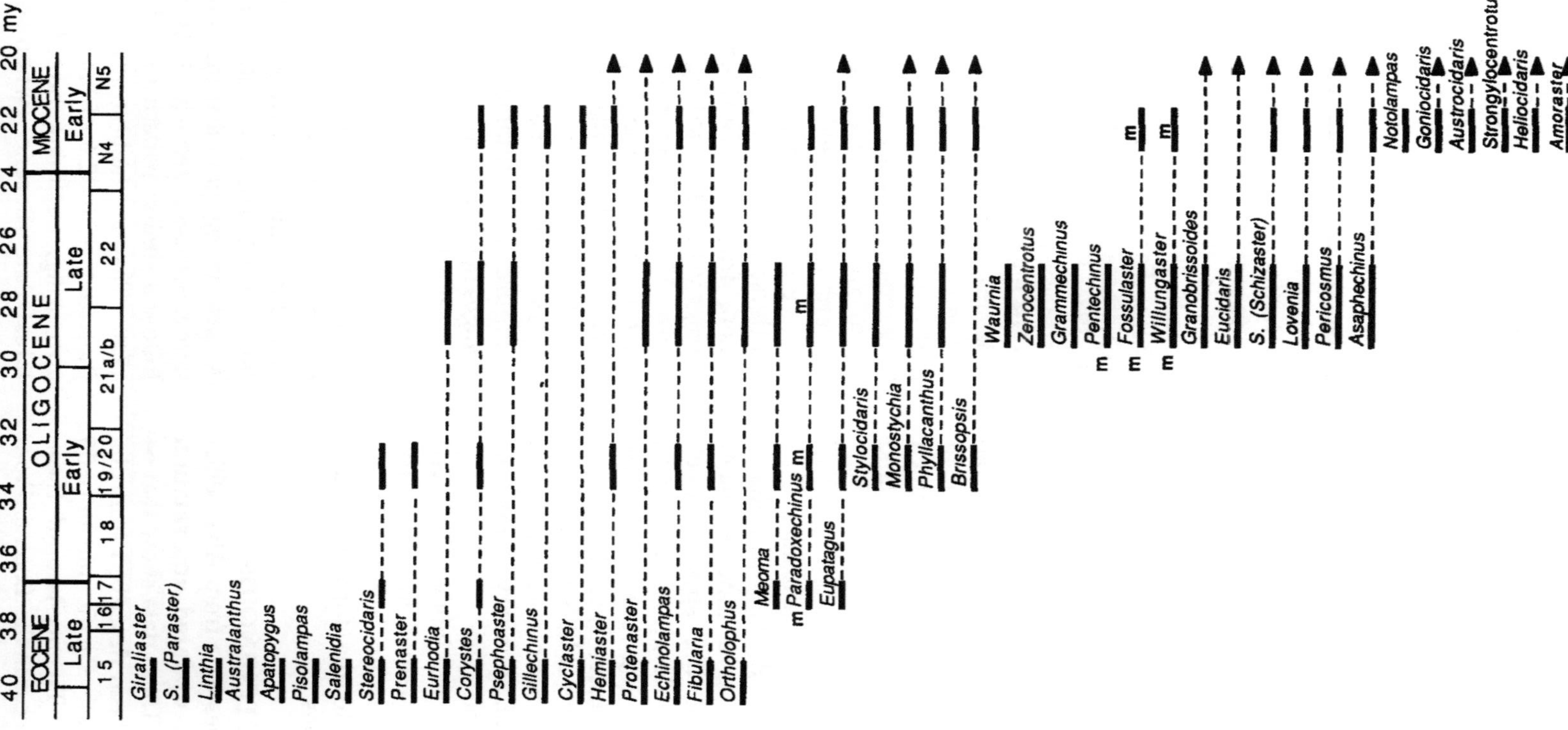

FIGURE 17.7. Stratigraphic ranges of all known echinoid genera from southern Australia; m=marsupiate. See text for discussion.

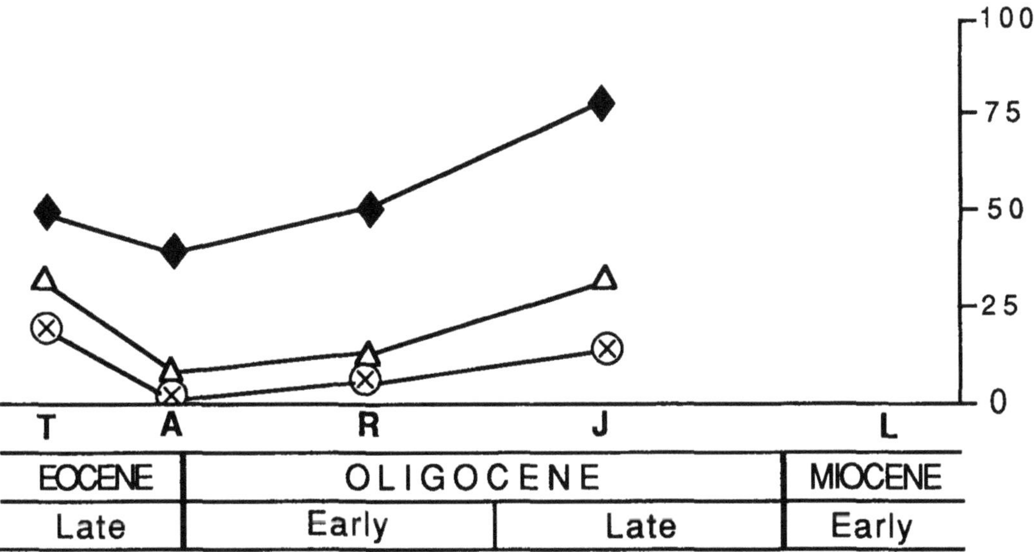

♦ = generic diversity, expressed as percentage of total late Paleogene (=late Eocene and Oligocene) genera

⊗ = extinctions, expressed as percentage of total late Paleogene genera becoming extinct at the end of each fossiliferous horizon

Δ = percentage of total late Paleogene genera appearing at each horizon

FIGURE 17.8. Summary of diversity changes of echinoid genera seen in Figure 17.7. See text for discussion.

Eocene level. Generic diversity (Figure 17.8) shows parallel trends to both extinction and origination. Late Paleogene genera go from 50% in the early late Eocene to a low of 39% at the Terminal Eocene, rising to a high of 76% in the late Oligocene.

In summary, the following major patterns emerge.

1) A high extinction rate occurs before the Terminal Eocene.

2) The latest Eocene shows lower extinction and origination rates and a reduction in overall generic diversity.

3) The Oligocene shows a trend of increasing origination and diversity. Unlike patterns 1) and 2), which are generally concordant with patterns in the U.S. and Caribbean, this increasing diversity is in stark contrast with them. The reason for this seems to be Australia's northerly motion into a warmer climate since the Eocene. This brings us to the topic of what caused the Australian diversity drop at the Terminal Eocene Event.

Cooling temperature: A major control on diversity

The high rate of early late Eocene extinctions and the low levels of origination

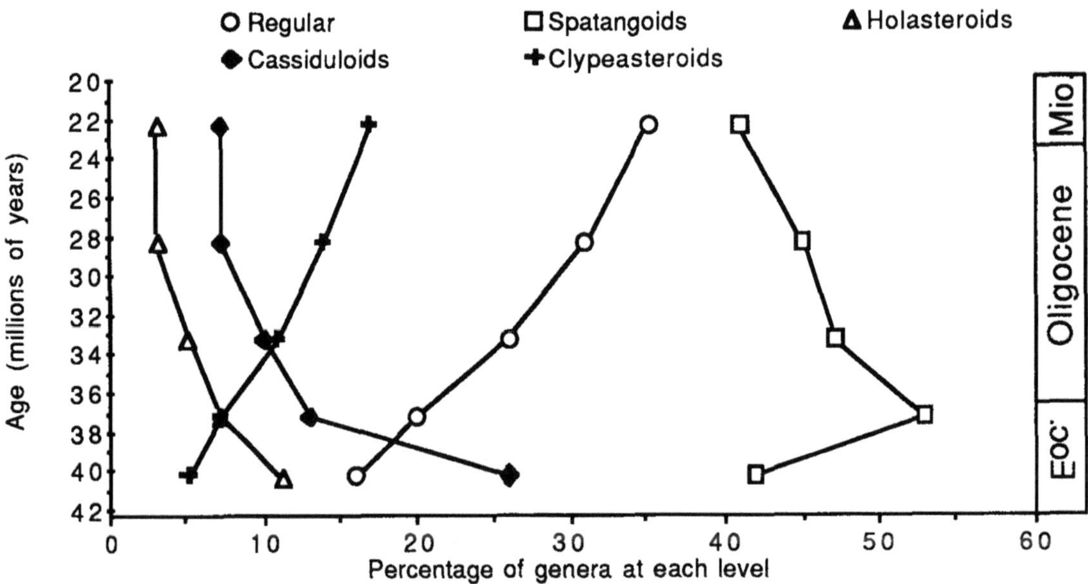

FIGURE 17.9. Changing Australian echinoid generic diversity, grouped by order. See text for discussion.

and overall lower diversity in the latest Eocene in southern Australia have been attributed to marked drops in sea temperatures prior to the Terminal Eocene boundary in this area (Jenkins, 1974; Lindsay, *in* Daily et al., 1976). Key echinoid evidence substantiating this is the influx at this time of marsupiate echinoids (see Foster, 1974, for discussion). These brood-protecting forms are predominant in cooler waters (in today's Antarctic, 23 of 28 species are marsupiate). They first appear in southern Australia in the latest Eocene (Zone 17), leading Foster (1974) to suggest that this influx was caused by the opening of the Drake Passage between South America and Antarctica just prior to the Eocene/Oligocene boundary. This created a circumpolar current, causing much cooler water to sweep by southern Australia. The influx of cold-adapted species is also seen in the selective demise of older genera. Up to the end of the Eocene, 36% of Paleocene genera were still present in the area. By the early Oligocene, only 9% (1 genus, *Hemiaster*) remained.

Marsupiate echinoids gradually decline in abundance and diversity in southern Australia throughout the Oligocene and Miocene (Philip and Foster, 1971), until only a few marsupiates remain today (such as *Eupatagus valenciennesi*). This steady decline, when coupled with the observation of increasing overall diversity, shows that Australia's increasing echinoid diversity is due to the northerly movement of the continent into warmer waters. Adding to this argument is that warmer water faunas in northwestern Australia contain no marsupiate echinoids.

Echinoid Orders

Generic diversity, when grouped by orders, reveals some marked variations in the Australian late Paleogene, as shown in Figure 17.9. (Regular echinoids were grouped together owing to their general ecologic similarity relative to the irregulars.) The cassiduloids and holasteroids show strong reductions through the late Eocene and early Oligocene, with slower reductions later, to their current diversity of less than 10%. However, some single genera within these orders, such as *Echinolampas*, have had continued success in both species diversity and abundance throughout this time. The two groups undergoing increased generic diversity (Figure 17.9) were the clypeasteroids and regular echinoids. Finally, the fifth group shown, the spatangoids, have fluctuated, but never dropped below 40% of all genera, far more than the other groups. They have thus been consistently dominant over the other groups in generic percentage. Indeed, even apparent decreases of spatangoid dominance are misleading; the Oligocene decline is due to a large increase in regular echinoid diversity. The number of spatangoid genera actually increased from 8 to 13 through the Oligocene.

Role of dispersal

By comparing time of generic origination in the northwestern Indian Ocean eastward through to Australasia, patterns of east-west dispersal in the late Paleogene can be documented. As shown in Table 17.3, a number of cosmopolitan genera first appear in the Mediterranean, spread northeastward through the Indian Ocean, to then go eastward. No doubt some eastward movement did not involve the Indian Ocean. The old Tethyan Seaway would have provided a route for such dispersal. For instance, *Brissopsis* and *Lovenia* first appear in the Eocene Mediterranean, and only in the Oligocene become common in Australasia.

Perhaps of most interest is the clear increase in dispersal from the Indian Ocean to Australasia in latest Eocene to early Oligocene time. Thus, whereas 42% (8/19) of the early late Eocene genera are endemic to southern Australia, only 27% were endemic just prior to the Eocene/Oligocene boundary. This low endemism persisted into the early Oligocene as influx continued, including such important genera as *Meoma*, *Eupatagus*, and *Brissopsis* (these being ultimately from the Mediterranean). By the late Oligocene the level of endemic genera had again risen, to 44%.

Indo-Pacific Echinoids: Conclusions

Indo -Pacific echinoid genera show declines in diversity and origination that generally track those seen at the ordinal and group level where cassiduloids and holasteroids declined after the Eocene and regulars and clypeasteroids increased. The appearance of marsupiate echinoids in the late Eocene provides important independent evidence that markedly cooler waters from the Antarctic region caused the faunal declines.

Echnoid patterns in the Australian region deviate from much of the rest of the world in the apparent increase in overall diversity after the Terminal Eocene. This may be due to the northerly motion of Australia into warmer, more benign waters. As is well known in ecology, diversity of most groups tends to increase toward the equator (e.g, Ricklefs, 1990). The Australian increase was supplied by dispersal from the west, such as the Mediterranean Sea and Indian Ocean.

MEDITERRANEAN ECHINOIDS

The Mediterranean region provides perhaps the best evidence for Eocene echinoid evolutionary trends for two reasons. First, it has very high diversity compared to other regions, and second, it has the only well-represented Ypresian echinoid fauna. Mediterranean echinoids are most common in southern Europe and Egypt, but occur throughout North Africa and southern and western Europe. Fossil European and African echinoids have been greatly "oversplit" taxonomically. The resultant descriptive literature is presently both vast and outdated. Therefore the following discussion focuses only on spatangoids, which one of us (BC) is now studying. However, because this order was among the most consistently diverse and abundant throughout the Paleogene world, they provide much information on echinoids in general. Genera discussed herein were taken from the *Treatise on Invertebrate Paleontology*.

The Mediterranean region is dichotomous relative to Paleogene spatangoid biogeography, with North Africa and Europe having significantly different genera: 42% (20/48) of Mediterranean Eocene and Oligocene spatangoid genera are restricted to Europe; 25% (12/48) to Africa; only 33% (16/48) are shared. Of the 28 genera known from Africa, nine (32%) are endemic. Of the 36 known European genera, 13 (36%) are endemic.

Eocene

In the European Eocene, there exist roughly 350 nominal spatangoid species belonging to 36 genera. However, only 184 species can be assigned to a stage. (The Priabonian is assumed to include the Tongrian local stage.) The diversity, by stage, reveals a fairly even distribution, except for the Ypresian, which shows a dramatic paucity, with only about 2% of Eocene species:

Ypresian: 4/184 = 2%
Lutetian: 58/184 = 32%
Bartonian: 64/184 = 34%
Priabonian: 58/184 = 32%

These data are consonant with all the global and regional patterns discussed above: very low early Eocene diversity giving rise to very high diversity in mid-late Eocene times, with the decline beginning before the Priabonian.

North Africa has about 150 nominal spatangoid species belonging to 29 genera. In Egypt and Libya, the stratigraphic distributions are fairly well tied into those of nummulitids and planktonic foraminifera. The Ypresian fauna is the most diverse with 38 of 87 nominal species. The Priabonian has the fewest species, with only six. Thus, as shown in Figure 17.10, Eocene echinoid diversity in northeastern Africa follows trends different from those of Europe. The post-Bartonian diversity decline is still apparent but the Ypresian maximum varies dramatically from that inferred from other parts of the world. This has major implications for inferences of echinoid diversity patterns because the low Ypresian diversity elsewhere in the world may be due more to global poor preservation of Ypresian sediments than an actual diversity decline.

Eocene African echinoids outside Egypt and Libya are less rich and more poorly studied. Algeria and Tunisia have a few known species as does Somalia. The best-estimates for spatangoid diversity for these areas is shown in Figure 17.10, but interpretation is difficult and tentative.

Oligocene Decline

Europe has about 60 nominal species of Oligocene echinoids, belonging to 25 genera. This represents only about 17% of the specific diversity and 71% of the generic diversity of the Mediterranean region. Eight genera (32%) first appear in the Oligocene. Ten genera (40%) are endemic to Europe. North Africa has only about 20 Oligocene species in ten genera, two of which (20%) are endemic.

Overview: Mediterranean Eocene -Oligocene

From the above discussion, the following information can be tabulated to estimate the loss of spatangoid diversity at the Terminal Eocene.

Europe	# Species	# Genera
Eocene	350	36
Oligocene	60	25
Oligocene Loss	290/350 = 83%	11/36 = 31%

NorthAfrica	# Species	# Genera
Eocene	150	29
Oligocene	20	10
Oligocene Loss	130/150 = 87%	19/29 = 66%

Mediterranean	# Genera
Eocene	48
Oligocene	28
Oligocene Loss	20/48 = 42 %

These data show that European and the less diverse North African fauna suffered similar, profound losses of over 80% of species diversity between the Eocene and Oligocene. However, the loss of genera was strikingly less (about half) in Europe. In terms of specific genera, 13 European genera (36%) are not found in post-Eocene rocks while 22 African genera (76%) are not. The high disparity between European and African generic loss (versus the similarity of species loss) is at least partly because the greater absolute species diversity of Europe has a sampling effect on the number of genera represented. Overall, the Mediterranean Oligocene has less than half the genera of that in the Eocene.

The profound species loss of over 80% for both Europe and North Africa is somewhat higher than comparable data on global spatangoid diversity (Table 17.1), where Oligocene diversity (163) is a reduction by only 56% of the Eocene. However, the Mediterranean loss is similar to that of spatangoid species in the United States (Table 17.2), where a reduction of 81% occurs (from 27 to 5). It is likely that marine non-deposition plays a large role in the apparently depauperate Mediterranean Oligocene. Marine Oligocene rocks are largely absent in North Africa, such that virtually all known species are from Tunisia and Algeria. The famous

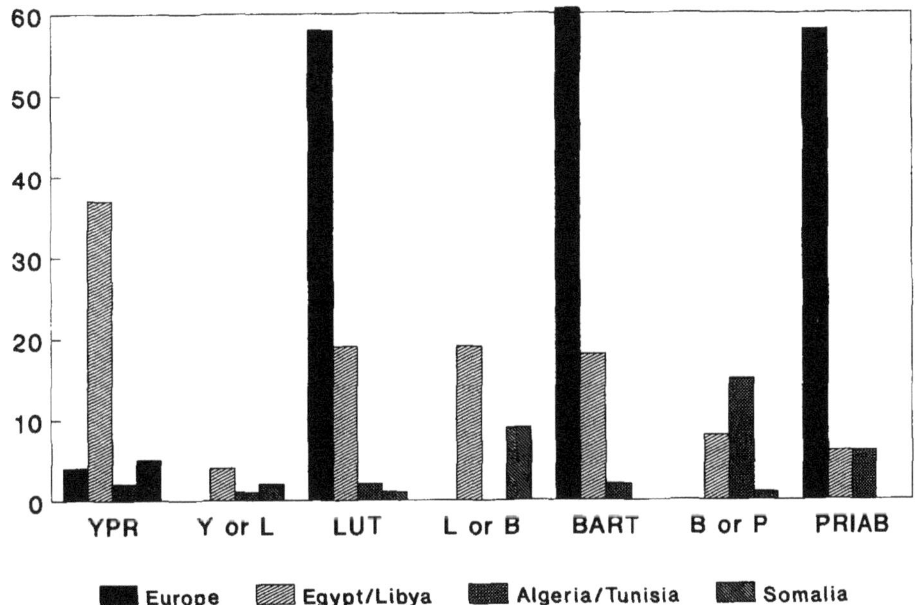

FIGURE 17.10. Histogram of spatangoid generic diversity, grouped by order. See text for discussion.

Fayum terrestrial deposits are more characteristic of the Oligocene in the area.

Two final patterns are worth mentioning. One, North African diversity seems to show a gradual decline throughout the Eocene, in contrast to the typical pattern elsewhere in the world (including Europe) which peaks around the middle Eocene. Whether this decline is real or preservational is a key question for future work. Two, patterns of endemism within the Mediterranean are not dramatically affected by the Eocene/Oligocene boundary. In the Eocene, 36% of spatangoid genera are endemic to Europe while 31% are endemic to North Africa. In the Oligocene, comparable figures are 40% and 20%, respectively.

GENERAL CONCLUSIONS

Echinoid global diversity shows an Eocene peak bounded by much lower diversities in the Paleocene and Oligocene. The low Paleocene diversity is thought to represent a lag phase, as speciation gradually replaced diversity lost at the end of the Mesozoic. The low Oligocene diversity is thought to result from a global-scale deterioration of the physical environment.

The pattern of the deterioration is better seen via a regional view of echinoids, with stage-level resolution. The general pattern, seen in most regions, is that the Eocene peak occurred rather early in the middle Eocene. This was followed by a small to moderate loss of diversity in the middle Eocene, possibly corresponding to the Lutetian-Bartonian boundary. A second, much larger diversity drop occurred during the Terminal Eocene when massive reductions of over 50% occur in all orders. These fine-scale patterns are visible in the United States, Caribbean, Indo-Pacific and Europe. They may directly correspond to Prothero's (1989) Lutetian-Bartonian and Terminal Eocene events, which were inferred from independent, largely terrestrial evidence. North African echinoids deviate from the pattern in peaking in the early Eocene. This is at least partly due to better preservation of early Eocene sediments in North Africa than other regions. The major implication is that the low global early Eocene echinoid diversity may be an artifact of poorer preservation of this time than, say, the generally well-preserved late Eocene.

After the Eocene, the data indicate a general loss of diversity caused by a deterioration of previously more favorable conditions. The ultimate cause of deterioration is inferred to be climatic cooling, probably associated with the isolation of Antarctica discussed elsewhere in this book. Evidence for this in the echinoid record is twofold: 1) a strong statistical correlation of southeastern United States echinoid diversity with temperature, and 2) the appearance of (cold-adapted) marsupiate echinoids in Australia in the late Eocene. Both of these also indicate that the deterioration continued through at least most of the Oligocene, in the: 1) continuing low diversity in the Oligocene (especially throughout the early Oligocene), and 2) continuing presence of marsupiate echinoids during the Oligocene. An important regional exception to the prolonged low Oligocene diversity is Australia, where northerly plate motion apparently acted to locally counteract the global cooling trend. Much of this diversity origination is documented to have come from migration from the Mediterranean, moving eastward through the Indian Ocean and other intermediate areas.

In inferring physical deterioration as a cause of late Eocene and Oligocene diversity loss, we do not exclude a partial role for biotic events. Most notable is the disproportionate loss of cassiduloid and holectypoid echinoids. Thus, competitive exclusion, such as caused by the early Cenozoic origin of clypeasteroids, may have played a role in the extinction of species within some groups. Alternatively, and perhaps more likely (Sherman Suter, personal communication), the increasing clypeasteroid diversity simply represents invasion of new habitats. Also, we do not wish to oversimplify the case, to say that temperature alone can explain all diversity changes. Sedimentation and sea level are but two of many environmental parameters that often vary with temperature; these will also affect diversity. Nevertheless, temperature seems to be the predominant factor.

ACKNOWLEDGEMENTS

We thank Anita Argabright for typing the manuscript onto the computer. Gordon Hendler and Sherman Suter provided helpful reviews.

REFERENCES

Ali, M.S.M. 1983. Tertiary echinoids and the time of collision betwen Africa and Eurasia. *N.Jb. Geol. Palaont. Mtg.* 1983: 213-227.

Arnold, B.W. and Clark, H.L. 1927. Jamaican fossil echini. *Mem. Mus. Comp. Zool.* 50: 1-75.

Arnold, B.W. and Clark, H.L. 1934. Some additional fossil echini from Jamaica. *Mem. Mus. Comp. Zool.* 54: 139-156.

Carter, B.D., Beisel, T.H., Branch, W.B., and Mashburn, C.M. 1989. Substrate preferences of late Eocene (Priabonian/Jacksonian) echinoids of the eastern Gulf Coast. *Jour. Paleont.* 63: 495-503.

Cooke, C.W. 1961. Cenozoic and Cretaceous echinoids from Trinidad and Venezuela. *Smithson. Misc. Coll.* 142(4): 1-35.

Cutress, B.M. 1980. Cretaceous and Tertiary Cidaroids (Echinodermata: Echinoidea) of the Caribbean area. *Bull. Amer. Paleont.* 77(309): 1-221.

Dockery, D.T. 1986. Punctuated succession of Paleogene mollusks in the north Gulf Coast Plain. *Palaios* 1: 582-589.

Donovan, S.K. 1988. A preliminary biostratigraphy of the Jamaican fossil Echinoidea. In R.D. Burke, P.V. Mladenov, P. Lambert, and R.L. Parsley, eds., *Echinoderm Biology; Proceedings of the Sixth International Echinoderm Conference,* Rotterdam, A.A.Balkema, pp. 125-131.

Donovan, S.K., Gordon, C.M., Schickler, W.F., and Dixon, H.L. 1989. An Eocene age for an outcrop of the 'Montpelier Formation' at Beecher Town, St. Ann, Jamaica, using echinoids for correlation. *J. Geol. Soc. Jamaica* 26: 5-9.

Foster, R.J. 1974. Eocene echinoids and the Drake Passage. *Nature* 249: 751.

Foster, R.J. and Philip, G.M. 1976a. *Corystas dysasteroides,* a Tertiary holasteroid echinoid formerly known as *Duncaniaster australasia. Trans. R. Soc. S.Aust.* 100: 113-116.

Foster, R.J. and Philip. G.M. 1976b. Statistical analysis of the Tertiary holasteroid *Corystas dysasteroides* from Australia. *Thalassia Jugoslavica* 12: 129-144.

Foster, R.J. and Philip, G.M. 1978. Tertiary holasteroid echinoids from Australia and New Zealand. *Palaeontology* 21: 791-822.

Frakes,L. 1979, *Climate Evolution Through Geologic Time.* Elsevier, Amsterdam.

Hansen, T.A. 1987. Extinction of late Eocene to Oligocene molluscs: relationship to shelf area, temperature changes, and impact events. *Palaios* 2: 69-75.

Jenkins, F.J.F. 1974. A new giant penguin from the Eocene of Australia. *Palaeontology* 17: 291-310.

Kier, P.M. 1967. Revision of the oligopyoid echinoids. *Smithson. Misc. Coll.* 152(2): 1-149.

Kier, P.M. 1974. Evolutionary trends and their functional significance in the post-Paleozoic echinoids. *Paleont. Soc. Mem.* 5: 1-95.

Kier, P.M. 1977. The poor fossil record of the regular echinoids. *Paleobiol.* 3: 168-174.

McGowran, B. and Beecroft, A. 1986a. Neritic, southern extratropical Foraminifera and the terminal Eocene event. *Palaeogeogr., Palaeoclimat., Palaeoecol.* 55: 23-3

McGowran, B. and Beecroft, A. 1986b. Foraminiferal biofacies in a silica-rich neritic sediment, late Eocene, South Australia. *Palaeogeog., Palaeoclimat., Palaeoecol.* 52: 321-345.

McKinney, M.L. 1984. Allometry and heterochrony in an Eocene echinoid lineage. *Paleobiology* 10: 207-219.

McKinney, M.L., and Oyen, C.W. 1989. Causation and non-randomness in biological and geological time series: Temperature as a proximal control of extinction and diversity: *Palaios* 4: 3-15.

McNamara, K.J. 1985a. Taxonomy and evolution of the Cainozoic spatangoid echinoid *Protenaster. Palaeontology* 28: 311-330.

McNamara, K.J. 1985b. The spatangoid echinoid *Linthia* from the Late Eocene of southern Australia. *Trans. R. Soc. S. Aust.* 109: 161-165.

McNamara, K.J. and Ah Vee, C. 1989. A new genus of brissid echinoid from the Miocene of Australia. *Geol. Mag.* 126: 177-186.

McNamara,K.J. and Philip, G.M. 1980a, Australian Tertiary schizasterid echinoids. *Alcheringa* 4: 47-65.

McNamara, K.J. and Philip, G.M. 1980b. Tertiary species of *Echinolampas* (Echinoidea) from southern Australia. *Mem. Nat. Mus. Vict.* 41: 1-14.

McNamara, K.J. and Philip, G.M. 1984. A revision of the spatangoid echinoid *Pericosmus* from the Tertiary of Australia. *Rec. West. Aust. Mus.* 11: 319-356.

McNamara, K.J., Philip, G.M. and Kruse, P.D. 1986. Tertiary brissid echinoids of southern Australia. *Alcheringa* 10: 55-84.

Mooi, R. 1989. Living and fossil genera of the Clypeasteroidea (Echinoidea: Echinodermata): an illustrated key and annotated checklist. *Smithson. Contr. Zool.* 488:1-51.

Patterson, C. and Smith, A.B. 1987. Periodicity of extinction: a taxonomic artefact? *Nature* 330: 248-251.

Philip, G.M. 1963a. The Tertiary echinoids of south-eastern Australia. I. Introduction and Cidaridae (1): *Proc. R. Soc. Vict.* 76: 181-236.

Philip, G.M. 1963b. Two Australian Tertiary neolampadids, and the classification of cassiduloid echinoids. *Palaeontology* 6: 718-726.

Philip, G.M. 1964. The Tertiary echinoids of south-eastern Australia. II. Cidaridae (2). *Proc. R. Soc. Vict.* 77: 433-477.

Philip, G.M. 1965. The Tertiary echinoids of south-eastern Australia. III. Stirodonta, Aulodonta, and Camarodonta (1). *Proc. R. Soc. Vict.* 78: 181-196.

Philip, G.M. 1969. The Tertiary echinoids of south-eastern Australia. IV. Camarodonta (2). *Proc. R. Soc. Vict.* 82: 233-275.

Philip, G.M. and Foster, R.J. 1971. Marsupiate Tertiary echinoids from south-eastern Australia and their zoogeographic significance. *Palaeontology* 14: 124-134.

Prothero, D.R. 1989. Stepwise extinctions and climatic decline during the later Eocene and Oligocene, In S.K. Donovan, ed., *Mass Extinctions: Processes and Evidence.* New York, Columbia Univ. Press, pp. 211-234.

Ricklefs, R. 1990. *Ecology*. New York, W.H. Freeman.

Wolfe, J. 1978. A paleobotanical interpretation of Tertiary climates in the northern hemisphere. *Amer. Sci.* 66: 694-703.

18. CETACEAN EVOLUTION AND EOCENE/OLIGOCENE ENVIRONMENTS

By R. Ewan Fordyce

ABSTRACT

Marine mammal evolution during the Eocene-Oligocene involved mainly two groups, the Sirenia and Cetacea. Sirenians were largely restricted to low latitudes, and did not show the taxonomic and structural diversity seen amongst Cetacea. Primitive archaeocete whales appeared about the Ypresian-Lutetian, and diversified in the eastern Tethys during the Lutetian (10-12 described species, 1-3 families). Bartonian archaeocetes (8-11 species, 2 families), mostly Basilosauridae, radiated into mid temperate latitudes. A lower diversity in the Priabonian (3-7 species, 1-2 families) may reflect Bartonian/Priabonian boundary extinctions, but there is no evidence for any major extinction at the end of the Eocene. By this time some species occupied polar latitudes. The early Oligocene saw some of the most dramatic changes in cetacean history. Poorly-known early Oligocene taxa (8-11 species, $\geq$4 families) reveal the evolution of Mysticeti by the early Rupelian, and Odontoceti by the late Rupelian. Chattian Cetacea are diverse (36-37 to perhaps >50 species, $\geq$11 families), marking an explosive radiation of modern Cetacea that continued through much of the Neogene. Food resources and, ultimately, the physical structure of the oceans, seem to be the key driving forces in cetacean evolution and extinction. In particular, cetacean evolution during the earlier Oligocene seems linked strongly to progressive cooling and, by inference, new feeding opportunities offered by increased vertical and horizontal differentiation of the oceans.

OVERVIEW OF MARINE MAMMAL HISTORY

Marine mammal history starts in the Eocene, long after the demise of large marine reptiles at the end of the Cretaceous. Two main groups, the Cetacea (whales, dolphins, porpoises), and Sirenia (dugongs, manatees and relatives) dominate the Eocene to Oligocene interval, although other taxa are known. This review concentrates on the evolutionary history of Cetacea and the possible role of environmental changes in their evolution.

The oldest whale-like animal is of uncertain early to middle Eocene age; other Cetacea provide evidence of a Lutetian and later radiation of the rather conservative archaeocetes. The Oligocene saw rapid evolution amongst Cetacea, with the derivation of both odontocetes and mysticetes from amongst the archaeocetes, so that by the end of the Oligocene (Figures 18.1, 18.2) the Cetacea encompassed species close to the ancestry of some extant groups. The Eocene and Oligocene record of Cetacea is considered further below. Early Sirenia occur in the Lutetian, and the group may have an earlier record than this. Domning's ongoing and authoritative contributions (e.g. Domning, 1977, 1981, 1982, 1989; Domning et al., 1982, 1986) obviate the need here for a detailed account of Sirenia. Furthermore, there is no suggestion of any evolutionarily significant interaction between cetaceans and sirenians.

Other marine mammal groups appeared later than Cetacea and Sirenia. Seals, sea lions and relatives (Pinnipedia) appear about the Oligocene/Miocene boundary (Berta and Ray, 1990; Mitchell and Tedford, 1973). The extinct herbivorous Desmostylia are known from the Oligocene and Miocene (Domning et al., 1986) while marine otter-like forms (Carnivora: Mustelidae) appear by late Miocene (Berta and Morgan, 1985). These and some other minor groups were reviewed by Barnes et al. (1985).

None of the groups mentioned above are

closely related to each other, and each had a separate terrestrial ancestry. Thus, there have been repeated invasions of the oceans. Why should mammals, animals with a long terrestrial history, return to the sea? The availability of food otherwise not cropped by homeothermic large vertebrates has probably been a key driving force in marine mammal evolution and extinction (Barnes, 1977; Fordyce, 1980, 1989b; Lipps and Mitchell, 1976). The rise of new marine mammal taxa is commonly marked by innovative feeding structures, reinforcing the idea of a link between evolution and food resources and, through that, changes in oceanic environments. Thus, there is a role for extrinsic influences in cetacean evolution. Furthermore, living marine mammals are regarded by some as K strategists (e.g. Estes, 1979), and as such are potentially sensitive indicators of environmental change in terms of expanding or contracting food resources. Although there is only a broad picture of Cenozoic oceanic changes (circulation, thermal stratification) that may have influenced marine mammal evolution, there are rough correlations. Changing food resources probably have been a major factor in extinctions, but there may also be a place for extinction through competitive ecological displacement.

Because of their homeothermy, marine mammals may be less directly influenced by absolute temperature fluctuations than poikilotherms or ectotherms. Some extant Cetacea are cosmopolitan, but why most species are more restricted is not clear. It is tempting to interpret distribution simply as governed by adaptations to feeding on particular prey, but many Cetacea are opportunistic feeders. In the case of Sirenia, past and present subtropical to tropical distributions probably reflect the occurrence of seagrasses, a favoured food (Domning et al., 1986); a dramatic exception is provided by gigantic extinct kelp-feeding *Hydrodamalis* species (late Miocene to Quaternary, North Pacific, Domning, 1977). Furthermore, although the ranges of many cetacean species are well established, few have clearly identifiable structural adaptations to specific water masses. Absolute temperature probably governs minimum body size, through heat losses in species with relatively high surface area to volume ratios (Downhower and Blumer, 1988).

THE STRATIGRAPHIC RECORD OF CETACEA

The Eocene and Oligocene stratigraphic record of cetacean species is outlined in Figure 18.1. This was compiled from published literature, particularly the summaries given by Barnes and Mitchell, 1978; Fordyce, 1989b; Kellogg, 1936; Kumar and Sahni, 1986; Mchedlidze, 1984; Okazaki, 1988; Rothausen, 1970; Russell, 1982; and Whitmore and Sanders, 1977, as well as other articles cited below. Conventional binomina are used, other than the new combinations suggested by Gingerich et al. (1990). Some unpublished observations were used to refine age ranges and taxonomy. Listed names include apparently distinct taxa that are published but not named formally to species level (e.g. *Basilosaurus* sp. of Halstead and Middleton, 1972, from Britain), while names of uncertain application are excluded (e.g. *Babiacetus indicus* Trivedy and Satsangi, 1984). Recognised but unpublished new species amongst museum collections are also omitted. These include Cetacea from the late Oligocene of South Carolina (Barnes and Sanders, 1990; Sanders et al., 1982; Whitmore and Sanders, 1977), ?early and late Oligocene of Oregon and Washington states (Barnes, 1977; Ray, 1977), and late Oligocene of New Zealand (Fordyce, 1987, 1989b).

The time scale is that of Berggren et al., 1985. Sources used in correlation include: Australia -- Abele et al., 1976; New Zealand -- Hornibrook et al., 1989; Japan -- Tsuchi et al., 1987; India -- Mohan and Soodan, 1970; Singh and Singh, 1986; Egypt -- Abdel-Kireem, 1985; Allam et al., 1988, Haggag, 1985; western Europe -- Rothausen, 1986; Russell, 1982; Britain -- Hooker, 1986; Atlantic Coastal Plain and Gulf Coast -- Gohn, 1988; Popenoe et al., 1987; Westgate, 1988; Oregon and Washington states -- Barnes, 1977; Domning et al., 1986; Ray, 1977. For most species, stratigraphic distributions are cited imprecisely in the literature, e.g. often only to stage level, without mention of associated age-diagnostic fossils such as planktic foraminifera. Bars in Figures 18.1 and 18.2 indicate likely limits of error (effectively, maximum ranges) rather than precisely established ranges for different species. Nowhere does a published species history have clear lower and upper limits based on a stratigraphic

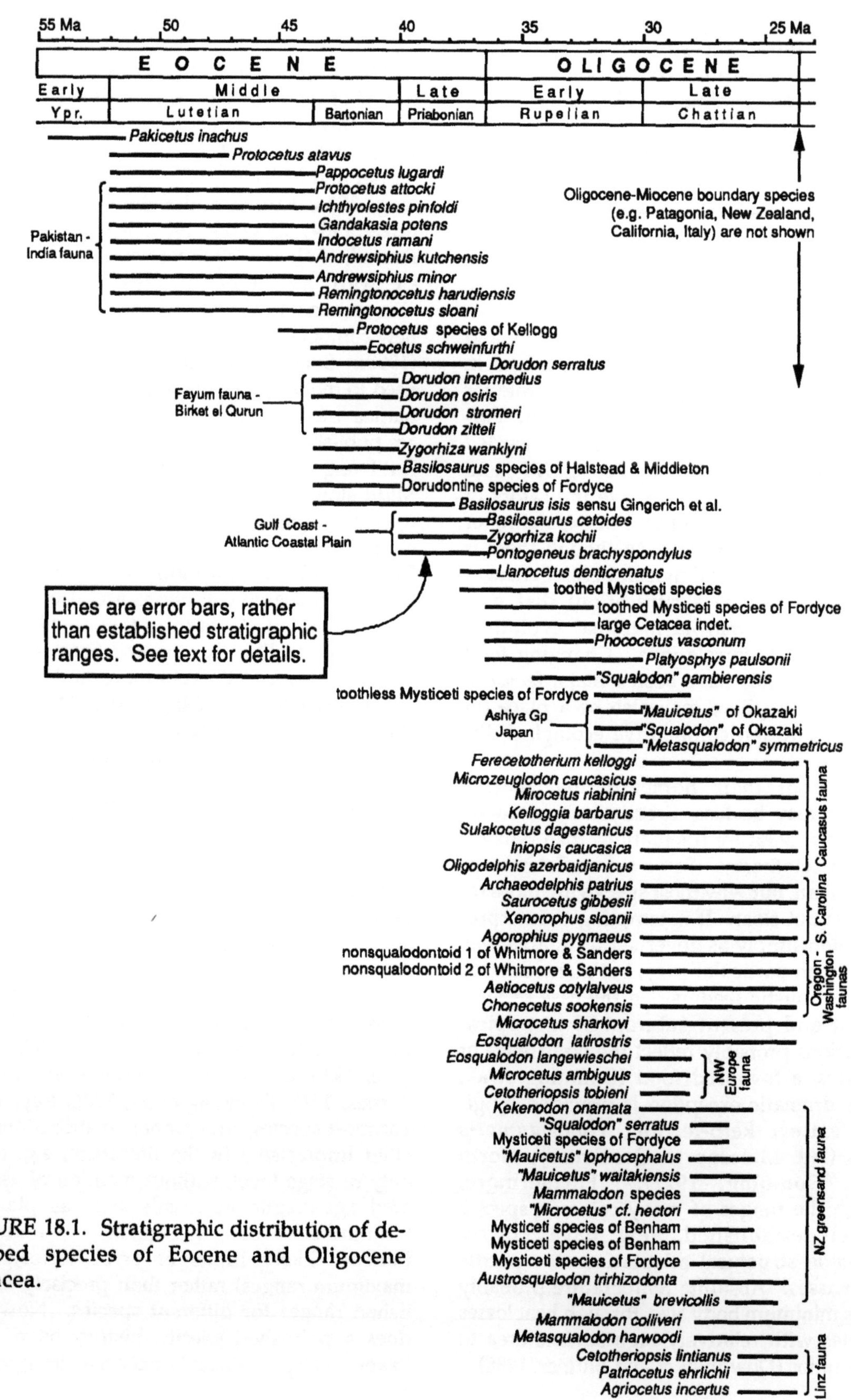

FIGURE 18.1. Stratigraphic distribution of described species of Eocene and Oligocene Cetacea.

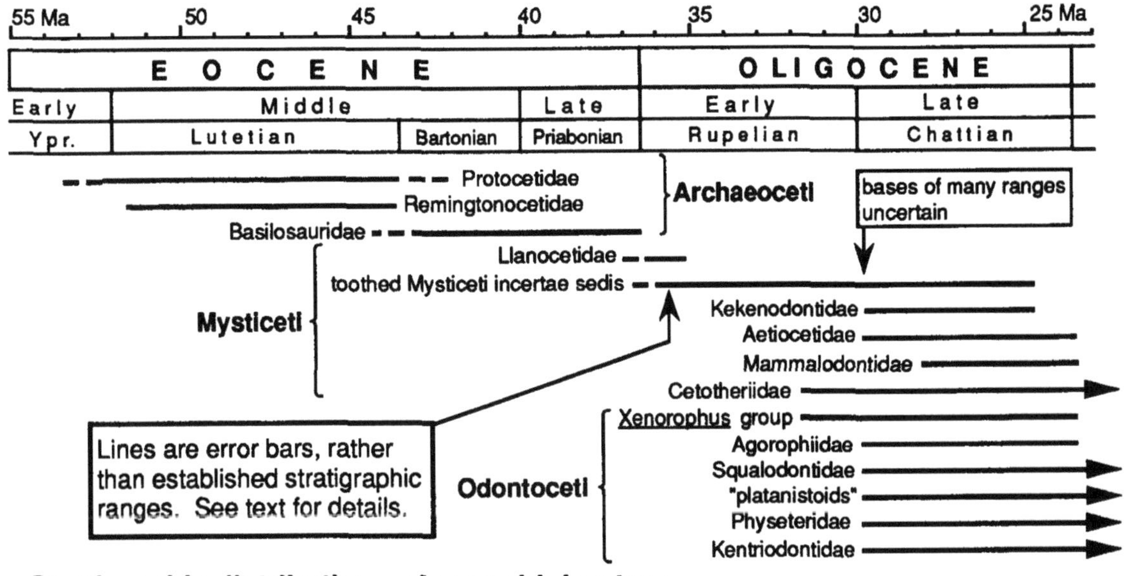

FIGURE 18.2. Stratigraphic distribution of described higher taxa of Eocene and Oligocene Cetacea.

succession of specimens. There is no clear evidence that any Eocene or Oligocene cetacean species ranged through more than one stage, and, as noted below, it is possible that most species had durations of no more than a million years.

This review concentrates on taxa and events before the end of the Oligocene. It is noteworthy that the very latest Oligocene reveals yet more taxa barely considered below, taxa perhaps better viewed as the start of a Neogene radiation of "modern" whales and dolphins than as part of an Eocene - Oligocene story. For example, rocks of the New Zealand Waitakian Stage, which have yielded significant Cetacea, are partly of Oligocene age, and the Argentine "Patagonian Formation" is close to the Oligocene/Miocene boundary. Age determinations around the Oligocene/Miocene boundary, especially in the south, seem plagued by ambiguous occurrences of planktonic microfossils.

There are other limits to the study of broader trends, though these are not unique to marine mammals. Many long-established species are poorly defined and diagnosed by modern standards, and shaky taxonomic foundations make it difficult to interpret broader patterns. Most higher taxa need to be reviewed in light of changing concepts of taxonomy, and many groups are acknowledged to be paraphyletic. Marine mammals are large animals that, although often locally conspicuous in outcrop, are less abundant in terms of numbers of fossilised individuals than many smaller organisms. Perhaps it is not surprising that amongst Eocene and Oligocene Cetacea many species are represented by only one published specimen.

TAXONOMIC TRENDS AND THEIR IMPLICATIONS

Broad stratigraphic patterns at species level are revealed in Figure 18.1. This review assumes that all the species in a higher taxon (e.g. a family) have some sort of ecological unity. Thus, trends at family level and above are potentially more revealing about broader environmental changes than are those at species level.

In the text below, a taxon - author - comma - date combination indicates the first publication of a taxonomic name. Otherwise, authors and dates are cited merely for the most appropriate source.

Archaeoceti

Archaeocetes are defined conventionally as primitive toothed Cetacea that lack features of the extant Odontoceti and Mysticeti. They include Protocetidae, Remingtonocetidae, and Basilosauridae, all of which are Eocene only. Reports of supposed Oligocene Archaeoceti refer either to species that belong elsewhere (e.g. *Mammalodon colliveri* - Mysticeti) or to those that are too incomplete to place conclusively (e.g. *Kekenodon onamata*, late Oligocene, New Zealand; *Platyosphys paulsonii*, early Oligocene, Ukraine). Archaeocetes were possibly similar in habits to living pinnipeds, with an amphibious lifestyle, and probably did not use the feeding techniques of either modern Odontoceti (echolocation-assisted predation) or Mysticeti (cropping by filter-feeding).

The ancient, primitive Protocetidae derive their identity from *Protocetus atavus* (Lutetian, Egypt - Tethys), a small and probably long-beaked fish-eating species. The family as used conventionally is a diverse paraphyletic group that includes the oldest reported cetacean, *Pakicetus inachus* (Ypresian or Lutetian, Pakistan - Tethys) (Gingerich et al., 1983). Other possible protocetids include taxa based on teeth of reported Lutetian age from India and Pakistan (e.g. Kumar and Sahni, 1986; West, 1980), skull remains from Egypt and Nigeria (Barnes and Mitchell, 1978; Kellogg, 1936), and fragments from the Gulf Coast (Kellogg, 1936). Interspecific variation in tooth form suggests a diversity of feeding modes not yet substantiated by the description of skulls or skeletons. Geographic ranges mimic those of contemporaneous Sirenia (Domning et al., 1982), with no evidence of spread outside the Tethys-Caribbean. If this absence is real and not preservational, it must support the idea of a Tethyan presumably warm water origin for cetaceans. This constrasts with the Bartonian radiation of cetaceans into temperate latitude waters, and polar waters by the end of the Eocene.

The bizarre remingtonocetids (Lutetian, India - Tethys) were a short-lived archaeocete group characterised by a long, narrow rostrum. Such a rostrum was presumably an adaptation for more rapid, forceps-like feeding than likely to be used by other Eocene Cetacea. Kumar and Sahni (1986) suggested that remingtonocetids are ancestral to odontocetes, but the evidence is not convincing.

Basilosaurids (Basilosaurinae, Dorudontinae), characteristic of the Bartonian and Priabonian, are more derived than protocetids and remingtonocetids in their more complex teeth and earbones (Barnes and Mitchell, 1978; Kellogg, 1936). More advanced feeding strategies, and different underwater hearing capabilities are indicated, but whether these features arose in response to changing external environments is uncertain. Fossils are geographically widespread (Gulf Coast, Atlantic Coastal Plain, and Britain - North Atlantic; New Zealand - southwest Pacific), and indicate the presence of, thus dispersal, of Cetacea beyond the Tethys well before the end of the Eocene. Such geographic spread suggests tolerance for waters cooler than tropical. The smaller dorudontines were perhaps dolphin-like in habit, while the larger basilosaurines include the strikingly large zeuglodon, *Basilosaurus cetoides* (Priabonian), of the Gulf Coast and Atlantic Coastal Plain. The latter, which was probably the largest late Eocene marine carnivore, has become a tacit index fossil for the North American Jacksonian Stage. Regretfully, it seems that there are no accounts of the precise stratigraphic distribution of basilosaurids within the Priabonian.

Odontoceti

Extant odontocetes (toothed whales, dolphins and porpoises) have a characteristic skull and jaw structure (Miller, 1923) which probably reflects the ability to transmit and receive echolocation sounds used in navigation and hunting prey. All fossil odontocetes for which skulls are known have the same basic structure as extant species, which suggests that fossil odontocetes may also have echolocated, and that the evolution of echolocation was critical in the origin of odontocetes (Fordyce, 1980). The oldest reported odontocete fossils are rather incomplete and, as is often the case with taxa at the base of a radiation, are difficult to place. Even some of the late Oligocene species known from skulls are of uncertain relationships. For years it was thought that many of the isolated heterodont teeth from Oligocene strata were teeth of odontocetes, and some species are based on such specimens; examples include *"Squalodon" gambierensis* (?Rupelian, Australia) and *"Squalodon" serratus* (Chattian, New Zealand). It now seems that some of these heterodont teeth, including the latter examples, are those of early Mysticeti.

The oldest described odontocetes appear to be "*Metasqualodon*" [= *Xenorophus*?] *symmetricus* and "*Squalodon*" sp. (Okazaki, 1982, 1987, 1988) from the Waita Formation, uppermost Ashiya Group, Japan. These are probably as old as basal Zone P21 (correlations after Tsuchi et al., 1987). Most other primitive odontocetes are identified only as late Oligocene, zone unspecified, because of lack of published information about age in terms of planktonic microfossil zones. Of these, the Agorophiidae are widely identified as the most primitive odontocetes. However, the family Agorophiidae is known strictly only from the holotype of *Agorophius pygmaeus* (late Oligocene, South Carolina - west Atlantic). Other primitive odontocetes sometimes referred to the family include, for example, species of *Xenorophus, Archaeodelphis, Microzeuglodon, Atropatenocetus,* and *Mirocetus* (see comments by Barnes et al., 1985, Fordyce, 1981), but it is likely that all belong elsewhere. Undescribed agorophiid-like specimens are also known from South Carolina, Oregon and Washington (Ray, 1977; Sanders et al., 1982; Whitmore and Sanders, 1977). These "agorophiids" suggest a plethora of low-diversity but high rank (perhaps family level) early taxa, again a common feature of early phases of evolutionary radiations. Apart from the record of "*Metasqualodon*" *symmetricus*, none of these "agorophiids" is known clearly to be older than late Oligocene. "Agorophiids" are overall of more primitive grade than Squalodontidae, and give some idea of the likely appearance of archaic odontocetes

"Shark-toothed dolphins," Squalodontidae, which were a rather widespread largely Miocene group (Rothausen, 1968), appear first sometime in the late Oligocene (e.g. species of *Eosqualodon, Kelloggia,* and perhaps *Patriocetus*). Earlier records of "*Squalodon*" are not clearly squalodontids. Few of these early squalodontids are stratigraphically well placed (e.g. to within part of a stage) and, furthermore, the taxonomic limits of the family are not clear. Most squalodontids are medium-sized odontocetes with narrow beaks. Presumably they were fairly agile fish-eaters; perhaps austral species ate coexisting small penguins. There are no significant analyses of functional morphology and life modes. The Austral short-beaked *Prosqualodon australis* appears about the latest Oligocene or earliest Miocene around the Southern Ocean (Tasmania, New Zealand, Patagonia) and provides an early example of Circum-Antarctic distribution.

Sperm whales, Physeteridae, appear sometime in the late Oligocene (e.g. *Ferecetotherium*, discussed by Barnes, 1985a, and Mchedlidze, 1984; and an undescribed specimen from New Zealand). Specimens are too incomplete to tell whether they were squid-eating deep divers like the extant *Physeter*, but such a lifemode is possible.

Marine dolphins apparently related to the extant river dolphins *Platanista* spp. occur in the Paleogene. New Zealand late Oligocene and latest Oligocene or earliest Miocene specimens formerly assigned to *Microcetus* and *Prosqualodon* are currently under study (Fordyce, 1990). An undescribed long-beaked late Oligocene "agorophiid" with platanistoid-like features is known from Oregon, and judging from the reconstruction by Pilleri (1986) *Sulakocetus* (late Oligocene, Caucasus; Mchedlidze, 1984) could be a platanistoid.

Small early delphinoids are represented by fossils in the grade family Kentriodontidae, including species of *Oligodelphis* (late Oligocene, Caucasus, Mchedlidze, 1984) and *Kentriodon* (late Oligocene or early Miocene, New Zealand, Fordyce, 1989b; late Oligocene or early Miocene, Patagonia, Cozzuol, 1986). The kentriodontids, which have a significant Miocene history (Barnes, 1978), seem related to the living Delphinidae, and perhaps also had pelagic habits. Other small odontocetes not obviously related to the delphinoids include the problematic eurhinodelphid-like *Iniopsis* (late Oligocene, Caucasus, Mchedlidze, 1984).

Mysticeti

The oldest known species of Mysticeti, filter-feeding whales, are toothed species from about the Eocene/Oligocene boundary in middle and high latitudes of the Southern Hemisphere. One, *Llanocetus denticrenatus*, was based on a fragment of mandible (Mitchell, 1989) of uncertain late Eocene or early Oligocene age from Seymour Island, Antarctica. The rest of the relatively complete holotype skull, still under preparation (Fordyce, 1989b), supports the idea of mysticete relationships, but much remains to be done on this mainly undescribed specimen before relationships within the Mysticeti can be assessed properly. *Llanocetus denticrenatus* may be related to a basal Oligocene "protosqualodont" (Keyes, 1973) from New Zealand and to the enigmatic "*Squalodon*" *serratus* (late Oligocene, New Zealand). I also regard the supposed archaeocete, *Phococetus vasconum* (early Oligocene?, France), and "*Squalodon*" *gambierensis* (early Oligocene, Australia), and an undescribed species close to "*Squalodon*" *gambierensis* (late Oligocene, New Zealand) as early toothed species of Mysticeti. An early Oligocene toothed mysticete seemingly unrelated to any of the above also occurs in New Zealand (Fordyce, 1989a).

Of better known species, the small *Mammalodon colliveri* has a short rostrum and heterodont teeth (about Oligocene/Miocene boundary, Australia; Fordyce, 1984). A second undescribed species occurs in Australia, and *Mammalodon* sp. occurs in the late Oligocene of New Zealand. It is not clear how *Mammalodon* is related to *Aetiocetus cotylalveus* and *Chonecetus sookensis*, both late Oligocene toothed mysticetes from the northeast Pacific (Oregon, Washington, British Columbia; Barnes, 1987; Emlong, 1966; Russell, 1968).

Toothless baleen-bearing mysticetes, conventionally placed in the grade family Cetotheriidae, appeared sometime late in the early Oligocene. This marks a major ecological shift from tooth-assisted filter-feeding to baleen-assisted filter-feeding. The early records are of undescribed specimens reported from New Zealand (Fordyce, 1989b) and from the Ashiya Group of Japan (Okazaki, 1987). The few species formally described are assigned to *Mauicetus* (Duntroonian, late Oligocene, New Zealand; Marples, 1956) and *Cetotheriopsis* (Chattian, northern Europe; Rothausen, 1971). In addition, undescribed late Oligocene baleen whales are known from South Carolina (Barnes and Sanders, 1990), Oregon and Washington, and New Zealand (at least 12 new species; Fordyce, 1987, 1989b). It is unlikely that all are related closely, and revision of the Cetotheriidae, a family based ultimately on a Miocene Eurasian species, may see other supra-generic taxa established.

Morenocetus (Patagonia - southwest Atlantic), which is the earliest described species assigned to the extant family Balaenidae, is usually regarded as early Miocene. Correlations in this part of the column are difficult in Austral middle to high latitudes, and it is just possible that, like other supposed early Miocene Austral taxa, *Morenocetus* could be close to the Oligocene/Miocene boundary. The advent of balaenids marks the evolution of a highly derived method of filter-feeding, that of slow skimming (Pivorunas, 1979).

None of the extant families Neobalaenidae, Eschrichtiidae, or Balaenopteridae is known from the Paleogene.

DIVERSITY TRENDS–THE SPECIES LEVEL

Marine mammals are potentially useful in the study of species-level evolutionary patterns, but the published Paleogene record is currently of uncertain value. Apparent species diversity per stage, as derived from the literature and summarised in Figure 18.1, shows a continued increase over the Ypresian to Chattian interval, with an apparent drop in the Priabonian. There is no clear diversity drop in the Rupelian. Totals are: Ypresian, minimum of 0, maximum of 1 species; Lutetian, 10-12 species; Bartonian, 8-11 species; Priabonian, 3-7 species; Rupelian, 8-11 species; Chattian, 36-37 species. Undescribed specimens known to reside in museum collections will add to these totals, especially for the late Oligocene (e.g. the more than 12 new species known from New Zealand alone).

Raw diversity figures should be interpreted cautiously, especially in terms of ecological implications. The Rupelian figure is probably far below the actual total for reasons discussed below. The Chattian total, of around 50 species, seems high, but it does not measure instantaneous ecological diversity. Not all 50 species were contemporaneous. Spread over an interval of some 5 million years, this Chattian total is modest compared to the modern diversity of over 70 obviously contemporaneous species.

For Paleogene Cetacea there are no clearly demonstrated cases of ancestor to descendant relationships. Thus it is difficult to assess species level trends over time in the manner used in stratophenetic analyses of terrestrial mammal sequences. Published records indicate that no Paleogene species is known to occur in more than one stage, thus suggesting rather short species durations. The notion of short ranges is supported further by known fossil records for living species of Cetacea. Such records go back conclusively to the late Pleistocene, with dubious early Pleistocene or Pliocene occurrences, and species durations of less than a million years seem likely. The clear evidence that Pliocene species became extinct attests to rapid recent ecological turnover, and further supports the idea of short species durations. The same might apply to the Paleogene.

Species level diversity changes are potentially more informative than trends at genus level, since some Paleogene genera have been used rather broadly as grades rather than clades (e.g. *Squalodon, Mauicetus*) and the generic position of other species is uncertain. Orr and Faulhaber (1975) plotted diversity changes at the genus level, and suggested that there was a marked drop in diversity from late Eocene (10 genera) to early Oligocene (2 genera) with a rise to 3 genera in the late Oligocene. Given the uncertain generic nomenclature of Rupelian Cetacea indicated in Figure 18.1, it seems likely that diversity increased from perhaps 3-6 Priabonian genera to 6-8 in the Rupelian. The apparent drop in generic diversity identified by Orr and Faulhaber seems unlikely, given trends at species level.

At the family level, the expanding record of early Cetacea, especially the early Odontoceti and Mysticeti, continues to reveal short-lived, early branched, and thus high ranked clades containing a few species. There is no firm evidence of high diversity families analogous to the modern Delphinidae, Ziphiidae or Balaenopteridae. Such low diversity, high rank clades might be viewed as representing varied experiments in cetacean lifestyles which in retrospect appear bizarre; examples include Remingtonocetidae, Mammalodontidae, and the *Xenorophus* group. These and others are potential candidates for plesion rank, or perhaps with careful justification, for formal high level taxa. Taxonomic rank aside, these higher level taxa do give an index of ecological diversity. Thus, for purposes of comparison with other literature, suprageneric taxa likely to be accepted by a consensus of paleocetologists are plotted in Figure 18.2. This suggests a drop during the Priabonian, a modest increase in the Rupelian, and a dramatic increase in the Chattian. A proviso to this trend is that the grades Protocetidae and Basilosauridae could be subdivided further upon cladistic analysis. The Chattian family diversity parallels that seen at the species level (Figure 18.1), and further attests to the Oligocene radiation of the modern Cetacea in response to new ecological opportunities afforded by increasingly differentiated oceans.

PALEOZOOGEOGRAPHY AND THE EFFECTS OF CLIMATE

The earliest Cetacea are Tethyan species from the Ypresian or Lutetian of India and Pakistan. Waters were probably subtropical, and the early species were probably warm water stenotherms. Bartonian archaeocetes are known from Britain and New Zealand as well as lower latitudes, suggesting geographic spread but not clear adaptation to cooler and/or more seasonal climates. Cetacea reached high southern latitudes by about the end of the Eocene (Fordyce, 1989b) and have probably occupied these waters since then, despite the poor fossil record beyond about 45° South. In the north, odontocete-like teeth are known from possible Oligocene sediments in Alaska, but details are not published. Since adaptations to cool climates have not been noted in the skeletons of extant Cetacea, they cannot be identified in any of the higher latitude fossils. However, high latitude organisms must deal with extreme seasonality whatever the temperature range encountered. Most extant Mysticeti are well adapted to high latitudes, employing a strategy of high latitude feeding and low latitude breeding. Such behaviour may have been a feature of the earliest mysticetes, and strengthens suggestions (Fordyce, 1980, 1989b) of high latitude (Austral?) origins for mysticetes. The implications of migration and seasonal feeding are considered elsewhere (Fordyce, 1980, 1989b).

Tropical waters today, in contrast to inferences about the middle Eocene, do not have a peak diversity of Cetacea (Barnes, 1977: 337). Davies (1963), in an analysis of historical zoogeography, suggested that modern warm waters might limit the distribution of many extant species (although a few species are restricted to the tropics). Davies further suggested that equatorial waters could act as a variable filter, allowing interchange between the hemispheres at times of global cooling, but acting as barriers to mixing during times of global warmth. Rothausen (1967) speculated that equatorial warm waters probably limited the distribution of some fossil odontocetes (squalodontids). Barnes (1977) briefly mentioned the broader evolutionary implications of antitropical distributions, and later (Barnes, 1985b) discussed these issues in relation to Neogene Phocoenidae. There has been no significant discussion of Paleogene Cetacea. Although some Paleogene cetacean species could have been cosmopolitan, little affected by latitudinal water masses, there is no evidence of this yet; nor is there clear evidence of north-south species pairs of the sort seen (Davies, 1963) amongst extant species.

PHYLETIC SIZE CHANGES AND THEIR IMPLICATIONS

Whereas the earliest Cetacea were small, some late Eocene taxa were large. For example, the skull of *Pakicetus inachus* (Ypresian?) is about 145 mm in zygomatic width, that of *Protocetus atavus* is about 240 mm, and that of *Basilosaurus cetoides* (Priabonian) is about 576 mm (Kellogg, 1936, Gingerich and Russell, 1981). Fischer and Arthur (1977) suggested that the presence of large *Basilosaurus cetoides* indicates equable late Eocene conditions, with this phase of late Eocene gigantism apparently followed by times of smaller and/or less diverse species. Since species at the base of a radiation are often small, this might also apply to Oligocene early Odontoceti and Mysticeti. Although there are no species demonstrably as long as *B. cetoides*, however, the late Eocene or early Oligocene *Llanocetus denticrenatus* is a mysticete with a skull larger than that of *B. cetoides* (zygomatic width estimated at about 800 mm, maximum cheek-tooth length about 47 mm at crown base). Furthermore, an undetermined cetacean from the New Zealand lower Whaingaroan Stage (early Oligocene) has large vertebrae over 100 mm in diameter. Thus, there is no evidence for simple phyletic size changes through the Priabonian-Rupelian.

INTERPRETATIONS: TIMING, MAGNITUDE, CAUSES

Ypresian-early Lutetian

The age of the most primitive described cetacean, *Pakicetus inachus*, is uncertain, but it is clear that there was a rapid radiation probably early in the Lutetian that resulted in diverse protocetids and remingtonocetids. These primitive whales, which were probably warm water stenotherms, did not attain a diversity comparable to that seen amongst extant mysticetes. Perhaps this is not surprising given their restricted distribution in a homogeneous subtropical to tropical epicontinental remnant of the Tethys sea. Gingerich et al. (1983) reasonably suggested that the initial radiation of cetaceans was linked to the availability of food resources in warm epicontinental seas, but ultimate causes are uncertain.

Later Lutetian

Despite the lack of clear ancestor to descendant relationships, it is likely that species of protocetid grade gave rise to the basilosaurids, a Bartonian-Priabonian group. It would be incorrect, therefore, to identify the Lutetian/Bartonian boundary simply as a time of protocetid extinction. It is more difficult to interpret the history of remingtonocetids; were these Cetacea really only Lutetian taxa restricted to a small area, or was there a wider and longer unrecorded history? The simplest explanation is that remingtonocetids became extinct about the late Lutetian, perhaps as a result of climate change or from reduction in habitat with continued closure of the eastern Tethys. Extinction of remingtonocetids through competition or ecological displacement seems unlikely in the absence of ecological equivalents. The rise of basilosaurids in the Bartonian, and the absence at that time of protocetids supports the idea of a major ecological turnover about the end of the Lutetian. Furthermore, the Bartonian radiation of basilosaurids marks increasingly sophisticated feeding, and reveals geographic spread from subtropical and tropical regions to middle latitudes with the presumed abandonment of warm water stenothermy. These changes may have been initially in response to new food resources, but it is uncertain whether they reflect increased temperature gradients from the poles to the tropics.

Bartonian-Priabonian

In contrast to other elements of pelagic ecosystems (Keller, 1983; Prothero, 1989), there is no evidence of a major event involving Cetacea at the end of the Bartonian. Species of *Dorudon, Zygorhiza* and *Basilosaurus* are known from both the Bartonian and Priabonian. Individual cetacean species do not clearly span two stages here, but stratigraphic and taxonomic information is too uncertain to support a claim of 100% turnover in species at the end of the Bartonian.

End Priabonian - Eocene/Oligocene boundary

On a broad scale, the end of the Priabonian was a time of profound ecological change. Basilosaurids, both small and large, are not known certainly to range beyond the Eocene, and the earliest filter-feeding mysticetes appeared in middle to high south latitudes at about this time. I have suggested elsewhere (Fordyce, 1980, 1989b) that the initial evolution of mysticetes might be linked ultimately to changes in oceanic food resources associated with the progressive isolation of Antarctica, the development of Antarctic ice and psychrosphere, and increased horizontal and vertical stratification of the Southern Ocean. The advent of filter-feeding mysticetes seems to be a consequence of increased horizontal and/or vertical partitioning of the oceans and could well have arisen concomitant with increased thermal gradients from poles to tropics. If this model is correct, early mysticetes may indeed show high latitude southern heterochroneity like that of some invertebrates (Zinsmeister and Feldmann, 1984). Much better stratigraphic and taxonomic resolution of both Priabonian and Rupelian taxa are needed before details can be put in sequence; for example, the significance of ecological changes will be revealed further once it is determined whether mysticete origins lie within the basilosaurines or dorudontines. Did basilosaurines become extinct because of climate changes, or through competition with equally large primitive mysticetes (*Llanocetus*)?

End Rupelian-Chattian - mid Oligocene

The few Rupelian Cetacea, poorly known in terms of taxonomy and stratigraphy, hint at ongoing ecological partitioning amongst early toothed mysticetes. It is not until the end of the Rupelian that the first toothless (baleen-feeding) mysticetes and odontocetes are reported. The Chattian is marked by a profound increase in taxa at species and family level, and an increase in ecological diversity that surely must attest to an earlier (late Rupelian?) unrecorded history. Eochattian (*sensu* Rothausen, 1986) records from Europe, for example, include certain cetotheres and squalodontids, and there can be little doubt about a high cetacean diversity by the end of the Chattian. It seems unlikely that this diversity arose through rapid radiation after the Rupelian, so there must be a significant unrecorded early Oligocene history. Simpson (1945) suggested that "missing" phases of cetacean evolution like this perhaps occurred in open oceans remote from sedimentary environments likely to be preserved on land, but oceanic species are elsewhere preserved in the nearshore record. If the early Oligocene record as a whole was poor, the erosion of fossiliferous proximal marine sediments by middle Oligocene sea level changes (Popenoe et al., 1987) could be invoked to account for the poor record (Fordyce, 1980). Figure 18.1 reveals that only 2 of the 8-11 Rupelian Cetacea are likely to be Odontoceti, but, in light of the possible biasing role of erosion events, it is difficult to explain why mysticetes should dominate.

Field-based stratigraphic studies are needed to better refine the events during this interval. In New Zealand, there is an explosive increase in cetacean abundance above a conspicuous "mid" Oligocene disconformity. This horizon widely marks an abrupt change from distal deepwater sediments to more proximal, glauconitic sediments shortly after the extinction of *Subbotina angiporoides*. It probably indicates the large sea level change reported worldwide in the middle of the Oligocene (Keller, 1983; Prothero, 1989). A glacioeustatic cause for the sea level drop might also explain the dramatic increase in cetacean diversity, through continued stratification of the oceans and oceanic food reserves. Ironically, the unconformity probably eroded the proximal shelf sediments that contained critical cetacean fossils, and the climate changes probably induced the extinction of many planktonic foraminifera that would have otherwise usefully calibrated the history of the marine mammals that remained as fossils. Stratigraphic studies on deep water sediments now exposed on land would seem to have the most promise to resolve the later Rupelian history of Cetacea.

The Chattian marks the appearance of other marine mammal groups (Desmostylia, Pinnipedia) in mid latitudes. By the end of the Oligocene, then, the radiation of the modern Cetacea was well established and the oceans were partitioned by marine mammals in manner little different from today.

ACKNOWLEDGEMENTS

I thank Larry Barnes and Jerry Hooker for comments on the manuscript, Frank Whitmore for a decade of discussion, and Don Prothero for enabling me to attend the Penrose Conference that led to this article. Research leading to results reported here was supported variously by postdoctoral fellowships from the Smithsonian Institution and Monash University, by an NSF subcontract through M.O. Woodburne, and particularly by grants from the University of Otago and the National Geographic Society.

REFERENCES

Abdel-Kireem, M. R. 1985. Planktonic foraminifera of Mokattam Formation (Eocene) of Gebel Mokattam, Cairo, Egypt. *Rev. Micropal.* 28: 77-96.

Abele, C., Gloe, C. S., Hocking, J. B., Holdgate, G., Kenley, P. R., Lawrence, C. R., Ripper, D., Threlfall, W. R. 1976. Tertiary. In J.G. Douglas and J.A. Ferguson, eds., *Geology of Victoria.* Melbourne, Geological Society of Australia, pp. 177-274.

Allam, A., Bassiouni, M. A. and Zalat, A. 1988. Calcareous nannoplankton from Middle and Upper Eocene rocks at Gebel Mokattam, East Cairo, Egypt. *Jour. Afr. Earth Sci.* 7: 201-211.

Barnes, L. G. 1977. Outline of eastern North Pacific fossil cetacean assemblages. *Syst. Zool.* 25: 321-343.

Barnes, L. G. 1978. A review of *Lophocetus* and *Liolithax* and their relationships to the delphinoid family Kentriodontidae (Cetacea: Odontoceti). *Nat. Hist. Mus. Los Angeles Cty. Sci. Bull.* 28: 1-35.

Barnes, L. G. 1985a. Review [of Mchedlidze 1984, General features of the paleobiological evolution of Cetacea]. *Mar. Mamm. Sci.* 1: 90-93.

Barnes, L. G. 1985b. Evolution, taxonomy and antitropical distributions of the porpoises (Phocoenidae, Mammalia). *Mar. Mamm. Sci.* 1: 149-165.

Barnes, L. G. 1987. *Aetiocetus* and *Chonecetus,* primitive Oligocene toothed mysticetes and the origin of baleen whales. *Jour. Vert. Paleontol.* 7 (3, suppl.): 10A.

Barnes, L. G., Domning, D. P. and Ray, C. E. 1985. Status of studies on fossil marine mammals. *Mar. Mamm. Sci.* 1: 15-53.

Barnes, L. G. and Mitchell, E. D. 1978. Cetacea. In V.J. Maglio, and H.S.B. Cooke, eds., *Evolution of African mammals.* Cambridge, Mass., Harvard University Press, pp. 582-602.

Barnes, L. G. and Sanders, A. E. 1990. An archaic Oligocene mysticete from South Carolina. *Jour. Vert. Paleontol* 10 (3, suppl.): 14A.

Berggren, W. A., Kent, D. V. and Flynn, J. J. 1985. Paleogene geochronology and chronostratigraphy. *Geol. Soc. Lond. Mem.* 10: 141-195.

Berta, A. and Morgan, G. S. 1985. A new sea otter (Carnivora: Mustelidae) from the late Miocene and early Pliocene (Hempillian) of North America. *Jour. Paleontol.* 59: 809-819.

Berta, A. and Ray, C. E. 1990. Skeletal morphology and locomotor capabilities of the archaic pinniped *Enaliarctos mealsi. Jour. Vert. Paleontol.* 10: 141-157.

Cozzuol, M. A. 1986. Primer registro de la Familia Kentriodontidae (Delphinoidea: Cetacea) para el Oligoceno Tardio-Mioceno temprano de Patagonia. *Bol. Inf., Asoc. Pal. Argentina* 15: 14.

Davies, J. L. 1963. The antitropical factor in cetacean speciation. *Evolution* 17: 107-116.

Domning, D. P. 1981. Sea cows and sea grasses. *Paleobiology* 7: 417-420.

Domning, D. P. 1982. Evolution of manatees: a speculative history. *Jour. Paleontol.* 56: 599-619.

Domning, D. P. 1989. Kelp evolution: a comment. *Paleobiology* 15: 53-56.

Domning, D. P., Morgan, G. S. and Ray, C. E. 1982. North American Eocene sea cows (Mammalia: Sirenia). *Smithsonian Contrib. Paleobiol.* 52: 1-69.

Domning, D. P., Ray, C. E. and McKenna, M. C. 1986. Two new Oligocene desmostylians and a discussion of Tethytherian systematics. *Smithsonian Contrib. Paleobiol.* 59: 1-56.

Downhower, J. F. and Blumer, L. S. 1988. Calculating just how small a whale can be. *Nature* 335: 675.

Emlong, D. R. 1966. A new archaic cetacean from the Oligocene of Northwest Oregon. *Bull. Mus. Nat. Hist., Univ. Oregon* 3: 1-51.

Estes, J. A. 1979. Exploitation of marine mammals: r-selection of K-strategists? *Jour. Fish. Res. Board Can.* 36: 1009-1017.

Fischer, A. G. and Arthur, M. A. 1977. Secular variations in the pelagic realm. In H.E. Cook and P. Enos, eds., *Deep-water carbonate environments*. Tulsa, Oklahoma, Soc. Econ. Paleont. Mineral., pp. 19-50.

Fordyce, R. E. 1980. Whale evolution and Oligocene Southern Ocean environments. *Palaeogeog., Palaeoclimat., Palaeoecol.* 31: 319-336.

Fordyce, R. E. 1981. Systematics of the odontocete *Agorophius pygmaeus* and the family Agorophiidae (Mammalia: Cetacea). *Jour. Paleontol.* 55: 1028-1045.

Fordyce, R. E. 1984. Evolution and zoogeography of cetaceans in Australia. In M. Archer and G. Clayton, eds., *Vertebrate zoogeography and evolution in Australasia*. Carlisle, Hesperian, pp. 929-948.

Fordyce, R. E. 1987. New finds of whales from the Kokoamu Greensand. [Abstract.]. *Geol. Soc. N. Z. Misc. Publ.* 37A: [unpaginated].

Fordyce, R. E. 1989a. Problematic Early Oligocene toothed whale (Cetacea, ?Mysticeti) from Waikari, North Canterbury, New Zealand. *N.Z. Jour. Geol. Geophys.* 32: 385-390.

Fordyce, R. E. 1989b. Origins and evolution of Antarctic marine mammals. *Spec. Publ. Geol. Soc. Lond.* 47: 269-281.

Fordyce, R. E. 1990. Squalodelphid-like dolphins from the Duntroonian and Waitakian Stages (Late Oligocene - earliest Miocene) of North Otago, New Zealand [abstract]. *Geol. Soc. N. Z. Misc. Publ.* 50A: 53.

Gingerich, P. D. and Russell, D. E. 1981. *Pakicetus inachus*, a new archaeocete (Mammalia, Cetacea) from the Early-Middle Eocene Kuldana Formation of Kohat (Pakistan). *Contr. Mus. Paleo. Univ. Mich.* 25: 235-246.

Gingerich, P. D., Smith, B. H. and Simons, E. L. 1990. Hind limbs of Eocene *Basilosaurus*: evidence of feet in whales. *Science* 249: 154-156.

Gingerich, P. D., Wells, N. A., Russell, D. E. and Shah, S. M. 1983. Origin of whales in epicontinental remnant seas: new evidence from the Early Eocene of Pakistan. *Science* 220: 403-406.

Gohn, G. S. 1988. Late Mesozoic and early Cenozoic geology of the Atlantic Coastal Plain: North Carolina to Florida. In R.E. Sheridan and J.A. Grow, eds., *The geology of North America, volume* I-2. *The Atlantic coastal margin, U.S.* Boulder, Colo., Geological Society of America, pp. 107-130.

Haggag, M. A. Y. 1985. Middle Eocene planktonic foraminifera from Fayum area, Egypt. *Rev. Esp. Micropal.* 17: 27-40.

Halstead, L. B. and Middleton, J. A. 1972. Notes on fossil whales from the Upper Eocene of Barton, Hampshire. *Proc. Geol. Assoc.* 83: 185-190.

Hooker, J. J. 1986. Mammals from the Bartonian (middle/late Eocene) of the Hampshire Basin, southern England. *Bull. Brit. Mus. (Nat. Hist.), Geol. Ser.* 39: 191-478.

Hornibrook, N. de B., Brazier, R. C. and Strong, C. P. 1989. Manual of New Zealand Permian to Pleistocene foraminiferal biostratigraphy. *N. Z. Geol. Surv. Paleontol. Bull.* 56: 1-175.

Keller, G. 1983. Paleoclimatic analyses of Middle Eocene through Oliogocene planktonic foraminiferal faunas. *Palaeogeog., Palaeoclimat., Palaeoecol.* 43: 73-94.

Kellogg, A. R. 1936. A review of the Archaeoceti. *Carnegie Inst.Washington Publ.* 482: 1-366.

Keyes, I. W. 1973. Early Oligocene squalodont cetacean from Oamaru, New Zealand. *N.Z. Jour. Mar. Freshw. Res.* 7: 381-390.

Kumar, K. and Sahni, A. 1986. *Remingtonocetus harudiensis*, new combination, a middle Eocene archaeocete (Mammalia, Cetacea) from western Kutch, India. *Jour. Vert. Paleontol.* 6: 326-349.

Lipps, J. H. and Mitchell, E. D. 1976. Trophic

model for the adaptive radiations and extinctions of pelagic marine mammals. *Paleobiology* 2: 147-155.

Marples, B. J. 1956. Cetotheres (Cetacea) from the Oligocene of New Zealand. *Proc. Zool. Soc. Lond.* 126: 565-580.

Mchedlidze, G. A. 1984. *General features of the paleobiological evolution of Cetacea.* New Delhi, Amerind, 136 pp.

Miller, G. S. 1923. The telescoping of the cetacean skull. *Smithsonian Misc. Coll.* 76 (5): 1-70.

Mitchell, E. D. 1989. A new cetacean from the Late Eocene La Meseta Formation, Seymour Island, Antarctic Peninsula. *Can. Jour. Fish. Aquatic Sci.* 46: 2219-2235.

Mitchell, E. D. and Tedford, R. H. 1973. The Enaliarctinae, a new group of extinct aquatic Carnivora and a consideration of origin of the Otariidae. *Bull. Amer. Mus. Nat. Hist.* 151: 201-284.

Mohan, M. and Soodan, K. S. 1970. Middle Eocene planktonic foraminiferal zonation of Kutch, India. *Micropaleontology* 16: 37-46.

Okazaki, Y. 1982. A lower Miocene squalodontid from the Ashiya Group, Japan. *Bull. Kitakyushu Mus. Nat. Hist.* 4: 107-112.

Okazaki, Y. 1987. Additional materials of *Metasqualodon symmetricus* (Cetacea: Mammalia) from the Oligocene Ashiya Group, Japan. *Bull. Kitakyushu Mus. Nat. Hist.* 7: 133-138.

Okazaki, Y. 1988. Oligocene squalodont (Cetacea/Mammalia) from the Ashiya Group, Japan. *Bull. Kitakyushu Mus. Nat. Hist.* 8: 75-80.

Orr, W. N. and Faulhaber, J. 1975. A middle Tertiary cetacean from Oregon. *Northw. Sci.* 49: 174-181.

Pilleri, G. 1986. *Beobachtungen an den fossilen Cetaceen des Kaukasus.* Hirnanatomisches Institut, Ostermundigen. 40 pp.

Pivorunas, A. 1979. The feeding mechanisms of baleen whales. *Amer. Scient.* 67: 432-440.

Popenoe, P., Henry, V. J. and Idris, F. M. 1987. Gulf trough - the Atlantic connection. *Geology* 15: 327-332.

Prothero, D. 1989. Stepwise extinctions and climatic decline during the later Eocene and Oligocene. In S.K. Donovan, ed., *Mass extinctions: processes and evidence.* London, Belhaven, pp. 217-234.

Ray, C. E. 1977. Fossil marine mammals of Oregon. *Syst. Zool.* 25: 420-436.

Rothausen, K. 1967. Die Klimabindung der Squalodontoidea (Odontoceti, Mamm.) und anderer mariner Vertebrata. *Sonderveroff. Geol. Inst. Univ. Köln* 13: 157-166.

Rothausen, K. 1968. Die systematische Stellung der europaischen Squalodontidae (Odontoceti: Mamm.). *Paläont. Z.* 42: 83-104.

Rothausen, K. 1970. Marine Reptilia and Mammalia and the problem of the Oligocene-Miocene boundary. *Giorn. Geol.* (series 2) 35: 181-189.

Rothausen, K. 1971. *Cetotheriopsis tobieni* n. sp., der erste paläogene Bartenwale (Cetotheriidae, Mysticeti, Mamm.) nördlich des Tethysraumes. *Abh. Hess. L.-Amtes Bodenforsch.* 60: 131-148.

Rothausen, K. 1986. Marine Tetrapoden im tertiären Nordsee-Becken. 1. Nord- und mitteldeutscher Raum ausschließlich Niederrheinische Bucht. In *Nordwestdeutschland im Tertiär, Teil 1 (Beitr. Reg. Geol. Erde, 18).* Berlin-Stuttgart, Gebrüder Borntraeger, pp. 510-557.

Russell, D. E. 1982. Tetrapods of the Northwest European Tertiary basin. *Geol. Jahrb.* A60: 5-74.

Russell, L. S. 1968. A new cetacean from the Oligocene Sooke Formation of Vancouver Island, British Columbia. *Can. Jour. Earth Sci.* 5: 929-933.

Sahni, A. and Mishra, V. P. 1975. Lower Tertiary vertebrates from western India. *Paleont. Soc. India, Monogr.* 3: 1-48.

Sanders, A. E., Weems, R. E. and Lemon, E. M. 1982. Chandler Bridge Formation - a new Oligocene stratigraphic unit in the lower coastal plain of South Carolina. *U.S. Geol. Surv. Bull.* 1529-H: 105-124.

Simpson, G. G. 1945. The principles of classification, and a classification of mammals. *Bull. Amer. Mus. Nat. Hist.* 85: 1-350.

Trivedy, A. N. and Satsangi, P. P. 1984. A new archaeocete (whale) from the Eocene of India. *27th Int. Geol. Congr., Moscow, abstrs* 1: 322-323.

Tsuchi, R., Shuto, T. and Ibaraki, M. 1987. Geologic ages of the Ashiya Group, North Kyushu from a viewpoint of planktonic foraminifera. *Rept. Fac. Sci., Shizuoka Univ.* 21: 109-119.

West, R. M. 1980. Middle Eocene large mammal assemblage with Tethyan affinities, Ganda Kas region, Pakistan. *Jour. Paleontol.* 54: 508-533.

Westgate, J. W. 1988. Biostratigraphic implications of the first Eocene land mammal fauna from the North American coastal plain. *Geology* 16: 995-998.

Whitmore, F. C. and Sanders, A. E. 1977. Review of the Oligocene Cetacea. *Syst. Zool.* 25: 304-320.

Zinsmeister, W. J. and Feldmann, R. M. 1984. Cenozoic high latitude heterochroneity of Southern Hemisphere marine faunas. *Science* 224: 281-283.

19. PALEOSOLS AND CHANGES IN CLIMATE AND VEGETATION ACROSS THE EOCENE/ OLIGOCENE BOUNDARY

by Gregory J. Retallack

ABSTRACT

Fossil soils in alluvial sequences, like those of Badlands National Park, South Dakota, can be evidence for changes in climate and vegetation during Tertiary geological time, supplementing other geological records of paleoclimate, such as eolian sediments, fossil leaves and the isotopic composition of foraminifera. In the Badlands of South Dakota, the depth of calcic horizons in paleosols when compared with depths in soils of the present Great Plains of North America, indicates late Eocene (38 Ma) mean annual rainfall in excess of 1000 mm, declining by Oligocene time (32 Ma) to 500-900 mm, then later in that epoch to 450-500 mm by 30.5 Ma and to 250-450 mm by 29.5 Ma. Drier climate of late Oligocene time also may explain the decreased abundance of kaolinite and increased abundance of smectite and then illite in paleosols higher in the sequence, as well as changes in hue from red to brown and then yellow, and changes in texture from clayey to silty. Climatic deterioration across the Eocene/ Oligocene boundary is prominently displayed in the appearance of these colorfully banded badlands.

Although few fossil plants are preserved in the Badlands of South Dakota, the paleosols contain abundant drab-haloed root traces and their profile form and soil structures also are evidence of changing vegetation: moist forests of 38 Ma giving way to dry forests by 34 Ma, to dry woodland by 33 Ma, to wooded grassland with streamside gallery woodland by 32 Ma and large areas of open grassland by 30 Ma. Some changes in assemblages of fossil mammals of Badlands National Park can be related to the progressively drier conditions and more open vegetation. The long term evolutionary trend toward the kinds of cursorial limbs and high crowned teeth well displayed in modern mammalian faunas of grasslands was initiated during this time, but in these respects Eocene and Oligocene faunas remained more like modern faunas of woodland than those of grassland. Mid-Tertiary paleoclimatic cooling and drying was an impetus for evolution of the grassland biome in continental interiors.

Local volcanism and tectonic uplift in the Black Hills and Rocky Mountains to the west also have affected this record in paleosols of mid-Tertiary paleoclimatic change, but not to the extent of obscuring it. A more critical limitation on this sequence as a paleoclimatic record is non-deposition during long periods of dry climate, as indicated by abrupt changes in the nature, development and spacing of paleosols. These limitations may be overcome when more is learned about other mid-Tertiary sequences of paleosols in Oregon, Argentina, Egypt, Mongolia and Kazakhstan.

INTRODUCTION

The demonstration by Dokuchaev (1883) that soil types such as the Russian Chernozem (or Mollisols in modern terminology) were similar over many different kinds of parent rocks marked the liberation of soil science as a discipline, independent of geology and agronomy. In this example and others gathered by this pioneering Soviet school of soil geography (Glinka, 1931), climate and vegetation were seen as more important factors in determining the nature of soils, than parent material, topographic setting or time for formation. These other factors are known to be significant, even limiting in certain cases, so that each factor must be considered in trying to understand how a particular soil may have formed (Jenny, 1941). There is now a great body of published information on how surface soils form (Birkeland, 1984). This can be applied to understanding the paleoenvironmental significance of soils buried in sedimentary sequences, including many sequences of paleosols across the Eocene/Oligocene boundary.

Late Eocene and Oligocene terrestrial sequences with mammalian fossils have been known for more than a century, but paleosols in them have been reported only in the last few decades (Schultz et al., 1955; Pomerol, 1964; Pettyjohn, 1966; Morand et al., 1968). The Badlands of South Dakota are typical of such paleosol sequences: subtly banded, colorfully variegated, red and green, bioturbated, clayey rocks. Paleosols in one part of the Badlands of South Dakota have now been studied in detail (Retallack, 1983a,b), and feature prominently in this chapter.

In addition to such red and variegated beds, paleosols also are common in coal measures that accumulated in swamps and marshes (Fastovsky and McSweeney, 1987). However, these are less useful guides to ancient climate because their formation is dominated by waterlogging that isolates them from climatic influences. Paleosols also are found in many sequences of evaporites. Precipitation of salts in exposed tidal flats and playa lakes results in a kind of soil called a Salorthid (of Soil Survey Staff, 1975), and is a very different process than the accumulation of salts within lake or ocean water (Warren, 1989). Because so little research has been reported on Eocene and Oligocene paleosols beyond the Badlands of South Dakota, a preliminary assessment of paleosols in many parts of the world can only be made by attempting to read between the lines of existing geological reports on red beds, coal measures and evaporites. A brief discussion of this kind is offered in this chapter, not to present realistic paleoclimatic interpretations, but to suggest the potential of such studies elsewhere in the world.

Paleopedological studies of Tertiary non-marine rocks now are becoming commonplace (Singer and Nkedi-Kizza, 1980; Singler and Picard, 1981; Andreis, 1981; Allen, 1986; Retallack, 1983a,b, 1985, 1986b; Fastovsky and McSweeney, 1987; Bown and Kraus, 1981, 1987), but their recent publication dates and lack of worldwide coverage reflect past problems with expertise and interest in paleosols. Few geologists have been trained in soil science, and few soil scientists have examined sedimentary rock sequences. Many paleosols have been altered substantially during burial, and many of the sensitive analytical approaches of soil science are not appropriate for them. Much remains to be done, but these problems now are being addressed on a wide front. Paleosols can be recognized within sedimentary rocks by three broad classes of diagnostic features: root traces, soil horizons and soil structure (Retallack, 1988a). They are best characterized by a combination of petrographic, mineralogical and geochemical techniques (Retallack, 1983a, 1986b). The alteration of paleosols after burial is in many ways like that of other rocks (Nesbitt and Young, 1989; Retallack, 1990, 1991). There are a variety of approaches for measuring sections, mapping and naming of individual paleosols (Retallack, 1983a), deep weathering zones of paleosols (Senior and Mabbutt, 1979) and natural assemblages of paleosols (Bown and Kraus, 1987). Interpretation of paleosols may be made in two distinct ways: by identification of a paleosol in a classification of modern soils and by study of particular paleoenvironmentally significant features of paleosols (Retallack, 1983a, 1990; Fastovsky and McSweeney, 1987). The second interpretive approach is emphasized in this chapter.

CLIMATICALLY SENSITIVE FEATURES OF PALEOSOLS

A variety of features of surface soils show a clear and significant relationship, or climofunction, to climatic variables. In general, tropical soils are more red in hue, more clayey, less rich in organic matter, and more deeply and thoroughly weathered than soils of temperate climates (Birkeland, 1984). Unfortunately however, none of these generalizations are very useful for interpreting paleoclimate from paleosols for two reasons. Some of these features (reddening, increased clay content, greater depth and intensity of weathering) are also related to the time available for soil formation and mean annual rainfall, as well as former temperature. Some of these features (reddening, depletion of organic matter) also are known to be imposed during burial alteration of paleosols

(Retallack, 1983a, 1990). From the large number of climate–soil relationships now known, a few are sufficiently independent of other soil–forming factors and unaffected by burial to be useful in interpreting paleoclimate from paleosols (Retallack, 1990). Only a few of these can be outlined here.

One useful climofunction for estimating mean annual rainfall from paleosols is the depth to the top of the calcareous nodular (calcic or Bk) horizon, which is deeper in wetter climates. This has been demonstrated for surface soils in several different parts of the world: North American Great Plains (Jenny, 1941; Ruhe, 1984), Mojave Desert of the North American southwest (Arkley, 1963), Serengeti Plains of Tanzania (de Wit, 1978) and Indo–Gangetic Plains of India and Pakistan (Sehgal et al., 1968). This general relationship is remarkably consistent among soils of different climate, parent material and time for formation (Ruhe, 1984). The depth to the calcic horizon should not to be confused with the depth of leaching of carbonate in non–calcareous soils of humid climates (Birkeland, 1984), nor with the depth to calcareous stringers and horizons of very old soils (Kubiena, 1970), nor with the depth to pore–filling simple cements of the kind produced in groundwater calcretes (Retallack, 1991). Carbonate nodules within paleosols are robust in the face of alteration after burial even up to greenschist-grade metamorphism (Retallack, 1985). The main problems with using depth to the nodular horizon as an indicator of paleoclimate are compaction of the upper part of paleosols during burial and erosion of the upper part of paleosols prior to burial. Corrections for compaction can be made using geological information on depth of burial and standard compaction curves (Baldwin and Butler, 1985). An assessment of surface erosion can be based on preservation of root traces and soil structures (especially granular peds) typical of surface horizons of paleosols. A greater limitation is the variance of the relationship in surface soils. In soils of the midwestern United States the relationship between depth to the calcic horizon (D in cm) and mean annual precipitation (P in cm) is $D = 2.5(P-30.5)$. Slightly different regressions of this kind of data have been found elsewhere. Until these studies are extended and integrated into a coherent whole for a variety of different conditions, the interpreted ranges of former precipitation given here and previously (Retallack, 1983a, 1986a) correspond to the total range of values observed in postglacial soils with calcic horizons at comparable depths.

Another potential indicator of mean annual rainfall is the nature of clay minerals in paleosols. In general, sepiolite and palygorskite are found in desert climates, smectite in dry climates and kaolinite in wet climates (Weaver, 1989). The kinds of clay mineral produced also depend on the nature of the parent material and the duration of soil formation among other factors, but most of these variables can be controlled by examining the changing proportions of clay minerals within paleosol profiles or between paleosol profiles of comparable degree of development. Far more serious for the paleoclimatic interpretation of paleosol clays is the diagenetic alteration of smectite to illite during burial, especially beyond about 2 km and 200°C (Eberl et al., 1990).

In extremely dry climates, there is so little rain that soils and lakes become salty from the evaporation of groundwater. Gypsum is a common soil evaporite, and may also form horizons whose depth from the surface of the soil reflects mean annual rainfall (Dan and Yaalon, 1982). Unfortunately, such evaporite minerals are readily leached from rocks during burial, and so not always preserved in paleosols, although pseudomorphs of them are sometimes found (Warren, 1989).

Strongly seasonal wet–dry climates induce characteristic soil features in clayey soils: the hummock and swale surface (gilgai microrelief) and cracking and deformation of surface horizons (mukkara structure of Paton, 1974) found in Vertisols. These kinds of structures are very robust during burial, and are known in very ancient paleosols, in some cases metamorphosed to within the greenschist facies (Allen, 1986; Retallack, 1986b).

PALEOSOLS IN BADLANDS NATIONAL PARK, SOUTH DAKOTA

One place where paleosols spanning the Eocene/Oligocene boundary have been studied in detail is in the Pinnacles area of Badlands National Park, in southwestern South Dakota, U.S.A. (Figure 19.1). These colorful, clayey,

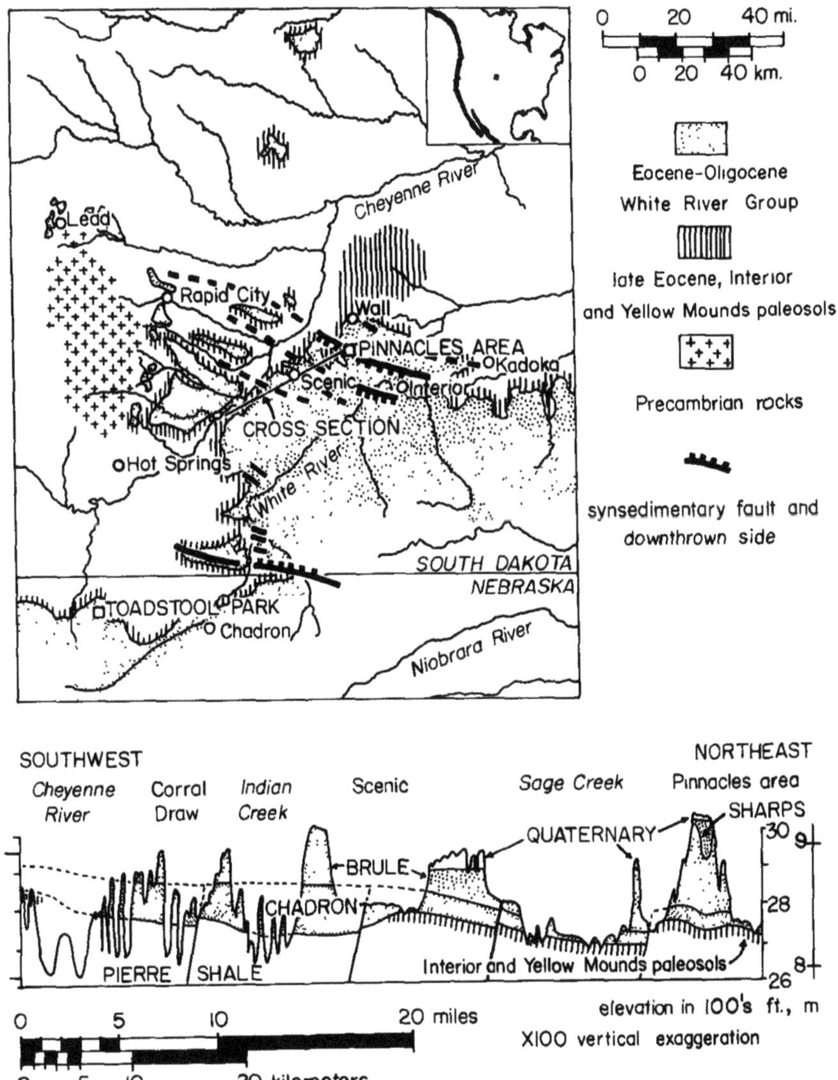

FIGURE 19.1. Map and cross section of Eocene–Oligocene alluvium of the White River Group, of extensive late Eocene Interior and Yellow Mounds paleosols developed in Cretaceous marine sediments, and of Precambrian rocks of the Black Hills uplift, in northeastern Nebraska and southwestern South Dakota (adapted from Retallack, 1988b).

non-marine rocks are exposed in a long ragged wall between high prairie to the north and the valley of the White River to the south. They have long been famous for their great variety of well preserved fossil mammals (Emry et al., 1987) and as the foremost example of badlands weathering (Schumm, 1962). It is now known that their subtle color banding reflects a large number of successive buried soil horizons (Figure 19.2). These paleosols have been described in detail, and used as guides to local Oligocene paleoenvironments (Retallack, 1983a,b), the evolution of ancient grassland ecosystems (Retallack, 1982, 1984b, 1988, 1990), completeness of stratigraphic sections (Retallack, 1984a) and factors in deposition of this volcanic–alluvial sedimentary sequence (Retallack, 1986a). This chapter looks beyond these issues to the record in paleosols of global climate change.

A particularly impressive break in the nature of these buried soils in the Badlands is between

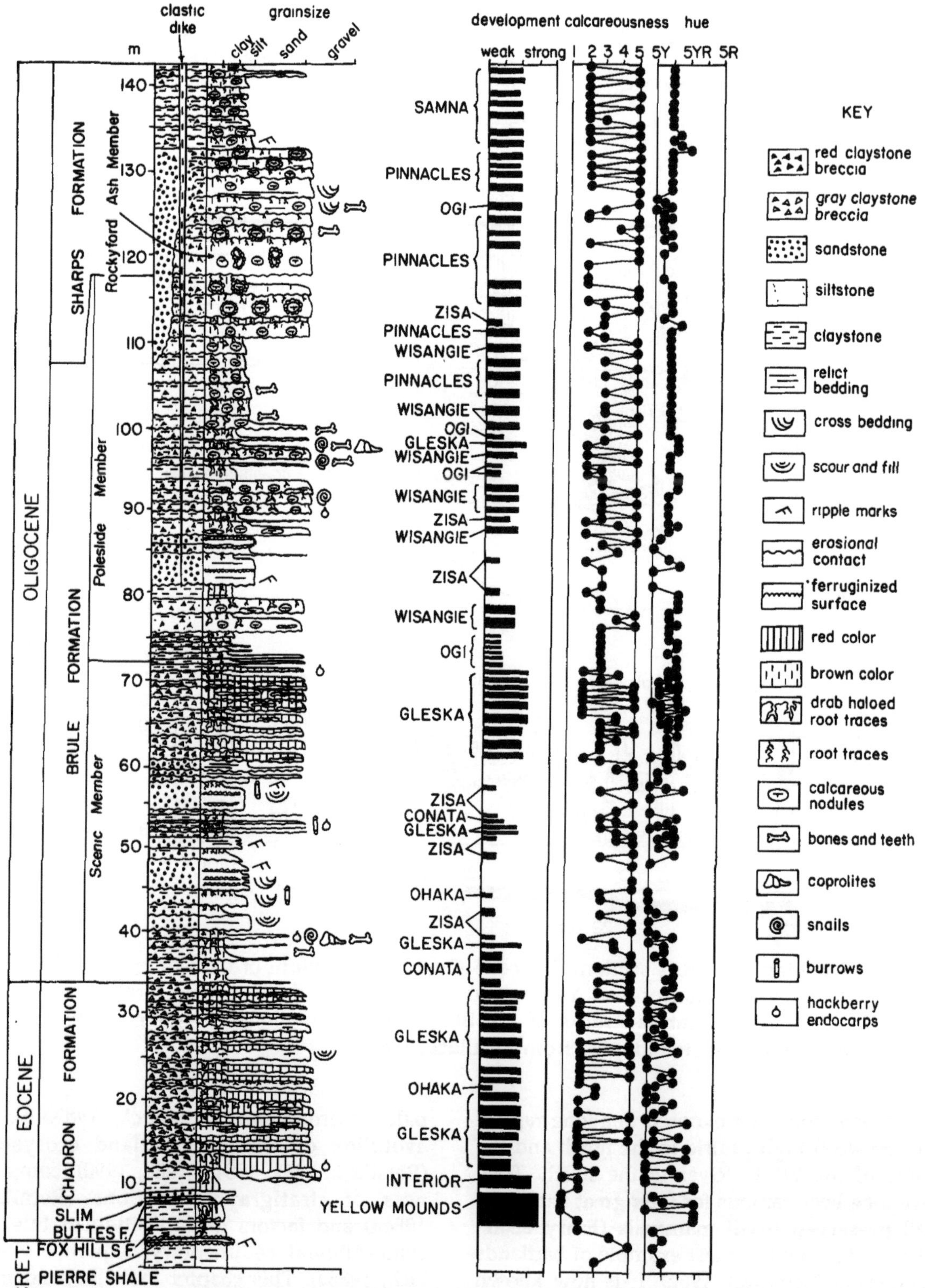

FIGURE 19.2. A measured section of paleosols in the Pinnacles area of Badlands National Park, South Dakota (from Retallack, 1990). Black boxes indicate position of paleosols and their width is their degree of development (in scale of Retallack, 1988a). Calcareousness by acid testing in field (using scale of Retallack, 1988a). Hue from olive to red after Munsell color charts.

the pink and green banded, clayey Chadron Formation and the brown to white, nodular, silty Brule Formation (Figure 19.3). This lithological contact was until recently considered mid–Oligocene in age (Prothero et al., 1983), but reassessment of radiometric dating and magnetostratigraphy of Oligocene continental deposits in the western United States now indicate that it is the local Eocene/Oligocene boundary and some 34 million years old (Swisher and Prothero, 1990). This and other boundaries between rock units in the Badlands are now known to reflect paleoclimatic changes, as revealed by studies of the depth of carbonate nodules, clay content, clay minerals and other features of paleosols in the sequence.

The depths to calcareous nodular horizons and layers within paleosols were recorded in the field during section measuring (Figure 19.4), and show a decline in paleosols higher within the section. Linear regressions of these data were only done within rock units, because of evidence from paleosols (Retallack, 1986a) and magnetostratigraphy (Prothero et al., 1983) for profound disconformities between rock units. The tabular calcareous (petrocalcic) horizons low in the sequence are of a kind that are not necessarily related to climate in surface soils (Kubiena, 1970), but they do continue the relationship of shallowing with time. The relationship also extrapolates back to the oldest paleosol on the unconformity with Cretaceous marine rocks. This paleosol has been leached of fossiliferous marine, calcareous nodules to a depth of 511 cm. For this paleosol and the non–calcareous profile above it, mean annual rainfall was probably in excess of 1000 mm, by comparison with surface soils in the Great Plains (Jenny, 1941; Ruhe, 1984) and elsewhere (Arkley, 1963; de Wit, 1978; Sehgal et al., 1968). The depth of petrocalcic horizons in the late Eocene Chadron Formation should not be trusted as a paleoclimatic indicator, but these paleosols probably received less than 1000 mm and more than 500 mm. For the Scenic Member of the early Oligocene Brule Formation, there also are petrocalcic horizons, indicating mean annual rainfall from about 900 mm to 500 mm. Rainfall then declined to 450 to 500 mm for the Poleslide Member of the Brule Formation and 350 to 450 mm for the Sharps Formation. These estimated ranges are the total range of precipitation found in surface soils of the Great Plains with nodules at comparable depths to the paleosols.

These results do not appear compromised by erosion of the paleosols prior to deposition. Surface horizons with fine root traces and granular ped structure are well preserved in many of the paleosols (Retallack, 1983a, 1990). Nor were these results compromised by compaction. The maximum thickness of the Eocene–Oligocene White River Group is 195 m, and the cumulative maximum thicknesses of other Oligocene and younger formations in this region total 361 m (Martin, 1983). If a marine, non–calcareous shale were buried to a depth of 500 m it would be compacted to about 90% of its former thickness (Baldwin and Butler, 1985). This seems a very unlikely maximal compaction at the base of the Chadron Formation, because these younger sediments did not form a single layer, but filled local paleovalleys created as the nearby Black Hills continued to rise, as has been especially well documented in nearby northwestern Nebraska (Schultz and Stout, 1980; Swinehart et al., 1985).

Another indicator of climatic change is the nature of clay minerals in the paleosols, as determined by x–ray diffraction studies (L.G.

FIGURE 19.3. (next page) The Eocene–Oligocene boundary is currently recognized in the Pinnacles area of Badlands National Park, South Dakota, between the clayey, pink, mound–like paleosols of the Chadron Formation with their titanothere–dominated fauna, and the silty, brown, cliff–forming paleosols and paleochannels of the Brule Formation with their oreodon–dominated fauna: A–C, are Late Cretaceous molluscs, *Hoploscaphites nicolleti* (A), *Tenuipteria fibrosa* (B), and *Discoscaphites cheyennensis* (C); D–F are late Eocene mammals, *Archaeotherium mortoni* (D), *Hyaenodon horridus* (E), and *Brontops robustus* (F); G–L are early Oligocene tortoise (G) and mammals, *Stylemys nebrascensis* (G), *Ischyromys typus* (H), *Merycoidodon culbertsoni* (I), *Hoplophoneus primaevus* (J), *Mesohippus bairdi* (K) and *Metamynodon planifrons* (L); M–P are mid–Oligocene mammals, *Palaeolagus haydeni* (M), *Leptomeryx evansi* (N), *Leptauchenia decora* (O) and *Protoceras celer* (P; from Retallack, 1983b).

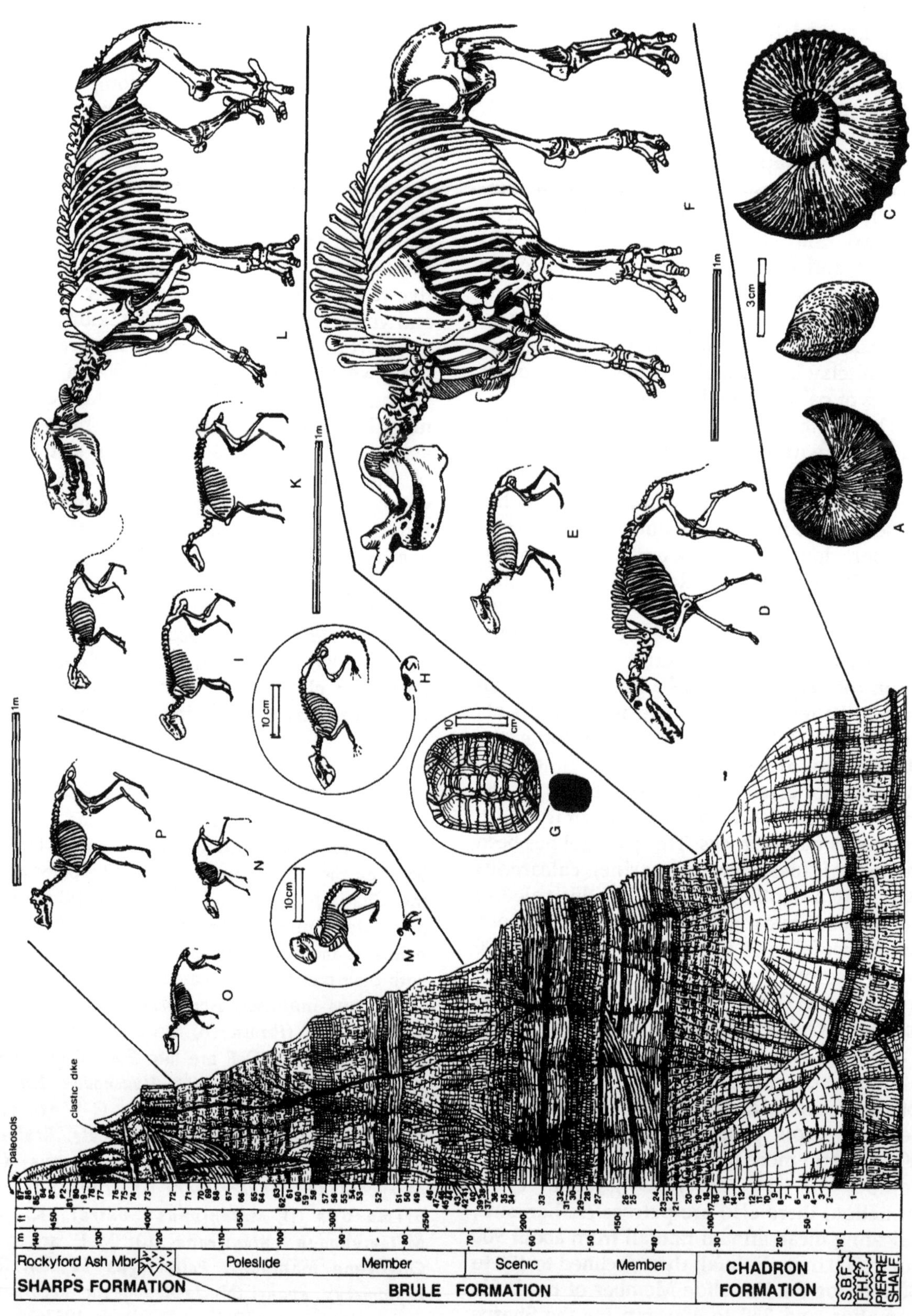
paleosols
clastic dike
Rockyford Ash Mbr
Poleslide Member
Scenic Member
SHARPS FORMATION
BRULE FORMATION
CHADRON FORMATION
S.B.F.
F.H.F.
PIERRE SHALE
m
ft
1m
10 cm
3cm
A
B
C
D
E
F
G
H
I
J
K
L
M
N
O
P

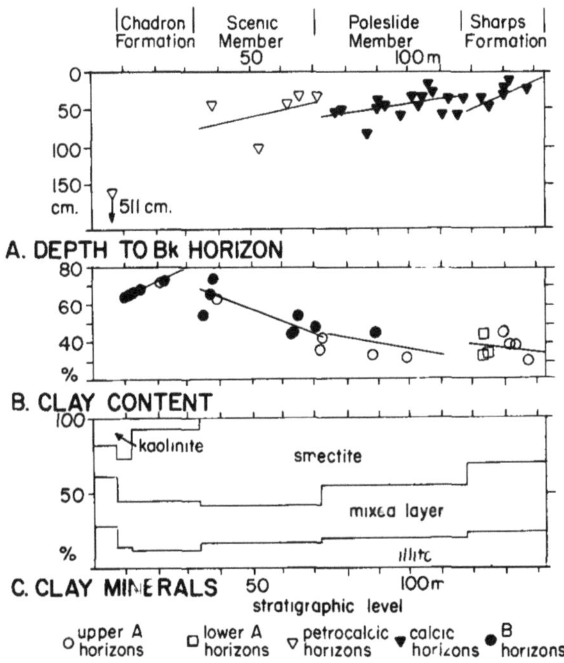

FIGURE 19.4. Indices of stepwise climatic drying in a measured section of paleosols in the Pinnacles area, Badlands National Park, South Dakota (from Retallack, 1990; data from Retallack, 1983a, 1986a): A, depth to calcareous horizons, both nodular (calcic) and continuous (petrocalcic), as measured in the field: B, clay content of paleosol horizons, determined by point counting petrographic thin sections: C, proportion of clay minerals in each recognized rock unit, by averaging many x-ray diffractometer traces for each unit. Correlation coefficients (r) for regression lines in A, are right to left, 0.72, 0.40, 0.41.

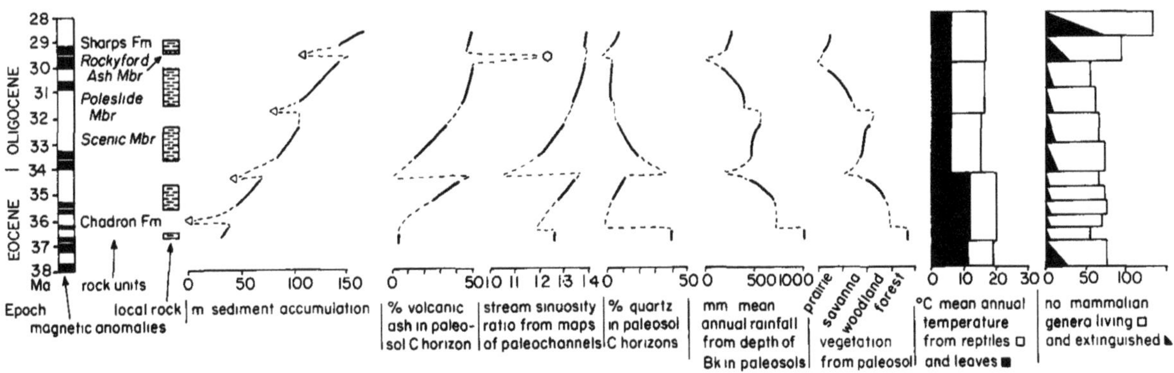

FIGURE 19.5. Estimated controls on sedimentation and erosional episodes in a sequence of paleosols in the Pinnacles area, Badlands National Park, South Dakota (adapted from primary data of Wolfe, 1978; Hutchison, 1982; Retallack, 1986a, Stucky, 1990): changing mammalian generic diversity and extinctions (from taxonomic compilations), mean annual temperature (from reptile diversity and percentage entire-margined leaves), vegetation (from paleosol morphology), rainfall (from depth to paleosol calcareous horizons), stream sinuosity (measured from maps of paleochannels), far travelled alluvium (from point counting for quartz in paleosol C horizons), volcanic ash (from point counting for volcanic shards in paleosol C horizons), and sediment accumulation (from degree of development of paleosols), plotted for current understanding of the Oligocene time scale (Swisher and Prothero, 1990).

Schultz, 1961; Retallack, 1983a, 1986a). The two lowest paleosols in the sequence contain appreciable amounts of kaolinite, as would be expected in soils of humid climate. Paleosols higher in the sequence contain largely smectite, with increasing amounts of illite up-section (Figure 19.4). These less intensely weathered clays were formed in a relatively drier climate during Oligocene time.

These results are not compromised greatly by effects of parent material. The lowest two paleosols developed on smectitic marine shale. The rest of the sequence developed on smectitic alluvium and redeposited soil derived ultimately from rhyolitic airfall ash. This material is of very uniform composition chemically and mineralogically, and relic shards are found throughout the sequence above the basal paleosol (Retallack, 1983a). Nor are these results compromised by illitization of the clay minerals in this sequence. The shallow depth of burial of these deposits is well short of that required for illitization, and the up-profile increase in illite abundance is the reverse of that seen in studies of illitization in boreholes (Weaver, 1989). These results are compromised somewhat by differences in development of the paleosols. The basal two profiles are much better developed than those of the Chadron to Sharps interval, but many of these younger paleosols show comparable degree of development (within the categories of Retallack, 1988a). Stepwise changes in clay composition above the basal paleosols remain as a climatic signal.

Evidence of soil cracking was seen in many of the paleosols in the form of soil clods (peds in the terminology of soil science) surrounded by clay skins (argillans), but deep cracking was uncommon. The zeolite clinoptilolite was common in several paleosols high within the exposed sequence, but this rare associate of evaporitic soils of very dry climates could have formed during burial of the paleosols (Retallack, 1983a). Evidence for very dry and highly seasonal climates was localized to boundaries between the rock units in the Badlands. For example, the Chadron Formation immediately below the Brule Formation is riddled with chalcedony-replaced pseudomorphs of gypsum and chalcedony-filled deep cracks (Honess, 1923; Lawler, 1923; Retallack, 1983a). Both these features have been documented in saline playa lakes and surrounding soils (Eugster, 1969). The prominent clastic dikes that penetrate as deeply as 60 m from the top of the exposed sequence of Sharps Formation (Figure 19.3), are most like the giant desiccation cracks seen in playa lakes and soils of the desert southwestern United States (Retallack, 1983a). A less extreme climatic drying may have created calcareous rhizoconcretions with partial chalcedony filling, rare for the sequence as a whole, within the upper part of the second paleosol above the base of the sequence in the Pinnacles area. Such features are found in desert soils (Retallack, 1983a). Magnetostratigraphic (Prothero et al., 1983) and completeness studies of these paleosol sequences (Retallack, 1984a, 1986a), have shown that the boundaries between recognized rock units are disconformities reflecting periods of erosion at least as long as the periods of net deposition that created each rock unit. Thus some of these periods of dry climate were also times erosion.

Other features of the paleosols such as clay content and hue may also reflect climatic change. Clay content was determined by point counting petrographic thin sections (Figure 19.4), and hue by comparison of freshly excavated rock chips with a Munsell color chart (Figure 19.2). More clayey and red paleosols were formed in humid to subhumid paleoclimates of the lower part of the Badlands sequence, but more silty and brown paleosols in semiarid paleoclimates of the upper part of the sequence.

Paleoclimatic interpretation of clay content and hue of these paleosols must remain imprecise, for a variety of reasons. Estimation of clay content from petrographic thin sections systematically overestimates clay abundance of paleosols and soils (Murphy and Kemp, 1984) and the color of rock chips is not directly comparable with the hue of either wet or dry soils. Both clay content and color reflect paleotemperature, duration of soil formation, and former vegetation of the paleosols, as well as climatic drying (Birkeland, 1984). Indeed there is evidence from fossil floras and reptiles in the western United States for cooling at the Eocene/Oligocene boundary (Wolfe, 1978; Hutchison, 1982). The color of these paleosols has certainly been altered during burial by the subsurface decomposition of most soil organic matter (Stevenson, 1969) and by the dehydration and recrystallization to coarser grainsize of ferric hydroxide minerals (Walker, 1967). Nevertheless, relative changes in clay content within this sequence, which is uniformly altered in this way, may be interpreted as a muted signal of relative paleoclimatic change, in support of other more robust paleoclimatic indicators.

Also broadly compatible with indications of paleoclimatic change are likely changes in vegetation and animals with time in this sequence. The two, thick, red, kaolinitic basal paleosols contain a large, deeply-penetrating root traces and well-differentiated clayey subsurface (Bt) horizons, as in surface soils under forests. Silicified wood has been found at this stratigraphic level (Clark et al., 1967), and it contains growth rings as evidence of some climatic seasonality.

Moderately developed, green-pink, calcareous, clayey paleosols of the late Eocene Chadron Formation also contain abundant, large, drab-haloed root traces and blocky, angular peds. The drab haloes were probably produced by bacterial reduction of organic matter associated with the roots during shallow burial (Retallack, 1990), but their size and distribution, as well as the overall profile form of these paleosols, are of kinds now found under dry woodland vegetation. A few paleosols in the Chadron Formation are thin and gray, with only fine root traces and some relict bedding, as in soils of herbaceous vegetation early in the ecological succession to colonize disturbed ground (Retallack, 1983a). Fossil logs, walnuts (*Juglans siouxensis*) and hackberry (*Celtis hatcheri*) endocarps from the Chadron Formation provide additional evidence of dicot trees (Chaney, 1925; Manchester, 1987), but the rarity and exceptional preservation of these remains limit their usefulness in paleoecological reconstruction (Retallack, 1984a).

Grey-pink paleosols with large root traces also persist in the Brule Formation, though closely associated with paleochannels. Thin paleosols with fine root traces and relict bedding also are found in the Brule Formation, but here they are red as well as gray in color. The distinctive brown, nodular, silty appearance of much of the Brule Formation is due to the appearance of new kinds of paleosols with abundant fine root traces, as well as scattered large ones. These paleosols have granular ped structure, and a subsurface (Bk) horizon of calcareous nodules. These paleosols are most like those of wooded grassland or wooded shrubland in the modern southern Great Plains of North America (Aandahl, 1982). Vegetation during deposition of the Brule Formation has been reconstructed as a mosaic of streamside early successional vegetation and gallery woodland, with more open wooded grassland away from streams. Hackberry endocarps (*Celtis hatcheri*) also are found in the Brule Formation as evidence of trees (Retallack, 1984a,b). Fossil grasses (*Stipidium* sp.) were not found here, but have been recorded as rare fossils from the Brule Formation in nearby Colorado (Galbreath, 1974).

A single paleosol with large, drab-haloed root traces has been found interbedded with a deeply incised sandstone paleochannel in the Sharps Formation at the top of the exposed sequence, but most of the paleosols at this stratigraphic level lack large root traces. They are thin profiles, with abundant fine root traces and granular peds, over a very shallow and moderately developed (Bk) horizon of calcareous nodules, as in dry parts of the modern Great Plains and intermontane west (Aandahl, 1982). By this time open rangelands were probably widespread. No plant fossils of any kind have yet been found from the Sharps Formation.

Stepwise changes are seen in the Badlands also in fossil vertebrate faunas. Studies involving careful sampling of many levels have shown that fossil mammalian faunas and individual mammalian species are relatively uniform from the bottom to the top of the recognized rock units, but there are profound changes in the fossil mammals between units (Clark et al., 1967; Prothero, 1985). Some of these faunal overturns may be related to climatic change. The boundary between the two basal paleosols in the Pinnacles area may correlate with the boundary between the Duchesnean and Chadronian North American land mammal "ages" (about 37 Ma old). This was a time, less significant than the Uintan/Duchesnean boundary (about 40 Ma), when many archaic forest mammals of the Eocene again became extinct and modern groups of mammals became established (Hutchison, 1982; Emry, et al., 1987; Stucky, 1990). The relevance of paleosols for this faunal shift would be better established by their study in localities that contain more abundant fossil mammals of this age than found in the Badlands. The Pinnacles section is however, relevant for interpreting the faunal change between the Chadron and Brule Formation, which corresponds to the boundary of the Chadronian and Orellan North American land mammal ages and the Eocene/Oligocene boundary (about 34 Ma). At this time alligators, many species of turtles, and the large titanotheres became extinct: only one specimen of a titanothere and a single small alligator have been found in the basal Brule Formation (Clark et al., 1967), which has an unusually low diversity of fossil turtles (Hutchison, 1982). Fossil land snails also were much reduced in diversity at this time (Evanoff et al., this volume). All these creatures would have been more dependent than other vertebrates on consistent supplies of water, which became scarce here at the Eocene/Oligocene boundary. In addition, reduced reptilian diversity could be explained by a drop from a mean annual temperature of 20° C to 15° C (Hutchison, 1982). A temperature drop of comparable magnitude at this time can be seen in montane floras of the Rocky Mountains: from 11° C to 6° C (Wolfe, 1978; Hutchison, 1982). Higher within the Brule Formation and within the Sharps Formation a similar assemblage of animals persists, but there are increases in the abundance and diversity of burrowing mammals and decreases in the overall body size of the mammals within the most prominent evolutionary lineages. Both trends may reflect a drier, cooler and less productive ecosystem through time.

Stepwise, episodic, long term, paleoclimatic change is clearly evident in this sequence of paleosols. By controlling the density of vegetation cover and landscape stability, climatic deterioration can also control rates of sediment accumulation, episodic periods of erosional downcutting and fluvial style. A variety of other factors, including tectonic uplift of the source terrane and volcanic input of ash also can determine the timing and nature of alluvial deposition (Retallack, 1986a). Accumulation of these sequences was due in part to an inbalance between tectonic subsidence and sediment influx from volcanic ash and stream flooding (Swinehart et al., 1985). However, exceptional uplift of the source region should introduce far-travelled components to the alluvium such as quartz and unusually voluminous volcanism should be evident from thick deposits of airfall ash unaltered by Oligocene soil formation. There is evidence that each of these occurred during at least one period of erosional downcutting, but the constant association of each erosional episode with evidence for climatic deterioration (Figure 19.5), implicates paleoclimate as an overiding extrinsic control on erosional episodes now dated at 40, 37, 34, 32 and 29.5 million years ago.

PALEOSOLS IN OTHER REGIONS

Paleosols are known now in several other Eocene–Oligocene sequences, and suspected in many more. This brief discussion of other areas for the purposes of illustrating the global scope of changes in paleosols at that time, is limited to areas in which at least a few paleosols have reported in more than cursory fashion.

Paleosols in the Chadron and Brule Formations change in appearance west from the Pinnacles area of Badlands National Park to Sheep Mountain Table in southwestern South Dakota, to Toadstool Park National Monument in northwestern Nebraska, and to the Lusk and Douglas areas of southeastern Wyoming and northeastern Colorado, northwest to Flagstaff Rim Wyoming and north to Slim Buttes, South Dakota (Schultz et al., 1955; Schultz and Stout, 1980; Singler and Picard, 1981; Retallack, personal observations). Proceeding westward, paleosols at the same stratigraphic level become more calcareous, silty and thinner, so that the Chadron Formation near Douglas and Slim Buttes is more like the Brule Formation in the Badlands National Park. A fossil flora found in the Chadron Formation at Flagstaff Rim is dominated by small-leaved plants that indicate dry climatic conditions (Wing, 1987). At any given time it was drier to the west from South Dakota into Wyoming and Colorado, and climatic drying through Eocene and Oligocene time lengthened the rain shadow cast by the Rocky Mountains and their active silicic volcanoes.

Abundant paleosols and fossil mammals of Eocene and Oligocene age also are known in the John Day Country of central Oregon (Retallack, 1981; Smith, 1988; Pratt, 1988). This thick sequence of volcanic and volcaniclastic rocks also has preserved a superb fossil record of fossil plants (Manchester, 1981; Manchester and Meyer, 1987). The thick, red, deeply weathered paleosols of the Eocene Clarno Formation are compatible with evidence from fossil plants for rainforest vegetation and climate (Smith, 1988). Red paleosols also are found in the overlying Oligocene John Day Formation, but are discernably less deeply weathered, and associated with a variety of brown and green, calcareous paleosols generally like those thought to have supported wooded grassland in the Pinnacles area of Badlands National Park. Fossil plants in the John Day Formation are of cool temperate climatic affinities (Wolfe, 1978), but do not show indications of dry climate, perhaps in part because they lived around lakes in whose deposits they are found. Nevertheless, Oregon probably became drier across the Eocene/Oligocene boundary, though remaining wetter than South Dakota at this time.

In southwestern Oregon (Bestland, 1987), northwestern California (Singer and Nkedi-Kizza, 1981) and central Washington (Gresens, 1981), there are thick deeply weathered paleosols of likely late Eocene age, and red paleosols with large root traces of Oligocene age. Much remains to be done on these paleosols and their climatically sensitive features.

In Argentina, grassland paleosols have been identified within volcaniclastic alluvium of Eocene age, or Mustersan in the local mammalian scheme (Spalletti and Mazzoni, 1978; Andreis, 1981). These are the oldest likely grassland paleosols known anywhere, and presumably reflect expansion over Patagonia of a rain shadow from the volcanic Andean Cordillera. Calcareous paleosols have been found in upper Oligocene, or local Deseadean, alluvial rocks in this region, but their paleoclimatic significance has not yet been established, and this sequence is not complete enough to preserve the Eocene/Oligocene boundary (MacFadden et al., 1985).

In the Paris Basin of France, there is a well-preserved late Eocene (probably about 40 Ma), podzolic paleosol (Pomerol, 1964). Similar Orthods now form under oligotrophic forest in humid climates (Retallack, 1990). This paleosol is followed by latest Eocene marine deposits and lacustrine evaporites, which latter indicate significant climatic drying. Oligocene coastal dune deposits of the Paris Basin include thin, humic paleosols (Morand et al., 1968). Such soils of swampy areas are relatively insensitive to climate because of their isolation by groundwater. Late Oligocene to early Miocene paleosols of Aquitaine have both clayey and calcareous (Bt and K) horizons of a kind found in strongly developed soils of dry woodland (Meyer et al., 1980).

In the Fayum Depression of Egypt there is a spectacularly well exposed sequence of paleosols that may span the Eocene/Oligocene boundary (Bown and Kraus, 1988; Rasmussen et al., this volume). This boundary may be better placed within the Jebel Qatrani Formation, rather than at its base as in the past [considering recent recalibration of the time scale by Swisher and Prothero (1990), and radiometric dates of Fleagle et al. (1986); see Rasmussen et al., this volume]. There are differences between the paleosols in the upper and lower part of this formation and those of underlying formations, but none yet documented that were obviously related to paleoclimatic change. There are marine intercalations in this sequence and many of the paleosols may have supported mangal vegetation buffered by marine influence from regional climate.

Near Lake Zaisan in Soviet Kazakhstan, inner Asia, there is a sequence of paleosols and lake beds rich in fossil plants (Akmetiev et al., 1986). Drab-sideritic and pink-mottled Eocene paleosols are overlain by a sequence with calcareous nodular Oligocene paleosols. In nearby Mongolia, there are red and green, nodular upper Eocene paleosols and reddish-brown Oligocene paleosols (Dashvezeg and Devyatkin, 1986).

In southwestern Queensland, Australia, coal measures and sandstones of Paleocene to Eocene age have been deeply weathered by a thick lateritic paleosol of Oligocene age (Senior and Mabbutt, 1979). This change may reflect improved drainage across the Eocene/Oligocene boundary, but this area formerly at high latitude probably remained covered in wet forests.

OTHER LINES OF EVIDENCE FOR PALEOCLIMATIC CHANGE

Eocene and Oligocene paleosols in most parts of the world are little known compared to those of South Dakota, but they do demonstrate a potential for further studies of global climatic change. Although a global data base from paleosols is not yet available, it is instructive to compare the timing of paleoclimatic changes indicated by paleosols in Badlands National Park with those proposed on the basis of other lines of evidence in different parts of the world.

There is evidence of climatic drying from the amounts of eolian quartz in deep sea cores of the Pacific Ocean (Rea et al., 1985). The late Eocene (about 37 Ma) and Eocene–Oligocene transition (about 34 Ma) were times of significantly increased mass accumulation and weight percent of eolian quartz, although there were much more significant influxes of eolian quartz during early Miocene time. The Eocene–Oligocene transition was also a time when the mean grain size of eolian quartz in these cores began to climb to near modern values, which were attained during early Miocene time.

Also in Pacific Ocean cores, oxygen–isotopic studies of foraminifera have shown declines in ocean temperature during the late Eocene (about 37 Ma), followed by a very profound drop at the Eocene/Oligocene boundary (now dated about 34 Ma) and smaller drops during Oligocene time (32 Ma: Miller, this volume). Climatic cooling at similar times has been postulated to explain the declining percentage of entire-margined leaves in fossil floras (Wolfe, 1978) and diversity of turtles and crocodilians (Hutchison, 1982) in the western United States, where the Eocene–Oligocene cooling was much more profound than the other episodes. In northwest Europe, fossil palynofloras show a decline in land temperature at the Eocene/Oligocene boundary (Boulter, 1984). In Europe this was also a time of major extinctions of forest–adapted archaic species (Hartenberger, 1988), and their replacement with a more modern fauna with a size distribution suggestive of dry, open rangeland (Legendre, 1987).

At these critical times, there were also major marine regressions (Haq et al., 1987). The circum–Antarctic current was becoming well established as Australia drifted away from Antarctica and the central Tethys Ocean was much constricted as Africa began to make full contact with Europe. The ability of oceans to absorb and circulate heat was thus dramatically curtailed, as was hydrothermal activity and the generation of carbon dioxide along mid–ocean ridges (Owen and Rea, 1985). Antarctic ice caps presumably began to grow at these times, although they did not reach sea level until late Oligocene time (Kennett, 1982). These were times when the world stepped out of the torrid and more equable paleoclimate of middle Eocene times, closer to the highly zoned, glacial climates of the past two million years.

CONCLUSIONS

Evidence of stepwise climatic deterioration during late Eocene time (40 and 37 Ma), the Eocene/Oligocene boundary (34 Ma) and Oligocene time (32 and 29.5 Ma) is provided by a variety of paleoclimatically–sensitive features of paleosols in a detailed measured section in Badlands National Park, South Dakota. The most informative of these features were the changing proportions of clay minerals and the depth to the calcareous nodular horizons in the paleosols. This latter line of evidence by comparison with surface soils in the North American Great Plains indicates mean annual precipitation of more than 1000 mm for the early part of the late Eocene (local Duchesnean), somewhat drier for the latest Eocene (Chadronian), 500 to 900 mm for the early Oligocene (Orellan), 450 to 500 mm for the mid–Oligocene (Whitneyan) and 350 to 450 mm for the late Oligocene (Arikareean). Deep clastic dikes and pseudomorphs of gypsum at the Eocene/Oligocene and the mid/late Oligocene boundaries are evidence that climate was at times even drier than these estimates.

These rainfall estimates are compatible with evidence from fossil root traces and profile form of the paleosols for late Eocene forest, latest Eocene dry woodland, early Oligocene wooded grassland with gallery woodland and mid–Oligocene open grassland with very few trees along watercourses. Paleoclimatic deterioration at 40 Ma is not well understood in Badlands National Park. Although the other steps at 37, 34, 32 and 29 Ma appear to have been of equal magnitude, the paleoclimatic change with the greatest biological effects was the one at 34 Ma, which heralded the earliest known wooded grasslands in North America.

Climatic drying also could explain the stepwise decrease in clay content and change from red to yellow hue of the paleosols from the base to the top of the section, although climatic cooling and other paleoenvironmental factors may also have played a role in these features of the paleosols.

A limitation on the paleosol sequence in Badlands National Park as a record of paleoclimate is its profound disconformities, separating each of the major rock units. These long periods of non-deposition and erosion may also have been caused by climatic deterioration. Tectonically determined subsidence rates were critical in allowing sediment to accumulate at all, but periodic variations in accumulation rate do not all appear to be due to exceptional tectonic or volcanic activity, but rather to thinner vegetation of dry periods less able to stabilize the landscape against erosion.

Much more could be done to improve the quality of paleoclimatic data from paleosols of Badlands National Park, and to extract similar information from other sequences of paleosols of this age, especially in central Oregon, southern Argentina, Kazakhstan, and Mongolia. These paleoclimatic records are useful additions to those based on fossil reptiles in paleosols, on fossil leaves in lake deposits, on isotopic composition of marine foraminifera in deep sea cores and on amounts of eolian dust in the deep sea. Not only was the stepwise, mid-Tertiary paleoclimatic deterioration a matter of cooling, but also of drying in the interiors of large mid-latitude continents. Both these climatic coolings and desiccations may have been due to a redistribution of water into polar ice caps. Away from the poles, paleoclimatic changes ushered in early grasslands and pronounced modernization of mammalian faunas. Neither the polar ice caps, grasslands nor grazers were as extensive or highly evolved as they are today, but a series of climatic deteriorations around the Eocene/Oligocene boundary were important turning points in their geological history.

ACKNOWLEDGEMENTS

Don Prothero convened several conferences on Eocene/Oligocene boundary events that did much to clarify my thinking, especially on persistent stratigraphic problems. Rob Lander, Jan Hardenbol and Margaret Collinson offered useful discussion on a variety of related topics. Also helpful was a field trip with Peter Birkeland, Emmett Evanoff and Penny Patterson. Careful reviews were provided by David Fastovsky and Mary Kraus. This work was funded by NSF grant EAR-7900898.

REFERENCES

Aandahl, A.R. 1982. *Soils of the Great Plains*, Lincoln, University of Nebraska Press, 282 pp.

Akhmetiev, M.A., Borisov, B.A., Erofeev, V.S., and Tsekhovsky, Y.G. 1986. The Kiin-Kerish section (U.S.S.R, south-eastern Kazakhstan, Lake Zaisan Basin). In C. Pomerol, ed., *Terminal Eocene Events*, Amsterdam, Elsevier, pp. 141–145.

Allen, J.R.L. 1986. Pedogenic calcretes in the Old Red Sandstone facies (Late Silurian–Early Carboniferous) of the Anglo-Welsh area, southern Britain. In V.P. Wright, ed., *Paleosols: their recognition and interpretation*, Oxford, Blackwell, pp. 58–86.

Andreis, R.R. 1981. *Identificacion e importancia geologica de los paleosuelos*, Porto Alegre, Editora do Universidade Federal do Rio Grande do Sul, 67 pp.

Arkley, R.J. 1963. Calculation of carbonate and water movement in soil from climatic data. *Soil Sci.* 96: 239–248.

Baldwin, B. and Butler, C.O. 1985. Compaction curves. *Bull. Amer. Assoc. Petrol. Geol.* 69: 622–626.

Bestland, E.A. 1987. Volcanic stratigraphy of the Oligocene Colestin Formation in the Siskiyou Pass area of southern Oregon. *Oregon Geology* 49: 79–86.

Birkeland, P.W. 1984. *Soils and geomorphology*, New York, Oxford University Press, 372 pp.

Boulter, M.C. 1984. Palaeobotanical evidence for land-surface temperature in the European Paleogene. In P.J. Brenchley, ed., *Fossils and climate*, Chichester, John Wiley and Sons, 35–47.

Bown, T.M., and Kraus, M.J. 1987. Integration of

channel and floodplain suites in aggrading fluvial systems. I. Developmental sequence and lateral relations of lower Eocene alluvial paleosols, Willwood Formation, Bighorn Basin, Wyoming. *Jour. Sed. Petrol.* 57: 587–601.

Bown, T.M., and Kraus, M.J. 1988. Geology and paleoenvironment of the Oligocene Jebel Qatrani Formation and adjacent rocks, Fayum Depression, Egypt. *Prof. Pap. U.S. Geol. Surv.* 1452: 64 pp.

Chaney, R.W. 1925. Notes on two fossil hackberries from the Tertiary of the western United States. *Publ. Carnegie Inst. Wash.* 349: 51–56.

Clark, J., Beerbower, J.R., and Kietzke, K.K. 1967. Oligocene sedimentation in the Big Badlands of South Dakota. *Fieldiana Geol. Mem.* 5: 158 pp.

Dan, J., and Yaalon, D.H. 1982. Automorphic saline soils in Israel. In D.H. Yaalon, ed., *Aridic soils and geomorphic processes, Catena Suppl.* 1: 103–115.

Dashvezeg, D. and Devyatkin, E.V. 1986. Eocene–Oligocene boundary in Mongolia. In C. Pomerol and I. Premoli-Silva, eds., *Terminal Eocene Events*, Amsterdam, Elsevier, pp. 153–157.

de Wit, H.A. 1978. *Soils and grassland types on the Serengeti Plains, Tanzania: their distribution and interrelationships*, Wageningen, Agricultural Institute, 260 pp.

Dokuchaev, V.V. 1883. Russian chernozem (Russkii chernozem). In N. Kaner, transl., *Collected works (Socheniye), v.1*, Jerusalem, Israel Program for Scientific Translations (1967), 419 pp.

Eberl, D.D., Srodon, J., Kralik, M., Taylor, B.E. and Peterman, Z.E. 1990. Ostwald ripening of clays and metamorphic minerals. *Science* 248: 474–477.

Emry, R.J., Bjork, P.R. and Russell, L.S. 1987. The Chadronian, Orellan and Whitneyan land-mammal ages. In M.O. Woodburne, ed., *Cenozoic Mammals of North America*, Berkeley, University of California Press, pp. 118–152.

Eugster, H.P. 1969. Inorganic bedded cherts from the Magadi area, Kenya. *Contrib. Mineral. Petrol.* 22: 1–31.

Fastovsky, D.E., and McSweeney, K. 1987. Paleosols spanning the Cretaceous–Paleogene transition, eastern Montana and western North Dakota. *Bull. Geol. Soc. Amer.* 99: 66–77.

Fleagle, J.G., Bown, T.M., Obradovich, J.D., and Simons, E.L. 1986. Age of the earliest African anthropoids. *Science* 234: 1247–1249.

Galbreath, E.C. 1974. Stipid grass "seeds" from the Oligocene and Miocene deposits of northeastern Colorado. *Trans. Illinois Acad. Sci.* 67: 366–368.

Glinka, K.D. 1931. *Treatise on soil science (Pochvovedeniye)*, transl. A. Gourevitch, Jerusalem, Israel Program for Scientific Translation (1963), 674 pp.

Gresens, R.L. 1981. Extension of the Telluride erosion surface to Washington State and its regional and tectonic significance. *Tectonophysics* 79: 145–164.

Haq, B.U., Hardenbol, J., and Vail, P.R. 1987. Chronology of fluctuating sea levels since the Triassic. *Science* 245: 1156–1167.

Hartenberger, J.–L. 1988. Etudes sur la longevite des genres mammiferes fossils du Paleogene d'Europe. *C.R. Acad. Sci. Paris* 306: 1197–1204.

Honess, P. 1923. Some interesting chalcedony pseudomorphs from the Big Badlands, South Dakota. *Amer. Jour. Sci.* 205: 173–174.

Hutchison, J.H. 1982. Turtle, crocodilian and champsosaur diversity changes in the Cenozoic of the north–central region of the western United States. *Palaeogeogr., Palaeoclimat., Palaeoecol.* 37: 149–164.

Jenny, H. 1941. *Factors of soil formation*, New York, McGraw–Hill, 281 pp.

Kennett, J.P. 1982. *Marine geology*, Englewood Cliffs, Prentice–Hall, 813 pp.

Kubiena, W.L. 1970. *Micromorphological features of soil geography*, New Brunswick, Rutgers University Press, 254 pp.

Lawler, T.B. 1923. On the occurrence of sandstone dikes and chalcedony veins in the White River Oligocene. *Amer. Jour. Sci.* 205: 160–172.

Legendre, S. 1986. Analysis of mammalian communities from the late Eocene and Oligocene of southern France. *Palaeovertebrata* 16: 191–212.

MacFadden, B.J., Campbell, K.E., Cifelli, R.L., Siles, O., Johnson, N.M., Naeser, C.W., and Zeitler, P.K. 1985. Magnetic polarity stratigraphy and mammalian faunas of the

Deseadean (late Oligocene–early Miocene) Salla Beds of northern Bolivia. *Jour. Geol.* 93: 223–250.

Manchester, S.R. 1981. Fossil plants of the Eocene Clarno nut beds. *Oregon Geol.* 43: 75–81.

Manchester, S.R. 1987. The fossil history of the Juglandaceae. *Monog. Syst. Bot. Missouri Bot. Gard.* 21: 137 pp.

Manchester, S.R. and Meyer, H.W. 1987. Oligocene fossil plants of the John Day Formation, Fossil, Oregon. *Oregon Geol.* 49: 115–127.

Martin, J.E. 1983. Composite stratigraphic section of the Tertiary deposits of western South Dakota. *Dakoterra* 2(1): 1–8.

Meyer, R., Guillet, B., Burtin, G., and Montanari, R. 1980. Facies differencies d'origine pedologique dans la molasse Oligo–Miocene d'Aquitaine centrale. *Bull. Sci. Geol. Strasbourg* 33: 67–80.

Morand, F., Riveline–Bauer, J., and Trichet, J. 1968. Etudes sedimentologiques, paleogeographiques et geomorphologiques de la Butte Chaumont (Champlan, Essone). *Bull. Soc. Geol.France* 7(10): 627–638.

Murphy, C.P. and Kemp, R.A. 1984. The overestimation of clay and underestimation of pores in soil thin sections. *Jour. Soil Sci.* 35: 481–495.

Nesbitt, H.W., and Young, G.M. 1989. Formation and diagenesis of weathering profiles. *Jour. Geol.* 97: 129–147.

Owen, R.M. and Rea, D.K. 1985. Sea–floor hydrothermal activity links climate to tectonics. *Science* 227: 166–169.

Paton, T.R. 1974. Origin and terminology for gilgai in Australia. *Geoderma* 11: 221–242.

Pettyjohn, W.A. 1966. Eocene paleosols in the northern Great Plains. *Prof. Pap. U.S. Geol. Surv.* 550C: 61–65.

Pomerol, C. 1964. Decouverte de paleosols de type podzol au sommet de l'Auversien (Bartonien inf)rieur) de Moisselles (Sein–et–Oise). *C.R. Acad. Sci. Paris* 258: 974–976.

Pratt, J.A. 1988. Paleoenvironments of the Eocene/Oligocene Hancock mammal quarry, upper Clarno Formation, Oregon. Unpublished MSc thesis, University of Oregon, Eugene, 104 pp.

Prothero, D.R. 1985. North American mammalian diversity and Eocene–Oligocene extinctions. *Paleobiology* 11: 389–405.

Prothero, D.R., Denham, C.R., and Farmer, H.G. 1983. Magnetostratigraphy of the White River Group and its implications for Oligocene geochronology. *Palaeogeogr., Palaeoclimat., Palaeoecol.* 42: 151–166.

Rea, D.K., Leinen, M., and Janacek, T.R. 1985. Geologic approach to the long–term history of atmospheric circulation. *Science* 227: 721–725.

Retallack, G.J. 1981. Preliminary observations on fossil soils in the Clarno Formation (Eocene to early Oligocene), near Clarno, Oregon. *Oregon Geol.* 43: 147–150.

Retallack, G.J. 1982. Paleopedological perspectives on the development of grasslands during the Tertiary. In B. Mamet and M.J. Copland, eds., *Proceedings of the 3rd North American Paleontological Convention,* v.2, Toronto, Business and Economic Services, 417–421.

Retallack, G.J. 1983a. Late Eocene and Oligocene paleosols from Badlands National Park, South Dakota. *Geol. Soc. Amer. Spec. Pap.* 193: 82 pp.

Retallack, G.J. 1983b. A paleopedological approach to the interpretation of terrestrial sedimentary rocks: the mid–Tertiary fossil soils of Badlands National Park, South Dakota. *Bull. Geol. Soc. Amer.* 94: 823–840.

Retallack, G.J. 1984a. Completeness of the rock and fossil record: some estimates using fossil soils. *Paleobiology* 10: 59–78.

Retallack, G.J. 1984b. Trace fossils of burrowing beetles and bees in an Oligocene paleosol, Badlands National Park, South Dakota. *Jour. Paleont.* 58: 571–592.

Retallack, G.J. 1985. Fossil soils as grounds for interpreting the advent of large plants and animals on land. *Phil. Trans. R. Soc. Lond.* B309: 105–142.

Retallack, G.J. 1986a. Fossil soils as grounds for interpreting the long term controls on ancient rivers. *Jour. Sed. Petrol.* 56: 1–18.

Retallack, G.J. 1986b. Reappraisal of a 2200–Ma–old paleosol from near Waterval Onder, South Africa. *Precambrian Res.* 32: 195–232.

Retallack, G.J. 1988a. Field recognition of paleosols. In J. Reinhardt and W.R. Sigleo, eds., *Paleosols and weathering through geologic time: techniques and applications. Geol. Soc. Amer. Spec. Pap.* 216: 1–20.

Retallack, G.J. 1988b. Down to earth approaches to vertebrate paleontology. *Palaios* 3: 335–344.

Retallack, G.J. 1990. *Soils of the past: an introduction to paleopedology*, London, Unwin–Hyman, 520 pp.

Retallack, G.J. 1991. *Miocene paleosols and ape habitats of Pakistan and Kenya*, New York, Oxford University Press, 346 pp.

Ruhe, R.V. 1984. Soil–climate system across the prairies in mid–western U.S.A. *Geoderma* 34: 201–219.

Schultz, C.B., and Stout, T.M. 1980. Ancient soils and climatic changes in the central Great Plains. *Trans. Nebraska Acad. Sci.* 8: 187–205.

Schultz, C.B., Tanner, L.G. and Harvey, C. 1955. Paleosols of the Oligocene of Nebraska. *Bull. Nebraska State Museum* 4(1): 1–15.

Schultz, L.G. 1961. Preliminary account on the geology and mineralogy of clays on the Pine Ridge Indian Reservation, South Dakota. *Open File Rept. U.S. Geol. Surv.* 61–153: 61 pp.

Schumm, S.A. 1975. Erosion in miniature pediments in Badlands National Monument, South Dakota. *Bull. Geol. Soc. Amer.* 73: 718–724.

Sehgal, J.L., Sys, C. and Bhumbla, D.R. 1968. A climatic soil sequence from the Thar Desert to the Himalayan Mountains in Punjab (India). *Pedologie* 28: 351–373.

Senior, B.R. and Mabbutt, J.A. 1979. A proposed method of defining deeply weathered rock units based on regional mapping in southwest Queensland. *Jour. Geol. Soc. Aust.* 26: 237–254.

Singer, M.J., and Nkedi-Kizza, P. 1980. Properties and history of an exhumed Oxisol in California. *Jour. Soil Sci. Soc. Amer.* 44: 587–590.

Singler, C.R. and Picard, M.D. 1981. Paleosols in the Oligocene of northwest Nebraska. *Contrib. Geol. Univ. Wyoming* 20: 57–68.

Smith, G.S. 1988. Palaeoenvironmental reconstruction of Eocene fossil soils from the Clarno Formation in eastern Oregon. Unpublished MSc thesis, University of Oregon, Eugene, 167 pp.

Soil Survey Staff. 1975. Soil taxonomy. *Hdbk. U.S. Dept. Agric.* 436: 754 pp.

Spalletti, L.A. and Mazzoni, M.M. 1978. Sedimentologia del Grupo Sarmiento en el perfil ubicado al sudeste del Lago Colhue Huapi, Provincia de Chubut. *Obra Centen. Mus. La Plata* 4: 261–283.

Stevenson, F.J. 1969. Pedohumus: accumulation and diagenesis during the Quaternary. *Soil Sci.* 107: 470–479.

Stucky, R.K. 1990. Evolution of land mammal diversity in North America during the Cenozoic. In H.H. Genoways, ed., *Current Mammalogy*, 2: 375–432.

Swinehart, J.B., Souders, V.L., DeGraw, H.M., and Diffendahl, R.F. 1985. Cenozoic paleogeography of western Nebraska. In R.M. Flores and S.S. Kaplan, eds., *Cenozoic paleogeography of the west–central United States*, Denver, Rocky Mountain Section of the Society of Economic Paleontologists and Mineralogists, 209–229.

Swisher, C.C. and Prothero, D.R. 1990. Single crystal $^{40}Ar/^{39}Ar$ dating of the Eocene–Oligocene transition in North America. *Science 249*: 760–762.

Walker, T.R. 1967. Formation of red beds in modern and ancient deserts. *Bull. Geol. Soc. Amer.* 78: 353–368.

Warren, J.K. 1989. *Evaporite sedimentology: importance in hydrocarbon accumulation*, Englewood Cliffs, Prentice–Hall, 285 pp.

Weaver, C.E. 1989. *Clays, muds and shales*, Amsterdam, Elsevier, 819 pp.

Wing, S.L. 1987. Eocene and Oligocene floras and vegetation of the Rocky Mountains. *Ann. Missouri Bot. Gard.* 74: 748–784.

Wolfe, J.A. 1978. A paleobotanical interpretation of Tertiary climates in the northern hemisphere. *Amer. Scientist* 66: 694–703.

20. LOW-BIOMASS VEGETATION IN THE OLIGOCENE?

by Estella B. Leopold, Gengwu Liu, and Scott Clay-Poole

ABSTRACT

In this paper we evaluate evidence of desert scrub, grassland, or savanna vegetation types in key mid-continent areas of the Northern Hemisphere during the mid Cenozoic. Low-biomass (plant weight per hectare) vegetation such as these types are linked with climates having moderate to low rainfall that is seasonally distributed. The pollen rain, though it varies greatly by region within each vegetational type, is in many instances diagnostic, first by the tendancy to have somewhat low arboreal pollen percentages compared to those of forest areas, and second by indicator genera.

Fossil floras from the mid-continent United States clearly indicate a trend toward greater seasonality of rainfall beginning in the late Eocene, and suggest a kind of woody (as opposed to grassy or herbaceous) savanna vegetation. Paleosol evidence strongly suggests the development of arid and seasonally dry climates during the Oligocene of the Great Plains, but the lack of botanical identifications does not tell us what vegetation occurred there. Fossil fruits of grass and prairie herbs become very common regionally during the mid and late Miocene, and suggest a grassy savanna developed about the time that specialized grazing mammals became common too.

In the Oligocene of the USSR, forest vegetation suggested by pollen data, was widespread regionally, except in Kazakhstan (south-central USSR) where pollen data indicate a more open vegetation, perhaps a woodland with abundant ferns and shrubs, with only rare grasses and herbs. Grassy savanna and steppe clearly begin to be important during the Miocene in south-central USSR.

In China an extensive semiarid belt existed across China from the southeast (present) coast northwestward to the Xinjiang Region during the Paleocene through mid-Eocene. A great change of climate and vegetation, possibly related to early uplift in the Xizang (Tibetan) Plateau, occurred between mid and late Eocene when the arid area retreated northwestward; during the Oligocene it became restricted mainly to northwest China and western north China. Based on pollen evidence from inland sites, a low-biomass vegetation occupying this arid area consisted mainly of the shrubs, *Ephedra*, cf. *Nitraria*, and salt bush with either scattered trees or woodlands. Herbs and grasses were rare or absent. By contrast, pollen data from the east coastal area of China show that no low-biomass vegetation developed there during mid Tertiary.

Interior regions of China and North America record the occurrence of a kind of woody savanna in the late Eocene, and in western China this persisted through the Oligocene. Miocene data show grassy savanna becoming prominent in the Great Plains of United States and in southern USSR; so far no convincing botanical evidence indicates grassy savanna before Neogene time in mid-continents of the Northern Hemisphere. Open grasslands probably did not develop until late in the Quaternary.

INTRODUCTION

Compelling evidence from paleosols and fossil root traces in the Oligocene deposits of the South Dakota Badlands documents the spacing of trees through time; indications are that the late Eocene forest stands were characterized by trees spaced rather closely together, while early Oligocene vegetation had widely spaced trees as in a savanna, and trees may have become eliminated in the final phases of the Oligocene by a drying climate. The extensive work of Retallack (1982, 1983, 1988) in the Badlands deposits documents the trend toward arid soil types and toward smaller root diameters during the Oligocene. Such climates and vegetation structures are in strong contrast with those typical of the humid subtropical forests in the middle or early Eocene of the North American mid-continent (MacGinitie, 1974; Dorf, 1960; Hickey, 1977).

The battery of evidence from the Great Plains region of North America strongly suggests that a savanna-like vegetation existed during much of the Oligocene in this part of the midcontinent. The absence of traces suggesting large roots and the presence of small root traces in local paleosols of the final Oligocene phases has indicated to Retallack (1983; this volume) that the region became treeless and may have become grassland.

There is as yet no direct evidence from the Great Plains of North America regarding the botanical makeup of the Oligocene vegetation. Leaf floras have not been found, and fossil fruit and seed floras seem to be limited, recording only hackberry (*Celtis*) at a wide number of sites. Oligocene pollen are apparently poorly preserved, as samples attempted by a number of workers proved sterile. However midcontinent floras of the Old World have been documented and are pertinent to the broader questions: did savanna-like vegetation occur in Asia, and if so, what was its botanical composition? Did grassland occur at the end of the Oligocene?

The ecological relevance of these questions relates first of all to the importance of discerning the nature and seasonality of climate during the late Eocene and Oligocene intervals. But the questions are also critical for discerning what kind of primary production supported the biota of late Paleogene ecosystems.

The purpose of this paper is to evaluate available Oligocene pollen floras of the Northern Hemisphere focusing particularly on evidence from midcontinent areas where savanna or other low-biomass vegetation occurs today. We will compare modern and fossil pollen evidence regarding savanna, grassland, steppe or desert in the mid Cenozoic.

CHARACTERISTICS OF SAVANNA, AND OTHER LOW BIOMASS VEGETATION

Low-biomass or non-forest vegetation typically exists under relatively dry climates with strongly seasonal distribution of rainfall; the total amount of rainfall is generally low, e.g. from about 150 to 5 cm. The severity or length of the dry season which is linked with a diminished annual precipitation, creates a graded series of vegetation types. These range from tropical or subtropical savanna with 4-5 months of drought climate (150-90 cm annual rainfall), temperate grassland with a dry season about 3-4 months long, (100-40 cm annual rainfall), and woody steppe bordering on desert conditions with ca. 5-6 months of water deficits (40-20 cm annual rainfall), and desert scrub with a dry season from about 6 to 9 or 10 months (45 to ~7 cm of precipitation annually). Some examples and their key features are summarized in Table 20.1.

Savanna

The term savanna has been variously used but typically is defined as a tropical or subtropical grassland containing scattered trees and drought-resistant undergrowth (Webster's 7th Dictionary). We use it in the structural sense of Dansereau (1957) to describe a mixed formation of grasses and woody plants in any geographical area, temperate or tropical. The critical factor controlling savanna vegetation is a "dry season occurring in the cooler half of the year" (MacGinitie, 1969, p. 47). In general, Richards (1957) states that grassy savanna with woodland (open forest) occupies the zone between the deciduous forest (mesic extreme) and thorn woodland and desert (xeric extreme). Thorn forest of the Sahel may be a kind of woody savanna, for it is dominated by spiny trees with finely divided leaves as in *Acacia;* grasses or herbaceous ground cover may be absent or discontinuous (Richards, 1957, p. 315). Grassy savanna has been shown to be a water-limited vegetation where competition between the grass-herb ground cover with the trees provides a sort of equilibrium state controlled by soil moisture seasonally through the year (Eagleson and Segarra, 1985).

Tropical savanna soils may be lateritic, deep red in color, while temperate savanna may have soils like those of grasslands. Nodules of carbonates or other salts form in the subsurface when annual precipitation is in the lower ranges (Buol et al., 1980). The range of monthly temperatures may be up to 14 °C.

TABLE 20.1. Defining characteristics for three types of low biomass vegetation (from Richards, 1957; Bonnefille, 1984; Spaulding and Graumlich, 1986; Sarmiento, 1984).

	SAVANNA	GRASSLAND tropical, temperate	WOODY STEPPE	DESERT
Annual Precipitation	90-150 cm	40-100 cm	20-40 cm	2-20 cm
Length of the Dry Season	4-5 months cool season	3-4 months winter	5-6 months summer	5-6 months
Seasonal Range of Temperature, °C.	5-14	20-30	20-30	25-30

Grassland

Technically, grassland only includes vegetation dominated by herbaceous graminoids, but is sometimes loosely applied to any herb-dominated vegetation (Daubenmire, 1978). In central United States, it refers to grass-dominated Great Plains or High Plains, that are seasonally dry, too dry for successful growth of trees. Sauer's (1950) argument that grassland only exists as a result of periodic fire and grazing, may be true at the prairie areas of the forest/grassland ecotone, if not in the grassland region itself. Grassland soils are mollisols (chernozems) with black A horizons. In semi arid areas white bands of caliche are typical in the subsurface, and the thickness of the caliche is inversely proportional to the precipitation/evaporation ratio (Buol et al., 1980). The range of temperature for temperate grasslands is generally greater than for savanna, and is about 20 to 30 °C in central United States.

Steppe

We use the term steppe in the American sense referring to shrub-dominated vegetation, though grasses may be present or even abundant. Europeans have long used the term steppe for extratropical grassland areas where the zonal soils are too dry for trees, but herbaceous perennials and grasses are well represented. (Daubenmire, 1970, 1978). The climatic difference between steppe in this sense and grassland is that the former has a summer-dry climate with relatively low precipitation during the growing season, while most grasslands are summer-moist. The soils of steppe and grassland are notably different. While the prairie has a thick, dark A horizon and occasional calcium carbonate nodules in the subsurface, steppe soils have a poorly developed A horizon and often more highly developed caliche accumulations in the subsurface (Buol et al., 1980).

Desert

In North America deserts range from the high-latitude steppe of the Great Basin to the subtropical scrub of the Sonoran Desert of northwestern Mexico; seasonality of rainfall varies, for the northern deserts receive 25% of their precipitation in the summer, while the southerly desert has a distinct summer monsoon accounting for 40-60% of annual rainfall (Spaulding and Graumlich, 1986). Vegetation occurring below the steppe shrub zone includes shrubs and succulents of desert scrub, xeric woodland or interior chaparral. Perennial grasses are fairly rare; annual plants are abundant in the spring of rainy years. Ground cover is usually limited as bare ground represents from 50 to nearly 100% of the area. Though there are many other kinds of deserts, the North American examples illustrate the common vegetation structures. Usually desert soil profiles are poorly developed (Buol et al., 1980).

TABLE 20.2. Some examples of modern pollen rain, selected types, low biomass vegetation. Percent total pollen and spores in tallies. Data indicate range of values.

	SAVANNA	GRASSLAND		STEPPE	DESERT
	Ethiopia (1)	Ethiopia (1)	Great Plains S. Dak. (2)	Wyoming & Idaho (2,4)	California (3)
Gramineae	20-55%	30-55%	10-48%	1-25%	+
Compositae total			7-75%	10-80%	10-20
Liguliflorae	ca. 2%	-	-	0-3%	
Tubuliflorae	ca. 9%	-	3-30%	2-12%	
Artemisia	-	-	5-70%	8-75%	4-10
Ambrosia	-	-	0-10%	1-3%	3-10
Chenopodiineae					
undiff	0-30%	10-50%	10-70%	10-80%	8-32
Sarcobatus	-	-	0- 2%	0-10%	0- 4
Dodonea	1%	-	-	-	
Legumes undet.	1%	-	-	-	+
Mimosoidae	0.5%	-	-	-	
Acalypha	0-40%	-	-	-	
Combretaceae	0- 5%	-	-	-	
Herbaceous families					
Caryophyllaceae				+	+
Cruciferae				+	+
Labiatae				+	+
Polemoniaceae				+	+
Non Tree Pollen	25-62	22-30	5-25	2-25	70-90
Tree Pollen					
Angiosperms	38-75	18-70	0-5	+	+
Pinaceae	-	-	3-22	2-30	8-18

1. Bonnefille (1984)
2. McAndrews and Wright (1969)
3. Leopold (1967)
4. Leopold and Wright (1985)

Pollen Rain

An important question is to ask is what do pollen rain signals from low-biomass vegetation types look like? At least concerning modern savanna, grassland, steppe, and desert, we may be able to recognize them in the fossil pollen record. As indicated by the few examples given in Table 20.2, savanna and grassland tend to have high percentages of non-tree pollen. The chief non-tree pollen types are grasses (20-60% but typically above 25%), or salt bushes (Chenopodiineae, 0-80%), asters (Compositae , 0-17%). Other special drought-resistant shrubs appear in the tropical savanna (legumes, Combretaceae, *Dodonea*). Compared to these, steppe from the High Plains tends to have more tree pollen than non-tree pollen (2-25% NAP), perhaps because with the low productivity of the region, drift of tree pollen from the mountains may be relatively important (McAndrews and Wright, 1929). Steppe of the Wyoming basins tends to be

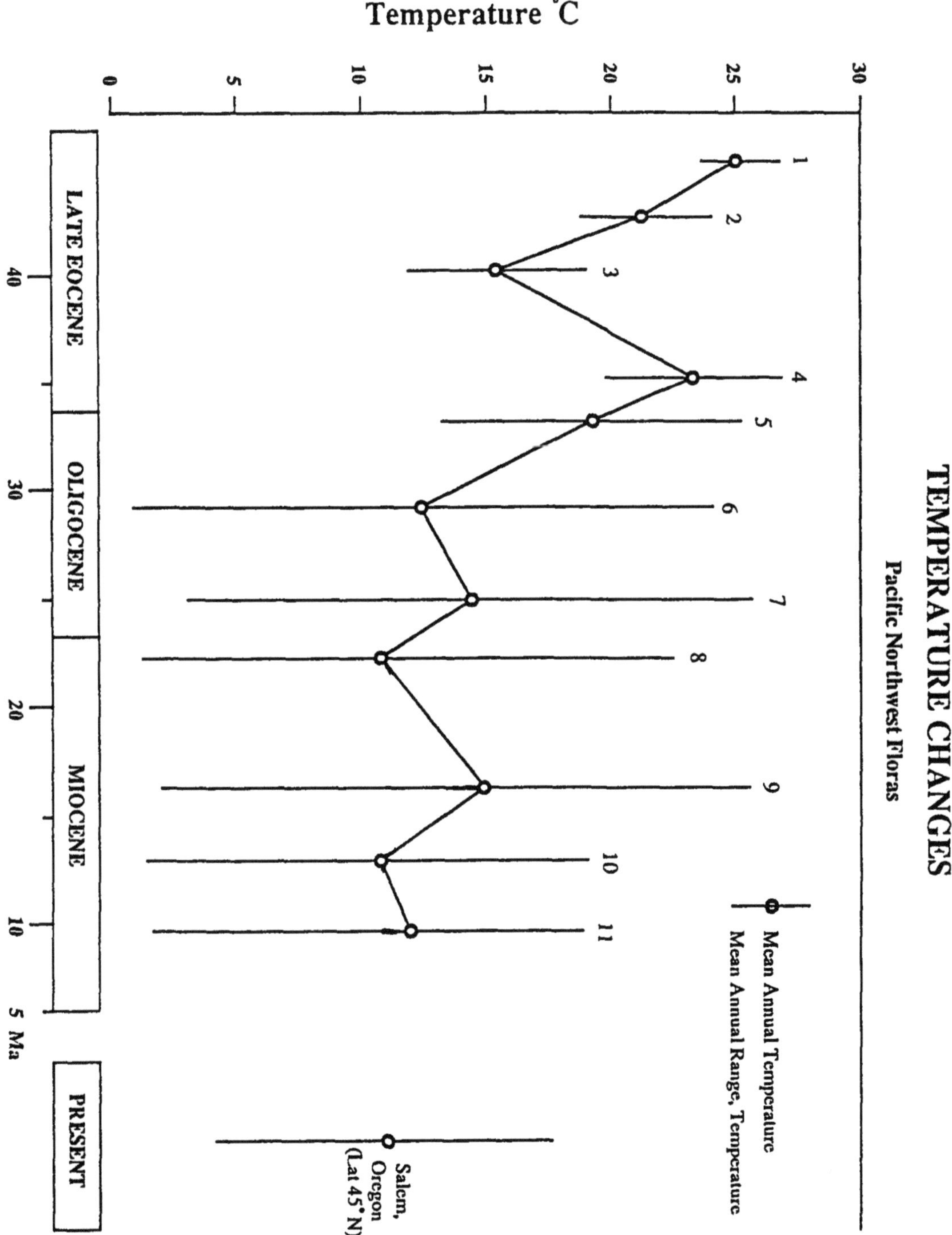

FIGURE 20.1. Estimated average annual temperatures based on Cenozoic floras of western Washington and western Oregon (Wolfe, 1978,1981; this volume) are shown by circles. Estimated range of temperatures (warmest month minus coldest month values) for the same floras shown by vertical bars; plot assumes the range is symmetrical around the mean. (see text). Localities 1 - 5 are from Wolfe's Puget Group sites. Localities 6 through 11 are near Salem, Oregon : 6. Bridge Creek, Willamette; 7. Yaquina; 8 Eagle Creek; 9, Latah and other sites; 10. Faraday; 11. Troutdale.

characterized by large amounts of Compositae pollen, especially that of sagebrush (*Artemisia*). Xeric shrubs, such as buffalo berry (*Shepherdia/Elaeagnus*), greasewood (*Sarcobatus*), and pollen of herbaceous groups such as the buckwheats (Polygonaceae), pinks (Caryophyllaceae), phlox (Polemoniaceae), mustards (Cruciferae), and mints (Labiatae) typically are present in trace amounts.

Pollen rain in areas of desert scrub is rather similar to that in steppe, except the types of shrubs are even more drought resistant. While the steppe is typically rich in sagebrush, desert scrub may feature fairly large amounts of *Ambrosia*, saltbush including greasewood, and other types of Compositae pollen. Perennial grasses are uncommon, but adventive annual grasses such as cheatgrass that are now common seem to contribute little to the pollen rain as their flowers are cleistogamous.

FOSSIL RECORDS

Our summary of available pollen data to infer Cenozoic vegetation types focuses on midcontinental areas in order to seek records of low-biomass vegetation. For western North America we include fossil sites from areas periferal to the Great Plains. From Asia we summarize selected pollen records from USSR and China.

Western United States

Based on local floras from the Pacific Northwest, Wolfe (1981; this volume) has shown that the single largest climatic shift of the Cenozoic took place during the late Eocene and early Oligocene, about 33 Ma. During that interval, he estimates that at mid latitudes the average annual temperature dropped ~12°C while the range of seasonal temperature increased from an equitable 7-8°C to a highly continental range of ~25°C. In Figure 20.1 the circles connected by the heavy line trace the average annual temperature, and the vertical bars assume that the range is symmetrically distributed around the mean. Actually, since the early Oligocene winter temperatures may have warmed more than the summer temperatures dropped (Wolfe, 1981, p.95), our chart is not quite representative in that respect. During the Miocene the average annual temperature (which was cooler than in the Oligocene) may have been similar to that of the present at Salem, Oregon, but the seasonal extremes continued to be greater than now.

Late Eocene

One would expect that if such changes had important impacts on the coastal climate and vegetation during mid Cenozoic (as described by Wolfe, this volume), that the effects on the midcontinent areas now characterized by highly seasonal climates would have been greater. In the Rocky Mountain area Wyoming and Colorado floras confirm that expectation, as they show a major drop in annual precipitation and a shift toward more open woodland and savanna (Figure 20.2). The trend in Figure 20.2 begins with a subtropical mid-Eocene environment indicated by broadleaved evergreen and deciduous forest in Wyoming; the pollen and leaf evidence from the Bridgerian Kisinger flora (~48 Ma ; MacGinitie, 1974, and other sites) connote a moist climate with wet summers and dry winters. An interval of "savanna woodland" followed in the early late Eocene represented by the Green River flora (~42 Ma) that MacGinitie (1969) described as representing an open-structured woody savanna, under a dry subtropical climate with highly seasonal rainfall (below 75 cm/yr). The botanical evidence for this was the wide number of identified woody plants having subhumid adaptations, including *Celtis*, *Ephedra* (pollen) *Mahonia*, *Astronium*, certain legumes and others of his "group 2" (MacGinitie, 1969, p.50). Grass pollen and fruits are present but they are extremely rare.

The next phase brought an impoverished warm-temperate woody flora and a vegetation that was also savanna like or woodland. At the end of a long sequence of Eocene pollen from the Washakie Basin of southern Wyoming, the Uinta B and C pollen show that a great impoverishment of the rich mid Eocene flora had occurred (Leopold and MacGinitie, 1972). Pine pollen is common to abundant with occasional hardwood pollen, e.g. *Platanus*, *Pterocarya*, *Caesalpinia*, *Eucommia* , oak and elm types. Fern spores, *Lygodium* and polypod types, are occasionally very abundant. These and small increases in diverse and sometimes abundant Rosaceae and occasional pollen of shrubs such as *Ephedra* suggest a more open landscape. Several cordilleran floras that are mainly

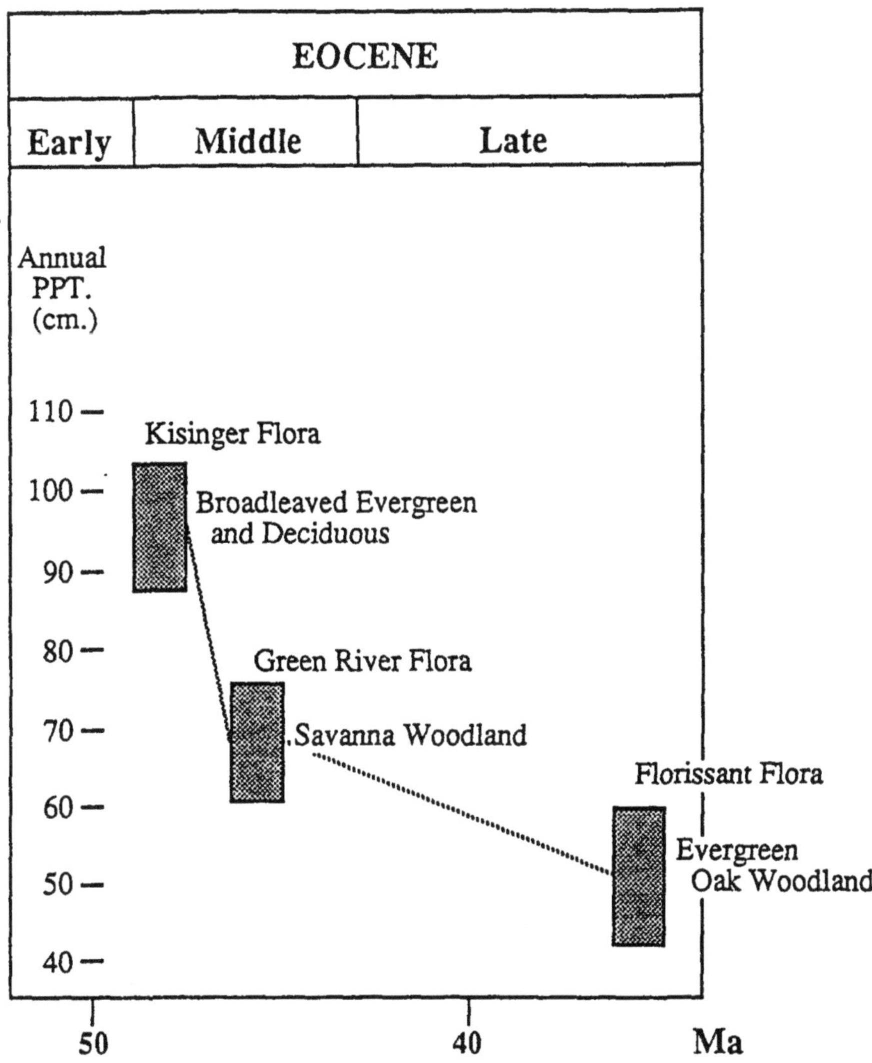

FIGURE 20.2. Trend of annual precipitation estimated from Rocky Mountain floras of middle and late Eocene age (after MacGinitie, 1953, 1969, 1974; Becker,1960)

coniferous and of temperate aspect, are recorded in the central Rocky Mountains during the Uintan, though farther south in the lowlands of New Mexico, a modified Green River pollen flora of subtropical character suggests a much warmer climate that may have been seasonally dry (Leopold and MacGinitie, 1972)

Late Eocene vegetation of Chadronian age is recorded at several localities and suggest open woodland and temperate savanna in the Rocky Mountain region. The Florissant lake beds (~34 Ma) from central Colorado, where the flora, rich in woody xerophytes, portends a seasonally dry savanna-type climate. The area of western Texas and northeastern Mexico where some related species live, is the nearest modern climatic analogue for the Florissant flora (MacGinitie, 1953, p. 57). The pollen flora at Florissant is dominated by a wide mixture of wind-pollinated plants (Figure 20.3) and supports MacGinitie's inferred environments: a lowland riparian and lakeside environment in which *Sequoia*, Juglandaceae and Ulmaceae dominated, and dry upper hillslopes on which grew open stands of "evergreen oak and pine woodland" (Leopold and MacGinitie, 1972). Precipitation is thought to have been below 60 cm/yr (Figure 20.2). We found fossil pollen assemblages very similar to the upland complexes from Florissant in West Yellowstone, Montana, and in the Chadronian White River Group of Wyoming (Hamilton and Leopold, 1962).

The pollen rain of the late Eocene Chadronian and middle Eocene Uintan sites in the Rocky Mountains is not comparable to that of any modern low-biomass vegetation we know. Riparian and conifer tree pollen counts are moderately high; herbs, including grasses, are virtually lacking except for the aquatics such as *Sparganium*. Modern desert or steppe genera are largely absent, though many shrubby groups are present--the Rosaceae, *Fremontodendron*, *Sarcobatus* *Ilex*, and *Ephedra* (1-4%). The pollen evidence suggests an odd kind of woody savanna lacking many modern genera that characterize temperate or subtropical savanna of the Texas and northwestern Mexico analogues. The Chadronian savanna was especially novel because herbaceous elements are virtually absent or undiverse and rare.

The shift toward more temperate floras indicates a cooling, but by the late Uintan or the Chadronian the range of temperatures was not extreme. Only scattered pollen grains in occasional samples of Oligocene sediments (Orellan, Whitneyan) have been found in the mid continent, though many workers have repeatedly attempted to prepare these sediments for pollen analysis. Consequently we have no pollen data for this interval from the central United States.

Based on Retallack's (1982, 1983, 1988) paleosol data, including caliche deposits, seasonally dry climates became fairly arid toward the end of the Oligocene in South Dakota. Field evidence from haloed root traces suggests widely spaced tree bases that may have represented an open woody vegetation, savanna-like in character. Without plant identifications for the Orellan and Whitneyan, the prediction that grassy savanna and finally grassland occupied the interfluves cannot very well be ascertained. The records of other mid continent areas discussed later will be of interest because of their richer fossil records for the Oligocene.

<u>Miocene</u>

The late Oligocene and Miocene deposits of the Arikaree and Ogalalla groups on the High Plains over a five-state region are chiefly volcanic ashfalls, and the Miocene sediments contain a rich paleobotanical record pertinent to our discussion of low-biomass vegetation (see reviews by Leopold and Denton, 1987; Axelrod, 1985). Pollen from a key leaf locality indicate the development of grassy savanna by 15 Ma at Valentine, Nebraska (Kilgore locality; MacGinitie, 1962). Fossil grass fruits and seeds documented in the major work of Elias (1942) and Thomasson (1980) demonstrate the following sequence for the High Plains:

a) an absence of grass remains in late Oligocene sediments,
b) appearance of rare fossils and a few undiverse extinct genera of grasses in the early Miocene;
c) abundant fossils in the middle and late Miocene representing many species of extinct grass genera with first appearances of extant genera of grasses. Prairie herbs which begin in this fossil record also become more diverse toward the end of the Miocene (e.g. members of Boraginaceae).

Clearly this taxonomic radiation of the grasses (Stipidae and Poidae groups) was accompanied by increasing abundance of grass fruits in the fossil record over a five-state region on the Plains. Together with the evidence from fossil leaves and pollen, the paleobotanical record points toward the development of grassy savanna or woodland with a mosaic of local prairies in the mid and late Miocene (Leopold and Denton, 1987; Axelrod 1985). Climatic indications from leaf floras indicate a lowering of annual precipitation under a temperate climate during the late Miocene from ~15-5 Ma. We deduce

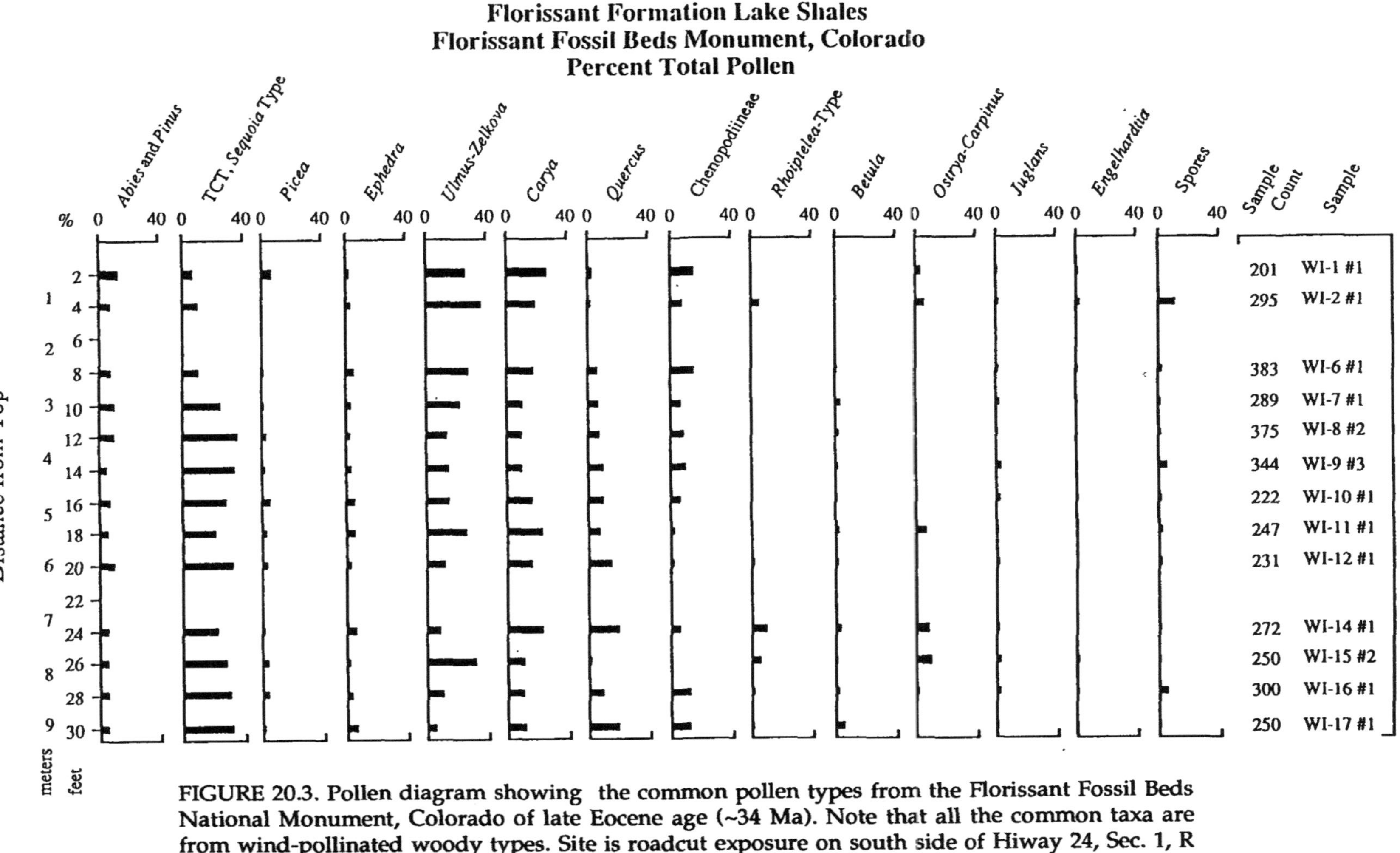

FIGURE 20.3. Pollen diagram showing the common pollen types from the Florissant Fossil Beds National Monument, Colorado of late Eocene age (~34 Ma). Note that all the common taxa are from wind-pollinated woody types. Site is roadcut exposure on south side of Hiway 24, Sec. 1, R 71W, T 13 S; locality 5 of E. Leopold.

that some summer precipitation must have been present in order to explain the abundance of grasses and broad leaved trees.

Oligocene Pollen Records of the USSR

In pursuit of our purpose to examine Oligocene pollen evidence from mid-continental areas, it is critical that we study the patterns of mid-Cenozoic vegetation inferred from central Asia. Summary vegetation maps of USSR during the various phases of the Tertiary are provided by Vinodradov et al. (1967), who also indicate outlines of the continent and paleogeography features including relief. For the central and northeastern parts of Asia, Pokrovskaya (1956a) edited a classic Oligocene compendium with many chapters on pollen floras by a wide number of palynologists. This volume contains profuse illustrations (drawings) of the pollen taxa, descriptions of the floras and vegetation, master tables showing the range of percentage values for pollen-spore composition, and maps indicating proposed distribution of vegetation types for early and late Oligocene. A similar compendium gives an analysis of the Miocene deposits in the USSR (Prokovskaya, 1956b) reviewed by Leopold (1969).

As the map in Figure 20.4 portrays, the outlines of the Asian continent were greatly altered by the Tethys Seaway that connected the Mediterranean with the Black and Caspian Seas and Sea of Aral. The Oligocene Asian continent was turned so that western Europe lay in a southerly position relative to the eastern or Pacific coast. The 60° latitude crossed through the middle of Asia and through the northern part of Japan, putting European Russia farthest south.

Most of eastern Siberia (east of 90° E. longitude) was a highland plateau (except the Lena River valley), and the Urals were low mountains. West Siberia, i.e. the entire Ob River basin, was a saline depression as was the southern Kazakhstan subprovince, but a highland and low mountains (northern Kazakhstan) lay in between. The Pacific coast was rimmed with low hills and mountains. The north and eastern seashore outlines of Asia were about in their present position.

The eastern European portion of USSR is mapped as broadleaved forest with a mixture of conifers and a very significant portions of evergreen subtropical plants (Figure 20.4; Vinogradov, 1967). Pokrovskaya (1956a) showed that there were widespread cypress (*Taxodium*) swamps along the coasts during the Oligocene, and in the late Oligocene conifer forest grew on the highlands west of Kiev and in the Caucasas highlands. In the north, forests of the eastern European area were rich in umbrella pines (*Sciadopitys*) in the early Oligocene.

Central USSR

During the early Oligocene the interfluves of the Ob River basin in the Urals/West Siberian province (area 4, Figure 20.4) were broadleaved hardwood and pine forests. The northern part of the Ob River basin was extensive cypress (*Taxodium*) swamp during the late Oligocene, while the southern part of the drainage was broadleaved forest with subtropical and possibly tropical elements. Oak (*Quercus*), beech (*Fagus*), and chestnut (*Castanea*), were the chief pollen taxa, and a wide variety of temperate hardwoods were present.

In the south-central district, the

FIGURE 20.4. (opposite page). Plant geography showing vegetation zones and outlines of coastal areas during the Oligocene epoch in the USSR (from Pokrovskaya, 1956a). Proposed positions of 60° and 30° N latitude lines are also shown.

1. Cis-Baltic subprovince; broadleaved coniferous forest with a great amount of umbrella pine (*Sciadopitys*).

2. South-Russian subprovince; broadleaved and coniferous forest with a great amount of evergreen subtropical plants.

3. Caucasas subprovince. Broadleaved forest with a great amount of evergreen subtropical plants; in mountain regions there are coniferous woods; cypress (*Taxodium*) swamps occurr in lowlands near the coasts.

4. Kazakhstan subprovince. Broad leaved and coniferous forest with a great amount of subtropical plants and grassy cenose.

5. West-Siberian subprovince. Broadleaved and coniferous forest.

6. Maritime subprovince. Broadleaved forest.

7. Cis-Ohotsk subprovince. Broadleaved forest with a thick fern cover.

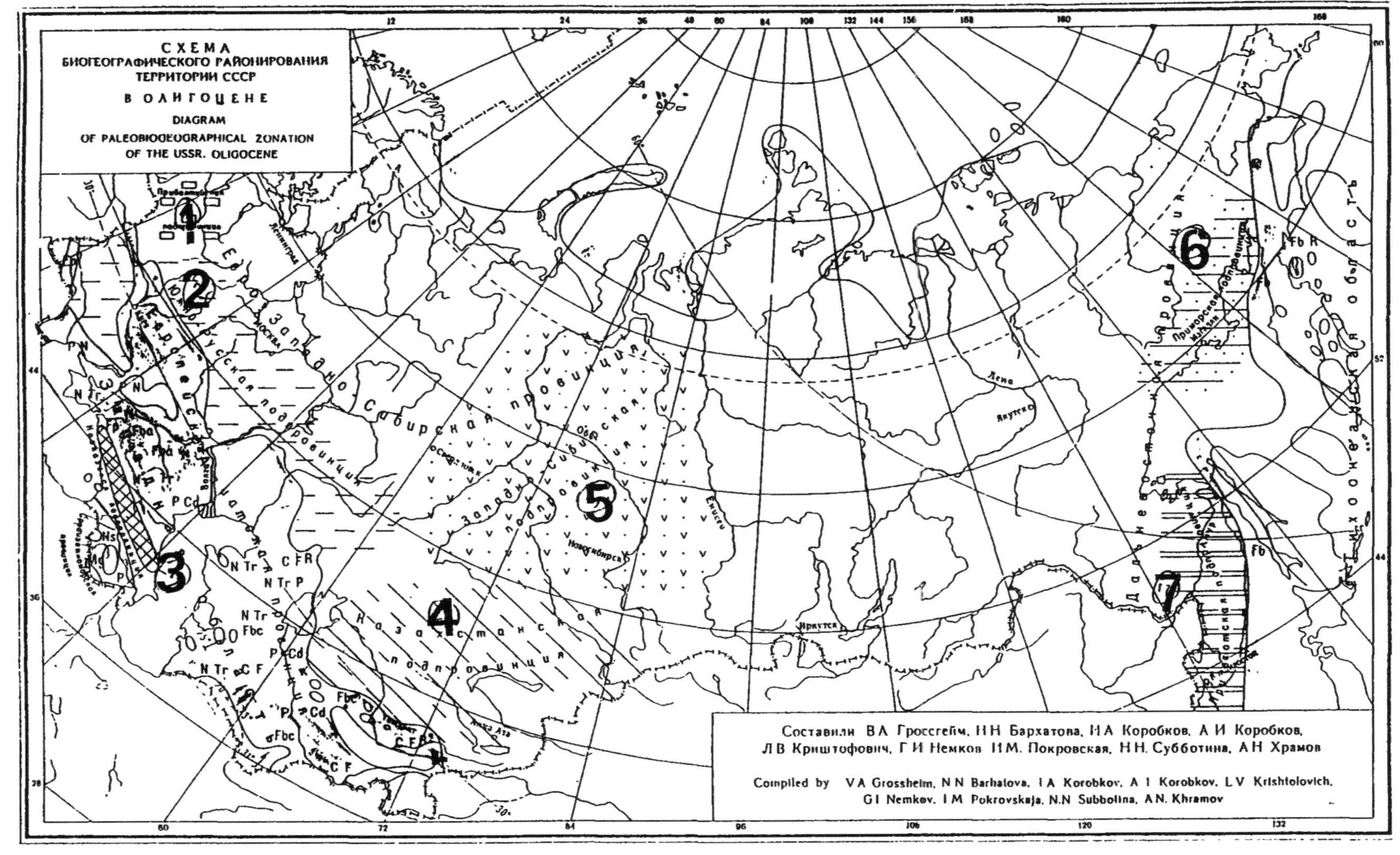
СХЕМА
БИОГЕОГРАФИЧЕСКОГО РАЙОНИРОВАНИЯ
ТЕРРИТОРИИ СССР
В ОЛИГОЦЕНЕ
DIAGRAM
OF PALEOBIOGEOGRAPHICAL ZONATION
OF THE USSR. OLIGOCENE
1
2
3
4
5
6
7
Москва
Ленинград
Лена
Якутск
Енисей
Обь
Новосибирск
Иркутск
Алма Ата
Западно Сибирская провинция
Западно Сибирская подпровинция
Казахстанская подпровинция
Южно Русская подпровинция
Приморская подпровинция
Дальневосточная провинция
N Tr C F R
N Tr P
N Tr Fbc
N Tr C F
P Cd
Fbc
C F
Fb R
Fb
Составили В.А. Гроссгейм, Н.Н. Бархатова, Н.А. Коробков, А.И. Коробков, Л.В. Криштофович, Г.И. Немков, Н.М. Покровская, Н.Н. Субботина, А.Н. Храмов
Compiled by V.A. Grossheim, N.N. Barhatova, I.A. Korobkov, A.I. Korobkov, L.V. Krishtolovich, G.I. Nemkov, I.M. Pokrovskaja, N.N. Subbotina, A.N. Khramov

Kazakhstan subprovince (area 4, Figure 20.4) was broadleaved forest with a mixture of needle-leaved conifers and great amounts of subtropical trees "and grassy cenose". The "early Oligocene" pollen spectra (which now might be considered late Eocene if the end of the Eocene is placed at 34 Ma; Swisher and Prothero, 1990) look like a suite from a mixed hardwood-conifer forest in which pine, fir and cypress are prominent, and *Cedrus* and *Gingko* are present. The hardwoods are members of the birch, walnut, elm and oak families. Warm subtropical taxa are rare (*Engelhardtia, Anemia).* The map legend "and grassy cenose" suggests grass associations. However the pollen spectra from that district for early mid and late Oligocene are fairly consistent in containing only single grains of grass in occasional samples, not enough to consider it important in any sense. Newly evolved herbaceous groups such as mustards and pinks (Cruciferae, Caryophyllaceae) are present but rare, and *Lygodium* ferns, *Osmunda,* polypod ferns, become very common, so these represented an herbaceous understory. Shrubby taxa become fairly abundant at the early part of the Oligocene in this region, e.g. *Rhus* (to 14%), Elaeagnaceae (ca. 1%) . This vegetation may have included areas of woody subtropical savanna, but considering the importance of so many temperate hardwood trees, it was more probably an open forest or woodland. "Middle Oligocene" spectra (which now might be considered early Oligocene if the end of the Eocene is placed at 34 Ma) are less diverse but also seem to represent forest with occasional openings dominated by *Osmunda* (up to 22%) and polypod ferns (2%). The late Oligocene assemblages are even less diverse and look like a simpler mixed conifer forest. Newly evolved taxa include the shrub, *Diervilla,* and herbs of the mint, carrot and pink families (Labiatae, Umbelliferae, Caryophyllaceae). Grasses are absent or not found. Climate appears temperate as the subtropical elements are gone.

<u>Eastern USSR</u>

In the Far East province along the Pacific coast (area 7, Figure 20.4), Oligocene conifer forest grew on the highlands, and broadleaved forest rich in tropical and subtropical elements occurred in the lowland, along with with extensive cypress swamps. Comparison of early and late pollen spectra (Pokrovskaya,1956a, e.g. increasing importance of conifer pollen types associated with a continuing prevalence of the rich broadleaved forest pollen assemblage) suggests that uplift along this coast was progressing during the Oligocene. The chief pollen types in lower, mid and upper Oligocene were *Picea, Larix,* Taxodiaceae types, *Fagus* and *Ulmus.* The lower Oligocene sediments (now probably considered upper Eocene) show large abundances locally of the fern spores, Polypodiaceae, *Onoclea, Lygodium,* and especially *Osmunda.* Consistently present in small quantities are pollen of hardwood trees such as *Quercus, Castanea,* various members of the walnut and birch families, and single grains of *Platycarya, Sterculia, Sapindus.* Shrubby groups such as *Rhus, Ilex,* are rare as are semiarid types such as *Ephedra* . This vegetation was probably a forest with fern-rich openings. Area 7, which was the most northerly site (Figure 20.4), reflects a very simple conifer forest dominated by pine.

In short, the Oligocene pollen spectra appear to represent chiefly forest vegetation with the possible exception of the south central Kazakhstan district. That vegetation may have been woody savanna or woodland in which ferns may have been an important ground cover, and shrubs were relatively important, confirming an open vegetation. Grass pollen was rare or absent in the Oligocene of USSR, however it did become locally abundant in the Miocene of Kazakhstan. Pokrovskaya (1956b) maps grassy steppe and forest for the interfluves of Kazakhstan and central Siberia for early to middle Miocene, and grassy steppes became widespread in central and southern USSR in the late Miocene. These low-biomass vegetation types which record a mix of deciduous broadleaved trees, conifers and nontree types suggest a Miocene mosaic of woodland, steppe and savanna.

Mid Cenozoic pollen spectra of China

The Gobi Desert region in northwestern China is a mid-continental area that today has extreme seasonality of rainfall. The modern vegetation is desert, semi-desert with scattered riparian woodland in northwestern China and mainly steppe and grassland in the western

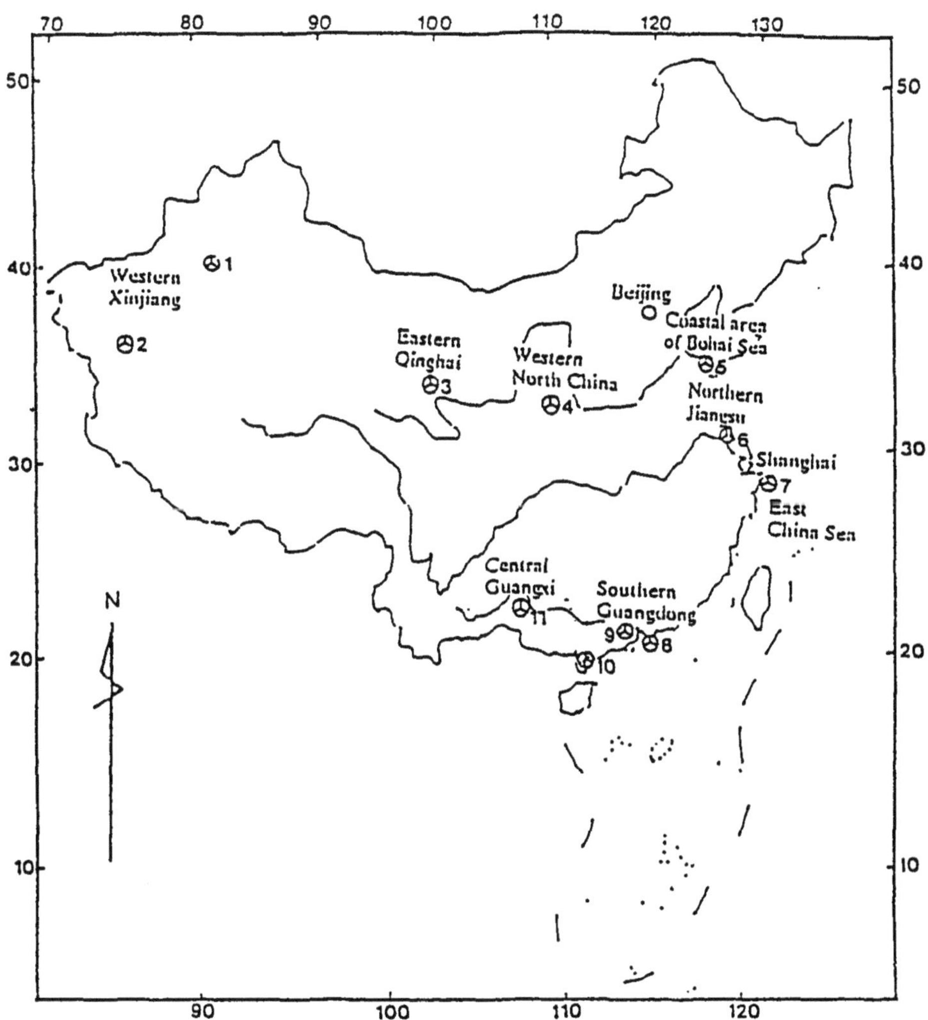

FIGURE 20.5. Map showing selected mid-Cenozoic pollen localities of China mentioned in text. Localities 1 through 5 comprise our west-east transect shown in Figure 20.6 (data base listed in Appendix). 1. Kuche, western Xinjiang (Zhao et al., 1982); 2. Kashi, western Xinjiang (Zhao et al., 1982); 3. Xining-Minhe Basin, eastern Qinghai (Sun et al., 1985); 4. Weihe Basin, western North China (Sun et al., 1980); 5. Coastal areas of Bohai Sea (Song et al., 1978); 6. Northern Jiangsu (Song et al., 1981); 7. East China Sea (Song et al., 1985); 8. Mouth of Zhujiang River, southern Guangdong (Lei,1985; Sun et al., 1985); 9. Sanshui, southern Guangdong (Song et al., 1986); 10. Leiqong Peninsula, southern Guangdong (Sun et al., 1985); 11. Bose Basin, central Guangxi (Liu, in press).

part of North China (Figure 20.5). The mean annual temperature for these areas is below 10°C. and precipitation is between 2 and 65 cm (Hou, 1983; Wu, 1980). This region is of special interest because it was an interior area in the Oligocene, and because low-biomass vegetation probably developed relatively early here.

The fossil pollen data available are scattered, mostly from core material, and the data themselves have some limitations in reconstruction of vegetation types. Some of the Oligocene sediments are dated by fossil vertebrates, and for other units the palynologists established relative age by comparisons with sections where dating is available. We have corrected the estimated ages to allow for the extended definition of late Eocene. We will summarize some of the main characteristics of regional pollen rain and vegetation during the mid Cenozoic by using a west-east transect to show the compositional changes in space and time. For comparison we also include mention of data from a north-south transect along the eastern coast of China.

The west-east transect starts with western Xinjiang, extends eastward approximately at 40° N. Latitude and ends at the coast of the Bohai Sea (localities 1-5, Figure 20.5). The north-south transect extends from the Bohai Sea coast southward to southern Guandong and Guangxi (localities 6-11, Figure 20.5). Tables showing percentages of key forms are included in the Appendix. Figure 20.6 gives histograms for percentages of key groups as a function of space and time.

A great change of climate and vegetation occurred between mid- and late Eocene in China; during the early Tertiary (through mid-Eocene) when the expanded coastline connected Japan and Taiwan with the mainland, arid and subarid climate dominated vast areas of northwestern, central and most of southern China (Hsu, 1983; Guo, 1980), typified by pollen suites with large amounts of *Schizaeoisporites*, arid-adapted types such as *Ephedra*, cf. *Nitraria*, as well as occurrence of gypsum deposits. The arid area had a kind of low-biomass vegetation, woody savanna and probably desert riparian deciduous forest with some arid-type shrubs. After the middle Eocene this arid climate area retreated northwestward and became restricted to northwestern China and western North China, although it expanded somewhat again in late middle Miocene (Sun et al., 1980; Liu 1988). The changes probably involved basic shifts in atmospheric circulation patterns; considering the modeling data of Ruddiman and Kutzbach (1990, 1991), a shift from a dry to a summer-moist climate in southeastern and eastern China could have resulted in part from uplift in the Tibetan (Xizang) Plateau. That Plateau is thought to have only been about 1000 m high during the Oligocene (Wang et al., 1975), but even a low Plateau might have forced the beginning of a monsoonal climate for the southeastern coast of China.

In northwestern China in all late Paleogene assemblages, pollen of arid-adapted shrubs occur commonly and their values decrease eastward toward the coast. Conversely the tree pollen counts show an increasing value toward the east. As we show, the rise of grass pollen and herbaceous groups is a major feature of the Miocene spectra but does not occur earlier. The overall change suggests that a Paleogene woody low-biomass vegetation was replaced in the Neogene by herbaceous savanna and steppe with woodland.

The late Eocene pollen spectra from northwestern China are typified by the data from western Xinjiang (Zhao et al., 1982) and Qinghai (Sun et al., 1985) (left hand two columns in Figure 20.6 and Appendix A). Arid-adapted shrubs such as *Ephedra* and cf.*Nitraria* are prominent (up to 55% of pollen tallies), and trees are chiefly broadleaved deciduous (chiefly Ulmaceae, Juglandaceae, Fagaceae) with a few subtropical elements, e.g. rare Bombacaceae. Pine pollen is rare, and tree pollen is less than 20% of the count. We deduce that the vegetation was probably of subtropical woody savanna type or open deciduous forest with a lot of shrubby areas dominated by cf. *Nitraria* and *Ephedra*. Herbs and grasses were rare or absent in the record.

In contrast, the coastal area of the Bohai Sea had late Eocene pollen spectra dominated by angiosperm tree pollen (on right of Figure 20.6), especially that of the beech family (Fagaceae; see Appendix). The vegetation was deciduous broadleaved forest, and cypress swamp probably occurred along the estuaries. In South China, pollen data from the coast suggest evergreen and deciduous forest with mangrove trees along the coastal area (Lei, 1985; Sun et al., 1985). Arid-adapted shrubs which are comparatively rare, may have grown at some distance or been transported eastward from the more arid regions of China upwind (see Maher,1964, for modern examples).

The early Oligocene pollen spectra along the transect show the same patterns as those described for the late Eocene with a continued strong representation of arid-adapted shrubs in the west. For example in western Xinjiang and in eastern Qinghai the average pollen values of xerophyte shrubs *Ephedra, cf. Nitraria* and saltbushes (Chenopodiineae) represent a total of about 40% of the counts, and tree pollen is low (ca 20-30% of counts). The lowest diversity

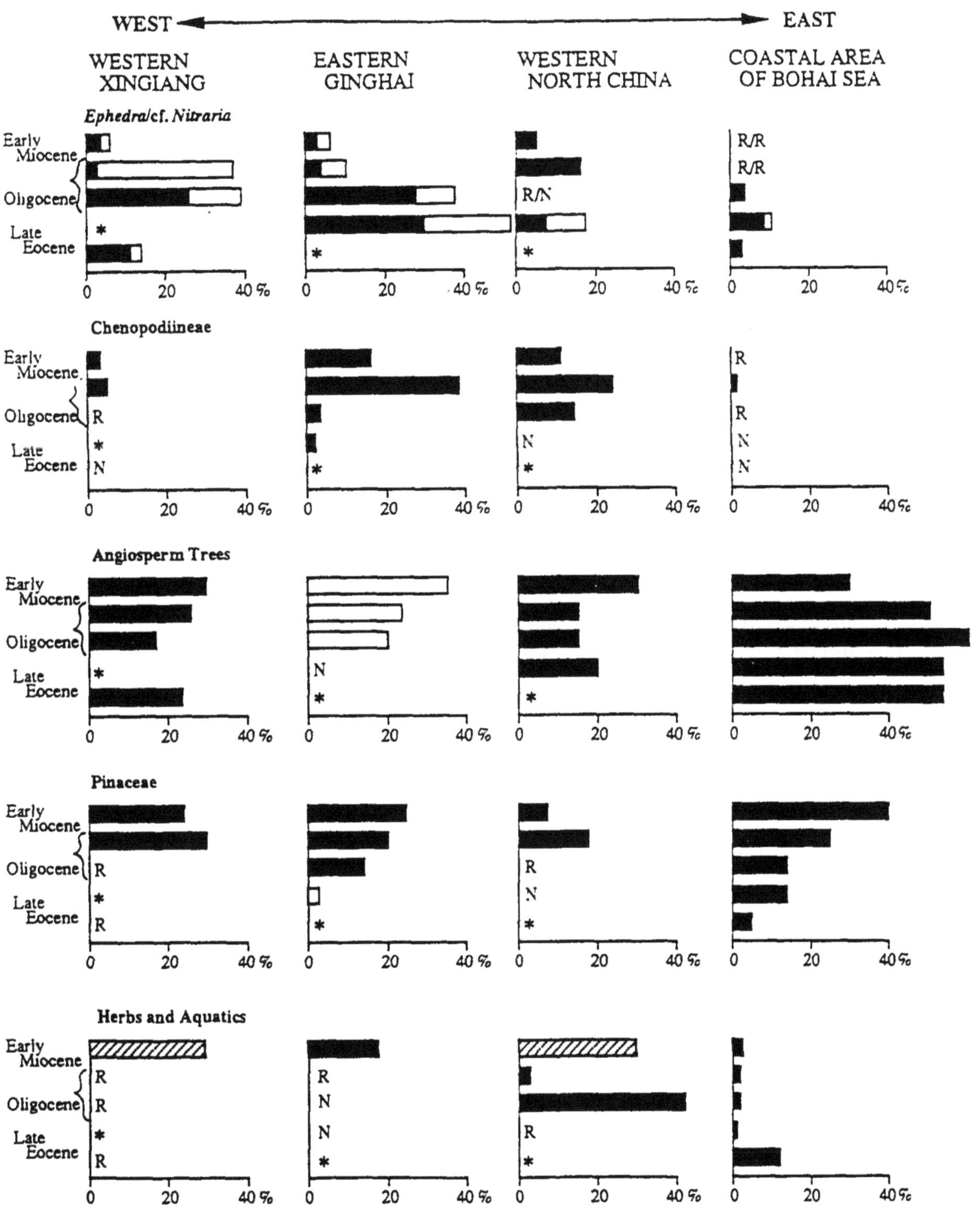

FIGURE 20.6. Selected pollen taxa from mid-Cenozoic sites along a west-east transect in Northwest and North China. Sites are listed in Figure 20.5. Data listed here from Zhao et al., 1982 (western Xingjiang), Sun et al., 1985 (eastern Qinghai), Sun et al., 1980 (western North China) and Song et al., 1978 (coastal area of Bohai Sea).

of tree vegetation occurred in the early Oligocene at the interior stations of our west-east transect (Figure 20.6). A strong representation of aquatic plant pollen accounts for the large herb count in western North China at this time.

In the late Oligocene at interior stations the arid-adapted shrubs continue with saltbush family (Chenopodiineae) becoming more important, i.e. at eastern Qinghai and western North China. Members of the aster family (Compositae) begin here as rare pollen types becoming common in Miocene sediments (Sun et al.,1985). Grass pollen is still lacking, and herbaceous taxa are still rare and undiverse. We deduce that a low-biomass woody savanna existed here under a temperate climate. Modern steppe taxa (saltbushes) were becoming important at the end of the Paleogene.

By contrast in eastern China along the Bohai Sea deciduous and evergreen forest still prevailed, in the late Oligocene this forest showed elevated participation of birch (Betulaceae) and elm (Ulmaceae) families. Highland conifer forests contained more hemlock and spruce (*Tsuga* and *Picea*), which led Song et al. (1978) to suggest the coastal climate cooled and became more humid in the late Oligocene. Pollen spectra from along the north-south transect (e.g., along the eastern seaboard), show two aspects of change during the Oligocene: one an increased role of temperate deciduous trees and conifers, and the other is the mangrove-related pollen types which became common along the coast of southeastern China. The climate remained humid subtropical to tropical in aspect (Sun et al., 1985; Lei, 1985; Liu, in press).

The Miocene assemblages differ from the Paleogene ones in containing more herb and grass pollen, less xerophyte shrub pollen, and generally pine family (Pinaceae) pollen becomes more important (Figure 21.6). In western Xinjiang temperate hardwood trees represent 30% of the pollen count. The vegetation of the Gobi Desert and interior stations was probably grassy savanna. Along the coast, in eastern North China pollen assemblages contain a large amount of Pinaceae pollen, with some hardwoods, especially elms, suggesting a mixed conifer and deciduous broadleaved forest under a humid temperate climate (Liu, 1988). Areas of southeastern China from northern Jiangsu and the shelf of East China Sea record broadleaved deciduous and evergreen forests, and indicate an early Miocene climate that was warm-temperate to subtropical in aspect (Song et al., 1981, 1985).

Though pollen data of mid-Cenozoic are from scattered localities and some are poorly dated, certain general trends and consistencies are striking. Low-biomass vegetation that had existed in widespread areas of China in the middle Eocene, had become more restricted in the late Eocene, i.e. the climate of southeastern China became more mesic but that of northwestern China remained semiarid (Liu, in press). In the subtropical woody savanna that developed in the northwestern interior of China during the late Eocene, arid-adapted shrubs became prominent or dominant members of the vegetation. Terrestrial herbaceous elements including grasses, were lacking. A similar woody savanna but with decreased floral diversity is recorded in the early Oligocene of interior stations in north China. Though the same regions show some increase in tree cover during the late Oligocene, the woody savanna continued with a growing importance of salt bushes. This low-biomass vegetation does not seem to have an analogue among the modern non-forest vegetation types we are aware of. The composition of the pollen rain does not resemble that of present day desert, steppe grassland or savanna in China.

Low-biomass vegetation did not develop along the coast of China during the mid-Cenozoic. Vegetation of the east coastal region maintained a forest structure, with minor amounts of arid zone shrub pollen appearing, perhaps as result of long distance pollen transport. During the Miocene the interior region of northern China developed a savanna vegetation that had a diversity of herbaceous groups and grasses as well as deciduous and coniferous trees. New groups such as the aster family begin to appear consistently in the Miocene sediments of China.

DISCUSSION

In the Rocky Mountain area the late Eocene pollen and leaf floras suggest the vegetation was open-structured with diverse shrubs many of which were drought adapted, and a climate that was seasonally dry. MacGinitie described two key floras of this interval (Green River and Florissant) as representing a "woody savanna", and based on these he commented: "The definition of savanna must be modified to fit conditions of the early Tertiary . The Tertiary (Paleogene) savanna can be described as a region of low or moderate relief in a tropical or subtropical climate characterized by a marked dry season, occupied by an open forest of shrubs and small trees. . . He concluded: "It appears certain that grass was not an important member of the vegetation ..." (MacGinitie, 1969, p. 49).

The pollen data we have summarized here from the Paleogene of the Rocky Mountain region fit well with MacGinitie's concept of the "woody savanna". Though there may be some structural similarity with the thorn forests of Africa, the same taxonomic groups are not represented. Further, the woody savannas of the Paleogene are unlike their floristic analogues in Mexico and West Texas because there the semi arid scrub with dwarf oaks, juniper, thorny legumes are accompanied by cactus, grasses, asters and many herbaceous perennials, all lacking or in the case of grasses and asters, rare, in the Paleogene. The diversity of Rosaceae that may represent shrubs in the Uintan is reminiscent of the radiation of roses described for the later Eocene in other mountainous areas of the western United States (Wolfe and Wehr, 1988). Other families that became important in savanna and steppe and desert vegetation did not appear until later or did not radiate till the Neogene, e.g. the grass and borage families.

The first direct evidence of grassy savanna in the Cenozoic record of the United States appears in the mid-Miocene. The increasingly abundant grass fossil remains together with the appearance of diverse grazing ungulates having specialized molars (Webb, 1989, 1983) suggests that a savanna biome emerged full-blown in the middle and late Miocene.

In summary the plant evidence from the mid continent United States leads us to propose that Retallack's (1982, 1983, 1988) paleosol records suggesting increasing distance between trees and trends toward smaller root sizes might be a record of woody savanna in MacGinitie's (1969) sense, not grassy savanna in Dansereau's (1957) sense. Perhaps the small-sized root traces were from small ferns or mosses, or even smaller roots of larger plants rather than grasses. We also note that while some A1 horizons have been identified (Retallack, 1983), the typical black or dark A-horizon features of grassland chernozems are not present; their organic contents may be altered by weathering.

The pollen data from the Oligocene of the USSR indicate that most of the plant cover types are forests suggesting relatively high biomass vegetation. The palynologists do not seem to describe them as open-structured, except in the case of the Kazakhstan area just north of the Sea of Aral, where they specifically indicate "and grassy cenose". Though grass pollen are absent, numbers of fern spores are occasionally very high, and pollen of shrubs such as *Rhus* become abundant (~14% of the count) in late Oligocene sediments. These data suggest a more open habitat and a somewhat lower biomass vegetation for the Sea of Aral district in the late Oligocene. The Soviet data clearly describe the establishment of grassy steppes in the Miocene of the Kazakhstan subprovince and central Siberia. Eastern Siberia developed herbaceous understory in conifer forest in the Miocene (Pokrovskaya 1956b), which parallels the western United States sequence just described.

Wolfe's (1985) world vegetation maps recognize the work of Guo (1980), who described a subhumid, semideciduous vegetation of China south of latitude 45° N, Guo's "Paratropical Arid Floral Province." The position of the Kazakhstan subprovince (loc. 4 , Figure 20.4) may represent the westward extension of Guo's Oligocene arid area. Though Wolfe suggests that the arid conditions may have occurred only in the early Oligocene, the pollen data (Appendix) indicate that aridity was a long-standing condition in northwestern China and central northern China during the Oligocene and late Eocene (if not Paleogene).

The Chinese pollen data provide a picture of temporal changes within a woody savanna of the late Paleogene and associated landscape change; this low-biomass vegetation occupied the interior district of western North China and the Gobi Desert at a time when the Pacific coast had deciduous broadleaved and broadleaved evergreen forest. It lasted until the early Miocene when more favorable conditions permitted somewhat greater forest cover to develop. Grassy savanna and steppe in a mosaic with woodland then became widespread in the Gobi Desert region.

CONCLUSIONS

Low-biomass vegetation in the form of woody savanna developed in the late Eocene of the Rocky Mountain region. Precipitation became more seasonal as annual temperatures lowered and seasonality increased at the end of the Eocene (Chadronian). The savanna was not comparable to those in areas where floristic analogues now grow, as herbaceous taxa and grasses were either absent or rare and undiverse. We lack Orellan and Whitneyan plant records for the Oligocene of the Great Plains of the United States, but Miocene pollen and seed floras clearly show the beginning of grassy savannas and steppe over a wide region on the High Plains. Open grasslands were probably absent in the Neogene, based on the pollen and leaf records. Axelrod (1985) makes a case for the first extensive open grasslands developing during the Holocene.

In eastern Asia, the late Eocene brought about development of extensive semiarid vegetation which continued into the Oligocene; it extended from western North China westward across the Gobi Desert area and included the Kazakhstan subprovince south of 45° N. latitude during the late Oligocene. This vegetation represented various kinds of woody savanna, with high representation of arid-adapted shrubs, and an open woodland of mostly temperate hardwoods. In the Miocene, there developed a new kind of low-biomass vegetation in Asia, one characterized by grassy and herbaceous steppes, and savanna.

Because in the Oligocene vegetation of Asia grasses and herbaceous plants were unimportant or lacking, the role of these plants in the late Oligocene low-biomass vegetation of the High Plains of the United States can be questioned, especially as direct evidence of those plant identifications is not yet at hand.

ACKNOWLEDGMENTS

We gratefully acknowlege the help of Kate Roudybush, Rita Stockwell, and Kay Suiter for their assistance in typing. Jack A. Wolfe and Greg J. Retallack were very helpful in giving us their comments on the manuscript.

REFERENCES

Axelrod, D. I. 1966. The Eocene Copper Basin flora of northeastern Nevada. *Univ. Calif. Pubs. Geol. Sci.* 59:1-125.

Axelrod, D. I. 1985. Rise of the grassland biome, central North America. *Bot. Rev.* 51:164-201.

Becker, H. F. 1960. The Tertiary Mormon Creek flora from the Upper Ruby River Basin in southwestern Montana. *Palaeontographica* B, 107:83-126.

Bonnefille, R. 1984. Palynological research at Olduvai Gorge. *Nat. Geogr. Soc. Research Report* 17: 227-243.

Buol, S.W., Hole, F.D., and McCracken, R.J. 1980. *Soil Genesis and Classification.* 2nd ed. Ames, Iowa; Iowa State University Press.

Dansereau, P. 1957. *Biogeography. An ecological perspective.* New York, Ronald Press.

Daubenmire, R. 1970. *Steppe Vegetation of Washington. Wash. Agric. Exp. Stn., Tech. Bull.* 62, 131 pp.

Daubenmire, R. 1978. *Plant Geography.* New York, Academic Press, 338 pp.

Dorf, E. 1960. Tertiary fossil forests of Yellowstone National Park, Wyoming. *Billings Geol. Soc. Eleventh Ann. Field Conf.* pp.253-260.

Eagleson, P. S. and Segarra, R. I. 1985. Water-limited equilibrium of savanna vegetation systems. *Water Resources Res.* 21(10):1483-1493.

Elias, M. K. 1942. Tertiary grasses and other herbs from the High Plains. *Geol. Soc. Amer. Spec. Pap.* 41.

Guo Shuangxin. 1980. Late Cretaceous and Eocene floral provinces. *Report Nanjing Inst. of Geol. and Palaeontol.*, Nanjing, Academia Sinica.

Hamilton, W. and Leopold, E.B. 1962. Volcanic rocks of Oligocene age in the southern part of the Madison Range, Montana and Idaho. *U.S. Geol. Surv. Prof. Paper* 450-B: 26-30.

Hickey, L. J. 1977. Stratigraphy and Paleobotany of the Golden Valley Formation (early Tertiary) of western North Dakota. *Geol. Soc. Amer. Mem.* 150, 177 pp.

Hou Hsiohyu. 1983. Vegetation of China with reference to its geographical distribution. *Ann. Missouri Bot. Garden* 70:509-549.

Hsu Jen. 1983. Late Cretaceous and Cenozoic vegetation in China, emphasizing their connections with North America. *Ann. Missouri Bot. Garden* 70:490-508.

Lei Zuo-qi. 1985. Tertiary Sporo-pollen assemblage of Zhujiangkou (Pearl River Mouth) Basin and its stratigraphical significance. *Acta Bot. Sinica*, 27(1):94-105.

Leopold, E. B. 1967. Summary of Palynological Data from Searles Lake. In G.I. Smith and E.B. Leopold. *Pleistocene Geology and Palynology, Searles Valley California.* Guidebook for Friends of the Pleistocene, Pacific Coast Section. pp. 52-66.

Leopold, E. B. 1969. Late Cenozoic palynology. In Tschudy, R. H. and Scott, R. A. eds., *Aspects of Palynology.* New York, Wiley-Interscience, pp. 377-438.

Leopold, E. B. and Denton, M. F. 1987. Comparative age of grass and steppe east and west of the northern Rocky Mountains. *Ann. Missouri Bot. Gard.* 74:841-867.

Leopold, E. B. and MacGinitie, H. D. 1972. Development and affinities of Tertiary floras in the Rocky Mountains. In A. Graham, ed., *Floristics and Paleofloristics of Asia and Eastern North America.* Amsterdam, Elsevier, pp. 147-200.

Leopold, E. B. and Wright, V. C. 1985. Pollen profiles of the Plio-Pleistocene transition in the Snake River Plain, Idaho. In C.J. Smiley, ed., *Late Cenozoic History of the Pacific Northwest.* pp. 323-348.

Liu Gengwu. 1988. Neogene climatic features and events of northern China. In *Neogene biotic evolution and related events*, IGCP-246 Symposium special publication, p. 21-30, Osaka Museum of Natural History, Osaka.

Liu Gengwu. (in press). Early Tertiary palynological assemblages from the Bose Basin, Guangxi.

MacGinitie, H. D. 1953. Fossil plants of Florissant beds, Colorado. *Carnegie Inst. Washington Publ.* 599, 198 pp.

MacGinitie, H. D. 1962. The Kilgore flora--a late Miocene flora from northern Nebraska. *Univ. Calif. Publ. Geol. Sci.* 35:67-158.

MacGinitie, H. D. 1969. The Eocene Green River flora of northwestern Colorado and northeastern Utah. *Univ. Calif. Publ. Geol. Sci.* 83, 140 pp.

MacGinitie, H. D. 1974. An early middle Eocene flora from the Yellowstone-Absaroka volcanic Province, northwestern Wind River Basin, Wyoming. *Univ. Calif. Publ. Geol. Sci.* 108. 103 pp.

Mack, R. N. and Bryant, V. M. Jr. 1974. Modern pollen spectra from the Columbia Basin, Washington. *Northwest Sci.* 48:183-194.

Maher, L. J. 1964. *Ephedra* pollen in sediments of the Great Lakes Region. *Ecology* 435 (2): 391-395.

McAndews, J. H. and Wright, H. E. Jr. 1969. Modern pollen rain across the Wyoming Basin and the northern Great Plains (U.S.A.). *Rev. Palaeobot. Palynol.* 9:17-43.

Pokrovskaya, I.M., ed. 1956a. *Atlas of Oligocene Spore-pollen Complexes of Various Areas, USSR.* Vses. Geol. Inst., Materialy, new ser., no. 16, 312 pp. Moscow.

Pokrovskaya, I. M., ed. 1956b. *Atlas of Miocene Spore-pollen Complexes of Various Areas, USSR.* Vses. Geol. Inst., Materialy, new ser., no.13, 460 pp. Moscow.

Retallack, G. J. 1982. Paleopedological perspectives on the development of grassland during the Tertiary. In *Third North Paleont. Conv., Proc.* 2:417-421.

Retallack, G. J. 1983. A paleopedological approach to the interpretation of terrestrial sedimentary rocks: The mid-Tertiary fossil soils of Badlands National Park, South Dakota. *Geol. Soc. Amer. Bull.* 94:823-840.

Retallack, G. J. 1988. Down-to-Earth approaches to vertebrate paleontology. *Palaios* 3:335-344.

Richards, P. W. 1957. *The Tropical Rain Forest.* Cambridge, Cambridge Univ. Press. 450 pp.

Ruddiman, W.F. and Kutzbach, J.E., 1990. Late Cenozoic plateau uplift and climate change. *Trans. Royal Soc. Edinburgh: Earth Sciences,* 81: 301-314.

Ruddiman, W.F. and Kutzbach, J.E., 1991. Plateau uplift and climatic change. *Scientific American* 264: 66-75.

Sarmiento, G. 1984. *The Ecology of Neotropical Savannas.* Cambridge, Massachusetts, Harvard University Press.

Sauer, C. O. 1950. Grassland, climax, fire and man. *Jour. Range Mgt.* 3:16-21.

Song Zhichen, Cao Liu, Zhou Heyi, Guan Xueting, and Wang Kede. 1978. *Early Tertiary Spores and Pollen Grains from the Coastal Region of Bohai.* Beijing, Science Press, 177 pp.

Song Zhichen, Guan Xueting, Zheng Yahui, Li Zengrui, Wang Weiming, and Hu Zhongheng. 1985. *A Research on Cenozoic Palynology of the Longjing Structure Area in the Shelf Basin of the East China Sea(Donghai) Region*. Hefei, China, Anhui Science and Technology Publishing House, 209 pp.

Song Zhichen, Li Manying, and Zhong Lin. 1986. *Cretaceous and Early Tertiary Sporo-Pollen Assemblages from the Sanshui Basin, Guangdong Province*. Palaeontologica Sinica, whole number 171, New Series A, Number 10, 170 pp., Beijing, Science Press.

Song Zhichen, Zheng Yahui, Liu Jinglin, Ye Pingyi, Wang Confeng, and Zhou Shanfu. 1981. *Cretaceous-Tertiary Palynological Assemblages from Jiangsu*. Peking, Geological Publishing House, 268 pp.

Spaulding, W. G. and Graumlich, L. J. 1986. The last pluvial climatic episodes in the deserts of southwestern North America. *Nature* 320: 441-444.

Sun Xiangjun et al. 1985. Pollen and Spore Fossils. In *Paleontological Atlas from the Shelf area of South China Sea*, pp. 1-30, Guangzhou, Guangdong Science and Technology Publishing House.

Sun Xiuyu, Fan Yongxiu, Deng Cilan, and Yu Zhengqing. 1980. Cenozoic sporo-pollen assemblage of the Weihe Basin, Shaanxi. *Bull. Geol. Inst., Chinese Academy of Geological Sciences* 1:84-109.

Sun Xiuyu, Zhao Yingniang, and He Zhuosheng. 1985. The Oligocene-Miocene palynological assemblages from the Xining-Minghe Basin, Qinghai Province. *Geol. Rev.* 30(3):207-216.

Swisher, C.C. III and Prothero, D.R. 1990. Single-crystal $^{40}Ar/^{39}Ar$ dating of the Eocene-Oligocene transition in North America. *Science* 249: 760-762.

Thomasson, J. R. 1980. *Paleoeriocoma* (Gramineae, Stipeae) from the Miocene of Nebraska: taxonomic and phylogenetic significance. *Syst. Bot.* 5:233-240.

Vinogradov, A. P. ed. 1967. *Atlas of the Lithological-Paleogeographical maps of the USSR*. Academy of Sciences of the USSR. Moscow, p.4-11.

Wang Kaifa, Yang Jiaowen, Li Zhe, and Li Zengrui. 1975. On the Tertiary sporo-pollen assemblages from Lunpola Basin of Xizang, China and their paleogeographic significance. *Sci. Geol. Sinica* 366-374.

Webb, S.D. 1983. The rise and fall of the late Miocene ungulate fauna in North America. In M.H. Nitecki, ed. *Coevolution*. Chicago, University of Chicago Press, pp. 267-306.

Webb, S.D. 1989. The fourth dimension in North American terrestrial mammal communities. In D.W. Morris, Z. Abramsky, B.J. Fox and M.R.Willig, eds. *Patterns in the structure of mammalian Communities. Spec. Publ. Mus., Texas Tech Univ. No.* 28, Lubbock, pp.181-203.

Wolfe, J. A. 1978. A paleobotanical interpretation of Tertiary climates in the Northern Hemisphere. *Amer. Sci.* 66:694-703.

Wolfe, J. A. 1981. Paleoclimatic significance of the Oligocene and Neogene floras of northwestern United States. In Niklas, K. J., ed., *Paleobotany, Paleoecology, and Evolution*, v. 2. New York, Praeger, pp.79-101.

Wolfe, J. A. 1985. Distribution of major vegetational types during the Tertiary. *Amer. Geophys. Union Monogr.* 32:357-375.

Wolfe, J. A. and Schorn, H. E. 1989. Paleoecologic, paleoclimatic, and evolutionary significance of the Oligocene Creek flora, Colorado. *Paleobiol.* 15(2):180-198.

Wolfe, J.A. and Wehr, W. 1988. Rosaceous *Chamaebatiaria*-like Foliage from the Paleogene of Western North America. *ALISO* 12(1), pp. 177-200.

Wu Zhenyi (chief editor). 1980. *Vegetation of China*. Beijing, Science Press, 1375 pp.

Zhao Yingniang, Sun Xiuyu, Wang Daning. 1982. Tertiary palynological assemblages from Shache and Kuche Basins, Xinjiang District. *Bull. Geol. Inst. Chinese Academy of Geological Sciences*, 4:95-125.

APPENDIX

Selected pollen taxa from sites along W-E transect in Northwest and North China

	WESTERN XINJIANG (Zhao et al., 1982)	EASTERN QINGHAI (Sun et al., 1985)	WESTERN NORTH CHINA (Sun et al., 1980)	COASTAL AREA BOHAI SEA (Song et al., 1978)	
Spores	1 (0.5-2)	1 (0-1	R	10 (1-20)	MIOCENE
Pinaceae	25 (2-49)	25 (14-40	8 (5-12)	40 (35-60)	
Ephedra	4 (3-6)	3 (1-6)	5 (3-8)	R	
Juglandaceae	18 (15-20)	2 (1-3)	4 (2-6)	10 (5-15)	
Betulaceae	2 (2-3)	1 (1-2)	2 (1-4)	9 (3-15)	
Fagaceae	4 (1-10)	12 (5-15)	6 (3-8)	8 (5-13)	
Ilmaceae	10 (8-10)	18 (16-30)	15 (10-20)	10 (5-15)	
Nitraria/Melia	2 (1-2)	3 (1-5)	N	R	
Chenopodiaceae	3 (1-5)	16 (5-28)	11 (10-20)	R	
Compositae	3 (1-5)	R	3 1-5)	R	
Angiosperm Trees	30 (15-35)	35 (23-50)	30 (25-35)	30 (20-35)	
Angiosperm Herbs	G,30 (20-35)	18 (5-30)	G,30 (25-36)	3 (2-5)	
T-C-T	N	N	R	R	
Spores	2 (0-4)	R (0-1)	2 (0-3)	10 (1-30)	LATE OLIGOCENE
Pinaceae	30 (25-35)	20 (12-25)	18 (9-15)	25 (15-40)	
Ephedra	3 (3-4)	4 (2-8)	17 (15-20)	R	
Juglandaceae	5 (3-8)	1 1-2)	1 (1-2)	7 (5-10)	
Betulaceae	6 (5-7)	2 0-6)	R	5 (3-7)	
Fagaceae	2 (2-3)	5 (3-10)	5 (4-7)	10 (3-50)	
Ilmaceae	9 (8-10)	13 (9-17)	7 5-10)	25 (8-32)	
Nitraria/Melia	34 (30-38)	6 (5-7)	N	R	
Chenopodiaceae	5 (3-13)	39 (24-50)	23 (20-25)	1 (0-2)	
Compositae	2 (1-3)	R	N	R	
Angiosperm Trees	26 (20-35)	23 (15-40)	15 (10-20)	50 (40-55)	
Angiosperm Herbs	R	R	3 (2-5)	2 (0-3)	
T-C-T	R	R	4	4 (2-5)	
Spores	2 (0-4)	1 (0-1)	8	2 (0-4)	EARLY OLIGOCENE
Pinaceae	R	14	R	13 (12-14)	
Ephedra	26 (23-35)	28 (10-48)	R	3 (1-6)	
Juglandaceae	R	2 (1-3)	R	5 (3-10)	
Betulaceae	R	1 (0-1)	N	6 (3-12)	
Fagaceae	10 (7-21)	8 (1-18)	R	39 (35-43)	
Ilmaceae	7 (4-12)	3 (0-8)	13 (10-20)	17 (13-21)	
Nitraria/Melia	13 (5-36)	10 (6-12)	N	R	
Chneopodiaceae	R	3 (0-6)	14 (10-20)	R	
Compositae	N	R	R	N	
Angiosperm Trees	18 11-35)	20 (7-35)	15 (11-20)	65 (60-70)	
Angiosperm Herbs	1 (1-2)	N	42	2 (1-3)	
T-C-T	4 (3-7)	R (0-1)	N	5 (3-8)	

	WESTERN XINJIANG (Zhao et al., 1982)	EASTERN QINGHAI (Sun et al., 1985)	WESTERN NORTH CHINA (Sun et al., 1980)	COASTAL AREA BOHAI SEA (Song et al., 1978)
Spores		1 (1-2)	33	6 (2-13)
Pinaceae		3 (0-10)	N	13 (6-20)
Ephedra		30 (15-50)	8	9 (7-12)
Juglandaceae		1 (0-3)	R	4 (4-5)
Betulaceae		R (0-1)	R	6 (4-10)
Fagaceae		8 (1-17)	9	28 (25-35)
Ilmaceae		3 (0-8)	R	14 (10-16)
Nitraria/Melia		24 (4-49)	10	1 (0-2)
Chenopodiaceae		2 (0-3)	N	N
Compositae		N	N	N
Angiosperm Trees		13 (5-27)	20	55 (40-62)
Angiosperm Herbs		N	R	1 (1-2)
T-C-T		R	R	8 (3-17)
LATE EOCENE				
Spores	10 (4-23)			6 (1-20)
Pinaceae	R			5 (4-8)
Ephedra	11 (9-15)			2 (1-3)
Juglandaceae	2 (1-3)			5 (2-9)
Betulaceae	2 (1-3)			8 (4-10)
Fagaceae	7 (6-8)			50 (35-60)
Ilmaceae	12 (11-13)			14 (13-15)
Nitraria/Melia	3 (1-6)			R
Chenopodiaceae	N			N
Compositae	N			N
Angiosperm Trees	24 (19-31)			55 (50-60)
Angiosperm Herbs	R			R
T-C-T	7 (5-13)			12 (5-21)

R = Rare
N = None
G = Grass pollen included

Selected pollen taxa from mid-Cenozoic sites along the west-east transect in Northwest and North China; values are percent total pollen and spores. Samples are arranged from oldest at the bottom to youngest at the top. Note that "T-C-T" refers to a large group including pollen of Taxaceae, Cupressaceae and Taxodiaceae. Only a few groups are shown here for the purpose of comparison.

21. CLIMATIC, FLORISTIC, AND VEGETATIONAL CHANGES NEAR THE EOCENE/OLIGOCENE BOUNDARY IN NORTH AMERICA

By Jack A. Wolfe

ABSTRACT

Multivariate analysis of the physiognomy of leaf assemblages in western North America indicates that a marked decline in mean annual temperature occurred near the Eocene/Oligocene boundary; this deterioration was also accompanied by an increase in mean annual range of temperature. Radiometric ages suggest that this deterioration occurred about one million years following the end of the Eocene. Palynological data from eastern North America indicate that a marked palynofloral change occurred within calcareous nannoplankton Zone NP21 and above the Eocene/Oligocene boundary.

Although extinctions in the land flora were probably major in the megathermal vegetation, they were less severe in the mesothermal flora, and least pronounced in the microthermal flora. Regional extinction at high latitudes was particularly marked. The early Oligocene deterioration resulted in a major physiognomic change in the microthermal vegetation, which was dominantly coniferous before the deterioration and dominantly broad-leaved deciduous following the deterioration.

INTRODUCTION

The climatic deterioration near the Eocene/Oligocene boundary was the single most important global climatic event that occurred between events near the Cretaceous/Tertiary boundary and the glaciation of the late Pliocene. Although the deterioration is recognized in marine environments, the effects of the deterioration were more marked in nonmarine environments, from which the deterioration was also first inferred.

MacGinitie (1953) was the first to suggest that a major, abrupt climatic deterioration might have occurred in the Oligocene. He discussed the stratigraphic evidence that placed the Goshen (now considered of earliest Oligocene age) and Bridge Creek floras (now considered of late early Oligocene age) close in time and concluded (1953, p. 66) that, if they were, "...a sharp climatic change must have occurred...since the Bridge Creek floras show marked differences from the Goshen" and (p. 67) "...that period must have witnessed an almost complete revolution in the flora of the region, and more critical events must be crowded into [that period]...than paleontologists have hitherto been willing to concede."

A decade after publication of MacGinitie's (1953) analysis, the application of potassium-argon dating to Tertiary leaf assemblages (Evernden and James, 1964) provided significant data for evaluating rapidity and timing of climatic changes. Wolfe and Hopkins (1967), using the univariate analysis of percentage of entire-margined dicot species, confirmed MacGinitie's floristic analysis and, using the newly available radiometric ages, concluded that (p. 73) "...the cooling took place within an interval no longer than two million years and possibly as short as one million."

Oxygen isotope data from New Zealand (Devereux, 1967) also indicated a marked temperature decline near the Eocene/Oligocene boundary, and Devereux suggested that the New Zealand decline was the same event documented by Wolfe and Hopkins (1967). Oxygen-isotope data from the Deep Sea Drilling Project (Shackleton and Kennett, 1975; Keigwin, 1980) also indicated that both surface and bottom water temperatures lowered significantly in this same interval in the subantarctic.

Axelrod and Bailey (1969) disputed the rapidity and magnitude of the Oligocene deterioration, but their analysis largely involved (1) assigning older ages than had Axelrod (1966) to critical pre-deterioration floras and (2) assigning lower temperature values than had Axelrod (1966) to these same assemblages (Wolfe, 1971). The net result of these unexplained alterations was that the temperature decline appeared to be gradual over several millions of years.

The intensity of the early Oligocene deterioration was also questioned based on paleobotanical data from the British Paleogene section (Collinson et al., 1981). Numerous late Eocene taxa were cited that ranged into the Oligocene and thus seemingly negated a significant temperature change. These taxa, however, are all aquatics: most extant aquatics have very broad geographic ranges and are, compared to most extant woody terrestrial plants, relatively insensitive to climate (cf. Szafer, 1954, p. 195-198). Investigations of terrestrial plants (and especially physiognomic analyses of them) across the Eocene/Oligocene boundary in Europe would be more productive in inferring climatic changes or stability.

The ages assigned in epochal terminology to the Oligocene deterioration have varied. MacGinitie (1953), using the then-accepted molluscan chronology, considered the deterioration to have occurred between the middle and late Oligocene, whereas Wolfe and Hopkins, using a revised molluscan chronology, assigned a middle Oligocene age to the deterioration. I later (1978) considered the deterioration to fall near the Eocene/ Oligocene boundary as defined in the rapidly developing planktonic biochronology and referred to the deterioration as the "terminal Eocene event." Because both the deterioration (~33 Ma; see below) and the Eocene/ Oligocene boundary (~34 Ma; Swisher and Prothero, 1990) have now been radiometrically dated and, as discussed below, the deterioration occurred in NP21, the deterioration occurred about ~1 m.y. after the Eocene/ Oligocene boundary.

In this report I will review evidence for the deterioration in North America and discuss some biotic consequences relative to the North American flora and vegetation. Paleoclimatic inferences for fossil-leaf assemblages in this report are largely based on CLAMP (Climate-Leaf Analysis Multivariate Program), which involves the correspondence analysis of >30 rigorously defined foliar character states scored for each species in modern vegetation samples (Wolfe, 1990a, 1990b). Almost all samples were collected proximal (<5 km) to stations where meteorological data have been recorded. The multivariate analysis produces scores for each sample along two principal axes that account for ~70% of total variance of the physiognomic characters. Ranking on the first axis largely corresponds to mean annual temperature, while ranking on the second axis corresponds to water stress. Calibration of mean annual temperature of the samples is valid (accurate) to <1°C; water-stress calibration, being dependent on factors such as soil, only generally reflects precipitation, although calibration of samples from areas of <100 cm precipitation is valid to ± 3-4 cm. The score for a fossil-leaf assemblage can thus produce an accurate and reproducible estimate of mean annual temperature and a general but also reproducible estimate of precipitation. The CLAMP database only includes modern samples that contain at least 20 species; if a fossil-leaf assemblage of <20 species was subjected to CLAMP analysis, I will so state. In this report, CLAMP estimates of mean annual temperature are given to the nearest half degree Celsius.

EVIDENCE FOR THE DETERIORATION

The most convincing evidence for the early Oligocene deterioration comes from the Pacific Northwest, where a number of stratigraphic sequences contain both pre- and post-deterioration floras, some radiometrically dated. Other areas, however, also have assemblages that provide important data relative to the biotic consequences of the deterioration or data relative to the timing of the deterioration.

Main part of Alaska

No sequence of floras in the main part of Alaska north of the Alaska Panhandle is known to span the late Eocene-early Oligocene interval. Indeed, from this main part of Alaska, only one probable pre-deterioration assemblage falls close to the Eocene/ Oligocene boundary. The Rex Creek assemblage from the central part of the Alaska Range (paleolatitude ~72°N; Figure 21.1) is from a unit that was deposited on granitic rocks that are ~36 Ma (Wolfe and Tanai, 1987).

Floristic composition of the Rex Creek assemblage is strongly indicative of a pre-

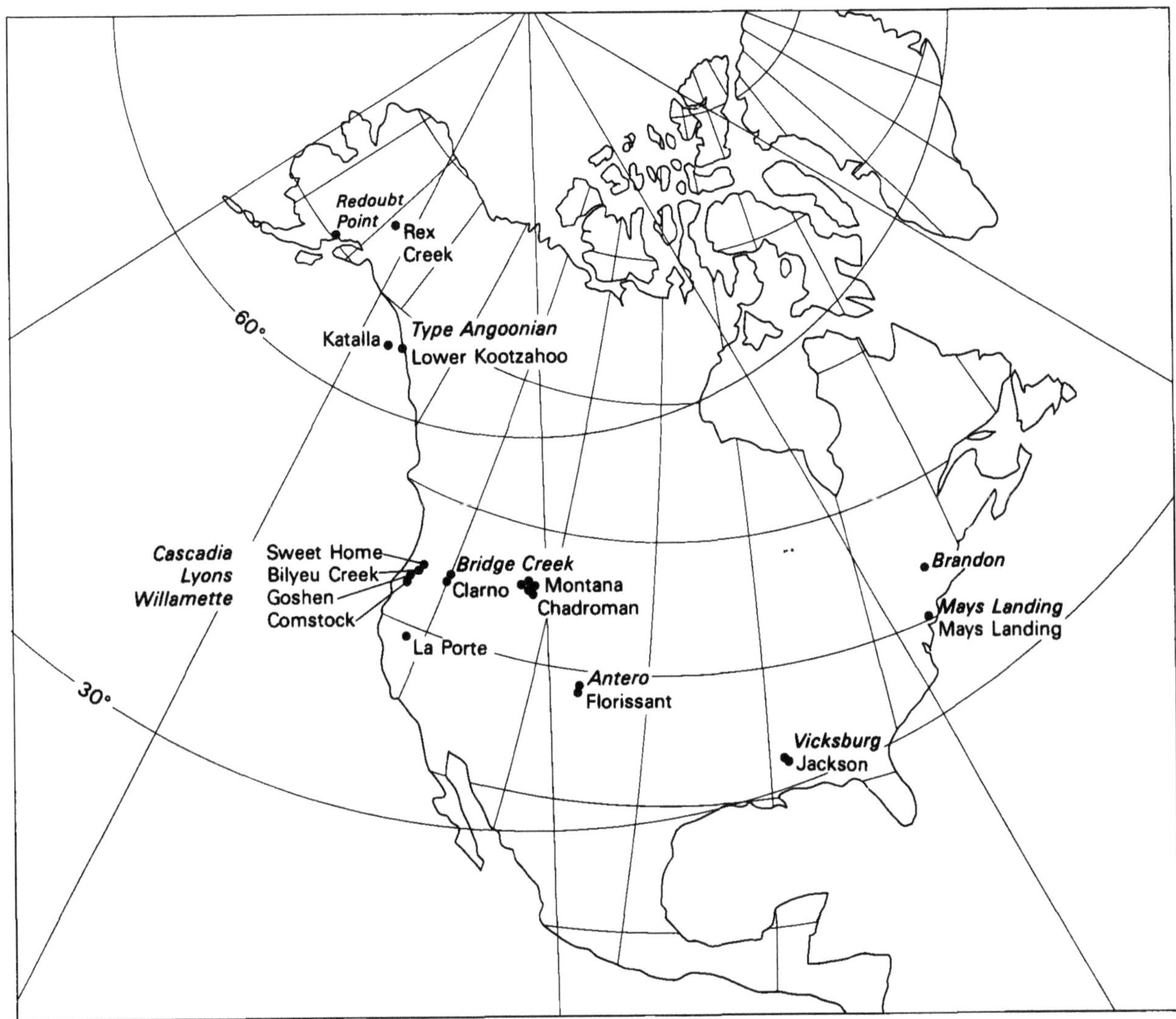

FIGURE 21.1. Map of North America showing location of critical late Eocene (roman) and early Oligocene (italic) plant assemblages. Paleolatitudes follow Smith et al.'s (1981) map 15 for the late Eocene.

deterioration age. Among the well-preserved palynomorphs are *Engelhardtia*-type, *Cedrela-Melia*, and the extinct *Aquilapollenites* (Leopold, *in* Wahrhaftig et al., 1969). The plant megafossils include *Engelhardtia* fruits (as *Palaeocarya*; Manchester, 1987) and *Ilex* (holly) leaves, which are coriaceous and thus were probably evergreen; broad-leaved evergreens, except for nanophylls such as *Ledum* (Labrador tea), are unknown in the main part of Alaska in floras of certain post-deterioration age. Available evidence indicates that the Rex Creek assemblage is between 33 and 36 Ma, i.e., latest Eocene or earliest Oligocene. For their size, the collections of plant megafossils contain a moderately diverse (>35 species) flora, and the palynofloras contain a high diversity of woody broad-leaved plants (E. B. Leopold, unpubl. data).

Floras of Angoonian (~Oligocene) age occur widely in the main part of Alaska (Figure 21.1; Wolfe, 1972). In the Cook Inlet region, Angoonian plant-bearing rocks crop out extensively along the western shore of Cook Inlet at and between Harriet and Redoubt Points. At Redoubt Point, several horizons contain abundant, well-preserved plants, and large talus blocks densely covered with impressions were examined for about 1 km on either side of Redoubt Point. This examination resulted in the collection of only 12 species; by far the most abundantly represented taxa are *Metasequoia* and *Alnus evidens* (alder). This low diversity assemblage is typical of Angoonian assemblages throughout Alaska; no broad-leaved evergreens are represented.

The stratigraphic and/or age separation between the Rex Creek and Angoonian assemblages is not well controlled, but the differences between the Rex Creek and all known younger assemblages are marked. The CLAMP MAT estimate for the Rex Creek assemblage is ~15°C; although the Angoonian assemblages have only 14 species, they produce an estimate of ~4.5°C. No post-Rex Creek assemblage in Alaska has a MAT estimate of >7°C. This indicates a decline of MAT of at least 8°C, and, if the Angoonian analysis is accepted, the decline was 10.5°C. No difference in mean annual precipitation between the Rex Creek and Angoonian assemblages is indicated by CLAMP; both pre- and post-deterioration precipitation was mesic (~150 cm/yr).

Southeastern Alaska

Although now situated in the main part of Alaska, the floras of the predominantly marine Poul Creek Formation (including the Katalla assemblage; Wolfe, 1977) were, during the Eocene, located on the Yakutat terrane seaward of the Alaskan panhandle and northern Queen Charlotte Islands (Figure 21.1; Plafker, 1987). The Katalla flora was collected from rocks that are now at the western end of the terrane, which would have been at the northern end during the Eocene. Thus, the Katalla floras would have been at approximately the same paleolatitude as the flora of the lower part of the Kootznahoo Formation at Kootznahoo Inlet on Admiralty Island (Wolfe, *in* Lathram et al., 1965).

Both the Katalla and lower Kootznahoo floras contain diverse evergreen Lauraceae and Fagaceae. Broad-leaved deciduous taxa (e.g., *Fagus* [beech], *Carya* [hickory]) are present but are a minor element. The CLAMP MAT estimates for both the Katalla (17 spp.) and lower Kootznahoo (15 spp.) assemblages are ~16°C.

No leaf assemblages of Oligocene, post-deterioration age are known on the Yakutat terrane. In the Kootznahoo Formation, however, are the floras of the type section of the Angoonian Stage (Wolfe, 1977), which are stratigraphically about 300 m above the lower Kootznahoo flora. As in the main part of Alaska, the Angoonian assemblages are broad-leaved deciduous and of low diversity (Wolfe, 1977); again, analysis of the small Angoonian assemblages could be questioned, but they produce a CLAMP MAT estimate of ~6°C. As in the instance of the assemblages to the north, the Katalla, lower Kootznahoo, and the type Angoonian assemblages indicate mesic conditions both before and after the deterioration.

Pacific Northwest

The physiognomic and floristic differences between leaf assemblages of the Pacific Northwest that are >33 Ma and those that are <32 Ma were the original basis for inferring a major, abrupt decline in mean annual temperature near the Eocene/ Oligocene boundary. Based only on the univariate analysis of percentage of entire-margined dicot species, I (1978) inferred a decline in MAT of ~12°-13°C. Estimates from CLAMP for the latest Eocene Comstock (Sanborn, 1935) and earliest Oligocene Goshen (Chaney and Sanborn, 1933) floras of western Oregon (Figure 21.1) are both ~20°C; the CLAMP estimate for the stratigraphically higher Willamette flora (cf. Wolfe, 1981b), which is dated at ~32 MA, is ~12°C. The decline is thus less than inferred from leaf-margin analysis, but still substantial, i.e., from marginally megathermal to warm microthermal climate and from broad-leaved evergreen to dominantly broad-leaved deciduous vegetation. Relative to precipitation, no significant change is apparent from the mesic Comstock-Goshen interval to the mesic Willamette interval.

Other leaf assemblages in western Oregon that are stratigraphically correlative with the Goshen are the Sweet Home (Brown, 1950) and the Bilyeu Creek/Scio (Sanborn, 1949). These produce CLAMP MAT estimates of ~19°-20°C. Stratigraphically above these assemblages are the Cascadia and Lyons floras (Meyer, 1973), respectively. Both these younger assemblages are broad-leaved deciduous, although they have yet to be analyzed with CLAMP.

In east-central Oregon, the uppermost assemblage of the Clarno Formation near Iron Mountain (Wolfe, unpubl. data) comes from a stratigraphic interval radiometrically dated at ~35 Ma; this interval also contains probable Chadronian mammals. Stratigraphically above, the many assemblages from the lower member of the John Day Formation are collectively referred to the Bridge Creek flora(s) (e.g., Chaney, 1927; Manchester and Meyer, 1987). The oldest radiometric ages from the Bridge Creek plant-bearing horizons are 32.8 Ma. I have not analyzed any of these assemblages with CLAMP, but the Iron Mountain is dominantly broad-leaved evergreen in contrast to the broad-leaved deciduous Bridge Creek vegetation.

California

The only leaf assemblage in California that bears directly on the deterioration is the La Porte (Potbury, 1935), which comes from tuffs radiometrically dated at 33.2 Ma (Evernden and James, 1964). The La Porte leaf assemblage produces a CLAMP MAT estimate of ~20°C, which is the same as for assemblages such as the Comstock and Goshen 8° latitude to the north. Even allowing for a lower latitudinal temperature gradient than now during the earliest Oligocene, the La Porte temperature should be at least 2°C higher than these more northern assemblages if the La Porte represents sea-level vegetation. The most logical inference is that the La Porte tuffs were deposited at an altitude of at least several hundred meters (they are currently in the Sierra Nevada at an altitude of 1,600 m).

Rocky Mountains

Based on presumed Chadronian and Orellan leaf assemblages from southwestern Montana (Becker, 1960, 1961, 1969, 1972, 1973), Wing (1987) inferred no significant climatic deterioration in the Rocky Mountain region near the Chadronian/Orellan boundary. Wing suggested that these leaf assemblages are all indicative of a generally similar climate, but also suggests that the dating of these assemblage was not well controlled and thus equivocal. CLAMP MAT estimates for these assemblages vary from ~11° to 12°C, in agreement with Wing's inference that no deterioration is evidenced.

In South Park, Colorado, however, the famous Florissant lake beds have a radiometric age of ~35 Ma (Epis and Chapin, 1974) and also contain a Chadronian oreodont (MacGinitie, 1953), whereas the younger Antero lake beds have a radiometric age of ~34.5 Ma and also contain Orellan mammals (MacGinitie, 1953). Because many of the radiometric ages obtained from the Tertiary volcanic rocks in South Park were recognized to contain contaminants (Epis and Chapin, 1974) and because the Chadronian/ Orellan boundary is dated by single-crystal method at ~34 Ma, both the Florissant and Antero radiometric samples may also be slightly contaminated; both the Florissant and Antero lake beds should be redated by the single-crystal method. The significant point, however, is that the Florissant and Antero floras are apparently separated in time by ~0.5 m.y.

CLAMP MAT estimate for the Florissant leaf assemblage is ~12.5°C. The Antero flora is, unfortunately, small but is indicative of a very cool coniferous forest; MacGinitie (1953) noted that the Antero flora was basically similar to the late Oligocene Creede flora, for which the CLAMP MAT estimate is ~4.5°C. Included in the Antero is *Pinus* cf. *P. crossii* (the common Creede pine and the probable ancestor of the subalpine bristle-cone pine). The 15 dicot leaves include, in decreasing abundance, an evergreen *Quercus* (oak), *Vaccinium* (blueberry), *Cercocarpus* cf. *C. henricksonii* (mountain mahogany), *Eleopoldia* cf. *E. lipmanii* (related to the extant alpine-subalpine *Luetkea*), and an indeterminate nanophyllous taxon. Of the six taxa, three are apparently conspecific with Creede taxa. The evergreen oak is microphyllous to nanophyllous and non-spinose, as in extant low-growing, subalpine species such as *Q. vaccinifolia*. The assemblage is far too small to obtain a valid CLAMP estimate, but the small leaf size is consistent with subalpine vegetation. Small leaf size also characterizes sclerophyll woodland and chaparral vegetation, which, however, have high percentages of obovate and/or spinose leaves, unlike the few

Antero species and extant subalpine vegetation. Significant uplift between time of deposition of the Florissant lake beds and time of deposition of the Antero lake beds is highly improbable, and thus the marked temperature difference between the Florissant and Antero assemblages is inferred to reflect the deterioration closely following the Eocene/Oligocene boundary.

The southwestern Montana assemblages, some of which are certainly of Chadronian age (all could be Chadronian), and the Florissant assemblage all indicate that they were deposited at a significant altitude. Using univariate physiognomic analyses, Meyer (1986), for example, inferred an altitude of 2,450 m for the Florissant (present altitude is 2,600 m). Although CLAMP produces somewhat lower MAT estimates for the assemblages Meyer dealt with, the relative altitudinal estimates are approximately the same as Meyer's and indicate that the Rocky Mountains in central Colorado and southwestern Montana were near (or even above in southwestern Montana) their present altitudes by the latest Eocene.

Eastern North America

Although late Eocene (~Jackson time) leaf floras are known from the Mississippi embayment, leaf material from early Oligocene (~Vicksburg) time is too sparse for valid comparisons. Palynological data from two areas, however, indicate that a major vegetational change occurred near the Eocene-Oligocene bounday. In Mississippi and western Alabama, palynological studies (Frederiksen, 1980) indicate that extinctions occurred near the end of Jackson deposition, and a marked vegetational change occurred by Vicksburg time, as indicated by a major increase in pollen of *Quercoidites*, a pollen taxon that may represent *Quercus*. Regionally, the boundary between the Jackson and Vicksburg approximates the Eocene/ Oligocene boundary.

A core from Mays Landing, New Jersey, was taken from marine rocks (Owens et al., 1988), and thus the age of floristic and vegetational changes can be related directly to the marine chronology (cf. Poore and Bybell, 1988). Calcareous nannofossils and planktonic foraminifera indicate that the interval from 175 to 210 m represents Zone NP21 and the Eocene/ Oligocene boundary occurs at 192 m within Zone NP21. Ager (in Owens et al., 1988) found the typical Jackson-type, pre-deterioration palynoflora in samples at 189 m (and below), i.e., the pre-deterioration palynoflora extends into the earliest Oligocene part of the core. The next highest sample at 186 m, however, records "...a dramatic increase in the *Quercus* and quercoid pollen..." (Owens et al., 1988, p. 14). Extinction of only a few palynomorph genera are associated with this vegetational change. Ager concluded that this marked vegetational change noted by him and by Frederiksen (1980) resulted from the climatic deterioration evidenced by megafloras in western North America. Most significantly, the vegetational change occurred a short time after the end of the Eocene as defined in the planktonic chronology.

PHYSICAL ASPECTS OF THE DETERIORATION

Age and rapidity

From radiometric ages on pre- and post-deterioration assemblages, the deterioration occurred between 33.2 and 32.8 Ma, i.e., at about 33 Ma. Most of these radiometric ages are based on older techniques of potassium-argon dating and could be considerably improved with newer techniques. Further, undated floral assemblages such as the Goshen, Sweet Home, Bilyeu Creek, Cascadia, and Lyons occur in sequences containing much volcanic debris and could also be dated.

Such improvement of previous dates and securing of new dates will be critical to determining rapidity of the deterioration. The two ages cited above are experimentally indistinguishable and suggest a very rapid deterioration. Rapidity is also indicated by proximity in the New Jersey core of the pre-deterioration and post-deterioration samples, which are separated stratigraphically by only 3 m of sediment.

Change in mean annual range of temperature Some investigators (e.g., Collinson et al., 1981; Kennett et al., 1985) have argued that the "terminal Eocene event" was simply the culmination of a cooling trend that began late in the middle Eocene, if not earlier. The marine record of climatic change during the later Eocene is, however, subject to differing interpretations. Based on both isotopic and nannofossil

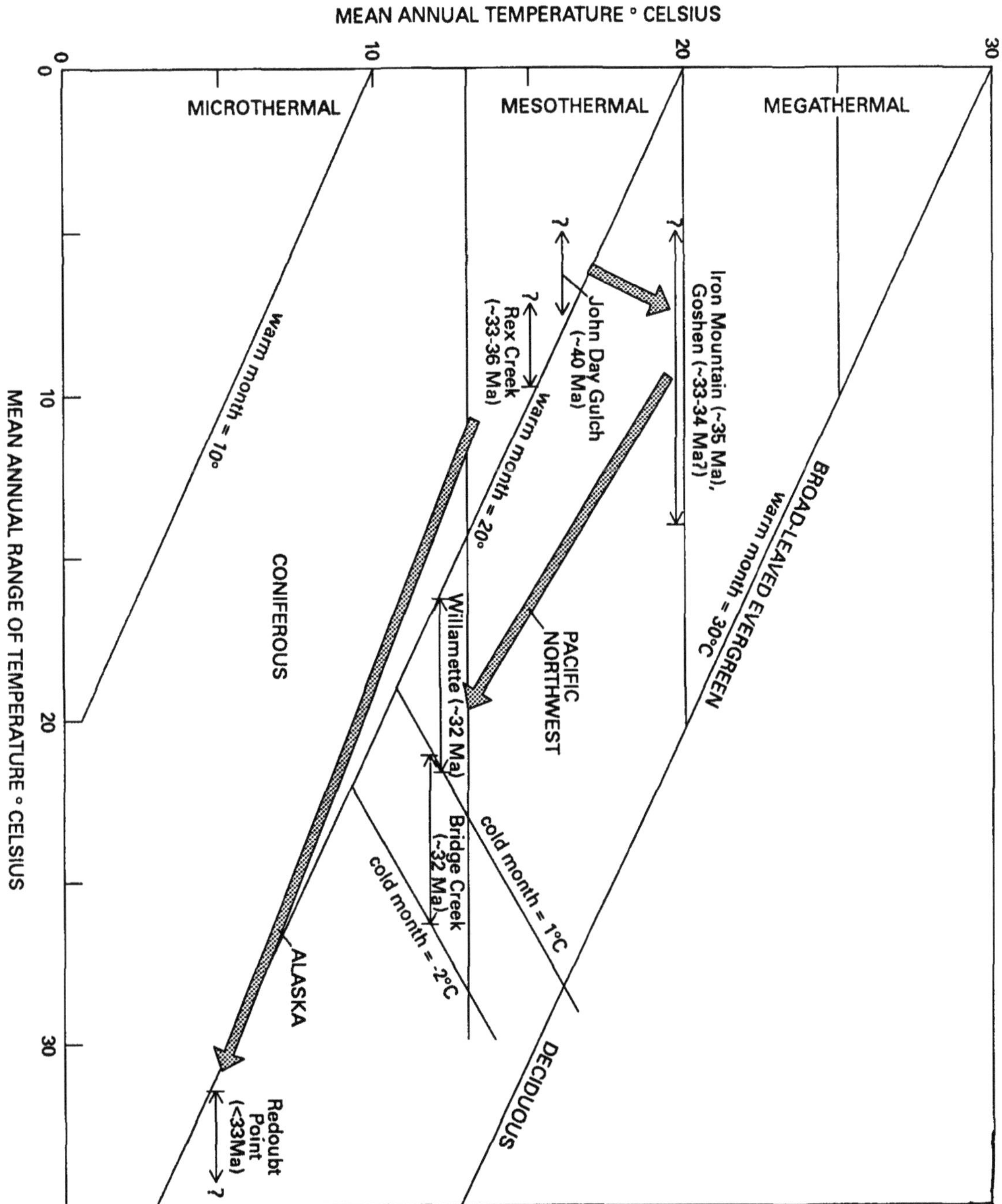

FIGURE 21.2. Nomograph showing suggested directions of temperature changes in the Eocene-Oligocene boundary interval. Base nomograph and vegetational units adapted from Wolfe (1979).

data from the Atlantic, Poore (in Wolfe and Poore, 1982) showed a significant temperature decrease near the middle/late Eocene boundary followed by a moderate temperature increase approaching the Eocene/Oligocene boundary. Poore's interpretation of the Atlantic data is paralleled by physiognomic interpretations from the Mississippi Embayment leaf assemblages (Wolfe and Poore, 1982). Leaf assemblages of latest Eocene-earliest Oligocene age are consistently indicative of warmer climate than leaf assemblages slightly older or slightly younger. The leaf assemblages agree with a general cooling trend from the early Eocene on, but the cooling trend was reversed during the middle and again during the late Eocene.

Comparison of leaf assemblages from the cool interval near the middle/late Eocene boundary with those from the post-deterioration early Oligocene emphasize that the climates of the two intervals were significantly different, i.e., that the early Oligocene cool climate was not simply a cooler version of the middle/ late Eocene cool climate.

I previously emphasized (Wolfe, 1971, 1978) that the early Oligocene deterioration was also characterized by a marked shift to a higher mean annual range of temperature (MART). The last valid "fix" on MART in the Eocene of the Pacific Northwest is at ~40 Ma, when MART was certainly <14°C and was probably <10°C. This estimate is based on the John Day Gulch assemblage of the Clarno Formation of east-central Oregon, which represents Microphyllous Broad-leaved Evergreen forest and has a CLAMP MAT estimate of ~16°C. By 32-33 Ma, however, MART was >19°C, because the Bridge Creek and isochronous assemblages represent broad-leaved deciduous forest; except in anomalous vegetation subjected to intense Arctic cold fronts (as in central and eastern United States today), broad-leaved deciduous forest naturally lives under MART that exceeds 19°C (Figure 21.2; Wolfe, 1979). In the intervening ~7 m.y. from 40 Ma to 33 Ma, paleobotanical evidence from the Pacific Northwest is equivocal relative to MART; whether MART gradually, although still markedly, increased during this interval or abruptly increased at ~33 Ma is uncertain. The Ruby (Becker, 1961), which is the youngest of the probable Chadronian floras of southwestern Montana, however, has a significant coniferous element (*Chamaecyparis, Abies, Picea, Pinus* spp.), indicating that coniferous forest was present. Combined with a CLAMP MAT estimate for the Ruby of ~12°C, MART had to be <16°C in this interior region (see Figure 21.2), and a somewhat lower MART (<14°C?) would be expected in the Clarno region during the late Chadronian. Thus, available data would appear to indicate an abrupt increase in MART near 33 Ma.

POSSIBLE CAUSES

In relation to the marked temperature deterioration in the subantarctic near the Eocene/ Oligocene boundary, at least three factors could be involved: an increase in intensity of bottom-water circulation in the subantarctic, significant expansion of montane glaciation on Antarctica (Kennett et al., 1975), or a major increase in volcanic activity (Kennett et al., 1985). All three factors and the temperature deterioration itself present a chicken or egg conundrum, and all are also capable of producing feedback mechanisms.

Increased volcanism is possibly the least significant factor. Whether the increase is real or apparent on a world-wide basis has yet to be demonstrated, although volcanism certainly increased in some areas of the Pacific Basin (Kennett et al., 1985); needed are analyses that estimate the actual volumes of volcanic debris erupted into the atmosphere during different intervals. The late Eocene increase, however, started at least a few millions of years prior to the end of the Eocene, and, if volcanism played a crucial role, some unstated threshhold effect must be assumed to cause a major, abrupt temperature decline a few millions of years later. Further, if volcanism was the critical forcing factor, when this forcing factor apparently subsided in the Oligocene (Kennett et al., 1985), a return to conditions that prevailed prior to the presumed increase in volcanism might be expected.

An increase in bottom-water circulation similarly began during at the end of the middle Eocene, but the circum-Antarctic current itself was not fully developed until the middle or late Oligocene (Kennett et al., 1975). Again, an assumption is required regarding the level of intensity of bottom-water circulation necessary to produce the temperature deterioration, which occurred a few millions of years after increased circulation was initiated and a few millions of years before the circum-Antarctic current was in place.

An increase in montane glaciation on Antarctica is more likely to result from temperature deterioration than the reverse. That is, air temperatures must cool (especially in the summer) to allow glaciers to grow and finally reach sea level. The North American leaf assemblages, however, indicate that air temperatures had actually warmed during the latest Eocene/earliest Oligocene, and, within current age resolution, the marked decline in bottom-water temperature in the subantarctic and the temperature deterioration on land in North America are isochronous. The continuing growth of ice on Antarctica during the Oligocene (Kennett et al., 1975) should have, if this were a major factor in causing the temperature deterioration, continued to cause a temperature decline in North America during the Oligocene; the leaf assemblages in western North America, however, indicate a warming of ~3-4°C during the later part of the Oligocene (Wolfe, 1978).

Comparisons of the early Oligocene deterioration to the Cretaceous/Tertiary (K/T) boundary events were made by Kennett et al. (1985), who noted that, unlike the K/T events, the Oligocene deterioration was not accompanied by "catastrophic" extinctions. As discussed below, however, regional extinctions in the land flora at high latitudes could be classed as catastrophic, with 80-90% of the genera disappearing from high latitudes in North America. Like the K/T events, moreover, the Oligocene deterioration was abrupt. Unless some unknown threshholds are invoked, I find it difficult to explain the abruptness of the Oligocene deterioration in relation to plate tectonic factors and subsequent changes in oceanic circulation.

This is not to say that I would resort to a bolide impact to explain the abruptness of the Oligocene deterioration. The K/T bolide(s) resulted in a brief impact winter immediately followed by a 0.5-1.0 m.y. wet greenhouse (Wolfe, 1990a), and, moreover, whether the Oligocene deterioration was as abrupt as climatic change at the K/T boundary has yet to be demonstrated. Nevertheless, I suggest that explanations proposed for the Oligocene deterioration are unsatisfactory and that the search for explanations should continue.

BIOTIC CONSEQUENCES

Megathermal

Biotic changes in the broad-leaved evergreen vegetation and flora are difficult to infer, especially relative to megathermal vegetation. Late Eocene megathermal vegetation in western North America extended into the coastal lowlands of northern California and possibly into part of the Pacific Northwest; the deterioration must have restricted megathermal vegetation to the southern part of North America (<25°N paleolatitude), where no Oligocene megafossil assemblages are known. The pre-deterioration earliest Oligocene La Porte flora of northern California has many boreotropical families, tribes, and genera that are either totally extinct or survive only in southeastern Asia. Whether some or even most of these lineages survived the deterioration in southern North America is uncertain, but, considering the inferred great geographic restriction of megathermal climate in North America following the deterioration and the fact that the refugia were already occupied (although many of these refugial occupants may have been related to the more northern megathermal lineages), major regional extinction probably occurred in the megathermal flora.

Mesothermal

Areas of mesic mesothermal climate must have occurred along the Pacific Coast, especially in California, during the post-deterioration, early Oligocene, but no assemblages are known. Renewed warming into the late Oligocene allowed mesothermal vegetation to extend north along the coastal plain of Oregon. In coastal Oregon, the late Oligocene Yaquina flora (McClammer, 1978) has a CLAMP MAT estimate of ~15°C. This flora and correlative assemblages in Oregon (unpubl. data) have numerous lineages of broad-leaved evergreens that are known in pre-deterioration floras of the Pacific Northwest, e.g., *"Phoebe" oregonensis* (Lauraceae), *Diploclisia* (Menispermaceae), *Prunus pristina* (Rosaceae), *Meliosma* (Meliosmaceae), and *"Tetracera" oregona* (Fagaceae).

Unknown in the Yaquina and coeval floras (or Neogene mesothermal assemblages, e.g., the Weaverville flora; MacGinitie, 1937), however, are other taxa of the pre-deterioration floras of the Pacific Northwest. Conspicuous among these absences are Tribe Spondieae of Anacardiaceae, Mastixiaceae, and Tribe Phytocreneae of Icacinaceae, which are now exclusively paleotropical. Although certainly data from Oligocene assemblages in lowland California are needed, available data suggest that the deterioration resulted in considerable extinction in the mesothermal flora.

A similar level of extinction can be inferred for mesothermal vegetation in eastern North America, despite the lack of pre-deterioration assemblages that have been studied by modern techniques. The many tribes and genera shared by Eocene floras of Europe and western North America strongly indicate some direct interchange of taxa between the Laurasian continents. Tiffney (1985a, 1985b) emphasized that the North Atlantic land bridge was the most southerly migration route available to megathermal and mesothermal lineages during the early Tertiary, and presumably eastern North America (at least the coastal plain) was part of the migratory pathway. The pre-deterioration boreotropical flora of eastern North America should have shared many of the taxa with both Europe and western North America.

The Oligocene Brandon Lignite in Vermont has produced a fruit and seed assemblage that, on floristic criteria, represents mesothermal vegetation. The Brandon flora includes some typical boreotropical taxa that are now endemic to the Old World (e.g., *Euodia* and *Phellodendron* of Rutaceae; Tiffney, 1980) and that survived the early Oligocene deterioration. However, many boreotropical taxa that were presumably present prior to the deterioration (and many of which, such as Mastixiaceae, persisted in Europe) are unknown in the Brandon flora. Thus, at least moderate extinction in the mesothermal flora is also inferred for eastern North America.

Microthermal

Biotic changes in the microthermal vegetation and flora in western North America are better documented than in mesothermal and megathermal vegetation. The close relationships between many lineages in the early Oligocene, post-deterioration Bridge Creek flora of Oregon and the montane late Eocene pre-deterioration floras of the Rocky Mountains was first pointed out by MacGinitie (1953). In the framework of the older hypothesis of an "Arcto-Tertiary Geoflora" (Chaney, 1959), the Bridge Creek taxa should have resulted from southward migration of lineages from high latitudes, but MacGinitie suggested that these taxa were closest to Florissant taxa. More recent work has confirmed MacGinitie's suggestion (e.g., Manchester and Crane, 1987; Wolfe and Tanai, 1987) of the derivation of Bridge Creek (and later) microthermal taxa from Eocene montane lineages.

These data, however, should not be interpreted as substituting an "Eocene-Montane Geoflora" for an "Arcto-Tertiary Geoflora." First, the Eocene montane vegetation at middle latitudes of western North America was coniferous forest and dominated by Pinaceae, Cupressaceae, and evergreen Taxodiaceae; the Bridge Creek assemblages represent broad-leaved deciduous forest accompanied by deciduous Taxodiaceae. Second, although many montane microthermal lineages survived the deterioration by apparent migration into the more coastal lowlands, many montane lineages are unknown in western North America following the deterioration. In western North American *Acer* (maples), for example, the later Eocene montane floras had 15 sections and 24 species, but the deterioration reduced *Acer* to 10 sections and 13 species in the lowlands (Wolfe and Tanai, 1987). Microthermal extinctions also occurred at the generic level; the characteristic *Fagopsis* (Fagaceae?), for example, has yet to be found in post-deterioration floras.

Survival of the deterioration by a microthermal lineage did not guarantee survival into the Miocene. Numerous early Oligocene taxa are unknown in the late Oligocene or younger (and better documented) Miocene floras, including *Macginitiea* (Platanaceae), *Cedrelospermum* (Ulmaceae), *Asterocarpinus* (Betulaceae), *Plafkeria* (Tiliaceae?), *Bohlenia* (Sapindaceae), and sections *Glabroidea* and *Manchesteria* of *Acer*. The environmental and biotic pressure of the deterioration apparently continued long after the deterioration; a lineage may have survived for a few million year

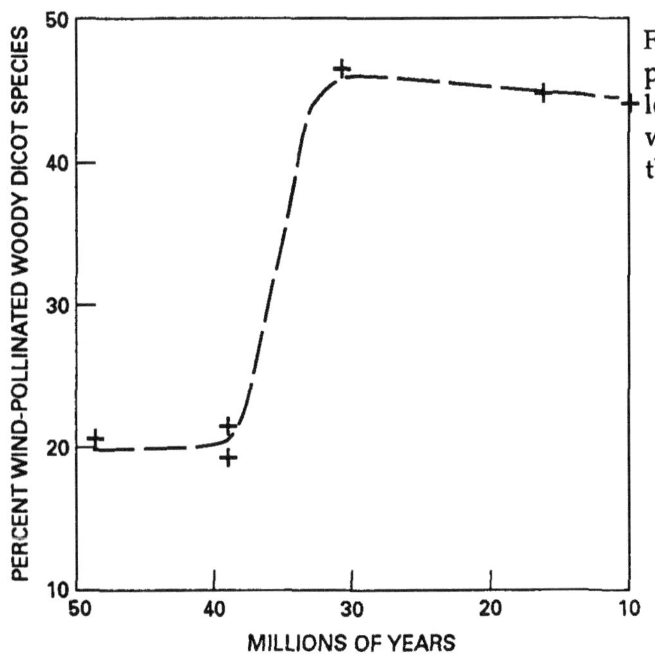

FIGURE 21.3. Change in percentages of probable wind-pollinated woody dicotyledons in microthermal vegetation of western United States from the later Eocene through the middle Miocene.

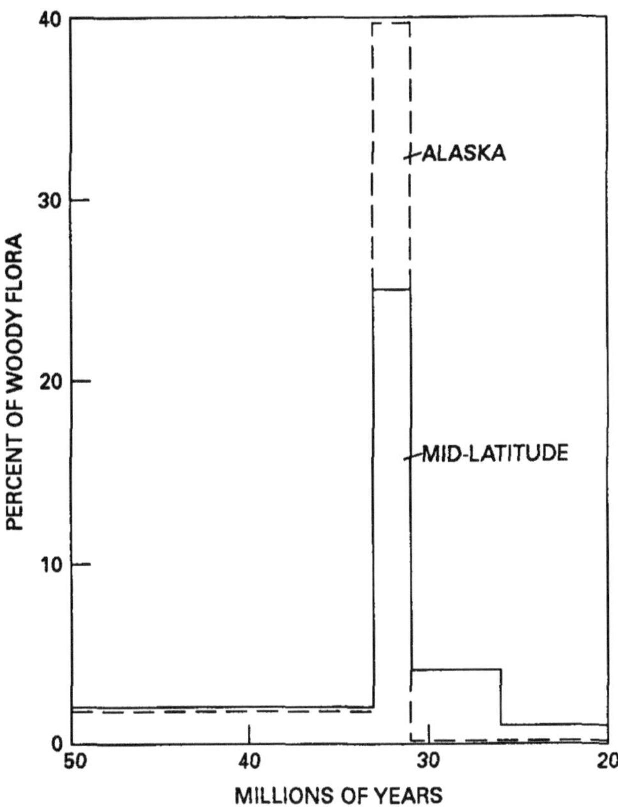

FIGURE 21.4. Rate of inferred regional extinction of genera during the later Eocene and Oligocene in western North America. Extinction during the early Oligocene deterioration is conservatively extended over 2 m.y.; if the deterioration were shorter, the rate would be increased. Based on compilations of published and unpublished floral lists. A genus is included in the regionally extinct category whether extinction was world-wide or the genus survived in regions outside of western, mid-latitude North America or Alaska. Genera that reappeared in western, mid-latitude North America or Alaska during the late Oligocene warming are not included in the regionally extinct category.

but was finally eliminated by better-adapted lineages.

The stream-side community of the late Eocene montane coniferous forest primarily comprised *Alnus* (alders) and/or *Fagopsis* (MacGinitie, 1953), which were in places accompanied by a low diversity of *Populus* (cottonwoods), *Salix* (willows), and *Metasequoia*. Following the deterioration, the stream-side community in the lowland, broad-leaved deciduous forest was overwhelmingly dominated by *Alnus* and *Metasequoia* from Oregon north into Alaska. In the late Oligocene of Alaska, Salicaceae were diversifying and became the dominant stream-side element; they apparently spread southward into middle latitudes during the late Oligocene and were a major stream-side element by the early Miocene. *Liquidambar* (sweet-gum) and *Platanus* (sycamore), on the other hand, were largely in mesothermal to megathermal stream-side vegetation during the Eocene. Species of both genera were generally rare or absent in most lowland, post-deterioration assemblages of the early Oligocene, but both genera were an important element in the stream-side community of the early Miocene at middle latitudes and spread north into southern Alaska by the mid-Miocene (Wolfe and Tanai, 1980); lineages in *Liquidambar* and *Platanus* gradually adapted to the cooler and more extreme post-deterioration climates.

Another effect of the deterioration in microthermal vegetation in western North America was a marked increase of the proportion of taxa that are wind-pollinated (Figure 21.3). This increase is a corollary of the change from a conifer-dominated canopy to a broad-leaved canopy. In coniferous forests, where the canopy is of course dominated by the wind-pollinated conifers, most broad-leaved, sub-canopy trees and shrubs are insect-pollinated (except in open spaces as along streams). Insect-pollinated groups such as *Acer* and Rosaceae were highly diverse in the Eocene montane forests; in *Acer* and possibly also in Rosaceae, diversity in western North America declined following the deterioration. In the early Oligocene broad-leaved deciduous forests, however, the wind-pollinated Juglandaceae, Betulaceae, Fagaceae, and Ulmaceae were probably canopy dominants, leading to an over-all increase in the flora and vegetation of woody broad-leaved plants that are wind-pollinated.

Regional diversity would, of course, decline at middle latitudes (Figure 21.4) because of the extinction or southward restriction of many megathermal and mesothermal lineages. If the discussion is restricted to microthermal ecosystems (Figure 21.5), diversity also declined, i.e., the montane Eocene flora was more diverse than the lowland early Oligocene flora. However, microthermal diversity then increased to a high level by the mid-Miocene; this is especially true for genera that include large, canopy-forming trees (e.g., *Fagus* [beech], *Quercus* [oak], *Carya* [hickory], *Ulmus* [elm]).

At high latitudes, the deterioration resulted in even more regional extinction than at middle latitudes (Figure 21.4). The Oligocene microthermal flora was very depauperate. By the end of the epoch, diversity again increased, and, as at mid-latitudes, reached a maximum in the mid-Miocene. The very low diversity of the high-latitude flora following the deterioration may have been a significant biotic factor in the evolution of groups such as Salicaceae. That is, the reduction by the deterioration in numbers of lineages would result in the remaining lineages being freed to radiate and to occupy different habitats; as lineages in the south gradually adapted to the more northern light regime and spread northward, they would find little competition in many habitats and could undergo explosive (for plants) radiation. This appears to be the case for willows of subgenus *Vetrix*, which first appeared in North America and Alaska during the Oligocene (presumably an immigrant from Asia) and underwent extensive diversification.

Elimination of many mesothermal and warm microthermal genera from Alaska (and presumably other high-latitude areas) by the deterioration also resulted in a vicariant distribution for many of these genera. For example, the pre-deterioration flora of the main part of Alaska included genera such as *Itea* (Virginia willow), *Pachysandra* (Appalachian spurge), and *Cedrela*, which are unknown in Alaskan post-deterioration floras and many of which today are disjunct between eastern Asia and eastern North America. On the other hand, some now-vicariant genera (e.g., *Liquidambar*, *Platanus*) that were eliminated from Alaska by

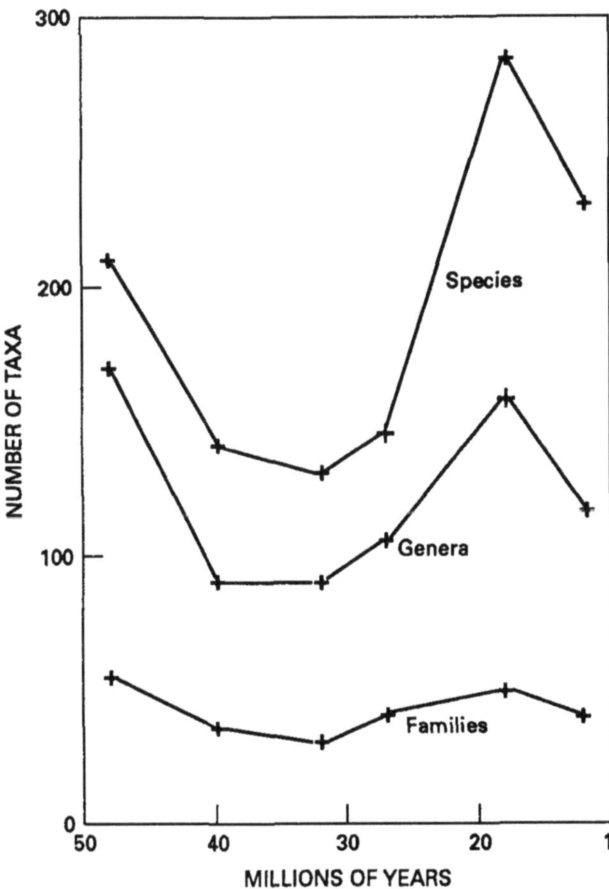

FIGURE 21.5. Known diversity of microthermal vegetation during the Eocene through Miocene in middle-latitude western North America. Based on mostly unpublished floral lists of Eocene to earliest Oligocene microthermal assemblages (see Wolfe, 1987, Figure 1, for locations and repositories) and on published and unpublished floral lists of post-deterioration microthermal assemblages (see Wolfe, 1987, Figure 2, for locations, citations, and repositories; compilation excludes taxa occurring only in mesothermal assemblages (11, 12, and 17).

the deterioration apparently returned to Alaska during the mid-Miocene warm interval, and vicariance in these genera again developed following the mid-Miocene.

CONCLUSIONS

The climatic deterioration near the Eocene/Oligocene boundary is well indicated by floristic and/or vegetational changes throughout North America. Where radiometric data are available, the best estimate for the deterioration is ~33 Ma, and, where independent biostratigraphic data are available, the deterioration occurs in an interval correlative with calcareous nannoplankton Zone NP21 and above the Eocene/Oligocene boundary. Regional extinction resulting from the deterioration was greatest at high latitudes and was at least moderate at middle latitudes. However, because many lineages at middle latitudes survived the deterioration, either by migrating from montane to lowland areas or by surviving in southern refugia and thus reappearing during later warm intervals, floristic criteria based on middle latitude assemblages do not as clearly delineate the deterioration as do physiognomic changes in leaf assemblages or changes in relative abundances in palynological spectra.

A shift to considerably higher mean annual ranges of temperature can be inferred: microthermal vegetation at middle latitudes was altered from dominantly coniferous to dominantly broad-leaved deciduous. Concomitantly, the proportion of wind-pollinated broad-leaved trees markedly increased. Some broad-leaved deciduous tree genera that presumably occupied the canopy of microthermal vegetation following the deterioration, although initially of low diversity, were of much greater diversity in the Neogene.

ACKNOWLEDGEMENTS

I have benefited greatly by discussions with H. E. Schorn (University of California, Berkeley), B. H. Tiffney (University of California, Santa Barbara), and S. L. Wing (Smithsonian Institution). E. B. Leopold (University of Washington) generously allowed access to some unpublished palynological data. T. A. Ager (U.S. Geological Survey) and Tiffney also provided comments on the manuscript.

REFERENCES

Axelrod, D. I. 1966. The Eocene Copper Basin flora of northeastern Nevada. *Univ. California Pubs. Geol. Sci.* 59:1-125.

Axelrod, D. I. and Bailey, H. P. 1969. Paleotemperature analysis of Tertiary floras. *Palaeogeogr., Palaeoclimatol., Palaeoecol.* 6: 163-195.

Becker, H. F. 1960. The Tertiary Mormon Creek flora from the upper Ruby River Basin in southwestern Montana. *Palaeontographica* B107:83-126.

Becker, H. F. 1961. Oligocene plants from the upper Ruby River Basin, southwestern Montana. *Geol. Soc. Amer. Mem.* 82:1-127.

Becker, H. F. 1969. Fossil plants of the Tertiary Beaverhead Basins in southwestern Montana. *Palaeontographica* B127:1-142.

Becker, H. F. 1972. The Metzel Ranch flora of the upper Ruby River Basin, southwestern Montana. *Palaeontographica* B141:1-61.

Becker, H. F. 1973. The York Ranch flora of the upper Ruby River Basin, southwestern Montana. *Palaeontographica* B143:18-93.

Brown, R. W. 1950. An Oligocene evergreen cherry from Oregon. *Wash. Acad. Sci. Jour.* 40:321-324.

Chaney, R. W. 1927. Geology and paleontology of the Crooked River Basin with special reference to the Bridge Creek flora. *Carnegie Inst. Wash. Pub.* 346:45-138.

Chaney, R. W. 1959. Miocene floras of the Columbia Plateau, Part I, composition and interpretation. *Carnegie Inst. Wash. Pub.* 617: 1-134.

Chaney, R. W. and Sanborn, E. I. 1933. The Goshen flora of west-central Oregon. *Carnegie Inst. Wash. Pub.* 439:1-103.

Collinson, M. E., Fowler, K. and Boulter, M. C. 1981. Floristic changes indicate a cooling climate in the Eocene of southern England. *Nature* 291:315-317.

Devereux, I. 1967. Oxygen isotope paleotemperature measurements on New Zealand Tertiary fossils. *New Zealand Jour. Sci.* 10:988-1011.

Epis, R. C. and Chapin, E. E. 1974. Stratigraphic nomenclature of the Thirtynine Mile volcanic field, central Colorado. *U.S. Geol. Surv. Bull.* 1395-C:1-23.

Evernden, J. F. and James, G. T. 1964. Potassium-argon dates and the Tertiary floras of North America. *Amer. Jour. Sci.* 262:945-974.

Frederiksen, N. O. 1980. Sporomorphs from the Jackson Group (upper Eocene) and adjacent strata of Mississippi and western Alabama. *U.S. Geol. Surv. Prof. Paper* 1084:1-75.

Keigwin, L. D. 1980. Palaeoceanographic change in the Pacific at the Eocene-Oligocene boundary. *Nature* 287:722-725.

Kennett, J. P., et al. 1975. Cenozoic paleoceanography in the southwest Pacific Ocean, Antarctic glaciation, and the development of the Circum-Antarctic Current. In J. P. Kennett et al., eds., *Init. Rept. Deep Sea Dril. Proj.* Washington, D.C., U.S. Govt. Printing Office, 29:1155-1169.

Kennett, J. P., et al. 1985. Palaeotectonic implications of increased late Eocene-early Oligocene volcanism from South Pacific DSDP sites. *Nature* 316:507-511.

Lathram, E. H., Pomeroy, J. S., Berg, H. C. and Loney, R. A. 1965. Reconnaissance geology of Admiralty Island, Alaska. *U.S. Geol. Surv. Bull.* 1181-R:R1-R48.

MacGinitie, H. D. 1937. The flora of the Weaverville beds of Trinity County, California. *Carnegie Inst. Wash. Pub.* 465:83-151.

MacGinitie, H. D. 1953. Fossil plants of the Florissant beds, Colorado. *Carnegie Inst. Wash. Pub.* 599:1-198.

Manchester, S. R. 1987. The fossil history of the Juglandaceae. *Missouri Bot. Gard. Monog. Syst. Botany* 21:1-137.

Manchester, S. R. and Crane, P. R. 1987. A new genus of Betulaceae from the Oligocene of western North America. *Bot. Gaz.* 148:263-273.

Manchester, S. R. and Meyer, H. W. 1987. Oligocene fossil plants of the John Day Formation, Fossil, Oregon. *Oregon Geology* 49: 115-127.

McClammer, J. U. 1978. Paleobotany and stratigraphy of the Yaquina flora (latest Oligocene-earliest Miocene) of western Oregon. Univ. Maryland, unpubl. M.A. dissertation.

Meyer, H. W. 1973. The Oligocene Lyons flora of northwestern Oregon. *Ore Bin* 35:37-51.

Meyer, H. W. 1986. An evaluation of the methods for estimating paleoaltitudes using Tertiary floras from the Rio Grande Rift vicinity, New Mexico and Colorado. Univ.

California (Berkeley), unpubl. Ph.D. dissertation.

Owens, J. P., Bybel, L. M., Paulachok, G., Ager, T. A., Gonzalez, V. M. and Sugarman, P. J. 1988. Stratigraphy of the Tertiary sediments in a 945-foot-deep corehole near Mays Landing in the southeastern New Jersey Coastal Plain. *U.S. Geol. Surv. Prof. Paper* 1484:1-39.

Plafker, G. 1987. Regional geology and petroleum potential of the northern Gulf of Alaska continental margin. In D. W. Scholl, A. Grantz, and J. G. Vedder, eds., *Geology and Resource Potential of the Continental Margin of Western North America and Adjacent Ocean Basins*, Circum-Pacific Council for Energy and Mineral Resources Earth Science Series 6:229-268.

Poore, R. Z. and Bybell, L. M. 1988. Eocene to Miocene biostratigraphy of New Jersey core ACGS #4: implications for regional stratigraphy. *U.S. Geol. Surv. Bull.* 1829:1-22.

Potbury, S. S. 1935. The La Porte flora of Plumas County, California. *Carnegie Inst. Wash. Pub.* 465:29-82.

Sanborn, E. I. 1935. The Comstock flora of west central Oregon. *Carnegie Inst. Wash. Pub.* 465:1-28.

Sanborn, E. I. 1949. The Scio flora of western Oregon. *Oregon State Univ. Monogr., Studies in Geology* 4:1-29.

Shackleton, N. J. and Kennett, J. P. 1975. Paleotemperature history of the Cenozoic and the initiation of Antarctic glaciations: oxygen and carbon isotope analyses in DSDP sites 277, 279, and 281. In J. P. Kennett et al., eds., *Init. Rept. Deep Sea Drill. Proj.* Washington, D.C., U.S. Govt. Printing Office, 29:743-755.

Smith, A. G., Hurley, A. M. and Briden, J. C. 1981. *Phanerozoic Paleocontinental World Maps*, Cambridge, Cambridge Univ. Press.

Swisher, C. C. and Prothero, D. R. 1990. Single-crystal $^{40}Ar/^{39}Ar$ dating of the Eocene-Oligocene transition in North America. *Science* 249:760-762.

Szafer, W. 1954. Pliocene flora from the vicinity of Czorsztyn (west Carpathians) and its relationship to the Pleistocene. *Inst. Geol. Prace, Warsaw* 11:5-238.

Tiffney, B. H. 1980. Fruits and seeds of the Brandon lignite, V. Rutaceae. *Jour. Arnold Arboretum* 61:1-40.

Tiffney, B. H. 1985a. Perspectives on the origin of the floristic similarity between eastern Asia and eastern North America. *Jour. Arnold Arboretum*, 66:73-94.

Tiffney, B. H. 1985b. The Eocene North Atlantic land bridge: its importance in Tertiary and modern phytogeography of the Northern Hemisphere. *Jour. Arnold Arboretum* 66:243-273.

Wahrhaftig, C., Wolfe, J. A., Leopold, E. B. and Lanphere, M. A. 1969. The coal-bearing group in the Nenana coal field of the central Alaska Range. *U.S. Geol. Survey Bull.* 1274-D:D1-D30.

Wing, S. L. 1987. Eocene and Oligocene floras and vegetation of the Rocky Mountains. *Annals Missouri Bot. Gard.* 74:748-784.

Wolfe, J. A. 1971. Tertiary climatic fluctuations and methods of analysis of Tertiary floras. *Palaeogeogr., Palaeoclimat., Palaeoecol.* 9:27-57.

Wolfe, J. A. 1972. An interpretation of Alaskan Tertiary floras. In A. Graham, ed., *Floristics and Paleofloristics of Asia and eastern North America*. Amsterdam, Elsevier, pp. 201-233.

Wolfe, J. A. 1977. Paleogene floras from the Gulf of Alaska region. *U.S. Geol. Surv. Prof. Paper* 997:1-108.

Wolfe, J. A. 1978. A paleobotanical interpretation of Tertiary climates in the Northern Hemisphere. *Amer. Scientist* 66:694-703.

Wolfe, J. A. 1979. Temperature parameters of the humid to mesic forests of eastern Asia and relation to forests of other regions of the Northern Hemisphere and Australasia. *U.S. Geol. Surv. Prof. Paper* 1106:1-37.

Wolfe, J. A. 1981a. A chronologic framework for Cenozoic megafossil floras of northwestern North America and its relation to marine geochronology. *Geol. Soc. Amer. Spec. Paper* 184:39-47.

Wolfe, J. A. 1981b. Paleoclimatic significance of the Oligocene and Neogene floras of the northwestern United States. In K. J. Niklas, ed., *Paleobotany, Paleoecology, and Evolution*. New York, Praeger, pp. 79-101.

Wolfe, J. A. 1987. An overview of the origins of the modern vegetation and flora of the Northern Rocky Mountains. *Ann. Missouri Bot. Garden* 74:785-803.

Wolfe, J. A. 1990a. Palaeobotanical evidence for a marked temperature increase following the Cretaceous/Tertiary boundary. *Nature* 343: 153-156.

Wolfe, J. A. 1990b. Estimates of Pliocene precipitation and temperature based on multivariate analysis of leaf physiognomy. *U.S. Geol. Surv. Open-File Rept.* 90-94:39-42.

Wolfe, J. A. and Hopkins, D. M. 1967. Climatic changes recorded by Tertiary land floras in northwestern North America. In K. Hatai, ed., *Tertiary Correlation and Climatic Changes in the Pacific.* Tokyo, Sasaki Printing and Publishing. pp. 67-76.

Wolfe, J. A. and Poore, R. Z. 1982. Tertiary marine and nonmarine climatic trends. In W. Berger and J. C. Crowell, eds., *Climate in Earth History.* Washington, D.C., Natl. Acad. Sci., pp. 154-158.

Wolfe, J. A. and Tanai, T. 1980. The Miocene Seldovia Point flora from the Kenai Group, Alaska. *U.S. Geol. Surv. Prof. Paper* 1105:1-52.

Wolfe, J. A. and Tanai, T. 1987. Systematics, phylogeny, and distribution of *Acer* (maples) in the Cenozoic of western North America. *Hokkaido Univ. Jour. Fac. Sci.*, ser. 4 22:1-246.

22. VEGETATIONAL AND FLORISTIC CHANGES AROUND THE EOCENE/OLIGOCENE BOUNDARY IN WESTERN AND CENTRAL EUROPE

by Margaret E. Collinson

ABSTRACT

Northwest Bohemia (Czechoslovakia) and the Weisselster Basin (Germany) yield macrofloral evidence of late Eocene and early Oligocene forest vegetation. This changes from dominantly evergreen subtropical (late Eocene) to mixed evergreen and deciduous with a warm but seasonal climate (early Oligocene). Unfortunately more precise dating of this transition is not possible. Palynological evidence from these areas and also from Belgium, the Paris Basin (France) and southern England indicates that the transition is marked by incoming of temperate elements; loss of tropical and subtropical elements and an increase in conifer pollen. These changes occur below MP21 mammal level. In the absence of radiometric dates and with limited biological corelation the Eocene/Oligocene boundary cannot be precisely located in any of the areas considered. Southern England and the Paris Basin have the best stratigraphic control and a major perturbation in pollen floras (sudden increase in proportion of temperate forms) in the Paris Basin may be a reflection of the cooling event observed in the marine realm. The marsh and swamp floras in southern England cross the boundary unchanged and only minor changes are observed in pollen floras at one level, possibly coeval with the event in the Paris Basin. In the context of the extensive sequence of Paleogene macrofloras in southern England the changes described here at the Eocene/Oligocene boundary in Europe are seen as the culmination of floristic change, resulting from cooling climate, which began in the early late Eocene. Palynological evidence from southern England, the Paris Basin and the Massif Armoricain supports this conclusion. Understanding of the extent to which any sharp perturbation near the boundary influenced future floral develop-ment must await improved stratigraphic resolution and further studies outside the Paris Basin.

INTRODUCTION

A critical assessment of vegetational change around the Eocene/Oligocene boundary requires two major sources of evidence. Firstly a complete, or more or less complete, sequence of strata through this interval or, alternatively, a series of isolated exposures with good local correlation. Secondly these strata should contain plant macrofossils and microfossils in comparable depositional settings or in settings where the differential taphonomic bias can be reliably assessed. In practice very few geological circumstances combine these features to provide ideal information. Around the Eocene/Oligocene boundary in Europe the situation is particularly problematic. Areas which yield diverse plant macrofossils (e.g. Germany and Czechoslovakia) tend to lack both radiometric dates and biostratigraphically useful fossils. Furthermore good plant-bearing sites are often in geologically isolated basins. Alternatively, the more continuous, well-dated sequences have been deposited in marine to marginal marine or coastal floodplain settings which mostly yield only plant microfossils (e.g. the Paris Basin) or a more restricted, local assemblage of macrofossils (e.g. southern England). In this paper I summarise current evidence from these disparate sites and indicate possibilities for future work. I deal with evidence from each geographical area separately. The summary attempts to integrate this evidence to form an overall picture of vegetational and floristic change. The main area of Western and Central Europe under consideration includes England, and the major western European landmass (e.g. Mai, 1989, Figure 2; Mai and Walther, 1985, Figure 2, p. 145) but excludes the Fennoscandinavian platform and the eastern land masses. The area was situated between about 45-55° north palaeolatitude (Mai and Walther, 1985; Ziegler et al., 1983).

ENGLAND: LONDON AND HAMPSHIRE BASINS

The strata of the Solent Group (Headon Hill; Bembridge Limestone and Bouldnor Formations in ascending order; Figure 22.2) provide a relatively complete sequence through late Eocene and early Oligocene strata. Local correlation and correlation with the Paris Basin sequence can be established on the basis of lithostratigraphy and zonations for mammals, calcareous nannoplankton, dinoflagellates, and charophytes (For references see explanations to Figures 22.1 and 22.2; see also Hooker, this volume).

In view of the importance of the pollen *Boehlensipollis hohli* in continental Europe its occurrence in England must be considered. The species was recorded by Collinson et al. (1981) from the Ramnor Borehole (London Basin -- M.C. Boulter record) and from Whitecliff Bay, Isle of Wight, (Hampshire Basin -- K. Fowler record) from the uppermost Barton and Headon Beds (the latter = Headon Hill Formation; Figure 22.2). According to Vinken (1988, Figure 144) this pollen is diagnostic of zone SP7 of Oligocene age (see discussion below under Czechoslovakia and Germany). However, also in Vinken (1988, Figure 145) Fowler did not record *Boehlensipollis* on his Hampshire Basin chart even though the chart coverage extends into the lower part of the Headon Beds. The stratigraphic range of this chart does imply that in compiling it Fowler only considered mainland Hampshire Basin and not the Isle of Wight. Hence it is unclear whether he revised his earlier determination or merely excluded it from consideration. Boulter and Hubbard (1982) repeated the range from upper Barton Beds into Headon Beds for Hampshire Basin *Boehlensipollis* which was therefore included in Hubbard and Boulter (1983) and Boulter (1984). Gruas-Cavagnetto (1976) only recorded this pollen from her upper Hamstead Beds sample (= Bouldnor Formation above the top of Figure 22.1; Cranmore Member Figure 22.2). Her next sample in descending order was from the middle Headon Beds (= Colwell Bay Member Figure 22.2) of Colwell Bay, Isle of Wight. *Boehlensipollis* was not recorded by Collinson (1983) or Machin (1971) who studied the Bembridge Marls member and the entire Solent Group of the Isle of Wight respectively. Boulter (pers. comm., 1991) confirms the late Eocene occurrence of *Boehlensipollis* in southern England and considers that this pollen occurs earlier in England than elsewhere in Europe. For the purposes of this paper I am following correlations established by a variety of other biostratigraphical indicators (Figures 22.1 and 22.2).

Floral changes during this interval have been discussed using palynological and macrofossil evidence (Collinson et al., 1981; Collinson, 1983, 1990; Collinson and Hooker, 1987; Boulter and Hubbard, 1982; Hubbard and Boulter, 1983; Boulter, 1984; Machin, 1971). Collinson et al. (1981) noted a gradual decline in pollen and macrofossils whose nearest living relatives are tropical or subtropical, along with a general change in floristsic composition, beginning in the early late Eocene. Boulter and Hubbard (1982), Hubbard and Boulter (1983) and Boulter (1984) expanded the palynological aspect of that work and utilised principal components analysis of a large number of species rich palynofloras to recognise groups of pollen and spore taxa. On the basis of diagnostic grains and nearest living relatives these groups were considered to represent "Eocene deciduous angiosperm forest" which included water plant associations, "Eocene fern and conifer forest" and "Eocene paratropical forest." In their analysis, Boulter and Hubbard (1982, Text-Figure 6, p. 65) the Eocene paratropical forest dominates late early Eocene (London Clay) and the early middle Eocene (Bracklesham group) whereas "Eocene fern and conifer forest" (precedes and) succeeds this in the late middle Eocene (Barton Beds). The "Eocene deciduous angiosperm forest" dominates in the late Eocene Headon Beds

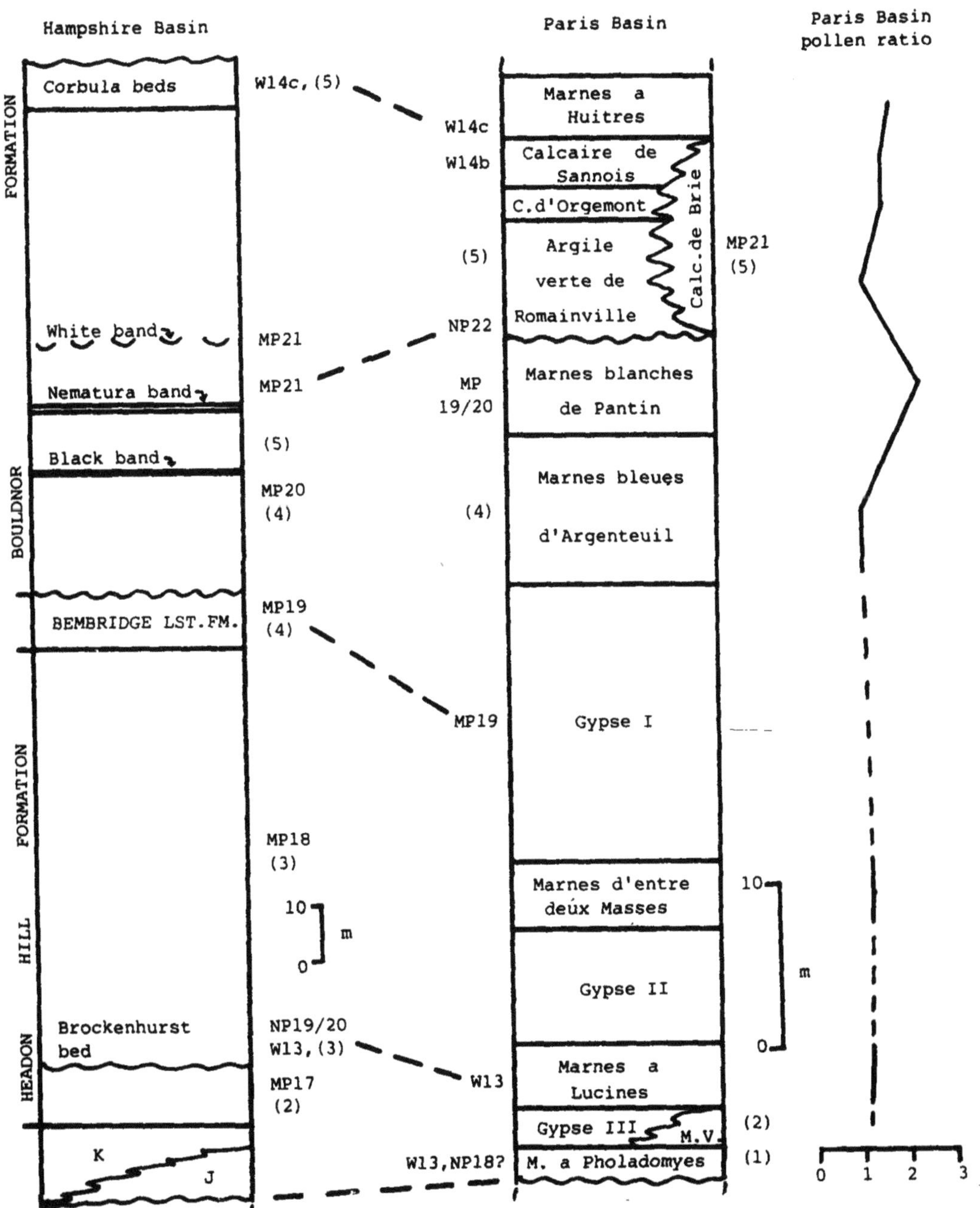

FIGURE 22.1. Lithic logs of the late Eocene (Priabonian) and early Oligocene (Rupelian) strata in the Hampshire (mainly from Bristow et al., 1889; Insole and Daley, 1985) and Paris Basins (mainly from Blondeau et al., 1968; Pomerol, 1973); and ratio of tropical to subtropical/temperate pollen forms in the Paris Basin, numbers representing the ratio scale (redrawn from Chateauneuf, 1980 with gaps inserted for the barren gypsum levels). J and K are beds of the Barton Clay and Becton Sand Formations respectively (Hooker, 1986). Mammal levels (MP) folow Brunet et al. (1987), Cavelier (1979) and Hooker (this volume); calcareous nannoplankton zones (NP) follow Aubry (1985); dinocyst zones (w) in the Paris Basin follow Chateauneuf and Gruas-Cavagnetto in Vinken (1988, Figure 167) and are interpreted in the Hampshire Basin from taxa recorded by Liengjarern et al. (1980); charophyte zones are from Riveline (1984) and Riveline and Cavelier (1987) and are referred to by numbers :- 1 *Psilochara repanda*; 2 *Gyrogona tuberosa*; 3 *Harrisichara vasiformis-tuberculata*; 4 *Stephanochara vectensis*; 5 *S. pinguis*. The Eocene/Oligocene boundary is likely to be within the Marnes bleues d'Argenteuil in the Paris Basin and in the lower part of the Bouldnor Formation in the Hampshire Basin. Abbreviations :- Calc = Calcaire; C = Caillasses; M = Marnes; M.V. = Marnes de Verzenay.

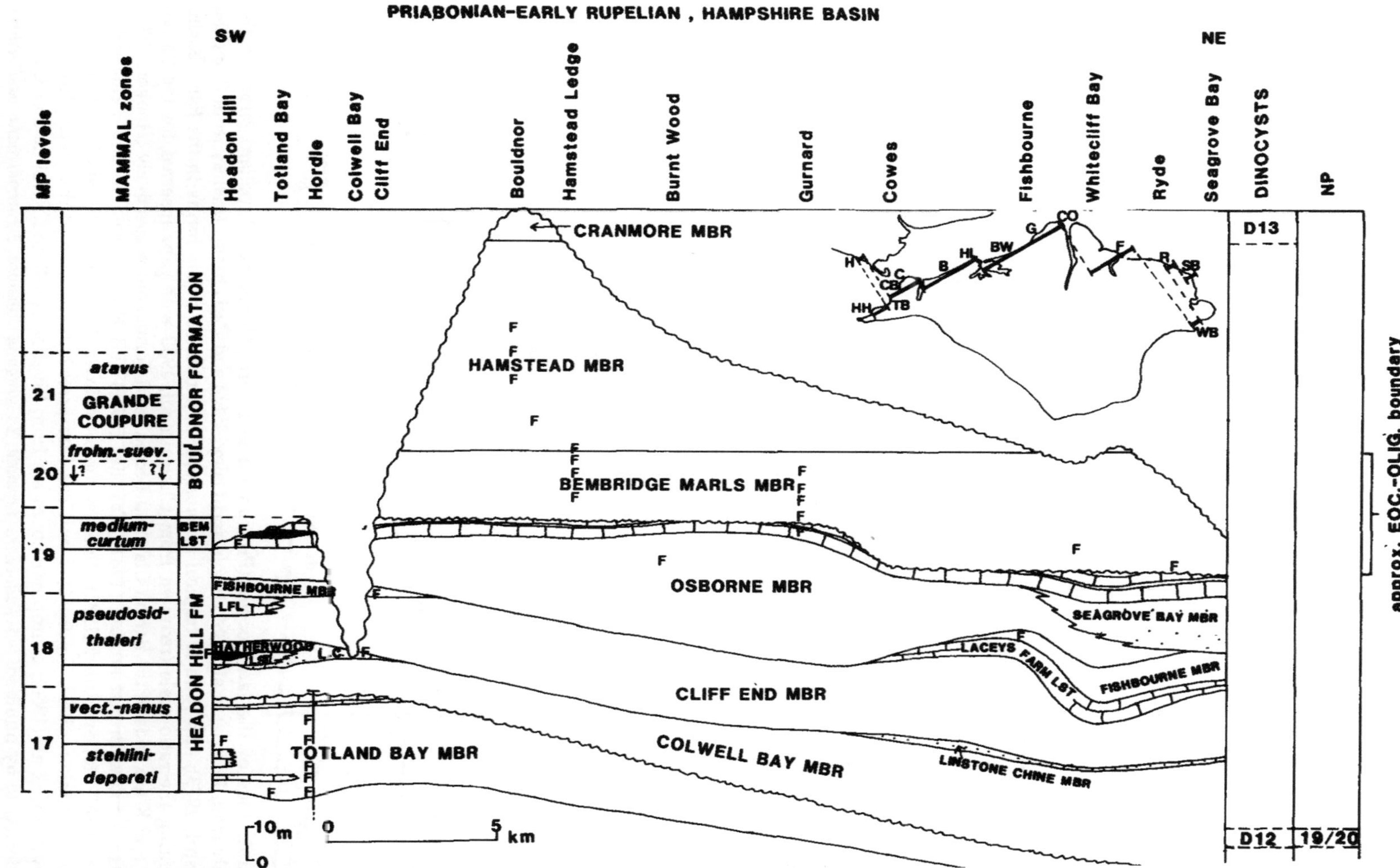

FIGURE 22.2. Section through late Eocene (Priabonian) to early Oligocene (Rupelian) strata in the Hampshire Basin. Lithostratigraphic and thickness data from Insole and Daley (1985) with personal observations by J.J. Hooker. Lithological symbols :- brickwork = limestone; solid black = dominantly organic; stipple = sand; unshaded = dominantly mud. Mammal zones from Hooker (1987) and Bosma (1974); MP levels from Brunet et al. (1987); NP occurrences from Aubry (1985); Dinocysts from Meyer (in Vinken 1988). Wavy lines indicate major unconformities and the levels of modern erosion in the Isle of Wight. T-shaped bar indicates the extent of strata at Hordle. Inset shows location of logs, mainly on the Isle of Wight. Major levels at which plant macrofossils have been collected are indicated with F.

(=Headon Hill Formation, Figure 22.2) and in a single sample from the upper Hamstead Beds (= Cranmore Member, Figure 22.2). They state (p.65 explanation to Text-Figure 6) "The same stratigraphic variations are shown in the results from data published on material from France (Chateauneuf, 1980 [Paris Basin] and Ollivier-Pierre, 1980 [Massif Armoricain]) and Nigeria." Their correlation to these regions is not clarified. Boulter and Hubbard's work strengthens the earlier conclusions that floral change, with climatic cooling, occurred during the late Eocene in Southern England. However, they do not provide any evidence for changes within the latest late Eocene or early Oligocene. Ollivier-Pierre et al. (1987) also recorded floristic change and the gradual loss of pollen with tropical or subtropical nearest living relatives, from the early middle Eocene onwards in pollen floras from southern England, the Paris Basin, the Massif Armoricain and Belgium. Details of the French and Belgian studies are considered below.

A late Eocene marker, NP19/20 occurs in the Brockenhurst Bed (Figure 22.1) in the Colwell Bay Member (Figure 22.1). The hantkeninid extinction occurs within NP21 (Pomerol and Premoli-Silva, 1986) so the Eocene/Oligocene boundary must be some distance above the Colwell Bay Member. Hooker (this volume; see also Figs 22.1 and 22.2) argued that the likely position of the Eocene/Oligocene boundary in southern England is within the lower part of the Bouldnor Formation which probably correlates with the Marnes bleues d'Argenteuil and the Marnes Blanches de Pantin in the Paris Basin (see Figure 22.1). A fluctuation in pollen occurrences in the Marnes Blanches de Pantin was interpreted by Chateauneuf (1980) and Cavelier et al. (1981) as a sharp climatic deterioration (see France below). A cooling event is recognised above the hantkeninid extinction in the marine realm from physical evidence such as the oxygen isotope record (Pomerol and Premoli-Silva, 1986). The floras of the Lower Bouldnor Formation (Bembridge Marls and Hamstead Members) in southern England may thus encompass both the Eocene/ Oligocene boundary and the climatic event; constrained above by the occurrence of mammal Zone MP21 (Hooker, this volume) above the Nematura Band (Figure 22.1).

These floras were examined in detail by Collinson (1983) and Machin (1971). Whereas the Bembridge Marls crop out from East to West Isle of Wight, the Hamstead Member can only be studied at Hamstead Ledge and Bouldnor Cliff having been removed by erosion elsewhere (Figure 22.2). Pomerol (1989) recognised a major unconformity "at the Eocene/Oligocene boundary" situated in England "between the Bembridge and Hamstead Beds at the level of the Black Band" (For Black Band, see Figure 22.1). This positioning of the unconformity in England seems to have been based on correlation with the Marabet Gypsum (underlying the Argile Verte, at the top of the Marnes Blanches [Figure 22.1]) in the Paris Basin. Liengjarern et al. (1980) in contrast suggest a correlation of the Nematura Band and the Argile Verte on the basis of dinoflagellates (Figure 22.1).

Machin (1971) and, to a lesser extent Collinson (1983), noted a palynofloral change in the Black Band compared with Bembridge Marls below. However, Collinson (1983) recorded no significant variation in the macroflora of marsh and swamp elements. The Black Band and its underlying green clay were considered to be the uppermost representative of a sequence of nine such alternations in the Bembridge Marls. In each case, olive-brown clays rich in plant material overlie olive-green clays largely lacking plant macrofossils. The similarities of the Black Band to the others include a transitional boundary between the two lithologies marked by minor intertonguing and the same micro- and macro-floral and faunal (mollusc and ostracod) associations in all the green clays and all the brown clays. The only significant differences in the Black Band were an absence of *Pediastrum* in the immediately underlying green clays; an absence of palm pollen and a marked increase in *Inaperturopollenites magnus* (a conifer pollen possibly of Araucariaceae or Pinaceae; cf. *Larix* or *Pseudotsuga*). The Black Band is the thickest of the nine organic-rich levels reported by Collinson (1983) and it contains least clastic input. It is also the most lateraly extensive having been recorded in old boreholes over much of the Isle of Wight (Bristow et al., 1889). It is thus possible that this band may represent a prolonged period of very slow sedimentation (although probably not an unconformity). Varving within the organic rich levels was noted by Collinson (1983) as a possible indication of annual couplets; thus a detailed palynological analysis through this band could prove very informative. Pomerol's hiatus HP11, if one can be recognised in this part of the succession, may exist in the lower part of the Hamstead Member between the Black Band and the Nematura Band according to biostrati-graphic correlation (Figure 22.1). Further study here would be of considerable value for clarification of Eocene/Oligocene boundary events.

Palynological analyses by Machin (1971) suggested a "more marked floral change at the Bembridge: Hamstead border than at any other horizon" in her study of Headon to Hamstead Beds (i.e. the Solent Group) on the Isle of Wight. Machin noted in particular a disappearance of pollen assigned to *Thrinax* (a palm) and the incoming of *Tsuga, Keteleeria, Pterocarya* and *Alnus;* Ulmaceae pollen were also more abundant and the increase in inaperturate conifer pollen was most striking (from 10-20%). *Ephedra* (resembling *E. distachya*--personal observations of type slides) was recorded by Machin in the Bembridge Marls and Hamstead Beds but not earlier. As the nearest living relatives are largely plants of dry areas this pollen may indicate increasing aridity. It has recently been encountered (R. Singer, pers. comm.) in the clastic sequence of the Bembridge Limestone Formation (Figure 22.2) i.e., within MP 19. Additional evidence for seasonal aridity during deposition of the Bembridge Limestone Formation is provided by silcretes and calcretes (Daley, 1989) and by mammalian faunas indicative of an open environment (Hooker, this volume; in press). Machin (1971) considered that the pollen changes generally indicated more temperate conditions in the Hamstead Beds compared with underlying strata. None of the pollen diagnostic of Oligocene zone SP6 or 7 (Vinken 1988, Figure 144) were recorded by Machin (1971) or Collinson (1983). It would be unwise to consider this as conclusively negative evidence in view of the early date of Machin's work, her treatment of pollen mainly as modern genera and the fact that Collinson concentrated on the olive green/olive brown alternations.

Vegetational change prior to the late Eocene in England and comparison with the main European landmass.

Buzek et al. (1990) remark that changes in forest floras (such as those seen in Germany and Czechoslovakia) are not likely to be recognised in the largely water and swamp associations of the coastal floodplain settings of southern England. Whilst this is to some extent correct, palynological assemblages ought to record some change in regional floras even if macrofloras do not. Furthermore, Collinson (1990) has documented the overall changes in the coastal wetlands for southern England and has indicated very clear transition from a *Nipa*-dominated mangrove in the late early and early middle Eocene to increasing freshwater marshlands until these dominate in the latest late Eocene and early Oligocene. These changes are paralled by a gradual loss of fruit and seed macrofossils whose nearest living relatives are tropical, paratropical or subtropical, from a maximal diversity in the early and early middle Eocene to almost absent in the latest late Eocene and early Oligocene (Collinson et al., 1981; Collinson and Hooker, 1987). The late Eocene and early Oligocene floras which have contributed to this study are indicated by (F) on Figure 22.2.

Early middle Eocene assemblages in England do show some similarity with those of Messel and Geiseltal in Germany (Collinson and Hooker, 1987; Collinson, 1988; Mai, 1976) and the Grès à Sabals in France (Vaudois-Miéja, 1985). Middle Eocene leaf floras in southern England still await detailed study. Most are impression fossils only, but a physiognomic treatment based on the methods of Wolfe (1990, this volume) would be helpful. The leaf floras of Messel suggest warm humid and equable climate but with possible seasonality (Wilde, 1989); but they may reflect influence of the local lake microclimate rather than representing the regional climate. Similarly in the Geiseltal the middle Eocene floras (pollen zone Pg 15 of Krutzsch, 1976; see Haubold, 1989, for mammal zonation) reflect the unusual depositional setting in a "salt dome" complex (Krumbiegel et al., 1983). The upper Eocene of Geiseltal (placed in Pg Zone 18 by Krutzsch, 1976, and see Franzen and Haubold, 1987) is considered to have a similar flora to that of the Weisselster Basin (see below) (Mai and Walther, 1985). There is an unconformity between the middle and upper Eocene at Geiseltal and no higher strata are represented. Messel is an isolated lake basin. In the Weisselster Basin and Bohemia rare floral assemblages may represent late middle Eocene (Pg16-17) vegetation. They include dominant Lauraceae and typical tropical elements but are generally poorly understood (Kvacek et al., 1989). Earlier strata are lacking. There is thus currently no sequence in continental Europe which is sufficiently extensive to enable comparison with the spectrum of changes observed in macrofloras in southern England.

FRANCE AND BELGIUM

Major studies of palynofloras around the Eocene/Oligocene boundary in France and Belgium have been undertaken by Chateauneuf (1980), and also reported in Cavelier et al. (1981), Ollivier-Pierre et al. (1987) and Ollivier-Pierre et al. (1988). These studies all indicate changes in palynofloras at this time, particularly losses of taxa with tropical or subtropical nearest living relatives. In some cases different taxa, but still with tropical or subtropical affinities, replace those lost (e.g. in the Paris Basin during the middle Eocene). In other cases a gradual decline occurs without replacement e.g., the Paris Basin and Massif Armoricain during the late Eocene (Ollivier-Pierre et al., 1987). Note that Boulter and Hubbard (1982) recognised floristic changes in the Paris Basin and the Massif Armoricain comparable to those in southern England (see above).

In the Paris Basin Ollivier-Pierre et al. (1987) document the coincidental loss of 8 "tropical" taxa above the Marnes bleues d'Argenteuil (see Figure 22.1) with no new tropical taxa replacing them. Thus a fall from 23 to 20 then to only 12 "tropical" taxa characterises the late Eocene. This "event" is apparently coincident with that recorded by Chateauneuf (1980) and Cavelier et al (1981) represented on Figure 22.1 by a sharp fluctuation in the tropical and subtropical to temperate pollen ratio in the Marnes blanches de Pantin. Chateauneuf's detailed study (1980) examined numerous sections in the Paris Basin, hence the curve and pollen distributions (Chateauneuf, 1980, Figures 67 and 68; Cavelier et al., 1981, Figure 1) are a summation of a large body of data. The ratio remains more or less constant during the late middle and early late Eocene. It is marked by a major rise in proportions of temperate taxa in the Marnes blanches de Pantin. It is then followed by recovery to previous state in the overlying Argile Verte de Romainville, followed by a subsequent gradual increase in proportions of temperate taxa (Figure 22.1). The pollen assemblage also shows a marked reduction in diversity falling from about 270 species in the Marnes bleues to only about 50 in the Marnes blanches. The diversity only recovers to about 130 species after this event. The apparent contradiction between the recovery of tropical taxa (Chateauneuf, 1980) and survival only of persistent forms with no incoming tropical taxa (Ollivier-Pierre et al., 1987) may be explained either by different concepts of "tropical and subtropical" or by the larger data set analysed by Chateauneuf. Chateauneuf also remarked upon a temporary decline in swamp and aquatic vegetation (a total which included pollen related to mangroves, freshwater aquatic and marsh herbs, wetland shrubs and trees, and riparian forest elements (Chateauneuf, 1980, pp. 333-334)) in the Marnes blanches from a previous average of about 65 species (peak 90 species) to only 12 species. This is only a slightly larger fall than might be expected on the basis of the overall drop in diversity but it was considered to indicate a possible fall in humidity (Chateauneuf, 1980, p. 338). Aridity in the Paris Basin area during the late Eocene is indicated by numerous gypsum deposits (Figure 22.1) and also by silcretes (Koeniguer, 1987). Pollen of *Ephedra* appears prior to the sequence illustrated on Figure 22.1 and extends into the Oligocene (Chateauneuf, 1980, pp. 338-9). One additional feature of the sequence was a major increase in conifer pollen (including *Tsuga, Picea* and *Sciadopitys*) in the Marnes blanches de Pantin.

In Belgium (Roche, in Ollivier-Pierre et al., 1987) originations and extinctions of "tropical" taxa through the Eocene indicate an overall pattern of gradual floristic change but do not show any major perterbations of abundance. However, the Sables de Neerrepen show an increase in bisaccate conifer pollen along with the expansion of temperate taxa such as Betulaceae, *Ulmus*, *Fraxinus* and *Pterocarya*. The overlying Henis Clay is dominated by tropical to warm temperate elements, while the temperate elements return above this. From the recent study by Gullentops (1990) the Hoogbutsel mammal bed (MP 21, Brunet et al., 1987) is placed between the Neerrepen Sands and the Henis Clay. The Grimmertingen Sands, which underlie the Neerrepen Sands, contain calcareous nannofossils of Zone NP21 (Gramann, 1990). [See also Figure 153 by Roche in Vinken (1988)]. These pollen fluctuations may therefore possibly reflect the same event recorded by Chateauneuf (1980) in the Paris Basin (see above).

In the Massif Armoricain (Ollivier-Pierre, in Ollivier-Pierre et al., 1987) and Ollivier-Pierre et al. (1988) discuss changes around the late Eocene. Ollivier-Pierre et al. (1988) studied a borehole from the area of Céaucé. Charophyte occurrences permit correlation of part of this sequence with the Marnes bleues in the Paris Basin and the Bembridge Marls in southern England. The Céaucé pollen assemblage is similar to that from the Paris Basin Marnes bleues. Overlying Oligocene pollen assemblages show a decrease in tropical pollen taxa and an increase in conifers (Ollivier-Pierre in Ollivier-Pierre et al., 1987).

GERMANY: WEISSELSTER BASIN

The floras of the Zeitz complex (for discussion of the use of the term floral complex see Walther, 1990, p.150) in Germany are similar to those of the late Eocene in Bohemia whilst those at Seifhennersdorf and the Haselbach complex in Germany are similar to the early Oligocene of Bohemia (Buzek et al 1990, Kvacek et al 1989). The Zeitz complex (Weisselster Basin) is the type section for pollen zone Pg 18 (Kvacek et al., 1989) of the late Eocene (Priabonian). The Haselbach flora is considered to be middle Oligocene (Kvacek et al., 1989; Walther, 1990; Mai and Walther, 1978) according to local stratigraphic correlation within the Basin where the index pollen *Boehlensipollis hohli* (Zone Pg 20) is recorded (Mai and Walther, 1978, p.169). Intervening sediments are absent in the Weisselster Basin due either to a hiatus or subsequent erosion (Kvacek et al., 1989).

The late Eocene Zeitz flora (typified by elements termed palaeotropical by authors cited here) consists mainly of broad-leaved evergreen elements, interpreted as representing a subtropical mesophytic forest, which existed under nearly frostless perhumid climate with indistinct dry winter periods. The most important forest community was an *Eotrigonobalanus*/ Lauraceae/ *Rhodomyrtophyllum* forest. Vegetation indications are supported by foliar physiognomy (e.g. many large leaves with entire margins and thick coriaceous texture typical of evergreens) as well as by climatic tolerances of nearest living relatives (Mai and Walther, 1985; Walther, 1990). Lists of dominant taxa are given by Kvacek et al. (1989) and Walther (1990) while a full palaeobotanical monograph is given by Mai and Walther (1985). This floral complex was termed early mastixioideen flora by Mai (1967).

The Haselbach floral complex contains deciduous elements (termed Arctotertiary by authors cited here) such as *Taxodium*, *Alnus*, *Ulmus*, *Carya*, *Acer*, *Carpinus* and *Populus* but these grew alongside persisting "Palaeotropical" elements such as *Dryophyllum*, Lauraceae, Mastixiaceae, Symplocaceae and *Myrica longifolia*. The "Arcto-tertiary" elements dominate in terms of numbers of specimens but the "Palaeotropical" forms dominate the species list. The vegetation is interpreted as mixed mesophytic forest (transitional to broad-leaved evergreen forest) existing under a warm temperate humid climate with distinct seasons (tending towards a subtropical climate with summer rain). Again the climatic and vegetational inferences are supported by foliar physiognomy (many leaves thin and deciduous and with serrate margins for example) as well as climatic tolerances of nearest living relatives (Mai and Walther, 1978, 1985). A summary of the flora is given by Kvacek et al. (1989) and by Walther (1990). Full palaeobotanical treatment is given by Mai and Walther (1978). These floras are known as late Mastixioideen floras (Mai, 1967).

The Weisselster Basin deposits accumulated in a nearshore lowland in a deltaic area of a complex river system entering the German Polish sea. Deposits include channel fills, lake sediments and oxbow lake infills, the latter often producing lignitic sediments which are most productive of plants (Walther, 1990; Kvacek et al., 1989). The Zeitz complex includes plant communities representing water and swamp associations and riparian forests but is dominated by the subtropical laurel forests. Details of more varied plant communities have been recognised within the Haselbach complex. Water and swamp associations include a floating community; a submerged community and a fern-monocotyledonous riparian community; an *Arthrotaxus* swamp forest and a *Nyssa/ Taxodium* humid forest, the latter is the first ocurrence of an association which later becomes very important in forming some brown coal deposits of the European Miocene (Teichmuller, 1989). A *Populus/Liquidambar/ Ulmus* riparian forest, dominated by *Populus*, is also recognised. The swamp and riparian forests have affinities with modern North America whereas the major mixed mesophytic forests have affinities with east Asia. The recognition of these various wetland sub communities provides tentative evidence that the Eocene/Oligocene floristic change also involved increased partitioning of habitats, resulting in patchiness (perhaps due to increased seasonality) and generally more open vegetation. There is however, apparently no indication of significant herbaceous communities other than those of marsh or riparian associations.

CZECHOSLOVAKIA

The northwest Bohemian area contains sediments of late Eocene age (Stare Sedlo Formation and equivalents) which are mainly sands, sandstones and quartzites; overlain unconformably in most areas (most complete in the Cheb Basin) by volcanodetritic strata including coals and coaly clays of the early Oligocene. These were deposited in an inland setting. Precise corelation is not possible. The amount of missing strata, the position within the late Eocene or early Oligocene, and the exact position of the Eocene/Oligocene boundary are not known.

Fossil floras are known from northwest Bohemia in the Cheb and Sokolov Basins, the Doupovské Hory Mountains, the north Bohemian Basin, the Ceské Stredohorí Mountains (from west to east; see Buzek et al., 1990, map, Figure 2). In the north Bohemian Basin the floras are known mainly from early Miocene strata (Buzek et al., 1988).

In the Doupovské Hory Mountains volcanic complex, from Dvérce and Detan near Podborany, mammalian faunas document the occurrence of MP 21 an Oligocene, post- "Grande Coupure," fauna (Buzek et al., 1988; Buzek et al., 1990). According to Buzek et al. (1990) this represents the only safely dated mammal zone MP 21 in Central Europe and it is also the most easterly occurrence of this zone. Two pollen taxa are thought significant stratigraphically: *Boehlensipollis hohli* and *Aglaoreidia cyclops* (Buzek et al., 1988); and the associated macroflora includes a high ratio of "Arcto-tertiary" elements (Buzek et al., 1988). *Boehlensipollis hohli* was considered an index pollen for zone SP7 which begins within the early Oligocene and terminates at the end of the middle Oligocene (Range chart Figure 144, Vinken, 1988). The first appearance of *B. hohli* in the Paris Basin is in the Argiles Vertes de Romainville (see Figure 22.1; Chateauneuf and Gruas-Cavagnetto, Figure 147, in Vinken, 1988). The underlying sands, sandstones and clays contain pollen spectra and macrofossils similar to those of the Staré Sedlo Formation in the Cheb and Sokolov Basins, but their dating is uncertain.

In the Ceske Stredohorí mountains two localities have yielded MP21, Oligocene, post "Grande Coupure" mammals but an extensive time span may be represented and local correlation between isolated sites is difficult in this area (Buzek et al., 1990). The main flora here is from Kuclin near Bilina (Buzek et al., 1988, Buzek et al., 1990) but it occupies an isolated position not enabling correlation. Several elements in the Kuclin flora also occur in the Bembridge Marls of southern England (*Acrostichum, Hooleya, Palaeocarya, Doliostrobus*) but it is unwise to attach much significance to these in the absence of independent correlation.

In the Sokolov Basin the Staré Sedlo Formation is overlain by the Nové Sedlo Formation. The latter contains a single mammalian record of *Entelodon* sp. (Buzek et al., 1990). in a tuffaceous clay overlying the Josef Seam. This is certainly no older than MP21. The index pollen *Boehlensipollis hohlii* is found in the

Josef Seam (Buzek et al., 1988). However, the time difference between the two formations is unclear. In the Cheb Basin a more continuous depositional sequence is present with the lower sands grading through sandy clay facies into the overlying coaly facies.

Palynological evidence from both the Cheb and Sokolov Basins places the Stare Sedlo formation in the zones Pg 16-18 of Krutzsch suggesting a late Eocene age., most probably Priabonian (Konzalova, 1990). Stratigraphically diagnostic pollen include the *Bombacacidites* group and *Porocolpopollenites vestibulum* group (Fig.144; Vinken, 1988). The base of the Stare Sedlo formation may be late middle Eocene or early late Eocene (Buzek et al., 1990). Similar pollen assemblages are found in presumed equivalents of the Staré Sedlo Formation in the Ceské Stredohori Mountains.

The late Eocene floras suggest humid and equable paratropical to subtropical climate (Buzek et al., 1990). Typical elements (which are termed Palaeotropical in the papers cited here) include *Rhodomyrtophyllum, Steinhauera,* species of *Litsea* and *Laurophyllum* and *Eotrigonobalanus furcinerve* (Buzek et al., 1990; Knobloch, 1990). The leaves are mainly entire margined, thick and coriaceous suggesting an evergreen habit (Knobloch, 1990). This flora is very similar to that termed the "Zeitz floristic complex" in Germany (Knobloch, 1990; Buzek et al., 1990) (see above). In contrast, typical elements (which are termed Arctotertiary in papers cited here) of the early Oligocene floras include more serrate margined, deciduous taxa with Betulaceae, Juglandaceae, *Acer, Platanus,* and so on, mixed with some of the previous evergreen forms, suggesting a mixed forest development following a climatic decline (Buzek et al., 1990). However, the transition is not total because the Josef seam contains mainly *Eotrigonobalanus furcinerve* a laurophyllous element with an *Alnus-Taxodium* alluvial forest assemblage overlying this (Buzek et al., 1990). Clearly some facies control is indicated; perhaps pockets of the older floras were surviving in certain microhabitats. Suggestions of floral similarity with the Bembridge Marls (Kuclin macroflora and aquatic and marshy palynoflora from Dvercé) are interesting (Buzek et al., 1990). Kvacek et al. (1989, p. 395) also refer to fresh water associations with *Stratiotes* and *Brasenia* in lignites of the Josef seam and lateral lithological correlates. These genera are typical of the upper part of the Solent Group (Figure 22.2) in England. However, the Czech records come from strata including MP 21 fauna which does not occur in England until the Lower Hamstead Beds overlying the Bembridge Marls (Figures 22.1 and 22.2; Hooker, this volume). Collinson (1983) noted similar macrofloral assemblages, dominated by aquatic and marsh elements, in both the Bembridge Marls and Lower Hamstead Beds and many of the pollen taxa are in common despite some changes (Machin, 1971; Collinson, 1983). It is possible that a similar aquatic and marsh plant community existed in various parts of Europe at this time.

The MP21 mammal fauna in northwest Bohemia is associated with numerous "Arctotertiary" deciduous elements. Precise timing is unknown but late Eocene humid equable evergreen forests seem to have been replaced by mixed mesophytic deciduous forests at some time within or below MP21. According to Kvacek et al. (1989), the locality at Valec (in the Doupov mountains) may represent the Eocene/Oligocene boundary interval. This is a massive limestone containing *Doliostrobus* and two "Palaeotropical" leaf taxa. Buzek et al. (1990) emphasise the uncertain dating of this level while noting that it may represent pre "Grande Coupure" time. The statement by Kvacek et al. (1989) that this locality "corresponds obviously to" the Bembridge Marls in England, Haring (Austria) and partly with the Tard Clays in Hungary should not be taken as proven.

OTHER AREAS

The Bovey Basin in the western part of England probably includes Eocene and Oligocene strata (Chandler, 1957; Wilkinson and Boulter, 1981; Edwards, 1976) but dating is uncertain. This is an isolated basin and the lignites, clays and sands are devoid of faunal remains although rich in flora in places. The deposits are exposed during temporary working of clays so prospects for future study are limited.

Schuler (in Ollivier-Pierre et al., 1987) discusses the sequence in the Rhine Graben,

Germany where nannoplankton control exists with recognition of NP22 in the early Oligocene middle and upper Pechelbronn Formation. The upper Pechelbronn contains MP21 and the lower Pechelbronn MP19/20 (Tobien, 1987, modified by Hooker, this volume). Several tropical taxa disappear below the Lower Pechelbronn Formation. The middle and upper Pechelbronn are characterised by temperate taxa, many conifers and the occurrence of *Boehlensipollis hohli* (see also Figure 149 by Schuler, in Vinken, 1988).

Tertiary macrofloras from Poland have recently been reviewed by Lancucka-Srodoniowa et al. (1983). They list several floras from the Eocene of the Tatra mountains but these have not yet been studied in detail and stratigraphic position is unclear. Very few Oligocene floras are known from Poland. According to the charts (Figures 158 and 159 in Vinken, 1988) areas in northern Poland might provide data pertinent to Eocene/ Oligocene boundary events but I am not aware of relevant publications.

According to Givulescu (1990) macrofloras do not occur in Roumania between the middle Eocene and the middle Oligocene. Cernjavska et al. (1988) describe extensive macro and microfloras from Paleogene sediments in Bulgaria including ecological and physiognomic interpretations. Unfortunately no independent dating is available for this material.

Mihajlovic (1990) gives a summary of Eocene and Oligocene floras from Yugoslavia which show changing character and indicate the influence of aridity especially in floras dated by means of NP 23 nannoplankton as late early Oligocene. Unfortunately independent dating for the earlier floras is not yet available.

Teslenko (1990), Pulatova (1990) and Iljinskaya (1988) indicate floral changes around the Eocene/Oligocene boundary in Ukraine and the USSR which involves, in the lower Oligocene, the incoming of deciduous taxa including Betulaceae, Juglandaceae and *Alnus* and an increase in conifers, but with some evergreen taxa persisting. Also, pollen indicating "half-open" ... "steppe-meadow plant communities" was recorded in the Oligocene by Pulatova (1990). *Ephedra* was apparently widespread in the late Eocene at least in the Tadzik depression (Pulatova, 1990). Undoubtedly these eastern areas can potentially provide data relevant to interpretation of Eocene/Oligocene boundary events but they are outside the scope of this paper. It will probably be necessary for the publication of the floristic monographic revision "Magnoliophyta Fossilia URSS" to be completed before a thorough study can be made of floral changes in this region.

ACKNOWLEDGEMENTS

I wish to thank M. Boulter, C. Buzek, J. Hooker, M. Konzalová, Z. Kvacek, H. Walther, J. Wolfe and S. Wing for helpful comments and discussion; Jerry Hooker for invaluable assistance with stratigraphic questions and for critically reading the manuscript; and Don Prothero for inviting me to participate in the Penrose symposium and for his patience while awaiting this manuscript. This work was undertaken during the tenure of a Royal Society 1983 University Research Fellowship which is gratefully acknowledged.

REFERENCES

Aubry, M.-P. 1985 Northwest European Palaeogene magnetostratigraphy, biostratigraphy, and paleogeography: Calcareous nannofossil evidence. *Geology* 13: 198-202.

Blondeau, A., Cavelier, C. and Pomerol, C. 1968. *Colloque sur l'Eocene Milan, Nice, Paris, Reims 1968. Livret Guide des Excursions dans le Bassin de Paris.* 25-26-28 Mai 1968.

Bosma, A.A. 1974. Rodent biostratigraphy of the Eocene-Oligocene transitional strata of the Isle of Wight. *Utrecht micropaleont. Bull. spec. Publs.* 1: 1-128.

Boulter, M.C. 1984. Palaeobotanical evidence for land-surface temperature in the European Palaeogene. In Brenchley, P., ed., *Fossils and Climate.* Chichester, John Wiley and Sons, pp. 35-47.

Boulter, M.C. and Hubbard, R.N.L.B. 1982. Objective paleoecological and biostratigraphic interpretation of Tertiary palynological data by multivariate statistical analysis. *Palynology* 6: 55-68.

Bristow, H.W., Reid, C. and Strahan, A. 1889. The Geology of the Isle of Wight. *Mem. Geol. Surv. U.K.*

Brunet, M., Franzen, J.L., Godinot, M., Hooker, J.J., Legendre, S., Schmidt-Kittler, N. and

Vianey-Liaud, M., coordinators. 1987. European reference levels and correlation tables. *Münchner Geowiss. Abh. (A)* 10: 13-31.

Buzek, C., Ctyroky, P., Fejfar, O., Konzalova, M. and Kvacek, Z. 1988. Biostratigraphy of Tertiary coal-bearing deposits of Bohemia and Moravia (C.S.R.). In Pesek, J and Vozar, J., eds., *Coal-bearing formations of Czechoslovakia.* Bratislava, Dionyz Stur Institute of Geology, pp. 291-305.

Buzek, C., Fejfar O., Konzalova, M. and Kvacek, Z. 1990. Floristic changes around Stehlin's Grande Coupure in Central Europe. In Knobloch, E. and Kvacek, Z., eds., *Proceedings of the Symposium Paleofloristic and paleoclimatic changes in the Cretaceous and Tertiary.* Prague, Geological Survey Publisher, pp.167-181.

Cavelier, C. 1979. La limite Eocène-Oligocène en Europe Occidentale. *Sci. Géol. Strasb., Mém.* 54: 1-280.

Cavelier,C., Chateauneuf, J-C., Pomerol, C., Rabussier, D., Renard, M. and Vergnaud-Grazzini, C. 1981. The geological events at the Eocene/Oligocene boundary. *Palaeogeogr., Palaeoclimat., Palaeoecol.* 36: 223-248.

Cernjavska, S., Palamarev, E. and Petkova, A. 1988. Micropaleobotanical and macropalaeobotanical characteristics of the Paleogene sediments in the Hvojna Basin, (Central Rhodopes). *Palaeontology, Stratigraphy and Lithology* 26: 26-36; Bulgarian Academy of Sciences, Sofia.

Chandler, M.E.J. 1957. The Oligocene flora of the Bovey Tracey Lake Basin, Devonshire. *Bull. Br. Mus. nat. Hist. (Geol.)* 3: 71-123.

Chateauneuf, J-J, 1980. Palynostratigraphie et paléoclimatolgie de l'Eocène superieur et de l'Oligocène du Bassin de Paris (France). *Mém. B.R.G.M.* 116: 1-357.

Collinson, M.E. 1983. Palaeofloristic assemblages and palaeoecology of the lower Oligocene Bembridge Marls, Hamstead Ledge, Isle of Wight. *Bot. Jour. Linn. Soc.* 86: 177-225.

Collinson, M. E. 1988. The special significance of the middle Eocene fruit and seed flora from Messel, West Germany. *Cour. Forsch-Inst. Senckenberg* 115: 187-197.

Collinson, M.E. 1990. Vegetational change during the Palaeogene in the coastal wetlands of southern England. In Knobloch, E. and Kvacek, Z., eds., *Proceedings of the Symposium Paleofloristic and paleoclimatic changes in the Cretaceous and Tertiary.* Prague, Geological Survey Publisher, pp.135-139

Collinson, M.E., Fowler, K. and Boulter, M.C. 1981. Floristic changes indicate a cooling climate in the Eocene of southern England. *Nature* 291: 315-317.

Collinson, M.E. and Hooker, J.J. 1987. Vegetational and mammalian faunal changes in the early Tertiary of southern England. In Friis, E.M., Chaloner, W.G. and Crane, P.R., eds. *The origins of angiosperms and their biological consequences.* Cambridge, Cambridge University Press, pp. 259-304.

Daley, B. 1989. Silica pseudomorphs from the Bembridge Limestone (Upper Eocene) of the Isle of Wight, southern England and their palaeoclimatic significance. *Palaeogeogr., Palaeoclimat., Palaeoecol.* 69: 233-240.

Edwards, R.A 1976. Tertiary sediments and structure of the Bovey Basin, South Devon. *Proc. Geol. Assoc.* 87: 1-26.

Franzen, J.L. and Haubold, H. 1987. The biostratigraphic and palaeoecologic significance of the middle Eocene locality Geiseltal near Halle (German Democratic Republic). *Münchner Geowiss. Abh.* 10: 93-99.

Givulescu, R. 1990. Paleofloral changes in Romania between the Eocene and the Pliocene. In Knobloch, E. and Kvacek, Z., eds., *Proceedings of the Symposium Paleofloristic and paleoclimatic changes in the Cretaceous and Tertiary.* Prague, Geological Survey Publisher, pp. 113-114.

Gramann, F. 1990. Eocene/Oligocene boundary definitions and the sequence of strata in N.W. Germany. *Tert. Res.* 11: 73-82.

Gruas-Cavagnetto, C. 1976. Etude palynologique du Paléogène du sud de l'Angleterre. *Cah. Micropaléont.* 1: 5-49.

Gullentops, F. 1990. Sequence stratigraphy of the Tongrian and early Rupelian in the Belgian type area. *Tert. Res.* 11: 83-96.

Haubold, H. 1989. Die referenzfauna des Geiseltalium, MP levels 11 bis 13 (Mitteleozän, Lutetium). *Palaeovertebrata* 19: 81-93.

Hooker, J.J. 1986. Mammals from the Bartonian (middle/late Eocene) of the Hampshire

Basin, southern England. *Bull. Br. Mus. nat. Hist. (Geol.)* 39: 191-478.

Hooker, J.J. 1987. Mammalian faunal events in the English Hampshire Basin, (late Eocene-early Oligocene) and their application to European biostratigraphy. *Munchner Geowiss. Abh.* 10: 109-116.

Hooker, J.J. in press. Mammalian palaeoecology of the Bembridge Limestone Formation (late Eocene, S. England). *Historical Biology.*

Hubbard, R.N.L.B. and Boulter, M.C. 1983. Reconstruction of Palaeogene climate from palynological evidence. *Nature* 301: 147-150.

Iljinskaja, I.A. 1988. Contributions to the characterisation and origin of the Turgai flora of the U.S.S.R.. *Tert. Res.* 9: 169-180.

Insole, A.N. and Daley, B. 1985. A revision of the lithostratigraphical nomenclature of the late Eocene and early Oligocene strata of the Hampshire Basin, southern England. *Tert. Res.* 7: 67-100.

Koeniguer, J.-Cl. 1987. Paléoécologie de quelques gisements a végétaux fossiles du Bassin Parisien. *Bull. Inf. Geol. Bass. Paris* 24: 23-32.

Knobloch, E. 1990. The flora of the Staré Sedlo Formation in West Bohemia, upper Eocene. In Knobloch, E. and Kvacek, Z., eds., *Proceedings of the Symposium Paleofloristic and paleoclimatic changes in the Cretaceous and Tertiary.* Prague, Geological Survey Publisher, pp. 159-166.

Knobloch, E. and Kvacek, Z. 1990. *Proceedings of the Symposium Paleofloristic and paleoclimatic changes in the Cretaceous and Tertiary.* Prague, Geological Survey Publisher.

Konzalová, M. 1990. Palynological investigation of the basal Tertiary sediments (Palaeogene) in western Bohemia and their correlation with the Carpathian region. *Západné Karpaty, ser. Pal.* 14: 73-102.

Krumbiegel, G. Rüffle, L. and Haubold, H. 1983. *Das Eozäne Geiseltal.* Wittenberg Lutherstadt, A. Ziemsen.

Krutzsch, W. 1976. Die Mikroflora der Braunkohle des Geiseltals, Teil IV : Die stratigraphische Stellung des Geiseltalsprofils im Eozän und die sporenstratigraphische Untergliederung des mittleren Eozäns. *Abh. zentr. geol. Inst.* 26: 47-92

Kvacek, Z., Walther, H. and Buzek, C. 1989. Palaeogene floras of W. Bohemia (C.S.S.R.) and the Weisselster Basin (G.D.R.) and their correlation. *Casopis pro miner a geol,* 34: 385-401.

Lancucka-Srodoniowa, M., Zastawniak, E. and Guzik, J. 1983. Macroscopic plant remains from the Tertiary of Poland. *Acta Palaeobot.* 23: 21-76.

Liengjarern, M., Costa, L and Downie, C. 1980. Dinoflagellate cysts from the upper Eocene-lower Oligocene of the Isle of Wight. *Palaeontology* 23: 475-499.

Machin, J. 1971. Plant microfossils from the Tertiary deposits of the Isle of Wight. *New Phytol.* 70: 851-872.

Mai, D.H. 1967. Die Florenzonen, der Florenwechsel und die Vorstellungen uber den Klimaablauf im Jungtertiär der Deutschen Demokratischen Republik. *Abh. zentr. geol. Inst. Berlin* 10: 55-81.

Mai, D.H. 1976. Fossile Fruchte und samen aus dem Mitteleozän des Geiseltales. *Abh. zentr. geol. Inst.* 26: 93-149.

Mai, D.H. 1989. Development and regional differentiation of the European vegetation during the Tertiary. *Pl. Syst. Evol.* 162: 79-91.

Mai, D.H. and Walther, H. 1978. Die Floren der Haselbacher Serie im Weisselster-Becken (Bezirk Leipzig, DDR). *Abh. St. Mus. Miner. Geol. Dresd.* 28: 1-200 + 1-102.

Mai, D. H. and Walther, H. 1985. Die obereozänen Floren des Weisselster-Beckens und seiner Randgebiete. *Abh. St. Mus. Miner. Geol. Dresd.* 33: 1-260.

Mihajlovic, D. 1990. Palaeogene flora of Yugoslavia, a review. In Knobloch, E. and Kvacek, Z., eds. *Proceedings of the Symposium Paleofloristic and paleoclimatic changes in the Cretaceous and Tertiary.* Prague, Geological Survey Publisher, pp. 141-146.

Ollivier-Pierre, M-F., Gruas-Cavagnetto, C., Roche, E. and Schuler, M. 1987. Eléments de flore de type tropical et variations climatiques au Paléogene dans quelques bassins d'Europe nord-occidentale. *Mém. Trav. E.P.H.E. Inst. Montpellier* 17: 173-205.

Ollivier-Pierre, M-F., Riveline, J., Lautridou, J.P. and Cavelier, C. 1988. Le fosse de Céaucé (Orne) et les bassins ludiens (Eocene supérieur) de la partie orientale du Massif armoricain: Sedimentologie, paléontologie Intéret stratigraphique et tectonique. *Géologie de la France* 1: 51-60.

Pomerol, C. 1973. *Stratigraphie et Paleogeographie Ere Cenozoique (Tertiaire et Quaternaire)*. Paris, Doin.

Pomerol, C. 1989. Stratigraphy of the Palaeogene: hiatuses and transitions. *Proc. Geol. Assoc.* 100: 313-324.

Pomerol, C. and Premoli-Silva, I., eds., 1986. *Terminal Eocene Events*. Amsterdam, Elsevier.

Pulatova, M.S. 1990. The main stages in development of the Eocene and Oligocene floras of south-eastern middle Asia. In Knobloch, E. and Kvacek, Z., eds., *Proceedings of the Symposium Paleofloristic and paleoclimatic changes in the Cretaceous and Tertiary*. Prague, Geological Survey Publisher, pp.147-148.

Riveline, J. 1984. Les gisements à charophytes du Cénozoique (Danien à Burdigalien) d'Europe Occidentale. Lithostratigraphie, biostratigraphie, chronostratigraphie. *Bull. Inf. Geol. Bass. Paris (1984) Mém. h.s. no.* 4, 523pp.

Riveline, J. and Cavelier, C. 1987. Les charophytes du passage Eocène moyen-Eocène supérieur en Europe occidentale. Implications stratigraphique. *Bull. Soc. géol. France* III (2): 307-315.

Teichmuller, M. 1989. The genesis of coal from the viewpoint of coal petrology. *Int. J. Coal. Geol.* 12: 1-87.

Teslenko, Ju.V. 1990. Floristic and palaeoclimatic changes in Palaeogene and Neogene times over the territory of the Ukraine. In Knobloch, E. and Kvacek, Z., eds., *Proceedings of the Symposium Paleofloristic and paleocli-matic changes in the Cretaceous and Tertiary*. Prague, Geological Survey Publisher, pp.115-118.

Tobien, H. 1987. The position of the "Grande Coupure" in the Palaeogene of the Upper Rhine Graben and the Mainz Basin. *Münchner Geowiss. Abh.* 10: 197-202.

Vaudois-Miéja, N. 1985. La flore des Grès à Palmiers de l'ouest de la France. *Bull Sect Sciences, vol. jubilaire* 8: 259-273

Vinken, R. (compiler) 1988. The Northwest European Tertiary Basin. Results of the International Geological Correlation Programme Project No 124. *Geol. Jb. (A)* 100: 7-508; Figs. 1-267; 3 tabs; 7 maps.

Walther, H. 1990. The Weisselster Basin (GDR)-- an example of the development and history of Paleogene forest vegetation in Central Europe. In Knobloch, E. and Kvacek, Z., eds., *Proceedings of the Symposium Paleofloristic and paleoclimatic changes in the Cretaceous and Tertiary*. Prague, Geological Survey Publisher, pp.149-158.

Wilde, V. 1989. Untersuchungen zur Systematik der Blattreste aus dem Mitteleozan der Grube Messel bei Darmstadt (Hessen, Bundesrepublik Deutschland). *Cour. Forsch-Inst. Senckenberg* 115: 1-213.

Wilkinson, G.B. and Boulter, M. C. 1981. Oligocene pollen and spores from the western part of the British Isles. *Palaeontographica* B, 175: 27-83.

Wolfe, J. A. 1990. Palaeobotanical evidence for a marked temperature increase following the Cretaceous/Tertiary boundary. *Nature* 343: 153-156.

Ziegler, A.M., Scotese, C.R. and Barrett, S.F. 1983. Mesozoic and Cenozoic paleogeographic maps. In Brosche and Sundermann, eds., *Tidal friction and the earth's rotation* II. Berlin, Springer, pp.239-252.

23. WESTERN NORTH AMERICAN REPTILE AND AMPHIBIAN RECORD ACROSS THE EOCENE/ OLIGOCENE BOUNDARY AND ITS CLIMATIC IMPLICATIONS

By J. Howard Hutchison

ABSTRACT

The changes in herpetofaunas across the Eocene/Oligocene boundary in the western interior of North America appear to have begun as early as the Uintan, with a general decrease in diversity of aquatic amphibians and reptiles. The interval between the Chadronian and Orellan (Eocene/Oligocene boundary) marks a sharp terminal decline in aquatic forms, but only a modest decrease in diversity of terrestrial forms. The decline in the aquatic forms appears to be related to a general increase in aridity, decrease in permanent rivers and streams, and filling of the major lacustrine basins. The pattern of reptile and amphibian diversity and size changes is less dramatically correlated with temperature.

INTRODUCTION

The Eocene-Oligocene interval marks a dramatic transformation and modernization of the non-marine herpetofauna of western North America, a modernization substantially achieved by the end of the Oligocene. The record of amphibian and reptile diversity (richness) across the interval of the Eocene/ Oligocene boundary is only adequately represented in the western part of North America, west of the Mississippi River and from southern Canada to northern Mexico, specifically along the southern and eastern slopes of the Rocky Mountain province south to the Big Bend region of Texas. The eastern non-marine faunas, especially during the Eocene-Oligocene interval, are very poorly known and are therefore excluded. Another reason for limiting the study area is that eastern and western North America were divided during long intervals in the Cretaceous by the interior seaway and it is likely that faunal remnants of this division persisted until the late Eocene or Oligocene. It is also likely that significant latitudinal faunal provincialism existed within the study area. Much of this current record is still undocumented but a survey of several major collections provides additional data on the large reptiles (turtles and crocodilians). The smaller vertebrates are sporadically represented and are to an unknown extent a reflection of the activity of paleomammalogists and the nature and thoroughness of their sampling of the nonmammalian part of microvertebrate collections.

Many previous studies of selected parts of the herpetofauna history noted striking changes in the stratigraphic record. In discussing the shift from predominantly aquatic to terrestrially dominated turtle faunas in the northern Rockies and Plains, Hay (1908, p. 39) noted that "There can be no doubt that at this time [Orellan] the climate of the Plains region had become arid and the streams few."

Tihen (1964) did not discuss the Eocene-Oligocene transition in detail but noted that "there appears to be an abrupt transition from the archaic to the basically modern faunas, probably beginning in the late Oligocene [Whitneyan], but certainly centered around the Lower Miocene Arikareean."

Estes (1970a), in discussing the origins of the Recent North American lower vertebrate fauna summarized the temporal distributions of the families and the appearance of extant genera. For the Amphibia, he tabulated that 29% of the extant families appeared during the Oligocene and 64% of the extant families are represented by the first appearance of an extant genus in the Oligocene (but note that his Oligocene included the Chadronian). For the reptiles, he tabulated that only 7% of the modern families appeared in the Oligocene but that this included 50% of the extant genera of extant families. He notes both the rarity of Oligocene lower vertebrates and that the change from an "Austrariparian type of lower vertebrate fauna so common earlier is replaced by one quite different." He cited Dorf (1959) regarding a general cooling trend initiated in the Oligocene and speculates that "as midcontinental aridity began to develop in the late Oligocene [Orellan and Whitneyan], a number of derivatives of forms associated with the more moist Arctotertiary Geoflora seem to have moved into and became adapted for drier conditions, thereby becoming associated with an essentially Madrotertiary flora". Estes and Baez (1985, p. 156) again noted "in the Oligocene (including Early Arikareean...), there is a considerable increase in the number of Recent genera represented in the fossil record."

In a survey of diversity changes in turtles and crocodilians throughout the Cenozoic of North-Central region of western United States, I noted a sharp reduction in large aquatic reptiles during the Orellan - Arikareean interval (Hutchison, 1982). I also attributed this decline to a decrease in availability of permanent surface waters. The less complete record in southwestern Untied States shows a similar pattern. The previous data indicated that the greatest magnitude in the declines leading toward the Orellan-Arikareean reduction in diversity occurred in steps between the Bridgerian and Uintan (50% of 19 genera) and between the Chadronian and Orellan (38% of 8 genera). As will be noted below, these stepped reductions persist, although the relative proportions of the declines have altered owing to additions of new records, expanded taxonomic coverage, and some changes in correlations. While in 1982, I noted a rough correlation of the sea level stands (see Vail et al., 1977) with the aquatic reptile diversity curves, the major drop in sea level stand appears to follow rather than precede the major decline in reptile diversity in the Oligocene (30 Ma).

Rage (1984b) provided a detailed tabulation of temporal range data of amphibians and reptiles across the Eocene/Oligocene boundary (the Grande Coupure, between the Escamps and Hoogbutsel reference levels) in France. Of the 49 taxa (excluding one questioned taxon), 53 percent disappear at the boundary, although 8 percent reappear later in the record. He noted only one appearance at the boundary, but several new taxa and a few older taxa appear in the next higher interval. In a shorter summary paper, Rage (1986) reported that "...more than three fourths of the uppermost Eocene fauna died out by the Eocene-Oligocene transition" and the "Grande Coupure" ... "represents the major break in the amphibian and reptile Cenozoic history." He noted that the "extinctions may be the consequence of the deterioration of climatic conditions" and that "some oriental immigrants settled in Western Europe and their arrival probably contributed to the extinction of the Eocene autochthonous forms."

Finally, Schleich (1986) provided a summary of reptile and amphibian diversity in the Tertiary of Europe at the level of Epoch. He illustrated a peak of diversity in the Eocene followed by a sharp decline in the Oligocene, and suggested that this trend may indicate a shift from "tropical" to "?colder/?drier" climates (Schleich, 1986, Figure 4).

WESTERN RECORD

Compilation of taxa over a wide region and temporal span is significantly influenced by a number of factors that can not be readily controlled or evaluated. The reader is referred to Donovan (1989) regarding the many potential pitfalls with regard to taxonomy and phylogeny, geography, preservation, sampling, sedimentation, correlation, and succession. It would be desirable to tabulate only lineage extinctions and appearances but these are not readily available or known for many of the groups and may have occurred beyond the geographic area of study. When a single lineage is divided into several taxa on a morphologic basis, the individual taxa will appear as a series of appearances and disappearances. This may fatally flaw a pure diversity compilation. The object of this survey, however, is to attempt to assess the changes in the herpetofaunas across the Eocene/ Oligocene boundary in the study area (western United States). These changes may be owing to either true extinctions and originations, immigrations or emigrations, or shifts in morphology (punctuations) distinctive enough to be regarded as taxa by systematists. The latter may also be a simple artifact of a spotty record with gaps in an otherwise gradual temporal morphocline. It is beyond the scope of this survey to assess which of these patterns is in play for particular taxa, although these should be kept in mind and are worth further exploration. In a strict sense, the relationships of the taxa (morphotaxa) are irrelevant because it is only necessary to identify, by some means, different morphologic units and look at their distributions in the temporal record. Sorting out of lineage relationships is a second order of analysis that is not attempted here.

Although I only have confidence in the relatively better sampling of the turtles which contribute large and easily preserved elements, the other groups are included for a more complete view of the changing herpetofauna. The questions to be addressed here are whether the changes between the early Eocene and late Oligocene reflect a gradual and temporally correlated modernization of the fauna, an episodic change, or a catastrophic change.

Although the raw data is plotted and tabulated, I have taken the liberty of interpolating taxonomic occurrences (questioned and higher level taxa usually excluded) when there is a gap in the record of the study area. This can only be justified as a recognition of the poor sampling or study (which I think is great) of the affected groups for some of the time intervals. The gaps may be real and the reappearance of a taxon later in the record (the Lazarus effect) may indicate an immigration following extinction in the study area. The degree that the curves with and without the interpolated data differ may provide a crude indication of uncertainties in the data and thus the conclusions drawn from them.

Because of the lumping together of faunas of the same Land Mammal "Age" for tabulation purposes, latitudinal differences are likely to be submerged in the attempt to assess the general regional trend. This will also tend to minimize the stratigraphic differences within local sections. For example the genus *Baptemys* appears to vanish from the record in the northern states by the beginning of the Uintan but is present in later beds in Texas and its possible descendant or sister taxon *Dermatemys* survives today in southern Mexico.

The following discussion builds upon from the stratigraphic range summaries of Estes (1970a, 1981, 1983a) for amphibians, lizards, amphisbaenians, and salamanders and Rage (1984a) for the snakes with additions or modifications from subsequent primary literature. I attempted little or no reevaluation of these summaries, and I made no attempt to reconfirm the data in the summaries other than to refine the stratigraphic occurrence, thus errors in these summaries are likely to be incorporated here in addition to those I may have introduced. The turtle and crocodilian ranges, however, are based upon my own observations and collections, the primary literature, and compilations from museum collections and thus represent a single, if not correct, point of view with regard to taxonomy and synonymy. For this reason, some turtle and crocodilian curves are presented separately as well as in combination with other groups.

Most of the herpetofaunas in the West occur with or in close proximity to land mammal faunas and their age assignment is thus subject to the same spectrum of correlation problems as for North American Land Mammal "Ages" (Woodburne, 1987). The herpetofaunas of the Wasatchian and Bridgerian intervals are generally better and more finely sampled, and studied, than those of the Uintan to Whitneyan. Even more significant is the patchy geographic nature of the record which may overemphasize local and provincial differences (latitudinal, elevation, etc.) and much additional collecting and faunal analysis is needed to support the generality of the pattern.

Only by including as wide a variety of depositional environments as possible does a general picture of herpetofauna diversity begin to emerge. The breadth of included habitats depends on a number of important factors, including the original environmental habitat diversity, diagenetic history, current accessibility of appropriate exposures, and intensity of collecting and study. For all of these factors, assessment or adjustment is difficult or impossible. The present summary is influenced both stratigraphically and numerically by the preponderance of localities

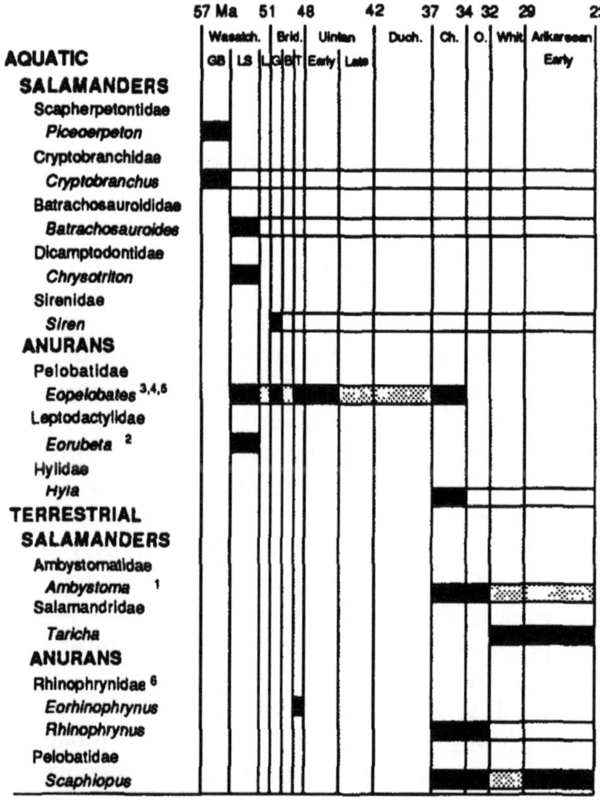

[1] Holman, 1987. Estes (1983a:137) notes that the ? *Batrachosauroides* reported by Setoguchi (1978) may be an ambystomatid.
[2] Hecht, 1960.
[3] Estes, 1970b.
[4] Grande, 1984.
[5] Golz and Lillegraven, 1977.
[6] Rhinophrynids - genus indet. are also known from the early Uintan (Golz and Lillegraven, 1977).

TABLE 23.1. Record of Eocene-Oligocene amphibians in western North America (Canada, USA). Basic data from Estes (1970a). Abbreviations B = Blackbuttean; Bridg. = Bridgerian; C = Chadronian; Duch. = Duchesnean; G = Gardnerbuttean; GB = Graybullian; L = Lostcabinian; LS = Lysitean; O = Orellan; Ma = Mega-annum; Wasatch. = Wasatchian; Whit. = Whitneyan. Bars: Black = record; shaded = interpolated (filled gap); unshaded = unfilled gap (extralimital or at least one age unit removed from the intervals under study).

from the Rocky Mountain states of Montana, South Dakota, Wyoming, and Utah and to a lesser extent additions from Texas and California. In spite of these problems, the fossil record of the herpetofauna of western North America exhibits previously noted general trends that have been strengthened with additional collecting.

The stratigraphic distributions used in the analysis are presented in Tables 23.1-23.4 and partitioned by major taxonomic groupings and by estimated habitat preference (aquatic or terrestrial). The stratigraphic range tables are compiled arbitrarily at the level of genus because of the many problems in determining the number and synonymy of the species but, in general, additions of species would tend to accentuate the amplitude of the Bridgerian and Chadronian curves. The biostratigraphic ranges are recorded only to the level of North American Land Mammal "Ages" and "Subages" (Woodburne, 1987) and the bars extended to span the interval regardless of whether narrower intra-age ranges are known, a bias that will tend to force a more stepped appearance to the data. The year time span of the Land Mammal "Ages" correlations are rounded to the nearest whole number. The Land Mammal "Ages" span the Eocene to late Oligocene, and are subdivided as follows: Wasatchian ("subages": Graybullian, Lysitian, Lostcabinian), Bridgerian ("subages": Gardnerbuttean, Blackbuttean, Twinbuttean), Uintan (subdivisions: early, late), Duchesnean, Chadronian, Orellan, Whitneyan, and early Arikareean.

The raw plots (Figure 23.1A-C) exhibit peaks in generic diversity in the Wasatchian, Bridgerian, and Chadronian and a low point in the Duchesnean. The addition of interpolated points (infilling of range gaps in the record) increases the peaks in the Wasatchian and Bridgerian but adds little to the Chadronian and reduces the overall amplitude of the fluctuations. The addition of the interpolated data notably elevates the Duchesnean (and Gardnerbuttean) diversity above those of the Whitneyan and Arikareean. Despite the long temporal range of the Duchesnean, it is the most poorly sampled of the age intervals. The general pattern is one of general decline in diversity through the later Eocene and Oligocene.

Dividing the taxa by habitat preference into

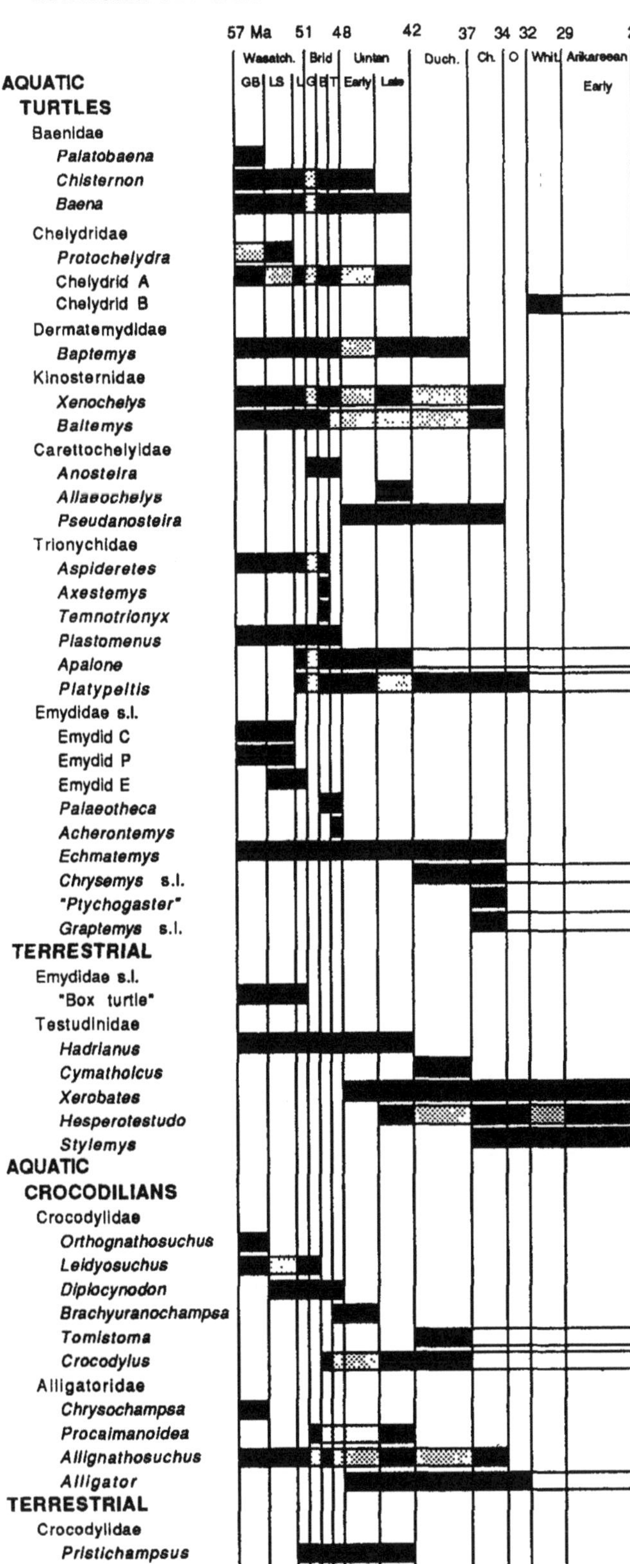

TABLE 23.2. Record of Eocene-Oligocene turtles and crocodilians in western North America (Canada, USA). See Table 23.1 for definition of abbreviations and symbols.

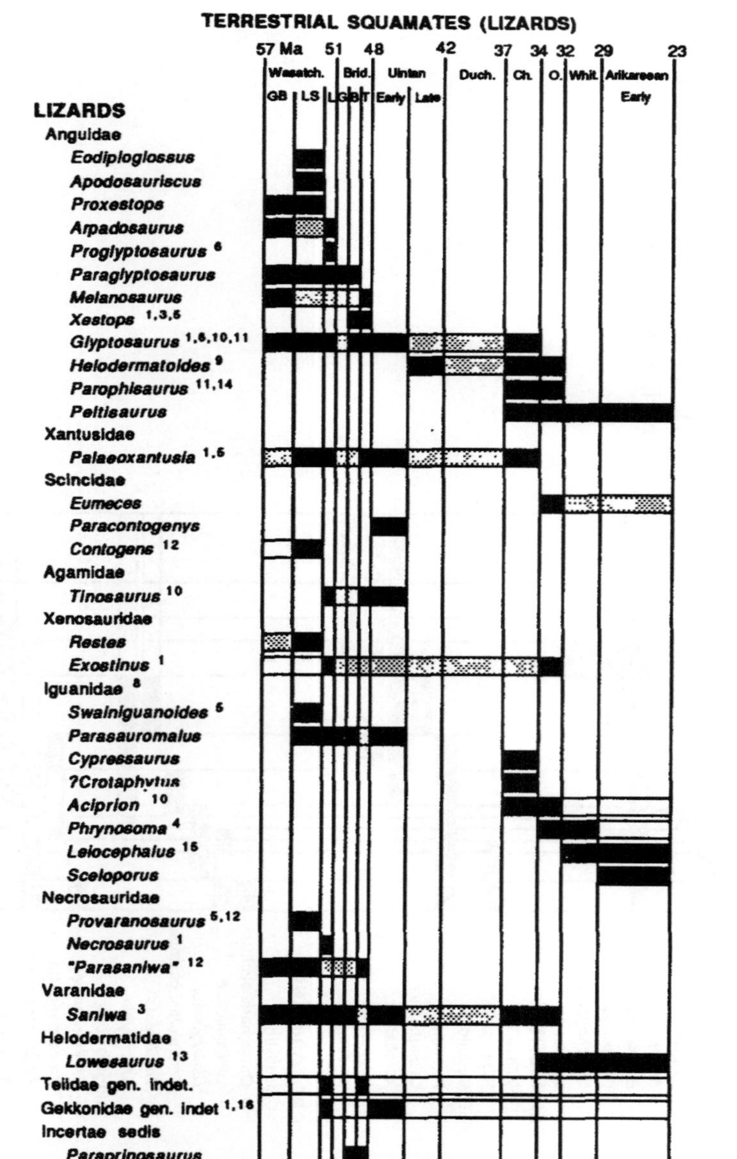

[1] Hirsch et al., 1987, list a Lostcabinian *?Ophisaurus* (not recorded here - possibly referable to *Parophisaurus*.

[2] Sullivan, 1987.

[3] Estes, 1988.

[4] Holman, 1987.

[5] Sullivan and Lucas, 1988, recorded *Swainiguanoides* and *Provaranosaurus* as "cf.".

[6] Sullivan, 1989.

[7] Torres, 1985.

[8] Eaton, 1982, record of cf. *Lepidophyma* ignored.

[9] Estes, 1983a, lists the Uintan record of *Helodermatoides* as questioned.

[10] Emry, 1973, recorded a questioned ? *Tinosaurus*. None of Emry's records are mentioned by Estes,1983a.

[11] Ostrander, 1985.

[12] Gauthier, pers. com., 1990. He regards the Eocene records of *Parasaniwa* as referable to a new genus.

[13] Pregill et al., 1986.

[14] Both pre-Chadronian and post Whitneyan records of *Peltisaurus* are questioned but only the former are not used.

[15] Setoguchi, 1978.

[16] Hirsh et al., 1987, on egg shell and Estes, 1983a on dentary fragments recognize gekkotan records but question the family.

TABLE 23.3. Record of Eocene-Oligocene lizards in western North America (Canada, USA). Basic data from Estes (1983a). See Table 23.1 for definition of abbreviations and symbols.

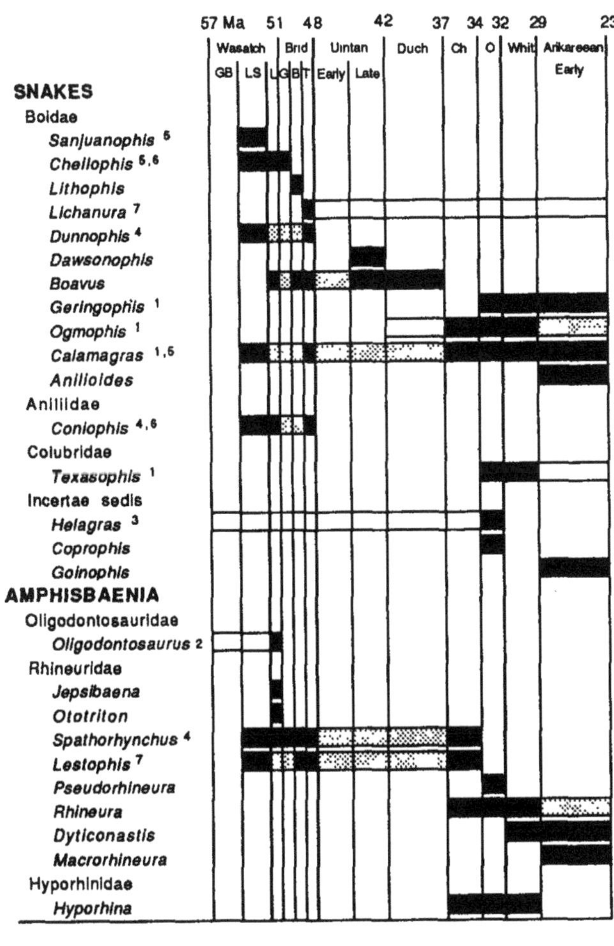

[1] Holman, 1987.
[2] Hirsch et al., 1987.
[3] Holman, 1983.
[4] Gauthier, pers. com., 1991.
[5] Sullivan and Lucas, 1988, Lysitian *Calamagras* listed as "cf.".
[6] Hecht in Robinson, 1966.
[7] Kluge, 1988.

TABLE 23.4. Record of Eocene-Oligocene snakes and amphisbaenians in western North America (Canada, USA). Basic data from Rage (1984a) and Estes (1983a). See Table 23.1 for definition of abbreviations and symbols.

two groups, terrestrial and aquatic, provides further insight into the alternation of reptile and amphibian diversity through the Eocene-Oligocene transition (Figure 23.1B-C). Through the Eocene, the herpetofaunas are approximately equally or predominantly aquatic, but by the early Oligocene (Orellan) and into the late Oligocene (early Arikareean), the faunas are overwhelmingly terrestrial. The terrestrial element reaches its peak of diversity in the Orellan but this record exhibits greater fluctuation than for aquatics and there is little overall trend in its diversity. The great departure from the approximately middle Eocene equilibrium of aquatic and terrestrials is evident by the sharp decline in aquatic diversity after the Chadronian and culminates in the virtual absence of aquatic forms in the Oligocene. Although two salamanders are recorded in the latter intervals, these genera contain extant species capable of existing on surficially dry land for many months provided moist microhabitats (burrows, under logs, etc.) are available and only need seasonal bodies of water for breeding purposes. A subset of this general diversity of which I am more familiar, the large-bodied reptiles (turtle and crocodilians), clearly shows this decline (Figure 23.1D). The aquatic forms while exhibiting a gradual decline through the later Eocene, experience a precipitous decline in the vicinity of the Eocene/ Oligocene boundary. The terrestrial turtles conversely show no change in diversity through this interval despite their low diversity. Aside from a trionychid and an alligatorid that just extend in the lower Orellan, only one specimen of an aquatic turtle, a chelydrid from the John Day of Oregon, is known from the Oligocene. No doubt as the record is expanded to include southern coastal areas, additional aquatic taxa will be found, but their general rarity in continental deposits is nonetheless an indication of their marginal distribution.

Size in large-bodied reptiles may be a useful tool in assessing environmental changes. For aquatic reptiles, large size is generally correlated with large bodies of water. Temperature also plays a role in body size of aquatic turtles, but it appears that the presence of a warm summer is more important than that of a cold winter, which they may escape by burrowing or subaqueous hibernation. Water availability

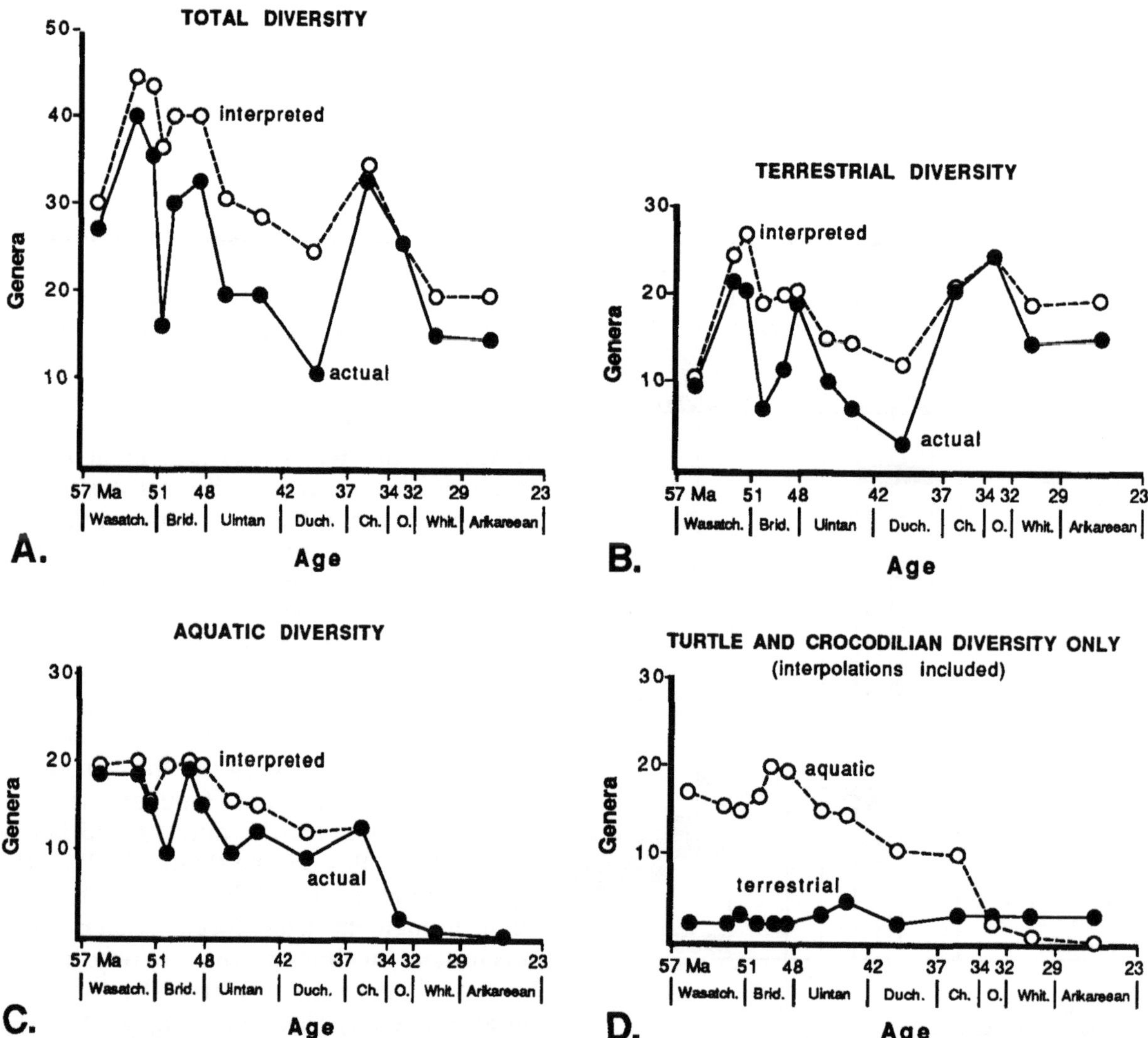

FIGURE 23.1. Patterns of change of generic diversity of the herpetofauna of western North America (richness). A. Total diversity. B. Diversity of terrestrial genera. C. Diversity of aquatic genera. D. Aquatic and terrestrial diversity for turtles and crocodilians. "Actual" indicates an uninterpreted count; "interpreted" indicates and interpreted count with range gaps filled (see Tables 23.1-23.4). See Table 23.1 for other abbreviations.

Land Mammal Age	Major lakes	Major drainages	Aquatic genera	Terrestrial genera
Wasatchian-Bridgerian	4	6	19	36
Early Uintan	1	5	9(16)	11(15)
Early Chadronian	0	3	12	21(22)

TABLE 23.5. Comparison of Land Mammal "Age" to habitat diversity as indicated by lakes, drainage divides, and number of aquatic and terrestrial genera (interpolated estimates in parentheses). Geomorphic data from Lillegraven and Ostresh (1980, Figures10-12, Table 2).

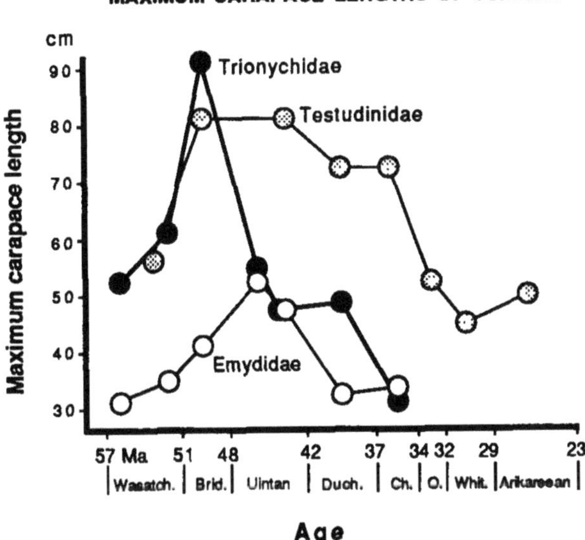

FIGURE 23.2 . Relationship between maximum carapace lengths of selected turtle groups through the Eocene and Oligocene of western North America. See Table 23.1 for other abbreviations.

and temperature function independently and may complicate the analysis. Temperature is also a significant factor in both distribution and size of terrestrial tortoises (Hibbard, 1960; Holman, 1976; Hutchison, 1982) with large-bodied non-burrowing tortoises being more susceptible to cold winters than are aquatic reptiles. The comparison of aquatic and terrestrial turtles size, therefore, provides a means of assessing whether the decline in aquatic turtle diversity is owing to lack of aqueous habitats or owing to temperature. As shown in Figure 23.2, the emydids (pond turtles) and trionychids (soft-shelled turtles) exhibit a sharp increase in size in the early and middle Eocene. Peak size is reached in the Bridgerian for soft-shelled turtles with a sharp decline thereafter. This peak in size for emydids is reached in the early Uintan and then decreases. Both groups vanish from the record by or in the earliest Orellan. Both groups, however, are broadly distributed in North America today indicating a southern or eastern refugium during the Oligocene and early Miocene (Estes, 1970a).

While exhibiting a similar increase in size into the Bridgerian, the tortoises (already exceeding 50 cm in carapace length when they first appear in the record) only begin to show a strong decrease in size between the Chadronian and Orellan, although they still may exceed 50 cm in length. The turtles thus demonstrate that the principal decline in turtle (and crocodilian) diversity was the result of an increase in aridity rather than an decrease in temperature. The above-noted decline in tortoise size may well point to a drop in temperature but not so severe as to limit their range within the study area. Crocodilians also follow a pattern of size decline shown by the aquatic turtles but their size estimates are more difficult to measure from fragmentary material. The absence of crocodylids and the persistent presence of the more cool tolerant alligatorids in the Chadronian and Orellan and the eventual disappearance of the latter from the record may indicate a decline in temperature but is readily explained by a lack of permanent water supplies. The alligator recorded in the Orellan occurs just above the base and further records are unknown in the main part of the Orellan. The presence of large-bodied lizards appear to indicate mild climates (at least comparable to Recent southwestern United States) well into the northern part of the United States during the Oligocene.

The herpetofauna record in the South Dakota Badlands agrees well with the soil formation and vegetational scenarios proposed by Retallack (1983, 1986) but the presence of large tortoises and virtual absence of aquatic forms in the Orellan and Whitneyan suggest climates (at least seasonally) nearer the drier and warmer extremes of ranges that he proposed respectively for these time intervals.

The distribution of various reptiles and amphibians is significantly influenced by general environmental parameters such as temperature (annual and seasonal), rainfall (annual and seasonal), surface water availability, and topographic diversity (Hutchison, 1982; Kiester, 1971; Pianka, 1977). While the former three have been addressed above to some extent, the latter is difficult to assess on faunal evidence alone.

Lillegraven and Ostresh (1988) provide a detailed, topographic drainage and volcanic history of Wyoming through the Paleogene at the scale of the Land Mammal "Ages." They note that during the early Uintan that volcanic activity intensified and that volcanic debris

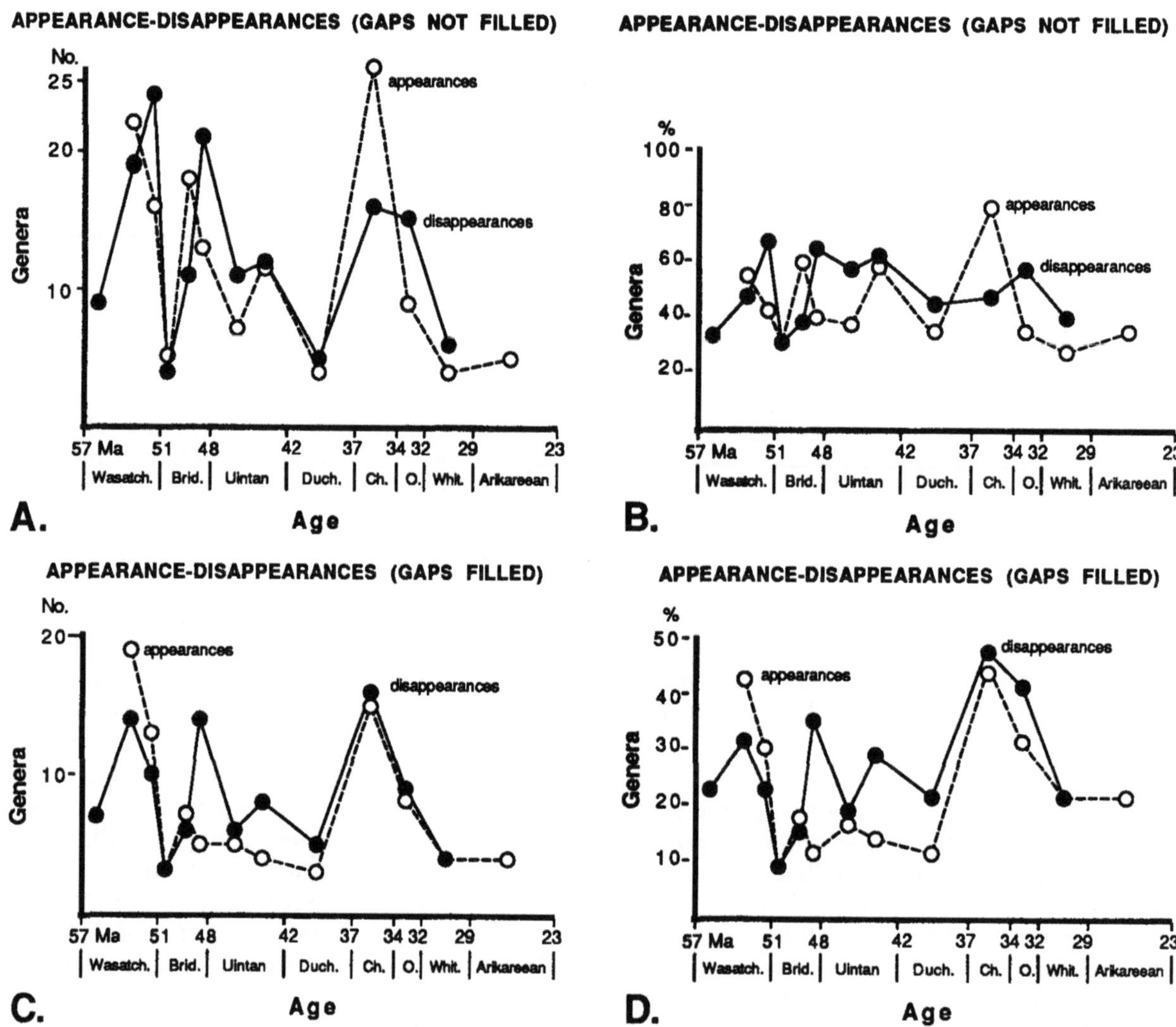

FIGURE 23.3. Patterns of change in the appearance and disappearance of the amphibians and reptiles of western North America. A. Uninterpreted record based on number of genera (range gaps not filled). B. Uninterpreted record based on percentage genera per total number of genera within an interval (range gaps not filled). C. Interpreted record based on number of genera (range gaps filled). D. Interpreted record based on percentage genera per total number of genera within an interval (range gaps filled). Appearances indicate first record occurs within the interval; disappearances indicates that the last record occurs within the interval. See Table 23.1 for other abbreviations.

and erosion buried mountain ranges and contributed to a generally southward integration of the drainages. The massive intermountain lakes of the Bridgerian and Wasatchian interval were reduced and possibly eliminated. They note the extreme rarity of Duchesnean deposits and suggest a period of continental-wide erosion with a reduction in volcanism. The drainage patterns are unknown. By early Chadronian, colossal volumes of tephras derived from widespread volcanic centers in the far west largely complete the filling of Wyoming's intermountain basins. The filling of the basins resulted in a basic eastward grain to the drainage and an increasing topographic monotony dominated by a vast savannah shrublands. Although their analyses were derived from direct geologic interpretations, there is a distinct but not unexpected correlation in the timing and habitat inferences between Lillegraven and Ostresh's (1988) geomorphic and hydrological conclusions and the record of the aquatic/terrestrial herpetofauna (Table 23.5). The general patterns they developed for Wyoming can be projected well into adjacent states (Flores and Kaplan, 1985). The patterns they provide indicate a progressive change, abet episodic, rather than a single or even a few discrete catastrophic changes related to the tectonic and volcanic history of the west.

Analysis of the appearances and disappearances is complicated by unexplored lineage relationships as noted above and by exclusion of the geographic and temporally extralimital records (excluding immediately adjacent ages), such records are indicated by the clear bars in the tables. Thus, these analyses should not be interpreted as general origination and extinction curves. Both immigrations and emigrations and morphotype appearances and disappearances are counted as appearances and local extinctions in the study area.

Whether analyzed as counts (Figure 23.3A) or as percentage of total genera (Figure 23.3B), the raw data do not show a clear trend with the notable exception of the Chadronian. The very high appearance point and low disappearance point suggest a differential expansion of the fauna during the Chadronian that is reversed in the Orellan (i.e., across the Eocene/ Oligocene boundary). I think that the high appearance peak in the Chadronian is a reflection of the poor sampling of the Duchesnean. When the gaps in the record are filled (Figure 23.3C-D), the data show that until the Bridgerian, appearances closely track or exceed disappearances; just the opposite action occurs during the latest Bridgerian through Whitneyan. There is still an expansion of the faunal diversity and decline into the Orellan, but the differential in appearances and disappearances reduces to near equilibrium. Both the raw and interpreted data indicate significant changes across the Eocene/ Oligocene boundary and into the Oligocene. The dynamic nature of these changes, however, is also indicated by similar, although of frequently lesser amplitude, fluctuations in the pre-Chadronian record. The long record of disappearances exceeding appearances indicates a general decline in diversity with a notable quickening of turnover near the boundary interval.

SUMMARY

The late Eocene through early Oligocene represents an interval of great change in the herpetofaunas of western North America. The available evidence indicate that these changes were initiated by the Uintan and completed by Orellan-Whitneyan. The overriding factor appears to be the integration of the main drainages into ocean directed systems, filling of the early Tertiary lacustrine basins, general increase in aridity and decrease in the number or even absence of permanent rivers and streams. These changes in freshwater regimes may also have been accompanied by a mean decrease in temperatures but not so severe as to exterminate tortoises and large-bodied lizards. Although the record is much less complete than is desirable, the available evidence suggests that diversity declines occurred, whether episodic or progressively, over a period of approximately 16 m.y. and are not attributable to a single catastrophic event, although the tempo of the change appears to have quickened across the Eocene/ Oligocene boundary.

ACKNOWLEDGEMENTS

I thank the University of California Museum of Paleontology A. M. Alexander endowment and the Geological Society of America Penrose Conference for financial support toward this study. I thank the late Richard Estes, Jean-Claude Rage, and an unknown reviewer for constructive criticisms and J. A. Gauthier for information on unpublished data. Contribution No. 1538 of the University of California Museum of Paleontology.

REFERENCES

Donovan, S. K. 1989. Palaeontological criteria for the recognition of mass extinctions. In S. K. Donovan, ed., *Mass Extinctions: Processes and Evidence*. New York, Columbia Univ. Press, pp. 19-36.

Dorf, E. 1959. Climatic changes of the past and present. *Contrib. Mus. Paleontol. Univ. Michigan* 13:181-210.

Eaton, J. E. 1982. Paleontology and correlation of Eocene volcanic rocks in the Carter Mountain area, Park County, southern Absaroka

Range, Wyoming. *Univ. Wyoming Contrib. Geol.* 21:153-194.

Emry, R. J. 1973. Stratigraphy and preliminary biostratigraphy of the Flagstaff Rim area, Natrona County, Wyoming. *Smithsonian Contrib. Paleobiol.* 18:1-43.

Estes, R. 1970a. Origin of the Recent North American lower vertebrate fauna: an inquiry into the fossil record. *Forma et Functio* 3:139-163.

Estes, R. 1970b. New fossil pelobatid frogs and a review of the genus *Eopelobates*. *Bull. Mus. Comp. Zool.* 139:293-340.

Estes, R. 1981. Gymnophiona, Caudata. *Encyclopedia of Paleoherpetology*. Stuttgart, Gustav Fischer Verlag. Part 2: xv+115 pp.

Estes, R. 1983a. Sauria terrestria, Amphisbaenia. *Encyclopedia of Paleoherpetology*. Stuttgart, Gustav Fischer Verlag. Part 10A: xxii+249 pp.

Estes, R. 1988. Lower vertebrates from the Golden Valley Formation, early Eocene of North Dakota (U. S. A.). *Acta Zool. Cracov.* 31: 541-562.

Estes, R. and Baez, A. 1985. Herpetofauna of North and South America during the Late Cretaceous and Cenozoic: evidence for interchange? In F. G. Stehli and S. D. Webb., eds., *The Great American Biotic Interchange*. New York and London, Plenum Press, pp. 139-197.

Flores, R. M., and Kaplan, S. S. (eds.), 1985. *Cenozoic paleogeography of the west-central United States. Rocky Mountain Paleogeography Symposium 3, Rocky Mountain Section, Soc. Econ. Paleontol. Mineral.* 460 pp.

Golz, D. J., and Lillegraven, J. A. 1977. Summary of known occurrences of terrestrial vertebrates from Eocene strata of southern California. *Univ. Wyoming Contrib. Geol.* 15:43-64.

Grande, L. 1984. Paleontology of the Green River Formation, with a review of the fish fauna. *Bull Geol. Surv. Wyoming.* 63: xviii+333 pp.

Hay, O. P. 1908. The fossil turtles of North America. *Carnegie Inst. Washington Publ.* 75:568 pp.

Hecht, M. K. 1960. A new frog from an Eocene oil-well core in Nevada. *Amer. Mus. Novitates* 2006:1-14.

Hibbard, C. W. 1960. An interpretation of Pliocene-Pleistocene climates in North America. *Ann. Rep. Mich. Acad. Sci., Arts Lett.* 62:5-30.

Hirsch, K., Kristalka, L., and Stucky, R. K. 1987. Revision of the Wind River faunas, early Eocene of central Wyoming. Part 8. First fossil lizard egg (?Gekkonidae) and list of associated lizards. *Ann. Carnegie Mus.* 56:223-230.

Holman, J. A. 1976. Cenozoic herpetofaunas of Saskatchewan. In C. S. Churcher, ed., *Athlon, Essays on Paleontology, in Honor of Loris Shano Russell. Royal Ont. Mus. Life Sci. Misc. Publ.* pp. 80-92.

Holman, J. A. 1983. New species of *Helagras* (Serpentes) from the middle Oligocene of Nebraska. *J. Herpetol.* 17:417-419.

Holman, J. A. 1987. Some amphibians and reptiles from the Oligocene of northeastern Colorado. *Dakoterra.* 3:16-21.

Hutchison, J. H. 1982. Turtle, crocodilian, and champsosaur diversity changes in the Cenozoic of the north-central region of western United States. *Palaeogeogr., Palaeoclimat., Palaeoecol.* 37:149-164.

Kiester, A. R. 1971. Species density of North American amphibians and reptiles. *Syst. Zool.* 20:127-137.

Kluge, A. 1988. Relationships of the Cenozoic boine snakes *Paraepicrates* and *Pseudoepicrates*. *Jour. Vert. Paleo.* 8: 229-230.

Lillegraven, J.A.; and Ostresh, L. M., Jr. 1988. Evolution of Wyoming's Early Cenozoic topography and drainage patterns. *Nat. Geogr. Res.* 4:303-327.

Ostrander, G. E. 1985. Correlation of the early Oligocene (Chadronian) in northwestern Nebraska. *Dakoterra.* 2:205-231.

Pianka, E. R. 1977. Reptilian species diversity. In C. Gans and D. W. Tinkle, eds., *Biology of the Reptilia*. London, Academic Press, 7:1-34.

Pregill, G. K., Gauthier, J., and Greene, H. W. 1986. The evolution of helodermatid squamates, with description of a new taxon and an overview of Varanoidea. *Trans. San Diego Nat. Hist.* 21:167-202.

Rage, J.-C. 1984a. Serpentes. *Encyclopedia of Paleoherpetology*. Stuttgart, Gustav Fischer Verlag. Part 11: xi+80 pp.

Rage, J.-C. 1984b. La "Grande Coupure" éocène/

oligocène et les herpétofaunes (Amphibians et Reptiles): problèmes du synchronisme des événements paléobiogéographiques. *Bull. Soc. geol. France.* ser. 7. 26:1251-1257.

Rage, J.-C. 1986. The amphibians and reptiles of the Eocene-Oligocene transition in western Europe: an outline of the faunal alterations. In C. Pomerol and I. Premoli-Silva, ed., *Terminal Eocene Events.* Amsterdam, Elsevier Science Publ., pp. 309-310.

Retallack, G. J. 1983. A paleopedological approach to the interpretation of terrestrial sedimentary rocks: The mid-Tertiary fossil soils of the Badlands National Park, South Dakota. *Geol. Soc. Amer. Bull.* 94:823-840.

Retallack, G. J. 1986. Fossil soils as grounds for interpreting long-term controls on ancient rivers. *Jour. Sed. Pet.* 56:1-18.

Robinson, P. 1966. Fossil Mammalia of the Huerfano Formation, Eocene of Colorado. *Bull. Peabody Mus. Nat. Hist.* 21:viii+95 pp.

Schleich, H. H. 1986. Reflections upon the changes of local Tertiary herpetofaunas to global events. In O. Walliser, ed., *Lecture Notes in Earth Sciences, vol. 8. Global Bio-Events*, Heidelberg, Spring-Verlag Berlin. pp. 429-442.

Setoguchi, T. 1978. Paleontology and geology of the Badwater Creek area, central Wyoming. Part 16. The Cedar Ridge Local Fauna (late Oligocene). *Bull. Carnegie Mus. Nat. Hist.* 9:1-16.

Sullivan, R. M. 1987. *Parophisaurus pawneensis* (Gilmore, 1928) new genus of anguid lizard from the middle Oligocene of North America. *Jour. Herpetol.* 21:115-133.

Sullivan, R. M. 1989. *Proglyptosaurus huerfanensis*, new genus new species: glyptosaurine lizard (Squamata: Anguidae) from the early Eocene of Colorado. *Amer. Mus. Novitates* 2949:1-9.

Sullivan, R. M., and Lucas, S. G. 1988. Fossil Squamata from the San Jose Formation, early Eocene, San Juan Basin, New Mexico. *Jour. Paleont.* 62:631-639.

Tihen, J. A. 1964. Tertiary changes in the herpetofaunas of temperate North America. *Senckenb. Biol.* 45:265-279.

Torres, V. 1985. Stratigraphy of the Eocene Willwood, Aycross, and Wapiti Formations along the North Fork of the Shoshone River, north-central Wyoming. *Univ. Wyoming Contrib. Geol.* 23:83-97.

Vail, P. R., Mitchum, R. M., Jr., Thompson, S., III. 1977. Seismic stratigraphy and global changes of seal level. Part 4: Global cycles of relative changes of seal level. *Amer. Assoc. Petrol. Geol. Mem.* 26:83-97.

Woodburne, M. O., ed. 1987. *Cenozoic mammals of North America.* Berkeley, Univ. California Press, 336 pp.

24. MAMMALIAN FAUNAS IN NORTH AMERICA OF BRIDGERIAN TO EARLY ARIKAREEAN "AGES" (EOCENE AND OLIGOCENE)

by Richard K. Stucky

ABSTRACT

Diversity patterns of mammalian faunas from the Bridgerian through Arikareean Land Mammal "Ages" (middle Eocene to Oligocene, ca. 50-24 million years ago) are discussed in relation to intercontinental migration, paleoclimates and paleofloral patterns. Diversity is analyzed in terms of the number of taxa (species richness) and their relative abundance and geographic distribution. The highest levels of local species richness occur near the early/middle Eocene boundary after which it declines to levels consistent with modern faunas of North America. Paleocene through early middle Eocene local faunas show "even" abundance patterns whereas later Eocene and Oligocene local faunas have higher relative abundances of the most common taxa. Provinciality increased during the middle Eocene. These patterns of diversity are consistent with global warming near the early/middle Eocene boundary and global cooling and a reduction in climatic equability through the Oligocene. This climatic change resulted in a shift from generally subtropical closed forest habitats to more open, savanna-like habitats in the Western Interior. Faunal turnover through the latter half of the Eocene (Uintan to Chadronian) resulted in the extinction of many primitive mammalian groups and the appearance of many modern families. High rates of extinction and origination occurred during the middle Chadronian but are generally consistent among the Land Mammal "Ages" and "Subages". Conclusions regarding stepwise or catastrophic extinction during the Eocene and Oligocene are currently premature. Much more detailed stratigraphic correlations and more refined radioisotopic ages are necessary to outline in detail the patterns of extinction. Lists of mammalian genera known from North America during the Eocene and Oligocene are presented for each Land Mammal "Age" and "Subage".

INTRODUCTION

Fossil mammals from the middle Eocene through Oligocene (~50 to 24 million years ago, Ma) are well represented in the geological record of North America. A number of sedimentary basins and regions preserve thick sequences of strata of these ages with fossil mammals, especially in the Western Interior (central Rocky Mountains and Great Plains) and coastal southern California. New data and more refined biostratigraphies and chronologies now allow an updated assessment of the patterns of mammalian diversification and faunal change (Prothero, 1985, 1989; Krishtalka et al., 1987; Emry et al., 1987; Tedford et al., 1987; Aubry et al., 1988; Emry, this volume).

The patterns of first and last appearances (origination and extinction, respectively), and alpha, beta and continental diversity in middle Eocene through Oligocene mammalian faunas in North America suggest some concordance with paleofloral, paleoclimatic and continental relations (Webb, 1977; Prothero, 1985, 1989; Stucky, 1990).

Diversity is used as a general term that applies to the occurrence, distribution and abundance of taxa. It is a composite of richness, the number of taxa in a specified area (locality, region, state, continent); abundance, the relative representation (percent frequency, number) of each taxon in the specified area; and distribution, the occurrences of taxa from area to area. Diversity is considered here at three scales of resolution: alpha diversity (measurements at a restricted locality, or within habitat); beta diversity (distributional changes across landscapes, or turnover from area to area; provinciality) and continental richness. Diversifi-

FIGURE 24.1. Land mamal "ages," "subages," and faunal zones used in this paper. Modified from Krishtalka et al. (1987), Emry et al. (1987), Tedford et al. (1987), and Prothero (1990).

AGE (Ma)	EPOCH	LAND MAMMAL "AGE"	FAUNAL ZONE/ LAND MAMMAL "SUBAGE"
	MIOCENE		
24.0	OLIGOCENE	ARIKAREEAN	Early Arikareean
28.0		WHITNEYAN	Whitneyan
31.0		ORELLAN	Orellan
34.0	EOCENE	CHADRONIAN	Late Chadronian
			Middle Chadronian
			Early Chadronian
37.0		DUCHESNEAN	Late Duchesnean
			Early Duchesnean
42.5		UINTAN	Late Uintan
			Middle Uintan
			Shoshonian
48.0		BRIDGERIAN	Twinbuttean
			Blacksforkian
			Gardnerbuttean
50.5		WASATCHIAN	Lostcabinian

cation is the pattern of evolutionary radiation within a clade and should not be confused with diversity. The scope of diversity studies in the mammalian fossil record and the methods employed here have been discussed in detail elsewhere (Stucky, 1990). Mammalian higher level classification follows McKenna (in Stucky and McKenna, in press).

The purposes of this paper are to (1) review the changes in mammalian diversity from the middle Eocene through Oligocene (Bridgerian through early Arikareean Land Mammal "Ages") in North America, (2) present a list of mammalian genera during each time interval, and (3) review briefly the major climatic, paleofloral and migration patterns bearing on the dynamics of mammalian diversity.

STRATIGRAPHIC FRAMEWORK

The chronology of middle Eocene through Oligocene mammalian faunas in North America has been reviewed recently in the Woodburne (1987) volume on the North American Cenozoic mammalian record by Krishtalka et al. (1987, Eocene), Emry et al. (1987, late Eocene and Oligocene) and Tedford et al. (1987, late

Oligocene). These three summary papers enhance and expand the Eocene and Oligocene framework of the North American Land Mammal "Ages" (Wood et al., 1941) and provide up-to-date information on their chronologies (Figure 24.1). Data presented here are analyzed within the context of the North American Land Mammal "Ages" (LMA) and "Subages" (LMS).

Since the publication of the Woodburne volume in 1987, Wilson (1986), Westgate (1987), Storer (1987, 1988, 1990), Novacek et al. (1987), Gingerich (1989), Stucky et al., (1989, 1990), Emry (1990, this volume), Swisher and Prothero (1990), Evanoff (1990) and Beard and Tabrum (1991) have added new information bearing on the chronology and composition of Eocene and Oligocene North American mammalian faunas. Taxonomic revisions of such groups as the rodents (Korth, 1987, 1989; Emry and Korth, 1989; Ivy, 1990), perissodactyls (various papers in Prothero and Schoch, 1989), archontans (Gunnell, 1989; Doi, 1990; McKenna, 1990) and carnivorous mammals (Gingerich and Deutsch, 1989), among others, have added new taxa to the record as well as revised the occurrences of genera during different time intervals.

Analysis of middle Eocene through Oligocene faunas is restricted to those from the western part of North America encompassing the region from Texas north to Saskatchewan, and west to California. The reader is referred to the correlation charts and maps in Woodburne (1987) and Savage and Russell (1983) for primary stratigraphic data and geographic coverage.

Recent work has brought evidence that bears on the calibration of North American LMAs with Eocene and Oligocene Epoch boundaries and stages. Beard and Tabrum (1991) report the discovery of an omomyid primate from the Bashi Formation of Mississippi which is within nannoplankton Zone NP10 and planktonic foramiferal Zone P6b (see Aubry et al., 1988). The mammalian fauna from the Bashi Formation is equivalent to that from Wasatchian 0 (here included in Wasatchian 1) according to the format of Gingerich (1989), establishing that either the Wasatchian begins at the beginning of the Eocene or that the Wasatchian encompasses the Paleocene/Eocene boundary (see Wing, 1984). This also confirms that the base of the Wasatchian is essentially equivalent to that of the Ypresian (Berggren et al., 1985; also see Gingerich, 1989).

Radioisotopic dates from early Bridgerian stratigraphic sequences surround an age of 50.5 Ma which corresponds approximately to the beginning of the Lutetian (Berggren et al., 1985; Krishtalka et al., 1987). Golz and Lillegraven (1977) have reviewed the correlations of terrestrial and marine sections in California. Most authors have concluded that the faunas from the Friars and Mission Valley formations are earlier than typical faunas from the Uintan LMA but younger than the Bridgerian LMA in their characterizing areas in the Western Interior. New and revised information on fossil sites in Wyoming (Flynn, 1986) and Colorado (Stucky, in preparation; see West and Dawson, 1975) have identified Western Interior sites that are equivalent to those in California. Flynn (1986) defined the Shoshonean LMS (early Uintan) for these faunas and has based the correlations on radioisotopic, biostratigraphic and magnetostratigraphic data. Importantly, the Mission Valley and Friars formations are closely associated with the Ardath Shale that contains a calcareous nannoplankton assemblage correlative with the *Discoaster sublodoensis* Zone. This indicates that the Uintan faunas are Lutetian or middle Eocene in age (?NP14, see Berggren et al., 1985).

Wilson's (1986) recognition of early and late Duchesnean is followed here. A date of 42.5 Ma tentatively marks the beginning of the Duchesnean (Krishtalka et al., 1987), although this date may be too old and unreliable (based on Badwater Locality 20, Swisher and Stucky, in prep). Early Duchesnean faunas include those from Badwater Locality 20, Wyoming (Dawson, 1980; Maas, 1985), the Duchesne River Formation, Utah (Emry 1981), the Sespe Formation localities, California (Golz and Lillegraven, 1977), and the Skyline faunas, Texas (Wilson, 1986). Late Duchesnean faunas include those of the Porvenir local fauna, Texas (Wilson, 1986), and the Lac Pelletier of Saskatchewan (Storer, 1987, 1988, 1990).

The beginning of the Chadronian is approximately 37 Ma (Swisher and Prothero, 1990; Prothero and Swisher, this volume). Montanari et al. (1988), Swisher and Prothero (1990) and Evanoff (1990) have argued convincingly that the Eocene/Oligocene boundary should be placed at approximately 34 Ma or near the Chadronian/Orellan LMA boundary in North America. Previously, this boundary was thought to coincide with the Duchesnean/Chadronian boundary. New radioisotopic and revised magnetostratigraphic calibrations with the standard European sequence for the Eocene and Oligocene appear to require a shift to 34 Ma. The Eocene/Oligocene boundary may lie within the Chadronian, depending on the definition of the termination of this LMA relative to the chronostratigraphy of this boundary (e.g., extinction of Brontotheriidae). The Chadronian correlates in part with the later Priabonian; early Chadronian may be equivalent to the late Bartonian (Berggren et al. 1985; see Hooker, 1986).

The Chadronian LMA is subdivided into three units based on the work of Emry (Emry et al., 1987; Emry, this volume) and others (Prothero, 1985; Ostrander, 1985). Further studies of faunas from Orellan, Whitneyan and Arikareen sequences, especially those based on firm stratigraphic grounds (such as, for example, the work of Emry, this volume) will improve the resolution of the data presented here.

Tables 24.1 and 24.2 provide a listing of the stratigraphic occurrences of mammalian genera for each Land Mammal "Age" or "Subage" considered. These data have been compiled using faunal lists, taxonomic revisions, site reports and personal observations of collections (primarily those of Wasatchian, Bridgerian, Uintan, and Duchesnean age). References to the tables provide the basis for major correlations among faunas. Such data are prone to error in age assignment; future stratigraphic and systematic studies will modify these ranges. The data are compiled to ensure more refinement and improvement in our knowledge of occurrences of different mammalian genera.

INTERCONTINENTAL CORRELATIONS

Correlations of Bridgerian to Arikareean LMA faunas with land mammal assemblages from Europe and Asia is more ambiguous than with the standard marine stages or epochs. An earliest Eocene correlation is well established for the Wasatchian of North America, Ypresian of Europe and the Bumbanian of Asia on the basis of co-occurring genera of mammals (McKenna, 1975; Dashzeveg, 1988). Although Asian and North American middle Eocene through Oligocene faunas can also be correlated on the same basis, European and North American faunas can not because of their isolation after the early Eocene. Based on radioisotopic calibration of Bartonian mammal localities in England, this stage may correspond with the late Uintan, Duchesnean and early Chadronian (Hooker, 1986; Krishtalka et al., 1987). In Europe, the Grande Coupure has been classically correlated with the Duchesnean/Chadronian boundary in North America (Savage and Russell, 1983; Prothero, 1989). North American Duchesnean and Chadronian faunas are not associated with marine sediments, while critical land mammal sites in Europe documenting the Grande Coupure have little hope for either radioisotopic or magnetostratigraphic calibration (Prothero, personal communication, 1991; see Berggren and Prothero, this volume). Although late Eocene and early Oligocene are generally equivalent in the Holarctic (Legendre and Hartenberger, this volume; Wang, this volume), precise correlations of the North American record with the Grande Coupure is equivocal at present, requiring much more future study (see Berggren and Prothero, this volume).

TEMPORAL RESOLUTION AND STRATIGRAPHIC COMPLETENESS

The durations of middle Eocene through Oligocene range from estimates of 0.8 to 2.75 million years. These differences in duration can influence greatly the recorded diversity of an interval simply because of an aliquot effect (McKenna, 1983a). Intervals with longer durations may have more genera than shorter ones simply because they are longer. In addition, a miscalculation of time interval length can alter the calculations of rates of evolution geometrically. The radioisotopic ages of boundaries between middle Eocene through Oligocene Land Mammal "Ages" are well documented (see Krishtalka et al., 1987; Emry et al., 1987; Tedford et al., 1987; Swisher and Prothero, 1990), such that the estimates of extinction and origination rate are robust in a relative sense. New $^{40}Ar/^{39}Ar$ dating may modify these estimates as they become available (Prothero and Swisher, this volume). Estimates for the duration of the LMS used here are provisional. Subdivisions of the Bridgerian, Duchesnean and Chadronian are considered to be equal in length. The earliest Uintan (Shoshonean) is estimated at 1.0 million years, and the middle and late Uintan as 2.75 million years.

The stratigraphic completeness and geographic coverage of the North American fossil record varies considerably through the middle Eocene and Oligocene. This results both from poor stratigraphic representation and a lack of published correlations of biostratigraphic, lithostratigraphic and chronostratigraphic data. Among the faunas discussed here, the late Bridgerian, Duchesnean and late Chadronian are represented by only a handful of sites that are well-documented. Relatively complete, localized, stratigraphic sequences are known for each LMA but detailed stratigraphic studies have only been accomplished for the early Bridgerian of Wyoming (McGrew and Sullivan, 1970; West, 1973a; Stucky 1984; Bown, 1982), the early Uintan of Wyoming (Flynn, 1986), the Uintan/Duchesnean rocks of California (Golz and Lillegraven, 1977; Flynn, 1986), the Uintan through Chadronian rocks of Texas (Wilson, 1986), the Chadronian of Wyoming (Emry, this volume), the Orellan and Whitneyan of South Dakota (in part, Lillegraven, 1970) and the Arikareean of Oregon (Fischer and Rensberger, 1973). In these well documented stratigraphic sequences, last appearances of taxa can at present only be characterized to the level of LMA or LMS. Many sites are also isolated and cannot be placed into an overall stratigraphic context. As shown by Badgley (1990) during the early Eocene, variability in samplng from well to poorly sampled horizons can be erroneously identified as an extinction event. Until detailed stratigraphic studies have been conducted in other well-exposed sequences, the characterization of the nature and kinds of extinctions that occurred cannot be determined for the terrestrial record of North America with great confidence.

Lack of stratigraphic control can lead to premature conclusions regarding diversity patterns or extinction among mammalian faunas. For example, some authors imply that extinction may have been major at the end of the Eocene whereas others suggest that it was stepwise through the late Eocene into the early Oligocene among terrestrial faunas (e.g., Alvarez et al., 1980; Prothero, 1989; Stanley, 1990). The supposition that extinction was catastrophic or stepwise is based on treating each temporal subdivision at the scale of Land Mammal "Subage" or "Age" as a single event (see McKenna 1983a, p. 474). Clearly this cannot be assumed until detailed stratigraphic studies have been completed, especially across the Eocene/Oligocene boundary in all representative strata. In addition, Duchesnean faunas are often restricted to single localities in geographically disparate areas. The principal records of the Duchesnean in the Western Interior are known only from Badwater locality 20 of the northern Wind River Basin, Wyoming (Maas, 1985) and Lac Pelletier, Saskatchewan (Storer, 1987, 1988, 1990). These two sites appear to represent early and late Duchesnean, respectively, and account for most of the record of small mammals during the Duchesnean in the Western Interior which lasted for ~5.5 million years from 42.5 to 37 Ma (Krishtalka et al., 1987). In North America, rates of extinction can be calculated and inferences on increases or decreases in extinction can be proposed, but characterizations of extinction as catastrophic or stepwise are beyond the scope of data resolution at this time (especially for the late Uintan through Chadronian).

PALEOGEOGRAPHY, PALEOCLIMATOLOGY, PALEOBOTANY

Intercontinental relations and migration

Holarctic mammalian interchange and dispersal was episodic during the Paleogene. North America apparently was a corridor for interchange with Asia and Europe during the latter Paleocene and early Eocene (McKenna, 1975, 1983a, 1983b; Rose, 1981; Dashzeveg, 1988; reviewed in Krause and Maas, 1990). Paleogene Holarctic dispersal, as currently known, involved 16 genera in the orders Mesonychia, Artiodactyla, Perissodactyla, Cimolesta, Carnivora, and Rodentia that are common to North America, Europe and Asia. This dispersal was apparently in several pulses during the late Paleocene (*Dissacus* and Rodentia), early Eocene (*Diacodexis, Hyracotherium, ?Hyrachyus* [European record very questionable], *Coryphodon, Hyopsodus, Pachyaena,* and *Miacis*) and late Eocene/early Oligocene

(*Pterodon, Hyaenodon, Nimravus, Eusmilus, Bothriodon, Elomeryx,* and *Plesiosminthus*). Significantly, some of the late Paleocene and early Eocene immigrants--rodents, artiodactyls and perissodactyls--were all relatively small, generalist herbivores which diversified substantially during the later Paleogene and became the principle elements of the herbivore guild in North America as well as Europe.

Intercontinental connections between Europe and North America existed during the late Paleocene, and at the beginning of the Ypresian/ Wasatchian (~57 Ma) resulted in the migration and interchange of hyaenodontid creodonts, euprimates, artiodactyls, perissodactyls and other mammal groups (McKenna, 1975; 1983a, 1983b; Rose, 1981). The initial continents of origin for some of the early Eocene groups may have been Africa (euprimates, hyaenodontid creodonts; Gheerbrant, 1987) and Asia (perissodactyls, rodents and ?artiodactyls). As a result, approximately 40 genera known in the Wasatchian of North America also occur in the Ypresian of Europe (see, e.g. Godinot, 1982; McKenna, 1975, 1983a, 1983b; Russell et al., 1983; Krause and Maas, 1990; Stucky, 1990). Co-occurring Ypresian and Wasatchian genera are scattered across nearly the full spectrum of mammalian orders (see Table 24.1). After the earliest Eocene, exchange between Europe and North America did not take place again during the Paleogene. Co-occurring North American and European middle Eocene through Oligocene genera appear to have dispersed independently to these continents via Asia.

Connections between North America and Asia occurred during the Paleocene and continued through much of the Eocene and Oligocene. Indeed, many first appearances in North America that mark the beginning of LMAs or LMSs are Asian immigrants.

Rodents, tillodont cimolestans and *Arctostylops* dispersed to North America from Asia in the late Paleocene (Clarkforkian). Genera in common between Asia and North America during the early Eocene include several perissodactyls (*Homogalax, Hyracotherium,* and *Heptodon*) and mesonychids (*Hapalodectes, Dissacus,* and *Pachyaena*), and single genera of Cimolesta (*Coryphodon*), Carnivora (*Miacis*), Condylarthra (*Hyopsodus*) and Dinocerata (*Prodinoceras*). Perissodactyls were common immigrants into North America through the Eocene and included brontotheriids (*Eotitanops, Palaeosyops,* Bridgerian), ceratomorphs (*Heptodon,* Wasatchian 4; *Hyrachyus,* early Bridgerian; *Triplopus, Forstercooperia,* Shoshonean; *Amynodontopsis, Metamynodon, Colodon,* late Uintan), and chalicotheres (*Grangeria,* Uintan). Different lineages of cimolestan tillodonts also dispersed to North America from Asia in the latest Wasatchian and early Bridgerian (Stucky and Krishtalka, 1983). Among the Asian immigrants, the rodents and perissodactyls underwent North American endemic diversification. The strong bias toward the immigration of perissodactyls during the latter two-thirds of the Eocene suggests that some type of ecological filter across high latitudes was involved as suggested by McKenna (1983a, 1983b).

The overall effects of immigration were variable through the Paleogene. Immigration into North America during the early Eocene increased local species richness and overall continental generic richness (Stucky, 1990), but later immigrations seem to have had little effect on diversity patterns as a whole. Rodents, artiodactyls, perissodactyls and euprimates diversified substantially during different times, each showing its own unique history. The apparently endemic evolution of European faunas through the Eocene and Oligocene and the evolution of American and Asian faunas will provide an experimental arena for understanding global controls on diversity and morphological change (patterns of parallel evolution). Critically, future studies of immigration should be placed within a cladistic framework to provide a firmer basis for recognizing the Holarctic migration patterns.

Rocky Mountain orogeny

Several authors have reviewed Eocene and Oligocene changes in geography of the Western Interior that resulted from the origin and development of the modern Rocky Mountains (e.g., Robinson, 1972; Lillegraven and Ostresh, 1988). This affected the distribution of plant associations (Wing, 1987) and the intracontinental relations of mammals (Lillegraven, 1979). Rocky Mountain uplift was strongest during the late Paleocene through the early Eocene when many of the modern basins were being defined. The result was increasingly accentuated geographic barriers that influenced migration and led to greater biotic provinciality. Orographic effects also resulted in drier climates east of the Rocky Mountains during the middle Eocene (Wing, 1987). During the late Paleocene and earliest Eocene, the western basins may have been drier than the eastern ones (K. Johnson, oral communication, 1991). During the latest early Eocene, Laramide tectonism shifted to Absaroka volcanism in the state of Wyoming (Lillegraven and Ostresh, 1988). From the Bridgerian through the duration of the Paleogene and into the Neogene, volcanic materials are a major component of sedimentary rocks in the Western Interior.

Global temperature and paleoenvironments

Oxygen isotope data from marine organisms and analyses of fossil plants suggest that, during the early and early middle Eocene (~55-50 Ma), global temperatures were higher, differences between seasons in temperature were lower, and differences between low and high latitude temperatures were less compared to the rest of the Cenozoic (Savin, 1977; Savin and Douglas, 1984; Wolfe, 1978; Wing, 1987, 1991; Upchurch and Wolfe, 1988; Rea et al., 1989). Some authors (e.g., Owen and Rea, 1985; Rea et al., 1989) suggest that this global warming coincides with a carbon dioxide greenhouse. Peak warming for the entire Cenozoic occurs during the latter part of the early Eocene and earliest middle Eocene (Savin, 1977; Savin and Douglas, 1984) with the trend toward warming beginning in the late Paleocene (Hickey, 1980; Rea et al., 1989). Although there are some discrepancies in the actual timing of peak warming, all models agree that after the early Eocene worldwide temperatures declined continuously with a relatively rapid decline at the Eocene/Oligocene boundary to a temperature minimum in the late Oligocene (Wolfe, 1978) for the Paleogene.

Wing (1987) has identified three stages in the development of Eocene and Oligocene floras of the Rocky Mountain region. First, during the Wasatchian and early Bridgerian, subtropical and paratropical forest migrated northward; paleofloras of "tropical", "subtropical", "paratropical" or temperate aspect (broad-leaved, evergreen and deciduous forests) existed across the North American continent to 60°N in apparent response to global warming. Second, during the later Bridgerian and Uintan, subtropical and paratropical forests in the Rocky Mountain region were replaced by "more open subtropical vegetation of lower stature, probably as a result of local or regional orographic effects." Elevational differences also were a factor in floral variability. Finally, in the late Eocene and Oligocene, floras were dominated by a "mixture of conifers and broad-leaved deciduous forms" (Wing, 1987, pp. 754-755).

The apogee in "tropical"-aspect floras in coastal regions and the central part of the continent apparently occurred near the early/middle Eocene boundary at 51-50 Ma (Leopold and MacGinnitie, 1972; Wolfe, 1978; Wing, 1987, 1991). This suggests that terrestrial productivity and biomass were probably at their highest levels for the Cenozoic at this time in North America. The high abundance of obligatory arboreal taxa at this boundary, especially primates, is indicative of a closed canopy forest architecture that may have characterized most forest environments since the middle (?) Paleocene. A similar pattern has been recorded in Europe (Collinson and Hooker, 1986).

According to isotopic and paleobotanical data, temperatures decreased at a relatively constant rate after the early middle Eocene as seasonal differences and differences between latitudes in temperature increased. Paleofloral data suggests that greater provinciality developed in North America (Leopold and McGinnitie, 1972; Wing, 1987). Paleosol data and more cursorial adaptations among mammals suggests that habitats became more open by the late Eocene (Chadronian-- Webb, 1977; Retallack, 1983; Stucky, 1990). Arboreal taxa (principally primates) also decrease in richness and abundance through the late Eocene, becoming extinct in North America prior to the beginning of the Oligocene. The first North American macrofossil grasses are of late Eocene age (Chadronian, *Stipa*, MacGinitie, 1953), but grasses are apparently not abundant in the fossil record until the Miocene. By the early part of the Miocene, continental interiors were occupied by upper floodplain grass, sedge and composite grassland/parklands and riverine gallery forests

(Hunt, 1985, 1990).

ALPHA, BETA AND CONTINENTAL DIVERSITY

Alpha diversity

During the Cenozoic, species richness for individual fossil horizons in North America was at peak levels across the early/middle Eocene boundary (late Wasatchian to early Bridgerian) (Stucky, 1990). Local species richness declined during and after the Bridgerian and dropped dramatically in the Chadronian to levels consistent with modern grassland and woodland communities of North America (Figure 24.2). Lac Pelletier, Saskatchewan (67 species; Storer, 1988) and Raben Ranch, Nebraska (63 species; Ostrander, 1975) have high species richness for comparably aged Duchesnean and Chadronian localities, respectively. The richness estimates for Lac Pelletier and Raben Ranch are inconsistent with other Duchesnean (e.g., Badwater Locality 20, Maas, 1985; California and Texas sites, Golz and Lillegraven 1977; Wilson, 1986) and Chadronian localities (Clark et al., 1965; Stucky 1990), indicating that either taphonomic and sampling factors may be involved or that, during the Duchesnean, northern faunas were more diverse than southern ones, and during the Chadronian, there was much greater variability in species richness within a region.

Abundance distributions for middle Eocene faunas are generally even with less variance in the abundance of common taxa compared to Chadronian and later faunas. The abundance of high-ranking species among Chadronian and later faunas is very variable from one locality to the next. Quarry sites with one species or only a few taxa are more common among Chadronian through Arikareean faunas (for example Figgins, 1934; Schlaikjer, 1935; Hunt, 1985, 1990). Similarly, relatively diverse Chadronian through Arikareean sites (greater than 15 genera) are often dominated by one or a few species such as rabbits, small to medium sized artiodactyls, and/or rodents (e.g., Clark et al., 1965; Figure 24.3).

Less even abundance distributions coupled with a lower species richness imply that habitats and resource abundance were less predictable during the late Eocene and Oligocene than during the earlier Paleogene. This may be associated as well with more open savanna-like habitats, greater seasonality, and/or development of local habitat patchiness.

Beta diversity and provinciality

Geographic turnover in mammalian faunal composition across the continent is relatively low during the Bridgerian and early Uintan showing an increasing pattern of provinciality during the late Uintan and Duchesnean that is also reflected in the floras (Lillegraven, 1979; Storer, 1984, 1988, 1990; Wilson, 1986; Stucky, 1990; Wing, 1987). During the late Eocene and Oligocene (Chadronian, Orellan, and Whitneyan), mammalian faunas are again more similar to one another in species composition, or more continental in aspect.

Continental generic richness

North American continental generic richness from the middle Eocene to Oligocene averages approximately 96 genera during each time interval and varies from a low of around 70 genera in the late Bridgerian to highs surrounding 105 genera during the late Uintan and middle Chadronian (Figure 24.4). Late Bridgerian faunas are very poorly known, based principally on sites from southwestern Wyoming, suggesting that this low in richness is a sampling artifact (Gazin, 1976). Although the late Uintan represents a relatively long interval of time, high continental generic richness may be a phenomenon related to the development of provinciality discussed above.

Decreasing local richness (alpha diversity) coupled with increased turnover between habitats (beta diversity) appears to have generally maintained continental richness during the Eocene. This suggests that any stability or equilibrium in continental richness may be an epiphenomena of underlying patterns and not directly related to any overall process of extinction/origination equilibria on a continental scale.

ORIGINATION, EXTINCTION AND FAUNAL CHANGE

Origination

The number of originating genera is generally consistant within each time interval during the middle Eocene through Oligocene (Figures 24.4 and 24.5). During the Bridgerian,

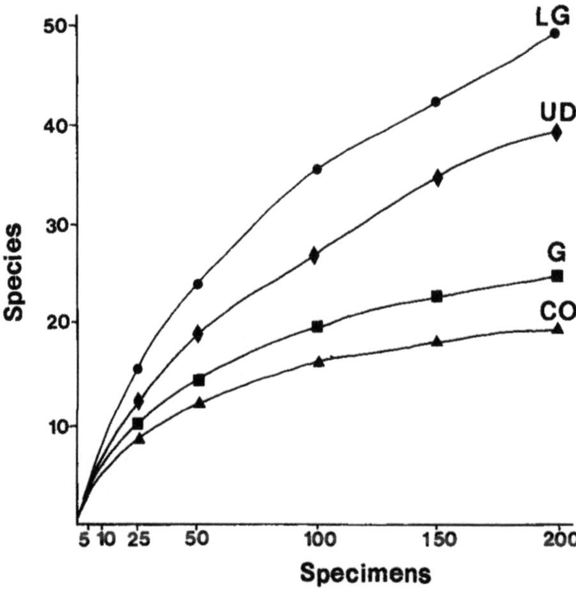

FIGURE 24.2. Composite rarefaction curves for early Wasatchian (G, Greybullian LMS), late Wasatchian-early Bridgerian (LG, Lostcabinian-Gardnerbuttean LMSs), Uintan-Duchesnean (UD), and Chadronian-Orellan mammals from single fossiliferous horizons of North America. The X-axis represents the number of specimens found at a site, and the Y-axis predicts the number of species for different specimen sample sizes during the different age intervals. The Chadronian-Orellan curve predicts that species richness was much lower than earlier Paleogene localities (after Stucky, 1990).

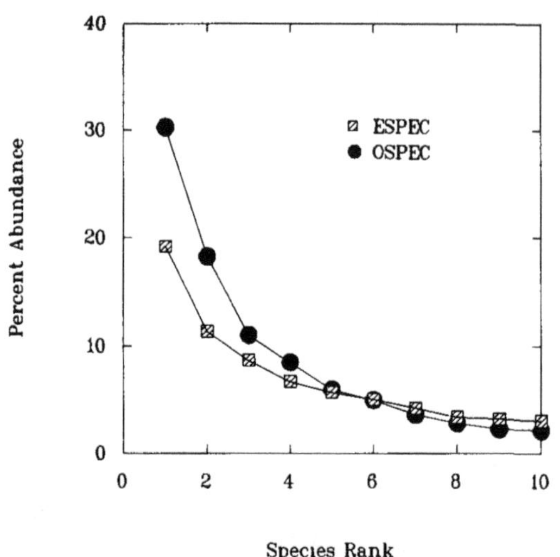

FIGURE 24.3. Composite percentage rank abundance curve among mammals for the first ten most abundant species in Chadronian-Orellan local faunas (OSPEC, n=13) and Uintan-Duchesnean local faunas (ESPEC, n=6). The first two most abundant mammals in the Chadronian-Orellan faunas represent nearly 50 percent of all specimens. The skewed abundance distribution for faunas of these ages suggests much less predictable resources or greater habitat patchiness.

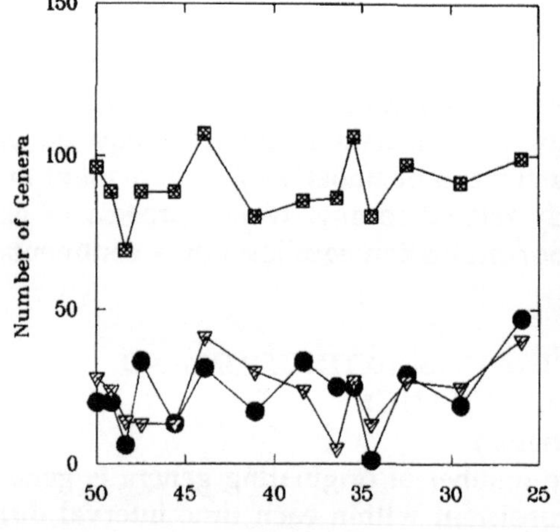

FIGURE 24.4. Total genera (TOTAL), number of last appearances (EXTINCT), and number of first appearances (ORIGIN) of mammals of middle Eocene through Oligocene age in North America. Data points represent intervals of time identified in Figure 1 and are based on information provided in Tables 24.1 and 24.2.

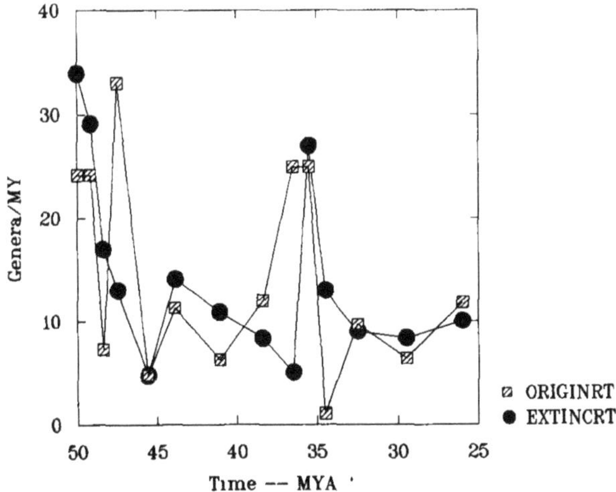

FIGURE 24.5. Number of Genera per million years (MY) appearing (ORIGINRT) and disappearing during different intervals of the middle Eocene through Oligocene of North America. See explanation to Figure 24.4.

FIGURE 24.6. Numbers of genera of rodents (RO), perissodactyls (PE) and artiodactyls (ART) present in each time interval. See text for explanation.

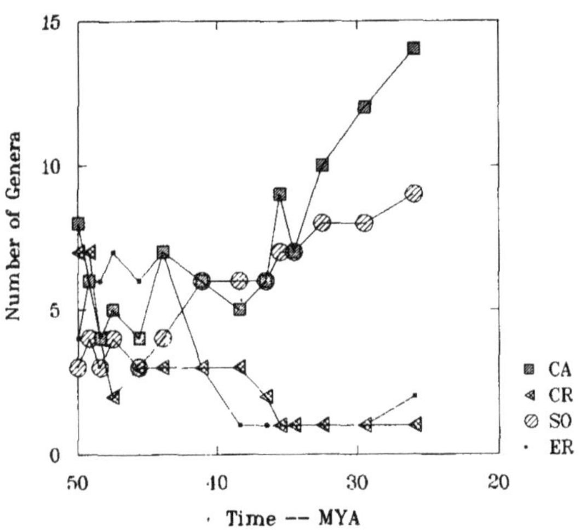

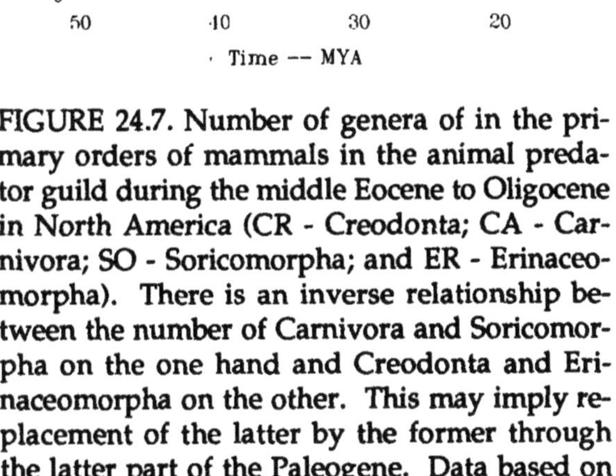

FIGURE 24.7. Number of genera of in the primary orders of mammals in the animal predator guild during the middle Eocene to Oligocene in North America (CR - Creodonta; CA - Carnivora; SO - Soricomorpha; and ER - Erinaceomorpha). There is an inverse relationship between the number of Carnivora and Soricomorpha on the one hand and Creodonta and Erinaceomorpha on the other. This may imply replacement of the latter by the former through the latter part of the Paleogene. Data based on occurrences of taxa as outline in Tables 24.1 and 24.2.

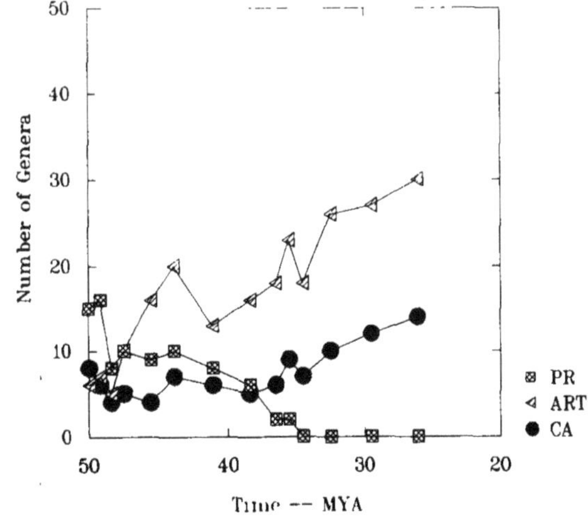

FIGURE 24.8. Number of genera of Primates (PR), Artiodactyla (ART) and Carnivora (CA) during the middle Eocene to Oligocene in North America. The increasing abundance of Carnivora genera follows very closely the abundance of Artiodactyla, while the primates decline and are extinct by the late Chadronian. The negative relationship between Artiodactyla (terrestrial herbivores) and Primates (arboreal herbivores and omnivores) generic richness may be a function of habitat change from "tropical-subtropical" community associations to more temperate-savanna-like community associations.

first appearances are spread across thirteen different orders of mammals, but during the early Uintan, perissodactyls, artiodactyls and rodents dominate first appearances. Among these groups, the perissodactyls show low diversification after the early Chadronian and through the Oligocene.

The middle Eocene through Oligocene witnessed the first appearances of many extant mammal families (Black and Dawson, 1966; Lillegraven, 1972; Stucky and McKenna, in press). In North America, the modern non-volant mammal families which first appeared are: Bridgerian - dipodids; Uintan - leporids, aplodontids, geomyids, soricids, camelids and rhinocerotids; Duchesnean - murids and canids; and Chadronian - castorids, sciurids and tayassuids. In Europe and/or Asia, the first ochotonids, cervids and ursids appeared during the late Eocene, and during the Oligocene, the first procyonids, felids and tapirids appeared. Most other extant mammal families currently found in North America first appear in the early and middle Miocene.

Temporal scaling of generic first appearances results in high rates of originations (genera/my) during the early and middle Bridgerian, early Uintan and early and middle Chadronian.

Extinction

The number of extinctions during each interval shows much greater fluctuations than do those for originations (Figures 24.4 and 24.5). During most intervals extinctions were apparently non-selective in terms of ordinal affinities. During the middle Bridgerian, extinctions of primates were especially high and during the late Uintan and Whitneyan, both perissodactyls and artiodactyls had high numbers of generic extinctions. High numbers of extinction occurred among the rodents during the middle Chadronian and Orellan.

Many clades of mammals which originated and diversified during the late Cretaceous, Paleocene, and early Eocene became extinct by the end of the Oligocene in North America. These include the multituberculates, leptictids, cimolestans, dermopterans, creodonts, euprimates, mesonychids, arctocyonids, condylarths, uintatheres and palaeandonts. Generic extinctions of these groups is attritional through the middle Eocene and Oligocene, representing a steady reduction in generic richness among them through this time. During the early Bridgerian, these groups compose nearly 50 percent of generic richness, whereas during the early Arikareean, they compose only two percent.

High rates of extinction occurred during the early and middle Bridgerian and middle Chadronian. Otherwise the rates of extinction are similar for each time interval.

New records of taxa for the Chadronian (see references to Table 24.2) and refined radiometric date control (Swisher and Prothero, 1990) has modified an earlier conclusion that most extinctions were concentrated in the North American record between 44 and 39 Ma (late Uintan to Duchesnean) (Stucky, 1990). High numbers of extinctions do occur through this time, but a high rate of extinctions is also recorded for the middle Chadronian. My prior estimates for low numbers of extinctions during the Chadronian were based on the supposed duration of the Chadronian of 5 m.y., and Prothero's (1985) identifications for extinctions for three intervals which are here combined into a single middle Chadronian interval. Based on the primary literature, the middle Chadronian is considered as a single interval of approximately one million years duration, rather than three intervals, each one million years in duration. Much future stratigraphic clarification is needed to determine if this high level of extinction was catastrophic or attritional in nature, whether or not it extended into the late Chadronian (a poorly sampled interval), and whether or not it correlates with the Grande Coupure in Europe. Importantly, the middle Chadronian is also characterized by high origination, suggesting a faunal turnover or replacement sequence allied with climatic or habitat change.

Faunal change

The decline and decimation of mammalian orders and families that first appeared in the Cretaceous, Paleocene and early Eocene and the appearance and increase in modern-aspect orders and families is the rule for the latter half of the Eocene and Oligocene (Black and Dawson, 1966b; Lillegraven 1972). Several trends in the composition and representation among ecological guilds in North American communities can be recognized through this time.

Herbivores increase almost continuously through the middle Eocene and Oligocene. Among these, the browsing herbivores, especially perissodactyls and artiodactyls, increase substantially during the Uintan (Figure 24.6; Stucky, 1990). Rodents increase in richness as well through the latter Paleogene. During the early Bridgerian, perissodactyls, artiodactyls and rodents compose approximately one-third of the fauna. By the middle Uintan, over one-half of the known genera are members of these clades, and by the late Eocene (Chadronian) these three orders represent nearly two-thirds of generic richness. A parallel increase in true Carnivora of lesser magnitude also occurs.

The diversity of mammalian animal predators is virtually stable from the middle Eocene through the Oligocene. During the latter part of the Eocene, soricomorph insectivores replace erinaceomorph ones, and similarly, true Carnivora replaces the creodonts among predators of vertebrates (Figure 24.7).

Obligatory arboreal taxa (primates and dermopterans) decline through the Eocene and are virtually extinct by the Oligocene (Figure 24.8). Among Bridgerian faunas arboreal taxa compose approximately one-fifth of the genera, declining to one-tenth in the Uintan to less than four percent during the Chadronian. Obligatory arboreal taxa are extinct in North America prior to the beginning of the Neogene.

HERBIVORE MORPHOLOGIC CHANGE

Collinson and Hooker (1987), Hooker (this volume) and Legendre and Hartenberger (this volume) have documented the principal evolutionary changes among Eocene and Oligocene European herbivorous mammals. These changes dramatically parallel those for the North American mammals although their evolution and diversification was entirely independent. North American late Eocene and Oligocene herbivores have much higher crowned and lophodont dentitions than do early and early middle Eocene ones (with the exception of a few, generally rare taxa; see Stucky, 1990). Late Wasatchian and early Bridgerian herbivores show incipient patterns of higher molar crowns and greater loph lengths that occurs at the same time that floras are tropical in aspect, preceding climatic deterioration and the opening of habitats. Among these taxa, many still retain dentitions that suggest broad-spectrum herbivory or omnivory (seeds, fruit, leaves and possibly insects) rather than a focus on leaves as is typical of later Eocene and Oligocene herbivores. By the Oligocene, most perissodactyls, artiodactyls and rodents have higher tooth crowns and longer loph lengths. This implies a reliance on browse and a generally progressive trend in many different lineages towards more efficient oral food processing. Most large-bodied Arikareean herbivores have hypsodont, or nearly hypsodont, dentitions.

Additional changes among middle Eocene and Oligocene herbivores involve more elongate limb segments, more robust or co-ossified metapodials, and other features which imply more efficient and faster cursorial locomotion (compare data in Rose, 1990, with that of Scott and Jepsen, 1936-1941). Many independent clades (some within the same order) show similar patterns in the evolution of limb and dental morphology (Stucky, 1990).

The overall body size spectrum among mammals expands during the middle Bridgerian and by the Uintan the maximum size of the largest member of each mammalian order has increased. Predator selection resulting in a size dichotomy in prey species may also have been important during the late Eocene and Oligocene (Legendre, 1987; Stucky 1990). These patterns may be related to the occurrences of more open habitats, which in turn may be related to climatic deterioration.

CONCLUSIONS

Two climatic episodes during the Eocene may have affected the patterns of diversity among the mammals of North America: Global warming across the early/middle Eocene boundary and climatic deterioration during the Uintan and into the Oligocene. Global warming (potentially due to a carbon dioxide greenhouse) resulted in high rates of extinction and origination and high levels of species richness for the Cenozoic. Climatic deterioration beginning during the late Bridgerian and early Uintan apparently resulted in decreased local species richness, but increased provinciality among mammals and plants by the late Uintan

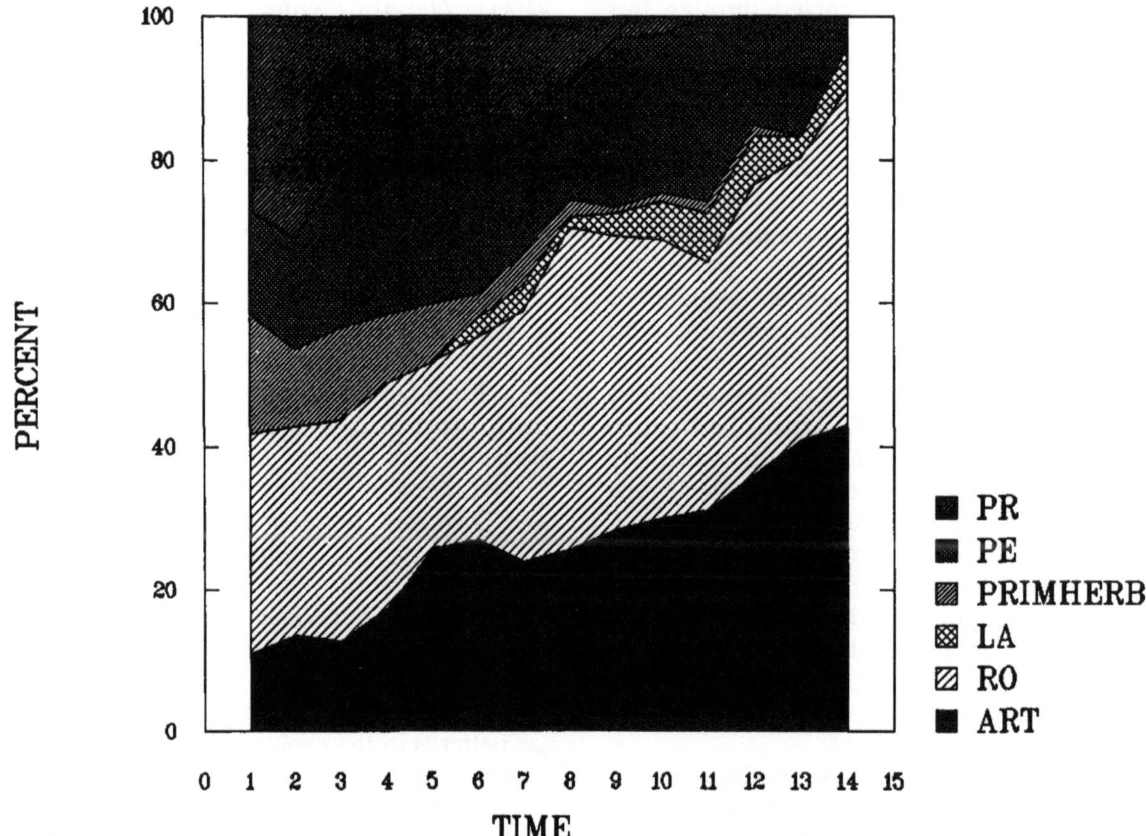

FIGURE 24.9. Relative percentages of Primates (PR), Perissodactyla (PE), primitive herbivores (PRIMHERB, including Condylarthra, Arctocyonia, Dinocerata, and tillodont, taeniodont, and pantodont Cimolesta), Lagomorpha, Rodentia, and Artiodactyla during the middle Eocene to Oligocene in North America. Time intervals along the X-axis are: Bridgerian - 1, Gardnerbuttean; 2, Blacksforkian; 3, Twinbuttean. Uintan - 4, Shoshonean; 5, middle Uintan; 6, late Uintan. Duchesnean - 7, early Duchesnean; 8, late Duchesnean. Chadronian - 9, early Chadronian; 10, middle Chadronian; 11, late Chadronian. Orellan - 12. Whitneyan - 13. Early Arikareean - 14.

and Duchesnean. By the Chadronian local faunas were much less diverse and show an abundance distribution that is less even than earlier Eocene local faunas. The rise of herbivores and the extinction of primates implies a link between the patterns of diversification and habitat change (cf. Figure 24.9). Changes in morphology among the herbivores that suggest more efficient food processing appear prior to climatic deterioration and the supposed origin of more open habitats. By the end of the Eocene, more cursorial adaptations appear at a time coincident with the opening up of habitats. These changes in morphology are most conceivably the result of diffuse competition among guild members and the coevolution of predator and prey. Climatic constraints on extinction/origination and alpha/beta/continental diversity among mammals may be related to productivity (=energy availability) and habitat structure. Patterns of dental morphological evolution on the other hand may be related to biotic interactions and diffuse competition among guild members (Stucky, 1990).

ACKNOWLEDGEMENTS

I thank D.R. Prothero and W.A. Berggren for the opportunity to contribute to this volume and to participate in the GSA theme session on the Eocene and Oligocene. Kirk R. Johnson and Mary R. Dawson made valuable suggestions on the manuscript, and an anonymous reviewer critically changed the direction of the manuscript. Any errors in this paper are entirely the responsibility of the author. Research support from the National Science Foundation, NASA, The Carnegie Museum of Natural History, and the Denver Museum of Natural History for studies related to this research is gratefully acknowledged.

REFERENCES

Alvarez, W., Asaro, F., Michel, H.V., and Alvarez, L.W. 1982. Iridium anomaly approximately synchronous with terminal Eocene extinctions. *Science* 216:886-888.

Archibald, J.D., Clemens, W.J., Gingerich, P.D., Krause, D.W., Lindsay, E.H., and Rose, K.D. First North American Land Mammal Ages of the Cenozoic Era. In M.O. Woodburne, ed., *Cenozoic Mammals of North America: Geochronology and Biostratigraphy,* Berkeley, Univ. California Press, pp. 24-76.

Aubry, M.-P., Berggren, W.A., Kent, D.V., Flynn, J.J., Klitgord, K.D., Obradovich, J.D., and Prothero, D.R. 1988. Paleogene geochronology: An integrated approach. *Paleoceanography* 3:707-742.

Badgley, C. 1990. A statistical assessment of last appearances in the Eocene record of mammals. *Geol. Soc. Amer. Spec. Pap.* 243: 153-168.

Beard, K.C. and Tabrum, A.R. 1991. The first early Eocene mammal from eastern North America: An omomyid primate from the Bashi Formation, Lauderdale County, Mississippi. *Miss. Geol.* 11(2):1-6.

Berggren, W.A., Kent, D.V., and Flynn, J.J. 1985. Jurassic to Paleogene: Part 2, Paleogene geochronology and chronostratigraphy. In N.J. Snelling, ed., *The Chronology of the Geological Record, Geol. Soc. Mem.* 10:141-195.

Bjork, P.R. 1967. Latest Eocene vertebrates from northwestern South Dakota. *Jour. Paleont.* 41: 227-236.

Black, C. C. 1967. Middle and late Eocene mammal communities: a major discrepancy. *Science* 156:62-64.

Black, C. C. 1969. Fossil vertebrates from the late Eocene and Oligocene, Badwater Creek area, Wyoming, and some regional correlations. *21st Field Conf., Wyo. Geol. Assoc. Guidebook*, pp. 43-47.

Black, C. C. 1970. Paleontology and geology of the Badwater Creek area, central Wyoming. Part 5. The cylindrodont rodents. *Ann. Carnegie Mus.* 41:201-214.

Black. C. C. 1971. Paleontology and geology of the Badwater Creek area, central Wyoming. Part 7. Rodents of the Family Ischyromyidae. *Ann. Carnegie Mus.* 43:197-217.

Black. C. C. 1974. Paleontology and geology of the Badwater Creek area, central Wyoming. Part 9. Additions to the cylindrodont rodents from the late Eocene. *Ann. Carnegie Mus.* 45:151-160.

Black. C. C. 1978. Paleontology and geology of the Badwater Creek area, central Wyoming. Part 14. The artiodactyls. *Ann. Carnegie Mus.* 47:223-259.

Black. C. C. 1979. Paleontology and geology of the Badwater Creek area, central Wyoming. Part 19. Perissodactyla. *Ann. Carnegie Mus.* 48:391-407.

Black, C. C. and Dawson, M.R. 1966. Paleontology and geology of the Badwater Creek area, central Wyoming. Part 1. History of field work and geologic setting. *Ann. Carnegie Mus.* 38:297-307.

Black, C.C. and Dawson, M.R. 1966. A review of late Eocene mammalian faunas from North America. *Amer. Jour. Sci.* 264:321-349.

Bown, T.M. 1979. Geology and mammalian paleontology of the Sand Creek facies, Lower Willwood Formation (Lower Eocene), Washakie County, Wyoming. *Mem. Geol. Survey Wyoming* 2:1-151.

Bown, T.M. 1980. The Willwood Formation (lower Eocene) of the southern Bighorn Basin, Wyoming, and its mammalian fauna. In P.D. Gingerich, ed., *Early Cenozoic Paleontology and Stratigraphy of the Bighorn Basin, Wyoming, Univ. Michigan Pap. Paleont.* 24:51-68.

Bown, T.M. 1982. Geology, paleontology, and correlation of Eocene volcaniclastic rocks, southeast Absaroka Range, Hot Springs County, Wyoming. *U.S. Geol. Survey Prof. Pap.* 1201-A:1-75.

Clark, J. 1937. The stratigraphy and paleontology of the Chadron Formation in the Big Badlands of South Dakota. *Ann. Carnegie Mus.* 25:261-350.

Clark, J., Beerbower, J. R., and Kietzke, K.K. 1967. Oligocene sedimentation, stratigraphy, paleoecology and paleoclimatology of the

Big Badlands of South Dakota. *Fieldiana: Geol. Mem.* 5:1-158.

Collinson, M.E. and Hooker, J.J. 1987. Vegetational and mammalian faunal changes in the early Tertiary of southern England. In E.M. Friis, W.G. Chaloner, and P.R. Crane, eds., *The Origins of Angiosperms and their Biological Consequences,* Cambridge, Cambridge Univ. Press, pp. 259-304.

Dashzeveg, D. 1988. Holarctic correlation of non-marine Palaeocene-Eocene boundary strata using mammals. *Jour. Geol. Soc. London* 145:473-478.

Dawson, M. R. 1970. Paleontology and geology of the Badwater Creek area, central Wyoming. Part 6. The leporid *Mytonolagus* (Mammalia, Lagomorpha). *Ann. Carnegie Mus.* 41:215-230.

Dawson, M. R. 1973. Paleontology and geology of the Badwater Creek area, central Wyoming. Part 8. The rodent *Microparamys. Ann. Carnegie Mus.* 45:145-150.

Dawson, M. R. 1980. Paleontology and geology of the Badwater Creek area, central Wyoming. Part 20. The late Eocene Creodonta and Carnivora. *Ann. Carnegie Mus.* 49:79-91.

Dawson, M.R., Krishtalka, L., and Stucky, R.K. 1990. Revision of the Wind River faunas, early Eocene of central Wyoming. Part 9. The oldest known hystricomorphous rodent (Mammalia: Rodentia). *Ann. Carnegie Mus.* 59:135-147.

Dawson, M. R., Stucky, R.K., Krishtalka, L., and Black, C.C. 1986. *Machaeroides simpsoni,* new species, oldest known sabertooth creodont (Mammalia) of the Lost Cabin Eocene. *Univ. Wyoming Contrib. Geol., Spec. Pap.* 3:177-182.

Delson, E. 1971. Fossil mammals of the Wasatchian Powder River local fauna, Eocene of northeast Wyoming. *Bull. Amer. Mus. Nat. Hist.* 146:309-364.

Doi, K. 1990. Geology, [sic] and paleontology of two primate families of the Raven Ridge, northwestern Colorado and northeastern Utah. Unpubl. M. Sc. Thesis, University of Colorado, Boulder, 215 pp.

Eaton, J.G. 1980. Preliminary report on paleontological exploration of the southeastern Absaroka Range, Wyoming. In P.D. Gingerich, ed., *Early Cenozoic Paleontology and Stratigraphy of the Bighorn Basin, Wyoming, Univ. Michigan Pap. Paleont.* 24:139-142.

Eaton, J. G. 1982. Paleontology and correlation of Eocene volcanic rocks in the Carter Mountain area, Park County, southeastern Absaroka Range, Wyoming. *Contrib. Geol. Univ. Wyoming* 21(2):153-194.

Emry, R. J. 1981. Additions to the mammalian fauna of the type Duchesnean, with comments on the status of the Duchesnean "Age". *Jour. Paleont.* 55:563-570.

Emry, R.J. 1990. Mammals of the Bridgerian (middle Eocene) Elderberry Canyon Local Fauna of eastern Nevada. *Geol. Soc. Amer. Spec. Pap.* 243:187-210.

Emry, R.J., Bjork, P.R., and Russell, L.S. 1987. The Chadronian, Orellan, and Whitneyan North American Land Mammal Ages. In M.O. Woodburne, ed., *Cenozoic Mammals of North America: Geochronology and Biostratigraphy,* Berkeley, University of California Press, pp. 118-152.

Emry, R.J. and Korth, W.W. 1989. Rodents of the Bridgerian (middle Eocene) Elderberry Canyon Local Fauna of eastern Nevada. *Smithsonian Contrib. Paleobiol.* 67:1-14.

Evanoff, E. 1990. Late Eocene and early Oligocene paleoclimates as indicated by the sedimentology and nonmarine molluscs of the White River Formation near Douglas, Wyoming. Unpublished Ph.D. dissertation, University of Colorado, Boulder, 440 pp.

Figgins, J.D. 1934. New material for the study of individual variation, from the lower Oligocene of Colorado. *Proc. Colo. Mus. Nat. Hist.* 13:7-17.

Fisher, R.V., and Rensberger, J.M. 1972. Physical stratigraphy of the John Day Formation, central Oregon. *Univ. Cal. Pub. Geol. Sci.* 101:1-45.

Flynn, J.J. 1986. Correlation and geochronology of middle Eocene strata from the western United States. *Palaeogeogr., Palaeoclimat., Palaeoecol.* 55:335-406.

Galbreath, E.C. 1953. A contribution to the Tertiary geology and paleontology of northeastern Colorado. *Univ. Kansas Paleont. Contrib. Vertebrata* 4.

Gazin, C. L. 1952. The Lower Eocene Knight Formation of western Wyoming and its mammalian faunas. *Smithsonian Misc. Coll.* 117:1-82.

Gazin, C.L. 1955. A review of upper Eocene Artiodactyla of North America. *Smithsonian Misc. Coll.* 128:1-96.

Gazin, C.L. 1956. The geology and vertebrate paleontology of upper Eocene strata in the northeastern part of the Wind River Basin, Wyoming. Part 2. The mammalian fauna of the Badwater area. *Smithsonian Misc. Coll.* 131:1-35.

Gazin, C.L. 1962. A further study of the Lower Eocene mammal faunas of southwestern Wyoming. *Smithsonian Misc. Coll.* 144:1-98.

Gazin, C.L. 1976. Mammalian faunal zones of the Bridger middle Eocene. *Smithsonian Contrib. Paleobiol.* 26:1-25.

Gheerbrant, E. 1987. Les vertebres continentaux de l'Adrar Mgorn (Maroc, Paleocene): une dispersion de mammiferes transtethysienne aux environs de la limite Mesozoique-Cenozoique. *Geodynamica Acta, Paris*, 1, 233-246.

Gingerich, P.D. 1980. History of early Cenozoic vertebrate paleontology in the Bighorn Basin. In P.D. Gingerich, ed., *Early Cenozoic Paleontology and Stratigraphy of the Bighorn Basin, Wyoming, Univ. Michigan Pap. Paleont.* 24:7-24.

Gingerich, P.D. 1983a. Paleocene-Eocene faunal zones and a preliminary analysis of Laramide structural deformation in the Clark's Fork Basin, Wyoming. *34th Annual Field Conference, Wyo. Geol. Assoc. Guidebook*, pp. 185-195.

Gingerich, P.D. 1983b. Systematics of early Eocene Miacidae (Mammalia, Carnivora) in the Clark's Fork Basin, Wyoming. *Contrib. Mus. Paleont. Univ. Mich.* 26:197-225.

Gingerich, P.D. 1989. New earliest Wasatchian mammalian fauna from the Eocene of northwestern Wyoming: Composition and diversity in a rarely sampled high-floodplain assemblage. *Pap. Paleont. Univ. Mich.* 28:1-97.

Gingerich, P.D., and Deutsch, H.A. 1989. Systematics and evolution of early Eocene Hyaenodontidae (Mammalia, Creodonta) in the Clark's Fork Basin, Wyoming. *Contrib Mus. Paleont., Univ. Michigan* 27, 327-391.

Gingerich, P.D., Rose, K.D., and Krause, D.W. 1980. Early Cenozoic mammalian faunas of the Clark's Fork Basin-Polecat Bench area, northwestern, Wyoming. In P.D. Gingerich, ed., *Early Cenozoic Paleontology and Stratigraphy of the Bighorn Basin, Wyoming, Univ. Michigan Pap. Paleont.* 24:51-68.

Godinot, M. 1982. Aspects Nouveaux des Echanges entre les faunes Mammaliennes d'Europe et d'Amerique du Nord a la base de l'Eocene. *Geobios, Mém. Spec.* 6:403-412.

Golz, D.J., and Lillegraven, J.A. 1977. Summary of known occurrences of terrestrial vertebrates from Eocene strata of southern California. *Contrib. Geol. Univ. Wyoming* 15, 43-60.

Gunnell, G.F. 1989. Evolutionary history of Microsyopoidea (Mammalia, ?Primates) and the relationship between Plesiadapiformes and Primates. *Pap. Paleont. Univ. Mich.* 27:1-157.

Hickey, L.J. 1980. Paleocene stratigraphy and flora of the Clark's Fork Basin. *Pap. Paleont. Univ. Mich.*, 24:33-49.

Hooker, J.J. 1986. Mammals from the Bartonian (middle/late Eocene) of the Hampshire Basin, southern England. *Bull. Brit. Mus. (Nat. Hist.), Geology* 39:191-478.

Hough, J. and Alf, R. 1956. A Chadron mammalian fauna from Nebraska. *Jour. Paleont.* 30:132-140.

Hunt, R.M., Jr. 1985. Faunal succession, lithofacies and depositional environments in Arikaree rocks (lower Miocene) of the Hartville Table, Nebraska and Wyoming. *Dakoterra* 2:155-204.

Ivy, L.D. 1990. Systematics of late Paleocene and early Eocene Rodentia (Mammalia) from the Clark's Fork Basin, Wyoming. *Contrib. Mus. Paleont. Univ. Mich.* 28:21-70.

Kay. J.L. 1934. The Tertiary formations of the Uinta Basin, Utah. *Ann. Carnegie Mus.* 23: 357-371.

Kay, J.L. 1953. Faunal list of vertebrates from the Uinta Basin, Utah. *6th Ann. Field Conf., Soc. Vert. Paleont.* pp. 20-25.

Kay, J.L. 1957. The Eocene vertebrates of the Uinta Basin, Utah. *Intermountain Assoc. Petrol. Geol. Guidebook, 8th Ann. Field Conf.* pp. 110-114.

Kihm, A. J. 1984. Early Eocene mammalian faunas of the Piceance Creek Basin, northwestern Colorado. Unpublished Ph.D. Dissert., Univ. Colorado, Boulder, 381 pp.

Kihm, A.J. 1987. Mammalian paleontology and geology of the Yoder Member, Chadron Formation, East-central Wyoming. In J.E. Martin and Ostrander, G.E., eds., *Papers in Vertebrate Paleontology in honor of Morton Green. Dakoterra* 3:28-45.

Kihm, A.J. and Lammers, G.E. 1986. Vertebrate biochronology of Oligocene sediments in southwest North Dakota. *Proc. N. Dak. Acad. Sci.* 40:18.

Korth, W. 1984. Earliest Tertiary evolution and radiation of rodents in North America. *Bull. Carnegie Mus. Nat. Hist.* 24:1-71.

Korth, W.W. 1987. Sciurid rodents (Mammalia) from the Chadronian and Orellan (Oligocene) of Nebraska. *Jour. Paleont.* 61:1247-1255.

Korth, W.W. 1989. Stratigraphic occurrence of rodents and lagomorphs in the Orella Member, Brule Formation (Oligocene) northwestern Nebraska. *Contrib. Geol., Univ. Wyoming* 27:15-20.

Krause, D.W. and Maas, M.C. 1990. The biogeographic origins of late Paleocene-early Eocene mammalian immigrants to the Western Interior of North America. *Geol. Soc. Amer. Spec. Pap.* 243:71-106.

Krishtalka, L. 1978. Paleontology and geology of the Badwater Creek area, central Wyoming. Part 15. Review of the late Eocene primates from Wyoming and Utah, and the Plesitarsiiformes. *Ann. Carnegie Mus.* 47:335-360.

Krishtalka, L. 1979. Paleontology and geology of the Badwater Creek area, central Wyoming. Part 18. Revision of late Eocene *Hyopsodus*. *Ann. Carnegie Mus.* 48:377-389.

Krishtalka, L. and Black, C.C. 1975. Paleontology and geology of the Badwater Creek area, central Wyoming. Part 12. Description and review of late Eocene Multituberculata from Wyoming and Montana. *Ann. Carnegie Mus.* 45:287-297.

Krishtalka, L. and Setoguchi, T. 1977. Paleontology and geology of the Badwater Creek area, central Wyoming. Part 13. The late Eocene Insectivora and Dermoptera. *Ann. Carnegie Mus.* 46:71-99.

Krishtalka, L. and Stucky, R.K. 1983a. Revision of the Wind River faunas, early Eocene of central Wyoming. Part 3. Marsupialia. *Ann. Carnegie Mus.* 52:205-228.

Krishtalka, L. and Stucky, R.K. 1983b. Paleocene and Eocene marsupials of North America. *Ann. Carnegie Mus.* 52(10): 229-263.

Krishtalka, L. and Stucky, R.K. 1984. Middle Eocene marsupials from northeastern Utah, and the mammalian fauna from Powder Wash. *Ann. Carnegie Mus.* 53(2): 31-45.

Krishtalka, L. and Stucky, R.K. 1985. Revision of the Wind River faunas, early Eocene of central Wyoming. Part 7. Revision of *Diacodexis* (Mammalia, Artiodactyla). *Ann. Carnegie Mus.* 54 (14): 413-486.

Krishtalka, L., Stucky, R.K., West, R.M., McKenna, M.C., Black, C.C., Bown, T.M., Dawson, M.R., Golz, D.J., Lillegraven, J.A. and Turnbull, W.D. 1987. Eocene (Wasatchian through Duchesnean) chronology of North America. In Woodburne, M.O., ed., *Cenozoic Mammals of North America: Geochronology and Biostratigraphy.* Berkeley, University of California Press, pp. 77-117.

Legendre, S. 1987. Les communautes de mammiferes d'Europe occidentale de l'Eocene superieur et Oligocene: structures et mileux. *Munchner Geowiss. Abh.* 10:301-310.

Leopold, E.B. and MacGinitie, H.D. 1972. Development and affinities of Tertiary floras in the Rocky Mountains. In A. Graham, ed., *Floristics and paleofloristics of Asia and Eastern North America,* Amsterdam, Elsevier Publ. Co., pp. 147-200.

Lillegraven, J.A. 1970. Stratigraphy, structure, and vertebrate fossils of the Oligocene Brule Formation, Slim Buttes, northwestern South Dakota. *Bull. Geol. Soc. Amer.* 81:831-850.

Lillegraven, J.A. 1972. Ordinal and familial diversity of Cenozoic mammals. *Taxon* 21: 261-274

Lillegraven, J.A. 1979. A biogeographical problem involving comparisons of later Eocene terrestrial vertebrate faunas of western North America. In J. Gray and A.J. Boucot, eds., *Historical Biogeography, Plate Tectonics, and the Changing Environment,* Corvallis, Oregon State University Press, pp. 333-347.

Lillegraven, J.A. and Ostresh, L.M., Jr. 1988. Evolution of Wyoming's early Cenozoic topography and drainage patterns. *Nat. Geogr. Res.* 4:303-327.

Lucas, S.G. 1982. Vertebrate paleontology, stratigraphy and biostratigraphy of Eocene Galisteo Formation, north-central New Mexico. *New Mexico Bur. Mines Min. Res.* 186:1-34.

Lucas, S.G., Schoch, R.M., Manning, E., and Tsentas, C. 1981. The Eocene biostratigraphy of New Mexico. *Bull. Geol. Soc. Amer.* 92:951-

967.

Maas, M.C. 1985. Taphonomy of a late Eocene microvertebrate locality, Wind River Basin, Wyoming (U.S.A.). *Palaeogeog., Palaeoclimat., Palaeoecol.* 52:123-142.

MacDonald, J.R. 1963. The Miocene faunas from the Wounded Knee area of western South Dakota. *Bull. Amer. Mus. Nat. Hist.* 125:139-238.

MacDonald, J.R. 1970. Review of the Miocene Wounded Knee faunas of southwestern South Dakota. *Los Angeles Co. Mus. Nat. Hist. Bull Sci.* 8.

MacGinitie, H.D. 1953. Fossil plants from the Florissant Beds, Colorado. *Carn. Inst. Wash. Publ.* 599:1-188.

McGrew, P.O., ed. 1959. The geology and paleontology of the Elk Mountain and Tabernacle Butte area, Wyoming. *Bull. Amer. Mus. Nat. Hist.* 117(3):117-176.

McGrew, P.O. and Sullivan, R. 1970. The stratigraphy and paleontology of Bridger A. *Contrib. Geol. Univ. Wyoming* 9(2):66-85.

McKenna, M.C. 1960. Fossil Mammalia from the early Wasatchian Four Mile local fauna, Eocene of northwest Colorado. *Univ. Calif. Publ. Geol. Sci.* 37:1-130.

McKenna, M.C. 1975. Fossil mammals and Early Eocene North Atlantic land continuity. *Ann. Missouri Bot. Gard.* 62:335-353.

McKenna, M. C. 1980. Late Cretaceous and early Tertiary vertebrate paleontological reconnaissance, Togwotee Pass area, northwestern Wyoming. In L. Jacobs, ed., *Aspects of Vertebrate History*, Flagstaff, Museum of Northern Arizona Press,pp. 321-343.

McKenna, M.C. 1983a. Holarctic landmass rearrangement, cosmic events, and Cenozoic terrestrial organisms. *Ann. Missouri Bot. Gard.* 70: 459-489.

McKenna, M.C. 1983b. Cenozoic paleogeography of North Atlantic land Bridges. In Bott et al., eds., *Structure and Development of the Greenland-Scotland Ridge*, New York, Plenum, pp. 351-399.

McKenna, M.C. 1990.Plagiomenids (Mammalia: ?Dermoptera) from the Oligocene of Oregon, Montana, and South Dakota, and middle Eocene of northwestern Wyoming. *Geol. Soc. Amer. Spec. Pap.* 243: 211-234.

Montanari, A., Deino, A.L., Drake, R.E., Turrin, B.D., DePaolo, D.J., Odin, G.S., Curtis, G.H., Alvarez, W., and Bice, D.M. 1988. Radioisotopic dating of the Eocene-Oligocene boundary in the pelagic sequence of the northeast Apennines. In I. Premoli-Silva, R. Coccioni., and A. Montanari, eds., *The Eocene-Oligocene Boundary in the Marche-Umbria Basin (Italy), Int. Union Geol. Sci. Comm. Strat.* Ancona, Italy, pp. 195-208.

Novacek, M.J., Flynn, J.J., Ferrusquia-Villafranca, I., and Cipolleti, R.M. 1987. An early Eocene (Wasatchian) mammal fauna from Baja California. *Nat. Geogr. Res.* 3:376-388.

Ostrander, G.E. 1985. Correlation of the early Oligocene (Chadronian) in northwestern Nebraska. *Dakoterra* 2:205-231.

Owen, R.M. and Rea, D.K. 1985. Sea-floor hydrothermal activity links climate to tectonics: The Eocene carbon dioxide greenhouse. *Science* 227:166-169.

Patton, T.H. 1969. An Oligocene land vertebrate fauna from Florida. *Jour. Paleont.* 43:543-546.

Peterson, O. A. 1919. Report upon the material discovered in the Upper Eocene of the Uinta Basin by Earl Douglass in the years 1908-1909 and by O. A. Peterson in 1912. *Ann. Carnegie Mus.* 12:40-168.

Peterson, O. A.. 1931. New species of the genus *Teleodus* from the upper Uinta of northeastern Utah. *Ann. Carnegie Mus.* 20: 307-312.

Peterson, O. A. 1931. New mesonychids from the Uinta. *Ann. Carnegie Mus.* 20:333-339.

Peterson, O. A. 1931. New species from the Oligocene of the Uinta. *Ann. Carnegie Mus.* 21:51-78.

Peterson, O. A. 1934. List of species and description of new material from the Duchesne River Oligocene, Uinta Basin, Utah. *Ann. Carnegie Mus.* 23:373-389.

Peterson, O. A. and J. L. Kay. 1931. The Upper Uinta Formation of northeastern Utah. *Ann. Carnegie Mus.* 20:293-306.

Prichinello, K.A. 1971. Earliest Eocene mammalian fossils from the Laramie Basin of southeast Wyoming. *Contrib. Geol. Univ. Wyoming* 10:73-88.

Prothero, D.R. 1985. North American mammalian diversity and Eocene-Oligocene extinctions. *Paleobiology* 11:389-405.

Prothero, D.R. 1989. Stepwise extinctions and climatic decline during the later Eocene and Oligocene. In S.K. Donovan, ed., *Mass Extinc-*

tions: Processes and Evidence, London, Belhaven Press, pp. 217-234.

Prothero, D.R., and Schoch, R.M., eds. 1989. *The Evolution of Perissodactyls,* New York, Oxford Univ. Press, 537 pp.

Rasmussen, D.L. 1977. Geology and mammalian paleontology of the Oligocene-Miocene Cabbage Patch Formation, central-western Montana. Unpublished Ph. D. Dissertation, Univ. Kansas.

Rea, D.K., Zachos, J.C., Owen, R.M., and Gingerich, P.D. 1990. Global change at the Paleocene-Eocene boundary: climatic and evolutionary consequences of tectonic events. *Palaeogeogr., Palaeoclimat., Palaeoecol.* 79: 117-128.

Retallack, G. 1983. Late Eocene and Oligocene fossil Paleosols from Badlands National Park, South Dakota. *Geol Soc. Amer. Spec. Pap.* 193.

Rensberger, J.M. 1973. Pleurolicine rodents (Geomyoidea) of the John Day Formation, Oregon. *Univ. Calif. Publ. Geol. Sci.* 102:1-130.

Rensberger, J.M. 1983. Succession of meniscomyine and allomyine rodents (Aplodontidae) in the Oligo-Miocene John Day Formation, Oregon. *Univ. Calif. Publ. Geol. Sci.* 124:1-157.

Robinson, P. 1966a. Paleontology and geology of the Badwater Creek area, central Wyoming. Part 3. Late Eocene Apatemyidae (Mammalia, Insectivora) from the Badwater area. *Ann. Carnegie Mus.* 38:317-320.

Robinson, P. 1966b. Fossil Mammalia of the Huerfano Formation, Eocene of Colorado. *Bull. Yale Peabody Mus. Nat. Hist.* 21:1-95.

Robinson, P. 1968. Paleontology and geology of the Badwater Creek area, central Wyoming. Part 4. Late Eocene primates from Badwater, Wyoming, with a discussion of material from Utah. *Ann. Carnegie Mus.* 39:207-326.

Robinson, P. 1972. Tertiary history. In W.M. Mallory and others, eds., *Geologic Atlas of the Rocky Mountain Region.* Denver, Rocky Mtn. Assoc. Geologists, pp. 233-242.

Rose, K. D. 1981. The Clarkforkian Land-Mammal Age and mammalian faunal composition across the Paleocene-Eocene boundary. *Univ. Mich. Pap. Paleont.* 26: 1-197.

Rose, K.D. 1990. Postcranial skeletal remains and adaptations in early Eocene mammals from the Willwood Formation, Bighorn Basin, Wyoming. *Geol. Soc. Amer. Spec. Pap.* 243:107-133.

Rose, K.D., Krishtalka, L., and Stucky, R.K. 1991. Revision of the Wind River faunas, early Eocene of central Wyoming. Part 11. Palaeanodonta (Mammalia). *Ann. Carnegie Mus.* 60:63-82.

Russell, D.E., Hartenberger, J.-L., Pomerol, C., Sen, S., Schmidt-Kittler, N., and Vianey-Liaud, M. 1982. Mammals and stratigraphy: The Paleogene of Europe. *Palaeovertebrata, Mem. Extraord.* 77 pp.

Savage, D.E. and Russell, D.E. 1983. *Mammalian Paleofaunas of the World.* London, Addison-Wesley, 432 pp.

Savin, S.M. 1977. The history of the Earth's surface temperature during the past 100 million years. *Ann. Rev. Earth Planet. Sci.* 5: 319-355.

Savin, S.M. and Douglas, R.G. 1984. Sea level, climate, and the Central American land bridge, In F.G. Stehli and S.D. Webb, eds., *The Great American Biotic Interchange,* New York, Plenum, pp. 303-324.

Schankler, D. M. 1980. Faunal zonation of the Willwood Formation in the central Bighorn Basin, Wyoming. In P. D. Gingerich, ed., *Early Cenozoic Paleontology and Stratigraphy of the Bighorn Basin, Wyoming, Univ. Michigan Pap. Paleont.* 24:99-114.

Schlaikjer, E.M. 1935. Contributions to the stratigraphy and paleontology of the Goshen Hole area, Wyoming. IV. New vertebrates and the stratigraphy of the Oligocene and early Miocene. *Bull. Mus. Comp. Zool., Harvard* 76:97-189.

Scott, W. B. 1945. The Mammalia of the Duchesne River Oligocene. *Trans. Amer. Phil. Soc.* 34:209-253.

Scott, W.B. and Jepsen, G.L., eds. 1936-1941. The mammalian fauna of the White River Oligocene, parts I-V. *Trans. Amer. Phil. Soc.* 28:1-980.

Setoguchi, T. 1975. Paleontology and geology of the Badwater Creek area, central Wyoming. Part 11. Late Eocene marsupials. *Ann. Carnegie Mus.* 45:263-275.

Setoguchi, T. 1978. Paleontology and geology of the Badwater Creek Area, central Wyoming. Part 16. The Cedar Ridge Local Fauna (Late Oligocene). *Bull. Carnegie Mus.* 9:1-61.

Skinner, M.F., Skinner, S.M., and Gooris, R.J. 1968. Cenozoic rocks and faunas of Turtle Butte, south-central South Dakota. *Bull. Amer. Mus. Nat. Hist.* 138:381-436.

Stanley, S.M. 1990. Delayed recovery and the spacing of major extinctions. *Paleobiology* 16:401-414.

Storer, J.E. 1984. Mammals of the Swift Current Creek local fauna (Eocene: Uintan), Saskatchewan. *Nat. Hist. Contrib. Sask. Mus. Nat. Hist.* 7:1-158.

Storer, J.E. 1987. Dental evolution and radiation of Eocene and early Oligocene Eomyidae (Mammalia, Rodentia) of North America, with new material from the Duchesnean of Saskatchewan. *Dakoterra* 3:108-117.

Storer, J.E. 1988. Thc rodents of the Lac Pelletier lower fauna, late Eocene (Duchesnean) of Saskatchewan. *Jour. Vert. Paleont.* 8:84-101.

Storer, J.E. 1990. Primates of the Lac Pelletier lower fauna (Eocene: Duchesnean), Saskatchewan. *Can. Jour. Earth Sci.* 27:520-524.

Stucky, R.K. 1984. The Wasatchian-Bridgerian land mammal age boundary (early to middle Eocene) in western North America. *Ann. Carnegie Mus.* 53:347-382.

Stucky, R.K. 1990. Evolution of land mammal diversity in North America during the Cenozoic. *Curr. Mammal.* 2:375-432.

Stucky, R.K. and Krishtalka, L. 1982. Revision of the Wind River Faunas, early Eocene of central Wyoming. Part l. Introduction and Multituberculata. *Ann. Carnegie Mus.* 51(3): 39-56.

Stucky, R.K. and Krishtalka, L. 1983. Revision of the Wind River faunas, early Eocene of central Wyoming. Part 4. The Tillodontia. *Ann. Carnegie Mus.* 52:375-391.

Stucky, R.K. and Krishtalka, L. 1990. Revision of the Wind River faunas, early Eocene of central Wyoming. Part 10. *Bunophorus* (Mammalia, Artiodactyla). *Ann. Carnegie Mus.* 59:149-171.

Stucky, R.K., Krishtalka, L., and Dawson, M.R. 1989. Paleontology, geology and remote sensing of Paleogene rocks in the northeastern Wind River Basin, Wyoming, USA. In J.J. Flynn and M.C. McKenna, eds., *Mesozoic/Cenozoic Vertebrate Paleontology: Classic Localities, Contemporary Approaches. 28th Int. Geol. Cong., Field Trip Guidebook* T322:34-44.

Stucky, R.K., Krishtalka, L., and Redline, A.D. 1990. Geology, vertebrate fauna, and paleoecology of the Buck Spring Quarries (early Eocene, Wind River Formation), Wyoming. *Geol. Soc. Amer. Spec. Pap.* 243:169-186.

Stucky, R.K. and McKenna, M.C. in press. Mammalia. In M.J. Benton and M.A. Whyte, eds., *The Fossil Record,* London, Chapman and Hall.

Swisher, C.C., and Prothero, D.R. 1990. Single-crystal $^{40}Ar/^{39}Ar$ dating of the Eocene-Oligocene transition in North America. *Science* 249: 760-762.

Tabrum, A.R. and Fields, R.W. 1980. Revised mammalian faunal list for the Pipestone Springs Local Fauna (Chadronian, early Oligocene), Jefferson County, Montana. *Northwest Geol.* 9:45-51.

Tedford, R.H., Galusha, T., Skinner, M.F., Taylor, B.E., Fields, R.W., MacDonald, J.R., Rensberger, J.M., Webb, S.D., and Whistler, D.P. 1987. Faunal succession and biochronology of the Arikareean through Hemphillian interval (late Oligocene through earliest Pliocene Epochs) in North America. In M.O. Woodburne, ed., *Cenozoic Mammals of North America: Geochronology and Biostratigraphy,* Berkeley, Univ. California Press, pp. 153-210.

Torres, V. and Gingerich, P.D. 1983. Summary of Eocene stratigraphy at the base of Jim Mountain, North Fork of the Shoshone River, Northwestern Wyoming. *34th Ann. Field Conf. Wyom. Geol. Assoc. Guideb.* pp. 205-208.

Turnbull, W.D. 1972. The Washakie Formation of Bridgerian-Uintan ages and the related faunas. *Guideb. Field Conf. Tert. Biostrat. of Southern and Western Wyoming,* pp. 3-19.

Upchurch, G.R., Jr. and Wolfe, J.A. 1987. Mid-Cretaceous to Early Tertiary vegetation and climate: evidence from fossil leaves and woods. In E.M. Friis, W.G. Chaloner, and P.R. Crane, eds., *The Origins of angiosperms and their biological consequences.* Cambridge, Cambridge Univ. Press, pp. 75-106.

Webb, S.D. 1977. A history of savanna vertebrates in the New World. Part I: North America. *Ann. Rev. Ecol. Syst.* 8:355-380.

West, R. M. 1973a. Geology and mammalian paleontology of the New Fork-Big Sandy area, Sublette County, Wyoming. *Fieldiana, Geol.* 29:1-193.

West, R.M. 1973b. New records of fossil mammals from the early Eocene Golden Valley Formation, North Dakota. *Jour. Paleont.* 54: 749-750.

West, R.M. 1982. Fossil mammals from the Lower Buck Hill Group, Eocene of Trans-Pecos Texas: Marsupicarnivora, Primates, Taeniodonta, Condylarthra, bunodont Artiodactyla, and Dinocerata. *Texas Mem. Mus. Pearce-Sellards Ser.* 35:1-20.

West, R.M. and Dawson, M.R. 1973. Fossil mammals from the upper part of the Cathedral Bluffs tongue of the Wasatch Formation (early Bridgerian), northern Green River Basin, Wyoming. *Contrib. Geol. Univ. Wyoming* 12(1):33-41.

West, R.M. and Dawson, M.R. 1975. Eocene fossil Mammalia from the Sand Wash Basin, northwestern Moffat County, Colorado. *Ann. Carnegie Mus.* 45(11):231-253.

Westgate, J.W. 1987. First Eocene land mammal fauna from the North American coastal plain. *Jour. Vert. Paleont.* 7(Supp. to 3):29A.

Wilson, J. A. 1977. Stratigraphic occurrence and correlation of early Tertiary vertebrate faunas, Trans-Pecos Texas. *Bull. Texas Mem. Mus.* 25:1-42.

Wilson, J.A. 1986. Stratigraphic occurrence and correlation of early Tertiary vertebrate faunas, Trans-Pecos Texas: Agua Fria-Green Valley areas. *Jour. Vert. Paleont.* 6:350-373.

Wing, S.L. 1984. A new basis for recognizing the Paleocene/Eocene boundary in Western Interior North America. *Science* 226, 439-441.

Wing, S.L. 1987. Eocene and Oligocene floras and vegetation of the Rocky Mountains. *Ann. Missouri Bot. Garden* 74:176-212.

Wing, S.L. 1991. Comment on "'Equable' climates during Earth history?" *Geology* 19: 539-540.

Wolfe, J.A. 1978. A paleobotanical interpretation of Tertiary climates in the northern hemisphere. *Amer. Scientist* 66:694-703.

Wood, H.E., Chaney, R.W., Clark, J., Colbert, E.H., Jepsen, G.L.;, Reeside, J.B. Jr., and Stock, C. 1941. Nomenclature and correlation of the North American continental Tertiary. *Bull. Geol. Soc. Amer.* 52:1-48.

Woodburne, M.O., ed. 1987. *Cenozoic Mammals of North America: Geochronology and Biostratigraphy.* Berkeley, University of California Press, 336 pp.

TABLE 24.1. Genera occurring in the Paleocene (PAL: Pu, Puercan LMA; To, Torrejonian LMA; Ti, Tiffanian, LMA), and Clarkforkian (CLF), Wasatchian (WA), Bridgerian (BR), Uintan (UN), and early Duchesnean (DU) Land Mammal Age faunas of North America. An asterisk (*) indicates a common occurrence in the European record and a plus (+) indicates a common occurrence in the Asian record; a queried record is identified by a question mark (?) following the symbol. Data abstracted from the following principle sources (see references in these papers for additional sources): **Paleocene** - All areas: Archibald et al., 1987. **Clarkforkian** - All areas: Rose, 1981; Archibald et al., 1987. **Wasatchian** - All areas: Krishtalka and Stucky, 1983; Korth, 1984; Stucky, 1984; Krishtalka et al. 1987. Colorado: McKenna, 1960; Robinson, 1966b; Kihm, 1984. New Mexico: Lucas et al., 1981. North Dakota: West, 1973b. Wyoming: Gazin, 1952, 1962; Delson, 1971; Prichinello, 1971; West, 1973a; Bown, 1979, 1980; Gingerich, 1980, 1983a, 1983b, 1989; Gingerich et al., 1983; Schankler, 1980; Krishtalka and Stucky, 1985; Stucky and Krishtalka, 1982, 1983, 1990; Stucky et al., 1990; Dawson et al., 1986, 1990; Rose et al., 1991. **Bridgerian** - All areas: Stucky, 1984; Krishtalka et al., 1987. Wyoming: McGrew, 1959; McGrew and Sullivan 1970; West, 1973a; West and Dawson, 1973; Gazin, 1976; Eaton 1982; Bown 1982; Torres and Gingerich, 1983. Colorado: Robinson, 1966b. Utah: Krishtalka and Stucky, 1984. Nevada: Emry and Korth, 1989; Emry, 1990. **Uintan** - All areas: Black and Dawson, 1966a; Flynn, 1986. California: Golz and Lillegraven 1977. Colorado: West and Dawson, 1975. New Mexico: Lucas, 1982. Saskatchewan: Storer, 1984. Texas: West, 1982; Wilson, 1977, 1986. Utah: Kay, 1934, 1953, 1957; Peterson, 1919, 1931a, 1931b, 1931c, 1934; Peterson and Kay, 1931; Gazin, 1955. Wyoming: Black and Dawson, 1966a; Gazin, 1956; Black, 1967, 1969, 1970, 1971, 1974, 1978, 1979; Dawson, 1970, 1973, 1980; Turnbull, 1972; Krishtalka, 1978, 1979; Krishtalka and Black, 1975; Setoguchi, 1975; Krishtalka and Setoguchi, 1977; Robinson, 1966a, 1968; McKenna, 1980; Eaton, 1980. **Early Duchesnean** - Montana: Tabrum, personal comm.; Wyoming: Maas, 1985. Texas: Wilson, 1986. Utah: Emry, 1981; Scott, 1945.

	PAL	CLF	WA1	WA2	WA3	WA4	WA5	BR1	BR2	BR3	UN1	UN2	UN3	DU1
MULTITUBERCULATA														
*Ectypodus**	Pu^2	X	X	X	X	-	X	-	-	-	-	-	X	X
Parectypodus	Pu^2	X	X	X	-	-	X							
*Neoplagiaulax**	Pu^2	-	X											
*Microcosmodon**	Pu^2	X^2												
Neoliotomus	Ti^2	X	X	X	-	X								
Prochetodon	Ti^3	X^2												
MARSUPIALIA														
*Peradectes**	To^3	X	X	X	X	-	X	X	X	X	X	-	X	X
Mimoperadectes		X^3	X	X	X									
*Peratherium**			X	X	X	X	X	X	X	X	X	-	X	X
Armintodelphis							X	X	X					
LEPTICTIDA														
Prodiacodon	Pu^3	X	X	X	X	X								
Palaeictops						X	X	X	-	X	X	-	X	-
ERINACEOMORPHA														
Diacodon	Ti^1	X	-	-	X	X	-	-	-	-	-	-	X	
Leipsanolestes		X^1	X											
Wyonycteris	X	X	X	X	X									
Alveojunctus							X	-	X					
Megadelphus								X						
Tarka											X			
Thylacaelurus													X	X
Craseops														X
SORICOMORPHA														
Palaeoryctes	To^2	X	X											
*Leptacodon**	To^3	X	-	-	X									
Pararyctes	Ti^2	-	X											
Pontifactor		X^2	X	X	X	-	-	-	-	X				
Plagioctenodon		X^2	X	X	X	X	X							
Centetodon			X	X	X	-	X	X	X	X	X	-	X	X
Parapternodus			X											
Plagioctenoides			X											
Nyctitherium						X	X	X	X	X	X	X	X	X
Myolestes									X					
Domnina													X	X
Apternodus														X
Micropternodus														X
CIMOLESTA														
Aphronorus	To^1	X												
Titanoides	Ti^1	X^1												
Cyriacotherium	Ti^3	X^3												
Barylambda	Ti^3	X	-	-	-	X								
Haplolambda	Ti^4	X^1												
*Palaeosinopa**	Ti^4	X	X	X	X	X	X	X						
Ectoganus	Ti^5	X	X	X	X	X	X							

	PAL	CLF	WA1	WA2	WA3	WA4	WA5	BR1	BR2	BR3	UN1	UN2	UN3	DU1
*Coryphodon**+		X[1]	X	X	X	X	X	X						
*Apatemys**		X	X	X	X	X	X	X	-	X	X	X	X	X
Azygonyx		X	X	X	X									
*Esthonyx**			X	X	X	X	X	X						
Protentomodon		X												
Wyolestes			X	X	X									
*Didelphodus**			X	X	X	X	X	X	-	X	-	-	X	-
Amaramnus				X	-	-	X							
Stylinodon							X	X	X	X	X	X		
Megalesthonyx							X							
Pantolestes								X	X	X	X			
Trogosus								X	X					
Simidectes											X	X	X	X
PALAEANODONTA														
Palaenodon		X	X	X	X	X								
Alocodontulum					X									
Dipassalus						X	X	X						
Tubulodon							X	X						
Pentapassalus							X							
Metacheiromys								X	X	X				
Tetrapassalus									X	X				
CHIROPTERA														
*Icaronycteris**							X	X						
*Ageina**							X							
Wallia													X	
PRIMATES														
Ignacius	To[3]	X	X	X	X	-	X	-	-	-	-	-	X	
*Plesiadapis**	Ti[1]	X	X											
*Chiromyoides**	Ti[3]	X[2]												
*Phenacolemur**	Ti[5]	X	X	X	X	X	X	X	-	-	X	-	X	X
Carpolestes	Ti[5]	X	X											
*Cantius**			X	X	X	X	X							
Tetonoides			X	X	X	X								
Tetonius			X	X	X	X								
*Teilhardina**			X											
Pseudotetonius			X	X	X									
Copelemur				X	X	X	X	X						
Uintanius				X	-	X	-	X	X	-	X	X		
Absarokius					X	X	X	X	X					
Pelycodus					X	X	X							
Loveina						X	X							
Anemorhysis						X	X	X						
Arapahovius						X								
Jemezius						X								
Trogolemur							X	X	X	X	X	-	X	
Hemiacodon							X	-	-	X	X	X		
Notharctus							X	X	X	X	X	X		
Smilodectes							X	X	X					
Gazinius							X	-	X					
Shoshonius							X	X						
Uintalacus							X							
Chlororhysis							X							
Omomys								X	X	X	X	X	X	
Washakius								X	X	X	X			
Aycrossia								X	X					
Utahia									X					
Elwynella									X					
Anaptomorphus									X					
Macrotarsius											X	-	X	-
Ourayia											X	X	X	
Yaquius													X	
Dyseolemur													X	X
Chumashius													X	X
Mytonius													X	X
Mahgarita														X
LAGOMORPHA														
Mytonolagus													X	X
Procaprolagus+													X	-

	PAL	CLF	WA1	WA2	WA3	WA4	WA5	BR1	BR2	BR3	UN1	UN2	UN3	DU1
RODENTIA														
Apatosciurus		X												
Acritoparamys		X[1]	-	-	-	-	X	X						
Franimys		X[2]	X	X	X	X								
*Paramys**		X[1]	X	X	X	X	X	X	X	X	-	-	-	-
*Microparamys**			X	X	X	X	X	-	X	X	X	X	X	X
Knightomys			X	X	X	X	X	X						
Dawsonomys			X	X	X	-	X							
Lophiparamys			X	X	X	X	X	X	X	X	X			
Pseudotomus					X	X	X	-	X	X	X	X	X	X
Mytonomys						X	X	-	-	-	-	X	X	X
Reithroparamys						X	X	X	X	X	X	-	X	
Thisbemys						X	X	X	X	X	X	X	X	
Pauromys						X	X	X	X	X	-	-	-	-
Notoparamys						X								
Sciuravus							X	X	X	X	X	X		
Mattimys							X	X						
Leptotomus								X	X	X	X	X	X	X
Mysops								X	X	X	X	X	X	
Tillomys								X	X	X	X			
Quadratomus								X	X					
Armintomys								X						
Elymys									X					
Taxymys									X					
Protadjidaumo											X	-	X	X
Protoptychus											X	X		
Pareumys											X	X	X	X
Eohaplomys											X	X	X	X
Simimys											X	X	X	X
Prolapsus											X	-	X	
Manitsha													X	X
Pseudocylindrodon+													X	X
Griphomys													X	X
Namatamys													X	X
Spurimus													X	X
Rapamys													X	X
Janimus													X	
Presbymys														X
CREODONTA														
Dipsalodon	Ti	X												
*Oxyaena**	Ti[5]	X	X	X	X	X	X							
*Palaeonictis**		X[3]	X	X										
Prolimnocyon?*			X	X	X	X	X	X						
Tritemnodon			X	X	X	X	X	X	X					
*Prototomus**			X	X	X	X	X	X						
*Arfia**			X	X	X	-	-	X						
Dipsalidictes?*			X											
Acarictis			X											
Galecyon				X	X									
Pyrocyon		X												
Ambloctonus					X	X	X							
*Patriofelis**							X	X	X	X				
Machaeroides							X	X	X					
Limnocyon								X	X	-	-	X	X	
Proviverra									X	X	?			
Thinocyon+									X	X				
Proviveroides									X					
Apataelurus												X	X	
Oxyaenodon												X		
Hemipsalodon													X	-
Hyaenodon+*														X
Pterodon+*														X
CARNIVORA														
Protictis	To[1]	-	-	-	-	-	-	-	-	-	X			
*Viverravus**	Ti[5]	X	X	X	X	X	X	X	X	X	X	X	X	
*Didymictis**	Ti5	X	X	X	X	X	X	X						
Uintacyon?*		X[2]	X	X	X	X	X	X	X	X	X	X	X	
Miacis+*			X	X	X	X	X	X	X	X	X	X	X	X
Oodectes			X	X	X	X	-	X	X					
Vassacyon			X	X	X	X	X	X						
Vulpavus?*				X	X	X	X	X						

	PAL	CLF	WA1	WA2	WA3	WA4	WA5	BR1	BR2	BR3	UN1	UN2	UN3	DU1
Palaearctonyx									X					
Plesiomiacis											X	-	-	X
Tapocyon													X	X
Prodaphaenus													X	-
*Procyonodictis**													X	
Eosictis														X
Daphoenus														X
MESONYCHIA														
Dissacus+*	To[2]	X	X	X	X									
Pachyaena+*			X	X	X	X	X							
Hapalodectes+						X	X	X						
Mesonyx+								X	X	X	-	X	X	
Harpagolestes+									X	X	X	X	X	X
Synoplotherium										X				
Hessolestes											X	-	X	
ARCTOCYONIA														
Chriacus?*	Pu[3]	X	X	X	X	X	X							
Thryptacodon	Ti[1]	X	X	X	X	X	X	X						
Anacodon	Ti[3]	X	X	X	X	X								
Princetonia	Ti	X	X											
Lambertocyon	Ti[3]	X												
CONDYLARTHRA														
*Phenacodus**	Ti[1]	X	X	X	X	X	X	X	X	X				
Ectocion	Ti[1]	X	X	X	X	-	-	X						
Phenacodaptes	Ti[4]	X[1]												
Aletodon	Ti[3]	X[3]												
Hyopsodus+*		X[3]	X	X	X	X	X	X	X	X	X	X	X	X
Meniscotherium		X	X	X	X	X	X							
Apheliscus		X[1]	X	X	X	X								
Haplomylus		X[1]	X	X	X									
Copecion		X	X											
ARCTOSTYLOPIDA														
Arctostylops+	Ti[4]	X[3]												
DINOCERATA														
Prodinoceras+	Ti[5]	X	-	-	-	X								
Bathyopsis							X	X	X					
Uintatherium+										X	X	X		
Eobasileus											X	X		
PERISSODACTYLA														
Hyracotherium+*			X	X	X	X	X	X	X					
Homogalax+			X	X	X	X	X							
Xenicohippus					X	X								
Palaeomoropus					X									
Heptodon+						X	X	X	X					
Orohippus							X	X	X	X	X			
Helaletes+							X	X	X	X				
Lambdotherium							X	X						
Hyrachyus?+*								X	X	X	X	X	X	
Eotitanops+?								X						
Palaeosyops+								X	X	X				
Selenaletes								X						
Isectolophus									X	X	X	X	X	
Fouchia									X					
Dilophodon										X	X	X	X	X
Telmatherium+										X	X	X	X	X
Manteoceras										X	X	X	X	
Mesatirhinus										X				
Triplopus+											X	X	X	X
Amynodon+											X	X	X	X
Epihippus											X	X	X	
Forstercooperia+											X	X	X	
Metarhinus											X	X	X	
Sthenodectes											X	-	X	
Eomoropus+											X	X		
Amynodontopsis+												X	-	X
Eotitanotherium												X		
Sphenocoelus												X		
Grangeria+													X	
Diplacodon													X	
Protitanotherium													X	
Metamynodon+													X	X

	PAL	CLF	WA1	WA2	WA3	WA4	WA5	BR1	BR2	BR3	UN1	UN2	UN3	DU1
Colodon+													X	-
Epitriplopus													X	-
Duchesneodus														X
Hyracodon														X
Mesamynodon														X
ARTIODACTYLA														
*Diacodexis**+?			X	X	X	X	X	X	X					
*Bunophorus**			X	X	X	X	X	X						
*Simpsonodus**				X	X	X								
Hexacodus						X	X	X						
Antiacodon							X	X	X	X				
Ibarus								X	-	-	X	-	X	
Homacodon								X	-	X				
Helohyus									X	X	X	X		
Microsus									X	X				
Ithygramon									X					
Leptoreodon											X	X	X	X
Protoreodon											X	X	X	X
Protylopus											X	X	X	X
Hylomeryx											X	X	X	
Malaquiferus											X	-	X	
Achaenodon+?											X	X		
Parahyus											X			
Merycobunodon											X			
Oromeryx												X	-	-
Pentacemylus												X	X	
Mytonomeryx												X	X	
Diplobunops												X	X	-
Leptotragulus												X	X	-
Bunomeryx												X	X	
Mesomeryx												X	X	
Auxontodon												X	X	
Texodon													X	
Simimeryx													X	X
Toromeryx													X	
Poebrodon													X	
Apriculus													X	
Hendryomeryx													X	X
Eotylopus														X
Poabromylus														X
Tapochoerus														X
Brachyhyops														X
Agriochoerus														X
MAMMALIA, *incertae sedis*														
Aethomylos									X	-	X			

TABLE 24.2. Genera occurring in "Late" Duchesnean (DU2), Chadronian(CH1-CH3), Orellan (OR), Whitneyan (WH), and early Arikareean (EA) Land Mammal Age Faunas of North America. An asterisk (*) indicates a common occurrence in the European record and a plus (+) indicates a common occurrence in the Asian record. An "o" indicates that zone of occurrence is unknown. Data abstracted from the following sources (see references in these papers for additional sources): **Late Duchesnean** - South Dakota: Bjork, 1967. Saskatchewan: Storer, 1987, 1988, 1990. Texas: Wilson, 1977, 1986. **Chadronian** - All areas: Prothero, 1985; Emry et al., 1987. Colorado: Galbreath, 1953; Stucky, unpublished. Montana: Tabrum and Fields, 1980. Nebraska: Hough and Alf, 1956; Ostrander, 1985. North Dakota: Kihm and Lammers, 1986. South Dakota: Clark, 1937; Scott and Jepsen, 1936-1941; Clark et al., 1967. Texas: Wilson, 1986. Wyoming: Emry, this volume; Kihm, 1987. **Orellan**- All areas: Savage and Russell, 1983; Prothero, 1985; Korth 1989; see references under Chadronian. Wyoming: Setoguchi, 1978. South Dakota: Lillegraven, 1970. **Whitneyan**- See above references; principally from Savage and Russell, 1983. **Arikareean** - All areas: Savage and Russell, 1983; Tedford et al., 1987. South Dakota, Nebraska: MacDonald, 1963, 1975; Skinner et al., 1965. Oregon: Fisher and Rensberger, 1972; Rensberger, 1973, 1983.

	LD	CH1	CH2	CH3	OR	WH	EA
MULTITUBERCULATA							
Ectypodus	X	X	X				
MARSUPIALIA							
*Peratherium**/*Herpetotherium*	X	X	X	X	X	X	X
*Peradectes**	X	X	X	X	X	X	-
Didelphidectes			X				
LEPTICTIDA							
Leptictis	X	X	X	X	X	X	
Palaeictops	X						
CIMOLESTA							
Sinclairella	?	X	X	-	X		
Didelphodus	X						
Apatemys	X						
Chadronia			X	-	X		
SORICOMORPHA							
Micropternodus	-	-	X	-	X	-	X
Domnina	X	-	X	-	X	X	X
Centetodon	X	X	X	X	X	X	X
Oligoryctes	X	-	X	X	X		
Apternodus	X	X	X	X			
Nyctitherium	X						
Proscalops		X	-	-	X	X	X
Oligoscalops			X	-	X	X	
Pseudotrimylus					X	-	X
*Trimylus**					X	X	
Quadrodens							X
Domninoides							X
ERINACEOMORPHA							
Ankylodon	?	-	X	X	X		
Proterix					X	X	
Ocajila							X
DERMOPTERA							
Thylacaelurus	X	-	X				
Tarkadectes			?				
Ekgmowechashala							X
PRIMATES							
Macrotarsius	X	-	X				
Chumashius	-	-	X				
Trogolemur	X						
Phenacolemur	X						
Omomys	X						
Rooneyia	X						
PALAEANODONTA							
Epoicotherium		X	X	X			
Xenocranium			X	-	X		
CHIROPTERA							
Wallia	X						
Chadronycteris			X				
*Stehlinia**			X				
PHOLIDOTA							
Patriomanis		X	X				

	LD	CH1	CH2	CH3	OR	WH	EA
RODENTIA							
Heliscomys	X	-	X	-	X	X	X
Eutypomys	X	X	X	-	X	X	X
Prosciurus	?	X	X	X	X	X	X
Omegodus (=*Adjidaumo*)	X	X	X	-	X	X	
Eumys	X	-	X	-	X	X	
Ischyromys	X	-	X	-	X	X	
Microparamys	X	-	-	-	X		
Namatomys	?	X	X	-	?		
Aulolithomys	X	-	X	-	X		
Cylindrodon	X	X	-	-	X		
Manitsha	X	X	-	-	X		
Yoderimys	X	X	X	-	X		
Paramys	-	-	-	X			
Jaywilsonomys	X	-	X				
Pauromys	-	-	X				
Nonomys	X	X	X				
Leptotomus	X	-	X				
Pseudocylindrodon+	X	X	X				
Ardynomys+	X	X	X				
Simimys	X	-	X				
Protadjidaumo	X	X					
Leptotomus	X	X					
Anonymus	X						
Pseudotomus	X						
Janimus	X						
Presbymys	X						
Viejadjidaumo	X						
Mytonomys	X						
Titanotheriomys		X	X	X	X	X	
Paradjidaumo		X	X	-	X	X	
Cedromus		*	*	*	X		
Eoeumys		X	-	-	X		
Oligospermophilus			X	-	X	X	
Agnotocastor			X	X	X	X	X
Meliakrouniomys			X	X			
Pipestoneomys			X				
Montanamus			X				
Centimanomys			X				
Cupressimus			X				
Scottimus				X	X	X	X
Protosciurus					X	X	X
Proheteromys					X	X	X
Tenudomys					?	-	X
Downsimus					X	-	X
Pelycomys					X	X	
Wilsoneumys					X	X	
Metadjidaumo					X	X	
Diplolophus					X	X	
Coloradoeumys					X		
Ecclesimus					X		
Orelladjidaumo					X		
Miospermophilus						X	X
Palaeocastor						X	X
Protospermophilus						X	X
Leidymys						X	X
Paciculus						X	X
Geringia						X	X
Kirkomys						X	X
Alwoodia							X
Parallomys							X
Haplomys							X
Mylogaulodon							X
Capatanka							X
Capacikala							X
Promylagaulus							X
Nototamias							X
Allomys							X

	LD	CH1	CH2	CH3	OR	WH	EA
Meniscomys							X
Crucimys							X
Niglarodon							X
Entoptychus							X
Pleurolicus							X
Grangerimus							X
Sanctimus							X
Florentiamys							X
Hitonkala							X
Zetamys							X
*Plesiosminthus**+							X
Sespemys							X
LAGOMORPHA							
Procaprolagus	-	-	-	-	X		
Palaeolagus		X	X	X	X	X	X
Megalagus			X	X	X	X	X
Chadrolagus			X	X	X		
Litolagus					X		
Desmatolagus					X	-	X
CREODONTA							
*Hyaenodon**+	X	X	X	X	X	X	X
Hemipsalodon	X	X					
Ischnognathus	X						
CARNIVORA							
Hesperocyon	?	X	X	X	X	X	X
Daphoenus	X	?	X	X	X	X	
Parictis	?	X	X	X	X	?	
*Miacis**+	X	X	X				
Prodaphaenus	X						
Daphoenocyon		X	X	X			
Daphoenictis		X	X				
Dinictis			X	X	X	X	X
*Palaeogale**+			X	X	X	X	X
Hoplophoneus+?			X	X	X	X	
Mesocyon					X	X	X
*Eusmilus**+?					X	X	X
Brachyrhynchocyon					X	X	X
Drassonax					X		
Nothocyon						X	X
Oxetocyon						X	X
*Nimravus**+						X	X
*Temnocyon**							X
Pogonodon							X
Dinaelurus							X
Paradaphaenus							X
Ekgmoiteptecela							X
PERISSODACTYLA							
Hyracodon	X	X	X	X	X	X	X
Colodon+	X	X	X	-	X	X	
Metamynodon+	X	-	-	-	X	X	
Mesohippus	X	X	X	X	X	X	
Telataceras	X						
Trigonias	?	X	X	X	X		
*Menops**?	X	o	o	o			
Toxotherium	X	X	X				
Duchesneodus	X						
Haplohippus	X						
Miohippus		o	o	o	X	X	X
Teleodus		o	o	o			
Brontops		o	o	o			
Protitanops		o	o	o			
Megacerops		o	o	o			
Amphicaenopus		o	o	o	X	X	
Penetrigonias		o	o	o	X	X	
Amphicaenopus		o	o	o	X	X	
Subhyracodon			X	-	X	X	
Triplopides			X				
*Protapirus**					X	X	
Diceratherium						X	X

	LD	CH1	CH2	CH3	OR	WH	EA
ARTIODACTYLA							
Archaeotherium	X	X	X	-	X	X	X
Agriochoerus	X	X	-	-	X	X	X
Hypertragulus	X	-	-	-	X	X	X
Heptacodon	X	X	X	-	X	X	
Pseudoprotoceras	X	X	X	X			
Leptotragulus	?	X	X	X			
Eotylopus	X	X	X	X			
Aclistomycter	X	X	X				
Hendryomeryx	X	-	X				
Brachyhyops	X	X					
Protoreodon	X						
Heteromeryx	X						
Diplobunops	X						
Oromeryx	X						
Poabromylus	X						
Leptomeryx+?		X	X	X	X	X	X
Poebrotherium		X	X	X	X	X	
Hypisodus		X	X	X	X	X	
Stibarus		X	X	-	X	X	
Merycoidodon		X	X	X	X	X	
Bathygenys		X	X	X	X	X	
Parvitragulus		X	X				
Montanatylopus		X					
Miniochoerus			X	X	X	X	
Perchoerus			X	-	X	X	
Oreonetes			X	-	X	X	
Aepinacodon			X	-	X		
Limnenetes			X				
Bothriodon+*			X				
Leptochoerus					X	X	X
Leptauchenia					X	X	X
Pseudolabis					X	X	
Paratylopus					X	X	
Entelodon+*					?		
Megachoerus					X		
Thinohyus					X		
Elomeryx+*						X	X
Protoceras						X	X
Nanotragulus						X	X
Chaenohyus						X	X
Pithecistes						X	X
Daeodon						?	X
Promesoreodon						X	
Mesoreodon							X
Megoreodon							X
Paramerycoidodon							X
Arretotherium							X
Hypsiops							X
Merycoides							X
Cyclopidius							X
Sespia							X
Pronodens							X
Oxydactylus							X
Dyseotylopus							X
Kukusepasutanka							X
MAMMALIA, incertae sedis							
Ideogenomys			X				

25. BRITISH MAMMALIAN PALEOCOMMUNITIES ACROSS THE EOCENE-OLIGOCENE TRANSITION AND THEIR ENVIRONMENTAL IMPLICATIONS

by J.J. Hooker

ABSTRACT

Faunal turnovers, changes in species numbers and in ecological diversity in the southern English mammal faunas from the late middle Eocene to early Oligocene are used to interpret environmental changes at this important time. Integration with similar ecological information from the Franco-Swiss Province demonstrates latitudinal differences in mammalian habitats early on, with forests in the north, more open but mosaic habitats in the south. Both areas became dominantly more open and uniform from the latest Eocene onwards.

Extrapolation from continental mammalian zonations to the standard marine scale via the few available points of facies interdigitation in Europe allows more precise calibration than hitherto of one of the mammalian dispersal events and of community changes. The latter especially can be linked with cooling climate as indicated by the marine oxygen isotope record. The 'Grande Coupure,' the major late Palaeogene mammalian faunal turnover event in Europe, took place in the early Oligocene.

INTRODUCTION

This paper investigates the changes in diversity and ecological composition of mammal faunas in southern England from the late middle Eocene to the early Oligocene. It puts forward available evidence for the dating of these changes in terms of the global time scale. It then attempts to relate them to southern European faunas of the same age and to major contemporaneous events like changes in climate and in land and sea distribution.

Because of a long sequence of superposed mammaliferous deposits, southern England is a key area for tracing the changes in diversity and ecological composition, which mirror and interrelate with those in the rest of Europe. These deposits are notable in containing other biota with a high content of ecological information, i.e. plants and invertebrate animals such as molluscs and ostracods. Figures 25.1 and 25.2 show the relationships of the strata of the late middle Eocene and late Eocene-early Oligocene respectively in the Hampshire Basin, with positions of the main mammal occurrences.

DIVERSITY CHANGES

Changes in diversity including origination and extinction events over this timespan have already been addressed (Hooker, 1989). The summary herein is based on that study and includes a minor update of the data (Figure 25.4A). The 1989 study included for consideration bats, but excluded didelphid marsupials as all except those from Creechbarrow (Hooker, 1986) are essentially undescribed. Didelphids are included here; bats, on the other hand, are excluded to facilitate comparison across Europe to include karst sites whose species numbers would be disproportionately increased solely by inclusion of the bats.

Species numbers

Figure 25.4B shows species numbers at key localities/levels from the late middle Eocene to early Oligocene in southern England. There is an overall reduction from beginning to end of the timespan, a pattern similar to that obtained by Legendre (1989) for the sequence of Quercy fissure fillings in southern France.

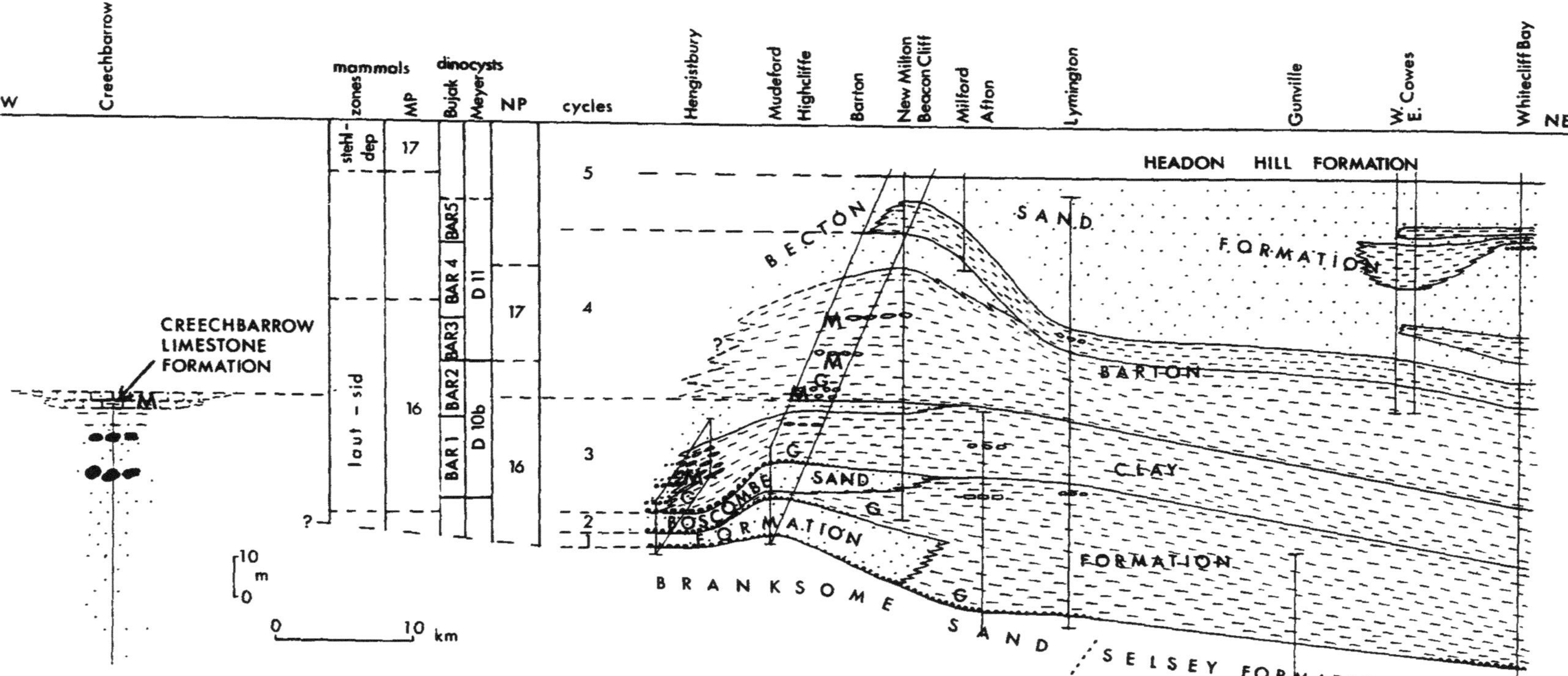

FIGURE 25.1. Section through late middle Eocene (Bartonian) strata in the Hampshire Basin, modified from Hooker (1986, Figure 68), q.v. for details of lithology and literature sources. Lithostratigraphic modification largely follows Edwards and Freshney (1987). Mammal zones are from Hooker (1986, 1987); MP levels from Brunet et al. (1987); dinocyst zones from Bujak et al. (1980) and extrapolated from Meyer (1988); nannoplankton (NP) zones from Aubry (1983). Logs of each locality are aligned as if lying on a straight SW-NE line, which is the direction of basinward facies change. Principal mammal-bearing units are indicated with an M.

However, there is an intermediate peak in southern England in the Bembridge Limestone organic and marl facies of Headon Hill, whereas the main trough is in the Osborne/ Seagrove Bay Members. The significance of these is discussed below.

Faunal turnover

In southern England there are two major phases of faunal turnover. One is between the *lautricense-siderolithicum* Zone (equivalent to Reference Level MP16) and the *stehlini-depereti* Zone (c.= lower part of MP17) (approximately the middle/late Eocene boundary--see below). The other is at the 'Grande Coupure' between the *frohnstettense-suevicum* Zone (c.= MP20) and the *atavus* Zone (=MP21-23) (not far from the Eocene/ Oligocene boundary). For definition of the zones see Hooker (1986, 1987) and Bosma (1974); for definition of the MP levels see Hooker, Legendre, Brunet and Vianey-Liaud in Brunet et al. (1987).

Both turnovers are significant at species and family level but differ in the area of origination of the incoming families. At the middle-late Eocene turnover they comprise the Theridomyidae, Gliridae, Viverravidae, Hyaenodontidae and Talpidae. Except for the last, these are all known from elsewhere in Europe in the late middle Eocene, where close relatives of most of the species or genera occur. A notable exception is *Hyaenodon*, which appears suddenly in Europe at the beginning of the late Eocene (Lange-Badré, 1979). The scale of the turnover in the Franco-Swiss Province (FSP) was similar at genus level (Hartenberger, 1986) and a number of originations seem to derive from south England (Hooker, 1986, p. 450). At family level the scale of extinctions was similar, although there were no originations. At the 'Grande Coupure' all the incoming families (Erinaceidae, Plesiosoricidae, Castoridae, Cricetidae, Eomyidae, Entelodontidae, Anthracotheriidae and Rhinocerotidae) are unknown earlier in Eocene mainland Europe (with the unique and enigmatic exception of the anthracothere *Elomeryx crispus* from the late Eocene of La Débruge). The pattern and scale of the turnover was similar throughout the main European landmass and the incoming families are thought to have emanated from southwest Asia via the Balkan archipelago and mid Alpine rise (Heissig, 1979). The exact timing and pattern of the middle-late Eocene turnover in southern England are uncertain as there is a gap in the mammal record in the vicinity of the boundary (timespan of deposition of the Becton Sand Formation and upper tongue of Barton Clay Formation; see Hooker, 1986) although the scale is not in doubt. There is a better record, however, across the 'Grande Coupure'. An increase in the rate of extinction had begun one zone earlier by the beginning of Bembridge Marls time and the rate was maintained at the time of immigration of the post-'Grande Coupure' fauna.

A smaller turnover occurred at the base of the Bembridge Limestone Formation, where one new family, the Cainotheriidae (originating in the FSP), made a temporary appearance. In addition, a number of species, nearly all (*Plagiolophus minor*, *Palaeotherium medium*, *Xiphodon gracilis*, *Dichobune leporina* and *Pterodon dasyuroides*) originating from the same area except one (*Ectropomys exiguus*) which was apparently from Germany East of the Rhine (Hooker, in press b). Details of this dispersal event along with extinctions that coincided with or followed it are related to the palaeoecological changes discussed in the next section.

FIGURE 25.2. (opposite page) Section through late Eocene (Priabonian) to early Oligocene (Rupelian) strata in the Hampshire Basin. Lithostratigraphic and thickness data mainly from Insole and Daley (1985) with additional personal observations. Lithological symbols: brickwork = limestone; solid black = dominantly organic; stipple = sand; unshaded = dominantly mud. Biostratigraphic sources largely as for Figure 25.1. Alignment of locality logs as for Figure 25.1 (see inset of Isle of Wight map for scheme followed). Wavy lines indicate major unconformities and the levels of modern erosion in the Isle of Wight. T-shaped bar indicates extent of strata at Hordle. Mammal biozonation after Hooker (1987) and Bosma (1974). Principal mammal-bearing units are indicated with an M.

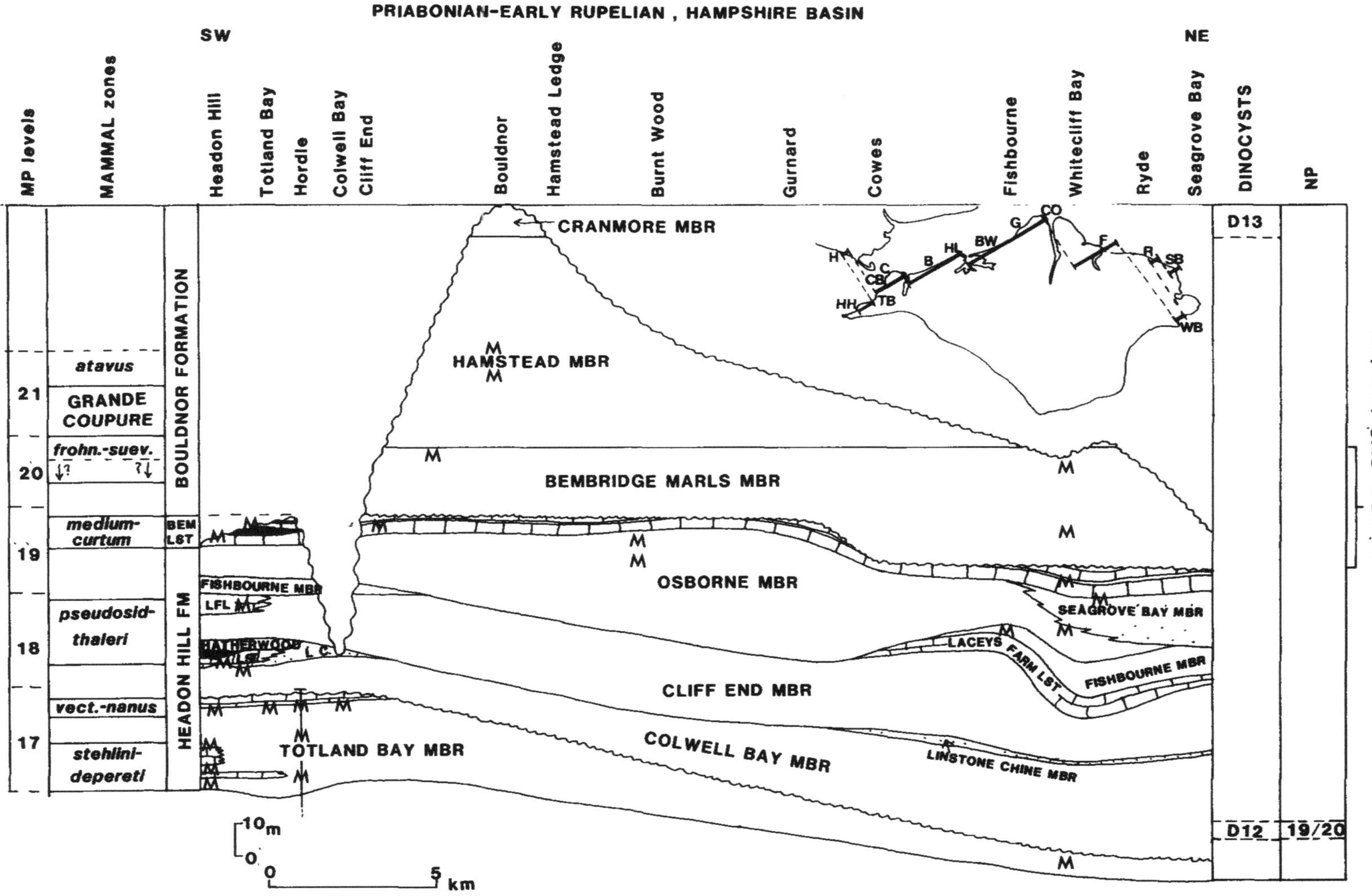
PRIABONIAN-EARLY RUPELIAN, HAMPSHIRE BASIN
SW
NE
MP levels
MAMMAL zones
Headon Hill
Totland Bay
Hordle
Colwell Bay
Cliff End
Bouldnor
Hamstead Ledge
Burnt Wood
Gurnard
Cowes
Fishbourne
Whitecliff Bay
Ryde
Seagrove Bay
DINOCYSTS
NP
21
20
19
18
17
atavus
GRANDE COUPURE
frohn.-suev.
medium-curtum
pseudosid-thaleri
vect.-nanus
stehlini-depereti
BOULDNOR FORMATION
BEM LST
HEADON HILL FM
CRANMORE MBR
HAMSTEAD MBR
BEMBRIDGE MARLS MBR
OSBORNE MBR
FISHBOURNE MBR
LFL
SEAGROVE BAY MBR
LACEYS FARM LST
CLIFF END MBR
COLWELL BAY MBR
LINSTONE CHINE MBR
TOTLAND BAY MBR
D13
D12
19/20
10 m
0
0
5 km
approx. EOC.-OLIG. boundary

ECOLOGICAL CHANGES

Changes in composition of rodent and perissodactyl faunas

Some of the most striking faunal changes are among rodents and perissodactyls. They are important markers of the ecological changes discussed below. They have already been covered in detail (Collinson and Hooker, 1987) and will only be summarised here. Major changes in rodent and perissodactyl faunas accompany the faunal turnovers at the middle/ late Eocene boundary and at the 'Grande Coupure.' At the middle/late Eocene boundary, the ailuravine 'paramyid' rodents and the rhinocerotoid and lophiodontid perissodactyls became extinct whilst the manitshine 'paramyids' and pseudosciurid rodents as well as the low crowned hippomorph perissodactyls suffered a major decline. Although the pseudosciurid decline was followed by a radiation (Hooker, in press a), their individual numbers compared to other rodent types remained small. Some developed cresting on the teeth in the course of the late Eocene and shifted their diet from dominantly frugivorous to dominantly herbivorous. The new late Eocene rodent faunas were dominated in both species and specimen numbers by the related but higher crowned and more herbivorous theridomyids and to a lesser extent by glirids. The new perissodactyl faunas were dominated by new radiations of the palaeotheriid genera *Palaeotherium* and *Plagiolophus* which, with their higher crowned selenolophodont teeth, were adapted for eating coarser types of browse than their predecessors the lophiodonts and primitive rhinocerotoids (*Hyrachyus*) with their low crowned bilophodont tapir-like teeth.

At the 'Grande Coupure,' the taxonomic changes were even more marked. All the hippomorphs had become extinct, although three species survived in the FSP. They were replaced by a rhinocerotid *Ronzotherium*, whose dental specialisations appear to be broadly similar to those of some of the earlier palaeotheres. The incoming rodents mainly added to, rather than replaced, those of the late Eocene, although the pseudosciurids entirely disappeared. The newcomers belonged to the families Castoridae, Eomyidae and Cricetidae.

Community changes

An earlier study (Collinson and Hooker, 1987) traced the changes in both vegetation and mammal communities from the late Palaeocene to early Oligocene. Analysis of the mammalian changes was accomplished using the ecological diversity method of Andrews et al. (1979). The conclusions were that mammal habitats from early Eocene to early Oligocene had changed from complex forests to a more open environment with forest patches. An apparent reversal of the trend within the late Eocene raised the question of a local versus more widespread effect. A more detailed study (Hooker, in press b) of five levels within the Bembridge Limestone Formation, documents a critical time for the Hampshire Basin area in the late Eocene. The following sections examine the late middle Eocene to early Oligocene part of the sequence in southern England, in more detail than previously, to try and elucidate the pattern of change. Comparison is made with the southern Franco-Swiss Province (FSP) (see Schmidt-Kittler and Vianey-Liaud, 1975 for definition).

Caveats and community differences in the early Tertiary

Details of the ecological diversity method as applied to these early Tertiary assemblages are given by Collinson and Hooker (1987). In summary, Andrews et al. (1979) differentiated five main modern habitat types in the African tropics: two closed (lowland forest and montane forest) and three open (floodplain, woodland/ bushland and short grass plains) according to the patterns exhibited by ecological diversity histograms. Three categories of ecological diversity are used here: size, locomotor adaptation and diet. Abbreviations for the different classes included in each category are explained in Figure 25.3. Various problems in interpreting fossil assemblages have been discussed by Andrews et al. (1979), Collinson and Hooker (1987) and Hooker (in press b). Some of these relate to collecting and tapho-nomic biases. Others question what degree of habitat discrimination can be achieved; for instance the modern East African localities Budongo and Semliki have near identical size and locomotor plots (although slightly different dietary plots) but represent respectively lowland semideciduous forest and montane evergreen forest; similarly, the floodplain habitat may vary greatly with the density of trees and can overlap extensively with the woodland/bushland habitat (Andrews et al., 1979). Note that the different ecological diversity categories have slightly different information content. Size and locomotor adaptation are giving evidence of how close together or otherwise the trees grew; the arboreal percentage may also go some way to explaining how complex the forest structure is, since today large numbers of truly arboreal mammals are only found in high canopy lowland tropical forest (Fleming, 1973). Diet is instead giving information about food availability, which is partly dependent on the vegetation structure via climate, partly on the resource partitioning of the mammals and other animals.

Following on from this point, there is yet another problem in interpreting fossil assemblages, which appears to relate specifically to the early Tertiary and is likely to result from different community structures then and now. Different ecological diversity categories, whilst each showing the same trend through a sequence, give slightly different individual signals. In particular, the dietary category tends to indicate a more closed environment than do the other two. The dietary spectra differ from most modern ones in two principal ways: they lack grazers and they show a high percentage of frugivores. These two characteristics appear to be achieved today only in tropical rainforests. In cooler climates and/or more open environments, the near ubiquity of areas of grassland (however small) allows grazers to enter the communities, which also support a smaller number of mammalian frugivores.

It is clear that there has been a change in the dominant locomotor adaptations of mammalian frugivores between the early Tertiary and the present day. In the Eocene they were mainly ground dwellers (either fully terrestrial, termed large ground mammals (LGM) or capable of climbing into bushes, termed small ground mammal (SGM) by Andrews et al., 1979). There were relatively few arboreal and scansorial and almost no aerial types. Today there are very few LGM frugivores although SGM's, especially rodents, exploit herbaceous seeds in open environments; arboreal and scansorial types are much more abundant and in the tropics there are also aerial types in the form of megachiropteran fruit bats (Fleming, 1973). There are no authenticated records of megachiropterans until the early Miocene (Sigé and Aguilar, 1987), although their level of evolution suggests an unsampled earlier record. The shift in locomotor adaptation probably represents the evolution through competition of a feeding strategy towards obtaining fruit at source especially in the trees, rather than only near ground level or after having fallen to the ground. Scansoriality and arboreality impose limitations of size and ability to live in a non-forested environment; and the specialised dietary requirements of megachiropterans restrict them to the tropics today. This evolution of feeding strategy does not explain the greater mid-latitude abundance of frugivores in the early Tertiary unless there was also a greater abundance and less seasonally restricted availability of fruit, approaching the situation in the modern tropics. By latest Eocene time, there was already marked wet-dry seasonality in southern England (Daley, 1989) and temperatures had fallen considerably from their late early Eocene maximum (Shackleton, 1986). However, it was not until the early Oligocene that a major temperature fall occurred, introducing cold ocean bottom waters linked to Antarctic glaciation (Shackleton, 1986), and only at the end of the Tertiary that major glaciation intensified seasonality to its present level.

Bearing in mind these points, how can habitat be interpreted from the dietary histograms? If one adds the plot of grazers to that of browsers in modern faunas, the histograms bear a closer resemblance to the majority of Eocene ones. This has the effect of shifting the emphasis of the Eocene dietary histograms from a more closed to a more open habitat, thus corresponding more closely to those of size and locomotion. However, the problem of frugivore percentages is less easily overcome. Apart from recognising trends within the early Tertiary sequence, it seems best to rely on the non-frugivore plots (mainly the insectivore-herbivore proportions) in a given dietary histogram for habitat interpretation.

ECOLOGICAL DIVERSITY ANALYSIS
(Figure 25.3)

The next section documents and interprets the changes in ecological diversity in southern England and the southern part of the Franco-Swiss Province (FSP). Only two levels/sites (Creechbarrow and the Bembridge Limestone organic and marl facies of Headon Hill) have so far been considered for collecting and taphonomic bias (Hooker, 1986; in press b) and the potential effect of these biases on the following analyses is discussed under each site account. Some sites/levels in southern England are the same as or similar to those treated earlier (Collinson and Hooker, 1987), but as well as additions and corrections to the faunal lists (Table 25.1), attempts have been made to narrow the stratigraphic and/or geographic range of each assemblage to improve resolution.

Ecological diversity analysis, southern England (Figure 25.3)

Creechbarrow Limestone Formation, Creechbarrow

The size plot shows nearly twice as many AB as C class, very few D and no larger animals. Screenwashing may have undersampled the large end of the spectrum but the sample was large and it is unlikely that the AB:C proportions would reverse. This indicates forest. The locomotor plot is somewhat contradictory. Only one of the forest histograms of Andrews et al. (1979) show more LGM's than SGM's. This is Sokoke-Gedi described as 'lowland deciduous forest and woodland' thus at the more open end of the spectrum. The large arboreal percentage at Creechbarrow excludes the floodplain and woodland/bushland habitats and implies a complex forest structure. It seems that only in lowland tropical forest today does one find such a high arboreal percentage. By adding grazers to browsers and for the moment ignoring the frugivores in the modern plots as advocated above, the slightly greater percentage of HB over I in the Creechbarrow dietary histogram fits better with forest than more open habitats. Thus Creechbarrow can best be regarded as representing a tropical-type high stature forest with glades.

Mammal bed, lower Totland Bay Member, Hordle, Hampshire

This is a well sampled fauna from a thin horizon collected by extensive prospecting and screenwashing. There are more large mammals than at Creechbarrow, probably partly a feature of better balanced collecting. However, the smaller C size percentage compares better with modern histograms for either montane or less tropical forest types than Creechbarrow. The locomotion histogram has equal LGM and SGM and significantly fewer arboreal types than Creechbarrow. It compares closely with that of modern Rwenzori, which is 'lowland semideciduous forest' of Budongo type (Andrews et al., 1979) and therefore rather open. The near equal I/HB plots indicate forest. Although it is not helpful to compare the F class with modern histograms, the plot is lower than at Creechbarrow suggesting less fruit productivity. Thus the Hordle Mammal Bed appears to represent a rather open forest or one with glades, but of a less tropical type than Creechbarrow. The arboreal percentage, however, implies a fairly complex structure.

How Ledge Limestone green clay, upper Totland Bay Member, west Isle of Wight; Lignite Bed, Hatherwood Limestone Member, Headon Hill; Lacey's Farm Limestone Member, Headon Hill outlier

These three assemblages are discussed together. The How Ledge green clay has been extensively sampled by screenwashing; the Hatherwood Lignite Bed has been both screenwashed and prospected (but is undersampled compared with the Hordle Mammal Bed); whilst the screenwash samples of thin clay and sand seams in the Lacey's Farm Limestone (Bosma and Insole, 1972) are supplemented by a small number of prospected finds in the main limestone facies. All the assemblages show a very high percentage of AB, suggesting some bias from screenwashing. The Hatherwood Lignite Bed is the least extreme and is close to the Hordle Mammal Bed. All have far fewer LGM's than SGM's but the arboreal percentage remains similar to that of the Hordle Mammal Bed. Again there may be some screenwashing influence here but it probably cannot entirely account for the LGM/SGM ratio. There are consistently more frugivores than in the Hordle Mammal Bed; the How Ledge green clay and the Lacey's Farm Limestone have more and fewer insectivores respectively. Andrews et al. (1979) found that particularly high percentages of SGM's typify montane forest. This is unlikely to have been the habitat of the faunas discussed here in view of the coastal floodplain/lagoonal environment judged from various lines of evidence for the Hampshire Basin at this time (e.g. Keen, 1977). Moreover, a similarly high percentage of SGM's occur in the species poor faunas of lowland temperate rainforests of Tasmania (data from Grainger et al., 1987). The indications are of a slightly more closed forest than in the Hordle Mammal Bed, but of similar complexity and structure (i.e. less tropical than Creechbarrow). Collinson and Hooker (1987) suggested that the Hatherwood Limestone Member fauna as a whole could represent a local forest patch because of the restricted extent of the unit). The Lacey's Farm Limestone assemblage is also almost as local, but in contrast, the How Ledge green clay extends for at least 4km with no sign of change in lithofacies or faunal content. Moreover, it seems to be a lateral equivalent of the Hordle Rodent Bed, some 7 km distant, where the lithofacies and fauna are slightly different but the latter still dominated by small animals. More closed forest in the *vectisensis-nanus* Zone may thus be more than just a phenomenon of local or mosaic vegetation. Study of faunas lateral to those already known would provide a test of this idea.

Osborne and Seagrove Bay Members

These members interfinger laterally. The Osborne Member has been collected mainly by screenwashing, the Seagrove Bay Member entirely by prospecting and so they complement each other. The large mammals found by screenwashing are the same as those found by prospecting and the small mammal component within the Osborne Member varies little with locality. The fauna is, however, small and of low diversity. The relative importance of the larger size categories (D and EF each = nearly 50% of AB) plus the small percentage of C is not like any of Andrews et al.'s (1979) modern histograms, but the size histogram overall is closest to the more open habitats, e.g. Gabiro ('wooded grassland') or Rukwa ('floodplain grassland with bordering woodland and lake swamps'). An open habitat is clearly indicated by the locomotor histogram. The percentage of LGM's being rather less than twice that of SGM's, together with absence of the aquatic class, fits woodland-bushland rather than floodplain. The HB percentage is more than twice that of the I percentage which is indicative of one of the open habitats. The F percentage is less than that of any earlier English assemblage but it is uncertain how to interpret this. The casting vote on floodplain versus woodland-bushland should perhaps go to the lithofacies. The Osborne Member is a sequence of dominantly colour-mottled clays with mudcrack horizons and occasional bonebeds, suggestive of the overbank facies of a floodplain. The Seagrove Bay Member appears to be a higher energy fluvial deposit prograding over the Osborne Member. Perhaps this assemblage represents a floodplain with woodland patches. If so it could explain the small size of the fauna by the normal low numbers of small species on modern floodplains (Andrews et al., 1979) and by some collecting bias against the large ones.

Lower Bembridge Limestone Formation, west Isle of Wight

This is discussed in more detail by Hooker (in press b). It has been collected largely by prospecting, but included is one small screen-washed assemblage from a local clay seam. The relatively high percentage of large mammals indicates open habitat. The LGM percentage being less than twice that of the SGM together with absence of aquatics fits best with woodland-bushland. The very low I yet very high HB percentages suggests some collecting bias but in a less extreme state these would still indicate a relatively open environment. Together with the sedimentological evidence of seasonal drying out in the form of silcretes and calcretes (Daley, 1989), a woodland-bushland habitat seems most likely.

Bembridge Limestone Formation, organic and marl facies, Headon Hill

This has been collected entirely by screen-washing and involves several horizons of al-ternating organic and marly deposits here

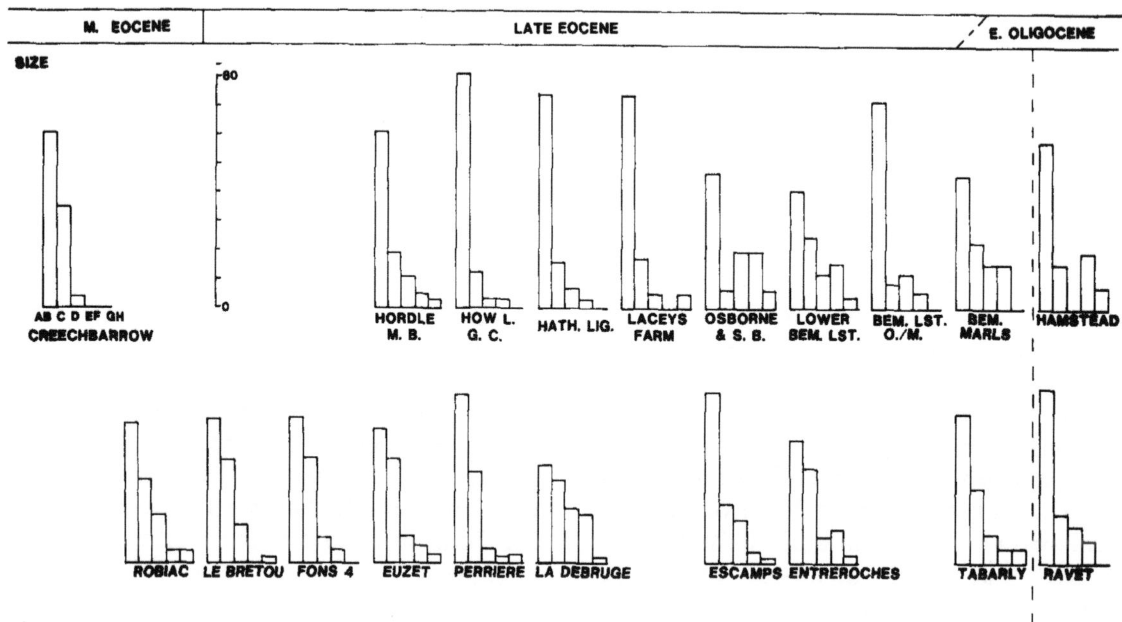
M. EOCENE
LATE EOCENE
E. OLIGOCENE
SIZE
80
0
AB C D EF GH
CREECHBARROW
HORDLE M. B.
HOW L. G. C.
HATH. LIG.
LACEYS FARM
OSBORNE & S. B.
LOWER BEM. LST.
BEM. LST. O./M.
BEM. MARLS
HAMSTEAD
ROBIAC
LE BRETOU
FONS 4
EUZET
PERRIERE
LA DEBRUGE
ESCAMPS
ENTREROCHES
TABARLY
RAVET

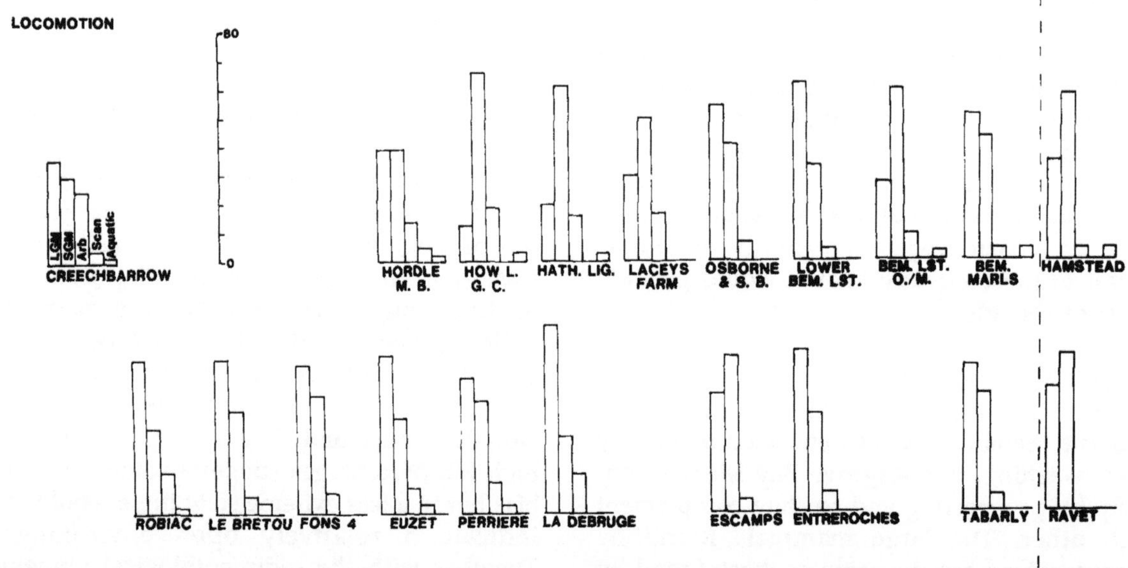
LOCOMOTION
80
0
LGM
SGM
Arb
Scan
Aquatic
CREECHBARROW
HORDLE M. B.
HOW L. G. C.
HATH. LIG.
LACEYS FARM
OSBORNE & S. B.
LOWER BEM. LST.
BEM. LST. O./M.
BEM. MARLS
HAMSTEAD
ROBIAC
LE BRETOU
FONS 4
EUZET
PERRIERE
LA DEBRUGE
ESCAMPS
ENTREROCHES
TABARLY
RAVET

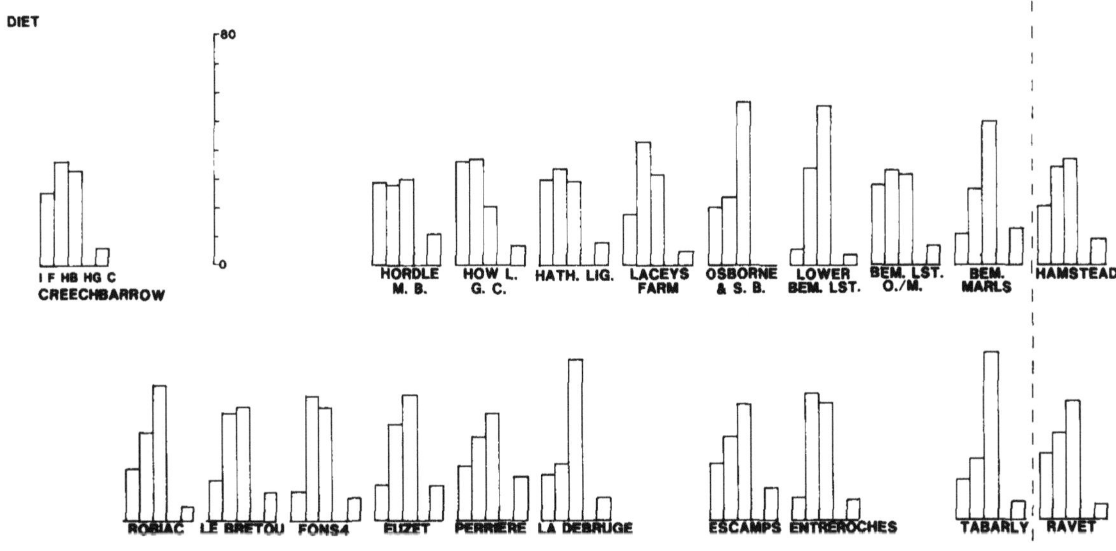

FIGURE 25.3. Ecological diversity histograms for late middle Eocene to early Oligocene mammal faunas of southern England (upper row) and the Franco-Swiss Province (lower row). A, Size; B, Locomotion; C, Diet. Scale indicates percentages of each category. Size, locomotory and diet classes follow Andrews et al. (1979) and Collinson and Hooker (1987): AB = less than 1kg; C = 1-10kg; D = 10-45kg; EF = 45-180kg; GH = more than 180kg. LGM = large ground mammal; SGM = small ground mammal; Arb = arboreal; Scan = scansorial; Aquatic (unabbreviated). I = insectivore; F = frugivore; HB = herbivore browser; HG = herbivore grazer; C = carnivore. Dashed line indicates position of 'Grande Coupure'. Abbreviations of localities/levels: M.B. = Mammal Bed in lower Totland Bay Member at Hordle; How L.G.C. = green clay beneath How Ledge Limestone, in upper Totland Bay Member at Headon Hill, Totland and Colwell Bays; Hath. Lig. = Lignite Bed in Hatherwood Limestone Member, Headon Hill; Laceys Farm = Laceys Farm Limestone Member at Laceys Farm Quarry and Headon Hill; Osborne and S.B. = Osborne and Seagrove Bay Members at Burnt Wood, Whitecliff Bay, Binstead and Seafield; Bem. Lst. = Bembridge Limestone Formation; O./M. = organic and marly beds about the middle of the latter formation on Headon Hill; Bem. Marls = Bembridge Marls Member; Hamstead = Hamstead Member at Bouldnor.

lumped together but analysed separately by Hooker (in press b). There is a return to the type of histogram of How Ledge green clay and Lacey's Farm Limestone, but with more D and EF and fewer C size classes, indicating a less complex forest. The locomotor and dietary histograms are very similar to those of the Hatherwood Lignite Bed but have a slightly lower arboreal percentage and a slightly higher HB percentage. As concluded by Hooker (in press b), the assemblages indicate alternations of a more or less densely forested area fringing a lake. It was probably a relatively low complexity forest because of the low arboreal percentage and was probably local in extent.

Bembridge Marls Member, Isle of Wight

This has been collected by both prospecting and screenwashing. The screenwashing has been mainly at one horizon and locality which is different in level from that of most of the prospected finds (see Hooker, 1987). There is a relatively high percentage of large animals indicating open habitat, but the histogram is not exactly like any in Andrews et al. (1979). The LGM plot is only slightly higher than the SGM. This together with a very low arboreal percentage suggests woodland-bushland, but an aquatic presence weights more in favour of floodplain. The much higher percentage of HB versus I indicates an open habitat. The apparent contradictions over the type of open environment may result from the necessary lumping of several small assemblages within the 20-30 m thickness of the Bembridge Marls, which according to sedimentology and flora were deposited in a lake/lagoon with extensive fringing reedmarsh (Collinson, 1983).

Hamstead Member, Bouldnor Cliff

This has been collected extensively by both prospecting and screenwashing, mainly from one or two thin units at one locality. There is a relatively high percentage of EF size class and no D. There is no modern histogram available exactly like this, but it suggests open habitat. The locomotor histogram is anomalous because there are significantly fewer LGM's than SGM's yet a very low arboreal percentage. Of the modern African open habitats in Andrews et al.(1979), only Jebel Mara has a higher SGM than LGM percentage and here the difference is small. Jebel Mara is 'woodland with rocky outcrops and many glades'. The I/HB proportions are like those of the Lacey's Farm Limestone, the plots being more similar to each other than in most of the modern African open habitat histograms. The assemblage seems to be at the more wooded end of the woodland-bushland spectrum.

Ecological diversity analysis, Franco-Swiss Province

Robiac (Calcaire de Fons inférieur); Le Bretou (fissure filling); Fons 4 (Calcaire de Fons supérieur); and Euzet (Marnes d'Euzet)

With the C size class being approximately two thirds of the percentage of AB, all these assemblages show variations on the theme of lowland tropical forest. All have a higher percentage of LGM's than SGM's ranging from about twice (Robiac) to scarcely higher (Fons 4); and their arboreal percentages are low except at Robiac. Thus on locomotor percentages, they seem to be intermediate between the open and closed modern habitat histograms. They are closest to extreme versions of each: e.g. Sokoke-Gedi which is 'lowland deciduous forest and woodland' and Tana which is 'floodplain grassland with patches of deciduous forest' (Andrews et al., 1979). All show a much higher percentage of HB than I, suggesting a rather open habitat. The conflict and intermediate character of some of the histograms suggests a rather mixed, possibly mosaic habitat consisting of rather open areas with patches of tropical type forest. Robiac emerges as the most wooded/forested on the basis of its arboreal percentage. It is worth noting that the histograms are similar despite potential collecting bias.

Perrière (fissure filling)

The C size class is about half the percentage of the AB class, which is more consistent with a less than tropical forest type, resembling some modern histograms for montane and deciduous forest. The locomotor histogram fits within the range of the earlier localities and is close to the modern histogram of Sokoke-Gedi. The less widely different I and HB plots suggest a more closed habitat than do earlier localities. It seems to represent a more forested mosaic version of the earlier localities.

La Débruge (lignites)

The relatively large percentage of larger animals gives it an open habitat aspect. The very large percentage of LGM more than twice that of the SGM is also indicative of an open habitat. The relatively large arboreal percentage makes it most like floodplain with forest patches, but this is slightly contradicted by the absence of aquatics, which are consistently present in Andrews et al. (1979) floodplain localities. The much higher percentage of HB than I is indicative of an open habitat. It thus probably represents floodplain or woodland/ bushland with forest patches. The virtual absence of screenwashing at this site (S. Legendre, pers. comm., 1991) may bias it slightly towards the open end of the habitat scale.

Escamps (fissure filling)

The size histogram is very similar to that of the Hordle Mammal Bed, thus indicating forest. The higher SGM than LGM percentage supports this, but the very low arboreal percentage indicates woodland-bushland. The relative percentages of I/HB indicate something somewhat intermediate between closed and open habitat. It is not clear why there is a conflict between the size and locomotor histograms.

Entreroches, Tabarly and Ravet (fissure fillings)

All the plots are similar to their age equivalents in southern England. The main differences are: fewer large mammals in the Ravet size histograms, although the proportions of the AB:C:GH are the same; no aquatics in any of the FSP locomotor histograms; and fewer HB and more F in the Entreroches dietary histogram. These minor differences have little effect on the habitat interpretations. There may, however, be some collecting bias as Entreroches was not screenwashed and as Tabarly is as yet poorly sampled (pers. comm. S. Legendre).

Habitat trends in western Europe.

In summary (see Figure 25.4A), the southern English habitats were forested from the late middle Eocene to at least the *vectisensis-nanus* Zone of the late Eocene. In the succeeding *pseudosiderolithicus-thaleri* Zone, these habitats may have been only local in extent. There was a reduction in structural complexity inferring change from tropical to less than tropical type (or more equable to more seasonal) forests after the late middle Eocene and a slight fluctuation towards more closed forest in the *vectisensis-nanus* Zone. In the southern FSP there existed a more open habitat varying in its degree of colonization by forested patches which may have maintained their tropical type complexity longer than the north, although the evidence is unclear. Just before the *medium-curtum* Zone, more open floodplain conditions affected southern England. The *medium-curtum* Zone was typified by the establishment of a drier more open habitat in both areas, at which time a number of species dispersed from the south to the north (see faunal turnover section). They include more cursorial forms like *Xiphodon* and *Palaeotherium medium* which had probably already adapted to the more open mosaic environments of the FSP. A brief, wetter, forested interlude in southern England was accompanied by the return of most of the species that were displaced (?further north) by the preceding dispersal event. The *frohnstettense-suevicum* Zone saw a return to the more open conditions in both areas which was accompanied in southern England by final disappearance of the 'Lazarus' species. Immediately after the 'Grande Coupure,' both areas had slightly less open habitats, i.e. more wooded, than just before.

Legendre (1989) using his cenogramme method (curve of distribution of body weight across the fauna) interpreted Le Bretou and other fissure fillings of *lautricense-siderolithicum* and *stehlini-depereti* Zones age as 'forêt tropicale humide.' In contrast, he interpreted Perrière and Escamps as 'savanes arborées' and Ravet (and others) as 'relativement ouverts, ou correspondent à des forêts armées.' His conclusions differ little from those which would be derived solely from the above size histograms; but those of locomotion and diet give very different interpretations except in the early Oligocene. Moreover, Legendre (1989) concluded that southern England was more open and more arid from Hatherwood Limestone to Bembridge Marls times than age equivalents in southern France. This is largely the opposite to what is concluded here. The study of ecological diversity from three of the categories of Andrews et al. (1979) as conducted here coincides more closely with the palaeobotanical data (Collinson and Hooker, 1987). Also, by covering a greater diversity of parameters it should provide a better balanced environmental interpretation than that relying solely on body weight distribution and should tend to 'buffer' against potential collecting and taphonomic biases.

DATING THE BIOTIC EVENTS

(Figures 25.1-2)

The position of the middle/upper Eocene boundary in the British sequence

Hooker (1986) has argued on 'nearest fit' grounds, backed up by biostratigraphy, that the transgression of the Marnes à Pholadomyes of the Paris Basin (base Ludian) correlates with that at the base of Barton Clay bed J (the upper tongue of the Barton Clay Formation). This correlation is supported by Riveline and Cavelier (1987) on the basis of charophytes. Aubry (1985) has tentatively recognised Zone NP18 in the Marnes à Pholadomyes, thus equating base Ludian with base Priabonian (in the sense of nannoplankton zonation) and therefore with the middle/upper Eocene boundary. Correlation with the stratified

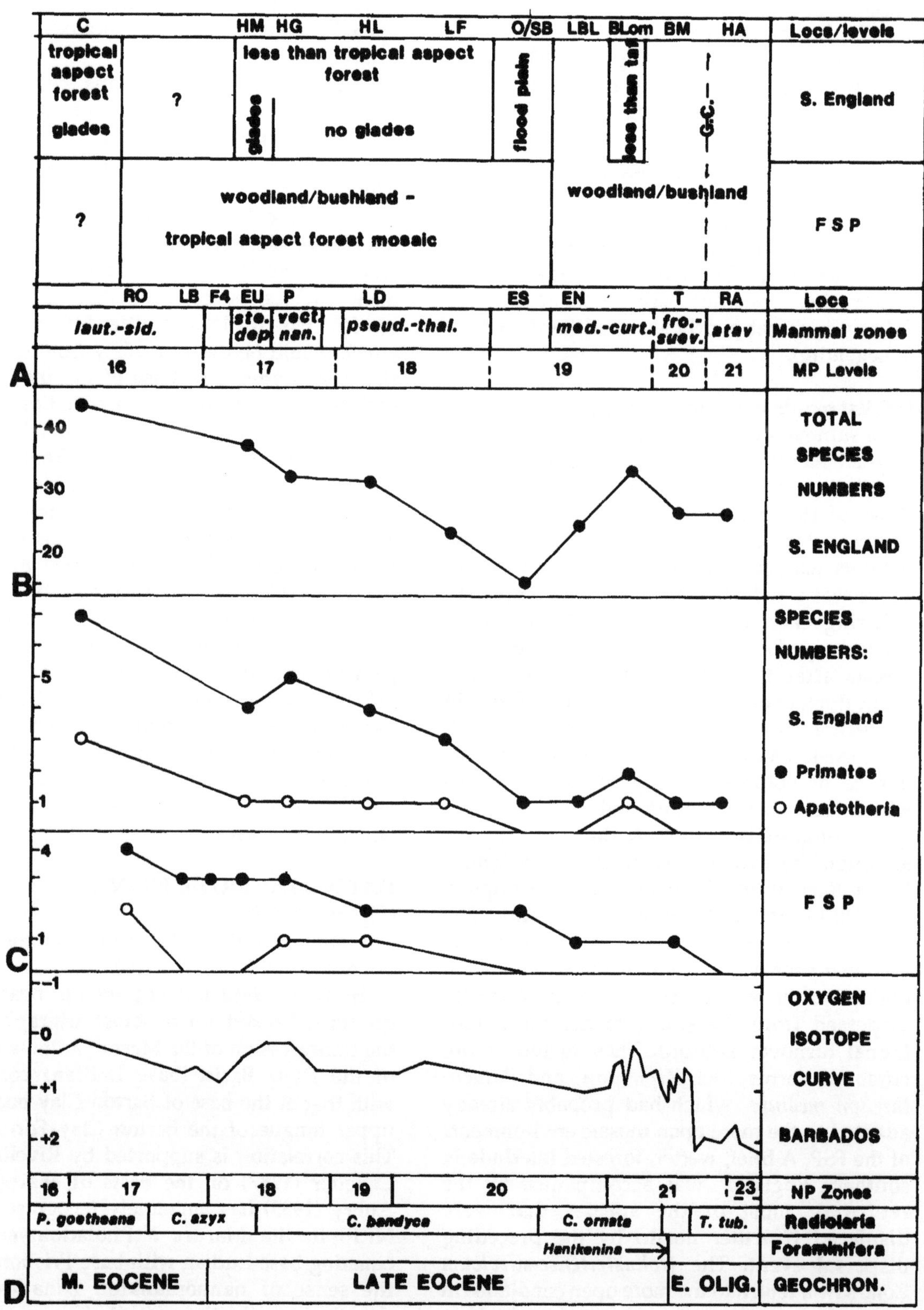
C
HM HG
HL
LF
O/SB
LBL
BLom
BM
HA
Locs/levels
tropical aspect forest
glades
?
less than tropical aspect forest
glades
no glades
flood plain
less than tab
G.C.
S. England
?
woodland/bushland - tropical aspect forest mosaic
woodland/bushland
F S P
RO
LB
F4
EU
P
LD
ES
EN
T
RA
Locs
laut.-sid.
ste. dep.
vect. nan.
pseud.-thal.
med.-curt.
fro.-suev.
atav
Mammal zones
16
17
18
19
20
21
MP Levels
A
TOTAL SPECIES NUMBERS
S. ENGLAND
B
SPECIES NUMBERS:
S. England
Primates
Apatotheria
F S P
C
OXYGEN ISOTOPE CURVE
BARBADOS
NP Zones
P. goetheana
C. azyx
C. bandyca
C. ornata
T. tub.
Radiolaria
Hantkenina
Foraminifera
M. EOCENE
LATE EOCENE
E. OLIG.
GEOCHRON.
D

southern French localities discussed here is also by means of charophytes (Riveline, 1986). One occurrence of Zone NP19/20 markers in the Brockenhurst Bed (base Colwell Bay Member) helps the standard calibration of this otherwise almost entirely nonmarine sequence (Aubry, 1985).

The position of the Eocene/Oligocene boundary in the British sequence

Following the recommendations of Pomerol and Premoli-Silva (1986) that the Eocene/Oligocene boundary coincide with the hantkeninid extinction (=top of Zone P17; within Zone NP21), this section attempts to relate the main mammal event, the 'Grande Coupure' to this boundary. The planktonic foraminifers which are the main basis for the boundary in offshore areas scarcely occur in the dominantly continental areas of Eocene Europe. It is therefore to the scattered nannoplankton occurrences that one must turn to correlate with the marine standard sequence. None occur close to the boundary in south England, so it is necessary to extrapolate from other European areas. The meagre northwestern European records of markers of the *Turborotalia cerroazulensis* Zone (P17) provide little help but do not contradict the nannoplankton evidence; they occur in the Gehlberg Formation which underlies the Silberberg Formation with Zone NP21 in the Helmstedt region of northern Germany (Spiegler et al., 1988). The meagre dinoflagellate evidence also tends to support the correlations (Chateauneuf, 1986).

FIGURE 25.4. (opposite page) Biotic and environmental events from the late middle Eocene to early Oligocene, calibrated to available biostratigraphy. A, Summary of the mammal habitats concluded from the EDS's in S. England and the Franco-Swiss Province (FSP). Abbreviations: G.C. = 'Grande Coupure'; taf = tropical aspect forest; C = Creechbarrow; HM = Hordle Mammal Bed; HG = How Ledge green clay; HL = Hatherwood Lignite Bed; LF = Laceys Farm Limestone Member; O/SB = Osborne/Seagrove Bay Members; LBL = lower Bembridge Limestone; BLom = Bembridge Limestone organic and marly beds; BM = Bembridge Marls; HA = Hamstead Member; RO = Robiac; LB = Le Bretou; F4 = Fons 4; EU = Euzet; P = Perrière; LD = La Débruge; ES = Escamps; EN = Entreroches; T = Tabarly; RA = Ravet; *laut.-sid.* = *Lophiodon lautricense - Lophiotherium siderolithicum; ste.-dep.* = *Palaeotherium magnum stehlini - 'Paradelomys' depereti; vect.-nan.* = *'Paradelomys' quercyi vectisensis - Heterohyus nanus; pseud.-thal.* = *Isoptychus pseudosiderolithicus - Palaeotherium muehlbergi thaleri; med.-curt.* = *Palaeotherium medium medium - P. curtum curtum; fro.-suev.* = *Palaeotherium curtum frohnstettense - P. medium suevicum; atav* = *Eucricetodon atavus.* B, Total numbers of species in the southern English faunas in A. C, Numbers of species of primates and apatotheres in the S. English and FSP faunas in A. D, Oxygen isotope curve from the benthic foraminifer *Oridorsalis umbonatus* in the Bath Cliff section, Barbados, calibrated to standard biostratigraphies (redrawn from Müller-Merz and Saunders, 1986, Figure 1); scale indicates parts per thousand of δ^{18}O PDB.

Rhine Graben and Mainz Basin, Germany

In the Rhine Graben, the earliest known post-'Grande Coupure' fauna (attributed to level MP21) at Pechelbronn in the Upper Pechelbronn Formation occurs above a distinctly pre-'Grande Coupure' fauna at Rot-Malsch in the Lower Pechelbronn Formation. The Rot-Malsch fauna consists of *Palaeotherium magnum magnum, P. medium, Plagiolophus fraasi* and *Anchilophus* sp. and Tobien (1987) attributed it to level MP20 (= *frohnstettense-suevicum* Zone). However, this fauna is not diagnostic of this level or zone (Legendre, *in* Brunet et al., 1987; Hooker, 1987), only MP19 or 20 (*medium-curtum* or *frohnstettense-suevicum* Zones). Between these two faunas in the middle Pechelbronn Formation of Bodenheim are a few teeth attributed to *'Theridomys (Th.) pseudosiderolithicus'* by Tobien (1987). According to a personal communication from Mr. Gad to him 'this taxon also occurs in the Bembridge Marls of the Isle of Wight', which belong to the *frohnstettense-suevicum* Zone (=MP20). *Theridomys* (or *Isoptychus*) *pseudosiderolithicus* defines an earlier zone, but is part of a lineage leading to *Theridomys* (or *Isoptychus*) *aquatilis* (type assemblage in the MP21 locality Ronzon) and it is assumed that what is meant here is that the same stage in the lineage has been attained by those from Bodenheim as from the Bembridge Marls. Crucially, nannoplankton indicative of Zone NP22 occur in the middle Pechelbronn Formation (Tobien, 1987). In the nearby Mainz Basin, another small fauna of mammals occurs in the Melanienton of Neustadt. It has been attributed to the '*Frohnstetten niveau*' by Tobien (1986) which is equivalent to level MP20 or the *frohnstettense-suevicum* Zone. Once again, the fauna is no more distinctive of level or zone than that of Rot-Malsch, but the Melanienton contains nannoplankton indica-tive of Zone NP22 (Gramann and Spiegler, 1986, p. 670). All this, however, does indicate that at least the *frohnstettense-suevicum* Zone (level MP20) and the 'Grande Coupure' postdate Zone NP21 and therefore must be Oligocene.

Belgium

A post-'Grande Coupure' fauna attributed to level MP21 (Brunet et al., 1987, Table 2) occurs at Hoogbutsel in a nonmarine stratum which is below the Oude Biesen (Alde Biezen) Sand containing Zone NP22 and above the Grimmertingen Sand containing Zone NP21; it also underlies the Rupelian transgression (Steurbaut, 1986; Gramann, 1990; Gullentops, 1990).

Paris Basin

Zone NP22 is recorded from the Glaises à Cyrènes at the base of the Argile Verte de Romainville by Aubry (1985). A lateral equivalent of the Argile Verte: the Calcaire de Brie contains in its lower part an early post-'Grande Coupure' mammal fauna (Cavelier, 1979, p. 158) with taxa appropriate for level MP21. This would correlate well with the post 'Grande Coupure' mammal fauna in the Hamstead Member of southern England.

Conclusions

Authors (e.g. Gramann, 1990) often correlate the Grimmertingen Sand (Zone NP21) with the Bembridge Oyster Bed (basal Bembridge Marls) although no nannoplankton has been recorded from the latter. It is the transgressions that have been matched. The *Nematura* Band in the Hamstead Member has been correlated with the Argile Verte de Romain-ville (with Zone NP22) by Liengjarern et al. (1980) on the basis of dinocysts. This implies that the NP21/22 Zone boundary is somewhere between the Bembridge Oyster Bed and the *Nematura* Band (shortly above which is the earliest post-'Grande Coupure' fauna). The Eocene/Oligocene boundary could thus be as low as the Bembridge Oyster Bed (possibly though improbably even lower). Pomerol's (1989) assessment of the major hiatus (HP11) marking the Eocene/Oligocene boundary as occurring at the Black Band was based on its correlation by him with the Marabet gypsum band underlying the Argile verte. If instead the correlation by Liengjarern et al. (1980) is correct, the hiatus in England would have to underlie the *Nematura* Band. If such a hiatus can be demonstrated and if of sufficient duration to have removed a significant interval of NP21 zone, the Eocene/Oligocene boundary could be higher in the sequence that spans Bembridge Oyster Bed to below *Nematura* Band. However, in the Paris Basin, Chateauneuf (1986) has claimed recognition through pollen data of the earliest Oligocene cooling event in the Marnes blanches (underlying the Marabet gypsum). This implies that HP11 postdates the boundary. It also suggests that the highest estimated position for the boundary in England would be a significant distance below the *Nematura* Band.

MECHANISM FOR THE DIVERSITY AND ECOLOGICAL CHANGES AND FOR THE FAUNAL TURNOVERS

Figure 25.4B shows the species diversity curve for southern England alongside a summary of ecological events in both southern England and the southern FSP. According to the dating of these events as discussed above, an oxygen isotope curve derived from benthic foraminifers closely linked to the standard marine biostratigraphy at Bath Cliff, Barbados (Muller-Merz and Saunders, 1986), is shown adjacent (Figure 25.4D). Similar data from several sites were the basis for Shackleton's (1986) more generalised curve which is used to infer global temperature changes. That of Muller-Merz and Saunders was chosen because of its close link to standard biostratigraphies, omitted from Shackleton's curve. There is a reasonable correlation between the oxygen isotope curve and the southern English species diversity curve. Legendre (1989, Figure 47) similarly obtained a good fit for the southern

French Quercy localities, although time resolution is low in the vicinity of the main temperature fall, enhancing the steepness of the drop in species numbers. The loss of the complex tropical type forest in southern England corresponds to the middle/late Eocene boundary trough on the oxygen isotope curve (Figure 25.4A,D). The development of more open and apparently drier conditions over the entire area and the northward dispersal of more open country species at the beginning of *medium-curtum* Zone time coincides approximately with the pre-Eocene/Oligocene boundary trough. The temporary return of displaced species in the forest assemblages of the Bembridge Limestone organic and marl facies matches the brief warm peak near the Eocene/Oligocene boundary. The return of lower Bembridge Limestone type habitats in the Bembridge Marls fits with the beginning of the sharp cooling event just after the boundary. The slight increase in tree density by post-'Grande Coupure' times fits the slight recovery in temperatures after the earliest Oligocene major fall. These changes are also closely mirrored by the species diversity of primates and apatotheres (Figure 25.4C), the two orders which provide nearly all the arboreal representatives in the faunas considered here (see Koenigswald and Schierning, 1987, for evidence of arboreality in the Apatotheria). There is a lower overall diversity of arboreal mammals in the FSP than in southern England for the FSP localities chosen. These localities are reasonably reresentative and the data support the lower degree of tree cover in the FSP as demonstrated by the ecological diversity histograms. An exception, the MP17 site of La Bouffie which has 3 species of apatothere and 5 species of primate, suggests localised greater tree cover but does not affect the overall pattern (Remy et al., 1987).

The major turnover at the middle/late Eocene boundary coincides with a major fall in sea level (Plint, 1988) and might represent exploitation of the ability to disperse more easily in both directions across the English Channel area (Hooker, 1986). The 'Grande Coupure' not only followed a major retreat of the sea from the main Eocene European area approximately coincidental with the major earliest Oligocene cooling event (Pomerol, 1989), but also with bridging of the seaways between Eocene Europe, the mid Alpine Rise and the Balkan archipelago, allowing the Asian faunas to migrate (Heissig, 1979).

Thus the mammalian faunas in southern England from the late middle Eocene to the early Oligocene provide patterns of change which give evidence of a strong link with changes in the vegetation, global climate, regional tectonics and sea level change.

ACKNOWLEDGEMENTS

I would like to thank the National Trust and Mr. and Mrs. J. Smith for permission to collect from important sites. Drs. P. Andrews, M.E. Collinson, J.-L. Hartenberger, S. Legendre, and D.E. Russell contributed helpful discussion and critical reading of the manuscript.

REFERENCES

Andrews, P., Lord, J.M. and Nesbit Evans, E.M. 1979. Patterns of ecological diversity in fossil and modern mammalian faunas. *Biol. Jour. Linn. Soc.* 11:177-205.

Aubry, M.P. 1985. Northwestern European Paleogene magnetostratigraphy, biostratigraphy, and paleogeography: calcareous nannofossil evidence. *Geology* 13:198-202.

Bosma, A.A. 1974. Rodent biostratigraphy of the Eocene-Oligocene transitional strata of the Isle of Wight. *Utrecht Micropaleont. Bull. Spec. Publs.* 1: 1-128.

Bosma, A.A. and Insole, A.N. 1972. Theridomyinae (Rodentia, Mammalia) from the Osborne Beds (late Eocene), Isle of Wight, England. *Proc. K. Ned. Akad. Wet.* (B)75(2) : 133-144.

Brunet, M., Franzen, J.L., Godinot, M., Hooker, J.J., Legendre, S., Schmidt-Kittler, N. and Vianey-Liaud, M. (coordinators) 1987. European reference levels and correlation tables. *Münchner Geowiss. Abh.* (A)10 :13-31.

Cavelier, C. 1979. La limite Eocène-Oligocène en Europe occidentale. *Sci. Géol. Inst. Géol. Strasbourg, (Mém.)* 54 :1-280.

Chateauneuf, J.J. 1986. Evolution of the microflora and dinocysts at the Eocene-Oligocene boundary in western Europe. In Pomerol, C. and Premoli-Silva, I., eds., *Terminal Eocene Events*. Amsterdam, Elsevier, pp. 289-293.

Collinson, M.E. 1983. Palaeofloristic assemblages and palaeoecology of the Lower Oligocene Bembridge Marls, Hamstead Ledge, Isle of Wight. *Bot. Jour. Linn. Soc.* 86: 177-205.

Collinson, M.E. and Hooker, J.J. 1987. Vegetational and mammalian faunal changes in the early Tertiary of southern England. In Friis, E.M., Chaloner, W.G. and Crane, P.R., eds., *The Origins of Angiosperms and their Biological Consequences.* Cambridge, Cambridge University Press, pp. 259-304.

Daley, B. 1989. Silica pseudomorphs from the Bembridge Limestone (Upper Eocene) of the Isle of Wight, southern England and their palaeoclimatic significance. *Palaeogeogr., Palaeoclimat., Palaeoecol.* 69: 233-240.

Fleming, T.H. 1973. Numbers of mammal species in north and central American forest communities. *Ecology* 54 :555-563.

Grainger, M., Gunn, E. and Watts, D. 1987. *Tasmanian Mammals, a Field Guide.* Hobart, Tasmanian Conservation Trust Inc., 111pp.

Gramann, F. 1990. Eocene/Oligocene boundary definitions and the sequence of strata in N.W. Germany. *Tert. Res.* 11 :73-82.

Gramann, F. and Spiegler, D. 1986. The Northwest German Tertiary Basin - Oligocene. In Tobien, H., ed., Nordwestdeutschland im Tertiär. *Beitr. Reg. Geol. Erde* 18: 669-687.

Gullentops, F. 1990. Sequence stratigraphy of the Tongrian and early Rupelian in the Belgian type area. *Tert. Res.* 11 :83-96.

Hartenberger, J.L. 1986. Crises biologiques en milieu continental au cours du Paléogène: exemple des mammifères d'Europe. *Bull. Centr. Rech., Expl.-Prod. Elf-Aquitaine* 10: 489-500.

Heissig, K. 1979. Die hypothetische Rolle Sudosteuropas bei den Säugetierwanderungen im Eozän und Oligozän. *Neues Jb. Geol. Paläont. Monatsh.* 1979: 83-96.

Hooker, J.J. 1986. Mammals from the Bartonian (middle/late Eocene) of the Hampshire Basin, southern England. *Bull. Br. Mus. Nat. Hist. (Geol.)* 39: 191-478.

Hooker, J.J. 1987. Mammalian faunal events in the English Hampshire Basin (late Eocene-early Oligocene) and their application to European biostratigraphy. *Münchner Geowiss. Abh.* (A)10 :109-116.

Hooker, J.J. 1989. British mammals in the Tertiary Period. *Biol. Jour. Linn. Soc.* 38: 9-21.

Hooker, J.J. In press a. Two new pseudosciurids (Rodentia, Mammalia) from the English late Eocene and their implications for phylogeny and speciation. *Bull. Br. Mus. Nat. Hist. (Geol.)* 47: 35-50.

Hooker, J.J. In press b. Mammalian palaeoecology of the Bembridge Limestone Formation (late Eocene, S. England). *Historical Biol.*

Insole, A.N. and Daley, B. 1985. A revision of the lithostratigraphical nomenclature of the late Eocene and early Oligocene strata of the Hampshire Basin, southern England. *Tert. Res.* 7 : 67-100.

Keen, M.C. 1977. Ostracod assemblages and the depositional environments of the Headon, Osborne, and Bembridge Beds (upper Eocene) of the Hampshire Basin. *Palaeontology* 20: 405-445.

Koenigswald, W.v. and Schierning, H.-P. 1987. The ecological niche of an extinct group of mammals, the early Tertiary apatemyids. *Nature* 326 : 595-7.

Lange-Badré, B. 1979. Les créodontes (Mammalia) d'Europe occidentale, de l'Eocène supérieur à l'Oligocène supérieur. *Mém. Mus. Natn. Hist. Nat. Paris* (C)42 :7-249.

Legendre, S. 1989. Les communautés de mammifères du Paléogène (Eocène supérieur et Oligocène) d'Europe occidentale: structures, milieux et évolution. *Münchner Geowiss. Abh.* (A)16 :1-110.

Liengjarern, M., Costa, L.I. and Downie, C. 1980. Dinoflagellate cysts from the Upper Eocene-Lower Oligocene of the Isle of Wight. *Palaeontology* 23 :475-499.

Muller-Merz, E. and Saunders, J.B. 1986. The Eocene-Oligocene boundary in the Bath Cliff section, Barbados, West Indies. In Pomerol, C. and Premoli-Silva, I., eds., *Terminal Eocene Events.* Amsterdam, Elsevier, pp. 193-198.

Plint, A.G. 1988. Global eustacy and the Eocene sequence in the Hampshire Basin, England. *Basin Res.* 1 : 11-22.

Pomerol, C. 1989. Stratigraphy of the Palaeogene: hiatuses and transitions. *Proc. Geol. Assoc.* 100 :313-324.

Pomerol, C. and Premoli-Silva, I., eds., 1986. *Terminal Eocene events.* Amsterdam, Elsevier, pp. 1-414.

Remy, J.A., Crochet, J-Y., Sigé, B., Sudre, J., Bonis, L. de, Vianey-Liaud, M., Godinot, M., Hartenberger, J-L., Lange-Badré, B. and Comte, B. 1987. Biochronologie des phosphorites du Quercy: mise à jour des listes fauniques et nouveaux gisements de mammifères fossiles. *Münchner Geowiss. Abh.* (A)10: 169-188.

Riveline, J. 1986. *Les Charophytes du Paléogène et du Miocène inférieur d'Europe Occidentale.* Cahiers de Paléontologie, CNRS, Paris, 227pp., 38pls.

Riveline, J. and Cavelier, C. 1987. Les charophytes du passage Eocène moyen-Eocène supérieur en Europe occidentale. Implications stratigraphiques. *Bull. Soc. géol. Fr. Paris* (8)3:307-315.

Schmidt-Kittler, N. and Vianey-Liaud, M. 1975. Les relations entre les faunes de rongeurs d'Allemagne et du Sud de France pendant l'Oligocène. *C. r. hebd. Séanc. Acad. Sci. Paris,* (D)281: 511-514.

Shackleton, N.J. 1986. Paleogene stable isotope events. *Palaeogeogr., Palaeoclimat., Palaeoecol.* 57: 91-102.

Sigé, B. and Aguilar, J.P. 1987. L'extension stratigraphique des megachiroptères dans le Miocène d'Europe méridionale. *C.r. hebd. Séanc. Acad. Sci. Paris* (2)304 : 469-474.

Spiegler, D., Gramann, F. and Daniels, C.H. von. 1988. Planktonic foraminifera. In Vinken, R. (compiler), The Northwest European Tertiary Basin. Results of the International Geological Correlation Programme, Project 124. *Geol. Jb.* (A)100: 208-213, figs 96, 101-106.

Steurbaut, E. 1986. Late middle Eocene to middle Oligocene calcareous nannoplankton from the Kallo Well, some boreholes and exposures in Belgium and a description of the Ruisbroek Sand Member. *Meded. Werkgr. Tert. Kwart. Geol.* 23(2): 49-83.

Tobien, H. 1986. Continental tetrapods in the Northwest German Tertiary Basin. In Tobien, ed., Nordwestdeutschland im Tertiär, Teil 1. *Beitr. Reg. Geol. Erde,* 18: 558-566.

Tobien, H. 1987. The position of the "Grande Coupure" in the Paleogene of the Upper Rhine Graben and the Mainz Basin. *Münchner Geowiss. Abh.* (A)10: 197-202.

TABLE 25.1. Mammalian faunal lists for the southern English and Franco-Swiss Province (FSP) localities treated herein. Those from southern England are updated from Hooker (1989), those from the FSP are largely from Legendre (1989) and sources quoted therein, modified according to personal observations. Size, locomotor and diet classes are indicated following the procedure of Collinson and Hooker (1987). See explanation to Figure 25.3 for abbreviations. **Southern English localities:** 1=Creechbarrow; 2=Mammal Bed, Hordle; 3=How Ledge Limestone green clay, W.

	S.England										FSP												
	1	2	3	4	5	6	7	8	9	10	1	2	3	4	5	6	7	8	9	10			
Amphiperatherium minutum												+		+	+	+	+			+	AB	SGM	I/F
A. bourdellense											+	+	+								AB	SGM	I/F
A. fontense/ambiguum	+										+	+	+				+	+	+	+	AB	SGM	I/C/F
A. exile																	+	+			AB	SGM	I/F
A. goethei/lamandini	?																+				AB	SGM	I/F
A. sp.A		+	+	+	+		+	+	+	+											AB	SGM	I/F
A. sp.B		+	+	+	+			+	+	+											AB	SGM	I/C/F
A. sp.C		+	+			+		+													AB	SGM	I/F
A. sp.D					+		+	+	+												AB	SGM	I/F
A. sp.3																+					AB	SGM	I/F
Peratherium cuvieri			+	+		+				+					+	+	+		+		AB	SGM	I/F
P. sudrei											+										AB	SGM	I/F
P. bretouense												+									AB	SGM	I/F
P. lavergnense												+	+	+	+		+				AB	SGM	I/F
P. perrierense/cayluxi															+	+					AB	SGM	I/C/F
P. elegans																				+	AB	SGM	I/F
P. sp.1										+								+			AB	SGM	I/F
P. sp.2																	+				AB	SGM	I/F
Opsiclaenodon major		+	+																		AB	Aq	C
'Dyspterna'				+				+	+	+											C	Aq	C
Pantolestidae indet.	+																				AB	Aq	C
Pseudorhynchocyon		+													+		+				AB	SGM	I/C
Eotalpa anglica		+	+	+				+													AB	SGM	I
Talpidae undet.										+											AB	SGM	I
Saturninia beata	?	+										+	+		+						AB	SGM	I
S. hartenbergeri											+										AB	SGM	I
S. grisollensis			+					+			?		+								AB	SGM	I
S. tobieni?	?	?	+	+	+	+		+	+					?		+	+				AB	SGM	I
S. gracilis			+	+				+				+			+	+	+		+	+	AB	SGM	I
S. grandis											+										AB	SGM	I
S. mamertensis											+										AB	SGM	I
Scraeva woodi		+	+	+																	AB	SGM	I
S. hatherwoodensis	?			+																	AB	SGM	I
Paradoxonycteris soricodon																		+			AB	SGM	I
Amphidozotherium cayluxi								+									+		+		AB	SGM	I
Amphidozotheriinae sp.1																	+				AB	SGM	I
Amphidozotheriinae sp.2																	+				AB	SGM	I
Amphidozotheriinae sp.3																+					AB	SGM	I
Darbonetus aubrelongensis																				+	AB	SGM	I
Nyctitheriidae indet.																				+	AB	SGM	I
Gesneropithex figularis	+																				AB	SGM	I/F
G. grisollensis		+	+																		AB	SGM	I/F
G. sp.1											+										AB	SGM	I/F
G. sp.2											+										AB	SGM	I/F
Tetracus sp.										+										+	AB	SGM	I
Butselia biveri										+											AB	SGM	I
Pseudoloris parvulus		+	+			+						+		+	+						AB	Arb	I
P. crusafonti	+																				AB	Arb	I
Nannopithex quaylei	+																				AB	Arb	I/F
N. sp.1	+																				AB	Arb	I/F
Necrolemur sp.											+	+	+	+							AB	Arb	F
Microchoerus wardi	+																				AB	Arb	F/HB
M. creechbarrowensis	+	+																			AB	Arb	F/HB
M. erinaceus		+	+	+	+		+	+					+								AB	Arb	F/HB
M. edwardsi					?				?												AB	Arb	F/HB

Isle of Wight; 4=Lignite Bed, Hatherwood Limestone Member, Headon Hill; 5=Lacey's Farm Limestone Member, Headon Hill outlier; 6=Osborne/Seagrove Bay Members, Isle of Wight; 7=Lower Bembridge Limestone, W. Isle of Wight; 8=Bembridge Limestone organic and clastic unit, Headon Hill; 9=Bembridge Marls Member, Isle of Wight; 10=Hamstead Member, Bouldnor Cliff. **Franco-Swiss Province localities:** 1=Robiac; 2=Le Bretou; 3=Fons 4; 4=Euzet; 5=Perrière; 6=La Débruge; 7=Escamps; 8=Entreroches; 9=Tabarly; 10=Ravet

M. ornatus																+		+			AB	Arb	F/HB
M. sp.C											+										AB	Arb	F/HB
M. sp.D											+										AB	Arb	F/HB
M. sp. indet.																	+				AB	Arb	F/HB
M. sp.2															+						AB	Arb	F/HB
Protoadapis ulmensis			+	+				+													AB	Arb	I/F
Europolemur collinsonae	+																				C	Arb	F/HB
Anchomomys quercyi												+									AB	Arb	I/F
Leptadapis magnus	+	+	+	+									+	+	+						C	Arb	HB
L. assolicus			+	+	+					+											C	Arb	HB
Adapis parisiensis																+					C	Arb	HB
A. betillei																	+		+		C	Arb	HB
A. sudrei											+										C	Arb	HB
Plesiarctomys curranti	+																				AB	SGM	F
P. hurzeleri/gervaisii	+										+	+				+		+			C	SGM	F
P. sp.													+								AB	SGM	F
Manitshinae undet.	+																				AB	SGM	F
Ailuravus stehlinschaubi	+										+										C	Arb	HB
Plesispermophilus																				+	AB	SGM	HB
Gliravus robiacensis											+	+	+	+							AB	SGM	I/F
G. meridionalis													+	+	+						AB	SGM	I/F
G. daamsi		+	+	+	+			+													AB	SGM	I/F
G. minor			+	+																	AB	SGM	I/F
Glamys priscus/fordi		+	+	+	+		+	+	+	+					+	+	+	+			AB	SGM	I/F
'Branssatoglis' bahloi			+	+	+			+	+	+											AB	SGM	F
Oligodyromys																				+	AB	SGM	F
Sciuroides rissonei	+																				AB	SGM	F
S. siderolithicus											+	+									AB	SGM	F
S. ehrensteinensis			+											+							AB	SGM	F
Suevosciurus authodon	+																				AB	SGM	F
S.minimus																		+			AB	SGM	F
S. sp. nov.		+	+	+																	AB	SGM	F
S. fraasi								+									+				AB	SGM	F
S. ehingensis					+			+										?			AB	SGM	F
Treposciurus preecei	+																				AB	SGM	F
T. mutabilis			+	+	+		+						+		+						AB	SGM	F
T. intermedius															+						AB	SGM	F
T. sp. nov.		+	+	+				+	+				+					+			AB	SGM	F
Paradelomys crusafonti											+	+	+	+			?				AB	SGM	F
P. rutimeyeri																		+			AB	SGM	F
'P.' depereti/quercyi		+	+	+	+	+	+	+					+	+		+	+	+	+		AB	SGM	F-F/HB
Tarnomys quercinus																				+	AB	SGM	HB
Estellomys cansouni													+								AB	SGM	F
Elfomys parvulus													+	+	+						AB	SGM	F
E. tobieni											+										AB	SGM	F
E. sp.												+					+				AB	SGM	F
Pseudoltinomys				+					+	+		+	+	+	+		+			+	AB	SGM	F/HB
P. sp.															+						AB	SGM	F/HB
Oltinomys platyceps																	+		+		AB	SGM	F/HB
Remys minimus											+	+									AB	SGM	F
R. garimondi													+								AB	SGM	F
R. sp.												+									AB	SGM	F
Ectropomys exiguus							+	+	+												AB	SGM	F/HB
Patriotheridomys																	+		+		AB	SGM	F/HB
Blainvillimys																	+		+	+	AB	SGM	F/HB
Thalerimys		+	+	+	+	+	+	+													AB	SGM	F/HB
Isoptychus aquatilis			+	+	+	+	+	+	+	+			+	+	+	+		+			AB	SGM	F/HB
I. sp.1								+		+											AB	SGM	F/HB
I. sp.2																		+			AB	SGM	F/HB

Theridomys perrealensis																+					AB	SGM	F/HB
T. sp.																	+				AB	SGM	F/HB
Eomys sp.										+											AB	SGM	I/F
Eucricetodon atavus										+										+	AB	SGM	I/F
Steneofiber										+											AB	SGM	F/HB
Desmatolagus																				+	AB	LGM	HB
Heterohyus nanus	?		+	+	+			+			+				+	+					AB	Arb	I
H. sudrei	?										+										AB	Arb	I
H. morinionensis	+																				AB	Arb	I
H. sp.		+																			AB	Arb	I
Prototomus bulbosus															+						C	LGM	C
P. minor												+									AB	LGM	C
Pterodon dasyuroides							+		+							+	+				C	LGM	C
Quercytherium													+	+							C	LGM	C
Cynohyaenodon magnus												+									C	LGM	C
Hyaenodon minor		+	?										+	+	+						C	LGM	C
H. requieni													+	+	+	+					D	LGM	C
H. brachyrhynchus				+											+			+			C	LGM	C
H. sp.										+							+				C	LGM	C
Quercygale angustidens		+									+	+		+	+						C	Scan	C
Paramiacis	+	+																			AB	Scan	I/C/F
P. sp.	+											?									AB	Scan	I/C/F
Simamphicyon helveticus											+	+									D	LGM	C
Cynodictis lacustris					+			+	+							+	+	+	+		C	SGM	C
C. compressidens																	+				C	SGM	C
Amphicynodon brachyrostris																				+	C	SGM	C
Vulpavoides cooperi	+																				AB	Scan	I/C/F
Entelodon magnus										+											GH	LGM	F
Cebochoerus 'helveticus'												+									C	LGM	F
C. robiacensis/fontensis	+										+		+	+							C	LGM	F
Acotherulum pumilum															+			+			C	LGM	F
A. campichii/saturninum	+	+			+						+	+				+	+	+			C	LGM	F
A. quercyi							+											+			C	LGM	F
Choeropotamus	+	+				+		+	+		+		+	+		+	+	?			D(C)	LGM	F/HB
Haplobunodon venatorum	+																				C	LGM	F/HB
H. lydekkeri		+																			C	LGM	F/HB
Rhagatherium					+		+											+			C	LGM	F/HB
Anthracobunodon											+										C	LGM	F/HB
Amphirhagatherium									+												C	LGM	F/HB
Diplopus aymardi		+																			EF	LGM	HB
Elomeryx crispus																+					D	LGM	HB
E. porcinus										+											EF	LGM	HB
Bothriodon aymardi										+											EF	LGM	HB
B. velaunus										+											EF	LGM	HB
B. leptorhynchus										+											EF	LGM	HB
A. alsaticum										+											EF	LGM	F/HB
Mixtotherium gresslyi	+																				C	LGM	HB
M. cuspidatum															+						C	LGM	HB
M. sp.	+																				AB	LGM	HB
'M'. sp.	+																				C	LGM	HB
Robiacina minuta											+	+									AB	LGM	HB
Oxacron							+									+	+	+	+		AB	LGM	HB
Paroxacron																	+				AB	LGM	HB
Plesiomeryx cadurcensis																				?	AB	LGM	HB
Tapirulus									+	+	+	+			+	+				+	C	LGM	F/HB
Catodontherium robiacense											+	+									D	LGM	HB
Dacrytherium elegans	+											+									C	LGM	HB
D. ovinum		+		+									+	+	+		+				C	LGM	HB
D. saturninii																+					C	LGM	HB

Robiatherium cournovense											+	+									D	LGM	HB
Anoplotherium commune						+	+	?	+							+	+	?			EF	LGM	HB
A. latipes																+					EF	LGM	HB
A. laurillardi							?		?							+					D	LGM	HB
Diplobune secundarium								?								+	+	+			D	LGM	HB
D. quercyi																	+				D	LGM	HB
D. sp.																				+	D	LGM	HB
Leptotheridium											+	+	+	+	+						C	LGM	HB
Xiphodon							+	+	+		+	+	+	+		+	+		+		C	LGM	HB
Paraxiphodon													+								C	LGM	HB
Dichodon cervinus	+	+	?			+		+					+	+	+						C	LGM	HB
D. cuspidatus	?	+		+		?										+					D	LGM	HB
D. frohnstettensis																+	+				C	LGM	HB
D. vidalenci												+									C	LGM	HB
D. sp.												+									C	LGM	HB
Haplomeryx					+				+		+	+	+	+	+		+		+		AB	LGM	HB
Dichobune leporina							+									+	+		+		C	LGM	F
Mouillacitherium												+	+		+						AB	LGM	F
Hyperdichobune spinifera																		+			C	LGM	F
H. spectabilis																		+			C	LGM	F
H. sp.	+																				AB	LGM	F
Dichobunidae indet.																				+	C?	LGM	F
Pseudamphimeryx pavloviae											+	+			+						AB	LGM	HB
P. renevieri											+	+	+	+	+						AB	LGM	HB
P. hantonensis	?			+	+																AB	LGM	HB
Amphimeryx murinus																+	+	+	+		C	LGM	HB
Gelocus communis																				+	C	LGM	HB
Pachynolophus												+	+								C	LGM	F/HB
Anchilophus dumasii											+	+	+	+							C	LGM	F/HB
A. gaudini/radegondensis		+					+				+	+	+	+		+		+			C	LGM	F/HB
A desmaresti											+	+									C	LGM	F/HB
Lophiotherium	+										+		+	+							C	LGM	F/HB
Propalaeotherium parvulum A	+											+	+								C	LGM	F/HB
P. parvulum B	+											+									C	LGM	F/HB
Leptolophus stehlini											+	+									D	LGM	HB
Plagiolophus annectens		+	+	+	+	+		+			+	+	+	+	+						D	LGM	HB
P. cartailhaci											+										D	LGM	HB
P. curtisi	+																				D	LGM	HB
P. minor							+		+							+	+	+	+	+	D	LGM	HB
P. oweni																+					D	LGM	HB
P. fraasi									+								?		+	?	D	LGM	HB
Palaeotherium magnum		+			+	+	+							+	+	+	+	+	+		GH	LGM	HB
P. medium							+		+				+	+	+	+	+		+		EF	LGM	HB
P. siderolithicum											+	+	+			+					D	LGM	HB
P. ruetimeyeri											+										EF	LGM	HB
P. muehlbergi		+	+			+	+	+	+							+					EF	LGM	HB
P. curtum				+		+	+		+				+	+		+		+			EF	LGM	HB
P. crassum																+		+			EF	LGM	HB
P. pomeli											+										EF	LGM	HB
P. castrense											+										GH	LGM	HB
P. duvalii		+					+	?										+			D	LGM	HB
***P.* sp.** indet.																				+	EF?	LGM	HB
Lophiodon lautricense											+	?									GH	LGM	HB
Chasmotherium cartieri											+										D	LGM	HB
Ronzotherium										+											GH	LGM	HB

26. EVOLUTION OF MAMMALIAN FAUNAS IN EUROPE DURING THE EOCENE AND OLIGOCENE

by Serge Legendre and Jean-Louis Hartenberger

ABSTRACT

An overview of the Eocene and Oligocene mammalian history in Europe shows three main events: (i) the Paleocene-Eocene event, the less documented, corresponds to the disappearance of primitive mammals and the arrival of modern orders; (ii) the Lutetian-Bartonian event corresponds to a small immigration wave; (iii) the Eocene-Oligocene event, known as the "Grande Coupure," corresponds to a major turnover in mammalian faunas, and to an important change in mammalian community structure. The "Grande Coupure" is contemporaneous with a major ecoclimatic change.

INTRODUCTION

During the two last decades, our knowledge of Paleogene mammalian faunas has increased, but data are still expected for the late Cretaceous and Paleocene history of mammals in Europe. In contrast, major progress and results have been obtained for the periods following the Paleocene. Eocene and Oligocene faunas are now well documented.

Thus, the overview proposed here does not concern the Paleocene, but will focus on Eocene and Oligocene mammalian evolution and faunal changes in western Europe. Tempo, patterns and processes in evolutionary changes in mammalian faunas are studied using different methods:

—origination/extinction analysis and cohort analysis using survivorship curves highlight major biotic events affecting European mammalian faunas;

—diversity analysis within faunas and major adaptive events in mammalian evolution are tentatively related to ecological and climatic changes;

—community analysis shows structural changes in local faunas related to drastic climatic changes;

—finally, some major events as revealed by the preceding analysis are linked to major global changes, and are used for correlation of faunal sequences within continents, between continents, and with the marine realm.

THE BACKGROUND: TAXONOMIC TREATMENT, FAUNAL DATA SET, AND THE TIME SCALE

The basic taxonomic data used here result from study started about 30 years ago and in which all mammalian orders were included. Updated faunal lists and a bibliographic compilation of these most recent studies are given in Hartenberger (1987), Legendre (1987a, 1989), Remy et al. (1987), Russell et al. (1982), Savage and Russell (1983), Vianey-Liaud and Legendre (1986). Some unpublished data were provided by specialists of different groups and regions, and they were added in our study. The specialists and groups studied are: J.-Y. Crochet (marsupials, Oligocene insectivores, Eocene creodonts and carnivores), L. de Bonis (carnivores, Oligocene perissodactyls), M. Godinot (primates), J.-L. Hartenberger (Eocene rodents), B. Lange-Badré (creodonts), B. Marandat (Eocene mammals), J.A. Remy (Eocene perissodactyls), B. Sigé (bats, Eocene insectivores), J. Sudre (artiodactyls), M. Vianey-Liaud and B. Comte (Oligocene rodents).

The genus and species concepts used correspond to specific lineages, and consequently does not depend strictly on taxonomic treatment (i.e., the generic or specific name). Thus, as far as possible we avoid introducing biases into faunal analysis by eliminating pseudoextinctions, which are artifacts due to the definition of chronospecies (see also Legendre, 1987a).

The time scale used for faunal analysis is

Age		Marine Stages		MP	Reference Levels
- 23	OLIGOCENE	STAMPIAN	CHATTIAN	30	Coderet
				29	Rickenbach
				27-28	Pech du Fraysse
				26	Mas de Pauffié
			RUPELIAN	24-25	Garouillas
				23	Itardies
				22	Villebramar
				21	Soumaille
- 34	EOCENE	PRIABONIAN		20	St-Capraise d'Eymet
				19	Escamps
				18	La Débruge
				17 a	Perrière
				17 b	Fons 4
		BARTONIAN		16	Robiac
- 41				15	La Livinière
				14	Egerkingen α et β
- 45		LUTETIAN		12-13	Geiseltal oMK
				11	Geiseltal UK
		YPRESIAN		10	Grauves
				9	Avenay
				8	Mutigny
				7	Dormaal
- 55	PAL	THANETIAN		6	Cernay
- 65				1-5	Hainin

TABLE 26.1.- Subdivision of the European continental Paleogene, based on mammals, as proposed at the Mainz Symposium (Schmidt-Kittler et al. 1987), with minor modifications explained in Legendre (1989). Calibration and correlation given here follow Hartenberger (1987) and for the Eocene/Oligocene boundary Pomerol & Premoli-Silva (1986) and Premoli-Silva et al. (1988).

that defined in the *Symposium on mammalian biostratigraphy and paleoecology of the European Paleogene* held at Mainz in 1987 (Schmidt-Kittler *et al.*, 1987).

The basis for this time scale is the use of a rich and diversified local mammalian fauna as the basic unit. The choice of the reference faunas, and their relative position along the time scale is based on evolutionary stages of component specific lineages (i.e. successive chronospecies). In addition to these basic data, but as second order of interest, first and last appearances of taxa, as well as association of taxa can be defined for correlation purposes. The whole local fauna can be used to represent the reference level; the principles underlying this biochronological time scale are discussed by Schmidt-Kittler (1987).

A consequence of this time scale is that it is defined by points, which are real, and it never represents intervals with implicit boundaries; these boundaries could never be defined--and thus tested--and induce endless discussions. Moreover the lack of developed continental deposits and the high provinciality in Europe during the middle to late Paleogene are the main reasons for using such units (Franzen, 1968; Hartenberger, 1969; Jaeger and Hartenberger, 1975; Schmidt-Kittler et al., 1987). Table 26.1 summarizes the correlation table used in this paper. Comments on the biochronological time scale and on its calibration against the geochronological time scale can be found in Hartenberger (1989, p. 123) and Legendre (1987a, 1989).

ORIGINATION/EXTINCTION, SURVIVORSHIP CURVES

About 300 genera have been described in the Paleogene of Europe (see Russell et al., 1982, and Schmidt-Kittler, 1987). Extinction and origination rates were calculated using the biochronological duration of genera (see Hartenberger, 1987 for more details). Survivorship curves concern 30 faunal strata considered as cohorts of genera (see Hartenberger, 1988 for more details).

Three main extinction/origination events are observed (Figures 26.1-26.2: E1, E2, E3). The first one of earliest Eocene age (Figures 26.1-26.2: E1), corresponds to the arrival in Europe of the Perissodactyla, Artiodactyla, and Rodentia. At the same period, the diversity of Condylarthra and Multituberculata decreased dramatically.

The second event of Lutetian age (Figures 26.1-26.2: E2) is not well understood at present. A minor immigration wave is noted for Primates (Franzen, 1988), Chiroptera (Sigé, 1977), Artiodactyla (Sudre, 1980) and Rodentia (Hartenberger, 1990).

The third event (Figures 26.1-26.2: E3) is situated at the Eocene/Oligocene boundary. It corresponds to the "Grande Coupure." [But see Hooker, this volume, who argues that the "Grande Coupure" is early Oligocene]. This "Grande Coupure" event, which results from the combination of an important immigration event with an extinction event, was first descibed in the Paris Basin by Stehlin (1910), and it is now recognized on a continental scale in Europe (Lopez and Thaler, 1974; Heissig, 1978, 1987; Russell et al., 1982; Hooker, 1987).

Regional studies at the species level show that the turnover in faunas is very high, and, in the Quercy area for example, up to 60% of species locally disappeared (Figure 26.3 and Legendre, 1987b).

The survivorship curves emphasize two aspects concerning the "Grande Coupure": (1) all cohorts are involved so that there is no selectivity; (2) some cohorts are affected before the main event (occurring after MP 20), and so the event is not geologically instantaneous but has an undoubted duration.

During the Oligocene, a minor extinction event occurs, which involves only the "old Eocene cohorts".

DIVERSITY AND ADAPTIVE CHANGES

This part of the study does not concern the lower Eocene: the fossil record is still too poor for this kind of approach. Thus, it begins with the middle Eocene (see Legendre et al., in press).

Evolution of the diversity in faunas from France (Figure 26.4) features three main phases. First, a diversification during the Lutetian, followed by relatively high richness during the Bartonian, and a reduction in diversity during the Oligocene.

In contrast to the Eocene, "insectivores", primates, large herbivores, theridomyid and ischyromyid rodents, and creodonts are less

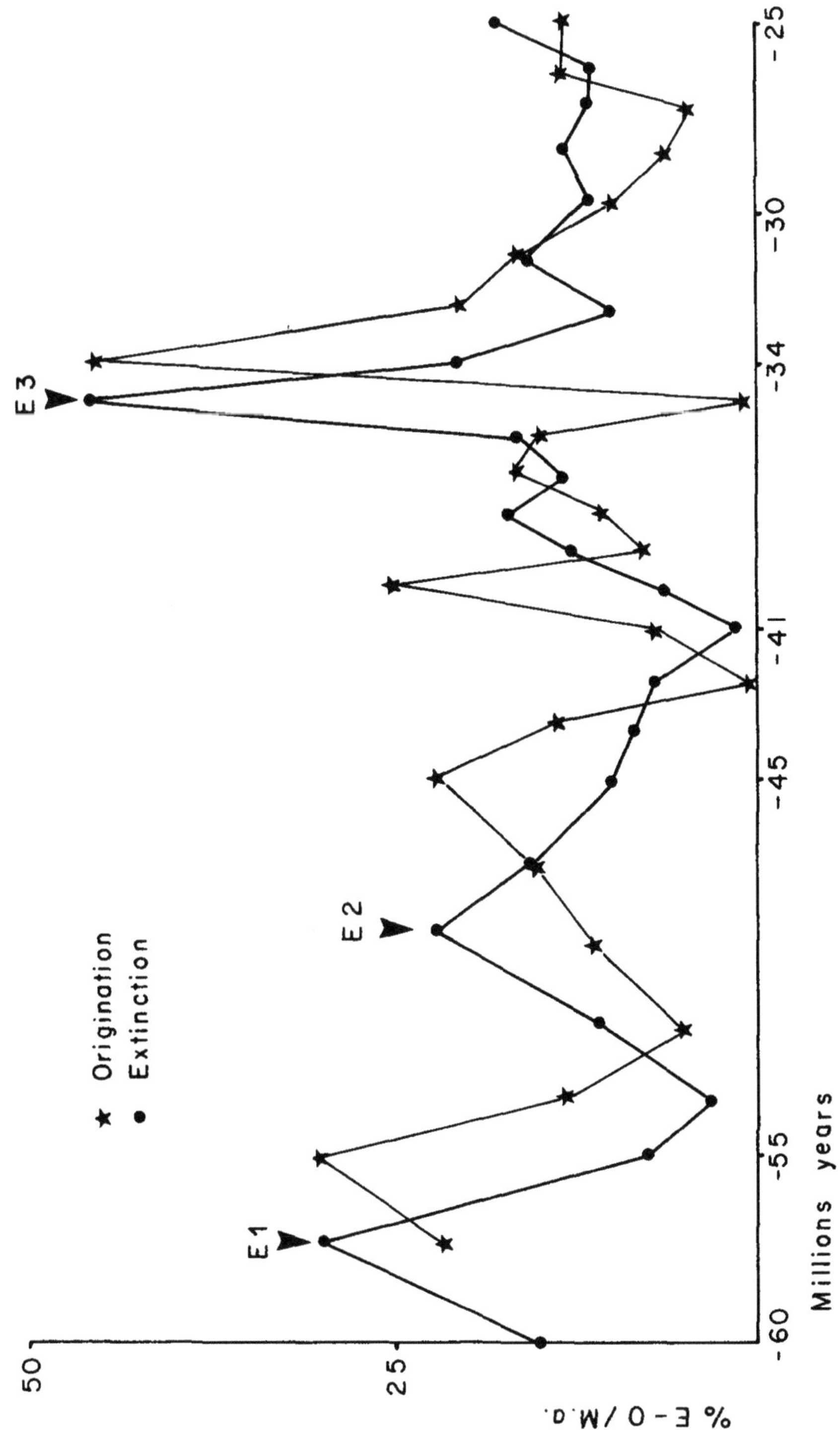

FIGURE 26.1.- Extinction and origination rates (percent genera per million years) in the Paleogene mammalian record (after Hartenberger, 1987). The three main events are noted E1, E2, E3.

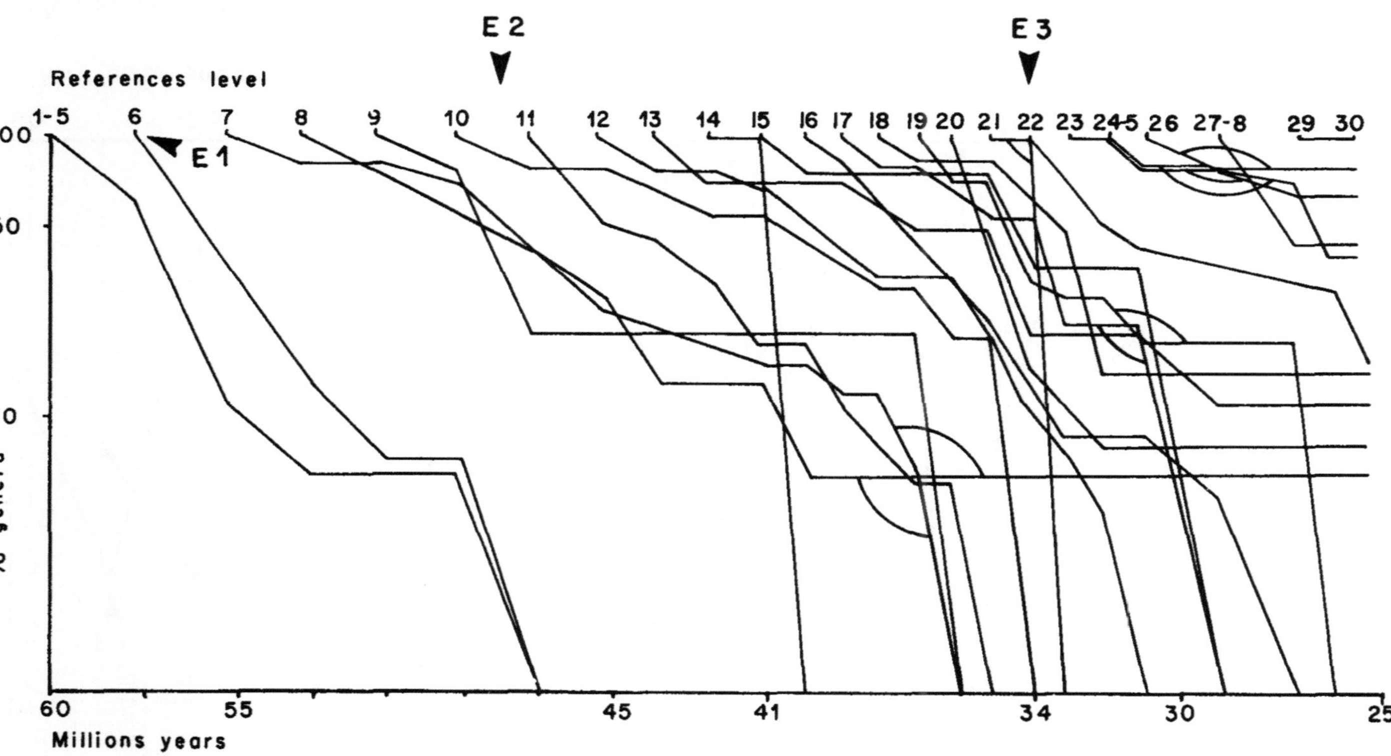

FIGURE 26.2.- Survivorship curves of mammalian cohorts from the European Paleogene record. E1, E2, E3 show the three main extinction events.

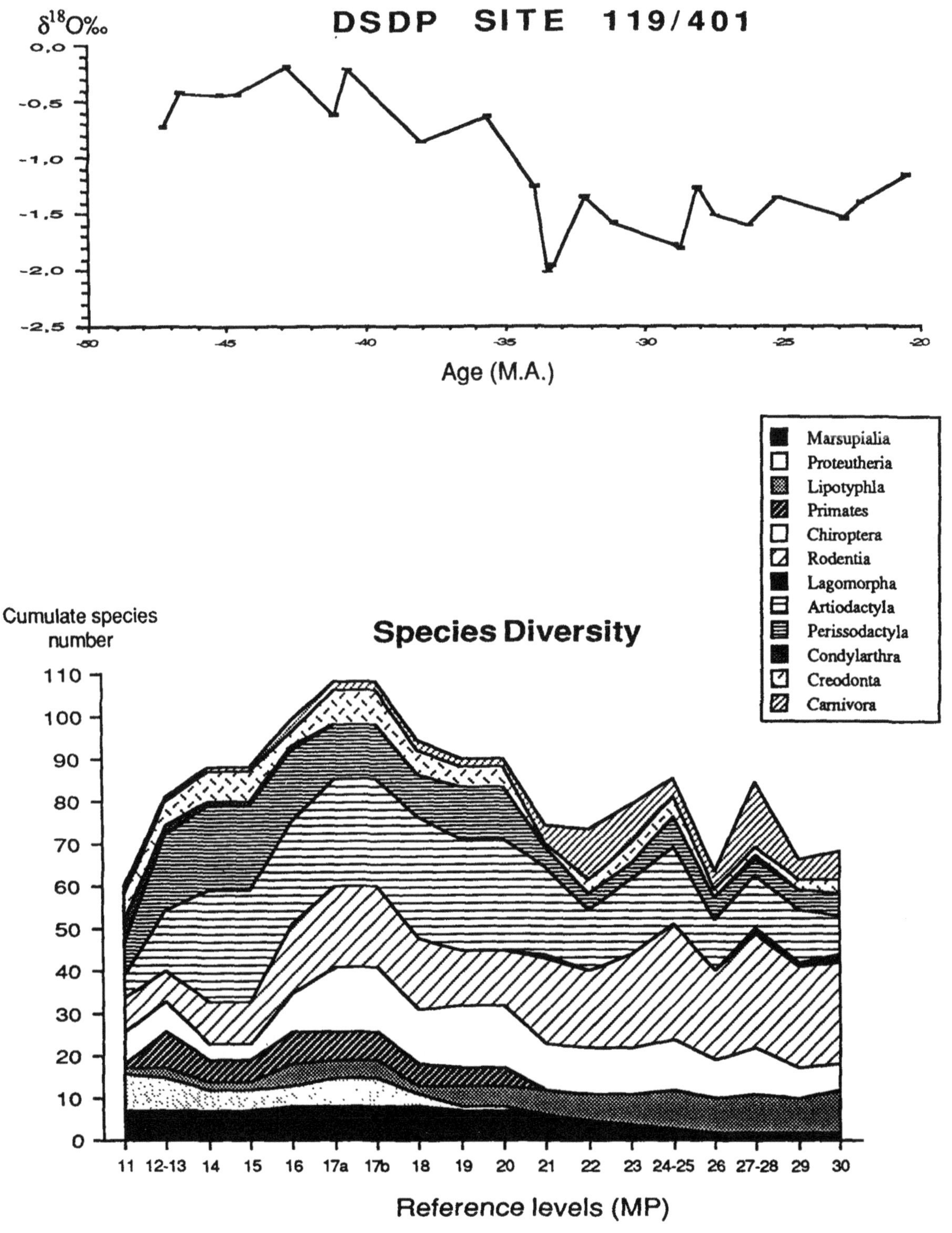

FIGURE 26.3.- Evolution of species diversity (BOTTOM) in mammalian faunas from Europe during the Eocene and the Oligocene (after Legendre et al., in press) and oxygene isotopic curves (TOP) from Site 119/401 in the Bay of Biscaye (after Miller and Curry, 1982). Decreases in mammalian diversity are coincident with decreases in oceanic temperature.

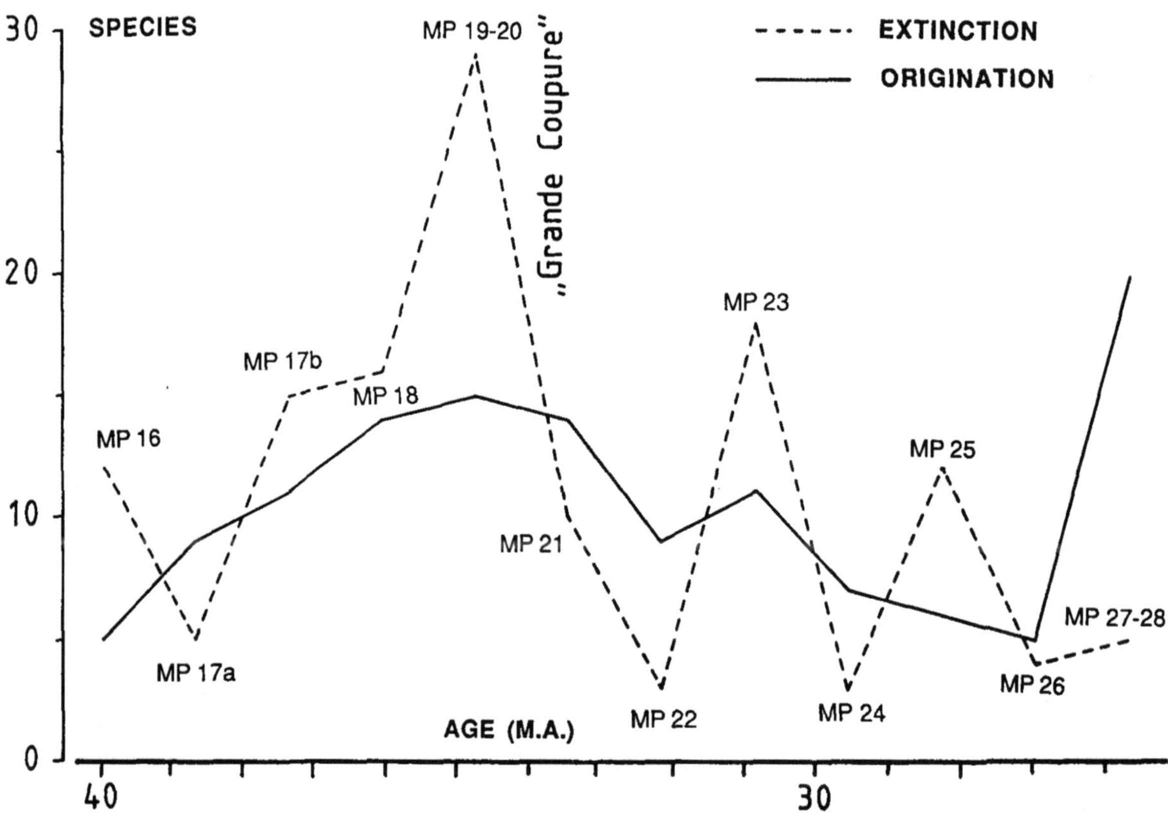

FIGURE 26.4.- Number of extinctions in mammalian species from the Quercy area (after Legendre, 1987b). The main extinction event corresponds to the "Grande Coupure" of Stehlin (1910), occurring at the Eocene-Oligocene transition.

abundant or absent during the Oligocene. Only carnivores and other rodents (except the two families mentioned above) are more diverse during the Oligocene. Diversity seems stable in bats.

A first observation concerning herbivore diversity is the apparent inverse relationship between the decline of medium to large herbivores and the increased diversity of small ones. This involves mainly artiodactyls and rodents.

Otherwise, there seems to be a correlation between predators and rodents. In particular, diversity in creodonts seems to be linked with variations undergone by theridomyids, whereas the number of species of carnivores seems to follow the variation in modern rodent families (Legendre et al., in press).

Some of the faunal events can be correlated with the acquisition of adaptive features in different mammalian groups. In artiodactyls, the diversification phase which seems to end at the beginning of the Bartonian (MP 14-15) corresponds with the development of selenodonty and premolar molarization (or premolar lengthening) in some lineages (*Pseudamphimeryx, Dichodon, Haplomeryx*). Selenodonty is also noted in rodents at the Eocene/Oligocene boundary (*Pseudosciurus*).

Among late Eocene primates the Adapinae progressively replace more primitive Adapiformes; these new lineages develop adaptations to a more folivorous diet and to harder food.

Hypsodonty begins to develop in some theridomyid rodents at the MP 15 reference-level (lower Bartonian) whereas it started at the MP 13 level (late Lutetian) in perissodactyls, with the appearance of the Palaeotheriidae (*sensu* Franzen, 1989). Some rodents are completely hypsodont by the end of Oligocene time. A clear faunal break can be observed in perissodactyls between the MP 16 and

	Marine Stages		MP	Immigration event	Extinction event	Characteristic extinction	Total diversity	Intrafaunal changes	Morphological innovation	Physical environmental changes
OLIGOCENE	STAMPIAN	CHAT-TIAN	30						Hypsodonty complete in rodents	
			29							
			27-28		Minor		Medium-High			
			26				Lowest			Maximum aridity ?
		RUPE-LIAN	24-25			Eocene genera		Small herbivores more diverse than large herbivores		
			23							
			22							
			21	Major			Low		Selenodonty in rodents Cement in perissodactyls	Major cooling
EOCENE	PRIABONIAN		20		Major	Primates				
			19				High			
			18							Seasonality
			17 a			Condylarthres			Cuboid and navicular fused in Artiodactyls	
			17 b						Semi-hypsodonty in rodents	
	BARTONIAN		16			*Lophiodon* + some other perissodactyls	Highest			
			15					Medium and large herbivores more diverse than small herbivores	Selenodonty in Artiodactyls Molarisation of premolars	
			14							
	LUTETIAN		12-13	Minor			Medium-High		Hypsodonty in perissodactyls	
			11		Major					Pyrenean phase Tropical (warm/humid) climate
	YPRESIAN		10							
			9			Multituberculata	Low			
			8							
			7	Major						
			6		Major					
PAL	THANETIAN		1-5							

TABLE 26.2.- Main events occurring in mammalian faunas during the Eocene and the Oligocene of Europe.

MP 17 levels (Bartonian/Priabonian boundary): all the large forms disappear by the end of the Bartonian and are progressively replaced by new lineages of the genus *Palaeotherium*. This latter group declines during the late Priabonian.

In the artiodactyls, fusion between the cuboid and navicular bones of the tarsus occurs in some lineages near the MP 16/MP 17 boundary (Bartonian/Priabonian); this structure is a key-character in the ruminants (see Geraads et al., 1987).

An important faunal turnover occurs between the MP17 and MP18 reference-levels (early/late Priabonian transition); it is characterized by the decline of proteutherian insectivores, the increase in familial diversity in artiodactyls, and the appearance of new groups within the creodonts (Proviverrinae).

The main event is the "Grande Coupure" which marks the transition between the Eocene and Oligocene epochs (MP 20/MP 21). This event is characterized by the disappearance of apatemyids and primates (if one excepts the occurrence of an adapid in the very early Oligocene of England; Hooker, 1986), and by the beginning of the diversification of modern insectivores, rodents and fissiped carnivores. The arrival of new perissodactyl families must also be noted, although this group does not show any clear diversification. On the other hand, the artiodactyl decline does not exactly coincide with this event; it seems to be delayed (MP 22).

During the Oligocene, the MP 26 reference level shows a minimum in diversity marking a transition between the Stampian and the late Oligocene faunas.

COMMUNITY ANALYSIS

Mammalian communities are analysed using the cenogram method (Legendre, 1986, 1989). A cenogram is constructed by plotting the natural logarithm of the body weight of each component species on the Y-axis ranked by decreasing size on the X-axis (Figure 26.5). A schematic cenogram shows the slope and gaps in the size distribution of species in the community.

In extant faunas, environmental factors affect characteristics of cenograms:

—in open environments, medium-sized species (i.e., species with body weight ranging from 500 g to 8 kg) are rare or absent, whereas in forest environments they are normally present. In cenograms this corresponds to a gap for the open environment and continuous distribution for forest;

—in arid environments, large-sized species (weighing over 8 kg) are rare, whereas they are abundant in humid environments. In a cenogram, this corresponds to the left part of the diagram, and the slope is steep under arid conditions, or smooth in humid environments.

These characteristics of cenograms have been found in faunas from different continents, and they have therefore been shown to be independent of the taxonomic composition of the community (Legendre, 1989). Thus, the method was extended to fossils, and it was applied to fossil communities from the late Eocene and Oligocene of western Europe.

Regional sequences of faunas from the Phosphorites of Quercy (Figure 26.6), southern France, Germany and southern England, show similarities in shape for the cenogram of faunas of the same age as well as pattern of change during the Eocene and Oligocene (Legendre, 1987c, 1989).

The older faunas, corresponding to the middle and late Eocene, are characterized by the abundance of large-sized species, indicating humid conditions (Legendre, 1986, 1989). Generally, medium-sized species are well represented in the faunas for this period, which can be related to forest environments (Legendre, 1986, 1989).

At the beginning of the Oligocene, a drastic change occurs in the communities. Large species become rare in local faunas, and medium-sized species are absent. This indicates arid and open environments. A lower species diversity found in each local fauna during the Oligocene could be related to a general decrease of the mean annual temperature (Legendre, 1987a, b; Legendre et al., in press). Some regional differences occur and are interpreted as resulting from latitudinal differences: faunas from England are characterized by a gap in the medium-sized part of the distribution of species showing more open environments during the end of the Eocene.

The main event in faunas occurred at the Eocene/Oligocene boundary, where large and medium sized species are less diverse. This reflects a drastic change in climate and environment: from humid, warm, forested conditions,

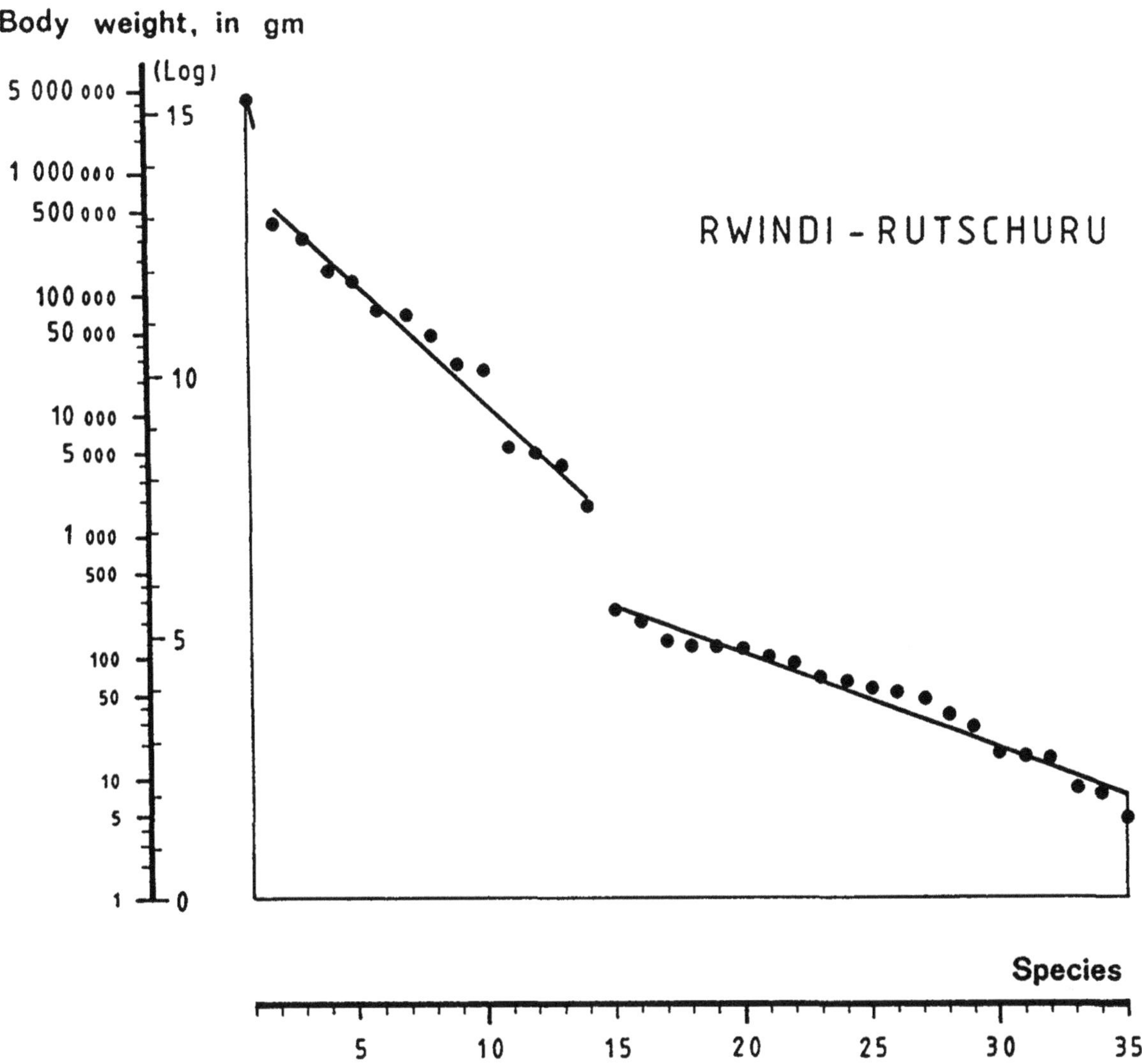

FIGURE 26.5.- Cenogram of the extant fauna of Rwindi-Rutschuru (Zaire). The average body weight (expressed in natural logarithms) of all species is plotted on the Y-axis and the rank of the species in decreasing order of size on the X-axis. The carnivores and the bats are omitted. Lines schematize the cenogram.

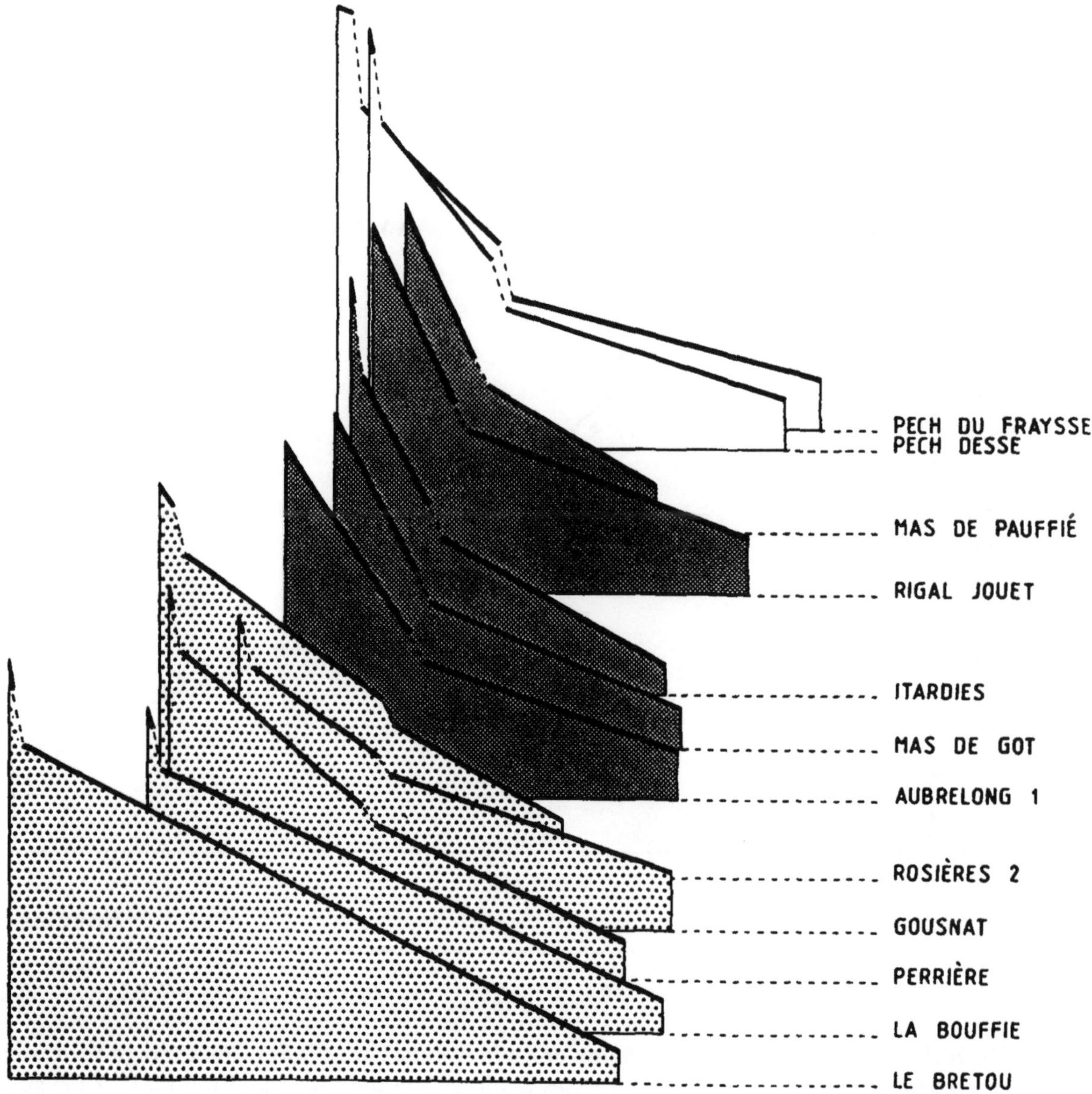

FIGURE 26.6.- Cenogram sequence of fossil faunas from the Phosphorites of Quercy, France (after Legendre, 1987c, 1989). The oldest faunas (Le Bretou to Rosières 2) are characterized by a large number of medium to large-sized species indicating forest and humid environments. The youngest faunas (Aubrelong 1 to Pech du Fraysse) show less abundant large-sized species and rare (or absent) medium-sized ones, indicating more open and arid environments.

near a tropical rain forest type during the Eocene, to arid, colder, more open environments, near a savanna type during the Oligocene (Legendre, 1986, 1987a-c, 1989).

ACKNOWLEDGEMENTS

We address many thanks to all paleontologists who collected the fossils and increased our knowledge on European mammals during the last decades, especially to our colleagues from Montpellier. We thank R.K. Stucky, J.J. Hooker, and an anonymous reviewer for helpful comments on a previous draft of the manuscript. This is contribution C.N.R.S.-I.N.S.U.-D.B.T. number 210.

REFERENCES

Franzen, J.L. 1968. Revision der Gattung *Palaeotherium* Cuvier 1804 (Palaeotheriidae, Perissodactyla, Mammalia). Inaugur.-Dissert. Albert Ludwigs-Universität, Freiburg, 1-181.

Franzen, J.L. 1988. Ein weiterer Primatenfund aus der Grube Messel bei Darmstadt. *Cour. Forsch.-Inst. Senckenberg* 107: 279-289.

Franzen, J.L. 1989. Origin and systematic position of the Palaeotheriidae. In D.R. Prothero and R.M. Schoch, eds., *The Evolution of Perissodactyls*, New York, Oxford Univ. Press, pp. 102-108.

Geraads, D., Bouvrain, G., and Sudre, J. 1987. Relations phylétiques de *Bachitherium* Filhol, ruminant de l'Oligocène d'Europe occidentale. *Palaeovertebrata* 17: 43-73.

Hartenberger, J.-L. 1969. Les Pseudosciuridae (Mammalia, Rodentia) de l'Eocène moyen de Bouxwiller, Egerkingen et Lissieu. *Palaeovertebrata* 3: 27-61.

Hartenberger, J.-L. 1987. Modalités des extinctions et apparitions chez les mammifères du Paléogène d'Europe. *Mém. Soc. géol. Fr.* 150: 133-144.

Hartenberger, J.-L. 1988. Etudes sur la longévité des mammifères du Paléogène d'Europe. *C.R. Acad. Sci. Paris*, Sér. 2, 306: 1197-1204.

Hartenberger, J.-L. 1989. Summary of Paleogene rodents in Europe. In C.C. Black and M.A. Dawson, eds., *Papers on fossil rodents in honor of A.E. Wood, Sciences Ser., Nat. Hist. Mus. Los Angeles County* 33: 119-124.

Hartenberger, J.-L. 1990. Données nouvelles et hypothèses sur l'origine des Theridomyidae (Mammalia, Rodentia). *C.R. Acad. Sci. Paris*, sér. 2, 311: 1017-1023.

Heissig, K. 1978. Fossilführende Spaltenfüllungen Süddeutschlands und die Ökologie ihrer Huftiere. *Mitt. Bayer. Staatsslg. Paläont. hist. Geol.* 18: 237-288.

Heissig, K. 1987. Change in the rodent and ungulate fauna in the Oligocene fissure fillings of Germany. *Münchner geowiss. Abh.* (A) 10: 101-108.

Hooker, J.J. 1986. Mammals from the Bartonian (middle/late Eocene) of the Hampshire Basin, southern England. *Bull. Brit. Mus. Nat. Hist. (Geol.)* 13: 191-478.

Hooker, J.J. 1987. Mammalian faunal events in the English Hampshire Basin (late Eocene - early Oligocene) and their application to European biostratigraphy. *Münchner Geowiss. Abh.*, (A), 10: 109-116.

Jaeger, J.-J. and Hartenberger J.-L. 1975. Pour l'utilisation systématique de niveaux-repères en biochronologie mammalienne. *3ème Réun. ann. Sci. Terre*, Montpellier: 201.

Legendre, S. 1986. Analysis of mammalian communities from the late Eocene and Oligocene of southern France. *Palaeovertebrata* 16: 191-212.

Legendre, S. 1987a. Concordance entre paléontologie continentale et les événements paléocéanographiques: exemple des faunes de mammifères du Paléogène du Quercy. *C.R. Acad. Sci. Paris*, Sér. 3, 304: 45-50.

Legendre, S. 1987b. Mammalian faunas as paleotemperature indicators: concordance between oceanic and terrestrial paleontological evidence. *Evol. Theory* 8: 77-86.

Legendre, S. 1987c. Les communautés de mammifères fossiles d'Europe occidentale de l'Eocène supérieur et Oligocène: structures et milieux. *Münchner Geowiss. Abh.* (A) 10: 301-312.

Legendre, S. 1989. Les communautés de mammifères du Paléogène (Eocène supérieur et Oligocène) d'Europe occidentale: structures, milieux et évolution. *Münchner Geowiss. Abh.* (A) 16: 1-110.

Legendre, S., Crochet, J.-Y., Godinot, M., Hartenberger, J.-L., Marandat, B., Remy, J.A., Sigé, B., Sudre, J., and Vianey-Liaud, M. In press. Evolution de la diversité des faunes de mammifères d'Europe occidentale au Paléogène (MP 11 à MP 30). *Bull. Soc.*

géol. Fr.

Lopez, N. and Thaler, L. 1974. Sur le plus ancien lagomorphe européen et la "Grande Coupure" oligocène de Stehlin. *Palaeovertebrata* 6: 243-251.

Miller, K.S. and Curry, W.B. 1982. Eocene to Oligocene benthic foraminiferal isotopic record in the Bay of Biscay. *Nature* 296: 347-350.

Pomerol, C. and Premoli-Silva, I. 1986. The Eocene-Oligocene transition: events and boundary. In Pomerol, C. and Premoli-Silva. I., eds., *Terminal Eocene Events*. Amsterdam, Elsevier, pp. 1-24.

Premoli-Silva, I., Coccioni, R., and Montanari, A. 1988. *The Eocene-Oligocene boundary in the Marche-Umbria Basin (Italia)*. Ancona, Italy, *Intern. Union Geol. Sciences, Paleog. Strat. spec. Publ.* 1: 1-268.

Remy, J.A., Crochet, J.-Y., Sigé, B., Sudre, J., Bonis, L. de; Vianey-Liaud, M., Godinot, M., Hartenberger, J.-L., Lange-Badré, B., and Comte, B. 1987. Biochronologie des phosphorites du Quercy: mise à jour des listes fauniques et nouveaux gisements de mammifères fossiles. *Münchner Geowiss. Abh.* (A) 10: 169-188.

Russell, D.E., Hartenberger, J.-L., Pomerol, C., Sen, S., Schmidt-Kittler, N., and Vianey-Liaud, M. 1982. Mammals and stratigraphy: the Paleogene of Europe. *Palaeovertebrata, Mém. extra.* 1-77.

Savage, B. and Russell, D.E. 1983. *Mammalian palaeofaunas of the world,* Reading, Mass, Addison-Wesley, 432 pp.

Schmidt-Kittler, N. 1987. Comments of the editor. *Münchner Geowiss. Abh.* (A) 10: 15-16.

Schmidt-Kittler, N. et al. 1987. European reference levels and correlation tables. *Münchner Geowiss. Abh.* (A) 10: 13-31.

Sigé, B. 1977. Les insectivores et chiroptères du Paléogène moyen d'Europe dans l'histoire des faunes de mammifères sur ce continent. *Jour. Palaeont. Soc. India* 20: 178-190.

Sudre, J. 1980. *Aumelasia gabineaudi* n. g. n. sp. nouveau Dichobunidae (Artiodactyla, Mammalia) du gisement d'Aumelas (Hérault) d'âge lutétien terminal. *Palaeovertebrata, Mém. jubil. R. Lavocat* 197-211.

Stehlin, H.G. 1910. Remarques sur les faunules de mammifères de l'Eocène et de l'Oligocène du Bassin de Paris. *Bull. Soc. géol. Fr.* 4 (9): 488-520.

Vianey-Liaud, M. and Legendre, S. 1986. Les faunes des phosphorites du Quercy: principes méthodologiques en paléontologie des mammifères fossiles. *Eclogae Geol. Helv.* 79: 917-944.

27. THE CHINESE OLIGOCENE: A PRELIMINARY REVIEW OF MAMMALIAN LOCALITIES AND LOCAL FAUNAS

by Banyue Wang

ABSTRACT

The Chinese Oligocene is here subdivided into 11 horizons representing 3 subepochs: early Oligocene: 1. Caijiachong l.f., 2. Urtyn Obo l.f., 3. Baron Sog l.f., 4. Houldjin l.f.; middle Oligocene: 5. Kekeamu l.f., 6. Early Wulanbulage l.f., 7. Late Wulanbulage l.f.; late Oligocene: 8. Shargaltein l.f., 9. Taben Buluk l.f., 10. Yikebulage l.f., and 11. Suosuoquan l.f. The terminal Eocene event of the mammalian faunas in Asia is not so abrupt as in Europe and North America. There is also an extinction event at the boundary between early and middle Oligocene in the Asian mammalian faunas. During the Oligocene the exchange of mammalian faunas between both Asia and Europe and Asia and North America proceeded.

INTRODUCTION

The study of the Oligocene mammals of China began in 1922, when the Third Asiatic Expedition led by R. C. Andrews made the first discovery of Tertiary mammals from Houldjin beds near Iren Dabasu, Nei Mongol (Inner Mongolia), China. The beds were considered to be Oligocene by Osborn one year later. Very soon some other Oligocene localities, such as Ulan Gochu, Baron Sog and Urtyn Obo were discovered. At about the same time, exploring along the middle reach of the Yellow River, P. Teilhard de Chardin and E. Licent found Oligocene mammals near Saint Jacques, Nei Mongol. From the 1930s through the 1940s work that added significantly to our knowledge of the Chinese Oligocene was done by C. C. Young and M. Bien in Yunnan Province and the Sino-Swedish Expedition in Gansu Province. Since then, only scattered work on the Oligocene had been done until the late 1970s when another wave of intensive research started, resulting in finding a number of new Oligocene mammal-bearing localities. The major progress in this respect has been the discovery of micromammals by screenwashing, and recognition of a more or less complete Oligocene faunal sequence. The newly discovered taxa have provided new evidence on the subdivision and correlation of the Oligocene in China.

PRELIMINARY ANALYSIS OF OLIGOCENE MAMMALIAN FAUNAS

The Oligocene mammal-bearing sediments are mostly found in North China, i.e., in Nei Mongol, Xinjiang, Gansu, Ningxia, Shaanxi and Shanxi, and partly in south-western China (Yunnan, Guizhou, and Guangxi). All of them are continental. Most of them are composed of red mudstone, siltstone and partly of grayish green marl and sandstone of lacustrine and fluvial origin (Figure 27.1).

So far, 30 Oligocene mammalian local faunas (l.f.) have been reported. According to the assemblages of fossil mammals, the Oligocene of China can be subdivided into 11 horizons representing early, middle and late Oligocene.

Early Oligocene

The early Oligocene of China is characterized both by retention of ancient forms, such as Anagalidae, Coryphodontidae, Mesonychidae, Yuomyidae, Deperetellidae, Lophialetidae, Brontotheriidae, *Amynodon* and *Sianodon*, and occurrence of new taxa: primitive zapodids, *Ardynia, Cadurcodon, Entelodon, Miomeryx, Lophiomeryx*, etc. It may be subdivided into 4 horizons:

Early early Oligocene: Caijiachong local fauna

The Caijiachong l.f. was found in the Caijiachong Formation near Caijiachong village,

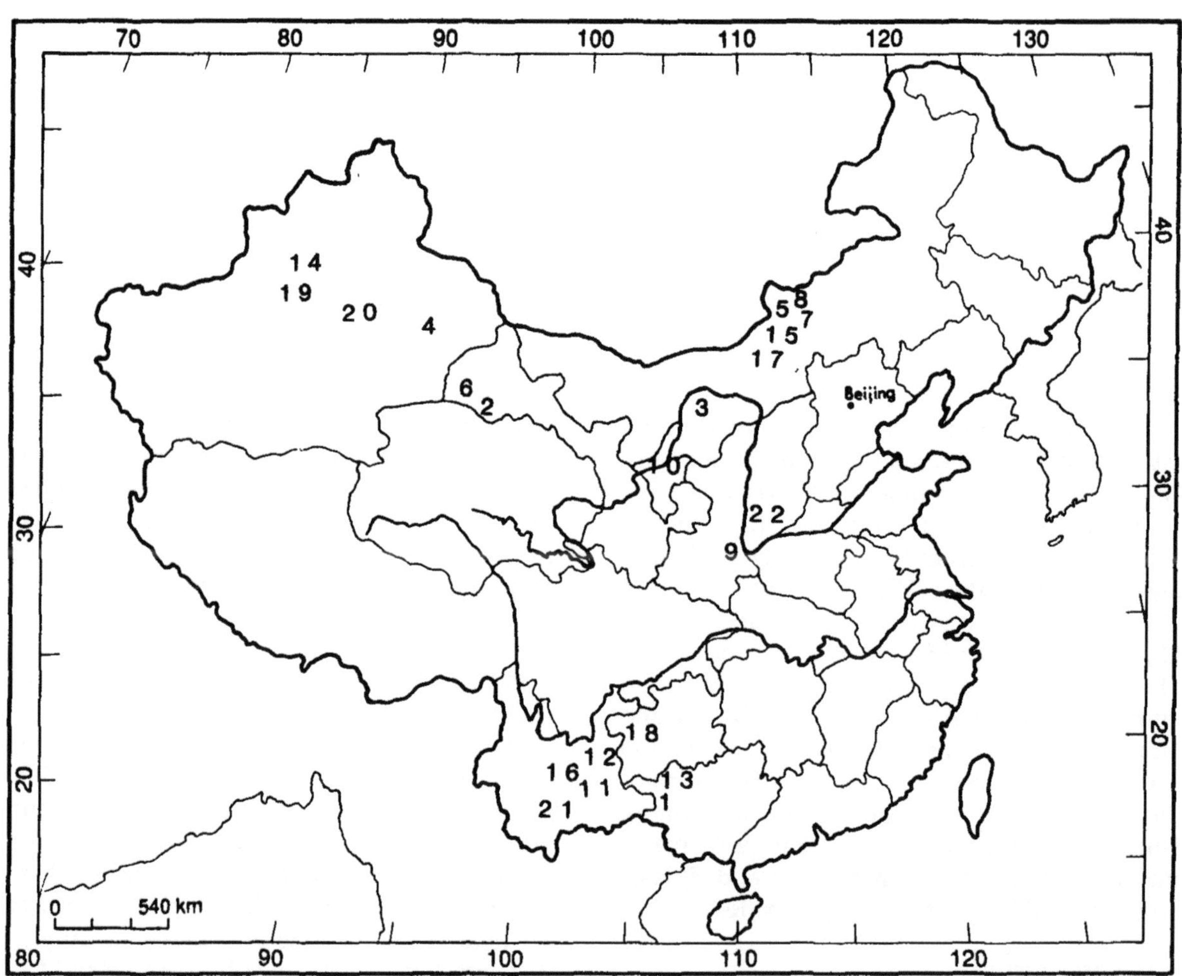

FIGURE 27.1. Principal Oligocene mammal-bearing localities in China: 1. Bose and Yongle Basin, 2. Danghe Area, 3. Dengkou, 4. Hami Basin, 5. Haosibuerdu Basin, 6. Hui-hui-p'u Area, 7. Irenhot, 8. Jilantai Basin, 9. Lantian Area, 10. Lingwu County, 11. Lunan Basin, 12. Luoping Basin, 13. Nanning Basin, 14. NorthZunggar Basin, 15. Qianlishan District, 16. Qujing County, 17. Shara Murun Area, 18. Shinao Basin, 19. South Zunggar Basin, 20. Turpan Basin, 21. Xuanwei Basin, 22. Yuanqu Basin.

Qujing county, Yunnan Province (Wang and Zhang, 1983). The fauna was considered to be equivalent to the Ardyn Obo l.f., Mongolian People's Republic (MPR), in age, on the basis of 7 shared genera. Having carefully examined the faunas, we found that in fact only one species, *Cadurcodon ardynensis*, was common to both faunas, all the others were different. Some species from the Caijiachong Formation show more primitive features than those of the Ardyn Obo local fauna. For example, *Lophiomeryx* sp. of the Caijiachong fauna is similar to *L. shinaoensis* and *L. gracilis* of the Shinao local fauna, Guizhou Province, but smaller and more primitive than *L. gobiae* and *L. angarae* from Ardyn Obo. *Entelodon* sp. from Caijiachong seems to be more primitive than *E. gobiensis* from Ardyn Obo. *Indomeryx* and *Prohyracodon*, two Eocene survivors in the Caijiachong fauna, have not been found in the Ardyn Obo l.f. In addition, *Eucricetodon meridionalis* and *E. leptaleos* represent the most primitive species in *Eucricetodon* known, and are more primitive than *E. atavus* of MP 21 in Europe. The zapodids from Caijiachong (*Heosminthus, Sinosminthus* and *Allosminthus*) are known to represent the most primitive members of the Zapodidae. It is possible that the Caijiachong local fauna is older than both the Ardyn Obo local fauna in Asia and MP 21 of Europe in age.

It may either be equivalent to the base of the Stampian or the top of the Priabonian. In view of the fact that it is more similar to the other early Oligocene than late Eocene faunas of Eurasia in sharing first appeared Oligocene genera (*Eucricetodon, Entelodon, Lophiomeryx* and *Bothriodon* for Europe and *Indricotherium, Cadurcodon* and *Miomeryx* in addition for Asia), it seems that the first alternative is more reasonable. The Caijiachong fauna is here considered the first horizon of the Asian early Oligocene fauna.

The Shinao local fauna of the Shinao Basin, Panxian county, Guizhou Province (Miao, 1982), the Xiaotun local fauna of the Lunan Basin, Yunnan Province (Zheng et al., 1978) and the Baishuicun local fauna of the Yuanqu Basin, Shanxi Province(Russell and Zhai, 1987, Wang and Hu, 1963), may be of the same age as the Caijiachong local fauna.

Middle early Oligocene: Urtyn Obo local fauna

This horizon is represented by the Ardyn Obo local fauna in MPR and by the Urtyn Obo local fauna in Shara Murun area, Nei Mongol in China(Russell and Zhai, 1987). The Urtyn Obo Fauna has 5 genera and species (*Cadurcodon ardynensis, Parabrontops gobiensis, Ardynia mongolensis, Schizotherium* cf. *avitum* and *Entelodon* sp.) in common with the Ardyn Obo local fauna. Since *Lophiomeryx gobiae* and *L. angarae* are smaller and more primitive than *L. mouchelini* of MP 22 of Europe, the Ardyn Obo and Urtyn Obo faunas are considered to be earlier than MP 22, possibly equivalent to MP 21 of Europe. Other formations or localities which may belong to this horizon are the Ulan Gochu Fm. in the Shara Murun Area(Osborn, 1929) and the Chaganbulage Fm. in the Haosibuerdu Basin, Nei Mongol(Qi, 1975, Qi, 1981), the Bailuyuan Fm. of the Lantian Area, Shaanxi Province (Zhang et al., 1978), the Yemaquan locality in the Hami Basin, Xinjiang (Hu, 1961), and, probably, the lower part of the Taoshuyuanzi Formation in the Turpan Basin, Xinjiang(Xu and Wang,1978, Zhai, 1978).

Late early Oligocene: Baron Sog local fauna

The Baron Sog Fauna was collected from the Baron Sog Formation of the Shara Murun Area, Nei Mongol (Osborn, 1929), and is composed of 3 species (*Embolotherium ultimum, Schizotherium avitum* and *Schizotherium* sp.). Since the Baron Sog Formation overlies the Ulan Gochu Formation, and *Embolotherium ultimum* represents a species more advanced than other species of *Embolotherium* (*E. andrewsi, E. grangeri* and *E. loucksii*) from the Ulan Gochu and Ardyn Obo formations, the age of the Baron Sog local fauna may be younger than that of the Ulan Gochu and Ardyn Obo faunas.

Latest early Oligocene: Houldjin local fauna

This fauna comes from the Houldjin Formation near Iren Dabasu, Irenhot, Nei Mongol(Matthew and Granger, 1923). Osborn (1923) considered the Houldjin Fauna as Oligocene. Later, some fossils collected by the Third Central Asiatic Expeditions (1930) and the Sino-Soviet Paleontological Expedition (1959) from the sandstone of Camp Margetts were also attributed to the Houldjin beds (Chow and Rozhendestrensky, 1960; Li and Ting, 1983; Osborn and Granger, 1923; Russell and Zhai, 1987). The mixture of the two samples has caused some confusion as to the age of the Houldjin Formation. Its Oligocene age assignment was once doubted. An Eocene age (equivalent to Irdin Manhan) was even suggested for it (Russell and Zhai, 1987). However, Radinsky (1964) stated, "From Granger and Morris' field notes, stratigraphic sequences at the various localities in the Camp Margetts area do not appear similar enough to one another or to the type section at the Houldjin and Irdin Manha escarpment to allow definite correlation on the basis of lithology." I agree with Radinsky. In fact, the fossils with Eocene hallmarks were all collected from Camp Margetts, 25 miles south-southwest of Iren Dabasu (Osborn and Granger, 1923; Qi, 1980, 1987; Tong, 1989). Therefore, the so-called "Houldjin beds" of Camp Margetts are in fact to be equivalent to the Irdin Manha Formation. The fossils from the typical Houldjin locality near Iren Dabasu are more advanced. Recently more fossil mammals were collected from the yellow gravel, the same level of the Houldjin Formation in the type locality area, Iren Dabasu, Nei Mongol. The new material includes *Paraceratherium* sp. nov., *Cadurcodon* sp., *Aprotodon* sp. nov., *Entelodon dirus*, Ctenodactyloidea gen. et sp. nov., Rhinocerotidae gen. et sp. indet. and Brontotheriidae gen. et sp. indet. Among the fauna

Cadurcodon, Ctenodactyloidea gen. et sp. nov. and brontotheriids support that the Houldjin Fauna is of early Oligocene age. On the other hand, *Entelodon dirus* is more progressive than *Entelodon gobiensis* of the Ardyn Obo, and both *Paraceratherium* sp. nov. and *Aprotodon* sp. nov. represent later and more advanced forms. One may doubt that the Houldjin fauna near Iren Dabasu is also a mixture of different ages. However, as far as the new material is concerned, except the ctenodactyloid and some specimens of *Cadurcodon* sp., all were collected from the same level of the same locality. It seems that the Houldjin Fauna near Iren Dabasu is of the same age and may represent the latest horizon of the early Oligocene.

Some other localities or formations, the Luoping locality (Chiu, 1962), and Xuanwei locality, Yunnan Province (Zhang, 1980), the Yongning Fm. in the Nanning Basin, Guangxi (Zhao, 1981), and, perhaps, the Shih–ehr–ma–ch'eng locality of the Hui–hui–p'u area, Gansu Province (Bohlin, 1951), may be of early Oligocene age. However their materials are too scarce for further precision.

The Gongkang local fauna collected from the Gongkang Formation in the Bose and Yongle Basins, Guangxi (Tang and Qiu, 1979), was considered as early Oligocene or late Eocene–early Oligocene. It consists of Nimravidae gen. nov., *Hoplophoneus*? sp., *Schizotherium nabanensis*, *Schizotherium* sp., *Forstercooperia* sp. nov., *Huananodon hypsodonta, Guixia youjiangensis, Anthracokeryx kwangsiensis, A. gungkangensis, Anthracokeryx* sp., "*Bothriodon*" *tientongensis, Heothema media, H. chengbiensis, H. angusticalxia* and *Odoichoerus unicornus*. As pointed out by Savage, Zhai, Tong and Ciochon (manuscript, 1983) and Russell and Zhai (1987, p. 238–241) the Gongkang local fauna is seriously confused with the Naduo local fauna collected from the underlying Naduo Fm., which is thought to be late Eocene. The Gongkang local fauna may be composed of some early Oligocene members. For example, *Schizotherium* was thought to be scarcely distinguishable from early Oligocene *Schizotherium avitum* of the Ardyn Obo fauna. However it is too difficult to distinguish further the Oligocene members from those of the Eocene ones.

Middle Oligocene

In this subepoch all the Eocene survivors still lingering in the early Oligocene became extinct, while some advanced forms, such as *Karakoromys, Tataromys, Tsaganomys, Cyclomylus, Parasminthus* and *Amphechinus*, appeared. The taxa restricted in this time span are *Ordolagus, Desmatolagus gobiensis, Cyclomylus lohensis, C. minutus, Cricetops dormitor, Karakoromys decessus* and *Selenomys mimicus*.

The Hsanda Gol local fauna of MPR is known to be one of the classic middle Oligocene faunas in Asia owing not only to its abundant fossils but also to its isotopic dating (Devyatkin and Smelov, 1979; Evernden et al., 1964). However, the specimens of the Hsanda Gol Fauna were collected from different sites and levels of the Hsanda Gol Formation. Mellett (1968, p. 9) suggested that there were three fossiliferous levels at Hsanda Gol. Kowalski (1974, p. 149) also said, "It seems that the remains found in this formation may be referred to separate periods."

The middle Oligocene of China was correlated with the Hsanda Gol local fauna as a whole and had not been subdivided for a long time. In 1981 Wang et al. discovered an Oligocene section in the Qianlishan District, Nei Mongol, that included the middle Oligocene Wulanbulage Formation and the late Oligocene Yikebulage Formation. The Wulanbulage Formation consists of upper and lower members containing two mammalian faunas. It reveals that the middle Oligocene of Asia can be subdivided into two horizons at least. Recently, Wang et al. (1991) found a new fauna (Kekeamu local fauna) in Alxa Juoqi (=Alxa Left Banner), Nei Mongol, which may be slightly older than the Wulanbulage Fauna. It seems that the middle Oligocene of China can be subdivided into three horizons. The Hsanda Gol Fauna may be equivalent to the whole Wulanbulage Fauna.

Early middle Oligocene: Kekeamu local fauna

This fauna was collected from the base of the Ulantatal Formation in the Jilantai Basin, Alxa Juoqi, Nei Mongol, just below the levels yielding the Ulantatal local fauna. It includes the typical middle Oligocene forms, *Karakoromys decessus, Prosciurus* sp., *Tupaiodon* sp., etc. Among them *Karakoromys decessus* is abundant. The fauna is distinguished from other Asian middle Oligocene faunas by retention of some early Oligocene mammals, such as *Heosminthus* sp. and *Ardynia mongoliensis*, as well as in lacking *Tataromys*, which is abundant in the Hsanda Gol, Wulanbulage, Ulantatal and Saint Jacques local faunas. It seems that the Kekeamu Fauna is older than those faunas.

Middle middle Oligocene: Early Wulanbulage local fauna

The fauna was collected from the lower member of the Wulanbulage Formation in the Qianlishan District, Nei Mongol (Wang et al., 1981). In this horizon *Cadurcodon ardynensis* continued to exist. This fauna contains typical middle Oligocene taxa, such as *Desmatolagus gobiensis*, *Cyclomylus lohensis* and *C. minutus*. It differs from the Kekeamu local fauna in having *Tataromys*, which made its first appearance, and *Parasminthus* replacing *Heosminthus*.

The Qingshuiying (Young and Chow, 1956) and Shuidonggou (Teilhard de Chardin, 1926) local faunas, Lingwu county, Ningxia, may be equivalent either to the whole Wulanbulage fauna or a part of it in age.

Late middle Oligocene: Late Wulanbulage local fauna

The late Wulanbulage Fauna was collected from the upper member of the Wulanbulage Formation of the Qianlishan District, Nei Mongol (Wang et al., 1981). The other main localities are Ulantatal of the Jilantai Basin, Alxa Zuoqi, Nei Mongol (Huang, 1982), Saint Jacques near Dengkou, Nei Mongol (Teilhard de Chardin, 1926) and Wu-tao-ya-yu in the Danghe Area, Gansu Province (Bohlin, 1946). In this horizon, besides retention of most of the middle middle Oligocene taxa, some advanced forms occur: *Amphechinus*, *Ordolagus*, *Sinolagomys*, *Leptotataromys*, *Tataromys sigmodon*, *Selenomys mimicus*, and *Parasminthus asiae-centralis*. *Desmatolagus pusillus* and *D.* cf. *gobiensis*, are more progressive than *D. gobiensis* of middle middle Oligocene in having the premolar foramen and mental foramen situated more posteriorly and more hypsodont cheek teeth.

Kowalski (1974) recognized two rodent faunas from the lower and upper parts of the profile in Tatal Gol, the richest locality of the Hsanda Gol Fauna. If it is true it seems that the fauna from the lower part may be mostly equivalent to the late Wulanbulage local fauna and the fauna from the upper one may be equivalent to the later period of the late Wulanbulage Fauna or may be even later than the latter.

Late Oligocene

The mammal assemblages of this subepoch can be distinguished from those of the middle Oligocene by the extinction of the primitive and typical middle Oligocene forms, such as *Desmatolagus gobiensis*, *Ordolagus*, *Prosciurus*, *Cyclomylus*, *Ardynomys*, *Pseudocylindrodon*, *Cricetops*, *Selenomys*, and *Karakoromys*. At the same time *Amphechinus minimus*, *Sinolagomys gracilis*, Sciuridae, *Yindirtemys* and *Dzungariotherium* occur.

Early late Oligocene: Shargaltein local fauna

The Shargaltein local fauna was collected from Shargaltein Gol in the Danghe Area, Gansu Province (Bohlin, 1937). While differing from the middle Oligocene in the characters already mentioned above it shares with the middle Oligocene fauna such animals as *Palaeoscaptor* cf. *acridens*, *Desmatolagus pusillus*, *D. robustus*, *Leptotataromys*. They made their last appearance in this horizon.

Middle late Oligocene: Taben Buluk local fauna

The typical locality yielding the fauna is Taben Buluk in the Danghe Area, Gansu Province (Bohlin, 1942, 1946). In this fauna the middle Oligocene survivors which were seen in the Shargaltein Fauna as mentioned above disappeared completely, and some new taxa, including *Parasminthus parvulus*, *Amphechinus minimus*, *Yindirtemys* and *Tataromys grangeri*, took their place. The two latter ones are restricted to this horizon. However, no Miocene or later forms occur during this time. The fauna collected from the upper part of the Taoshuyuanzi Formation in the Turpan Basin, Xinjiang, may be equivalent to or earlier than the Taben Buluk Fauna in age.

Late late Oligocene: Yikebulage local fauna

The represented fauna was collected from

the Yikebulage Formation in the Qianlishan District, Nei Mongol (Wang et al., 1981). In addition to the typical late Oligocene taxa in common with the Taben Buluk Fauna, it is marked by the occurrence of some more advanced and Miocene forms, *Tachyoryctoides kokonorensis, Tataromys suni* and *Distylomys qianlishanensis*, which have been also found in the early Miocene of China (Xiejia and Urtu).

Latest late Oligocene: Suosuoquan local fauna

The Suosuoquan local fauna was collected in the North Zunggar Basin, Xinjian (Tong et al., 1990). This fauna consists mainly of micromammals. So far only three species have been reported: *Sinolagomys ulungurensis*, *Prodistylomys xinjiangensis* and *Tachyoryctoides* sp. Among them *Sinolagomys ulungurensis* represents the latest and most progressive species of *Sinolagomys*, more advanced than *S. kansuensis* of Taben Buluk and Yikebulage, and continued to live in the early Miocene (MN 3). *Distylomys* is also a genus persisting into the medial Miocene (Tung Gur). Besides, *Desmatolagus* disappeared in this period. It seems that the Suosuoquan Fauna is later than the Taben Buluk and Yikebulage faunas in age and may be latest Oligocene or even later.

The "Brown" Formation in the South Zunggar Basin, Xinjiang (Chiu, 1965, 1973), yielded two macromammals, *Dzungariotherium turfanensis* and *Lophiomeryx* sp. It is difficult to correlate the Suosuoquan Fauna with the Fauna from the "Brown" Formation owing to the lack of common animals. However, both *Dzungariotherium turfanensis* and *Lophiomeryx* sp. are more progressive. *Lophiomeryx* sp. is the largest and morphologically very progressive in comparison with all the *Lophiomeryx* species known, *L. chalaniati* of MP 28 in Europe included. The Fauna from the "Brown" Formation may be later than MP 28 and equivalent to MP 29 or MP 30 of Europe in age.

CONCLUSION

So far more than 80 genera representing 44 families of 11 orders of the Oligocene mammals from 30 faunas have been reported in China. The Oligocene mammalian local faunas in China are chronologically ordered. The Chinese Oligocene is further subdivided into 11 horizons, representing three subepochs.

At present the viewpoints on the Eocene/Oligocene boundary are different among the paleontologists and geologists. Some thought it is between the Latdorfian and the Bartonian *sensu lato* just in the lower part of the Priabonian; while others placed the Latdorfian in the late Eocene, in which case the Oligocene would begin with base of the Stampian (=Rupelian), after the Priabonian and the "Grande Coupure" (Berggren et al., 1985; Russell et al., 1982; Schmidt-Kittler, 1987). Recently the latter viewpoint has been adopted during the 28th International Geological Congress in Washington, D. C. (1989).

As far as the Asian Oligocene mammal faunas (including faunas not only from China, but also from MPR and Central Asia) are concerned, the Caijiachong, Urtyn Obo, Baron Sog, Houl-djin and Ardyn Obo faunas are correlative with the lower Rupelian (=lower Stampian) of Europe in having a series of common first appearances (*vide supra*). Middle Oligocene of Asia (Kekeamu, early Wulanbulage and late Wulanbulage faunas) may be equivalent to the upper Stampian of Europe, while late Oligocene of Asia to the Chattian of Europe.

Primarily based on the isotopic dating, Swisher and Prothero recently (1990) placed the Eocene/Oligocene boundary between the Chadronian and Orellan in North America. However, the chronometric calibration of the geologic time scale is still a difficult and much debated problem. Geological opinions differ about the dating of the Eocene/Oligocene boundary, with estimates ranging from 34–37 Ma (Berggren et al., 1985; Russell et al., 1982; Schmidt-Kittler, 1987). Even in the Apennine Mountains of Italy, the data are still controversial. For example, in the Contessa section an age of 35.4 Ma has been obtained from the base of Oligocene, while the date of the top of the Eocene in the Massignano section is younger (34 Ma) (Deino et al., 1988; Montanari et al., 1988; Odin and Montanari, 1988). If we choose 36 or 37 Ma as the boundary age most of the Chadronian should still remain in the early Oligocene. Second, the correlation of the pelagic Massignano section with the shallow water sequence of the type Priabonian is still problem. Furthermore, some elements of the Chadronian fauna, such as *Plesictis, Eusmilus, Trigonias, Adjidaumo*

(=*Eomys*) and *Plesispermophilus*, occurred in the early Stampian rather than in the Priabonian in Europe. Judging from the mammals, it seems better now to keep the major part of the Chadronian in the early Oligocene. [*Editors' note:* for different conclusions, see Chapters 1 and 2 of this volume].

The Asian Oligocene mammalian faunas, like the European and North American ones, have much more advanced appearance than those of the Eocene. During the transition from the late Eocene to the early Oligocene a great number of the archaic forms became extinct and many new taxa appeared. The Terminal Eocene Event occurs in Asia as well. However, the change of the Asian mammalian faunas is not so abrupt as in Europe and North America. It has such characteristics as following:

1) The extinction of the archaic mammal forms occurs step by step. It is known that one order (Tillodonta) and 3 families (Eurymylidae, Arctocyonidae and Helohyidae) and 45 genera became extinct in the late Eocene. Among them one family and 13 genera became extinct at the end of the Naduan age and others disappeared earlier, at the end of the Sharamurunian age. Seven archaic families of mammals (see below) were not extinct until the end of the early Oligocene. Some of them, such as Condylarthra, survived in Asia longer than in Europe and North America.

2) Some archaic families, such as the Brontotheriidae, not only lingered but also diversified in the early Oligocene. Five new genera of the Brontotheriidae are known to occur in that period.

3) The advanced forms appear gradually. Some new families including Entelodontidae and the living Cricetidae, Ochotonidae, etc., made their first appearance in the late Eocene in Asia, earlier than in Europe and North America. One living family (Castoridae) appeared in the early Oligocene. Among 31 genera, 11 occurred in the early early Oligocene, 18 in the middle early Oligocene and 2 in the latest early Oligocene.

4) The Asian Oligocene mammals are mostly endemic, few appear to have migrated from other continents. It seems probable that the replacement of the largely endemic Asian mammalian faunas from the Eocene to Oligocene is more a response to the change of the climate than to the barriers between the continents (Cavelier et al., 1981).

An extinction event in the Asian mammalian faunas appeared at the boundary between the early Oligocene (Houldjinian) and middle Oligocene (Kekeamuan). The faunal change across this boundary was the most conspicuous in the Oligocene. By the beginning of the middle Oligocene a number of archaic mammals, including Anagalida (Anagalidae), Condylarthra (Mesonychidae), Brontotheriidae, Deperetellidae, Lophialetidae and Yuomyidae, were extinct. At first, the Ctenodactylidae with more crested cheek teeth appeared. Then, the high crowned tsaganomyids and more lophodont tachyoryctoidids made their first appearance in the middle Oligocene. There was some diversification of cricetids and ctenodactylids. The two cricetid genera appearing at that time, *Cricetops* and *Selenomys*, had more crested and higher crowned cheek teeth. The ctenodactylids had split into 5 genera (*Karakoromys, Tataromys, Leptotataromys, Woodomys* and *Terarboreus*).

The changes of the paleogeographic configuration and the paleoclimatical conditions affected the mammalian relationships between continents. From the beginning of the Oligocene, with the retreat of the Turgai Strait, the barrier of exchange of the terrestrial mammals between Asia and Europe gradually disappeared. More and more mammals migrated from Asia into Europe. These include Lagomorpha, Cricetidae, Zapodidae, Rhinocerotidae, Chalicotheriidae, Entelodontidae, and others. It caused the great change in the composition of the mammalian faunas in Europe, known as the "*Grande Coupure.*" More genera common to Asia and Europe occurred. *Eucricetodon, Hyaenodon, Entelodon* and *Lophiomeryx* continued to inhabit both continents throughout Oligocene. The genera common to Asia and Europe during the early Oligocene include *Ronzotherium* and *Bothriodon*. In the middle Oligocene Asia and Europe shared *Eggysodon, Aceratherium, Schizotherium, Prodremotherium* and *Amphicynodon*. During the late Oligocene both continents possessed such common genera as *Aceratherium, Prodremotherium, Steneofiber, Cephalogale, Plesictis, Brachypotherium* and *Amphitragulus*.

As far as the relationships between Asia

and North America are concerned, it was previously thought that the exchange of the Oligocene mammalian faunas between the two continents declined. However, recent discoveries have revealed that their exchange went on throughout Oligocene. For example, apart from *Hyaenodon* and *Colodon*, which were shared by Asia and North America during the whole Oligocene, genera common to both areas in the early Oligocene include *Ardynomys*, *Metamynodon* and *Bothriodon*. *Amynodontopsis*, a late Eocene genus of North America, has been cited by Wall (1980) in the early Oligocene Urtyn Obo and Ulan Gochu formations. Its occurrence in Asian Oligocene deposits is rather problematic and should be further verified. In the middle Oligocene the number of the genera common to both continents increased. They were *Desmatolagus*, *Palaeocastor*, *Agnotocastor*, *Prosciurus* and *Paleogale*. *Desmatolagus* may have migrated from Asia into North America, while *Palaeocastor*, *Agnotocastor* and *Prosciurus* migrated in the opposite direction. *Pseudocylindrodon* of the early Oligocene of North America may have come to Asia in the middle Oligocene as well. In the late Oligocene, mammals shared by the two areas were *Agnotocastor*, *Nimravus* and *Bothriodon*. In addition, the late Oligocene *Haplomys* and *Meniscomys* of North America may have derived from Asian middle Oligocene *Haplomys* and *Promeniscomys*. It seems that the mammalian fauna exchange between both Asia and Europe, and Asia and North America proceeded throughout Oligocene.

The study of Chinese Oligocene mammals has just begun, although our knowledge of Oligocene has been greatly increased. New localities and abundant faunas are much needed. Intensive investigation of the localities and taxa already known is also needed. It is hoped that this paper will offer a few generalizations that can be drawn from the work that has been done to date.

ACKNOWLEDGEMENTS

The author is gratefully indebted to Dr. Robert J. Emry of National Museum of Natural History, Smithsonian Institution, for his manifold help during her stay at the Smithsonian Institution and for his revising this manuscript. The author is much obliged to Dr. Richard H. Tedford of American Museum of Natural History and Dr. Zhan–xiang Qiu of the Institute of Vertebrate Paleontology and Paleoanthropology, Academia Sinica, and Dr. Mary R. Dawson of the Carnegie Museum for their valuable remarks and information during her preparation of the manuscript. The author feels especially grateful to Dr. Malcolm C. McKenna of AMNH, Prof. Su–yin Ting of IVPP and Dr. Spencer G. Lucas of the New Mexico Museum of Natural History for their critical review of the manuscript. Special thanks are also due to Mr. Dan Chaney and Mrs. Mary Parrish (NMNH, Smithsonian Institution) for their help in preparing the tables and map.

REFERENCES

Berggren, W. A., Kent, D. V., and Flynn, J. J. 1985. Jurassic to Paleogene: Part 2: Paleogene geochronology and chronostratigraphy. In: Snelling, ed. *The Chronology of the Geological Record. Geol. Soc. London Memoir* 10: 141–177.

Bohlin, B. 1937. Oberoligozäne Saugetiere aus dem Shargaltein–tal (Western Kansu). *Palaeont. Sinica N. S. C*, 3: 1–66.

Bohlin, B. 1942. The fossil mammals from the Tertiary deposit of Taben–Buluk, western Kansu. Part 1: Insectivora and Lagomorpha, *Palaeont. Sinica N. S. C* 8a: 1–113.

Bohlin, B. 1946. The fossil mammals from the Tertiary deposit of Taben–Buluk, Western Kansu. Part 2: Simplicidentata, Carnivora, Perissodactyla and Primates. *Palaeont. Sinica N. S. C.* 8b: 1–259.

Bohlin, B. 1951. Some mammalian remains from Shih–ehr–ma–ch'eng, Hui–hui–p'u area, Western Kansu. Reports from the Scientific Expedition to the north–western Provinces of China under leadership of Dr. Seven Hedin. *The Sino–Swedish Expedition, Publ.* 35, VI Vertebrate Paleontology, 5: 1–47. Stockholm.

Cavelier, C., Chateauneuf, J.–J., Pomerol, C., Rabussier, D., Renard, M., and Vergnaud-Grazzini, C. V. 1981. The geological events at the Eocene/Oligocene boundary. *Palaeogeogr., Palaeoclimat., Palaeoecol.* 36: 223–248.

Chiu, C. 1962. Giant rhinoceros from Loping, Yunnan, and discussion on the taxonomic characters of *Indricotherium grangeri*. *Vert.*

PalAs. 6(1): 57–71.

Chiu, C. 1965. First discovery of *Lophiomeryx* in China. *Vert. PalAs.* 9(4): 395–398.

Chiu, C. 1973. A new genus of giant rhinoceros from Oligocene of Dzungaria, Sinkiang. *Vert. PalAs.* 11(2): 182–191.

Chow, M. and Rozhdestrensky, A. K. 1960. Exploration in Inner Mongolia--A preliminary account of the 1959 field work of the Sino-Soviet Paleontological Expedition (SSPE). *Vert. PalAs.* 4(1): 1–10.

Deino, A. L., Drake, R. E., Curtis, G. H., and Montanari, A. 1988. Preliminary laser-fusion $^{40}Ar/^{39}Ar$ dating results from Oligocene biotites of the Contessa Quarry section (Gubbio, Italy). In Premoli-Silva, I., Coccioni, R., and Montanari, A., eds., *The Eocene–Oligocene boundary in the Marche-Umbria Basin (Italy).* Intern. Subcom. Paleogene Stratig. Ancona(Italy), pp. 229–238.

Devyatkin, Y. V. and Smelov, S. B. 1979. Position of basalts in the Cenozoic sedimentary sequence of Mongolia. *AN SSSR Izvestya ser. geol.* 1: 16–29.

Evernden, J. F., Savage, D. E., Curtis, G. H., and James, G. J. 1964. Potassium–argon dates and the Cenozoic mammalian chronology of North America. *Amer. Jour. Sci.* 262: 145–198.

Hu, C. 1961. The occurrence of *Parabrontops* in Hami, Sinkiang. *Vert. PalAs.* 5(1): 41–42.

Huang, X. 1982. Preliminary observations on the Oligocene deposits and mammalian fauna from Alashan Zuoqi, Nei Monggol. *Vert. PalAs.* 20(4): 337–349.

Kowalski, K. 1974. Middle Oligocene rodents from Mongolia. Results of the Polish–Mongolian Palaeontological Expeditions, Part V. *Palaeont. Polonica* 30: 1–178.

Li, C. and Ting, S. 1983. The Paleogene mammals of China. *Bull. Carnegie Mus. Nat. Hist.* 21: 1–93.

Matthew, W. D. and Granger, W. 1923. The fauna of the Houldjin Gravel. *Amer. Mus. Novitates* 97: 1–6.

Mellett, J. S. 1968. The Oligocene Hsanda Gol Formation, Mongolia: A revised faunal list. *Amer. Mus. Novitates* 2318: 1–16.

Miao, D. 1982. Early Tertiary fossil mammals from the Shinao Basin, Panxian county, Guizhou Province. *Acta Palaeont. Sinica* 21(5): 526–536.

Montanari, A., Deino, A. L., Drake, R. E., Turrin, B. D., DePaolo, D. J., Odin, G. S., Curtis, G. H., Alvarez, W., and Bice, D. M. 1988. Radioisotopic dating of the Eocene–Oligocene boundary in the Pelagic sequence of the Northeastern Apennines. In Premoli-Silva, I., Coccioni, R., and Montanari, A., eds. *The Eocene–Oligocene boundary in the Marche–Umbria Basin (Italy),* Intern. Subcom. Paleogene Stratig. Ancona (Italy), pp. 195-208.

Odin, G. S. and Montanari, A. 1988. The Eocene–Oligocene boundary at Massignano (Ancona, Italy): A potential boundary stratotype. In Premoli-Silva, I., Coccioni, R., and Montanari, A., eds. *The Eocene–Oligocene boundary in the Marche–Umbria Basin (Italy),* Intern. Subcom. Paleogene Stratig. Ancona (Italy), pp. 253–263.

Osborn, H. F. 1923. *Cadurcotherium* from Mongolia. *Amer. Mus. Novitates* 92: 1–2.

Osborn, H. F. 1929. *Embolotherium*, gen. nov., of the Ulan Gochu, Mongolia. *Amer. Mus. Novitates* 353: 1–20.

Osborn, H. F. and Granger, W. 1923. Coryphodonts and uintatheres from the Mongolian Expedition of 1930. *Amer. Mus. Novitates* 552: 1–16.

Qi, T. 1975. An early Oligocene mammalian fauna of Ningxia. *Vert. PalAs.* 13(4): 215–224.

Qi, T. 1980. Irdin Manha Upper Eocene and its mammalian fauna of Huhebolhe Cliff in Central Inner Mongolia. *Vert. PalAs.* 18(1): 28–32.

Qi, T. 1981. New materials of the early Oligocene Chaganbulage fauna from Alxa Zuoqi, Inner Mongolia. *Vert. PalAs.* 19(2): 145–148.

Qi, T. 1987. The Middle Eocene Arshanto fauna (Mammalia) of Inner Mongolia. *Ann. Carnegie Mus.* 56(1): 1–73.

Radinsky, L. B. 1964. Notes on Eocene and Oligocene fossil localities in Inner Mongolia. *Amer. Mus. Novitates* 2180: 1–11.

Russell, D. E., Hartenberger, J. -L., Pomerol, Ch., Sen, S., Schmidt-Kittler, N., and Vianey-Liaud, M. 1982. Mammals and stratigraphy: The Paleogene of Europe. *Paleovertebrata Mém. extr:* 1–77.

Russell, D. E. and Zhai, R. 1987. The Paleogene of Asia: mammals and stratigraphy. *Mem.*

Mus. Nat. Hist. Natur. Sci. Terre 52: 1–488.

Schmidt-Kittler, N., ed. 1987. International Symposium on mammalian biostratigraphy and paleoecology of the European Paleogene. Mainz, February 18th–21st 1987. *Münchner Geowiss. Abh. Reihe* A, 10: 1–312.

Swisher, C. C. and Prothero, D. R. 1990. Single–crystal $^{40}Ar/^{39}Ar$ dating of the Eocene–Oligocene transition in North America. *Science* 249: 760–762.

Tang, Y. and Qiu, Z. 1979. Vertebrate faunas of Baise, Guanxi. In Inst. Vert. Paleont. Paleoanthr. and Nanking Inst. Geol. Paleont. Acad. Sinica, eds., *Mesozoic–Cenozoic red beds of South China*, Science Press, pp. 407–415.

Teilhard de Chardin, P. 1926. Descriptions des mammiferes Tertiares de Chine et de Mongolie. *Ann. Paleont.*15: 1–52.

Tong, Y. 1989. A review of middle and late Eocene mammalian faunas from China. *Acta Palaeont. Sinica* 28(5): 663–682.

Tong, Y., Qi, T., Ye, J., Meng, J., and Yan, D. 1990. Tertiary stratigraphy of the North of Junggar Basin, Xinjiang. *Vert. PalAs.* 28(1): 59–70.

Wall, W. P. 1980. Cranial evidence for a proboscis in *Cadurcodon* and a review of snout structure in the family Amynodontidae (Perissodactyla, Rhinocerotoidea). *Jour. Paleont.* 54 (5): 968–977.

Wang, B., Chang, J., Meng, X., and Chen, J. 1981. Stratigraphy of the upper and middle Oligocene of Qianlishan district, Nei Mongol (Inner Mongolia). *Vert. PalAs.* 19(1): 26–34.

Wang, B. and Wang P. 1991. Discovery of early middle Oligocene mammalian fauna from Kekeamu, Alxa Left Banner, Nei Mongol. *Vert. PalAs.* 29(1): 64–71.

Wang, B. and Zhang, Y. 1983. New finds of fossils from Paleogene of Qujing, Yunnan. *Vert. PalAs.* 21(2): 119–128.

Wang, T. and Hu, C. 1963. An Oligocene mammalian horizon in the Yuanchu Basin, South Shansi. *Vert. PalAs.* 7(4): 357–351.

Xu, Y. and Wang, J. 1978. New materials of giant rhinoceros. *Mem. Inst. Vert. Paleont. Paleoanthr.* 13: 132–140.

Young, C. C. and Chow, M. 1956. Some Oligocene mammals from Lingwu, N. Kansu. *Acta Palaeont. Sinica* 4(4): 447–459.

Zhai, R. 1978. Late Oligocene mammals from the Taoshuyuanzi Formation of Eastern Turfan Basin. *Mem. Inst. Vert. Paleont. Paleoanthr. Acad. Sinica* 13: 126–131.

Zhang, Y. 1980. A small indricothere from Xuanwei, Yunnan. *Vert. PalAs.* 18(4): 348.

Zhang, Y., Huang, W., Tang, Y., Ji. H., You, Y., Tong, Y., Ding, S., Huang, X., and Zheng J. 1978. The Cenozoic of the Lantian Region, Shaanxi Province. *Mem. Inst. Vert. Paleont. Paleoanthr. Acad. Sinica* 14: 1–64.

Zhao, Z. 1981. The vertebrate fossils and lower Tertiary from Nanning Basin, Guangxi. *Vert. PalAs.* 19(3): 218–227.

Zheng, J., Tang, Y., Zhai, R., Ding, S., and Huang, X. 1978. Early Tertiary strata of Lunan Basin, Yunnan. *Prof. Papers Stratigr. Paleont.* 7: 22–29.

TABLE 27.1. List of early Oligocene mammals of China.

	Cj	Sn	Xt	Bc	Uo	Ug	Cg	Ym	Bl	Ts	Bs	Hd	Lp	Xw	Nn	Sh
Insectivora																
Dormaaliidae indet.	X															
Erinaceidae indet.	X															
Insectivora indet.	X															
Chiroptera																
Vespertilionoidea indet.	X															
Primates? indet.	X															
Lagomorpha																
Leporidae																
Gobiolagus andrewsi						X										
Gobiolagus? major					X											
Ochotonidae																
Desmatolagus vetustus						X										
Largomopha indet.	X						X		X							
Lagomorpha?																
Mimotonidae?																
Mimolagus rodens																X
Rodentia																
Paramyidae																
Hulgana ertnia						X										
Paramyidae? indet.						X										
Cylindrodontidae.																
Ardynomys sp.	X					X										
Cricetidae																
Eucricetodon meridionalis	X															
E. leptaleos	X															
Yuomyidae.																
Dianomys obscuratus	X															
Dianomys qujingensis	X															
Ctenodactiloidea																
gen. et sp. n.												X				
Dipodidae																
Allosminthus ernos	X															
Allosminthus majusculus	X															
Sinosminthus inapertus	X															
Heosminthus primiveris	X															
Anagalida																
Anagalidae.																
Anagale gobiensis						X										
Anagalopsis kansuensis																X

TABLE 27.1. (continued)

	Cj	Sn	Xt	Bc	Uo	Ug	Cg	Ym	Bl	Ts	Bs	Hd	Lp	Xw	Nn	Sh
Condylarthra																
Mesonychidae																
Mongolestes hadrodens						X										
Harpagolestes alxaensis							X									
Mesonychidae indet.					X											
Perissodactyla																
Brontotheriidae																
Metatitan progressus						X										
Metatitan primus						X										
Parabrontops gobiensis					X											
Parabrontops sp.								X								
Embolotherium loucksii						X										
Embolotherium grangeri						X	X									
Embolotherium andrewsi						X										
Embolotherium ultimum											X					
Pygmaetitan panxianensis		X														
Brontotheriidae indet.	X											X				
Equidae																
Qianohippus magicus		X														
Chalicotheriidae																
Schizotherium avitum											X					
Schizotherium cf. *S. avitum*					X											
Schizotherium sp.											X					
Deperetellidae																
Teleolophus magnus							X									
Teleolophus cf. *T. medius*							X									
Lophialetidae																
Simplaletes xianensis									X							
Hyracodontidae																
Prohyracodon sp.	X															
Caenopinae indet.	X											X				
Hyracodontidae indet.		X	X													
Indricotheriidae																
Urtinotherium incisivum					X											
Indricotherium parvum			X													
Indricotherium intermedium	X												X			
Indricotherium qujingensis	X															
Indricotherium sp.	X													X		
Paraceratherium sp. n.												X				
Rhinocerotidae																
Aprotodon sp. n.												X				
Rhinocerotidae indet.	X											X				

TABLE 27.1. *(continued)*

	Cj	Sn	Xt	Bc	Uo	Ug	Cg	Ym	Bl	Ts	Bs	Hd	Lp	Xw	Nn	Sh
Amynodontidae																
Cadurcodon ardynensis	X				X					X						
Cadurcodon sp.	X				X	X						X				
Gigantamynodon giganteus	X															
Gigantamynodon cf. *G. giganteus*	X															
cf. *Gigantamynodon giganteus*			X													
Gigantamynodon sp.	X															
Amynodon alxaensis					X		X									
Amynodontopsis parvidens					X											
Amynodontopsis sp.						X										
Sianodon bahoensis									X							
Sianodon sp.							X		X							
Zaisanamynodon? sp.						X										
Paracadurcodon suhaituensis							X									
cf. *Metamynodon* sp.	X															
Amynodontidae indet.						X	X									
Artiodactyla																
Entelodontidae.																
Entelodon dirus												X				
Entelodon sp.	X				X											
Entelodontidae indet.							X									
Anthracotheriidae																
Heothema nanningensis															X	
Heothema youngi															X	
Bothriogenys hui			X	X												
Bothriodon chowi	X															
Leptomerycidae																
Miomeryx sp.	X															
Cervidae indet.							X									
Gelocidae.																
Lophiomeryx shinaoensis		X														
Lophiomeryx gracilis		X														
Lophiomeryx gracilis?		X														
Lophiomeryx sp.	X															
cf. *Indomeryx* sp.	X															
Bovidae indet.							X									
Family indet.																
Lantianius xiehuensis									X							
Artiodactyla indet.			X													

Bc: Baishuicun; Bl: Bailuyuan; Bs: Baron Sog; Cg: Chaganbulage; Cj: Caijiachong; Hd: Houldjin; Lp: Luoping; Nn: Nanning; Sh: Shih-ehr-ma-ch'eng; Sn: Shinao; Ts: Taoshuyuanzi; Ug: Ulan Gochu; Uo: Urtyn Obo; Xt: Xiaotun; Xw: Xuanwei; Ym: Yemaquan.

TABLE 27.2. List of middle Oligocene mammals of China.

	Kk	Ew	Lw	Ul	Sj	Wt	Qs	Sd
Insectivora								
Erinaceidae								
Palaeoscaptor acridens				X				
Amphechinus rectus				X	X			
Amphechinus cf. *A. rectus*				X				
Tupaiodon sp.	X							
Insectivora indet.			X					
Lagomorpha								
Leporidae								
Ordolagus teilhardi				X	X			
Leporidae indet.		X	X					
Ochotonidae								
Desmatolagus gobiensis		X	X		X			
Desmatolagus cf. *D. gobiensis*				X				
Desmatolagus pusillus				X	X	X		
Desmatolagus robustus					X			
Desmatolagus sp.	X							
Sinolagomys major				X				
Sinolagomys cf. *S. major*					X			
Sinolagomys kansuensis				X				
Rodentia								
Tsaganomyidae								
Tsaganomys altaicus					X	X		
Tsaganomys sp.			X					X
Cyclomylus lohensis		X	X	X	X		X	
Cyclomylus minutus		X						
cf. *Cyclomylus minutus*				X				
Cyclomylus sp.		X						
Cylindrodontidae								
Ardynomys sp.	X			X				
Anomoemys lohiculus					X			
Aplodontidae								
Haplomys arboraptus					X			
Prosciurus ordosicus					X			
Prosciurus sp.	X							
Promeniscomys sinensis					X			
Cricetidae								
Cricetops dormitor			X		X			
Cricetops minor					X			
Eucricetodon caducus					X			
Eucricetodon sp.	X							
Selenomys mimicus				X	X			
Cricetidae indet.				X				

TABLE 27.2. (continued)

	Kk	Ew	Lw	Ul	Sj	Wt	Qs	Sd
Rodentia cont.								
Ctenodactylidae.								
Tataromys sigmodon			X		X			
Tataromys cf. *T. sigmodon*				X				
Tataromys cf. *T. plicidens*					X			
Tatatomys ulantatalensis				X				
Tataromys bohlini				X				
Tataromys deflexus			X		X			
Tataromys sp.		X	X			X		
Karakoromys decessus	X	X		X	X			
Karakoromys sp.		X						
Leptotataromys gracilidens				X	X	X		
Leptotataromys cf. *L. gracilidens*				X				
Leptotataromys minor				X				
Dipodidae.								
Parasminthus tangingoli		X						
cf. *Parasminthus tangingoli*				X				
Parasminthus asiae-centralis			X					
Parasminthus sp.			X					
Heosminthus sp.	X							
Carnivora								
Miacidae indet.		X						
Canidae								
Cynodictis? sp.				X				
Amphicyonidae								
Amphicyon sp.			X					
Amphicyon? sp.					X			
Mustelidae.								
Palaeogale ulysses				X				
Palaeogale parvulus				X				
Carnivora indet.		X	X			X		
Hyaenodontidae.								
Hyaenodon sp.			X		X			
Hyaenodon? sp.		X	X	X				
Perissodactyla								
Chalicotheriidae.								
Schizotherium cf. *S. avitum*					X			
Schizotherium sp.	X		X				X	
Helaletidae								
"Hyrachyus" sp.					X			

TABLE 27.2. (continued)

	Kk	Ew	Lw	Ul	Sj	Wt	Qs	Sd
Perissodactyla cont.								
Hyracodontidae								
Ardynia cf. *A. mongoliensis*	X							
Hyracodontidae indet.		X						X
Indricotheriidae								
Indricotherium transouralicum					X		X	X
Indricotherium sp.						X		
Paraceratherium sp.			X		X			
Rhinocerotidae								
Aprotodon sp.			X					
Aceratherium sp.				X				
Aceratherium? sp.					X			
Amynodontidae								
Cadurcodon ardynensis		X						
Cadurcodon sp.				X				
Artiodactyla								
Entelodontidae								
Entelodon ordosius							X	
Cervidae								
Eumeryx culminus				X				
Eumeryx cf. *E. culminus*					X			
Eumeryx sp.		X	X			X		X
"*Eumeryx*" sp.							X	
Cervidae indet.		X			X			
Gelocidae								
Lophiomeryx gobiae		X						
Lophiomeryx sp.		X	X					
Tragulidae indet.		X						
Bovidae								
Palaeohypsodontus asiaticus			X					
Palaeohypsodontus cf. *P. asiaticus*				X				
Hanhaicerus qii				X				
Order incertae sedis								
Didymoconidae								
cf. *Didymoconus berkeyi*				X				
Didymoconus sp.				X				

Ew: Early Wulanbulage; Kk: Kekeamu; Lw: Late Wulanbulage; Qs: Qingshuiying; Sd: Shuidonggou; Sj: Saint Jacques; Ul: Ulantatal; Wt: Wu-tao-ya-yu

TABLE 27.3. List of late Oligocene mammals of China.

	Sg	Tb	Ts	Yk	Ss	Br
Insectivora						
Erinaceidae						
Palaeoscaptor cf. *P. acridens*	X					
Amphechinus minimus		X		X		
Amphechinus cf. *A. rectus*		X	X	X		
Amphechinus sp.	X			X		
Amphechinus? sp.			X			
Erinaceidae indet.	X					
Erinaceidae? indet.		X				
Soricidae indet.		X				
Talpidae? indet.		X				
Order incertae sedis						
Didymoconidae						
Didymoconus berkeyi			X			
Didymoconus sp.	X					
Didymoconus? sp.		X				
Lagomorpha						
Ochotonidae						
Desmatolagus pusillus	X					
Desmatolagus robustus	X					
Desmatolagus sp.				X		
Sinolagomys kansuensis	X	X	X	X		
Sinolagomys major	X	X		X		
Sinolagomys gracilis	X			X		
Sinolagomys ulungurensis					X	
Sinolagomys sp.				X		
Rodentia						
Sciuridae?						
"Sciurus" sp.		X				
Sciuridae indet.	X					
Tsaganomyidae						
Tsaganomys altaicus	X					
Tsagalomys sp.				X		
Castoridae indet.				X		
Cricetidae						
Eucricetodon sp.		X				
Tachyoryctoididae						
Tachyoryctoides obrutschewi	X			X		
Tachyoryctoides pachygnathus	X					
Tachyoryctoides intermedius	X					
Tachyoryctoides kokonorensis				X		
Tachyoryctoides sp.		X			X	

Br: "Brown" Formation; Sg: Shargaltein; Ss: Suosuoquan; Tb: Taben Buluk; Ts: Taoshuyuanzi; Yk: Yikebulage.

TABLE 27.3. (continued)

	Sg	Tb	Ts	Yk	Ss	Br
Ctenodactylidae						
Tataromys cf. *T. plicidens*	X	X		X		
Tataromys deflexus				X		
Tataromys sigmodon		X				
Tataromys cf. *T. sigmodon*			X	X		
Tataromys suni				X		
Tataromys grangeri		X				
Tataromys cf. *T. grangeri*				X		
cf. *Tataromys* sp.	X					
Leptotataromys gracilidens	X					
Yindirtemys woodi		X				
Distylomyidae						
Distylomys qianlishanensis				X		
Prodistylomys xinjiangensis					X	
Dipodidae						
Parasminthus parvulus		X		X		
Parasminthus tangingoli		X		X		
Parasminthus asiae-centralis		X				
Sicistinae indet.	X	X				
Carnivora indet.	X	X				
Creodonta						
Hyaenodontidae						
Hyaenodon? sp.			X			
Perissodactyla						
Chalicotheriidae						
Schizotherium sp.			X			
Schizotherium? sp.		X				
Rhinocerotidae						
Aceratherium sp.		X	X			
Rhinocerotidae indet.		X				
Indricotheriidae.						
Paraceratherium tienshanensis			X			
Paraceratherium lipidus			X			
Indricotherium sp.	X					
Dzungariotherium turfanensis			X			
Dzungariotherium orgosensis						X
Artiodactyla						
Cervidae						
Eumeryx sp.		X				
Eumeryx? sp.	X					
Cervulinae indet.	X					
Gelocidae.						
Lophiomeryx sp.						X
Bovidae indet.	X					

TABLE 27.4. Correlation of Oligocene mammal faunas in China.

Ages		Horizons	Typical localities or formations	Other localities or formations	
OLIGOCENE	LATE	11	Suosuoquan Fm.	"Brown" Fm.	
		10	Yikebulage Fm.		
		9	Taben Buluk loc.	upper Taoshuyuanzi Fm.	
		8	Shargaltein loc.		
	MIDDLE	7	U. Mbr. Wulanbulage Fm.	Saint Jacques loc., Ulantatal loc., Wu-tao-ya-yu loc.	Qingshuiying FM.
		6	L. Mbr. Wulanbulage Fm.		Shuidonggon loc.
		5	Kekeamu loc.		
	EARLY	4	Houldjin Fm.		Shih-ehr-ma-ch'eng loc.
		3	Baron Sog Fm.		
		2	Urtyn Obo Fm.	Ulan Gochu Fm., Chaganbulage Fm., Bailuyuan Fm., Yemaquan Fm., lower Taoshuyuanzi Fm.	Yongning Fm. Xuanwei loc.
		1	Caijiachong Fm.	Shinao Fm., Xiaotun Fm., Baishuicun Fm.	Luoping loc.

28. THE EOCENE-OLIGOCENE TRANSITION IN CONTINENTAL AFRICA

by D. Tab Rasmussen, Thomas M. Bown, and Elwyn L. Simons

ABSTRACT

Knowledge of African terrestrial mammals during the Paleogene is limited to 12 sites, most of which are in North Africa. Only one of these, the Fayum region of Egypt, has produced an extensive mammalian record. The Eocene/Oligocene boundary has been difficult to identify in Africa because of the high proportion of unique endemic taxa, the lack of radiometrically datable rocks at appropriate stratigraphic positions, and other problems. To obtain an estimate of the boundary, the geology and mammalian fauna of the Fayum is analyzed, especially with respect to: (1) faunal change through time; (2) faunal correlations between the Fayum and other sites in Africa; and (3) the stratigraphic positions of major erosional unconformities and inferred regressive events of the Tethys Sea. When integrated, these data suggest that the most likely position of the Eocene/Oligocene boundary (34 Ma) in the Fayum is near an erosional unconformity situated at the 157 m level of measured section of the Jebel Qatrani Formation, between Fayum Faunal Zones 2 and 3. Diverse assemblages of terrestrial mammals recovered from above and beneath this hypothetical boundary level provide information about faunal change across the Eocene/ Oligocene boundary. Faunal and climatic change is less dramatic in Africa than in the northern continents, with consequently minor faunal shifts and no major extinction events. In Africa, wet, tropical conditions prevailed into the Oligocene, supporting a relatively stable, tropical fauna and flora.

INTRODUCTION

The Eocene and Oligocene fossil record of African terrestrial mammals is regrettably sparse. Knowledge of the mammalian fauna has been based almost exclusively on fossils from several localities in one region, the Fayum Depression, Egypt. Fossiliferous rocks there have generally been viewed as spanning a time period from about the middle Eocene to the middle Oligocene. Even as knowledge of the Fayum fauna and geology has grown, it has proven difficult to establish precise biostratigraphic correlations between the Fayum fauna and those elsewhere. Correlation of Eocene and Oligocene Afro-Arabian vertebrates with non-African fossil assemblages has been a long-standing problem for several reasons: (1) the high proportion of mammals that were endemic to Africa or had only rather distant relatives elsewhere; (2) floras and lower vertebrate faunas characterized by long-ranging taxa and taxa of unknown ranges; (3) absence of a good record of marine invertebrates stratigraphically close to the boundary interval; (4) lack of radiometrically datable rocks at appropriate stratigraphic positions; and (5) the inability to directly correlate African rocks with sections elsewhere that have reliable biostratigraphic or radiometric dates. For the purposes of this paper, we accept an approximate date of 34 Ma for the Eocene/Oligocene boundary (Odin and Montanari, 1989; Swisher and Prothero, 1990).

We are now in a position to overcome, or at least to address, some of the major problems listed above. The past decade has heralded a burst of new fossil finds (Figure 28.1), as well as more sophisticated geological studies in the African Paleogene. During the 1980s, important new mammal-bearing sites reported to be of Eocene age were discovered in Algeria and Tunisia (Coiffait et al., 1984; Mahboubi et al., 1986; Hartenberger et al., 1985). A fauna from Malembe, Angola, that had been described earlier as Miocene (Dartevelle, 1935; Hooijer, 1963) was properly recognized as being approximately contemporaneous with the Fayum fauna (Pickford, 1986). Terrestrial mammals were also discovered in the Ashawq Formation

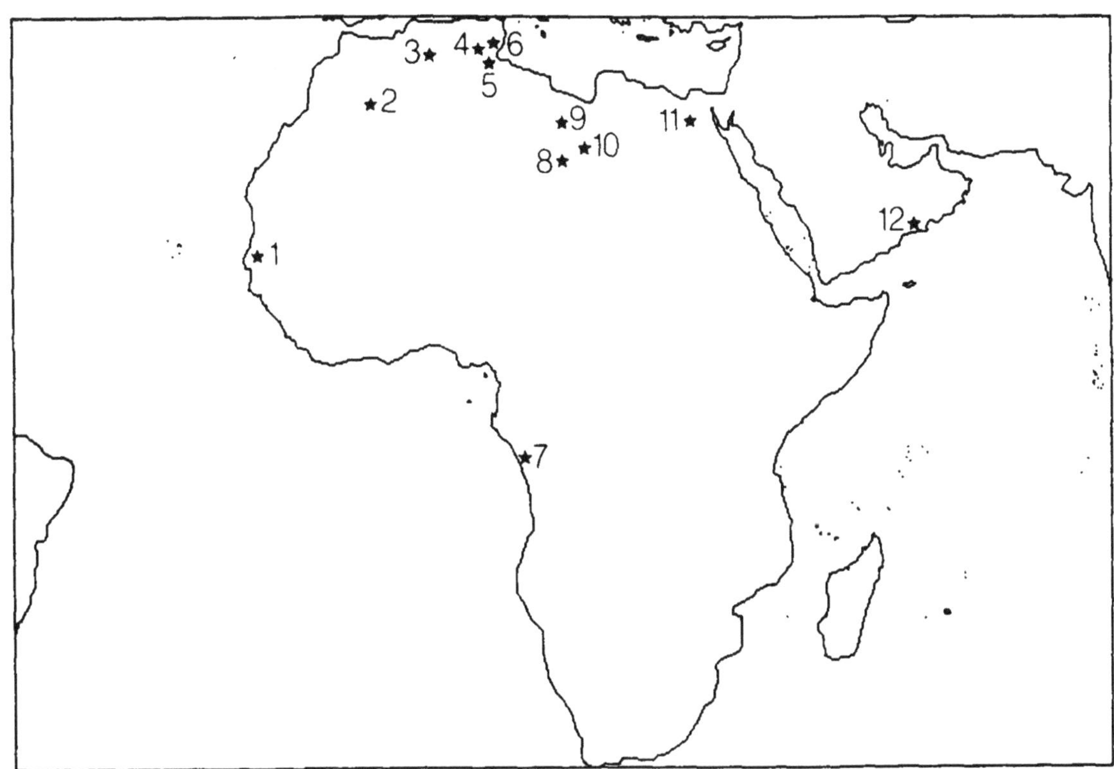

FIGURE 28.1. Geographic positions of the Afro-Arabian Eocene and Oligocene localities that have yielded terrestrial mammals: (1) M'Badione Dadere, Senegal; (2) Glib Zegdou, Algeria; (3) El Kohol, Algeria; (4) Nementcha, Algeria; (5) Gebel Bou Gobrine, Tunisia; (6) Chambi, Tunisia; (7) Malembe, Angola; (8) Gebel Hasawnah, Libya; (9) Zella, Libya; (10) Dor el Talha, Libya; (11) Fayum, Egypt; (12) Thaytiniti and Taqah, Oman.

of Oman on the Arabian Peninsula (Thomas et al., 1988, 1991), which was continuous and confluent with Africa during the Paleogene (Brown et al., 1989; Morgan, 1990). The first paleomagnetic dates associated with African Paleogene terrestrial mammals were derived from study of those sediments (Thomas et al., 1989). Finally, in the ongoing Fayum field project, detailed geological studies and correlations of the entire stratigraphic section (Bown and Kraus, 1988), and the recent discovery of a rich, vertebrate-bearing locality at a much lower stratigraphic level in the terrestrial beds than hitherto sampled (Simons, 1989; Rasmussen and Simons, in press) have made significant new contributions towards an understanding of the African Eocene-Oligocene transition.

Despite the increasing number of discoveries, terrestrial mammals from the new sites outside Egypt remain scarce. Most finds from these new sites, as well as from non-Fayum sites discovered earlier involve only a very few mammal specimens, many of which are unidentified below the family or even ordinal taxonomic level (Arambourg and Magnier, 1961; Arambourg and Burollet, 1962; Savage, 1969; Gevin et al., 1975; Sudre, 1979). Nor do these other sites provide a record spanning the great length of time represented in the Fayum Depression. The Fayum therefore remains the centerpiece of any attempt to address the question of the Eocene-Oligocene transition in Africa.

During the first decade of Fayum fieldwork (1901-1907; see Simons and Rasmussen, 1990) researchers believed that the age of the terrestrial Jebel Qatrani Formation and its associated fauna was late Eocene. After this work, intensive collecting was not resumed until 1961 under the direction of E.L. Simons, by which time it had been concluded that the Jebel Qatrani Formation and its mammalian fauna was Oligocene in age (e.g., Simons, 1968; Bowen and Vondra, 1974). More recent geological investigations of the contact between the Qasr el Sagha Formation, putatively of middle to late Eocene age, and the Jebel Qatrani Formation, putatively of Oligocene age, have demonstrated greater sedimentological continuity across the formation boundary than was hitherto recognized (Bown and Kraus, 1988). It is now evident that there is no sudden, dramatic break between the lower, largely marine Qasr el Sagha Formation and the upper, largely terrestrial Jebel Qatrani Formation. The results presented in this paper, although not definitive, suggest that the Eocene/Oligocene boundary is probably encompassed by a major erosional unconformity within the Jebel Qatrani Formation. The occurrence of terrestrial mammals on both sides of this unconformity allows the construction of a preliminary mammalian biostratigraphy for Africa, and an analysis of faunal and climatic change during the Eocene-Oligocene transition.

GEOLOGY

Terrestrial vertebrates are currently known from 102 Fayum localities: 94 distributed throughout the Jebel Qatrani Formation, and eight in the underlying Dir Abu Lifa Member of the Qasr el Sagha Formation (Figure 28.2). Eight of the localities in the Jebel Qatrani Formation may be considered to be the principal Fayum sites because it is from them that about 90% of the terrestrial mammals have been found (quarries A, B, E, G, I, M, V and L-41; Figure 28.2). Quarry L-41 is a remarkably productive site only recently discovered, and it is the only significant site that occurs this low in the section; the fauna from L-41 is only beginning to be described (Simons, 1989; Rasmussen and Simons, in press).

The Jebel Qatrani Formation consists of about 340 m of sandstones, granule and pebble gravels, gravelly sandstones, mudrocks, shales, and dolomitic limestones of principally alluvial (meandering streams) and associated backswamp origin (Bowen and Vondra, 1974; Bown and Kraus, 1988). These rocks overlie the Dir Abu Lifa Member of the Qasr el Sagha Formation, and are overlain by the Widan el Faras Basalt, both with pronounced erosional unconformity. The Jebel Qatrani Formation is separated into upper and lower sequences by the "barite sandstone," a widespread multistorey-multilateral sandbody locally containing an appreciable volume of evaporites and barite, and which lies atop a major intraformational erosional unconformity (Figure 28.2).

The barite sandstone also divides the formation into two sedimentologic units that evince two distinct alluvial depositional styles. The lower part is coarse sand and gravel-dominated coalesced channel deposits of meandering streams that accumulated under essentially quiescent tectonic conditions, thus permitting a considerable amount of lateral scouring of floodplain sediment and affording a high degree of sandbody interconnectedness (*sensu* Allen, 1978). In contrast, the upper part of the Jebel Qatrani Formation has a higher ratio of mudrock to sand, preserves a greater volume of floodplain deposits, and shows lesser sandbody interconnectedness. These attributes, combined with the presence of a thin marine sandstone near the top of the formation, indicate a greater rate of net sediment accumulation and a relatively more mobile strandline than occurred during deposition of the lower part of the formation. Both greater sediment accumulation and strand movements were influenced by a relatively less stable tectonic setting (Bown

FIGURE 28.2. (opposite page). Rock- and time-stratigraphic correlations of the Jebel Qatrani Formation and adjacent rocks showing: the geologic formations and their subdivisions; relative stratigraphic positions (in meters above the base of Jebel Qatrani Formation) of principal fossil mammal quarry sites (A,B,E,G,I,M,V,L-41); Fayum biostratigraphic zones (FFZ 1 - FFZ 4); major and minor sequence-bounding erosional unconformities (horizontal hatched lines – thickness and length of hatching denote magnitude of lacunae -- lacunae generally concident with the bases of significant bodies of coarse gravelly sandstone); and probable correlations with other fossil mammal sites of the Paleogene Afro-Arabian continent. In the Fayum Depression, the Eocene-Oligocene boundary probably coincides with a time encompassed by the major erosional unconformity between FFZ 2 and FFZ 3 (at 157 m along measured lines of section).

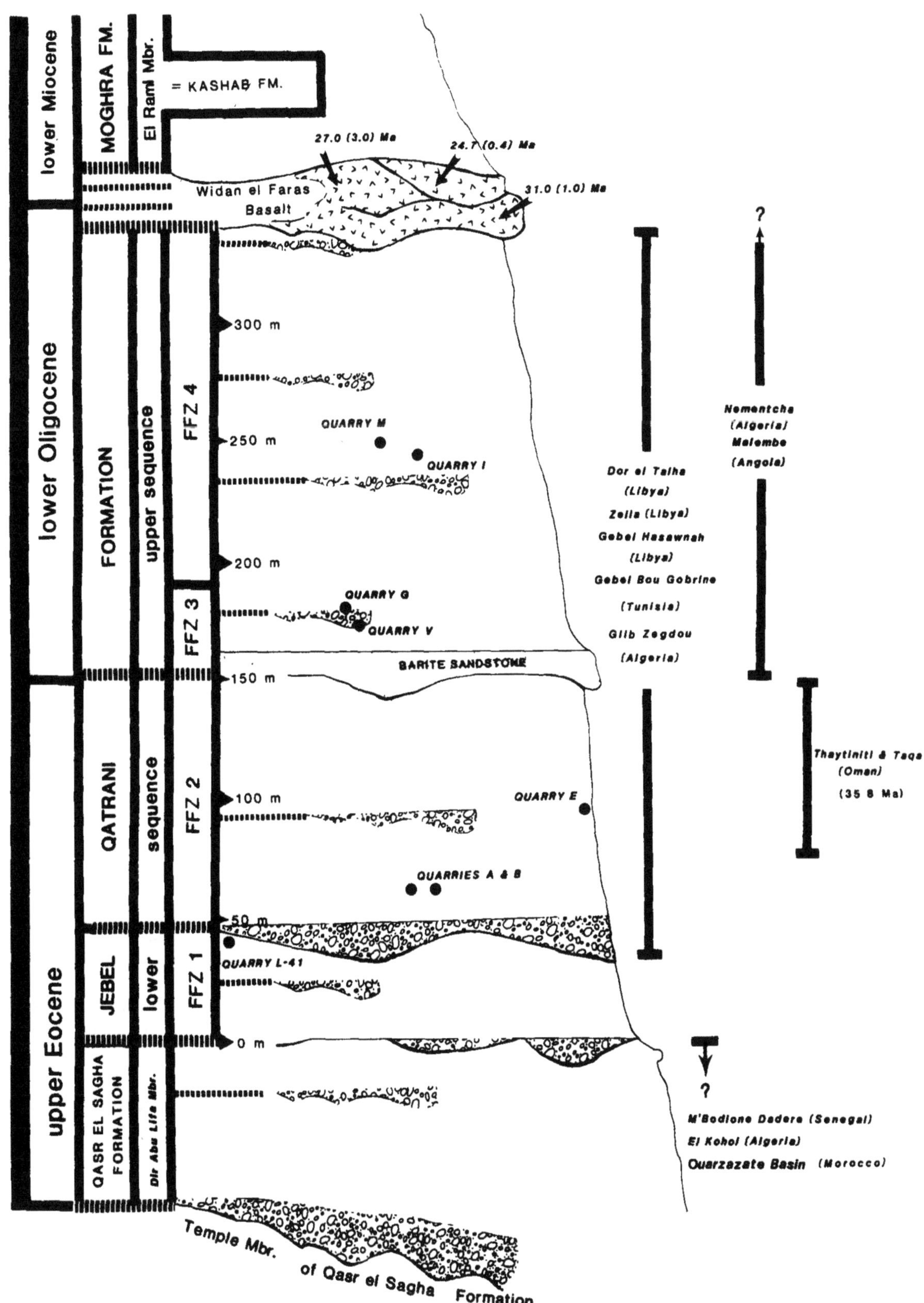
lower Miocene
MOGHRA FM.
El Raml Mbr.
= KASHAB FM.
27.0 (3.0) Ma
24.7 (0.4) Ma
31.0 (1.0) Ma
Widan el Faras
Basalt
lower Oligocene
QATRANI
FORMATION
upper sequence
FFZ 4
300 m
QUARRY M
250 m
QUARRY I
200 m
QUARRY G
QUARRY V
FFZ 3
BARITE SANDSTONE
150 m
sequence
FFZ 2
100 m
QUARRY E
QUARRIES A & B
50 m
QUARRY L-41
upper Eocene
JEBEL
lower
FFZ 1
0 m
QASR EL SAGHA FORMATION
Dir Abu Lifa Mbr.
Temple Mbr. of Qasr el Sagha Formation
Dor el Talha (Libya)
Zella (Libya)
Gebel Hasawnah (Libya)
Gebel Bou Gobrine (Tunisia)
Glib Zegdou (Algeria)
Nementcha (Algeria)
Malembo (Angola)
Thaytiniti & Taqa (Oman)
(35 8 Ma)
M'Bodione Dadere (Senegal)
El Kohol (Algeria)
Ouarzazate Basin (Morocco)

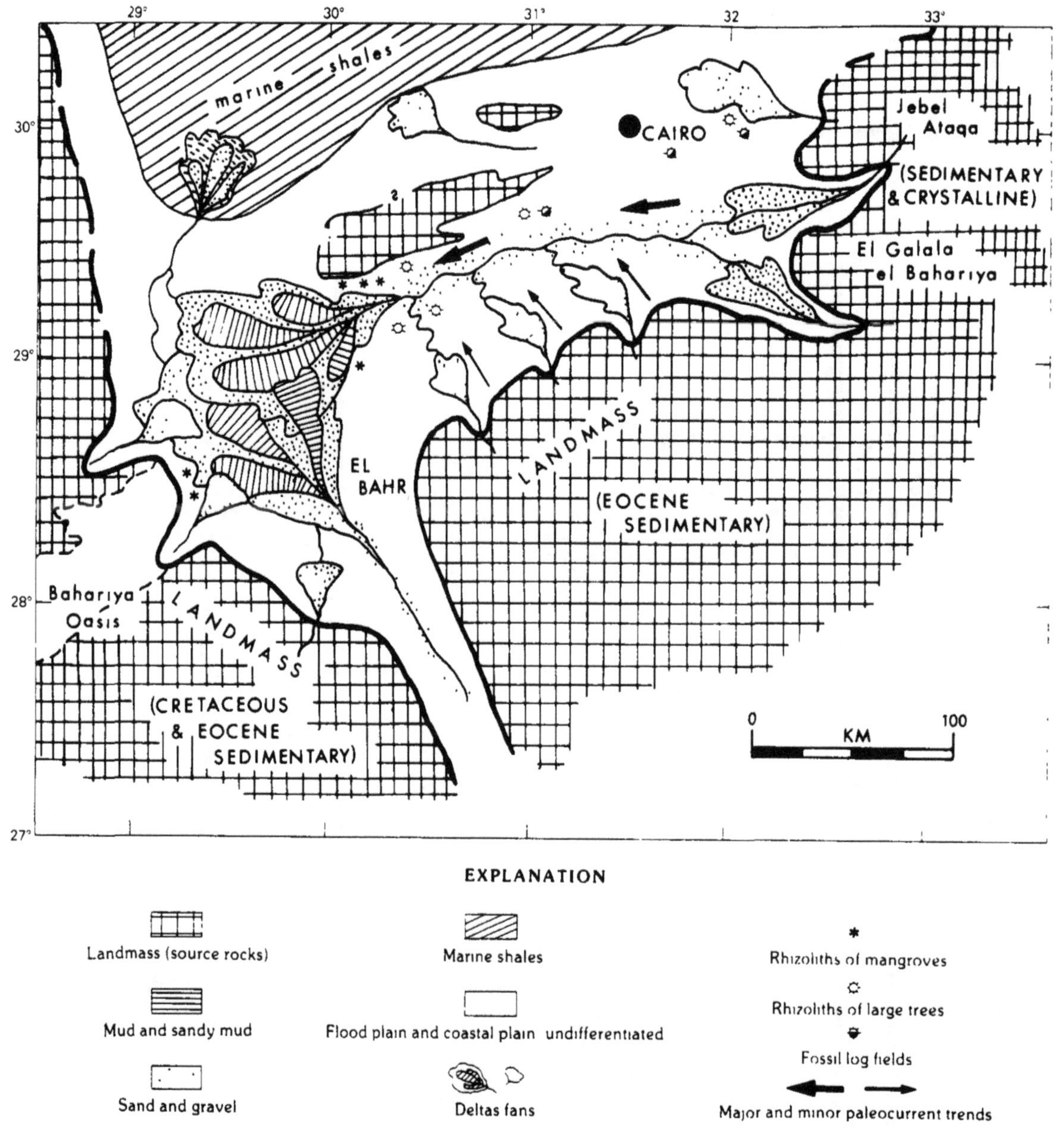

FIGURE 28.3. Hypothetical paleogeographic map of Fayum Depression and adjoining areas of northern Egypt during deposition of the Jebel Qatrani Formation (from Bown and Kraus, 1988; in turn, modified from Salem, 1976, but with considerable revision).

and Kraus, 1988).

Paleocurrent determinations indicate that lower Jebel Qatrani streams flowed principally to the west and southwest, a depositional pattern inherited from the time of accumulation of the underlying, largely alluvial Dir Abu Lifa Member of the Qasr el Sagha Formation (Figure 28.3). Upper Jebel Qatrani stream deposits indicate paleocurrent azimuths to the west but with recurrent northwestern or northern components that probably reflect gradual filling of the Fayum tectonic basin and more direct access by streams to the southern coast of Tethys (Bown and Kraus, 1988). Dir Abu Lifa and lower Jebel Qatrani paleocurrent azimuths and large-scale deposit geometry of the Jebel Qatrani Formation reflect buried yet periodically reactivated west/southwest to east/northeast late Cretaceous structural trends identified by Shata (1953) and Bayoumi and el-Gamili (1970; see also Bown and Kraus, 1988). These trends are principally those of an elongate synclinal basin bounded by antiform highs that may be fault controlled at depth. Therefore, the sedimentary tectonic setting of the Fayum area during Dir Abu Lifa through Jebel Qatrani times was not one of a simple unstable south Tethyan shelf influenced directly by the vagaries of a varying Eocene-Oligocene strandline. Rather, the controlling Jebel Qatrani stream systems indirectly entered a southern embayment of the Tethys from the east or northeast and their depositional styles were correspondingly affected only indirectly by Tethyan eustatic fluctuations (Bown and Kraus, 1988; Figure 28.3). The effect of this structurally isolated position with respect to the south coast of Tethys was to permit continuance of more-or-less sporadic alluvial deposition at times when marine deposition or major erosion obtained in truly coastal areas farther north.

Most of Africa's other Paleogene mammal-bearing sites are also situated near the south coastline of the Tethys Sea. Arambourg (1963) divided the North African sites from this time period into two major geographic "zones," which he called the Nilotic-Saharian Zone and the Atlas Zone. The former includes marine Eocene and overlying continental beds deposited on or near the Tethyan coastal plain to the east of the Atlas Mountains (the Fayum Depression and three Libyan sites, Dor el Talha, Zella, and Gebel Hasawnah; Figure 28.1). The sediments at Dor el Talha, the best studied of the Libyan sites, record a general Tethyan regression; Eocene marine beds occur at the base of the exposures which then grade upwards through a variable series of lagoonal, deltaic, channel, interchannel, levee and overbank deposits (Wight, 1980).

Localities of Arambourg's Atlas Zone are associated with the uplifts and foldings of the Saharan Atlas and High Atlas in Morocco, Algeria, and Tunisia (Ouarzazate Basin, Glib Zegdou [= Gour Lazib], El Kohol, Nementcha, Chambi, and Gebel Bou Gobrine; Figure 28.1). At least some of the terrestrial mammals from these sites, notably those from the Paleocene Ouarzazate Basin and the early to late Eocene El Kohol Formation, appear to be significantly older than the mammals from the continental beds of Libya and Egypt (Cappetta et al., 1978, 1987; Mahboubi et al., 1986). Other Atlas Zone sites that are apparently Eocene in age on the basis of charophyte correlations (e.g., Glib Zegdou, Nementcha; Gevin et al., 1974) may actually be contemporaneous with parts of the Fayum terrestrial sequence (see below).

The only two sites outside North Africa that have yielded any terrestrial mammals (aside from two small, broken, condylarth-like tooth fragments from Senegal; Sudre, 1979) are of special interest because of their geographic positions away from the immediate Tethyan region (Figure 28.1). The Malembe mammals, the only ones of Paleogene age known from south of the equator, come from estuarine and coastline deposits near the present Atlantic (Pickford, 1986). The Omani mammals come from relatively thin terrestrial beds representing regressive episodes bound above and below by thick marine sequences near the present Indian Ocean coastline (Thomas et al., 1989).

MAMMALIAN FAUNA

The Fayum mammalian fauna shows notable changes in taxonomic composition and diversity from the oldest to the youngest strata. Four relatively distinct faunal or biostratigraphic zones have been recognized within the Jebel Qatrani section, which are also tectono-stratigraphically defined (Rasmussen, 1989; Rasmussen and Simons, in press). It is possible that the Eocene/ Oligocene boundary corresponds to one of the breaks between faunal zones, and so the fauna at each of these levels will be reviewed. Hyracoids, the most taxonomically diverse of the endemic African herbivores, and primates are currently the two mammalian orders most useful for detailed faunal correlations among these levels, in large part simply because they have been the best studied taxa to date. It should be possible to place well-studied hyracoids and primates found elsewhere in Africa in a temporal framework relative to the Fayum faunal sequence.

The lowest Fayum faunal zone (FFZ 1) is known principally from quarry L-41 (Figure 28.2). FFZ 1 may be readily distinguished from younger faunal zones by unique genera and species that occur there, as well as by the distinctive assemblage of taxa taken as a whole.

The most common of the eight hyracoid species is a new species of *Saghatherium* (Rasmussen and Simons, in press) that occurs alongside two structurally distinctive species of *Thyrohyrax*, and two undescribed forms related to *Titanohyrax* that differ generically from *Titanohyrax* higher in the section. The primates comprise two very primitive genera and species of early anthropoids, *Catopithecus* and *Proteopithecus*, that are unique to this level (Simons, 1989, 1990). Despite having many primitive dental features, these primates are taxonomically lumped at the family level with the much different propliopithecids of higher Fayum levels largely on the strength of dental similarities and the derived catarrhine dental formula of 2-1-2-3 evident in *Catopithecus*. Anthracotheres are relatively rare in this faunal zone. Several small mammalian taxa that are rare higher in the formation are relatively frequent here (e.g., marsupials, macroscelideans and "insectivores"; these groups are currently under study by Simons, P. Holroyd, M. Gagnon and others).

Fayum faunal zone 2 (FFZ 2) includes the classic sites of the "Lower Fossil Wood Zone" (quarries A, B and C), and also the slightly higher quarry E, which has yielded a diverse fauna of small mammals. The hyracoids of FFZ 2 are characterized by the well-known species *Saghatherium antiquum*, the distinctive *Titanohyrax andrewsi*, and several species of the bunodont taxa *Geniohyus* and *Bunohyrax*. Primates have been recovered only from quarry E, where two anthropoids (*Oligopithecus savagei*, *Qatrania wingi*) and an unnamed omomyid prosimian are present. The omomyid and *Oligopithecus* are unique to this level. *Oligopithecus* is closely related to *Catopithecus* of FFZ 1, whereas *Q. wingi* shows affinities to a species of *Qatrania* found higher in the section (Simons and Kay, 1988).

FFZ 3, known primarily from quarries G and V, represents the first appearance of propliopithecine primates (*Propliopithecus ankeli*) and typical parapithecids (*Apidium moustafai*) which differ from lower sequence primates in having a distinct distal fovea and the development of a wear facet x on the lower molars (Kay, 1977). The small hyracoids of FFZ 3 (*Saghatherium*, *Thyrohyrax*) differ from earlier species in the same genera by showing independent evolutionary trends towards increasing molarization of the premolars, and greater molar selenodonty or lophodonty (Figure 28.4). The most distinctive hyracoid taxon of FFZ 3 is *Selenohyrax*, which is unknown at other levels (Rasmussen and Simons, 1988a).

FFZ 4 includes several very productive and well-studied mammal localities, especially quarries I and M. The FFZ 4 fauna is typified by the common hyracoid *Thyrohyrax domorictus*, the unique genus *Pachyhyrax*, and by a variety and abundance of anthropoid primates (*Apidium*, *Parapithecus*, *Aegyptopithecus*, *Propliopithecus*). Among the herbivores, the larger bodied hyracoids are much less common than the abundant anthracotheres, a reversal of the situation in the lowest faunal zones.

Meaningful quantitative data on total faunal abundance at different Fayum quarries is currently unavailable. However, it is clear that the number of specimens of hyracoids, *Arsinoitherium*, and *Palaeomastodon* declines upwards through the section (Meyer, 1978). In contrast, the abundance of anthracotheres increases dramatically; specimens of this family (which is in need of revision) are relatively rare or infrequent at quarries below 100 to 150 m, but they become abundant and widespread in the upper part of the sequence. Similarly, primates are very rare low in the section, but are one of the most commonly collected mammals in the uppermost quarries.

This information about the Fayum faunal sequence may now be used in comparison with other Paleogene sites of Africa. Table 28.1 presents a list of mammalian taxa identified from putative Eocene or Oligocene sites in other parts of Africa with a comparison to closely related Fayum taxa. Because most of the non-Fayum localities have yielded few mammalian taxa, precise quantitative comparisons or correlation matrices among sites cannot be produced with any confidence.

As noted earlier, two African Paleogene sites apparently fall well outside the time interval represented by the Fayum sequence and are not included in Table 28.1: (1) the Ouarzazate Basin, Morocco, is early to late Paleocene in age (Cappetta et al., 1978, 1987; Gheerbrant, 1988, 1990); and (2) the small molar fragments from M'Badione Dadere, Senegal, are said to be "condylarth" (Sudre, 1979), which presumably

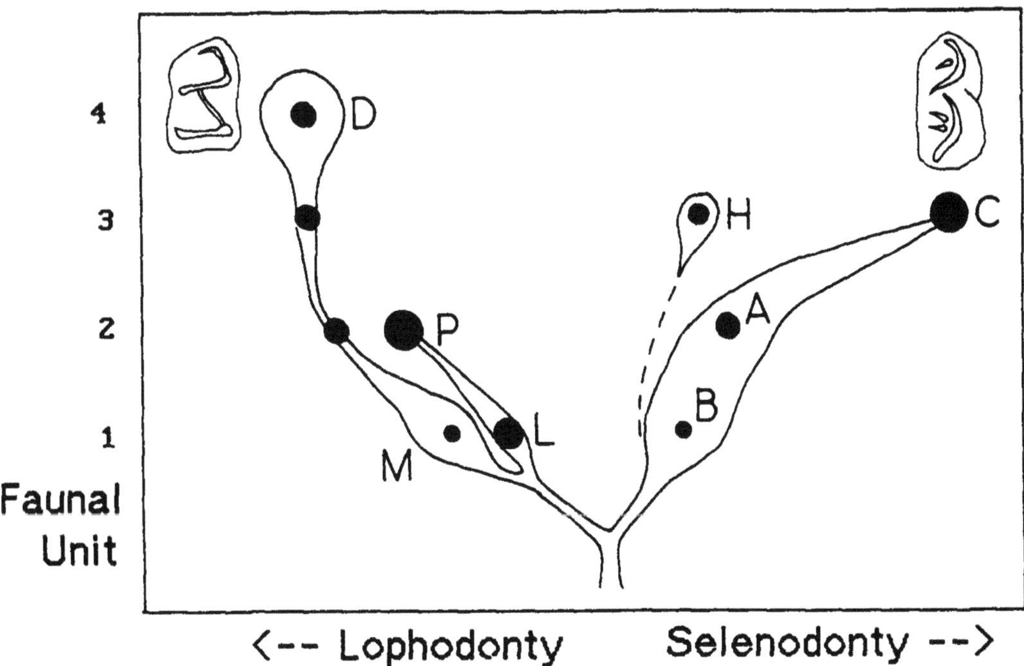

FIGURE 28.4. Evolutionary changes in the small saghatheriine hyracoids across biostratigraphic zones in the Fayum (ordinate axis). Each labeled dot represents a species (see Rasmussen and Simons, in press); the position along the abscissa indicates a gradient from very lophodont lower molars on the left to very selenodont molars on the right, while primitive, unspecialized molars lie near the middle. The size of the dots is proportional to body size; the thickness of the branches is proportional to relative abundance. *Thyrohyrax* lies on the left side of the chart; *Saghatherium* and *Selenohyrax* lie on the right. The occlusal surfaces of the most lophodont right M_2 (*T. domorictus*) and the most selenodont M_2 (*S. chatrathi*) are illustrated in the upper corners. (From Rasmussen and Simons, in press.)

indicates that they are significantly older than any Fayum sites. The poorly known site of Gebel Hasawnah is also omitted from the table (Spinar, 1980).

The remaining Paleogene sites may be placed in two categories: those that are generally believed to be Oligocene in age, and roughly contemporaneous with the Jebel Qatrani Formation, and those that are Eocene, which hitherto have been believed to be older than the Fayum fauna. Among the former are the Libyan sites of Dor el Talha, Zella, and Gebel Bou Gobrine. The faunal comparisons in Table 28.1 substantiate the hypothesis that these sites are similar in age to the Fayum mammals, as do the geological and faunal comparisons of researchers who have worked on those sites (Arambourg, 1963; Savage, 1969, 1971; Wight, 1980).

Among the sites that may be Eocene in age are El Kohol, Glib Zegdou, and Nementcha, in Algeria, and Chambi in Tunisia. El Kohol contains a unique fauna including a proboscidean, *Numidotherium*, and a hyracoid, *Seggeurius*, that both appear to be more primitive morphologically than the Fayum proboscideans and hyracoids (Mahboubi et al., 1986). Also found at El Kohol are the unique genera *Garatherium*, a marsupial, and *Koholia*, a creodont (Crochet, 1988). Glib Zegdou also has some unique genera (*Azibius*, *Helioseus*) but these are divergent and thereby not obviously more generalized or primitive than the Fayum taxa. Unlike *Seggeurius* from El Kohol, the small Glib Zegdou hyracoid, *Microhyrax*, does generally resemble Fayum taxa; efforts should be made to

TABLE 28.1. Comparison of African Paleogene faunas to the Fayum fauna.

Site and Mammalian Taxa[1]	L-41	A,B	E	G,V	I,M
THAYTINITI, OMAN					
Omomyidae ?	-	-	+	-	-
Arsinoitherium	-	+	+	+	+
Phiomys cf. *P.andrewsi*	*	+	+	+	-
Phiomys cf. *P.lavocati*	*	-	+	-	-
cf. *Metaphiomys* spp. 1&2	-	+	+	+	+
TAQAH, OMAN					
cf. *Oligopithecus savagei*	-	-	+	-	-
Propliopithecus markgrafi	-	-	?	?	-
Phiomys cf. *P.lavocati*	*	-	+	-	-
cf. *Metaphiomys* sp.	-	+	+	+	+
EL KOHOL, ALGERIA					
Geratherium mahboubii	-	-	-	-	-
Seggeurius amourensis	-	-	-	-	-
Numidotherium koholensis	-	-	-	-	-
Koholia atlasense	-	-	-	-	-
GLIB ZEGDOU, ALGERIA					
Helioseus insolatus	-	-	-	-	-
Azibius trerki	-	-	-	-	-
Microhyrax lavocati	-	-	-	-	-
Titanohyrax mongereaui	*	*	*	*	*
Megalohyrax gevini	*	*	*	*	*
NEMENTCHA, ALGERIA					
Biretia piveteaui	-	-	-	@	@
Bunohyrax sp.	+	+	-	+	+
Anthracotheriidae	+	+	+	+	+
Phiomyidae	+	+	+	+	+
Anomaluridae	-	-	-	-	-
CHAMBI, TUNISIA					
Kasserinotherium tunisiense	-	-	-	-	-
Paleoryctidae	-	-	-	-	-
Chambius kasserinensis	-	-	-	-	-
cf. *Pachyhyrax*	-	-	-	-	+
cf. *Saghatherium*	+	+	+	+	-
Ischyromidae	-	-	-	-	-
Anomaluridae	-	-	-	-	-
GEBEL BOU GOBRINE, TUNISIA					
Phiomia (=*Palaeomastodon*)[3]	-	+	+	+	+
? anthracothere	+	+	+	+	+
DOR EL TALHA, LIBYA					
Arsinoitherium	-	+	+	+	+
Palaeomastodon	-	+	+	+	+
Phiomia (=*Palaeomastodon*)[3]	-	+	+	+	+

Site and Mammalian Taxa[1]	L-41	A,B	E	G,V	I,M
ZELLA, LIBYA					
Palaeomastodon	-	+	+	+	+
Titanohyrax "palaeotherioides" (=*T. angustidens*?)[4]	-	-	-	+	+
Brachyodus (=*Bothriogenys*)[5]	+	+	+	+	+
MALEMBE, ANGOLA					
Arsinoitherium	-	+	+	+	+
Palaeomastodon	-	+	+	+	+
Pachyhyrax[6]	-	-	-	-	+
Bunohyrax/Geniohyus[6]	+	+	+	+	-

+ Indicates a faunal match

- Lack of a faunal match

* Asterisk indicates that the named species is not recognized in the Fayum, but that a closely related congener is present.

@ The genus is not present in the Fayum, but similar primate genera with bunodonty, distal foveae, and wear facet x occur only at the levels of FFZ 3 and FFZ 4.

1 Includes only mammalian taxa identified to the level of family or better.

2 Thomas et al. (1991) have resurrected the genus *Moeripithecus* Schlosser for this species; we retain it as a subgenus of *Propliopithecus*. The stratigraphic position of the type and only specimen from the Fayum is unknown, but on structural grounds it may lie between quarries E and V.

3 Synonymy of *Phiomia* and *Palaeomastodon* follows Coppens et al. (1978).

4 For discussion of the *nomen* used here see Rasmussen (1989).

5 Generic designation of anthracotheres follows Black (1978).

6 For comments on genus and species identification see Rasmussen (1989).

---

compare these directly. A large species of *Titanohyrax* from Glib Zegdou is similar in size and structure to *T. ultimus* of FFZ 4. Quarry L-41 contains two undescribed species related to *Titanohyrax* that are more primitive than the species of the upper sequence; for example, these L-41 species lack distinct metastylids, a derived character which occurs in younger species of *Titanohyrax*. The L-41 species may also be more primitive than the teeth attributed to this genus from Glib Zegdou. This would suggest that Glib Zegdou is approximately contemporaneous with that part of the Jebel Qatrani Formation stratigraphically above quarry L-41.

The sediments at Nementcha have also been described as older than the Fayum deposits, but the Nementcha fauna suggests an age approximately contemporaneous with the time interval represented by the Jebel Qatrani Formation. Three of the five Nementcha taxa are widespread through the Fayum section. *Biretia* is known only from one lower molar, but this tooth is bunodont, has a distinct wear facet x, and a discrete distal fovea (de Bonis et al., 1988). Primates with these attributes are notably absent below the level of quarry V in the Fayum (FFZ 3), but are abundant above this level, suggesting that the Nementcha deposits may be similar in age to FFZ 3 or 4. The rodent family Anomaluridae is not recorded from the Fayum, despite the recovery of hundreds of rodent jaws and teeth (Lavocat's 1978 reference to the occurrence of Anomaluridae in the upper sequence of the Fayum is incorrect). Despite relative dates on charophytes to the contrary (Coiffait et al., 1984), we can only infer a relatively young age for the Nementcha mammals, corresponding to the upper sequence of the Jebel Qatrani Formation.

Chambi, a site said to be early Eocene in age (Hartenberger et al., 1985; Hartenberger, 1986) is difficult to place relative to the Fayum faunal zones. Chambi has yielded undescribed paleoryctid insectivores and ischyromyid rodents, both apparently old and primitive families not represented in the Fayum, along with a didelphid, *Kesserinotherium*, and a macroscelidid, *Chambius*, that are generically distinct from Fayum didelphids and macroscelidids. However, the relatively young pliohyracid *Pachyhyrax* has also been tentatively identified from Chambi; this taxon occurs in the Fayum only at the uppermost levels (FFZ 4; Rasmussen and Simons, 1988a). Another pliohyracid from Chambi, *Saghatherium*, occurs in FFZ 1-3. The reported presence of anomalurids at Chambi may indicate that this rodent family was indeed present in Africa before the Miocene.

The sites of Thaytiniti and Taqah in the Ashawq Formation, Oman have also been described as predating the Fayum fauna (Thomas et al., 1989). Table 28.1 shows that a very close faunal match between the Omani sites and the Fayum occurs at quarry E (6 of 7 taxa). *Oligopithecus savagei* is known only from this level (Rasmussen and Simons, 1988b). Older, more primitive anthropoids occur at a considerably lower Fayum level at quarry L-41 (Simons, 1989), whereas younger, more derived propliopithecines occur at higher levels. Specimens of teeth and jaws allocated to the primate *Propliopithecus (Moeripithecus) markgrafi* represent the one species from Oman that does not fit comfortably at the quarry E level (Thomas et al., 1991). *P. markgrafi* is of unknown provenance within the Jebel Qatrani Formation, but structurally, this species is more primitive than other species of *Propliopithecus*, including *P. ankeli*, suggesting that it may have come from a level beneath quarry V. The Omani fauna thus indicates that Thaytiniti and Taqah are near contemporaries of quarry E. This strong faunal correlation is of particular interest because of the paleomagnetic analyses that have been conducted on the Omani sites. These suggest an age of about 35.8 Ma, or late Eocene.

IDENTIFYING THE EOCENE-OLIGOCENE TRANSITION

Because of how epochs and series were originally defined and have come to be used subsequently, precise location of epochal (time) and series (rock) boundaries within any body of rocks must ultimately rely upon correlations of faunas and floras contained within them (differences in European and North American time- and rock-stratigraphic usage notwithstanding). Radiometric and other means of dating rocks may be used in correlation of the rocks and their contained faunas; however, they technically play no primary role in defining epochs, series, or their boundaries. How much simpler geologic correlation would be had it been possible from the outset to define boundaries by fiat in terms of millions of years! Instead, temporal boundaries were (and are yet to most European workers) tied to the boundaries of type series (rocks) within the Tertiary System (also rocks). What was unknown then is that these divisions were largely based upon significant faunal change that was either due to, or was magnified by, the effects of intervening transgressions, regressions, or hiatuses. Thus, geologists find that they have inherited a system that obliges them to undertake the ponderous task of ascertaining exactly what is meant by a specific epochal or series boundary in cases in which an adequate, temporally complete boundary does not actually exist within the series type areas. Just how onerous this chore is can be visualized by examining the works of Pomerol and Premoli-Silva (1986) and Cavelier (1972, 1979) with respect to the Eocene/Oligocene boundary.

In European continental stratigraphy, the "Grande Coupure" (Stehlin, 1909) denotes a significant faunal change among land mammals and one that is still believed by many current workers to span, if not record, the Eocene/Oligocene boundary (see, e.g., Lopez and Thaler, 1974, for a contrasting view). However, refined faunal resolution and the discovery of temporally more complete sections spanning this interval indicate that both marine and continental faunal change across it was temporally staggered, and perhaps so over some millions of years (Crochet et al., 1981; Prothero, 1989). The magnitude of faunal separation on either side of the boundary in different sections is, of course, increased by the magnitude of intervening lacunae or by the amount of time represented by transgressive and regressive facies changes.

In the Fayum, the boundary between upper Eocene and lower Oligocene rocks was originally placed somewhere within the "fluvio-marine series" (=Jebel Qatrani Formation; Beadnell, 1905). Subsequent workers placed the boundary at the convenient contact of the Jebel Qatrani Formation with the underlying Qasr el Sagha Formation (Said, 1962; Simons, 1968; Bowen and Vondra, 1974; Bown and Kraus, 1988). The Qasr el Sagha Formation was, for a long time, believed to be wholly marine. Several new lines of evidence necessitate a reappraisal of the probable stratigraphic position of the Eocene/Oligocene boundary in the Fayum Depression: (1) new radiometric work on dating the Eocene/Oligocene boundary that suggests a younger age than previously accepted (about 34 Ma; Odin and Montanari, 1989; Swisher and Prothero, 1990); (2) combined new knowledge regarding the sedimentology of the upper part of the Qasr el Sagha Formation and the Jebel Qatrani Formation (Bown and Kraus, 1988); (3) a radiometric date of 31.0 ± 1.0 Ma on basalts lying unconformably on the uppermost Jebel Qatrani Formation (Fleagle et al., 1986); (4) the newly discovered primitive primate, hyracoid and other mammal faunas from quarry L-41 (Simons, 1989; Rasmussen and Simons, in press); and (5) recent paleomagetic analyses of the Fayum sequence, and (6) paleomagnetic correlations of mammalian faunas from the Ashawq Formation of Oman with those from the Fayum (Thomas et al., 1989). In this section, we will integrate data on major erosional unconformities, faunal change, and the Omani paleomagnetic dates, to arrive at a multi-disciplinary estimate of the position of the Eocene/ Oligocene boundary in the Fayum Depression.

Correlation with other localities

To the extent that the Omani paleomagnetic correlations are accurate and the faunal correlation of Omani sites with FFZ 2 is reliable, then the 92 meter level of the Jebel Qatrani Formation (quarry E) is late Eocene, approximately 35.8 Ma. The L-41 primate fauna is more primitive than the Omani primates, while the L-41 hyracoids are more primitive than those of some other Eocene African sites, suggesting that L-41 is Eocene. The marine and in part alluvial Qasr el Sagha Formation is Eocene (Bowen and Vondra, 1974; Bown and Kraus, 1988). Thus, the Eocene-Oligocene transition almost certainly lies somewhere within the Jebel Qatrani Formation, probably above quarry E.

Unconformities and marine regression

A major regression of the Tethys Sea took place at a time more-or-less coincident with the Eocene/ Oligocene boundary and is recorded in marine and continental sections throughout the Tethyan region (e.g., Plaziat, 1981). It is reasonable to assume that this regression was accompanied by the advance of continental conditions in coastal areas of Egypt and by erosion in these and more landward areas. A pronounced unconformity exists between Eocene and Oligocene rocks in some subsurface sections in northern Egypt landward to the Fayum basin, but it is not recorded in other sections farther north and closer to the south shore of the Tethys Sea (Sheikh and Faris, 1985). A relatively more complete section is present in the Fayum Depression, but there the rocks pertinent to the position of the Eocene-Oligocene boundary are continental, were deposited in an isolated structural basin, do not contain foraminifera, and therefore cannot be confidently tied to any of the subsurface sections.

The first regressive conditions in the Fayum area during the Tertiary are evinced by emergent offshore bars supporting mangroves in the middle and upper Eocene Birket Qarun Formation (Bown and Kraus, 1988). The overlying Temple Member of the Qasr el Sagha Formation (probably upper but not uppermost Eocene) witnessed a return to shallow marine conditions (Vondra, 1974; Bown and Kraus, 1988) and, although fluvial rocks dominate the 77 m of the Dir Abu Lifa Member, three separate minor transgressions punctuate its record of alluvial deposition (Figure 28.2). The thin marine intercalations of the Dir Abu Lifa Member record very minor transgressive events and have thus far yielded no planktonic foraminifera or indicative macroinvertebrates. Though regressive, the Qasr el Sagha Formation/Jebel Qatrani Formation contact does not herald the advent of continental conditions in the Fayum as was earlier believed. It is clear, therefore, that the Eocene/ Oligocene boundary in the Fayum cannot be picked with confidence based solely on a change to more continental conditions.

If major Tethyan regressions occurred during a time interval that embraces the 34 Ma boundary, and if that time is included in an unconformity in Egypt (as opposed to somewhere within the section of preserved rock), then it might be possible to demonstrate that one unconformity is more likely to contain a time interval spanning 34 Ma than is another. Unfortunately, the unconformable contact of the Qasr el Sagha and Jebel Qatrani Formations is only one of five major and as many as seven moderate to minor unconformities of unknown temporal magnitudes contained in the Fayum beds (Figure 28.2). All of these unconformities are erosional, all are quite extensive throughout the Fayum basin, and all were followed by initiation of deposition of coarse alluvial sand and gravel forming large-scale multistorey-multilateral sandbodies.

Paleocurrents and clast size, composition and resiliancy indicate that the source area for coarse clastic materials in the Jebel Qatrani gravels was almost certainly to the east or northeast, up the Fayum Basin trough (Bown and Kraus, 1988; Figure 28.3). This evidence suggests that erosion and subsequent deposition of coarse clastic materials was probably instigated by lowered marine strandlines in the embayment at the lower (west end) of the Fayum Basin trough; the magnitude of the individual unconformities being directly related to the magnitude of eustatic changes. Along the lines of measured sections (Figure 28.2), the Fayum unconformities occur: (1) at the base of the Dir Abu Lifa Member (-77 m, a major unconformity); (2) in the Dir Abu Lifa Member at the -26 m level (moderate); (3) at the contact of the Dir Abu Lifa Member and the lower part of the Jebel Qatrani Formation (0 m, major); (4) at the base of the lower green sandstone below quarry L-41 (unit names from Bown and Kraus; 1988) (23 m, minor); (5) at the base of the middle gravelly sandstone, just above quarry L-41 (48 m, major); (6) at the base of the upper gravelly sandstone (87 m, moderate); (7) at the base of the barite sandstone (152 m, major); (8) in the middle of the lower variegated sequence of the upper sequence of the Jebel Qatrani Formation (176 m, minor); (9) at the base of the sheet gravel sequence (235 m, moderate); (10) in the lower part of the upper variegated sequence (275 m, minor); (11) in the upper part of the upper variegated sequence (330 m, minor); and (12) between the Jebel Qatrani Formation and the Widan el Faras Basalt (340 m, major). The units bounded by these unconformites are therefore classic *sequences* in the sense of Sloss et al. (1949) and Wheeler (1958). The major unconformities in the lower part of the Jebel Qatrani Formation (#3, 5, and 7) bound two major sequences that are faunally distinguishable as FFZ 1 and FFZ 2. Drawing from the sedimentological evidence, the two unconformities most likely to include the 34 Ma boundary are probably the major erosional breaks at 48 m (#5) and 157 m (#7). The unconformity at 157 m is above the quarry E fauna that correlates best with the 35.8 Ma Thaytiniti and Taqah localities in Oman; therefore, the erosional unconformity at 157 m, above quarry E, is the one most likely to contain a 34 Ma Eocene/Oligocene boundary.

Shifts in faunal composition

The documented shifts in species diversity and faunal composition as one moves upwards through the Jebel Qatrani Formation involve changes in the number of species per family, or in the appearance and disappearance of species, genera, or occasionally, poorly sampled subfamilies or families represented by small, rare species. Never are there wholesale, correlated changes in the representation of families or higher level taxa.

The most notable faunal changes take place among the primates (Figure 28.5). A species of the Holarctic primate subfamily Anaptomorphinae, otherwise Eocene in age, occurs at quarry E, but not higher (Simons et al., 1987). The world's earliest occurrences of the extant primate families Lorisidae and Tarsiidae appear in FFZ 4 but not lower (Simons and Bown, 1985; Simons et al., 1987). These occurrences support a general changeover from archaic Eocene prosimians to modern prosimians; however, all three taxa are represented in the Fayum by very few isolated teeth (in the case of Tarsiidae, one jaw) and therefore little confidence can be placed on the known temporal distributions of those taxa. The propliopithecid primates in the Fayum show a change from the primitive oligopithecines of FFZ 1 and 2, which have teeth closely resembling Eocene

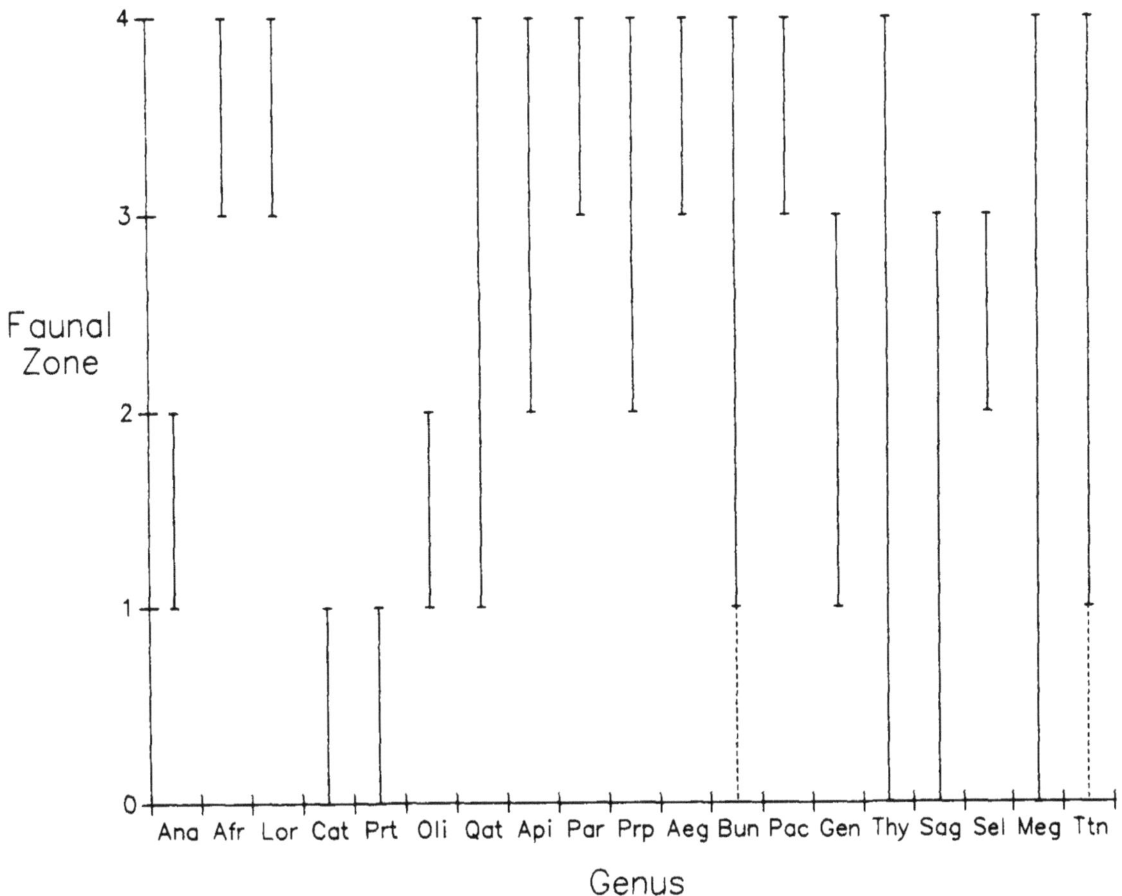

FIGURE 28.5. Known stratigraphic ranges of genera of primates (left) and hyracoids (right) within the Jebel Qatrani Formation. There is no significant difference among faunal zones in the number of appearances or dissappearances of genera; thus, there is no evidence of a major extinction event. A dashed line indicates the occurrence of an undescribed genus closely related to the genus of the solid line to which it is connected (Rasmussen, unpublished data). Taxa from left to right are as follows. Prosimians: Ana, Anaptomorphinae, gen. indet.; Afr, *Afrotarsius*; Lor, Lorisidae, gen. indet. Oligopithecines: Cat, *Catopithecus*; Prt, *Proteopithecus*; Oli, *Oligopithecus*. Parapithecids: Qat, *Qatrania*; Api, *Apidium*; Par, *Parapithecus*. Propliopithecines: Prp, *Propliopithecus*; Aeg, *Aegyptopithecus*. Pliohyracid hyracoids: Bun, *Bunohyrax*; Pac, *Pachyhyrax*; Gen, *Geniohyus*; Thy, *Thyrohyrax*; Sag, *Saghatherium*; Sel, *Selenohyrax*; Meg, *Megalohyrax*; Ttn, *Titanohyrax*.

adapid prosimians (Rasmussen and Simons, 1988b; Simons, 1989, 1990) to the more derived propliopithecines of FFZ 3 and 4, which have teeth closely resembling Miocene hominoids (Simons, 1965, 1974). Thus, on the basis of propliopithecids and prosimians, the greatest faunal break would appear to occur between FFZ 2 and 3. However, another family of Fayum primates, the parapithecids, shows some generic continuity across this level, with species of *Qatrania* known from FFZ 2 and FFZ 4 (Simons and Kay, 1988).

Another well-studied mammalian order, the Hyracoidea, also shows much generic continuity across faunal zones (Figure 28.5). *Selenohyrax* and *Pachyhyrax* are the only described genera that are currently known from a single faunal zone; the remaining genera occur in two or more zones. However, the pattern of zone overlap varies from genus to genus. There is no place in the stratigraphic section where numerous appearances or dissappearances of hyracoid genera coincide. The hyracoid fauna thus offers few clues as to a possible Eocene/Oligocene boundary.

Faunal correlations outside Africa

Many African Paleogene mammals (and, traditionally, those of the Fayum), belong to endemic groups unknown outside Africa in the early Tertiary (e.g., Tenrecoidea, Chrysochloridae, Macroscelidea, Proboscidea, Embrithopoda, Hyracoidea, phiomyid rodents, anthropoid primates). Among the taxa that are shared with other continental areas are some that show only subfamilial or higher level systematic affinities (e.g., omomyids, manids). Several of the most precise faunal matches between Africa and the northern continents occur during the Paleocene, and are thus not useful in dating the Eocene/Oligocene boundary (e.g., the paleoryctids *Paleoryctes* and *Cimolestes*; Gheerbrant, 1990). Only a small handful of Eocene and Oligocene taxa are shared between Africa and other continents. Among these, many are long-lived and therefore of limited value for faunal correlation (*Peratherium*, *Pterodon*, anthracotheres, anaptomorphine primates). A recent comparison of species of *Apterodon* in Europe and the Fayum support an early Oligocene age for the uppermost Fayum levels (FFZ 3 and 4; Tilden et al., 1990). The anthracotheres probably offer the greatest potential for broader faunal correlation but, to be useful the Paleogene members of this group require detailed systematic revision.

Conclusion

At present, the relative age sequence outlined above on the basis of mammalian faunal correlations can be anchored to a numerical time scale only by the paleomagnetic data from the Ashawq Formation of Oman. As outlined above, the Ashawq mammals, which are about 35.8 Ma in age, appear to correlate well only with the Fayum's quarry E fauna (Table 28.1). The anthropoid primates from quarry L-41 are clearly more primitive than those from the Ashawq Formation. Quarry L-41 is separated from quarry E by 45 m of deposits and by a major erosional unconformity. This evidence suggests that the age of L-41 must be significantly older than 35.8 Ma (perhaps as old as 38 to 40 Ma) or late Eocene at the the very youngest. Therefore, the 34 Ma Eocene/ Oligocene boundary in the Fayum continental sequence is probably contained within or is stratigraphically near to the extensive erosional unconformity between FFZ 2 and 3, near the 157 m mark of measured section, just below the barite sandstone. Though by no means definitive, we believe this placement of the Eocene/Oligocene boundary is a plausible and realistic solution because it utilizes all the published paleontologic, geologic, and radiometric data. If this placement of the boundary is correct, then the entire lower sequence (Bown and Kraus, 1988; formerly known as the "Lower Fossil Wood Zone") of the Jebel Qatrani Formation is Eocene. The second most likely placement of the boundary, if the faunal correlations with Oman or the radiometric dates from sediments there are not reliable, would probably be at the major unconformity between quarries L-41 and E, near the 48 m level, between FFZ 1 and 2.

John Kappelman (1991) has conducted paleomagnetic sampling in the Fayum Depression, and he has constructed a magnetic stratigraphy for the Fayum that can be correlated with the magnetic polarity time scale. Kappelman concludes that most fossil mammal localities in the Fayum are within Chrons C12 to C13R (about 32-37 Ma). Quarry L-41 may be as old as Chron C15R (37.7 to 38.1 Ma). An Eocene/ Oligocene boundary at 34 Ma may lie even higher in the Jebel Qatrani Formation than the 157 m unconformity suggested here.

CHANGES ACROSS THE EOCENE/ OLIGOCENE BOUNDARY

Unlike on the Holarctic continents, where the Eocene/Oligocene boundary has been traditionally associated with apparent significant shifts in faunal composition and climatic deterioration, the data available from Africa indicate little association between the inferred boundary and any dramatic extinction events or faunal turnovers. To the contrary, there is evidence of notable continuity for several taxa across any hypothesized Eocene/Oligocene boundary above the level of quarry L-41. Several genera persist from quarry L-41 to quarries I and M, including the hyracoids *Thyrohyrax* and *Megalohyrax* (Rasmussen and Simons, in press), and based on unpublished data, the anthracothere *Bothriogenys* and the stork *Palaeoephippiorhynchus*. Rodents and creodonts also show some degree of continuity up through the section, and these and other groups are currently under study by M. Gagnon and P. Holroyd. Older African sites such as Ouarzazate and El Kohol have faunas unlike those of the Fayum, but because of the small mammalian samples from these sites, and the great geographic and temporal ranges involved in comparisons of these sites to the Fayum fauna, it is impossible to attribute these apparent faunal differences to an Eocene/ Oligocene boundary event.

The climate of the Fayum and, by inference, that of the broader Tethyan Afro-Arabian area apparently remained relatively stable across the Eocene/Oligocene boundary. In contrast to the Oligocene climate of the northern continents, Jebel Qatrani sediments were deposited under warm and wet tropical to subtropical conditions, as indicated by fossil vertebrates, charophytes, ostracodes, leaves, paleosols, plant steinkerns, and trace fossils of social insects (Bown, 1982; Bown et al., 1982; Wing and Tiffney, 1982; Olson and Rasmussen, 1986; Rasmussen et al., 1987; Bown and Kraus, 1988). Many of the floral elements most closely resemble tropical to subtropical species of modern day Indomalaysia, such as the fruit of *Epipremnum*, which occurs in the Fayum both above and below the hypothesized Eocene/ Oligocene boundary (Wing and Tiffney, 1982; and in Bown et al., 1982). Among the birds, species that inhabit warm, wet, tropical, freshwater environments, such as jacanas (family Jacanidae) are also known from both above and below the hypothesized boundary (Olson and Rasmussen, 1986; Rasmussen et al., 1987). Africa thus contrasted with northern continental areas in maintaining a relatively stable, warm, wet, tropical climate across the Eocene/ Oligocene boundary. This relatively stable, equable climate allowed for the maintenance of taxonomic diversity among tropical African mammals, birds, and plants (e.g., the primates and the jacanas), whereas the northern continents experienced a more temperate climatic regime.

ACKNOWLEDGEMENTS

We thank M. Gagnon, P. Holroyd, A. Nelson, S. Gauld, R.J.G. Savage and G.E. Meyer for their technical reviews of the manuscript. We thank Gagnon and Holroyd for informative discussion about unpublished components of the Fayum fauna, including the rodents and creodonts. Field work in Egypt was done in cooperation with the Egyptian Geological Survey and Mining Authority, and the Director and staff of the Cairo Geological Museum. Primary funding for field and lab work was provided by National Science Foundation. Additional laboratory research funds were provided by the Academic Senate, University of California. This is Duke Primate Center publication no. 498.

REFERENCES

Allen, J.R.L. 1978. Studies in fluviatile sedimentation: an exploratory quantitative model for the architecture of avulsion-controlled alluvial suites. *Sed. Geol.* 21:149-197.

Arambourg, C. 1963. Continental vertebrate faunas of the Tertiary of North Africa. In F.C. Howell and F. Bouliere, eds., *African Ecology and Human Evolution*, London, Methuen, pp. 55-64.

Arambourg, C. and Burollet, P.F. 1962. Restes de vertebres oligocenes en Tunisie centrale. *C.R. Soc. Geol. Fr.* 2:42.

Arambourg, C. and Magnier, P. 1961. Gisements de vertebres dans le bassin tertiaire de Syrte (Libye). *C.R. Acad. Sci. Paris* 252:1181-1183.

Bayoumi, A.I. and el-Gamili, M.M. 1970. A geophysical study on the Fayum-Rayan area, with reference to its subsurface structures. *Arab Petroleum Congress, 7th, Kuwait, Proc.* 2 (35, B-2):25-34.

Beadnell, H.J.L. 1905. *The Topography and Geology of the Fayum Province, Egypt*. Cairo, Survey Department of Egypt.

Black, C.C. 1978. Anthracotheriidae. In V.J. Maglio and H.B.S. Cooke, eds., *Evolution of African Mammals*, Cambridge, Mass., Harvard Univ. Press, pp. 423-434.

Bonis, L. de, Jaeger, J.-J., Coiffait, B., and Coiffait, P.-E. 1988. Decouverte du plus ancien primate Catarrhinien connu dans l'Eocene superieur d'Afrique du Nord. *C.R. Acad. Sci. Paris*, II, 306: 929-934.

Bowen, B.E. and Vondra, C.F. 1974. Paleoenvironmental interpretations of the Oligocene Gabal Qatrani Formation, Fayum Depression, Egypt. *Annals Geol. Survey Egypt* 4:115-138.

Bown, T.M. 1982. Ichnofossils and rhizoliths of the nearshore fluvial Jebel Qatrani Formation (Oligocene), Fayum Province, Egypt. *Palaeogeogr., Palaeoclimat., Palaeoecol.* 40: 255-309.

Bown, T.M. and Kraus, M.J. 1988. Geology and paleoenvironment of the Oligocene Jebel Qatrani Formation and adjacent rocks, Fayum Depression, Egypt. *U. S. Geol. Surv. Prof. Paper* 1452:1-64.

Bown, T.M., Kraus, M.J., Wing, S.L., Fleagle, J.G., Tiffney, B.H., Simons, E.L. and Vondra, C.F. 1982. The Fayum primate forest revisited. *Jour. Human Evol.* 11:603-632.

Brown, G.F., Schmidt, D.L. and Huffman, A.C., Jr. 1989. Geology of the Arabian Peninsula. *U.S. Geol. Surv. Prof. Paper*, 560-A:1-188.

Cappetta, H., Jaeger, J.-J., Sabatier, M., Sige, B., Sudre, J. and Vianey-Liaud,M. 1978. Decouverte dans le Paleocene du Maroc des plus anciens mammiferes eutheriens d'Afrique. *Geobios* 11:257-263.

Cappetta, H., Jaeger, J.-J., Sabatier, M., Sige, B., Sudre, J. and Vianey-Liaud, M. 1987. Complements et precisions biostratigraphiques sur la faune paleocene a mammiferes et selaciens du bassin d'Ouarzazate (Maroc). *Tertiary Research* 8: 147-157.

Cavelier, C. 1972. L'age Priabonien superieur de la "zone a *Ericsonia subdisticha*" (nannoplancton) en Italie et l'attributiondes Latdorf Schichten allemands a l'Eocene superieur. *Bull. B.R.G.M.* 2:15-24.

Cavelier, C. 1979. La limite Eocene-Oligocene en Europe occidentale. *Universite L. Pasteur Strasbourg, Sci. geol.* 54:1-280.

Coiffait, P.E., Coiffait, B., Jaeger, J.J., and Mahboubi, M. 1984. Un nouveau gisement a mammiferes fossiles d'age Eocene superieur sur le versant sud des Nementcha (Algerie orientale): decouverte des plus anciens rongeurs d'Afrique. *C.R. Acad. Sc. Paris*, II, 299: 893-898.

Coppens, Y., Maglio, V.J., Madden, C.T. and Beden, M. 1978. Proboscidea. In V.J. Maglio and H.B.S. Cooke, eds., *Evolution of African Mammals*, Cambridge, Mass., Harvard Univ. Press, pp. 336-367.

Crochet, J.-Y. 1988. Le plus ancien creodonte africain: *Koholia atlasense* nov. gen., nov. sp. (Eocene inferiur d'El Kohol, Atlas saharien, Algerie). *C. R. Acad. Sc. Paris*, II, 307:1795-1798.

Crochet, J.-Y., Hartenberger, J.-L., Rage, J.-C., Remy, J.A., Sige, B., Sudre, J. and Vianey-Liaud, M. 1981. Les nouvelles faunes de vertebres anterieures a la "Grande Coupure" decouvertes dans les phosphorites du Quercy. *Bull. Mus. nat. Hist. natur. Paris* 3:245-265.

Dartevelle, E. 1935. Premiers restes de mammiferes du Tertiaire du Congo: La faune miocene de Malembe. *2nd Cong. Nat. Sci., Brussells, C.R.I.* 1935:715-720.

Fleagle, J.G., Bown, T.M., Obradovich, J.D. and Simons, E.L. 1986. Age of the earliest African anthropoids. *Science* 234:1247-1249.

Gevin, P., Feist, M. and Mongereau, N. 1974. Decouverte de charophytes au Glib Zegdou (frontiere algero-marocaine). *Bull. Soc. Hist. Nat. Afrique du Nord, Alger*, 65:371-375.

Gevin, P., Lavocat, R., Mongereau, N. and Sudre, J. 1975. Decouverte de mammiferes dans la moitie inferieure de l'Eocene continental du Nord-Ouest du Sahara. *C.R. Acad. Sci. Paris* 280 D:1539-1542.

Gheerbrant, E. 1988. *Afrodon chleuhi* nov. gen., nov. sp., "insectivore" (Mammalia, Eutheria) lipotyphle (?) du Paleocene marocain: donnees preliminaires. *C. R. Acad. Sc. Paris*, II, 307:1303-1309.

Gheerbrant, E. 1990. On the early biogeographical history of the African placentals. *Historical Biology* 4: 107-116.

Hartenberger, J.-L. 1986. Hypothese paleontologique sur l'origine des Macroscelidea (Mammalia). *C. R. Acad. Sc. Paris*, II, 302:247-249.

Hartenberger, J.-L., Martinez, C. and Ben Said, A. 1985. Decouverte de mammiferes d'age Eocene inferieur en Tunisie centrale. *C. R. Acad. Sc. Paris*, II, 301:649-652.

Hooijer, D. 1963. Miocene Mammalia of Congo. *Ann. Mus. Roy. Afr. Centr., Tervuren (Belg.) ser. in-8, Sci. Geol.* 46:1-77.

Kappelman, J. 1991. Paleomagnetic stratigraphy and age estimates for the Fayum primates. *Amer. Jour. Phys. Anthro. Suppl.* 12: 102-103 [abstract].

Kay, R.F. 1977. The evolution of molar occlusion in the Cercopithecidae and early catarrhines. *Am. Jour. Phys. Anthropol.* 46:327-

352.

Lavocat, R. 1978. Rodentia and Lagomorpha. In V.J. Maglio and H.B.S. Cooke, eds., *Evolution of African Mammals*, Cambridge, Mass., Harvard University Press, pp. 69-89.

Lopez, N. and Thaler, L. 1974. Sur les plus ancien lagoporphe europeen et la "Grande Coupure" Oligocene de Stehlin. *Palaeovertebrata* 6:243-251.

Mahboubi, M., Ameur, R., Crochet, J.-Y. and Jaeger, J.-J. 1986. El Kohol (Saharan Atlas, Algeria), a new Eocene mammal locality in northwestern Africa: stratigraphic, phylogenetic and paleobiogeographical data. *Palaeontographica Abteilung* A, 192:15-49.

Meyer, G.E. 1978. Hyracoidea. In V.J. Maglio and H.B.S. Cooke, eds., *Evolution of African Mammals*, Cambridge, Mass., Harvard University Press, pp. 284-314.

Morgan, P. 1990. Egypt in the framework of global tectonics. In R. Said, ed., *The Geology of Egypt*, Rotterdam, A.A. Balkema, pp. 91-111.

Odin, G.S. and Montanari, A. 1989. Age radiometrique et stratotype de la limite Eocene-Oligocene. *C. R. Acad. Sc. Paris*, II, 309:1939-1945.

Olson, S.L. and Rasmussen, D.T. 1986. Paleoenvironment of the earliest hominoids: new evidence from the Oligocene avifauna of Egypt. *Science* 233:1202-1204.

Pickford, M. 1986. Premiere decouverte d'une faune mammalienne terrestre paleogene d'Afrique sub-saharienne. *C.R. Acad. Sci. Paris*, II, 302:1205-1210.

Plaziat, J.-C. 1981. Late Cretaceous to late Eocene palaeogeographic evolution of southwest Europe. *Palaeogeogr., Palaeoclimat., Palaeoecol.* 36:263-320.

Pomerol, C. and Premoli-Silva, I., eds. 1986. *Terminal Eocene Events*. Amsterdam, Elsevier, pp. 1-414.

Prothero, D.R. 1989. Stepwise extinctions and climatic decline during the later Eocene and Oligocene. In S.K. Donovan, ed., *Mass Extinctions, Proesses and Evidence*, London, Belhaven, pp. 211-234.

Rasmussen, D.T. 1989. The evolution of the Hyracoidea: a review of the fossil evidence. In D.R. Prothero and R.M. Schoch, eds., *The Evolution of Perissodactyls*, New York, Oxford Univ. Press, pp. 57-78.

Rasmussen, D.T., Olson, S.L. and Simons, E.L. 1987. Fossil birds from the Oligocene Jebel Qatrani Formation, Fayum, Egypt. *Smithsonian Contrib. Paleobiol.* 62:1-20.

Rasmussen, D.T. and Simons, E.L. 1988a. New Oligocene hyracoids from Egypt. *Jour. Vert. Paleontol.* 8:67-83.

Rasmussen, D.T. and Simons, E.L. 1988b. New specimens of *Oligopithecus savagei*, early Oligocene primate from the Fayum, Egypt. *Folia Primatol.* 51:182-208.

Rasmussen, D.T. and Simons, E.L. in press. The oldest Egyptian hyracoids (Mammalia: Pliohyracidae): new species of *Saghatherium* and *Thyrohyrax* from the Fayum. *N. Jb. Geol. Palaont. Abh. Stuttgart.*

Said, R. 1962. *The Geology of Egypt*. Amsterdam, Elsevier, 377 pp.

Salem, R. 1976. Evolution of Eocene-Miocene sedimentation patterns in parts of northern Egypt. *Amer. Assoc. Petrol. Geol. Bull.* 60:34-64.

Savage, R.J.G. 1969. Early Tertiary mammal locality in southern Libya. *Proc. Geol. Soc. London* 1657:167-171.

Savage, R.J.G. 1971. Review of the fossil mammals of Libya. In C. Gray, ed., *Symposium on the Geology of Libya*, Tripoli, Univ. of Libya, pp. 215-226.

Shata, A. 1953. New light on structural developments of the Western Desert of Egypt. *Institute Desert Egypte Bulletin* 3:101-106.

Sheikh, H.A. and Faris, M. 1985. The Eocene/Oligocene boundary in some wells of the Western Desert, Egypt. *Neu. J. Geol. Palaontol., Stuttgart*, 1985:23-28.

Simons, E.L. 1965. New fossil apes from Egypt and the initial differentiation of the Hominoidea. *Nature* 205:135-139.

Simons, E.L. 1968. Early Cenozoic mammalian faunas, Fayum Province, Egypt. Part I. African Oligocene mammals: introduction, history of study, and faunal succession. *Bull. Peabody Mus. Nat. Hist., Yale Univ.*, 28:1-21.

Simons, E.L. 1974. The relationships of *Aegyptopithecus* to other primates. *Ann. Geol. Survey Egypt* 4:149-156.

Simons, E.L. 1989. Description of two genera and species of Late Eocene Anthropoidea from Egypt. *Proc. Natl. Acad. Sci. USA* 86:9956-9960.

Simons, E.L. 1990. Discovery of the oldest known anthropoidean skull from the Paleogene of Egypt. *Science* 247:1507-1509.

Simons, E.L. and Bown, T.M. 1985. *Afrotarsius chatrathi*, first tarsiiform primate (?Tarsiidae) from Africa. *Nature* 313:475-477.

Simons, E.L., Bown, T.M. and Rasmussen, D.T. 1987. Discovery of two additional prosimian primate families (Omomyidae, Lorisidae) in the African Oligocene. *Jour. Human Evol.* 15:431-437.

Simons, E.L. and Kay, R.F. 1988. New material of *Qatrania* from Egypt with comments on the phylogenetic position of the Parapithecidae (Primates, Anthropoidea). *Amer. Jour. Primatol.* 15:337-347.

Simons, E.L. and Rasmussen, D.T. 1990. Vertebrate paleontology of Fayum: history of research, faunal review and future prospects. In R. Said, ed., *The Geology of Egypt*. Rotterdam, A.A. Balkema, pp. 627-638.

Sloss, L.L., Krumbein, F.C. and Dapples, E.C. 1949. Integrated facies analysis. *Geol. Soc. Amer. Mem.* 39:91-124.

Spinar, Z.V. 1980. The discovery of a new species of pipid frog (Anura, Pipidae) in the Oligocene of central Libya. In M.J. Salem and M.T. Busrewil, eds., *The Geology of Libya*, Vol. I, London, Academic Press, pp. 327-348.

Stehlin, H.G. 1909. Remarques sur les faunules de mammiferes des couches eocenes et oligocenes du Bassin de Paris. *Bull. Soc. Geol. France* 9:488-520.

Sudre, J. 1979. Nouveaux mammiferes eocenes du Sahara occidental. *Palaeovertebrata* 9:83-115.

Swisher, C.C., III, and Prothero, D.R. 1990. Single-crystal $^{40}Ar/^{39}Ar$ dating of the Eocene-Oligocene transition in North America. *Science* 249:760-762.

Thomas, H., Roger, J., Sen, S. and Al-Sulaimani, Z. 1988. Decouverte des plus anciens "Anthropoides" du continent arabo-africain et d'un Primate tarsiiforme dans l'Oligocene du Sultanat d'Oman. *C. R. Acad. Sc. Paris*, II, 306:823-829.

Thomas, H., Roger, J., Sen, S; Bourdillon-de-Grissac, C. and Al-Sulaimani, Z. 1989. Decouverte de vertebres fossiles dans l'Oligocene inferieur du Dhofar (Sultanat d'Oman). *Geobios* 22:101-120.

Thomas, H., Sen, S., Roger, J. and Al-Sulaimani, Z. 1991. The discovery of *Moeripithecus markgrafi* Schlosser (Propliopithecidae, Anthropoidea, Primates), in the Ashawq Formation (early Oligocene of Dhofar Province, Sultanate of Oman). *Jour. Human Evol.* 20:33-49.

Tilden, C.D., Holroyd, P.A., and Simons, E.L. 1990. Phyletic affinities of *Apterodon* (Hyaenodontidae, Creodonta). *Jour. Vert. Paleont.* 10:46A [abstract].

Vondra, C.F. 1974. Upper Eocene transitional and near-shore marine Qasr el Sagha Formation, Fayum Depression, Egypt. *Annals Geol. Surv. Egypt* 4:79-94.

Wheeler, H.E. 1958. Time-stratigraphy. *Amer. Assoc. Petrol. Geol. Bull.* 42:1047-1063.

Wight, A.W.R. 1980. Paleogene vertebrate fauna and regressive sediments of Dur at Talhah, southern Sirt Basin, Libya. In M.J. Salem and M.T. Busrewil, eds., *The Geology of Libya*, Vol. I, London, Academic Press, pp. 309-325.

Wing, S.L. and Tiffney, B.H. 1982. A paleotropical flora from the Oligocene Jebel Qatrani Formation of northern Egypt: a preliminary report. *Misc. Series Botanical Soc. Amer.* 162:67.

INDEX

www.ingramcontent.com/pod-product-compliance
Ingram Content Group UK Ltd.
Pitfield, Milton Keynes, MK11 3LW, UK
UKHW051935140225
455139UK00002B/6